Charged Particle and Photon
Interactions with Matter

Charged Particle and Photon Interactions with Matter

Chemical, Physicochemical, and Biological Consequences with Applications

edited by

A. Mozumder
University of Notre Dame
Notre Dame, Indiana, U.S.A.

Y. Hatano
Tokyo Institute of Technology
Tokyo, Japan

CRC Press
Taylor & Francis Group
Boca Raton London New York

CRC Press is an imprint of the
Taylor & Francis Group, an **informa** business

CRC Press
Taylor & Francis Group
6000 Broken Sound Parkway NW, Suite 300
Boca Raton, FL 33487-2742

First issued in paperback 2019

© 2004 by Taylor & Francis Group, LLC
CRC Press is an imprint of Taylor & Francis Group, an Informa business

No claim to original U.S. Government works

ISBN-13: 978-0-8247-4623-0 (hbk)
ISBN-13: 978-0-367-39486-8 (pbk)

Library of Congress Cataloging-in-Publication Data
A catalog record for this book is available from the Library of Congress.

Visit the Taylor & Francis Web site at
http://www.taylorandfrancis.com

and the CRC Press Web site at
http://www.crcpress.com

Preface

The purpose of this work is to present a coherent account of high-energy charged particle and photon interactions with matter, in vivo and in vitro, that will be of use to both students and practicing scientists and engineers. The book encompasses not only radiation chemistry and photochemistry, but also other aspects of charged particle and photon interactions, such as radiation physics, radiation biochemistry, radiation biology, and applications to medical and engineering sciences and to radiation synthesis and processing and food irradiation. Any phenomenon of ionization and excitation induced by charged particle and photon interactions with matter is considered of interest, since information on the interactions of photons with matter help us understand the interactions of charged particles with matter. Further, the study of the interactions of high-energy photons, particularly in the vacuum ultraviolet–soft X-ray (VUV-SX) region, is of great importance in bridging the area between photochemistry and radiation chemistry (see Chapter 5). Throughout the book a major aim has been to elucidate the physical and chemical principles involved in applications of significance to practicing scientists and engineers. We have secured the foremost workers in their respective areas to contribute chapter articles in a manner that ensures contact with adjacent disciplines, keeping in mind the needs of the general readership.

Since its inception, radiation effects have had ramifications in various fields, which sometimes use different technical language, units, etc. Reviews condense the subject matter while amplifying important points and bring these up-to-date for students and researchers. In a field that is developing as rapidly as radiation effects in vivo and in vitro, there is a need for periodic summaries in the form of a book. In the present millenium, a great need has developed for in-depth studies with a balance of experiment, theory, and application. Yet, chapters written by several authors in the same book may use different styles, outlooks, or even notations. We have paid particular attention to these factors and have striven to make the presentations uniform.

By its very nature this book is interdisciplinary. The first eleven chapters delineate the fundamentals of radiation physics and radiation chemistry that are common to all irradiation effects. Chapters 12 and 13 deal with specific liquid systems, while Chapter 14 is concerned with LET effects. Chapters 15 to 18 describe biological and medical consequences of photon and charged-particle irradiation. The rest of the book is much more applied in character, starting with irradiated polymers in Chapter 19 and ending with applications of heavy ion impact in Chapter 27.

Our aim has been to provide a self-contained volume with sufficient information and discussion to take the reader to the frontier of investigation. The level of presentation is at the advanced undergraduate level, so that undergraduates, graduate students, researchers, and practicing professionals may all benefit from it. It is assumed that the reader has some knowledge of chemistry or biology or physics but is not necessarily versed in the properties of high-energy radiation.

It is our pleasure and privilege to thank our friends and colleagues all over the world who have contributed to our understanding of the high-energy charged particle and photon interactions with matter.

A. Mozumder
Y. Hatano

Contents

v

Contributors

Andrew D. Bass University of Sherbrooke, Sherbrooke, Québec, Canada

Jacqueline Belloni UMR CNRS–UPS, Université Paris-Sud, Orsay, France

William A. Bernhard University of Rochester, Rochester, New York, U.S.A.

G. V. Buxton University of Leeds, Leeds, England

David M. Close East Tennessee State University, Johnson City, Tennessee, U.S.A.

John F. Dicello Johns Hopkins University School of Medicine, Baltimore, Maryland, U.S.A.

József Farkas Szent István University, Budapest, Hungary

Mitsuhiro Fukuda Japan Atomic Energy Research Institute, Takasaki, Gunma, Japan

Reinhard Gahbauer Ohio State University, Columbus, Ohio, U.S.A.

John Gueulette Université Catholique de Louvain, Brussels, Belgium

Y. Hatano Tokyo Institute of Technology, Tokyo, Japan

Richard A. Holroyd Brookhaven National Laboratory, Upton, New York, U.S.A.

Tsuneki Ichikawa Hokkaido University, Sapporo, Japan

Hisayoshi Itoh Japan Atomic Energy Research Institute, Takasaki, Gunma, Japan

Charles D. Jonah Argonne National Laboratory, Argonne, Illinois, U.S.A.

Dan T. L. Jones iThemba Laboratory for Accelerator Based Sciences, Somerset West, South Africa

Yosuke Katsumura The University of Tokyo, Tokyo, Japan

Katsumi Kobayashi High Energy Accelerator Research Organization, Tsukuba, Japan

Noriyuki Kouchi Tokyo Institute of Technology, Tokyo, Japan

T. Kozawa   Osaka University, Osaka, Japan

Jay A. LaVerne   University of Notre Dame, Notre Dame, Indiana, U.S.A.

Tetsuro Majima   Osaka University, Osaka, Japan

Mehran Mostafavi   UMR CNRS–UPS, Université Paris-Sud, Orsay, France

A. Mozumder   University of Notre Dame, Notre Dame, Indiana, U.S.A.

Hideki Namba   Japan Atomic Energy Research Institute, Takasaki, Gunma, Japan

Hooshang Nikjoo   Medical Research Council, Harwell, Oxfordshire, England

Takeshi Ohshima   Japan Atomic Energy Research Institute, Takasaki, Gunma, Japan

Simon M. Pimblott   University of Notre Dame, Notre Dame, Indiana, U.S.A.

Masahiro Saidoh   Japan Atomic Energy Research Institute, Takasaki, Gunma, Japan

*Léon Sanche**   University of Sherbrooke, Sherbrooke, Québec, Canada

Myran C. Sauer, Jr.   Argonne National Laboratory, Argonne, Illinois, U.S.A.

S. Seki   Osaka University, Osaka, Japan

Ilya A. Shkrob   Argonne National Laboratory, Argonne, Illinois, U.S.A.

M. Tachiya   National Institute of Advanced Industrial Science and Technology (AIST), Tsukuba, Japan

S. Tagawa   Osaka University, Osaka, Japan

Atsushi Tanaka   Japan Atomic Energy Research Institute, Takasaki, Gunma, Japan

Larry H. Toburen   East Carolina University, Greenville, North Carolina, U.S.A.

Shuzo Uehara   Kyushu University, Fukuoka, Japan

Masatoshi Ukai   Tokyo University of Agriculture and Technology, Tokyo, Japan

André Wambersie   Université Catholique de Louvain, Brussels, Belgium

Mariusz Wojcik   Technical University of Lodz, Lodz, Poland

L. Wojnarovits   Hungarian Academy of Sciences, Budapest, Hungary

Marco Zaider   Memorial Sloan-Kettering Cancer Center, New York, New York, U.S.A.

---

*Canada Chair in the Radiation Sciences.

Charged Particle and Photon Interactions with Matter

1

Introduction

A. Mozumder
University of Notre Dame, Notre Dame, Indiana, U.S.A.

Y. Hatano
Tokyo Institute of Technology
Tokyo, Japan

1. EARLY INVESTIGATIONS

The effect of visible and UV-light on matter, in vitro and in vivo, has a long history. In this sense photochemistry and photobiology are forerunners, respectively, of radiation chemistry and radiation biology. The involvement of visible light on the discoloration of dyes and on the growth and transformation of plants has been known since time immemorial. In the 18th century the action of light on silver salts was definitely established, culminating in rudimentary photography in the mid-19th century (see Chapter 20). Some qualitative understanding of photochemical action was obtained by Grottus (1817) and by Draper (1841) in which it was stated that only the absorbed light can be effective in bringing out chemical transformation and that the rate of chemical change should be proportional to the light intensity. With the advent of the quantum theory, Einstein (1912) formalized what has since been called the *first law of photochemistry*, i.e., only one quantum of light is absorbed per reacting molecule. This actually applies to the primary process and at relatively low light intensity to which all early studies were confined. Later work at high intensity provided by lasers has necessitated some modification of this important law. As not all primary activations are observable, the idea of the quantum yield or the number of molecules transformed per quantum of light absorbed was introduced by Bodenstein (1913) who also noticed chain reactions in the combination of hydrogen and chlorine giving quantum yields many orders of magnitude greater than 1.

It is clear that, along with the discovery of x-rays in 1895, Roentgen also found the chemical action of ionizing radiation. He drew attention to the similarity of the photographic effect induced by light and x-rays. Application to medicine appeared very quickly, followed by industrial applications. However, this field of investigation remained nameless until Milton Burton, in 1942, christened it "radiation chemistry" to separate it from radiochemistry which is the study of radioactive nuclei. Historical and classical work in radiation chemistry has been reviewed by Mozumder elsewhere [1]. Here we will only make a few brief remarks.

Lind [2] has defined radiation chemistry as the science of the chemical effects brought about by the absorption of ionizing radiation in matter. It should be distinguished from radiation damage which refers to structural transformation induced by irradiation, particularly in the solid state. The distinction is not always maintained, perhaps unconsciously, and sometimes both effects may be present simultaneously. Following a suggestion of M. Curie around 1910, that ions were responsible for the chemical effects of radioactive radiations, the symbol M/N was introduced to quantify the radiation chemical effect, where M is the number of molecules transformed (created or destroyed) and N is the number of ion pairs formed. Later, Burton [3] and others advocated the notation G for the number of species produced or destroyed per 100 eV ($= 1.602 \times 10^{-17}$ J) absorption of ionizing radiation. It was purposely defined as a purely experimental quantity independent of implied mechanism or assumed theory.

Bethe [4] first pointed to the quantitative similarity of the excitation of a molecule by light and by fast charged particle impact, thereby underscoring the importance of *dipole oscillator strength*. Despite modifications required for slower charged particles, this *optical approximation* (see Sec. 3 and Chapters 2 and 5) remains a cornerstone in the understanding of the primary processes in radiation chemistry. In the totality of the various effects brought about by irradiation, the radiation chemical effects occupy a central position. In a similar fashion electrons and photons are the most important agencies for bringing the radiation effects. Electrons are always produced as secondary irradiation irrespective of the nature of the primary ionizing particle. Photons are ubiquitous and, in any case, understanding of charged particle impact requires a knowledge of photon impact processes.

Generally speaking, industrial applications of irradiation have a relatively short history except for some medical applications. The field, however, has gained considerable momentum since the 1960s. Spinks and Woods [5] have summarized radiation-induced synthesis and processing; Mozumder [1] has also given a brief account of various applications to science and industry including dosimetry, food irradiation, waste treatment, sterilization of medical equipment, etc. Salient features of radiation chemistry, both in the gaseous and condensed phases, have been discussed in the *CRC Handbook of Radiation Chemistry* edited by Tabata et al. [6]. Special topics of the interactions of excess electrons in dielectric media have been adequately reviewed in a CRC handbook edited by Ferradini and Jay-Gerin [7]. Further, specific applications to atomic and molecular processes in reactive plasmas have been summarized by Hatano [8]. Chapters 15 through 27 of this book are involved with radiation applications including radiobiological, medical, and industrial applications. In addition, Chapters 16, 19, and 20 discuss, in considerable details, the effects and applications of photon irradiation. An important role of synchrotron radiation (SR) as a new photon source was theoretically pointed out for the first time about 40 years ago by R.L. Platzman and U. Fano. Recent progress in SR research has been remarkable, as reviewed by Hatano elsewhere [9]. See also Chapter 5 for the newly established dedicated SR facilities combined with new experimental methods for spectroscopy and dynamics.

2. TIME SCALE OF RADIATION ACTION

In the action of charged particle and photon irradiation on condensed matter, it is instructive to consider several stages of overlapping time scales. The stages, in ascending orders of time, are called the physical, physicochemical, and chemical stages, respectively, for radiation chemical studies. To these, biochemical and biological stages have been added for application to radiobiology [10,11]. Within each stage various events take place occupying their respective time scales. For example, within the biochemical stage ($\sim 10^{-3}$

to $\sim 10^{+4}$ sec), secondary radicals form, evolve, and react, while the biological stage covers a wide range of time from mitosis to late biological effect. The earliest discernible time, in the physical stage, is obtained from the uncertainty principle, $\Delta E \Delta t \sim \hbar$, as $\sim 10^{-17}$ sec, which corresponds to the production of secondary electrons with energy ≥ 100 eV. Times such as these or shorter are just computed values. The longest time scale in radiobiology may exceed several years or generations if genetic effects are included. Morrison [12] first gave an approximate time scale of events for application to radiobiology. Several authors have periodically updated the concept for different types of application. More recently, Lentle and Singh [10] and Mozumder [11] have discussed this topic from the points of view of radiation biology and radiation chemistry, respectively.

Table 1 summarizes some of the events that occur through the various stages. It has been suggested that, following ionization in liquid water, the "dry" hole H_2O^+ can

Table 1 Approximate Time Scale of Events in Radiation Chemistry, for Example, of Liquid Water

	$-\log_{10} t(sec) \equiv pt$	Event	Stage
	18	Fast electron traverses molecule	
	17	MeV proton traverses molecule; energy loss to fast secondary electrons.	
fs	16	Energy loss to electronic states, (Vertical excitation)	Physical
	14	Fast ion-molecule reaction (H-atom transfer). H_3O^+, H and OH formed.	
	13	Solvated electron formed in water. Longitudinal dielectric relaxation in water. Molecular vibration. Fast dissociation.	Physico-Chemical / Chemical
ps	12	Electron thermalized or trapped. Spur formed. Self-diffusion time scales in simple liquids.	
	11	Transverse dielectric relaxation time in water.	
ns	9	Spur reactions continued.	
	8	Charge neutralization in media of low viscosity. Secondary reactions including intertrack reactions.	
	6	Electron escape time in low-viscosity media. Intratrack reactions completed. Secondary radical formation and reaction.	Biochemical
	0	Radiative lifetime of triplets. Biochemical effects	Biological
	-3	Neutralization time for media at very high viscosity. Time for mitosis.	
		Biochemical effects of metastables.	
		DNA synthesis time	
		Biological effects	
		Late biological effects	

move by exact resonance until the ion-molecule reaction $H_2O^+ + H_2O \rightarrow H_3O^+ + OH$ localizes the hole. The earliest chemical transformation, indicated by H-atom transfer in reactions of the type $ROH^+ + ROH \rightarrow ROH_2^+ + RO$, then occurs in $\sim 10^{-14}$ sec in water and alcohols [13]. The time needed for electron hydration is theoretically computed to be ~ 0.2 psec [14] while experiment gives an upper limit of 0.3 psec [15]. The information presented in Table 1 is not to be taken too literally. Each stage may contain many orders of magnitude of time, but the usefulness of the time scale picture rests on the perception that within an order of magnitude relatively few processes compete. As the species that exist at the end of one stage serve as the input to the next stage, the qualification "early" is relative. In radiobiological systems, the time of appearance of a change may be subjective depending on the cell cycle, which is due to several different kinds of generations of biochemical and biological transformations that must take place to render the radiation-induced transformation "visible." Also, these processes can occur both in the directions of repair and amplification. Certain biochemical reactions, for example, that of OH with sugars, may occur in the nanosecond time scale while others, such as those giving the O_2 effect, may take microseconds. DNA strand breakage may be considered an early biological effect while other damages can take ~ 1 day to ~ 40 years when genetic effects are considered. This should be contrasted with ~ 1 μsec needed for track dissolution in liquid water [16].

In the current millennium there is a strong need for interdisciplinary research involving an integrated outlook. In this sense various disciplines connected with radiation action all start with radiation physics and radiation chemistry. From that point on one sees a bifurcation. One way goes to radiation biochemistry to radiation biology and to medical applications. Another way goes to engineering and industrial applications including, but not limited to, polymerization, waste management, food preservation, etc.

3. FUNDAMENTAL PROCESSES IN THE PHYSICAL, PHYSICO-CHEMICAL, AND CHEMICAL STAGES OF INTERACTIONS OF HIGH-ENERGY CHARGED PARTICLES AND PHOTONS WITH MATTER

In the interactions of high-energy incident particles, i.e., photons, electrons, heavy charged particles (or positive and negative ions), and other particles, with matter, the succession of events that follow the absorption of their energies has been classified into three characteristic temporal stages: physical, physico-chemical, and chemical stages [17] (see Sec. 2). The physical stage consists of the primary activation of molecules in matter (in the case, for example, it is composed of molecular compounds), due to the collision of high-energy incident particles to form electronically excited or ionized states of molecules and ejected electrons. The electrons thus formed may have sufficient energy to further ionize the surrounding molecules. At the end of the interactions, electrons are formed in a wide energy range via cascading electron–molecule collision processes. These secondary electrons further decrease their energies to the subexcitation energy range, leading to the formation of reactive species, i.e., ions, excited molecules, free radicals, and low-energy electrons. These species interact with each other or with stable molecules. Subexcitation electrons, which are electrons below the first excitation potential, degrade their energies by vibrational, rotational, and/or elastic collisions to reach thermal energy. Electrons in the thermal and epithermal ranges disappear predominantly by recombination, attachment, or diffusion. It is concluded therefore that a decisive step in the physical and physico-chemical stages of the interactions of high-energy incident particles with matter is the collision of secondary electrons with molecules in a wide energy range (see Chapters 2–5, 9, 17). The entire feature of the fundamental processes of the interactions of high-energy incident particles with molecules, i.e., the molecular processes in the physical and physico-chemical stages (and further, the chemical stage which mainly consists of free radical reactions), are summarized in Table 2 [18] (see Chapter 6).

Table 2 shows that molecules AB in collisions with electrons in a wide energy range are directly ionized and excited into superexcited states [9] above their first ionization potentials (IP) and into excited states below IP. Superexcited states AB′ are competitively autoionized or dissociated into neutral fragments, i.e., free radicals or stable product molecules (see Chapter 5). Electronically excited states AB* are also dissociated to neutral fragments. Parent ions directly formed via ionization, or indirectly via autoionization, are collisionally (and/or unimolecularly) stabilized or dissociated into fragment ions. Such ions, particularly in the condensed phase, may be quickly neutralized with electrons by geminate recombination or competitively converted to other ions via ion–molecule reactions (see Chapter 6), while in the gas phase such conversion is predominant compared with recombination, depending on the pressure. In case of the addition of electronegative solute molecules, negative ions are produced in electron attachment processes. Electrons with characteristic energies are selectively captured by molecules to form negative ions (see Chapter 6). It is generally accepted that molecular clusters, large aggregates of molecules, and molecules in the condensed phase can capture electrons with large cross sections due to electron attachment dynamics different from those in the gas phase for isolated single molecules [19,20]. For polar molecules, electrons or ions are solvated with molecules to form solvated electrons and ions, respectively (see Chapter 7). The recombination of positive ions with electrons, particularly in the condensed phase, is classified as geminate and bulk recombination processes (see Chapter 10). A relative importance of these two processes is closely related with the transport mechanism of electrons in it, i.e., the mag-

Table 2 Fundamental Processes in the Physical, Physicochemical, and Chemical Stages of the Interactions of High-Energy Charged Particles and Photons with Molecules AB

$AB \longrightarrow AB^+ + e^-$	Direct ionization
$\longrightarrow AB'$	[Superexcitation (direct excitation)
$\longrightarrow AB^*$	Excitation]
$AB' \longrightarrow AB^+ + e^-$	Autoionization
$\longrightarrow A + B$	Dissociation
$AB^+ \longrightarrow A^+ + B$	Ion dissociation
$AB^+ + AB$ or $S \longrightarrow$ Products	Ion–molecule reaction
$AB^+ + e^- \longrightarrow AB^*$	Electron–ion recombination
$AB^+ + S^- \longrightarrow$ Products	Ion–ion recombination
$e^- + S \longrightarrow S^-$	Electron attachment
$e^- + nAB \longrightarrow e^-_s$	Solvation
$AB^* \longrightarrow A + B$	Dissociation
$\longrightarrow AB$	Internal conversion and intersystem crossing
$\longrightarrow BA$	Isomerization
$\longrightarrow AB + h\nu$	Fluorescence
$AB^* + S \longrightarrow AB + S^*$	Energy transfer
$AB^* + AB \longrightarrow (AB)_2$	Excimer formation
$2A \longrightarrow A_2$	Radical recombination
$\longrightarrow C + D$	Disproportionation
$A + AB \longrightarrow A_2B$	Addition
$\longrightarrow A_2 + B$	Abstraction

nitudes of the electron drift mobility or the electron conduction band energy in it. Phenomenologically, it is more dependent on the properties of the medium molecules (see Chapter 8). In some cases, particularly in admixed systems both in the gas and condensed phases, energy or charge transfer processes may predominantly occur from molecules of the major component in the system to added solute molecules (see Chapter 6).

As described above, a crucial step of the physical and physicochemical stages is the collision of secondary electrons with molecules in a wide energy range. Therefore the information on electron–molecule collision cross-section data, which must be correct, absolute, and comprehensive, is of great importance and is helpful for understanding the essential features of the fundamental processes in these two stages [21]. Such information is available both experimentally and theoretically in the gas-phase (see Chapter 3), but in some cases also in the condensed-phase as well (see Chapter 9). Cross sections for the ionization and excitation of molecules in collisions with electrons in the energy range greater than about 10^2 eV are well elucidated quantitatively by the Born–Bethe theory [22]. This theory is also helpful for calculating, at least roughly, these cross-section values further in the lower energy range.

According to the optical approximation, which was shown by Platzman [23] to be based on the Born–Bethe theory, a radiation chemical yield G may be estimated from optical data, viz., photoabsorption cross sections, oscillator strength distributions, photoionization cross sections, and photodissociation cross sections of molecules in a wide range of the photon energy or wavelength, particularly in VUV–SX region [9] (see Chapter 5). Basically, this approximation should be applied only for a rough estimation of G values. Such a limitation may be attributed to a superficial comparison of radiation chemical yields for atomic systems with electron- atom cross-section data in a limited collision-energy range which were available in earlier stages of the related research fields in the 1960s. Recently, however, its applicability has been examined by comparing electron-collision cross sections for molecules in a wide collision energy range between optically allowed and forbidden transitions, providing the conclusion, at least tentatively, that this approximation may be applicable to the estimation of G values more precisely than previously considered [24]. There is a great need to examine the applicability more systematically because of recent progress in the experimental and theoretical investigations of electron–molecule collision processes [21].

The experimental and theoretical investigations of the physical, physico-chemical, and chemical stages of the interactions of high-energy incident particles with matter have made a remarkable contribution to recent progress in fundamental studies of the static and dynamic behavior of reactive species, i.e., electrons, ions, excited atoms and molecules, and free radicals. Conversely, these are needed for a comprehensive understanding of recent progress in the fundamental studies of reactive species, where a scientifically reasonable analysis of the interesting but complex mechanism of the interactions of high-energy incident particles with matter is involved. Thus it often provides new findings of reactive species as well as their static or dynamic behavior. For that purpose, the matrix representation is helpful for summarizing or surveying the diverse information on such fundamental studies of a variety of reactive species [18].

4. FIELDS OF STUDY: RELEVANCE TO BASIC AND APPLIED SCIENCES

The fields of study undertaken in the present book are rather broad. However, the contributors, recognized experts in their respective areas, have striven to ensure continuity by connecting important concepts to adjacent areas. Thus, in one example, the logical flow would be from physics to chemistry to biology and medicine, and in another example it could be from physics to chemistry to industrial applications, including polymerization, waste, and flue-gas treatment, etc. In choosing the subject matters it has been the objective of the editors that comparable importance is given to fundamentals and applications. While it has been well recognized by almost everyone that radiation effect starts with a physical interaction, very often followed by chemical reactions, etc., the chemical stage is sometimes understressed in radiobiological discussions [25]. In a similar vein, the applied radiation scientist may have a right to think that aqueous radiation chemistry has been overstressed by the physical chemists who claim it to be a better understood topic in chemical kinetics [26]. In any case the editors have taken care so that each significant area of investigation would receive comparable attention.

Studies in radiation science and technology may be viewed from three angles, namely, life, industry, and basic knowledge. According to some estimates [27], the continual exposure to environmental radiation over geological periods is important and must be taken into account. There is now a widespread belief that aging and, indeed, the genetic evolution of man may be significantly influenced by ambient radiation. Understanding of these effects duly belongs to radiation biology, which depends on radiation physics and radiation chemistry for basic information. Radiation is applied in industry in various ways—for example, as reaction initiators, sustainers, and also as control mechanisms. Some of the important ones are included in this book, while others, such as curing of paints, materials for textile finishing, applications in mining and metallurgy, etc., are left out for obvious space limitation. Finally, the important contribution made by radiation chemistry to experimental basic science should be appreciated in making available excitations to states which would otherwise be inaccessible by thermal activation.

REFERENCES

1. Mozumder, A. *Fundamentals of Radiation Chemistry*; Academic Press: San Diego, 1999.
2. Lind, S.C. *Radiation Chemistry of Gases*; Reinhold Publishing Corporation: New York, 1961.
3. Burton, M. J. Phys. Colloid. Chem. 1947, *51*, 611.
4. Bethe, H.A. Ann. Physik. 1930, *5*, 325.
5. Spinks, J.W.T.; Woods, R.J. *An Introduction to Radiation Chemistry*. 3rd ed. Wiley-Interscience: New York, 1991.
6. *Handbook of Radiation Chemistry*; Tabata, Y., Ito, Y., Tagawa, S., Eds.; CRC Press: Boca Raton, 1991.
7. *Excess Electrons in Dielectric Media*; Ferradini, C., Jay-Gerin, J.-P., Eds.; CRC Press: Boca Raton, 1991.
8. Hatano, Y. Adv. At. Mol. Opt. Phys. 2000, *43*, 231.
9. Hatano, Y. Phys. Rep. 1999, *313*, 109.
10. Lentle, B.; Singh, H. Radiat. Phys. Chem. 1984, *24*, 267.
11. Mozumder, A. Radiat. Res. 1985, *104*, S-33.
12. Morrison, P. *Symposium in Radiobiology: The Basic Aspects of Radiation Effects on Living Systems*; In Nickson, J.J., Ed.; Wiley: New York, 1952; 1–12.
13. Magee, J.L.; Chatterjee, A. *Radiation Chemistry: Principles and Applications*; Farhataziz, Rodgers, M.A.J., Eds.; Chap. 5. VCH Publishers: New York, NY, 1987.
14. Mozumder, A. J. Chem. Phys. 1969, *50*, 3153.
15. Wiesenfeld, J.M.; Ippen, E.D. Chem. Phys. Lett. 1980, *73*, 47.
16. Turner, J.E.; Magee, J.L.; Wright, H.A.; Chatterjee, A.; Hamm, R.N.; Ritchie, R.H. Radiat. Res. 1983, *96*, 437.

17. Platzman, R.L. Vortex 1962, *23*, 327.
18. Hatano, Y. In: *Handbook of Radiation Chemistry*; Tabata, Y. Ito, Y., Tagawa, S., Eds.; CRC Press: Boca Raton, 1991.
19. Hatano, Y. In: *Electronic and Atomic Collisions*; Lorents, D.C. Meyerhof, W.E., Peterson, J.R., Eds.; Elsevier: Amsterdam, 1986; 153 pp.
20. Hatano, Y. Aust J Phys 1997, *50*, 615.
21. Inokuti, M., Ed. In: *Atomic and Molecular Data for Radiotherapy and Radiation Research;* IAEA-TECDOC-799 IAEA: Vienna, 1995.
22. Inokuti, M. Rev. Mod. Phys. 1971, *43*, 297.
23. Platzman, R.L. Radiat. Res. 1962, *17*, 419.
24. Hatano, Y. Radiat Phys Chem, to be published in 2003.
25. Lea, D.E. In: *Actions of Radiations on Living Cells*; Second ed. University Press: Cambridge, 1956.
26. Allen, A.O. In: *The Radiation Chemistry of Water and Aqueous Solutions*; D. Van Nostrand: Princeton, 1961.
27. Vereshchinskii, I.V.; Pikaev, A.K. In: *Introduction to Radiation Chemistry (translated from Russian)*; Daniel Davey: New York, 1964.

2

Interaction of Fast Charged Particles with Matter

A. Mozumder
University of Notre Dame, Notre Dame, Indiana, U.S.A.

1. ENERGY TRANSFER FROM FAST CHARGED PARTICLES

1.1. General Features

Any momentum change of an incident particle constitutes a collision. In the interaction of an incident particle with an atom or a molecule, the collisions are classified as *elastic* or *inelastic*. In *elastic* collisions, energy and momentum are conserved between the external degrees of freedom (e.g., translation) of the colliding partners. In *inelastic* collisions, energy is transferred into, or from, the internal degrees of freedom of the struck molecule, designated as *collisions of the first and the second kind*, respectively. Because collisions of the second kind are important for epithermal and thermal particles alone, we will only be concerned with the collisions of the first kind in this chapter. Furthermore, akin to the Grottus principle (see the Introduction) in photochemistry, only the energy absorbed by a molecule can bring about radiation–chemical change, or the ensuing radiobiological transformation. Therefore the internal molecular excitation (electronic, vibrational, etc.) is a necessary precursor of radiation-induced transformation. In this context, ionization is included as an extreme form of electronic excitation.

Particles and waves are distinguished only in classical mechanics. In quantum mechanics they are interchangeable, giving rise to *first and second quantizations*, respectively. However, it is customary to group the incident radiations as light (electrons, muons, x- and γ-rays, etc.) or heavy (protons, α-particles, fission fragments, etc.) particles. Photons, classified as a separate group, can cause excitation and ionization in a molecule. Ionization by photons can proceed by any of the three following mechanisms: *photoelectric effect, Compton effect, and pair production*. In the photoelectric effect, the photon is absorbed and the emergent electron inherits the photon energy reduced by the sum of the binding energy of the electron and any residual energy left in the resultant positive ion (Einstein equation). In the Compton effect, the photon is scattered with significantly lower energy by a medium electron, which is then ejected with the energy difference. In the pair production process, the photon is annihilated in a nuclear interaction, giving rise to an electron–positron pair, which together carries the photon energy lessened by twice the rest energy of the electron. This process therefore has an energetic threshold. With increase in photon energy, the dominant interaction changes from photoelectric to Compton to pair production. For

9

^{60}Co-γ rays, almost the entire interaction is induced by the Compton effect which, by the Klein–Nishina theory, gives a nearly flat electron energy spectrum. Thus the 1.2-MeV photon generates a wide spectrum of Compton electrons having an average energy of ~0.6 MeV; consequently, the medium molecules respond as if being showered by these fast electrons. Pair production is a characteristic of very high energy photons; again, to the medium molecules, the effect is the same as a spectrum of fast electrons and positrons of appropriate energy. Electrons of various energy are the most important sources as primary radiations in the laboratory and in the industry [1], and also as secondary radiations in any form of ionizing event. Other often-used irradiations include x-rays, radioactive radiations (α, β, or γ), protons, deuterons, various accelerated stripped nuclei, and fission fragments. X-rays differ from γ-rays operationally, i.e., x-rays are generated by machines whereas γ-rays are produced in nuclear transformations.

At ultrarelativistic speeds ($v \sim c$, the speed of light), a radiative process, called the nuclear *bremsstralung*, becomes significant. Bethe and Heitler [2] estimated the ratio of energy losses of a charged particle due to radiative (bremsstralung) and collisional (electronic) processes at an energy of E MeV approximately as $EZ/800$, where Z is the nuclear charge. For electrons in water, this radiative process only becomes significant above 100 MeV. For the vast majority of cases with incident velocity ~$0.99c$ down to about the speed of a least-bound electron, the dominant energy transfer mechanism is by electrostatic and electromagnetic interactions between the incident charged particle and the medium electrons. Of these, the electromagnetic interaction becomes important only at relativistic speeds [3]. Therefore in most cases of importance to radiation science, the principal interaction of an incident-charged particle with the medium electrons is of electrostatic origin.

For fast charged particles impinging on matter, the very first discernible effect is the electronic excitation of molecules or atoms. Because electron is the lightest particle, its excitation occurs at the earliest time scale consistent with the uncertainty principle, i.e., in ~10^{-16} sec. Later, transformation of the deposited energy in the various internal degrees of freedom occurs in their respective time scale, e.g., in ~10^{-12} sec for vibration, etc. Some details were presented elsewhere [4]. Molecular dissociation and low-grade heat appear at longer time scales and, in some cases involving exothermal reactions, the level of low grade heat can terminate above that of absorbed electronic energy. There is no evidence of local rise of temperature for fast incident charged particles.

1.2. Radiation Physics and Radiation Chemistry

Although in nature there is no distinction between physical and chemical effects, in practice it is profitable, depending on the investigator's choice of discipline, to view the radiation effect either from the point of view of the incident particle or from that of the medium molecules, designated as radiation physics and radiation chemistry, respectively. Thus the study of the charge, rate of energy loss, range, and penetration, etc. of the incident particle, any of which may change under the interaction, constitutes radiation physics. Study of the matter receiving the absorbed energy to produce chemical changes, charge separation, luminescence, etc. essentially belong to radiation chemistry. It is evident that radiation chemistry is the link between radiation physics and radiation biology (see Ch. 1, Sec. 2). Some investigators recognize this link explicitly [5,6], while others short circuit the chemical part, working directly from the physics of energy deposition to biological effects [7]. That necessitates the introduction of additional parameters, such as a critical dose or concepts such as microdosimetry [8,9].

1.3. Condensed Matter: Localization of Deposited Energy

Fano [10] drew attention to the fact that, derived from the uncertainty relation, an energy loss ~15 eV from a high-speed (~c) particle cannot be localized within ~90 nm. That such energy losses in a condensed medium would give rise to collective excitations (plasmons) was a source of some concern until recently. On the other hand, understanding of radiation-induced transformations, operation of detectors, etc. would require that the deposited energy be localized in a molecule before dissociation or ionization can occur. Because no reasonable explanation has yet been proposed for the localization of energy after the initial delocalization, rationalization has been advanced as the delocalization applying only to the first interaction [4]. After that, all energy losses are localized and correlated on a track [11].

Nevertheless, certain collective excitations can occur in the condensed phase. These may be brought about by longitudinal coulombic interaction (plasmons in thin films) or by transverse interaction, as in the 7-eV excitation in condensed benzene, which is believed to be an exciton [12]. Special conditions must be satisfied by the real and imaginary parts of the dielectric function of the condensed phase for collective excitations to occur. After analyzing these factors, it has been concluded that in most ordinary liquids such as water, collective excitations would not result by interaction of fast charged particles [13,14].

2. STOPPING POWER, LET, AND FLUCTUATIONS

2.1. Bohr's Theory and the Bragg Rule

The rate of energy loss of a charged particle per unit pathlength is called the *linear energy transfer* (LET). The corresponding energy received by the medium, which remains in the vicinity of the particle track, is designated as the *stopping power*. Therefore LET and stopping power refer to the penetrating particle and the medium, respectively. The difference between these quantities arises because, even in the shortest time scale (~10^{-15} s), some of the deposited energy may be removed from the track vicinity by fast secondary electrons, by Cerenkov radiation, and, in the case of ultrarelativstic particles, by bremsstralung. Cerenkov radiation is seen as a faint bluish light when the speed of the penetrating particle (usually an electron) exceeds the group velocity of light in the medium. Although it contributes negligibly to the LET, it has importance as a time marker, because it is emitted almost instantaneously. The term LET was first coined by Zirkle et al. [15] in a radiobiological context, replacing some earlier nebulous terminology. Both LET and stopping power are average concepts and considerable fluctuations are to be expected around these values. The physical theories of energy loss rate should refer to LET, but these are often called stopping power theories. The International Commission on Radiation Units and Measurements recommends the use of the symbol L_A to denote energy transfers below a limit A. Thus L_{100} would denote LET for energy losses of less than 100 eV and L_∞ would include all energy losses.

The first successful theory of stopping power is attributed to Bohr [16,17]. He takes the simplest case of a fast, yet nonrelativistic, heavy particle of velocity v and charge ze, where e is the electronic charge. Under this condition, the basic electrostatic interaction can be treated as a perturbation, that is, the particle path can be considered to be virtually undeflected. In addition, the energy losses occur through quasi-continuous inelastic collisions with medium electrons, so that the average stopping power is a good representation of the statistical process of energy loss. All stopping power theories recognize these simplifications. Neglecting the binding energy of the electron and calling the distance of closest approach of the incident particle as the impact parameter b, it may be shown that the component of momentum transfer parallel to the particle trajectory vanishes by symmetry and that, perpendicular to the trajectory, is given by $-2ze^2/bv$ [18]. The same expression is obtained by multiplying the peak force $-ze^2/b^2$ with the so-defined *duration of collision* $2b/v$. Thus the energy transfer is given by $Q = (-2ze^2/bv)^2/2m = 2z^2e^4/mb^2v^2$, where m is the electron mass. The differential cross-section for this process is given geometrically as:

$$d\sigma = -\pi d(b^2) = (2\pi z^2 e^4/mv^2)(dQ/Q^2),$$

which is just the Rutherford cross-section for scattering by free charges receiving the energy transfer Q. For application to bound electrons which can be excited (or ionized), for the same momentum transfer, to the nth state with energy E_n, Bohr surmised the sum rule $\sum_n f_n E_n = ZQ$, where Z is the atomic number and f_n is the oscillator strength of that transition (see later for proper definition). The stopping power $-dE/dx$ is the energy loss to all the medium electrons per unit path of the incident particle, over all permissible energy transfers. Therefore

$$-dE/dx = N \int d\sigma \sum_n f_n E_n = (2\pi z^2 e^4 NZ/mv^2)\ln(Q_{max}/Q_{min}). \qquad (1)$$

For the maximum energy transfer, binding effect may be ignored, giving in the non-relativistic case $Q_{max} = 2mv^2$. For the minimum energy transfer, Bohr argued that collisions must be sudden so that an effective energy loss may occur. That is, the collision time $2b/v$ must be $\lesssim \hbar/E_1$, where E_1 is atypical atomic transition energy and $\hbar$ is Planck's constant divided by 2π. This gives $b_{max} = \hbar v/2E_1$ and $Q_{min} = 8z^2e^4E_1/[(mv^2)(\hbar^2 v^2)]$. Substituting in Eq. (1), one obtains

$$-(dE/dx) = (4\pi z^2 e^4 N/mv^2)B, \qquad (2)$$

where the expression within the parentheses in Eq. (2) is called the kinematic factor and B, *denoted as the stopping number*, is given in Bohr's theory as $Z \ln[(2mv^2/E_1)(\hbar v/4ze^2)]$. Although the impact parameter is not an observable quantity, Bohr's theory has a wide range of validity if the typical transition energy E_1 is properly chosen. Equation (2) also shows the typical structure of a stopping power equation as a product of a kinematic factor and a stopping number. The kinematic factor is free from any target property, which is solely contained in the stopping number. In most stopping power expressions, the kinematic factor therefore remains the same as in Eq. (2).

As it stands, Bohr's theory is only applicable to atoms. To extend it to molecules and mixtures, one invokes the *Bragg rule*, originally used for charged particle ranges in ionization chambers. Simply stated, the Bragg rule equates the stopping number of the molecule (or the mixture) to the weighted sum of the stopping numbers of the constituting atoms. Bragg's additivity rule applies impressively within a few percent for many compounds. However, the contribution of an atomic stopping number in different compounds is not necessarily the same as that for the free atom. That this contribution remains the same in each case indicates the similarity of that atomic participation in various compounds. The apparent success of Bragg's rule has been traced to two factors: (1) similarity of atomic binding in different molecules, and (2) electronic transitions that have the most oscillator strengths involve excitation energies far in excess of chemical binding. Where such is not the case as, for example, in H_2, NO and compounds of the lightest elements, the rule does not hold well. There is evidence that the Bragg rule starts to break down at low incident energies. The quantum mechanical theories of stopping power are also amenable to this additivity rule in a similar fashion (see Sec. 2.2). However, this does not mean that the ensuing chemical response of the molecule is necessarily related to those of the constituent atoms. There are definite effects of chemical binding and state of aggregation which alter the energy levels and the oscillator strengths of transitions. The Bragg rule refers only to the stopping of the incident particle.

2.2. Bethe's Theory and Extensions

The first (and still the foremost) quantum theory of stopping, attributed to Bethe [19,20], considers the observables energy and momentum transfers as fundamental in the interaction of fast charged particles with atomic electrons. Taking the simplest case of a heavy, fast, yet nonrelativistic incident projectile, the excitation cross-section is developed in the first Born approximation; that is, the incident particle is represented as a plane wave and the scattered particle as a slightly perturbed wave. Representing the Coulombic interaction as a Fourier integral over momentum transfer, Bethe derives the differential Born cross-section for excitation to the nth quantum state of the atom as follows.

$$d\sigma_n = (2\pi z^2 e^4 / mv^2)|F_n(q)|^2 dQ/Q^2 \qquad (3)$$

In Eq. (3), q is the magnitude of the momentum transfer, $Q = q^2/2m$ is the energy of a free electron having that momentum q, and other kinematic parameters have the same significance as in the Bohr theory. The quantity, $F_n(q) = \Sigma_j \langle n|\exp(2\pi i hq \cdot r_j)|0\rangle$, called *the inelastic form factor*, is the sum over all the target electron coordinates, of the matrix elements of the exponential operator involving the vector momentum transfer q between the ground and the nth excited state. Obviously, the stopping power would be obtained from Eq. (3) as $-dE/dx = N\Sigma_n \int_{Q_{min}}^{Q_{max}} E_n d\sigma_n$, where N is the atomic number density, E_n is the transition energy, and Q_{min} and Q_{max} stand for the minimum and maximum values of Q, respectively, consistent with a given excitation energy. It is convenient to divide the range of Q into small and large momentum transfers [3]. In collisions involving small momentum transfers, sometimes called soft or *glancing collisions*, Bethe shows that $|F_n(q)|^2 \rightarrow Qf_n/E_n$, where f_n is the corresponding *dipole* (optical) *oscillator strength* (*vide infra*). This allows the contribution of such collisions to the stopping power to be written (cf. Eq. (3)) as $(2\pi z^2 e^4 N/mv^2)\Sigma_n f_n \ln \frac{Q_1}{(E_n^2/2mv^2)}$, where Q_1 is the rather inconsequential upper limit of Q for small momentum transfers, and Q_{min}, for excitation energy E_n, is given kinematically as $E_n^2/2mv^2$. For large momentum transfer collisions, sometimes designated as hard or *knock-on collisions*, Bethe directly proves the sum rule $\Sigma_n E_n|F(q)_n|^2 = ZQ$. Inserted into Eq. (3), one then obtains the contribution of hard collisions to the stopping power as $(2\pi z^2 e^4 N/mv^2)Z \ln(2mv^2/Q_1)$, where Q_1 is taken as the lower limit of Q for hard collisions and the kinematic maximum $2mv^2$ is used for Q_{max}. Noting the Thomas–Kuhn sum rule $\Sigma_n f_n = Z$, the total number of electrons in the atom, and adding the contributions of soft and hard collisions, the total stopping power is then given by

$$-dE/dx = (4\pi z^2 e^4 NZ/mv^2)\ln(2mv^2/I), \qquad (4)$$

where we used Bethe's definition of the *mean excitation potential I* of the atom through Eq. (5).

$$Z \ln I = \sum_n f_n \ln E_n \qquad (5)$$

The basic stopping power formula of Bethe has a structure similar to that of Bohr's classical theory [cf. Eq. (2)]. The kinematic factor remains the same while the stopping number is given by $B = Z \ln(2mv^2/I)$ for incident heavy, nonrelativistic particles. The Bethe theory gives a meaning to the mean excitation potential I in terms of strengths and energies of atomic transitions and underscores the importance of the dipole oscillator strength. It should be noted that, in the definition of the mean excitation potential [Eq. (5)], as well as in the derivation of the Thomas–Kuhn sum rule, excitation to energetically inaccessible states are also formally included. At a sufficiently high incident velocity for the Born approximation to be valid, it usually does not entail much error. However, it is a separate approximation. Another approximation, called *Bethe's asymptotic cross-section*, derives from expanding the interaction cross-section at high incident energies and retaining only the lead term [21]. A simplified derivation of Bethe's formula has been given by Magee [22].

Several extensions of the basic stopping power theory of Bethe can be made. For compounds and mixtures, Bragg's additivity rule may be applied with the resultant value of I given by the geometrical average of the mean excitation potentials of the constituent atoms over their electron numbers. In the case of incident electrons, Q_{max}, ignoring atomic binding, would be given by $(1/4)mv^2$ rather than by $2mv^2$, because the distinction between the incident and the secondary electron can only be made on the basis of energy. Incorporating this modification, Bethe gives $B(\text{electron}) = Z \ln(mv^2/2\sqrt{e/2I})$, where e is the base of natural logarithm. Thus the electron stopping power at the same velocity is always a little less than that of a heavy particle, sometimes by as much as 20%.

At relativistic speeds, the maximum energy transfer increases to $2mv^2/(1-\beta^2)$, where $\beta = v/c$. Also, electromagnetic effects have to be considered in addition to electrostatic interaction. These result in the addition of a term $-Z[\beta^2+\ln(1-\beta^2)]$ to the stopping number. This extra stopping, called the relativistic rise, is proportional to β^4 for small values of β. Its effect is $< 0.1\%$ for $v < 5 \times 10^9$ cm sec^{-1}, but it can be very significant when $v \sim c$. Fermi showed that the electronic stopping power would diverge were it not for the mutual polarization screening of medium electrons. This effect, therefore, is more important in the condensed phase and is called a density correction. This corrective factor, added to the stopping number, is denoted by $-Z\delta/2$, where $\delta = \ln[\hbar\omega_p^2/I^2(1-\beta^2)]-1$ and $\omega_p = (4\pi Ne^2Z/m)^{1/2}$, denoted as *the plasma frequency* [3]. Another correction is usually needed when the incident particle is not faster than the speed of atomic electrons, as required for the validity of the Born approximation. This is more important for the K electrons, but it is also sometimes needed for the L electrons in heavy elements. Denoting the total so-defined *shell correction* as C, the result is a little reduction of the stopping number by that amount. Combining the effects of relativity, density, and shell corrections, the Bethe stopping number of a heavy particle is given by

$$B = Z(\ln 2mv^2/I - \beta^2 - \ln(1 - \beta^2) - C/Z - \delta/2). \tag{6}$$

The corresponding stopping number for the electron is given by

$$B(\text{electron}) = (Z/2)\left[\ln mc^2\beta^2 E/2I^2(I - \beta^2) - \left(2\sqrt{1 - \beta^2} - 1 + \beta^2\right)\ln 2 \right.$$
$$\left. + (1 - \beta^2) + (1/8)\left(1 - \sqrt{1 - \beta^2}\right)^2 - 2C/Z - \delta\right] \tag{7}$$

where E is the electron energy at velocity βc. Because electrons can acquire relativistic speed at moderate energies, say > 20 keV, the required correction is more important for electrons.

Slow positive ions tend to capture electrons from the medium when their speeds fall below $z_1 v_0$, where z_1 is the bare nuclear charge and v_0 is Bohr's velocity. At first, the captured electron is promptly lost in a subsequent charge-changing collision. The cycle of capture and loss continues for a while and an equilibrium charge is established. The incident charge in the stopping number should then be changed from z to z_{eff}, given by $\langle z^2 \rangle^{1/2}$, where the angular brackets indicate averaging with respect to charge-changing collision cross-sections. After the first electron is firmly captured, another cycle of electron capture and loss may continue until a second electron is firmly captured, and so on. The net effect of charge changing collisions is to lessen the stopping power a little because z_{eff} is always $< z$. Even so, electron capture and loss constitute minor energy loss processes by themselves.

Despite the apparent similarity of the Bohr and the Bethe stopping power formulae, the conditions of their validity are rather complimentary than the same. Bloch [23] pointed out that Born approximation requires the incident particle velocity $v \gg ze^2/h$, the speed of a 1s electron around the incident electron while the requirement of Bohr's classical theory is exactly the opposite. For heavy, slow particles, for example, fission fragments penetrating light media, Bohr's formula has an inherent advantage, although the typical transition energy has to be taken as an adjustable parameter.

2.3. Dipole and Generalized Oscillator Strengths: Sum Rules

From the *inelastic form factor* (see Sec. 2.2), Bethe [19] defines a *generalized oscillator strength* $f_n(q) \equiv E_n F_n(q)^2/Q$, for excitation with an energy loss E_n, simultaneously with a momentum transfer q, in close association with the (optical) *dipole oscillator strength*. The latter may be obtained experimentally from the absorption coefficient $\alpha(v)$, viz. $f_n = (mc/\pi n' c^2) \int \alpha(v) dv$, where n' is the medium refractive index and the integration is carried over the frequency v for the same transition. Theoretically, it is related to the dipole moment matrix element for the transition from the ground to the nth state through the Einstein A—coefficient as $f_n = (8\pi m v/e^2 h) g_n \langle n|e \sum_j x_j|0\rangle^2$. Here g_n is the degeneracy of the excited state (the ground state is taken as nondegenerate), x_j is the x component of the position of the jth electron, and the summation is carried over all the atomic electrons. Expanding the exponential in the inelastic form factor [see Eq. (3) et seq.] for a small momentum transfer, and noting the orthogonality of the wave functions, one obtains $\lim_{q \to 0} f_n(q) = f_n$, thus establishing a relationship between the generalized and dipole oscillator strengths. Fig. 1 shows such a procedure for the electron impact excitation in helium at an incident energy of 500 eV [24]. Platzman [25] advocated the construction of a complete *optical spectrum* (f_n), and therefrom, that of the *excitation spectrum* (f_n/E_n) [cf. Eq. (3) and the definition of the generalized oscillator strength], from a few experimental determinations augmented by the various sum rules. However, the results in the cases of methane and water were not very satisfactory, probably because of the unavailability of a sufficient number of dipole sums of adequate accuracy.

Both generalized and dipole oscillator strengths satisfy sum rules, which can be used to unravel their character, although their calculation from first principles is not easy except for the lightest elements. For the dipole oscillator strength, a sum may be defined by

$$S(\mu) = \sum_n E_n^\mu f_n, \tag{8}$$

where the sum is over all the states of excitation, including integration over continuum states (here and in later discussion). $S(0) = Z$, the number of electrons in the atom or molecule. This is called the Thomas–Kuhn sum rule. Bethe [19] has shown that the same rule is satisfied by the generalized oscillator strength for *any* momentum transfer,

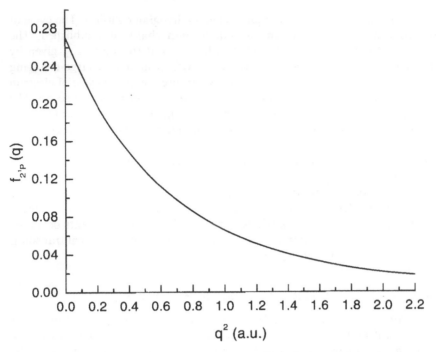

Figure 1   Generalized oscillator strength for the $1^1S \rightarrow 2^1P$ transition in helium obtained by Lassettre et al. [24], with 500-eV incident electrons. The abscissa is the square of the momentum transfer in atomic unit (1.99×10^{-19} g cm/s). Note that, according to the Lassettre procedure, $\lim_{q \rightarrow 0} f_n(q) = f_n$, the optical oscillator strength, is attained irrespective of the validity of the Born approximation. However, if that approximatiom remains valid, then the curve of $f_n(q)$ would be independent of incident energy.

i.e., $\Sigma_n E_n^\mu F_n(q) = Z$, for *any* q. Considering the oscillator strength as a normalized distribution, $S(\mu)$ in Eq. (8) may be taken as the μth moment of the transition energy. While $S(\mu)$ diverges for $\mu \geq 3$ in systems where the ground state electron density is nonvanishing at the nucleus [21], for $-4 \leq \mu \leq 2$, the sums are related to important physical properties [26]. $S(-4)$ may be experimentally obtained from the refractive index and the Verdet constant, the latter referring to the rotation of the plane of polarization per unit thickness per unit magnetic field parallel to the propagation direction. $S(-2) = \alpha/4$, where α is the polarizability. $S(2) = (16\pi Z/3)$times the average electron concentration at system center. $S(1)$ is related, apart from a constant ground state energy, to momentum correlation of electrons [26]. $S(-1)$ equals the total dipole matrix element squared, $M_{tot}^2 = \Sigma_{i,j}\langle x_i x_j \rangle$, which appears as a lead term in the total inelastic collision cross-section [27].

Another class of sum rules may be defined as logarithmic moments, or by formally differentiating $S(\mu)$ with respect to μ [see Eq. (8)], yielding

$$L(\mu) = \sum_n E_n^\mu f_n \ln E_n = S(\mu)\frac{d}{d\mu}S(\mu). \qquad (9)$$

$L(0) = Z \ln I$, where I is the mean excitation potential appearing in Bethe's stopping power equation [Eq. (4)]. $L(2)$ is proportional to the logarithm of average excitation energy, which is also involved in the Lamb shift [26]. $L(-1)$ has been shown to be an optical quantity [27]. Oscillator strength sum rules have been used with success, limited chiefly to the lightest atoms, in amending uncertain oscillator strengths and in calculating a range of physical properties including the refractive index and the Verdet constant [26]. Thus, for example, the dipole oscillator strength of the transition $2^1P \leftarrow 1^1S$ in helium was corrected by Miller and Platzman [28] to 0.277, against an earlier calculated value of 0.19.

2.4. Special Features of the Condensed Phase

One aspect of the condensed phase regarding the delocalization of the deposited energy has been alluded to in Sec. 1.3. Here we will consider the modifications on the oscillator strength and the mean excitation potential, due to condensation, which would enter in Bethe's stopping power theory [see Eqs. (4–7)].

For polyatomic molecules such as water, the oscillator strength is, in general, continuously distributed because of any combination of dissociation, ionization, etc. The stopping power is then determined by the *differential oscillator strength distribution* (DOSD). Phase effects occur naturally through the DOSD. In the gas phase, light absorption and/or inelastic scattering experiments can be used to obtain the DOSD, and a fairly complete determination has been made for water vapor [29]. In the condensed phase (solid or liquid), absorption measurement in the far-UV and beyond is experimentally very difficult. In such cases, an indirect method, depending on electromagnetic relationships, can be sometimes used if reflectance measurements are available at the vacuum–liquid (or vacuum–solid) interface over a wide range of energies. Denoting the reflectance at energy E as $R(E)$, the phase angle is obtained from the Kramers–Kronig relation, $\phi(E) = (E/\pi)P\int_0^\infty \ln R(E')dE'(E'^2 - E^2)^{-1}$, where P is Cauchy's principal value for the integral. From these, the real and imaginary parts of the refractive index are obtained as $n = (1-R)/(1 + R - 2R^{1/2} \cos \phi)$ and $k = (-2R^{1/2} \sin \phi)/(1-R-2R^{1/2} \cos \phi)$, respectively. The real and imaginary parts of the dielectric function are then given by $\varepsilon_1 = n^2 - k^2$ and $\varepsilon_2 = 2nk$, with $\text{Im}(-1/\varepsilon) = \varepsilon_2/(\varepsilon_1^2 + \varepsilon_2^2)$. The DOSD $f'(E)$ is finally calculated from the equation $\text{Im}(-1/\varepsilon) = (h^2e^2NZ/2m) f'(E)/E$. Here $f'(E)dE$ is the oscillator strength contained for transitions with energy between E and $E + dE$. In a certain sense within the context of the Bethe theory, the function $\text{Im}(-1/\varepsilon)$ is more fundamental for the interaction in the condensed phase.

There is an upper limit of experimental reflectance measurement, usually around 26 eV. With present-day synchrotron sources, this limit is being progressively extended. Nevertheless, in the global integration of the Kramers–Kronig relation, a long extrapolation is needed. In the original experiments of Heller et al. [30] on liquid water, an exponential and a power-law extrapolation beyond the upper limit of experiments were used. Later, LaVerne and Mozumder [31], requiring transparency in the visible region and the correct number of valence electrons (8.2 for water), opted for the power-law with index 3.8, close to the theoretical limit of 4.0 for valence electrons. The contribution of the K electrons remains the same as in the gas phase and are, therefore, simply added. The so-determined DOSD for liquid water is shown in Fig. 2, and compared with that of the gas phase given by Zeiss et al. [29], and with that of gaseous cyclohexane, obtained by Koizumi et al. [32] by using synchrotron UV-absorption measurement. In water, there are two main effects of condensation. First, there is a loss of structure in the condensed phase. Second, there is an upward shift of the excitation energy. The peak in the DOSD changes from ~18 eV in the gas phase to ~21 eV in the liquid phase. This means that collisions in the liquid phase are more difficult. Using the oscillator distribution, the computed values of I, the mean excitation potential is 74.9 and 71.4 eV for liquid and gaseous water,

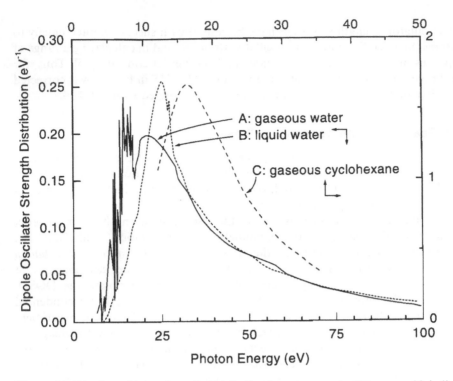

Figure 2 Dipole oscillator strength distribution in gaseous water [29, curve, A], in liquid water [31, curve, B] and in gaseous cyclohexane [32, curve, C]. Data in liquid water are obtained from an analysis of UV-reflectance and that in cyclohexane, from synchrotron–UV absorption. The Thomas–Kuhn sum rule is satisfied approximately in each case.

respectively [31]. This difference may not seem critical in determining the stopping power. However, the nature and energy of the excited states are different in the two phases, yielding different chemical outcomes (see later).

2.5. Range–Energy Relation

The range of an incident particle is defined as the average crooked path length between the initial and final energies, while penetration refers to the vector distance between the starting and ending points. The difference between these quantities is attributed to scattering, mostly elastic. Obviously, range increases with initial energy. The range–energy relation is important for the identification and energy measurement of charged particles, a tool that is routinely used in nuclear physics. In radiation chemistry, it provides a basis for charged particle tracks and the reactions on them. Various tables of ranges of light and heavy particles are now available, of which an earlier collection [33] and a later compilation by Ziegler [34] are noteworthy.

Range, obtained by (numerical) integration of the stopping power as, $R = \int_{E_0}^{E_i} dE/(-dE/dx)$, is called the continuous slowing down approximation (CSDA) range. Here E_i is the initial energy, E_0 is the energy where the particle is considered stopped, and the approximation refers to the stopping power being a continuous function of energy, ignoring random fluctuations. For electronic stopping, E_0 should be the energy of the first excited state, but a higher value is often adopted for fast incident particles, because, at low energies, the stopping power formula is not reliable and, in any case, it is superseded by subelectronic processes. Such an arbitrary value of E_0 becomes inconsequential if the incident energy is sufficiently high.

Often, the range–energy relation is displayed as a power law, $R \propto E^m$, where the index m would approach 2 if the slow variation of the stopping number B with particle velocity is ignored [see Eq. (4) et seq.]. For protons and He ions of a few hundred MeV, m approaches 1.8, decreasing somewhat with decreasing energy. The index m also depends on the medium traversed. For electrons in water, it diminishes gradually from ~1.7 to ~0.9 as the initial energy is decreased from a few thousand to a few hundred eV. Low-energy electron range in water is of great importance to the spur theory of radiation chemistry, but it is also rather difficult to ascertain and often greatly influenced by elastic scattering. This field is being continually improved and track simulation techniques have been employed [35], which give not only the mean range, but also its distribution (see the next section). However, very little is known about elastic scattering in liquid water, forcing many investigators to use gas phase data for that purpose. Typical values of computed ranges in water are ~550 nm for 5-keV electrons, ~0.5 mm for 5-MeV protons, and ~0.04 mm for 5-MeV α-particles. Ranges of slow heavy ions and fission fragments are nearly proportional to energy, because the increase of stopping power in relation to slowing down is largely compensated by a lowering of effective charge as a result of electron capture, thus providing an almost constant stopping power. The typical density-normalized range of Ne ions of 10 MeV/amu in Al is 40 mg/cm^2.

2.6. Discussion of Stopping Power, Range, and Fluctuations

The most important factor in Bethe's stopping power equation, which embodies the aggregate effect of transition energies and oscillator strengths, is the mean excitation potential, I [see Eqs. (4) and (5)]. Its direct calculation requires the knowledge of the ground and all excited state wave functions. Such a calculation for atomic hydrogen gives $I = 15$ eV, whereas in atomic binding in different molecules, a value of 18 eV is preferred to satisfy the Bragg rule [3]. For other atoms, application of the Thomas–Fermi model by Bloch shows that the spectral distribution of the oscillator strength has universal shape in all atoms, if the transition energy is scaled by Z. This means $I \propto Z$, with the constant of proportionality being ~12 eV. A detailed examination of the Thomas–Fermi–Dirac model gives $I/Z = a + bZ^{-2/3}$, with $a = 9.2$ and $b = 4.5$ as best adjusted values, and where I is expressed in eV. This equation agrees rather well with range experiments. However, Fano [3] points out that it should not be construed as a validation of the statistical model of the atom. I, being a logarithmically averaged quantity, can be well approximated without an accurate knowledge of the distribution. An error δI in determining I only appears as a relative error of ~$(1/5)\delta I/I$ in the evaluation of range. In practice, I is often treated as an adjustable parameter to be fixed by range measurement of fast protons or α-particles. With known I values of constituent atoms, the mean excitation potential of a molecule may be obtained by the application of the Bragg rule (see Sec. 2.1). Thus if the molecule has n_I atoms of atomic number Z_I, and mean excitation potential I_I, then the overall mean excitation potential of the molecule would be given by the relation $Z \ln I = \Sigma n_i Z_i \ln I_i$, where $Z = \Sigma n_i Z_i$ is the total number of electrons in the molecule.

The stopping power of water, over a wide span of energy, for various incident particles has been discussed elsewhere [Ref. [4], Sec. 6], with special consideration for effects at low and very high energies. Generally speaking, the electronic stopping power shows a peak at a relatively low energy (~100 eV for incident electrons and ~1 MeV for incident protons), as a result of the combined effect of the velocity denominator in the kinematic factor and the logarithmic velocity dependence in the stopping number [c.f., Eq. (4) et seq.].

The usual stopping power formula for low-energy electrons (Born–Bethe approximation) requires corrections on two accounts. First, the integrals defining the total oscillator strength and the mean excitation potential must be truncated at the maximum energy transfer ($\approx (1/4)mv^2$), resulting in the replacement of Z and I by Z_{eff} and I_{eff}, respectively[36]. These quantities are lower than their respective asymptotic values and their effects partially cancel each other. Second, the effect of the nonvanishing of the momentum transfer for inelastic collisions becomes significant. When a quadratic extension of the generalized oscillator strength is made in the energy-momentum plane still using the DOSD, the net effect is expressible as a reduction of the electron stopping number by an amount $Z_{eff}(\bar{\varepsilon}/2E)\ln(4\tilde{e}E/\bar{I})$, where $\tilde{e}$ is the base of natural logarithm, E is incident electron energy, ε is average energy loss in an encounter, and $\bar{I}$ is a mean energy defined by $Z_{eff}\tilde{e}\ln\bar{I} = \int_0^{\varepsilon_{max}} \varepsilon \ln \varepsilon f(\varepsilon)d\varepsilon$ [36]. Here ε_{max} is the maximum transferable energy. In track simulation procedure, it is more convenient to work with basic cross-sections (with which the stopping power is calculated) and keep a detailed account of energy loss, distance traversed, etc. The stopping power may then be computed from definition by averaging over a large number (typically $\sim 10^4$) of simulations. The results of such a calculation by Pimblott et al. [35] is shown in Fig. 3 for water in the gaseous and liquid phases. The difference, of course, is directly traceable to that in the DOSD as discussed earlier.

Even if the statistical fluctuation of energy loss is ignored, there would be a difference between the CSDA range and the penetration, because of large-angle elastic scattering (see Sec. 2.5). The distribution of the final position vector of the penetrating particle can be obtained in terms of its energy-dependent stopping power and the differential cross-section of scattering [36]. Actual calculation in the general case is complex, but it simplifies considerably when the number of scatterings are $\gg 1$. The distribution can then be shown as gaussian and spheroidal, i.e., $W(r) = (2\pi A)^{-1}(2\pi B)^{-1/2} \exp[-\{(x^2+y^2)/2A + (z-\langle z\rangle)^2/2B\}]$, where $A = \langle x^2\rangle$ and $B = \langle z^2\rangle - \langle z\rangle^2$, and the angular brackets refer to mean values. The mean square radial penetration is then given by $\langle r^2\rangle = 2\pi\int_0^\infty r^4 dr\int_0^\pi W(r)\sin\theta d\theta$, where θ is the angle of the position vector relative to the z-axis, the initial direction of motion. The root mean square (rms) range, defined by $\langle r^2\rangle^{1/2}$ is often the more significant quantity. Fig. 4 shows the CDSA and rms ranges in gaseous water as functions of electron energy according to LaVerne and Mozumder [36]. In these calculations, stopping power was computed by the procedure detailed above. The scattering cross-section was fitted to swarm and beam data by adopting Moliere's modification of screened Rutherford scattering. It is obvious that the rms range is always shorter than the CSDA range, the difference becoming more significant at lower energies. The following general observations can be made for the penetration distribution. (1) The memory of the initial direction is maintained by the electron until it has lost $\sim 80\%$ of the starting energy, after which the distribution becomes spherical. (2) The electron penetrates a certain distance with only a few scatterings and then diffusive motion sets in. (3) The number of elastic scatterings needed to give a spherical distribution increases with energy, being ~ 15 and ~ 74, respectively, for electrons of 1 and 10 keV energy in water.

Because of the statistical nature of collision processes, charged particles starting with the same energy do not travel the same distance after losing a fixed amount of energy, nor do they have the same energy after traveling a fixed distance, respectively giving rise to *pathlength straggling* and *energy straggling*. Fano [3] adds a third kind of distribution of distance for particles that have dropped below a certain energy at the last collision, which

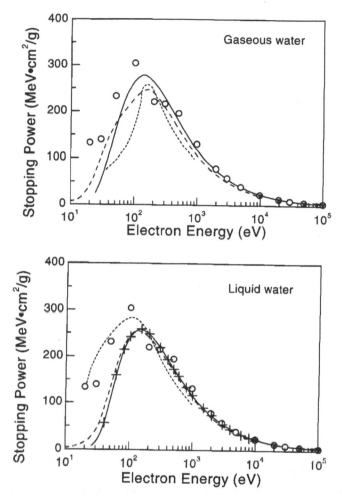

Figure 3 Density-normalized stopping power (MeV•cm^2/g) of gaseous and liquid water as a function of electron energy according to track simulation by Pimblott et al. [35]. There is a noticeable phase effect, while a peak is seen at ~100 eV in both phases.

is not necessarily the same as pathlength straggling, because of the discrete nature of the energy loss process. While all these distributions are related, the most important distribution is *range straggling* when the final energy is reduced to insignificance. Bohr [17] considers a group of energy losses and a large number of loss processes in each group, thereby obtaining a gaussian distribution by virtue of the central limit theorem. This means $(d/dx)\langle \Delta E^2 \rangle = 4\pi e^4 z^2 NZ$, where $\langle \Delta E^2 \rangle$ is the mean square fluctuation in energy loss of particles after traversing a distance x. The corresponding pathlength and range distributions are also gaussian. These considerations apply mainly to heavy particles of relatively high energy penetrating virgin media. The ratio of the root mean square range dispersion to the mean range is typically a few percent for a heavy particle, but may be a few tens of percent for electrons of moderate energy. Straggling is always more important for electrons of any energy because of the large fractional energy that an electron can lose in a single encounter. Also, if the sample penetrated is thin, the resultant energy

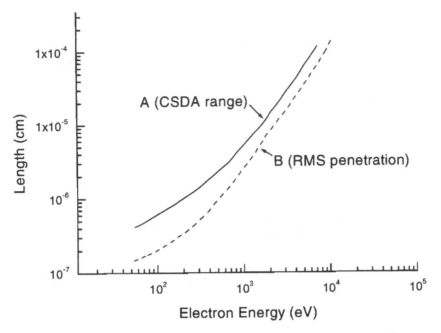

Figure 4 Continuous slowing down (CSDA) range and root mean square (rms) penetration in gaseous water according to LaVerne and Mozumder [36]. See text for details.

distribution of emergent particles of any kind will not be gaussian, but will show a high energy tail. Such *Landau–Vavilov distribution* has been extensively tabulated [33], having experimental justification. Its application to radiation biology derives from the fact that the radiation effect in a cell depends on the energy deposited in it, while the cell size is often small compared to the particle range [37].

A collision-by-collision approach was introduced by Mozumder and LaVerne [38] for range straggling of low-energy electrons without the consideration of large-angle elastic scattering. The *probability density* of pathlength x for the first collision is given by $P_1(x) = \Lambda^{-1} \exp(-x/\Lambda)$, where the mean free path Λ is computed from the same differential cross-section of energy loss as used in the stopping power (vide supra). The integral of the energy-loss weighted differential cross-section $\langle \varepsilon \rangle$ gives the mean energy loss in a single inelastic encounter. The ratio, $\langle \varepsilon \rangle / \Lambda$, gives the spatial rate of energy loss for that encounter, while the true stopping power is the mean value of the ratio of energy loss to path length. To this extent, the procedure is an approximation. To have a pathlength distribution $P_n(x)$ in n collisions, the electron must have had a distribution $P_{n-1}(x-y)$ in $(n-1)$ collisions and a distribution $P_1(y)$ in the last collision. This generates a convolution, $P_n(x) = \int_0^x P_{n-1}(x - y) P_1(y) dy$. By repeatedly using the Laplace transform method, one can evaluate the integral with the result: $P_n(x) = \Sigma_{i=1}^n \Lambda_i^{-1} S_i^n \exp(-x/\Lambda_i)$, where $S_i^n = \Pi_{j=1}^n (1 - \Lambda_j/\Lambda_i)^{-1} (j \neq 1)$. Note that the mean free paths are energy-dependent. Therefore a concurrent use of the stopping power is required to give the electron energy after each collision. To obtain the range distribution, the total number of collisions is so adjusted that the mean final energy becomes immaterial. Thus, as shown in Fig. 5, Mozumder and LaVerne [38] give the range distribution of 200-eV electrons in N_2 by convoluting the pathlength distributions between energies 200 to 110 eV and between energies 110 to 31 eV. In this case, the final energy has been taken to be twice the ionization potential. Notice

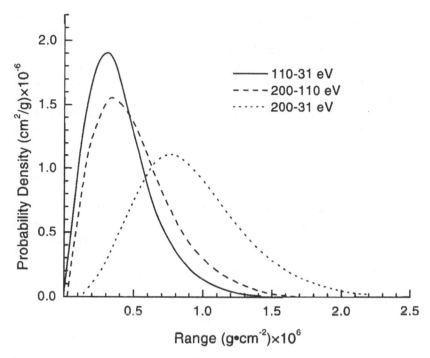

Figure 5 Range distribution in nitrogen for 200-eV electrons obtained via the convolution method at 110 eV [38]. The final energy is about twice the ionization potential. See text for details.

that, in these calculations, the fundamental input still remains to be the DOSD. The following remarks may be made about the range distribution of low-energy electrons. (1) For electron energies below about 5 keV, CSDA > mean > median range. CSDA range is the least reliable because it considers energy losses in a continuous manner within the mean free path. (2) At low energies, the distribution is skewed toward longer ranges. The skewness parameter approaches the gaussian limit of unity above ~ 10 keV, almost independent of the material traversed. (3) The most probable density-normalized range for a 1-keV electron in H_2O, N_2 and O_2 is 6.1, 7.1, and 8.4 $\mu g/cm^2$, respectively. However, relative straggling, given by the ratio of FWHM of the range distribution to the most probable value, is nearly independent of the medium, being 1.4, 0.7, and 0.4 at energies of 100, 500, and 2000 eV, respectively.

So far, the discussion presented here is based either on penetration deriving from large angle elastic scattering, or on the statistical nature of inelastic collisions giving straggling. As yet, there has not been any analytic theory that combines both, but Monte Carlo simulation, including some form of realistic differential cross-section of elastic scattering, shows promise [35].

3. SECONDARY IONIZATION

3.1. Cross-Sections

Ionizations produced by electrons generated in the primary interaction of the incident fast charged particle are called secondary. The definition may be extended to all successive generations if sufficient energy is available. In this sense, secondary ionization is very important and it is a ubiquitous feature of all fast primary charged particles. Typically, ~25% of all ionizations are produced in secondary processes. Of course, experiments count all ionizations, irrespective of whether they are primary or secondary.

The importance of *total* ionization yield derives from dosimetry and from providing a standard against which other radiation–chemical yields may be discussed [25]. For the complete description of impact ionization, a fivefold differential cross-section is required in principle; one for ejected electron energy and two each for the angular distributions of the ejected electron and the incident particle. Such a detailed description is rarely needed. In radiation chemical track structure, a cross-section doubly differential in the energy and angle of the ejected electron is sufficient. Moreover, the Born approximation, against which measurements and numerical calculations may be compared, starts to break down as the incident energy is decreased. The onset for this breakdown occurs at a greater energy if higher order differential cross-sections are considered. A special issue of Radiation Research [39] contains articles written by several experts on the experimental and theoretical aspects of impact ionization, including higher order differential cross-sections. The calculation of the impact ionization cross-section for incident velocity v proceeds, in the Born–Bethe approximation, in the same manner as that for energy loss (see Sec. 2.2). In effect, energy loss is written as $W_i = E + I_i$, where E is the secondary electron energy and I_i is the ionization potential for the ith orbital from which the electron is ejected. Combining the contribution from all the orbitals in the atom or molecule, the differential ionization cross-section for ionization may be given as follows [40].

$$d\sigma/dE = (4\pi a_0^2 z^2 R^2/T) \sum_i W_i^{-1}(df/dW_i)\ln(4c_iT/R) \qquad (10)$$

In Eq. (10), R is the Rydberg energy (13.6 eV), a_0, the Bohr radius (0.053 nm), $T = mv^2/2$, df/dW_i is the differential dipole oscillator strength for the ionization process, and c_i is a defined additional contribution of the generalized oscillator strength [21]. Note that this cross-section is basically given by the dipole oscillator strength, while there is a weak dependence on the generalized oscillator strength through c_i. Platzman has advocated the use of the ratio of ionization cross-section to the Rutherford cross-section, $Y(T,W) = (d\sigma/dE)(T/4\pi a_0^2 z^2)(W/R)^2$, and its plot vs. R/W is called the Platzman plot [see Eq. (19) et seq. of Chap. 3]. Here $W = E + B_1$ and B_1 is the lowest binding energy of the electron in the molecular orbital. The usefulness of such a plot rests on: (1) $\lim_{W \to \infty} = Z$, the total number of electrons, (2) autoionization and Auger processes show up as dips or peaks, (3) for small values of E, Y resembles the shape of $W(df/dW)$, and (4) in proton impact, a peak is seen at ejected electron speed equal to the speed of the proton. Furthermore, the area under the Platzman plot is proportional to the total ionization cross-section, the constant of proportionality being $T/4\pi a_0^2 R$. Kim [40] has made repeated use of the Platzman plot in a careful analysis of secondary electron spectra, in some cases correcting some experimental values. Another plot, $(T/R)(d\sigma/dE)$ vs. $\ln(T/R)$, called the Fano plot, is useful in extrapolation and interpolation of experimental data. It approaches a straight line at high T.

The above considerations need relativistic correction at $v \sim c$, which may be performed in a straightforward manner. More importantly, Eq. (10) assumes that the ionization process is direct, i.e., once a state above the ionization potential is reached, ionization occurs with a certainty. Platzman [25] points out that in molecules, this is not necessarily so and superexcited states with energy exceeding the ionization potential may exist, which will dissociate into neutral fragments with a certain probability. For example, in water in the gas phase, ionization occurs with a sharp threshold at the ionization potential (I.P.) = 12.6 eV, but only with an efficiency of 0.4. Beyond the I.P., the ionization efficiency, determined by the ratio of ionization to the photoabsorption cross-section, varies with the energy until ~18 eV, beyond which it approaches unity [41]. In such cases, the cross-section formula given in Eq. (10) must be suitably modified to account for the ionization efficiency. There is a very definite phase effect on the ionization efficiency, because of which the existence of superexcited states in liquid water has recently been questioned [42]. The typical values of proton impact ionization cross-section in the gas phase, for H, He, H_2, N_2, CO_2, NH_3, and CH_4 are 1.12, 0.93, 1.96, 5.33, 8.11, 4.62 and 6.84 A^2, respectively, at 100 keV, and 0.19, 0.21, 0.38, 1.46, 2.13, 1.30, and 1.66, respectively, at 1 MeV. Rudd etal. [43] have given an empirical formula for the total ionization cross-section due to a heavy ion at energy U, viz. $\sigma_t^{-1} = \sigma_l^{-1} + \sigma_h^{-1}$, where $\sigma_h(A^2) = 3.52(R/U)(A \ln[1 + U/R] + B)$ and $\sigma_l(A^2) = 3.52C(U/R)^D$. This form has the correct asymptotic behavior given by the Born–Bethe cross-section at high energy. The parameters A, B, C, and D have been tabulated, and a similar formula for incident electrons is also available. The corresponding secondary electron spectrum involving some 10 adjustable parameters has been empirically found to be very useful. It is discussed in some detail in Chap. 3 [Eqs. (29)–(35)].

3.2. The Degradation Spectrum

The complexity of electrons of different energy and generation actually present in an irradiated system, together with their primary counterparts, if different from the electron, is represented by the degradation spectrum, which is defined to be the normalized energy spectrum of these particles flowing through the surface of a small cavity inside the medium. Spencer and Fano [44] show that a convenient measure of this spectrum is given by $y(T_0, T)$, such that $y(T)dT$ is the total mean distance traveled by electrons of all generations (primary, secondary, etc.) between energies T and $T + dT$, with T_0 being the initial energy at unit flux. The degradation spectrum serves a purpose similar to the distribution function in statistical mechanics, in the sense that the yield of *any* primary species x is given by $N_x(T_0) = N\int_{I_x}^{T_0} \sigma_x(T)y(T_0, T)dT$, where σ_x is the cross-section for the production of x with an energetic threshold I_x (e.g., for ionization). Note that y needs to be evaluated only once for each initial energy, after which it may be used for all products. Although y depends on T_0, some approximate scaling property of the electron degradation spectrum has been noted [45].

Calculation of the degradation spectrum has proved to be difficult in all but the simplest cases of H, He, and H_2, because of the lack of accurate energy-loss cross-sections (mainly nonionizing) over a wide span of energy. Furthermore, to calculate the yield of a particular species, its own production cross-section must be known over a wide interval of energy, which is rarely available. In the continuous slowing down approximation (CSDA), the contribution to the degradation spectrum of a single electron is the reciprocal of the stopping power. Therefore after knowing the ionization cross-section and the electron stopping power, one can compute the partial degradation spectrum generation by generation. Finally, the total degradation spectrum may be obtained by summing up. Such a procedure has been followed by Kowari and Sato [46] for gaseous water. Apart from demonstrating the methodology, the actual results are of limited use because of various dubious cross-sections used. More sophisticated methods employ the Spencer–Fano equation [45], or the Fowler equation [47]; however, the problem of accurate cross-sections still remain in most cases. Another numerical approach by Turner et al. [48] relies on Monte Carlo simulation. Their computed spectra show considerable difference between the gas and liquid phases of water, especially in the low energy region, which is a direct consequence of the peculiar kinds of cross-sections used by them in these phases. One feature is common to all electron degradation spectra. It has large values at the high starting energy and also at very low energies. The former is attributed to low stopping power and the latter is attributed to the accumulation of electrons of different generations. Therefore the spectrum is expected to show a drop at some intermediate energy, often near where the stopping power goes through a broad peak. It should be noted that, like the stopping power and the range, the degradation spectrum also refers to mean values, and considerable fluctuation may be expected, especially at higher energies.

3.3. The *W* Value

The reciprocal of the ratio of the mean number of ionizations n_i produced by the complete absorption of a primary particle of energy T is defined as the (integral) W value. Ionizations produced by all secondary electrons are counted in it. A differential, or track segment ω value, may be similarly defined, so that $1/\omega = (d/dT)(T/W)$. If W is independent of T, so is ω and vice versa. Generally, W depends weakly on T, unless the initial energy is very low. Its dependence on the nature of the incident radiation (electrons, alpha-particles, etc.) is also minor. These factors constitute the basis of dosimetry, by collecting and measuring the total ionization. It also means that the Bragg curve, or the variation of total ionization with incident particle energy, is a reliable measure of the relative stopping power. Thus the W value is mainly a property of the medium, determined by the cross-sections of ionization and of other nonionizing energy-loss processes. As such, the W value often depends on the phase.

Platzman's [49] analysis for the W value may given as $W = \langle E_{\text{ion}} \rangle + v_{\text{ex}} \langle E_{\text{ex}} \rangle + \langle E_s \rangle$, where $\langle E_{\text{ion}} \rangle$, $\langle E_{\text{ex}} \rangle$, and $\langle E_s \rangle$ are the average energies for ionization, excitation, and subexcitation electrons, respectively, and v_{ex} is the relative number of excitations to ionizations. It provides a basis for understanding the insensitivity of W on particle energy, because the quantities involved depend not so much on absolute cross-sections for ionization and nonionizing energy loss, as it depends on their ratio.

An early theory of the W value was proffered by Spencer and Fano [44], based on the degradation spectrum. Another method, the Fowler equation, was employed by Inokuti [47] for electron irradiation, based on the approximation that there is only one ionization potential and that the ionization efficiency is unity. These restrictions can be relaxed. The main result of Inokuti's analysis may be given as follows.

$$W(T) = W_\infty (1 - U/T)^{-1}, \text{ where } U = E' - I. \tag{11}$$

In Eq. (11), W_∞ is the asymptotic W value at very high energy, I is the ionization potential, U is a so-defined energy parameter, and E' is interpreted as the average energy transfer in ionizing collisions where the ejected electron is subionization, i.e., incapable of further ionization. Thus, U is the excess of E' over the ionization potential. An expression of E' may be given in terms of the differential probabilities of ionization with energy loss between E and $E + dE$, viz. $E' = \int_I^{2I} [E dp_i(E, T)/dE] dE / \int_I^{2I} [dp_i(E, T)/dE] dE$, where these probabilities are simply the ratios of the relevant ionization cross-sections to the total energy loss cross-section. Eq. (11) impressively applies for various media over a wide span of energy, except perhaps at very low energy.

Another numerical method devised by LaVerne and Mozumder [50] has been applied to gaseous water under electron and proton irradiation. Considering a small section of the track, *the W value due to the primary particle only*, may be written as $\omega_P = S(E)/N\sigma_i(E)$, where $S(E)$ is the stopping power, N is the molecular density, and $\sigma_i(E)$ is the total ionization cross-section at energy E. The number of secondary electrons produced, in all generations, per unit track length is given by $N\int_I^{\varepsilon_m} (d\sigma_i/d\varepsilon)[\varepsilon/W(\varepsilon)]d\varepsilon$, where $W(\varepsilon)$ is the integral W value for electrons at energy ε, $d\sigma_i(\varepsilon)/d\varepsilon$ is the differential cross-section for producing secondary electrons of energy ε, and ε_m is the maximum secondary electron energy often well approximated by $(E-I)/2$. Combining, the differential ω value at electron energy E is given by $\omega(E) = S(E)/N[\sigma_i + \int_I^{\varepsilon_m}(d\sigma_i/d\varepsilon)\varepsilon W^{-1}(\varepsilon)d\varepsilon]$. From this equation, the overall W value may be calculated through integration by using the stopping power, ionization cross-section, and the integral W values obtained at lower energies. *A building-up principle* has been used by dividing the incident energy into several intervals, I–3I, 3I–7I, 7I–15I, and so on. In the first interval, no secondary ionization is possible and $\omega(E) = \omega_P(E)$, although an integration is necessary to compute $W(E)$. In subsequent intervals, the $W(E)$ values of previous intervals may be used together with stopping power and ionization cross-section obtained either from experiments or tabulations. By careful analysis of experiments and compilations, LaVerne and Mozumder [50] made the following conclusions. (1) Various experiments on the ionization cross-section are basically in agreement with each other, but those on the excitation cross-section are not, which is the major source of uncertainty in the calculation of W value. (2) For electron irradiation, an asymptotic W_∞ is reached at 1 keV, but the computed value (34.7 eV) exceeds measurement [51] (29.6 eV) by 15%, attributable to errors in inelastic cross-sections. (3) For proton irradiation, W_∞ is reached at 500 keV and the computed value (28.9 eV) is in good agreement with experiment [51] (30.5 eV). In both cases, ω_P greatly exceeds W, underscoring the importance of secondary ionization.

Cole [52] measured the W value in air for electron energies from 5 to 20 eV, while the electrons were completely absorbed in the ionization chamber. Later, Combecher [53] extended the measurements of $W(E)$ to several gases including water vapor. Fig. 6 shows the variation of $W(E)$ with electron energy in water vapor, as measured by Combecher,

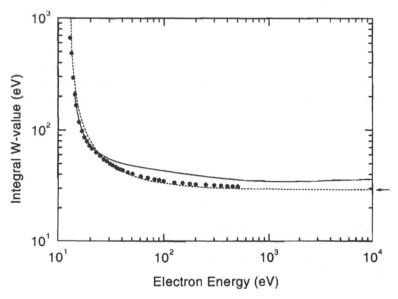

Figure 6 Comparison of experimental (...) variation of integral W value for water vapor [50] with theoretical calculation (___) and with an empirical model (---). Note that the accuracy of the theoretical calculation is limited by an inherent error in inelastic collision cross-sections, while in the empirical model, $U = 12.6$ eV has been adjusted to obtain best agreement with the experiment. The experimentally determined asymptotic limit, $W_\infty = 29.6$ eV is indicated by the arrow.

and compares it with the calculation of LaVerne and Mozumder and also with Inokuti's result, obtaining U = 12.6 eV, close to the ionization potential, for best adjustment.

The effect of condensation on the W value has been treated elsewhere in some detail [54]. Except for liquefied rare gases, where the W value in the liquid is somewhat less than that in the gas phase, there is an operational problem in the definition of W—thus requiring long extrapolation. In some hydrocarbons such as cyclohexane, the admittedly long extrapolation of scavenging yield gives a W value not far from the level obtained in the gas phase ($\sim$26 eV for cyclohexane). In most polar liquids, W_{liq} (22 $\pm$ 1 eV for water) is considerably less than W_{gas} (31 $\pm$ 1 eV for water). The exact reason for the additional yield of ionization in the liquid phase is not clear, but a significant contributing factor could be the conversion of excited states into ionization by interaction with neighboring ground state molecules in the liquid phase.

As for the stopping power and the degradation spectrum, the W value is also a mean quantity and ionization fluctuation is expected because of the various ways a deposited energy can be partitioned into ionizing and nonionizing events. The Fano factor represents such fluctuations, and it is defined as the ratio of mean square fluctuation of ionization to the average number of ionizations, i.e., $F = (\langle n_i^2 \rangle - \langle n_i \rangle^2)/\langle n_i \rangle$, where n_I is the number of ionizations detected in a sample. For a Poisson distribution, $F = 1$; for other distributions, it is < 1. For the operation of a radiation detector, a smaller Fano factor means better energy resolution. With solid state detectors using Si or Ge, Fano factors of $\sim$0.05 can be achieved. Other liquefied rare gas detectors have Fano factors in the range $\sim$0.1 to $\sim$0.2.

4. CONCLUSION

In this chapter, we have attempted to describe the interaction of fast charged particles with matter from the point of view of the incident radiation, i.e., basically radiation physics. We have discussed such topics as energy transfer from incident radiation to ionized and excited states of medium molecules, as well as theories of stopping power and secondary ionization effects. In many of the succeeding chapters, topics which are more chemical in nature will be found, i.e., the response of the matter receiving the transferred energy. In this sense, this chapter may be considered as preparatory groundwork for the succeeding contributions.

REFERENCES

1. Tabata, Y.; Ito, Y.; Tagawa, S., Eds. CRC Handbook of Radiation Chemistry. Boca Raton, FL: CRC Press, 1991, chaps. 2 and 3.
2. Bethe, H.A.; Heitler, W. Proc. Roy. Soc. A 1934, *146*, 83.
3. Fano, U. Annual Review Nuclear Science 1963, *13*, 1.
4. Mozumder, A. In *Advances of Radiation Chemistry*; Burton, M.; Magee, J.L., Eds.; Wiley-Interscience: New York, 1969, chap. 1.
5. Ward, J.F. In *Advances in Radiation Biology*; Lett, J.T.; Adler, H., Eds; Academic Press: New York, Vol. 5,181–239.
6. Land, L.L.; Hanrahan, R.J. In *Environmental Applications of Ionizing Radiation*; Cooper, W.J.; Curry, R.D.; O'Shea, K.E.; Eds.; John Wiley: New York, 1998, chap. 24.
7. Woods, R.J.; Pikaev, A.K. Applied Radiation Chemistry. John Wiley: New York, 1994, chap. 10.
8. Rossi, H.H. In *Radiation Dosimetry*, Attix, F.H.; Roesch, W.C., Eds.; Academic Press: New York, 1968, 43–92.

9. Kellerer, A.M.; Rossi, H.H. In *Proceedings of 2nd Symposium on Microdosimetry*, Ebert, H.G., Ed.; Euratom: Brussels, 1969, 843–853.
10. Fano, U. *Comparative Effects of Radiations*; Burton, M.; Kirby-Smith, J.S.; Magee, J.L., Eds.; John Wiley: New York, 1960, 14–21.
11. Mott, N.F. Proc. Roy. Soc. A 1930, *126*, 79.
12. Killat, U. J. Phys. C 1974, *7*, 2396.
13. Fano, U. Revs. Mod. Phys. 1992, *64*, 313.
14. LaVerne, J.A.; Mozumder, A. Radiat. Res. 1993, *133*, 282.
15. Zirkle, R.E.; Marchbank, D.F.; Kuck, K.D. J. Cell. Comp. Physiol. 1952, *39* (suppl. 1), 75.
16. Bohr, N. Philos. Mag. 1913, *25*, 10.
17. Bohr, N. Kgl. Dan. Vied. Selskab. Math.-Fys. Medd. 1948, *18*, 9.
18. See, for example, Kuppermann, A. J. Chem. Ed. 1959, *36*, 279.
19. Bethe, H.A. Ann. Physik. 1930, *5*, 325; 1933, *24*, 273.
20. Bethe, H.A. Z. Physik. 1932, *76*, 293.
21. Inokuti, M. Rev. Mod. Phys. 1971, *43*, 297.
22. Magee, J.L. Annu. Rev. Phys. Chem. 1961, *12*, 389.
23. Bloch, F. Ann. Physik. 1933, *16*, 285.
24. Lassettre, E.N.; Skerbele, A.; Dillon, M.A. J. Chem. Phys. 1969, *50*, 1829.
25. Platzman, R.L. In *Radiation Research*, Silini, G., ed. Amsterdam: North Holland, 1967, 20.
26. Nicholls, R.W.; Stewart, A.L. In *Atomic and Molecular Processes*, Bates, D.R., Ed.; Academic Press: New York, 1962, 47.
27. Inokuti, M.; Kim, Y.-K.; Platzman, R.L. Phys. Rev. 1967, *164*, 55.
28. Miller, W.F.; Platzman, R.L. Proc. Phys. Soc. (London) A 1957, *70*, 299.
29. Zeiss, G.D.; Meath, W.J.; Macdonald, J.C.F.; Dawson, D.J. Radiat. Res. 1975, *63*, 64.
30. Heller, J.M.; Hamm, R.N.; Birkhoff, R.D.; Painter, L.R. J. Chem. Phys. 1974, *60*, 3483.
31. LaVerne, J.A.; Mozumder, A. J. Phys. Chem. 1986, *90*, 3242.
32. Koizumi, H.; Sinsaka, K.; Hatano, Y. Radiat. Phys. Chem. 1989, *34*, 87.
33. Studies in Penetration of Charged Particles in Matter. National Academy of Sciences–National Research Council: Washington, DC, 1964, (publication 1133).
34. *The Stopping and Ranges of Ions in Matter*; Ziegler, J.F., Ed.; Pergamon Press: Elmsford, New York, 1980–1985, *1–6*.
35. Pimblott, S.M.; LaVerne, J.A.; Mozumder, A. J. Phys. Chem. 1996, *100*, 8595.
36. LaVerne, J.A.; Mozumder, A. Radiat. Res. 1983, *96*, 219.
37. Maccabee, H.D.; Raju, M.R.; Tobias, C.A. Phys. Rev. 1968, *165*, 469.
38. Mozumder, A.; LaVerne, J.A. J. Phys. Chem. 1985, *89*, 930.
39. Various, Radiat. Res. 1975, *64*, 1–204.
40. Kim, Y.-K. Radiat. Res. 1975, *64*, 96.
41. Haddad, G.N.; Simpson, J.A.R. J. Chem. Phys. 1986, *84*, 6623.
42. Mozumder, A. Phys. Chem. Chem. Phys. 2002, *4*, 1451.
43. Rudd, M.E.; Kim, Y.-K.; Madison, D.H.; Gallagher, J.W. Rev. Mod. Phys. 1985, *57*, 965.
44. Spencer, L.V.; Fano, U. Phys. Rev. 1954, *93*, 1172.
45. Fano, U.; Spencer, L.V. Int. J. Radiat. Phys. Chem. 1975, *7*, 63.
46. Kowari, K.; Sato, S. Bull. Chem. Soc. Jpn. 1978, *51*, 741.
47. Inokuti, M. Radiat. Res. 1975, *64*, 6.
48. Turner, J.E.; Paretzke, H.G.; Hamm, R.N.; Wright, H.A.; Ritchie, R.H. Radiat. Res. 1982, *92*, 47.
49. Platzman, R.L. Int. J. Appl. Radiat. Isot. 1961, *10*, 116.
50. LaVerne, J.A.; Mozumder, A. Radiat. Res. 1992, *131*, 1.
51. Christophorou, L.G. *Atomic and Molecular Radiation Physics*; Wiley-Interscience: London, 1971.
52. Cole, A. Radiat. Res. 1969, *38*, 7.
53. Combecher, D. Radiat. Res. 1980, *84*, 189.
54. Mozumder, A. *Fundamentals of Radiation Chemistry*; Academic Press: San Diego, 1999.

3

Ionization and Secondary Electron Production by Fast Charged Particles

Larry H. Toburen
East Carolina University, Greenville, North Carolina, U.S.A.

1. INTRODUCTION

From the first studies of cathode rays leading to the discovery of the electron in 1897 by J.J. Thomson, there has been intense interest in understanding the characteristics of the interactions of fast charged particles with matter. Early studies of electron impact excitation of gases led to important constraints on evolving models of atomic structure as well as contributing to the discovery of x-rays and natural radioactivity. With the discovery of energetic alpha particles emanating from natural radioactivity, Rutherford and his coworkers were able to investigate heavy charged-particle scattering in materials and gain new insights into the physical nature of atomic structure and of charged particle scattering. These studies also stimulated new questions regarding the mechanisms of energy loss by heavy charged particles in matter.

The first decades of the 20th century saw great advances in our knowledge of charged particle interactions with matter. With the discovery of nuclear fission came enhanced availability of radionuclides and nuclear radiations, including fast fission fragments, neutrons, alpha and beta particles. Coincidently, with the development of nuclear technology came the need to better understand the physical, chemical, and biological interactions of radiation with matter. The potential exposure of materials and people to nuclear radiation emphasized a need to understand the underlying mechanisms leading to damage by radiations of different types. The early part of the 20th century was also a period of rapidly expanding technology enabling one to produce and accelerate electrons and ions and to explore new applications of radiation in science and industry. Understanding of the interaction of radiation with matter has important applications in numerous research and applied fields of endeavor, including the sterilization of food and medical instruments, medical diagnosis and treatment of disease, and developments of nuclear energy and radioactive waste management (see the Preface and Chap. 26). In this chapter, we will explore the fundamental mechanisms for the interaction of fast charged particles with matter, with focus on their significance to the subsequent chemistry initiated by the absorption of ionizing radiation.

Ionizing radiation, as the term implies, defines those radiations that interact with matter by the production of charged particles, namely electrons and residual positive ions.

Energetic electromagnetic radiation, i.e., x-and γ-rays, interact with matter predominately via production of photoelectrons and Compton electrons, and, if the photon energy is greater than about 1 MeV, the production of electron–positron pairs. These primary electrons can have a broad spectrum of energies from subexcitation, or very slow electrons, to what are generally called fast charged particles. Here "fast" implies particles with velocities much larger than bound electrons of the medium. As we will see, electrons and fast heavy charged particles, e.g., protons and alpha particles, predominantly interact with the bound electrons of the medium producing secondary electrons via target ionization. For completeness, we note that the interaction of neutrons with matter also leads to local energy deposition by charged particles; neutrons primarily interact through elastic nuclear scattering, resulting in the production of energetic recoil charged particles characteristic of the absorbing medium. Ionization resulting from the slowing of energetic ions and electrons leads to the production of charged and/or neutral molecular fragments, reactive radicals, and other excited chemical species that foster subsequent chemical reactivity in the absorbing medium. Because all ionizing radiation leads to electron production, the different chemical yields observed for different types of radiation must depend on the detailed spatial distributions of initial events produced in the physical process of electron slowing. All radiation-induced chemistry begins with the stochastic physical processes involved in energy deposition by charged particles.

Investigation of the interactions of charged particle with matter has traditionally been viewed as comprised of studies falling into two fundamental classes: Class I studies focus on the fate of the particle, and class II studies focus on the fate of the absorbing medium [1] (cf. Sec. 2.1.2). The former is most commonly represented by the studies of charged-particle stopping powers, whereas the latter includes studies of media quantities such as the average energy per ion pair (W value), or more specifically, the fundamental processes of ionization and excitation that contribute to spatial structure in the patterns of energy deposition. In the following, we will first discuss the stopping of fast charged particles through the general theory of stopping power, its strengths and limitations, and then look more closely at the individual processes that contribute to energy loss and are incorporated within the general concepts of stopping power.

2. STOPPING POWER

When a fast charged particle passes through matter, it loses energy by interactions involving momentum transfer to the bound electrons of the media. If the particle slows to the point that insufficient momentum can be transferred to excite these bound electrons to vacant states, energy loss will then occur through elastic collisions with the target atom, or the target matrix, as a whole. This latter energy-loss mechanism, termed "nuclear scattering," dominates for ion energies less than a few keV per atomic mass unit (keV/u), i.e., for ions near the extreme end of their range. For high-energy or fast particles, i.e., particles with velocities large relative to the velocities of bound electrons, inelastic collisions involving electron excitation and ionization dominate the energy loss process. At intermediate energies, the competition between elastic and inelastic processes depends on the atomic, molecular, and condensed-phase properties on the absorbing medium and, for heavy charged particles, additional channels of electron capture and loss can become important mechanisms for energy loss.

Stopping power has been the subject of study for more than a century beginning with the pioneering work of Thompson and Rutherford. With the discovery of nuclear fission and subsequent developments in accelerator technology, energetic heavy ions with relatively high energies became readily available and the experimental and theoretical study of stopping power made steady advances (see, e.g., Refs. 2–4; and the excellent review of stopping power for protons and alpha particles published in 1992 by the International Commission on Radiation Units and Measurements [5]). These studies have been an important means of testing our theoretical understanding of charged particle interactions with matter as well as being of practical use in many scientific and applied fields of application. Because stopping power can be measured with relatively high accuracy using modern charged particle detection techniques, of order 1%, we can refine our theoretical understanding of the penetration of charged particles in matter to a high degree by comparisons of theoretical and experimental results.

Many textbooks describe derivations of the dominant features of energy loss incorporated in stopping power (see, e.g., Refs. 6 and 7). The primary energy loss mechanism for a fast charged particle is the interaction between the coulomb charge of the moving charged particle and that of electrons of the absorbing medium. This interaction is generally approximated as a transverse momentum impulse delivered to the bound electron by the coulomb charge of the passing charged particle integrated over the interaction times and impact parameters. A detailed derivation of stopping power has been presented elsewhere (see Sec. 2.2). The important concept from stopping power is that the energy loss by the fast particle is to the bound electrons of the medium. Increasingly complete theories of stopping power, from detailed first-order approximations to full quantum calculations, are described in a number of reviews (see, e.g., Ref. 5 and references therein). The expression for the mass stopping power for a charged particle of velocity v and atomic number z in a medium of atomic number Z and atomic mass A is given in ICRU report 49 [5] as

$$-\frac{1}{\rho}\frac{dE}{dx} = \frac{0.30708}{\beta^2} z^2 \frac{Z}{A} L(\beta) \left(\frac{\text{MeV cm}^2}{\text{g}}\right),$$

(1)

where $L(\beta)$ is obtained from the sum of three terms

$$L(\beta) = L_0(\beta) + zL_1(\beta) + z^2 L_2(\beta).$$

(2)

The parameters in Eq. (1) illustrate the primary dependence of stopping power on the square of the projectile charge z, the velocity of the projectile β, and the density of target electrons Z/A. The mean excitation energy I, a major target parameter, is incorporated in the first term of $L(\beta)$ given as

$$L_0(\beta) = \frac{1}{2}\ln\left(\frac{2mc^2\beta^2 W_m}{1-\beta^2}\right) - \beta^2 - \ln I - \frac{C}{Z} - \frac{\delta}{2}$$

(3)

with

$$W_m = \frac{2mc^2\beta^2}{1-\beta^2} \times [1 + 2(m/M)(1-\beta^2)^{1/2} + (m/M)^2]^{-1}.$$

(4)

Equation (3) incorporates relativistic effects, effects of target density, and corrections to account for binding of inner-shell electrons, as well as the mean excitation energy; C/Z is determined from the shell corrections, $\delta/2$ is the density correction, W_m accounts for the maximum energy that can be transferred in a single collision with a free electron, m/M is the ratio of the electron mass to the projectile mass, and mc^2 is the electron rest energy. If the value in the bracket in Eq. (4) is set to unity, the maximum energy transfer for protons is only overestimated by 0.1% at 1 MeV, and by 0.23% at 1000 MeV [5]. The shell corrections are included to reflect the fact that, as the ion velocity decreases, the assumption of the particle being "fast" can fail for interactions involving inner shell electrons; thus contributions from inner-shell electrons decrease and corrections to the Bethe theory are required. The density correction accounts for the fact that a moving charge tends to polarize the transport medium thereby reducing the stopping power slightly. The shell corrections are predominately a low-energy phenomenon whereas the density correction occurs for high energies. The second term in Eq. (4) is required to account for the observation that stopping power for negatively charged particles is smaller that that for positively charged particles at the same velocity. This correction is often called the Barkas correction after his measurements determined that the range of negative pions was somewhat larger than that for positively charged pions. This correction is often simply referred to as the z^3 correction because it depends on the cube of the particle charge. The last term in Eq. (4) is needed to bring agreement between the first-order theory of Bethe and a full quantum mechanical result of Bloch valid for high-energy particles. All of these additional "correction" factors have come about because of approximations made in the theoretical development of stopping power theory and because the ability to make accurate measurements of stopping power enables one to identify subtle differences between theory and experiment. The reader is directed to ICRU Report 49 [5] and to Chap. 2 for details including the full description of stopping power and the terms identified in Eqs. (1)–(4).

The most significant target parameters for the determination of stopping power are the electron density and the mean excitation energy I found in the logarithmic term of Eq. (3). The mean excitation energy is related to the oscillator strength distributions of the absorber and is, in general, determined from fitting experimental data after all corrections to the first-order theory have been included. Once values of I have been determined for elemental constituents of molecules or materials, one can obtain I for a composite material by

$$\ln I = \frac{\sum_i n_i Z_i \ln I_i}{\sum_i n_i Z_i}, \tag{5}$$

where n_i, Z_i, and I_i represent the atomic concentration, the atomic number, and the mean excitation energy of the ith component of the target molecule or compound. This technique has been extensively used by Porter et al. (see Refs. 5–9 and references therein) to determine stopping power for composite materials such as mylar and Al_2O_3. It is this formula for determining the mean excitation energy for compounds from individual elements that also leads to additivity among stopping powers. One can obtain the stopping power for a compound from a weighted sum of the stopping powers of the constituent elements.

The discussion of stopping power above has focused on interactions by bare fast charged particles. As fast positively charged particles slow, they can capture and subsequently lose electrons; thus they can move through material with a somewhat reduced "average" charge. Electrons bound to the projectile screen the projectile charge, reducing the interaction strength, thereby reducing the stopping power. This reduction is generally incorporated in stopping power theory by assuming a reduced value of the projectile charge, z_{eff} in Eq. (1). This reduced "effective" charge is defined as the square root of the ratio of the stopping power for the screened projectile to that of the bare projectile. It must be emphasized that this effective charge of stopping power theory is not the same as the average charge state of the moving ion. The difference exists because interactions leading to energy loss occur at a somewhat different range of impact parameters than those associated with the capture and loss of electrons, and the screened nuclear charge of the moving particle is strongly dependent on the impact parameter. The effective charge of stopping power theory is an average over all individual stochastic events contributing to the stopping power, each of which is weighted by an effective charge $z_{eff}(E)$ that depends on the energy loss E occurring in that event [10]; this is discussed in more detail later in this chapter. The difference between the effective and average charge is illustrated in Fig. 1 where the effective/average charge of a proton is plotted vs. the ratio of the projectile velocity to the bound target electron velocity. Data shown in Fig. 1 were derived from stopping power measurements of Yarlagadda et al. [11], the effective charge formula of Barkas derived for stopping power [12], and the average charge obtained from charge-transfer measurements [13]. Note that only for large velocities of the projectile, where the projectile in essentially bare, are the average and effective charges the same.

The importance of the various interactions that contribute to stopping power is illustrated for a range of energies involved in stopping a fast proton in Fig. 2. These data, taken from the compilation by Uehara et al. [14], clearly illustrate the importance of target ionization as the leading contributor to energy loss by protons: ionization by protons occurs at the higher energies and ionization by H^{o} at the lower energies. Energy loss processes involving electron capture by protons and electron loss by H^{o} are also important at particle energies less than about 100 keV; or energies less than a few hundred keV/u for heavier charged particles (not shown). Excitation is observed to be a small contribution to

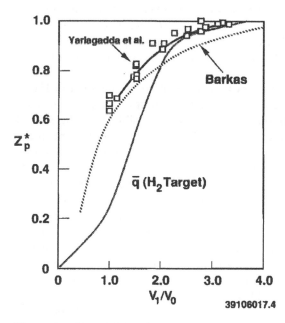

Figure 1 Comparison of the *effective* charge of a proton obtained from the work of Yarlagadda et al. [11], the *effective* charge obtained from the empirical formula of Barkus [12], and the *average* charge $\bar{q}$ determined from charge transfer cross section tabulated by McDaniel et al. [13]. The solid line is a guide through the data of Yarlagadda et al.

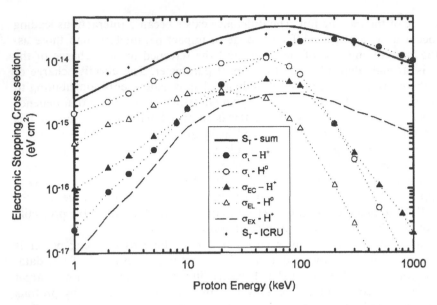

Figure 2 Contributions to the total stopping cross section S_T from ionization σ_i by H^+ and H°, electron capture σ_{EC} by H^+, electron loss σ_{EL} by H°, and excitation σ_{EX} by protons in water. The data are from a compilation by Uehara et al. [14]. The ICRU recommendation S_T-ICRU is from Ref. 5.

total energy loss at all energies and never exceeds about 10% even at its maximum contribution. However, it should be emphasized that contributions to energy loss attributed to excitation are relatively uncertain having been estimated from an empirical model based on limited data [15]; there is little actual data available for excitation by protons or other charged, or neutral, heavy particles.

There are also many channels of energy loss that have not been included in Fig. 2 because they either contribute only a small amount to the total energy loss, or because we lack reliable data to assess their contribution. However, it is possible that the special nature of some of these less-probable interactions might have weighted important in the evolution of subsequent chemical reactions. For example, multiple ionization of the target atom or molecule involving the simultaneous ejection of two or more electrons in a single event has been documented for ion collisions with rare gas atoms (see, e.g., Refs. 16 and 17, and references therein). Although these interactions are relatively rare events, the emission of two or more electrons correlated in time and space might have important consequences on the subsequent chemistry [18]. Unfortunately, there are few data available to guide our understanding of the relevance of such interactions in chemical systems. Energy loss by ionization of inner-shell electrons is also missing from Fig. 2. Ionization of the K-shell of oxygen in the water molecule happens only rarely compared to outer shell interactions, but inner shell ionization followed by Auger electron emission can lead to multiply ionized targets with unique chemical properties. When estimating the importance of energy deposition events, the subsequent chemical reactions under consideration can offer the defining criteria as to whether the rare events have weighted importance. Also lacking from Fig. 2 is information on the contributions leading to different molecular end points such as dissociative excitation, dissociative ionization, autoionization, etc. Surely such processes occur and are relevant to understanding the subsequent chemical activity following energy deposition by fast charged particles, but again there is little information regarding these processes for heavy charged particles that can be used to define their importance. On the other hand, a great deal is known regarding the interactions of electrons with molecules and the production of various final molecular states (see, e.g., Ref. 19), this will be addressed below.

The discussion of stopping power presented above was not meant to be rigorous, but to illustrate that extensive study has been devoted to both experimental and theoretical features of stopping power, and that these results provide benchmarks against which stochastic models of energy loss can be tested. Stopping powers are known with extreme accuracy and any alternative full description of the stopping of charged particles, e.g., stochastic track–structure models, *MUST* be consistent with known stopping powers (cf. Chap. 4). Second, we note that the stopping power gives essentially no information on the fate of the energy deposited, it is a particle-centered parameter. Thus stopping power, no matter how accurate, is of limited use in generating detailed information as to what chemistry might follow energy deposition by ionizing radiation.

To model the potential chemical reactions that might follow energy deposition, it is necessary to better understand the local spatial characteristics of the initial products of ionization and excitation, e.g., are products within reaction distances, is recombination likely prior to chemical reaction, etc. This information is contained in the collection of many small events clustered along the path of the particle, and to understand the subsequent chemistry, one must describe the basic interactions that accompany energy deposition on a spatial scale relevant to the reaction of interest (see Chap. 4). For *fast* protons and electrons, the important fundamental interactions are fairly straightforward. Fast ions deposit about 90% of their energy in ionizing collisions with about 10% of the energy deposited as excitation. Thus if we understand ionization, the production and slowing down of secondary electrons, we can understand how *most* of the energy is deposited; but be aware that determining the fate of the energy in the absorbing medium requires information on the *differential* ionization cross sections. For fast electrons and other bare particles, e.g., protons and alpha particles, there are many sources of experimental and theoretical data; the same is not true for heavier particles or for slow particles that carry bound electrons.

3. *W* VALUES

As has been emphasized, energy deposition by energetic charged particles is a stochastic process involving large numbers of events, each event involving small quantities of energy transfer. All detection devices used in the laboratory provide some form of averaging, whether we are counting numbers of target ions created, the number of electrons or photons emitted, yields of radicals, etc. Stopping power is the stochastic average over all energy loss processes of the projectile. On the other hand, the *W* value is a media-centered, class II, quantity representing the average energy required to produce an ion pair by the passage of a charged particle (cf. Sec. 2.3.3). Thus *W* is given by

$$W = \frac{E}{J},$$ (6)

where E is the total amount of energy absorbed in a volume of gas and J is the total number of ions produced in that volume. This quantity has been, as was stopping power, the subject of extensive investigation for more than a century (see, e.g., Ref. 20 and references therein) and provides a media-centered quantity of relatively high accuracy for testing of quantities obtained from stochastic averages of energy deposition events. As was the case for stopping power, this average provides little information on the actual events produced during the absorption of radiation energy and thereby is of little guidance in estimating the chemical reactions that might follow energy deposition. On the other hand, the study of W values provided the first approach to the analysis of energy deposition in terms of a description of the target as a mixture of oscillators whose frequencies are characteristic of the spectral absorption frequencies of the target [21]. In addition, as an accurately measurable target property, it provides another benchmark that any stochastic calculation must return when averaged over sufficient numbers of interactions.

4. CHARGED PARTICLE TRACK STRUCTURE

Both stopping power and W values give averages over all energy deposition events making up the path of the charged particle in the medium of interest. As averages, they tell us little of the individual processes that can affect the subsequent chemical reactions initiated by the absorption of ionizing radiation. Historically, the term linear energy transfer (LET) was defined to help describe information on the "local" densities of energy transferred to the medium per unit length of particle track (class II), as distinct from the energy lost defined by stopping power (class I). This was a first step in better understanding the effect of local energy density on chemical yields. The quantity LET differs from stopping power in that excitation or ionization resulting in emission of photons that might carry energy from the local area of interest are not included in LET. However, the use of LET as a radiation parameter has been found inadequate for many applications because particles of the same LET can produce quite different yields of chemical products [22]. An extension of the LET concept to define a restricted LET_Δ to either (1) restrict the energy deposited to be within a certain radius of the charged particle track, or (2) to restrict the magnitude of the energy deposited in any single event, was made to further investigate the effects of local energy deposition. These definitions both effectively restrict the energy of secondary electrons that are included in the analysis of events along the charged particle track. Although restricted LET has had some success in radiation biology [23,24], it has found little application in radiation chemistry.

The importance of the stochastic nature of energy deposition in nonhomogeneous reaction kinetics was emphasized in early studies of molecular yields by Mozumder and Magee [25,and,referentherein,therein]. They realized that the density of energy deposition would randomly vary along a charged particle track and that these variations would be a function of the primary electron energy. Their first effort to incorporate the statistical distribution of energy along charged particle tracks was to define regions called spurs in which energy deposition of up to 100 eV would occur stochastically distributed along a particle track. Later, these entities were augmented with regions called "blobs," in which energy depositions from 100 to about 500 eV were defined, and with "short tracks," which would include energy deposition from 500 eV to about 5 keV. These entities represent regions of significantly different chemical reactivity brought about by the statistical nature of energy deposition. The probabilities for energy deposition associated with each of these track entities were either determined analytically or by the use of Monte Carlo techniques. Their estimates of the frequency of such entities were somewhat crude at the time owing to the lack of detailed information on the interaction of electrons in the medium of interest, but their results were successful in illustrating the importance of spatial variations in energy deposition on the subsequent radiation chemistry. These pioneering studies have led to a large number of new Monte-Carlo-based models of energy deposition that maintain, to varying degrees, the stochastic nature of the energy deposition process (see, e.g., Sec. 4.4 and Refs. 26–34). The basic premise in the application of charged particle track structure models is that one must include accurate information on the initial spatial distribution of events involving ionization and excitation occurring along the path of a charged particle if accurate models of the subsequent chemistry are to be developed. A motivating force in the development of these stochastic models has been their application in radiation biology where the chemistry and biochemistry following energy deposition by ionizing radiation ultimately leads to biological damage.

To develop complete descriptions of the process of energy deposition along a charged particle track requires an extensive database of interaction cross sections for the primary particle as well as those initiated by secondary electrons produced in ionizing collisions. Some of the more important interaction processes for heavy charged particles were shown in Fig. 2. Not explicit in the depiction of cross sections shown in Fig. 2 is the importance of *differential* cross sections for electron production (ionization) that enable one to derive a three-dimensional description of the spatial patterns of energy deposition. To develop detailed models of the spatial characteristics of charged particle tracks requires detailed information on the energy and angular distributions of electrons produced by primary electrons or ions, along with similar cross sections for all secondary electrons. In the case of fast heavy charged particle track structure, differential cross sections for electron production for partially screened projectile nuclei are also needed as they capture electrons during slowing. The initial positions of all products of excitation, ionization, and, where appropriate, charge transfer, will determine whether secondary electrons and the corresponding residual media ions will recombine or proceed to initiate chemical reactions.

In the following sections, the interaction mechanisms contributing to the stochastic description of the energy loss process are discussed, as well as the trends in cross sections as a function of the incident particle energy and charge state. To develop a full stochastic description of a charged particle track useful for chemical and radiological applications, it is important that the available cross section data fulfill what has been called by Inokuti [35] as the "trinity of requirements"; that the cross sections be "absolute, correct, and comprehensive." Where gaps in data exist, one must often use theoretical techniques to bridge the gaps, in which case it is important to understand the limitations of the theories and/or models being used and to strive to test the methodology with experimental data wherever possible.

5. IONIZATION

The primary mechanisms for energy loss by ionizing radiation lead to ionization of the atoms and molecules of the absorbing medium. The subsequent slowing of the electrons from these primary ionizing events leads to the spatial relationships among the final physical products, e.g., ions, excited states, molecular fragmentation, etc., that are of utmost importance to subsequent chemical reactivity. For sufficiently fast ions and electrons, there are many similarities in the interaction process and some obvious differences. Electrons lose their energy primarily through coulomb interactions with bound target electrons, in much the same way as heavy ions. However, electrons can lose all their initial kinetic energy in a single collision with a bound electron whereas a heavy ion is restricted to a maximum energy loss to a free electron E_{max} given by

$$E_{max} = \frac{4m_1 m_e}{(m_1 + m_e)^2} E,$$ (7)

where m_1 is the incident particle mass, m_e is the electron mass, and E is the energy per atomic mass unit u of the incident particle. Furthermore, electrons can undergo large angle scattering in collisions with media electrons, leading to extremely torturous paths, whereas heavy ions move in a relatively straight line until very near the end of their range. For heavy ions or neutral particles with energies less than a few keV/u, elastic scattering from the atom as a whole, referred to as nuclear scattering, can dominate energy loss. At the other extreme, very fast electrons can also lose energy by emission of radiation, bremsstrahlung, if they are abruptly accelerated in collisions with target nuclei. The ratio of energy loss by radiation, $(dE/dx)_{Rad}$ relative to that by ionization and excitation $(dE/dx)_{Col}$ is given approximately by

$$\frac{(-dE/dx)_{Rad}}{(-dE/dx)_{Col}} = \frac{ZE}{800},$$ (8)

where Z is the atomic number of the target nucleus and E is the electron energy in MeV. Slowing in water, $Z \approx 8$, a 10-MeV electron would lose somewhat less than 10% of the energy loss per unit path length via radiation. In principle, heavy charged particles can also lose energy through radiation; however, the probability of radiative energy loss is proportional to the inverse of the square of the mass of the accelerated charge. Thus the probability of radiative energy loss by a proton is a few million times smaller than that for an electron, and for heavier ions the probability of radiative energy loss is even smaller.

Both electrons and heavy charged particles lose energy by coulomb interactions with target electrons in basically the same manner. Collisions can be divided into two general categories, those representative of distant, or soft, collisions and those characteristic of close, or hard, collisions. For soft collisions, the energy transfer can be described as occurring essentially with the entire atom or molecule leading to excitation of bound and continuum states of the target. In hard collisions, the energy is directly transferred to a bound electron in close "knock-on" collisions. Soft collisions resemble excitation and ionization produced by photons; that is, the moving charged particle can be viewed as the source of a continuum spectrum of photons with excitation and ionization of the target produced by virtual photon exchange to the "oscillators" of the target. Thus characteristics of soft collisions are similar to photo-excitation resulting in ejected electron spectra determined from the oscillator strengths of the target and with angular distributions of ejected electrons exhibiting dipole-like characteristics. On the other hand, the characteristics of hard collisions are normally described by binary-encounter models of momentum and energy transfer. Models of excitation and ionization also differ for electron and heavy charged particle impact because electron exchange must be included in models for incident electrons—this occurs because of the indistinguishability of incident and ejected electrons. Ionization models specific to electron and heavy particle impact will be independently discussed in the following sections of this chapter.

5.1. Ionization by Electrons

The fundamental mechanisms through which electrons interact with atoms and molecules have been reviewed is detail by several authors; for additional details, see Märk et al. [19],

ICRU Report 55 [36], and references therein. Fast electrons predominantly interact via the coulomb force with the bound electrons of the medium resulting in ionization, i.e., leading to formation of free electrons and residual positive ions. The residual ions can be left in any of a wide variety of final states ranging from their ground ionic state, to states resulting from simultaneous ionization and excitation, and/or dissociation of molecular constituents of the target material. The interaction of a primary electron e_p with the diatomic molecule AB can precede via many channels, e.g., ionization can occur by any of the following reaction pathways

$$e_p + AB \rightarrow$$

$AB^+ + e_p + e_s$	single ionization	(9.1)
$AB^{n+} + ne_s + e_p$	n times multiple ionization	(9.2)
$AB^{**} + e_p \rightarrow AB^+ + e_p + e_s$	autoionization	(9.3)
$AB^{+*} + e_p + e_s \rightarrow$	Ionization plus $\rightarrow$	
$\quad A^+ + B + e_p + e_s$	fragmentation	(9.4)
$\quad AB^{2+} + e_p + 2e_s$	autoionization	(9.5)
$\quad AB^+ + e_p + e_s + h\nu$	radiation	(9.6)
$A^+ + B + e_p + e_s$	dissociative ionization	(9.7)
$A^+ + B^- + e_p$	ion pair formation	(9.8)

where e_s is the secondary, or ionized, electron. For fast electrons, the most probable reaction is that described by Eq. (9.1) with ionization occurring predominately with valence electrons; however, in general, ionization can involve electrons bound in either inner or outer electronic shells of the molecule or constituent atoms. There are also a number on nonionizing channels that can be excited by an incident electron. These include

$$e_p + AB \rightarrow$$

$AB^* + e_p$	simple excitation	(9.9)
$A + B + e_p$	neutral dissociation	(9.10)
$A^* + B + e_p$	dissociation into excited neutrals	(9.11)
AB^-	electron attachment	(9.12)
$A^- + B$	dissociative attachment	(9.13)

where the last two are most likely to occur for very low energy electron impact.

For collisions involving fast electrons, most of the relevant reactions given by Eqs. (9.1)–(9.13) occur with the primary and secondary electrons leaving the target molecule promptly, in about 10^{-16} sec. One the other hand, autoionization and dissociative ionization channels can result in a "secondary" electron being delayed relative to the primary, and in the case of resonant electron attachment, there may be a measurable delay in the exit of the primary electron. These processes are described in considerable detail by Märk et al. [19].

Total Cross Sections

Conceptually, the cross section is used as a measure of the probability of interaction between charged particles (and energetic neutrals) and atomic and molecular "targets."

The concept of a cross section can be defined by considering a beam of fast particles traversing a media of sufficiently low density that there is a negligible probability of one of the particles in the beam making more than one interaction. Under these conditions, the number of interactions N is proportional to the number of incident beam particles passing through the target N_o, the number of target particles per unit volume n, and the length of beam path through the target material L. Mathematically, this can be described under these conditions as [36]

$$N = N_o n L \sigma, \tag{10}$$

where σ is the constant of proportionality, i.e., the probability of interaction. From the dimensions of the parameters in Eq. (10), we note that σ has units of area and we call this measure of the probability of interaction the "cross section." We hasten to note that this is not a physical size of the target, but an area representing the target for the particular type of event under consideration.

Numerous reviews of interaction cross sections for elastic and inelastic collisions have been published for electron impact with atoms and molecules for electron energies from a few meV (milli-electron volts) to a few keV (see, e.g., Secs. 4.2 and 4.3, as well as Refs. 19 and 36, and references therein). For fast electron impact, simple ionization and electronic excitation represent the most likely avenues of energy loss, with ionization occurring in about 90% on the interactions. This is illustrated in Fig. 3, where recommended values of inelastic and elastic cross sections presented by Märk et al. [19] are shown for electron impact excitation of water molecules. Here σ_T is the total interaction cross section determined as the sum of cross sections for elastic and inelastic processes. The inelastic channels shown in Fig. 3 include the vibrational modes σ_{v1} (the bending mode with threshold 0.198 eV), σ_{v2} (sum of two stretching modes with thresholds 0.453 and 0.466 eV), and σ_{v3} (a lump-sum of other vibrational excitation modes including higher hormonics and combinational modes with an assigned threshold of 1 eV). The electronic excitations σ_{e1} and σ_{e2} are shown with threshold energies of 7.5 and 13.3 eV. In addition to excitation, the open circles show contributions to inelastic energy loss from ionization for incident electron energies from near the ionization threshold to 150 eV, and the closed circles represent ionization energy losses for electron energies from 50 to 1000 eV; these data are attributed to Djuric et al., and Bolorizadeh and Rudd, respectively, and were taken from the review by Märk et al. [19].

From the asymptotic high-energy behavior of the excitation cross sections shown in Fig. 3, one observes, as discussed above, that the primary means of energy loss for electron energies greater than 1 keV will be target ionization ($\sim$90%). Of course, at very high energies, e.g., many MeV, bremsstrahlung will become an important factor in the energy loss by electrons; as was also noted above, energy loss by bremsstrahlung can reach nearly 10% of the total energy loss for electrons of 10 MeV in water.

It should be emphasized that the data shown in Fig. 3 do not provide information on the fate of the excited or ionized molecule; these data constitute a sum over final molecular states. Generally, measurements of excitation and ionization cross sections for fast electrons do not provide information on subsequent target relaxation modes. However, this information is often available from separate measurements that focus on state-selected partial cross sections and molecular fragmentation [19].

The total cross sections shown in Fig. 3 can be used to provide information on the relative distances between collisions, i.e., the mean free path of an electron moving in the target medium. The mean free path between collisions is simply obtained from the reciprocal of the cross section $1/\sigma_f$, where f designates the interaction channel of interest. For

Energy Loss by Electrons

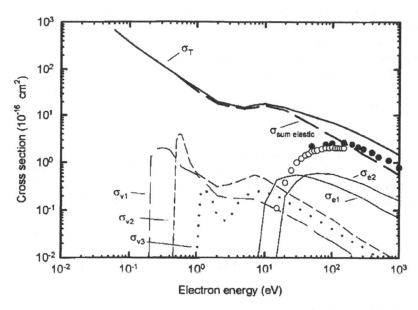

Figure 3 Inelastic and elastic cross sections for electron impact excitation of the water molecule; the data are from the review by Märk et al. [19]. The total interaction cross section σ_T was determined from the sum of cross sections for all elastic and inelastic processes. Inelastic channels include the vibrational modes σ_{v1} (the bending mode with threshold 0.198 eV), σ_{v2} (the sum of two stretching modes with thresholds 0.453 and 0.466 eV), and σ_{v3} (a lump sum of other vibrational excitation modes including higher hormonics and combinational modes with an assigned threshold of 1 eV). The electronic excitations σ_{e1} and σ_{e2} have threshold energies of 7.5 and 13.3 eV. Ionization cross sections are those of Djuric et al. (○), and Bolarizadah and Rudd (●). (From Ref. 19.)

example, the mean free path for ionization would be $1/\sigma_i$ and would be quite different from the mean free path for excitation $1/\sigma_{ex}$. The total ionization, or total scattering, cross section provides information on the distance between collisions, but gives only limited information of the spatial distribution of products produced along the electron track. A full three-dimensional distribution of events requires the additional detail provided by energy and angular *differential* inelastic and elastic scattering cross sections for both the primary and secondary electrons. Obviously, if differential cross sections are available, total cross sections can be obtained by integration over the appropriate variables. Historically, measurements of total ionization cross sections have been obtained from both the total charge measurements and by the integration of measured differential cross sections. For a review of the differential cross section measurements, see Refs. 19 and 36, and references therein.

Differential Ionization Cross Sections for Electrons

Differential ionization cross sections, differential in ejected electron energy and emission angle, were the subject of intense study during the 1970s and 1980s. Considerable progress was made in both experimental and theoretical methodologies needed to describe differential cross sections; these have been reviewed in IAEA TECDOC-799 [19] and ICRU-

Report 55 [36]. Because of experimental constraints, most differential cross-section measurements are undertaken in electron beam experiments with the energy spectra of ejected electrons being measured at various emission angles. This type of measurement results in the determination of doubly (single-electron spectrometer) or triply (two-electron spectrometers in coincidence) differential cross sections. Although the triple differential cross sections, which provide a complete kinematic description of the collision, have been the subject of much study in physics (for a review, see Ref. 37), the major studies of radiological and radiochemical importance for ionization by fast electrons have been confined primarily to doubly differential cross sections (see Sec. 4.4.2 for the use of these cross sections for track simulation). The measured quantity is generally presented as $\sigma(\varepsilon_p, \varepsilon_s, \theta_s)$, or simply $\sigma(\varepsilon_s, \theta_s)$, where ε_p is the energy of the incident or "primary" electron, ε_s is the energy of the secondary or ejected electron, and θ_s is the emission angle of the secondary electron relative to the direction of the incident primary electron. From doubly differential cross sections, one can readily obtain singly differential cross sections $\sigma(\varepsilon)$, differential in electron energy ε, by integration with respect to the emission angle

$$\sigma(\varepsilon) = 2\pi \int_0^\pi \sigma(\varepsilon, \theta)\sin\theta \, d\theta, \tag{11}$$

or single differential cross sections $\sigma(\theta)$, differential in emission angle, by integration over ejected electron energy

$$\sigma(\theta) = 2\pi \int_0^{\varepsilon_{max}} \sigma(\varepsilon, \theta)\sin\theta \, d\varepsilon. \tag{12}$$

In the same manner, one can determine the total ionization cross sections σ_T by integration of the doubly differential cross sections over both ejected electron energy and emission angle

$$\sigma_T = 2\pi \int_0^{\varepsilon_{max}} d\varepsilon \int_0^\pi \sigma(\varepsilon, \theta)\sin\theta \, d\theta \tag{13}$$

Doubly differential cross sections for ionization of water vapor by 1-keV electrons from the work of Bolorizadeh and Rudd [38] are shown in Fig. 4. Here the spectrum of electrons emitted with emission angles ranging from 10° to 150° are shown. Because electrons are indistinguishable, the low-energy electrons, $\varepsilon < 1/2 \, \varepsilon_p$, are commonly designated as secondary electrons, and electrons with energies $\varepsilon > 1/2 \, \varepsilon_p$ are considered the primary electrons that have undergone inelastic collisions; the very narrow line profile of the elastically scattered 1-keV electrons is not shown in Fig. 4. The spectra shown in Fig. 4 illustrate that the probability, as reflected in the cross sections, of producing low-energy secondary electrons is essentially independent of emission angle, whereas the probability of scattering a primary electron to a large angle rapidly decreases with angle.

The angular distribution of ejected electrons of selected energies from ionization of water vapor by 1-keV electrons is shown in Fig. 5. This illustration explicitly shows the isotropic nature of the cross sections for ejection of electrons with low energies. These low-energy secondary electron cross sections are representative of the angular distributions resulting from distant, or soft, collisions. At an ejected energy of 100 eV, we begin to see a "peak" forming at about 70°, and this peak moves to smaller angles for secondary electron energies of 350 and 500 eV. The relatively broad peaks kinematically correspond to close collisions between the incident primary and target electrons.

Kim and Rudd [39] have provided an excellent review of the theoretical techniques used to describe ionization of atoms and molecules by fast electrons and that have been

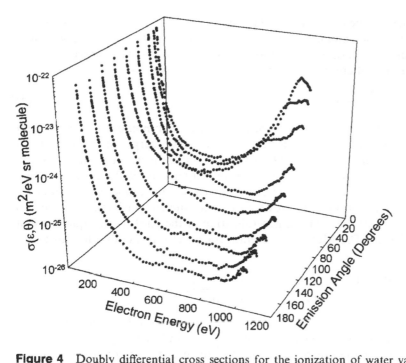

Figure 4 Doubly differential cross sections for the ionization of water vapor by 1-keV electron impact. (From Ref. 38.)

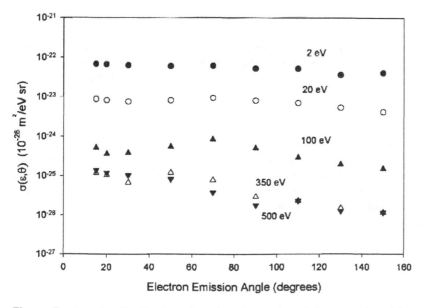

Figure 5 Angular distributions for selected energies of electrons ejected from water vapor by 1-keV electron impact. (From Ref. 38.)

used to develop models for determining the differential ionization cross sections used in numerous applications. The interaction of an electron with a free electron was first investigated by Rutherford as a simple coulomb interaction between two charged particles. The Rutherford equation, differential in energy loss W, was given as

$$\frac{d\sigma(W,T)}{dW} = \frac{4\pi a_o^2 Z_1^2 R^2}{T} \frac{1}{W^2},$$ (14)

where W is the energy of the ejected electron, Z_I is the atomic number of the incident charged particle ($Z_1 = 1$ for incident fast electrons), R is the Rydberg unit of energy (13.6 eV), a_o is the classical Bohr radius (0.0529 nm), and T is the incident particle's scaled kinetic energy ($T = 1/2 \, m_e v^2$, where v is the particle velocity and m_e is the electron mass). However, the Rutherford formula fails for electron impact because it does not account for electron exchange, a process brought about because of the indistinguishability of the two colliding electrons. The Mott formula is a generalization of the Rutherford equation that includes electron exchange. The Mott cross section for an electron collision with a free electron is given by

$$\frac{d\sigma(W,T)}{dW} = \frac{4\pi a_o^2 R^2}{T} \left[\frac{1}{W^2} - \frac{1}{W(T-W)} + \frac{1}{(T-W)^2} \right],$$ (15)

where the first term in the brackets corresponds to the Rutherford direct collision term, the second term results from interference between the direct and exchange collision terms, and the last term represents the exchange collision term. We should emphasize that the Mott cross section still applies to a collision with a free electron. To account for the fact that the electron is bound with energy B in an atom or molecule, one can introduce the binding energy through $E = W + B$ and rewrite the Rutherford formula, Eq. (14), with a simple replacement of E for W. The corresponding Mott formula of Eq. (15) can be rewritten as

$$\frac{d\sigma(W,T)}{dW} = \frac{4\pi a_o^2 R^2 N}{T} \left[\frac{1}{E^2} - \frac{1}{E(T-W)} + \frac{1}{(T-W)^2} \right],$$ (16)

where N is now the number of electrons with binding energy B. However, this replacement to account for a binding energy comes at a cost because both the modified Rutherford and Mott formulas are good approximations only for energy loss E much larger than the binding energy B.

These classical formulas still do not account for the motion of the bound electrons in the atom or molecule. To be more appropriate to the interaction of an incident electron with the bound target electron, one must recognize that the velocity vector of the bound electron can be randomly oriented with respect to the incident electron providing a broadening of the energy of the secondary electron as calculated by the modified Mott cross section. If one integrates over the velocity distribution of the bound electron, the more familiar binary encounter approximation is derived that, in its simplest form, is given by Kim and Rudd [39] as

$$\frac{d\sigma(E,T)}{dE} = \frac{4\pi a_o^2 R^2 N}{T+U+B} \left[\frac{1}{E^2} - \frac{1}{E(T-W)} + \frac{1}{(T-W)^2} \right.$$

$$\left. + \frac{4U}{3} \left(\frac{1}{E^3} + \frac{1}{(T-W)^3} \right) \right]$$ (17)

where U is the average kinetic energy of an electron with binding energy B given by $U = \langle \vec{p}^2 \rangle / 2m$ and $\vec{p}$ is the momentum operator for the electrons in the subshell under consideration.

Although these semiclassical approaches to electron impact ionization tend to be reasonably accurate at predicting the cross sections for large momentum transfer, when integrated with respect to energy loss, the total ionization cross section derived suffers from having an incorrect asymptotic high-energy dependence on the incident electron energy. The energy dependence of the total cross section varies as T^{-1}, whereas experimental evidence, and more complete quantum mechanical descriptions of the energy loss process, suggests that the asymptotic dependence should be $T^{-1} \ln T$.

The first quantum mechanical approach to describe the interaction of charged particles with atomic systems was that of Bohr, who considered the interaction in terms of an impulse approximation resulting in the transfer of energy and momentum from the incident charged particle to the atom or molecule. This perturbation approach was further developed by Bethe in the 1930s; an extensive review of this work has been published by Inokuti [40]. The Bethe theory provides the basis for a quantum mechanical understanding of collisions of fast charged particles with atomic and molecular targets. Within the Bethe theory, the differential cross section for energy loss can be expressed [41] as

$$\frac{d\sigma}{dW} = \frac{4\pi a_0^2 Z^2}{T} \sum_{k=1}^{N} \frac{R^2}{E_k} \frac{df_k}{dW} \ln\left(\frac{4RT}{E_k^2}\right) + b_k(W) + O\left(\frac{E_k}{T}\right), \tag{18}$$

where df_k/dW is the partial optical oscillator strength for ejection of an electron with energy W by photoionization of the kth subshell of the target, and E_k is the total energy transfer given by $W + B_k$. Expressed in this way, the first term in Eq. (18) represents the contributions to the cross sections from soft collisions, the second term represents the hard collisions contributions, and the last term includes contributions of higher order in powers of T^{-1}. The optical nature of soft collisions is clearly illustrated by the relationship to optical oscillator strengths. As expressed in Eq. (18), the contributions from exchange to electron impact excitation and ionization are not included. A more complete description of the use of the Bethe theory for the study of energy loss by electrons, including electron exchange, is given by Dingfelder et al. [42].

The strength of the Bethe approximation is that it properly treats soft collisions, thereby giving the proper energy dependence $T^{-1} \ln T$ for the total ionization cross sections. A weakness of the Bethe approximation is the computational difficulty, including the need for accurate wave functions for all initial bound and final continuum states. Still it can be a powerful tool for the calculation of cross sections representative of the target structure through the use of appropriate measured target oscillator strengths to evaluate matrix elements otherwise difficult to obtain (see Ref. 42 and references therein).

By careful inspection of the relationships of the semiclassical close collision approximations and Bethe formula, one can obtain simple and accurate information on ionization cross sections. By the method first proposed by Platzman, and used extensively by others, it is instructive to form the ratio of the differential cross sections [measured $\sigma(W, T)$ or calculated $d\sigma(W, T)/dW$] to the Rutherford cross section. This ratio, called Y, is mathematically defined as

$$Y = \frac{d\sigma(W, T)}{dW} \frac{T}{4\pi a_0^2} \frac{E^2}{R^2} \tag{19}$$

and has the property that if plotted as a function of R/E, the integral is proportional to the total ionization cross section

$$\int Y(T, W)\mathrm{d}(R/E) = \sigma_{\text{ion}}(T)\frac{T}{4\pi a_0^2 R}. \tag{20}$$

Therefore when experimental data are plotted as Y vs. R/E, the relative importance of electrons of different energies as contributors to the total ionization cross section is clearly apparent and the relative magnitude of the differential cross sections must be such that the area under the curve is equal to the total ionization cross section [43]. The latter point is particularly important because total ionization cross sections can generally be obtained more accurately than differential cross sections. Because Eq. (20) is a sum over energy loss to all electrons, in the limit of large secondary energy the ratio Y tends to a value equal to the number of electrons in the target. If $d\sigma(W,T)/dW$ given by the Bethe formula of Eq. (18) is inserted in Eq. (19) and terms in higher-order T are considered negligible, one can focus on the low-energy electrons, soft collisions, dominated by the first term in the Bethe theory

$$Y \cong \sum_{k=1}^{N} E_k \frac{\mathrm{d}f_k}{\mathrm{d}W} \ln\left(\frac{4RT}{E_k^2}\right). \tag{21}$$

If we now plot Y as a function of R/E, owing to the small variation of the logarithm term with low-energy ejected electron energy, the ratio Y is expected to have the same general structure with ejected electron energy as $\sum_{k=1}^{N} E_k \frac{\mathrm{d}f_k}{\mathrm{d}W}$, i.e., as the optical oscillator strength multiplied by the energy loss.

An example of the utility of the "Platzman Plot" from the work of Kim [44] is shown in Fig. 6 where single differential cross sections for ionization of N_2 by electrons are shown. Here the low-energy ejected electron spectra have the characteristic shape of the optical oscillator strengths multiplied by the energy loss, Edf/dW, and the magnitude of the cross sections are determined such that the areas under the curves are proportional to the total ionization cross section. At the lowest incident electron energies, the optical nature of the spectra tends to disappear as close, nondipole, interactions dominate. Note also that as the ejected electron energy increases, the Rutherford ratio approaches a value equal to the number of active, outer shell, electrons in the molecule. This type of plot provides an excellent means to evaluate the internal consistency and accuracy of experimental cross sections. The cross sections for ionization of water vapor by electron impact measured by Bolorizadeh and Rudd [38] are shown in Fig. 7 along with the distribution obtained from the dipole oscillator strengths [45]. Note that the low-energy electrons do not decrease as would be expected by the shape of Edf/dW, indicating that some extra electrons are probably being detected from scattering events in the experimental system. However, the high-energy electrons ejected from fast collisions do approach a value near 10, in agreement with the number of electrons in the water molecule. Note that the very rapid rise in the ratio at the extreme high energy end of each spectra is an effect of the indistinguishability of electrons and results from the elastic scattering contribution to the spectrum.

Because the semiclassical theories can be used to calculate differential cross sections with relative ease for close collisions between the incident charged particle and the bound electron, and the Bethe theory provides a straightforward method to describe low-energy electrons ejected in distant collisions, it is only natural to combine the best characteristics of the two approaches to derive a comprehensive description of electron impact ionization.

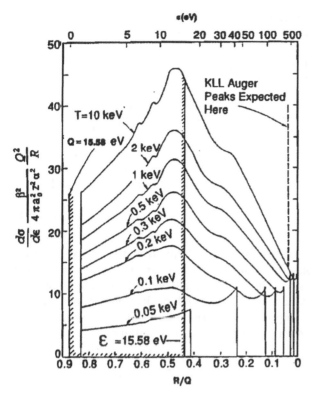

Figure 6 A "Platzman" plot of the ratio of the secondary electron cross sections for ionization of N_2 by electrons to the comparable Rutherford cross sections. Q is the energy loss, $Q = W + B$, and is equivalent to the parameter E used in the text. (From Ref. 44.)

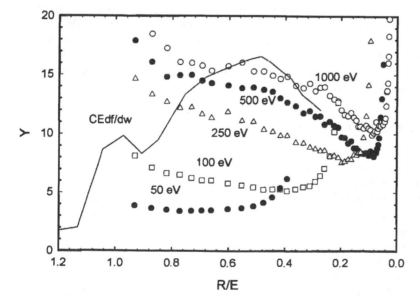

Figure 7 The ratio of measured single differential secondary electron production cross sections to the corresponding Rutherford cross sections for ionization of water vapor by electrons. (From Ref. 38.)

Kim and Rudd [39] have developed such a model, the binary encounter dipole (BED) model, in which the differential ionization cross sections are given by

$$
\frac{\mathrm{d}\sigma(W, T)}{\mathrm{d}W} = \frac{4\pi a_0^2 N}{B(t + u + 1)} \left(\frac{R}{B}\right)^2 \left\{ \frac{(N_i/N) - 2}{t + 1} \left(\frac{1}{w + 1} + \frac{1}{t - w}\right) \right.
$$
$$
\left. + [2 - (N_i/N)] \left(\frac{1}{(w + 1)^2} + \frac{1}{(t - w)^2}\right) + \frac{\ln t}{N(w + 1)} \frac{\mathrm{d}f(w)}{\mathrm{d}w} \right\}
$$

(22)

where $t = T/B$, $w = W/B$, $u = U/B$, N is the number of electrons in the subshell, $\mathrm{d}f/\mathrm{d}W$ is the differential dipole oscillator strength of the electron in this subshell with binding energy B, and N_i is obtained from the differential oscillator strength distribution by

$$
N_i = \int_0^\infty \frac{\mathrm{d}f(w)}{\mathrm{d}W} \mathrm{d}W.
$$

(23)

This relationship becomes very useful because dipole oscillator strengths are available for a wide selection of atoms and molecules. The authors also provide a means of modifying this equation for use when oscillator strengths are not available; the interested reader is directed to their work [39] for additional details. The strength of their work is also evident in that it provides a functional relationship for the determination of total cross section as well. By integration of the single differential cross sections, one obtains

$$
\sigma_i = \frac{4\pi a_0^2 N}{t + u + 1} \left(\frac{R}{B}\right)^2 \left[D(t)\ln t + \left(2 - \frac{N_i}{N}\right)\left(\frac{t - 1}{t}\right) - \left(\frac{\ln t}{t + 1}\right) \right],
$$

(24)

where

$$
D(t) = N^{-1} \int_0^{(t-1)/2} \frac{1}{w + 1} \frac{\mathrm{d}f(w)}{\mathrm{d}w} \mathrm{d}w.
$$

(25)

Both the single differential and total ionization cross sections derived from these models have been very successful in reproducing reliable experimental results. Because of the dominance of the dipole term for low-energy ejected electron energies, one would expect that the normal dipole distribution, as a function of angle for these electrons and the high-energy electrons, will have an angular distribution described by the kinematic constraints of the binary encounter theory.

5.2. Ionization by Protons and Other Bare Ions

As with the study of cross sections for electron impact discussed above, the decades of the 1970s and 1980s saw rapid advances in our understanding of ionization by fast protons and other heavy charged particles; see, e.g., reviews presented in Refs. 10, 36, and 46. Experimental studies have focused on the measurement of doubly differential cross sections, differential in ejected electron energy and emission angle, for protons with energies from a few tens of keV to several MeV. Measurements have also been conducted for He^+ and He^{2+} impact; however, these data are not nearly as comprehensive, and a few measurements were made for ions of C, O, Ne, and Ar. As with electron impact, measurements of electron emission cross sections are conducted for atoms and molecules in gas phase owing to the very short range of low-energy electrons and residual ions in condensed phase material.

The process of ionization by protons, and other bare ions, is essentially the same as that described above for incident electrons, except that heavy ions and their secondary electron are distinguishable, removing the need to consider exchange processes, and these projectiles, being much heavier than the secondary electrons, will travel predominantly in a straight line. One also must be more careful in describing ionization for heavy charged particles because ionization of the target atom or molecule can occur with either the release of a free electron, as with electron impact, or by the charged particle capturing an electron from the target. In the latter case, a residual ion is produced, but no free electron is released; thus one needs to specify whether ionization measurements are measuring free electrons or residual ions—for ionization by heavy ions, these cross sections are different.

As with electron impact, one can see a division of energy loss events occurring between either from soft or hard collisions. The spectra of electrons ejected from H_2 and N_2 at 20° with respect to an incident 1-MeV proton is shown in Fig. 8; these data are from the work of Toburen et al., Refs. 47 and 48, respectively. Note the rather clear separation between the low-energy peak in the cross sections at near zero ejected electron energy and the high-energy peak near 3000 eV; the high-energy peak is the so-called binary encounter peak. The electron energy W and emission angle θ observed for the high-energy peak are related to the kinematics of the system as

$$W = \frac{4MmE_p}{(M+m)^2}\cos^2\theta, \tag{26}$$

where M is the incident ion mass, m is the electron mass, and E_p is the incident ion energy in MeV/u. The shape and width of the binary encounter peak is determined by the velocity distribution of the initially bound electron, i.e., the Compton profile. The largest differences between the two spectra shown in Fig. 8, other than the N_2 cross sections being

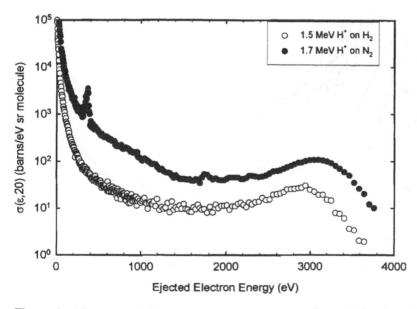

Figure 8 Electron emission at 20° with respect to the forward direction of the proton for ionization of H_2 and N_2 by fast proton impact. The data for ionization of H_2 are from Ref. 47 and those for N_2 are from Ref. 48.

about an order of magnitude larger, are the contributions to the spectra at about 360 eV from Auger electrons emitted following inner-shell ionization of the nitrogen atom, and the larger contribution of electrons in the energy region from about 200 to 1500 eV in the N_2 spectrum relative to the H_2 spectrum. Cross sections in this intermediate-energy region are somewhat enhanced by contributions from electrons ejected from inner shells of the nitrogen molecule, in contrast to the spectrum from hydrogen where only a single shell is involved. Because the experiments generally do not determine the initial state of the ejected electron, these spectra are an average over all initial bound states.

An example of the extent of the data obtained in typical measurements of doubly differential cross sections is shown in Fig. 9 for ionization of CH_4 by 1-MeV protons. This example from the work of Lynch et al. [49] also serves to illustrate the features of the spectra of electrons ejected from molecules by fast protons. This three-dimensional display shows the isotropic nature of the ejection of low-energy electrons and the sharply defined angular dependence of the binary encounter peaks in the ejection of high-energy electrons. The ridge formed by the binary encounter electrons across the different angles is commonly referred to as the Bethe ridge [40] and the peak energies are approximately given by the expression provided by Eq. (26). One can also see a ridge running across the surface for electron energies of approximately 250 eV. This ridge results from Auger electron emission following ionization of the K shell of carbon in the CH_4 molecule. For

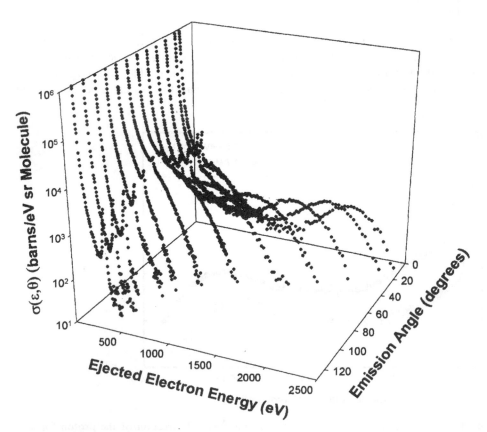

Figure 9 Energy and angular distribution of cross sections for electron emission from CH_4 following 1.0-MeV proton impact. (From Ref. 49.)

low atomic number elements, such as carbon, the fluorescence yield is very small ($\omega_K \approx$ 0.0024 for carbon), meaning that the cross sections for Auger electron emission are essentially identical to the cross sections for K-shell ionization. Thus secondary electron spectra, such as these shown in Fig. 9, provide a good source of information on inner-shell ionization cross sections.

The theoretical descriptions of the ejected electron spectra for heavy ion impact are basically the same as that for electron impact discussed above, except that the theory is simplified for heavy ions because exchange forces are not an issue. One can write the equivalent of Eq. (17) for the binary encounter approximation to the single differential ionization cross sections for bare heavy ion impact [36] as

$$\frac{d\sigma(E, T)}{dE} = \frac{4\pi a_0^2 Z_1^2 R^2}{TE^2} \left(1 + \frac{4U}{3E}\right) \tag{27}$$

for $E_- \geq E \geq 0$, and

$$\frac{d\sigma(E, T)}{dE} = \frac{4\pi a_0^2 Z_1^2 R^2}{TE^2} \frac{U}{E} \left\{ \frac{4}{3} \left(\frac{T}{U}\right)^{3/2} - \frac{1}{6} \left[\left(\frac{E}{U} + 1\right)^{1/2} - 1\right]^3 \right\} \tag{28}$$

for $E_+ \geq E \geq E_-$, where $E_\pm = 4T \pm 4(TU)^{1/2}$, Z_1 is the charge of the incident ion, and the other parameters were defined above for Eq. (17). For heavy ions, as defined above for electrons, the kinetic energy T of the incident ion is given as $T = m_e v_p^2/2$, where m_e is the electron mass and v_p is the incident particle velocity; thus T is the energy of the incident particle in terms of the equivalent velocity electron. The agreement between experiment and results of binary encounter theory can be somewhat improved by integrating over a Fock distribution of orbital electron velocities, rather than a simple hydrogenic distribution, as was shown by Rudd et al. [50]; however, for light atoms, the hydrogenic distribution is also a good approximation.

More detailed information on the characteristics of secondary electron emission can be obtained by consideration of the angular distributions of the double differential cross sections for electron production by fast ions. A comparison of the results of the binary encounter approximation, the quantum mechanical Born approximation, and experimental results is shown in Fig. 10 for ionization of helium by 2-MeV protons. These cross sections clearly illustrate the isotropic nature of low-energy electron emission compared with the sharply peaked angular distributions for high-energy electrons. The theoretical results shown in Fig. 10 include a plane wave Born approximation (PWBA) of Madison [51] in which the author used Hartree–Fock initial discrete and final continuum wave functions, and the binary encounter approximation (BEA) of Bonsen and Vriens [52] in which they average over the initial velocity distribution of the bound electrons. Notice that the BEA is more sharply peaked in angle than the quantum mechanical result. It should be noted that the PWBA results using hydrogenic wave functions provides essentially the same results as the BEA. Although the PWBA calculations of Madison are in much better agreement with experimental data than the BEA, there are still discrepancies for small angles and ejected electron velocities less than and near that of the proton velocity, i.e., equivalent to electron energies less than or equal to $E_e = m_e v_p^2/2$. For the particle energy shown in Fig. 10, this occurs at electron energies of approximately 1 keV and less. This enhancement of the cross sections for small angles of emission has been attributed to an ionization mechanism called charge transfer to continuum states of the proton, an additional ionization channel not included in the PWBA calculation [53].

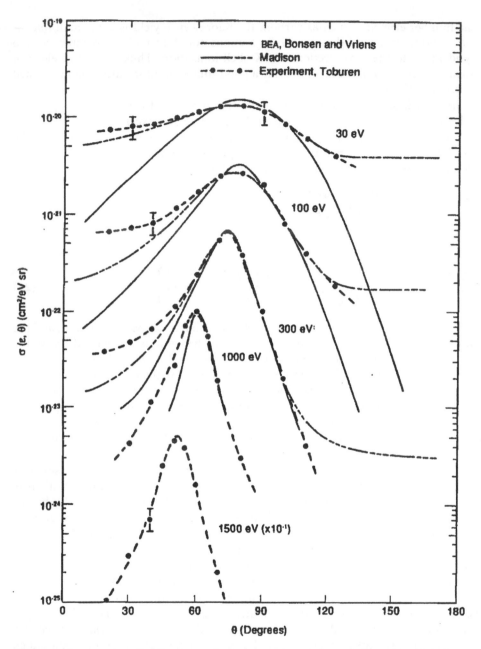

Figure 10 Angular distributions of electrons ejected from helium by 2-MeV protons. The experimental data are from Ref. 54; theoretical results are the binary encounter approximation of Bonsen and Vriens [52]; and the plan wave is the Born calculation of Madison [51].

As was discussed for electron impact, the classical Rutherford formula, Eq. (14), and Bethe theory, Eq. (18), provide excellent tests of the asymptotic features of the singly differential cross sections for electron emission in fast proton, or other bare ions, in collisions with atoms and molecules. The Platzman plot, a plot of the ratio of measured single differential cross sections to the Rutherford cross section as a function of the R/E, is shown in Fig. 11 for ionization of helium by protons [54]. Note that from small values of R/E (energetic ejected electrons), the ratio tends to a value of 2 as expected from the number of electrons in the atom, and the spectra of low-energy ejected electrons has the general shape expected from $E df/dW$ as predicted by the Bethe theory. The use of the Rutherford ratio was also proposed by Kim [55] as an excellent method to test the absolute accuracy of ionization cross sections by plotting the Rutherford ratio as a function of E/R, in which case the ratio should reach a asymptotic value for ejection of high-energy electrons equal to the number of electrons in the atom. This use of the quantity Y is shown in Fig. 12 for ionization of helium by protons. The dashed line in Fig. 12 is the predicted ratio for emission of high-energy electrons. The experimental results presented by Manson et al. [54] are measured *absolute* cross sections, and they are in excellent agreement with the expected cross sections based on the Rutherford theory. There is some evidence from this ratio that the experimental results might actually be as much as 10% larger than expected from Rutherford theory—this difference is well within the stated 25% uncertainties of the absolute values of the experimental data, suggesting that a renormalization to the asymptotic Rutherford value might be advised. From the data shown in Fig. 12, we see that, for the asymptotic form of the ratio Y to be reached requires incident proton energies greater than about 500 keV. At lower energies, the Rutherford plateau is not reached by the most energetic of the ejected electrons.

Doubly differential cross sections for electron emission following proton impact have been measured for a wide variety of different atomic and molecular target systems (for

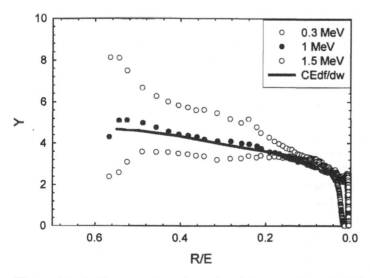

Figure 11 A Platzman plot, the ratio of the experimental single differential cross section for electron emission from helium by 1-MeV protons to the corresponding Rutherford cross sections plotted as a function of R/E. The experimental cross sections are from Ref. 54 and the differential oscillator strength is taken from Ref. 43.

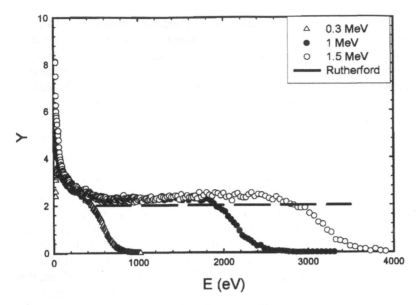

Figure 12 The ratio of the measured single differential cross section for ionization of helium by protons to the corresponding Rutherford cross sections plotted as a function of the ejected electron energy. The solid line represents the expected high-energy behavior of the ratio; it should approach the number of electrons in the atom. The measurements are from Manson et al. [54].

details, see reviews of these data in Refs. 10, 36, 46, and 56) providing an opportunity to explore the effects of target and projectile parameters on the interaction cross sections. For molecules composed of second-row elements, where only K and L shells are involved, the cross sections for ejection of fast electrons is observed to be approximately proportional to the number of electrons in the atom or molecule. This is illustrated in Fig. 13 where the single differential cross sections for ionization of a number of molecules are plotted per "effective" target electron for ionization by 1-MeV protons; hydrocarbon data are from Refs. 49 and 57 and the oxygen-containing molecular data are from Ref. 58. The number of effective electrons used here is simply the number of outer-shell electrons, and this is determined by subtracting the number of K-shell electrons from the total number of electrons in the molecule. Note that the differences in the spectra scaled in this manner occur only in the regions of Auger electron emission and for electron energies less than about 50 eV. The Auger electron peak energies depend on the K-shell binding energies and occur at about 250 eV for carbon, 365 eV for nitrogen, and 540 eV for oxygen; for the energy resolution used in these measurements, about 4%, the fine structural details of the spectra are not observed [59]. The differences in the spectra for low-energy electron emission reflect the differences in oscillator strength distributions for the different molecules as is expected from the Bethe theory.

It is also interesting to examine the angular distributions of these doubly differential cross sections for low-Z molecules. Angular distributions of cross sections scaled by the number of outer-shell electrons for selected energies of electrons ejected by 1-MeV protons are shown in Fig. 14. Scaled in this manner, cross sections for these molecules are in excellent agreement with one another except for those of molecular hydrogen. The hydrogen cross sections are much more sharply peaked in emission angle, and in this

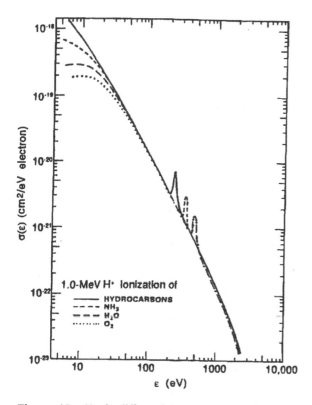

Figure 13 Singly differential cross sections for ionization of several molecular targets by 1-MeV protons are plotted as a function of the ejected electron energy. The cross sections are scaled by the effective number of target electrons in each molecule; the effective number of electrons is defined as the total number of molecular electrons minus those of the K-shell. (From Refs. 49, 56, and 58.)

way reflect the results of theoretical calculations that employ hydrogenic wave functions, as was also observed in the data shown in Fig. 10.

The relative success of the binary encounter and Bethe theories, and the relatively well established systematic trends observed in the measured differential cross sections for ionization by fast protons, has stimulated the development of models that can extend the range of data for use in various applications. It is clear that the low-energy portion of the secondary electron spectra are related to the optical oscillator strength and that the ejection of fast electrons can be predicted reasonable well by the binary encounter theory. The question is how to merge these two concepts to predict the full spectrum.

Miller et al. [60,61] examined the structure of the Bethe–Born formula and observed that the second term of Eq. (18), the hard collisions term, is independent of the projectile properties. In addition, for many target atoms and molecules, the optical oscillator strength has been measured, making evaluation of the first term readily accessible. Thus if the higher-order terms in E/T can be ignored and experimental data are available for single differential cross sections, even for a very limited incident ion energy range, it should be possible to extract the hard collisions term from the experimental data. Because the hard collision contribution is independent of particle energy, it can be used along with the first term in the Bethe formula to obtain differential cross sections throughout the range of validity of the Bethe formula. This model was shown to be very useful in predicting the

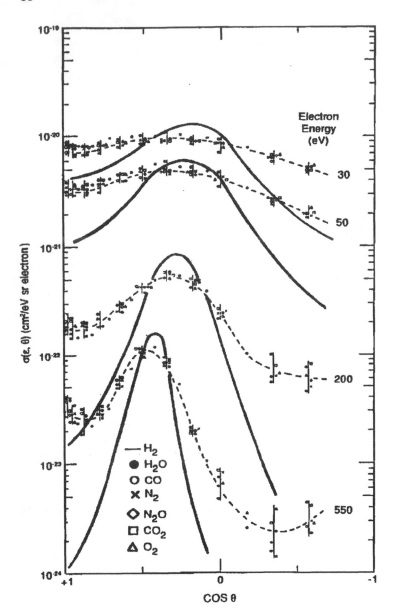

Figure 14   Doubly differential cross sections for ionization of several low-Z molecular targets by 1-MeV protons plotted for selected ejected electron energies as a function of the emission angle. (From Refs. 47, 48, 56–58.)

single differential cross sections where limited data existed, and in improving the accuracy of low-energy electron cross sections where experimental artifacts can render measurements unreliable. Their model uses the optical oscillator strengths to determine the shape of low-energy electron cross sections where experimental accuracy can be lacking. However, the model was limited to fast protons by the high-energy nature of the theory.

The most comprehensive model of secondary electron emission by bare charged particles has been developed by Rudd [62] for secondary electron production by protons;

this model is described in detail in ICRU Report 55 [36] and in an excellent review by Rudd et al. [63]. Rudd's model has the advantage of being applicable to the full range of incident proton and ejected electron energies. It has the proper asymptotic dependence on proton energy at both low and high energies. The major shortcoming is that it must be fit to sufficient experimental data to fix the model's 10 parameters. For those target atoms and molecules where parameters have been fixed, it is a powerful tool because of its analytic capabilities for application in fields such as radiation chemistry and biology that need secondary electron spectra to evaluate subsequent chemical reactivity. Rudd's formula is given as

$$\frac{\mathrm{d}\sigma(W,T)}{\mathrm{d}W} = \frac{(S/B)(F_1 + F_2 W)(1 + W)^{-3}}{1 + \exp[\alpha(W - W_c)/v]}, \tag{29}$$

where $S = 4\pi a_o^2\, N(R/B)^2$ and W_c is the cutoff energy given by

$$W_c = 4v^2 - 2v - R/4B, \tag{30}$$

N is the number of electrons initially bound with energy B, v is the ion velocity, and W is the ejected electron energy. The functions F_1 and F_2 are given as

$$F_1(v) = L_1 + H_1 \tag{31}$$

and

$$F_2(v) = L_2 H_2/(L_2 + H_2), \tag{32}$$

with

$$H_1 + A_1 \ln(1 + v^2)/(v^2 + B_1/v^2), \tag{33}$$

$$L_1 = C_1 v^{D_1}/[1 + E_1 v^{D_1+4}], \tag{34}$$

$$H_2 = A_2/v^2 + B_2/v^4, \tag{35}$$

and

$$L_2 = C_2 v^{D_2}. \tag{36}$$

The 10 parameters A_1, B_1, C_1, D_1, E_1, A_2, B_2, C_2, D_2, and α are given in both Refs. 36 and 63 for 10 different atomic and molecular targets. An indication of the accuracy of the cross section provided by the Rudd model is illustrated in Fig. 15, where cross sections for ionization of N_2 by protons is compared to measurements [48,64–66] over a wide energy range. The major advantage of this model is that it is accurate for both low-energy and high-energy protons where most models have been developed for only "high"-energy ions.

5.3. Ionization by Dressed Ions

Perhaps the greatest challenge for the detailed investigation of energy deposition by heavy charged particles occurs at particle velocities where the capture of electrons become probable and cross sections must be determined for the interaction of "dressed" ions with target electrons. As an ion captures electrons, these electrons can partially shield the ion nucleus from coulomb interactions with the electrons of the medium. For example, a proton with a bound electron (a moving hydrogen atom) looks like a neutral particle from a large distance and might be expected to have little probability of interacting with

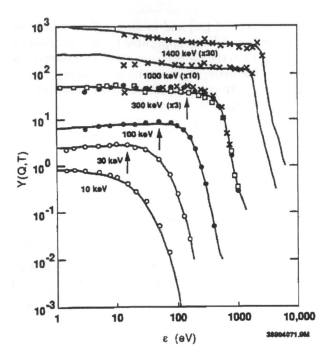

Figure 15 The ratio of calculated and measured single differential cross sections for electron emission from N_2 following proton impact to the corresponding Rutherford cross sections. The solid line is the result of the model of Rudd [62]; experimental data are from Rudd (O) [64], Crooks and Rudd (●) [65], Toburen (×) [48], and Stolterfoht (□) [66].

electrons of the target medium. However, this is far from what is actually observed; the moving hydrogen atom can actually be as effective as a proton in producing ionization of the target. To explain this phenomenon, we recall that the hydrogen electron in its ground state is bound with a spherically symmetric probability distribution having its maximum at the Bohn radius. As we approach the hydrogen atom on a microscopic scale, this charge distribution provides somewhat less than a full charge screening of the nucleus—the unit of electronic charge is distributed over the orbital volume of the ground state of the atom. In fact, if we were inside the Bohr radius, we would expect little or no screening of the nucleus by the bound electron. This distance-dependent shielding of the nucleus by bound electrons occurs for all dressed particles and results in ejection of low-energy ionization electrons, those produced in large impact parameter (soft) collisions, with a reduced probability owing to a more effective screening of the projectile charge at large distances. On the other hand, the ejection of high-energy electrons, those produced in small impact parameter (hard) collisions, will show little effect of screening by the projectile electron(s) because the impact parameter can be inside of some, or all, of the distributions of bound electrons. In addition, the electrons bound to the moving particle must be considered as active participants in the collision. They can also interact with target electrons via their coulomb charge and can either be ionized in the collision leading to free electrons in the media frame of reference with velocity comparable to the projectile (elastic scattering), or they can interact with the bound electrons of the media producing excitation or ionization (inelastic scattering) of the target. The interactions of projectile electrons with target electrons is complicated by the convolution of the initial velocity distributions resulting from their binding, with the velocity of the moving projectile, and conversion of these vectors to the medium (laboratory) frame-of-reference.

An example of the features of the spectrum of secondary electrons emitted in H° impact on water molecules from the work of Bolorizadeh and Rudd [67] is shown in Fig. 16. Compared to the simple spectrum of electrons emitted by proton impact shown as the solid line in Fig. 16 the spectrum from H° impact has an additional peak centered at an electron energy of approximately 82 eV. This broad peak is from the superposition of the spectrum of electrons stripped (elastically scattered) from the projectile on the spectrum of electrons ejected from the target. Because the stripped projectile electrons originate as bound electrons in the rest frame of the moving projectile, their laboratory energy is given approximately by $W = m_e E_p / M_p$ and the width of the peak is determined by the Compton profile of electrons in the projectile frame, but also transformed to the laboratory frame-of-reference. The results shown in Fig. 16 clearly illustrate that the cross-sections for

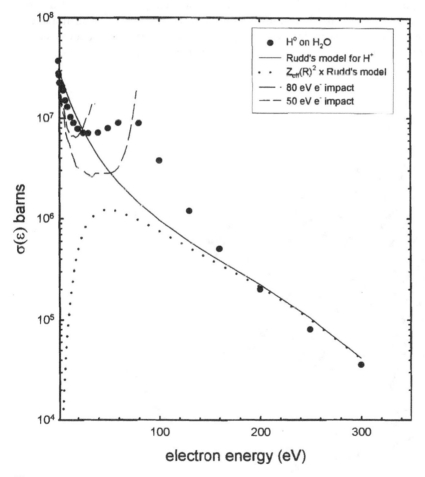

Figure 16 Ionization of water vapor by 150-keV H° particles. The e⁻ and H° impact data are from Refs. 38 and 67, and the parameters for the Rudd model are from Ref. 36. The dotted line is discussed in the text under the section on effective charge.

electron production by "fast" neutral hydrogen atoms can actually be larger than those for the comparable velocity proton. Owing to screening by the projectile electron, we might expect low-energy electrons emitted from the target in soft collision to be suppressed relative to the bare proton (as predicted by the dashed line in Fig. 16). However, for ionization by $H°$, this effect of screening is compensated by contributions from elastic scattering of projectile electrons (the electron loss peak at 82 eV in this case) and by the very low-energy electrons produced by inelastic scattering of projectile electrons from target electrons resulting in ionization of the target. The dashed lines in Fig. 16 are from the cross sections for inelastic energy loss by electrons with energies representative of those in the elastic peak, the data come from the work of Bolorizadeh and Rudd [38]. To predict the full extent of the low-energy inelastic scattering contribution to the spectra requires integration of the inelastic cross sections over the weighted distribution of projectile electron velocities.

For more complex projectiles and targets, additional features become apparent in the spectra of ejected electrons. Doubly differential cross sections for ionization of CH_4 by 0.3 MeV/u C^+ ions (3.6 MeV total projectile energy), taken from some of the author's unpublished work, are shown in Fig. 17 to illustrate the increasing complexity of dressed-ion collisions compared to that for bare ions represented in Fig. 9 by proton impact. In Fig. 17, the cross sections are seen to have their maximum at very low energy ejected electrons as was the case for proton impact, and, although their shape is not as sharp and

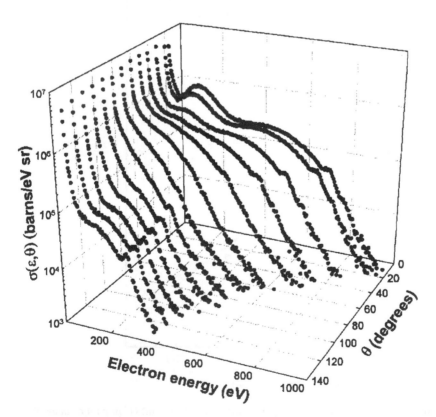

Figure 17   Ionization of CH_4 by 0.3-MeV/u C^+ ions. (These are unpublished data of Toburen discussed in Ref. 56.)

discrete for this intermediate velocity ion as for the higher velocity proton impact, there is a broad distribution of binary encounter electrons at the kinematically predicted forward angles. In addition, we see evidence of K-Auger emission following the inner-shell ionization of the target carbon atom as a ridge in the surface at approximately 250 eV; this is most clearly seen for large angles. At small angles, the Auger electron yield becomes insignificant relative to electrons directly ejected from outer shells. The broad peak at about 160 eV, most strongly seen for small angles, results from electrons stripped from the projectile, and Auger electrons resulting from excitation of the K-shell of the projectile are observed as the Doppler-shifted Auger peaks superimposed on the continuum spectrum at the high-energy edge of the binary encounter peaks. Although electrons stripped from the projectile are predominantly observed ejected into small angles, there is also evidence of their presence in the spectra at large angles.

Most of the features of the secondary electron spectra induced by impact of dressed ions can be simply explained by the combined effects of the individual interactions of the incident ion and its bound electrons with the target. However, a quantitative description is complicated by the effects of electron binding and nuclear screening by bound electrons. As discussed above for the calculation of stopping power for dressed ions, the effects of bound projectile electrons on target ionization can be included through use of an effective projectile charge. However, it is clear that a useful effective charge must be a function of energy loss to be able to adequately describe the effects of electron screening on the secondary electron emission spectra induced by dressed ions. To simply multiply the spectrum of electrons induced by bare ions by a constant factor, such as the effective charge determined from stopping power, would not provide a realistic estimation of the spectra of electrons ejected in collisions involving dressed ions. This will be explored in detail in the next section of this chapter.

6. EFFECTIVE CHARGE

As we have seen, the spectra of electrons ejected from a target atom or molecule by dressed ions and neutrals are quite different from those induced by bare projectiles. These differences cannot be explained by use of a simple effective charge as was defined by stropping power theory. For fast relatively simple dressed ions, such as H° or He^{+}, the separation of target and projectile ionization is relatively straightforward. The single differential cross sections for ionization of water vapor by 2-MeV He^{+} and He^{2+} ions from the work of Toburen and Wilson [68] are shown in Fig. 18. For comparison, the data for He^{2+} ion impact has been scaled by z^2 (divided by 4) to represent the spectrum expected of a bare singly charged ion (e.g., similar to proton impact at the same velocity as the He^{+} ion); this is shown as the solid line. The cross sections for He^{+} impact are in good agreement with the singly charged ion impact for electrons ejected with low energies (soft, or distant, collisions), whereas the high-energy ejected electron cross sections are in excellent agreement with the He^{2+} results showing that screening is inefficient in these close collisions. The dashed line in Fig. 18 is simply a "free-hand" interpolation of *target* ionization in the energy region containing the peak from electrons stripped from the projectile. From experimental data such as these, one can determine the effective nuclear charge as a function of energy loss from the formula

$$z_{\text{eff}}(E) = z \left[\frac{d\sigma(E, q)}{dE} \bigg/ \frac{d\sigma(E, z)}{dE} \right]^{1/2}, \tag{37}$$

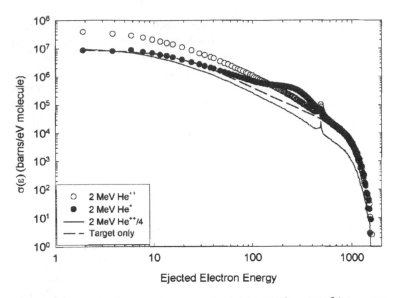

Figure 18 Ionization of water vapor by 2-MeV He$^+$ and He^{2+} ions. The dashed line is an estimate of the contribution of target ionization by He$^+$ obtained by interpolation from regions beyond the vicinity of electrons contributed by electron loss by the He$^+$ ion. (From Ref. 68.)

where $d\sigma(E,q)/dE$ is the measured differential cross section for target ionization by an incident ion of charge q and $d\sigma(E,z)/dE$ is the measured differential cross section for energy loss E by the bare ion of charge z. An example of $z_{\text{eff}}(E)$ as a function of energy loss obtained for ionization of He by He$^+$ ions is shown in Fig. 19. We must take care that the portion of the electron spectrum that includes electrons from projectile stripping is not included in the ratio, or that the stripping contribution is not included because they can otherwise dominate the ratio. One can estimate the target ionization as was performed in the case of the interpolated target contribution shown as the dashed line in Fig 18, or simply use the portion of the electron spectra beyond the energy limitations of the stripped projectile electrons. The origins of the theoretical curves shown in Fig. 19 will be discussed in some detail below.

The theoretical basis for an effective charge $z_{\text{eff}}(E)$ associated with collisional energy loss has been investigated by several authors (see, e.g., Refs. 69–72). Within the Born approximation, the doubly differential cross section for ejection of an electron with energy W by a He$^+$ ion can be written as

$$\frac{\partial^2\sigma}{\partial W \partial\Omega} = \int_{K_{\min}}^{K_{\max}} \left\{ 2\frac{1}{[1+(Ka_0/4^2)^2]^2} \right\}^2 I(K)dK, \tag{38}$$

where K is the momentum transfer, a_0 is the Bohr radius, and $I(K)$ is a complicated function of radial matrix elements and phase shifts [70]. The second term in the brackets accounts for the effective screening of the nuclear charge by the bound helium electron. Manson and Toburen have evaluated this expression using Hartree–Slater wave functions for initial discrete and final continuum states for ionization of He by He$^+$ ions. Notice that, within the Born approximation, the screening is a function of momentum transfer and not energy transfer. The effect on energy transfer is obtained by an appropriate

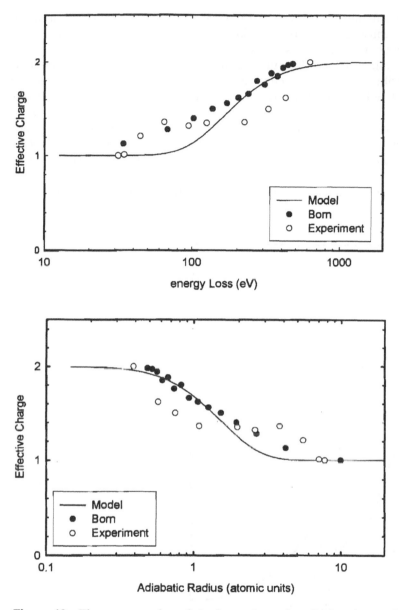

Figure 19 The upper portion of the figure shows the effective charge of the He⁺ projectile as a function of the ejected electron energy that is obtained from the experiment [68], Born theory [70], and the model of Toburen et al. [71]. The lower portion of the figure is the same data, but plotted in terms of the impact parameter. The impact parameter is obtained using the relationship between projectile velocity, energy loss, and impact parameter defined by the Massey criterion. (See the text and Ref. 71 for details.)

integral over momentum transfer. However, the effective charge for target ionization calculated for electron emission at 60° using the Born approximation is in good agreement with the measured single differential cross sections shown in Fig. 19.

An important finding of the calculations using the Born approximation was that there is a sizable contribution of target excitation (including ionization) that accompanies projectile ionization. Doubly differential cross sections for emission of 219 eV electrons from He by 2 MeV He ions are shown in Fig. 20 with the results of PWBA calculations. The peak in the distribution of electrons ejected at 0° in the laboratory frame of reference is dominated by target ionization simultaneously occurring with projectile ionization. This is in contrast to the binary encounter peak at about 70° that is attributed to target ionization with projectiles left in the ground state, and to electrons ejected at large angles (backward angles) where simultaneous excitation appears only of minor importance.

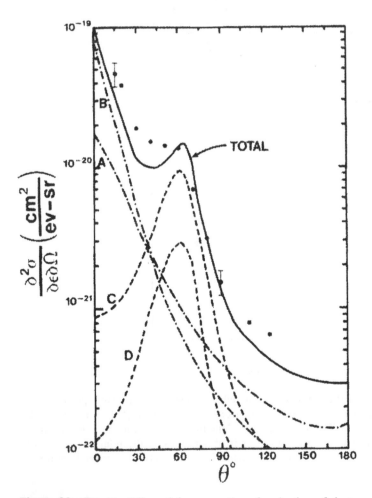

Figure 20 Doubly differential cross sections for ejection of electrons of 219 eV (16 Ry) from He by 2-MeV He⁺ ions. The points are measured cross sections and the calculated results are: line A—projectile ionization, target remains in the ground state; line B—projectile ionization with simultaneous target excitation; line C—target ionization, projectile remains in the ground state; and line D—target ionization with simultaneous projectile excitation. (From Ref. 70.)

The evaluation of the $I(K)$ term in Eq. (38) becomes increasingly difficult for more complex target and projectile systems. This encouraged the development of a phenomenological approach for use with such systems. In the method proposed by Toburen et al. [71], one considers the screening of the projectile by its bound electrons as a function of the distance R from the nucleus as

$$z_{\text{eff}}(R) = z - S(R), \tag{39}$$

with

$$S(R) = \sum_i N_i \int_0^R |\psi_i(r)|^2 r^2 dr, \tag{40}$$

where ψ_i is the normalized radial wave function for the N_i electrons residing in the ith subshell of the projectile. To relate the energy loss E with the radial distance R, the Massy adiabatic criterion $R_{\text{ad}} = v_p/E$ is used, where v_p is the incident ion or neutral particle velocity and all parameters are in atomic units. Using this model, we obtain the solid line in Fig. 19 in good agreement with both the experiment results and the Born calculation. This model was also used to obtain the dotted line in Fig. 16 for target ionization of water by atomic hydrogen impact. In the case of the neutral projectile, the model predicts zero cross section for ionization involving zero-energy ejected electrons, i.e., for electrons ejected by very distant collisions; this reflects the lack of coulomb field from the neutral particle as observed at sufficiently large distances.

As noted above, the calculations including effects of screening by heavy ions and interactions with complex targets, i.e., molecular targets, are difficult to accomplish by fundamental theories such as is represented by the Born approximation. However, it is possible to determine approximate cross sections by combining the model of Rudd, described above for ionization by protons, with the screening model presented in Eqs. (39) and (40). Single differential cross sections for ionization of neon by protons and C^+ ions of 0.1 MeV/u are shown in Fig. 21 along with these model predictions. The dotted line is the result of Rudd's model for ionization of neon by 0.1-MeV protons; it is in excellent agreement with measured proton impact ionization cross sections for neon of Crooks and Rudd [65] shown as the filled circles. A simple z^2 multiplication of the proton cross sections, i.e., classical z^2 scaling, is shown by the solid line with filled points; notice the large overestimation of the low-energy electron cross sections by this naive scaling technique, although the high-energy portion of the spectrum is in relatively good agreement with the measured cross sections. Applying the screening model for z_{eff} given by Eqs. (39) and (40) above, evaluated using hydrogenic wave function for the electrons bound to the carbon projectile, one obtains the solid line shown in Fig. 21. This application of the screening model returns cross sections in much better agreement with the unpublished measurement of Toburen throughout the energy range of the ejected electrons. There are still differences between measured and predicted cross sections in the intermediate electron energy region, but the differences have been reduced from factors of 30 or more at the lowest ejected electron energies to a factor of about two at intermediate energies. The remaining differences might be due to overestimates of the amount of contribution from continuum electron capture derived from scaling of the bare proton spectrum, or to simple inaccuracies in the model.

From the discussion of projectile charge and its effect on the spectra of secondary electron production, it is clear that the effective charge of the projectile is a function of the ejected electron energy, i.e., projectile energy loss. From this it is evident that the single

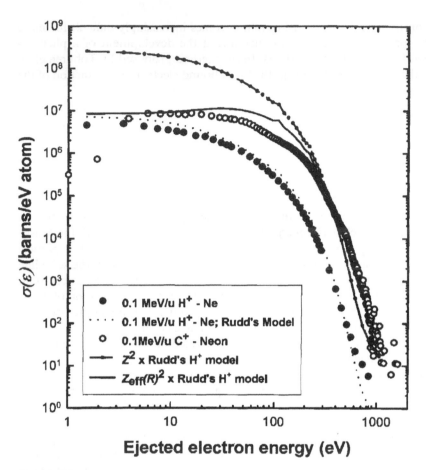

Figure 21 Cross sections form ionization of neon by 0.1-MeV/u protons and singly-charged carbon ions. The experimental data for protons in from the work of Crooks and Rudd [65] and the C^+ results are from some unpublished work of Toburen [56]. The results of Rudd's model were obtained using the parameters given in Ref. 36.

value for effective charge obtained from stopping power theory cannot predict the distribution of secondary electrons that, through subsequent collisions in slowing, produce the spatial patterns of energy deposition important to subsequent radiation chemistry.

7. CHARGE TRANSFER

As positively charged particles slow and reach velocities comparable to bound electron of the medium, they capture and/or lose electrons. These processes can result in ionization of the medium, with no free electron released (electron capture), or release of a free electron from the projectile, with no concomitant target ion being produced (electron loss); however, as pointed out above, these processes often occur with simultaneous ionization of both the projectile and the target. The capture and lose of electrons by the projectile can be an important energy loss mechanism for the incident particle (see Fig. 2) accounting for as much as 15–20% of the total energy-loss cross sections at their maximum. Although the state-to-state electron transfer from target to projectile might not involve significant energy loss by the projectile (in some cases, an energy gain can be experienced), the kinetic energy that must be gained by the captured electron to reach the velocity of the moving ion can be significant. For electron loss by a moving particle, the energy loss by the projectile is simply the binding energy of the stripped electron, assuming energy is not contributed by the velocity distribution of the target electrons that might aid in stripping the projectile electron. With regard to the effects on subsequent chemistry, pure electron capture leaves a residual ion, but no free electron and electron loss contributes a relatively energetic electron to the track of the charged particle, with no residual target ion; again, in reality, simultaneous excitation of the target or projectile often accompanies these processes.

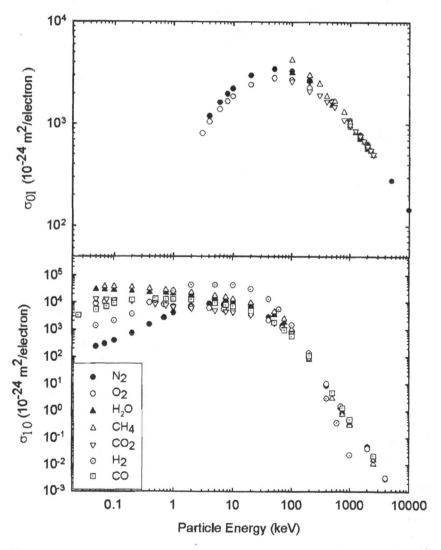

Figure 22   Cross sections for electron capture by protons σ_{10} and electron loss by hydrogen atoms σ_{01} for a number of molecular targets are shown scaled by the number of outer-shell target electrons. (From Ref. 73.)

Because of the importance of electron capture and loss in a number of applied fields, such as fusion plasma, atmospheric physics, and radiological physics, there has been substantial effort contributed to the study of these events for simple collision systems, such as protons and neutral hydrogen particles in simple gas targets. Less is known of the cross sections for helium ions and neutrals, and only a few data exist for heavier ions (for a recent review of the charge transfer data, see Ref. 73 and references therein). Data for electron capture and loss cross sections for proton collisions with a few selected molecular targets are shown in Fig. 22; these data are taken from the review of Toburen [73]. In this example, the cross sections have been scaled by the number of target electrons using the relatively naive assumption that the molecule simply provides a "cluster" of electrons from which the moving ion can either capture or be ionized. Notice that this cross section scaling is reasonably appropriate for fast collisions, i.e., ion energies larger than about 100 keV (and perhaps somewhat lower energies for electron loss), although target structure is seen to play an important role for electron capture by low-energy protons. From the data shown in Fig. 22, one can obtain an appreciation for the broad range of particle energies and target species that have been investigated for protons and neutral hydrogen. A similar plot for other ions would be much less complete; fewer targets have been investigated and a more limited range of particle energies explored.

8. MULTIPLE IONIZATION

Interactions involving electrons and heavy ions can lead to multiple ionization of the target atom or molecule. For example, ionization of an inner shell of an atom can result in subsequent emission of one or more Auger electrons as the excited atom relaxes to the ground state. This leads to two or more electrons, correlated in time and space, being emitted in the immediate region of the original point of ionization. The correlation involving multiple electrons produced by a single ionizing event can have a strong influence on subsequent chemical reactivity of the products of energy deposition. Although inner-shell ionization cross sections are well known, the consequence of multiple ionization is generally neglected in models of chemical kinetics because inner-shell ionization cross sections are generally several orders of magnitude smaller than outer shell ionization, i.e., multiple ionization via inner-shell ionization is a "rare" event.

On the other hand, the study of Auger electron spectra has been a productive means of gaining information on the extent that multiple ionization occurs simultaneous with the production of inner-shell ionization, i.e., the simultaneous ejection of outer-shell electrons accompanying inner-shell ionization. When outer-shell vacancies exist in an atom or molecule that is also ionized in an inner shell, the Auger transitions initiated in the inner shells are shifted in energy by the presence of these outer-shell vacancies and these shifts can be detected in the Auger spectra [74,75]. In fact, for collisions of high-energy heavy ions, the degree of multiple ionization can provide such extreme shifts in the Auger line energies that the normal spectral shapes are replaced with a statistical distribution of transition energies [76]. Although Auger spectra provide indirect information on the amount of multiple outer-shell ionization that accompanies inner-shell ionization, it cannot give a good indication of the overall yields of multiple ionization because a sizable fraction of the multiple ionization does not involve inner-shell ionization processes, but occurs from direct ionization of the outer shell electrons.

As discussed above, heavy charged particles, particularly multiply charged heavy ions, have a relatively large probability for inducing multiple ionization involving outer, as well as inner, shells. For ions that carry several electrons interacting with a target atom, or molecule that also have several bound electrons, multiple ionization may be the rule rather than the exception. Such a collision can be represented as

$$A^{q+} + B \rightarrow A^{(q-j)+} + B^{i+} + (i-j)e^-, \qquad (41)$$

where the collision results in the capture of j electrons by the projectile and ionization of i electrons from the target. Because of the multiple interactions between the nuclei and electrons of the projectile and target, the theoretical treatment of such systems is generally approached statistically [77].

The most detailed experimental information describing multiple ionization has been obtained through the study of recoil target ion states in coincidence with selected projectile states. To fully investigate the system would also require measurements of the electron spectra in coincidence with post collision projectile and target charge states; however, such experiments are extremely difficult. Therefore the general experimental focus has been on the measurements of recoil charge state in coincidence with the selected projectile charge state. Results of a coincidence study by DuBois [17,78] of the multiple ionization of a neon target by proton and helium ions are shown in Fig. 23. The significantly greater degree of double and triple ionization of neon by He ions relative to that by protons is clearly evident. At a few hundred keV, the contribution of double ionization reached more than 25% of that for single ionization; this is much greater than could be explained by contributions from Auger cascades following inner shell ionization. For higher Z target, the contributions from multiple ionization are even greater, reaching 50% for ionization of Kr by He$^+$. Only limited data exist for studies of multiple ionization of molecular targets, and these studies are further complicated by the fact that molecules undergo a many

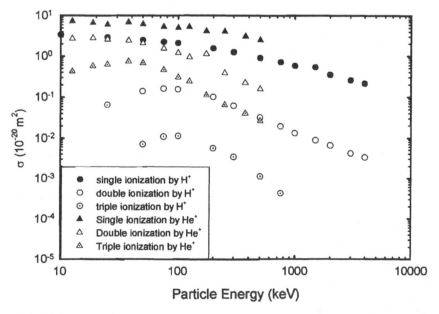

Figure 23 Cross sections for single, double, and triple ionization of neon by singly charged hydrogen and helium ions. (From Refs. 17 and 78.)

possible avenues of dissociation following ionization. In addition, coulomb "explosions" can occur following multiple ionization that can disrupt molecular bonds and have severe consequences on the subsequent radiation chemistry. Clearly more data is needed to evaluate this important mechanism of ionization for heavy charged particles.

There is a wealth of fundamental information available on the interaction of fast charged particles with atoms and molecules. Secondary electron distributions resulting from ionizing collisions are well understood for incident bare projectiles and a few simple dressed ions and neutrals. Even where a quantitative understanding of the collision is not feasible, the basic features of the interactions are understood, making it possible to make educated guesses as to the important parameters of the processes of energy loss. Still our understanding and application of many of the known processes is limited. We are particularly challenged to quantitatively describe the secondary electron spectra for dressed ions and neutrals. We have only limited information on multiple ionization and its consequences in radiation chemistry. There is a wealth of information in hand and a plethora of questions remaining to answer.

REFERENCES

1. Inokuti, M. In *Electronic and Atomic Collisions*; Oda, N., Takayanagi, K., Eds.; North Holland Publishing Co.: Amsterdam, 1980; p 31.
2. Bohr, N. Phys. Rev. 1940, *58*, 654.
3. Bohr, N. Phys. Rev. 1941, *59*, 270.
4. Bohr, N. *Det Kgl. Danske Videnskabernes Selskab Mathematisk-fysiske Meddelelser XVIII, 8.* Hafner Publishing: New York, 1948.
5. ICRU Report 49. *Stopping Powers and Ranges for Protons and Alpha Particles*; International Commission on Radiation Units and Measurements: Bethesda, MD, 1992.
6. Evans, R.D. *The Atomic Nucleus*; McGraw-Hill Book Company, Inc.: New York, 1955.
7. Turner, J.E. *Atoms, Radiation, and Radiation Protection*; John Wiley & Sons, Inc.: New York, 1995.
8. Porter, L.E. Int. J. Quantum Chem. 1997, *65*, 997.
9. Porter, L.E. Nucl. Instrum. Methods 2000, *170*, 35.
10. Toburen, L.H. *IAEA-TECDOC-799;* International Atomic Energy Agency: Vienna, 1995; p 47.
11. Yarlagadda, B.S.; Robinson, J.E.; Brandt, W. Phys. Rev. B 1978, *17*, 3473.
12. Barkas, W.H. *Nuclear Research Emulsions-I. Techniques and Theory*; Academic Press: New York, 1963.
13. McDaniel, E.W.; Flannery, M.R.; Thomas, E.W.; Ellis, H.W.; McCann, K.T.; Manson, S.T.; Gallagher, J.W.; Rumble, J.R.; Beaty, E.C.; Roberts, T.G. *Compilation of Data Relevant to Nuclear Pumped Lasers, Technical Report H-78-1*; US Army Missile Research an Development Command: Redstone Arsenal, Alabama, 1979; Vol. 5, p 1917.
14. Uehara, S.; Toburen, L.H.; Wilson, W.E.; Goodhead, D.T.; Nikjoo, H. Radiat. Phys. Chem. 2000, *59*, 1.
15. Miller, J.H.; Green, A.E.S. Radiat. Res. 1973, *54*, 343.
16. DuBois, R.D.; Toburen, L.H. Phys. Rev. A 1988, *38*, 3960.
17. DuBois, R.D. Phys. Rev. A 1989, *39*, 4440.
18. Ferradini, C.; Jay-Gerin, J.-P. Radiat. Phys. Chem. 1997, *51*, 263.
19. Märk, T.D.; Hatano, Y.; Linder, F. *IAEA-TECDOC-799*; International Atomic Energy Agency: Vienna, 1995; p 163.
20. Srdoč, D.; Inikuti, M.; Krajcar-Bronić, I. *IAEA-TECDOC-799*; International Atomic Energy Agency: Vienna, 1995; p 547.

21. Fano, U. Phys. Rev. 1946, *70*, 44.
22. Miller, J.H.; Wilson, W.E. Radiat. Phys. Chem. 1989, *34*, 129.
23. Bartels, E.R.; Harder, D. Radiat. Prot. Dosim. 1990, *31*, 211.
24. Blohm, R.; Harder, D. Radiat. Prot. Dosim. 1985, *13*, 377.
25. Mozumder, A.; Magee, J.L. J. Chem. Phys. 1966, *45*, 3332.
26. Miller, J.H. Radiat. Res. 1981, *88*, 280.
27. Paretzke, H.G.; Turner, J.E.; Hamm, R.N.; Ritchie, R.H.; Wright, H.A. Radiat. Res. 1991, *127*, 121.
28. Chatterjee, A.; Holley, W.R. Int. J. Quantum Chem. 1991, *39*, 709.
29. Kraft, G.; Krämer, M.; Scholz, M. Radiat. Environ. Biophys. 1992, *31*, 161.
30. Pimblott, S.M.; LaVerne, J.A.; Mozumder, A. J. Phys. Chem. 1996, *100*, 8595.
31. Cobut, V.; Frongillo, Y.; Patau, J.-P.; Goulet, T.; Fraser, M.-J.; Jay-Gerin, J.P. Radiat. Phys. Chem. 1998, *51*, 229.
32. Wilson, W.E.; Nikjoo, H. Radiat. Environ. Biophys. 1999, *38*, 97.
33. Emfietzoglou, D.; Papamichael, G.; Kostarelos, K.; Moscovitch, M. Phys. Med. Biol. 2000, *45*, 3171.
34. Friedland, W.; Bo, W.; Jacob, P.; Paretzke, H. Radiat. Res. 2001, *155*, 703.
35. Inokuti, M. *Physical and Chemical Mechanisms in Molecular Radiation Biology*; Glass, W.A., Varma, M.N., Eds.; Plenum Press: New York, 1991; p 29.
36. ICRU Report 55. *Secondary Electron Spectra from Charged Particle Interactions*; International Commission on Radiation Units and Measurements: Bethesda, MD, 1995.
37. Weigold, E. Aust. J. Phys. 1990, *43*, 543.
38. Bolorizadeh, M.A.; Rudd, M.E. Phys. Rev. A 1986, *33*, 882. Bolorizadeh, Doctoral Dissertation; University of Nebraska, Lincoln, 1984.
39. Kim, Y.-K.; Rudd, M.E. Phys. Rev. 1994, *50*, 3954.
40. Inokuti, M. Rev. Mod. Phys. 1971, *43*, 297.
41. Miller, J.H.; Wilson, W.E.; Manson, S.T.; Rudd, M.R. J. Chem. Phys. 1987, *86*, 157.
42. Dingfelder, M.; Hantke, D.; Inokuti, M.; Paretzke, H.G. Radiat. Phys. Chem. 1998, *53*, 1.
43. Kim, Y.-K. Phys. Rev. A 1983, *28*, 656.
44. Kim, Y.-K. *RADIATION RESEARCH Biomedical Chemical and Physical Perspectives*; Academic Press, Inc.: New York, 1975.
45. Paretzke, H.G. *Simulation von Elektronenspuren im Energiebereich 0.01–10 keV in Wasserdampf*; Gesellschaft für Strahlen-und Umweltforschung mbH: München, 1989.
46. Stolterfoht, N.; DuBois, R.D.; Rivarola, R.D. In *Electron Emission in Heavy Ion-Atom Collisions, Springer Series on Atoms + Plasmas*; Toennies, J.P., Ed.; Springer: Berlin, 1997.
47. Toburen, L.H.; Wilson, W.E. Phys. Rev. A 1972, *5*, 247.
48. Toburen, L.H. Phys. Rev. A 1971, *3*, 216.
49. Lynch, D.J.; Toburen, L.H.; Wilson, W.E. J. Chem. Phys. 1976, *64*, 2616.
50. Rudd, M.E.; Gregoire, D.; Crooks, J.B. Phys. Rev. A 1971, *3*, 1635.
51. Madison, D.H. Phys. Rev. A 1973, *8*, 2449.
52. Bonsen, T.F.M.; Vriens, L. Physica 1970, *47*, 307.
53. Rudd, M.E.; Macek, J. Case Stud. At. Phys. 1972, *3*, 47.
54. Manson, S.T.; Toburen, L.H.; Madison, D.H.; Stolterfoht, N. Phys. Rev. A 1975, *12*, 60.
55. Kim, Y.-K. Radiat. Res. 1975, *61*, 21.
56. Toburen, L.H. In *Physical and Chemical Mechanisms in Molecular Radiation Biology*; Glass, W.A., Varma, M.N., Eds.; Plenum Press: New York, 1991; p 51.
57. Wilson, W.E.; Toburen, L.H. 0.3-2.0 MeV. Phys. Rev. A 1975, *11*, 1303.
58. Toburen, L.H.; Wilson, W.E. J. Chem. Phys. 1977, *66*, 5202.
59. Siegbahn, K.; Nordling, C.; Johansson, G.; Hedman, J.; Hedén, P.F.; Gelius, U.; Bergmark, T.; Werme, L.O.; Manne, R.; Baer, Y. *ESCA Applied to Free Molecules*; North Holland Publishing Co.: Amsterdam, 1969.
60. Miller, J.H.; Toburen, L.H.; Manson, S.T. Phys. Rev. A 1983, *27*, 1337.
61. Miller, J.H.; Wilson, W.E.; Manson, S.T.; Rudd, M.E. J. Chem. Phys. 1987, *86*, 157.

62. Rudd, M.E. Phys. Rev. A 1988, *38*, 6129.
63. Rudd, M.E.; Kim, Y.-K.; Madison, D.H.; Gay, T.J. Rev. Mod. Phys. 1992, *64*, 441.
64. Rudd, M.E. Phys. Rev. A 1979, *20*, 789.
65. Crooks, J.B.; Rudd, M.E. Phys. Rev. A 1971, *3*, 1628.
66. Stolterfoht, N. Z. Phyzik 1971, *248*, 92.
67. Bolorizadeh, M.A.; Rudd, M.E. Phys. Rev. A 1986, *33*, 893.
68. Toburen, L.H.; Wilson, W.E.; Popowich, R.J. Radiat. Res. 1980, *82*, 27.
69. Briggs, J.S.; Taulbjerg, K. Top. Curr. Phys 1978, *5*, 105.
70. Manson, S.T.; Toburen, L.H. Phys. Rev. Lett. 1981, *46*, 529.
71. Toburen, L.H.; Stolterfoht, N.; Ziem, P.; Schneider, D. Phys. Rev. A 1981, *24*, 1741.
72. McClure, J.H.; Stolterfoht, N.; Simony, P.R. Phys. Rev. A 1981, *24*, 97.
73. Toburen, L.H. Radiat. Environ. Biophys. 1998, *37*, 221.
74. Watson, R.L.; Toburen, L.H. Phys. Rev. A 1973, *7*, 1853.
75. Stolterfoht, N.; Schneider, D. Phys. Rev. A 1975, *11*, 721.
76. Jamison, K.A.; Woods, C.W.; Kauffman, Robert; Richard, P. Phys. Rev. A 1975, *11*, 505.
77. Olson, R.E.; Ullrich, J.; Schmidt-Böcking, H. Phys. Rev. A 1989, *39*, 5572.
78. DuBois, R.D.; Toburen, L.H.; Rudd, M.E. Phys. Rev. A 1984, *29*, 70.
79. DuBois, R.D. Phys. Rev. A 1989, *39*, 4440.

4

Modeling of Physicochemical and Chemical Processes in the Interactions of Fast Charged Particles with Matter

Simon M. Pimblott and A. Mozumder
University of Notre Dame, Notre Dame, Indiana, U.S.A.

1. INTRODUCTORY REMARKS

In Chapter 2 of this book, the energy transfer interactions of fast charged particles with matter were discussed from the viewpoint of the incident particle. This chapter will describe and examine the consequences of energy deposition in matter with the goal of providing an understanding of the physico-chemical and chemical processes involved.

Energy deposition in a molecule is a necessary precursor to radiation-induced transformations (see Sec 1.1 for a similar statement for photochemistry). In most situations, especially in condensed media, the deposited energy is localized and available for physical and chemical transformations. A sequence of correlated and localized energy loss events, frequently in electronic form, may be termed a track. The incident particle generating the track may suffer occasional changes in direction, due mainly to elastic scatterings. This effect becomes increasingly important as the particle slows down. At first, the lost energy appears in electronic form then generations of transformations may take place before a particular radiation effect is "observable." When there is sufficient deposited energy, ionization ensues and, frequently, secondary ionization and excitation will also occur. These correlated events are localized in the vicinity of the primary interaction. For low linear energy transfer (LET) radiations, the primary energy loss events are well separated. The cluster of ionizations and excitations associated with each primary event is called a spur, and on a certain timetable and in a certain chemical sense these clusters develop independently. For high-LET radiations, the spurs of a track coalesce to form a more continuous, cylindrical distribution. The thrust of this chapter is concerned mainly with spur and track reactions. However, in conformity with the general aim and purpose of this book, it should be remembered that the arguments are not limited to radiation chemistry but can and should be extended to other applications (such as radiation biology and certain industrial applications, etc.). One restriction is that the discussion will be chiefly concerned with condensed phase phenomena, although sometimes reference will be made to gas-phase effects for the sake of comparison.

The organization of this chapter is as follows. In the following section, Sec. 4.2, the elastic and inelastic interaction cross sections necessary for simulating track structure (geometry) will be discussed. In the next section, ionization and excitation phenomena and some related processes will be taken up. The concept of track structure, from historical idea to modern track simulation methods, will be considered in Sec. 4.4, and Sec. 4.5 deals with nonhomogeneous kinetics and its application to radiation chemistry. The next section (Sec. 4.7) describes some application to high temperature nuclear reactors, followed by special applications in low permittivity systems in Sec. 4.8. This chapter ends with a personal perspective. For reasons of convenience and interconnection, it is recommended that appropriate sections of this chapter be read along with Chapters 1 (Mozumder and Hatano), 2 (Mozumder), 3 (Toburen), 9 (Bass and Sanche), 12 (Buxton), 14 (LaVerne), 17 (Nikjoo), and 23 (Katsumura).

2. INTERACTION CROSS SECTIONS

In the interaction of fast charged particles with matter, the averaged macroscopic quantities related to radiation physics can be evaluated from cross-sectional information that is of limited complexity. For instance, the stopping power, which is the energy-weighted integral over the inelastic collision cross section, can be computed fairly accurately using fairly approximate treatments (see Sec. 4.2.2). Bohr's theory is a prime example of such an approach. On the other hand, chemical and other effects may depend on a specific excited or ionized state of a molecule arising from a particular energy deposition event and on the local track geometry. Elucidation of such effects therefore requires more accurate cross-sectional information. Furthermore, because of the generation of copious low-energy secondary electrons, the elastic and inelastic collision cross sections for these interacting particles are always significant, even if the incident primary particle is highly energetic.

As an incident radiation particle slows down, various "track" processes occur with sometimes several processes occurring simultaneously. Initially, electronic excitation dominates inferior vibrational and rotational excitation. All the time, elastic collisions delineate the track geometry. For the purpose of track simulation, it is often convenient to think in terms of a collision mean free path, which is inversely related to the cross section. The basic requirement for realistic simulation is then a comprehensive set of inelastic and elastic collision cross sections for energetic (heavy) ions and for electrons over a wide span of energy. In most cases, a doubly differential cross section in energy and angle is sufficient, but in certain cases, where detailed track geometry is not critical, a cross-section differential in energy is adequate.

2.1. Cross-Sections for Track Structure Simulation

Despite the fact that Bohr's stopping power theory is useful for heavy charged particles such as fission fragments, Rutherford's collision cross section on which it is based is not accurate unless both the incident particle velocity and that of the ejected electron are much greater than that of the atomic electrons. The quantum mechanical theory of Bethe, with energy and momentum transfers as kinematic variables, is based on the first Born approximation and certain other approximations [1,2]. This theory also requires high incident velocity. At relatively moderate velocities certain modifications, shell corrections, can be made to extend the validity of the approximation. Other corrections for relativistic effects and polarization screening (density effects) are easily made. Nevertheless, the Bethe–Born approximation gradually breaks down at electron energies of about 500 to 1000 eV in most media of low atomic number.

The basic input information used in most cross-section calculations is the differential dipole oscillator strength distribution (DOSD) introduced in Sec. 2.2.2 and 2.2.3. For polyatomic molecules, this is a continuous density distribution $f(\varepsilon)$ such that the dipole oscillator strength for transition energies between ε and $\varepsilon + d\varepsilon$ is given by $f(\varepsilon)d\varepsilon$. In terms of the DOSD, the total number of electrons receiving any energy transfer, the effective mean excitation potential, the mean energy loss during a collision, and the mean excitation energy for straggling [3] are given by

$$Z_{\text{eff}} = \int_{\varepsilon_0}^{\varepsilon_{\max}} f(\varepsilon)d\varepsilon$$

$$Z_{\text{eff}} \ln(I_{\text{eff}}) = \int_{\varepsilon_0}^{\varepsilon_{\max}} f(\varepsilon)\ln \varepsilon \, d\varepsilon$$

$$Z_{\text{eff}} \bar{\varepsilon} = \int_{\varepsilon_0}^{\varepsilon_{\max}} \varepsilon f(\varepsilon)d\varepsilon \tag{1}$$

$$Z_{\text{eff}} \bar{\varepsilon} \ln(\bar{I}) = \int_{\varepsilon_0}^{\varepsilon_{\max}} \varepsilon f(\varepsilon)\ln \varepsilon \, d\varepsilon$$

respectively [4]. In the above integrals ε_0 is the lowest excitation potential of the molecule and $\varepsilon_{\max}$ is the maximum energy transfer that is kinematically permissible. Because the incident and ionized electrons are indistinguishable, $\varepsilon_{\max}$ virtually equals one-half of the incident energy E, if that is large compared with the binding energy B. Otherwise, it should be taken as $\sim (E + B)/2$. With known values of Z_{eff}, I_{eff}, $\bar{\varepsilon}$, and I, the electron stopping power at relatively low incident energies may be expressed in the form [5]:

$$S(\text{electron}) = Z_{\text{eff}} \left[\ln \left(\frac{mv^2}{I_{\text{eff}}} \sqrt{\tilde{e}/8} \right) - \frac{\bar{\varepsilon}}{2E} \ln \frac{4\tilde{e}E}{I} \right] \tag{2}$$

When E is sufficiently large that almost all the oscillator strength is contained within $\varepsilon_{\max}$, Z_{eff} and I_{eff} approach the limiting values of Z and I, respectively, in which case Bethe's nonrelativistic formula is recovered. Comparing Eq. (2) with Eq. (2.2.7), which includes relativistic, shell, and density corrections, reveals that, as is stands, Eq. (2) does correct for the energetically inaccessible states in the Bethe theory and therefore may be used at intermediate electron energies. However, it is still limited to nearly zero momentum transfer in collision kinematics. Ashley [6,7] has extended the use of the DOSD spectrum to account for the nonvanishing momentum transfer by a quadratic approximation in the energy–momentum plane. It is better represented by the imaginary part of the reciprocal of the dielectric response function $\varepsilon(q,\omega)$, where $\hbar q$ and $\hbar \omega$ are the momentum and energy losses, respectively, in the collision process. Ashley's procedure involves replacing $\text{Im}(-1/\varepsilon(q,\omega))$ by $\text{Im}(-1/\varepsilon(0,\omega - \hbar q^2/2m))$. The relationship between the reciprocal dielectric response function and the DOSD has been discussed in Sec. 2.2.4 and is reproduced here with a slight and obvious change in notations, viz. $\text{Im}(-1/\varepsilon(0,\omega)) = (\hbar^2 e^2 NZ/2m)f(\varepsilon)/\varepsilon$, with $\varepsilon = \hbar \omega$. Ashley's approximation has been well documented in a number of cases, including C and the metals Al and Cu, and has been found to be accurate. Its most frequent application has been to liquid water for relatively low energy electrons and protons [8,9].

Following Ashley's procedure and using the DOSD, $f(\varepsilon)$, the inelastic mean free path Λ_i for electronic excitation and ionization may be given by [8]

$$\Lambda_i^{-1} = \chi N \int \varepsilon^{-1} f(\varepsilon) L(\varepsilon/E) d\varepsilon \tag{3}$$

here $\chi = 2\pi e^4/mv^2$, $s = (1-2a)^{1/2}$ and

$$L(a) = a^{-1} \ln \left[\frac{(1+a-s)(1-a+s)}{(1-a-s)(1+a+s)} \right] - \frac{2}{(1+a)} F\left[\arcsin\left(\frac{s}{1-a}\right), \frac{(1-a)}{(1+a)} \right]$$

Here F denotes an elliptic integral of the first kind in Legendre's normal form [10]. The corresponding stopping power equation may also be expressed in a closed form [8] and when expanded in powers of ε/E, the lead term generates Bethe's nonrelativistic result for the stopping power for an electron. For the purpose of track simulation, the desired quantity is the cumulative inelastic cross section of an event with energy loss less than ε, which is given by

$$\sigma(E,\varepsilon) = N^{-1} \left[\Lambda_i^{-1} + (\chi/E) \int_{\varepsilon_0}^{\varepsilon} f(\varepsilon') M(\varepsilon/E, \varepsilon'/E) d\varepsilon' \right] \tag{4}$$

in Ashley's approximation, where

$$M(b,a) = \frac{1}{a} \ln \left[\frac{(b-a)(1+a-b)}{b(1-b)} \right] + \frac{2}{1+a} F\left[\arcsin\left(\frac{1+a-2b}{1-a}\right), \frac{1-a}{1+a} \right]$$

The density of the actual energy loss is, of course, the derivative of the expression in Eq. (4) The Ashley approximation greatly enhances the validity of the computed cross section for electronic excitation. However, as only the DOSD is used, excitations to states that are dipole forbidden are not included. This may not be a serious limitation unless the incident energy is very low. An additional advantage of the approximation is that it is conveniently used in the condensed phase.

2.2. Vibrational and Elastic Collision Cross Sections

For vibrational excitation under electron impact in the gas phase, experimental measurements are generally available and are usually more reliable than calculation. In the case of gaseous water, an almost complete compilation by Hayashi [11] is available in the all-important lower energy regime. Hayashi decomposed the total vibrational collision cross section into three components: σ_{v_1,v_3} for the v_1 and v_3 stretching modes, which are virtually indistinguishable; σ_{v_2} for the v_2 bending mode; and σ_w for the balance. For polyatomic molecules, rotational excitation involves very little energy relative to electronic or vibrational excitation. In any case, experimentally these events cannot be distinguished from elastic collision and are included with them in cross sections. Cross-sectional data for vibrational excitation are of very limited availability in the condensed phase. Because internuclear forces are much stronger than the intermolecular forces, it is generally believed that condensation does not effect vibrational cross sections a great deal and gas-phase values are frequently used to represent the liquid in simulations. A detailed description of the cross sections for electrons in liquid water is presented in Ref. 9, where the relevant cross sections have been quantified for incident electron energy in the interval 10 eV to 100 keV.

Elastic collision cross sections are important for track simulation and differential cross sections are needed to calculate angular deviation in the track trajectory. Pimblott et al. [9] have given an elaborate analysis for gaseous water and compared the results with the experiments of Katase et al. [12]. In brief, the total elastic cross section

$$\sigma_{el} = 2\pi \int_0^\pi [d\sigma/d\theta]\sin\theta d\theta$$

is well represented by the Moliere form for the differential cross section, viz. $d\sigma/d\theta = (Z^2 + Z)e^4/(p^2v^2(1-\cos\theta + 2\eta^2))$, where Z is the number of electrons in the molecule, p is the electron momentum at velocity v, and the screening parameter, η, is a carefully chosen parameter. Grosswendt and Waibel's [13] choice for η is

$$\eta = Z^{2/3}[1.64 - 0.0825 \ln(mc^2 T)]/[T(T+2)]$$

where T is electron kinetic energy in units of mc^2. This represents the total elastic cross section well but, unfortunately, does not match the experimental differential cross section for water. The analytical formula gives too little scattering in the forward direction, and it results in a nearly isotropic distribution. Therefore Pimblott et al. [9] chose to fit the differential experimental elastic cross section with a fourth-order polynomial $a_0 + a_1\theta + a_2\theta^2 + a_3\theta^3 + a_4\theta^4$, in which the a-coefficients themselves were fitted, as a function of energy, with another fourth-order polynomial. The finally fitted cross section agreed with experiment quite well. It should be emphasized that the fitting was done for the *gas phase* of water. The use of these cross sections to represent liquid water is questionable, but there are no experimental measurements available. The approximation has been justified by the belief that condensation does not affect the elastic collision cross section a great deal as these processes are

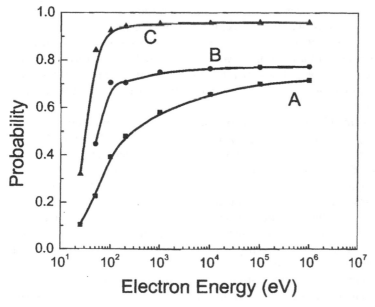

Figure 1 Electron energy dependence of the inelastic collision probability, $\sigma_{inelastic}/(\sigma_{inelastic} + \sigma_{elastic})$, in liquid water (A) and the probability of an inelastic collision causing ionization, $\sigma_{ion}/(\sigma_{ion} + \sigma_{ex})$, in gaseous (B) and liquid (C) water.

dominated by interaction of the electron with nuclear charges. While this may be true for polyatomic molecules, it is well known that the elastic collision cross sections of atomic liquids can be very different from those in the gas phase due to correlation effects represented by the structure factor. In any event, it is customary to use the gas-phase elastic cross sections in the condensed phases of molecular compounds for track simulation purposes.

For the purpose of track simulation, it is necessary to know whether the next realized collision would be inelastic, and if it is, then either the collision would result in ionization or in the production of a molecular excited state. The probabilities of these processes are given by $\sigma_{inelastic}/(\sigma_{elastic} + \sigma_{inelastic})$, where $\sigma_{elastic}$ and $\sigma_{inelastic}$ are the cross sections for elastic and inelastic collisions, and $\sigma_{ion}/(\sigma_{ex} + \sigma_{ion})$, where the cross sections for molecular excitation and ionization are denoted by σ_{ex} and σ_{ion}, respectively. These probabilities for water in the gaseous and liquid phases are shown as a function of electron energy in Fig. 1. The probability of an inelastic process is virtually independent of the phase, although there is a strong dependence on electron energy. On the other hand, the ionization probability depends both on the phase and on the electron energy, with more ionization occurring in the liquid phase at any electron energy.

3. IONIZATION, EXCITATION, AND RELATED PHENOMENA

The first excitation potential of a polyatomic molecule is generally unambiguous. For ordinary molecules it usually corresponds to transitions in the UV or in the far UV. The excited molecule, either in the first or in a higher excited state, may undergo prompt dissociation, or fast internal conversion to states of the same multiplicity or convert, relatively slowly, to states of different multiplicity by intersystem crossing. Kahsa's rule applies in most cases, signifying that the conversion to the state of lowest energy, singlet or triplet, is very fast and further transformations occur relatively slowly. Occasionally, the excited molecule will undergo a radiative transition resulting in the emission of a photon. This process generally has a low probability, except for aromatic molecules where the probability can be significant. (Of course, the probabilities of the various outcomes, once the excited molecule is formed, must add up to unity.) When the excited state is formed by the absorption of a single photon, the probabilities are referred to as quantum yields. If the state results from the absorption of n photons then the quantum yield is, by definition, that respective probability divided by n.

Excited states can be formed by a variety of processes, of which the important ones are photolysis (light absorption), impact of electrons or heavy particles (radiolysis), and, especially in the condensed phase, ion neutralization. To these may be added processes such as energy transfer, dissociation from super-excited and ionized states, thermal processes, and chemical reaction. Following Brocklehurst [14], it is instructive to consider some of the direct processes giving excited states and their respective inverses. Thus luminescence is the inverse of light absorption, super-elastic collision is the inverse of charged particle impact excitation, and collisional deactivation is the inverse of the thermal process, etc.

The term ionization may refer to different processes depending on the context. For radiation effects in the gas phase, it usually implies the removal of the least bound electron to infinity. Such a theoretical definition is not feasible in the condensed phase and it is necessary use a heuristic or operational procedure. Thus, in liquid hydrocarbons, one may use the electron scavenging reaction or a conductivity current to quantify the electrons liberated from molecules. It has only been possible to extrapolate the conductivity current at a low irradiation dose and at a relatively low external field to saturation in the cases of liquefied rare gases (LAr, LKr, and LXe) so that an unequivocal meaning to ionization may be accorded [15].

The ionization potential is the minimum energy required to remove an electron from a molecule, and as such it may be defined uniquely in the gas phase. Even so, as first pointed out by Platzman, ionization is not a certainty when that amount of energy is supplied to the molecule. In "super-excited" states, ionization for polyatomic molecules is in effective competition with dissociation into neutral fragments. At the same time, a small amount of ionization may occur at energies a little below the ionization potential by a variety of secondary processes. This blurring of the boundaries gives rise to the concept of ionization efficiency, which is the experimental ratio of the photoionization cross section to the photoabsorption cross section. This efficiency can be given a probabilistic interpretation if it is assumed that once an excited state is formed its eventual fate is independent of the process by which it was formed. In this sense, once the ionization efficiency η_i of a given state is known, the cross sections for ionization and excitation may be given by $\eta_i\sigma$ and $(1-\eta_i)\sigma$, respectively, where σ is the cross section of production of that state either by photoabsorption or by the impact of charged particles. This, of course, is true only when there is no other channel for energy disposal, such as nonradiative transition to the ground state. In the case of water in the gas phase, where the ionization potential is well established at 12.6 eV, the ionization efficiency, as a function of photon energy, shows a series of maxima and minima until it asymptotically approaches unity at ~ 20 eV [16]. This dependence implies that the efficiency of ionization depends not only on energy but also on the quantum nature of the state reached by the absorption of a specified amount of energy.

Closely associated with the ionization potential, but distinct from it, are the W-value and the appearance potential. The minimum energy necessary to observe the production of a particular ion out of a given molecule is called its appearance potential (see Ref. 2, p. 72). As this is an experimental quantity without reference to the state originally produced, there has been considerable confusion in the literature between the ionization potential and the appearance potential in the liquid phase. In Sec. 4.3.2, a clarification is attempted for the special case of liquid water. The W-value refers to the gross average energy needed to create a single ionization in the medium, whether primary or secondary. It is greater than the ionization potential, because of energy wastage in various nonionizing events, viz., production of excited states, delivering energy to subexcitation electrons and to the resultant positive ions, etc. A more detailed discussion will be found in Sec. 2.3.3.

3.1.   Photoionization and Electron Impact Ionization

Photoionization is the process by which an electron is removed to generate a positive ion by the absorption of light. A related process in which an electron is removed from a negative ion to produce a neutral entity by light absorption is called photo detachment. The inverse of photoionization is radiative capture of an electron. Applying microscopic reversibility to the direct and inverse processes, the respective cross sections of photoionization (σ_i) and radiative capture σ_r may be related by detailed balance as, $\sigma_i/\sigma_r = (2\omega_f/\omega_i)(mc^2 T/h^2\nu^2)$, where ν is the frequency of light, and ω_f and ω_i are the statistical weights of the final and initial states, respectively. The factor 2 represents the spin degeneracy. A similar relationship may be expected between the photo detachment and electron attachment cross sections. These are very useful when one or another process is not easily amenable to experimental investigation. It should be noted that the "ionization cross sections" are usually smaller than the respective "photoabsorption cross sections" to which they are related by the ionization efficiency.

Ionization by a charged particle or by electron impact is a special inelastic collision process in which molecular ionization occurs and the ejected electron carries the residual energy. The basic cross section is still given by the Bethe–Born theory as was used in Sec. 2.2 for stopping power and in Sec. 2.3 for secondary ionization. It should be noted that the complete description of the impact ionization process requires a fivefold differential cross section: one in the ejected electron energy and two each for the angular distributions of the ejected electron and the primary particle. Such detailed information is rarely available experimentally. Furthermore, the approximations used in the theory break down more easily when higher order differential cross sections are used. In radiation science, one is frequently concerned with the total number of ionizations. As such, a singly differential cross section in ejected electron energy is sufficient. For track simulation, a cross-section further differential in the angle of the ejected electron is necessary. However, in the more important case when the energy of the secondary electron is much greater than the ionization potential, a nearly classical relation between the ejected electron energy and angle can be invoked. In any case, it should be emphasized that the ionization cross section, differential in energy, such as represented by Eq. (2.10), implies that ionization is a certainty once the deposited energy exceeds the ionization potential. Because of the existence of super-excited states, these cross sections must be multiplied by the ionization efficiency factor. Presently, the only useful source for this ionization efficiency is experiment in the far UV. Even then, it is still necessary to assume that the efficiencies remain valid for the dipole-allowed states reached by charged particle impact.

Under electron impact ionization, when the energy transfer greatly exceeds the orbital ionization energy, the process resembles a Rutherford-type collision between two nearly free electrons. On this Mott [17] imposed the condition of indistinguishability of the outgoing electrons and obtained in the nonrelativistic limit

$$(d\sigma/dE)_{\text{Mott}} = \left(4\pi a_0^2 R^2/T\right) \sum_i n_i \left[W_i^{-2} - W_i^{-1}(T - E)^{-1} + (T - E)^{-2} \right] \tag{5}$$

where the same symbols are used as in Eq. (2.10). The second term within the brackets on the right-hand side of Eq. (5) represents the interference between direct and exchange scatterings. A manifestly relativistic treatment of the process has been given by Moller [18], however, with the caveat that in the extreme relativistic case and for a large angle scattering, energy loss in the radiative process may not be negligible [19].

A wide variety of chemical reactions can occur following ionization or excitation of a molecule in both gaseous and condensed phases. These may be of uni-molecular or bi-molecular nature, initiated by electrons, ions or by the transformations of excited or ionized molecules. These reactions include, but are not limited to, dissociation, elimination of atoms and smaller molecules (H, H_2, etc.), transfer of H^+, H, H_2, H^- and H_2^-, fragmentation, ion-molecule reaction, luminescence and energy transfer, neutralization, chain reaction, condensation, and polymerization, etc. These reactions will not be reviewed in this chapter but may be found elsewhere in this book. A brief summary is also found in Chapters 4 and 5 of Ref. 2. In the next section, some features of yields and mechanisms following excitation and/ or ionization in the liquid phase are discussed with special reference to water.

3.2. Condensed Phase: Yields and Mechanisms

The difficulty of providing a theoretical definition of ionization in the liquid akin to that in the gas phase has already been discussed. In practice, any available reaction that mimics that of a free electron can be taken as an operational definition of ionization. These include scavenging, attachment (dissociative or nondissociative), and other reactions of the electron, as well as generation of solvated electrons in polar liquids and of free electrons in nonpolar liquids. In some hydrocarbon liquids, careful extrapolation of the yield of scavenged electrons gives a total ionization yield which is not much different from the corresponding ionization yield in the gas phase; however, this is not always true. Extrapolation of the free-ion yield at infinite external electric field strength is generally uncertain, depending on an assumed form of the initial electron-ion separation distance, although in many cases such extrapolated ionization yields have been rationalized with those obtained in scavenging experiments.

In highly polar liquids, such as water, the situation is altogether different and until recently, the mechanisms and the yields of ionization and excitation were controversial. Although it has been known for a long time that the fates of the lower excited states of water are the same in all phases, viz. dissociation into $H + OH$, or into $H_2 + O$, the calculated yields in the liquid phase exceeded experimental determinations by as much as a factor of four [20,21]. An ad interim rationalization in terms of fast cage recombination lacks realism in view of the open structure of liquid water. As for ionization in liquid water, a very small yield is seen at 6.5 eV [22,23] and this threshold has been variously called ionization potential [23], threshold potential [24], or appearance potential [21]. Sander et al. [24,25] have postulated that the mechanism of ionization by which low energy processes give e_{aq}^- is not due to direct or auto-ionization but to optical charge transfer or to photo-induced electron transfer. A concerted proton-coupled electron transfer mechanism was first proposed by Keszei and Jay-Gerin [26] with later experimental support up to 9.3 eV [27]. Sander et al. [24] suggest that quasi-free electrons may not be produced in the liquid until the band gap energy, estimated around ~10–12 eV is reached.

The early work on the photolysis of water was in the gas phase employing one photon. The branching ratio of the photodissociation into $H + OH$ and $H_2 + O$ was reported by McNesby et al. [28] as 3:1 at a photon energy of 10.03 eV. Ever since, that ratio has been consistently revised in favor of the $H + OH$ reaction with the final result of Stief et al. [29] giving 0.99:0.01 for 6.70–8.54 eV photon energy and 0.89:0.11 for the interval 8.54–11.80 eV. In the absence of direct determination these ratios often are assumed valid in the liquid phase. In the early work of Sokolev and Stein [30], mainly the photodissociation quantum yield in liquid water was measured, but a small photoionization yield of ~0.05 was attributed to the process

$$(H_2O)_{aq} \xrightarrow{h\nu} (H_2O)^{**}_{aq} \rightarrow e_{aq}^- + OH + H_3O_{aq}^+$$

In the sense that the hydrated electron may be formed by the reaction of an excited state with a ground state water molecule, the above reaction may have some validity [27]. In some of the earlier work the initial yields were directly determined from scavenger studies. Later works, using mostly two-photon photolysis, have been extended to higher energies and have yielded more accurate photodissociation and photoionization yields. These have been recently summarized in Ref. 31. Briefly, the time dependencies were determined in the pico- and subpicosecond time scales from which the initial and escape yields were found using either analytical or Monte Carlo methods for the recombination kinetics. Fig. 2 shows the variation of the quantum efficiencies of photodissociation (η_d) and photoionization (η_i) in liquid water as a function of photon energy in which some earlier data are incorporated. The quantum efficiency of nonradiative transition to the ground state, computed as

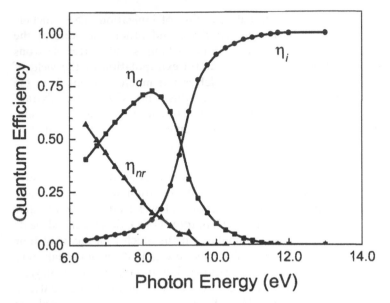

Figure 2 Variation of the quantum efficiencies of photodissociation (η_d), of photoionization (η_i), and of nonradiative transition to the ground state, $\eta_{nr} = 1 - \eta_d - \eta_i$, in liquid water as a function of photon energy.

$(\eta_{nr}) = 1 - \eta_d - \eta_i$, is also shown. It is significant below 9 eV but was ignored in earlier expositions. The photodissociation quantum yield is very low below 6.70 eV because of the high probability of nonradiative transition to the ground state and again very low above 10.0 eV because of the high ionization probability. It therefore exhibits a peak at ~8 eV. The ionization probability in liquid water has been postulated to reach a near certainty at a photon energy 11.7±0.2 eV (see Fig. 2). Therefore, in accordance with the analysis of Sander et al. [24,25], the ionization potential in liquid water may be defined as the minimum energy needed to generate a quasi-free electron in the band gap, and a value of 11.7 eV can be assigned to it [31]. Notice that, according to this definition, the existence of the superexcited state is denied in liquid water as the ionization potential is defined with respect to the ionization efficiency approaching unity and this quantity does not have maxima or minima. An approximate calculation of the ionization yield in liquid water for high energy electron radiolysis, using the data of Fig. 2 and a simplified Bethe theory, gives a G-value (yield for 100 eV absorbed energy) of 1.0 at first collision and a total G-value of 2 obtained from it by heuristic energy balance [31]. The latter is in good agreement with a recent experimental re-evaluation of Bartels and co-workers [32]. The computed dissociation yield, however, falls short of the experimental value by more than an order of magnitude. Therefore not only is there no need for invoking cage recombination, there is actually a need for an additional channel for dissociation into neutral fragments. Such a process, called dissociative electron attachment (DEA), is well known in the gas phase and appears as a compound resonance in low temperature ice in the energy interval ~5–15 eV [33]. The immediate product of DEA is $H^- + OH$, where H^- carries most of the excess energy and is likely to detach the electron quickly. The net result is an electron degraded in energy and a dissociated water molecule. An estimate of the total dissociation yield ($G \sim 0.8$) based on this hypothesis seems to be correct in the order of magnitude [31].

4. TRACK STRUCTURE

4.1. Historical Perspective

The spatial distribution of the energy loss events of a charged particle is usually referred to as a track. This conceptual picture of a track is the backbone of the theoretical description of radiation chemistry. Tracks are considered to have a transitory existence and exist so long as permitted by the diffusion and fast reactions of radiation-produced intermediates (ions, electrons, and radicals). A large body of radiation-physical and radiation-chemical phenomena requires track models for their elucidation, including (1) LET variation of product yields; (2) energy loss in primary excitations and ionizations; (3) yield of escaped ions; (4) radiation-induced luminescence; and (5) particle identification.

The statistics of the energy loss events and their spatial distribution along a track, and its branches, defines the structure of a radiation track. The ensuing chemistry depends on incident particle energy and particle quality (LET). In 1966, a rudimentary Monte Carlo method was used by Mozumder and Magee [35] for the statistics of energy loss events by high energy electrons in water, employing simplified Bethe cross section derived from a dipole oscillator distribution, which in turn was synthesized from then known moments and experimental constraints. Their calculation divided energy deposition events into three groups as follows: (1) spurs (spherical entities, up to 100 eV); (2) blobs (spherical or ellipsoidal, 100–500 eV); and (3) short tracks (cylindrical, 500–5000 eV). This arbitrary classification has proven to be tremendously useful and extremely robust in modeling radiation-chemical kinetics. The track structure calculation showed that the energy partition between the three track structure entities is a strong function of incident electron energy, dividing approximately as the ratio of 0.64:0.12:0.24 in the spur, blob, and short track fractions for ^{60}Co-γ irradiation. Recently, a more sophisticated calculation based on improved cross-section methodology employing a realistic dipole oscillator distribution for liquid water has provided an improved estimate of the ratio, 0.75:0.12:0.13, between the spur, blob, and short track fractions [36].

The reaction of the radiation-induced radicals and ions comprising a track is significant in the condensed phases and the observed chemistry reflects the competition between the diffusive relaxation of the track structure and the intra-track reaction. By contrast, in the gas phase the reactants are widely dispersed at any reasonable dose and LET, and the intermingling of inter- and intra-track reactants gives a homogeneous character to the reactions. The first nonhomogeneous diffusion–reaction treatment of the radiation chemistry of liquid water was proposed by Samuel and Magee [37] to explain the relative forward yield as a function of LET. (Forward yield means the yield of observable molecular product against radical yield, where the latter is estimated as the yield of scavenger reaction at a small scavenger concentration.) Their simple analysis used a one-radical model, which made no distinction between H and OH. Explicit consideration of reaction with homogeneously distributed scavengers within the context of a one-radical-species model was later introduced by Ganguly and Magee [38].

4.2. Track Structure Simulation

Track structure simulation has found application in many areas of radiation research since the pioneering studies of Mozumder and Magee [35]. These studies all employ essentially the same type of approach, a collision-to-collision modeling of the trajectory of the primary radiation particle and of its daughter secondary electrons, with the most significant difference between different calculations being the interaction cross sections used to describe the various inelastic and elastic processes. The attenuating medium is usually treated as a homogeneous continuum. The basic methodology for fast electrons is as follows: The primary electron has an energy E and is at point z traveling in a defined direction. The distance the radiation particle travels before its next collision, Δz, has a Poisson distribution with a mean free path, Λ_{total}, which is defined by the total cross section, σ_{total}, for elastic and inelastic processes, i.e.,

$$\Lambda_{total} = (\rho\sigma_{total})^{-1} = (\rho(\sigma_{elastic} + \sigma_{inelastic}))^{-1} \qquad (6)$$

where ρ is the number density of molecules. Consequently, the distance between two consecutive collisions is obtained by the inversion method sampling from the probability distribution function

$$P(\Delta z, \Lambda_{total}) = 1 - \exp(-\Delta z/\Lambda_{total}) \qquad (7)$$

and employing a uniformly distributed random number, U, in the range (0–1). Having determined the distance traveled between collisions and knowing the direction of travel, the position of the primary electron is modified to z_{new}. The elastic and inelastic cross sections ($\sigma_{elastic}$ and $\sigma_{inelastic}$) comprising the total cross section are now used to determine the nature of the collision by comparing a second uniformly distributed random number with the probability of an inelastic collision, i.e., $\sigma_{inelastic}/(\sigma_{elastic} + \sigma_{inelastic})$. If the event is inelastic, the type of inelastic event is determined in a similar manner from the cross sections for ionization ($\sigma_{ionization}$), excitation ($\sigma_{excitation}$), and the "inferior" energy loss processes ($\sigma_{vibration}$), i.e.

$$
\begin{aligned}
0 &< U < \sigma_{ionization}/\sigma_{inelastic} &&\Rightarrow \text{ionization} \\
\sigma_{ionization}/\sigma_{inelastic} &< U < (\sigma_{ionization} + \sigma_{excitation})/\sigma_{inelastic} &&\Rightarrow \text{excitation} \\
(\sigma_{ionization} + \sigma_{excitation})/\sigma_{inelastic} &< U < 1 &&\Rightarrow \text{vibration}
\end{aligned}
$$

The magnitude of the energy transfer, γ, is calculated and the primary energy is reduced to reflect the loss, $E_{new} = E - \gamma$. For elastic events and for inelastic events that do not result in ionization, the new trajectory of the electron is determined and the simulation proceeds in the same manner. Following an elastic event, the direction of travel is determined from the differential elastic cross section, $\sigma'_{elastic}(\theta, E)$, while excitation events are often assumed not to alter the trajectory of the electron. If an inelastic collision causes an ionization event, the ionization channel is determined using the cross sections appropriate for the different ionization processes and the energy lost to the sibling positive ion is calculated. The trajectories of the primary and the secondary electron are calculated from the kinematics. When the energy of the daughter electron is smaller than a pre-defined cut-off energy, γ_{stop}, this electron is not considered further, and the attenuation of the parent electron is continued. If the energy of the secondary electron is greater than γ_{stop}, then its energy is degraded until it is smaller than γ_{stop} before continuing the simulation of the parent electron trajectory. The trajectory of the primary electron is followed until its energy drops below a pre-defined cutoff, γ_{final}. Individual track simulations have many uses, for instance, in the modeling of radiation chemical kinetics or quantifying radiation-induced DNA damage. However, repeated realization of many different tracks is necessary to obtain statistically significant track averaged quantities such as stopping powers, ranges, and penetrations. The mean values for many of these "macroscopic" parameters are available from alternative calculations, but track structure simulation provides the best means for performing sensitivity analyses and for accessing the distribution of a radiation parameter about its mean value.

The simulation of a heavy ion track structure employs essentially the same methodology as described for energetic electrons; except that

- the effects of elastic and inferior inelastic collisions are not significant for energetic heavy ions,
- electron capture and loss by the primary ion must be considered, and
- large numbers of electronic interactions may occur in close proximity.

The first of these factors reduces the complexity of the simulation, but the second has entirely the opposite effect as charge cycling events affect both the energy of the primary ion and its inelastic collision cross section. While the proximity of energy loss events does not affect the details of track structure simulation (at reasonable LET), it may cause significant complications in subsequent diffusion-kinetic calculations due to the (potentially unphysically) high local concentration of radiation-induced reactants.

5. NONHOMOGENEOUS KINETICS

5.1. Transport and Reactions

As discussed, the passage of an ionizing radiation particle through a material leaves in its wake a spatially nonhomogeneous track of highly reactive radicals and ions, which undergo a characteristic nonhomogeneous chemistry. The ultra-fast chemistry of the radiation-induced reactants is due to the relaxation of the nonhomogeneous distribution by the diffusion of the reactive radicals and ions and their diffusion-limited reaction. Consequently, the observed kinetics contains direct information about not only the short-time chemical processes but also about the physical and physico-chemical processes that produce the spatially nonhomogeneous distribution of reactants. Describing the chemistry immediately following irradiation, and subsequent biological damage, requires a model that describes not only the positions at which energy is transferred from the radiation particle to the molecular medium, the physico-chemical consequences of each energy loss, and the local spatial distribution of the radiation-induced reactants resulting from each ionization/ excitation event, but also the fast diffusion-limited chemical reactions.

A number of different techniques have been developed for studying nonhomogeneous radiolysis kinetics, and they can be broken down into two groups, deterministic and stochastic. The former used conventional "macroscopic" treatments of concentration, diffusion, and reaction to describe the chemistry of a typical cluster or track of reactants. In contrast, the latter approach considers the chemistry of simulated tracks of realistic clusters using probabilistic methods to model the kinetics. Each treatment has advantages and limitations, and at present, both treatments have a valuable role to play in modeling radiation chemistry.

In the following two sections, deterministic (prescribed diffusion and FACSIMILE) and stochastic (random flights and IRT) approaches for the modeling of radiation chemical kinetics will be described. Then representative calculations for simple aqueous systems will be shown: The stochastic approach to modeling radiolysis kinetics is more physically realistic than the primitive deterministic models; however, it is also more conceptually advanced, requiring a more detailed (fuller) knowledge of the system under consideration.

5.2. Deterministic Kinetics

The first serious attempt to treat the recombination kinetics of radiolysis was made by Jaffe in 1913 [39] following Wilson's experimental demonstration of the columnar nature of a radiolysis track [40,41]. The general approach outlined by Jaffe provides the basis for most of the deterministic models used to model radiation chemistry. These models consider a hypothetical average system characterized in terms of a nonhomogeneous concentration of each species. Macroscopic laws are used to describe the diffusion of the radiation-induced particles and reaction is modeled using a deterministic rate equation dealing with net changes in the average concentration. The result is a set of coupled diffusion reaction equations that are used to describe the nonhomogeneous kinetics. These equations have the form [42],

$$\frac{\partial c_i}{\partial t} = D_i \nabla^2 c_i - \sum k_{ij} c_i c_j + \sum k_{mn} c_m c_n \qquad (8)$$

where D_i and c_i are the diffusion coefficient and spatially dependent concentration of species i. The first term of the right-hand side of the equation represents the diffusive contribution to the evolution of c_i, while the second and third terms represent removal and production of i by reaction. The set of coupled equations is solved either by analytic approximation or numerically.

The most frequently used analytic, approximate models for the fast kinetics in radiolysis are based on an approximation suggested by Jaffe and known as *prescribed diffusion*. The initial spatial distribution of the radiation-induced particles, their concentration profile, is assumed to be Gaussian, and the kinetic analysis invokes the approximation that reaction only affects the number of particles and not the form of the nonhomogeneous spatial profile, which is therefore always Gaussian. The prescribed diffusion approximation relies upon the Green's function solution for the diffusion equation being a Gaussian whose variance develops as a linear function of time. This assumption is incorrect if the kinetics is nonlinear, which is the case in real scenarios. In his pioneering study, Jaffe considered a cylindrically symmetric columnar system, appropriate for high-LET radiation. In conventional studies considering the kinetics of low-LET radiolysis, the system is usually described in terms of a spherically symmetric typical isolated spur. The concentration profile of species i is

$$c_i(r) = \frac{N_i}{(2\pi\sigma_i^2)^{3/2}} \exp(-r^2/2\sigma_i^2) \qquad (9)$$

where N_i is the number of i particles in the spur and σ_i^2 is the variance of the Gaussian profile and is time dependent, $\sigma = (\sigma^2(t = 0) + 2D't)^{1/2}$. Schwarz [42] extended the methodology to a fully functional multi-radical model. By considering the species e_{aq}^-, $H + OH$, H_3O^+, and H_2 and employing an appropriate reaction scheme with realistic diffusion and reaction rate coefficients, he was able to give a satisfactory (although not entirely accurate) description of the variation of molecular and radical yields with LET, scavenger concentration, and pH.

In addition to the prescribed diffusion approximation, several different numerical treatments have been developed for solving the set of coupled diffusion-reaction equations (Eq. 9) [43–45]. Most recent studies have been performed using a technique in which a "typical, average" spur is used as a representation of the electron track. The spur is divided into concentric shells as shown in Fig. 3. Each shell is sufficiently thin that the concentration of reactants within the shell can be regarded as constant. Diffusion takes place between adjacent shells and reactions within a shell are modeled using deterministic rates. The resulting set of coupled equations describing the diffusion and the kinetics is solved using the FACSIMILE implementation of the Gear algorithm [46]. The parameters defining the spur for aqueous solutions have been optimized to match measured experimental yields in simple scavenger studies [45]. The optimum energy of the spur, ε_{spur}, is 62.5 eV, which given the

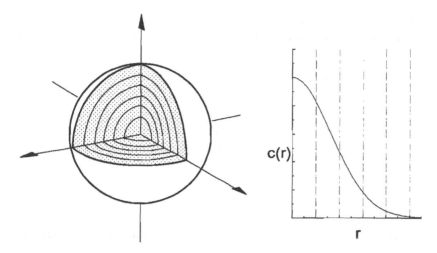

Figure 3 Division of the typical spur into concentric shells in the deterministic treatment of spur kinetics using the FACSIMILE algorithm.

approximate nature of this model is about the mean energy loss predicted by track structure calculations.

5.3. Stochastic Kinetics

Stochastic diffusion-kinetic treatments address the two fundamental aspect of the radiolysis, i.e., the structure of the radiation track and the chemistry of the resulting spatially nonhomogeneous distribution of radiation induced reactants. The track structure, describing the initial positions of the reactants, is usually prescribed by a simulation, as described above, employing a collision-to-collision methodology with cross sections appropriate for liquid water, then a diffusion-kinetic simulation using either a computer intensive random flight approach or the more elegant, but approximate, Independent Reaction Times methodology is employed to model the radiation chemical kinetics.

Random Flight Simulation

The modeling of the radiation chemistry by diffusion-kinetic simulation relies upon the generation of reaction times from initial coordinate positions. In a random flight simulation, this calculation is performed in a "brute force" manner by mimicking the diffusion and encounter of the reactive particles. Several variants of the random flight simulation method have been developed at different levels of sophistication, but all follow the same general approach [47–50]. Each simulation begins from a realizable configuration, generated at random from the assumed spatial profile, either a spur or an idealized track structure. The diffusive trajectories of the particles are then simulated using a discretization of the appropriate stochastic differential equations. Reaction is modeled by overlap at the end of a time step and/or by estimating the probability of encounter during a time step using a bridging process. Each realization is continued until reaction is complete or a pre-defined cut-off time is attained. Reaction kinetics is obtained by simulation of $\sim 10^4$ spatial configurations using different initial random number seeds. Each diffusive jump has two components. The first represents the motion caused by the buffeting of the solvent and the second reflects the drift caused by any force $\mathbf{F}$. The jump is calculated using an Euler-type discretization of the stochastic differential equation, which gives change in a particle's position of

$$\delta r_i = \sqrt{2D_i\delta t}N_3(0, 1) + \frac{D_i F_i}{k_B T}\delta t$$

where D_i is the diffusion coefficient of the particle, δt is the length of the time step, and N_3 is a three-dimensional normal random vector with unit variance. This simulation methodology has been validated, for several types of idealized system, including radical spurs and ionic spurs in both high and low permittivity solvents [47,48,51].

For neutral reactants the force $\mathbf{F} = 0$, and therefore [47]

$$\delta r_i = \sqrt{2D_i\delta t}N_3(0, 1)$$

When the particles are charged, the strong inter-ion forces modify the diffusion. The (Coulombic) force on ion i is then [48]

$$\mathbf{F}_i = -k_B T \sum_j \frac{z_i z_j r_c}{r_{ij}^3} r_{ij}$$

where z_i is the charge on ion i, and $\mathbf{r}_{ij}$ is the inter-ion vector. The Onsager distance, r_c, is the distance at which the potential energy of the Coulombic interaction of two unit charges has magnitude $k_B T$.

In the presence of an applied electric field, $\mathbf{E}$, the expression for the force on an ion is further modified to [52]

$$\mathbf{F}_i = -k_B T \sum_j \frac{z_i z_j r_c}{r_{ij}^3} r_{ij} + z_i e\mathbf{E}$$

where e is the electron charge.

The majority of simulations reported in the literature use a fixed time step, δt [48,50]; however, this treatment is computationally inefficient. When the particles are well separated the probability of encounter is small. Sophisticated methods have been developed, which allow more efficient computation by incorporating variable time steps [47,49,51]. In these treatments, the time step, δt, is determined by the proximity of the particles. The time step is selected by one of two methods:

(i) so that the pair with the shortest inter-particle separation has a very small encounter probability [47], or

(ii) so that the change in the inter-particle drift of every pair during the time step is small [49,51].

In both treatments, the minimum time step becomes increasingly small as two particles approach so a minimum time step, typically ~ 10 fsec, has to be employed to ensure that the simulation does not stall.

If a reaction is diffusion-controlled then reaction occurs on encounter of two particles. In a simulation employing finite time steps, a pair of reactants may encounter during a time step and then "separate" before the end of the time step. This pair should have reacted, but the reaction is not registered. Consequently, the modeled kinetics underestimates both the rate and the amount of reaction. This problem of encounter during a time step can be overcome using the conditional encounter probability for an interpolating "bridging process" [53]. Two forms of bridging process have been developed: the Bessel bridge, which assumes that the separation behaves as a Bessel process, and the Brownian bridge in which the separation between the two particles is approximated by a Wiener process.

While many of the important reactions in radiation and photochemistry are fast, not all are diffusion-limited. The random flight simulation methodology has been extended to include systems where reaction is only partially diffusion-controlled or is spin-controlled [54,55]. The technique for calculating the positions of the particles following a reflecting encounter has been described in detail, but (thus far) this improvement has not been incorporated in realistic diffusion kinetic simulations. Random flight techniques have been successfully used to model the radiation chemistry of aqueous solutions [50] and to investigate ion kinetics in hydrocarbons [48,50,56–58].

Independent Reaction Times Modeling

The IRT model has been developed in detail in a series of papers of Green, Pimblott and co-workers and has been validated by comparison with full random flight simulations [47,49,51]. The IRT treatment of the radiation chemistry relies upon the generation of random reaction times from initial coordinate positions from pair reaction time distribution functions. A simulation, such as a random flight calculation, starts with the initial spatial distribution of the reactants. The separations between all the pairs of particles are evaluated

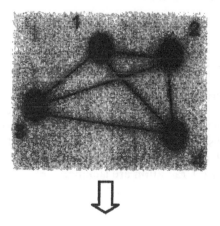

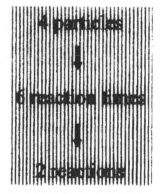

pair	sepn	time	reaction
1+2	1.2 nm	12 ps	x
1+3	1.4 nm	27 ps	x
1+4	1.9 nm	40 ps	2
2+3	2.4 nm	8 ps	1
2+4	1.2 nm	125 ps	x
3+4	2.5 nm	∞	x

Figure 4 Schematic representation of the IRT method for a four-radical spur.

and overlapping pairs are allowed to react. Reaction times for all the surviving pairs are calculated from the reaction time distribution functions for the pairs *as if they were in isolation*. (This approximation is an application of the independent pairs approximation that is implicit in the Debye–Smoluchowski treatment of diffusion limited kinetics.) The resulting ensemble of times is then used to determine the times of subsequent reactions. When a reaction occurs, any reactive products are positioned using an independent "diffusion" treatment [47]. New reaction times are determined from the appropriate first passage time distribution function and a modified ensemble of times is created. The simulation proceeds until a predefined time (usually 1 μsec) is reached or until no reactive particles remain. The kinetics of the system is obtained by repeated realization using a different initial configuration and different random number seeds. This technique is demonstrated schematically for a four-radical spur in Fig. 4.

The IRT method was applied initially to the kinetics of isolated spurs. Such calculations were used to test the model and the validity of the independent pairs approximation upon which the technique is based. When applied to real radiation chemical systems, isolated spur calculations were found to predict physically unrealistic radii for the spurs, demonstrating that the concept of a distribution of isolated spurs is physically inappropriate [59]. Application of the IRT methodology to realistic electron radiation track structures has now been reported by several research groups [60–64], and the excellent agreement found between experimental data for scavenger and time-dependent yields and the predictions of IRT simulation shows that the important input parameter in determining the chemical kinetics is the initial configuration of the reactants, i.e., the use of a realistic radiation track structure.

6. APPLICATION TO RADIATION CHEMICAL STUDIES

Investigation of the effects of radiation type and energy on radiation chemical kinetics using a combination of experimental and stochastic simulation methods provides the most direct insight into the nonhomogeneous diffusion and reaction processes occurring in a radiation track, which ultimately determine the damage caused by radiation. While the chemistry induced by low-LET radiation has received considerable experimental and theoretical attention, similar comparisons on the effects of high-LET radiation are lacking.

6.1. Gamma and Electron Radiolysis

The irradiation of water is immediately followed by a period of fast chemistry, whose short-time kinetics reflects the competition between the relaxation of the nonhomogeneous spatial distributions of the radiation-induced reactants and their reactions. A variety of gamma and energetic electron experiments are available in the literature. Stochastic simulation methods have been used to model the observed short-time radiation chemical kinetics of water and the radiation chemistry of aqueous solutions of scavengers for the hydrated electron and the hydroxyl radical to provide fundamental information for use in the elucidation of more complex, complicated chemical, and biological systems found in real-world scenarios.

Hydrated Electron Kinetics

There is a large amount of experimental data on the radiation chemical yield of e^-_{aq}, including direct absorption measurements of the time dependence of the yield in deaerated water [66–73] and a variety of scavenger studies that have been used to determine the effect of scavenging capacity, s (equal to the product of the scavenger concentration and the rate coefficient for the scavenging reaction, $k[S]$) on the scavenged yield of e_{aq}^- [45]. Extrapolation of the recently re-evaluated time dependence of e_{aq}^- measured experimentally to the pico-second time scale suggests a yield of e_{aq}^- of 4.0 ± 0.2 [32]. These data are compared with the simulated kinetics of a 1-MeV electron degraded by 10 keV in Fig. 5. The standard deviation of the "thermalization" distribution of e_{aq}^- that best reproduces the experimentally observed kinetics is ~ 5.2 nm. Notice that very little reaction occurs at times less than 0.1 nsec. The short-time yield is primarily determined by the ionization yield and the amount of ultra-fast recombination of the precursors to the hydrated electron, e_{pre}^-, with the molecular cation, H_2O^+. The precursors to the hydrated electron do not participate in significant intra-track chemistry in neat water.

In the IRT modeling of the radiation chemistry of water, the ionization yield is determined by the track structure simulation: it is *not* a variable. In contrast, the "thermalization" distribution for e_{aq}^-, which represents the degradation of the electron energy from $E = \gamma_{final}$ to solvation, is a variable in the diffusion-kinetic calculation. The value obtained is in agreement with estimates of Crowell and Bartels derived from picosecond laser study of the multiphoton ionization of liquid water [74]. Furthermore, the mean distance is essentially the same as the "average thermalization distance," ~ 8.3 nm, estimated by Monte Carlo simulation of the energy loss of subexcitation electrons in solid water using the cross sections for ice [75] and is similar to the mean thermalization lengths estimated by Konovalov et al. from electron photo-ejection experiments in water, $\langle l \rangle = 6.0$–8.0 nm [76]. (The thermalization distances of Konovalov depend upon the initial electron energy, whereas the radiation chemical spur width is for a distribution of energies.) The spur width derived by stochastic simulation and from these experiments is considerably larger than obtained from previous deterministic analyses of scavenger experiments, where $\sigma(e_{aq}^-) \sim 2$–3 nm [42].

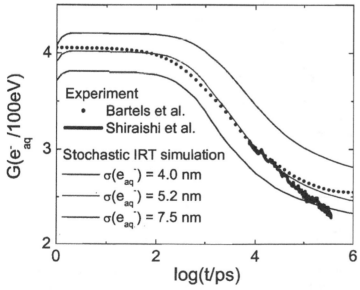

Figure 5 Decay kinetics of the hydrated electron. Experimental data: ($\bullet\bullet\bullet\bullet$) [32]; (— — —) [73]. IRT kinetics: (— — —) $\sigma(e_{aq}^-) = 4.0$ nm; (— — —) $\sigma(e_{aq}^-) = 5.2$ nm; (— — —) $\sigma(e_{aq}^-) = 7.5$ nm.

Precursors to the Hydrated Electron

The scavenging capacity dependence of the scavenged yield of e_{aq}^- is shown in Fig. 6 [77–82]. Comparison of the modeled chemistry with data from experimental scavenger studies shows excellent agreement. The calculations reproduce both the absolute yields and the relative variation as a function of scavenging capacity of the majority of the data. While Fig. 5 shows that e_{pre}^- is not significantly involved in the observed intra-track radiation chemistry of pure water, in concentrated aqueous solutions of e_{aq}^- scavengers, scavenging reactions may take place at early times. Two types of scavenging reaction may contribute to the observed chemistry, scavenging of e_{pre}^- and of e_{aq}^-. Scavenger systems are commonly used to estimate the yield of e_{aq}^- in the radiolysis of water. Consequently, understanding the effects of scavenging e_{pre}^- is important. Fig. 6 compares calculations for several e_{aq}^- scavengers. The yields of N_2 from N_2O solutions correspond to predictions from the Laplace transform of the decay kinetics of e_{aq}^-, and examination of the details of the IRT calculations clearly shows that the precursors of the hydrated electron do not influence the chemistry. In nitrate solutions, for e_{aq}^- scavenging capacities less than 10^8 sec^{-1}, the $(e_{pre}^- + NO_3^-)$ reaction does not have an effect on the amount of electrons scavenged. At higher concentrations, $k(e_{aq}^- + S)[S] \sim 10^9$ sec^{-1}, the scavenging of e_{pre}^- has a statistically (although probably not experimentally) significant effect on the amount of electrons scavenged. This difference becomes more distinct as the concentration of NO_3^- increases. The curve for the effect of selenate concentration on the amount of electrons scavenged is shifted considerably from those for the generic e_{aq}^- scavenger N_2O and for NO_3^-. The shift reflects the primary role that

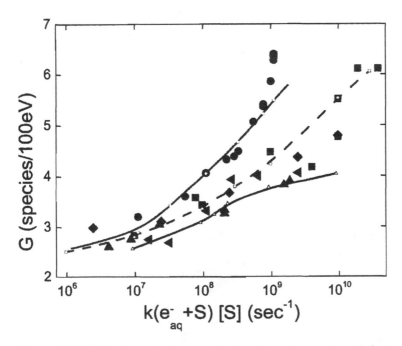

Figure 6 Effect of scavenging capacity for e_{aq}^- on the number of electrons scavenged in aqueous solutions of nitrous oxide, nitrate, and selenate. Experimental data: nitrate—(■) [77], (◆) [78]; selenate—(●) [79]; nitrous oxide—(▲) [80], (▼) [81], (◄) [82]. IRT simulation: nitrate—(— —); selenate—(— — —); nitrous oxide—(— — —).

the reaction of e_{pre}^- with SeO_4^{2-} plays in determining the amount of scavenging. Only in very dilute solution is the reaction of e_{aq}^- with SeO_4^{2-} dominant.

6.2. Light and Heavy Ion Radiolysis

Stochastic simulation methods have recently been developed for modeling heavy ion track structures using collision cross sections appropriate for liquid water [83]. Realistic diffusion-kinetic calculations on the effects of high-LET radiations such as alpha particles and accelerated light and heavy ions are scarce [84–86], and the only system that has been investigated thoroughly is the dependence of the Fricke dosimeter on radiation ion and energy [86]. Fig. 7 compares the results of stochastic IRT simulations for track segments with experimentally determined differential yields [87–89], showing excellent agreement for energetic electrons and for nonrelativistic ions, including ^1H, ^4He, ^{12}C, and ^{20}Ne. The data reveal a significant effect of particle type and energy, which reflects the competition between intra-track reaction of the radiation-induced radicals, diffusion, and scavenging. This competition is modified by changes in the ion track structure. Examination of the underlying Monte Carlo track structure simulations shows that the radial energy loss profiles are similar for ions with the same velocity/charge ratio and that ~40% of the energy is initially deposited within a water diameter of the track axis. The simulations demonstrate that LET of the ionizing radiation is a poor parameter for characterizing the Fricke dosimeter, and the

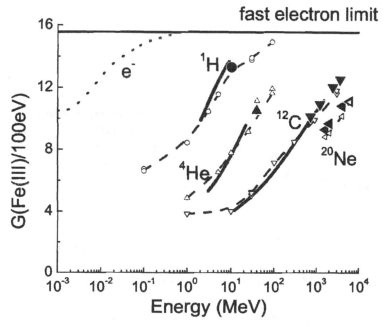

Figure 7   Effect of ion energy on the response of the Fricke dosimeter. The solid lines and filled points are the experimental data. (———) labeled curves for ^1H, ^4He, ^{12}C, and ^{20}Ne ions [87]; (●) ^2D(at $E/2$) [88]; (▲) $^4He^{2+}$ [88]; (▼) ^{12}C [89]; (◄) ^{20}Ne [89]. The dashed lines joining the open points represent stochastic IRT simulations for heavy ions, and the dotted line refers to calculations for the complete slowing down of energetic electrons.

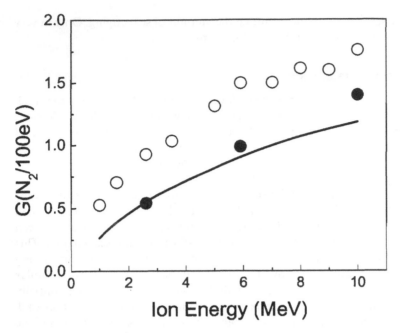

Figure 8 Effect of ^1H ion energy on the yield of N_2 from saturated N_2O solution. Experimental data: track average yield (●) [90]. IRT simulation: track segment yield (O); track average yield (——).

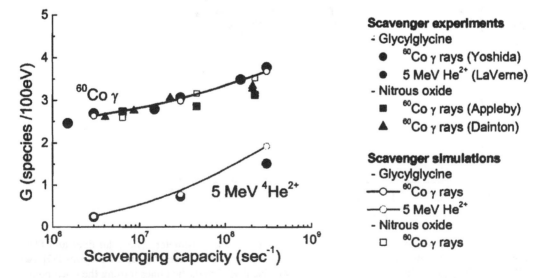

Figure 9 Effect of radiation type and of scavenging capacity for e_{aq}^- on the yield of NH_3 and N_2 from glycyl-glycine and N_2O solutions, respectively.

observed chemistry is predicted more precisely by the square of the ratio of the particle charge to its velocity.

There are almost no radiation chemical studies of the yields of the primary radiation-induced species made with ions of well-defined energy. To develop a realistic, predictive model for the effects of radiation it is important to chart the interplay between the various physical and chemical factors affecting track structure and track kinetics for all of the primary radiation-induced species, e_{aq}^-, OH, H_2, and H_2O_2. For instance, modeling of the track chemistry of a heavy ion necessitates the simulation of track segments for ions of different energy, followed by integration, as track segment yields are very different from the track average yield except at high ion energies. A typical calculation of this kind is shown in Fig. 8, which examines the production of N_2 from the ^1H ion radiolysis of aqueous nitrous oxide solutions [90]. A study of this type is necessary for each calculated result in a scavenger concentration study, such as Fig. 9, which compares recent experimental yields with stochastic IRT predictions for the effect of the concentration of glycyl-glycine on the yield of NH_3 and of nitrous oxide on the yield of N_2 from aqueous solutions irradiated with ^{60}Co gamma rays and 5 MeV He ions [91,92]. The potential interplay of calculation with experiment suggested by the data shown in Figs. 7–9 has many benefits. In addition to the accurate prediction of the effect of radiation quality on radiation-induced chemistry, realistic kinetic modeling will enable the elucidation of the underlying physical and chemical processes by analysis of the simulated track structures and the modeled interplay of diffusion and reaction.

7. APPLICATION TO HIGH-TEMPERATURE NUCLEAR REACTORS

There is a large amount of data on nonhomogeneous track chemistry of energetic electrons at room temperature, and the track structure and diffusion-limited kinetics are well parameterized. This wealth of knowledge contrasts with the limited information about the effects of radiation on aqueous solutions at elevated temperatures. The majority of the studies at elevated temperatures have been performed at AECL, Canada [93], or at the Cookridge Radiation Laboratory, University of Leeds, UK [94]. These two groups have focused on measuring the rate coefficients of the reactions of the radiation-induced radicals and ions of water. The majority of the temperature dependencies can be fitted with an empirical Arrhenius-type expression, $k = A \exp(-E_a/k_B T)$. This type of parameterization provides a satisfactory estimate of the rate coefficient, but it should not be taken to have any mechanistic implications. Radiation-induced radicals are very reactive, and so most of their reactions are close to diffusion-controlled at room temperature. At elevated temperature, however, deviations from diffusion-control are apparent. Consequently, the temperature dependence has to be described using two components, $k_{obs}^{-1} = k_{diff}^{-1} + k_{react}^{-1}$, one for the encounter process, k_{diff}, and one for the reaction process, k_{react}. The temperature dependencies of the diffusion coefficients of the radiation-induced ions and of the water molecule are partially known; however, the diffusion-coefficients of the neutral radicals are poorly characterized even at room temperature. In the diffusion-kinetic calculations, it has been necessary to assume that these species have diffusion coefficients with the same temperature dependence as the self-diffusion of water.

The temperature dependencies of the yields of e_{aq}^-, H, H_2, OH, and H_2O_2 following γ irradiation have been investigated using radical scavengers. The yields of the principal reducing, e_{aq}^-, and oxidizing, OH, radicals are quite well known and increase with increasing temperature. There are some discrepancies between different experimental studies for H_2;

however, the yield is either independent of temperature or increases slightly with increasing temperature (< 30% at 300°C). The yield of H atom is not accessible directly and can only be obtained in conjunction with H_2. The sum of the yields of H and H_2 is more or less independent of temperature. The yield of H_2O_2 appears to decrease with increasing temperature in contrast to H_2; however, there is no reliable information above 150°C. A number of deterministic and stochastic calculations have investigated the gamma and heavy ion radiolysis of water at elevated temperatures [95–99]. For the most part these studies have focused on the observable yields rather than the underlying track structures. The principal effect of elevated temperature on the track structure of energetic electrons in water is due to changes in the inter-event mean-free path. This property depends inversely upon the density of water so increasing temperature increases the separation between intra-track energy loss events. Consequently, (i) the dose distribution of the track is more diffuse at elevated temperatures than that at room temperature, and (ii) the spurs (clusters of radiation-induced reactants) comprising the nonhomogeneous spatial distribution of reactants are more isolated. Fig. 10 compares the dose distribution of a 1-keV electron at 25°C and at 300°C. The very apparent spreading of the track structure is significant as chemistry in clusters emphasizes stochastic effects and should therefore lead to increased recombination

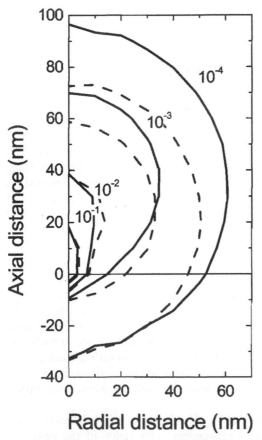

Figure 10 Comparison of the dose distribution of a 1-keV electron at 25°C (dashed contours) and at 300°C (solid contours). Dose in eV/nm^3.

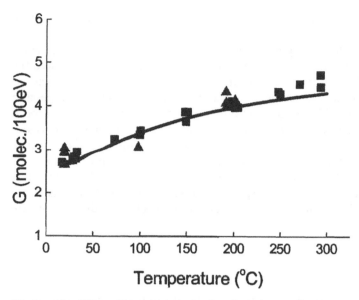

Figure 11 Effect of temperature on the yield of OH radicals from gamma radiolysis. Experiments: (▲) Kent and Sims [100], (■) Elliot et al. [101] Calculation: (solid line) IRT modeling using track structure simulation.

relative to molecular product formation. The reaction time for a pair of radiation-induced reactants depends upon their separation, their transport properties, and the rate of reaction upon diffusive encounter. In addition, physical parameters, such as the relative dielectric coefficient and the density of water, have to be considered in calculations as they affect the diffusion-reaction kinetics by modifying the inter-ionic forces and the rates of processes involving the solvent, i.e., water. Stochastic (and deterministic) radiation chemical kinetic calculations have been made for gamma radiolysis of water over the temperature range 25°C to 300°C, and the predictions have been compared with available experimental data for gamma radiolysis. Agreement is found for all of the radiation-induced species and for the combined yield of H and H_2. For the radicals e_{aq}^- and OH, the calculations quantitatively reproduce the measured yields and their temperature dependence. The predictions of stochastic IRT calculations, employing simulated fast electron tracks, for the yield of OH radicals are compared with the available experimental data in Fig. 11 [100,101].

8. LOW PERMITTIVITY SOLVENTS

In contrast to liquid water, a detailed mechanistic understanding of the physical and chemical processes occurring in the evolution of the radiation chemical track in hydrocarbons is not available except on the most empirical level. Stochastic diffusion-kinetic calculations for low permittivity media have been limited to simple studies of cation–electron recombination in aliphatic hydrocarbons employing idealized track structures [56–58], and simplistic deterministic calculations have been used to model the radical and excited state chemistry [102]. While these calculations have been able to reproduce measured free ion yields and end product yields, respectively, the lack of a detailed mechanistic model makes it very difficult to predict the effects of different types of radiation or to speculate on the effects of radiation on uninvestigated systems.

Differential dipole oscillator strengths are available for a large variety of gaseous hydrocarbons, and for a limited number of liquids and solids. These data have allowed the calculation of the condensed phase energy loss properties, including the cross sections for inelastic collisions, of electrons and positrons in n-hexane, cyclohexane and benzene, and in polyethylene and polystyrene [103]. Cross sections for elastic collisions have also been evaluated using partial wave methods [104]. Recently, these inelastic and elastic cross sections have been successfully used in the stochastic diffusion-kinetic simulation of the electric field dependence of the o-positronium yield in n-hexane [58], showing that positronium formation involves reaction of the thermalized positron with electrons produced along the last few hundred nanometers of its track-end, corresponding to several kiloelectronvolts of energy attenuation. More complex studies of radiation chemical effects in hydrocarbons are necessary.

9. PERSPECTIVE

Over the last 15 years, stochastic methodologies for modeling the radiation chemical kinetics of aqueous systems have made significant progress, advancing from isolated spur studies to ^{58}Ni track chemistry simulations. The IRT methodology for modeling radiation and photochemical kinetics is now used worldwide. Development of realistic models for track chemistry is crucial to predicting radiation effects in complicated practical applications and where experiments cannot be performed. However, the calculations for both low- and high-LET radiation still suffer from a number of deficiencies. The physical, physico-chemical, and chemical processes that occur in the radiation track on a subpicosecond time scale are not well characterized despite recent advances into this time domain in laser irradiation studies. A multi-faceted approach employing experiments in conjunction with Monte Carlo kinetic modeling and electronic structure calculations is clearly desirable to elucidate fundamental radiation chemical processes from their earliest stages.

REFERENCES

1. Inokuti, M. Revs. Mod. Phys. 1971, *43*, 297.
2. Mozumder, A. *Fundamentals of Radiation Chemistry*; Academic Press: San Diego, Ch.2.
3. Fano, U. Annu. Rev. Nucl. Sci. 1963, *13*, 1.
4. LaVerne, J.A.; Mozumder, A. J. Phys. Chem. 1985, *89*, 4216.
5. LaVerne, J.A.; Mozumder, A. Radiat. Res. 1983, *96*, 219.
6. Ashley, J.C.; Williams, M.W. Report RADC-TR-83-87, Rome Air Force Development Center; Grifiss Air Force Base: NY, 1983; 5 pp.
7. Ashley, J.C. J. Electron Spectrosc. Relat. Phenom. 1988, *46*, 199.
8. Pimblott, S.M.; LaVerne, J.A. J. Phys. Chem. 1991, *95*, 3907, and references therein.
9. Pimblott, S.M.; LaVerne, J.A.; Mozumder, A. J. Phys. Chem. 1996, *100*, 8595.
10. See, for example, Gradshteyn, I.S.; Ryzhik, I.M. *"Tables of Integrals, Series and Products"*; Academic Press: New York, 1980.
11. Hayashi, M. *"Atomic and Molecular Data for Radiotherapy"*; International Atomic Energy Agency: Vienna, 1989; 193 pp.
12. Katase, A.; Ishibashi, K.; Matsumoto, Y.; Sakae, T.; Maezono, S.; Murakami, E.; Watanabe, K.; Maki, H. J. Phys. B 1986, *19*, 2715.

13. Grosswendt, B.; Waibel, E. Nucl. Instrum. Methods 1978, *155*, 145.
14. Brocklehurst, B. Radiat. Res. Rev. 1970, *2*, 149.
15. Doke, T. Portugal Phys. 1981, *12*, 9.
16. Haddad, G.N.; Simpson, J.A.R. J. Chem. Phys. 1986, *84*, 6623.
17. Mott, N.F. Proc. R. Soc. A 1930, *216*, 259.
18. Moller, C. Z. Physik 1931, *70*, 786.
19. Mott, N.F.; Massey, H.S.W. *"The Theory of Atomic Collisions"*; Third Edition; O.U.P.: London Ch. XXII. 3, 1965.
20. Kaplan, I.G.; Miterev, A.M.; Sukhonosov, V.Ya. Radiat. Phys. Chem. 1990, *36*, 493.
21. Pimblott, S.M.; Mozumder, A. J. Phys. Chem. 1991, *95*, 7291.
22. Bernas, A.; Goulet, T.; Jay-Gerin, J-P.; Ferradini, C. Proc. 7th Tihany Symposium on Radiation Chemistry; Dobo, J., Nykos, L., Schiller, R., Eds.; Budapest, 1991; 17 pp. And the following discussion on p. 23.
23. Nikogosyan, D.N.; Oraevsky, A.A.; Rupasov, V.I. Chem. Phys. 1983, *77*, 131.
24. Sander, M.U.; Luther, K.; Troe, J. Ber. Bunsen-Ges. Phys. Chem. 1993, *97*, 953.
25. Sander, M.U.; Luther, K.; Troe, J. J. Phys. Chem. 1993, *97*, 11489.
26. Keszei, E.; Jay-Gerin, J.-P. Can. J. Chem. 1992, *70*, 21.
27. Bartels, D.M.; Crowell, R.A. J. Phys. Chem. A 2000, *104*, 3349.
28. McNesby, J.R.; Tanaka, I.; Okabe, H. J. Chem. Phys. 1962, *36*, 605.
29. Stief, L.J.; Payne, W.A.; Klemm, R.B. J. Chem. Phys. 1975, *62*, 4000.
30. Sokolov, U.; Stein, G. J. Chem. Phys. 1966, *44*, 2189 (ibid, 44, 3329 (1966)).
31. Mozumder, A. Phys. Chem. Chem. Phys. 2002, *4*, 1451.
32. Bartels, D.M.; Cook, A.M.; Mudaliar, M.; Jonah, C.D. J. Phys. Chem. A 2000, *104*, 1686.
33. Simpson, W.C.; Orlando, T.M.; Parenteau, L.; Nagesha, K.; Sanche, L. J. Chem. Phys. 1998, *108*, 5027.
34. Mott, N.F.; Frame, J.W. Proc. Camb Philos. Soc. 1931, *27*, 511.
35. Mozumder, A.; Magee, J.L. J. Chem. Phys. 1966, *45*, 3332.
36. Pimblott, S.M.; LaVerne, J.A.; Mozumder, A.; Green, N.J.B. J. Phys. Chem. 1990, *94*, 488.
37. Samuel, A.H.; Magee, J.L. J. Chem. Phys. 1953, *21*, 1080.
38. Ganguly, A.K.; Magee, J.L. J. Chem. Phys. 1956, *25*, 129.
39. Jaffe, G. Ann. Phys. IV 1913, *42*, 303.
40. Wilson, C.T.R. Proc. R. Soc. A 1911, *85*, 285.
41. Wilson, C.T.R. Proc. R. Soc. A 1912, *87*, 277.
42. Schwarz, H.A. J. Phys. Chem. 1969, *73*, 1928.
43. Trumbore, C.N.; Short, D.R.; Fanning, J.E.; Olson, J.H. J. Phys. Chem. 1978, *82*, 2762.
44. Burns, W.G.; Sims, H; Goodall, J.A.B. Radiat. Phys. Chem. 1984, *23*, 143–180.
45. LaVerne, J.A.; Pimblott, S.M. J. Phys. Chem. 1991, *95*, 3196.
46. Chance, E.M.; Curtis, A.R.; Jones, I.P.; Kirby, C.R. Report AERE-R 8775; AERE: Harwell, 1977.
47. Clifford, P.; Green, N.J.B.; Oldfield, M.J.; Pilling, M.J.; Pimblott, S.M. J. Chem. Soc., Faraday Trans. 1986, *82*, 2673–2689.
48. Bartczak, W.M.; Hummel, A. J. Chem. Phys. 1987, *87*, 5222.
49. Clifford, P.; Green, N.J.B.; Pilling, M.J.; Pimblott, M. J. Phys. Chem. 1987, *91*, 4417–4422.
50. Turner, J.E.; Hamm, R.N.; Wright, H.A.; Ritchie, R.H.; Magee, J.L.; Chatterjee, A.; Bolch, W.E. Radiat. Phys. Chem. 1988, *32*, 503–510.
51. Green, N.J.B.; Pimblott, S.M. J. Phys. Chem. 1990, *94*, 2922.
52. Pimblott, S.M. J. Chem. Soc., Faraday Trans. 1993, *89*, 3533.
53. Green, N.J.B. Mol. Phys. 1988, *65*, 1399.
54. Pimblott, S.M.; Green, N.J.B. J. Phys. Chem. 1992, *96*, 9338.
55. Bolton, C.E.; Green, N.J.B. J. Phys. Chem. 1996, *100*, 8807.
56. Siebbeles, L.D.A.; Bartczak, W.M.; Terrissol, M.; Hummel, A. J. Phys. Chem. A 1997, *101*, 1619.
57. Bartczak, W.M.; Hummel, A. Radiat. Phys. Chem. 1997, *49*, 675.

58. Alba García, A.; Pimblott, S.M.; Schut, H.; van Veen, A.; Siebbeles, L.D.A. J. Phys. Chem. B 1997, *106*, 1124.
59. Clifford, P.; Green, N.J.B.; Pilling, M.J.; Pimblott, S.M.; Burns, W.G. Radiat. Phys. Chem. 1987, *30*, 125.
60. Zaider, M.; Brenner, D.J. Radiat. Res. 1984, *100*, 245.
61. Green, N.J.B.; Pilling, M.J.; Pimblott, S.M.; Clifford, P. J. Phys. Chem. 1990, *94*, 251.
62. Hill, M.A.; Smith, F.A. Radiat. Phys. Chem. 1994, *43*, 265.
63. Cobut, V.; Frongillo, Y.; Patau, J.P.; Goulet, T.; Fraser, M.J.; Jay-Gerin, M.J. Radiat. Phys. Chem. 1998, *51*, 229; Frongillo, Y.; Goulet, T.; Fraser, M.J.; Cobut, V.; Patau, J.P.; Jay-Gerin, J.-P. Radiat. Phys. Chem. 1998, *51*, 245.
64. Pimblott, S.M.; LaVerne, J.A. J. Phys. Chem. A 1997, *101*, 5828.
65. Watanabe, R.; Saito, K. Radiat. Phys. Chem. 2001, *62*, 217.
66. Buxton, G.V. Proc. R. Soc. London A 1972, *328*, 9.
67. Wolff, R.K.; Bronskill, M.J.; Aldrich, J.E.; Hunt, J.W. J. Phys. Chem. 1973, *77*, 1350.
68. Jonah, C.D.; Hart, E.J.; Matheson, M.S. J. Phys. Chem. 1973, *77*, 1838.
69. Fanning, J.E. "Evidence for spurs in aqueous radiation chemistry". Ph.D. Thesis, University of Delaware, 1975.
70. Jonah, C.D.; Matheson, M.S.; Miller, J.R.; Hart, E.J. J. Phys. Chem. 1976, *80*, 1267.
71. Sumiyoshi, T.; Tsugaru, K.; Yamada, T.; Katayama, M. Bull. Chem. Soc. Jpn. 1985, *58*, 3073.
72. Chernovitz, A.C.; Jonah, C.D. J. Phys. Chem. 1988, *92*, 5946.
73. Shiraishi, H.; Katsumura, Y.; Ishigure, K. Radiat. Phys. Chem. 1989, *34*, 705.
74. Crowell, R.A.; Bartels, D.M. J. Phys. Chem. 1996, *100*, 17940.
75. Pimblott, S.M. unpublished results.
76. Konovalov, V.V.; Raitsimring, A.M.; Tsvetkov, Y.D. Radiat. Phys. Chem. 1988, *32*, 623; Konovalov, V.V.; Raitsimring, A.M. Chem. Phys. Lett. 1990, *171*, 326.
77. Hyder, M.L. J. Phys. Chem. 1965, *69*, 1858.
78. Draganic, I.G.; Draganic, Z.D. J. Phys. Chem. 1973, *77*, 765.
79. Pastina, B.; LaVerne, J.A. J. Phys. Chem. A 1999, *103*, 209.
80. Dainton, F.S.; Logan, S.R. Trans. Faraday Soc. 1965, *61*, 715.
81. Head, D.A.; Walker, D.C. Can. J. Chem. 1967, *45*, 2051.
82. Jha, K.N.; Ryan, T.G.; Freeman, G.R. J. Phys. Chem. 1975, *79*, 868.
83. Pimblott, S.M. unpublished results.
84. Brenner, D.J.; Zaider, M. Radiat. Prot. Dosim. 1985, *13*, 127.
85. Herve du Penhoat, M.-A.; Meesungnoen, J.; Goulet, T.; Filali-Mouhim, A.; Mankhetkorn, S.; Jay-Gerin, J.-P. Chem. Phys. Lett. 2001, *341*, 135.
86. Pimblott, S.M.; LaVerne, J.A. J. Phys. Chem. A 2002, *106*, 9420.
87. LaVerne, J.A.; Schuler, R.H. J. Phys. Chem. 1987, *91*, 5770.
88. Sauer, M.C.; Hart, E.J.; Naleway, C.A.; Jonah, C.D.; Schmidt, K.H. J. Phys. Chem. 1978, *82*, 2246.
89. Christman, E.A.; Appleby, A.; Jayko, M. Radiat. Res. 1981, *85*, 443.
90. Sims, H.E.; Ashmore, C.B.; Tait, P.K.; Walters, W.S. "Proceedings of the 1998 JAIF International Conference on Water Chemistry in Nuclear Power Plants"; Kashwazaki, 1998.
91. Yoshida, H. Radiat. Res. 1994, *137*, 145.
92. LaVerne, J.A.; Yoshida, H. J. Phys. Chem. 1993, *97*, 10720.
93. Elliot, A.J. "Rate constants and G-values for the simulation of light water over the range 0–300°C". Report AECL-11073, A.E.C.L. Chalk River, Canada, 1994.
94. Elliot, A.J.; McCracken, D.R.; Buxton, G.V.; Wood, N.D. J. Chem. Soc., Faraday Trans. 1990, *86*, 1539.
95. LaVerne, J.A.; Pimblott, S.M. J. Phys. Chem. 1993, *97*, 3291.
96. Swiatla-Wojcik, D.; Buxton, G.V. J. Phys. Chem. 1995, *99*, 11464.
97. Swiatla-Wojcik, D.; Buxton, G.V. J. Chem. Soc. Faraday Trans. 1998, *94*, 2135.
98. du Penhoat, M.A.H.; Goulet, T.; Frongillo, Y.; Fraser, M.J.; Bernat, P.; Jay-Gerin, J.-P. J. Phys. Chem. A 2000, *104*, 11757.

99.   Begusová, M.; Pimblott, S.M. Radiat. Prot. Dosim. 2002, *99*, 73.

100. Kent, M.C.; Sims, H.E. Report AEA-RS-2302; AERE: Harwell, 1992.

101. Elliot, J.A.; Chenier, M.P.; Ouellette, D.C.; Koslowsky, V.T. J. Phys. Chem. 1996, *100*, 9014.

102. LaVerne, J.A.; Pimblott, S.M.; Wojnarovits, L. J. Phys. Chem. A 1997, *101*, 1628.

103. Pimblott, S.M.; LaVerne, J.A.; Alba-Garcia, A.; Siebbeles, L.D.A. J. Phys. Chem. B 2000, *104*, 9607.

104. Salvat, F. Radiat. Phys. Chem. 1998, *53*, 247.

5

Interaction of Photons with Molecules: Photoabsorption, Photoionization, and Photodissociation Cross Sections

Noriyuki Kouchi and Y. Hatano
Tokyo Institute of Technology, Tokyo, Japan

SUMMARY

The absolute values of the photoabsorption, photoionization, and photodissociation cross sections are key quantities in investigating not only the interaction of photons with molecules but also the interaction of any high-energy charged particle with matter. The methods to measure these, the real-photon and virtual-photon methods, are described and compared with each other. An overview is presented of photoabsorption cross sections and photoionization quantum yields for normal alkanes, C_nH_{2n+2} ($n = 1-4$), as a function of the incident photon energy in the vacuum ultraviolet range and of the number of carbon atoms in the alkane molecule. Some future problems are also given.

1. INTRODUCTION

The interactions of photons with molecules are classified into absorption, scattering, and pair production. In this chapter, photons of moderate energies, particularly in the vacuum ultraviolet range, are discussed, and therefore only the absorption process is considered. The absorption of a single photon by a molecule in the electronically ground state changes its electronic state from the ground state 0 to a final excited or ionized state j. Its transition probability is expressed in terms of the optical oscillator strength f_j [1,2].

The optical oscillator strength f_j is expressed as

$$f_j = (E_j/R)M_j^2 \qquad (1)$$

where E_j is the excitation energy in the transition of $0 \rightarrow j$, R is the Rydberg energy ($R = (1/2)$(atomic units) $= 13.6$ eV), and M_j^2 is the dipole matrix element squared in the transition of $0 \rightarrow j$ in atomic units. Note that f_j is dimensionless. A set of E_j and f_j characterizes a discrete spectrum. To discuss a continuous spectrum, one expresses the oscillator strength in a small range of the excitation energy between E and $E + dE$ as $(df/dE)dE$, and calls df/dE the oscillator-strength distribution, or, more precisely, the spectral density of the oscillator strength. Thus df/dE has the dimension of the reciprocal of energy. The quantity df/dE is also defined for the discrete transition $0 \rightarrow j$ as

$$df/dE = \sum_j f_j \delta(E_j - E) \tag{2}$$

The integral of df/dE over a total space of E gives the total number Z of electrons in the molecule under consideration,

$$\int (df/dE) dE = Z. \tag{3}$$

Equation (3) is known as the Thomas–Kuhn–Reiche (TKR) sum rule.

The oscillator-strength distribution df/dE is proportional to the cross section σ for the absorption of a photon of energy E, the so-called photoabsorption cross section. Note that the excitation energy is equivalent to the photon energy. Explicitly, one may write

$$\sigma(E) = 4\pi^2 \alpha a_0^2 \frac{df}{d(E/R)} \tag{4}$$

where α is the fine structure constant and a_0 is the Bohr radius. Equation (4) is more conveniently written as

$$\sigma(E) = 1.098 \times 10^{-16} \left(\frac{df}{dE} \right) \tag{5}$$

where $\sigma(E)$ is expressed in cm^2 and df/dE in eV^{-1}.

The partial oscillator-strength distribution of a channel "ξ", $(df/dE)_\xi$, is also defined and is related to the cross section of channel ξ following the absorption of a photon of energy E, σ_ξ, by

$$\sigma_\xi(E) = 4\pi^2 \alpha a_0^2 \left(\frac{df}{d(E/R)} \right)_\xi \tag{4$'$}$$

and more conveniently by

$$\sigma_\xi(E) = 1.098 \times 10^{-16} \left(\frac{df}{dE} \right)_\xi \tag{5$'$}$$

where $\sigma_\xi(E)$ is expressed in cm^2 and $(df/dE)_\xi$ in eV^{-1}. The channel ξ refers to, for example, the ionization of a specified electron, formation of a specified product, and so on. The summations of $\sigma_\xi(E)$ and $(df/dE)_\xi$ over all channels give $\sigma(E)$ and df/dE, respectively, as

$$\sigma(E) = \sum_\xi \sigma_\xi(E) \tag{6}$$

$$\frac{df}{dE} = \sum_\xi \left(\frac{df}{dE} \right)_\xi \tag{7}$$

Thus the quantum yield of channel ξ, $\eta_\xi(E)$, the number of times of channel ξ per absorption of a photon of energy E, is expressed as

$$\eta_\xi(E) = \frac{\sigma_\xi(E)}{\sigma(E)} = \frac{(df/dE)_\xi}{(df/dE)} \tag{8}$$

The dipole matrix element squared for channel ξ in atomic units, M_ξ^2, is given by

$$M_\xi^2 = \int \left(\frac{R}{E}\right)\left(\frac{df}{dE}\right)_\xi dE \tag{9}$$

Note that E in the right hand side of Eq. (9) is not in atomic units.

As described in the Introduction of Chap. 1, a decisive step in the physical and physicochemical stages of the interactions of any high-energy incident particles with matter, in case it is composed of molecular compounds, is the collision of secondary electrons with molecules in a wide energy range. Cross sections for the ionization and excitation of molecules in collisions with electrons in the energy range greater than about 10^2 eV are well elucidated quantitatively by the Born–Bethe theory [3]. This theory is also helpful for calculating, at least roughly, these cross-section values further in the lower energy range.

According to the optical approximation, which was shown by Platzman [4] to be based on the Born–Bethe theory, a radiation chemical yield of channel ξ, G_ξ, may be estimated from optical data as shown in the following equation:

$$G_\xi = (100/W)\left(M_\xi^2/M_i^2\right) \tag{10}$$

where W is the mean energy (in eV) for the production of an ion pair, and M_ξ^2 and M_i^2 are the dipole matrix elements squared for channel ξ and ionization, respectively. Because this equation indicates that a radiation chemical yield is estimated from optical cross-section data, it is concluded, therefore, that photoabsorption, photoionization, and photodissociation cross-section data of molecules are of great importance in understanding not only the interactions of photons with molecules, but also those of any high-energy incident particles with molecules [5].

There have been remarkable advances in synchrotron radiation research and related experimental techniques in the range from the vacuum ultraviolet radiation to soft X-ray, where the most important part of the magnitudes of these cross-section values is observed, as shown below. Therefore, it is also concluded that synchrotron radiation can bridge a wide gap in the energy scale between photochemistry and radiation chemistry. Such a situation of synchrotron radiation as a photon source is summarized in Fig. 1 [5,6].

Fig. 1 shows the wavelengths of electromagnetic radiation from the infrared to the γ-ray ranges and corresponding photon energies. Characteristic X-rays, ^{60}Co and ^{137}Cs γ-rays, and vacuum ultraviolet light from discharge lamps are indicated by the arrows. The shaded areas indicate the ranges for which photon sources, apart from synchrotron radiation, are available and correspond to photochemistry and radiation chemistry.

In this chapter, we discuss the photoabsorption and photon-induced processes in terms of σ and σ_ξ (or η_ξ), respectively.

In Fig. 2, the photoabsorption (σ), photoionization (σ_i), and photodissociation (σ_d) cross sections of CH_4 measured by Kameta et al. [7] are shown as a function of the incident photon energy in the range 10–24 eV. The electron configuration of the ground electronic state of CH_4 in T_d symmetry is [8,9,10,11]:

$$\underbrace{(1a_1)^2}_{\text{inner shell}} \qquad \underbrace{(2a_1)^2}_{\text{inner valence}} \qquad \underbrace{(1t_2)^6}_{\text{outer valence}}$$

The vertical ionization potentials of the $(1t_2)^{-1}$ and $(2a_1)^{-1}$ ionic states are indicated in Fig. 2, while that of the $(1a_1)^{-1}$ ionic state is far away from the range in Fig. 2. Kameta et al. examined their photoabsorption cross sections σ in terms of the TKR sum rule for the oscillator-strength distribution df/dE, Eq. (3), following the conversion of σ to df/dE,

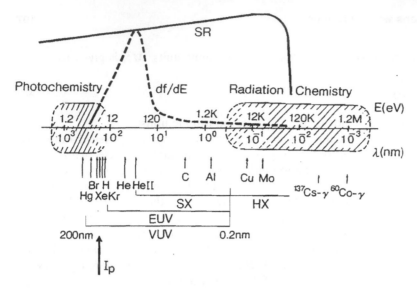

Figure 1 Synchrotron radiation (SR) chemistry as a bridge between radiation chemistry and photochemistry. The oscillator-strength distribution df/dE is shown as a function of wavelength λ and photon energy E. Note the relation $E \times \lambda = 1.24 \times 10^3$, where E is measured in eV and λ in nm. The intensity of synchrotron radiation is also roughly shown, as are the energies of photons from several line sources. VUV, EUV, SX, and HX stand for vacuum ultraviolet, extreme ultraviolet, soft X-rays, and hard X-rays, respectively. (From Refs. [5,6].)

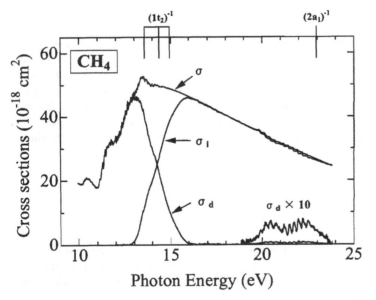

Figure 2 The photoabsorption (σ), photoionization (σ_i), and photodissociation (σ_d) cross sections of CH_4 as a function of the incident photon energy measured via the double ionization chamber and synchrotron radiation as mentioned in Section 2.1. The values of σ in the range below the first ionization potential were measured by the photon-beam attenuation method, using the ionization chamber as a conventional gas cell. The bandpass was 0.1 nm, which corresponds to the energy width of 32 meV at the incident photon energy of 20 eV. The vertical ionization potentials of the ionic states involved are also indicated by the vertical bars [11]. (From Ref. [7]. Reprinted with permission from Elsevier Science.)

Eq. (4). The result is shown in Table 1. The value of $\int (df/dE)dE$ should be equal to the number of electrons in CH_4 and the result yields 9.90, as shown in Table 1, only 1% less than 10. Let us stress from Fig. 2 and Table 1 that the maximum value of df/dE lies around the ionization potential for the removal of an outer-valence electron and the value of $\int_{11 \, eV}^{24 \, eV}(df/dE)dE$ amounts up to approximately 45% of the total integral. A similar feature is also seen for a wide range of molecules [13–16], and thus let us focus on the range of the energy of incident photons for the excitation and ionization of valence electrons as in Fig. 2, i.e., the vacuum ultraviolet range.

We note that the photoionization cross sections (σ_i) of CH_4 is much smaller than the photoabsorption cross sections (σ) just above the first ionization potential, as seen in Fig. 2. The values of σ_i then increase as the incident photon energy increases and become almost equal to those of σ around 16 eV. The small difference between σ and σ_i is seen around 22 eV, i.e., just below the $(2a_1)^{-1}$ ion state. This feature is shown in a more straightforward way in the photoionization quantum yields (η_i) of CH_4 as a function of the incident photon energy, as seen in Fig. 3 [$\eta_i = \sigma_i/\sigma$ following Eq. (8)]. The significant deviation of η_i from unity around the ionization potentials of the outer-valence electrons is seen for a wide range of molecules [13–16]. This experimental fact of interest is attributed to the excited states of molecules embedded in their ionization continua. The existence and importance of such excited states were first predicted by Platzman in the early 1960s in a pioneering theoretical work on the interaction of ionizing radiation with matter, and they were named "superexcited states" [4,18]. Let us consider a molecule, AB, absorbing a photon of the energy higher than the first ionization potential of AB. Then the following processes occur

$$AB + h\nu \rightarrow AB^+ + e^- : \text{ direct ionization} \tag{11}$$

$$\rightarrow AB^{**} : \text{ superexcitation} \tag{12}$$

$$\rightarrow AB^+ + e^- : \text{ autoionization} \tag{13}$$

$$\rightarrow A + B : \text{ neutral dissociation} \tag{14}$$

$$\rightarrow \text{ other processes} \tag{15}$$

where AB denotes not only a diatomic molecule but also, more generally, a polyatomic molecule. The most important feature in the dynamics of a superexcited molecule, AB^{**}, is the competition between the autoionization [process (13)] and neutral dissociation [process (14)], because it cannot be seen in a superexcited atom. The processes other than auto-ionization and neutral dissociation, such as emission of fluorescence and ion-pair for-

Table 1 Oscillator-Strength Distribution in CH_4

Energy Range (eV)	Integral of the Oscillator-Strength Distribution
≥23.8	5.15[a]
10.95–23.8	4.47[b]
8.55–10.95	0.28[c]
Total range	9.90

[a] From Ref. [12].
[b] From Ref. [7].
[c] From Ref. [13].
Source: Ref. [7].

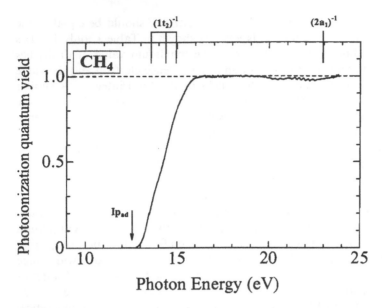

Figure 3 The photoionization quantum yields (η_i) of CH_4 as a function of the incident photon energy measured via the double ionization chamber and synchrotron radiation as mentioned in Section 2.1. The bandpass was 0.1 nm, which corresponds to the energy width of 32 meV at the incident photon energy of 20 eV. The vertical ionization potentials of the ionic states involved are indicated by the vertical bars [11] along with the first adiabatic ionization potential by the arrow [17]. (From Ref. [7]. Reprinted with permission from Elsevier Science.)

mation, are possible although their contribution is not really significant. Thus the difference between σ and σ_i gives the cross section for the neutral dissociation [process (14)] in the energy range above the first ionization potential, which is refereed to as the photodissociation cross section in this chapter, σ_d:

$$\sigma_d = \sigma - \sigma_i \tag{16}$$
$$= (1 - \eta_i)\sigma \tag{17}$$

The recent stage for the study of the spectroscopy and dynamics of the superexcited molecules was comprehensively discussed in Ref. [5], and let us stress that the formation of the superexcited molecules and their decay processes produce a wide range of interesting structures in the σ, σ_i, and σ_d curves as a function of the incident photon energy.

In this chapter, we aim at giving an overview of the photoabsorption (σ), photoionization (σ_i), and photodissociation (σ_d) cross sections of molecules in the vacuum ultraviolet range, where as stated the significant part of the photoabsorption cross sections lies.

2. EXPERIMENTAL METHODS TO MEASURE THE PHOTOABSORPTION, PHOTOIONIZATION, AND PHOTODISSOCIATION CROSS SECTIONS

2.1. Double Ionization-Chamber Method (Real-Photon Method)

The double ionization-chamber method [19] provides an excellent means of measuring the photoabsorption cross sections of atoms and molecules in the range of the incident photon energy higher than their first ionization potentials, and bears several advantages over the photon-beam attenuation method as follows (see also Fig. 4).

1. The light intensities need not be measured to obtain the photoabsorption cross sections (σ), as will be shown later using Eq. (20). Therefore we do not need to consider the absorption of light by gas that effuses through the entrance and exit apertures of the chamber.
2. Photoionization quantum yields (η_i) and, therefore, photoionization cross sections (σ_i) are obtained together with the photoabsorption cross sections (σ).

Fig. 4 shows the illustration of a double ionization chamber. We describe the process of measuring the photoabsorption cross sections as follows. I_0 denotes the incident photon flux coming into the chamber filled with atoms or molecules of the number density n, I_1 and I_1' denote the photon fluxes entering and leaving plate 1, respectively, and I_2 and I_2' denote the photon fluxes entering and leaving plate 2, respectively. The ion currents i_1 and i_2 collected by plates 1 and 2, respectively, are expressed as

$$i_1 = e\eta_i I_0 \exp(-\sigma n L_1)(1 - \exp(-\sigma n d)) \tag{18}$$

$$i_2 = e\eta_i I_0 \exp(-\sigma n L_2)(1 - \exp(-\sigma n d)) \tag{19}$$

where L_1, L_2, and d are lengths as defined in Fig. 4, and e is the elementary charge. Thus we obtain

$$\sigma = \frac{\ln(i_1/i_2)}{n(L_2 - L_1)} \tag{20}$$

According to Eq. (20), we measure only i_1, i_2, and the pressure of a gas in the chamber to obtain the absolute values of the photoabsorption cross sections (σ), and then we obtain the values of $\eta_i I_0$ from σ following Eq. (18) or Eq. (19). If we use a rare gas as reference, of which η_i is unity in the whole range above its first ionization potential, then I_0 is obtained. Thus the relation between I_0 and the signal from the incident photon detector,

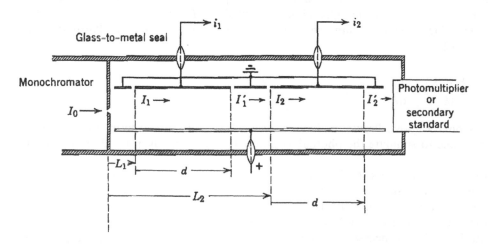

Figure 4   The outline of the double ionization chamber. (From Ref. [19].)

which is not shown in Fig. 4, is determined as a function of the incident photon energy by using a suitable rare gas as reference. When this relation is known, the quantities i_1, i_2, I_0, and n are obtained together to give the absolute values of both photoabsorption cross sections (σ) and photoionization quantum yields (η_i) following Eqs. (18–20). From σ and η_i, the photoionization cross sections (σ_i) and photodissociation cross sections (σ_d) are easily derived, as mentioned in Section 1: $\sigma_i = \eta_i \sigma$ [Eq. (8)] and $\sigma_d = (1 - \eta_i)\sigma$ [Eq. (17)].

We note that the relation between I_0 and the signal from the incident photon detector is influenced by absorption of photons due to gas effusing from the entrance aperture of the chamber, and thus the ionization chamber needs to be equipped with a suitable window to prevent gas effusion while measuring the accurate values of η_i [20,21]. The window is not needed when aiming to measure only the photoabsorption cross sections (σ) as stated above. However, the use of window is recommended even in measuring σ alone, because it prevents the pressure gradient inside the chamber caused by effusion of gas. It also suppresses the contribution from higher-order components that become mixed in the monochromatized photon beam from the continuous-wavelength light source such as synchrotron radiation.

The important element in the double ionization-chamber method is a photon source in the vacuum ultraviolet range. Discharge lamps were used in the 1960s and 1970s, and are still used today. Since 1980s, however, synchrotron radiation has been widely used in the field covered by this chapter because of its broadband tunability in the range from the vacuum ultraviolet radiation to soft X-ray, which is strongly required in measuring the cross sections as a function of the incident photon energy.

2.2. Fast Electron Impact Dipole-Simulation Method (Virtual-Photon Method) and Comparison with Real-Photon Method

In electron impact experiments of atoms and molecules for incident electron energies exceeding 1000 eV, for example, the generalized oscillator-strength distribution, $dF(E,K)/dE$, is derived from the electron scattering cross sections differential with respect to the energy loss of electrons, E, and the solid angle for scattered electrons according to the Born–Bethe approximation, where K is the momentum transfer [3]. The energy loss is equivalent to excitation energy or photon energy, and thus the symbol "E" is used to denote the energy loss as well. The most important characteristics of the generalized oscillator-strength distribution is [3]:

$$\lim_{K \to 0} \frac{dF(E,K)}{dE} = \frac{df}{dE} \tag{21}$$

It follows from the above discussion that the fast electron scattering experiments at forward direction is able to simulate the photoabsorption cross sections over a wide range of the energy loss.

Brion and co-workers [14,22–25] have extensively used and applied such an experimental approach by using fast electrons as the virtual-photon source. Brion et al. [14,25] pointed out some expected characteristics of the use of synchrotron radiation in comparison with virtual photons in studies of the photoionization and photoexcitation of molecules, and clarified some necessary assumptions to virtual photons instead of real photons. It should be noted here, as described below, that these two methods, real- and virtual-photon experiments, have complementary roles with each other in understanding the essential features of the interaction of photons with molecules and obtaining the absolute values of the cross sections (σ, σ_i, and σ_d) [7,5,26].

The fast electron scattering (dipole) approach to real-photon experiments using the virtual-photon source is called the electron impact dipole-simulation method. The dipole (e, e) method simulates total photoabsorption while coincidence techniques are used for measurements of the electronic partial photoionization [dipole (e, 2e)] and ionic photofragmentation [dipole (e, e + ion)] cross sections. These virtual-photon experiments are summarized in Table 2 [14] together with the corresponding real-photon experiments. These simulation experiments have provided a large body of data in comparison with real-photon experiments of absolute cross sections [14,25]. The real-photon experiments using synchrotron radiation, in which a lot of progress in experimental techniques is being made, have been comprehensively compared with the simulation measurements [5,16], which is briefly described below.

In most cases, except those in earlier comparative studies between the real-photon method and the dipole-simulation method, the absolute cross-section values obtained by both methods agree with each other [27]. Comparison of obtained cross-section values between the two methods were discussed in detail [27, 2, and references therein] and summarized in conclusion [5]. It should be noted, at least briefly, that it is essentially difficult to accurately obtain the absolute values of photoabsorption cross sections (σ) in the dipole-simulation experiments, and it is necessary to use indirect ways in obtaining those values as the application of the TKR sum rule, Eq. (3), to the relative values of the cross sections obtained partly with theoretical assumptions. Moreover, in some cases, in relatively earlier dipole-simulation experiments, particularly of corrosive molecules upon their electron optics with poorer energy resolutions, serious discrepancy from the real-photon experiments was clearly pointed out in the obtained absolute values of photoabsorption cross sections [5,20,25–28].

As for the absolute values of photoionization quantum yields (η_i), a situation of the dipole-simulation experiments in comparison with the real-photon experiments is much more serious and controversial because their absolute scales are determined by the assumption that the photoionization quantum yields should be unity around 20 eV photon energy in addition to the above-mentioned difficulty in obtaining absolute cross-section values in the dipole-simulation experiments.

In case of simple diatomic molecules such as H_2 and N_2, sharp peaks are observed with large photoabsorption cross sections, especially in the lower energy range. In real-photon experiments, if the peak shape is narrower in energy than the bandpass of the incident

Table 2 Photon and Electron Impact Experiments

Photon Experiment	Equivalent Electron-Impact Experiment
Total photoabsorption	Electron-energy-loss spectroscopy, dipole (e, e)
Total photoionization	Dipole (e, 2e) or (e, e + ion) (from sums of partials)
Photoelectron spectroscopy	Electron energy loss-ejected electron coincidence, dipole (e, 2e)
Photoionization mass spectrometry	Electron-ion coincidence, dipole (e, e + ion)

Source: Ref. [14].

photon beam, the measured photoabsorption cross section suffers from the serious "line saturation effect" [29]. In such cases, we should refer to the result by the virtual-photon experiments or the dipole-simulation experiments, because the integrated photoabsorption cross sections over a range of the incident photon energy for a transition give the correct value of the oscillator strength of that transition even if the measured shape is affected by the energy resolution [29,30]. However, it should be noted that the results of the dipole-simulation experiments are obtained with rather poor energy resolutions, i.e., poorer than several tens of meV, in comparison with the real-photon experiments, especially in the lower energy range. It means that some important spectral features in the cross-section curves may be sometimes missed in the dipole-simulation experiments.

The real-photon method is essentially more direct and easier compared to the dipole-simulation method in obtaining absolute values of photoabsorption cross sections (σ), photoionization cross sections (σ_i), and photoionization quantum yields (η_i). In the real-photon method, however, there is a practical need to use the big and dedicated facilities of synchrotron radiation—where, in many cases, one should change the beam lines equipped with different types of monochromators depending on used photon-wavelengths—and to develop some specific new experimental techniques in the range from the vacuum ultra-violet radiation to soft X-ray.

In conclusion, as summarized elsewhere [5,26], these two methods have complementary roles in understanding the interaction of photons with molecules, as well as in obtaining the absolute values of the cross sections (σ, σ_i, and σ_d).

3. PHOTOABSORPTION AND PHOTOIONIZATION OF MOLECULES OF CHEMICAL INTEREST, NORMAL ALKANES—AN OVERVIEW

In this section, we present an overview of the photoabsorption cross section (σ) and the photoionization quantum yields (η_i) for normal alkanes, C_nH_{2n+2} ($n = 1$–4), as a function of the incident photon energy in the vacuum ultraviolet range, and of the number of carbon atoms in the alkane molecule, because normal alkanes are typical polyatomic molecules of chemical interest. In Fig. 5, the vertical ionization potentials of the valence electrons, which interact with the vacuum ultraviolet photons, in each of these alkane molecules are indicated to show how the outer- and inner-valence orbitals associated with carbon 2p and 2s orbitals, respectively, locate in energy [7].

3.1. Photoabsorption Cross Sections of Normal Alkanes

In Fig. 6 [7], the photoabsorption cross sections (σ) of CH_4, C_2H_6, C_3H_8, and n-C_4H_{10} are compared, which were measured by our group by using the double ionization chamber and synchrotron radiation, as described in Section 2.1. Those in the range below the first ionization potentials were measured by the photon-beam attenuation method using the ionization chamber as a conventional gas cell. The following features can be noted [7].

1. Each cross section has a maximum around 13–16 eV, and the observed maxima shift to the higher energies with increasing the number of carbon atoms, i.e., from CH_4 to n-C_4H_{10}, while the first ionization potentials shift to the lower energies as shown in Fig. 5. This shift of the maxima seems to be amenable to the trends in the ionization potentials of the deepest outer-valence orbitals of each molecule.

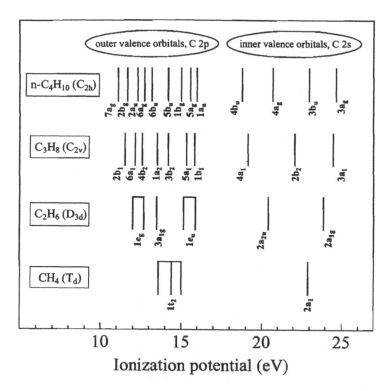

Figure 5 Ionization potentials of the valence electrons of CH_4, C_2H_6, C_3H_8, and n-C_4H_{10} [8,9, 10,11], which are the vertical ionization potentials of the ionic states produced with the removal of an electron from each valence orbital. The point groups for the molecules are also shown as well as the notation of the orbitals based on them. (From Ref. [7]. Reprinted with permission from Elsevier Science.)

2. In the energy range below the peak maximum, the σ curves show some undulations, which are attributed to either excitation, superexcitation, or ionization of the outer-valence electrons.

3. Around 18–23 eV, vibrational structures are clearly observed for CH_4, C_2H_6, and C_3H_8, which were assigned to the superexcited Rydberg states converging to the $(2a_1)^{-1}$, $(2a_{2u})^{-1}$, and $(4a_1)^{-1}$ ion states, respectively. These superexcited Rydberg states move to lower energies as the number of carbon atoms increases along the order of the ionization potentials of the $2a_1$ electron in CH_4, $2a_{2u}$ electron in C_2H_6, and $4a_1$ electron in C_3H_8 as shown in Fig. 5.

It is difficult to give a simple explanation of the gross features in the σ curves in Fig. 6 with existing theoretical knowledge. In this context, it should be noted that Nakatsukasa and Yabana [31] calculated the absolute values of the photoabsorption cross sections (σ) as a function of the energy of up to 40 eV for SiH_4 (silane), C_2H_2 (acetylene), and C_2H_4 (ethylene) based on the time-dependent local density approximation, and obtained good agreement with the results yielded by real-photon [20,32,33] and virtual-photon experiments [24,28,34]. It is also interesting, in terms of the gross features mentioned above, that the σ curves of cyclic alkanes seem to behave differently from those of normal alkanes, e.g., C_3H_8 (propane) vs. C_3H_6 (cyclopropane) [35].

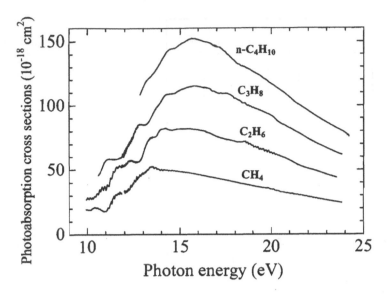

Figure 6 The photoabsorption cross sections (σ) of CH_4, C_2H_6, C_3H_8, and n-C_4H_{10} as a function of the incident photon energy measured via the double ionization chamber and synchrotron radiation, as mentioned in Section 2.1. Those in the range below the first ionization potentials were measured by the photon-beam attenuation method, using the ionization chamber as a conventional gas cell. The energy resolution of the incident photon energy at 20 eV is as follows: 32 meV for CH_4, 29 meV for C_2H_6, 29 meV for C_3H_8, and 116 meV for n-C_4H_{10}. (From Ref. [7]. Reprinted with permission from Elsevier Science.)

3.2. Photoionization Quantum Yields of Normal Alkanes

In Fig. 7 [7], we compared the photoionization quantum yields (η_i) of CH_4, C_2H_6, and C_3H_8, which were measured by our group using the double ionization chamber and synchrotron radiation, as described in Section 2.1. The photon energies are considered in two ranges, as follows, in terms of the behavior of the η_i curves as a function of the incident photon energy [7]:

1. The energy range of the outer-valence electrons, where the η_i curves rise up gradually and then reach unity.
2. The energy range of the inner-valence electrons, i.e., higher than approximately 17 eV, where the η_i curves show small but discernible deviations from unity. The superexcited Rydberg states mentioned in Section 3.1 play an important role in these deviations.

The η_i curves rise up from the first adiabatic ionization potentials and then reach unity at photon energies, slightly higher than the ionization potentials of the deepest outer-valence electrons. The energy differences between the onsets of the η_i curves and the photon energies where the η_i curves reach unity are approximately 3 eV for CH_4, 5 eV for C_2H_6, and 6 eV for C_3H_8, respectively, reflecting the difference between the ionization potentials of the shallowest and deepest outer-valence orbitals shown in Figs. 5 and 7.

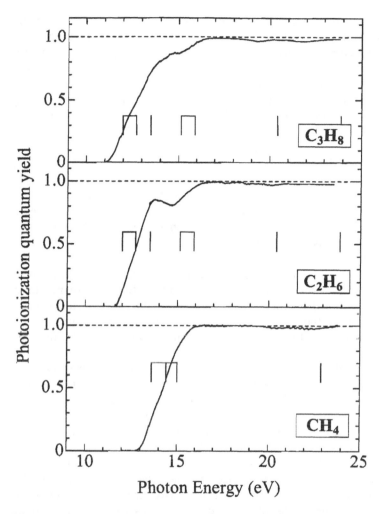

Figure 7 The photoionization quantum yields (η_i) of CH_4, C_2H_6, and C_3H_8 as a function of the incident photon energy measured via the double ionization chamber and synchrotron radiation as mentioned in Section 2.1. The energy resolution of the incident photon energy at 20 eV is as follows: 32 meV for CH_4, 29 meV for C_2H_6, and 29 meV for C_3H_8. The vertical ionization potentials of the ionic states involved are indicated by the vertical bars for each molecule [11]. (From Ref. [7]. Reprinted with permission from Elsevier Science.)

4. CONCLUSIONS AND FUTURE PROBLEMS

The features in C_1–C_4 normal alkanes discussed in Section 3 seem to be generalized to a wide range of molecules, and thus we conclude that the major part of the photoabsorption cross sections of molecules (σ) is associated with the ionization and excitation of the outer-valence electrons. Hence, there is a strong need to measure the absolute values of σ in the vacuum ultraviolet range, particularly in the range of the incident photon energy 10–30 eV, which is covered by the normal incidence monochromator used to monochromatize synchrotron radiation. The photoionization (σ_i) and photodissociation (σ_d) cross sections, which are partial photoabsorption cross sections, are also key quantities for a full under-standing of the interaction of photons with molecules in the vacuum ultraviolet range. The structures seen in the cross-section curves as a function of the incident photon energy are attributed to the superexcited states or shape resonances. In conclusion, the absolute values of σ, σ_i, and σ_d as a function of the incident photon energy in the vacuum ultraviolet range are quantities to be measured with a foremost priority in investigating the interaction of photons with molecules from a chemical point of view. However, they have been measured only for a limited number of molecules [13–16]; in particular, the quantity and quality of the data sets for polyatomic molecules are still insufficient. In this context, we note that it is more difficult to measure the correct and absolute values of σ_i and σ_d than σ, because the incident photon flux is needed in measuring the photo-ionization quantum yields (η_i) with the double ionization-chamber method, and the coincidence measurements should be carried out in the dipole-simulation experiments with the assumption that η_i is unity around 20 eV photon energy (see also Section 2).

Finally, some future problems that need to be solved from the view point of this chapter are summarized below.

1. The effect of temperature on the absolute values of σ, σ_i, and σ_d as a function of the incident photon energy. All the cross-section data shown in this chapter were measured for molecules in the gas phase at room temperature and thus, the target molecules do not lie in a single energy level as an initial level. This means that the measured cross sections seem to be dependent on gas temperature, which is important in various applications of the cross-section data.

2. The measurement of the absolute values of the partial photodissociation cross sections as a function of the incident photon energy. As mentioned in Section 1, the interesting structures due to superexcited states appear in the σ, σ_i, and σ_d curves as a function of the incident photon energy. They are much more noticeable in the σd curves than in the σ and σ_i curves because no contribution from ionization is involved in the σd curves (e.g., see the σ, σ_i, and σ_d curves around 22 eV in Fig. 2). It follows that the measurement of "partial" photodissociation cross sections provides us with a more powerful tool to investigate the superexcited molecules than that of σ_d. Most of the experiments along this idea, however, have yielded just the relative cross sections. We have recently been successful in measuring the absolute cross sections for the emission of fluorescences by excited CH radicals and hydrogen atoms in the photoexcitation of methane as a function of the incident photon energy in the range 12.65–41 eV, and found an unexpectedly large contribution from the doubly excited states in terms of the independent electron model [36]. The dynamics and spectroscopy of doubly excited molecules are the subject of current interest from the viewpoint of the few-body problem in quantum mechanics [37], and thus the absolute values of the partial photodissociation cross sections should be extensively measured.

3. The extension of target molecules from those in the gas phase to the condensed phase to investigate the effect of density on the cross sections [5]. The extension to clusters and transient species such as radicals is also important.

ACKNOWLEDGMENTS

The authors wish to thank Drs. Kosei Kameta and Masatoshi Ukai for their efforts in measuring the absolute values of the cross sections cited in this chapter.

REFERENCES

1. Inokuti, M. Photochem. Photobiol. 1986, *44*, 279.
2. Hatano, Y.; Inokuti, M. In *Atomic and Molecular Data for Radiotherapy and Radiation Research*. Inokuti, M., Ed.; IAEA: Vienna, 1995, (Chapter 5).
3. Inokuti, M. Rev. Mod. Phys. 1971, *43*, 297.
4. Platzman, R.L. Radiat. Res. 1962, *17*, 419.
5. Hatano, Y. Phys. Report 1999, *313*, 109.
6. Hatano, Y. In *Radiation Research*, Fielden, E.M.; Fowler, J.F.; Hendry, J.H.; Scott, D., Eds.; Taylor & Francis: London, 1987; 35 p.
7. Kameta, K.; Kouchi, N.; Ukai, M.; Hatano, Y. J. Electron Spectrosc. Relat. Phenom. 2002, *123*, 225.
8. Kimura, K.; Katsumata, S.; Achiba, Y.; Yamazaki, T.; Iwata, S. Handbook of He I Photoelectron Spectra of Fundamental Organic Molecules. Japan Scientific Societies: Tokyo, 1981.
9. Potts, A.W.; Streets, D.G. J. Chem. Soc., Faraday Trans. II 1974, *70*, 875.
10. Bieri, G.; Burger, F.; Heilbronner, E.; Maier, J.P. Helvetica. Chim. Acta 1977, *60*, 2213.
11. Bieri, G.; Åsbrink, L. J. Electron Spectrosc. Relat. Phenom. 1980, *20*, 149.
12. Samson, J.A.R.; Haddad, G.N.; Masuoka, T.; Pareek, P.N.; Kilcoyne, D.A.L. J. Chem. Phys. 1989, *90*, 6925.
13. Berkowitz, J. Photoabsorption, Photoionization, and Photoelectron Spectroscopy. Academic Press: New York, 1979; 1–469.
14. Gallagher, J.W.; Brion, C.E.; Samson, J.A.R.; Langhoff, P.W. J. Phys. Chem. Ref. Data 1988, *17*, 9.
15. Berkowitz, J. Atomic and Molecular Photoabsorption, Absolute Total Cross Sections. Academic Press: San Diego, 2002; 1–350.
16. Kameta, K.; Kouchi, N.; Hatano, Y. In *Landolt–Börnstein*, New series volume I/17C, *Photon and Electron Interactions with Atoms, Molecules and Ions — Photon- and electron-interactions with molecules: Ionization and dissociation*, Itikawa, Y., Ed.; Springer-Verlag: Berlin, 2003; 4-1–4-61, Chapter 4.
17. Rabalais, P.W.; Bergmark, T.; Werme, L.O.; Karlsson, L.; Siegbahn, K. Physica Scripta 1971, *3*, 13.
18. Platzman, R.L. Vortex 1962, *23*, 372.
19. Samson, J.A.R. Techniques of Vacuum Ultraviolet Spectroscopy. Pied Publications: Lincoln, 1980; 1–348; Samson, J.A.R. J. Opt. Soc. Am. 1964, *54*, 6.
20. Kameta, K.; Ukai, M.; Chiba, R.; Nagano, K.; Kouchi, N.; Hatano, Y.; Tanaka, K. J. Chem. Phys. 1991, *95*, 1456.
21. Kameta, K.; Machida, S.; Kitajima, M.; Ukai, M.; Kouchi, N.; Hatano, Y.; Ito, K. J. Electron Spectrosc. Relat. Phenom. 1996, *79*, 391.
22. Brion, C.E.; Hamnett, A. Adv. Chem. Phys. 1981, *45*, 1.
23. Brion, C.E. Comm. At. Mol. Phys. 1985, *16*, 249.
24. Cooper, G.; Burton, G.R.; Chan, W.F.; Brion, C.E. Chem. Phys. 1995, *196*, 293.
25. Olney, T.N.; Cann, N.M.; Cooper, G.; Brion, C.E. Chem. Phys. 1997, *223*, 59.
26. Hatano, Y. Dissociation dynamics of superexcited molecules. In *The Physics of Electronic and Atomic Collisions*, Dube, L.J.; Mitchell, J.B.A.; McConkey, J.W.; Brion, C.E. AIP Press: New York, 1995; 67 p.
27. Hatano, Y. In *Dynamics of Excited Molecules*, Kuchitsu, K., Ed.; Elsevier: Amsterdam, Chapter 6, 1994; 151–216.
28. Cooper, G.; Olney, T.N.; Brion, C.E. Chem. Phys. 1995, *194*, 175.
29. Chan, W.F.; Cooper, G.; Brion, C.E. Phys. Rev. 1991, *A44*, 186.
30. Chan, W.F.; Cooper, G.; Sodhi, R.N.S.; Brion, C.E. Chem. Phys. 1993, *170*, 81.
31. Nakatsukasa, T.; Yabana, K. J. Chem. Phys. 2001, *114*, 2550.
32. Ukai, M.; Kameta, K.; Chiba, R.; Nagano, K.; Kouchi, N.; Shinsaka, K.; Hatano, Y.; Umemoto, H.; Ito, Y.; Tanaka, K. J. Chem. Phys. 1991, *95*, 4142.

33. Holland, D.M.P.; Shaw, D.A.; Hayes, M.A.; Shpinkova, L.G.; Rennie, E.E.; Karlsson, L.; Baltzer, P.; Wannberg, B. Chem. Phys. 1997, *219*, 91.
34. Cooper, G.; Burton, G.R.; Brion, C.E. J. Electron Spectrosc. Relat. Phenom. 1995, *73*, 139.
35. Kameta, K.; Muramatsu, K.; Machida, S.; Kouchi, N.; Hatano, Y. J. Phys. 1999, *B32*, 2719.
36. Kato, M.; Kameta, K.; Odagiri, T.; Kouchi, N.; Hatano, Y. J. Phys. 2002, *B35*, 4383.
37. Kouchi, N.; Ukai, M.; Hatano, Y. J. Phys. 1997, *B30*, 2319.

6

Reactions of Low-Energy Electrons, Ions, Excited Atoms and Molecules, and Free Radicals in the Gas Phase as Studied by Pulse Radiolysis Methods

Masatoshi Ukai
Tokyo University of Agriculture and Technology
Tokyo, Japan

Y. Hatano
Tokyo Institute of Technology
Tokyo, Japan

SUMMARY

After a brief survey of recent advances in gas-phase collision dynamics studies using pulse radiolysis methods, the following two topics in our research programs are presented with emphasis on the superior advantages of the pulse radiolysis methods over the various other methods of gas-phase collision dynamics, such as beam methods, swarm methods, and flow methods. One of the topics is electron attachment to van der Waals (vdW) molecules. The attachment rates of thermal electrons to O_2 and to other molecules in dense gases, measured in wide ranges of both gas temperatures and pressures, present experimental evidence for electron attachment to van der Waals molecules. The results are compared with theories and discussed in terms of the effect of van der Waals interaction on the electron attachment resonance. The conclusions obtained are related with investigations of electron attachment, solvation, and localization in the condensed phase. Another topic is Penning ionization and its related processes. The rate constants for the de-excitation of $He(2^3S)$, $He(2^1P)$, $Ne(^3P_0)$, $Ne(^3P_1)$, $Ne(^3P_2)$, $Ar(^1P_1)$, and $Ar(^3P_1)$, by atoms and molecules measured in the temperature range from 100 to 300 K, provide the collisional energy dependence of the de-excitation cross sections. The results are compared in detail with theories classified according to the excited rare gas atoms in the metastable and resonant states, which present a systematic understanding of the de-excitation processes of excited rare gas atoms. For typical examples, the de-excitation process of metastable atoms, the de-excitation process of resonant atoms, and the de-excitation process by molecules containing a group IV element are discussed.

1. INTRODUCTION

Pulse radiolysis is of great importance in the understanding of gas-phase reactions [1–3]. The results obtained are also useful for understanding condensed phase reactions. The objectives of pulse radiolysis studies in the gas phase are divided into two parts. One is to understand the fundamental processes, in particular, early processes in radiolysis. The other is to make an important contribution, as one of the powerful experimental methods, to gas-phase collision dynamics studies. Recent advances in the latter studies are surveyed in this paper; those in the former studies are not included here. The above-mentioned objectives are, however, closely related with one another in terms of the following interface relationships. New information obtained from the latter studies is useful for understanding the fundamental processes in radiolysis, whereas that from the former studies is an important source of new ideas and information in collision dynamics studies.

Advances in pulse radiolysis studies in the gas phase have been summarized in several review papers. In a comprehensive review by Sauer [4], a review presented by Firestone and Dorfman [5] in 1971 was referred to as the first review on gas-phase pulse radiolysis. Experimental techniques and results obtained were summarized by one of the present authors [6], with emphasis on an important contribution of pulse radiolysis to gas-phase reaction dynamics studies. Examples were chosen by Sauer [7] from the literature prior to 1981 to show the types of species that were investigated in the gas phase using pulse radiolysis technique. Armstrong [8] reviewed experimental data obtained from gas-phase pulse radiolysis together with those from ordinary steady-state radiolysis. Advances in gas-phase pulse radiolysis studies since 1981 were also briefly reviewed by Jonah et al. [9], with emphasis on an important contribution of this technique to free radical reaction studies. One of the present authors reviewed comprehensively the gas-phase collision dynamics studies of low-energy electrons, ions, excited atoms and molecules, and free radicals by means of pulse radiolysis method [1–3]. An important contribution of pulse radiolysis to electron attachment, recombination, and Penning collision studies was also reviewed in Refs. 10–15.

In this chapter, firstly, a very brief survey is given of recent advances in such studies as classified according to the detection technique of transient species in pulse radiolysis. Secondly, examples are chosen from our recent investigations, with special emphasis on the important contributions of pulse radiolysis methods to gas-phase collision dynamics; one is electron attachment, the other is Penning ionization and related processes. The detection techniques and corresponding reaction processes, together with major references, are given below:

(A) Optical emission and absorption spectroscopy

(1) De-excitation of excited rare gas atoms such as Penning ionization and related processes (see Section 3)
(2) Formation and reaction of rare gas excimers and exciplexes [16–23]
(3) Reaction of atoms and free radicals (see references cited in Jonah et al. [9]).
(4) Electron–ion or ion–ion recombination [17,24]
(5) Ion–molecule reaction [25]
(6) Reaction of excited aromatic molecules [26]
(7) Electron thermalization [27]

(B) DC conductivity measurements

(1) Electron mobility [28–34]
(2) Electron detachment [35].

(3) Electron attachment (see Section 2)
(4) Electron–ion or ion–ion recombination [30,33,36]

(C) Microwave conductivity measurements

(1) Electron attachment (see Section 2)
(2) Electron–ion or ion–ion recombination [37]
(3) Electron thermalization [38–44]
(4) Penning ionization [45]

(D) Mass spectroscopy

(1) Ion–molecule reaction [46–48]

2. ELECTRON ATTACHMENT PROCESSES

Pulse radiolysis, as combined mainly with microwave conductivity techniques, has been applied to thermal electron attachment studies. The molecules studied are O_2, N_2O, NO, NO_2, SF_6, alkyl halides, and perfluorocarbons. Alkyl halides, fluorocarbons, and N_2O were studied in the early 1970s [46,49–51], whereas O_2, N_2O, and other simple molecules have been extensively studied with special emphasis on the important contribution of a pulse radiolysis method to electron attachment studies, in particular, to the studies on electron attachment to van der Waals molecules [10,11,52–70]. In the following, one may be a unique standpoint of the microwave technique combined with the pulse radiolysis method in electron attachment experiments. In the study of electron attachment mechanisms, ordinary swarm techniques may have some essential limitations in the experiment as it is almost unavoidable to use only a few environmental buffer gases for which the swarm parameters, electron energy distributions, etc., are well known. Therefore, in cases where the attachment mechanism is strongly dependent on the particular nature of the environmental gases, the technique may not give enough information to adequately evaluate the mechanism in detail. Usual beam techniques are evidently not suitable for the study of environmental effects on the attachment mechanism. On the other hand, the microwave technique has been used as an alternative means for such studies, with the main advantage of the technique being that it allows us to observe the behavior of thermal electrons, thus excluding any factor dependent on the electron energy distribution. For usual swarm and beam techniques, it has been difficult to study electron collision processes at very low energies such as thermal energy, whereas such studies are obviously of great importance in the understanding of not only a two-body problem such as electron interaction with molecules, but also various phenomena in ionized gases. The microwave technique, combined with the pulse radiolysis method, has shown a distinct advantage in studying thermal electron attachment to molecules. By employing the pulse radiolysis method, it is possible to perform a time-resolved observation of decaying electrons with a very fast response in a very wide range of pressure of an environmental gas, which is chosen with virtually no limitation. Thus, the mechanism of low-energy electron attachment to molecules has been discussed primarily in terms of the interaction of electrons with molecules. Recent studies of thermal electron attachment to O_2, N_2O, and other molecules have revealed that the electron attachment to pre-existing van der Waals molecules or neutral clusters plays a significant role in the overall mechanism. A significant development in such studies has been started in electron attachment studies using experimental techniques that originated from radiation chemistry; one is the microwave technique combined with pulse radiolysis (see references cited in Refs. 10 and 11), the other is the competition kinetics of steady-state γ-radiolysis [71–73]. Such studies have made an important contribution to advances in electron attachment studies, as summarized in the last part of this section, and have triggered a very recent development in beam experiments of electron–vdW molecule collisions [74–77].

Evidently, the existence of electron attachment to vdW molecules compels us to re-interpret more or less various experimental data obtained previously. Furthermore, because such processes must be more important in dense gases or in the condensed phase, the studies of those processes will provide insight as to the effect of the density on reactions involving electrons or, generally, on the electron–molecule interaction processes [78].

Because oxygen is probably the most extensively studied molecule in both experimental and theoretical investigations of low-energy electron attachment, the experimental results and detailed discussion are presented in this paper particularly for O_2. The only accepted mechanism has been the overall two-step three-body mechanism, which was originally suggested by Bloch and Bradbury [79] (referred to below as "BB") and was later modified by Herzenberg [80] to make it consistent with modern experimental data. The BB mechanism for O_2–M mixture, where M is a molecule other than O_2, is expressed as follows:

$$e^- + O_2 \xrightarrow{k_1} O_2^-* \tag{1}$$

$$O_2^-* \xrightarrow{k_2} O_2 + e^- \tag{2}$$

$$O_2^-* + O_2 \xrightarrow{k_3} O_2^- + O_2 \tag{3}$$

$$O_2^-* + M \xrightarrow{k_4} O_2^- + M \tag{4}$$

The vibrational characteristics of the negative ion O_2^-* are established from electron impact and scattering experiments, and the electron affinity of O_2 is 0.44 eV. For convenience, the potential energy diagrams for O_2 ($X^3\Sigma_g^-$) and O_2^- ($X^2\Pi_g$) are shown in Fig. 1. It can be seen that the lowest resonance involves the vibrational levels, O_2 ($v=0$) and O_2^- ($v'=4$), at low electron energies and the resonance energy is about 0.08 eV.

Because all electron decays for O_2–M mixtures in the above-mentioned experimental conditions show pseudo-first-order behavior, each decay curve gives an electron lifetime τ_0, which is related to molecular number densities [O_2] and [M] as:

$$\tau_0[O_2] = \frac{1}{k_1} + \frac{1}{k_M}[M] \tag{5}$$

when [O_2] $\ll$ [M], where k_M ($=k_1 k_4/k_2$) is the overall three-body rate constant. Based on Eq. (5), Shimamori and Fessenden [58] determined the value of $k_1 = 4.8 \times 10^{-11}$ cm^3/sec and the value of k_M for each stabilization partner, which is listed in Table 1 with the value obtained by other workers. The autoionization lifetime of O_2^-* (i.e., the value of $1/k_2$) was also estimated to be around 10^{-10} sec, which was comparable to the predictions of some theoretical treatments [80,94–97] (see Table 2).

Some inconsistencies existed between the results obtained by Shimamori and Hatano [61] and other data. Christophorou [98–100], Goans and Christophorou [101], and McCorkle et al. [102] investigated electron attachment in O_2–C_2H_4, O_2–C_2H_6, and O_2–N_2 mixtures at very high gas pressures (<30 atm), and observed that the attachment rate in the O_2–C_2H_4 system showed saturation at very high pressures, but the rates in O_2–C_2H_6 and O_2–N_2 mixtures continued to increase steeply with increasing pressures of C_2H_6 and

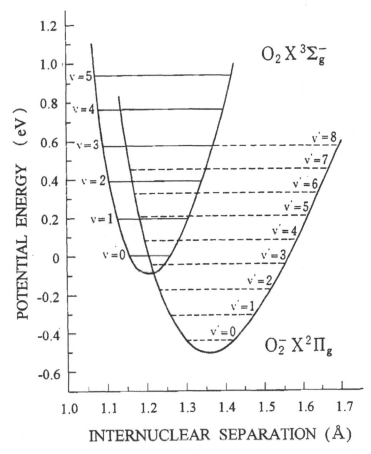

Figure 1 Potential energy curves for O_2 and O_2^-. (From Refs. 10 and 11.)

N_2, respectively. They reported the value of k_1 to be much larger than that obtained by Shimamori and Hatano [61] and, correspondingly, to have a considerably shorter lifetime of O_2^-* (see Table 2). Such inconsistencies have been demonstrated more clearly and analyzed by the recent work of Kokaku et al. [52,53]. Kokaku et al., in a joint research between Tokyo and Notre Dame, have studied, using the microwave technique combined with the pulse radiolysis method, O_2–CO_2 and several O_2–hydrocarbon systems at pressures from <100 to 1000 Torr and found that the BB mechanism can account for the data in the low-pressure range but fails to explain the results at higher pressures. The data for O_2–C_2H_4 mixtures are shown in Fig. 2 as an example, where the high-density part of the data agrees well with the data obtained by Goans and Christophorou [101] using a swarm technique. This research has clarified a significant discrepancy between the values of k_1 and $1/k_2$ obtained by the pulse radiolysis microwave technique and those obtained by the swarm technique. Thus, it has become obvious that a consistent interpretation in terms of only the BB mechanism is not possible in a wide range of pressures of M, and additional mechanisms must be considered to explain the high-pressure data. One of the strong

Table 1 Three-Body Attachment Rate Constants k_M for the Reaction $e^- + O_2 + M \rightarrow O_2^- + M$ at Room Temperature (From Refs. 10 and 11.)

M	10^{-30} (cm^3/sec)	Authors	Reference	M	10^{-30} (cm^3/sec)	Authors	Reference
He	0.033	Shimamori and Hatano	[61]	C_2H_2	1.3	Kokaku et al.	[53]
	0.03	Chanin et al.	[83]		1.5	Christophorou	[105]
	0.07	van Lind et al.	[88]		1.7	Shimamori and Hatano	[61]
Ne	0.023	Shimamori and Hatano	[61]		0.9	Toriumi and Hatano	[68]
Ar	0.05	Shimamori and Hatano	[61]	C_3H_8	3.2	Kokaku et al.	[52]
Kr	0.05	Shimamori and Hatano	[61]		3.3	Shimamori and Hatano	[61]
Xe	0.085	Shimamori and Hatano	[61]	n-C_4H_{10}	4.2	Shimamori and Fessenden	[58]
H_2	0.48	Shimamori and Hatano	[59]		4.5	Kokaku et al.	[53]
D_2	0.14	Shimamori and Hatano	[59]		5	Shimamori and Hatano	[61]
N_2	0.06	Chanin et al.	[83]	n-C_5H_{12}	7.9	Shimamori and Hatano	[61]
	0.085	Shimamori and Hatano	[60]	neo-C_5H_{12}	7	Kokaku et al.	[52]
	0.09	Shimamori and Fessenden	[58]		8.0	Shimamori and Hatano	[61]
	0.1	Crompton et al.	[84]	n-C_6H_{14}	8.1	Shimamori and Hatano	[61]
	0.11	Hegerberg and Crompton	[86]	C_6H_6	8.5	Shimamori and Hatano	[61]
		van Lind et al.	[88]		18	Bouby and Abgrall	[81]
	0.15	Goans and Christophorou	[101]	CO	1.31	Shimamori and Fessenden	[58]
		Hackam and Lennon	[85]	CO_2	3	Bouby and Abgrall	[81]
		McCorkle et al.	[102]			Pack and Phelps	[90]
	0.26	Young et al.	[93]			Warman et al.	[92]
	0.10	Toriumi and Hatano	[70]			Bouby et al.	[82]
O_2	1.7	Young et al.	[93]		3.2	Kokaku et al.	[53]
	2.0	Pack and Phelps	[90]		3.23	Crompton et al.	[84]
	2.1	van Lind et al.	[88]		3.5	Hegerberg and Crompton	[86]
		Nelson and Davis	[89]				
		Parlant and Fiquet-Fayard	[97]				
	2.15	Crompton et al.	[84]	H_2O	14	Bouby et al.	[82]
	2.2	Shimamori and Fessenden	[54]			Pack and Phelps	[90]
		Shimamori and Hatano	[59]			Stockdale et al.	[91]
		Warman et al.	[92]		15.2	Bouby and Abgrall	[81]
	2.3	Hackam and Lennon	[85]	H_2S	9	Bouby et al.	[82]
		Shimamori and Hatano	[60]		10	Bouby and Abgrall	[81]
	2.6	Hurst and Bortner	[87]	NH_3	6.8	Bouby et al.	[82]
	2.8	Chanin et al.	[83]		7.5	Bouby and Abgrall	[81]
	2.2	Hegerberg and Crompton	[86]				
CH_4	0.34	Shimamori and Hatano	[61]	CH_3OH	8.8	Bouby et al.	[82]
C_2H_4	1.5	Goans and Christophorou	[101]		9.6	Bouby and Abgrall	[81]
	1.7	Bouby and Abgrall	[81]		11	Shimamori and Hatano	[61]
	2.0	Kokaku et al.	[52]	C_2H_5OH	18	Shimamori and Hatano	[61]
	2.3	Hurst and Bortner	[87]	CH_3COCH_3	27	Bouby et al.	[82]
	2.5	Bouby et al.	[82]		35	Bouby and Abgrall	[81]
	3	Shimamori and Hatano	[61]				
	3.1	Stockdale et al.	[91]				
	3.4	Stockdale et al.	[91]				
	1.3	Toriumi and Hatano	[69]				

candidates for such a mechanism is electron attachment to vdW molecules, as was indicated by Kokaku et al. [52,53]:

$$O_2 + M \overset{K_{eq}}{\Longleftrightarrow} (O_2 \cdot M) \tag{6}$$

$$e^- + (O_2 \cdot M) \xrightarrow{k_5} (O_2 \cdot M)^{-*} \tag{7}$$

$$(O_2 \cdot M)^{-*} \xrightarrow{k_6} O_2 + M + e^- \tag{8}$$

$$(O_2 \cdot M)^{-*} + M \xrightarrow{k_7} (O_2 \cdot M)^- + M \tag{9}$$

Table 2　Lifetime of O_2^{-*} ($X^2\Pi_g$, $v'=4$) (From Refs. 10 and 11)

Lifetimes	$(10^{-12}$ sec)	Authors	Reference
Theory	300	Herzenberg	[80]
	170	Koike and Watanabe	[96]
	72	Koike	[94,95]
	88	Parlant and Fiquet-Fayard	[97]
Experiment	100	Shimamori and Hatano	[61]
	63	Shimamori and Fessenden	[58]
	2	Christophorou	[98]
		Goans and Christophorou	[101]
		McCorkle et al.	[102]
	91	Toriumi and Hatano	[68]
	66	Toriumi and Hatano	[70]

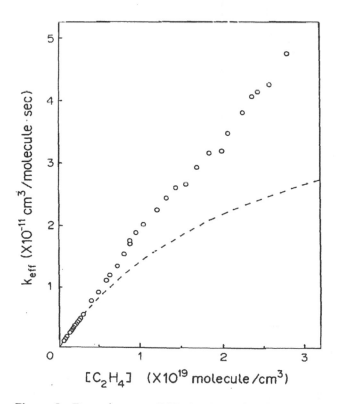

Figure 2　Dependence on C_2H_4 density of the effective two-body rate constant of thermal electron attachment in O_2–C_2H_4 mixtures at room temperature. (From Ref. 52.) The dashed curve represents the expected contribution from the BB mechanism.

where K_{eq} is the equilibrium constant for vdW molecule formation. One should note here that the density of vdW molecules is determined by K_{eq} [O$_2$][M], and the value of K_{eq} can be estimated by the theoretical treatment of Stogryn and Hirschfelder [103]. Several experiments have provided evidence for the existence of vdW molecules in the gas phase [78,104]. In the kinetic treatment made by Kokaku et al. [52,53], the estimated values of k_5 are (2–20) $\times$ 10^{-9} cm^3/sec depending on M, where it is highly attractive that all the values for k_5 are much larger than the value of k_1 ($= 4.8 \times 10^{-11}$ cm^3/sec). This result suggests that in the case of vdW molecules, the initial electron capture mechanism differs substantially from the case of isolated molecules. A recent study by Shimamori and Fessenden [58] has verified clearly the presence of the vdW mechanism. They have measured the temperature dependence of three-body rate constants in pure O$_2$, O$_2$–N$_2$, and O$_2$–CO mixtures. The result for O$_2$ is shown in Fig. 3. According to the theory of Herzenberg [80], the three-body rate constant, which corresponds to the experimentally obtained k_M, can be expressed as:

$$k_M = \frac{2}{3}\left(\frac{h_2}{2\pi m k_B}\right)^{3/2} \xi k_L \left(\frac{1}{T}\right)^{3/2} \exp\left(-\frac{E_0}{k_B T}\right) \tag{10}$$

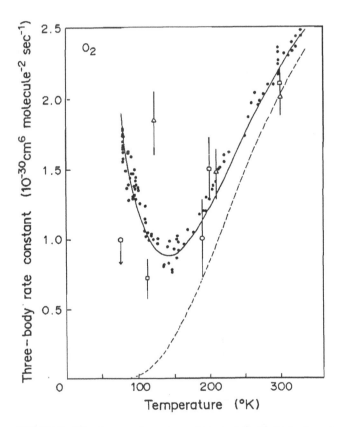

Figure 3 The temperature dependence of the three-body rate constant of O$_2$. (From Ref. 58.) The broken curve shows the temperature dependence of the rate constant calculated from Herzenberg's theory. The solid curve shows a calculated rate constant, which involves both the contributions from the broken curve and the rate constant due to electron attachment to van der Waals molecule (O$_2$)$_2$.

where h, m, and k_B have their usual meanings; k_L is the Langevin's rate constant; ξ is the stabilization efficiency; T is the absolute temperature; and E_0 is the resonance energy. Equation (10) predicts a simple decrease in the rate constant with reduced temperature. The expected curve for O_2 calculated from Eq. (10), assuming ξ to be unity, is drawn in Fig. 3. Because an extra contribution that increases with lowered temperature is evident, electron attachment to the vdW molecule $(O_2)_2$ has been proposed to account for this. Similarly, the importance of electron attachment to (O_2N_2) and (O_2CO) has also been demonstrated.

Because the temperature dependence of k_5 is given by:

$$k_5 = \frac{2}{3} \frac{2\pi h^2}{(2\pi m k_B)^{3/2}} \Gamma_5 \left(\frac{1}{T}\right)^{3/2} \exp\left(-\frac{E_r}{k_B T}\right) \tag{11}$$

where E_r and Γ_5 are, respectively, the energy and the width for the resonance attachment process (Eq. (7)), both pressure-dependent and temperature-dependent experiments [68–70] have given important rate parameters for the BB mechanism such as the rate constant for the initial electron attachment to $O_2(k_1)$, the lifetime τ of O_2^-* (i.e., the resonance width), and the overall three-body attachment rate constants for O_2–C_2H_6, O_2–C_2H_4, and O_2–N_2 mixtures. The values of τ are again in good agreement with those obtained by theories (see Table 2). Each three-body rate constant is, respectively, smaller than that obtained previously, without taking into consideration the vdW mechanism. The value of k_1 obtained from the O_2–N_2 system [70], which is selected as a convenient system to determine the value of k_1, is about 3×10^{-11} cm^3/sec. This value agrees within experimental error with those obtained from the O_2–C_2H_6 and O_2–C_2H_4 systems [68,69] and with the value 4.8×10^{-11} cm^3/sec, which was obtained previously by Shimamori and Hatano [61]. This value is also consistent with qualitative results of k_1 obtained by other groups, except for the value obtained by the extremely high-pressure swarm technique [101]. It should be noted here that the value 3×10^{-11} cm^3/sec, is in good agreement with the theoretical values 2.5×10^{-11} and 2.1×10^{-11} cm^3/sec [94–97].

The important rate parameters for the vdW mechanism, such as the rate constant for the initial electron attachment to (O_2M), where M $= C_2H_6$, C_2H_4, and N_2, and the lifetime τ of $(O_2 M)^-*$ (i.e., the resonance width), have been also obtained from this experiment and summarized in Table 3 [10]. The value of k_5 in Table 3 is again much larger than the above-mentioned k_1 values. The resonance energy for $e^- + (O_2M) \rightarrow (O_2M)^-*$ is much smaller than that for $e^- + O_2 \rightarrow O_2^-*$, whereas its width for the former process is much larger than that for the latter process. The large enhancement in the attachment rate constant from k_1 to k_5 has been discussed qualitatively as related to the decrease in the resonance energy and the increase in the resonance width. The reason for the decrease in the resonance energy has been ascribed to the fact that the resonance state is much

Table 3 Rate Constant k_5, Resonance Energy E_r, Resonance Width Γ_5, Electron Density $v_r f_r$. and Cross Section σ_h (From Refs. 10 and 11.)

$(O_2 \cdot M)$	E_r (meV)	Γ_5 (µeV)	$v_r f_r$	σ_h (Å^2)	k_5 (10^{-11} (cm^3/sec)
$(O_2 \cdot N_2)$	20	800	0.71	2500	3000
$(O_2 \cdot C_2H_6)$	30	450	0.89	1700	1100
$(O_2 \cdot C_2H_4)$	45	270	0.92	1100	380
O_2	88 (E_0)	10 (Γ_1)	0.47	570	3 (k_1)

stabilized by the polarization interaction between O_2^- and M. Such a situation is depicted in Fig. 4, where schematic potential energy curves are shown for the O_2–M and O_2^-–M systems [10,11]. Figure 4 shows that near the equilibrium intermolecular distance, the effective resonance energy of the O_2^-–M system is much reduced and even superimposed on the O_2–M curve. The existence of a number of vibrational states in both ion complex and neutral systems may be another major factor of the large transition probabilities.

When the resonance width is narrower than the energy distribution of thermal electrons, the attachment rate constant is expressed as:

$$k_5 = \int v f \sigma dv = v_r f_r \int \sigma dv \tag{12}$$

where v is electron velocity, f is a Maxwellian distribution function of electrons, σ is an attachment cross section, the suffix "r" means a value at resonance energy E_r. The factor $v_r f_r$ means the density of electrons with velocity v to attach to vdW molecules. The value of $v_r f_r$ is given in Table 3, which shows that $v_r f_r$ increases with decreasing resonance energy. The number of electrons that can attach to vdW molecules increases as compared with those that can attach to isolated O_2.

Because the cross section of a resonance process is expressed by the Breit–Wigner formula, the energy-integrated cross section is written as follows:

$$\int \sigma dv = \frac{4\pi^2}{3\kappa^2} \Gamma_5 \equiv \sigma_h \Gamma_5 \tag{13}$$

where κ is the wave number of incident electron, E_r and Γ_5 are assumed to be independent of energy, and σ_h means the effective magnitude of electron attachment cross section, which equals the geometrical cross section corresponding to the de Broglie wave length for

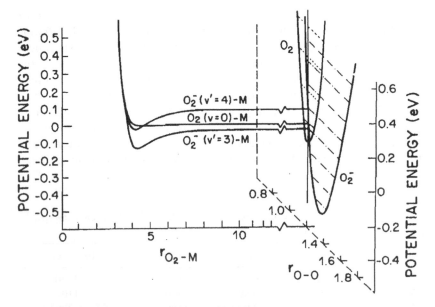

Figure 4 A model of variation of potential energies for O_2 ($v = 0$)–M and O_2^- ($v' = 4$)–M systems as a function of intermolecular distance. (From Refs. 10 and 11.)

the incident electron. The values of σ_h are also listed in Table 3. At extremely low-energy electron collision, such as electron attachment to vdW molecules, a "small" vdW molecule is supposed to collide with "large" electron clouds, of which the cross section is determined by the size of the de Broglie wave length of incident electrons. With decreasing resonance energy, therefore, the attachment cross section should increase. It should be noted that the maximum value of empirically obtained cross-section values for dissociative attachment processes at low energy is reasonably explained by the de Broglie wavelength of incident electron [105,106].

The resonance width Γ_5 is expressed by the Wigner's threshold rule. In the case of isolated O_2, the resonance state O_2^-* ($X^2\Pi_g$, $v' = 4$) can couple with only one electronic partial wave with an angular momentum $l = 2$. In the case of vdW molecules, intermolecular interaction may couple with additional partial waves such as p-wave and s-wave with low energy. If a third-body molecule distorts the orbital of O_2^- ($X^2\Pi_g$), new attachment channels can open with lower angular momentum of electrons and the resonance width may increase.

It has been necessary to make a quantitative calculation of these effects using precise wave functions of the O_2–M system. Huo et al. [107] made such calculations on the O_2–N_2 system and compared their result with experiments. They were successful in explaining the large enhancement in the attachment rate constant for vdW molecules using SCF wave functions corresponding to two geometries, T-shape and linear, for (O_2N_2) vdW molecules. The large enhancement in the attachment rate constant was clearly elucidated quantitatively in this theoretical calculation by the effect of additional vibrational structures of the vdW molecule on the attachment process, the symmetry breaking, which allows the molecule to attach a p-wave electron, and the lowering of resonance energy due to a deeper O_2–N_2 potential in comparison with O_2–N_2 potential, as shown schematically in Fig. 4.

An interesting approach [62,63] to this problem, the use of $^{18}O_2$ instead of $^{16}O_2$, further substantiated the electron attachment to vdW molecules. For the BB mechanism, the isotope effect may be expected to appear as a change in the rates of initial attachment and autoionization channels, which are caused by a decrease of the resonance energy for $^{18}O_2$ in comparison with $^{16}O_2$.

As mentioned in the beginning of this section, electron swarm data were reported [98–102] in the O_2–C_2H_6, O_2–C_2H_4, and O_2–N_2 systems up to about 10^{21} molecules/cm^3 (3×10^4 Torr) and simply elucidated only by the BB mechanism. An attempt [52,68,69] was made, therefore, to elucidate the high-pressure swarm data by the combination of the BB mechanism and the vdW mechanism. The electron swarm data are well explained up to about 4×10^{20} molecules/cm^3 by the combination of both mechanisms. It is obvious that the contribution from the vdW molecule is dominant in these density ranges. The large deviations in the data from the combination of the two mechanisms at densities higher than 4×10^{20} molecules/cm^3 may indicate electron attachment to large vdW molecules such as $O_2(C_2H_6)_2$, or may require additionally some collective properties of these hydrocarbon molecules to explain the density effect of electron attachment in this region. An attempt [108,109] was made to explain theoretically such density effects in the whole density range using the statistical model.

Electron attachment to O_2 has been investigated in supercritical hydrocarbon fluids at densities up to about 10^{22} molecules/cm^3 using the pulsed electric conductivity technique [110], and the results have been explained in terms of the effect of the change in the electron potential energy and the polarization energy of O_2^- in the medium fluids. In general, electron attachment to O_2 is considered to be a convenient probe to explore electron dynamics in the condensed phase.

A similar conclusion has been obtained also in the case of N_2O [54–58]. It may be plausible to extract from the results of O_2 and N_2O systems some general conditions under which one can predict the existence of electron attachment to vdW molecules [10,11]. The common feature to both O_2 and N_2O is that the rate constants of electron attachment to those isolated molecules are relatively small (10^{-11}–10^{-13} cm^3/sec) on an absolute scale. This is due to the presence of activation energy (i.e., the resonance energy) for electron attachment (0.08 eV for O_2 and 0.23 eV for N_2O). In contrast, there is virtually no activation energy in the electron attachment to vdW molecules containing O_2 or N_2O, thus yielding much larger rate constants for this process. The formation of vdW complexes appears to act just like it has an effect of lowering the activation energy or the resonance energy. Consequently, one may expect to observe the contribution of vdW molecules only for compounds that have activation energies for electron attachment, or for the molecules of which attachment cross section for electron energies near thermal energy increases with increasing electron energy.

One may expect generally that even in the case of molecules with negative electron affinities or with high threshold electron energies for attachment, some environmental effects or the effect of the vdW molecule formation bring about the large enhancement in the cross sections or the rate constants for the lower-energy electron attachment to these molecules. Based on the above discussions, the reasons for this expectation are summarized as follows [10,11]:

(1) The lowering of the resonance energy due to a deeper ion–neutral potential in comparison with neutral–neutral potential of the vdW molecule
(2) The additional vibrational structures of the vdW molecule
(3) The symmetry breaking due to the vdW interaction, which allows the molecule to attach electron with additional partial waves
(4) The deformation of the molecular structure, or the change of the vibrational modes due to the surrounding molecules
(5) The effective vibrational relaxation of the formed negative ion with excess energies due to the presence of a built-in third-body molecule in the vdW molecule.

The distinct features of the electron attachment to vdW molecules as summarized above may become a substantial clue to understand the fundamental nature of electron attachment not only in dense gases but also in the condensed phase [10,11]. It is also apparent that most of the electron attachment processes in the bulk system are no longer a simple process, as in the interaction of electron with isolated molecules. A definitely important role of pre-existing vdW molecules formed by weak intermolecular forces must be admitted. From this point of view, interesting phenomena in ionized gases, such as the attachment cooling effect [111–115] and the response time of the air-filled fast-response ionization chamber (J. W. Boag, 1984, private communication; R. W. Fessenden, 1985, private communication), should be analyzed by taking into account the important role of vdW molecules in the electron attachment mechanism.

Another recent aspect of the research of electron attachment processes, using the pulse radiolysis method combined with the microwave cavity technique, has been presented by a further combination of microwave electron heating [116]. This new technique has enabled to measure the electron attachment rate constants as a function of the mean electron energy in the range of 0.005–1 eV (Fig. 5). The new measurements have been made for various molecules, such as c-C_7H_{14}, CH_3I, and CH_2Br_2 [116,117], which have clearly revealed the disadvantage of the pseudo-free electron attachment by the collisional electron transfer using the higher-lying Rydberg atoms [117], which was

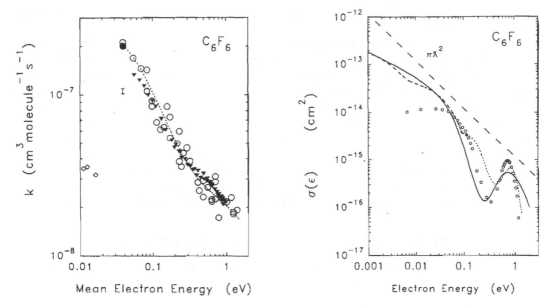

Figure 5   Left panel: Rate constants k for electron attachment to C_6F_6 as a function of the mean collisional energy. (From Ref. 118.) (O) The results obtained using the pulse radiolysis microwave cavity technique combined with microwave electron heating; (●) previous results by Shimamori et al. [J. Chem. Phys. 1993, *99*, 7789]; (▼) electron swarm [Spyrous and Christophorou, J. Chem. Phys. 1985, *82*, 1048]; and (◊) high-Rydberg atom beams [Marawar et al., J. Chem. Phys. 1988, *88*, 2853]. Right panel: Comparison of the cross section $\sigma(\varepsilon)$ for attachment to C_6F_6 as a function of the electron energy, (—) derived by unfolding the rate constants (From Ref. 118.) with the previous cross sections; (– – –) Chutijan and Alajajian [J. Phys. B 1985, *18*, 4159]; (⋯⋯⋯) Datkos et al. [J. Chem. Phys. 1993, *98*, 7875]; and (O) Grant and Christophrou [J. Chem. Phys. 1976, *65*, 2977].

assumed to be reliable for electron attachment in subthermal region. The pulse radiolysis microwave cavity technique, combined with microwave heating, has been further combined with the analytical "unfolding" technique by Shimamori et al. [118], which has made it possible to derive the attachment cross sections as a function of electron energy (i.e., the attachment cross sections in the region between the subthermal and subexcitation electron energy are continuously determined within the same experiment). The extremely low-energy electron attachment processes have been uncovered for the target molecules of C_6F_5X (X = F, Cl, Br, and I) [118], C_6H_5X (X = Cl, Br, and I) [119], various brominated ethanes and ethylenes [120], brominated methanes [121], and $CHCl_3$ [122]. The information of electron collision processes at such extremely low energies is not available by the resonant electron scattering method because of experimental difficulties. The new pulse radiolysis method will certainly provide the qualitatively brand-new information on the thermal electron attachment.

3. DE-EXCITATION OF EXCITED RARE GAS ATOMS

De-excitation processes of excited rare gas atoms in the lowest excited states have an important role in various phenomena in ionized gases. Recently, the importance has further been recognized in the control and modeling of reactive plasmas in industrial applications [123].

These atoms are produced both in the primary excitation processes in the radiation interaction with gases and in the electron–ion recombination followed by the collisional and optical relaxation processes. Penning ionization by long-lived metastable atoms has been studied experimentally using W-value methods, static afterglow methods, flowing afterglow methods, beam methods, and pulse radiolysis methods [1,2,10,124]. Comparative discussions on the methods concluded a superior advantage of the pulse radiolysis method [1,2], which was best demonstrated by Ueno and Hatano [125], over the other methods in determining absolute rate constant or cross-section values for this process. A further advantage of the pulse radiolysis method was presented for obtaining the absolute values of the rate constants or cross sections for short-lived excited rare gas atoms in optically allowed or resonant states by Hatano [2] and by Ukai and Hatano [12].

3.1. De-excitation of Excited Rare Gas Atoms in Metastable States

The main de-excitation processes for $He(2^3S)$ atoms are considered to be Penning ionization (Eq. (14)) or associative ionization (Eq. (15)):

$$He(2^3S) + M \rightarrow He + M^+ + e^- \tag{14}$$

$$He(2^3S) + M \rightarrow HeM^+ + e^- \tag{15}$$

The ionizing reaction occudrs over a certain range of intermolecular distance R, where $He(2^3S)$ and M can be regarded as intermediate quasi-molecules.

According to the theoretical investigations by Nakamura [126–128] and Miller [129], the collisional energy dependence of these processes can be calculated if the interaction potential $V(R)$ for the system $He(2^3S)$–M and the autoionization rate $\Gamma(R)/h$ from the intermediate quasi-molecule $[He(2^3S)M]$ to the resulting quasi-molecular ion $[HeM]^+ - e^-$ are known. For example, by the classical formula of Miller [129], we have theoretical cross sections for Penning ionization as:

$$\sigma(E) = 2\pi \int_0^\infty b \left(1 - \exp\left[-2\int_{R_0}^\infty \frac{\Gamma(R)}{\hbar} \frac{dR}{v} \right] \right) db \tag{16}$$

where R_0 is the classical turning point, v is the relative velocity, and μ is the reduced mass. The interaction potentials and the autoionization widths have been obtained in atomic scattering experiments only for some simple target molecules, such as rare gas atoms and hydrogen molecules. The proposed mechanism of autoionization via electron exchange is:

$$He(2^3S)(1) + M(2) \rightarrow He(2) + M^+ + e^-(1)$$

$$He(2^3S)(1) + M(2) \rightarrow He(2)M^+ + e^-(1)$$

where (1) and (2) denote the excited electron of the metastable He atom and the electron of M, respectively. The empirical forms of $V(R)$ and $\Gamma(R)$ commonly employed as the most reliable for theoretical analyses [130–133] are:

$$\Gamma(R) = A \, \exp(-\alpha R) \tag{17}$$

$$V^*(R) = B \, \exp(-\beta R) \tag{18}$$

where A, B, α, and β are constants and R is an intermolecular distance.

The de-excitation rate constants of $He(2^3S)$ and $Ne(^3P_0, {}^3P_1, \text{ and } {}^3P_2)$ by various atoms and molecules were obtained at room temperature using a pulse radiolysis method [125,134–136].

Experimental details for the cross-section measurements were presented in the literature. Briefly, after the irradiation by electron beam pulse for a few nanoseconds, the time-dependent absorption for the atomic line transition $Rg^{**} \to Rg^* + h\nu$ was measured to observe the time-dependent population of the excited rare gas atoms Rg^*. The population of excited Rg^* was determined using an absorption law for the atomic lines, where the broadening of the absorption profile due to the thermal Doppler effect and due to the attractive interatomic potentials was reasonably taken into consideration. The time-dependent optical emission from energy transfer products, such as:

$$He(2^3S) + N_2 \to He(1^1S) + N_2^+\left(B^2\Sigma_g^-\right),$$

$$N_2^+\left(B^2\Sigma_g^-\right) \to N_2^+\left(X^2\Sigma_g^-\right) + h\nu$$

was also measured to monitor the population of Rg^* [125,126].

An attempt has been made to correlate the rate constant values obtained with various molecular parameters such as ionization potentials and polarizabilities. A relatively good correlation has been obtained for $He(2^3S)$ [134], as shown in Table 4 between the de-excitation probability $P\,(=k_M/k_C)$ and the excess energy $\Delta E\,[= E(He^*) - IP_M]$, where k_M, k_C, $E(He^*)$, and IP_M are the experimentally obtained rate constant, the calculated gas kinetic rate constant, the excitation energy of $He(2^3S)$, and the ionization potential of the target atom or molecule M, respectively. However, the reason for the correlation has not been well understood. A similar experiment was made [135,136] also for the de-excitation of $Ne(^3P_0, {}^3P_1, \text{ and } {}^3P_2)$ by atoms and molecules. Two new features of the obtained results have been demonstrated in this experiment: one is the J-dependent de-excitation cross sections, the other is a comparison between a theory of the de-excitation of optically allowed resonant state and the experimental result of $Ne(^3P_1)$, which is partly allowed because of a weak spin–orbit coupling. A pulse radiolysis is very advantageous to obtain not only absolute de-excitation rate constants or cross sections but also their collisional energy dependence. A velocity-averaged absolute cross section σ_M is obtained as a function of mean collisional energy E from the temperature dependence of an absolute rate constant. The temperature dependence of the rate constants for the de-excitation of $He(2^3S)$ by atoms and molecules has been measured [122] in the temperature range from 133 to 300 K (Fig. 6). From the above classical formula of the Penning ionization cross section, the collisional

Table 4 Derived Parameters for De-excitation of $He(2^3S)$ (From Ref. 137.)

M	α/β	P	ΔE (eV)	$\sqrt{IP(M)/IP(He(2^3S))}$
N_2	1.4	0.05	4.24	1.81
CO	1.1	0.07	5.81	1.71
Ar	1.0	0.07	4.06	1.82
Kr	1.0	0.07	5.82	1.71
NO	0.6	0.16	10.56	1.39
O_2	0.6	0.18	7.75	1.59
C_2H_4	0.3	0.31	9.31	1.48
CO_2	0.1	0.39	6.04	1.70

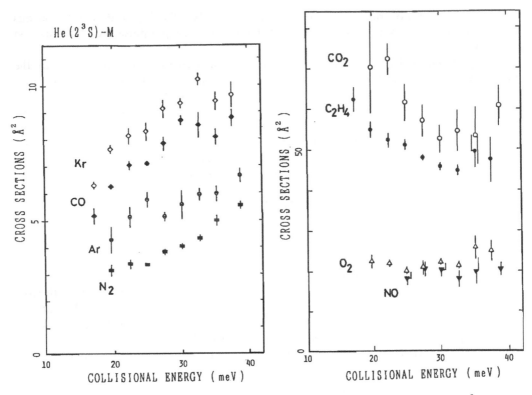

Figure 6 Collisional energy dependence of the cross sections for de-excitation of He(2^3S) by ($\blacksquare$) N_2; ($\ominus$) Ar; ($\blacklozenge$) CO; ($\Diamond$) Kr (left panel); ($\triangle$) O_2; ($\blacktriangledown$) NO; ($\bullet$) C_2H_4; and ($\bigcirc$) CO_2 (right panel). (From Ref. 137.)

energy dependence of its cross section is readily given, if the interaction potential $V^*(R)$ for He(2^3S)–M and the autoionization rate $\Gamma(R)/\hbar$ from He(2^3S)–M to He–(M^+–e^-) are obtained, by the following simple equation:

$$\sigma_T \propto E^{\alpha/\beta - 1/2} \quad \text{or} \quad k(T) \propto T^{\alpha/\beta} \tag{19}$$

where σ_T and $k(T)$ are the total Penning ionization cross section and the corresponding rate constant, respectively. The slope of log–log plots of $k(T)$ vs. T gives the values of α/β (Fig. 7) [137]. The obtained α/β value for each molecule M listed in Table 4 increases with decreasing value of P, where P is a de-excitation probability per collision, or $1/P$ is an effective collision number for energy transfer. Because P is not so relatively different for each M, Table 4 shows clearly that the bigger α (i.e., the shorter range interaction between He(2^3S) and M) gives the smaller P (i.e., the less efficient energy transfer from He(2^3S) to M). This conclusion satisfies the exterior electron density model by Ohno et al. [138].

3.2. De-excitation of Excited Rare Gas Atoms in Resonant States

De-excitation of the excited rare gas atoms in the resonant states has been studied less extensively than that of the metastable atoms. This is due to experimental difficulties caused by the short lifetimes of the resonant atoms. There have been reported, however, several theoretical formulations [139,140] based on a long-range dipole–dipole interaction

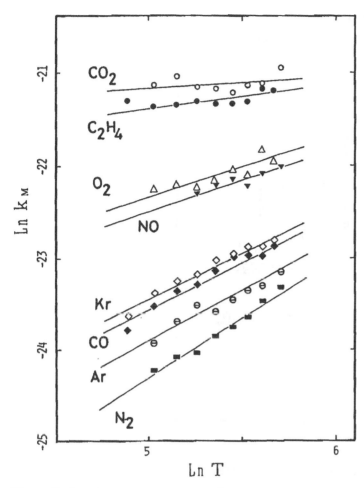

Figure 7 Log–log plot of k_M for $He(2^3S)$ vs. T. (From Ref. 137.) The same symbol indicates the same quenching gas as in Fig. 6.

for the de-excitation cross section of radiative atoms. It is, therefore, necessary to compare the theory with the results of the resonant or the lowest radiative state atoms. It was reported [141,142] that the collisional energy dependence of the de-excitation cross sections of $He(2^1P)$, $Ne(^3P_1)$, $Ar(^1P_1)$, and $Ar(^3P_1)$ by atoms and molecules, using a pulse radiolysis method, is very advantageous to obtain the absolute values of the de-excitation cross sections of the resonant atoms as well as the metastable atoms. The experimental technique of the time-dependent optical atomic line absorption is almost the same as that for the metastable states.

A comparison, as shown in Fig. 8, between the experimental results of the collisional energy dependence of the de-excitation cross section of $He(2^1P)$ by Ar and the theoretical ones calculated from the W–K theory [140] and the K–W theory [139] makes clearly possible for the first time to compare in detail the experimental results with the theoretical ones [141]. Previous comparisons between experiments and theories have been made only for a value of the rate constant or the cross section, respectively, at a particular temperature of collisional energy, usually at room temperature. The results of temperature-

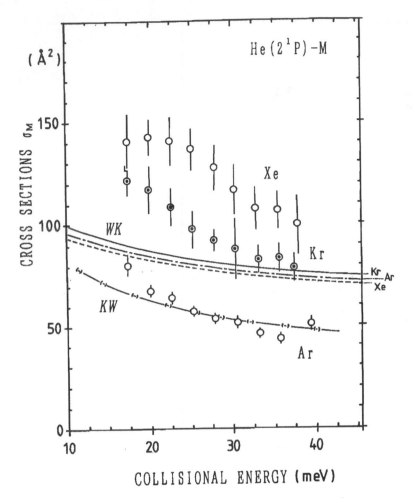

Figure 8 Experimental de-excitation cross sections of He(2^1P) by Ar, Kr, and Xe (denoted by circles). (From Ref. 144.) (———), (– – –), and (—— – —) represent σ_{WK} for Ar, Kr, and Xe. The —— ($\cdot$) —— curve shows σ_{KW} for Ar.

dependent measurement using the pulse radiolysis method supports evidently the modified new theory, the K–W theory [139], which means that the theory should take the bent trajectory into consideration.

A classical expression for the cross section for collisional de-excitation of He(2^1P) is also derived from the formula by Eq. (16). However, the autoionization widths $\Gamma(R)$ for Penning ionization by resonant atoms are not identical to the empirical form of Eq. (18) for electron exchange. Instead, a direct transition due to a dipole–dipole interaction is proposed to govern this Penning ionization [126,139,140,143], that is,

$$\text{He}(2^1\text{P})(1) + \text{M}(2) \rightarrow \text{He}(1) + \text{M}^+ + e^-(2)$$

$$\text{He}(2^1\text{P})(1) + \text{M}(2) \rightarrow \text{He}(1)\text{M}^+ + e^-(2).$$

According to the dipole–dipole interaction, $\Gamma(R)$ is given by the equation:

$$\Gamma(R) = \frac{2\pi\mu_{He}^2\mu_M^2}{R^6}(1 + 3\cos^2\theta) \tag{20}$$

where μ_{He}^2 and μ_M^2 are squared matrix elements for transitions of $He \rightarrow He(2^1P)$ and $M \rightarrow M^+ + e^-$ at 58.4 nm, respectively; θ is the angle between the intermolecular axis and the direction of the atomic $2p$ polarization. The values of μ_{He}^2 and μ_M^2 are obtained from optical oscillator strengths for He and the relevant atoms and molecules. Using Eq. (20) together with an appropriate treatment for the p-state polarization during a collision, Watanabe and Katsuura [140] (WK) proposed a formula of the Penning ionization cross section as:

$$\sigma_{WK} = 13.88\left(\frac{\mu_{He}^2\mu_M^2}{\hbar v}\right)^{2/5} \tag{21}$$

In the WK theory, the direction of the polarization axis of $He(2^1P)$ is carefully considered, but only rectilinear trajectories for relative motion are considered. The assumption of the rectilinear trajectories has been generally employed in the theories for higher-energy heavy- particle collisions, where the forward scattering is dominant. However, because the relative collisional energy is comparable to the depth energy of the interaction potential well in the thermal collisions, the assumption of the rectilinear trajectories can give rise to a considerable disagreement with experimental cross sections, which will be discussed in this section. The maximum limit for the cross section has been obtained by Katsuura [143] also by assuming linear trajectories together with the perturbed rotating atom approximation (i.e., the atomic polarization of $He(2^1P)$ is directed toward M throughout a collision).

The cross-section measurements using the pulse radiolysis method provided the opportunity to examine the applicability and inapplicability of the WK theory systematically. Fig. 8 shows the experimental de-excitation cross sections of $He(2^1P)$ by Ar, Kr, and Xe [144]. The values of the cross sections are, on the whole, extremely large in comparison with simple energy transfer cross sections between neutral atoms and molecules, and closely approach gas-kinetic close collision cross sections [125,134]. As compared with the results in Section 3.1, the cross sections are more than one order of magnitude larger than those for the metastable atoms. The cross section for Xe is larger than that for Kr in the present energy region, whereas the cross section for Ar is smaller than that for Kr.

The cross sections show a decrease with increasing collisional energy, which is in marked contrast to the de-excitation cross sections of $He(2^3S)$ atoms by the same quenching atoms, as presented in Fig. 6. However, this behavior is rather similar to other quenching molecules, such as CO_2 and C_2H_4, where the cross sections are fairly large, on the order of tens of squared angstroms.

Relatively good agreement between the experimental and theoretical cross sections is obtained for the absolute values and also for the collisional energy dependence of the cross sections. In detail, the experimental values for Kr and Xe are larger than σ_{WK} and the difference increases at lower energies. On the other hand, the experimental cross section for Ar is smaller than σ_{WK} and the difference increases at higher energies. It is known that the values of μ_M^2 are almost the same for Ar, Kr, and Xe, but the value for Xe is a little smaller than those for Ar and Kr, so that the cross-section values should be in the order of $\sigma_{Xe} < \sigma_{Ar} < \sigma_{Kr}$ (see σ_{WK} in Fig. 8). However, the experimental values are $\sigma_{Ar} < \sigma_{Kr} < \sigma_{Xe}$. The differences between σ_{Ar} and σ_{Kr} and between σ_{Kr} and σ_{Xe} are much greater than those calculated from Eq. (21). Thus, the absolute values, the order, and the energy dependence of the experimental cross sections are found to be different from those predicted by σ_{WK}. A similar consideration was also made for molecular target in He(2^1P)–H$_2$ collision [145].

To make the comparison a little more obvious, Fig. 9 shows the cross-section ratios of the experimental values of σ_M to σ_{WK} plotted against the collisional energy. The ratio for Xe is larger than unity and decreases with increasing collisional energy. The value for Kr is also larger but decreases almost to unity with increasing energy. The ratio for Ar is smaller than unity and continues to decrease with increase in energy, a result that is in good agreement with a classical trajectory calculation by Kohmoto and Watanabe [139] (σ_{KW}). The ratios presented in Fig. 9 suggest that the influence of the classical motion of the colliding particles is important in determining the absolute cross sections.

Fig. 10 shows the relation between the experimental values σ_M and the theoretical predictions σ_{KW}, together with the experimental cross sections previously reported for the de-excitation of He(3^1P) [146,147]. The difference between the values for He(2^1P) and He(3^1P) is explained by the difference in the optical oscillator strengths for the two states. The experimental values can generally be explained by σ_{WK} based on the long-range

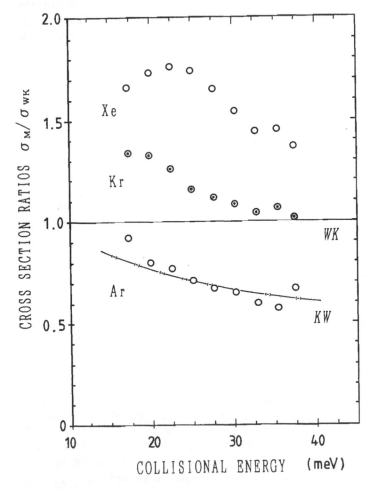

Figure 9 Ratios of experimental cross section to WK cross section for de-excitation of He(2^1P) by Ar, Kr, and Xe (circles). (From Ref. 144.) — (·) — represents the ratio of σ_{KW}/σ_{WK} for Ar.

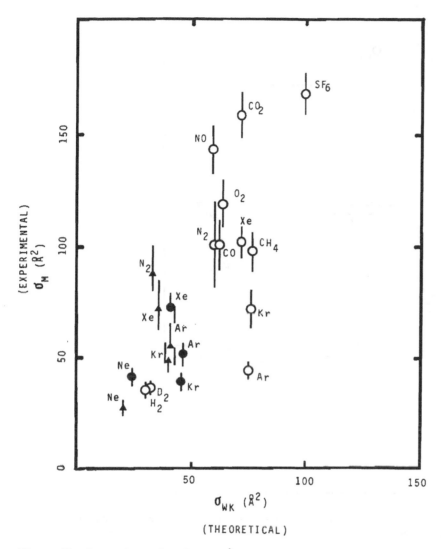

Figure 10   Comparison of σ_M for $He(2^1P)$ to σ_{WK} (From Ref. 144.): (O) present results; (●) cross section for $He(3^1P)$ at 300 K (From Ref. 146.); and (▲) cross section for $He(3^1P)$ at 600 K. (From Ref. 147.)

dipole–dipole interaction. However, each of the cross sections is characteristically different from the theory, which seems to be much beyond any possible systematic error.

The value of σ_M for Ar is smaller than the theory, but for CO_2, SF_6, or NO, σ_M exceeds σ_{WK}. The positive difference for diatomic molecules other than H_2 and D_2 is more than that for rare gas atoms, which is also observed for $He(3^1P)$.

The result for $He(2^1P)$–Ar in Fig. 9 shows that the long-range attractive part of the interaction potentials has a close relation with the polarizabilities of the molecules, M. Fig. 11 is a plot of σ_M/σ_{WK} vs. α_M, the polarizability of M. As for Ar, Kr, Xe, CH_4, and SF_6, some correlation is observed. From the slope of the σ_M/σ_{WK} vs. α_M plots, the effect of bent trajectories on the Penning ionization cross sections can be qualitatively evaluated

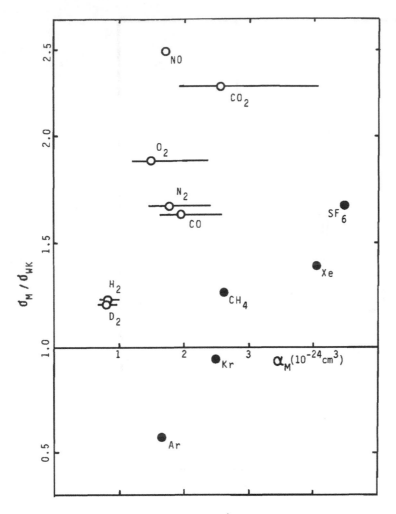

Figure 11 Cross section ratios of He(2^1P) vs. α_M. (From Ref. 142.) The horizontal lines show the limits of α_M values if seen in parallel or perpendicular direction to the molecular axis.

because, for molecules with large polarizabilities, the attractive potential extends to large intermolecular distances, whereas for molecules with smaller polarizabilities, the potential becomes repulsive sooner at relatively smaller intermolecular distances. On the other hand, for CO_2 and diatomic molecules, there is no same correlation between the cross-section ratios and the polarizabilities observed for the rare gas atoms and spherical molecules, although some other correlations characteristic of these molecules might be distinguished.

The effect of bent trajectories due to the attractive interaction on the absolute cross section is further clarified by the classical trajectory calculation shown in Fig. 12. The cross section was calculated for He(2^1P)–Ar, He(2^1P)–Kr, and He(2^1P)–Xe collisions assuming $\theta = 0$ (perturbed rotating atom approximation [143]) in Eq. (20) for the autoionization widths, employing the interaction potentials determined in cross-beams experiments for He(2^1S)–M collisions, and using Eqs. (14)–(16) for the general procedure [144]. The calculated cross sections in general are strongly dependent on the collisional energy. However,

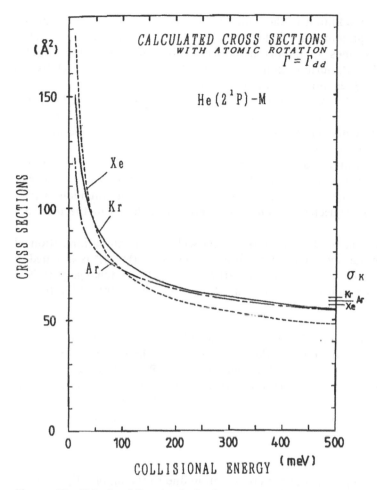

Figure 12 Calculated Penning ionization cross sections for collisions: (— – – —) He(2^1P)–Ar, (— —), He(2^1P)–Kr, and (——) He(2^1P)–Xe, assuming classical trajectories for relative motion. (From Ref. 144.) A perturbed rotating atom approximation is employed in the evaluation of the magnitudes of the dipole–dipole interaction ($\theta = 0$; see Eq. (20)). Theoretical values by Katsuura at 500 meV employing rectilinear trajectories are also indicated. (From Ref. 143.)

the energy dependence is different in each case, thus changing the relative order of the absolute magnitudes.

The order in the high-energy region is, as expected from the rectilinear calculations [125,128], $\sigma_{Xe} < \sigma_{Ar} < \sigma_{Kr}$. However, below 50 meV, the calculation reproduces the experimental order of $\sigma_{Ar} < \sigma_{Kr} < \sigma_{Xe}$. The result demonstrates that the effect of entangled trajectories decreases with the increase in the collisional energy. The behavior of the experimental cross sections has been further examined by a classical trajectory calculation employing the procedure of KW [139]. The problem of whether the direction of the polarization axis of He(2^1P) is conserved or not during a collision has also been examined. For Ar, an atomic target of small polarizability, the experimental cross section is in good agreement with the KW-type calculation for a spatially fixed orientation. For Xe, which has a large atomic polarizability, σ_M is more closely predicted by the perturbed rotating atom approximation. Therefore, in addition to the bent trajectories, the rotation of the polarization axis is also probably an important factor in determining the absolute magnitude of the cross sections. This includes an important collision physics, namely the angular momentum of polarized $He(2^1P)$ during a collision, which should be quantized relative to the interatomic axis or should remain in the initial fixed space.

As for a total decay width of the transient autoionizing state of [HeM]* formed during a collision, the direct transition due to the dipole–dipole interaction and the transition with the electron exchange should also be involved simultaneously. Because the outermost electron orbitals of the target rare gas atoms, CH_4 and SF_6, are fully occupied, the electron density outside their van der Waals sphere is supposed to be quite low. However, most of the linear molecules considered here have some amount of electron density outside their hard sphere, so that the contribution from a short-range interaction (i.e., electron exchange), which is neglected in WK, might also be included in the total de-excitation cross section.

In the present discussion, the total autoionization width for Penning ionization of He*–M or for decay of the transient molecular autoionizing state of [HeM]* is assumed to be divided into two components: a dipole–dipole part and an electron exchange part. The maximum width possible is assumed to be the sum of the two components [142], that is,

$$\Gamma_{\text{total}} = \Gamma_{\text{dd}} + \Gamma_{\text{ex}} \tag{22}$$

where Γ_{total} is the total decay width of [HeM]*, and Γ_{dd} and Γ_{ex} are the partial widths due to the dipole–dipole interaction and the electron exchange interaction, respectively. Substitution of Eq. (22) into the classical formulation of the Penning ionization cross section, as shown by Eq. (16), gives the theoretical cross section for simultaneous contributions from both the dipole–dipole interaction and the electron exchange.

First, in cases where the transition (ionization) rate is small, the total cross section will be given by:

$$\sigma_{\text{total}} = \sigma_{\text{dd}} + \sigma_{\text{ex}} \tag{23}$$

where σ_{dd} and σ_{ex} are the components of the cross section due to the dipole–dipole interaction and the electron exchange interaction, respectively. Fig. 13 gives plots of $(\sigma_M - \sigma_{WK})$ vs. σ_m, the de-excitation cross section of the metastable helium atoms.

The values of σ_m are assumed to represent the exchange part of the total de-excitation cross section for $He(2^1P)$. The values of $(\sigma_M - \sigma_{WK})$ are a measure of the cross section due to interactions other than the dipole–dipole interaction (i.e., the electron exchange interaction). The plots in Fig. 13 are shown for linear molecules and also for Xe, CH_4, and SF_6. The values of $(\sigma_M - \sigma_{WK})$ for Ar and Kr are negative and are not included in the figure. A good correlation is found for H_2, D_2, N_2, CO, O_2, and CO_2. A better, or rather more proportional, correlation is obtained for $\sigma(2^1S)$ than for $\sigma(2^3S)$, demonstrating the contribution from the electron exchange. The large deviation observed for NO is explained by the inapplicability of the WK theory for transitions other than s–p (σ–π) type. Xenon and SF_6 also seem to be within the correlation and may have a minor contribution from the electron exchange interaction.

Secondly, the cross section reflects the nature of the dependence of the partial widths on the intermolecular distance R. As expected from Eq. (20), Γ_{dd} is small for large R and increases gradually with decrease in R. On the other hand, Γ_{ex} in its empirical form (Eq. (17)) is extremely small at large distances and increases sharply near the repulsive wall. Therefore, the de-excitation probability due to Penning ionization is determined by the dipole–dipole part of the decay width in collisions with large impact parameter, whereas the probability for Penning ionization is already almost unity by the single contribution

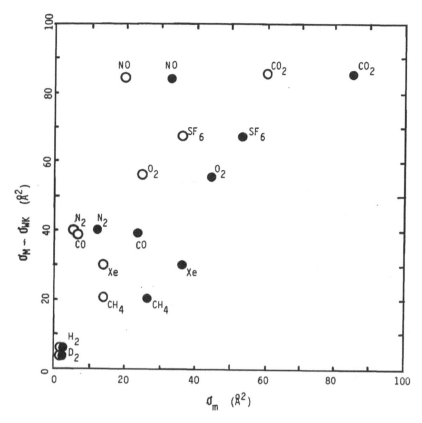

Figure 13 Plots of $(\sigma_M - \sigma_{WK})$ vs. σ_m for an analysis of the de-excitation cross section σ_M for He(2^1P), where the de-excitation cross section σ_m for the metastable helium atoms is assumed to represent the exchange part of the total de-excitation cross section. (From Ref. 142.) The σ_m values are for He(2^3S) (O) and He (2^1S) (●).

from the dipole–dipole part in collisions with a small impact parameter where the trajectory intersects the intermolecular region in which Γ_{ex} is comparably large relative to Γ_{dd}. The reason why the cross sections for a rare gas target can be explained almost completely by the dipole–dipole interaction originates from this relationship. As expected from the results for metastable atoms in Table 4, the molecules CO_2, O_2, and NO could have large Γ_{ex} at relatively large R, and the cross sections may have contributions from the exchange part even at large R, additional to the dipole–dipole part.

The above conclusion based on the cross-section measurements, using the pulse radiolysis method and with consideration of the Penning ionization mechanism including the classical trajectory analysis, is further rationalized by a quantum mechanical calculation for He(2^1P)–Ar, Kr, Xe, H_2, and N_2 collisions [148,149], where the radial wave function for relative motion is numerically solved by taking into account the partial widths for dipole–dipole interaction and electron exchange interaction. Fitting the calculated Penning ionization cross sections to the experimental cross sections enables to determine the interaction potentials for these collision systems for the first time (Fig. 14), which was not obtained by the scattering experiments. This is also recognized as an advantage of the pulse radiolysis method, which provides the insight of intermolecular interaction.

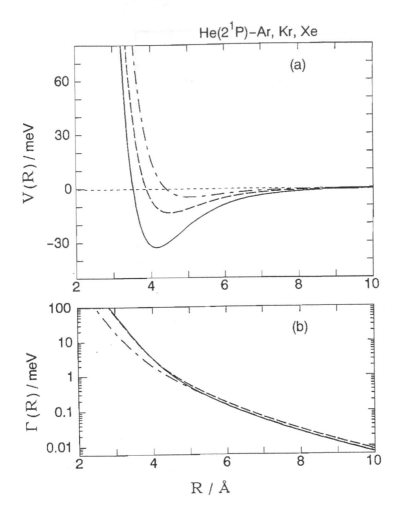

Figure 14 Interaction potential and autoionization decay width employed and determined in the calculations for He(2^1P)–M; (— · —) M = Ar; (– – –) M = Kr; and (——) M = Xe; (a) interaction potential: position of the repulsive potentials for He(2^1P)–Kr and He(2^1P)–Xe are determined by fitting to the experimental results; (b) autoionization decay width for Penning ionization.

It is also a fact that very few studies have been reported on simple excitation transfer, in which Penning ionization is energetically impossible, from resonant rare gas atoms to atoms and molecules. The temperature dependence of the de-excitation rate constants of the resonant states of Ar, ^{1}P$_1$, and ^{3}P$_1$ by SF$_6$ and N$_2$ has been measured in the temperature range from 133 to 300 K using a pulse radiolysis method [150], thus obtaining the collisional energy dependence of the de-excitation cross sections. The results of the cross sections for SF$_6$ are compared with the W–K theory and a good agreement is obtained. The results for N$_2$ agree with predictions of the cross sections for a nonresonant case. Even in the case of the de-excitation of the resonant state, such as Ar(^{1}P$_1$) or Ar(^{3}P$_1$), the cross-section value and its collisional energy dependence are very similar to those for the metastable state (i.e., Ar(^{3}P$_0$) or Ar(^{3}P$_2$)). This result is consistent with the W–K theory because of the fact that N$_2$ has almost no optical absorption in the energy region corresponding to the excitation energy of Ar($^1P_1$) and Ar($^3P_1$). It is concluded, therefore, that a long-range dipole–dipole interaction is important in the de-excitation processes of Ar($^1P_1$) and Ar($^3P_1$) by SF_6, but that a short-range interaction with curve crossing dominates in the de-excitation of Ar($^1P_1$) and Ar($^3P_1$) by N_2.

A little more complicated system is the de-excitation of He(2^1P) by Ne, where the de-excitation is dominated by the excitation transfer and only a minor contribution from the Penning ionization is involved. The experimental cross section obtained by the pulse radiolysis method, together with the numerical calculation for the coupled-channel radial Schrödinger equation, has clearly provided the major contribution of the following excitation transfer processes to the absolute de-excitation cross sections [151] (Fig. 15):

$$He(2^1P) + Ne \rightarrow He + Ne^*(6p_3, 5s_2, and\ 6s_4).$$

3.3. De-excitation of Excited Rare Gas Atoms by Molecules Containing a Group IV Element

De-excitation cross sections of He(2^3S, 2^1S, and 2^1P), Ne($^3P_2$, $^3P_1$, and $^3P_0$), and Ar($^3P_2$, $^3P_1$, $^3P_0$, and $^1P_1$) by molecules containing group IV elements were obtained by the pulse

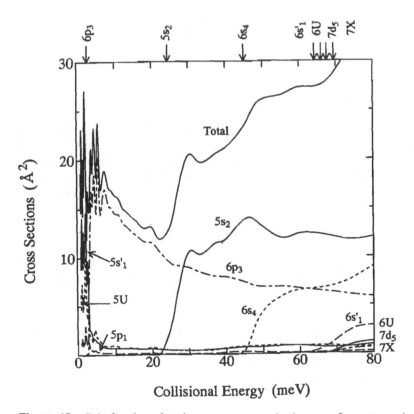

Figure 15 Calculated total and state-to-state excitation transfer cross sections in the de-excitation of He(2^1P)–Ne. (From Ref. 151.) Both electron exchange and dipole–dipole interactions are included in the coupling matrix elements. The threshold energy into each exit channel is shown on the upper axis.

radiolysis method [152–155]. The study was designed from the viewpoint of the control of reactive plasmas. As can be seen in the above, in reactive plasmas, the excited rare gas atoms in different excited states are produced simultaneously and react with atoms and molecules with different probabilities. Thus, the cross-section data for different excited atoms are strongly required.

Table 5 presents the de-excitation cross sections for different excited helium atoms. As demonstrated in the preceding sections, the de-excitation cross sections of $He(2^3S)$ and $He(2^1P)$ are the Penning ionization cross sections. It is natural that these cross sections are also understood as the Penning ionization cross sections. In general, the cross sections are extremely large compared with a simple energy transfer cross section between a neutral atom and a neutral molecule, and they approach the gas-kinetic collision cross sections. The values in Table 5 show an apparent dependence on the particular excited states of the helium atoms (i.e., the cross sections for the metastable atoms, $He(2^3S)$ and $He(2^1S)$, are much smaller than those for the resonant atom, $He(2^1P)$, whereas the cross sections for $He(2^1S)$ are larger than those for $He(2^3S)$). The behavior of the de-excitation cross sections, dependent both on the helium excited states and the target molecules, is ascribed to that of the Penning ionization cross sections. A systematic comparison in Table 6 between the total de-excitation cross sections and the partial cross sections for possible fragmentation confirms that the Penning ionization is the major de-excitation mechanism. For example, the optical emission cross sections in a UV–visible region from Si* atoms in several excited electronic states produced in $He(2^3S)$–SiH_4 collisions are reported to be $0.081\ \text{Å}^2$. Minor emission from other excited fragments H^*, SiH^*, SiH_2^*, and SiH_3^* does not increase the cross section significantly. The production of nonemitting fragments H, Si, SiH, SiH_2, and SiH_3 has been shown to give a minor contribution of such fragments to the formation of neutral products. The cross section of $0.081\ \text{Å}^2$ thus represents the total cross section for the production of emitting neutral fragments, or possibly the total cross section for the neutral dissociation of SiH_4 in $He(2^3S)$–SiH_4 collisions. The de-excitation cross section of $He(2^3S)$ by SiH_4 in Table 5 is, however, determined to be $18\ \text{Å}^2$ in the present experiment—a value that is more than about 200 times larger than that of the optical emission cross section. It is, therefore, concluded that a major part of the de-excitation processes in $He(2^3S)$–SiH_4 collisions is due to processes other than the formation of neutral dissociation fragments (i.e., due to Penning ionization). The Penning ionization here then

Table 5 De-excitation Cross Sections (σ_M) at a Mean Collisional Energy Corresponding to Room Temperature (295 K) and De-excitation Probabilities (P) of Excited Helium Atoms He* (He*= $He(2^3S)$, $He(2^1S)$, and $He(2^1P)$) and polarizabilities (α_M) of target molecules (From Ref. 154.)

	2^3S		2^1S		2^1P		
M	$\sigma_M\ (\text{Å}^2)$	P	$\sigma_M\ (\text{Å}^2)$	P	$\sigma_M\ (\text{Å}^2)$	P	$\alpha_M\ (\text{Å}^3)$
CH_4	14 ± 1	0.127	58 ± 3	0.461	98 ± 9	0.792	2.59
SiH_4	18 ± 1	0.127	60 ± 2	0.367	105 ± 2	0.663	5.44
GeH_4	26 ± 1	0.175	70 ± 4	0.415	109 ± 3	0.662	6.20
C_2H_6	28 ± 1	0.222	106 ± 4	0.472	152 ± 4	1.031	4.43
Si_2H_6	41 ± 1	0.228	58 ± 5	0.521	179 ± 7	0.904	11.1
CF_4	14 ± 3	0.109	58 ± 1	0.157	94 ± 6	0.669	3.84
SiF_4	50 ± 1	0.352	58 ± 4	0.439	147 ± 6	0.919	5.45
$SiCl_4$	59 ± 3	0.309	58 ± 5	0.503	172 ± 8	0.815	13.1

Table 6 De-excitation Cross Sections of He(2^3S) by CH_4, SiH_4, or GeH_4 in Comparison with the Respective Cross Sections for Reaction Products (in Å^2) (From Refs. 123 and 152.)

He(2^3S) + CH$_4$	$\xrightarrow{\sigma_M = 14}$	CH$_n^+$ ($n \leq 4$)	12
		C*	–
		CH*	0.051
		CH	–
		CH$_2$, CH$_3$	–
He(2^3S) + SiH$_4$	$\xrightarrow{\sigma_M = 18}$	SiH$_n^+$ ($n \leq 4$)	–
		Si*	0.081
		SiH*	0
		SiH	0
		SiH$_2$, SiH$_3$	0
He(2^3S) + GeH$_4$	$\xrightarrow{\sigma_M = 26}$	GeH$_n^+$ ($n \leq 4$)	–
		Ge*	0.44
		GeH*	0
		GeH	–
		GeH$_2$, GeH$_3$	–

means the formation of $SiH_n^+ + e^-$ ($n \leq 4$). Similar conclusions have also been reached about He(2^3S)–CH$_4$ and He(2^3S)–GeH$_4$ collisions [123,152].

It is seen that the individual de-excitation cross sections for a particular excited atom are governed by the individual interactions. The de-excitation cross sections of He(2^3S, 2^1S, and 2^1P) are reasonably interpreted as follows. Two important factors are readily extracted from the semiempirical formula for the Penning ionization cross section, so that the rational relation of the Penning ionization cross section is obtained:

$$\sigma_M \propto N_i \text{IP}^{-1/2}$$

where N_i is the number of equivalent electrons which can be ionized by the transfer of the excitation energy of He(2^3S) and the $\text{IP}^{-1/2}$ involves the electron densities of outer electrons extending out of the rigid surface of the target molecules, which penetrate into the inner hole of He($1s^{-1}$). Fig. 16 shows the excellent correlation between the experimental cross sections and the $N_i \text{IP}^{-1/2}$, thus confirming the mechanism of electron exchange interaction [154].

A systematic comparison of the de-excitation cross sections for He(2^1P) with the theoretical Penning ionization cross sections using the WK formula shows the importance of a long-range dipole–dipole interaction to determine the transition probabilities and the strongly attracted trajectories in ionizing collisions, both of which result in large values of de-excitation cross sections. Photoionization quantum yields of CH$_4$ and SiH$_4$ at $hv = 58.4$ nm (21.2 eV) are available for approximate estimation of the products resulting from He(2^1P)–CH$_4$ and He(2^1P)–SiH$_4$ collisions. The quantum yields of more than 95% on a single photoabsorption suggest that neutral fragmentation of SiH$_4$ is again much less important in comparison with Penning ionization in He(2^1P)–CH$_4$, He(2^1P)–SiH$_4$, and, presumably, He(2^1P)–GeH$_4$ collisions. Another feature of the cross sections for He(2^1P) is presented in Fig. 17. The ratios of the experimental cross sections to σ_{WK} values are much larger than unity for target molecules of smaller photoabsorption cross section, whereas the ratios approach unity with increasing photoabsorption cross sections. This behavior reveals the abundance of the dipole–dipole interaction relative to the electron exchange interaction as discussed above [154].

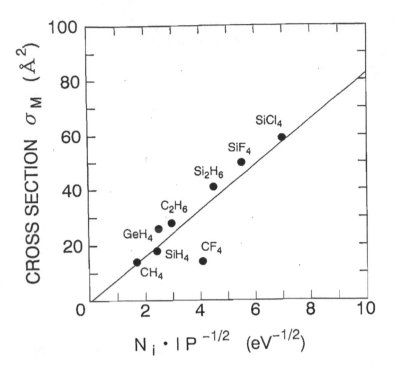

Figure 16 Relation of de-excitation cross sections (σ_M) for metastable helium atoms with the Hotop–Niehaus formula for (a): He(2^3S) and (b) He(2^1S). (From Ref. 154.)

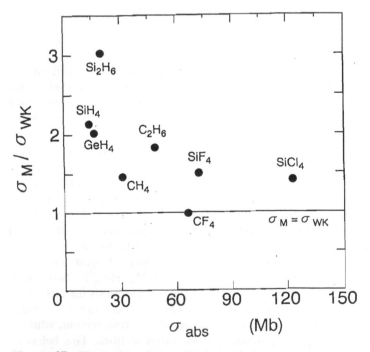

Figure 17 The ratios of the experimental cross sections to the theoretical ones (σ_M/σ_{WK}) for He(2^1P) as a function of photoabsorption cross section (σ_{abs}) of M at the corresponding energy to the excitation energy of He(2^1P). (From Ref. 154.)

A similar conclusion has been obtained for the de-excitation cross sections of Ne($^3P_2$, $^3P_1$, and $^3P_0$) by the molecules containing the group IV elements [155]. As shown, the cross sections are not so strongly dependent on the excited state of neon atoms. Instead, the dependence of the cross sections on the target molecule is more prominent. A systematic analysis of the cross sections with respect to the polarizability of target molecules has provided the physical insight of the mobile state of the σ, π, and nonbonding outermost electrons; the marked difference in the behavior of the cross section against the target polarizabilities in Fig. 18 presents the relation of the localized and delocalized electron characters of the σ, π, and nonbonding outermost electrons and the probabilities of electron exchange.

The cross sections of the de-excitation of Ar* (Ar*: Ar ($^3P_2$, $^3P_1$, $^3P_0$, and/or $^1P_1$)) by CH$_4$, SiH$_4$, and GeH$_4$ are extremely large [153] (Fig. 19). The difference in the magnitude of the cross sections is ascribed to the particular de-excitation mechanisms for the metastable and resonant excited argon atoms. However, in general, the Penning ionization is not a major de-excitation process. This is readily seen from the following. The excitation energy of Ar* is smaller than both of the vertical and adiabatic ionization potentials of CH$_4$, so that the Penning ionization of CH$_4$ is not accessible. In the case of Ar*–SiH$_4$ and Ar*–GeH$_4$ collisions, the excitation energy of Ar* is slightly greater than the adiabatic ionization potential of SiH$_4$ or GeH$_4$, and less than its vertical ionization potential. It is safely predicted that rendering the Penning ionization requires a strong structural deformation of a target molecule, which is not accessible with the thermal collisional energies. As a matter of fact, the ionic products such as CH$_n$ ($0 \leq n \leq 4$) or H$^+$ are not

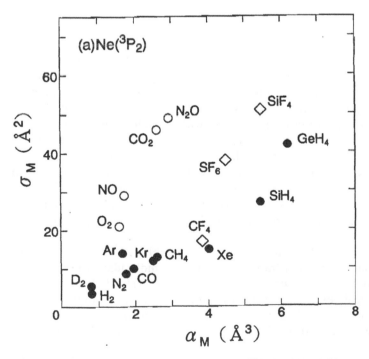

Figure 18   De-excitation cross sections of Ne($^3P_2$) by M (M = CH$_4$, SiH$_4$, GeH$_4$, CF$_4$, and SiF$_4$, Ar, Kr, Xe, H$_2$, D$_2$, N$_2$, O$_2$, CO, NO, CO$_2$, N$_2$O, and SF$_6$) as a function of the polarizability of M. (From Ref. 155.) The target molecules are classified by the type of their outermost molecular orbitals: σ-type ($\bullet$); π-type ($\circ$); and nonbonding ($\diamond$).

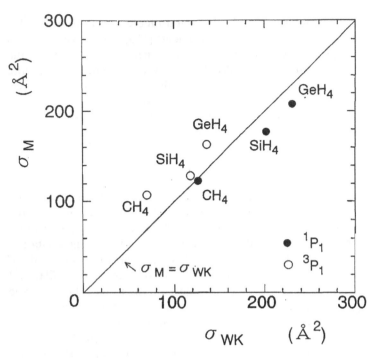

Figure 19 Comparison of the de-excitation cross sections of excited argon atoms Ar* [Ar* = $^3P_1$ (O) or $^1P_1$ (●)] by M (M = CH_4, SiH_4, or GeH_4) with the theoretical cross sections (σ_{WK}). (From Ref. 153.)

observed in mass spectrometric studies for Ar*–CH_4, Ar*–SiH_4, and Ar*–GeH_4, as seen in Table 7. At the same time, the above consideration on the energetics between the excitation energy of Ar* and the ionization potential of the target molecule leads to another insight of the de-excitation mechanism (i.e., the excitation transfer to form the molecular Rydberg state). The vertical excitation of a molecule with a slightly smaller excitation energy than the first ionization potential produces one of the molecular Rydberg states, which exist with a fairly high density of states in the region slightly below the vertical ionization potential. The density of the molecular Rydberg states increases with the increase in the excitation energy in the region slightly below the vertical ionization potential of a target molecule. The probability of de-excitation is enhanced with the increase in the density of molecular Rydberg states because the number of channels accepting the energy transferred from Ar* is increased.

Thus, it is concluded that the de-excitation of the metastable Ar($^3P_2$ and $^3P_0$) atoms is ascribed to the nonadiabatic excitation transfer at large intermolecular distance by the crossing of the intermolecular potential curves between the initial Ar*–M channel and the final Ar–M* channel, and the de-excitation of the resonant Ar($^3P_1$ and $^1P_1$) to the resonant excitation transfer by the dipole–dipole interaction [153]. This conclusion is compatible with the result of the above-mentioned conclusions for the de-excitation of He(2^1P) by Ne [135].

As a brief conclusion of this section, the cross-section measurements for the de-excitation of excited rare gas atoms have been best performed using the pulse radiolysis method. The pulse radiolysis method has provided not only the most reliable cross sections

Table 7 De-excitation Cross Sections of $Ar(^3P_{2,0})$ by CH_4, SiH_4, or GeH_4 in Comparison with the Respective Cross Sections for Reaction Products (in Å^2) (From Refs. 123 and 153.)

$Ar(^3P_{2,0}) + CH_4$	$\xrightarrow{\sigma_M=80}$	CH_n^+ $(n \leq 4)$	0
		C^*	–
		CH^* (A)	0
		$CH(X)$	4.1
		CH_2, CH_3	–
$Ar(^3P_{2,0}) + SiH_4$	$\xrightarrow{\sigma_M=101}$	SiH_n^+ $(n \leq 4)$	–
		Si^*	0.27
		SiH^* (A)	4.0
		$SiH(X)$	4.6
		SiH_2, SiH_3	–
$Ar(^3P_{2,0}) + GeH_4$	$\xrightarrow{\sigma_M=102}$	GeH_n^+ $(n \leq 4)$	–
		Ge^*	1.7
		GeH^* (A)	0.03
		$GeH(X)$	0
		GeH_2, GeH_3	–

both for metastable and the resonant atoms, but also several physical aspects of the collisional and molecular insight, so that the research of the collisional de-excitation proceeds into a new stage of investigation.

ACKNOWLEDGMENTS

The authors thank a number of collaborators for their excellent collaboration, particularly Dr. N. Kouchi. They also appreciate the early contribution to and the continuous development of the pulse radiolysis microwave conductivity/cavity technique by the late Dr. H. Shimamori.

REFERENCES

1. Hatano, Y. In *Proceedings of the 19th International Conference on Phenom. Ioniz. Gas., Belgrade, 1989* (Invited Papers); Zigman, V.J., Ed.; University of Belgrade Press: Belgrade, 1989; 242 pp.
2. Hatano, Y. Radiat. Phys. Chem. 1989, *34*, 675.
3. Hatano, Y. In *Pulse Radiolysis*; CRC Press: Boca Raton, Tabata, Y., Ed.; 1991. (Chap. 9).
4. Sauer, M.C., Jr. Adv. Radiat. Chem. 1976, *5*, 97.
5. Firestone, R.F.; Dorfman, L.M. In *Actions Chimiques et Biologiques des Radiations*; Haissinsky, M., Ed.; Masson: Paris, 1971; Vol. 15, 7 pp.
6. Hatano, Y. New Exp. Tech. Chem. 1978, *16*, 544. *in Japanese.*
7. Sauer, M.C., Jr. In *Study of Fast Processes and Transient Species by Electron Pulse Radiolysis*; Baxendale, J.H., Busi, F., Eds; Reidel: Dordrecht, 1982; 601 pp.
8. Armstrong, D.A. In *Radiation Chemistry*; Principles and Applications; Farhataziz, Rodgers, M.A.J., Eds; . VCH: Berlin, 1987 (Chap. 9).
9. Jonah, C.D.; Liu, A.; Mulac, W.A. In *Radiation Research*; Fielden, E.M., Fowlor, J.F., Hendry, J.H., Scott, D., Eds.; Taylor and Francis: London, 1987; Vol. 2, 60 pp.

10. Hatano, Y. In *Electronic and Atomic Collisions*; Lorents, D.C., Meyerhof, W.E., Peterson, J.R., Eds.; Elsevier: Amsterdam, 1986; 153 pp.

11. Hatano, Y.; Shimamori, H. In *Electron and Ion Swarms*; Christophorou, L.G., Ed.; Pergamon: Oxford, 1981; 103 pp.

12. Ukai, M.; Hatano, Y. In *Gaseous Electronics and Its Applications*; Crompton, R.W., Hayashi, M., Boyd, D.E., Makabe, T., Eds.; KTK Scientific: Tokyo, 1991; 51 pp.

13. Shinsaka, K.; Hatano, Y. Nucl. Instrum. Methods A 1993, *327*, 7.

14. Hatano, Y. In *Linking the Gaseous and Condensed Phases of Matter. The Behavior of Slow Electrons*, NATO-ASI B326; Christophrou, L.G., Illenberger, E., Schmidt, W.F., Eds.; Plenum: New York, 1994; 467 pp.

15. Hatano, Y. Aust. J. Phys. 1997, *50*, 615.

16. Cooper, R.; Denison, L.S.; Zeglinski, P.; Roy, C.R.; Gillis, H. J. Appl. Phys. 1983, *54*, 3053.

17. Cooper, R.; Mezyk, S.P.; Armstrong, D.A. Radiat. Phys. Chem. 1984, *24*, 545.

18. Doba, T.; Arai, S. J. Chem. Phys. 1981, *75*, 488.

19. Kasama, K.; Oka, T.; Arai, S.; Kurusu, H.; Hama, Y. J. Phys. Chem. 1982, *86*, 2035.

20. Loeb, D.W.; Chen, M.; Firestone, R.F. J. Chem. Phys. 1981, *74*, 3270.

21. Manzahares, E.R.; Firestone, R.F. J. Chem. Phys. 1982, *76*, 4475.

22. Manzahares, E.R.; Firestone, R.F. J. Chem. Phys. 1983, *79*, 1683.

23. Tanaka, M.; Sasaki, S.; Katayama, M. Bull. Chem. Soc. Jpn. 1985, *58*, 429.

24. Sauer, M.C., Jr.; Mulac, W.A. J. Chem. Phys. 1972, *56*, 4995.

25. Dreyer, J.W.; Perner, D. Chem. Phys. Lett. 1971, *12*, 299.

26. Ueno, T.; Kouchi, N.; Takao, S.; Hatano, Y. J. Phys. Chem. 1978, *82*, 2373.

27. Cooper, R.; Denison, L.; Sauer, M.C., Jr. J. Phys. Chem. 1982, *86*, 5093.

28. Freeman, G.R. In *Electron, Ion Swarms*; Christophorou, L.G., Ed.; Pergamon: Oxford, 1981; 93 pp.

29. Huang, S.S.-S.; Freeman, G.R. Phys. Rev. A 1981, *24*, 714.

30. Nakamura, Y.; Shinsaka, K.; Hatano, Y. J. Chem. Phys. 1983, *78*, 5820.

31. Nishikawa, M.; Holroyd, R.A.; Sowada, U. J. Chem. Phys. 1980, *72*, 3081.

32. Nishikawa, M.; Holroyd, R.A. J. Chem. Phys. 1982, *77*, 4678.

33. Shinsaka, K.; Codama, M.; Srithanratana, T.; Yamamoto, M.; Hatano, Y. J. Chem. Phys. 1988, *88*, 7529.

34. Wada, T.; Freeman, G.R. Phys. Rev., A 1981, *24*, 1066.

35. Hansen, D.; Jungblut, H.; Schmidt, W.F. J. Phys. D 1983, *16*, 1623.

36. Sennhauser, E.S.; Armstrong, D.A.; Wilkinson, F. J. Phys. Chem. 1980, *84*, 123.

37. Warman, J.M.; Sennhauser, E.S.; Armstrong, D.A. J. Chem. Phys. 1979, *70*, 995.

38. Scales, M.J.; Cooper, R.; Warman, J.M.; deHaas, M.P. Radiat. Phys. Chem. 1987, *29*, 365.

39. Shizgal, B.; Hatano, Y. J. Chem. Phys. 1988, *88*, 5980.

40. Suzuki, E.; Hatano, Y. J. Chem. Phys. 1986, *84*, 4915.

41. Suzuki, E.; Hatano, Y. J. Chem. Phys. 1986, *85*, 5341.

42. Warman, J.M.; deHaas, M.P. J. Chem. Phys. 1975, *63*, 2094.

43. Warman, J.M.; deHaas, M.P. Radiat. Phys. Chem. 1988, *32*, 31.

44. Warman, J.M.; Sauer, M.C., Jr. J. Chem. Phys. 1975, *62*, 1971.

45. Hatano, Y.; Kimizuka, Y.; Shimamori, H. Radiat. Phys. Chem. 1982, *19*, 265.

46. Fessenden, R.W.; Bansal, K.M. J. Chem. Phys. 1970, *53*, 3468.

47. Matsuoka, S.; Nakamura, H.; Tamura, T. J. Chem. Phys. 1981, *75*, 681.

48. Matsuoka, S.; Nakamura, H.; Tamura, T. J. Chem. Phys. 1983, *79*, 825.

49. Bansal, K.M.; Fessenden, R.W. Chem. Phys. Lett. 1972, *15*, 21.

50. Bansal, K.M.; Fessenden, R.W. J. Chem. Phys. 1973, *59*, 1760.

51. Warman, J.M.; Fessenden, R.W.; Bakale, G. J. Chem. Phys. 1972, *57*, 2702.

52. Kokaku, Y.; Hatano, Y.; Shimamori, H.; Fessenden, R.W. J. Chem. Phys. 1979, *71*, 4883.

53. Kokaku, Y.; Toriumi, M.; Hatano, Y. J. Chem. Phys. 1980, *73*, 6167.

54. Shimamori, H.; Fessenden, R.W. J. Chem. Phys. 1978, *68*, 2757.

55. Shimamori, H.; Fessenden, R.W. J. Chem. Phys. 1978, *69*, 4732.

56. Shimamori, H.; Fessenden, R.W. J. Chem. Phys. 1979, 70, 1137.
57. Shimamori, H.; Fessenden, R.W. J. Chem. Phys. 1979, 71, 3009.
58. Shimamori, H.; Fessenden, R.W. J. Chem. Phys. 1981, 74, 453.
59. Shimamori, H.; Hatano, Y. Chem. Phys. Lett. 1976, 38, 242.
60. Shimamori, H.; Hatano, Y. Chem. Phys. 1976, 12, 439.
61. Shimamori, H.; Hatano, Y. Chem. Phys. 1977, 21, 187.
62. Shimamori, H.; Hotta, H. J. Chem. Phys. 1983, 78, 1318.
63. Shimamori, H.; Hotta, H. J. Chem. Phys. 1984, 81, 1271.
64. Shimamori, H.; Hotta, H. J. Chem. Phys. 1986, 84, 3195.
65. Shimamori, H.; Hotta, H. J. Chem. Phys. 1986, 85, 887.
66. Shimamori, H.; Hotta, H. J. Chem. Phys. 1986, 85, 4480.
67. Shimamori, H.; Hotta, H. J. Chem. Phys. 1989, 90, 232.
68. Toriumi, M.; Hatano, Y. J. Chem. Phys. 1983, 79, 3749.
69. Toriumi, M.; Hatano, Y. J. Chem. Phys. 1984, 81, 3748.
70. Toriumi, M.; Hatano, Y. J. Chem. Phys. 1985, 82, 254.
71. Nagra, S.S.; Armstrong, D.A. Can. J. Chem. 1976, 54, 3580.
72. Nagra, S.S.; Armstrong, D.A. J. Phys. Chem. 1977, 81, 599.
73. Nagra, S.S.; Armstrong, D.A. Radiat. Phys. Chem. 1978, 11, 305.
74. Märk, T.D. In *Electronic and Atomic Collisions*; Gilbody, H.B., Newell, W.R., Read, F.H., Smith, A.C.H., Eds.; North-Holland: Amsterdam, 1988; 705 pp.
75. Märk, T.D.; Leiter, K.; Ritter, W.; Stamatovic, A. Phys. Rev. Lett. 1985, 55, 2559.
76. Stamatovic, A. In *Electronic and Atomic Collisions*; Gilbody, H.B., Newell, W.R., Read, F.H., Smith, A.C.H., Eds.; North-Holland: Amsterdam, 1988; 729 pp.
77. Matejcik, S.; Kiendler, A.; Stampfli, P.; Stamatovic, A.; Märk, T.D. Phys. Rev. Lett. 1996, 77, 3771.
78. Smirnov, B.M. Sov. Phys., Usp. 1984, 27, 1.
79. Bloch, F.; Bradbury, N.E. Phys. Rev. 1935, 48, 689.
80. Herzenberg, A. J. Chem. Phys. 1969, 51, 4942.
81. Bouby, L.; Abgrall, H. In *Proceedings of the 5th International Conference on Phys. Electron. Atom. Collision*, Leningrad, 1967; 584 pp.
82. Bouby, L.; Fiquet-Fayard, F.; LeCoat, Y. Int. J. Mass Spectrom. Ion Phys. 1970, 3, 439.
83. Chanin, L.M.; Phelps, A.V.; Biondi, M.A. Phys. Rev. 1962, 128, 219.
84. Crompton, R.W.; Hegerberg, R.; Skullerud, H.R. In *Proceedings of the International Seminar on Swarm Experiments in Atomic Collision Research*, Tokyo; Ogawa, I., Ed.; 1979; 18 pp.
85. Hackam, R.; Lennon, J.J. Proc. Phys. Soc. 1965, 86, 123.
86. Hegerberg, R.; Crompton, R.W. Aust. J. Phys. 1983, 36, 831.
87. Hurst, G.S.; Bortner, T.E. Phys. Rev. 1959, 114, 166.
88. van Lind, V.A.J.; Wikner, E.G.; Truedblood, D.L. Bull. Am. Phys. Soc. 1960, 5, 122.
89. Nelson, D.R.; Davis, F.J. Bull. Am. Phys. Soc. 1971, 16, 217.
90. Pack, J.L.; Phelps, A.V. J. Chem. Phys. 1966, 45, 4316.
91. Stockdale, J.A.; Christophorou, L.G.; Hurst, G.S. J. Chem. Phys. 1967, 47, 3267.
92. Warman, J.M.; Bansal, K.M.; Fessenden, R.W. Chem. Phys. Lett. 1971, 12, 211.
93. Young, B.G.; Johnson, A.W.; Garruthors, J.A. Can. J. Phys. 1963, 41, 625.
94. Koike, F. J. Phys. Soc. Jpn. 1973, 35, 1166.
95. Koike, F. J. Phys. Soc. Jpn. 1975, 39, 1590.
96. Koike, F.; Watanabe, T. J. Phys. Soc. Jpn. 1973, 34, 1022.
97. Parlant, G.; Fiquet-Fayard, F. J. Phys. B 1976, 9, 1617.
98. Christophorou, L.G. J. Phys. Chem. 1972, 76, 3730.
99. Christophorou, L.G. Chem. Rev. 1976, 76, 409.
100. Christophorou, L.G. Adv. Electr. Electron Phys. 1978, 46, 55.
101. Goans, R.E.; Christophorou, L.G. J. Chem. Phys. 1974, 60, 103.
102. McCorkle, D.L.; Christophorou, L.G.; Anderson, V.E. J. Phys. B 1972, 5, 1211.
103. Stogryn, D.E.; Hirschfelder, J.O. J. Chem. Phys. 1959, 31, 1531.

104. Blaney, B.L.; Ewing, G.E. Annu. Rev. Phys. Chem. 1976, 27, 553.
105. Christophorou, L.G. Environ. Health Perspect. 1980, 36, 3.
106. Christophorou, L.G.; Stockdale, J.A. J. Chem. Phys. 1968, 48, 1956.
107. Huo, W.M.; Fessenden, R.W.; Bauschlicher, C.W., Jr. J. Chem. Phys. 1984, 81, 5811.
108. McMahon, D.R.A. In Electron and Ion Swarms; Christophorou, L.G., Ed.; Pergamon: Oxford, 1981; 117 pp.
109. McMahon, D.R.A. Chem. Phys. 1982, 66, 67.
110. Nishikawa, M.; Holroyd, R.A. J. Chem. Phys. 1983, 79, 3754.
111. Crompton, R.W.; Hegerberg, R.; Skullerud, H.R. J. Phys. B 1980, 13, 603.
112. Koura, K. J. Chem. Phys. 1982, 76, 390.
113. Koura, K. J. Chem. Phys. 1983, 78, 604.
114. McMahon, D.R.A.; Crompton, R.W. J. Chem. Phys. 1983, 78, 603.
115. Skullerud, H.R. Aust. J. Phys. 1983, 36, 845.
116. Shimamori, H.; Tatsumi, Y.; Ogawa, Y.; Sunagawa, T. Chem. Phys. Lett. 1992, 194, 223.
117. Shimamori, H.; Tatsumi, Y.; Ogawa, Y.; Sunagawa, T. J. Chem. Phys. 1992, 97, 6335.
118. Shimamori, H.; Sunagawa, T.; Ogawa, Y.; Tatsumi, Y. Chem. Phys. Lett. 1994, 227, 609.
119. Shimamori, H.; Sunagawa, T.; Ogawa, Y.; Tatsumi, Y. Chem. Phys. Lett. 1995, 232, 115.
120. Sunagawa, T.; Shimamori, H. Int. J. Mass Spectrom. Ion Process 1995, 149/150, 123.
121. Sunagawa, T.; Shimamori, H. J. Chem. Phys. 1997, 107, 7876.
122. Sunagawa, T.; Shimamori, H. Int. J. Mass Spectrom. 2001, 205, 285.
123. Hatano, Y. Adv. At. Mol. Opt. Phys. 2000, 43, 231.
124. Yencha, A.J. In Electron Spectroscopy: Theory, Techniques and Applications; Brundle, C.R., Baker, A.D., Eds.; Academic Press: London, 1984; Vol. 5, 197 pp.
125. Ueno, T.; Hatano, Y. Chem. Phys. Lett. 1976, 40, 283.
126. Nakamura, H. J. Phys. Soc. Jpn. 1965, 26, 1473.
127. Nakamura, H. J. Phys. Soc. Jpn. 1971, 31, 574.
128. Nakamura, H. J. Phys. B 1975, 8, L489.
129. Miller, W.H. J. Chem. Phys. 1970, 52, 3563.
130. Niehaus, A. In Radiation Research; Nygaard, O.F., Adler, H.I., Sinclair, W.K., Eds.; Academic Press: London, 1975; 227 pp.
131. Niehaus, A. Adv. Chem. Phys. 1981, 45, 399.
132. Niehaus, A. In Electronic and Atomic Collisions; Datz, S., Ed.; North Holland: Amsterdam, 1982; 237 pp.
133. Illenberger, E.; Niehaus, A. Z. Phys., B 1975, 20, 33.
134. Ueno, T.; Yokoyama, A.; Takao, S.; Hatano, Y. Chem. Phys. 1980, 45, 261.
135. Yokoyama, A.; Takao, S.; Ueno, T.; Hatano, Y. Chem. Phys. 1980, 45, 439.
136. Yokoyama, A.; Hatano, Y. Chem. Phys. 1981, 63, 59.
137. Koizumi, H.; Ukai, M.; Tanaka, Y.; Shinsaka, K.; Hatano, Y. J. Chem. Phys. 1986, 85, 1931.
138. Ohno, K.; Matsumoto, S.; Harada, Y. J. Chem. Phys. 1984, 81, 4447.
139. Kohmoto, M.; Watanabe, T. J. Phys. B 1977, 10, 1875.
140. Watanabe, T.; Katsuura, K. J. Chem. Phys. 1967, 47, 800.
141. Ukai, M.; Tanaka, Y.; Koizumi, H.; Shinsaka, K.; Hatano, Y. J. Chem. Phys. 1986, 84, 5575.
142. Ukai, M.; Nakazawa, H.; Shinsaka, K.; Hatano, Y. J. Chem. Phys. 1988, 88, 3623.
143. Katsuura, K. J. Chem. Phys. 1965, 42, 3771.
144. Ukai, M.; Yoshida, H.; Morishima, Y.; Nakazawa, H.; Shinsaka, K.; Hatano, Y. J. Chem. Phys. 1989, 90, 4865.
145. Morishima, Y.; Ukai, M.; Shinsaka, K.; Kouchi, N.; Hatano, Y. J. Chem. Phys. 1991, 94, 2564.
146. Nayfeh, M.H.; Chen, C.H.; Payne, M.G. Phys. Rev., A 1976, 14, 739.
147. Penkin, N.P.; Devdariani, A.Z.; Ionih, W.Z.; Samson, A.V. In Abstract, Xth International Conference on Phys. Electron. Atom. Collision; Paris, Watel, G., Ed; 1977; 320 pp.
148. Morishima, Y.; Ukai, M.; Kouchi, N.; Hatano, Y. J. Chem. Phys. 1992, 96, 8187.
149. Morishima, Y.; Yohshida, H.; Ukai, M.; Kouchi, N.; Hatano, Y. J. Chem. Phys. 1992, 97, 3180.

150. Ukai, M.; Koizumi, H.; Shinsaka, K.; Hatano, Y. J. Chem. Phys. 1986, *84*, 3199.
151. Kitajima, M.; Hidaka, K.; Kusumori, H.; Ukai, M.; Kouchi, N.; Hatano, Y. J. Chem. Phys. 1994, *100*, 8072.
152. Yohshida, H.; Kawamura, H.; Ukai, M.; Shinsaka, K.; Kouchi, N.; Hatano, Y. Chem. Phys. Lett. 1991, *176*, 173.
153. Yohshida, H.; Kawamura, H.; Ukai, M.; Kouchi, N.; Hatano, Y. J. Chem. Phys. 1992, *96*, 4372.
154. Yohshida, H.; Ukai, M.; Kawamura, H.; Kouchi, N.; Hatano, Y. J. Chem. Phys. 1992, *97*, 3289.
155. Yohshida, H.; Kitajima, M.; Kawamura, H.; Hidaka, K.; Ukai, M.; Kouchi, N.; Hatano, Y. J. Chem. Phys. 1993, *98*, 6190.

7

Studies of Solvation Using Electrons and Anions in Alcohol Solutions*

Charles D. Jonah
Argonne National Laboratory, Argonne, Illinois, U.S.A.

1. INTRODUCTION

It has long been known that the solvent plays an important part in chemical reactivity. It can influence reactivity by altering the energy levels of the reactants and products, by removing energy from hot modes, by stabilizing intermediates, and by confining reactants. For these reasons, reactions in solution may be considerably different than in the gas phase.

Radiation chemistry highlights the importance of the role of the solvent in chemical reactions. When one radiolyzes water in the gas phase, the primary products are H atoms and OH radicals, whereas in solution, the primary species are e_{aq}^-, OH, and H^+ [1]. One can vary the temperature and pressure of water so that it is possible to go continuously from the liquid to the gas phase (with supercritical water as a bridge). In such experiments, it was found that the ratio of the yield of the H atom to the hydrated electron (H/e_{aq}^-) does indeed go from that in the liquid phase to the gas phase [2]. Similarly, when one photoionizes water, the threshold energy for the ejection of an electron is much lower in the liquid phase than it is in the gas phase. One might suspect that a major difference is that the electron can be transferred to a trap in the solution so that the full ionization energy is not required to transfer the electron from the molecule to the solvent.

One can consider two facets of the solvation process, the energetics and the kinetics. Clearly, the kinetics will not matter if reactions take place on a time scale that is much faster or much slower than the solvation process. However, if reaction and solvation occur on the same time scale, the considerable energy changes that the solvation process can engender will affect the reaction. In fact, exactly what solvent motions take place during the solvation process may well be important. Thus, it is of interest to understand the kinetics of the solvation process.

In a paper, Jortner stated that information about solvation dynamics can be obtained from (1) solvation of the electron, (2) solvation of a dipole, and (3) extracting information from the rates of electron transfer reaction [3,4]. We have added to these

*Work performed under the auspices of the Office of Science, Division of Chemical Science, US-DOE, under contract number W-31-109-ENG-38.

possibilities the study of the solvation of an anion that is created very quickly, and we show that some of the characteristics of electron solvation are specific to that species.

Solvation processes are generally studied by observing the shift of an absorption process or an emission process as a function of time [5,6]. For example, one might excite a molecule where the excited state is considerably more polar than the ground state. The solvent would then rearrange the solvent around the excited state to lower the total energy of the system. One can then observe the shift of the emission spectrum as a function of time. This is shown pictorially in Fig. 1, with the top showing the system and the bottom showing the "potential curves" against the solvent coordinate. This will then lead to a shift of the emission spectrum towards lower energy. These techniques have been used by many authors (see, e.g., Refs. [5,6]). The fluorescence measurements can make use of the single-photon-counting technique, and the signal-to-noise and dynamic range of those measurements are much better than for optical absorption measurements. In fact, such experiments are able to separate multiple-exponential processes, something that is not possible with the absorption measurements.

An alternative technique is to create a ground state species very quickly, by creating an anion or a cation, or possibly by decomposing a molecule and then observing the change of its optical spectrum. This technique has the advantage over observing the relaxation of the excited state in that one is not limited by the fluorescence lifetime of the probe molecule. In the experiments that we will discuss, we will attach an electron to a benzophenone molecule and then watch the change of the spectrum of the resulting anion. Fig. 2 shows a cartoon for this process. Note that the absorption will shift towards higher energies (the blue) for these experiments.

In this paper, we will limit ourselves to the discussion of two types of anions in alcohol solutions: the benzophenone anion [7–9] and the solvated electron. One can consider that these systems are both measurements of the solvation of an anion, if we consider the electron

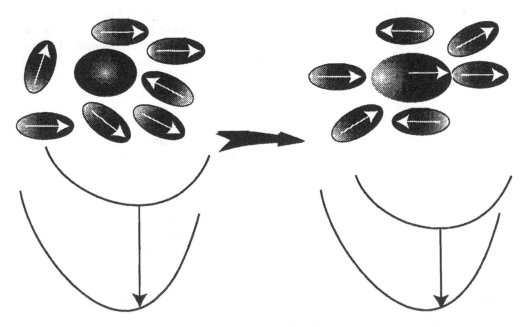

Figure 1   Cartoon showing solvation and how it will shift the spectrum of a dipolar excited state.

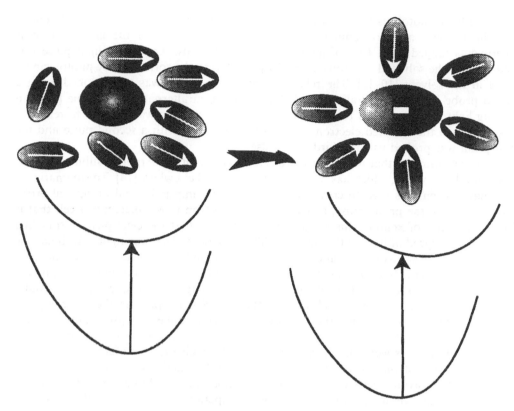

Figure 2   Cartoon showing solvation and how it will shift the spectrum of an anion.

as the simplest anion. We will show that while there are considerable similarities between the solvation of the two species, there are also some considerable differences. We will suggest that these differences arise out of the formation mechanism of the electron.

We will not discuss the solvation of the electron in water. While one might expect that the electron solvation in alcohols should be qualitatively the same as in water, with only quantitative changes, it turns out that the quantitative changes lead to qualitative changes in the mechanism. It is usually assumed that the solvated electron is formed in the excited state, which then relaxes to the ground state, and the solvent reorganizes around the ground-state electron. This picture seems to hold reasonably well for electrons in alcohols; however, in water, the solvent relaxation time is sufficiently fast that the relaxation of the excited state of the electron and the reorganization of the solvent occur on the same time scale, making interpretation difficult.

2. EXPERIMENTAL

The results in this discussion will include those from our laboratory and experiments on electron solvation from other laboratories. The experiments that were done at Argonne made use of the stroboscopic pulse radiolysis technique, which will be discussed below. Experiments from other laboratories have made use of pulse radiolysis and laser photolyis techniques for the measurement of electron solvation.

The laser photolysis techniques are discussed in detail in the appropriate references. We will just give a short summary here. The time resolution of the laser photolysis experiments ranged from 100s of femtoseconds to 20–30 ps. An intense light pulse was used to form the solvated electrons, either by photolyzing the solvent or by photolyzing a solute in the solvent [10–12]. The relaxation of the spectrum was then probed using a second probe light beam. Walhout and Barbara used an alternative technique [13]. An ionizing light pulse was used to form the electron. The electron was allowed to relax for 4 ns, and then the solvated electron was optically bleached with a second pulse and the absorption was probed using a third light pulse.

The experiments that we have carried out make use of the Argonne subnanosecond pulse radiolysis system, which has been described in detail elsewhere [14]. To summarize, a 30-ps electron pulse is used to create both an electron pump pulse and an optical probe pulse. To create the probe pulse, a portion of the electron pulse intercepts a cell that is filled with 1 atm of xenon. This pressure is sufficient so that the electrons are traveling faster than the speed of light in that medium. This generates Cerenkov light, which has the time profile of the electron pulse and which is a continuum light source where the intensity increases towards the blue. The portion of the electron beam that is not intercepted by the gas cell is delayed by a magnet (a 270° magnet or alpha magnet). The Cerenkov light pulse can be adjusted using an optical delay line so that it can arrive prior to the electron pulse or up to approximately 3 ns after the pulse (maximum time range of approximately 3.9 ns).

For the experiments that measure the solvation of the electron in an alcohol, different neat alcohols were placed in a flow cell and the kinetics at different wavelengths and the evolution of the spectra were observed [15–17]. Kinetics were determined to the blue side of the solvated electron spectrum. The temperature could be lowered to provide data over a range of temperatures. The lower-temperature experiments all utilized n-propanol or 2-propanol because they have a long liquid range [18].

Additional experiments were done in mixtures of alcohol–alkane [16,17]. The spectra and kinetics were measured in mixtures of 1-propanol–n-hexane. Some experiments were done in cyclohexane, where the behavior was qualitatively similar; however, the exact concentration where spectra and kinetics changed depended on the alkane [16]. Additional experiments observed the shift of the final spectrum of the solvated electron in supercritical ethane–methanol mixtures. These experiments were done using standard pulse radiolysis techniques and thus we were unable to observe the kinetics [19].

Anion-solvation experiments were done using benzophenone as a probe molecule. These experiments were suggested by Bernard Hickel and were based on work of Marignier and Hickel and Ichikawa et al. in low-temperature alcohols [7–9]. The concentration of benzophenone was in the range of 0.25 M. This concentration was shown to be sufficient so that it would react with all of the solvated-electron precursors; thus, virtually no solvated electrons would be formed under these conditions [20]. The data were analyzed by considering the time dependence of the spectra and the kinetics and evaluating a global fit for these data [20,21]. These experiments were done in a series of alcohols, as a function of temperature, and in alcohol–alkane mixtures.

The solvation processes for the anion systems were simulated using both simple ball models and more representational models of the solvent and solute [14,22].

3. RESULTS AND DISCUSSION

The measurement of the solvation of an aromatic anion and the measurement of the solvation of an electron in alcohols have quite a bit in common. They both observe the response of the solvent to the sudden creation of an anion in the bulk of the solution. Thus, one might expect that the final states would be similar. However, there may well be considerable differences in the initial states, which then may well lead to changes in the solvation process.

The electron will be solvated in a region where the solvent molecules are appropriately arranged. There must be a cluster of electrons of a size of 4–5 to support the formation of the solvated electron from the results of Gangwer et al., [23], Baxendale [24,25], and Kenney-Wallace and Jonah [16]. This behavior does not depend on the specific alcohol or alkane and even occurs in supercritical solutions, as has been shown in experiments done using mixtures of supercritical ethane–methanol mixtures [19]. Experiments have also shown that the thermodynamically lowest state might not be reached. For example, the experiments of Baxendale that measured the conductivity of the solvated electron in alcohol–alkane mixtures showed that when there was a sufficient concentration of alcohols to form dimers, there was a sharp decrease in the mobility of the electron [24,25]. This result showed that the electron was at least partially solvated. However, the conductivity was not as low as one would expect for the fully solvated electron, and the fully solvated electron was never formed on their time scale (many microseconds), a time scale that was sufficiently long for the electron–alcohol entity to encounter sufficient alcohols to fully solvate the electron. Similarly, the experiments of Weinstein and Firestone, in mixed polar solvents, showed that the electron that was observed depended on the initial mixture and would not relax to form the most fully solvated electron [26].

The situation for the benzophenone molecule would be expected to be considerably different. The major driving force for the attachment of the electron is the electron affinity of the benzophenone molecule. The arrangement of the solvent around that molecule will provide only a perturbation to that energy. Thus, one will observe an average of the solvent structures around the benzophenone and not just solvent arrangements that are optimal to trap the electron. Similarly, because the charge is more delocalized on the benzophenone molecule, the alcohol molecules might not be so strongly aligned, so that the alcohol molecules will be able to rearrange and reach the lowest energy configuration.

We shall discuss separately the results for electron solvation and anion solvation at room temperature in different alcohols to provide a basis for the discussion of solvation mechanisms. This will be followed by a discussion of solvation at lower temperatures and in alcohol–alkane mixtures to further highlight the similarities and differences between anion and electron solvation.

3.1. Electron Solvation at Room Temperature

There have been many experiments on electron solvation at very low temperatures, and the results depend on the temperature range and alcohol. For n-propanol, which has a very long liquid range, Baxendale and Wardman showed that the solvation behavior changes markedly from 110 to 120 K [27]. At the higher temperatures, kinetics and spectral changes are similar to what are seen at room temperature, while at lower temperatures, the results are similar to what are observed in glasses. The original experiments on electron solvation in alcohols near room temperature were carried out by Chase and Hunt [28]. In their experiments, they used their stroboscopic pulse radiolysis system to determine the kinetics of the electron solvation. Those experiments were followed by further pulse radiolysis experiments that obtained similar results [15,17]. With the advent of short-pulsed lasers, several groups explored the solvation of the electron in alcohols. A summary of the pulse radiolysis data was given by Kenney-Wallace and Jonah [16], and a summary of the laser data has been given by Walhout and Barbara [13].

If one observes the blue of the absorption maximum, the signal appears to consist of a fast rise followed by a slower growth. We are concerned here with the slower growth, because it arises from the relaxation of the solvent around the electron. This behavior was seen in both the photolytic and radiolytic formation of the electron. The data of Walhout and Barbara are a bit more complex in this time period, because one has a bleach followed by a recovery of the absorption of the electron (however, not all the absorption is recovered, suggesting that the electron reacts with the solvent after the photoexcitation) [13]. A summary of the measured kinetics from both laser and pulse radiolytic experiments are given in Fig. 3. As we can see from this figure, the rate of the slow growth does not appear to depend on the mode of formation. Early results suggested that there were differences between the rate of electron solvation for the different methods of electron formation, but it has been suggested by Sander et al. that the differences arose from the data analyses of some of the early laser data [11], because the earlier analyses of the laser-flash-photolysis data did not separate the slow growth process from the fast component, which would have been instantaneous on their time scale.

Fig. 3 displays the solvation time of the electron (using open symbols) plotted against the longitudinal relaxation time. The results were plotted against the longitudinal relaxation time because that is the time scale that some continuum models would have suggested. Note that the rates are comparable to those seen for the solvation of an anion, and considerably faster than those measured for dipole solvation. The anion data will be discussed below.

The shift of the spectrum has been well discussed previously. The early pulse radiolysis data suggested that the kinetics at 1300 nm [28] matched the kinetics in the blue [16], which led to the description of the kinetics as a two-state problem. However, the results measured by Chase and Hunt, where the kinetics at 1050 nm were considerably slower than the kinetics at 500 and 1300 nm, suggested that the kinetics were more

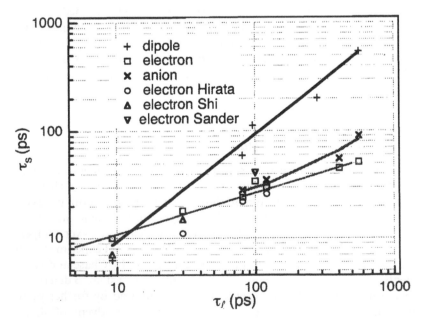

Figure 3 Solvation time in alcohols of the electron and the benzophenone anion, plotted versus the longitudinal relaxation time τ_1. (The dipole data are from Refs. [5,6].)

complicated than a two-state model would predict [28]. Some authors have suggested the existence of an approximate isosbestic point [12]. It seems to be clear that the response observed well to the blue of the peak does not depend on wavelength [16].

3.2. Anion Solvation

Anion solvation has been studied by observing the shift in the absorption spectrum of the benzophenone anion in various solvents and as a function of temperature. The benzophenone anion was formed from the reaction of the benzophenone molecule and a precursor to the solvated electron. Approximately 0.25 M benzophenone is put into the solution so that all the presolvated electrons will react with the benzophenone and virtually none will form the solvated electron. This process occurs much more quickly than the solvation processes that are observed [14,20].

The first experiments were done in a series of primary alcohols at room temperature. In Fig. 4, we show the spectra as a function of time at room temperature at different times in n-octanol. Similar data are seen in all the primary alcohols, and the final solvated spectrum is similar for all of the alcohols, suggesting that final solvated anion state is similar in all primary alcohols.

One of the models that has had considerable success for predicting solvation processes of dipoles in non-hydrogen-bonded solvents is the dielectric continuum model [5,14]. In this model, the amount of solvation will depend on the dipole density—that is, the molar concentration and strength of dipoles. While the position of the absorption maximum is not directly related to the energy of solvation that a molecule experiences, one would expect the two to be very strongly correlated. However, for the three different

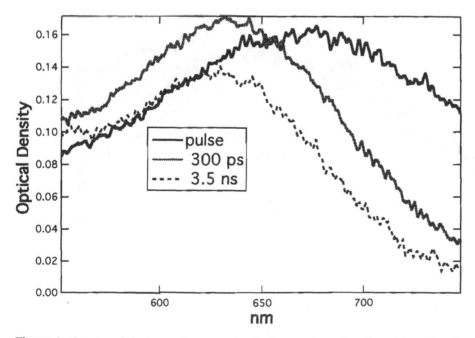

Figure 4 Spectra of the benzophenone anion in n-octanol as a function of time after the generation of the anion.

primary alcohols considered here, the dipole density changes by more than a factor of 2, with the highest density for the butanol alcohol and the lowest density for the decanol solvent, and the maxima of the spectra of the solvated anion do not change. We interpret these results to mean that it is the local concentration of dipoles around an anion that dominates the solvation and not the concentration of the dipoles in the bulk of the solution.

We have also carried out experiments in secondary alcohols. For example, Fig. 5 shows the spectra in primary and secondary octanol (1-octanol vs. 2-octanol). These data show that while the initial spectra are approximately the same in the two alcohols, the final spectra are considerably different. There is much less shift in the absorption spectrum in the 2-octanol, suggesting that 2-octanol is less effective at solvating the benzophenone molecule.

We suggest that in a secondary alcohol, there will be steric hindrance, which will limit the number of molecules that can get near the ion. This is shown, cartoon fashion, in Fig. 6. As one can see, the steric factors might well limit the ability of the alcohols to pack around an anion.

To test these suggestions, simple Monte Carlo and molecular dynamics simulations were carried out [22]. These explored the potential energy of a charge in the presence of solvent molecules of different forms. The form of these approximations is shown in the reference. These simple systems embody the primary characteristics of the present system. The results showed that the solvation energy (the field at the charge) does not depend on the length of the solvent molecule but on the orientation of the dipole. If the positive end of the dipole is at the end of the molecule, solvation will be strong. This agrees with the experimental results that show that the final spectra are very much the same in all primary alcohols.

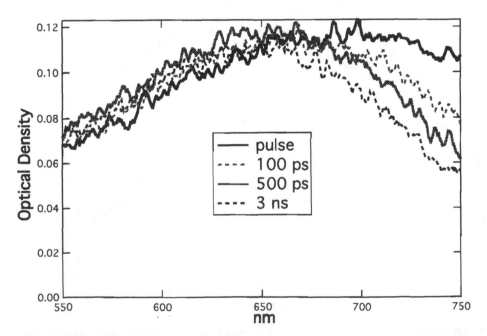

Figure 5 Spectra of the benzophenone anion in primary and secondary n-octanol.

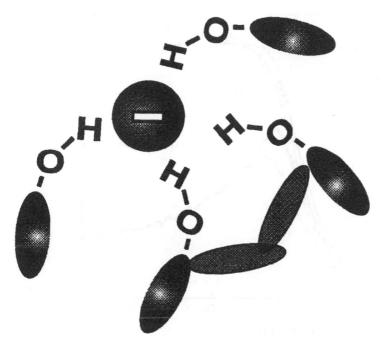

Figure 6 Cartoon of the effect of structure on benzophenone anion solvation.

Simulations were done where the dipole is reversed in the solvent molecule. This is equivalent to making measurements in acetonitrile. The calculations suggested that there would only be weak solvation in acetonitrile, despite the fact that acetonitrile is far more polar than any of the alcohols that we have measured. This fact is indeed borne out experimentally. The spectrum of the benzophenone anion is considerably to the red of the spectrum of the benzophenone anion in any of the alcohols that we have measured. In addition, there is no evidence for any shift of the spectrum in the time scale that we observe. This lack of shift may not be surprising, because experiments in acetonitrile suggest that the solvation time is very fast. Thus, any relaxation that is going to take place will have occurred on the time scale of the present experiments.

Typical kinetics for the solvation of an anion in 1-octanol are seen in Fig. 7. While an approximate isosbestic point has been seen for the solvation of the electron in an alcohol, this figure shows there is nothing approaching an isosbestic point for this process. The kinetics are different at different wavelengths, and even include both a growth and decay. The kinetics were determined by fitting the time profiles and spectra at different times. The results of these analyses are seen in Fig. 3, along with the data for the solvation of the electron in the same solvents, as discussed above. We see that the solvation time for the anion is longer but of the same order of magnitude as the solvation time for the electron, and it is much shorter than would be predicted for dielectric relaxation.

3.3. Temperature Dependence

Solvation takes place through multiple processes, each with different energy requirements. By making measurements at different temperatures, one can begin to separate the multiple processes and possibly shift which processes are dominant. For these reasons, temper-

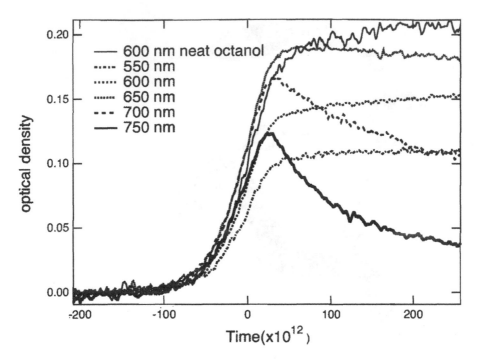

Figure 7   Time profiles of the absorption of the benzophenone anion as a function of wavelength.

ature-dependent measurements have always been an important technique for under-standing fundamental reaction pathways. As has been mentioned previously, the spectral and kinetic behavior of the solvation process alters drastically in n-propanol between 110 and 120 K [27]. There have also been studies of the activation energy of solvation for dipolar systems, which we will compare to the present data. We will first discuss the anion-solvation measurements, then the electron solvation measurements, and then compare the two different chemical systems.

Anion Solvation as a Function of Temperature

In Fig. 8 (top), we see the temperature dependence of the rate of solvation of the ben-zophenone anion in primary and secondary propanol and n-butanol as a function of temperature [21,29]. As one would expect, the solvation rate is slower in n-butanol than it is in n-propanol; however, as we can see from the slope of the Arrhenius plot, the acti-vation energies are about the same, both having an activation energy of approximately 22 kJ/mol. The activation energy for solvation of the anion in 2-propanol is considerably lower, −16 kJ/mol. These values are similar to the energy required to break a hydrogen bond in these systems, of 25±3 kJ/mol for the primary alcohols and 17±3 kJ/mol for 2-propanol [30]. The ordering of the activation energies is reasonable in that steric hindrance will decrease the hydrogen bonding in the secondary alcohol in comparison to the primary alcohol. The similarity of the solvation energy to the activation energy for hydrogen-bond breakage would suggest that hydrogen-bond breakage is important in the solvation process; however the correlation with τ_2, the second dielectric relaxation time, which is often attributed to the rate of rotation without breaking hydrogen bonds, is not consistent with that picture [31].

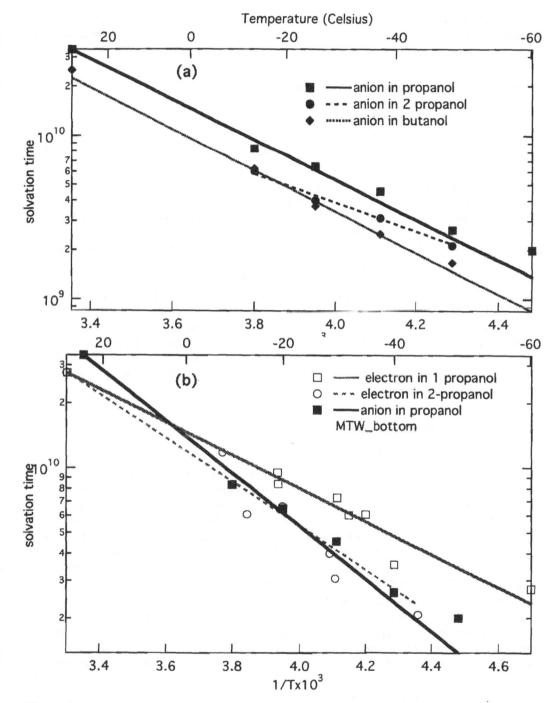

Figure 8   (a) Temperature dependence of the solvation of the benzophenone anion in primary and secondary propanol and n-butanol. (b) Temperature dependence of the solvation of the electron in n-propanol and 2-propanol.

If one compares the log of the solvation time to the log of viscosity, the two primary alcohols fall on the same straight line and this line has a slope of 1. The 2-propanol, while correlated with viscosity, does not show a linear dependence [21].

Temperature Dependence of Electron Solvation

These results can be compared to the temperature dependence that has been seen in electron solvation (see Fig. 8, bottom) [18,28]. We first note that the dependence of the solvation time on temperature is quite different for 1-propanol and 2-propanol. It appears as if the activation energy for 2-propanol is considerably higher than 1-propanol, which is the reverse of what was observed for the solvation of an anion or for the hydrogen-bond energetics. The activation energy for solvation in 1-propanol is 14.4 kJ/mol, which is less than the energy for hydrogen-bond breakage, while the activation energy for 2-propanol is 21 kJ/mol, which is considerably higher than the activation energy for anion solvation and considerably higher than the energy of a hydrogen bond.

It has been suggested that the electron will solvate in a preexisting structure in the liquid such as a cyclic tetramer or pentamer [16,23]. It is known that these types of structures are quite common in alcohol–alkane mixtures. Because the electron is dissolved in the preexisting structure that has an appropriate structural arrangement, it might not be necessary to break the hydrogen bonds to solvate the electron. In 2-propanol, experimental data suggest that in dilute alcohol solutions, the alcohol tends to form polymer chains rather than cyclic structures. This means that to create the final electron structure, one will need to break polymeric hydrogen bonds and allow the alcohols to rearrange to create a structure around the electron. This then requires the breaking of hydrogen bonds and the higher activation energy results from that fact.

3.4. Concentration Dependence of Solvation

Concentration Dependence of Anion Solvation

Figs. 9 and 10 show the shift of the spectra and the different kinetics that occur as a function of alcohol concentration in the benzophenone system in propanol–hexane mixtures. These results show that there is a fast absorption, which is due to the formation of the benzophenone anion. This absorption is then followed by a slow growth. This growth is much slower than what is seen in the neat alcohol, and the rate of the growth depends on the concentration of the alcohol in the propanol–hexane mixture. Fig. 11 displays the rate of growth as a function of concentration at two different wavelengths, 750 and 800 nm. The rate at 800 nm is approximately twice as fast as the rate at 750 nm. If the benzophenone anion is more solvated, the spectrum will be more blue-shifted. Presumably, the more deeply solvated anions will have more alcohols near them, and it should take longer to solvate. However, one does not expect this to be a simple function, because the spectra will be both homogeneously and heterogeneously broadened. The different rates of solvation as a function of wavelength are consistent with this result. The measured rate constants are consistent with what one would expect for the rate of an alcohol molecule diffusing to the anion and aligning itself ("reacting").

Concentration Dependence of Electron Solvation

The results for the solvation of the electron in mixed systems are greatly different from the results for the anion solvation. All evidence, which we summarize below, is consistent with

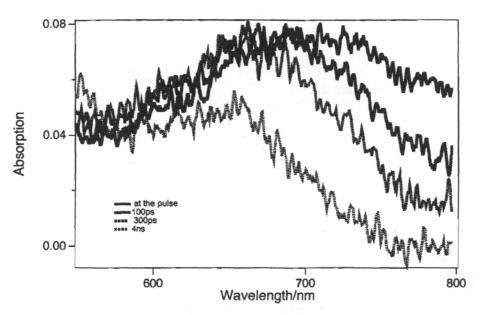

Figure 9   Spectral dependence of the benzophenone anion as a function of time at 0.15 M propanol in n-hexane.

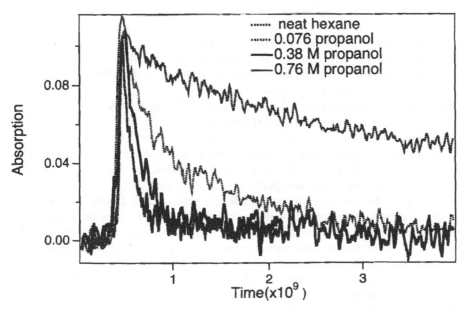

Figure 10   Solvation kinetics of the benzophenone anion in propanol–hexane mixtures measured at 800 nm.

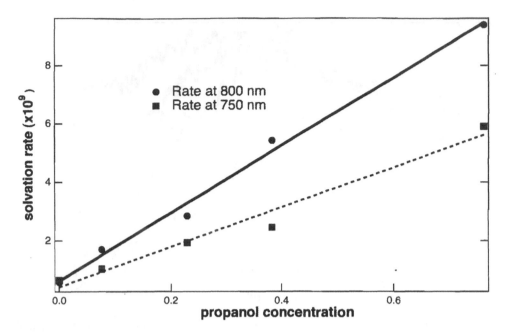

Figure 11 Rate of the solvation of the benzophenone. First-order rate constant at 800 nm is 1.15×10^{10} (Msec)$^{-1}$ and at 750 nm the rate is 6.8×10^9 (Msec)$^{-1}$.

the picture of an electron finding an appropriate location and then rearranging the solvent in the vicinity of it and not solvating further by bringing alcohol (or other solvent) molecules into the inner shell around the electron.

In the experiments of Weinstein and Firestone [26], it was shown that the spectrum of the electron in mixed solvents depends on the mixture of the two components, for example, an alcohol and an ether. If complete relaxation were possible, one would expect that the solvent molecules would interchange and the result would be the lowest energy configuration. This does not appear to occur.

In experiments done by Baxendale [24] and Baxendale and Sharpe [25] in mixed n-propanol–hexane systems, it was found that at low concentrations of alcohol, a solvated species was formed; however, it was not the solvated electron. The existence of the solvated species was determined by the decrease in mobility of the electron in the solvent. The mobility never decreased to the level that would be expected for the solvated electron and no evidence of the solvated electron spectrum was detected.

Experiments done by Kenney-Wallace and Jonah showed a fast initial solvation of the electron followed by a shift/increase of the spectrum towards the blue on a time scale that is similar to (but somewhat slower than) the rate of the solvation of the electron in the neat alcohol [16,17]. There was no long-term alteration of the spectrum on the time scale that would be needed to allow an alcohol molecule to diffuse to the solvated electron (a nanosecond or two, as seen in the kinetics for the benzophenone anion solvation).

The results of Baxendale and Wardman, Kenney-Wallace and Jonah, and Gangwer et al. are all consistent with the idea that there must be preformed traps in the solution [16,24–26]. Further evidence consistent with this view arises from experiments of Dimitrijevic et al., who measured the yield and the spectrum of the solvated electron as a function of temperature in mixtures of supercritical ethane–methanol [19]. As one raised the temperature, the maximum of the absorption spectrum of the solvated electron in the mixture shifted to the red the same amount as did the spectrum of the electron in neat alcohol. The amount of absorption decreased at higher temperatures. Above 110°C, the absorption spectrum disappeared in the alcohol–alkane mixture. In a solution, an electron will solvate where there exist sufficiently large clusters of alcohols to solvate the electron or it will react with the cations in the solution. The decrease in absorption as a function of temperature was interpreted to mean that the concentration of alcohol-molecule clusters capable of solvating the electron decreased at higher temperatures. At higher temperatures, the number of clusters that are sufficiently large to support the solvated electron decreased, and at 110°C, the probability of the electron finding such a cluster decreased to such a level that no solvated electrons were formed.

4. SUMMARY

While the solvated electron may be considered the smallest and simplest anion, its solvation behavior in alcohols is not the same as an anion. An anion can solvate by attaching solvent molecules from the bulk of the solution, the solvated electron does not do this. A solvated electron is formed making use of particular structures in the solvent so the arrangement of solvent molecules is not random. This means that we are observing a small subclass of the solvent structure when we observe electron solvation. Electron solvation can be an excellent probe for structure of the liquid because of this characteristic. This fact was utilized in estimating the size and structures of alcohol molecules in supercritical ethane–methanol mixtures [19]. However, the solvation behavior is not necessarily characteristic of the behavior that one would expect for an anion in the solvent. The driving force for the attachment of the electron to a large aromatic molecule will be the electron affinity of the solute molecule and not the solvent structure. This means that the anion that is quickly produced will be initially probing the average.

While noting the differences in behavior between electron solvation and anion solvation, we cannot forget the similarities. The activation energies for solvation seem to be related to the strength of the hydrogen bond in the solvent, when a solvent rearrangement will take place. In addition, the solvation times are closely related to the dielectric relaxation time τ_2, the time assigned to molecular rotation without breaking hydrogen bonds, rather than τ_1, the time for the rotation requiring hydrogen-bond breakage. The latter is the time that correlates with the rate of dipole solvation [5]. Whether this is a characteristic of the solvent or the long-range forces that occur with anion solvation is not clear at the present.

In this discussion, I have intentionally not discussed the solvation of the electron in water. These results are quite confusing because of the overlap of the electronic relaxation of the excited solvated electron and the rearrangement of the solvent. In an alcohol, these terms are well separated and so discussion is simplified. In addition, we are not aware of any study on the solvation of an anion in water.

There are many future roads for research that can greatly illuminate the physical processes. It would be interesting to observe the solvation of a neutral species that is formed from a negative species by photoionization. One might well expect solvation to be faster for this system because the solvent molecules near the anion/neutral could rearrange without breaking hydrogen bonds. The solvation of an anion primarily rearranges molecules that are hydrogen bonded to other solvent molecules.

Supercritical solutions also provide a large and interesting collection of phenomena that can be probed using these techniques. The greater fluctuations in these solvents may increase the rate of solvation because of the larger number of possible states that can exist.

Solvation is an important component of all reactivity and now it is possible to gain a clearer insight into these processes. Hopefully, this will continue in the years ahead.

REFERENCES

1. Spinks, J.W.T.; Woods, R.J. *An Introduction to Radiation Chemistry*; 3rd Ed; 1990.
2. Cline, J.; Takahashi, K.; Marin, T.W.; Jonah, C.D.; Bartels, D.M. J. Phys. Chem. A 2002, *106*, 12260.
3. Rips, I.; Klafter, J.; Jortner, J. J. Chem. Phys. 1988, *89*, 4288.
4. Rips, I.; Klafter, J.; Jortner, J. J. Chem. Phys. 1988, *88*, 3246.
5. Maroncelli, M. J. Mol. Liq. 1993, *57*, 1.
6. Simon, J.D. Acc. Chem. Res. 1988, *21*, 128.
7. Marignier, J.L.; Hickel, B. Chem. Phys. Lett. 1982, *86*, 95.
8. Marignier, J.L.; Hickel, B. J. Phys. Chem. 1984, *88*, 5375.
9. Ichikawa, T.; Ishikawa, Y.; Yoshida, H. J. Phys. Chem. 1988, *92*, 508.
10. Hirata, Y.; Murata, N.; Tanioka, Y.; Mataga, N. J. Phys. Chem. 1989, *93*, 4527.
11. Sander, M.; Brummund, U.; Luther, K.; Troe, J. Ber. Bunsen-Ges. 1992, *96*, 1486.
12. Shi, X.; Long, F.H.; Lu, H.; Eisenthal, K.B. J. Phys. Chem. 1995, *99*, 6917.
13. Walhout, P.K.; Barbara, P.F. Ultrafast Processes in Chemistry and Photobiology, 1995; 83 pp.
14. Lin, Y.; Jonah, C.D. Understanding chemical reactivity. 1994, *7*, 137.
15. Kenney-Wallace, G.A.; Jonah, C.D. Chem. Phys. Lett. 1976, *39*, 596.
16. Kenney-Wallace, G.A.; Jonah, C.D. J. Phys. Chem. 1982, *86*, 2572.
17. Kenney-Wallace, G.A.; Jonah, C.D. Chem. Phys. Lett. 1977, *47*, 362.
18. Zhang, X.; Jonah, C.D. Chem. Phys. Lett. 1996, *262*, 649.
19. Dimitrijevic, N.M.; Takahashi, K.; Bartels, D.M.; Jonah, C.D. J. Phys. Chem. A 2001, *105*, 7236.
20. Lin, Y.; Jonah, C.D. J. Phys. Chem. 1993, *97*, 295.
21. Zhang, X.; Jonah, C.D. J. Phys. Chem. 1996, *100*, 7042.
22. Lin, Y.; Jonah, C.D. Chem. Phys. Lett. 1995, *233*, 138.
23. Gangwer, T.E.; Allen, A.O.; Holroyd, R.A. J. Phys. Chem. 1977, *81*, 1469.
24. Baxendale, J.H. Can. J. Chem. 1977, *55*, 1996.
25. Baxendale, J.H.; Sharpe, P.H.G. Chem. Phys. Lett. 1976, *41*, 440.
26. Weinstein, J.B.; Firestone, R.F. J. Phys. Chem. 1975, *79*, 1322.
27. Baxendale, J.H.; Wardman, P. J. Chem. Soc. Faraday Trans 1973, *1*, *69*, 584.
28. Chase, W.J.; Hunt, J.W. J. Phys. Chem. 1975, *79*, 2835.
29. Zhang, X.; Lin, Y.; Jonah, C.D. Radiat. Phys. Chem. 1999, *54*, 433.
30. Levin, B.Y. Zhur. Fiz. Khim. 1954, *28*, 1399.
31. Garg, S.K.; Smyth, C.P. J. Phys. Chem. 1965, *69*, 1294.

8

Electrons in Nonpolar Liquids

Richard A. Holroyd
Brookhaven National Laboratory, Upton, New York, U.S.A.

1. INTRODUCTION

Excess electrons can be introduced into liquids by absorption of high-energy radiation, by photoionization, or by photoinjection from metal surfaces. The electron's chemical and physical properties can then be measured, but this requires that the electrons remain free. That is, the liquid must be sufficiently free of electron attaching impurities for these studies. The drift mobility as well as other transport properties of the electron are discussed here as well as electron reactions, free-ion yields, and energy levels.

Ionization processes typically produce electrons with excess kinetic energy. In liquids during thermalization, where this excess energy is lost to bath molecules, the electrons travel some distance from their geminate positive ions. In general, the electrons at this point are still within the Coulombic field of their geminate ions and a large fraction of the electrons recombine. However, some electrons escape recombination and the yield that escapes to become free electrons and ions is termed G_{fi}. Reported values of G_{fi} for molecular liquids range from 0.05 to 1.1 per 100 eV of energy absorbed [1,2]. The reasons for this 20-fold range of yields are discussed here.

Electrons in nonpolar liquids are either in the conduction band, trapped in a cavity in the liquid, or in special cases form solvent anions. The energy of the bottom of the conduction band is termed V_0. V_0 has been measured for many liquids and its dependence on temperature and pressure has also been measured. New techniques have provided quite accurate values of V_0 for the liquid rare gases. The energies of the trapped state have also been derived for several liquids from studies of equilibrium electron reactions. A characteristic of the trapped electron is its broad absorption spectrum in the infrared.

Electron attachment rates have been measured for numerous solutes. Many of these studies were limited to three solvents: cyclohexane, 2,2,4-trimethylpentane, and tetramethylsilane (TMS), and those rates are discussed here. What to expect in other liquids can be inferred from these results. Considerable insight has been gained into certain reactions. Equilibrium reactions of electrons are particularly interesting since they provide information not only on energy levels, as mentioned above, but also on the partial molar volume of trapped electrons. This has led to a better understanding of the mechanism of electron transport.

This chapter presents the current understanding of electrons in nonpolar liquids. Experimental as well as theoretical studies are discussed. For further detail than is provided here, the reader is referred to recent books on the subject [3,4] as well as references cited herein. Some questions still remain due largely to the theoretical difficulties of describing a quantum particle in a disordered environment. The future will hopefully bring new discoveries and revelations that will answer these questions. Finally, we discuss current applications of nonpolar liquids to indicate where future uses may develop.

2. ELECTRON ESCAPE AND RECOMBINATION

An important consideration in understanding a radiation chemical mechanism or predicting the outcome of a radiation experiment is knowing the yield of free electrons or G_{fi}, the number of ion pairs produced per 100 eV absorbed. The free ion yield is affected by the density of ionization along the track of the ionizing particle. The highest yield is observed for high-energy electrons, typically 1–2 MeV, which result in well-separated clusters of a few ionizations each. An alpha particle creates a dense column of ionization and ion yields are very low. X-rays show intermediate behavior. Other factors like molecular structure, temperature, applied field, pressure, and density also affect G_{fi}.

Making sense of free ion yield data requires first knowing what is meant by the free ion yield. Free electrons are those that escape initial spur or track recombination and are therefore diffusing in the bulk and can react with other species in a homogeneous fashion. The escape process can be considered to have two steps. First, the electrons released by ionizing the solvent molecules lose energy in scattering events and, in the process, travel some distance, r, from their positive ions. The electrons may lose energy to vibrational modes, but a significant fraction of the range occurs while the electron has lower, near thermal, energy. In general, there will be a distribution of ranges $D(r)$ and the mean thermalization range is designated by b. For most nonpolar liquids, the electrons after thermalization will still be within the Coulombic field of the positive ions, equal to $e^2/\varepsilon r$, where ε is the dielectric constant. Consequently, the fraction escaping is low.

In the second step, the thermalized electrons will either recombine with the ions in the track or spur or escape. The yield of free electrons is the integral of the product $D(r) \times P(r)$ times the number of electron–ion pairs formed initially in the spur or track, G_{tot}, where $P(r)$ is the probability of escape. For a single ion pair, $P(r)$ is given by [5]:

$$P(r) = \exp\left(-e^2/\varepsilon k_B T\right) \tag{1}$$

The initial yield, G_{tot}, is not known exactly for molecular liquids. However, the fraction of electrons that escape geminate recombination increases with the applied electric field and extrapolating such results to high field gives $G_{tot} = 4.0$ for neopentane [6], 3.1 ± 0.3 for TMS [7], and 4.5 for Ar [8]. A value for G_{tot} of 4.0 was obtained for cyclohexane by measuring the yields of methyl radicals formed in the reaction of electrons with methyl chloride and methyl bromide [9]. High concentrations of these solutes were used (up to 0.5 M) in order to compete with geminate recombination. An analysis by Jay-Gerin [10] of free ion yield data led to an average G_{tot} value of 3.7 ± 0.5 for typical hydrocarbons. Others have suggested that G_{tot} may vary considerably from liquid to liquid [11].

Fast pulse radiolysis studies have shown that geminate recombination occurs on the picosecond time scale [12,13]. Bartczak and Hummel [14] predicted that for n-dodecane, 82% of the geminate ions still remain at 5 psec for 1 MeV irradiation. Future accelerators, with pulses of a few picoseconds length, may soon provide experimental measurements of G_{tot} directly.

2.1. G_{fi} for Minimum Ionizing Radiation

For minimum ionizing radiation, where the ionization events are widely separated, the value of G_{fi} at room temperature correlates with the electron mobility, μ_D [1,10].

Fig. 1 illustrates this dependence for representative liquids. Typically, G_{fi} and μ_D are low for n-alkanes, n-alkenes, and aromatics and high for branched compounds with many methyl groups like neopentane. For $\mu_D > 0.1$ cm^2/Vs, the yields for alkanes follow reasonably well a relationship suggested by Jay-Gerin [10]:

$$G_{fi} = a(\mu_D)^x \tag{2}$$

Implicit in such a dependence is the recognition that scattering lengths of thermal and epithermal electrons are similar. A least-squares fit of the data for compounds containing only hydrogen and carbon leads to the solid line shown in Fig. 1 for which $a = 0.25$ and $x = 0.33$, for μ_D in cm^2/Vs.

Fig. 1 includes points for ethane and propane at 298 K for which $G_{fi} = 0.94$ and 0.43 ions/100 eV, respectively [16]. The electron mobilities for these liquids at this temperature are quite high. At even higher temperatures, in the supercritical fluid, G_{fi} and μ_D for ethane and propane are still higher [20,21]. On the other hand, at low temperatures near the boiling points, the yields of free ions as well as the mobilities in these liquids are quite low ($G_{fi} = 0.16$ for ethane for example), and points for these liquids then would be on the left side of the figure.

The data for compounds containing a silicon or germanium atom fall on a lower line (dashed) in Fig. 1 for which $a = 0.12$ and $x = 0.34$ in Eq. (2). That is, for any given mobility value, compounds containing a heavier atom have a lower free ion yield than that given by the line for alkanes [15]. A lower G_{fi} means a shorter mean thermalization range, suggesting that scattering of epithermal electrons is stronger when silicon or germanium

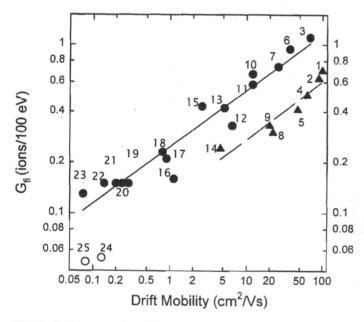

Figure 1 Log–log plot of free ion yield vs. electron mobility in nonpolar liquids. (From Refs. 2 and 15–19.) Numbers refer to liquids listed in Table 2.

atoms are present. The points for benzene and toluene are also below the line for alkanes. While Eq. (2) is a rough prediction of how G_{fi} changes with mobility, it fails in some cases. For example, when pressure is applied to 2,2-dimethylbutane and 2,2,4-trimethylpentane, μ_D increases yet G_{fi} decreases [22]. However, for n-pentane and TMS, Eq. (2) predicts the changes in G_{fi} with pressure quite well.

Free ion yields generally increase with increasing temperature, indicating that the mean thermalization distance, b, increases. Since the density, d, decreases with increasing temperature, several authors have examined how the product, bd, changes. In the case of n-alkanes and benzene, this product is fairly constant over a large temperature range [17]. This same study also found that at 296 K, bd is almost the same for the alkanes from C_4 to C_{14}. For the pressure study mentioned above, free ion yields were found to decrease with increasing pressure, yet the product, bd, remained quite constant for all six liquids studied [22].

2.2. G_{fi} for X-rays

Yields of free ions for exposure to x-rays are less than for high-energy electrons. Interaction of x-rays with nonpolar liquids occurs largely by the photoelectric effect, with Compton scattering becoming important as the photon energy increases. Both events release electrons. The photoelectron energy is given by the x-ray energy less the binding energy, which for carbon is 284 eV. For hydrocarbons, photoelectrons from the k-shell of carbon will dominate.

Results for x-rays are shown in Fig. 2. G_{fi} has been measured for three liquids for x-rays of 1.6- to 30-keV energy [23–25]. The figure also includes points for minimum ionizing

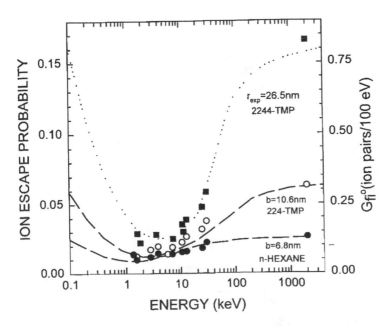

Figure 2   Free ion yields for n-hexane, 2,2,4-trimethylpentane, and 2,2,4,4-tetramethypentane as a function of x-ray energy. Points are experimental [23–25]. Dashed lines are theoretical [26]; dotted line is theory [27]. Reproduced by permission [28].

radiation at 2 MeV. There are very large changes in yield with energy. For 2,2,4,4-tetramethypentane, G_{fi} changes from 0.83 to 0.12 ions/100 eV. The probability of escape is only 3% in the 2–5 keV range.

The lower yields for x-rays come about as a result of the high rate of energy loss of the photoelectrons resulting in a high density of ionizations along the track of the electron. This rate of energy loss depends on the energy of the photoelectrons. For a 30-keV electron, ionizations occur on average 24 nm apart, while for a 2-keV electron, ionizations occur 2.9 nm apart. Thus there is overlap of ionizations because the thermalization ranges of the secondary electrons are comparable to the distance between ionizations. A secondary electron will sense several positive ions along the track, thus increasing the probability of recombination in the track.

Theoretical calculations of the free ion yields have been made as a function of the electron energy. These calculations start with a given track structure for the position of the positive ions and a distribution of distances for the secondary electrons. The charges are then allowed to diffuse in the electric field due to all the other charges. Electrons either recombine with positive ions or escape. The lines in Fig. 2 were calculated in this way. The dashed lines are for n-hexane and 2,2,4-trimethylpentane. In this case, a Gaussian distribution of electron separation distances was assumed, with the mean thermalization ranges, b, shown on the figure [26,29]. For the dotted line, an exponential distribution with an average thermalization range of 26.5 nm was assumed [27]. The computer simulations agree quite well with the experimental measurements.

2.3. G_{fi} for Alpha Particles

The few studies that have been made indicate that the free ion yield for exposure of liquids to alpha particles is quite small. For hydrocarbons, G_{fi} is very small, 0.005 per 100 eV [30,31]. Theoretically, a zero yield is expected for cylindrical geometry and alpha particles create such a track. The low yields in hydrocarbons can be attributed to those electrons on the tail of the distribution that thermalize some distance from the track; these are often called delta rays. For liquid rare gases, the yields are higher; for example, the zero field yield is 0.16 per 100 eV for Xe [32] because the thermalization ranges are much longer.

3. ENERGY OF THE QUASI-FREE ELECTRON

The existence of a band of states in which the electron is quite mobile and its wave function is extended is common to all nonpolar liquids. The energy of the lowest state in this band relative to vacuum is designated V_0. Values of V_0 for nonpolar molecular liquids range from $+0.2$ to -0.75 eV at room temperature [2]. Some representative values are given in Table 1. When V_0 is low in energy, there is usually little trapping and the electron mobility is high. Conversely, when V_0 is high, the trap state is likely to be favored and the mobility is low. When V_0 is positive, the electron favors the vacuum over the liquid energetically. Emission of electrons into the vacuum occurs readily for liquids for which V_0 is positive like n-hexane ($V_0 = 0.1$ eV) [35,36]. For liquid helium, V_0 is $+1.3$ eV and the electron resides in a bubble of radius 1.4 nm [37].

The energy of the V_0 state is usually considered to be the difference between two terms: an attractive polarization energy, U_P, and a kinetic energy term, T_0. Conceptually, this energy is like the energy of the ground state of an electron in a potential well, where the walls of the well are the impenetrable hard sphere surfaces of the molecules of the fluid.

Table 1 Energy Levels of Electrons at 298 K

Solvent	V_0 (eV)	$\Delta G_{soln}(e)$
n-Hexane	0.10	−0.33
Cyclohexane	0.01	−0.28
n-Pentane	0.01	−0.32
3-Methylpentane	0	−0.33
2-Methylbutane	–	−0.34
2,2,4-Trimethylpentane	−0.17	−0.40
2,2-Dimethylbutane	−0.26	−0.43
2,2-Dimethylpropane	−0.38	−0.44
Tetramethylsilane	−0.62	−0.62
Bis(trimethylsilyl)methane	−0.66	−0.66
Hexamethyldisiloxane	−0.70	−0.70

Source: Refs. 2, 33, and 34.

When these walls come closer, the energy T_0 will increase, which explains why V_0 increases with density for liquids. For example, V_0 increases with decreasing temperature for various hydrocarbons [38]. When pressures of 2.5 kbar are applied to hydrocarbons, increases in V_0 of a few tenths of an electron volt are observed [39]. Various theories have been used to calculate V_0 (see below).

Several methods have been employed to measure the energy of this state in nonpolar liquids. The methods fall into three categories: the change in work function of a metal when immersed in the liquid, photoionization, and field ionization. Of these, the latter, in which field ionization of high-lying Rydberg states is utilized to locate V_0, has in recent years provided what are considered to be the most accurate values of V_0 in fluid Ar, Kr, and Xe.

In the photoelectric method, V_0 is obtained as the change in work function, ϕ, of a metal when immersed in the liquid. Thus V_0 is given by the difference in work functions:

$$V_0 = \phi_{liq} - \phi_{vac} \tag{3}$$

This method was first used to determine V_0 in liquid Ar [40] and was later applied to liquid hydrocarbons [38,41–43] as well as to supercritical hydrocarbons like ethane [44–46]. Most of the data available on V_0 for nonpolar liquids were obtained by this method.

In a variation of the photoelectric method, V_0 can be determined by measuring the emission of electrons from the liquid into the vacuum. Even when V_0 is negative, electrons can penetrate this barrier and be collected in the gas phase [35,36,47]. Borghesani et al. [48] used this technique and from the time evolution of the current reaching the anode for a sample of liquid Ar at 87 K found V_0 to be −0.126 eV. This is in excellent agreement with the value of −0.125 eV given by Eq. (6) (see below) for this density (2.09×10^{22} cm^{-3}) using the field ionization technique.

When molecules are photoionized in a liquid, there is a lowering of the ionization threshold, E_{th}, due to both the sudden polarization of the liquid by the ion E_{pol}^+ and V_0. Thus:

$$E_{th} = IP + E_{pol}^+ + V_0 \tag{4}$$

A study of the photoionization of tetramethyl-*p*-phenylenediamine (TMPD) in solution [49] showed the dependence of the wavelength of ionization onset on the energy V_0 of the solvent. The data were used to evaluate V_0 for 18 different liquids. Typical results of this and other studies [50,51] are given in Table 1. Direct single-photon ionization of solvent molecules is also sensitive to the value of V_0 [52–55] and can, in principle, be used to determine V_0.

Recently, laser multiphoton ionization of solutes has been used. Defining the threshold of ionization, E_{th}, can be a problem in some of these methods. A recent multiphoton technique, utilizing femtosecond laser pulses, appears to give quite accurate thresholds [56]. In this work, a conductivity spectrum is measured at visible wavelengths and a sharp drop in current occurs as the mechanism changes from n-photon excitation to $(n + 1)$-photon excitation, where n is typically 3 to 4. The threshold is defined by fitting the current to an analytic function that defines the midpoint of this transition. E_{th} is then n times the energy at which the midpoint occurs. The thresholds are sensitive to V_0 and could be used for determination of this quantity.

Photodetachment from anions should also be mentioned [57–59]. In this case, the threshold, E'_{th}, given by:

$$E'_{th} = EA + V_0 - E^-_{pol} \tag{5}$$

is well defined since the electron escapes with high probability from the neutral molecule left behind, and the photodetachment yields as a function of photon energy follow a known power law. Thus V_0 values can be determined from such studies if values of the electron affinity, EA, are available.

Recently, V_0 has been measured in dense rare gas fluids by field ionization of Rydberg states of solutes lying close to the continuum. In this technique, Reininger et al. [60–63] utilized synchrotron radiation and measured photocurrent spectra in the vacuum ultraviolet (VUV) with a resolution of 7 meV. The photocurrent spectra were measured at several applied electric fields. A field ionization spectrum was obtained by subtracting the spectrum at low field from one obtained at high field. Such spectra typically showed one peak at threshold energies due to those high-lying states that ionized at the high voltage but not at the low voltage. Measurements were made in argon containing CH_3I [60] and H_2S [61] as solutes, and the density was varied from dilute gas to the triple point of argon. The position of the field ionization peak shifts as the density is changed. These shifts are the combined effect of changes in the ion-media polarization energy and V_0. Accurate calculation of the former allowed the evaluation of V_0 as a function of density, utilizing Eq. (4) above. The resulting values of V_0 are in good agreement with earlier photoelectric and theoretical values. Furthermore, the two solutes CH_3I and H_2S gave very comparable results. For argon, V_0 reaches a minimum value of -0.294 eV at a density of 12×10^{21} cm^{-3}, which is the density at which the electron mobility is a maximum.

The solid line in Fig. 3 is a best fit of Eq. (6) to the experimental data [61]. See cited reference for values of the parameters in Eq. (6).

$$V_0(N) = a_0 + a_1(N - a_2) + (a_3/a_4)\ln \cosh(a_4(N - a_2)) \tag{6}$$

The density dependence of V_0 in Kr was determined by field ionization of CH_3I [62] and $(CH_3)_2S$ [63]. Whereas previous studies found a minimum in V_0 at a density of 12×10^{21} cm^{-3} [66], the new study indicates that the minimum is at 14.4×10^{21} cm^{-3} (see Fig. 3). This is very close to the density of 14.1×10^{21} cm^{-3} at which the electron mobility reaches a maximum in krypton [67], a result that is consistent with the deformation potential model [68] which predicts the mobility maximum to occur at a density where V_0 is a minimum. The use of $(CH_3)_2S$ permitted similar measurements of V_0 in Xe because of its lower ionization potential. The results for Xe are also shown in Fig. 3 by the lower line,

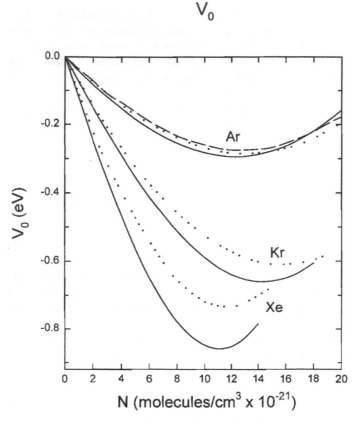

Figure 3 Energy of the quasi-free state in rare gas fluids as a function of density. Solid lines represent recent results obtained by field ionization for Ar [61], Kr, and Xe [see Eq. (6)] [63]. Recent theoretical calculations are shown for Ar by the dashed lines [64] and for Ar, Kr, and Xe by the dotted lines [65].

which represents the analytical equation (6) giving the best fit to the data. Parameter values, a_n, can be found in the reference cited.

The earliest theoretical model of the effect of a medium on the quasi-free electron energy was proposed by Fermi [69] to explain ionization of high Rydberg states. This model, which predicts a linear dependence on density, works well only at low densities. Springett et al. [37] introduced the Wigner–Seitz model to calculate V_0 in several liquids employing a simple electron–atom pseudopotential. This model was used by Stampfli and Bennemann to calculate V_0 for liquid Ar, Kr, and Xe, except they solved the eigenvalue problem numerically [64]. Their results for Ar, shown in Fig. 3 by the dashed line, are in good agreement with the experiment. Several other modifications to the theoretical treatments have been proposed [65,70–73] using various techniques. Simon et al. [74,75] used a classical percolation approach to predict V_0 for Ar, Kr, and Xe. The calculations of Plenkiewicz et al. [65], which use an accurate electron–atom pseudopotential, are shown by the dotted lines in Fig. 3. Their calculations also agree quite well with the experimental data and find the minima at the right densities. These newer studies are readily applicable to molecular liquids and have been applied to CH_4 and SiH_4 [76].

4. SOLVATED ELECTRONS

Evidence for the existence of a trapped state of the electron in some liquids comes from several sources. For one, pulse radiolysis studies have shown that for certain alkanes, there is a broad absorption in the infrared that can be attributed to the electron. Second, studies of the effect of pressure on the mobility of electrons have revealed information about cavity sizes and the role of electrostriction in the trapping process. Finally, conductivity studies of reversible reactions of the electron have given us the energetics of the trapping process. These latter equilibrium studies are described in detail in Sec. 5.1. Values of $\Delta G_{soln}(e)$ for selected liquids are given in Table 1. For many liquids, $\Delta G_{soln}(e)$ lies below the V_0 state making trapping energetically favorable.

Early pulse radiolysis studies of alkanes at room temperature showed that the solvated electron absorption begins around 1 μm and increases with increasing wavelength to 1.6 μm for *n*-hexane, cyclohexane, and 2-methylbutane [77]. More complete spectra for three liquid alkanes are shown in Fig. 4. The spectrum for methylcyclohexane at 295 K extends to 4 μm and shows a peak at 3.25 μm [78]. At the maximum, the extinction coefficient is 2.8×10^4 M^{-1} cm^{-1}. The spectrum for 3-methyloctane at 127 K, shown in Fig. 4, peaks around 2 μm. The peak for methylcyclohexane is also at 2 μm at lower temperature. Recently, the absorption spectra of solvated electrons in 2-methylpentane, 3-methylpentane, *cis*-decalin, and methylcyclohexane glasses have been measured accurately at 77 K [80]. For these alkanes, the maxima occur at 1.8 μm, where the extinction coefficient is 2.7×10^4 M^{-1} cm^{-1}.

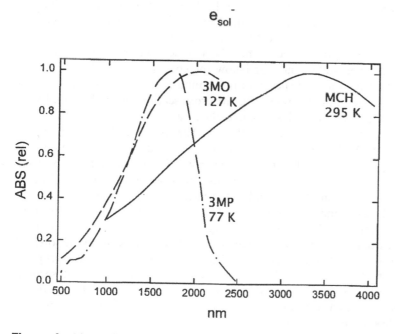

Figure 4 Absorption spectra of solvated electrons in alkanes vs. wavelength. Solid line is for methylcyclohexane at 295 K [78]. Dashed line is for 3-methyloctane at 127 K [79]. Dash–dot line is for 3-methylpentane at 77 K [80]. Spectra have been normalized to unity at the peaks.

The stronger absorption at lower temperatures can be attributed to several factors. One is that the equilibrium between quasi-free and trapped electrons shifts to favor trapped electrons as the temperature is lowered for these liquids. Another is that homogeneous recombination of electrons with positive ions is slower at the lower temperatures and therefore occurs to a lesser extent during the pulse, which, for the pulse radiolysis studies, was typically 10 to 20 nsec. The rate constant, k_r, for electron recombination with positive ions, in most nonpolar liquids, is given by:

$$k_r = \frac{4\pi e \mu_D}{\varepsilon} = 1.09 \times 10^{15}(\mu_e/\varepsilon) \text{ M}^{-1} \text{ sec}^{-1} \qquad \text{(for } \mu_e \text{ in cm}^2 \text{ Vs)} \qquad (7)$$

This equation is sufficiently valid that it is considered to be a law of electron dynamics. Exceptions exist only for very high values of μ_D (see Chap. 10).

It is important to consider the magnitude of the recombination rate in studies of this type. For methane, k_r is $1.7 \times 10^{17} \text{ M}^{-1} \text{ sec}^{-1}$ at 93 K [81]. Thus if a concentration of ions of 0.1 μM was formed in the pulse, the electrons would disappear with a first half-life of 50 psec. For 2,2,4-trimethylpentane, k_r is $3.6 \times 10^{15} \text{ M}^{-1} \text{ sec}^{-1}$ and for a similar concentration of electrons, the recombination lifetime would be a few nanoseconds. Where the electron mobility is lower, the recombination rate is slower. For methylcyclohexane, where $\mu_e = 0.07 \text{ cm}^2/\text{Vs}$ [18], k_r is $4 \times 10^{13} \text{ M}^{-1} \text{ sec}^{-1}$ and the first half-life should be about 250 nsec.

Electrons have not been detected by optical absorption in alkanes in which the mobility is greater than 10 cm²/Vs. For example, Gillis et al. [82] report seeing no infrared absorption in pulse-irradiated liquid methane at 93 K. This is not surprising since the electron mobility in methane is 500 cm²/Vs [81] and trapping does not occur. Geminately recombining electrons have, however, been detected by IR absorption in 2,2,4-trimethylpentane in a subpicosecond laser pulse experiment [83]. The drift mobility in this alkane is 6.5 cm²/Vs, and the quasi-free mobility, as measured by the Hall mobility, is 22 cm²/Vs (see Sec. 6). Thus the electron is trapped two-thirds of the time.

The nature of the absorption spectra has been discussed by several authors [78,80,84,85]. Since the trapped state is not far below the conduction band, it is at least reasonable to consider that the infrared spectrum is a bound-free transition. In the study of alkane glasses at 77 K, mentioned earlier [80], the cross sections near threshold were found to fit the Wigner [86] power law. This supports the idea that the spectra are due to a transition from a bound S-state to a continuum P-state. The threshold binding energy at 77 K was found to be 0.48 eV. From an analysis of the spectral distribution, the authors obtained ground state properties of the trapped electron. The experimental spectrum matched that derived from a simple spherical well model, suggesting that the electron resides in a cavity of radius 0.35 nm. Other studies have arrived at a similar value for the radius. For example, Ichikawa and Yoshida [84] obtained 0.36 nm for 3-methylpentane glass based on a model that included a short-range repulsive contribution and an attractive, Born-like, medium polarization. An electron spin resonance study of solvated electrons in deuterated 3-methylpentane glasses indicated that there were an average of three molecules in the first solvation shell. A combination of spin-echo and second-moment analysis gave electron to proton distances from 0.35 to 0.43 nm [87]. These studies have provided considerable insight into the nature of the trapped state and are consistent with the transport studies discussed below.

Studies of the effect of pressure on μ_D in a series of 10 hydrocarbons revealed more information about the trapped state [34]. The pressure data led to the conclusion that the partial molar volume of the electron, $\overline{V}_e$, is small but may be either positive or negative.

Values range from $+22$ cm^3/mol for m-xylene to -27 cm^3/mol for 1-pentene. This depends on the relative magnitude of two large volume terms that make up $\bar{V}_e$. One is the cavity volume, a positive term, and the other is the electrostriction of the solvent around the trapped electron. Whereas the electrostriction term varies considerably depending on the compressibility of the solvent, the cavity volume does not change much and the average value for the hydrocarbons is 96 cm^3/mol, corresponding to a cavity radius of 0.34 nm.

Thus from pulse radiolysis, mobility measurements, and electron reaction studies, we have information on the absorption spectra, the cavity volume, and the energy of the trapped or solvated state. The nature of this state seems to be an electron that is localized in a cavity in the liquid.

5. REACTIONS OF ELECTRONS

Electron attachment to solutes in nonpolar liquids has been studied by such techniques as pulse radiolysis, pulse conductivity, microwave absorption, and flash (laser) photolysis. A considerable amount of data is now available on how rates depend on temperature, pressure, and other factors. Although further work is needed, some recent experimental and theoretical studies have provided new insight into the mechanism of these reactions. To begin, we consider those reactions that show reversible attachment–detachment equilibria and therefore provide both free energy and volume change information.

5.1. Electron Equilibria

Electron equilibria of the type:

$$e^- + \text{solute} \underset{k_d}{\overset{k_a}{\rightleftharpoons}} \text{solute}^- \tag{8}$$

have been observed for solutes that do not dissociate on attachment and have gas phase electron affinities (EA) between -1.15 eV (benzene) [88,89] and 0.3 eV (phenanthrene) [90]. Application of high pressure facilitated the studies for solutes of very negative electron affinities like toluene, benzene, and butadiene. Equilibrium constants, K_{eq}, have been evaluated by conductivity methods from changes in electron mobility and from determination of both the attachment rate constants, k_a, and the detachment rate constants, k_d. Values of ΔG_r for various solutes are shown in Fig. 5 for tetramethylsilane, 2,2,4-trimethylpentane, n-hexane, and supercritical ethane as solvents. It has been shown that ΔG_r depends on the polarization energy of the product anion, $E_{pol}(S^-)$, and the solvation energy of the electron $\Delta G_s(e)$ according to:

$$\Delta G_r(l) = \Delta G_r^o(g) + E_{pol}(S^-) - \Delta G_s(e) \tag{9}$$

where $\Delta G_r^o(g)$ is derived from the electron affinity $\Delta G_r^o(g) = -\text{EA} - T\Delta S_r(g)$. For all the reactions, ΔG_r^o is lower in 2,2,4-trimethylpentane than in tetramethylpentane (TMS) partly because $\Delta G_s(e)$ is higher in the former solvent and also because the dielectric constant is higher in 2,2,4-trimethypentane, which makes the value of the solvation energy lower. The further lowering of ΔG_r for these reactions in n-hexane is due largely to the even higher value of $\Delta G_s(e)$ in this solvent.

Equilibria of this type have been used to determine the ground state of the electron in liquids [100]. Some of the values of $\Delta G_s(e)$ given in Table 1 were evaluated this way. The reaction of the electron with CO_2 was used to measure $\Delta G_s(e)$ for hexamethyldisiloxane

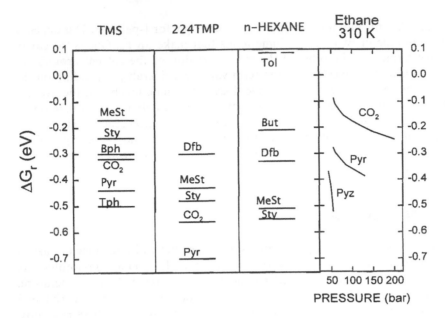

Figure 5 Free energies for attachment to solutes: MeSt—methylstyrene, Sty—styrene, Bph—biphenyl, CO_2, Pyr—pyrimidine, Tph—triphenylene, Dfb—p-difluorobenzene, Tol—toluene, But—1,3-butadiene, Pyz—pyrazine; in TMS, 2,2,4-trimethylpentane, and n-hexane at 298 K and in supercritical ethane at 310 K. (From Refs. 90–99.)

and bis(trimethylsilyl)methane. The electron mobility in these liquids is unusually high, and V_0 is therefore expected to be the same as $\Delta G_s(e)$. For hexamethyldisiloxane, $\Delta G_s(e)$ was found to be −0.70 eV and that for bis(trimethylsilyl)methane, it was found to be −0.66 eV (see Table 1).

These equilibrium reactions occur with large decreases in both volume and entropy. Volume changes range from −80 to −300 cm^3/mol depending on the solute and pressure. These volume changes, ΔV_r, are associated with the electrostriction of the solvent around the product anion, $V_{el}(ion)$, and, to some extent, with a contribution of the partial molar volume of the electron, $\overline{V}(e)$. Thus:

$$\Delta V_r = V_{el}(ion) - \overline{V}(e) \tag{10}$$

Values of $\overline{V}(e)$ are small compared to the overall volume changes (see Sec. 4). However, magnitudes of ΔV_r were found to be much larger than could be accounted for using the classical expression for $V_{el}(ion)$ [101]:

$$V_{el}(ion) = -\left(e^2/2R_{ion}\right)\left(1/\varepsilon^2\right)(d\varepsilon/dP) \tag{11}$$

To account for the difference between experiment and theory, Schwarz suggested that electrostriction includes the formation of a glassy shell of 7 to 9 solvent molecules around the ion [102]. This phase transition provides a substantial density increase, and the observed volume changes can then be accounted for.

These attachment–detachment equilibria [Eq. (8)] shift to the right with increasing pressure and to the left with increasing temperature. Thus the free energies decrease with pressure and increase with temperature. These effects are related to the solvent compressibility, which increases with pressure and decreases with temperature. The entropy contribution is less at high pressure but increases with temperature. Because both entropy and volume changes are due to electrostriction of the solvent, these quantities are related by:

$$\Delta S_{el}(\text{ion}) = (\alpha/\chi_T)\Delta V_{el}(\text{ion}) \tag{12}$$

where α is the coefficient of thermal expansion and χ_T is the isothermal compressibility.

Electron attachment equilibria have also been observed in supercritical ethane. The equilibrium constants are generally smaller than in liquids and consequently ΔG_r is higher. Thus, as shown in Fig. 5, ΔG_r for attachment to CO_2 is higher in ethane at all pressures than in either TMS or 2,2,4-trimethylpentane liquids. This is mainly a consequence of the lower density of the solvent that results in less polarization energy for the anion. However, the continuum model of Born is no longer correct for evaluating the polarization energy because of the significant density augmentation around the ions in supercritical nonpolar fluids. Instead, a compressible continuum model that takes this effect into account must be used [99,103]. In the case of a supercritical solvent like ethane, the free energy of attachment to a solute changes rapidly with pressure, particularly near critical densities where the compressibility of the fluid changes rapidly. Thus, as the pressure increases, the values of ΔG_r for attachment to CO_2, pyrimidine, and pyrazine decrease. The equilibria in supercritical ethane could only be observed over the pressure ranges indicated in the figure for these solutes.

The volume changes, ΔV_r, for attachment to solutes in supercritical ethane can be extremely large and they are always negative. For pyrazine as a solute, values of ΔV_r range from -1.0 to -45 L/mol, depending on temperature and pressure. The largest changes are found at the densities where the compressibility is the largest at each temperature. These changes are mostly due to electrostriction of the solvent around the ion formed and are predicted quite well by the compressible continuum model. Generally, the density augmentation around an ion in a supercritical fluid extends to a radial distance of 1 nm.

5.2. Attachment Rates

In nonpolar liquids, bimolecular electron attachment rate constants, k_a, are much larger than those for conventional reactions of ions or radicals. This is, in part, related to the high mobility of electrons in these liquids; but various other factors, like V_0, the kinetic energy of the electron, and dipole moment of the solute, are important as well. These and other factors are examined below; the dependence of k_a on the energy gap, ΔG_r, in representative liquids is also shown and discussed.

It is natural to conclude that the high rate constants for electron attachment reactions in nonpolar liquids are associated with the high mobility of electrons. Early studies [96,104,105] of attachment to biphenyl and SF_6 emphasized the dependence of k_a on mobility. This relationship is apparent if the expression for the rate constant for a diffusion-controlled reaction:

$$k_D = 4\pi R_e D_e \tag{13}$$

is combined with the Einstein relation:

$$D_e = \mu_D k_B T/e \tag{14}$$

then:

$$k_D = 4\pi R_e \mu_D k_B T/e \tag{15}$$

In the solvents ethane, propane, and hexane, a reaction radius, R_e, of 1.4 nm was found for attachment to SF_6. This is close to the theoretical maximum radius for electron attachment in the gas phase [106].

Modern theory suggests that rates of electron transfer reactions depend on the potential energy difference [107]. For electron attachment reactions in solution, that difference is given by the difference in energy of reactants, in this case, the electron plus solute, and the product anion [see Eq. (9)]. Since this theory works so well for electron transfer, it is interesting to examine the dependence of rates on $\Delta G_r(l)$ for electron attachment reactions; such a plot for cyclohexane as solvent is shown by Fig. 6. For one of the solutes, difluorobenzene, $\Delta G_r(l)$ is known from equilibrium studies. For the other solutes, $\Delta G_r(l)$ was calculated using Eq. (9). Values of rate constants and electron affinities are from references given in the figure legend. The continuum model of Born:

$$E_{\text{pol}}(S^-) = -\frac{e^2}{2R_S}\left(1 - \frac{1}{\varepsilon}\right) \tag{16}$$

was used to calculate $E_{\text{pol}}(S^-)$, where the radius of the anion, R_S, was calculated from the molar volume of the solute. Many of the rates for cyclohexane are close to 3×10^{12} M^{-1} sec^{-1} (dotted line in Fig. 6). Since the room temperature electron mobility in cyclohexane is 0.24 cm^2/Vs [19], this rate corresponds to a reaction radius of 0.7 nm, assuming that Eq. (15) applies. The lack of dependence on the exothermicity, $\Delta G_r(l)$, for most of the

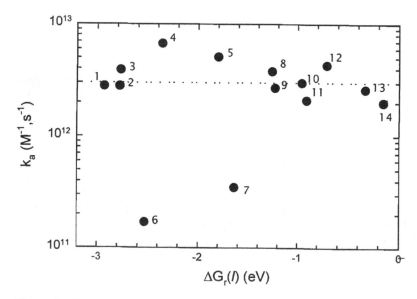

Figure 6 Rate constants for electron attachment to solutes in cyclohexane at 295 K. Solutes are: 1—CCl_4, 2—p-dinitrobenzene, 3—benzoquinone, 4—o-dinitrobenzene, 5—nitrobenzene, 6—O_2, 7—perfluoromethylcyclohexane, 8—pyrene, 9—anthracene, 10—biphenyl, 11—naphthalene, 12—CO_2, 13—p-difluorobenzene, 14—ethylbromide. Dotted line indicates calculated diffusion rate. References for rate data: [19,108–111]; references for electron affinities: [112–115].

reactions included in Fig 6 indicates that the rate is determined by the rate of diffusion of the electron to the solute.

A plot of the attachment rate in cyclohexane vs. simply the electron affinity of the solute, shown in a paper by Christophorou [116], is similar to Fig. 6 in appearance. Christophorou concluded that for solutes with EA > 0, k_a is close to the diffusion-controlled rate. The similarity in plots is not surprising since for many of the larger solute molecules, the polarization energy is nearly the same and thus changes in $\Delta G_r(l)$ are proportional to changes in EA. Christophorou suggests that the rate constant drops off to very low values when EA < −0.9 eV.

Some reactants have rate constants higher than 3×10^{12} M^{-1} sec^{-1}; examples are nitrobenzene and o-dinitrobenzene. These two compounds have large dipole moments of 4.1 and 6.1 Debye, respectively, and it has been shown [110] that the rate constants in cyclohexane increase with dipole moment because the reaction radius increases. That dependence is given by [117]:

$$R_e^{-1} = \int_{r_c}^{\infty} \frac{\exp[U(r)/k_B T]}{r^2} \, dr \tag{17}$$

where $U(r)$ is a function of the dipole moment and r_c is the hard-core radius of the reactant pair. Diffusion rates calculated with Eq. (15) agree with the experimental values for a series of nitrobenzenes when Eq. (17) is used to calculate R_e [110].

Many of these attachment reactions are also diffusion-controlled in other solvents of low electron mobility like, for example, n-hexane. It has been suggested that this is the case for all solvents for which $\mu_D < 1$ cm^2/Vs [118]. For this to be true, the rate constant k_a should scale as the mobility. For hexane, the rate constants for attachment to solutes like biphenyl, naphthalene, and difluorobenzene are close to 1×10^{12} M^{-1} sec^{-1} or one-third the value in cyclohexane. The mobility in n-hexane is approximately one-third that in cyclohexane [2]; thus k_a scales with μ_D for these two solvents.

An interesting example of a diffusion-controlled reaction is electron attachment to SF$_6$. Early studies showed that in n-alkanes, k_a increases linearly with μ_e over a wide range of mobilities from 10^{-3} to 1 cm^2/Vs [119]. Another study of the effect of electric field (E) showed that in ethane and propane, k_a is independent of E up to approximately 90 kV/cm, but increases at higher fields [105]. This field is also the onset of the supralinear field dependence of the electron mobility [120]. Thus over a wide range of temperature and electric field, the rate of attachment to SF$_6$ remains linearly dependent on the mobility of the electron, μ_D, as required by Eq. (15).

Diffusion is not always rate determining of course. Exceptions shown in Fig. 6 are oxygen and perfluoromethylcyclohexane for which the rate constants are below the diffusion rate; these are considered in more detail below. The dependence of k_a on electron mobility also breaks down if the mobility is >10 cm^2/Vs. This was noted early for the reaction of the electron with biphenyl [104,121].

Many attachment reactions have been studied in 2,2,4-trimethylpentane, a liquid for which μ_D is 6.6 cm^2/Vs at room temperature. Attachment rate constants for many solutes in 2,2,4-trimethylpentane are shown in Fig. 7 plotted vs. $\Delta G_r(l)$. The dotted line shows the diffusion rate for the radius of 0.72 nm, derived for cyclohexane. In 2,2,4-trimethylpentane, only a few solutes, like SF$_6$, C$_6$F$_6$, and the metal carbonyls, come close to the diffusion rate.

The attachment rate constants for the solvent tetramethylsilane (TMS), in which the electron mobility is 100 cm^2/Vs, are shown in Fig. 8. The solutes include those listed in a

2,2,4-TRIMETHYLPENTANE

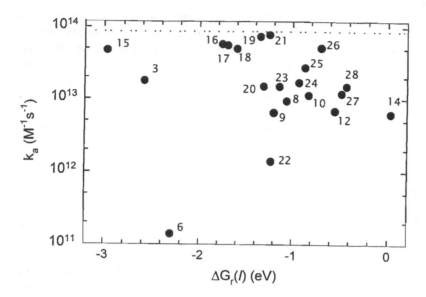

Figure 7 Rate constants for electron attachment to solutes in 2,2,4-trimethylpentane at 295 K. For solutes numbered 1–14, see legend to Fig. 6. Other solutes: 15—C_{60}, 16—SF_6, 17—C_6F_6, 18—$W(CO)_6$, 19—$Cr(CO)_6$, 20—perylene, 21—$Mo(CO)_6$, 22—t-stilbene, 23—benzperylene, 24—coronene, 25—pyrazine, 26—pyrimidine, 27—styrene, 28—α-methylstyrene. Dotted line is calculated diffusion rate. References for rate data: [19,58,108,109,122–124]; references for electron affinities: [112–115].

recent compilation by Nishikawa [2] plus pyrimidine and C_{60} from recent studies [93,122]. The values of k_a range over 4 orders of magnitude; nevertheless, some conclusions become apparent. For large energy gaps, k_a is close to 10^{14} M^{-1} sec^{-1}, with the exception of benzoquinone, duroquinone, and O_2, which are special cases (see below). The diffusion rate, shown by the dotted line, was calculated using the value of $R_e = 0.72$ nm suggested for cyclohexane. None of these reactions occurs at the diffusion rate in TMS. For moderate energy gaps ($\Delta G_r = -0.7\pm0.3$ eV), k_a tends to be around 10^{13} M^{-1} sec^{-1}. When ΔG_r is close to zero, much lower rate constants are observed.

The value of k_a for attachment to SF_6 given in Fig. 8 is 2.1×10^{14} M^{-1} sec^{-1}. This reaction has also been studied in other high mobility liquids including methane ($\mu_D = 400$ cm^2/Vs), argon ($\mu_D = 400$ cm^2/Vs), and xenon ($\mu_D = 2000$ cm^2/Vs) [127–129], and the rate constant is nearly constant at $3\pm1 \times 10^{14}$ M^{-1} sec^{-1}. This has been explained by Warman [106] and others as due to the fact that the residence time, τ_D, of an electron within a reaction radius, R_e, is short compared to the attachment time, τ_A. Thus rate constants would be expected to fall off with increasing mobility according to the equation:

$$k_a = \frac{4\pi R_e D}{1 + \tau_A/\tau_D} \tag{18}$$

At high-enough mobility, since τ_D is $R_e^2/2D$, k_a reaches a constant value of $2\pi R_e^3/\tau_A \cong 3 \times 10^{14}$ M^{-1} sec^{-1}, which explains the near constant rate for attachment to SF_6 observed in liquids of high mobility.

TETRAMETHYLSILANE

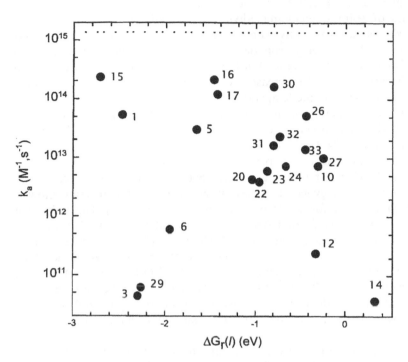

Figure 8 Rate constants for electron attachment to solutes in tetramethylsilane at 295 K. For solutes numbered 1–28, see legends to Figs. 6 and 7. Other solutes: 29—duroquinone, 30—CH$_3$I, 31—cycloC$_4$F$_4$, 32—C$_2$HCl$_3$, 33—phenanthrene. Dotted line is calculated diffusion rate. References for rate data: [18,19,58,90,93,122,123,125,126]; references for electron affinities: [112–115].

Figs. 7 and 8 show that the rate constants are mostly below the diffusion limit. Some rates are a few orders of magnitude lower, and these reactions have been interpreted as dependent on the energy gap. Attachment to O$_2$ is a case in point [123]. Like many of these reactions, this is a nondissociative attachment reaction. As pointed out by Henglein [130,131], attachment to O$_2$ is most favorable when the energy gap is small. As can be seen in Figs. 6–8, ΔG_r is between -2.6 and -2.0 eV for these solvents and thus the reaction is unfavorable at room temperature. Any decrease in the gap would be expected to increase the rate. The positive activation energy of approximately 2 kcal/mol for this reaction in most solvents can be accounted for in this way. As the temperature increases, the density decreases causing the energy level of the electron to decrease and the value of $E_{pol}(O_2^-)$ to increase thus narrowing the energy gap. The effect of increasing pressure on this reaction is of interest since increasing pressure causes the density to increase, causing the energy levels of the electron to increase and the value of $E_{pol}(O_2^-)$ to decrease. Thus the energy gap increases with increasing pressure. Preliminary studies show that the rate of attachment to O$_2$ in n-pentane as solvent decreases a factor of 2 in 1 kbar [132]. The attachment rate to O$_2$ has also been studied in liquid argon and xenon [127]. The rate decreases with increasing field or, since the field increases the average electron energy, the rate decreases as the kinetic energy of the electrons increases. Thus at higher fields, there is more energy available for the reaction and the rate slows down.

There is a class of solutes, including ethylbromide, N_2O, C_2HCl_3, and certain fluoroalkanes, that shows negative activation energies for attachment in liquids like TMS and neopentane [19,123]. For ethylbromide, the rate constant of electron attachment in TMS is 4.2×10^{10} at 23°C. This is point 14 in Fig. 8. Lowering the temperature causes the energy level of the electron, V_0, to increase and the polarization energy of the anion $E_{pol}(S^-)$ to decrease. Thus lowering the temperature increases the energy gap between the electron energy level and the energy of the ion in solution and the rate increases. For neopentane as solvent, the effect is similar except that the electron energy level is higher making the gap larger, and, at 24°C, k_a is 3.4×10^{11} M^{-1} sec^{-1}, an order of magnitude larger than in TMS. The rate reaches a maximum in 2,2,4-trimethylpentane where k_a is 6.3×10^{12} M^{-1} sec^{-1} at 23°C (point 14 in Fig. 7) because the electron level is even higher. Thus the rate increases as the energy gap increases. In the gas phase, this reaction shows a maximum in the attachment rate for an electron energy of 0.76 eV [133]. This value was used for $\Delta G_r(g)$ in calculating $\Delta G_r(l)$. In TMS, $\Delta G_r(l)$ turns out to be $+0.32$ eV, or this vertical attachment reaction is unfavorable. In 2,2,4-TMP, $\Delta G_r(l)$ is close to zero or, in terms of the electron redox level picture of Henglein [134], the occupied donor level (electron in the solvent) matches the unoccupied acceptor level (ethyl bromide). But the rate constant is less in cyclohexane, 2×10^{12} M^{-1} sec^{-1} at 23°C, where the electron level is even higher. This is explained by the fact that the reaction is limited by the diffusion rate, which should be somewhat less than the line in Fig. 6 because of a smaller reaction radius for this solute.

The explanation of the negative activation energies for attachment to N_2O, C_2HCl_3 [19], and perfluorocyclobutane [123] in TMS is similar to that for ethylbromide above. These molecules exhibit maxima for electron attachment in the gas phase at 2.2 eV [135], 0.4 eV [136], and 0.35 eV [137], respectively. Thus rates are expected to increase as the energy gap increases. However, in 2,2,4-TMP and n-hexane, the attachment rates show normal Arrhenius behavior. For these solvents, the ratio k_a/μ_D is constant over a range of temperatures indicating that the reaction becomes diffusion-limited as was the case for ethylbromide in cyclohexane. Other perfluoroalkanes like n-C_5F_{12} and n-C_6F_{14} are reported to show maxima as is seen for cyclo-C_4F_8 [138].

Carbon dioxide reacts in a manner similar to the solutes discussed above in that as the energy gap increases, the rate of attachment increases. However, unlike the other solutes, this reaction is reversible in solution and the equilibrium:

$$e^- + CO_2 \rightleftarrows CO_2^- \tag{19}$$

has been studied in several nonpolar liquids as well as in supercritical ethane. Thus not only the attachment rates, k_a, have been measured, but also values of the free energy of reaction, ΔG_r, are available. The rate constants, k_a, for this reaction, obtained at various temperatures and pressures in several fluids: TMS, 2,2,4-trimethylpentane, dimethylbutane, and supercritical ethane [126,139], are plotted in Fig. 9. In liquids, the rate constant is independent of temperature at any pressure but increases with pressure. The increase in rate with pressure can be explained by the increase in electron energy level [39] and the increase in stability of the ion with pressure that causes the energy gap to increase. As shown in Fig. 9, $\log k_a$ is linearly dependent on ΔG_r and k_a increases about 3 orders of magnitude as the energy gap increases by 0.4 eV.

The reaction of electrons with p-benzoquinone is an unusual case. A conductivity study showed that the reaction is slow at room temperature in TMS and neopentane and

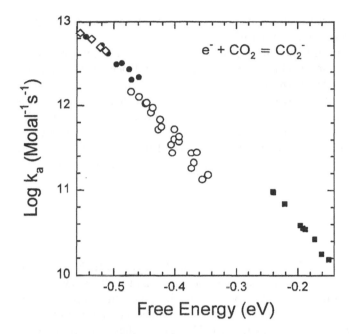

Figure 9 Rate constants for electron attachment to CO_2 vs. the free energy of reaction in different fluids: ◇—2,2,4-trimethylpentane [126], ●—2,2-dimethylbutane [139], ○—TMS [126], ■—super-critical ethane [99].

the activation energies are negative [140]. The energy gaps are large: $\Delta G_r(l)$ is -2.3 eV in TMS and -2.44 eV in neopentane, indicating possible inverted behavior. However, in 2,2,4-TMP, $\Delta G_r(l)$ is even lower, -2.58 eV, and the rate of attachment is fast. In cyclohexane, n-pentane, and n-hexane, where the electron levels are high, the reaction appears to be diffusion-limited. The results were explained by assuming an equilibrium with an excited state of the anion:

$$e^- + \text{benzoquinone} \rightleftarrows \text{benzoquinone}^-* \tag{20}$$

$$\text{benzoquinone}^-* \rightarrow \text{benzoquinone}^- \tag{21}$$

The excited state lies 2.07 eV above the ground state; thus there is sufficient energy available in all solvents to reach this state. The kinetic analysis indicated that the attachment rate is fast in all solvents. In TMS, neopentane, and 2,2,4,4-tetramethylpentane, the energy levels of the electron are low enough that the reverse reaction, autodetachment from the excited anion, occurs with small activation energy.

Recently, the excited state of benzoquinone anion was detected both by absorption and fluorescence [141]. The lifetime of the excited state in 2,2,4-trimethylpentane is ≈ 120 nsec or $k_{21} = 8 \times 10^6$ sec^{-1}. If it is assumed that the lifetime is similar in the other nonpolar solvents, then the results of the two studies may be combined to evaluate the detachment rate from the excited anion. At room temperature, this calculation gives $k_{-20} = 6 \times 10^8$ sec^{-1} in TMS; thus detachment readily competes with deactivation of the excited state of benzoquinone anion. The latter process includes internal conversion as well as fluorescence. In 2,2,4,4-tetramethylpentane, where the activation energy for detachment is 0.41 eV, k_{-20} is 4×10^6 sec^{-1} and some autodetachment occurs along with fluorescence. In solvents like n-pentane, n-hexane, and cyclohexane, the activation energies for detachment are much higher because the electron energy levels are higher, and the reaction proceeds directly to benzoquinone anion at the diffusion rate.

For attachment to CO_2, the rate constant clearly shows a dependence on $\Delta G_r(l)$. However, for attachment to other solutes like aromatics, the dependence on $\Delta G_r(l)$ is not yet clear. Questions remain like what is the role of reorganization energy, which can be large even in nonpolar solvents [142]; it was pointed out earlier that there is considerable density augmentation around the negative ions formed in attachment reactions. For solvents like TMS, rates maximize for values of $\Delta G_r(l)$ near -0.7 eV; what rate is to be expected for large values of $-\Delta G_r(l)$? Is the high rate observed for C_{60} typical or the low rate for quinones more typical? Further study is needed to resolve these questions.

6. TRANSPORT PROPERTIES

6.1. Quasi-Free Mobility

The drift mobility of electrons in nonpolar liquids ranges from high values such as that for liquid xenon of 2000 cm^2/Vs to low values like that for tetradecane of 0.02 cm^2/Vs. It has often been suggested that the mobility is high for symmetrical molecules and low for straight chain molecules like n-alkanes. Inspection of Table 2 shows that liquids with symmetrical molecules are indeed at the top of the list. However, other less symmetrical molecules like *bis*-trimethylsilylmethane and 2,2,4,4-tetramethylpentane also show high drift mobility. A more important factor may be the existence of many methyl groups in the molecule. In any case, for liquids for which $\mu_D > 10$ cm^2/Vs, the electron is considered to be quasi-free. This is supported by the Hall mobility studies, as discussed below.

The mobility of quasi-free electrons has recently been explained by the deformation potential theory. Originally from solid-state physics, this theory was applied by Basak and Cohen [68] to liquid argon. The theory assumes that scattering occurs when the electron encounters a change or fluctuation in the local density which results in a potential change. The potential is assumed to be given in terms of dV_0/dN, d^2V_0/dN^2, etc. The formula they derived for the mobility is:

$$\mu_D = \frac{2e\hbar^4\sqrt{2\pi}(m_e/m^*)^{5/2}}{3m_e^{5/2}(k_BT)^{1/2}k_BN^2\{V_0'^2T\chi_T + V_0''^2\beta/k_B + \ldots\}}, \tag{22}$$

where χ_T is the isothermal compressibility and m^* is the effective mass of the electron. A similar expression was derived by Berlin et al. [144]. Equation (22) was later shown to account for many of the features of the density dependence of the mobility in Ar, Kr, and Xe [66]. Namely, it predicts a minimum in the mobility near the critical density and it correctly predicts a maximum in the mobility at the density at which V_0 is a minimum, i.e., when $dV_0/dN = 0$ (compare Figs. 3 and 10). Scattering in these fluids is weakest at this density. At higher densities, a decrease in mobility is predicted as is observed experimentally. Although the general features of the experimental mobility are reproduced, the predicted values of μ_D around the minimum near N_C are much too small largely because χ_T for the fluids are very large in this region. Arguing that electron–medium interactions are relatively short-ranged, Nishikawa [147] suggested that χ_T be replaced by the adiabatic compressibility. This greatly improved the agreement at these densities for Ar.

Since these papers were published, accurate values of V_0 in argon have been obtained by the field ionization technique [60,61] as described in Sec. 3. Also, the effective mass, m^*,

Table 2 Free Ion Yields and Mobilities at 293–295 K

No.	Liquid	G_{fi} (electrons/100 eV)	μ_D (cm^2/Vs)
1	Tetramethylsilane	0.7	100
2	Tetramethylgermane	0.63	90
3	Neopentane	1.1	69
4	Bis-trimethylsilylmethane	0.5	63
5	Bis-trimethylsilylethane	0.41	47
6	Ethane	0.94	37
7	2,2,4,4-Tetramethylpentane	0.74	26
8	Hexamethyldisiloxane	0.30	22
9	Hexamethyldisilane	0.33	20
10	2,2-Dimethylbutane	0.58	12
11	2,2,5,5-Tetramethylhexane	0.67	12
12	2,2,4-Trimethylpentane	0.33	6.6
13	2,2,3,3-Tetramethylpentane	0.42	5.2
14	Polydimethylsiloxane	0.24	4.6
15	Propane	0.19	2.6
16	Cyclopentane	0.16	1.1
17	2-Methylbutane	0.17	0.93
18	3,3-Diethylpentane	0.23	0.76
19	2-Methylpentane	0.15	0.29
20	Cyclohexane	0.15	0.24
21	3-Methylpentane	0.15	0.20
22	n-Pentane	0.15	0.14
23	n-Hexane	0.13	0.074
24	Benzene	0.054	0.13
25	Toluene	0.051	0.08
26	Methylcyclohexane	0.12	0.068

Source: Refs. 2, 15–19, and 143.

of the electron in argon is now available from theory [148] (see Sec. 6.5). Previous workers had taken $m^* = m_e$. As a test of Eq. (22), the mobility in argon was recalculated here using these new data. The adiabatic compressibility was used and the value of β was obtained by making both sides of Eq. (22) equal at the density at which V_0 is a minimum. Fig. 10 shows that this calculated mobility compares well with the experimental data of Jahnke et al. [145]. This theory has also been shown to predict quite well the density dependence of the mobility in xenon [149].

This theory has also been used to predict mobility for molecular liquids. Neopentane and TMS are liquids that exhibit maxima in the electron mobility at intermediate densities [46]. These maxima occur at the same densities at which V_0 minimizes, in accordance with the Basak–Cohen theory. The drift mobility in TMS has been measured as a function of pressure to 2500 bar [150]. The observed relative experimental changes of mobility with pressure are predicted quite well by the Basak–Cohen theory; however, the predicted value of μ_D is 2.5 times the experimental value at 1 bar and 295 K. In this calculation, the authors used χ_T to evaluate the mobility. This is reasonable in this case since for liquids, there is little difference between the adiabatic and isothermal compressibilities. A similar calculation for neopentane showed that the Basak–Cohen theory predicted the Hall mobility of the electron quite well for temperatures between 295 and 400 K [151]. Itoh

ARGON

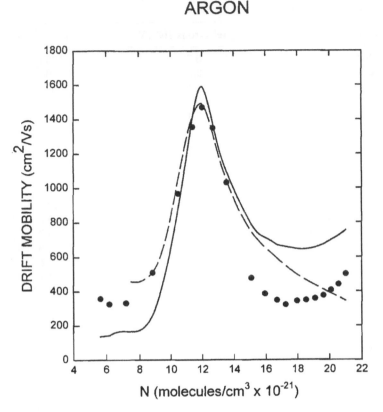

Figure 10 Drift mobility in argon as a function of density. ●: Experimental results at 55 bar [145]. —: Calculation using modified Basak–Cohen (see text). - - -: Calculation by Hsu and Chandler [146].

et al. [152] extended the theory to mixtures taking concentration fluctuations into account, but found that the data for TMS–neopentane mixtures were best fit if only density fluctuations were considered.

Other theories have been proposed in recent years to account for the mobility of quasi-free electrons. Borghesani et al. [153] studied the field dependence of the mobility in Ar–Kr and Ar–Xe mixtures and found that the results disagreed with the deformation potential model. Instead, the results could be described by a gas-kinetic model if concentration-dependent scattering cross sections were used. Atrazhev and Iakubov [154] used pseudopotential theory and similarly derived density-dependent scattering lengths, which varied from negative values at low density to positive values at high density. Their theory qualitatively describes the density dependence of the mobility in Ar, Kr, and Xe. Naveh and Laikhtman [155] introduced scattering from longitudinal acoustic phonons into the deformation potential framework. Their theory gave excellent agreement with experimental results for Ar for densities from 0.8 to 11.2×10^{21} cm^{-3}. A different approach was used by Hsu and Chandler [146] who used Feynman's polaron theory [156] for the mobility and a pseudopotential consisting of attractive and repulsive parts. These two terms counteract at some intermediate density and result in a peak in the mobility. Their calculation for Ar is shown as dashed line in Fig. 10.

The deformation potential model seems to provide a suitable framework to understand the quasi-free electron mobility in nonpolar liquids. Already several extensions or modifications on this theory have been proposed, and the dependence on temperature and pressure seems to be adequately explained. However, several authors have taken different approaches to the problem showing that a consensus in our understanding has not yet been reached.

6.2. Trapped State Transport

For many nonpolar liquids, the electron drift mobility is less than $10 \text{ cm}^2/\text{Vs}$, too low to be accounted for in terms of a scattering mechanism. In these liquids, electrons are trapped as discussed in Sec. 4. Considerable evidence now supports the idea of a two-state model in which equilibrium exists between the trapped and quasi-free states:

$$e^-_{qf} \rightleftarrows e^-_{tr} \tag{23}$$

The magnitude of the mobility then depends on the value of the quasi-free mobility in such liquids multiplied by the fraction of time the electron is quasi-free since the trapped electron is relatively immobile. Thus:

$$\mu_D = \mu_{qf} \frac{[e_{qt}]}{[e_{qf}] + [e_{tr}]} = \mu_{qf} \frac{1}{1 + K_{23}} \tag{24}$$

This equilibrium depends on many factors like V_0, $\Delta G_{soln}(e^-)$, temperature, and pressure. Differences in these and other factors are presumed to account for the wide range of mobilities observed (see Table 2).

Studies of the effect of pressure on μ_D for nonpolar liquids provided support for the two-state model. Pressure affects the position of the equilibrium [Eq. (23)] because of the volume change associated with trapping of the electron, ΔV_{tr}. These volume changes were deduced from changes in μ_D with pressure. For n-alkanes [157] as well as some alkenes [158], the mobility decreases with pressure, as shown in Fig. 11 for n-hexane and 1-pentene.

These changes in μ_D led to values of ΔV_{tr} equal to -22 and $-27 \text{ cm}^3/\text{mol}$ for n-hexane and 1-pentene, respectively. The negative volume changes are due to the role of electrostriction [160] of the solvent around the trapped electron, and the electrostriction volume, V_{el}, is a function of the isothermal compressibility χ_T of the liquid.

For certain liquids like cyclohexane [158], o-xylene, and m-xylene [159], the mobility increases with increasing pressure (see Fig. 11). These results provided the key to understand the two-state model of electron transport. In terms of the model, ΔV_{tr} is positive; for example, for o-xylene, ΔV_{tr} is $+21 \text{ cm}^3/\text{mol}$. Since electrostriction can only contribute a negative term, it follows that there must be a positive volume term which is the cavity volume, $V_{cav}(e)$. The observed volume changes, ΔV_{tr}, are the volume changes for reaction (23). These can be identified with the partial molar volume, $\overline{V}_e$, of the trapped electron since the partial molar volume of the quasi-free electron, which does not perturb the liquid, is assumed to be zero. Then the partial molar volume is taken to be the sum of two terms, the cavity volume and the volume of electrostriction of the trapped electron:

$$\Delta V_{tr} \cong \overline{V}_e = V_{cav}(e) + V_{el}(e) \tag{25}$$

Whether $\overline{V}_e$ is positive or negative depends on the relative value of the two terms in Eq. (25). Ten hydrocarbons were studied and the cavity volume term was found to be rela-

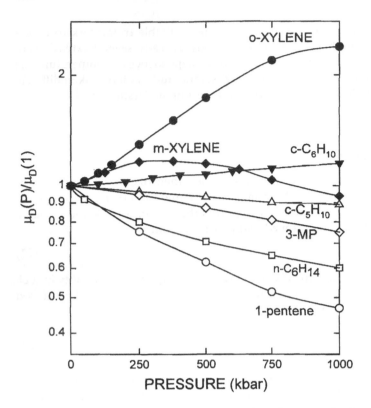

Figure 11   Relative drift mobility as a function of pressure for low mobility liquids. Mobility at 1 bar in parenthesis. $\bigcirc$: 1-pentene (0.048) [158]; $\square$: n-hexane (0.071) [19,150]; $\diamond$: 3-methylpentane (0.22) [2,157]; $\triangle$: cyclopentane (1.02) [157]; $\blacktriangledown$: cyclohexene (1.39) [158]; $\blacklozenge$: m-xylene (0.08) [159]; $\bullet$: o-xylene (0.019) [159].

tively constant at 96 ± 18 cm^3/mol. The electrostriction term varies, however, with the compressibility of the liquid. The compressibilities of the xylenes are approximately one-third that for 1-pentene. Thus the cavity volume term dominates for the xylenes and ΔV_{tr} is positive. The V_{el} term dominates for 1-pentene and n-alkanes and ΔV_{tr} is negative.

The main experimental effects are accounted for with this model. Some approximations have been made; a higher-level calculation is needed which takes into account the fact that the charge distribution of the trapped electron may extend outside the cavity into the liquid. A significant unknown is the value of the quasi-free mobility in low mobility liquids. In principle, Hall mobility measurements (see Sec. 6.3) could provide an answer but so far have not. Berlin et al. [144] estimated a value of $\mu_{qf} = 27$ cm^2/Vs for hexane. Recently, terahertz (THz) time-domain spectroscopy has been utilized which is sensitive to the transport of quasi-free electrons [161]. For hexane, this technique gave a value of $\mu_{qf} = 470$ cm^2/Vs. Mozumder [162] introduced the modification that motion of the electron in the quasi-free state may be in part ballistic; that is, there is very little scattering of the electron while in the quasi-free state.

Thus considerable support exists to support the two-state model of electron transport. The magnitude of the mobility is dependent on many factors including V_0, μ_{qf}, $\Delta G_{soln}(e)$, temperature, pressure, and other factors. Presumably, differences in these factors can explain the wide range of mobilities observed for nonpolar liquids. For example, Mozumder [163] has recently related μ_D to G_{fi} for a series of hydrocarbons through the thermalization distance. The fact that the mobility of *trans*-2-butene is approximately 100 times lower than that of *cis*-2-butene (2.2 cm^2/Vs) [164] is still quite surprising.

6.3. Hall Mobility

Measurements of the Hall mobility of electrons in nonpolar liquids are few in number, but those that have been made provide information about the transport processes that is not available from drift measurements alone. The Hall mobility, μ_H, is obtained by measuring the deflection of electrons by a magnetic field while they are drifting in an electric field. Since the deflection occurs only while the electrons are quasi-free, μ_H is a measure of μ_{qf}. Measurements of μ_H that have been done are for liquids of high drift mobility. The results for liquid argon [165] and xenon [166] show that μ_H is approximately equal to μ_D near the respective triple points. The results for TMS indicate that the ratio μ_H/μ_D is close to unity over a large temperature range from 295 to 437 K [167]. For neopentane, μ_H is quite comparable to μ_D at temperatures from 293 K to 413 K; however, near the critical temperature, 434 K, the ratio μ_H/μ_D is about 5, suggesting that there are localized states produced by density fluctuations in neopentane at this temperature [151,168] This effect is similar to the localization in clusters observed in xenon near the critical density (see above). For 2,2-dimethylbutane, μ_H is 12 cm^2/Vs, and for 2,2,4,4-tetramethylpentane, μ_H is 32 cm^2/Vs at 293 K [169]. These values are comparable to the values of the drift mobility for these compounds (see Table 2). For 2,2,4-trimethylpentane, μ_H is 3.5 times μ_D at room temperature, indicating that trapping occurs in this liquid, which is consistent with the observation of the solvated electron in this liquid (see Sec. 4).

6.4. Diffusion Coefficient

For molecular liquids, the diffusion coefficient of the electron can be obtained from the mobility and the Einstein relation Eq. (14). The electrons remain thermal in such liquids, and the temperature, T, of the liquid can be used in Eq. (14). In liquid rare gases, this is not the case if an electric field is applied, which causes the electrons to gain energy. The extent of this effect was measured for liquid argon by Shibamura et al. [170]. The energy was derived from the diffusion broadening of the electron cloud during drift of the electrons in an electric field. Values of D_e were derived and the energy, $k_B T$, of the electrons was calculated from Eq. (14). The electron energy was found to increase from 0.1 to 0.4 eV as the electric field increased from 2 to 10 kV/cm.

6.5. Effective Mass (m*)

Because of the lack of long-range order in liquids, the quasi-free electron is subjected to multiple scattering. This is taken into account by making an effective mass approximation in theoretical calculations of the mobility [see Eq. (22) and Sec. 3]. Previously, because of the lack of experimental data, the effective mass was put equal to the free electron mass. The effective mass has been evaluated [171] for some liquids from O_2^- ionization cross-section spectra [172]. This procedure gave $m^* = 0.26m_e$ for argon at 87 K and $m^* = 0.27m_e$ for TMS, 2,2-dimethylbutane, and 2,2,4-trimethylpentane at 296 K. From exciton spectra, Reininger et al. [173] determined m^* for argon to be $0.55m_e$ at the triple point. In a similar way, Reshotko et al. [174] determined m^* to be $0.28m_e$ for Xe at several densities near N_t.

Recently, the effective mass of the electron has been calculated [148] within a Wigner–Seitz framework [175] for Ar, Kr, and Xe. In all three liquids, m^* decreases with increasing density. At the triple point densities, $m^* = 0.6m_e$ in argon and $m^* = 0.3m_e$ in xenon.

6.6. Fast Negative Ions

In some liquids, the electron is trapped as a solvent negative ion, yet transport occurs much faster than expected for an ion. A well-documented example of this type of electronic transport is found in liquid perfluorobenzene where the electron attaches to form $C_6F_6^-$ [176]. Although the V_0 level is not known for this liquid, the anion is stable because the electron affinity of C_6F_6 is 1.1 eV and the anion is further stabilized by polarization. However, the negative ion has a mobility of 0.018 cm^2/Vs, much greater than that of the positive ion. Another example is the negative ion in liquid CS_2, which has a mobility approximately 10 times that of the positive ion [177]. In supercritical CO_2, the anion also has a high mobility. At 41°C, the solvent anion mobility is 0.01 cm^2/Vs near critical density and increases with density to 0.015 cm^2/Vs at 0.8 g/cm^3. This increase is expected for a hopping mechanism since the average distance between sites decreases as the density increases. Solute ions are reported to have a mobility of about 0.002 cm^2/Vs at this temperature in supercritical CO_2 [178].

Studies have shown that the electron mobility in aromatic liquids changes upon the application of external pressure but becomes constant above 2 kbar. At low pressures, the electron is trapped in a cavity as discussed for alkanes in Sec. 6.2. Increasing the pressure causes the equilibria

$$e^- + A \rightleftarrows A^- \tag{26}$$

to shift to the right. The mobility observed at high pressure is 0.06 cm^2/Vs for toluene [179], $\leq$0.08 cm^2/Vs for benzene [179], and 0.06 and 0.04 cm^2/Vs for m-xylene and o-xylene, respectively [159]. Transport at high pressure is believed to occur by hopping:

$$A^- + A \rightarrow A + A^- \tag{27}$$

Hopping occurs between neighboring molecules and the activation energies are small, from 0.12 to 0.15 eV for the hydrocarbons. Fast anions have also been suggested for liquid SF_6 [180]. Thus several examples of this type of transport for the negative charge exist and the mobilities are comparable. More examples of fast ions are expected to be found in the future.

6.7. Positron Mobility

It is interesting to compare the properties of positive electrons, positrons, with the properties of electrons in nonpolar liquids. Values of the mobility of positrons, μ_+, are now available for a few liquids. Early measurements for μ_+ in n-hexane ranged from 8.5 to 100 cm^2/Vs [181,182]. In a recent study, the Doppler shift in energy of the 511-keV annihilation gamma ray in an electric field was utilized to measure the drift velocity. This method led to $\mu_+ = 53$ cm^2/Vs in n-hexane and 69 cm^2/Vs in 2,2,4-trimethylpentane [183]. Interestingly, these values are comparable to the mobilities of quasi-free electrons in nonpolar liquids.

7. APPLICATIONS

This chapter would not be complete without mentioning the applications of electrons in nonpolar liquids. Scientists who are very interested in the properties of electrons in these liquids are the detector physicists, working in high-energy physics, gamma ray astronomy, cosmic radiation, or positron emission tomography [184]. Ionization chambers are used in these fields to detect particles by the current signals induced by the excess electrons produced. Ideal liquids for such applications would have high free ion yields, high drift velocities, and high density. Liquid purity is important as well so that the electrons live long enough to reach the electrode.

There are a variety of detectors currently in use to detect particles from neutrinos to weakly interacting massive particles (WIMPS). Table 3 is a partial list of some of these that are currently in use or under construction. As noted, some utilize multiton quantities of liquid. The first such detector was a sampling calorimeter containing 300 L of liquid argon [185]. Many of the detectors in use today in high-energy physics experiments are of this type. They are calorimeters because the particles are totally stopped within the detector and the energy of the particles can be determined by the ionization produced in the liquid. Heavy metal plates, each a few millimeters thick, are introduced and the ionization is "sampled" in the liquid between the plates. An electromagnetic calorimeter detects electrons and gamma rays. The latter create electron–positron pairs; these in turn produce lower-energy gammas that create more pairs, etc. This sequence of events is called an electromagnetic shower. The D0 detector at Femilab is an example that has been in operation since 1991 [186]. Calorimeters that record signals from hadrons (protons, neutrons, kaons, pions, etc.) are often incorporated in such detectors. These utilize thicker metal plates. Calorimeters have necessarily grown as the energy of the particles in high-energy physics experiments has increased. For example, the A Torcidal Large Hadron Collider Apparatus (ATLAS) detector at European Center for Nuclear Research (CERN) is a 200-kt detector. Specific details about these detectors are given in the cited references

Table 3 Ionization Detectors

Name	Liquid	Location	Type	Detects	Year
D0	42,000 L Ar	Fermilab	EM and hadronic sampling calorimeter	Electrons, photons, and hadrons	1991
Walic	TMP	UA1-CERN	EM sampling calorimeter		1988
—	210 L TMS	Karlsruhe	Calorimeter	Cosmic rays	1994
KEDR	24 t Kr	Novosibirsk	EM calorimeter		1996
ICARUS	600 t Ar	Gran Sasso Italy	Time projection chamber	Neutrinos p decay	2001
ATLAS	140 t Ar	CERN	EM and hadron calorimeters	Electrons, photons, hadrons, and Higgs boson	2007
LXeGRIT	7 L Xe	Balloon	Compton telescope	Gamma rays	1999
DAMA/Xe-2	2 L Xe	Gran Sasso	Scintillator	WIMPS	1998

[187]. To avoid the necessity of a cryogenic container, some calorimeters have utilized molecular liquids like TMS [188] and 2,2,4,4-tetramethylpentane [189].

Another type of liquid detector is the time projection chamber. Examples are Imaging Cosmic and Rare Underground Signals (ICARUS), a large liquid Ar detector [190], and LXeGRIT (gamma ray imaging telescope) [191]. These detectors determine both the energy and the direction of the incoming particle or photon [192]. Liquid xenon detectors are usually smaller due to the high cost of xenon. For the purposes of gamma-ray astronomy, the scintillation of the xenon provides a trigger for an event and the direction of the particle is obtained by measuring the time required for electrons to drift to the collecting electrodes. Balloon flights by E. Aprile's group from Columbia have demonstrated the feasibility of such gamma-ray telescopes [191].

ACKNOWLEDGMENTS

The preparation of this chapter was done at Brookhaven National Laboratory and supported under contract DE-AC02-98-CH10886 with U.S. Department of Energy and supported by its Division of Chemical Sciences, Office of Basic Energy Sciences.

REFERENCES

1. Freeman, G.R. *Kinetics of Nonhomogeneous Processes*; Freeman, R., Ed.;. Wiley: New York, 1987; 63 pp.
2. Nishikawa, M. *Handbook of Radiation Chemistry*; Tabata, Y. Ito, Y., Tagawa, S., Eds.;. CRC Press: Boca Raton, 1991; p 395.
3. Schmidt, W.F. *Liquid State Electronics of Insulating Liquids*; CRC Press: Boca Raton, 1997.
4. Mozumder, A. *Fundamentals of Radiation Chemistry*; Academic Press: San Diego, 1999.
5. Onsager, L. Phys. Rev. 1938, *54*, 554.
6. Schmidt, W.F. Radiat. Res. 1970, *42*, 73.
7. Hoshi, Y.; Higuchi, M.; Iso, H.; Sakamoto, M.; Ooyama, K.; Yuta, H.; Abe, K.; Hasegawa, K.; Suekane, F.; Kawamura, N.; Neichi, M.; Suzuki, K.; Masuda, M.; Kikuchi, R.; Miyano, K. IEEE Trans. Nucl. Sci. 1993, *40*, 532.
8. Huang, S.S.-S.; Freeman, G.R. Can. J. Chem. 1977, *55*, 1838.
9. Warman, J.M.; Asmus, K.D.; Schuler, R.H. *Advances in Chemistry Series 82*; Gould, R.F., Ed.; Washington: American Chemical Society, 1968; Vol. 82, p 25.
10. Jay-Gerin, J.-P. Can. J. Chem. 1993, *71*, 287.
11. Geer, S.; Holroyd, R.A. Phys. Rev. B 1992, *46*, 5043.
12. LeMotais, B.C.; Jonah, C.D. Radiat. Phys. Chem. 1989, *33*, 505.
13. Saeki, A.; Kozawa, T.; Yoshida, Y.; Tagawa, S. Radiat. Phys. Chem. 2001, *60*, 319.
14. Bartczak, W.M.; Hummel, A. Chem. Phys. Lett. 1993, *208*, 232.
15. Holroyd, R.A.; Itoh, K.; Nishikawa, M. Nucl. Instrum. Methods Phys. Res. A 1997, *390*, 233.
16. Gee, N.; Senanayake, C.; Freeman, G.R. J. Chem. Phys. 1988, *89*, 3710.
17. Schmidt, W.F.; Allen, A.O. J. Chem. Phys. 1970, *52*, 2345.
18. Allen, A.O.; Holroyd, R.A. J. Phys. Chem. 1974, *78*, 796.
19. Allen, A.O.; Gangwer, T.E.; Holroyd, R.A. J. Phys. Chem. 1975, *79*, 25.
20. Nishikawa, M.; Yamaguchi, Y.; Fujita, K. J. Chem. Phys. 1974, *61*, 2356.
21. Yamaguchi, Y.; Nishikawa, M. J. Chem. Phys. 1973, *59*, 1298.
22. Holroyd, R.A.; Chen, P.Y.; Stradowska, E.; Itoh, K. Radiat. Phys. Chem. 1996, *48*, 635.
23. Holroyd, R.A.; Sham, T.K. J. Phys. Chem. 1985, *88*, 2909.
24. Holroyd, R.A.; Sham, T.K.; Yang, B.X.; Feng, X.H. J. Phys. Chem. 1992, *96*, 7438.

25. Holroyd, R.A.; Sham, T.K. Radiat. Phys. Chem. 1998, *51*, 37.
26. Bartczak, W.M.; Hummel, A. Radiat. Phys. Chem. 1997, *49*, 675.
27. Siebbeles, L.D.A.; Bartczak, W.M.; Terrissol, M.; Hummel, A. J. Phys. Chem. 1997, *101*, 1619.
28. Holroyd, R.A.; Preses, J.M. *Chemical Applications of Synchrotron Radiation*; Sham, T.-K., Ed.; River Edge, NJ: World Scientific, 2002; Vol. 12B, p 987.
29. Bartczak, W.M.; Hummel, A. J. Phys. Chem. 1993, *97*, 1253.
30. Chybicki, M. Acta Phys. Pol. 1966, *30*, 927.
31. Munoz, R.C.; Cumming, J.B.; Holroyd, R.A. J. Chem. Phys. 1986, *85*, 1104.
32. Lindblad, T.; Bagge, L.; Engstrom, A.; Bialkowski, J.; Gruhn, C.R.; Pang, W.; Roach, M.; Loveman, R. Nucl. Instrum. Methods 1983, *215*, 183.
33. Mozumder, A. J. Phys. Chem. 1996, *100*, 5964.
34. Holroyd, R.A.; Nishikawa, M. Radiat. Phys. Chem. 2002, *64*, 19.
35. Balakin, A.A.; Boriev, I.A.; Yakovlev, B.S. Can. J. Chem. 1977, *55*, 1985.
36. Anderson, D.F.; Charpak, G.; Holroyd, R.A.; Lamb, D.C. Nucl. Instrum. Methods Phys. Res. 1987, *A261*, 445.
37. Springett, B.E.; Jortner, J.; Cohen, M.H. J. Chem. Phys. 1968, *48*, 2720.
38. Holroyd, R.A.; Tames, S.; Kennedy, A. J. Phys. Chem. 1975, *79*, 2857.
39. Holroyd, R.A.; Nishikawa, M.; Nakagawa, K.; Kato, N. Phys. Rev. B 1992, *45*, 3215.
40. Halpern, B.; Lekner, J.; Rice, S.A.; Gomer, R. Phys. Rev. 1967, *156*, 351.
41. Holroyd, R.A.; Allen, M. J. Chem. Phys. 1971, *54*, 5014.
42. Holroyd, R.A.; Tauchert, W. J. Chem. Phys. 1974, *60*, 3715.
43. Schiller, R.; Vass, S.; Mandics, J. J. Radiat. Phys. Chem. 1973, *5*, 491.
44. Yamaguchi, Y.; Nakajima, T.; Nishikawa, M. J. Chem. Phys. 1979, *72*, 550.
45. Nakagawa, K.; Ohtake, K.; Nishikawa, M. J. Electrostat. 1982, *12*, 157.
46. Holroyd, R.A.; Cipollini, N.E. J. Chem. Phys. 1978, *69*, 501.
47. Boriev, A.; Balakin, A.A.; Yakovlev, B.S. Khim. Vys. Energ. 1978, *12*, 20.
48. Borghesani, A.F.; Carugno, G.; Santini, M. IEEE Trans. Electr. Insul. 1991, *26*, 615.
49. Holroyd, R.A.; Russell, R.L. J. Phys. Chem. 1974, *78*, 2128.
50. Holroyd, R.A.; Dietrich, B.K.; Schwarz, H.A. J. Phys. Chem. 1972, *76*, 3794.
51. Nakato, Y.; Tsubomura, H. J. Phys. Chem. 1975, *79*, 2135.
52. Bottcher, E.-H.; Schmidt, W.F. J. Chem. Phys. 1984, *80*, 1353.
53. Holroyd, R.A.; Preses, J.M.; Bottcher, E.H.; Schmidt, W.F. J. Phys. Chem. 1984, *88*, 744.
54. Casanovas, J.; Grob, R.; Sabattier, R.; Gruelfucci, J.P.; Blanc, D. Radiat. Phys. Chem. 1980, *15*, 293.
55. Casanovas, J.; Grob, R.; Delacroix, D.; Guelfucci, J.P.; Blanc, D. J. Chem. Phys. 1981, *75*, 4661.
56. Greever, J.S.; Turner, J.B.M.; Kauffmann, J.F. J. Phys. Chem. A 2001, *105*, 8635.
57. Sowada, U.; Holroyd, R.A. J. Phys. Chem. 1980, *84*, 1150.
58. Sowada, U.; Holroyd, R.A. J. Phys. Chem. 1981, *85*, 541.
59. Lukin, L.V.; Yakovlev, B.S. Chem. Phys. Lett. 1976, *42*, 307.
60. Al-Omari, A.K.; Reininger, R. J. Chem. Phys. 1995, *103*, 506.
61. Al-Omari, A.K.; Altmann, K.N.; Reininger, R. J. Chem. Phys. 1996, *105*, 1305.
62. Al-Omari, A.K.; Reininger, R. J. Chem. Phys. 1995, *103*, 4484.
63. Altmann, K.N.; Reininger, R. J. Chem. Phys. 1997, *107*, 1759.
64. Stampfli, P.; Bennemann, K.H. Phys. Rev. A 1991, *44*, 8210.
65. Plenkiewicz, B.; Frongillo, Y.; Lopez-Castillo, J.-M.; Jay-Gerin, J.-P. J. Chem. Phys. 1996, *104*, 9053.
66. Reininger, R.; Asaf, U.; Steinberger, I.T.; Basak, S. Phys. Rev. B 1983, *28*, 4426.
67. Jacobsen, F.M.; Gee, N.; Freeman, G.R. Phys. Rev. A 1986, *34*, 2329.
68. Basak, S.; Cohen, M.H. Phys. Rev. B 1979, *20*, 3404.
69. Fermi, E. Nuovo Cim. 1934, *11*, 157.
70. Lopez-Castillo, J.-M.; Frongillo, Y.; Plenkiewicz, B.; Jay-Gerin, J.-P. Chem. Phys. 1992, *96*, 9092.

71. Lopez-Castillo, J.-M.; Jay-Gerin, J.-P. Phys. Rev. E 1995, *52*, 4892.

72. Space, B.; Coker, D.F.; Liu, Z.H.; Berne, B.; Martyna, G. J. Chem. Phys. 1992, *97*, 2002.

73. Boltjes, B.; deGraaf, C.; Leeuw, S.W. J. Chem. Phys. 1993, *98*, 592.

74. Simon, S.H.; Dobrosavljevic, V.; Stratt, R.M. Phys. Rev. A 1990, *42*, 6278.

75. Simon, S.H.; Dobrosavljevic, V.; Stratt, R.M. J. Chem. Phys. 1991, *94*, 7360.

76. Frongillo, Y.; Plenkiewicz, B.; Jay-Gerin, J.-P.; Jain, A. Phys. Rev. E 1994, *50*, 4754.

77. Baxendale, J.H.; Bell, C.; Wardman, P. J. Chem. Soc., Faraday Trans. I 1973, *69*, 776.

78. Atherton, S.J.; Baxendale, J.H.; Busi, F.; Kovacs, A. Radiat. Phys. Chem. 1986, *28*, 183.

79. Gillis, H.A.; Klassen, N.V.; Woods, R.J. Can. J. Chem. 1977, *55*, 2022.

80. McGrane, S.D.; Lipsky, S. J. Phys. Chem. 2001, *105*, 2384.

81. Nakamura, Y.; Shinsaka, K.; Hatano, Y. J. Chem. Phys. 1983, *78*, 5820.

82. Gillis, H.A.; Klassen, N.V.; Teather, G.G.; Lokan, K.H. Chem. Phys. Lett. 1971, *10*, 481.

83. Siebbeles, L.D.A.; Emmerichs, U.; Hummel, A.; Bakker, H.J. J. Chem. Phys. 1997, *107*, 9339.

84. Ichikawa, T.; Yoshida, H. J. Chem. Phys. 1980, *73*, 1540.

85. Kimura, T.; Ogawa, N.; Fueki, K. Bull. Chem. Soc. Jpn. 1981, *54*, 3854.

86. Wigner, E.P. Phys. Rev. 1948, *73*, 1002.

87. Kevan, L. Radiat. Phys. Chem. 1981, *17*, 413.

88. Itoh, K.; Holroyd, R. J. Phys. Chem. 1990, *94*, 8854.

89. Itoh, K.; Nishikawa, M.; Holroyd, R. J. Phys. Chem. 1993, *97*, 503.

90. Holroyd, R.A. Ber. Bunsenges. Phys. Chem. 1977, *81*, 28.

91. Sowada, U.; Holroyd, R.A. Private communication.

92. Chen, P.Y.; Holroyd, R.A. J. Phys. Chem. 1995, *99*, 14528.

93. Chen, P.Y.; Holroyd, R.A. J. Phys. Chem. 1996, *100*, 4491.

94. Holroyd, R.A.; Gangwer, T.E.; Allen, A.O. Chem. Phys. Lett. 1975, *31*, 520.

95. Holroyd, R.A.; McCreary, R.D.; Bakale, G. J. Phys. Chem. 1979, *83*, 435.

96. Warman, J.M.; deHaas, M.P.; Zador, E.; Hummel, A. Chem. Phys. Lett. 1975, *35*, 383.

97. Holroyd, R.A.; Nishikawa, M.; Itoh, K. J. Phys. Chem. B 1999, *103*, 9205.

98. Holroyd, R.A.; Nishikawa, M.; Itoh, K. J. Phys. Chem. B 2000, *104*, 11585.

99. Nishikawa, M.; Itoh, K.; Holroyd, R.A. J. Phys. Chem. A 1999, *103*, 550.

100. Holroyd, R.A.; Itoh, K.; Nishikawa, M. Chem. Phys. Lett. 1997, *266*, 227.

101. Hamann, S.D. Physico-Chemical Effects of Pressure. London: Butterworths, 1957.

102. Schwarz, H.A. J. Phys. Chem. 1993, *97*, 12954.

103. Itoh, K.; Holroyd, R.A.; Nishikawa, M. J. Phys. Chem. A 2001, *105*, 703.

104. Beck, G.; Thomas, J.K. Chem. Phys. Lett. 1972, *13*, 295.

105. Bakale, G.; Schmidt, W.F. Z. Naturforsch. 1981, *36a*, 802.

106. Warman, J.M. *The Study of Fast Processes and Transient Species by Electron Pulse Radiolysis*; Baxendale, J.H., Busi, F., Eds.; Reidel: Dordrecht, Holland, 1982; 433ff.

107. Tachiya, M. J. Phys. Chem. 1993, *97*, 5911.

108. Sauer, M.C.; Schmidt, K.H. Radiat. Phys. Chem. 1994, *43*, 413.

109. Bakale, G.; McCreary, R.D.; Gregg, E.C. Cancer Biochem. Biophys. 1981, *5*, 103.

110. Bakale, G.; Gregg, E.C.; McCreary, R.D. J. Chem. Phys. 1977, *67*, 5788.

111. Baxendale, J.H.; Keene, J.P.; Rasburn, E.J. J. Chem. Soc., Faraday Trans 1974, *70*, 718.

112. Chen, E.S.; Chen, E.C.M.; Sane, N.; Talley, L.; Kozanecki, N.; Shulze, S. J. Chem. Phys. 1999, *110*, 9319.

113. Christodoulides, A.A.; McCorkle, D.L.; Christophorou, L.G. Report No. DOE/EV/04703-39.

114. Gains, A.F.; Kay, J.; Page, F.M. Trans. Faraday Soc. 1966, *62*, 874.

115. Kebarle, P.; Chowdhury, S. Chem. Rev. 1987, *87*, 513.

116. Christophorou, L.G. Z. Phys. Chem. 1996, *195s*, 195.

117. Weston, R.E., Jr; Schwarz, H.A. Chemical Kinetics. Prentice-Hall: Englewood Cliffs, NJ, 1972.

118. Funabashi, K.; Magee, J.L. J. Chem. Phys. 1975, *62*, 4428.

119. Bakale, G.; Sowada, U.; Schmidt, W.F. J. Phys. Chem. 1975, *79*, 3041.

120. Schmidt, W.F.; Bakale, G.; Sowada, U. J. Chem. Phys. 1974, *61*, 5275.

121. Beck, G.; Thomas, J.K. J. Chem. Phys. 1972, *57*, 3649.

122. Holroyd, R.A. Radiat. Phys. Chem., submitted.
123. Holroyd, R.A.; Gangwer, T.E. Radiat. Phys. Chem. 1980, *15*, 283.
124. Kang, Y.S.; Holroyd, R.A. Radiat. Phys. Chem. 1982, *20*, 237.
125. Warman, J.M.; de Haas, M.P.; Hummel, A. *Conduction and Breakdown in Dielectric Liquid*; Goldschwartz, J.M., Niessen, A.K., Boone, W., Eds.; Delft Univ. Press: Delft, Netherlands, 1975; 70.
126. Nishikawa, M.; Itoh, K.; Holroyd, R. J. Phys. Chem. 1988, *92*, 5262.
127. Bakale, G.; Sowada, U.; Schmidt, W.F. J. Phys. Chem. 1976, *80*, 2556.
128. Bakale, G.; Sowada, U.; Schmidt, W.F. J. Phys. Chem. 1975, *79*, 3041.
129. Sowada, U.; Bakale, G.; Yoshino, K.; Schmidt, W.F. Chem. Phys. Lett. 1975, *34*, 466.
130. Henglein, A. Ber. Bunsenges. Phys. Chem. 1976, *79*, 129.
131. Henglein, A. Can. J. Chem. 1976, *55*, 2112.
132. Nakagawa, K.; Holroyd, R.A. Private communication.
133. Christophorou, L.G. *Atomic and Molecular Radiation Physics*; Wiley-Interscience: New York, 1971.
134. Henglein, A. Ber. Bunsenges. Phys. Chem. 1975, *79*, 129.
135. Wentworth, W.E.; Chen, E.; Freeman, R. J. Chem. Phys. 1971, *55*, 2075.
136. Blaunstein, R.P.; Christophorou, L.G. J. Chem. Phys. 1968, *49*, 1526.
137. Christophorou, L.G.; McCorkle, D.L.; Pittman, D. J. Chem. Phys. 1974, *60*, 1183.
138. Holroyd, R. *The Liquid State and Its Electronic Properties;* Kunhardt, E.E., Christophorou, L.G., Luessen, L.H., Eds.; Plenum Press. New York, 1987; Vol. 193, 221.
139. Ninomiya, S.; Itoh, K.; Nishikawa, M.; Holroyd, R. J. Phys. Chem. 1993, *97*, 9488.
140. Holroyd, R.A. J. Phys. Chem. 1982, *86*, 3541.
141. Cook, A.R.; Curtiss, L.A.; Miller, J.R. J. Am. Chem. Soc. 1997, *119*, 5729.
142. Matyushov, D.V.; Schmid, R. Mol. Phys. 1995, *84*, 533.
143. Astbury, A.; Keeler, R.K.; Li, Y.; Poffenberger, P.R.; Robertson, L.P. Nucl. Instrum. Methods Phys. Res. A 1991, *305*, 376.
144. Berlin, Y.A.; Nyikos, L.; Schiller, R. J. Chem. Phys. 1978, *69*, 2401.
145. Jahnke, J.A.; Meyer, L.; Rice, S.A. Phys. Rev. A 1971, *3*, 734.
146. Hsu, D.; Chandler, D. J. Chem. Phys. 1990, *93*, 5075.
147. Nishikawa, M. Chem. Phys. Lett. 1985, *114*, 271.
148. Plenkiewicz, B.; Frongillo, Y.; Plenkiewicz, P.; Jay-Gerin, J.-P. J. Chem. Phys. 1991, *94*, 6132.
149. Holroyd, R.A.; Nishikawa, M.; Itoh, K. J. Chem. Phys. 2003, *118*, 706.
150. Munoz, R.C.; Holroyd, R.A. J. Chem. Phys. 1986, *84*, 5810.
151. Munoz, R.C.; Ascarelli, G. J. Phys. Chem. 1984, *88*, 3712.
152. Itoh, K.; Nishikawa, M.; Holroyd, R.A. Phys. Rev. B 1991, *44*, 680.
153. Borghesani, A.F.; Iannuzzi, D.; Carugno, G. J. Phys. 5057, 1997, *9*, Condens. Matter 9.
154. Atrazhev, V.M.; Iakubov, I.T. J. Chem. Phys. 1995, *103*, 9030.
155. Naveh, Y.; Laikhtman, B. Phys. Rev. B 1993, *47*, 3566.
156. Feynman, R.P. Phys. Rev. 1955, *97*, 660.
157. Munoz, R.C.; Holroyd, R.A.; Itoh, K.; Nakagawa, K.; Nishikawa, M.; Fueki, K. J. Phys. Chem. 1987, *91*, 4639.
158. Itoh, K.; Holroyd, R.A.; Nishikawa, M. J. Phys. Chem. B 1998, *102*, 3147.
159. Itoh, K.; Nishikawa, M.; Holroyd, R.A. J. Chem. Phys. 1996, *105*, 5510.
160. Drude, P.; Nernst, W. Z. Phys. Chem. 1894, *15*, 79.
161. Knoesel, E.; Bonn, M.; Shan, J.; Heinz, T.F. Phys. Rev. Lett. 2001, *86*, 340.
162. Mozumder, A. Chem. Phys. Lett. 1993, *207*, 245.
163. Mozumder, A. J. Phys. Chem. A 2002, *106*, 7062.
164. Allen, A.O. Drift Mobilities and Conduction Band Energies of Excess Electrons in Dielectric Liquids, NSRDS-NBS 58, 1976.
165. Ascarelli, G. Phys. Rev. B 1989, *40*, 1871.
166. Ascarelli, G. Phys. Rev. Lett. 1991, *66*, 1906.
167. Munoz, R.C.; Holroyd, R.A. Chem. Phys. Lett. 1987, *37*, 250.

168. Munoz, R.C.; Ascarelli, G. Phys Rev. Lett. 1983, *51*, 215.

169. Itoh, K.; Munoz, R.C.; Holroyd, R.A. J. Chem. Phys. 1989, *90*, 1128.

170. Shibamura, E.; Takahashi, T.; Kubota, S.; Doke, T. Phys. Rev. A 1979, *20*, 2547.

171. Baird, J.K. J. Chem. Phys. 1983, *79*, 316.

172. Sowada, U.; Holroyd, R.A. J. Chem. Phys. 1979, *70*, 3586.

173. Reininger, R.; Steinberger, I.T.; Bernstorff, S.; Saile, V.; Laporte, P. Chem. Phys. 1984, *86*, 189.

174. Reshotko, M.; Asaf, U.; Ascarelli, G.; Reininger, R.; Reisfeld, G.; Steinberger, I.T. Phys. Rev. B. 1991, *43*, 14174.

175. Wigner, E.; Seitz, F. Phys. Rev. 1933, *43*, 804.

176. Ende, C.A.M.v.d.; Nyikos, L.; Warman, J.M.; Hummel, A. Radiat. Phys. Chem. 1982, *19*, 297.

177. Gee, N.; Freeman, G.R. J. Chem. Phys. 1989, *90*, 5399.

178. Itoh, K.; Holroyd, R.A.; Nishikawa, M. J. Phys. Chem. A 2001, *105*, 703.

179. Itoh, K.; Holroyd, R.A. J. Phys. Chem. 1990, *94*, 8850.

180. Gee, N.; Ramanan, G.; Freeman, G.R. Can. J. Chem. 1990, *68*, 1527.

181. Heinrich, F.; Schiltz, A. Presented at the 6th International Conference on Positron Annihilation, Arlington, TX, 1982. Unpublished.

182. Linderoth, S.; MacKenzie, I.K.; Tanigawa, S. Phys. Lett. A 1985, *107*, 409.

183. Wang, C.L.; Kobayashi, Y.; Hirata, K. Radiat. Phys. Chem. 2000, *58*, 451.

184. Lopes, M.I.; Chepel, V.; Solovov, V.; Marques, R.F.; Policarpo, A.J.P.L. Presented at the Proceedings of the 8th Internatl. Conf. on Calorimetry in High Energy Physics, Lisbon Portugal, 1999. Unpublished.

185. Willis, W.J.; Radeka, V. Nucl. Instrum. Methods 1974, *120*, 221.

186. Abachi, S.; Abolins, M.; Acharya, B.S.; Adam, I.; Ahn, S., et al. Nucl. Instrum. Methods Phys. Res. A 1994, *338*, 185.

187. Schinzel, D. Nucl. Instrum. Methods Phys. Res. A 1998, *419*, 217.

188. Engler, J. J. Phys. G. Nucl. Part. Phys. 1996, *22*, 1.

189. Albrow, M.G.; Apsimon, R.; Aubert, B.; Bacci, C.; Bezaguet, A., et al. Nucl. Instrum. Methods A 1988, *265*, 303.

190. Arneodo, F.; Badertscher, A.; Baibussinov, S.; Benetti, P.; Borio, A., et al. Nucl. Instrum. Methods A 2000, *455*, 376.

191. Aprile, E.; Curioni, A.; Egorov, V.; Giboni, K.L.; Oberlack, U.; Ventura, S.; Doke, T.; Takizawa, K.; Chupp, E.L.; Dunphy, P.P. Nucl. Instrum. Methods Phys. Res. A 2001, *461*, 256.

192. Lopes, M.I.; Chepel, V. Presented at the 2002 IEEE 14th Internatl. Conf. on Dielectric Liquids, Graz, Austria, 2002. Unpublished.

9

Interactions of Low-Energy Electrons with Atomic and Molecular Solids

Andrew D. Bass and Léon Sanche*
University of Sherbrooke, Sherbrooke, Québec, Canada

1. INTRODUCTION

Low-energy electron–molecule scattering has been studied in the gas phase for over 80 years, beginning with experiments by Franck and Hertz [1] in 1914 (for a review of gaseous electron–molecule collision processes, see Refs. 2–5). While such research has usually been performed on isolated targets, more recent work (commencing around 1980) has extended "electron-scattering" type measurements and theory into the condensed phase to determine the interactions of electrons with the atomic and molecular constituents of solids and liquids, thus providing new challenges to our understanding of the basic physics of electron scattering. Particular attention has been directed to the interactions of low-energy electrons (LEEs), that is, electrons with kinetic energies lower than 30 eV, since these species are implicated in a large number of phenomena and processes operating in the condensed phase systems, from radiation damage [6] and the electronic aging of dielectrics [7], friction induced damage to lubricants [8] and nanolithography [9], and to (potentially) the production of O_2 in extraterrestrial water ices [10] and depletion of the ozone [11].

Low-energy electrons are of special relevance to radiobiology since high-energy radiation interacting with condensed matter (i.e., liquids to solids) generates as the most abundant intermediate species electrons with energies lower than 70 eV [6,12,13], within nanoscopic volumes along ionization tracks. About four secondary electrons per 100 eV deposited are created with a most probable energy lying below 10 eV. At these energies, electrons have thermalization distances of the order of 1–10 nm [13], which effectively define the initial volume for energy deposition by high-energy radiation. In these volumes, termed "spurs," the highly excited atomic, molecular, and radical species ions and secondary electrons can induce nonthermal reactions within femtosecond times. A majority of these reactive species, including those that initiate further chemical reactions, are created by the secondary electrons. Thus to reach a comprehensive understanding of radiation energy deposition processes in condensed matter at the nanoscopic level, it is necessary to identify the interactions of these electrons.

---

* Canada Chair in the Radiation Sciences.

In the early 1990s, one of us (LS) reviewed the then available experimental data on LEE interactions with atomic and molecular thin films [14,15]. In the intervening years, considerable progress has been made, and it is now appropriate to review this more recent work. Thus in this contribution, we will largely restrict the discussion to research conducted since 1990, although reference to older material will also be given. Moreover, it should be noted that with the exception of materials of biological interest, discussion is further limited to measurements and processes occurring in the multilayer atomic and molecular solids. Particular attention will be paid to electron-driven processes that initiate molecular fragmentation and/or other chemical change and to the formation of transient negative ions (TNIs) which occur when an electron becomes temporarily bound to an atom or molecule and which can greatly enhance scattering cross sections.

The article is organized as follows. In Sec. 2, we describe some of the basic physics of the interactions of LEEs with isolated molecules, including an account of the DEA process. In Sec. 3, we review experimental investigations of LEE impact on simple atomic and molecular solids. Studies of electron impact on molecules of biological interest are reviewed separately in Sec. 4. Finally, in Sec. 5, we consider with examples the relevance of this work for application in areas other than radiobiology.

2. THEORETICAL BACKGROUND

Since electrons have a negligible mass compared to those of atoms and molecules, they can exchange very little energy by momentum transfer with such particles at low velocities. Hence to transfer considerable energy in a single "event," the electron must be coupled to a target by its electrodynamic interaction potential or by the constructive interference of its wave function. The latter is associated primarily with the internal structure (i.e., orbital configurations, internuclear distances, symmetry, etc.) of an atom or a molecule and, to a lesser extent, with the geometrical arrangement of the atomic or molecular constituents of a molecular solid or liquid. When intra-atomic and/or intramolecular constructive interference occurs, we refer to this phenomenon as the formation of an electron resonance. Constructive and destructive interference of electron waves, delocalized over many atoms or molecules of a solid or liquid, is known as electron diffraction and is highly dependent on the geometrical arrangement of the target and its band structure.

The electrodynamic interaction potential acting between a diatomic molecule and an electron outside a molecule can be written as

$$V = \frac{\mu_e}{r^2} P_1\left(\vec{r} \cdot \vec{R}\right) - \frac{Q_e}{r^3} P_2\left(\vec{r} \cdot \vec{R}\right)$$
$$-\frac{\alpha e^2}{2r^4} - \frac{\alpha' e^2}{2r^4} P_2\left(\vec{r} \cdot \vec{R}\right) - \dots \tag{1}$$

where r is the distance of the incident electron from the molecule and R is the internuclear separation; μ_e is the electric dipole moment. The term containing μ_e is necessarily absent in homonuclear diatomic molecules. The second term, involving Q_e, the quadrupole moment, characterizes the quadrupole interaction. These two terms are the "electrostatic" terms in that they pertain to the interaction of the incident electron and the unperturbed molecule. The next two terms involving α, the spherically symmetric, and α', the nonspherical part of the polarizability, are the "dynamic" terms. The latter involve the polarization of the molecule by the incident electron. In Eq. (1), $\vec{r}$ and $\vec{R}$ are unit vectors in the directions r

and R, respectively, and P_n are the Legendre polynomials. The parameters α, α', μ_e, and Q_e are functions of R. It can be seen from Eq. (1) that by estimating the magnitude of the various terms, it may be possible to sort out the dominant scattering mechanism. For example, the potential of Eq. (1) can be expanded around the equilibrium distance R_e along an internuclear coordinate of the molecule as

$$
V = V(R = R_e) + (R - R_e) \left. \frac{\partial V}{\partial R} \right|_{R = R_e} + \cdots \tag{2}
$$

When only these two terms are considered and the molecule is assumed to be a harmonic oscillator, solving the problem within the Born approximation (usually corresponding to small momentum transfer) leads to the optical selection rule $\Delta v = 1$ for vibrational transitions [16]. Thus within this most restrictive approximation, the electron behaves like electromagnetic radiation. From this analysis, we can expect the electron–molecule potential described by Eq. (2) to be responsible for the magnitude of the differential scattering cross sections which are large only for small scattering angles and the excitation of the first vibrational energy level ($v = 1$) from the ground state of the molecule.

Electron resonances occur when the scattered electron resides for a much longer time than the usual scattering time in the neighborhood of a target atom or molecule. From an atomic or molecular orbital perspective, a resonant state may be considered as a TNI formed by an electron in which it occupies an orbital of the target. This concept leads to the definition of two major types or categories of resonances or transient anions [17,18]. If the additional electron occupies a previously unfilled orbital of the target in its ground state, the transitory state is referred to as a single particle resonance. The term "shape" resonance applies more specifically when temporary trapping of the electron is due to the shape of the electron–molecule potential. When the transitory anion is formed by two electrons occupying previously unfilled orbitals, the resonance is called "core-excited" or may be referred to as a two-particle, one-hole state.

When an electron resonance occurs, the temporary capture of an electron at an atomic or molecular site increases the interaction time of the electron at that site in proportion to the resonance lifetime ($\tau_a \sim 10^{-16}$–10^{-13} sec) and the inverse of the electron transfer rate. This local interaction causes a distortion of the atom or molecule which accepts the additional electron. When the electron leaves the molecule, nuclear motion is initiated toward the initial internuclear distance, causing excitation of many overtones of the molecule, due to the strong overlap between the nuclear wave function of the resonant state and those of many vibrational states of the molecular ground state. On the other hand, when τ is much smaller than a typical vibrational period ($\tau 10^{-14}$ sec), the nuclei are not significantly displaced. In this case, overlap between the nuclear wave function of the resonant state and that of the vibrational levels of the ground state occurs only between the first few energy levels. Thus for short resonance times, only the lower vibrational levels become excited with considerable amplitude.

Dissociative electron attachment (DEA) occurs when the molecular transient anion state is dissociative in the Franck–Condon (FC) region, the localization time is of the order of or larger than the time required for dissociation along a particular nuclear coordinate, and one of the resulting fragments has positive electron affinity. In this case, a stable atomic or molecular anion is formed along with one or more neutral species. Dissociative electron attachment usually occurs via the formation of core-excited resonances since these possess sufficiently long lifetimes to allow for dissociation of the anion before autoionization.

Within a local complex potential curve crossing model, the cross section for the simple DEA reaction $e + AB \rightarrow AB^{-} \rightarrow A^{-} + B$, where AB is a diatomic molecule, may be expressed as [18]

$$\sigma_{DEA}(E_i) = \sigma_{cap}(E_i)P_S \tag{3}$$

where P_S represents the survival probability of the anion against autodetachment of the electron and E_i is the incident electron energy. The capture cross section σ_{cap} is given by

$$\sigma_{cap}(E_i) = \lambda 2g|\chi_v|^2 \left[\frac{\Gamma_a}{\Gamma_d}\right] \tag{4}$$

where λ is the de Broglie wavelength of the incident electron, g is a statistical factor, and χ_v is the normalized vibrational nuclear wave function. Γ_a is the *local energy* width of the AB^- state in the FC region and Γ_d is the *extent* of the AB^- potential in the FC region. The width of the transient anion state in the autodetaching region defines the lifetime τ_a towards autodetachment, $\tau_a(R) = \hbar/\Gamma_a(R)$, such that the survival probability of the temporary anion, after electron capture, is given by

$$P_S = \exp\left[-\int_{R_e}^{R_c} \frac{dt}{\tau_a(R)}\right] \tag{5}$$

where R_e is the bond length of the anion at energy E_i and R_c is that internuclear separation beyond which autodetachment is no longer possible. If we define an average lifetime τ_a and let $K \equiv \lambda^2 g|\chi_v|^2$, then Eq. (3) becomes

$$\sigma_{DEA}(E_i) = K\left[\frac{\Gamma_a}{\Gamma_d}\right]\exp\left[-\frac{\bar{\tau}_c}{\bar{\tau}_a}\right] \tag{6}$$

Here $\bar{\tau}_c(E) \equiv |R_c - R_e|/v$ where v is the average velocity of separation of the fragments A^- and B upon dissociation. Hence the DEA cross section depends exponentially on the product of the lifetime of the transient anion and the velocities of the fragments. Equation (6) defines most of the *intrinsic* characteristics of the DEA process. It may be seen from this equation that the magnitude of the DEA cross section depends on parameters which are influenced by the nature of the target, i.e., the attaching electron wavelength λ, the resonance lifetime $\bar{\tau}_a$, and curve crossing at R_c.

Equation (6) defines most of the *intrinsic* characteristics of the DEA process. For further information on the mechanism of transient anion formation and its effects in isolated electron–atom and electron–molecule systems, the reader is referred to the review articles by Schulz [17] and others [7,19,20]. Information on resonance scattering from single layer and submonolayer of molecules physisorbed or chemisorbed on conductive surfaces can be found in the review by Palmer et al. [21–23]. The present article provides information essentially on resonances in atoms and molecules condensed onto a dielectric surface or forming a dielectric thin film.

3. EXPERIMENTAL STUDIES OF RARE GAS AND MOLECULAR SOLID FILMS

3.1. Experimental Methods

A number of experimental techniques have been developed to investigate the interaction of low-energy electrons with atomic and molecular solids. To avoid surface contamination and to facilitate the use of low-energy electron beams, such experiments must be performed under ultrahigh-vacuum (UHV) conditions in chambers evacuated by cryogenic, ion, and turbo-molecular pumps to base pressures below 10^{-10} Torr. Molecular (and atomic) solid targets are usually formed by vapor deposition onto a clean substrate held at temperatures of between 15 and 100 K by a cryostat. The substrate is usually a polycrystalline metal foil, a metallic monocrystal, or a semiconductor crystal which can be cleaned by resistive heating and/or ion bombardment. Films of materials that exhibit a high vapor pressure are grown using a gas-volume expansion dosing procedure [24] that can be calibrated by monitoring the quantum size effect features observed for ultrathin films [25]. Films of materials that under ambient conditions might be described as low vapor pressure solids can be grown using an oven to generate a low-density molecular flux [26]. Films of larger organic and biomolecular materials can be prepared ex vacua as self-assembled monolayers on gold substrates [27,28]. Once formed, the sample films may be probed with a variety of electron beam techniques, the most common of which are now briefly described.

Low-Energy Electron Transmission Spectroscopy

In a typical low-energy electron transmission (LEET) experiment [24], a trochoidal monochromator [29] provides an electron current of between 1 and 10 nA, with an intrinsic resolution between 40 and 60 meV full width at half maximum (fwhm). The beam is incident normally on the film surface. A LEET spectrum is obtained by measuring the current I_t arriving at the substrate as a function of incident electron energy E_i. In LEET spectroscopy, and in the following other electron impact experiments, the absolute energy scale is calibrated to within ± 0.15 eV of the vacuum level by measuring the onset of electron transmission through the films. A description of the technique is given in Ref. 24.

Charge (Electron) Trapping

The LEET technique [24] can be modified to measure cross sections for charge (electron) trapping or stabilization by molecules condensed onto dielectric films [30,31]. Electronic charge trapped at the surface of a film, following exposure to electrons of a known E_i, produces a retarding potential (ΔV), which is apparent as a displacement of any subsequent LEET spectrum to higher incident energies. The observed rate of charging or *charging coefficient* A_s can be converted into a charging cross section (σ_{CT}) as follows:

$$A_s(E_i) \equiv d\Delta V(t)/dt|_{t=0} = (LI\mu_o/\varepsilon\pi r^2)\sigma_{CT} \qquad (7)$$

where L and ε are the spacer layer thickness and dielectric constant, respectively, μ_o is the surface density of molecular targets, and I and r are the total current and radius of the incident electron beam, respectively. Absolute trapping cross sections as small as 10^{-19} cm^2 can be measured with an uncertainty of $\pm 50\%$ using this technique. When combined with the mass spectrometric measurements of the electron-stimulated desorption (ESD) of anions, σ_{CT} cross sections can, under certain conditions, allow absolute cross sections for dissociative electron attachment (DEA) to be obtained.

In the form described above, charge-trapping measurements can only be performed on nanometer-thick films. However, analogous measurements on dielectric materials of macroscopic dimensions (thickness of 0.05 to 0.15 mm and area ~ 2 cm^2) have been obtained [32] with an instrument that employs a *pulsed* low-current electron beam (10^{-14}–10^{-12} C/pulse) to induce charging and a Kelvin probe to measure changes in the sample's surface potential $\Delta \overline{V}$. This latter can then be related to changes in the quantity of trapped charge ΔQ_t. Independently obtained electrometer measurements made during each electron pulse provide an instantaneous measure of the induced charge (e.g., ΔQ_t), which, with knowledge of the incident charge per pulse Q_0, allows a trapping probability ($\Delta Q_t/Q_0$) to be determined. The film is discharged with a UV lamp, mounted outside the vacuum chamber.

High-Resolution Electron Energy-Loss Spectroscopy

The energy losses of electrons scattered near the surface of thin films and their dependence on E_i can be measured with a high-resolution electron energy loss (HREEL) spectrometer of the type illustrated in Fig. 1. In addition to a cryogenically cooled substrate, such an instrument includes an electron monochromator and energy analyzer. Those shown in Fig. 1 employ hemispherical electrostatic deflectors to achieve optimal resolution (~ 5 to 30 meV fwhm). Typically, the monochromator produces a focused electron beam that strikes the film surface at an angle θ_0 from the film normal. Electrons scattered in a narrow pencil about angle θ_r relative to the surface normal are energy-analyzed with the second hemispherical deflector. Dependent on spectrometer design, the angles θ_0, θ_r, or both can be varied within limits set by available space. Energy-loss spectra are recorded by sweeping the potential of the monochromator or analyzer relative to the grounded target. The energy dependence of the magnitude of a given energy-loss event (i.e., *the excitation function*) is obtained by sweeping the energy of both deflectors with the potential difference between them corresponding to the probed energy.

The HREEL technique can be used in several ways: for detailed spectroscopic studies of vibrational and electronic excitation within the atomic and molecular films

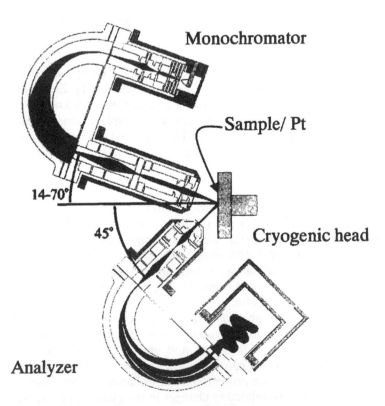

Figure 1 Schematic representation of a high-resolution electron energy loss (HREEL) spectrometer.

[33,34], for measuring effective elastic and absolute inelastic cross sections [35,36], and for determining effective cross sections for the production of fragment species [37,38]. This latter is achieved by measuring the amplitude of a signal associated with a particular degradation product (for example, a particular electronic or vibrational energy-loss feature) as a function of the integrated electron current at specific electron energies. Similar effective cross sections for the production of fragment species may be measured using x-ray photoelectron [39] and infrared spectroscopies [40].

Electron-Stimulated Desorption Experiments

Electron-induced dissociative processes and certain other reactions are perhaps most effectively detected when a fragment ion or neutral species is desorbed into vacuum. Generally, ion desorption is easier to observe than that of neutral species since desorbed charged particles can be immediately mass-analyzed. Fig. 2 illustrates schematically an experimental system for ion desorption studies of various types of target including large organic/biomolecular films. Note the load-lock section for the introduction sample slides and/or the degassing of organic molecules within an oven prior to their introduction into the analysis chamber. In this instrument [26], an electron gun produces an electron beam of variable energy having a current of between 5 and 300 nA with an energy resolution of ~ 250 meV fwhm. The electron beam is incident on a cryogenically cooled solid film. Ions desorbed during electron impact enter an ion lens (containing a set of retardation grids) which precedes a quadrupole mass spectrometer. The so-called *ion yield functions* are obtained by recording a particular ion signal as a function of E_i energy. In other ESD instruments, the electron gun is replaced with an electrostatic [41,42] or trochoidal [43] monochromator which produces an electron beam of a few nanoamperes at a resolution of < 80 meV. Pulsed electron sources have also been used [44] in conjunction with a time of flight mass spectrometer.

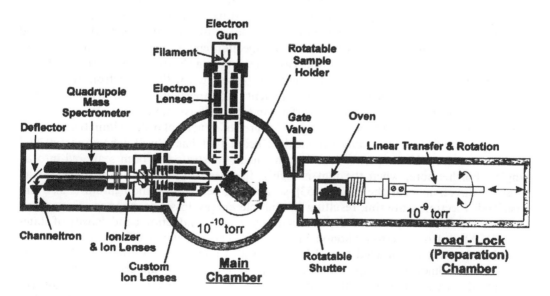

Figure 2 Schematic overview of an apparatus suitable for measurements of the electron-stimulated desorption of anions from vacuum deposited films. (From Ref. 26.)

In principle, the neutral desorbed products of dissociation can be detected and mass analyzed, if ionized prior to their introduction into the mass spectrometer. However, such experiments are difficult due to low *effective* ionization efficiencies for desorbed neutrals. Nevertheless, a number of systems have been studied in the groups of Wurm et al. [45], Kimmel et al. [46,47], and Harries et al. [48], for example. In our laboratory, studies of neutral particle desorption have been concentrated on self-assembled monolayer targets at room temperature [27,28]. Under certain circumstances, neutrals desorbed in electronically excited metastable states of sufficient energy can be detected by their de-excitation at the surface of a large-area microchannel plate/detector assembly [49]. Separation of the ESD signal of metastables from UV luminescence can be effected by time of flight analysis [49]; however, when the photon signal is small relative to the metastable yield, such discrimination is unnecessary and only the total yield of neutral particles (NP) needs to be measured.

3.2. Experimental Results

Elastic and Quasi-Elastic Scattering

Elastic scattering of LEEs by multilayer rare gas and molecular solid films has been investigated by low-energy electron transmission (LEET) spectroscopy [24,25,50–71], photoinjection [72–82], and elastic reflection [35,36,66,83–87] experiments. In the latter, the spectrometer shown in Fig. 1 is adjusted to measure electron elastically scattered from the film at particular incident and scattered angles θ_0 and θ_r. Then, the amplitude of the elastic peak is measured as a function of E_i. Since in these experiments both the incident and outgoing electron momenta are specified, features due to interferences of the electron waves are prominent in the spectra. In well-ordered films, the diffraction structure is dominated by long-range order [66], whereas in amorphous substances, variation in the structure factor, due to short-range order, can be detected in the energy dependence of the elastic reflectivity [83,84,87].

Both LEET and photoinjection experiments measure the electron current transmitted through a multilayer film deposited on a metal substrate. Usually in photoinjection experiments, electrons are injected in the film from the metal substrate with ill-defined energies and momenta, although some control of electron injection energy is now possible with synchrotron radiation sources [79]. In either case, the outgoing electrons which escape into the vacuum at a given energy and momentum can be selected with an electron analyzer or retarding electric field. This sort of measurement [72] is particularly sensitive to electron interactions near the substrate–film interface and, for self-assembled monolayers of long-chain organic molecules, to the direction and degree of structural order within the film [80].

In transmission, it is only the incident beam that has a well-defined energy and momentum since the current collected at the metal substrate has been scattered into all angles. Furthermore, when the film is highly disordered, electrons are scattered in all possible directions near the surface [53], so that the penetrating momentum is also unspecified. Electron scattering resulting in multiple energy losses to phonons (i.e., quasi-elastic scattering) is expected to depend on the electronic conduction band density of states (CB DOS) and on electron–phonon interactions. The calculated CB DOS of solid Ar (solid line) and that extracted from LEET spectra of multilayer (20–100 monolayers) Ar films (dashed line) are compared in Fig. 3 [60]. There exists an obvious relationship between the two sets of data, which indicates that it is essentially the CB DOS that governs quasi-elastic scattering of the LEEs. In HREEL experiments on rare gas (RG) solids [83,84], it has been shown that this correspondence arises because of multiple energy losses

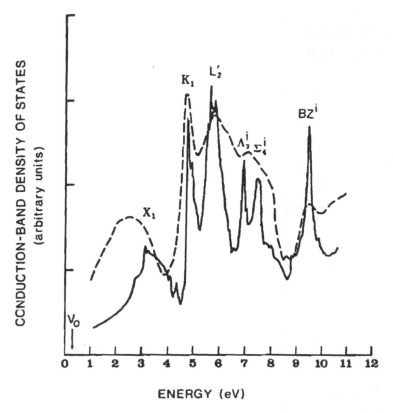

Figure 3 Electron conduction band density of states (CB DOS) for solid Ar; calculated (solid line) and determined from analysis of LEET data recorded at 20 K at energies below the first excitonic threshold (dashed line). The zero of energy is the vacuum level. V_0 is the energy of the bottom of the conduction band (0.25 eV). (From Ref. 60.)

to phonons by the scattered electrons. Taking again the example of an Ar multilayer film, this relationship is shown in Fig. 4. The curves in the figure were recorded with the HREEL spectrometer described in Sec. 3.1.3, set to measure the excitation function at an energy loss $\Delta E = 0.25$ eV and $\theta_r = 45°$ for several incident angles between 15° and 65°. These curves thus represent the probability that an electron, penetrating a 50-layer film of Ar deposited on Pt, loses 0.25 eV via multiple losses to phonons in the solid. Except for the measurement at $\theta_0 = 45°$ (i.e., in the specular direction), all features are found at essentially the same energy, independent of the incident angle. The similarity between these curves reveals an electron-scattering property of the solid that is averaged over various directions of electron propagation (i.e., various electron states) which may consequently reflect the CB DOS. In Fig. 4(b), the CB DOS of Ar as calculated by Bacalis et al. [88] is displayed with the bottom of the lowest conduction band fixed at the measured value [25] of 0.25 eV above the vacuum level. Clearly, the experimental curves of Fig. 4(a), especially those for large incident angles, bare a close resemblance with the calculated CB DOS. The progressive shift to higher energy of calculated features relative to their experimental counterparts (excluding those peaks at 9 and 12 eV) likely reflects a discrepancy between the face-centered cubic lattice parameter of Ar employed in the calculation (0.526 nm) and experimental values at 20 K (0.531 nm) [89].

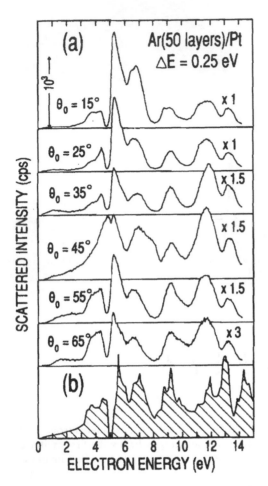

Figure 4   (a) Scattered electron intensity at fixed energy loss $\Delta E = 0.25$ eV as a function of incident electron energy E_i at several angles of incidence θ_0 on a 50-ML (monolayers) Ar film. (b) CB DOS for fcc structure of solid Ar as calculated by Bacalis et al. [88]. (From Ref. 83.)

The similarity between the experimental results and the calculated CB DOS can be explained by focusing on the electron transport properties in the bulk [83,84]. An electron propagating in a conduction band of an RG solid suffers scattering mainly from defects and lattice waves. This can be described by introducing the scattering probability per unit length $Q(E_{k0}, k_0, E_k, k)$ that a Bloch electron initially in a state $|\chi\rangle$ of momentum k_0 and energy E_{k0} is scattered into a final state $|\chi_k\rangle$ of momentum k and energy E_k, while the crystal changes from a state $|i\rangle$ of energy ε_i to a state $|f\rangle$ of energy ε_f. Then, by referring to the "golden rule" and solving the Boltzmann transport equation for a plane-parallel system in the "two-stream" approximation [35,36,55,90], one obtains for the electron mean-free path $\lambda(E_i)$ (i.e., the reciprocal of Q) the expression

$$\lambda(E_i) = \left\langle \left\{ \sum_k Q(E_{k0}, k_0, E_k, k) \right\}^{-1} \right\rangle_{E_i} \tag{8}$$

In this expression, the k summation extends over the first Brillouin zone, whereas $\langle \ldots \rangle_{E_i}$ stands for the average over the incident direction k_0 for a constant incident energy. If we replace the summations with integrations and assume for simplicity that the matrix element for calculating Q depends only on the momentum transfer (i.e., $k - k_0$), Eq. (8) yields

$$\lambda(E_i)^{-1} = D(E_i)\langle\{8\pi\hbar/\Omega S(E_i)\}\{1/\tau(E_i)\}\rangle \tag{9}$$

where $D(E_i)$ is the CB DOS of the solid at the energy E_i, $\tau(E_i)$ corresponds to a relaxation time (i.e., the time between scattering events) independent of the k_0 direction, $S(E_i)$ is the surface of constant energy E_i within the first Brillouin zone, and Ω is the volume of the crystal. Within the approximations of an electron effective mass and of an electron–phonon interaction described as a deformation-potential perturbation, one has $1/\tau(E_i) \propto k_0^2$ and $S(E_0) \propto k_0^2$ [91]. Consequently, the expression in parentheses in Eq. (9) is independent of E_i and the energy dependence of $\lambda(E_i)^{-1}$ or Q (i.e., the quasi-elastic scattering probability per unit length) becomes directly proportional to the CB DOS as shown experimentally in Fig. 4.

Influence of Temporary Electron Trapping on Quasi-Elastic Scattering and Phonon Creation

The formation of a TNI within an atomic or molecular solid has also been found to affect quasi-elastic scattering of incident electrons [85–87]. A TNI is formed when an electron of a particular E_i becomes trapped at a lattice site for a short time (10^{-15} to 10^{-13} sec). This trapping (and detrapping) process transfers energy to the lattice, which appears in HREEL measurements as an increase in phonon loss processes. The effects of a TNI can thus be identified by comparing HREELS spectra obtained at the "on-resonance" electron energy with those obtained "off resonance." The energy and dynamics involved in the temporary trapping of an electron in a thin solid film can be demonstrated from electron scattering experiments on RG solid films. In RG, an electron resonance occurs at an energy slightly below the threshold for creating the lowest bulk exciton. In solid Ar, for example, the lowest transient anion state of Ar^- occurs at 11.8 eV and has a lifetime of 2.2×10^{-14} sec (approximately 1 order of magnitude less than in the gas phase [85]). This Ar^- state is a core-excited resonance [i.e., a one-hole, two-electron state composed of a 4 s electron bound to the electronic excited state formed from the atomic orbitals ($\dots 3s^2, 3p^5, 4s$)]. It has the isolated (i.e., gas phase) configuration ($\dots 3s^2, 3p^5, 4s^2$) $^2P_{3/2}$ $^2P_{1/2}$. In the condensed phase, this resonance is 0.25 eV below the $n = 1$ bulk exciton; it is formed by the binding of an electron to the $n = 1, 1'$ (the prime denoting the $j = 1/2$ spin-orbit partner) excitons in the bulk. The apparent width of the resonance is greater than the spin-orbit splitting for the neutral exciton states of Ar (~ 0.18 eV [92,93]) so that the state, assigned as ^2P Ar^-, can be considered the unresolved condensed phase equivalent of the lowest two gas phase Feshbach resonances [94]. Since the anion state involves the coupling of an electron with an exciton, it is more properly termed as an *electron–exciton complex*.

Fig. 5 shows HREEL spectra obtained with a 12-layer film of polycrystalline Ar at $\theta_o = 15°$ and $\theta_r = 45°$, for incident energies slightly below the $n = 1, 1'$ bulk-exciton thresholds, located at 12.06 and 12.24 eV, respectively [95]. The shaded curve results from the calculation of the energy-loss distribution [85]. The peak near 0 eV corresponds to electrons scattered quasi-elastically from the film. At incident energies $E_0 = 11.45$ and 12.05 eV, the phonon tail of this latter is characteristic of "off-resonance" excitation and is similar to the data recorded at other energies where there are no resonances. However, at "on resonance" (i.e., at $E_0 = 11.75$), the Ar^- ^2P state is formed and a new phonon peak is superimposed on the phonon energy-loss tail. Subtracting the off-resonance data from the HREEL data recorded at $E_i = 11.75$ eV provides the contribution to phonon excitation in the Ar lattice due to the formation of a TNI in solid Ar. The position of the peak in Fig. 5 indicates that the energy released to the surrounding phonon modes by the localization of

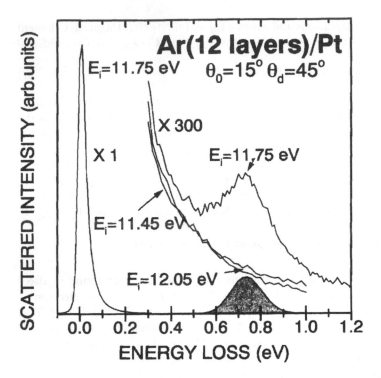

Figure 5 HREEL spectra of a 12-ML Ar film at three E_i with $\theta_0 = 15°$ and analysis angle $\theta_r = 45°$. At $E_i = 11.75$ eV, an Ar$^-$ is formed causing an additional peak in the phonon energy-loss tail. The shaded curve represents the calculated phonon excitation expected from a time-dependent charge at a lattice site. (From Ref. 85.)

an electron at a lattice site during its lifetime is equivalent to that released by about 200 inelastic "off-resonance" collisions.

The effect of a TNI on the surrounding medium can be described by considering the interaction between a time-dependent charge located on a lattice site and acoustical phonons. Using the approach of Mills [96], the result of such an analysis leads to an expression for the multiphonon loss feature whose intensity and mean energy loss depend mainly on the electron residence time τ_R and the polarizability of the surrounding atoms or molecules (i.e., dielectric constant ε of the medium). More specifically, it can be shown for a charge e, whose time dependency varies according to $e(t) = e \exp(-|t|/2\tau_R)$ on a lattice site, that the total probability to excite $1, 2, \ldots, m, \ldots$ phonons with an energy loss between Ω and $(\Omega + d\Omega)$ is given by

$$\frac{dP^{(\text{tot})}}{d\Omega} = \sum_{m=1}^{\infty} \frac{\gamma^m \exp(-\gamma)}{m!} \frac{\exp\left[-\dfrac{(\Omega - m\omega)^2}{2m(\Delta\omega)^2}\right]}{(2\pi m)^{1/2}\Delta\omega} \tag{10}$$

In this expression, $m\omega$ and $m^{1/2}\Delta\omega$ correspond to the mean frequency and width of the m phonon-loss process, respectively [96]. In the spirit of the Debye model for the longitudinal acoustic phonon at zero temperature, we have that $\omega = 4\omega_D/5$ and $\Delta\omega = \omega_D\sqrt{2/75}$ with ω_D the Debye frequency [96]. In Eq. (11), the term γ is referred to as the pseudo-Debye–

Waller factor because it depends on the transient character of the scattering event, whereas the usual Debye–Waller factor involves only the overall momentum transfer. It is given explicitly [96] by

$$\gamma = \frac{8}{3} \frac{e^4 (\varepsilon - 1)^2 \omega_D \tau_R^2}{\hbar M a_0^2 c_l^2 [1 + (\omega_D \tau_R)^2]^2} \qquad (11)$$

where M is the mass of the atom, a_0 is the nearest neighbor distance, c_l is the longitudinal sound velocity, and ε is the permittivity appropriate to the frequency range τ_R.

Calculations were performed with Eqs. (10) and (11) with τ_R as the sole adjustable parameter and experimentally determined values for a_0, ε, ω_D, and c_l [97]. The result of the calculation with $\tau_R = 2.2 \times 10^{-14}$ sec is displayed in Fig. 5. Note that the energy of the maximum is quite sensitive to τ_R; a variation of τ_R by 10% resulted in a 15% change for the energy of the maximum.

Vibrational Excitation

By scattering within molecular solids and at their surfaces, LEE can excite with considerable cross sections not only phonon modes of the lattice [35,36,83,84,87,90,98,99], but also individual vibrational levels of the molecular constituents [36,90,98–119] of the solid. These modes can be excited either by nonresonant or by resonant scattering prevailing at specific energies,* but as will be seen, resonances can enhance this energy-loss process by orders of magnitude. We provide in the next two subsections specific examples of vibrational excitation induced by LEE in molecular solid films. The HREEL spectra of solid N_2 illustrate well the enhancement of vibrational excitation due to a shape resonance. The other example with solid O_2 and O_2-doped Ar further shows the effect of the density of states on vibrational excitation.

High-Resolution Electron Energy-Loss Spectra of N_2. Vibrational excitation of ground-state N_2, induced by LEE impact on thin multilayer N_2 films, is illustrated in Fig. 6. These electron energy-loss spectra were recorded for electrons of energies $E_i = 2.9$, 10.8, and 19.8 eV, $\theta_o = 14°$ and $\theta_r = 45°$. The vertical gain in each curve or portion of a curve is referenced to the elastic peak. Each energy-loss peak can be ascribed to vibrational excitation of ground-state N_2 [100]. It can be seen in the figure that at certain impact energies ($E_i = 2.9$ and 19.8 eV), the intensities of vibrational energy losses are greatly increased (i.e., up to 2 orders of magnitude for overtones). In N_2 films, the production of overtones at $E_i = 2.9$ and 20 eV is attributable to $^2\Pi_g$ and $^2\Sigma_u$ shape resonances [100]. There is, however, no electron resonance at $E_i = 10.8$ eV in N_2 so that the strongest part of the interaction leading to vibrational excitation arises mainly from induced polarization (i.e., from direct scattering). This interaction is mainly effective to produce excitation of one vibrational quantum which amounts to an energy loss of less than 0.4 eV.

Excitation Functions of O_2 and O_2-Doped Ar Films. Resonances can be best identified by the structures they produce in excitation functions of a particular energy-loss process (i.e., the incident-electron energy dependence of the loss). Fig. 7 is reproduced from a recent study [118] of the electron-induced vibrational and electronic excitation of multilayer films of O_2 condensed on the Pt(111) surface and shows the incident electron energy dependence of major losses at the indicated film thickness and scattering angles. Also shown in this figure is the scattered electron intensity of the inelastic background

---

* For a theoretical description of the effects of a surface on resonant scattering, see Ref. 120.

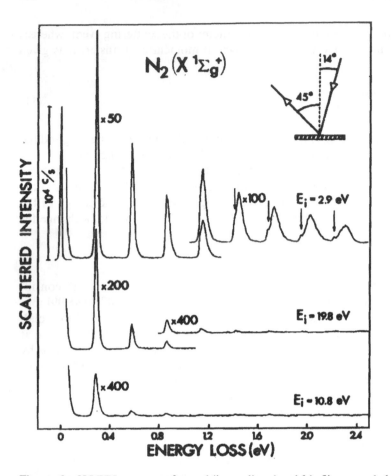

Figure 6 HREEL spectra of a multilayer disordered N_2 film recorded at the indicated E_i. (From Ref. 100.)

located just before the $v = 1$ peak onset at $\Delta E = 0.16$ eV along with that contributing to each energy loss (dashed lines).

With the exception of that of the $a^1\Delta_g$ ($v = 0$) loss, the vibrational excitation functions of O_2 in the solid phase differ significantly from their gas-phase equivalents. For example, the well-known sharp peaks of the $^2\Pi_g$ resonance, which dominate the vibrational cross sections up to 3 eV in the gas phase [121], are absent in the solid phase. This absence is understood as deriving from a lowering in energy of the $^2\Pi_g$ resonance by charge-image/charge polarization [101] of the O_2 film. At higher energies in the gas phase, excitation functions for the $X^3\Sigma_g^-$, $v = 1$, 2, and 3 vibrational losses display a single structure due to the $^2\Pi_u$ resonance near 6.5 eV. In contrast, in the condensed phase, the same vibrational losses are characterized by three broad overlapping bands, attributed to the excitation of the $^2\Pi_u$, $^2\Sigma_g^+(I)$, $^2\Sigma_u^+(I)$, and $^2\Sigma_{g,u}^+(II)$ anion states at 6.5, 8.5, 10.5, and 13.5 eV, respectively. These contributions from Σ^+ transient anion states, which are forbidden by the mirror plane symmetry selection rule in the gas phase (i.e., $\Sigma^- \leftarrow/\rightarrow \Sigma^+$), arise here due to the loss of the cylindrical symmetry of O_2 in the condensed phase [122].

The vibrational excitation is further altered when O_2 molecules are embedded within an Ar matrix [118,119]. Fig. 8 presents the results of experiments for 1% O_2 in 20 ML of Ar and shows excitation functions for the $v = 1$ and 2 vibrational losses of O_2 [Fig. 8(b) and (c),

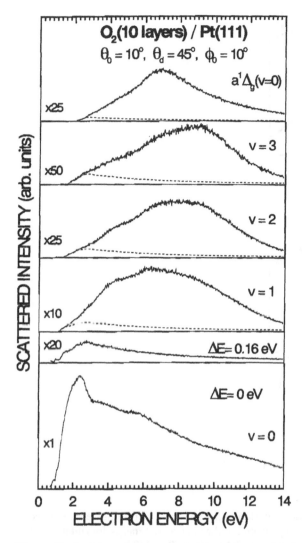

Figure 7   Incident electron energy dependence of the $X\,^3\Sigma_g^-$, $v = 0, 1, 2, 3$ vibrational and the $a^1\Delta_g$ ($v = 0$) electronic loss scattered intensities from a 10-layer film of O_2 condensed on Pt(111). θ_0 was set at 10° with θ_r at 45° and the azimuth ϕ_0 at 10°. Also shown is the energy dependence of the inelastic background intensity located just before the $v = 1$ loss peak onset at $\Delta E = 0.16$ eV along with that contributing to each energy-loss profile (dashed lines). (From Ref. 118.)

respectively] obtained after subtraction of a background of Ar phonon losses. Compared to the gas-phase result [121] [shown in Fig. 8(a) for $v = 1$], and those for solid O_2 in Fig. 7, the excitation functions for O_2 in Ar are much more structured, containing many narrow features. There is also a significant vibrational enhancement within the 2.5–5 eV incident energy range that is absent in the pure multilayer O_2 results (Fig. 7). Furthermore, except for the relative intensity changes, each maximum is located at the same energy independently of the angle of incidence and azimuthal orientation of the crystal. Such a behavior strongly suggests that the maxima arise primarily from an electron scattering property that is averaged over various directions of electron propagation within the crystal and thus depends on the CB DOS. In Fig. 8(d), we again reproduce the CB DOS of solid Ar [88]

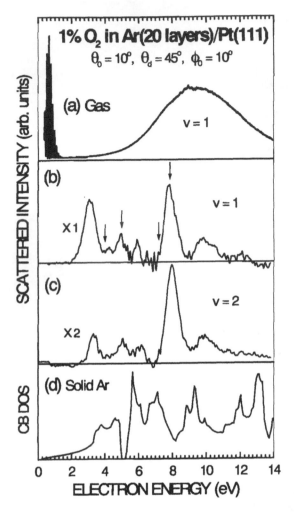

Figure 8 (a) Differential cross section for the excitation of $X^3 \Sigma_g^-$, $v = 1$ vibrational level of gaseous O_2. (b) and (c) are the incident electron energy dependence of the net $v = 1, 2$ energy-loss intensities for a 20-ML film of 1% O_2 in Ar. (d) is the CB DOS of solid Ar. (From Ref. 119.)

and it is easily seen that all *maxima* in Fig. 8(b) and (c) correlate respectively with *minima* in the CB DOS as opposed to the scattered intensity arising from multiple scattering on phonon (Fig. 4) [83,84], which was shown to follow the CB DOS.

This striking result can be qualitatively understood as related to CB DOS-influenced changes in the O_2^- anion lifetime [118]. For a diatomic molecule with R as the internuclear coordinate, a transient anion state is described in the fixed nuclei limit [123,124] by an energy and R-dependent complex potential: $V_{opt}(R,E_i) = V_d(R) + \Delta(E_i) - 1/2i\Gamma(E_i)$, where $V_d(R) \equiv \varepsilon_d(R) + V_G(R)$ is the potential energy curve of the discrete state, $V_G(R)$ is the potential energy curve of the ground state, and E_i is the kinetic energy of the scattered electron. The electronic decay width function $\Gamma(E_i)$ is given explicitly by

$$\Gamma(E_i) = \hbar/\tau(E_i) = 2\pi \sum_k |V_{d,k}|^2 \delta(E_i - E_k) \cong 2\pi |V_{d,E}|^2 D(E_i), \qquad (12)$$

where $V_{d,k}$ is the coupling matrix element responsible for the decay and E_k is the energy of the free electron continuum. Under sufficiently small variations of the matrix element with the wave vector k, the density of free electron states $D(E)$ appears explicitly with the summation over k [125].

For the short capture time limit, it can be shown [118] that the energy imparted to the nuclei ΔT_N can be approximated by

$$\Delta T_N \cong (2M_{red})^{-1}F_r^2(R_i)\tau^2[E_r(R_i)], \tag{13}$$

where M_{red} is the reduced mass of the molecule, τ is the anion lifetime, E_r is the energy of the resonance, R_i is the internuclear separation at which attachment occurs, and F_r is the internuclear force. It is apparent that the energy ΔT_N is a combination of two factors: the strength of the resonant vibrational coupling $F_r(R)$ (which depends on the shape of internuclear potential of the TNI) and the resonance lifetime $\tau[E_r(R)]$. In the absence of potential energy curve crossings *within the Franck–Condon region*, the factor $F_r(R)$ in Eq. (13) varies slowly as a function of the R. Moreover, as the transient anion potential energy curve is in a first approximation only slightly shifted down by the polarization of the surrounding medium, $F_r(R)$ is virtually the same in both gas and condensed phase.

On the other hand, the lifetime of the resonance $\tau[E_r(R)]$, whose inverse is explicitly given by $\tau^{-1}[E_r(R)] \cong (2\pi/\hbar)|V_{d,Er}|^2 D[E_r(R)]$ [cf., Eq. (12)], is directly linked to the CB DOS of the host medium, i.e., $D[E_r(R)]$. The remaining factor, which contains the matrix element responsible for the decay $V_{d,Er}$, is akin to a tunneling effect and yields essentially a monotonic behavior in energy. Owing to its relatively short-range character, it is also expected not to change significantly between gas and condensed phases. Consequently, the amount of energy transferred to the nuclei [cf., Eq. (13)] along with the corresponding overtone scattering probabilities (i.e., cross sections) varies mostly, via the modification of the resonance lifetime, as the opposite of the CB DOS of the host medium.

Electronic Excitation

Electron Spectroscopy. Electronic excitation induced by LEE in multilayer in atomic and molecular solids has been studied with LEET [24,51,52,57,62,126–131] and HREELS [104,131–147]. The slow-electron excitation of electronic states of atoms and nondissociative electronic states of molecules in dielectrics can result in localized energy deposits (e.g., Refs. 136 and 137) or into the formation of excitons [133] that propagate through the solid with well-defined wave vector. From HREELS measurements, one notes that electronic excitation in a molecular solids is similar and of similar magnitude to that observed for the isolated molecule, except for the cases of Rydberg and symmetry-forbidden transitions. It appears that a general characteristic of Rydberg states in condensed phases is poorly resolved since they broaden considerably into wide bands [148] due to autoionization by fast exchange of Rydberg electrons. In contrast, the strong exchange interaction in the low-energy regime causes singlet–triplet transitions to be more prominent in low-energy HREEL spectra than in measurements obtained with higher-energy electrons or photon excitation. Consequently, the low-energy HREELS has employed intensively to study the lowest triplet states of organic molecules [135–146].

Electron-Stimulated Desorption of Metastable Species. Electronic excitation of atomic and molecular solids is also demonstrated in the ESD yield of metastable atoms and molecules (i.e., lifetimes $>10^{-6}$ sec) and of UV photons from RG solids Ne and Ar [49,149–151], RG alloys [152], and from molecular solids of N_2 [153], N_2O [154], CO [153,155], and CO_2 [156,157], as well as N_2- and CO-covered Xe and Kr [158–160].

Desorption can proceed via several mechanisms. For solids with a negative electron affinity such as Ar [49,149–151] and N_2 [153], the extended electron cloud around a metastable center will interact repulsively with the surrounding medium and metastables formed at the film–vacuum interface will be expelled into vacuum (the so-called "cavity expulsion mechanism" [161]). Also permitted in solids with positive electron affinities (e.g., CO) is the transfer of energy intramolecular vibration to the molecule-surface bond with the resulting desorption of a molecule in lower vibrational level [153,155,158–160]. Desorption of metastables via the excitation of dissociative molecular (or excimer) electronic states is also possible [49,149–151,154,156,157]. A concise review of the topic can be found in Ref. 162.

Recent measurements have presented an interesting example of how metastable desorption can provide information on TNI electronic excitation in rare gas solids. Low-energy electron impact on solid Ar is known to induce the desorption of metastable Ar atoms (Ar*) [49,149] via the repulsive interaction of a surface-trapped exciton with the surrounding ground-state Ar atoms [161]. Fig. 9(b)–(e) shows NP yield functions from thin Ar films of varying preparation [163,164]. The signal at energies above 12 eV is correlated to the production of neutral excitons within the film and their decay by ejection and/or photon luminescence [151]. The maximum NP yield at ~ 15 eV coincides with the energy for optimum production of excited states in the gas [165] and condensed phases [151]. What is of note, however, in the data of Fig. 9 is that the ^2P Ar$^-$, shown in the HREEL data of Fig. 9(a), also contributes to the NP signal seen from the ~ 11.5 eV peak in each of the NP yield functions of Fig. 9(b), (d), and (e). This is surprising since the ^2P Ar$^-$ resonance *lies below* the energy of the first bulk and surface Ar excitons found at 12.03 and 11.71 eV [133], so that its observation in NP (and metastable) yield functions implies that the resonance has decayed into the higher-lying exciton energy levels [163]! In detailed studies of the ^2P Ar$^-$, as a function of Ar film thickness [163] and in Ar/n-hexane double-layer systems [164], it has been shown that Ar* desorption is initiated by electron emission from Ar$^-$ to the Pt(111) substrate at an energy of about 0.3 eV below the vacuum level. After electron emission at the Ar/Pt(111) interface, Ar* in the form of an exciton moves to the film–vacuum interface, where it can be ejected. Autoionization to the metal is inhibited when a spacer layer is introduced between the Ar layer and the substrate as evidenced in Fig. 9(c), where the resonance signal disappears for Ar condensed on a spacer of 3 ML of crystalline n-hexane.

The remaining results in Fig. 9 demonstrate that when a small amount of molecular oxygen is mixed in the Ar layer condensed on n-hexane [Fig. 9(e)] or deposited onto an isolated Ar layer [Fig. 9(d)], the ^2P Ar$^-$ resonance reappears in the Ar* desorption yield function. Since the n-hexane spacer inhibits Ar$^-$ decay by electron transfer to Pt(111), the presence of ^2P Ar$^-$ resonance in Fig. 9(c) and (d) was therefore interpreted [164] as due to electron transfer to O_2 leading to the formation of O_2^- in its ground-state $^2\Pi_g$. With the electron affinity of O_2 being of the order of the binding energy of the first electronically excited state of Ar, the decay of Ar$^-$ ^2P into lowest bulk excitons is possible by electron transfer to O_2.

Stable Anion Formation

The production of stable anions by low-energy electron impact has been studied by ESD [22,41,43,44,122,147,154,156,157,166–219] and charge-trapping [154,156,217,220–236] experiments. Below 20 eV, the desorption of stable anions from condensed systems is attributed to DEA, which produces oscillatory structures in the anion yield functions, and

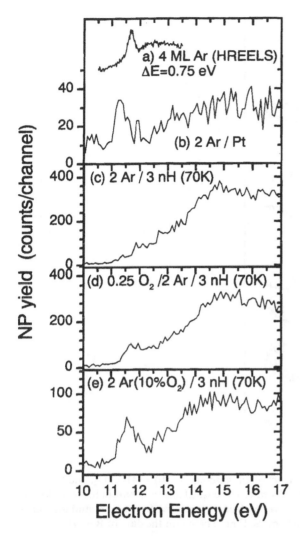

Figure 9 (a) High-resolution electron energy loss (HREEL) excitation function of 4 ML of Ar obtained at an energy loss $\Delta E = 0.75$ eV [85]. Neutral particle yield function produced by electron impact on: (b) 2 ML of Ar; (c) 2 ML of Ar condensed on a 3 ML crystalline n-hexane (nH) spacer layer; (d) same as in (c) but with 0.25 ML of O_2 adsorbed on the film surface; and (e) same as (c) but with 2 ML of Ar containing 10% volume of O_2 adsorbed on the nH film. In all cases, the substrate is Pt(111). (From Refs. 163 and 164.)

to dipolar dissociation (DD), which produces both anionic and cationic fragments. Typically, DD produces a featureless signal which increases linearly with electron energy from a threshold lying between 10 and 20 eV. Anion desorption data are often dominated by the DEA process. The first observations of DEA in a solid were made from the ESD of O^- from multilayer O_2 films [166]. Subsequently, the O_2 molecule has served as a model target to study factors affecting the physics of DEA in condensed phase systems [41,122,172–175,184,185,188–191,206,209,211]. The yield of O^- from electron impact on gas phase and multilayer-condensed O_2 is shown in Fig. 10(a) and (b), respectively. In the

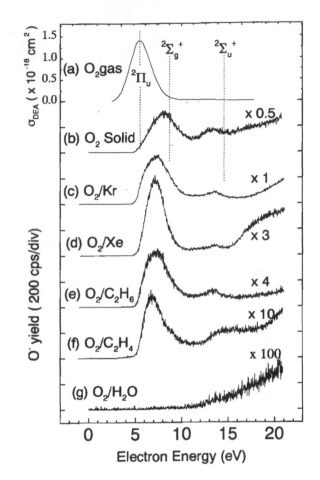

Figure 10 O⁻ yield functions for electron impact on (a) gaseous O_2, (b) 4 ML of O_2, and (c–g) 0.15 ML of O_2 condensed on 4 ML of the indicated substrate. The dashed lines and bars indicate the known positions of the O_2^- resonances. The gain factor over each curve is relative to curve (c), and the shifted baselines correspond to the zero anion intensity levels. Figure based on the data of Ref. 41.

gas phase, the O⁻ signal is dominated by DEA via the $^2\Pi_u$ state of O_2^- at ~6.7 eV [237]. In contrast, the anion yield function from multilayer O_2 shows in addition to DD above ~15 eV a much broader resonance feature between 5 and 10 eV and another structure at 13 eV. The "broadening" of the lower energy structure and the appearance of the 13-eV structure are attributed to DEA via Σ^+ states following a relaxation of the $\Sigma^- \nleftrightarrow \Sigma^+$ selection rule [122].

Also shown in Fig. 10(c)–(g) are the anion yield functions for submonolayer quantities of O_2 deposited onto various multilayer atomic and molecular solids. The data represent part of a study [41] on the environmental factors involved in the DEA process. As can be seen, the yield of desorbed ions can vary greatly with substrate composition. Such variations can be attributed to the so-called *extrinsic factors* that modify the ESD process at times before attachment and after dissociation, for example, electron energy-loss processes in the substrate and postdissociation interactions (PDI) of ions with the surrounding medium [41]. These processes can be contrasted with *intrinsic factors*, which modify the properties of the resonance (e.g., lifetime, energy, and decay channels). In general, the O^- yields per O_2 molecule are higher for O_2 deposited on RG solids Kr and Xe [Fig. 10(c) and (d)—see Ref. 211 for detailed comparison of O^- desorption from RG solids] than on a molecular solid where extrinsic effects such as electron energy-loss processes and PDI are more likely. The weakest O^- signal was observed from O_2 on H_2O [Fig. 10(g)] which, like the other substrate films, was condensed at temperature of 20 K. The figure shows that while the DD signal is reduced by approximately a factor of 100 relative to O_2 on Kr, the O^- ESD signal over the energy range associated with DEA is almost entirely absent. This dramatic effect (termed *quenching*) was initially attributed to some interaction of the transient O_2^- states with the dipole moment of the adjacent water molecules [175]. However, later work [206,238,239] has shown that amorphous water films formed at 20 K can be highly porous. When O_2 is deposited onto porous ice at 20 K (close to the sublimation temperature of oxygen), it diffuses rapidly over the film's extended surface, so that the effective density of O_2 at the film–vacuum interface is lowered considerably from that obtained when the same quantity is deposited onto a nonporous film. Oxygen ions formed inside the film have a much lower probability of desorption than those generated at the film–vacuum since they are more likely to scatter and lose kinetic energy [206]. Absolute cross sections for anion desorption and hence lower limits for DEA cross sections have been reported for O^- from O_3 [203] and CF_3I [200]. By combining ESD data with the results of charge-trapping experiments, DEA cross sections have also been from H_2O, [222,223], CF_4 [224,225], CF_3Cl [230,231,233], HCl [236], CF_2Cl_2 [235], CH_3Br [227], and the chloromethanes [226–229,231].

The E_i dependence of the surface charging cross section σ_{CT} for 0.1 ML of CH_3Cl condensed onto a 15-ML Kr film [226,227] is shown in Fig. 11(a). The two curves Fig. 11(b) and (c) show the H^- and Cl^- ESD yield functions for similar quantity of CH_3Cl deposited onto a 7-ML Kr film [182]. No ESD signal was detected below 5 eV. Curve Fig. 11(d) is the total anion yield from the gas phase as measured by Pearl and Burrow [240] and was obtained by subtracting a background signal at 0 eV arising from CCl_4. The 0.8-eV peak in this curve represents a cross section of $(2.0 \pm 0.4) \times 10^{-21}$ cm^2 for anion production, a value much smaller than that of $(13 \pm 2) \times 10^{-18}$ cm^2 for the 0.5-eV peak in Fig. 11(a). Moreover, the origin of this gas phase structure has been questioned, as its variation of magnitude with increasing temperature is consistent with the DEA of HCl [240].

The variation with Kr film thickness of the amplitude (full squares) and energy at maximum (open squares) of the lowest energy charge-trapping cross section feature is shown in Fig. 12. It is apparent that both these quantities are strongly dependent on Kr film thickness. The thickness dependence of the feature's energy is reminiscent of that observed by Michaud and Sanche [103] in the energy of the $^2\Pi_g$ resonance of N_2 on Ar, which was successfully described in terms of the polarizability of the Ar film and the image charge induced in the metal substrate. The lower dashed curve is a fit to the data of the same type of function used by Michaud and Sanche to describe the thickness-dependent changes in polarization energy V_p [103].

The structure near 0.5 eV in Fig. 11(a) has been interpreted [226,227] as due to the formation of Cl^- anions via DEA to the $^2A_1$ state of CH_3Cl^- and the subsequent dissociation of Cl^- and CH_3 along the strongly antibonding C–Cl^- σ^* orbital. Since no ESD signal is observed in Fig. 11(b) and (c) below 5 eV, the charge-trapping cross section represents an absolute DEA cross section at these low energies. The value reported for CH_3Cl on a 5-ML-thick Kr film (Fig. 12) thus represents an enhancement over gas-phase experimental [240] and theoretical values [241] of between 4 and 6 orders of magnitude,

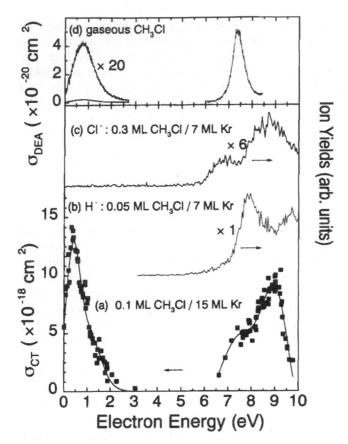

Figure 11 Charging cross section and anion yields induced by 0–10 eV electron impact on CH_3Cl. (a) Absolute surface charging cross section for 0.1 ML methyl chloride condensed on a 15-ML Kr film. (b) H^- and Cl^- desorption yields. (c) Total anion yield from gaseous methyl chloride. Figure taken from Ref. 227.

respectively. This remarkable enhancement has been attributed to image-charge-induced polarization energy V_p, which lowers the potential energy curve of the intermediate transient anion increasing the anion lifetime [242] and decreasing the distance the anion must stretch before dissociation is inevitable [226,227].

A quantitative understanding of the effect of V_p on electron attachment to CH_3Cl adsorbed on Kr [226,226] and within its bulk [229] has been attempted with a modified version of the R-matrix scattering model [241] which has been previously successful in describing LEE–CH_3Cl scattering in the gas phase [243]. Essentially, the model was modified to include the effects of V_p at short electron–molecule distances. The solid line in Fig. 12 represents the results of the R-matrix calculation. However, to obtain this result, it was also necessary to model in a semi-empirical fashion the electron's interaction with the metal substrate. The discontinuity represented by the film's surface complicates considerably this type of calculation, which is more straightforward for CH_3Cl embedded within bulk Kr. In this case [229], good agreement between theory and experiment is possible by adjusting V_p alone. However, it must be noted that the values of V_p required to fit the data

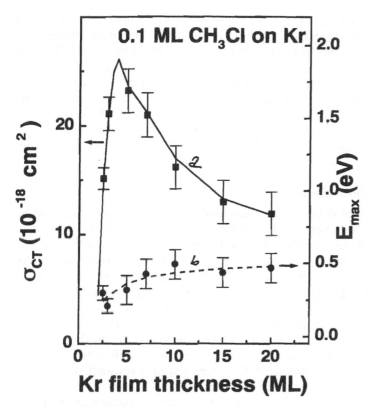

Figure 12 (a) Variation of maximum charge-trapping cross section (σ_{CT}) with Kr film thickness for low-energy electrons incident on 0.1 ML of CH_3Cl deposited on the surface of Kr films. The solid line drawn from the data represents the results of an R-matrix calculation to describe electron–CH_3Cl/Kr scattering. (b) Variation in energy E_{max} of the maximum in σ_{CT} with Kr film thickness. Dashed line is a parametric fit to E_{max}. (From Ref. 226.)

for CH_3Cl condensed on and within Kr are approximately 50% larger than those measured elsewhere in other experiments [103]. With these latter values, the calculated cross sections are about 1 order of magnitude too small. This disparity does not exist when comparing similar charging data for CF_3Cl [230] with an R-matrix calculation [231]. Comparison of the CH_3Cl and CF_3Cl data and calculations [231] suggest that the CH_3Cl^- PE curve may be modified in the Kr environment due to a change of symmetry and/or screening of the CH_3 and Cl^- interaction by the Kr medium at intermediate internuclear separations. Alternatively, Aflatooni and Burrow have suggested that the effective polarization energy seen by CH_3Cl on Kr could indeed be larger than that measured for N_2^- on Kr [103] due to the larger static dipole moment of this molecule.

Two further mechanisms are known to trap electronic charge in thin films: *intermolecular* and *resonance stabilization*. In resonance stabilization, electron attachment to a molecular center produces an anion in a vibrationally excited state that is then de-excited by energy exchange with neighboring molecules. When the initial anion ground state lies below the band edge or lowest conduction level of the dielectric, then the additional electron may become permanently trapped at the molecular site. In this case, a permanent anion is formed (e.g., the case of O_2 [220]). Intermolecular stabilization refers to the trapping of a LEE by an aggregate of molecules, typically unable to do so in the monomeric form. In this sense, the process is similar to solvation,* which usually requires the organization of polar molecules to form a suitable trap. Intermolecular stabilization has been observed for water clusters condensed on Kr and Xe surfaces at cryogenic temperatures [222] and in pure water films [223]. Both intermolecular and resonance stabilization occur at incident electron energies below 1 eV. Since DEA can, in some systems, occur at similar energies, it is not always easy to identify which process is responsible for charge accumulation. Fortunately, further information can sometimes be obtained from measurements of the ESD yield of anions from similar films and from comparison with gas phase results.

Electron Transfer from the Environment

In general, the properties of molecular TNI of condensed molecules are greatly affected by their local environment. It is unsurprising then to find that DEA can arise from electron transfer from an electron state of the substrate to a dopant molecule. Such a transfer has been observed between the electron–exciton complex of RGS and adsorbed molecules [179,180], between water films and adsorbed CF_2Cl_2 [235] and HCl [236], and between "image" states and CH_3Cl [232]. Fig. 13 presents recent charge-trapping data obtained on 10-ML-thick films of n-hexane [232]. Fig 13(a) represents the energy variation of the trapping coefficient for pure n-hexane in the glassy amorphous state "nH_g" that forms by vacuum deposition at 25 K [245].[†] The curve shows that below $E_i = 7$ eV, the threshold for charging via DEA to n-hexane [167], charging coefficients are very small. Annealing such films at 60 K produces crystalline n-hexane "nH_c", which is also resistant to charging below 7 eV [Fig 13(b)]. It is this property of both n-hexane film types (i.e., resistance to charging) which renders them suitable substrates to study electron attachment to surface dopant molecules at incident electron energies below 7 eV. Fig 13(c) presents $A_s(E_i)$ for 0.1 ML CH_3Cl on the surface of a 5-ML nH_g film. The sharp (50 meV fwhm) and intense peak near zero energy is shown on an extended energy scale in the inset [Fig 13(d)]. The maximum in A_s corresponds to a charge-trapping cross section of 2×10^{13} cm^2 which is 10^4 times larger than that for DEA to CH_3Cl on the Kr film surface [226] shown in Fig 13(e). Charging data for 0.1 ML CH_3Cl on the nH_c film surface are shown in Fig 13(f) and do not show the same intense charging at 0 eV seen on the glassy n-hexane substrate. The DEA peak observed near 0.5 eV for CH_3Cl on Kr [in Fig 13(e)] is hidden by the background signal in Fig 13(f).

Likewise, large charge-trapping cross sections, peaked sharply at 0 eV, were measured for submonolayer quantities of O_2 and H_2O deposited onto the nH_g substrate. Both of these molecules are known to trap near-zero electron volt electrons (as O_2^- [220] and by intermolecular stabilization within water assemblages [222]), but again, the cross sections on nH_g were several orders of magnitude larger than when measured on Kr or on nH_c films. Significantly, charging coefficients (and the cross sections for electron trapping) at higher energies on the nH_g surface were not enhanced, and no 0-eV peak was observed in charge-trapping measurements for CF_4 on nH_g (CF_4 on Kr *does not* trap near 0 eV electrons [224]). In light of these results, viz., a huge enhancement in $A_s(E_i)$ at 0 eV, independent of the mechanism of electron stabilization (but only for molecules that trap

---

* See, e.g., Ref. 244.

[†] More recent studies (Ref. 239) have shown n-hexane films formed at < 60 K to be highly porous, although this fact is not of great relevance to the present discussion.

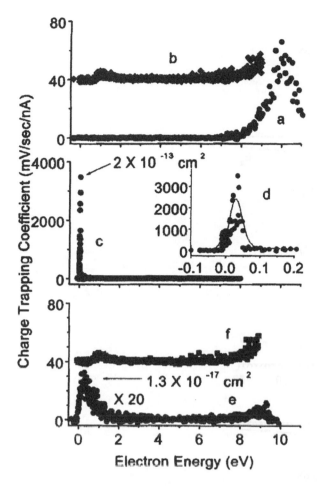

Figure 13 Charging coefficient $A_s(E_i)$ as a function of incident electron energy E_i for (a) a 10-ML film of glassy n-hexane nH_g; (b) a 13-ML film of crystalline n-hexane nH_c; (c) 0.1 ML CH_3Cl on a 5-ML nH_g surface; (d) the peak near zero energy in (c) shown on an extended scale; (e) 0.1 ML CH_3Cl on a 10-ML Kr surface; (f) 0.1 ML CH_3Cl on 13 ML nH_c surface. (From Ref. 232.)

electrons near 0 eV), it was argued that trapping proceeds via the initial capture of the electron into an "image state" of the substrate with subsequent resonant transfer of the surface electron to the molecular target [e.g., O_2, CH_3Cl, or $(H_2O)_n$]. Since in this case a large portion of the surface would be available for electron capture, the observed large trapping cross sections would likely result, provided that the lifetime of the electron in the image state near the surface was sufficiently long. Fig. 14(a) explains this process in more detail and shows schematically the band structure near the surface of the glassy n-hexane as deduced from the information contained in the LEET spectra of nH_g [55]. The nH_g film has a band gap at the vacuum level with the lowest conduction band edge lying at $V_0 = 0.8$ eV [55]. This latter, combined with the attractive interaction of an incident electron with its image charge, produces at the film–vacuum interface a potential well of a sufficient depth to support quasi-bound electron state(s). A realistic calculation for the nH_g system [232] predicts image states at 0.1 and 0.42 eV below the vacuum level with lifetimes greater than 3 nsec, assuming the decay of the image state via electron transfer through gap states.

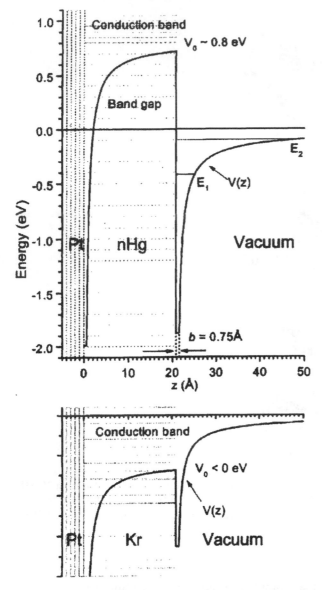

Figure 14 Schematic diagram of simplified one-dimensional band structure for an nH_g (top) and Kr (bottom) spacer layer film. (From Ref. 232.)

Normally incident near-zero electron volt electrons scattered into nonspecular directions will be captured into these states. While the image states are themselves too short-lived to contribute to charge coefficient measurements, the calculated lifetime is much longer than phonon vibrational periods, so that the probability that the excess electron transfer to any electron acceptor, be it O_2, CH_3Cl, or $(H_2O)_n$ clusters, would be large. In contrast, on Kr and nH_c films, the conduction band edge is lower [Fig. 14 (b)] [232] and the potential well at the film–vacuum interface is no longer capable of supporting an image state of sufficient lifetime to transfer to the adsorbate.

Electron-Induced Reactions

Impact of LEE on atomic and molecular solids has been observed to induce chemical changes other than dissociation. In the next two sections, we will review examples of electron-induced reactions that proceed via formation of a TNI.

Anion Desorption Measurements. Reactive scattering by DEA fragment ions in condensed media was first noted in the form of OH^- electron-stimulated desorption (ESD) yields from O_2 embedded in multilayer alkane films [246] and subsequently for aniline physisorbed on top of O_2 solids [215]. The O^- anion produced following DEA to N_2O has also been observed to react with other N_2O molecules within an Ar/N_2O matrix to generate a desorbed yield of NO^- and NO_2^-, among other products. Part of the H_2 ESD yield observed from multilayer films of H_2O, at incident electron energies below 10 eV, has also been attributed to proton abstraction by H^- fragments produced by DEA, viz., $H^- + H_2O \rightarrow \{H_3O\}^{*-} \rightarrow H_2 + OH^-$ [46], and has thus been associated with part of the unscavengeable H_2 yield formed in water radiolysis [247]. Another (post-DEA) ion–molecule reaction, directly observed in the condensed phase for ion kinetic energies well below 5 eV, is isotope exchange measured in $^{18}O_2/C^{16}O$ mixed solids [248], i.e., $^{18}O^- + C^{16}O \rightarrow \{C^{18}O^{16}O\}^{*-18} \rightarrow O^- + C^{18}O$. Formation of D_2O in synchrotron-irradiated films of N_2O has also been attributed to associative electron attachment reactions of DEA O^- fragments with matrix D_2 molecules, i.e., $O^- + D_2 \rightarrow D_2O^{*-} \rightarrow D_2O + e^-$ [249]. Measurements of electron-stimulated O_3 production in condensed O_2 films were attributed to postdissociation reactions of neutral $O(^3P,$ or $^1D)$ with adjacent molecules [250]. The threshold energy for this process being near 3.5 eV allowed DEA to be identified as the source of $O(^3P)$ at low incident electron energies. Similarly, Hedhili et al. [251], in their measurements of F^- and Cl^- desorption from condensed CF_2Cl_2, have observed the appearance of new features in the Cl^- yield function at long bombardment times. These changes were attributed to the synthesis of Cl_2 in the condensed phase by an as yet unidentified reaction pathway having DEA to CF_2Cl_2 as its initial step. Subsequent measurements of Cl^- desorption from LEE impact on films of 1,2-$C_2F_4Cl_2$ also show the production of Cl_2 and the cross section for formation of which exhibits maxima at near 0 and 10 eV. While Cl_2 production at the higher energy may involve either excited anionic or electronic molecular states, at energies below 2 eV, it can only be initiated by DEA. In other experiments, Tegeder and Illenberger [252] have shown, by comparison of anion ESD data and infrared–absorption–reflection spectroscopy, that DEA is the initial step in the production of N_2F_4 molecules within NF_3 films under bombardment with 0 to 5 eV electrons.

In early studies of anion desorption from solid O_2/hydrocarbon mixtures [246], several observations were made concerning the desorbed signal of OH^-. (1) The OH^- yield functions resembled more closely the yield functions for O^- with kinetic energies 1.5 eV from pure O_2 films than the H^- signal associated with DEA to the hydrocarbon molecule. (2) The OH^- signal from an O_2 film increased linearly with alkane surface coverage below 1 ML. (3) The threshold in energy for OH^- desorption coincided with that for O^- desorption, which is at least 2 eV below the H^- desorption onset. For these reasons, and the fact that O_2^- is the only charged product observed in the gas-phase collisions of H^- with O_2 [253], the OH^- desorption signal was attributed to hydrogen abstraction reactions of the type

$$O^- + C_nH_m \rightarrow C_nH_mO^{*-} \rightarrow OH^- + C_nH_{m-1}$$

The existence of such $C_nH_mO^{*-}$ intermediate anion collision complex had previously been suggested by Comer and Schulz [254], who measured the energies of electrons emitted during gas-phase collisions of O^- with C_2H_4. Further evidence for a complex anion formation and reactive scattering in the gas phase can be seen in the studies of Parkes [255] and Lindinger et al. [256].

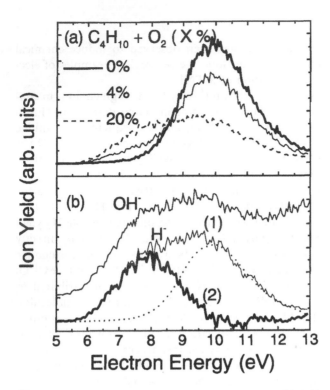

Figure 15 (a) The H⁻ yield function from 4-ML-thick films of an O_2/C_4H_{10} mixture of increasing percentage volume concentration of O_2. (b) The H⁻ signal from a 4-ML-thick mixture of 10% O_2 in C_4H_{10} can be resolved into two components: one centered at 10 eV, associated with H⁻ production via DEA to the alkane (dotted line), and another at 7.8 eV, associated with reactive scattering of O⁻ ions (bold line). Also shown is the yield of OH⁻ ions from the same film. (From Ref. 257.)

More recent measurements [257] reproduced in Fig. 15(a) show the H⁻ yield vs. E_i. An enhanced H⁻ signal *at low* E_i is observed from the O_2/C_4H_{10} mixed films. Two trends can be observed in Fig. 15(a) with increasing O_2 concentration. Firstly, there is a decrease in the yield of H⁻ at 10 eV, the peak energy of H⁻ production via direct DEA to the alkane molecule. Secondly, there is the appearance and the development of a second H⁻ ESD structure at lower energies, such that at an O_2 concentration of 20%, a substantial fraction of the H⁻ yield occurs at incident electron energies significantly below those required for DEA to C_4H_{10}. In Fig. 15(b), the H⁻ yield function from an n-butane film containing 10% O_2 by volume "curve (1)" is resolved into two separate components. One of these, with a maximum at 10 eV (dotted curve), is assumed identical in form to that of H⁻ desorption from pure C_4H_{10}. The other (bold curve 2) is obtained from curve (1) when the H⁻ signal from a pure film normalized to curve (1) at 10 eV (dotted line) is removed. This procedure yields a second quasi-Gaussian-shaped structure centered at 7.8 eV, close to the energy of maximum production of O⁻ via DEA and with a threshold at 5.5 eV, identical to that observed for OH⁻ production. From these observations, the increase in H⁻ production at incident electron energies below the threshold for DEA to C_4H_{10} was attributed to reactive scattering of O⁻ and atom-exchange reactions of the type

$$O^- + C_4H_{10} \rightarrow C_4H_{10}O^{*-} \rightarrow H^- + C_4H_9O$$

involving the same or similar transient anions as those responsible for OH^- production, as evidenced from the similarity of the OH^- and H^- yield functions in Fig. 15(b). Evidence for additional channels in the decay of such $C_nH_mO^{*-}$ anions was apparent in the anion yield functions of CH^- and CH_2^- from thin films of C_2H_4/O_2 mixtures [257] as a sub-DEA threshold component to the anion signals.

Electron-Induced Reactions—HREELS Measurements. Novel LEE-induced chemistry has also been observed in HREEL measurements of molecular solids and molecules physisorbed on the surface of RGS. For example, Lepage et al., building on the initial observations of Jay-Gerin et al. [141], have employed HREELS to measure in situ, neutral dissociation products arising from the impact of low-energy electrons on thin multilayer films of methanol [37] and acetone [38]. The technique is similar to that developed earlier by Martel et al. [258] for chemisorbed systems, in that the same electron beam is used for both the production and the detection of the neutral fragments. However, in the work of Lepage

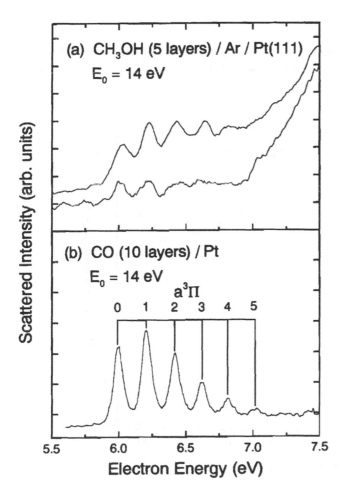

Figure 16 Electron energy-loss spectra with 14-eV electrons incident on: (a) a 5-ML film of methanol condensed on an Ar spacer after exposition to small (lower curve) and large (higher curve) electron doses and (b) a 10-ML film of CO condensed on a platinum substrate. The $a^3\Pi$ excited state of CO is characterized by a vibrational progression having a spacing of about 0.21 eV. (From Ref. 37.)

et al., neutrals are detected via their electronic excitation rather than by the spectroscopy of their vibrational levels. Fig. 16(a) shows HREEL spectra of multilayer methanol (CH_3OH) for energy losses in the range 5.5 to 7.5 eV. The incident electron energy was 14 eV. The lower curve in Fig. 16 was taken after 5 min of electron bombardment of the CH_3OH film. A weak structure is visible in the spectrum at energies below 6.7 eV. The upper curve was recorded after 20 min of exposure to the 14-eV electron beam and shows the evolution of the weak structure into a series of sharp peaks corresponding almost exactly to the vibrational progression in the lowest electronic state $a^3\Pi$ of condensed CO [shown as recorded for a 10-ML film of pure condensed CO in Fig. 16(b)].

In their first paper, Lepage et al. [37] demonstrated that the appearance of the CO electronic state derives exclusively from the electron-induced fragmentation of condensed methanol molecules. Assuming a uniform electron current density within the electron beam, it can also be shown that for induced CO concentrations below 2%, an effective cross section for CO production σ_p can be calculated via the formula

$$\sigma_p \cong \frac{n_{CO}S_0}{n_0 I_0 t} \tag{14}$$

where I_0 is the incident electron beam current and S_0 is its area; n_0 and n_{CO} are the initial number density of target molecules and the number density of CO molecules at time t, respectively.

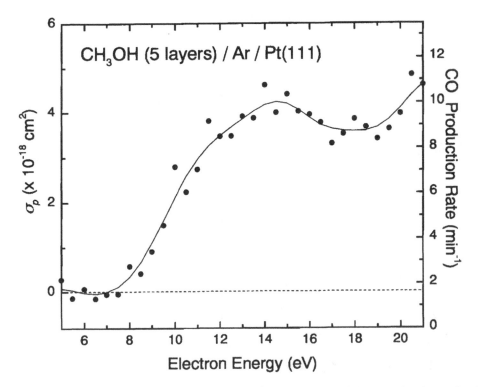

Figure 17 Total scattering cross section σ_p (left axis) and rate (right axis) for the production of CO by electron impact on a 5-ML film of methanol condensed on an Ar spacer as a function of the incident electron energy. The dashed line is the intrinsic minimum rate of CO production due to the measurement at $E_i = 14$ eV. (From Ref. 37.)

Fig. 17 exhibits the incident electron energy dependence of the σ_p and CO production rate as a function of incident electron energy. The horizontal dashed line represents the intrinsic minimum rate of CO production due to the measurement of the energy-loss spectra at 14 eV. The CO production rate was found to increase linearly with the quantity of deposited methanol indicating that the CO is generated by the interaction of an electron at a single molecular site, rather than through reactions of fragment species with surrounding (intact) molecules. The form of the CO yield function visible in Fig. 17 was attributed to the formation of multiple transient anions at these low incident energies. In addition to DEA via long-lived resonances, molecular dissociation is possible via shorter-lived states that decay to neutral repulsive states of CH_3OH. By comparison with H^- ESD yield functions [259] and vibrational excitation functions [116] for condensed CH_3OH, Lepage et al. ascribed the shoulder at 11.5 eV and broad maximum centered at 14.5 eV in Fig. 17 to the $...(6a')^1(3sa')^2$, ^2A,$...(1a)^1(3sa')^2$, ^2A, and $...(5a')1(3sa')^2$, $^2A'$ core-excited electron resonances, which decay into their parent repulsive states. The rising signal above 19 eV was attributed to direct ionization of CH_3OH.

The same experimental techniques were applied to measure effective cross sections for the electron-induced production of CO from condensed acetone [38], which again was attributed to the formation of TNI and their decay to neutral dissociative states. HREELS measurements have also been used to study electron-induced degradation of cyclopropane [260] and CH_3Cl on graphite [261].

4. RESULTS FOR LOW-ENERGY ELECTRON INTERACTIONS WITH DNA AND OTHER BIOMOLECULES

4.1. LEE Damage to Plasmid DNA

The DNA molecule is composed of two strands of repeated sugar-phosphate units hydrogen-bonded together by the bases (adenine, thymine, guanine, and cytosine) which are covalently linked to the sugar moiety of the backbone. Results obtained from electron bombardment of solid films of DNA and its constituents are presented in the following sections. These latter include H_2O, the bases, and the backbone dioxyribose analogs such as tetrahydrofuran.

Damage to pure dry samples of supercoiled DNA has been induced by bombardment with 3 eV to 1.5 keV electrons under a hydrocarbon-free 10^{-9} Torr residual atmosphere [262–265]. These experiments required the adoption of sample preparation and analytical techniques [263] unlike those already described for simple atomic and molecular solids. They are briefly discussed here. Supercoiled plasmid DNA [pGEM 3Zf(−), 3199 base pairs] is first extracted from *Escherichia coli* DH5α, purified, and resuspended in nanopure water into a clean N_2-filled glove box, where the following manipulations are performed. An aliquot of this pure aqueous DNA solution is deposited onto chemically clean tantalum substrates held at liquid nitrogen temperatures, lyophilized with a hydrocarbon-free sorption pump at 0.005 Torr, and transferred directly in a UHV chamber without exposure to air. After evacuation for ~24 hr, room-temperature DNA solid films of 5-ML average thickness and 6-mm diameter are irradiated with a monochromatic LEE beam of the same diameter. Each target is bombarded for a specific time at a fixed beam current density (2.2×10^{12} electrons sec^{-1} cm^{-1}) and E_i. After 24 hr under UHV, the DNA still contains 2.5 H_2O molecules/base pair as structural water [266]. After bombardment, the DNA is analyzed by agarose gel electrophoresis and quantified as supercoiled (undamaged), nicked circle (single strand break—SSB), full-length linear (double strand break—DSB), and short linear forms [264]. By measuring the relative quantities of the various forms of DNA in the sample as a function exposure to 10, 30, and 50 eV electrons, it was possible to obtain the total effective cross section ($\sim 4 \times 10^{-15}$ cm^2) and effective range (~ 13 nm) for destruction of supercoiled DNA at these energies [263]. Such experiments also allowed the identification of the regime in which the measured yields were linear to electron exposure. It is within this regime that the incident electron energy dependence of damage to DNA was recorded more continuously between 3 and 100 eV [262,263,265]. The most striking observations were that (1) these yields did not depend on the ionization cross section and (2) below 15 eV, they varied considerably with E_i. Above 20 eV, the yields of SSB and DSB did not exhibit pronounced energy variations nor did they increase considerably with increasing energy.

The incident electron energy dependence of these yields below 20 eV are displayed in Fig. 18 and show that LEE induce SSBs and DSBs, even at electron energies below the ionization limit of DNA (~ 8 eV) [267]. Also, the damage is highly dependent on the initial kinetic energy of the incident electron, particularly below 14 to 15 eV, where thresholds near 3 to 5 eV and intense peaks near 10 eV are observed. The results shown in Fig. 18 can be understood by investigating the fragmentation induced by LEEs to the various subunits of the DNA molecule, including its structural water.

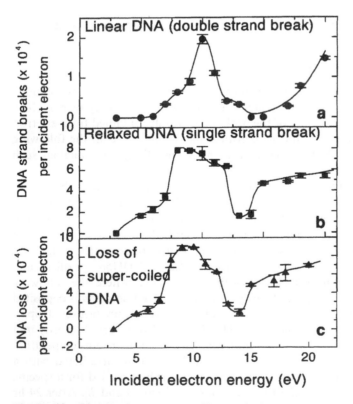

Figure 18 Measured quantum yields, per incident electron, (a) for the induction of DSBs, (b) SSBs, and (c) loss of the supercoiled DNA form, in DNA solids by low-energy electron irradiation as a function of incident electron energy; the curves are guides to the eye. (From Ref. 262.)

4.2. Electron-Stimulated Desorption from Water Ice Films

The damage induced by LEE impact on amorphous ice films has been measured by recording H[-] [223,268,269], H_2 [46,270], $D(^2S)$, $O(^3P)$, and $O(^1D_2)$ [47,271] desorption yield functions in the range 5–30 eV. Most of these functions exhibit resonance structures below 15 eV which are characteristic of TNI formation. From anion yields, DEA to condensed H_2O was shown to result principally into the formation of H[-] and the OH radical from dissociation of the $^2B_1$ state of H_2O^- located in the 7–9 eV region as shown in Fig. 19. Smaller contributions from the $^2A_1$ and $^2B_2$ anionic states formed near 9 and 11 eV, respectively [268,269], contribute to a lesser extent to the H[-] signal. At higher energies, nonresonance processes such as DD lead to H_2O fragmentation with the assistance of a broad resonance extending from 20 to 30 eV [268,269]. In some of the water ice experiments, the temperature of the substrate was modified before or after water condensation in order to investigate the effects of morphology and porosity [44,223,269].

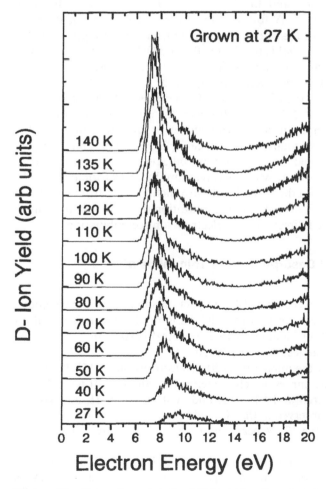

Figure 19 Dependence on E_i of D[-] yield collected at various temperatures from a 20-bilayer porous amorphous ice films grown at 27 K. Individual scans are offset vertically for display and are labeled with the temperature at which they were collected. (From Ref. 223.)

The temperature dependence of the D^- signal from 20 bilayers of porous amorphous ice, grown at 27 K [223], is shown in Fig. 19. Both the resonance energies and the ion yields change with temperature. An analysis of the D^- onset energy and the energy and intensity of the low-energy resonance peak indicates that there is a break in this behavior at around 60 K. The peak intensity increases fivefold between 27 and 60 K, remains roughly constant between 60 and 100 K, and then more than doubles in going from 100 to 140 K. These results were attributed to thermally induced movement of the hydrogen-bonding network which changes the orientation of the surface molecules and the lifetimes of the predissociative transient anions that lead to ESD.

Kimmel et al. measured the D_2 (X $^1\Sigma_g^+$), $D(^2S)$, $O(^3P_{j=2,1,0})$, and $O(^1D_2)$ products [47,271] desorbing from D_2O multilayers (~ 20 nm) grown on a platinum(111) crystal at 88 K under conditions which are known to produce amorphous ice. The ice samples were irradiated with a pulsed electron beam in an apparatus similar to the one shown in Fig. 2, and the desorbing neutral products were ionized by laser resonance-enhanced multiphoton absorption. State-specific time-of-flight distributions were obtained by varying the delay time between the electron-beam pulse and the laser pulse. The $D(^2S)$ yield function exhibited an apparent threshold at ~ 6.5 eV followed by a rapid increase in signal until a distinct plateau was reached for $\sim 14 \leq E_i \leq 21$ eV. The signal increased more gradually for electron energies above 21 eV. The thresholds for the $O(^3P_2)$ and $O(^1D_2)$ signals also lay in the range 6 to 7 eV, but the line shape of the yield function was featureless.

The low threshold energies for the production of $D(^2S)$, $O(^3P)$, and $O(^1D_2)$ show the importance of valence excited states in the ESD of neutral fragments [47]. The pathway for $D(^2S)$ desorption probably involves $D_2O^*D + OD$. However, the thresholds for producing $O(^3P_2)$ and $O(^1D_2)$, which are the same within experimental error, are lower than the 9.5- and 11.5-eV thermodynamic energies required to produce $O(^3P_2) + 2D(^2S)$ and $O(^1D_2) + 2D(^2S)$, respectively. The low threshold values therefore indicate that the formation of $O(^3P_2)$ and $O(^1D_2)$ must occur by a pathway which involves simultaneous formation of D_2. Kimmel et al. have in fact reported [46] a threshold for the production of D_2 from D_2O ice at ~ 6 to 7 eV, which supports this conclusion. Above the ionization threshold of amorphous ice, these excited states can be formed directly or via electron–ion recombination.

4.2. Anion Electron-Stimulated Desorption from Thin Films of Deoxyribose Analogs

To understand how breaks may occur via direct LEE impact on subunits of the backbone of DNA, ESD experiments were performed with solid films of the sugar-like analogs [272,273] tetrahydrofuran (**I**) and its analogs 3-hydroxytetrahydrofuran (**II**) and α-tetrahydrofuryl alcohol (**III**) (structures shown in Fig. 20).

The H^- ion yields desorbed by the impact of 1–20 eV electrons on 10-ML films of **I**, **II**, and **III** are shown in Fig. 20. The curves are characterized by an onset at 6.0, 5.8, and 6.0 eV and a yield maximum centered at 10.4, 10.2, and 10.0 eV for **I**, **II**, and **III**, respectively. A second feature is also observed in the H^- yield function for **II**; it appears as a shoulder on the low-energy side of the 10-eV peak and is characterized by a sharp onset and a peak maximum near 7.3 eV.

All features below 15 eV in Fig. 20 are characteristic of DEA to **I**, **II**, and **III**. The monotonic rising signal in the H^- yield with an energetic threshold near 14.5 eV is characteristic of nonresonant DD of C–H bonds in **I**, **II**, and **III**; it could also partially arise from DD of the O–H bond in **II** and **III**. However, the broad feature centered around 22 eV has been attributed to DEA and/or resonance decay into an electronically excited state dissociating into H^- and the corresponding cation. The formation of H^- via DEA

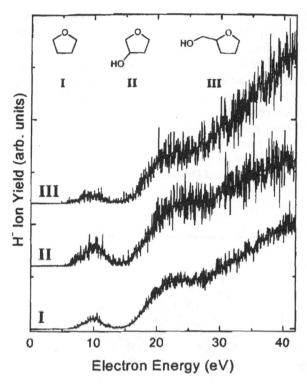

Figure 20 Comparisons of H$^-$ ESD yields produced by impact of 0- to 40-eV electrons on 6-ML-thick films of the DNA backbone sugarlike analogs tetrahydrofuran (**I**), 3-hydroxytetrahydrofuran (**II**), and α-tetrahydrofuryl alcohol (**III**). The smooth solid lines serve as guides to the eye. (From Ref. 272.)

from **I**, **II**, and **III** has been discussed in detail by Antic et al. [272,273]. These authors considered the possibility of H$^-$ arising from dissociation of the tetrahydrofuran ring, the OH, and the –CH$_2$OH group. Other decay channels of the transient anions could result in the formation of larger anion fragments, such as OH$^-$ and CH$_2$OH$^-$, and compete with H$^-$ production, but these heavier ions were not observed to desorb. Typically, large mass fragments do not possess sufficient kinetic energy to escape induced polarization and thus remain trapped within the film [147]. Owing to the strong similarity of the H$^-$ desorption profiles for **I**, **II**, and **III**, these authors concluded that "the majority of the anion yield for all three systems arises from at least one transient anion associated with electron attachment to the furan ring and located near 10 eV." Considering the largely Rydberg character of the excited states in **I** near the energy range of the observed resonance, they further suggested that this resonant state is of the core-excited type, possibly with dissociative valence σ* configurational mixing.

4.4. Anion Electron-Stimulated Desorption from Thin Films of DNA Bases

Electron impact dissociation of DNA bases has been investigated in the gas [274] and the condensed phase [26,39,275]. In both phases, a large variety of stable anions is produced via DEA to the bases for electron energies below 30 eV. The anions H$^-$, O$^-$, CN$^-$, CH$^-$, OCN$^-$, and OCNH$_2^-$ were observed from gaseous thymine, whereas electron impact on gaseous cytosine resulted in the formation of the anions H^-, C^-, O^-, and/or NH_2^-, CN^-, OCN^-, $C_4H_5N_3^-$, and/or $CH_4N_3O^-$ and $C_3H_3N_2^-$ [274]. All four bases were investigated in the form of thin multilayer films held at room temperature, but fewer anions of different masses were measured than in the gas phase [26,275]. The difference is principally due to the inability of the heavier anions to overcome the polarization potential they induce in the film, causing them to remain undetected in the target [147]. In fact, only the light anions H^-, O^-, OH^-, CN^-, OCN^-, and CH_2^- were found to desorb by the impact of 5–30 eV electrons on the physisorbed DNA bases adenine, thymine, guanine, and cytosine via either single or complex multibond dissociation [275]. As an example, the CN^- yield functions produced by 3–20 eV electron impact on thin films of each of the four bases are shown in Fig. 21. All yield functions exhibit maxima which are characteristic of DEA to the bases. Whereas anions

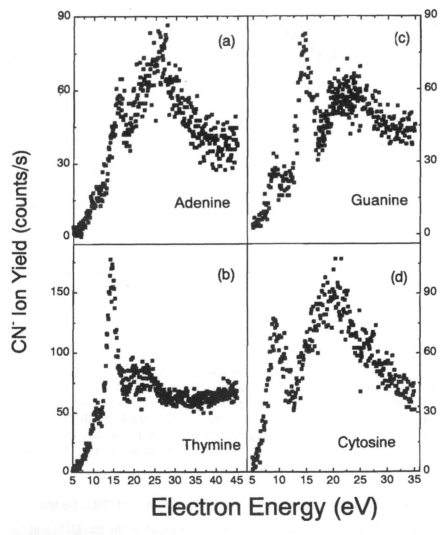

Figure 21 Dependence on E_i of CN^- ion yields desorbed under 500-nA electron bombardment of 8-ML-thick films of (a) adenine, (b) thymine, (c) guanine, and (d) cytosine. (From Ref. 275.)

such as H^- can be induced via a simple bond cleavage from the bases, the production of CN^- occurs under more complex ring dissociation. It should be noted that the extraction of a CN^- anion from any DNA base must involve more than a single bond cleavage. In other words, during the lifetime of the negative ion, not only exocyclic single bond occurs, but also complex endocyclic multibond cleavages are involved via most likely either stepwise [276] or rearrangement reactions [277]. The similarity of the CN^-, OH^-, and O^- ion yield functions with resonant peaks at 9.0 and 10 eV from guanine or cytosine and thymine films, respectively, observed by Abdoul-Carime et al. [275], suggests that they arise from the formation of the same excited isocyanic anion intermediate $(OCNH)^{*-}$, via closely lying but distinct resonances, i.e., $(G^-, Cy^-,$ or $T^-)R + (OCNH)^{*-}$, where R represents the remaining radical; then $(OCNH)^{*-}$ further undergoes fragmentation into different possible dissociative channels: $CN^- + OH$, $OH^- + CN$, or $O^- + CNH$.

4.5. Neutral Species Electron-Stimulated Desorption from Short Single DNA Strands

In recent experiments, the damage produced by LEE impact on short single strands of DNA was measured by ESD. Work in this area has been largely performed by Dugal et al. [278,279] and Abdoul-Carime et al. [280–283], who measured the yields of neutral fragments induced by 1–30 eV electron impact on oligonucleotides consisting of 6 to 12 base units. The oligomers were chemisorbed on a gold surface via sulfur bounding. Their results showed that LEE impact dissociation of DNA bases led to desorption of CN, OCN, and/or H_2NCN neutral species as the most intense observable yields. No sugar moieties were detected; neither phosphorus-containing fragments nor entire bases were observed to desorb. These results were obtained from mass spectrometric measurements of the residual atmosphere near the target during its bombardment in UHV by a high current 10^{-8} A electron beam. Fig. 22 presents the electron energy dependence of neutral CN (solid square) and OCN (and/or H_2NCN) (open circle) fragments desorbed per incident electron from oligomers consisting of nine bases [281]. The upper, middle, and lower panels show the results from oligonucleotides consisting of nine cytosine bases, Cy_9, six cytosine bases and three thymine bases, Cy_6T_3, and six cytosine bases and three bromouracil bases, $Cy_6\text{-}BrU_3$, respectively. The latter base is not naturally found in DNA but can be substituted for thymine and be incorporated into cellular DNA to enhance the therapeutic effects of high-energy radiation [284]. Above 20 eV in Fig. 22, the neutral fragment production increases linearly with the E_i, which is indicative of molecular fragmentation governed mostly by nonresonant DD and/or dissociative ionization of the bases. Below 20 eV, base fragmentation may involve excitation to dissociative electronic neutral states and DEA [278–280]. The curves in Fig. 22 present broad peaks extending from 7 to 15 eV for all oligomers. At these high energies, these broad maxima are likely to reflect the formation of core-excited resonances that are dissociative in the Franck–Condon region. This interpretation is supported by: (1) the electron energy losses in solid phase DNA bases [285] in the 7–15 eV range, which are attributed to either π or σ orbitals; and (2) observation of resonant formation CN^-, O^-, and/or NH_2^- and H^- at 9–10, 7–9, and 12–13 eV, respectively, in electron-stimulated desorption yields from thin films of DNA bases [26,275]. Moreover, the threshold of neutral species production observed at 5 eV coincides with the threshold for electronic excitations, suggesting that the onset of the nucleic acid fragmentation involves dissociative electronic transitions.

   An extra peak is observed at 3 eV for BrU-substituted oligonucleotides in the inset at the bottom of Fig. 22. Since this peak lies at an energy too low to involve electronically

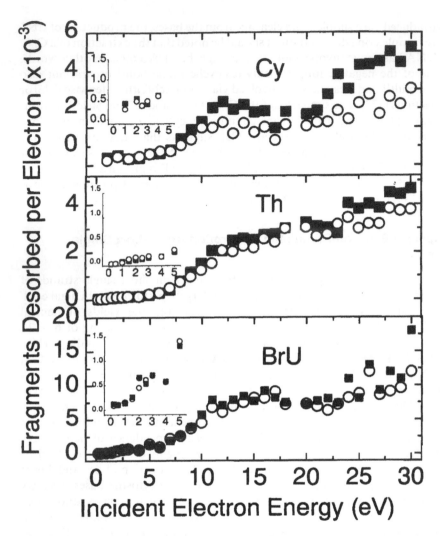

Figure 22 Dependence on E_i of neutral CN (solid square) and OCN (and/or H_2NCN) (open circle) fragment desorption yields per incident electron from Cy_6-$(Cy)_3$ oligos (upper panel), Cy_6-$(Th)_3$ oligos (middle panel), and C_6-$(BrU)_3$ oligos (bottom panel). The spread in the data is estimated to be 20%. (From Ref. 281.)

excited states, it probably arises from the formation of a shape resonance consisting of the BrU molecule in the ground state with an electron occupying a usually unfilled orbital. In XPS [39] investigations of LEE-induced damage to Bromouracil, it was found that Br production below 5 eV involves resonant electron capture by BrU followed by dissociation into a uracil-yl radical (U).

4.6. Discussion

In previous sections, we have seen that the dependence on E_i of damage to elementary constituents of DNA, probed in the form of desorbed anions, exhibits strong variations due to electron resonances. Maxima in the anion desorption yields are due to DEA, while those appearing in the neutral radical yields may result from either DEA or the decay of a transient anion into an electronically excited state which is dissociative. Thus from these data, it becomes quite obvious that the strong energy dependence of the DNA strand breaks below 15 eV in Fig. 18 can be attributed to the initial formation of transient anions, decaying into the DEA and/or dissociative electronic excitation channels. However, since the basic DNA components (i.e., the sugar and base units and structural H_2O) can all be fragmented via DEA between 5 and 13 eV, it is not possible to unambiguously attribute SSB and DSB to the initial dissociation of a specific component. For example, the maximum in the SSB yield lying near 8 eV is very close to the maximum in $H^- + OH$ production from H_2O, but the DNA bases also produce H^- and the corresponding radical with high efficiency near 9 eV. What appears to be more convincing is the coincidence in the 10-eV peak in H^- production in Fig. 20 from tetrahydrofuran and its analogs, with that in the DSB yield seen in Fig. 18. This is not surprising since two events are necessary to create a DSB, and the probability for such breaks is likely to be larger when the first hit breaks a strand (i.e., breaks the sugar-phosphate backbone). Another interesting aspect of the DSB yield curve is the absence of damage induced by direct (nonresonant) electronic transitions. Indeed, the two points in Fig. 18 around 14 and 15 eV lie at the zero base line, indicating that below ~16 eV, DSB occurs exclusively via the decay of transient anions. These observations suggest that below 16 eV, DSB occurs via molecular dissociation on one strand initiated by the decay of a transient anion, followed by reaction of at least one of the fragments on the opposite strand. This hypothesis is further supported by the observation of electron-initiated fragment reactions (such as hydrogen abstraction, dissociative charge transfer, atom and functional group exchange, and reactive scattering) occurring over distances comparable to the DNA's double-strand diameter (~2 nm) in condensed films containing water or small linear and cyclic hydrocarbons [246,257].

5. APPLICATIONS IN OTHER FIELDS

5.1. Dielectric Aging and Breakdown

Electrical insulation is often exposed to stresses which with time reduce the insulating properties of the dielectric material. This deterioration is referred to as "aging." Aging stresses include temperature, applied voltage, and mechanical force; these are affected by environmental factors such as the presence of water, oxygen, corrosive materials, and radiation. While "electronic" aging (i.e., that due to the application of high electric fields) produces macroscopic modifications of dielectric materials, it is probably initiated at the nanoscale level in the early life of the material under stress [72,286,287]. The accumulated "space charge" is both a precursor to and a contributing cause of electrical breakdown [72]. In fact, it has been shown that LEEs play a significant role in the processes leading to the breakdown of the dielectric materials used to insulate high-voltage power cables [162]. These so-called "hot" electrons can be released into the conduction band following energy absorption from phonons in the dielectric or from external particles penetrating the solid (e.g., UV photons and γ-rays) [7,162]. Despite being accelerated by the ambient electric field, initial hot electrons would rarely gain sufficient energy to break molecular bonds before becoming trapped. However, those accelerated in cavities of micrometer diameters could attain the higher kinetic energies of several eV, required to rupture molecular bonds and hence contribute to the aging process. Consequent changes in chemical nature or excessive charge accumulation at specific locations would eventually cause electrical breakdown [7,162].

To better understand the mechanisms responsible for charge accumulation in dielectrics, the technique for measuring induced charge in millimeter-thick molecular solids described in Sec. 3.1.2 has been applied to macroscopic samples of cross-linked polyethylene (XPLE) cut from industrial power lines [32]. As an example of the type of information obtainable with the new instrument, Fig. 23 shows both the instantaneous trapping probability $\Delta Q_t/Q_0$, measured with the electrometer (black dots), and $\Delta V/Q_0$ (open circles), obtained with the Kelvin probe, as functions of E_i. A geometrical factor α of 1.3, consistent with the experimental geometry [32], is required to bring these two type of data into quantitative agreement. The data presented in Fig. 23 reveal that charge trapping is most efficient at very low energies, as is evidenced by the peak near 1 eV. A smaller, second peak is observed near 10 eV. The negative data values obtained between 15 and 25 eV indicate that over this range, the sample charges *positively* with respect to the unbombarded sample, taken as the zero potential reference.

From comparison with the results of charge trapping and anion ESD experiments on pure and doped *n*-hexane films [32] (see as an example Fig. 13), it has been suggested that negative charging between 7 and 15 eV in XPLE is dominated by DEA to alkane chains. In contrast, charging below 5 eV is likely to be associated with molecular impurities (water, O_2, and antioxidant additives) since pure alkane molecular solids are unable to trap electrons at these energies (Fig. 13).

5.2. Electron Beam and Nanolithography

Intense interest exists in producing nanometer-scale structures for new and existing applications in the fields of electronics, optics, and molecular electronics. Among the most promising approaches to producing submicrometer structures are electron beam

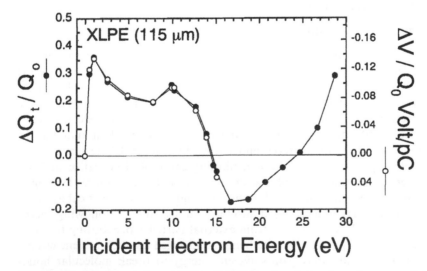

Figure 23 The trapping probability ($\Delta Q/Q_0$) and variation in surface potential per incident charge $\Delta \overline{V}/Q_0$, as functions of incident electron energy for the 115-mm-thick cross-linked polyethylene (XLPE) sample. (From Ref. 32.)

lithography and its newer "offspring," scanning-probe lithography, in which a scanning tunneling microscope (STM) or an atomic force microscope is used to deliver (or induce) a localized electron current [288,290].* One limitation to producing nanoscale etching masks is the large size of molecules used in the standard polymeric resists. Considerable interest lies then in producing new resists from smaller molecules and/or self-assembled mono-layers [9]. In other research, an STM is used to produce a resist by inducing reactions between a substrate and adsorbed molecules, for example, the oxidation of Si(111) and (100) surfaces under ambient conditions [290] or that between chlorobenzene and Si(111) [291]. Low-energy electrons are present in such situations, either as a primary source of energy or as secondary electrons produced by high-energy electron beams. Therefore the electron-scattering processes already described should play a role in the electron-induced modification and/or chemical changes. Indeed, LEEs via DEA or other processes might be usefully employed to induce particular chemical reactions.[†] In this case, it would appear advantageous to study the electron energy dependence of any chemical changes in the resist material.

As an example, XPS has been used to analyze modifications induced by 2 to 20 eV electrons incident on a hydrogen-passivated and sputtered Si(111) surface, onto which had been *physisorbed* thin films of H_2O [293,294] and CF_4 [295]. In both cases, following the electron-induced dissociation of the molecular adsorbate, a new XPS signal associated with the *chemisorption* of either O or F onto the Si surface was observed and an effective cross section for chemisorption was then calculated. This cross section for electron-induced chemisorption of oxygen from an H_2O bilayer onto a hydrogen-passivated Si(111) surface is shown in Fig. 24 as a function of E_i [293,294]. The low energetic threshold for the chemisorption process (i.e., 5.2 eV) has been interpreted as due to the formation of OH via the DEA process

$$H_2O[X] + e^- \leftrightarrow H_2O^{\cdot -}[^2B_1, {}^1A_1] \rightarrow H^-[^2S_0] + OH[^2\Pi_1]$$

and its subsequent reaction with the surface, viz.,

$$Si_3 - Si - H + OH \rightarrow Si_3 - Si - OH + H.$$

The chemisorption process has its maximum cross section at ~ 11 eV in contrast with cross sections for the radiolysis of bulk ice. This difference was understood as being dependent on the selective quenching of dissociative electronic states of water due to the resonant charge exchange between the substrate and adsorbate and the absence of multiple inelastic scattering in the H_2O bilayer on Si(111).

A similar XPS study has been performed for ~ 1 ML of CF_4 deposited onto the same hydrogen-passivated Si substrate [295] and clearly demonstrates that chemisorption of the F-containing species is dependent on the DEA reactions

$$CF_4 + e^- \rightarrow CF_4 \rightarrow F^- + CF_3$$

$$CF_4 + e^- \rightarrow CF_4 \rightarrow F + CF_3^-$$

as an initial step for further reactions of fragments with the Si substrate.

---

* For a review, see Ref. 289.
[†] For example, Ref. 292.

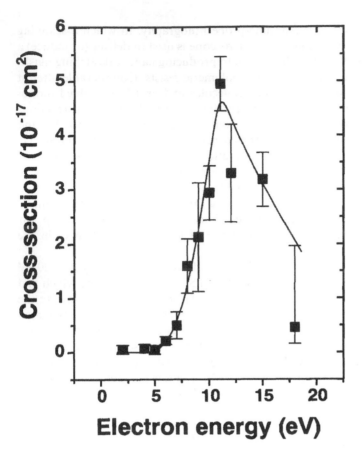

Figure 24 Experimental cross sections for the electron-induced oxidation of the a-H:Si(111) surface as a function of incident electron energy. The solid line is drawn to guide the eye. (From Ref. 294.)

5.3. Atmospheric Processes

In a recent review, Madey et al. [296] describe how in interplanetary space, energetic photons, ions, and electrons from the solar wind, together with galactic and extragalactic cosmic rays, constantly bombard surfaces of planets, planetary satellites, dust particles, comets, and asteroids. The authors suggest that much can be learned about the tenuous atmospheres surrounding these bodies by studying the interactions of energetic particles producing LEE secondary electrons with the molecular constituents of their surfaces. For example, Sieger et al. [10] have demonstrated the production of O_2 from ice films by electron impact at 200 eV and have posited this mechanism as a source of O_2 in the atmospheres of Ganymede and Europa.

The presence of ionizing radiation in the upper regions of the earth's atmosphere and the realization that "atmospheric" chemistry can occur on the surface of ice and dust particles have lead many authors to study on the interaction of LEE with molecular solids of ozone [203], HCl [236], and halogen-containing organic compounds [176,177,195–197,199–202,205,214,217,224–234] in an effort to shed new light on the problem of ozone depletion. In a recent series of experiments, Lu and Madey [297,298] found that the F^- and Cl^- yields produced by the impact of 250-eV electrons on CF_2Cl_2 adsorbed on a metal Ru surface were enhanced by several orders of magnitude when co-adsorbed with polar molecules such as H_2O and NH_3. Subsequent charge-trapping measurements by Lu and Sanche [235] also display an enhancement in stable anion formation at electron impact energies near 0 eV. Fig. 25(a) shows charging coefficient $A_s(E_i)$ for 10 ML of Kr deposited on Pt foil (solid triangles) and that for 0.1 ML of CF_2Cl_2 deposited on the Kr surface (open squares). These results can be compared to those for 5 ML of H_2O on 10 ML of Kr and for 0.1 ML of CF_2Cl_2 on 5 ML of H_2O on Kr [solid circles and open diamonds, respectively, in Fig. 25(b)]. In contrast to pure Kr films, which do not trap electrons, both H_2O- and CF_2Cl_2-covered films show significant charging. The results for H_2O are similar to earlier measurements [222,223]. Those for CF_2Cl_2 correspond to a maximum trapping cross section of 1.4×10^{-15} cm^2 [235] near 0 eV, attributed to charge stabilization as Cl^-, via the previously identified DEA reaction $CF_2Cl_2 + e(\sim 0\ \text{eV}) \rightarrow CF_2Cl + Cl^-$ [299], and here enhanced by approximately an order of magnitude by the effects of surface polarization (e.g., Ref. 226). However, when the same quantity of CF_2Cl_2 is deposited on the H_2O, the charging coefficient per CF_2Cl_2 molecule at

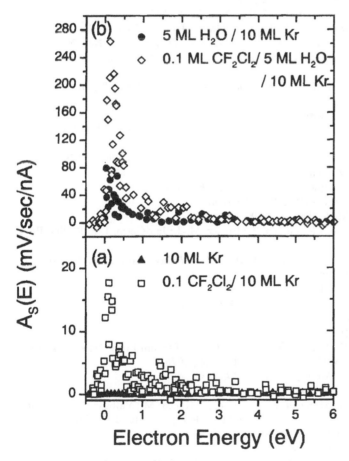

Figure 25 Charging coefficient A_s as a function of E_i for (a) 10 ML Kr condensed (solid triangles) and 0.1 ML CF_2Cl_2 on 10 ML Kr (open squares); (b) 5 ML H_2O on 10 ML Kr (solid circles) and 0.1 ML CF_2Cl_2 on 5 ML H_2O on 10 ML Kr (open diamonds). (From Ref. 235.)

0 eV increases approximately a factor 10 to 1.3×10^{-14} cm^2. A further order of magnitude enhancement is observed for CF_2Cl_2 on NH_3 [235].

It has been argued [235] by analogy with the case of molecules adsorbed on glassy n-hexane [232] that this enhancement is due to the electron transfer to CF_2Cl_2 of an electron previously captured in a precursor state of the solvated electron in the water layer, which lies at and just below the vacuum level [300,301] and the subsequent. Similar results have been reported for HCl adsorbed on water ice [236]. It has been proposed that enhanced DEA to CF_2Cl_2 via electron transfer from precursor-solvated states in ice [235] may explain an apparent correlation between cosmic ray activity (which would generate secondary LEE in ice crystals) and atmospheric ozone loss [11]. The same electron transfer mechanism may contribute to the marked enhancement in electron, and x-ray-induced dissociation for halo-uracil molecules is deposited inside water ice matrices [39].

REFERENCES

1. Franck, J.; Hertz, G. Verh. Dtsch. Phys. Ges. 1914, *16*, 512.
2. Brown, S.C. *Electron–Molecule Scattering*. Wiley: New York, 1979.
3. Hinze, J. *Electron–Atom and Electron–Molecule Collisions*. Plenum: New York, 1983.
4. Shimamura, I.; Takayangi, K. *Electron–Molecule Collisions*; Plenum Press: New York, 1984.
5. Christophorou, L.G. *Electron–Molecule Interactions and Their Applications, 1 and 2*. Academic: Orlando, Fl, 1984.
6. *International Commission on Radiation Units and Measurements, ICRU Report 31*; ICRU: Washington, DC.
7. Sanche, L. IEEE Trans. Electr. Insul. 1993, *28*, 789.
8. Lin, J.-L.; Singh Bhatia, C.; Yates, J.T. Jr. J. Vac. Sci. Technol., A 1995, *13*, 163; Lin, J.-L.; Yates, J.T. Jr. J. Vac. Sci. Technol., A 1995, *13*, 1867.
9. Frey, S.; Heister, K.; Zharnikov, M.; Grunze, M. Phys. Chem. Chem. Phys. 1979, 2002, 2; Hernandez, J.E.; Ahn, H.; Whitten, J.E. J. Phys. Chem. B. 2001, *105*, 8339.
10. Sieger, M.T.; Simpson, W.C.; Orlando, T.M. Nature 1998, *394*, 554.
11. Lu, Q.-B.; Sanche, L. Phys. Rev. Lett. 2001, *87*, 078501.
12. LaVerne, J.A.; Pimblott, S.M. Radiat. Res. 1995, *141*, 208.
13. Cobut, V.; Frongillo, Y.; Patau, J.P.; Goulet, T.; Fraser, M.-J.; Jay-Gerin, J.-P. Radiat. Phys. Chem. 1998, *51*, 229.
14. Sanche, L. In *Excess Electrons in Dielectric Media, Chap. I.* Jay-Gerin, J.-P., Ferradini, C. CRC Press: Boca Raton, FL, 1991.
15. Sanche, L. J. Phys. B. At. Mol. Opt. 1990, *23*, 1597.
16. Takayanagi, K. Prog. Theor. Phys. Japan Suppl. 1967, *40*, 216.
17. Schulz, G.J. Rev. Mod. Phys 1973, *45*, 378.
18. O'Malley, T.F. Phys. Rev. 1966, *150*, 14.
19. Shimamura, I.; Takayangi, K. *Electron–Molecule Collisions*; Plenum Press: New York, 1984.
20. Massey, H.S.W. *Negative Ions*; University Press: London, 1976.
21. Palmer, R.E.; Rous, P.J. Rev. Mod. Phys. 1992, *64*, 383.
22. Palmer, R.E. Prog. Surf. Sci. 1992, *41*, 51.
23. Franchy, R.; Bartolucci, F.; de Mongeot, F.B.; Cemic, F.; Rocca, M.; Valbusa, U.; Vattuone, L.; Lacombe, S.; Jacobi, K.; Tang, K.B.K.; Palmer, R.E.; Villette, J.; Teillet-Billy, D.; Gauyacq, J.P. J. Phys. Condens Matter 2000, *12*, R53–R82.
24. Sanche, L. J. Chem. Phys. 1979, *71*, 4860.
25. Perluzzo, G.; Bader, G.; Caron, L.G.; Sanche, L. Phys. Rev. Lett. 1985, *55*, 545.
26. Hervé du Penhoat, M.-A.; Huels, M.A.; Cloutier, P.; Jay Gerin, J.-P.; Sanche, L. J. Chem. Phys. 2001, *114*, 5755.

27. Rowntree, P.; Dugal, C.; Hunting, D.; Sanche, L. J. Phys. Chem. 1996, *100*, 4546.
28. Abdoul-Carime, H.; Dugal, P.-C.; Sanche, L. Radiat. Res. 2000, *153*, 23.
29. Stamatovic, A.; Schulz, G.J. Rev. Sci. Instrum. 1970, *41*, 423.
30. Marsolais, R.M.; Deschênes, M.; Sanche, L. Rev. Sci. Instrum. 1989, *60*, 2724.
31. Nagesha, K.; Gamache, J.; Bass, A.D.; Sanche, L. Rev. Sci. Instrum. 1997, *68*, 3883.
32. Bass, A.D.; Cloutier, P.; Sanche, L. J. Appl. Phys. 1998, *84*, 2740.
33. Jones, T.S. Vacuum 1992, *43*, 177.
34. Thiry, P.A. J. Electron Spectrosc. Relat. Phenom. 1986, *39*, 273.
35. Michaud, M.; Sanche, L. Phys. Rev. A 1987, *36*, 4672.
36. Michaud, M.; Sanche, L. Phys. Rev. A 1987, *36*, 4684.
37. Lepage, M.; Michaud, M.; Sanche, L. J. Chem. Phys. 2000, *113*, 3602.
38. Lepage, M.; Michaud, M.; Sanche, L. J. Chem. Phys. 1997, *107*, 3478.
39. Klyachko, D.V.; Huels, M.A.; Sanche, L. Radiat. Res. 1999, *151*, 177.
40. Olsen, C.; Rowntree, P.A. J. Chem. Phys. 1998, *108*, 3750.
41. Huels, M.A.; Parenteau, L.; Sanche, L. J. Chem. Phys. 1994, *100*, 3940.
42. Tronc, M.; Azria, R.; Le Coat, Y.; Illenberger, E. J. Phys. Chem. 1994, *100*, 14745.
43. Meinke, M.; Illenberger, E. J. Phys. Chem. 1994, *98*, 6601.
44. Sieger, M.T.; Simpson, W.C.; Orlando, T.M. Phys. Rev. B 1997, *56*, 4925.
45. Wurm, S.; Feulner, P.; Menzel, D. Surf. Sci. 1998, *400*, 155; Scheuer, M.; Menzel, D.; Feulner, P. Surf. Sci. 1997, *390*, 23.
46. Kimmel, G.A.; Orlando, T.M.; Vezina, C.; Sanche, L. J. Chem. Phys. 1994, *101*, 3282.
47. Kimmel, G.A.; Orlando, T.M. Phys. Rev. Lett. 1995, *75*, 2606.
48. Harries, T.D.; Lee, D.H.; Blumberg, M.Q.; Arumainaygam, C.R. J. Phys. Chem. 1995, *99*, 9530.
49. Leclerc, G.; Bass, A.D.; Mann, A.; Sanche, L. Phys. Rev. B 1992, *46*, 4865.
50. Perluzzo, G.; Sanche, L.; Gaubert, C.; Baudoing, R. Phys. Rev. B 1984, *30*, 4292.
51. Sanche, L.; Bader, G.; Caron, L. J. Chem. Phys. 1982, *76*, 4016.
52. Sanche, L.; Perluzzo, G.; Bader, G.; Caron, L.G. J. Chem. Phys. 1982, *77*, 3285.
53. Plenkiewicz, B.; Plenkiewicz, P.; Perluzzo, G.; Jay-Gerin, J.-P. Phys. Rev. B 1985, *32*, 1253; Plenkiewicz, B.; Plenkiewicz, P.; Jay-Gerin, J.-P. Phys. Rev. B 1986, *33*, 5744.
54. Ueno, N.; Sugita, K.; Seki, K.; Inokuchi, H. Phys. Rev. B 1986, *34*, 6386.
55. Caron, L.G.; Perluzzo, G.; Bader, G.; Sanche, L. Phys. Rev. B 1986, *33*, 3027.
56. Bader, G.; Perluzzo, G.; Caron, L.G.; Sanche, L. Phys. Rev. B 1984, *30*, 78.
57. Bader, G.; Perluzzo, G.; Caron, L.G.; Sanche, L. Phys. Rev. B 1982, *26*, 6019.
58. Goulet, T.; Jay-Gerin, J.-P. Radiat. Phys. Chem. 1986, *27*, 229; Goulet, T.; Pou, V.; Jay-Gerin, J.-P. J. Electron Relat. Phenom. 1986, *41*, 157; Goulet, T.; Jay-Gerin, J.-P.; Patau, J.-P. J. Electron Relat. Phenom. 1987, *43*, 17.
59. Goulet, T.; Keszei, E.; Jay-Gerin, J.-P. Phys. Rev. A 1988, *37*, 2176; Keszei, E.; Goulet, T.; Jay-Gerin, J.-P. Phys. Rev. A 1988, *37*, 2183.
60. Plenkiewicz, P.; Jay-Gerin, J.-P.; Plenkiewicz, B.; Perluzzo, G. Solid State Commun. 1986, *57*, 203.
61. Plenkiewicz, P.; Plenkiewicz, B.; Jay-Gerin, J.-P. Solid State Commun. 1988, *65*, 1227.
62. Hiraoka, K.; Nara, M. Bull. Chem. Soc. Jpn. 1984, *57*, 2243; Hiraoka, K.; Nara, M. Chem. Phys. Lett. 1983, *94*, 589.
63. Harrigan, M.E.; Lee, H.J. J. Chem. Phys. 1974, *60*, 4909.
64. Huang, J.T.J.; Magee, J.L. J. Chem. Phys. 1974, *61*, 2736.
65. Cheng, I.Y.; Funabashi, K. J. Chem. Phys. 1973, *59*, 2977.
66. Michaud, M.; Sanche, L.; Gaubert, C.; Baudoing, R. Surf. Sci. 1988, *205*, 447.
67. Yamane, H.; Setoyama, H.; Kera, S.; Okudaira, K.K.; Ueno, N. Phys. Rev. B 2001, *64*, 113407.
68. Azuma, Y.; Yokota, T.; Kera, S.; Aoki, M.; Okudaira, K.K.; Harada, Y.; Ueno, N. Thin Solid Films 1998, *329*, 303.
69. Ueno, N.; Azuma, Y.; Yokota, T.; Aoki, M.; Okudaira, K.K.; Harada, Y. Jpn. J. Appl. Phys. Part 1 1997, *36*, 5731.
70. Momose, M.; Kamiya, K.; Sugita, K.; Ueno, N. Jpn. J. Appl. Phys. Part 1 1994, *33*, 4754.

71. Ueno, N.; Suzuki, K.; Momose, M.; Kushida, M.; Sugita, K. Jpn. J. Appl. Phys Part 1 1994, *33*, 319.

72. Marsolais, R.M.; Cartier, E.A.; Pfluger, P. Jay-Gerin, J.-P., Ferradini, C., Eds. *Excess Electrons in Dielectric Media, Chap. 2.* CRC Press: Boca Raton, 1991.

73. Chang, Y.C.; Berry, W.B. J. Chem. Phys. 1974, *61*, 2727; Hino, S.; Sato, N.; Inokuchi, H. J. Chem. Phys. 1977, *67*, 4139; Grechov, V.V. Chem. Phys. Lett. 1983, *96*, 237.

74. Pfluger, P.; Zeller, H.R.; Bernasconi, J. Phys. Rev. Lett. 1984, *53*, 94.

75. Bernasconi, J.; Cartier, E.; Pfluger, P. Phys. Rev. B 1988, *38*, 12567; Cartier, E.; Pfluger, P. Phys. Scr. 1988, *T23*, 235.

76. Yasufuku, H.; Okumura, M.; Kera, S.; Okudaira, K.K.; Harada, Y.; Ueno, N. J. Electron Spectrosc. Relat. Phenom. 2001, *114*, 1025.

77. Kadyshevitch, A.; Naaman, R. Phys. Rev. Lett. 1995, *74*, 3443; Sanche, L. Phys. Rev. Lett. 1995, *75*, 2904.

78. Ray, K.; Shanzer, A.; Waldeck, D.H.; Naaman, R. Phys. Rev. B 1999, *60*, 13347.

79. Haran, A.; Naaman, R.; Ashkenasy, G.; Shanzer, A.; Quast, A.T.; Winter, B.; Hertel, I.V. Eur. Phys. J., B. 1999, *8*, 445.

80. Dimitrov, D.A.; Trakhtenberg, S.; Naaman, R.; Smith, D.J.; Samartzis, P.C.; Kitsopoulos, T.N. Chem. Phys. Lett. 2000, *322*, 587.

81. Naaman, R.; Haran, A.; Nitzan, A.; Evans, D.; Galperin, M. J. Phys. Chem. B. 1998, *102*, 3658.

82. Nitzan, A. Annu. Rev. Phys. Chem. 2001, *52*, 681 and references therein.

83. Michaud, M.; Sanche, L.; Goulet, T.; Jay-Gerin, J.-P. Phys. Rev. Lett. 1991, *66*, 1930.

84. Michaud, M.; Cloutier, P.; Sanche, L. Phys. Rev. B 1991, *44*, 10485.

85. Michaud, M.; Cloutier, P.; Sanche, L. Phys. Rev. B 1993, *47*, 4131.

86. Michaud, M.; Cloutier, P.; Sanche, L. Phys Rev. B 1994, *48*, 11336.

87. Michaud, M.; Cloutier, P.; Sanche, L. Phys. Rev. B 1994, *49*, 8360.

88. Bacalis, N.C.; Papaconstantopoulos, D.A.; Pickett, W.E. Phys. Rev. B 1988, *38*, 6218.

89. Peterson, O.G.; Batchelder, D.N.; Simmons, R.O. Phys. Rev. 1966, *150*, 703.

90. Michaud, M.; Sanche, L. Phys. Rev. B 1984, *30*, 6067.

91. Kittel, C. *Quantum Theory of Solids.* Wiley: New York, 1963; 137 p.

92. Schwentner, N.; Koch, E.-E.; Jortner, J. In *Electronic Excitations in Condensed Rare Gases.* Springer-Verlag: Berlin and references therein.

93. Michaud, M.; Sanche, L. Phys. Rev. B 1994, *50*, 4725.

94. Brunt, J.N.H.; King, G.C.; Read, F.H. J. Phys. B 1977, *10*, 1289.

95. Zimmerer, G. *Excited State Spectroscopy of Solids;* Grassano, U., Terzi, N., Eds. North Holland: Amsterdam, Netherlands, 1987.

96. Mills, D.L. Phys. Rev. B 1992, *45*, 36.

97. Goulet, T.; Jung, J.-M.; Michaud, M.; Jay Gerin, J.P.; Sanche, L. Phys. Rev. B 1993, *47*, 4131.

98. Michaud, M.; Sanche, L. Phys. Rev. Lett. 1987, *59*, 645; Sanche, L.; Michaud, M. J. Chem. Phys. 1984, *81*, 257; Bader, G.; Chiasson, J.; Caron, L.G.; Michaud, M.; Perluzzo, G.; Sanche, L. Radiat. Res 1988, *114*, 467.

99. Thiel, P.A.; Hoffmann, F.M.; Weinberg, W.H. J. Chem. Phys. 1981, *75*, 5556; Jacobi, K.; Bertolo, M.; Hansen, W. J. Electron Spectrosc. Relat. Phenom. 1990, *54/55*, 529; Jacobi, K.; Bertolo, M. Phys. Rev. B 1990, *42*, 3733; Hansen, W.; Bertolo, M.; Jacobi, K. Surf. Sci. 1991, *253*, 1 and citations therein.

100. Sanche, L.; Michaud, M. Chem. Phys. Lett. 1981, *84*, 497.

101. Sanche, L.; Michaud, M. Phys. Rev. Lett. 1981, *47*, 1008.

102. Sanche, L.; Michaud, M. *ACS Symp, Truhlar D.G., ed. 1984; Series No. 263* p. 211.

103. Michaud, M.; Sanche, L. J. Electron Spectrosc. Relat. Phenom. 1990, *51*, 237.

104. Sanche, L.; Michaud, M. Phys. Rev. B 1983, *27*, 3856.

105. Sanche, L.; Perluzzo, G.; Michaud, M. J. Chem. Phys. 1985, *83*, 3837.

106. Pireaux, J.J.; Thiry, P.; Caudano, R.; Pfluger, P. J. Chem. Phys. 1986, *84*, 6452; Rei Vilar, M.; Blatter, G.; Pfluger, P.; Heymann, M.; Schott, M. Europhys. Lett. 1988, *5*, 375.

107. Sakurai, M.; Okano, T.; Tuzi, Y. J. Vac. Sci. Technol A 1987, *5*, 431.

108. Hoffman, F.M.; Felter, T.E.; Thiel, P.A.; Weinberg, W.H. Surf. Sci. 1983, *130*, 173.
109. Demuth, J.E.; Schmeisser, D.; Avouris, P. Phys. Rev. Lett. 1981, *47*, 116; Schmeisser, D.; Demuth, J.E.; Avouris, P. Phys. Rev. B 1982, *26*, 4857.
110. Swiderek, P.; Winterling, H.; Ibach, H. Chem. Phys. Lett. 1997, *280*, 556.
111. Gootz, B.; Winterling, H.; Swiderek, P. J. Electron Spectrosc. Relat. Phenom. 1999, *105*, 1.
112. Gootz, B.; Krohl, O.; Swiderek, P. J. Electron Spectrosc. Relat. Phenom. 2001, *114*, 569.
113. Bartolucci, F.; Franchy, R.; Silva, J.A.M.C.; Moutinho, A.M.C.; Teillet-Billy, D.; Gauyacq, J.P. J. Chem. Phys. 1998, *108*, 2251.
114. Teillet-Billy, D.; Gauyacq, J.P.; Bartolucci, F.; Franchy, R.; Ramseyer, C.; Girardet, C. Surf. Sci. 2000, *465*, 138.
115. Marinica, D.C.; Teillet-Billy, D.; Gauyacq, J.P.; Michaud, M.; Sanche, L. Phys. Rev. B 2001, *64* (085408).
116. Wen, A.T.; Michaud, M.; Sanche, L. Phys. Rev. A 1996, *54*, 4162; Wen, A.T.; Michaud, M.; Sanche, L. J. Electron Spectrosc. Relat. Phenom. 1998, *94*, 23.
117. Lepage, M.; Michaud, M.; Sanche, L. J. Chem. Phys. 2000, *112*, 6707.
118. Michaud, M.; Lepage, M.; Sanche, L. Phys. Rev. B 1999, *59*, 15480.
119. Michaud, M.; Lepage, M.; Sanche, L. Phys. Rev. Lett. 1998, *81*, 2807–2811.
120. Gerber, A.; Hersenberg, A. Phys. Rev. B 1985, *31*, 6219; Teillet-Billy, D.; Gauyack, J.P. Surf. Sci. 1990, *239*, 343.
121. Allan, M. J. Phys. B 1995, *28*, 4329.
122. Azria, R.; Parenteau, L.; Sanche, L. Phys. Rev. Lett. 1987, *59*, 638.
123. O'Malley, T.F. Phys. Rev. 1965, *137*, A1668; Chen, J.C.Y. Phys. Rev. 1966, *148*, 66.
124. Domcke, W. Phys. Rep. 1991, *208*, 97; J. Phys. B 1981, *14*, 4889; Domcke, W.; Cederbaum, L.S. Phys. Rev. A 1977, *16*, 1465.
125. Mills, D.L. Phys. Rev. B 1992, *45*, 13221.
126. Steinberger, I.T.; Bass, A.D.; Shechter, R.; Sanche, L. Phys. Rev. B 1993, *48*, 8290.
127. Goulet, T.; Pou, V.; Jay-Gerin, J.P. J. Electron Relat. Phenom. 1986, *41*, 157.
128. Goulet, T.; Jay-Gerin, J.-P.; Patau, J.P. J. Electron Relat. Phenom. 1987, *43*, 17.
129. Ashby, C.I.H. Appl. Phys. Lett. 1983, *43*, 609.
130. Marsolais, R.M.; Sanche, L. Phys. Rev. B 1987, *38*, 11118.
131. Marsolais, R.M.; Michaud, M.; Sanche, L. Phys. Rev. A 1987, *35*, 607.
132. Merkel, P.B.; Hamill, W.H. J. Chem. Phys. 1971, *55*, 1409.
133. Michaud, M.; Sanche, L. Phys. Rev. B 1994, *50*, 4725.
134. Michaud, M.; Cloutier, P.; Sanche, L. Phys. Rev. A 1991, *44*, 5624.
135. Oeter, D.; Egalhaaf, H.-J.; Ziegler, Ch.; Oelkrug, D.; Gopel, W. J. Chem. Phys. 1994, *101*, 6344.
136. Sanche, L.; Michaud, M. Chem. Phys. Lett. 1981, *80*, 184.
137. Swiderek, P.; Michaud, M.; Holneicher, G.; Sanche, L. Chem. Phys. Lett. 1990, *175*, 667; Chem. Phys. Lett. 1991, *178*, 289; Chem. Phys. Lett. 1991, *187*, 583.
138. Michaud, M.; Fraser, M.-J.; Sanche, I. J. Chim. Phys. 1994, *91*, 1223.
139. Swiderek, P.; Michaud, M.; Sanche, L. J. Chem. Phys. 1995, *98*, 8397; J. Chem. Phys. 1995, *103*, 8424; J. Chem. Phys. 1997, *106*, 9403.
140. Swiderek, P.; Fraser, M.-J.; Michaud, M.; Sanche, L. J. Chem. Phys. 1994, *100*, 70.
141. Jay-Gerin, J.-P.; Swiderek, P.; Fraser, M.-J.; Michaud, M.; Ferradini, C.; Sanche, L. Radiat. Phys. Chem. 1997, *50*, 263.
142. Swiderek, P.; Schurfeld, S.; Winterling, H. Ber. Bunsenges. Phys. Chem. 1997, *101*, 1517.
143. Swiderek, P.; Winterling, H. Chem. Phys. 1998, *229*, 295.
144. Swiderek, P.; Gootz, B. Ber. Bunsenges. Phys. 1998, *102*, 882.
145. Swiderek, P.; Gootz, B.; Winterling, H. Chem. Phys. Lett. 1998, *285*, 246.
146. Krohl, O.; Swiderek, P. J. Mol. Struct. 1999, *481*, 237.
147. Huels, M.A.; Parenteau, L.; Michaud, M.; Sanche, L. Phys. Rev. A 1995, *51*, 337.
148. Robin, M.B. *Higher Excited States of Polyatomic Molecules, 1 and 2.* Academic: New York.
149. Arakawa, I.; Takahashi, M.; Takeuchi, K. J. Vac. Sci. Technol. A 1989, *7*, 2090.

150. Hirayama, T.; Hayama, A.; Koike, T.; Kuninobu, T.; Arakawa, I.; Mitsuke, K.; Sakurai, M.; Savchenko, E.V. Surf. Sci. 1997, *390*, 266.
151. Leclerc, G.; Bass, A.D.; Michaud, M.; Sanche, L. J. Electron Spectrosc. Relat. Phenom. 1990, *52*, 725.
152. Mann, A.; Leclerc, G.; Sanche, L. Phys. Rev. B 1992, *46*, 9683.
153. Shi, H.; Cloutier, P.; Sanche, L. Rev. B 1995, *52*, 5385.
154. Bass, A.D.; Lezius, M.; Ayotte, P.; Parenteau, L.; Cloutier, P.; Sanche, L. J. Phys., B At. Mol. Opt. Phys. 1997, *30*, 3527.
155. Shi, H.; Cloutier, P.; Sanche, L. Low Temp. Phys. 1998, *24*, 742.
156. Huels, M.A.; Parenteau, L.; Cloutier, P.; Sanche, L. J Chem. Phys. 1995, *103*, 6775.
157. Tronc, M.; Azria, R.; Le Coat, Y.; Cloutier, P.; Sanche, L. Chem. Phys. 2000, *254*, 69.
158. Mann, A.; Cloutier, P.; Liu, D.; Sanche, L. Phys. Rev. B 1995, *51*, 7200.
159. Shi, H.; Cloutier, P.; Gamache, J.; Sanche, L. Phys. Rev. B 1996, *53*, 13830.
160. Shi, H.; Cloutier, P.; Gamache, J.; Sanche, L. Surf. Sci. 1997, *380*, 385.
161. Cui, S.T.; Johnson, R.E.; Cummings, P.T. Phys. Rev. B 1989, *49*, 9580.
162. Sanche, L. IEEE Trans. Dielectr. Electr. Insul. 1997, *4*, 517.
163. Vichnevetski, E.; Bass, A.D.; Cloutier, P.; Sanche, L. Phys. Rev. B 1988, *57*, 14914.
164. Bass, A.D.; Vichnevetski, E.; Sanche, L. Phys. Rev. B 1999, *60*, 14405.
165. Borst, W.L. Phys. Rev. A 1974, *9*, 1195.
166. Sanche, L. Phys. Rev. Lett. 1984, *53*, 1638.
167. Rowntree, P.; Parenteau, L.; Sanche, L. J. Phys. Chem. 1991, *95*, 523; J. Phys. Chem. 1991, *95*, 4903.
168. Sambe, H.; Ramaker, D.E.; Parenteau, L.; Sanche, L. Phys. Rev. Lett. 1987, *59*, 236.
169. Akbulut, M.; Madey, T.E.; Parenteau, L.; Sanche, L. J. Chem. Phys. 1996, *105*, 6043.
170. Huels, M.A.; Parenteau, L.; Sanche, L. Nucl. Instrum. Methods Phys. Res. B 1995, *101*, 203.
171. Azria, R.; Parenteau, L.; Sanche, L. J. Chem. Phys. 1988, *88*, 5166.
172. Azria, R.; Sanche, L.; Parenteau, L. Chem. Phys. Lett. 1989, *156*, 606.
173. Azria, R.; Parenteau, L.; Sanche, L. Chem. Phys. Lett. 1990, *171*, 229.
174. Azria, R.; Le Coat, Y.; Ziesel, J.-P.; Guillotin, J.-P.; Mharzi, B.; Tronc, M. Chem. Phys. Lett. 1994, *220*, 417.
175. Huels, M.A.; Parenteau, L.; Sanche, L. Chem. Phys. Lett. 1993, *210*, 340.
176. Meinke, M.; Illenberger, E. J. Phys. Chem. 1994, *98*, 6601.
177. Meinke, M.; Parenteau, L.; Rowntree, P.; Sanche, L.; Illenberger, E. Chem. Phys. Lett. 1993, *205*, 213.
178. Parenteau, L.; Jay-Gerin, J.-P.; Sanche, L. J. Phys. Chem. 1994, *98*, 10277.
179. Rowntree, P.; Parenteau, L.; Sanche, L. Chem. Phys. Lett. 1991, *182*, 479.
180. Rowntree, P.; Sambe, H.; Parenteau, L.; Sanche, L. Phys. Rev. B 1993, *47*, 4537.
181. Parenteau, L.; Sanche, L. J. Chim. Phys. 1994, *91*, 1237.
182. Rowntree, P.; Sanche, L.; Parenteau, L.; Meinke, M.; Weik, F.; Illenberger, E. J. Chem. Phys. 1994, *101*, 4248.
183. Sambe, H.; Ramaker, D.E. Chem. Phys. Lett. 1987, *139*, 386.
184. Sambe, H.; Ramaker, D.E. Surf. Sci. 1992, *269*, 444.
185. Sanche, L. Phys. Rev. Lett. 1984, *53*, 1638.
186. Sanche, L. J. Vac. Sci. Technol. B 1986, *10*, 196.
187. Sanche, L.; Parenteau, L. J. Vac. Sci. Technol. A 1986, *4*, 1240.
188. Sanche, L.; Parenteau, L.; Cloutier, P. J. Chem. Phys. 1989, *91*, 2664.
189. Sanche, L.; Bass, A.D.; Parenteau, L.; Gortel, Z.W. Phys. Rev. B 1993, *48*, 5540.
190. Silva, L.A.; Palmer, R.E. Surf. Sci. 1992, *282*, 313.
191. Huels, M.A.; Parenteau, L.; Sanche, L. Phys. Rev. B 1995, *52*, 11343.
192. Le Coat, Y.; Azria, R.; Mharzi, B.; Tronc, M. Surf. Sci. 1995, *331*, 360.
193. Laitenberger, P.; Palmer, R.E. J. Phys., Condens. Matter 1996, *8*, L71–L78.
194. Azria, R.; Le Coat, Y.; Mharzi, B.; Tronc, M. Nucl. Instrum. Methods Phys. Rev. B 1995, *101*, 184.

195. Oster, T.; Ingolfsson, O.; Meinke, M.; Jaffke, T.; Illenberger, E. J. Chem. Phys. 1993, 99, 5141.
196. Weik, F.; Illenberger, E. J. Chem. Phys. 1995, 103, 1406; Ingolfsson, O.; Weik, F.; Illenberger, E. Int. Rev. Phys. Chem. 1996, 15, 133.
197. Tegeder, P.; Bruning, F.; Illenberger, E. Chem. Phys. Lett. 1999, 310, 79.
198. Tegeder, P.; Illenberger, E. Phys. Chem. Chem. Phys. 1999, 1, 5197.
199. Lecoat, Y.; Hedhili, N.M.; Azria, R.; Tronc, M.; Weik, F.; Illenberger, E. Int. J. Mass Spectrom. Ion Process. 1997, 164, 231.
200. Bruning, F.; Tegeder, P.; Langer, J.; Illenberger, E. Int. J. Mass Spectrom. 2000, 196, 507.
201. Tegeder, P.; Smirnov, B.M.; Illenberger, E. Int. J. Mass Spectrom. 2001, 205, 331.
202. Langer, J.; Matt, S.; Meinke, M.; Tegeder, P.; Stamatovic, A.; Illenberger, E. J. Chem. Phys. 2000, 113, 11063.
203. Tegeder, P.; Kendall, P.A.; Penno, M.; Mason, N.J.; Illenberger, E. Phys. Chem. Chem. Phys. 2001, 3, 2625.
204. Lachgar, M.; Le Coat, Y.; Azria, R.; Tronc, M.; Illenberger, E. Chem. Phys. Lett. 1999, 305, 408.
205. Lecoat, Y.; Azria, R.; Tronc, M.; Ingolfsson, O.; Illenberger, E. Chem. Phys. Lett. 1998, 296, 208.
206. Azria, R.; Le Coat, Y.; Lachgar, M.; Tronc, M.; Parenteau, L.; Sanche, L. Surf. Sci. 1999, 436, L671; Surf. Sci. 2000, 451, 91.
207. Tronc, M.; Azria, R. Int. J. Mass Spectrom 2001, 205, 325.
208. Bass, A.D.; Parenteau, L.; Weik, F.; Sanche, L. J. Chem. Phys. 2000, 113, 8746.
209. Bass, A.D.; Parenteau, L.; Weik, F.; Sanche, L. J. Chem. Phys. 2001, 115, 4811.
210. Siller, L.; Hedhili, M.N.; Le Coat, Y.; Azria, R.; Tronc, M. Chem. Phys. Lett. 1998, 288, 776; J. Chem. Phys. 1999, 110, 10554.
211. Hedhili, M.N.; Parenteau, L.; Huels, M.A.; Azria, R.; Tronc, M.; Sanche, L. J. Chem. Phys. 1997, 107, 7577.
212. Hedhili, M.N.; Azria, R.; Lecoat, Y.; Tronc, M. Chem. Phys. Lett. 1997, 268, 21.
213. Tronc, M.; Azria, R.; Lecoat, Y.; Illenberger, E. J. Phys. Chem. 1996, 100, 14745.
214. Tegeder, P.; Illenberger, E. Chem. Phys. Lett. 2001, 341, 401.
215. Huels, M.A.; Parenteau, L.; Sanche, L. Chem. Phys. Lett. 1997, 279, 223.
216. Huels, M.A.; Rowntree, P.; Parenteau, L.; Sanche, L. Surf. Sci. 1997, 390, 282.
217. Weik, F.; Illenberger, E.; Nagesha, K.; Sanche, L. J. Phys. Chem. B 1998, 102, 824.
218. Lu, Q.-B.; Madey, T.E.; Parenteau, L.; Weik, F.; Sanche, L. Chem. Phys. Lett. 2001, 342, 1.
219. Junker, K.H.; White, J.M. J. Vac. Sci. Technol. A 1998, 16, 3328.
220. Sanche, L.; Deschênes, M. Phys. Rev. Lett. 1988, 61, 2096.
221. Sambe, H.; Ramaker, D.E.; Deschênes, M.; Bass, A.D.; Sanche, L. Phys. Rev. Lett. 1990, 64, 523.
222. Bass, A.D.; Sanche, L. J. Chem. Phys. 1991, 95, 2910.
223. Simpson, W.C.; Orlando, T.M.; Parenteau, L.; Nagesha, K.; Sanche, L. J. Chem. Phys. 1998, 108, 5027.
224. Bass, A.D.; Gamache, J.; Parenteau, L.; Sanche, L. J. Phys. Chem. 1995, 99, 6123.
225. Nagesha, K.; Sanche, L. J. Appl. Phys. 2000, 88, 5211.
226. Sanche, L.; Bass, A.D.; Ayotte, P.; Fabrikant, I.I. Phys. Rev. Lett. 1995, 75, 3568.
227. Ayotte, P.; Gamache, J.; Bass, A.D.; Fabrikant, I.I.; Sanche, L. J. Chem. Phys. 1997, 106, 749.
228. Bass, A.D.; Gamache, J.; Ayotte, P.; Sanche J. Chem. Phys. 1996, 104, 4258.
229. Fabrikant, I.I.; Nagesha, K.; Wilde, R.; Sanche, L. Phys. Rev. B 1997, 56, R5725.
230. Nagesha, K.; Sanche, L. Phys. Rev. Lett. 1997, 78, 4725.
231. Nagesha, K.; Fabrikant, I.I.; Sanche, L. J. Chem. Phys. 2001, 114, 4934.
232. Nagesha, K.; Sanche, L. Phys. Rev. Lett. 1998, 81, 5892.
233. Weik, F.; Illenberger, E.; Nagesha, K.; Sanche, L. J. Phys. Chem. B 1998, 102, 824.
234. Weik, F.; Sanche, L.; Ingolfsson, O.; Illenberger, E. J. Chem. Phys. 2000, 112, 9046.
235. Lu, Q.B.; Sanche, L. Phys. Rev. B 2001, 63, 153403.
236. Lu, Q.B.; Sanche, L. J. Chem. Phys. 2001, 115, 5711.

237. Van Brunt, R.J.; Kieffer, L.J. Phys. Rev A 1970, 2, 1899.
238. Stevenson, K.P.; Gimmel, G.A.; Dohnalek, Z.; Smith, R.S.; Kay, B.D. Science 1999, 283, 1501; Ayotte, P.; Scott Smith, R.; Stevenson, K.P.; Dohnálek, Z.; Kimmel, G.A.; Kay, B.D. J. Geophys. Res. 2001, 106, 33387.
239. Vichnevetski, E.; Bass, A.D.; Sanche, L. J. Chem. Phys. 2000, 113, 3874.
240. Pearl, D.M.; Burrow, P.D. Chem. Phys. Lett. 1993, 206, 483.
241. Fabrikant, I.I. J. Phys., B At. Mol. Opt Phys. 1991, 24, 2213; J. Phys., B. At. Mol. Opt Phys. 1994, 27, 4325.
242. Pearl, D.M.; Burrow, P.D. J. Chem. Phys. 1994, 101, 2940; Aflatooni, K.; Burrow, P.D. J. Chem. Phys. 2000, 113, 1455.
243. Pearl, D.M.; Burrow, P.D.; Fabrikant, I.I.; Gallup, G.A. J. Chem. Phys. 1995, 102, 2737.
244. Jay-Gerin, J.-P., Ferradini, C., Eds. (1991). Excess Electrons in Dielectric Media. CRC Press, 1991: Boca Raton, FL; Kestner, N.R. Radiation Chemistry, Principles and Applications; Farhataziz, Rogers, M.J., Eds. . VCH: New York, 1987.
245. Norman, N.; Mathisen, H. Act Chem. Scand. 1961, 15, 1755; Firment, L.E.; Somorjai, G.A. J. Chem. Phys. 1977, 66, 1901; Hager, S.L.; Willard, J.E. J. Chem. Phys. 1975, 63, 942; Takeda, K.; Oguni, M.; Suga, H. J. Phys. Chem. Solids 1991, 52, 991.
246. Sanche, L.; Parenteau, L. Phys. Rev. Lett. 1987, 59, 136; J. Chem. Phys. 1990, 93, 7476.
247. Cobut, V.; Jay-Gerin, J.-P.; Frongillo, Y.; Patau, J.P. Radiat. Phys. Chem. 1996, 47, 247.
248. Azria, R.; Parenteau, L.; Sanche, L. Chem. Phys. Lett. 1990, 171, 229.
249. Hanson, D.M. Abstracts of the 46th Annual Meeting of the Radiat. Res. Soc. Louisville, KY, 1998; 86 p.
250. Lacombe, S.; Cemic, F.; Jacobi, K.; Hedhili, M.N.; Le Coat, Y.; Azria, R.; Tronc, M. Phys. Rev. Lett. 1997, 79, 1146.
251. Hedhili, M.N.; Lachgar, M.; Le Coat, Y.; Azria, R.; Tronc, M.; Lu, Q.B.; Madey, T.E. J. Chem. Phys. 2001, 114, 1844.
252. Tegeder, P.; Illenberger, E. Chem. Phys. Lett. 2001, 341, 401.
253. Huq, M.S.; Doverspike, L.D.; Champion, R.L. Phys. Rev. A 1983, 27, 785.
254. Comer, J.; Schulz, G.L. Phys. Rev. A 1974, 10, 2100.
255. Parkes, D.A. J. Chem. Soc. Faraday Trans. I 1972, 68, 613.
256. Lindinger, W.; Albritton, D.L.; Fehsenfeld, F.C.; Fergusson, E.E. J. Chem. Phys. 1975, 63, 3238.
257. Bass, A.D.; Parenteau, L.; Huels, M.A.; Sanche, L. J. Chem. Phys. 1998, 109, 8635.
258. Martel, R.; Rochefort, A.; McBreen, P.H. J. Am. Chem. Soc. 1994, 116, 5965.
259. Parenteau, L.; Jay-Gérin, J.-P.; Sanche, L. J. Phys. Chem. 1994, 98, 10277.
260. Winterling, H.; Haberkern, H.; Swiderek, P. Phys. Chem. Chem. Phys. 2001, 3, 4592.
261. Wilkes, J.; Palmer, R.E.; Lamont, C.L.A. Phys. Rev. Lett. 1999, 83, 3332.
262. Boudaiffa, B.; Cloutier, P.; Hunting, D.; Huels, M.A.; Sanche, L. Science 2000, 287, 1658.
263. Boudaïffa, B.; Cloutier, P.; Hunting, D.; Huels, M.A.; Sanche, L. Radiat. Res. 2002, 157, 227.
264. Boudaiffa, B.; Hunting, D.; Cloutier, P.; Huels, M.A.; Sanche, L. Int. J. Radiat. Biol. 2002, 76, 1209.
265. Boudaïffa, B.; Cloutier, P.; Hunting, D.; Huels, M.A.; Sanche, L. Méd. Sci. 2000, 16, 1281.
266. Swarts, S.G.; Sevilla, M.D.; Becker, D.; Tokar, J.C.; Wheeler, K.T. Radiat. Res. 1992, 129, 333.
267. Colson, A.O.; Besler, B.; Sevilla, M.D. J. Phys. Chem. 1992, 96, 9787.
268. Rowntree, P.A.; Parenteau, L.; Sanche, L. J. Chem. Phys. 1991, 94, 8570.
269. Simpson, W.C.; Sieger, M.T.; Orlando, T.M.; Parenteau, L.; Nagesha, K.; Sanche, L. J. Chem. Phys. 1997, 107, 8668.
270. Kimmel, G.A.; Orlando, T.M. Phys. Rev. Lett. 1996, 77, 3983.
271. Kimmel, G.A.; Cloutier, P.; Sanche, L.; Orlando, T.M. J. Phys. Chem. 1997, 101, 6301.
272. Antic, D.; Parenteau, L.; Lepage, M.; Sanche, L. J. Phys. Chem. B 1999, 103, 6611.
273. Antic, D.; Parenteau, L.; Lepage, M.; Sanche, L. J. Phys. Chem. B 2000, 104, 4711.
274. Huels, M.A.; Hahndorf, I.; Illenberger, E.; Sanche, L. J. Chem. Phys. 1997, 108, 1309.

275. Abdoul-Carime, H.; Cloutier, P.; Sanche, L. Radiat. Res. 2001, *155*, 625.
276. Andrieux, C.P.; LeGorande, A.; Saveant, J.-M. J. Am. Chem. Soc. 1992, *114*, 6892.
277. Stepanovic, M.; Pariat, Y.; Allan, M. J. Chem. Phys. 1999, *110*, 11376.
278. Dugal, P.; Huels, M.A.; Sanche, L. Radiat. Res. 1999, *151*, 325.
279. Dugal, P.; Abdoul-Carime, H.; Sanche, L. J. Phys. Chem. B 2000, *104*, 5610.
280. Abdoul-Carime, H.; Dugal, P.C.; Sanche, L. Radiat. Res. 2000, *153*, 23.
281. Abdoul-Carime, H.; Dugal, P.C.; Sanche, L. Surf. Sci. 2000, *451*, 102.
282. Abdoul-Carime, H.; Sanche, L. Radiat. Res. 2001, *156*, 151.
283. Abdoul-Carime, H.; Sanche, L. Int. J. Radiat. Biol. 2002, *78*, 89.
284. Zamenhof, S.; De Giovanni, R.; Greer, S. Nature 1958, *181*, 827.
285. Crewe, A.V.; Isaacson, M.; Johnson, D. Nature 1971, *231*, 262.
286. Zeller, H.R.; Baumann, Th.; Cartier, E.; Dersch, H.; Pfluger, P.; Stucki, F. Adv. Solid State Phys. (Festkörperprobleme) 1987, *27*, 223.
287. Zeller, H.R.; Pfluger, P.; Bemasconi, J. IEEE Trans. Electr. Insul. 1984, *19*, 200.
288. Ferry, D.K.; Khoury, M.; Pivin, D.P. Jr.; Connolly, K.M.; Whidden, T.K.; Kozicki, M.N.; Allee, D.R. Semicond. Sci. Technol. 1996, *11*, 1552.
289. Avouris, Ph. Acc. Chem. Res. 1995, *28*, 95.
290. Fontaine, P.A.; Dubois, E.; Stievenard, D. J. Appl. Phys. 1998, *84*, 1776.
291. Lu, P.H.; Polanyi, J.C.; Rogers, D. J. Chem. Phys. 1999, *111*, 9095.
292. Xu, C.; Koel, B.E. Surf. Sci. 1993, *292*, L803; Syomin, D.; Koel, B.E. Surf. Sci. 2001, *492*, L693; Syomin, D.; Kim, J.; Koel, B.E.; Ellison, G.B. J. Phys. Chem. B 2001, *105*, 8387.
293. Klyachko, D.; Rowntree, P.; Sanche, L. Surf. Sci. 1996, *346*, L49.
294. Klyachko, D.; Rowntree, P.; Sanche, L. Surf. Sci. 1997, *389*, 29.
295. Di, W.; Rowntree, P.; Sanche, L. Phys. Rev. B 1995, *52*, 16618.
296. Madey, T.E.; Johnson, R.E.; Orlando, T.M. Surf. Sci. 2002, *500*, 838.
297. Lu, Q.-B.; Madey, T.E. J. Chem. Phys. 1999, *111*, 2861.
298. Lu, Q.-B.; Madey, T.E. Phys. Rev. Lett. 1999, *82*, 4122; Surf. Sci. 2000, *451*, 238.
299. Christophorou, L.G.; Stockdale, J.A. J. Chem. Phys. 1968, *48*, 1956; Illenberger, E.; Scheunemann, H.-U.; Baumgartel, H. Chem. Phys. 1979, *37*, 21.
300. Migus, A.; Gauduel, Y.; Martin, J.L.; Antonetti, A. Phys. Rev. Lett. 1987, *58*, 1559; Silva, C.; Walhout, P.K.; Yokoyama, K.; Barbara, P.F. Phys. Rev. Lett. 1998, *80*, 1086..
301. Assel, M.; Laenen, R., Laubereau, A. J. Chem. Phys. 1999, *111*, 6869; Laenen, R.; Roth, T.; Laubereau, A. Phys. Rev. Lett. 2000, *85*, 50.

10

Electron–Ion Recombination in Condensed Matter: Geminate and Bulk Recombination Processes

Mariusz Wojcik
Technical University of Lodz, Lodz, Poland

M. Tachiya
National Institute of Advanced Industrial Science and Technology (AIST), Tsukuba, Japan

S. Tagawa
Osaka University, Osaka, Japan

Y. Hatano
Tokyo Institute of Technology, Tokyo, Japan

SUMMARY

A survey is given of the theoretical and experimental studies of electron–ion recombination in condensed matter as classified into geminate and bulk recombination processes. Because the recombination processes are closely related with the magnitudes of the electron drift mobility, which is largely dependent on molecular media of condensed matter, each recombination process is discussed by further classifying it to the recombination in low- and high-mobility media.

1. THEORY OF ELECTRON–ION RECOMBINATION IN CONDENSED MATTER

1.1. Introduction

Ionization of atoms or molecules is the main primary event induced by the interaction of radiations with condensed matter. The charged species produced by ionization, if not removed from the irradiated system, will naturally tend to recombine. The conventional theories of recombination treat the transport and reactions of charged species only after the electrons ejected from atoms or molecules become thermalized by dissipating their initially high kinetic energies to the surrounding medium and form a spatial distribution around their parent cations. The thermalization in condensed phases is fast and is usually completed on the timescale shorter than picoseconds. Following thermalization, electrons and their parent cations perform a diffusive motion in each other's Coulomb field, and may react on encounter to form neutral products. This phase of recombination, in which the ionized system may be described as a collection of independent electron–cation pairs, is called geminate *recombination*. In the course of this process, a fraction of electrons may be separated from their parent cations far away, so that the Coulomb interaction between the electron and the cation becomes very weak and they may be considered as free charges. Then, the initial correlations within the geminate pairs are lost, and the spatial distributions of positive and negative charges in the irradiated system become independent of each other and uniform. This sets the initial conditions for the next stage of recombination, called *bulk recombination*, in which all charges in the system will finally be neutralized in a homogeneous, second-order process.

Dividing the electron–ion recombination processes into two phases, well separated in the time domain, has provided a convenient basis for constructing theoretical models, in which the *geminate* phase is usually characterized by the pair reaction (or escape) probability, and the bulk phase by its reaction rate constant. However, one should realize that in real systems the general picture of the recombination processes is not in all cases so simple. The spatial distributions of ionization events in tracks of radiations are usually not homogeneous. Ionizations may occur in clusters (often called spurs or blobs) or form columns (sometimes called short tracks), where the distances between neighboring cations are shorter than the average electron thermalization distance. In such a case, in addition to geminate recombination between electrons and their parent cations, cross-recombination between electrons and nonparent cations belonging to the same cluster or column is also possible. Another complication arises from the fact that in many systems, the recombination processes are intercepted by chemical reactions involving electrons or cations.

The course of the recombination processes in a particular system depends on several factors. One of the most important ones is the polarity of the system. Both geminate and bulk recombination processes are strongly influenced by the Coulomb attraction between electrons and cations, and the range of this interaction in condensed matter is determined by the dielectric constant ε. The range of the Coulomb interaction in a particular system is usually represented by the Onsager radius, r_c, which is defined as the distance at which the electrostatic energy of a pair of elementary charges falls down to the thermal level $k_B T$.

$$r_c = \frac{e^2}{\varepsilon k_B T} \tag{1}$$

Here e is the electron charge, k_B is the Boltzmann constant, and T is temperature. The value of r_c in nonpolar liquids at room temperature may be as high as ~30 nm, while in water it is only 0.7 nm. Because the electron thermalization distances are usually on the order of a few nanometers, the effect of the Coulomb attraction between the ionization products in water will be much weaker than that in, e.g., liquid hydrocarbons. This will result in much lower probabilities of geminate recombination in polar liquids compared to those in nonpolar ones.

Another factor that strongly affects the recombination processes is the mobility of charges, or, more generally, the mechanism of charge transport in a particular system. It is normally assumed that the transport of both electrons and cations may be described as ideal diffusion. This assumption is justified in most polar systems, as well as in a part of nonpolar ones. However, in a number of nonpolar systems, usually characterized by high electron mobility, it is more appropriate to describe the electron motion as "ballistic" rather than Brownian. The transport of positive charges may be, in some cases, better described as a resonant charge hopping rather than the diffusion of cations. Different theoretical approaches are needed to describe the recombination processes in high-mobility systems.

The electron–ion recombination processes are also strongly dependent on the type of radiations that ionize a particular system. As it was already mentioned, track effects determine the initial spatial distributions of cations and electrons in irradiated systems.

In the following sections, we present the most important theoretical results concerning the electron–ion recombination in condensed matter. In Sec. 10.1.2, we concentrate on the geminate recombination, and in Sec. 10.1.3 on the bulk recombination. In both cases, we start with the diffusion theory (Secs. 10.1.2 "Diffusion-Controlled Geminate Ion Recombination" and 10.1.3 "Diffusion-Controlled Bulk Ion Recombination"), and present its results for the escape probability in the geminate phase, and the recombination rate constant in the bulk phase. We also discuss transient effects and the effects of an applied electric field on the recombination processes considering their experimental importance. In the case of the geminate recombination, we include the theory of electron scavenging and a brief discussion of the multipair effects in radiation tracks. Theoretical studies of the recombination processes in high-mobility systems require an approach that is quite different from that used in the theory of diffusion-controlled recombination. Therefore we discuss the recombination processes in high-mobility systems in separate subsections [Secs. 10.1.2 ("Geminate Ion Recombination in High-Mobility Systems") and 10.1.3 ("Bulk Ion Recombination in High-Mobility Systems")].

1.2. Geminate Ion Recombination

Diffusion-Controlled Geminate Ion Recombination

The central problem in the theory of geminate ion recombination is to describe the relative motion and reaction with each other of two oppositely charged particles initially separated by a distance r_0. If we assume that the particles perform an ideal diffusive motion, the time evolution of the probability density, $w(\mathbf{r}, t)$, that the two species are separated by $\mathbf{r}$ at time t, may be described by the Smoluchowski equation [1,2]

$$\frac{\partial w(\mathbf{r}, t)}{\partial t} = D \left[\nabla^2 w(\mathbf{r}, t) + \frac{1}{k_B T} \nabla w(\mathbf{r}, t) V(\mathbf{r}) \right] \tag{2}$$

where D is the sum of the diffusion coefficients of the two particles and $V(\mathbf{r})$ is the interaction potential. The initial condition is given by

$$w(\mathbf{r}, 0) = \frac{1}{4\pi r_0^2} \delta(\mathbf{r} - \mathbf{r}_0) \tag{3}$$

The probability density $w(\mathbf{r}, t)$ contains the complete information about the behavior of the geminate pair. However, it is difficult to solve Eq. (2) in most cases of interest. Moreover, the probability density $w(\mathbf{r}, t)$ itself is usually experimentally unobservable. The quantities of greater practical importance are the pair survival probability defined as

$$W(t) = \int w(\mathbf{r}, t) d\mathbf{r} \tag{4}$$

and its long-time limit

$$\varphi = \lim_{t \to \infty} W(t) \tag{5}$$

The latter quantity is called the escape probability, and describes the probability that the two oppositely charged particles of the geminate pair will never recombine with each other and become free ions.

By introducing an equation adjoint to Eq. (2), one can derive the following equation for the pair survival probability [3]

$$\frac{\partial W}{\partial t} = D\left(\nabla^2_{r_0} W - \frac{1}{k_B T} \nabla_{r_0} W \nabla_{r_0} V \right) \tag{6}$$

where the space derivatives are the ones with respect to the initial separation r_0. The escape probability is shown [3] to satisfy the following equation by taking the limit of $t \to \infty$ in Eq. (6)

$$\nabla^2_{r_0} \varphi - \frac{1}{k_B T} \nabla_{r_0} \varphi \nabla_{r_0} V = 0 \tag{7}$$

For the Coulomb potential given by

$$V(r) = -\frac{e^2}{\varepsilon r} = -k_B T \frac{r_c}{r} \tag{8}$$

Eq. (7) may be written in the spherical coordinates as

$$\frac{d^2\varphi}{dr_0^2} + \left(\frac{2}{r_0} - \frac{r_c}{r_0^2} \right) \frac{d\varphi}{dr_0} = 0 \tag{9}$$

The escape probability satisfies the following boundary condition at infinity

$$\varphi(r_0 \to \infty) = 1 \tag{10}$$

In the equations written so far, the reaction between the particles of the geminate pair is not accounted for. This can be performed by introducing a boundary condition at a distance R, which is called the reaction distance. If the reaction on encounter is very fast, one can use the totally absorbing boundary condition

$$w(R, t) = 0 \tag{11a}$$

Equation (11a) is sometimes called the absorption boundary condition. This boundary condition is equivalent to the following boundary condition on the escape probability [4]

$$\varphi(R) = 0 \tag{11b}$$

By solving Eq. (9) subject to boundary conditions (10) and (11b), the escape probability for the totally diffusion-controlled geminate ion recombination is calculated as

$$\varphi(r_0) = \frac{\exp\left(-\dfrac{r_c}{r_0}\right) - \exp\left(-\dfrac{r_c}{R}\right)}{1 - \exp\left(-\dfrac{r_c}{R}\right)} \tag{12}$$

Equation (12) is derived on the assumption of an instantaneous reaction of the opposite ions once they approach each other to the distance R. However, in many real systems, the rate of the reaction at the reaction distance is comparable to, or even slower than, the diffusive flux of the ions toward each other. This effect can be taken into account by modifying the boundary condition at the reaction distance R. Instead of the totally absorbing boundary condition (11a), one can use a partially reflecting boundary condition in the form [5]

$$D\left[\frac{\partial w(r,t)}{\partial r} + \frac{w(r,t)}{k_B T}\frac{\partial V(r)}{\partial r}\right]_{r=R} = pw(R,t) \qquad (13a)$$

where the left-hand side represents the flux of particles through the reaction surface, and the parameter p is a measure of the intrinsic rate constant at the reaction distance. Equation (13a) is sometimes called the radiation boundary condition. This boundary condition leads to the following boundary condition on the escape probability [3]

$$D\left[\frac{\partial \varphi(r_0)}{\partial r_0}\right]_{r_0=R} = p\varphi(R) \qquad (13b)$$

In the limit $p \to \infty$, Eqs. (13a) and (13b) reduce to Eqs. (11a) and (11b), respectively. The solution of Eq. (9) subject to the boundary conditions (10) and (13b) is given by [3]

$$\varphi(r_0) = \frac{\exp\left(-\dfrac{r_c}{r_0}\right) - \left(1 - \dfrac{Dr_c}{pR^2}\right)\exp\left(-\dfrac{r_c}{R}\right)}{1 - \left(1 - \dfrac{Dr_c}{pR^2}\right)\exp\left(-\dfrac{r_c}{R}\right)} \qquad (14)$$

This expression characterizes the escape probability for the partially diffusion-controlled geminate ion recombination.

In the case where the reaction distance may be assumed much shorter than the range of the Coulomb interaction ($R \ll r_c$), Eq. (14) reduces to the well-known Onsager formula [2]

$$\varphi(r_0) = \exp\left(-\frac{r_c}{r_0}\right) \qquad (15)$$

This simple expression, which gives the probability that a pair of charges initially separated by r_0 and performing a diffusive motion in the mutual Coulomb field will escape recombination, has found numerous applications in various fields of physics and chemistry.

Another method of accounting for the reaction between the electron and the ion, instead of imposing a specific boundary condition, is to introduce a sink term $-k_r(r)w(r,t)$ on the right-hand side of Eq. (2), where $k_r(r)$ is a first-order rate constant [6]. This approach is more general than that using the totally absorbing or the partially reflecting boundary conditions, in a sense that it is applicable to the case where the reaction occurs over a range of distances. Moreover, it does not even require introducing the reaction distance, which is a quantity difficult to define in a real system. If the sink term is used to represent the reaction, the escape probability satisfies

$$D\left[\frac{d^2\varphi}{dr_0^2} + \left(\frac{2}{r_0} - \frac{r_c}{r_0^2}\right)\frac{d\varphi}{dr_0}\right] - k_r(r_0)\varphi(r_0) = 0 \qquad (16)$$

When the reaction is represented by a sink term, a reflective boundary condition must be imposed at a contact distance d. This boundary condition leads to the following boundary condition on the escape probability

$$\left[\frac{\partial \varphi(r_0)}{\partial r_0}\right]_{r_0=d} = 0 \qquad (17)$$

The solution of Eq. (16) depends on the assumed rate constant. In many cases, an exponential form of the rate constant $k_r(r) \sim \exp(-\alpha r)$, where α is a parameter, provides a realistic model of the reaction. Unfortunately, it is usually not possible to find simple analytical solutions of Eq. (16), and the equation has to be numerically solved [7].

We have derived the escape probability for a pair of charges initially separated by a given distance r_0 for various cases. However, in real systems, the electron thermalization distance is distributed. If we denote the distribution of thermalization distances by $f(r)$, the total averaged escape probability, φ_{tot}, can be calculated from

$$\varphi_{tot} = \int_0^\infty \varphi(r)f(r)dr \tag{18}$$

The exact form of $f(r)$ is usually unknown, and the electron thermalization in irradiated systems still remains a subject of both theoretical and experimental scrutiny. Often, $f(r)$ is approximated by a simple analytical expression, e.g., the exponential distribution

$$f(r, r_0) = \frac{1}{r_0} \exp\left(-\frac{r}{r_0}\right) \tag{19}$$

or the Gaussian distribution

$$f(r, r_0) = \frac{4r^2}{\sqrt{\pi} r_0^3} \exp\left(-\frac{r^2}{r_0^2}\right) \tag{20}$$

where r_0 is a parameter.

One of the most important experimental methods of studying the electron–ion recombination processes in irradiated systems are measurements of the external electric field effect on the radiation-induced conductivity. The applied electric field is expected to increase the escape probability of geminate ion pairs and, thus, enhance the number of free ions in the system, which will result in an enhanced conductivity.

The escape probability of an electron in an external electric field F can generally be obtained by solving Eq. (7) or (16) with the potential energy V given by

$$V(r, \vartheta) = -\frac{e^2}{\varepsilon r} - eFr \cos \vartheta \tag{21}$$

Here ϑ is the polar angle. We assume that the field is directed along $\vartheta = \pi$. The boundary condition at infinity is given by Eq. (10), and the reaction may be described either by a specific boundary condition at the reaction distance [Eq. (7)], or by a sink term [Eq. (16)], as discussed before.

The solution of Eq. (7) with the potential energy described by Eq. (21), for the boundary condition (11b) with $R = 0$, was found by Onsager [2]. His result is given by

$$\varphi(r_0, \vartheta, F) = \exp[-\beta r(1 + \cos \vartheta)] \int_{r_c/r}^\infty J_0\left\{2[-\beta r(1 + \cos \vartheta)s]^{1/2}\right\} e^{-s} ds \tag{22}$$

where $\beta = eF/2k_B T$ and J_0 is a zeroth-order Bessel function of the first kind. Because in real systems, the angular dependence of the escape probability is usually unobservable, the quantity that is experimentally more significant is the escape probability $\varphi(r_0, F)$ averaged over the polar angle ϑ. Although no closed solution for $\varphi(r_0, F)$ is available, the field dependence of the angle-averaged escape probability can be described by an infinite series representation [8]

$$\varphi(r_0, F) = \exp(-a - b) \frac{1}{b} \sum_{l=1}^{\infty} l\left(\frac{b}{a}\right)^{l/2} I_l\left(2\sqrt{ab}\right) \tag{23}$$

where $a = r_c/r_0$, $b = eFr_0/k_BT$, and I_l are modified Bessel functions. A couple of different series representations of $\varphi(r_0, F)$ are also available [9,10].

It can be shown that the expansion of $\varphi(r_0, F)$ up to the first order in F gives the expression

$$\varphi(r_0, F) = \exp\left(-\frac{r_c}{r_0}\right)\left(1 + \frac{eFr_c}{2k_BT}\right) \tag{24}$$

This equation shows that at low electric fields, the escape probability is a linear function of F, and the slope-to-intercept ratio of this dependence is given by $er_c/2k_BT$. It is worth noting that this ratio is independent of r_0. Therefore plots of $\varphi(F)/\varphi(0)$ vs. F may be used to test the applicability of the presented theory to describe real systems, even if the distribution of electron thermalization distances is unknown.

Calculation of the electric field dependence of the escape probability for boundary conditions other than Eq. (11b) with $R = 0$ poses a serious theoretical problem. For the partially reflecting boundary condition imposed at a nonzero R, some analytical treatments were presented by Hong and Noolandi [11]. However, their theory was not developed to the level, where concrete results of $\varphi(r_0, F)$ for the partially diffusion-controlled geminate recombination could be obtained. Also, in the most general case, where the reaction is represented by a sink term, the analytical treatment is very complicated, and the only practical way to calculate the field dependence of the escape probability is to use numerical methods.

In Fig. 1, we plot the electron escape probability of geminate ion pairs as a function of the external electric field. The solid lines show the results calculated using Eq. (23) for various initial separations r_0. The broken lines represent the results of $\varphi(r_0, F)$ obtained by numerically solving Eq. (16), which includes the sink term $-k_r\varphi$ to represent the reaction [12]. The rate constant k_r is assumed in the form $k_r(r) = A\exp[-\alpha(r-d)]$, and the parameters are chosen in such a way that the recombination is partially diffusion-controlled. We see that the electric field dependence of the escape probability in the partially diffusion-controlled recombination is significantly different from that in the totally diffusion-controlled recombination. This indicates that the effects due to the finite rate of the reaction cannot be neglected in analyzing experimental data. We should also bear in mind that in real systems the electron thermalization distance r_0 is distributed. To calculate the escape probability for a given thermalization distance distribution $f(r)$, one has to average $\varphi(r, F)$ over $f(r)$.

In the preceding part of this section, we have concentrated on the electron escape probability, which is an important quantity in the geminate phase of recombination, and can be experimentally observed. However, modern experimental techniques also give us a possibility to observe the time-resolved kinetics of geminate recombination in some systems. Theoretically, the decay of the geminate ion pairs can be described by the pair survival probability, $W(t)$, defined by Eq. (4). One method of calculating $W(t)$ is to solve the Smoluchowski equation [Eq. (2)] for $w(\mathbf{r}, t)$ and, then, to integrate the solution over the space variable. Another method [4] is to directly solve Eq. (7) under relevant conditions.

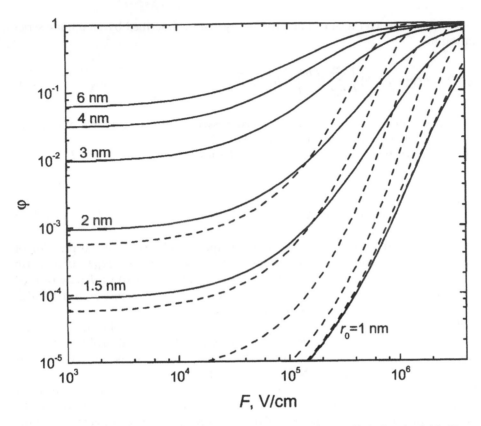

Figure 1 Electron escape probability as a function of the applied electric field. The solid lines are obtained from Eq. (23) for different values of the initial electron cation distance r_0. The broken lines are calculated for $r_0 = 1$ nm from the numerical solution of Eq. (16) with the sink term given by $k_r(r) = A \exp[-\alpha(r-d)]$, where $\alpha = 10$ nm^{-1} and $d = 0.6$ nm. Different lines correspond to different values of A from 10^{13} (the lowest broken curve) to 10^8, in decadic intervals. The parameter values were assumed as $\varepsilon = 4$, $T = 298$ K, and $D = 5 \times 10^{-3}$ cm^2/sec.

The analytical solution of the Smoluchowski equation for a Coulomb potential has been found by Hong and Noolandi [13]. Their results of the pair survival probability, obtained for the boundary condition (11a) with $R = 0$, are presented in Fig. 2. The solid lines show $W(t)$ calculated for two different values of r_0. The horizontal axis has a unit of r_c^2/D, which characterizes the timescale of the kinetics of geminate recombination in a particular system For example, in nonpolar liquids at room temperature $r_c^2/D \sim 10^{-8}$ sec. Unfortunately, the analytical treatment presented by Hong and Noolandi [13] is rather complicated and inconvenient for practical use. Tabulated values of $W(t)$ can be found in Ref. 14. The pair survival probability of geminate ion pairs can also be calculated numerically [15]. In some cases, numerical methods may be a more convenient approach to calculate $W(t)$, especially when the reaction cannot be assumed as totally diffusion-controlled.

It can be shown that for the totally absorbing boundary condition, and for $R \ll r_c$, the long-time behavior of the pair survival probability $W(t)$ is described by

$$W(t) = \exp\left(-\frac{r_c}{r_0}\right)\left(1 + \frac{r_c}{\sqrt{\pi D t}}\right) \tag{25}$$

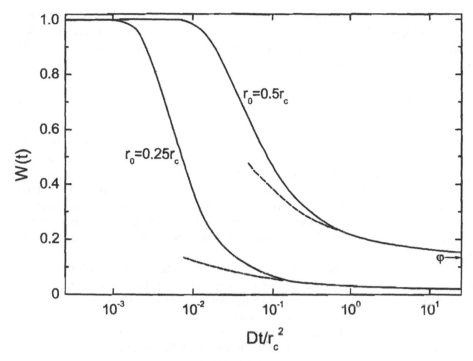

Figure 2 Survival probability of geminate ion pairs as a function of time. The two solid lines correspond to two different values of the initial electron–cation distance. The broken lines show the asymptotic kinetics calculated from Eq. (25). The value of the escape probability for $r_0 = 0.5r_c$ is indicated by φ. (From Ref. 13.)

This asymptotic form of $W(t)$ is also plotted in Fig. 2 (as dashed lines). However, we see that this approximation is valid only when the pair survival probability slightly exceeds the escape probability, so its practical importance is rather limited.

When a strong electron acceptor (electron scavenger) is added in the irradiated system, some of the electrons will react with the scavenger before they recombine with their parent cations. Therefore the measurement of the yield of the scavenging products as a function of the scavenger concentration can be used to monitor the geminate recombination, and the electron scavenging technique has proved to be an important tool in experimental studies.

The geminate recombination in the presence of a scavenger can be described by the Smoluchowski equation [Eq. (2)] with an additional term representing the loss of the geminate pairs by scavenging reactions

$$\frac{\partial w_s}{\partial t} = D\left[\nabla^2 w_s + \frac{1}{k_B T}\nabla w_s \nabla V\right] - k_s C_s w_s(\mathbf{r}, t, C_s) \tag{26}$$

Here k_s denotes the rate constant for the scavenging reaction and C_s is the scavenger concentration. The probability density for the geminate pairs, $w_s(\mathbf{r},t,C_s)$, may be related to the corresponding probability density in the absence of scavenger, $w(\mathbf{r},t)$, through

$$w_s(\mathbf{r}, t, C_s) = w(\mathbf{r}, t)\exp(-k_s C_s t) \tag{27}$$

which is easily verified by substitution. Experimentally important is the total scavenging probability $\varphi_s(k_s C_s)$, which is defined as the probability that an electron reacts with a scavenger before recombination with a positive ion. By making use of Eq. (27), $\varphi_s(k_s C_s)$ was shown to satisfy the relation

$$\varphi_s(k_s C_s) = k_s C_s \int_0^\infty W(t)\exp(-k_s C_s t)\mathrm{d}t = k_s C_s \overline{W}(k_s C_s) \tag{28}$$

where $\overline{W}(s)$ stands for the Laplace transform of $W(t)$. This equation shows that the total scavenging probability is proportional to the Laplace transform, $\overline{W}(k_s C_s)$, of the pair survival probability in the absence of scavenger $W(t)$ [16]. Eq. (28) has a great practical importance, as it allows us to calculate the lifetime distribution of the geminate pairs from measurements of the yield of the scavenging products by inverse Laplace transformation. This equation may also be considered as a generalization [17] of the Stern–Volmer relation widely used in photochemistry to the case where the survival probability $W(t)$ is not described by an exponential function.

The theory of geminate recombination presented so far is based on the solutions of the Smoluchowski equation for an isolated pair of a cation and an electron, which are assumed to perform a diffusive motion in each other's Coulomb field. The single-pair approach has proved to be very useful in modeling early stages of the electron–ion recombination processes in various systems, and was applied to interpret numerous experimental results. There are physical situations, especially in photochemical studies, where ionization is weak and the geminate electron–ion pairs are separated from one another by distances much larger than the Onsager radius. In such systems, the single-pair model of geminate recombination is directly applicable. On the other hand, ionization events in radiation tracks are known to form specific spatial patterns, in which only a fraction of the produced electron–cation pairs may be treated as independent of one another. The single-pair model is then only an approximation, and the "multipair effects" due to the interactions between the neighboring pairs of charges have to be taken into account.

Analytical treatment of the diffusion-reaction problem in a many-body system composed of Coulombically interacting particles poses a very complex problem. Except for some approximate treatments, most theoretical treatments of the multipair effects have been performed by computer simulations. In the most direct approach, random trajectories and reactions of several ion pairs were followed by a Monte Carlo simulation [18]. In another approach [19], the approximate Independent Reaction Times (IRT) technique was used, in which an actual reaction time in a cluster of ions was assumed to be the smallest one selected from the set of reaction times associated with each independent ion pair.

An important result of the theoretical studies of the multipair effects is that the recombination kinetics in a cluster of ions, in which the initial separation between neighboring cations is ~1 nm, is faster than the corresponding decay kinetics of a single ion pair [18]. Furthermore, the escape probability is lower than the Onsager value [Eq. (15)], and decreases with increasing number of ion pairs in the cluster (a relative decrease of about 30% for two ion pairs, and about 50% for five ion pairs). The average electron escape probability in radiation tracks obviously depends on the distribution of ionization events in the tracks, which is determined by the type of radiations and their energy.

Calculations of the external electric field effect on the escape probability in multipair clusters [18] have shown that significant deviations from the results of the single-pair theory are expected at high electric fields. At low fields, the escape probability is a linear function of the field, and the slope-to-intercept ratio of this dependence is almost the same as that obtained in the single-pair theory [20] [cf. Eq. (24)].

A study of electron scavenging in multipair clusters [21] has shown that the total scavenging probability decreases with increasing number of ion pairs in the cluster. However, the Laplace transform relationship [Eq. (28)] between the scavenging probability and the recombination kinetics was found to work reasonably well also in the multipair case.

In this section, a theory of geminate recombination of charged species is presented by fully taking into account the Coulomb interaction between the species. While for nonpolar systems, the theory of geminate ion recombination in its full form has to be used, some simple approximate theories can be used for systems of high polarity. An important example is water, where the Onsager radius is only 0.7 nm, and is shorter than typical electron thermalization distances of a few nanometers. In this case, the Coulomb interaction between the ionization products is not so important and, in the first approximation, it may even be neglected, as it was performed in numerous theoretical studies. The advantage of this approach is that the analytical expression describing the geminate recombination of uncharged species are much simpler than those for the geminate ion recombination. However, theoretical modeling of the geminate recombination processes in water is still complicated, even if one neglects the Coulomb interactions. This is because the primary ionization products in water undergo a series of chemical reactions, which proceed on the timescale shorter than that of the charge recombination.

Geminate Ion Recombination in High-Mobility Systems

The theory of geminate recombination presented in Sec. 10.1.2 ("Diffusion-Controlled Geminate Ion Recombination") is based on the assumption that the relative motion of electrons and cations that constitute geminate pairs is described by the diffusion equation. In a real system, the motion of a particle may be treated as ideal diffusion when its scattering by atoms or molecules of the system is very frequent, and its mean free path, λ, is much shorter compared with the typical length involved in the problem. In the case of the ion recombination, it is reasonable to take as the typical length involved in the problem the range of the Coulomb interactions in the considered system, so the applicability of the diffusion theory of geminate recombination is theoretically limited to systems where the scattering mean free path is much shorter than the Onsager radius ($\lambda \ll r_c$). In a simple model, the mean free path of charged particles is proportional to their mobility μ. Therefore the diffusion theory of geminate recombination will not be applicable, when the mobility of charges is too high. In fact, deviations from the diffusion theory were experimentally observed in systems where the electron mobility exceeds ~100 cm V^{-1} sec^{-1}.

Limitations of the diffusion approach also become apparent from the considerations based on energy. In the conventional diffusion approach, it is assumed that particles are always in thermal equilibrium with the system, and that their velocity distribution is close to the Maxwell–Boltzmann distribution. When the mean free path is long, and the scattering events not so frequent, electrons are no longer able to effectively dissipate the kinetic energy they acquire in the Coulomb field of the cations. The average electron energy is then significantly higher than the thermal level.

In this physical picture of geminate recombination in high-mobility systems, it is no longer valid to use the space criterion of reaction expressed in terms of the distance. Instead, the energy criterion of reaction expressed in terms of energy or, more preferably, the energy-dependent reaction cross section should be used.

Because of the inefficient dissipation of their kinetic energy, electrons in high-mobility systems are more likely to escape geminate recombination than predicted by the diffusion theory. Such an effect has been confirmed by the results of a computer simulation study [22] presented in Fig. 3. In this study, classical electron trajectories in the Coulomb field of a cation were calculated, starting at a distance r_0 from the cation, and the electron scattering was modeled in such a way that the electron velocity was randomized at each collision to restore the Maxwell–Boltzmann distribution. The times between the collisions were selected from an exponential distribution, and its mean value, τ, was used to characterize the strength of the dynamic interaction between the electron and the solvent. The mean free time τ may be related to the mean free path by $\lambda/r_c \approx 1.6\tau$. For each electron, the simulation was carried out until either the reaction occurred, or the electron was separated from the cation to a distance much larger than r_c. An energy criterion of reaction was used, namely, the electron was assumed to have reacted once its total energy in the Coulomb field of the cation was reduced below $-10k_BT$.

According to Fig. 3, the Onsager result [Eq. (15)], which shows the escape probability in the diffusion-controlled geminate recombination, gives the lower bound for the simulation results. The simulation results obtained for the lowest value of τ ($\tau = 0.05$) are

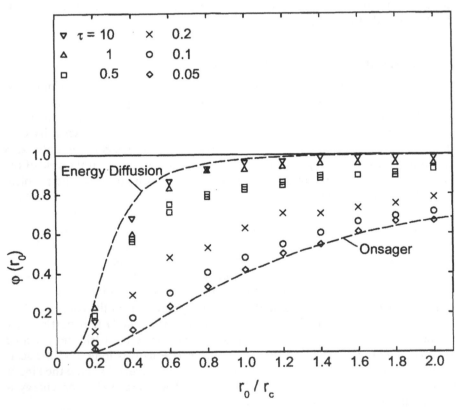

Figure 3 Escape probability as a function of the initial electron–cation distance. The lower broken curve is calculated from the Onsager equation [Eq. (15)]. The numerical results for different mean free times τ were taken from Ref. [22]. The unit of τ is $r_c/(k_BT/m)^{0.5}$, where m is the electron mass. The upper broken curve was calculated using the energy diffusion model. (From Ref. 23.)

very close to the Onsager result. At higher values of the mean free time, the escape probability exceeds the Onsager result and significantly increases with increasing τ.

The upper bound of the escape probability, corresponding to $\tau \to \infty$, can also be analytically calculated. In this limit, the electron is allowed to make many revolutions in the same orbit around the cation before it is scattered into another orbit. Under this condition, the electron motion may be described as diffusion in energy space [23]. The escape probability calculated by using the energy diffusion model is also included in Fig. 3. We see that the simulation results for finite τ properly approach the energy diffusion limit.

Computer simulation has also been used to calculate the external electric field effect on the geminate recombination in high-mobility systems [22]. For the mean free time τ exceeding ~0.05, the field dependence of the escape probability was found to significantly deviate from that obtained from the diffusion theory. Furthermore, the slope-to-intercept ratio of the field dependence of the escape probability was found to decrease with increasing τ. Unlike in the diffusion-controlled geminate recombination, this ratio is no longer independent of the initial electron-ion separation [cf. Eq. (24)].

The model of electron scattering in high-mobility systems applied in the simulations is rather simplified. Especially, the assumption that the electron velocity is randomized at each scattering to restore the Maxwell–Boltzmann distribution may be an oversimplification. If the dissipation of energy by electron collisions in a real system is less efficient than that assumed in the simulation, the escape probability is expected to further increase.

The electron–ion recombination in high-mobility systems has also been analyzed in terms of the fractal theory [24,25]. It was postulated that even when the fractal dimension of particle trajectories is not equal to 2, the motion of particles is still described by diffusion but with a distance-dependent "effective" diffusion coefficient $D'(r) = D(1 + l/r)^{-1}$, where the parameter l is proportional to the mean free path λ [24]. However, when the fractal dimension of trajectories is not equal to 2, the motion of particles is not described by orthodox diffusion.

1.3. Bulk Ion Recombination

Diffusion-Controlled Bulk Ion Recombination

When the motion of electrons and positive ions in a particular system may be described as ideal diffusion, the process of bulk recombination of these particles is described by the diffusion equation. The mathematical formalism of the bulk recombination theory is very similar to that used in the theory of geminate electron–ion recombination, which was described in Sec. 10.1.2 ("Diffusion-Controlled Geminate Ion Recombination"). Geminate recombination is described by the Smoluchowski equation for the probability density $w(\mathbf{r}, t)$ [cf. Eq. (2)], while the bulk recombination is described by the diffusion equation for the space and time-dependent concentration of electrons around a cation (or vice versa), $c(\mathbf{r}, t)$,

$$\frac{\partial c(\mathbf{r}, t)}{\partial \tau} = D \left[\nabla^2 c(\mathbf{r}, t) + \frac{1}{k_B T} \nabla c(\mathbf{r}, t) \nabla V(\mathbf{r}) \right] \tag{29}$$

The two equations have the same mathematical form, although they deal with different physical quantities. The initial and boundary conditions are also somewhat different. In the bulk recombination, the initial electron concentration is assumed to be uniform in the space and equal to the average concentration c_0

$$c(\mathbf{r}, 0) = c_0 \tag{30}$$

The boundary condition at infinite separation is given by

$$\lim_{r \to \infty} c(\mathbf{r}, t) = c_0 \tag{31}$$

where we assume that the average electron concentration changes little during the reaction.

The methods used in geminate recombination to account for the reaction are also applicable to bulk ion recombination. In the simplest case of fully diffusion-controlled recombination, the reaction can be represented by the totally absorbing boundary condition

$$c(R, t) = 0 \tag{32}$$

where R is the reaction distance. For partially diffusion-controlled recombination, we may use the boundary condition in the form

$$D\left[\frac{\partial c(r, t)}{\partial r} + \frac{c(r, t)}{k_B T} \frac{\partial V(r)}{\partial r}\right]_{r=R} = p\, c(R, t) \tag{33}$$

where the left-hand side represents the flux through the reaction surface, and the reactivity parameter p is a measure of the intrinsic rate constant at the reaction distance.

An example of numerical solutions of Eq. (29) for the attractive Coulomb potential, subject to the initial and the boundary conditions (30)–(32), is presented in Fig. 4 [26]. The figure shows how the initially uniform electron concentration gradually decreases in the vicinity of the cation until the steady state is established. The rate of reaction is determined by the diffusive flow of electrons from large distances toward the reaction sphere.

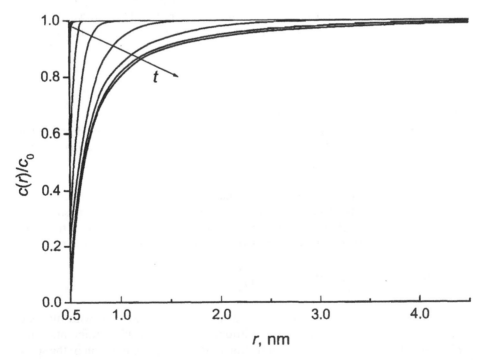

Figure 4 Time evolution of the normalized concentration of electrons around the cation $c(r,t)/c_0$ for $D = 10^{-4}$ cm^2/sec and $r_c = 0.7$ nm. Different curves correspond to different times from 10^{-14} to 10^{-7} sec, in decadic intervals. (From Ref. 26.)

The time-dependent rate coefficient for the bulk ion recombination may be obtained as the total current of particles flowing through the reaction surface at $r = R$, divided by c_0,

$$k(t) = 4\pi R^2 D \left[\frac{\partial c(r,t)}{\partial r} + \frac{c(r,t)}{k_B T} \frac{\partial V(r)}{\partial r} \right]_{r=R} \bigg/ c_0 \tag{34}$$

The long-time limit of $k(t)$ defines the steady state rate constant of the bulk ion recombination, which for the totally diffusion-controlled recombination is calculated as [27]

$$k = \frac{4\pi D r_c}{1 - \exp\left(-\dfrac{r_c}{R}\right)} \tag{35}$$

When the reaction distance is short in comparison with the range of the Coulomb interaction ($R \ll r_c$), Eq. (35) reduces to the so-called Debye–Smoluchowski expression

$$k_D = 4\pi D r_c \tag{36a}$$

This expression is a fundamental result of the theory of bulk ion recombination and has been extensively used in interpreting experimental results of diffusion controlled reactions. The Debye–Smoluchowski expression can also be written in terms of the mobility,

$$k_D = \frac{4\pi e}{\varepsilon} \mu \tag{36b}$$

where we have used the Einstein relation between the diffusion coefficient and the mobility $\mu/D = e/k_B T$.

In the case of partially diffusion-controlled recombination, which is described by Eq. (33), the recombination rate constant is calculated as [28]

$$k = \frac{4\pi D r_c}{1 - \left(1 - \dfrac{D r_c}{p R^2}\right) \exp\left(-\dfrac{r_c}{R}\right)} \tag{37}$$

The rate constants given by Eqs. (35)–Eqs. (37) give the steady state rate constants of bulk ion recombination. As seen from Fig. 4, it takes some time before the steady state is established, and, in general, the rate coefficient of the bulk ion recombination is time-dependent. Although no simple analytical expression is available, which describes $k(t)$ in the whole time domain, the expression for the asymptotic behavior of $k(t)$ at long times is easily obtained. For fully diffusion-controlled recombination, we have

$$k(t \to \infty) = \frac{4\pi D r_c}{1 - \exp(-r_c/R)} \left[1 + \frac{\exp(-r_c/R)}{1 - \exp(-r_c/R)} \frac{r_c}{\sqrt{\pi D t}} \right] \tag{38}$$

It was shown [26] that in water ($r_c = 0.7$ nm), the transient effect in the bulk ion recombination may be significant on the timescale shorter than a nanosecond. For nonpolar systems ($r_c \sim 30$ nm), the transient effect in Eq. (38) is much smaller, and the time dependence of the rate coefficient is not likely to be experimentally observed.

Bulk recombination is a many-body problem. It is not so obvious whether the approach based on Eqs. (29–34) correctly describes the many-body character of bulk recombination. Tachiya [29] formulated the rate of bulk recombination in terms of the pair survival probability of geminate recombination and showed that the approach described above is exact only when the minority reactants are fixed and the majority reactants are mobile (target problem). The above approach is not exact when the majority reactants are fixed and the minority reactants are mobile (trapping problem), or when the concentrations of two reactants are comparable, although it gives a very good approximation in most cases of interest.

Because of its importance for experimental studies, we will also discuss the effect of an applied electric field on the rate constant of bulk ion recombination. Within the framework of the diffusion theory, this effect can be calculated by solving the diffusion equation [Eq. (29)] with the potential energy given by Eq. (21) under relevant initial and boundary conditions. Analytical solutions of this problem, both for the totally absorbing [Eq. (32)] and the partially reflecting [Eq. (33)] boundary conditions, are found in Ref. 30. In the case of totally diffusion-controlled recombination, the rate constant is found to increase with increasing field strength. The general expression for the electric field dependence of the recombination rate constant, $k(F)$, obtained in Ref. 30 is quite complicated and involves an infinite series summation. An expansion of this expression in the Taylor series up to the first order in F gives [30]

$$k(F) = \frac{4\pi D r_c}{1 - \exp(-r_c/R)} \left[1 + \frac{\exp(-r_c/R)}{1 - \exp(-r_c/R)} \frac{eFr_c}{2k_B T} \right] \tag{39}$$

The increase of k with increasing field strength is explained by an enhanced rate of encounter between positively and negatively charged particles because of their drift in the opposite directions in the presence of an external field. However, in nonpolar systems, this effect is very weak and the recombination rate constant is almost independent of the electric field.

The results obtained in Ref. 30 for partially diffusion-controlled recombination show that the field dependence of the recombination rate constant is affected by both the reaction radius R and the reactivity parameter p [cf. Eq. (33)]. Depending on their relative values, the rate constant can be increased or decreased by the electric field. The latter effect predominates at low values of p, where the reactants staying at the encounter distance are forced to separate by the electric field.

Because of space limitation, we have discussed only a part of aspects of the theory of diffusion-controlled ionic processes, which we believe to be essential for understanding the bulk phase of ion recombination in irradiated systems. Discussions of other aspects of diffusion-controlled ionic processes are found in monographs on the diffusion-controlled reactions [31]. One important aspect that was only briefly touched in this section concerns the theoretical models to account for the reaction between ions (the final step of recombination). We have discussed the models, which introduce specific boundary conditions at the reaction distance [Eqs. (32) and (33)] to account for the reaction. In a more general approach, the reaction may be accounted for by introducing a sink term, as it was considered in geminate recombination [cf. Eqs. (16) and (17) in Sec. 10.1.2 ("Diffusion-Controlled Geminate Ion Recombination")]. Such an approach allows us to treat the case where the reaction occurs not only at the contact distance, but over a range of distances, as in the case of electron transfer reactions. Another important aspect concerns the influence of the solvent structure on the diffusive transport of ions. This is exhibited as the potential of mean force and hydrodynamic effects. However, these effects are usually overwhelmed by the Coulomb attraction between the ions, and may be safely neglected, especially in nonpolar systems. The competition effects in the diffusion-controlled bulk ion recombination are also expected to influence the recombination rate constant at high ion concentrations, although they may not be very important in normal experimental conditions [32].

The theory of diffusion-controlled recombination presented in Secs. 10.1.2 ("Diffusion-Controlled Geminate Ion Recombination") and 10.1.3 ("Diffusion-Controlled Bulk Ion Recombination") may not be applicable to a class of disordered solid state systems, including amorphous semiconductors and organic glasses, where the charge transport is strongly influenced by the processes of trapping and detrapping of charge carriers. The transport of charged particles in such systems is usually described as the continuous time random walk [33,34] or using the fractional diffusion approach [35]. These two approaches have also been used in the theoretical studies of both geminate and bulk ion recombination in the disordered systems [36,37]. However, in those studies, the Coulomb interaction between the charged particles was not taken into account.

Bulk Ion Recombination in High-Mobility Systems

The diffusion theory of bulk electron–ion recombination presented in Sec. 10.1.3 ("Diffusion-Controlled Bulk Ion Recombination") is not applicable to systems where the electron mobility is high. If the mobility exceeds ~ 100 cm^2 V^{-1} sec^{-1}, effects due to the finite mean free path for electron scattering become significant, and the electron motion can no longer be described as ideal diffusion. The general physical picture of electron transport processes and reactions in high-mobility systems is essentially the same both in the geminate and the bulk phases of recombination, and has already been discussed in Sec. 10.1.2 ("Geminate Ion Recombination in High-Mobility Systems"). In short, when the electron mean free path increases, the character of its motion gradually changes from diffusive to "ballistic." In place of the spatial diffusion of electrons and cations toward one another, the rate-determining process for their recombination becomes the dissipation of the kinetic energy the particles acquire in the mutual Coulomb field. In consequence, the electron–ion recombination rate constants in high-mobility systems are expected to be significantly lower than the Debye–Smoluchowski value k_D predicted by the diffusion theory [cf. Eq. (36)]. This was confirmed by the experimental results, which will be discussed later in this chapter.

An estimate of the electron–ion recombination rate constant in high-mobility systems based on an empirical model of energy dissipation processes was provided by Warman [38]. He related the rate constant to the field dependence of the electron mobility, and proposed an empirical formula

$$\frac{k}{k_D} = \left(1 + \frac{55T^4\varepsilon^2}{F_{10}^2}\right)^{-1} \tag{40}$$

Here F_{10} denotes the "critical" electric field strength expressed in units of V/m, at which the electron mobility deviates 10% from the thermal mobility, and is used as a measure of the rate of the electron energy dissipation in a particular system. Despite its simplicity, Eq. (40) is shown to give reasonable estimates of the electron–ion recombination rate constant for some of the experimentally studied high-mobility systems.

The breakdown of the diffusion theory of bulk ion recombination in high-mobility systems has been clearly demonstrated by the results of the computer simulations by Tachiya [39]. In his method, it was assumed that the electron motion may be described by the Smoluchowski equation only at distances from the cation, which are much larger than the electron mean free path. At shorter distances, individual trajectories of electrons were simulated, and the probability that an electron recombines with the positive ion before separating again to a large distance from the cation was determined. The value of the recombination rate constant was calculated by matching the net inward current of electrons obtained from the simulation results with that derived from the Smoluchowski equation at large distances. To simulate electron trajectories, he used a simplified model of electron scattering, in which the electron velocity was randomized at each collision to restore the Maxwell–Boltzmann distribution. An energy criterion of recombination was used, namely, an electron was assumed to have reacted once its total energy in the Coulomb field of the cation was reduced below a certain critical energy.

The simulation results of the electron–ion recombination rate constant obtained in Ref. 39 are plotted in Fig. 5. The figure shows that the rate constant becomes lower than the Debye–Smoluchowski value k_D when the electron mean free path exceeds ~0.01r_c. At higher values of λ, the ratio k/k_D further decreases with increasing mean free path. The simulation results are found to be in good agreement with the experimental data on the electron–ion recombination rate constant in liquid methane, which are also plotted in Fig. 5.

The dependence of the electron–ion recombination rate constant on the mean free path for electron scattering has also been analyzed on the basis of the Fokker–Planck equation [40] and in terms of the fractal theory [24,25,41]. In the fractal approach, it was postulated that even when the fractal dimension of particle trajectories is not equal to 2, the motion of particles is still described by diffusion but with a distance-dependent "effective" diffusion coefficient. However, when the fractal dimension of trajectories is not equal to 2, the motion of particles is not described by orthodox diffusion. For the

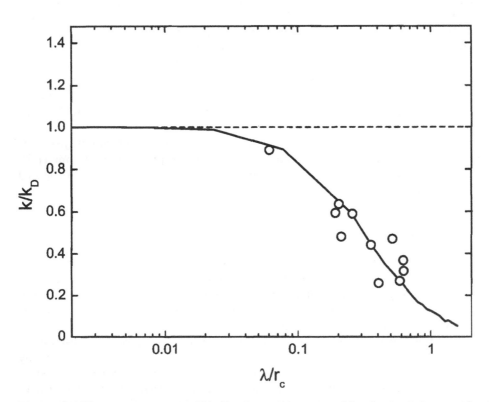

Figure 5 The rate constant of bulk electron–ion recombination, relative to the Debye–Smoluchowski value [Eq. (36)], as a function of the electron mean free path λ. The solid line represents the simulation results, and the circles show the experimental data for liquid methane [49]. (From Ref. 39.)

relation between the fractal dimensions of reactant trajectories and the transport-limited reaction rate, see Ref. 42.

In the limit of an infinitely long mean free path, the bulk electron–ion recombination may be described using the energy diffusion model [43,44]. This model is especially relevant to the electron–ion recombination processes in the gas phase.

The theoretical models of the bulk electron–ion recombination presented so far in this section use only one parameter, namely the electron mean free path, to characterize the electron transport processes in a wide class of high-mobility systems. This is undoubtedly an oversimplification, and more realistic models of electron scattering have to be used, if one wants to achieve a better understanding of the bulk recombination processes in a given system. Such models, which use electron scattering cross sections, were applied in the simulations of the electron–ion recombination in liquid methane [45] and liquid argon [46]. In the case of methane, the mean free path for elastic electron scattering was determined from the experimental electron mobility, and the inelastic scattering was simulated using model cross sections for vibrational and electronic excitations. In the simulation for argon [46], the electron motion was described using a model, which takes account of different rates of energy and momentum transfers in electron collisions, in accordance with the Cohen–Lekner theory [47] of hot electrons in liquids. In both

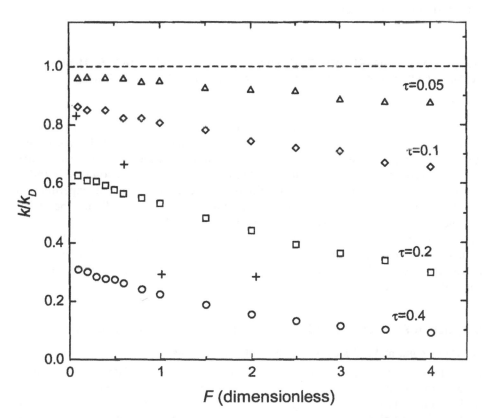

Figure 6 Dependence of the electron–ion recombination rate constant on the external electric field F, calculated for different values of the electron mean free time τ. The unit of F is $e/\varepsilon r_c^2$, and that of τ is $r_c/(k_B T/m)^{0.5}$. The simulation results of Morgan [45] for liquid methane at 120 K are shown by crosses. (From Ref. 48.)

simulations, a reasonable agreement of the calculated rate constants with the experimental results has been obtained.

Experimental data on the electron–ion recombination in high-mobility systems are usually obtained by measuring the transient conductivity of an ionized system in the presence of an external electric field. The applied field increases the average kinetic energy of charge carriers. As the dissipation of energy is believed to be the rate-determining process for the electron–ion recombination in high-mobility systems, the recombination rate constant in these systems should decrease with increasing field strength. Such an expectation has been confirmed by the simulations [48] in which the field dependence of the rate constant was calculated for systems characterized by different values of the electron mean free time τ (the mean free time may be related to the mean free path by $\lambda/r_c \approx 1.6\tau$). The results are presented in Fig. 6. We see that as the mean free time increases, the suppression of the electron–ion recombination by the electric field becomes stronger. In the model of the electron scattering assumed in Ref. 48, the electron velocity was randomized at each collision to restore the Maxwell–Boltzmann distribution. However, in real systems, the electron energy dissipation by collisions is usually not so efficient, and the decrease of the recombination rate constant with increasing electric field should be even stronger than calculated in Ref. 48. This expectation is supported by the simulation results for liquid methane [45], also included in Fig. 6, which were obtained by using a realistic model of electron scattering in that system.

2. PICOSECOND AND SUBPICOSECOND PULSE-RADIOLYSIS STUDIES OF GEMINATE ION RECOMBINATION IN LIQUID HYDROCARBONS

2.1. Introduction

The geminate ion recombination in liquid hydrocarbons has been studied both by stationary experiments and by kinetics data. Although the kinetics data of pulse radiolysis experiments give more detailed information about the geminate recombination than stationary data such as scavenger experiments, the very high time-resolved spectroscopy is indispensable for studies on the kinetics of the geminate ion recombination. The experimental studies on the kinetics of the geminate ion recombination have advanced with the development of picosecond and subpicosecond pulse radiolysis development.

Here the progress in the picosecond and subpicosecond pulse radiolysis is described first and then the experimental studies on the kinetics of the geminate ion recombination is explained in connection with their application to advanced technology such as the next generation nanolithography and nanotechnology.

2.2. Progress in Picosecond and Subpicosecosecond Pulse Radiolysis

History of Picosecond and Subpicosecosecond Pulse Radiolysis

With the development of the picosecond pulse radiolysis, the kinetics data of the geminate ion recombination have been directly obtained. The history of picosecond and subpicosecond pulse radiolysis is shown in Fig. 7. Very recently, the first construction of the femtosecond pulse radiolysis and the improvement of the subpicosecond pulse radiolysis started in Osaka University.

The first picosecond pulse radiolysis experiment was carried out in the late 1960s by the so-called stroboscopic method (generally pomp and probe method) at University of

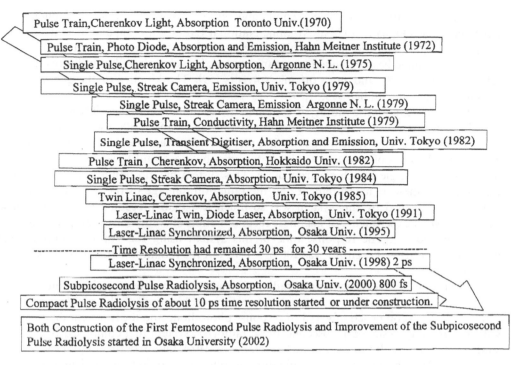

Figure 7 History of picosecond and subpicosecond pulse radiolysis and the start of construction of the first femtosecond pulse radiolysis.

Toronto [50]. Picosecond electron pulse trains generated with a S-band linear accelerator (linac) as an irradiation source and the Cerenkov light induced from the electron pulse trains as an analyzing light were used. The system was very sophisticated, but S/N ratio is not high. Therefore the operation of the system for obtaining small absorption and the detailed kinetics data is very difficult. Similar system was developed in Hokkaido University more than 10 years later [51]. The combination of picosecond electron pulse trains with high time-resolved detection systems was applied to the picosecond pulse radiolysis with optical [52] and conductivity [53] spectroscopy. Measurable time domain of these picosecond pulse radiolysis systems using picosecond electron pulse trains were limited from the time resolution of the system to 350 and 770 psec determined by the pulse interval in the pulse trains of S-band and L-band linacs, respectively. The overlapping of the correct signal with the signal of slow decaying intermediates species produced by the preceding picosecond electron pulses made difficult the detailed analysis of the kinetics of intermediates. These systems were unable to study the geminate ion recombination in irradiated liquid hydrocarbons.

The stroboscopic pulse radiolysis with the single bunch electron pulse instead of pulse trains started in Argonne National Laboratory in 1975 [54]. The research fields have been extended by the stroboscopic pulse radiolysis with the picosecond single electron bunch, although most of researches had been limited to hydrated and solvated electrons in the aqueous and alcoholic solutions. This system was unable to study the kinetics of the geminate ion recombination in liquid hydrocarbons until the modification of the Argonne linac in 1983, which made possible the quality measurements of the weak absorption.

The combination of the picosecond single electron bunch with streak cameras, independently developed in 1979 at Argonne National Laboratory [55] and at University of Tokyo by us [56], enabled the very high time resolution for emission spectroscopy. The research fields have been extended to organic materials such as liquid scintillators [55–57], polymer systems [58], and pure organic solvents [59]. The kinetics of the geminate ion recombination were studied [55,57,59].

The picosecond pulse radiolysis system in the combination of very fast response photodiode with the transient digitizer (the time resolution of about 800 psec) was developed by us in 1982 [60]. This system was the first picosecond pulse radiolysis system that could measure the kinetics of ions and triplet states on subnanosecond timescale by using only one picosecond single bunch electron pulse. This system has no limitation of nano- and microsecond timescale measurement and the radiation damage of the samples of the experiments performed by this system was so much less than preceding picosecond pulse radiolysis. Most experiments of geminate recombination of solute ions were performed by this system [60–66]. The response time of this system using the sampling oscilloscope instead of the transient digitizer was 200 psec [61], but the radiation damage of the samples was very large.

The important picosecond pulse radiolysis study on geminate recombination of solute ions in liquid alkanes [65] was carried out in the combination of the picosecond pulse radiolysis system composed of very fast response photodiode and the transient digitizer [60] with the specially designed picosecond pulse radiolysis system [67] composed of a streak camera (the time resolution of 50 psec) developed in 1984. Our experiment [65] showed for the first time that the geminate recombination of solute ions could be analytically explained by the extended Smoluchowski equation [13,68]. The S/N ratio was not high because of a limited dynamic range of the streak camera at that time, but the radiation damage of the samples was very low. The proof of the principle of the twin linac system, where one linac produced picosecond electron pulses as radiation source and the other linac produces the Cherenkov light as monitor light, was reported in 1985 [69] and the construction of the system was carried out in 1987 [70]. Although the twin linac system was very interesting system, the operation of the system was very difficult and it takes too much effort to obtain the data of radiation chemistry. The radiation damage of the samples was also very large. After the LL (Laser–Linac) twin picosecond pulse radiolysis was developed in 1991 [71], the twin linac system was not used any more. Very important results for radiation chemistry used by the twin linac system were the measurement of very rapid formation of alkyl radicals in irradiated liquid alkanes within the time resolution of the system and it indicates the importance of the excited radical cations in radiolysis of cyclohexane [66].

The LL twin picosecond pulse radiolysis system, which used picosecond semiconductor laser as monitor light instead of Cherenkov light of the twin linac pulse radiolysis, was successfully developed by us for the first time in 1991 [71]. It is the first picosecond pulse radiolysis in the combination of the single electron bunch with laser monitor light, although several trials of the picosecond pulse radiolysis with the single electron bunch and laser monitor light had attempted, but the system was not correctly operated [72]. Because the operation of the LL twin picosecond pulse system was much easier than that of the twin linac, pulse radiolysis and the laser diode as monitor light enabled the absorption spectroscopy from visible to infrared region in picosecond time range for the first time [71]. The kinetics of the geminate recombination of both electrons and cation radicals were first measured on a picosecond timescale in the same sample and agree with each other on picosecond timescale from 50 psec after a pulse [71]. However, it was very difficult to obtain the complete transient absorption spectra by this system because the laser diodes are not tunable light source.

We solved the lack of the tunability of the LL twin pulse radiolysis by the development of the laser synchronized picosecond pulse radiolysis in 1995 at Osaka University [73]. The time resolution of pulse radiolysis had remained about 30 psec for 30 years since the late 1960s (Fig. 7), although many modifications had been performed for picosecond pulse radiolysis techniques. The femtosecond monitoring laser pulse is synchronized with both the electron pulses from the electron gun and accelerating electron pulses through the accelerating tubes in the laser synchronized picosecond pulse radiolysis system, although the picosecond laser pulse is synchronized with only the electron pulse from the electron gun in the LL twin linac system. Although the time resolution of the laser synchronized pulse radiolysis at that time was 30 psec, this system had the possibility of the development of femtosecond pulse radiolysis.

The time resolution of 2 psec [74] for 2-mm optical path sample cell was obtained as shown in Fig. 8 by using laser synchronized subpicosecond pulse radiolysis with the femtosecond single electron bunch [76] and a jitter compensation system with 200-fsec streak camera [74]. In this case, the time difference between electron beam and light passing through a sample is 1.8 psec. Therefore the time resolution is mainly determined by the optical path length. The time resolution of 800 fsec [75] for 0.5-mm optical path sample cell was obtained as shown in Fig. 9, but the S/N ratio is very low. The S/N ratio of the laser synchronized subpicosecond pulse radiolysis was greatly improved by the double-pulse method as shown in Fig. 10 [77]. Now the kinetics of the geminate ion recombination of

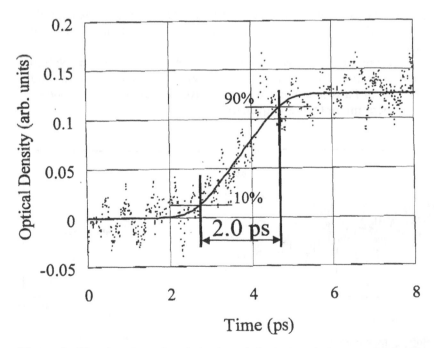

Figure 8 The time-dependent behavior of the hydrated electron obtained in the subpicosecond pulse radiolysis of neat water using 2-mm optical path sample cell, monitored at the wavelength of 780 nm.

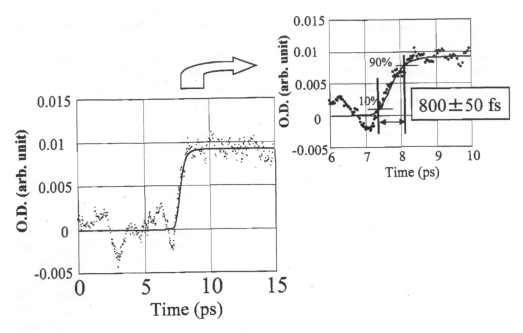

Figure 9 The time-dependent behavior of the hydrated electron obtained in the subpicosecond pulse radiolysis of neat water using 0.5-mm optical path sample cell, monitored at the wavelength of 780 nm.

electron and radical cations has been directly studied by the laser synchronized subpicosecond pulse radiolysis with the double-pulse method. The details of the double-pulse method are described in the next section.

Now several picosecond pulse radiolysis systems are operating in Argonne National Laboratory, Brookhaven National Laboratory and University of Tokyo and are under construction in University of Paris South and Waseda University, although only one subpicopulse radiolysis is operating in Osaka University.

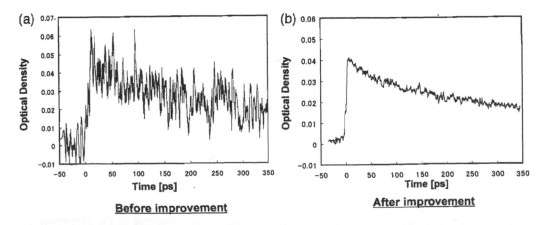

Figure 10 The geminate decay of the radical cation in irradiated neat n-dodecane observed by the subpicosecond pulse radiolysis before (a) and after (b) the improvement of the system by mainly using the double-pulse method.

Subpicosecond Pulse Radiolysis [74,77]

The subpicosecond pulse radiolysis [74,77] detects the optical absorption of short-lived intermediates in the time region of subpicoseconds by using a so-called stroboscopic technique as described in Sec. 10.2.2 ("History of Picosecond and Subpicosecosecond Pulse Radiolysis"). The short-lived intermediates produced in a sample by an electron pulse are detected by measuring the optical absorption using a very short probe light (a femtosecond laser in our system). The time profile of the optical absorption can be obtained by changing the delay between the electron pulse and the probe light.

Stroboscopic pulse radiolysis can achieve high time resolution because the time resolution of the stroboscopic pulse radiolysis does not depend on the time resolution of the detection systems such as photodiodes and oscilloscopes. The time resolution is principally determined by the lengths of the electron pulse and probe light and the timing accuracy (or jitter) between them as shown in Fig. 11. Assuming that the envelopes of both the electron pulse and probe light and the timing fluctuation have a Gaussian distribution, the response function, $g_1(t)$, can be expressed as follows,

$$g_1(t) = \frac{1}{\sqrt{2\pi}\sigma} e^{-\frac{t^2}{2\sigma^2}} \tag{41}$$

$$\sigma^2 = \sigma_i^2 + \sigma_b^2 + \sigma_j^2 \tag{42}$$

where σ_b is the length of the electron pulse (rms), σ_i is the length of the probe light (rms) and σ_j is the timing fluctuation (rms).

However, the time resolution is also degraded by the velocity difference between the light and the electron pulse in a sample. The time in which the electron pulse passes through a sample is given by $L/(\beta c)$, where L is the optical length of the sample, c is the velocity of light in vacuum, and β is the ratio of the velocity of the electron to c. On the other hand, in the case of light, the time in which it passes is given by Ln/c, where n is the refractive index of the sample. Therefore the time resolution is degraded with increasing

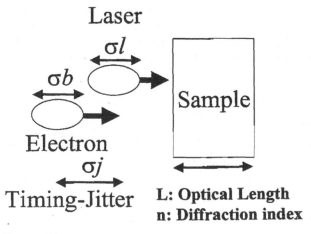

Figure 11 The components of the timing jitter of the laser synchronized subpicosecond pulse radiolysis. σ_b is the length of the electron pulse (rms), σ_i is the length of the probe light (rms), and σ_j is the timing fluctuation (rms).

thickness of the sample. The response function with the effect of velocity difference, $g_2(t)$, can be deduced from the definition of the response function as follows,

$$g_2(t) = \frac{c}{\sqrt{\pi}L\left(n - \frac{1}{\beta}\right)} \int_{\frac{2ct - L\left(n - \frac{1}{\beta}\right)}{2\sqrt{2}\sigma c}}^{\frac{2ct + L\left(n - \frac{1}{\beta}\right)}{2\sqrt{2}\sigma c}} e^{-x^2} dx \tag{43}$$

When the optical length of the sample is much shorter than the other factors such as the length of the electron pulse, the response function takes on a Gaussian shape. As the optical length increases, the shape of the response function becomes trapezoidal. Furthermore, a thick sample causes the prolongation of the electron pulse by electron scattering, which leads to the degradation of time resolution. Therefore the experiment to observe ultrafast phenomena requires the use of a thin sample.

Fig. 12 shows the subpicosecond pulse radiolysis system. The system consists of a subpicosecond electron linac as an irradiation source, a femtosecond laser as an analyzing light, and a jitter compensation system. A sample was irradiated by a subpicosecond electron single pulse. The subpicosecond electron single pulse was obtained by compressing a 30-psec electron single pulse from the ISIR L-band linac with a magnetic pulse compressor. This system can compress the width of electron pulse to approximately 125 fsec (FWHM) [76]. The time-resolved optical absorption was detected with a femtosecond laser. A mode locked ultrafast Ti:Sapphire laser (Tsunami, Spectra-Physics Lasers, Inc.) was synchronized to the ISIR L-band Linac using a commercially available phase lock loop. The width of the laser pulse was 60 fsec (FWHM). The intensity of the laser pulse was measured by a Si photodiode. The timing between the electron pulse and the laser pulse was controlled by radio frequency (RF) system. The time profile of the optical absorption could be obtained by changing the phase of the RF with an electrical phase shifter.

The jitter between the laser pulse and the electron pulse was estimated from the measurement using a streak camera (C1370, Hamamatsu Photonics Co. Ltd.), because the jitter is one of important factors that decide the time resolution of the pulse radiolysis. The jitter was several picoseconds. To avoid effects of the jitter on the time resolution, a jitter compensation system was designed [74]. The time interval between the electron pulse (Cerenkov light) and the laser pulse was measured by the streak camera at every shot. The Cerenkov radiation was induced by the electron pulse in air at the end of the beam line. The laser pulse was separated from the analyzing light by a half mirror. The precious time interval could be

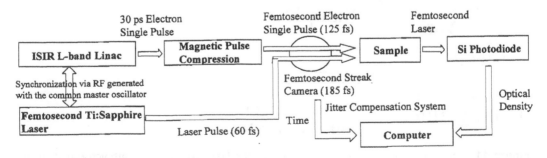

Figure 12 The subpicosecond pulse radiolysis system.

obtained by the analysis of the streak image. All equipment described above was controlled by a personal computer. The acquisition time was 1 sec per one shot. Fig. 8 shows the time-dependent behavior of the absorption due to hydrated electrons at the wavelength of 780 nm was observed in distilled and deaerated water in a quartz cell with the optical length of 2.0 mm [76]. Fig. 8 shows the time-dependent behavior of the hydrated electron obtained in the system. The solid line represents the fitting curve for the experimental data. The 10–90% rise time is 2.0 psec.

The time resolution of the stroboscopic pulse radiolysis depends on the width of electron pulse and that of laser. In this experiment, both widths are much shorter than 2.0 psec. Furthermore, the time resolution is limited by the difference between the velocity of the light and that of the electron pulse in the sample. The passing time of the electron pulse through a sample is given by $L/(\beta c)$, where L is an optical length of a sample, β is the ratio of the velocity of electron to that of light in the vacuum, and c is the velocity of light in the vacuum. On the other hand, in the case of light, the passing time is given by Ln/c, where n is a refractive index of a sample. Therefore the time resolution limited by the thickness of the sample is 1.8 psec (10–90% rise time) at $n = 1.33$ and $1 = 2$ mm. Therefore the 10–90% rise time of 2.0 psec is the time resolution of the system. In the present system, the time resolution is mainly limited by the thickness of the sample. The time resolution of 800 fsec can be obtained by a quartz cell with the optical length of 0.5 mm as shown in Fig. 9. The S/N ratio decreases with the optical length decreases. So the S/N ratio is not good for a 0.5-mm cell.

As discussed above, it is necessary to use a thin sample in high time resolution pulse radiolysis. The use of a thin sample leads to the degradation of the S/N ratio because the signal intensity is proportional to the optical length of the sample. In pulse radiolysis experiments, the optical density is generally calculated based on the intensity of the probe light passing through a sample with an electron beam present (I) and that of the probe light passing through a sample with the electron beam absent (I_0).

$$\text{O.D.} = \log \frac{I_0}{I} \tag{44}$$

If there is no fluctuation of laser intensity, we have to measure I_0 only once. Actually, the envelope of laser pulses changes in a relatively long time range (typically from several minutes to a few tens of minutes) because of the change of environmental factors such as room temperature and coolant temperature. There is also an intensity jitter caused by factors such as the mechanical vibration of mirrors and the timing jitter of electronics. Furthermore, in our system, the laser system is located about 15 m from the beam port to prevent radiation damage to the laser system. (Later, it was moved into a clean room, which was installed in the control room to keep the room temperature constant and to keep the laser system clean. The distance is about 10 m.) Therefore it is predicted that a slight tilt of a mirror placed upstream will cause a displacement of the laser pulse at the downstream position where the photodetector is placed.

From Eq. (44), to achieve a good S/N ratio measurement, it is necessary and sufficient to lower the fluctuation of I_0/I. Based on this information, we devised a new method to measure I and I_0 with a good S/N ratio. We named it the "double-pulse method." In the new method, I_0 is measured by the laser pulse passing just before the laser pulse used to measure I. Because the frequency of the Ti:Sapphire laser is 81 MHz, the time interval of two laser pulses is 12.3 nsec. The laser pulse for the measurement of I passes through the sample 12.3 nsec after the laser pulse for the measurement of I_0. An electron pulse enters the sample between two laser pulses. The two laser pulses are expected not to suffer from the vibration of optical elements because it is thought that the frequency of mechanical vibration mainly caused by the vibration of coolant and ventilator systems is lower than the MHz order. The fluctuation of I_0/I is measured at the position where the photodetector is located. The fluctuation of I_0/I is reduced from 5.34% to 0.80% using the newly developed method. The time-dependent behavior of geminate ion recombination in neat dodecane is shown in Fig. 10 as a typical example that demonstrates the effect of the newly developed system. The sample was deaerated by Ar bubbling in a quartz cell having an optical length of 2.0 mm and was irradiated at room temperature. The high S/N data obtained by the improved subpicosecond pulse radiolysis using the double-pulse method makes possible the detailed studies on the geminate recombination of electrons and radical cations. The details of the time-dependent behavior of the geminate recombination shown in Fig. 10 are discussed in the next section.

2.3. Experimental Studies of the Kinetics of Geminate Ion Recombination

Kinetics of Geminate Ion Recombination

The early work of the kinetics of geminate ion recombination studied by picosecond pulse radiolysis has been briefly reviewed [78]. The studies on the kinetics of geminate ion recombination in liquid hydrocarbons were difficult because the very high time-resolved spectroscopy is indispensable for monitoring the time behavior of the geminate ion recombination. The combination of the picosecond single electron bunch with streak cameras, independently developed in 1979 at Argonne National Laboratory [55] and at the University of Tokyo by us [56], enabled the time behavior of recombination fluorescence of the geminate ions. The direct measurement of the recombination fluorescence produced by geminate ion pairs [55,57] stimulated studies on the kinetics of the geminate ion recombination and gave us many kinds of the problems for understanding the geminate kinetics and the positive ions in irradiated hydrocarbons. But the recombination fluorescence is the indirect measurement of the geminate positive ions and is unable to make clear the details of both the identification and characteristics of positive ions in irradiated hydrocarbons, especially cyclohexane [79].

The picosecond pulse radiolysis system in the combination of very fast response photodiode with the transient digitizer (the time resolution of about 800 psec) was developed by us in 1982 [60]. This system was the first picosecond pulse radiolysis system that could measure the kinetics of ions and triplet states on subnanosecond timescale by using only one picosecond single bunch electron pulse. This system has no limitation of nano- and microsecond timescale measurement and the radiation damage of the samples by this system was much less than that by other preceding picosecond pulse radiolysis. The change of the ratio of the formation rate of the solute excited triplet state to the decay rate of the solute anion observed directly by this system showed the loss process of the spin correlation of geminate ion pairs very clearly in the time range between 5 and 20 nsec [60]. The decay of the geminate ion pairs in cyclohexane [62] was analytically analyzed by the Smoluchowski equation [13] and in various nonpolar solutions [63] on the basis of the absorption of solute ions in time range from 800 psec to several hundred nanoseconds by using the same system. In 1987 [64], the kinetics of the geminate pairs of solute ions was analytically analyzed in the time range from 50 psec to several hundred nsec by the extend Smoluchowski equation [68] on the basis of the combination of the picosecond pulse radiolysis system composed of very fast response photodiode and the transient digitizer [60] with the specially designed picosecond pulse radiolysis system composed of a streak camera [67]. Then, geminate ion recombination and the formation process of solute excited states in cyclohexane solutions of biphenyl observed by the combination of two types of picosecond pulse radiolysis with emission [56] and absorption [60] spectroscopy were analyzed by the expansion of the extended Smoluchowski equation to the formation of solute excited states [65]. The comparison of the theoretical treatment based on Smoluchowski equation with the absorption and emission data on the geminate recombination of solute ions and the formation of solute excited states on timescale from 50 psec to several hundred was successfully achieved [64,65].

Now we shall discuss about the geminate recombination of electrons and solvent cation radical in liquid hydrocarbons. The original stroboscopic pulse radiolysis in Argonne National Laboratory [54] was unable to make good quality measurements on geminate recombination of the electron and the positive ion in liquid hydrocarbons. In 1983, the modification of the Argonne linac made possible the first good quality measurements of the weak absorption at 600 nm in neat liquid n-hexane and cyclohexane. The transient absorptions at 600 nm are assigned to the positive ion and the excited state for neat n-hexane and cyclohexane, respectively [79]. The difference between broad absorption bands of the excited states and the radical cations become larger with increasing the number of carbon atoms of n-alkanes, although the peak wavelengths of both absorption bands become longer with increasing the number of carbon atoms of n-alkanes [66]. The lifetimes of alkane radical cations become shorter with decreasing the number of carbon atoms of n-alkanes. In irradiated neat cyclohexane, very broad visible absorption bands, mainly because of the excited state, were observed [66]. Although radiolysis of cyclohexane and n-hexane was heavily studied, both experimentally and theoretically, the kinetics of radical cations of both cyclohexane and n-hexane are difficult to be studied because of their very short lifetimes [66], the absorption overlapping of radical cations and excited states [66], and the complexity of their reactions and characters [80]. Because the complication of the absorption spectroscopy on radical cations of smaller hydrocarbons was clearly shown experimentally [66], the absorption spectroscopy of the geminate recombination of electrons and cation radicals in liquid hydrocarbons has mainly been carried out for higher n-alkanes such as n-dodecane instead of both cyclohexane and n-hexane.

Because the LL twin picosecond pulse system with the laser diode as monitor light enabled the absorption spectroscopy from visible to infrared region for the first time in picosecond pulse radiolysis, the kinetics of the geminate recombination of both electrons and cation radicals were first measured on picosecond timescale in the same sample and agree with each other on picosecond timescale from 50 psec after a pulse [71]. By using the laser synchronized picosecond pulse radiolysis [73] with the higher S/N ratio than that of the LL twin system, the good agreement of the diffusion theory with experimental data were confirmed on a picosecond timescale from 50 psec after a pulse, but a slight difference between the experimental data and the simulation based on Smoluchowski equation using the exponential function of the initial separation between electrons and radical cations [81]. Many factors from the time resolution of the system and simulation models are considered as the reason for the difference. Because the time resolution and S/N ratio of the laser synchronized subpicosecond pulse radiolysis was drastically improved by both the jitter compensation system [74] and the double-pulse method [77] as described above, the difference between the experimental data and the simulation based on Smoluchowski equation using the simple model was clearly observed on timescale within 50 psec. The details of the kinetics of the geminate ion recombination studied by the laser synchronized subpicosecond pulse radiolysis are described in the next section.

Subpicosecond Pulse-Radiolysis Studies of Geminate Ion Recombination

The kinetics data of the geminate ion recombination in irradiated liquid hydrocarbons obtained by the subpicosecond pulse radiolysis was analyzed by Monte Carlo simulation based on the diffusion in an electric field [77,81,82]. The simulation data were convoluted by the response function and fitted to the experimental data. By transforming the time-dependent behavior of cation radicals to the distribution function of cation radical–electron distance, the time-dependent distribution was obtained. Subsequently, the relationship between the space resolution and the space distribution of ionic species was discussed. The space distribution of reactive intermediates produced by radiation is very important for advanced science and technology using ionizing radiation such as nanolithography and nanotechnology [77,82].

The deposited energy of radiation is spent mostly for ionization of liquid hydrocarbon molecules. During ionization, a pair of a cation radical ($RH^+\cdot$) and an electron (e^-) is produced as follows,

$$RH \rightsquigarrow RH^+\cdot + e^- \tag{45}$$

After the ionization, the electrons with excess energy interact with surrounding molecules and become thermalized. The reaction from the ionization to thermalization is estimated to occur within about 1 psec. The initial separation length between the cation radical and the thermalized electron is several nanometers on average.

In nonpolar materials, most of the electrons recombine with the counter cation radicals as shown by Eq. (46). RH^* is one of the main products produced through the recombination.

$$RH^+\cdot + e^- \longrightarrow RH^* \tag{46}$$

This occurs because the initial separation length between the cation radical and the electron is smaller than the Onsager length (r_c) at which the thermal energy of an electron corresponds to the Coulomb field. This reaction is called the geminate ion recombination and has been investigated by many researchers globally.

The experimental method and apparatus, and a procedure of the Monte Carlo method that simulates the geminate ion recombination are described, and the time-dependent distribution is elucidated.

The experiments were carried out using the subpicosecond pulse radiolysis system [77] described in Sec. 10.2.2 ("Subpicosecond Pulse Radiolysis"). Considering the signal intensity and the degradation of the time resolution, a sample cell with the optical length of 2 mm was mostly used. The sample was saturated by Ar gas to eliminate the scavenging effect by the remaining O_2 gas.

At the initial stage of reactions, the produced intermediate species such as the cation radical and the electron exist in a narrow space, the so-called spur. After the electron thermalization process, a pair of a cation radical and a thermalized electron remain in a spur. The geminate ion recombination of the cation radical and the electron occurs before these ionic species diffuse and spread uniformly in the media. Therefore the geminate ion recombination takes place in the spur. On the condition of a so-called single pair model, the reaction, which is described by the following equation, can be solved in spherical co-ordinates with a radial component.

$$\frac{\partial w(\mathbf{r}, t)}{\partial t} = D\left[\nabla^2 w(\mathbf{r}, t) + \frac{1}{k_B T}\nabla w(\mathbf{r}, t)\nabla V(\mathbf{r})\right] \tag{2}$$

Hong and Noolandi [11,13,83] analytically solved Eq. (2) under the following initial and boundary conditions:

$$w(r, 0) = f(r) \quad \text{(initial condition)} \tag{47}$$
$$w(R, t) = 0 \quad \text{(boundary condition)} \tag{11a}$$

Equation 2 indicates that if the distance r between the cation radical and the electron becomes less than the reaction radius R, they recombine. The typical shape of $f(r, r_0)$, the initial distribution function, is expressed by the following exponential or Gaussian.

$$f(r, r_0) = \frac{1}{r_0}\exp\left(-\frac{r}{r_0}\right) \quad \text{(exponential)} \tag{19}$$

$$f(r, r_0) = \frac{4r^2}{\sqrt{\pi}r_0^3}\exp\left(-\frac{r^2}{r_0^2}\right) \quad \text{(gaussian)} \tag{20}$$

The solution of Eq. (2) can also be obtained by a numerical analysis similar to the calculus of finite differences. However, an analytical or semianalytical method based on Eq. (2) is not suitable for discussing the time-dependent distribution function because the calculation is lengthy.

On the other hand, the Monte Carlo method enables us to simultaneously obtain the time-dependent decay curve and the time-dependent distribution function. Therefore we adopted Monte Carlo simulation [18,21,84,85] for the analysis. The geminate ion recombination is also described as follows,

$$\Delta r_i = \sqrt{6D_i\Delta t n_i} + \mu_i\Delta t E_i \tag{48}$$

In Eq. (48), i represents the cation radical and electron. The D_i and μ_i are the diffusion constant and the mobility of the ith species, respectively. The relationship between them is described by the Einstein's relation. The calculation was carried out in the rectangular coordinates (x, y, z). For each time step Δt, the vector of movement Δr_i is calculated according to Eq. (48). The first term in Eq. (48) represents the diffusion component. It was calculated by uniform random variables n_i ranging from -1 to 1. The second term in Eq. (48) represents the drift in the electric field E_i, which is due to counter ionic species. At first, the cation radical and the electron were set at $(0, 0, 0)$ and $(x_0, 0, 0)$ in the coordinates, respectively. Therefore x_0 corresponds to r_0, which is the initial separation length between the cation radical and the electron. After that, the positions of the ith species were calculated by Eq. (48). If the distance between the cation radical and electron became less than the reaction radius (R), they recombine. The time taken for the recombination was recorded. The Δt was taken to be 0.1 psec. By carrying out calculations for 4000 pairs of the cation radical and the electron, the survival probability of the cation radicals was obtained. The time-dependent distribution function was also obtained by recording the distance between the cation radical and the electron. Before analyzing the experimental data, to check the validity of the simulation code, we compared the simulation data with the analytical solution given by Hong and Noolandi [11,13,83]. The calculation was completed in about 15 min by a personal computer. The result is shown in Fig. 13(a). The initial distribution function and the Onsager length (r_c) were taken as a delta function $\delta(r-r_0)$ of

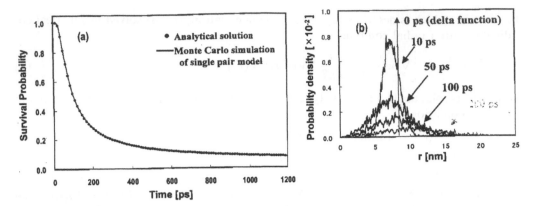

Figure 13 (a) Comparison of the simulation curve with the analytical solution given by Hong and Noolandi [11,13,83]. The initial distribution function and the Onsager length (r_c) were taken as a delta function $\delta(r-r_0)$ of $r_0 = 7.5$ nm and $r_c = 30$ nm, respectively. The dots and the solid line indicate the analytical solution and the simulation curve, respectively. (b) Time-dependent distribution function that was obtained from the simulation of (a). r indicates the distance between the cation radical and the electron.

$r_0 = 7.5$ nm and $r_c = 30$ nm, respectively. The dots and the solid line represent the analytical solution and the simulation curve, respectively. They perfectly agree with each other. Therefore the validity of the code was confirmed. At the same time, the time-dependent distribution function was obtained. It is shown in Fig.13(b). The vertical axis and the horizontal axis represent the probability density and the distance between the cation radical and the electron, respectively. The probability density is defined as the existing ratio of species per unit length. We can see that the electron, which existed at $r_0 = 7.5$ nm, diffuses and recombines with the cation radicals.

In the experiment, liquid n-dodecane was used due to the small overlap among intermediate species. At the wavelength of the Ti:Sapphire fundamental light, a large part of the absorption is assigned to the cation radicals in n-dodecane [66]. Thus the decay curve of the experimental result shows a decay of the cation radicals indicated in Eq. (46). Fig. 14(a) shows the time-dependent behavior of the cation radicals in liquid n-dodecane, which was obtained by the subpicosecond pulse radiolysis and monitored at 790 nm [82]. The vertical axis represents optical density, which is proportional to the density of the intermediate species. The dotted and the solid lines represent the experimental curve and the simulation curve, respectively. It was observed that the cation radicals began to be produced at time zero and decayed by the recombination. The rise time of the experimental data corresponded to the time resolution of the measurement system. In the simulation, the parameters of the electron diffusion coefficient $(D_e) = 6.4 \times 10^{-4}$ cm^2/sec, the cation radical diffusion coefficient $(D_+) = 6.0 \times 10^{-6}$ cm^2/sec, the relative dielectric constant $\varepsilon = 2.01$, the reaction radius $R = 0.5$ nm and the exponential function as shown in Eq. (19) with $r_0 = 6.6$ nm were used. These values are the same as those used in previous papers [81,86]. The exponential distribution function was realized by using the rejection method with two sets of uniform random variables. The shape and mean value of the initial distribution function are dependent on the materials used. The obtained simulation curve was convoluted by the response function [77] and fitted to the experimental curve so that the residual between two curves was minimized in the long timescale. Although the

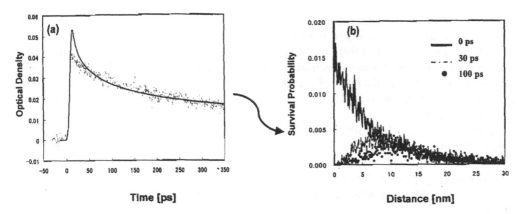

Figure 14   (a) Time-dependent behavior of cation radicals in liquid *n*-dodecane monitored at 790 nm. The dotted and the solid lines represent the experimental curve and the simulation curve, respectively. The parameters of the electron diffusion coefficient $(D_e) = 6.4 \times 10^{-4}$ cm²/sec, the cation radical diffusion coefficient $(D_+) = 6.0 \times 10^{-6}$ cm²/sec, the relative dielectric constant $\varepsilon = 2.01$, the reaction radius $R = 0.5$ nm, and the exponential function as shown in Eq. (19) with $r_0 = 6.6$ nm were used. (b) Time-dependent distribution function obtained from fitting curve of (a). *r* indicates the distance between the cation radical and the electron. The solid line, dashed line, and dots represent the distribution of cation radical–electron distance at 0, 30, and 100 psec after irradiation, respectively.

time range was terminated at 350 psec in Fig. 14(b), it was confirmed that the experimental curve is in accordance with the simulation curve in the nanosecond timescale [63,66,81,86]. As a result, the two curves corresponded with each other after a time interval of about 50 psec after irradiation. Some factors that may explain the disagreements within the time interval of about 50 psec are discussed but was not determined [81]. Fig. 14(b) shows the time-dependent distribution obtained by the simulation curve in Fig. 14(a). The solid line, dashed line, and dots represent the distribution of cation radical-electrons distance at 0, 30, and 100 psec after irradiation, respectively. At 0 psec, the distribution was in the shape of the exponential function. At 30 psec, a large part of the electrons, especially near the cation radical, have already recombined. At 100 psec, the shape of the distribution curve is highly similar to that of 30 psec.

The space distribution of ionic species, which move in several or a few nanometers scale, should not be ignored in future nanolithography and nanotechnology based on beam technology.

Further detailed kinetics of the geminate recombination of electrons and positive ions and their application to the advanced technology will be studied by higher time resolution of the femtosecond pulse radiolysis and both by the higher S/N ratio and the wider wavelength monitoring light of the improved subpicosecond pulse radiolysis shown in Fig. 7.

3. EXPERIMENTAL STUDIES OF ELECTRON–ION BULK RECOMBINATION IN CONDENSED MATTER

3.1. Introduction

The investigation of the dynamic behavior of low-energy electrons such as electron transport and reactivities in nonpolar dense molecular media is of essential importance in both fundamental and applied sciences [87–100]. The values of an electron mobility μ_e have been extensively measured in a variety of these media, and several theoretical models have been proposed for the transport mechanism. Electron reactivities with solute molecules or positive ions in these media have been also extensively investigated. The former reaction is electron attachment and the latter electron–ion recombination. The investigation of these electron reactivities has provided a new insight into that of the behavior of low-energy electrons in matter.

In the following, a brief survey is given of the experimental investigation of electron–ion recombination with particular emphasis placed on the comparison between the experimental and theoretical results as well as on that between the results in dense media and in low-density gases or beams.

A survey of electron attachment with a similar emphasis has been given elsewhere [92,100].

3.2. Comparative Aspects of Electron–Ion Recombination Processes in the Gas and Condensed Phases

It is of great interest to investigate the electron–ion recombination in dense molecular media in the density range from dense gases to liquids and solids [96,97,99,100] because the Coulombic interaction distance between an electron and an ion, as expressed by the Onsager length r_c, which is given by Eq. (1), is much bigger than the mean intermolecular distance of the nearest neighbor molecules in the medium.

Electron–ion recombination processes in isolated two-body collisions in beams or in low-pressure gases have been extensively studied mainly with a merged beam method [101], a pulsed afterglow method [102], a flowing afterglow method [103], and a recently developed method by means of a cooler ring [104,105]. Cross sections or rate constants for electron–ion recombination obtained by these experimental methods are theoretically explained by the mechanism including molecular superexcited states as collision complex or reaction intermediate species. Spectroscopy and dynamics studies of molecular super-excited states have been recently in great progress [106,107], and these states are assigned mainly to vibrationally, doubly, or inner-core excited molecular high Rydberg states converging to each of ion states.

In the gas phase at one to several atmospheric pressures, electron–ion recombination processes have been studied with a pulse-radiolysis method, and observed recombination rate constants k_r are expressed by,

$$k_r = k_2 + k_3 n \tag{49}$$

where k_2 and k_3 are the two-body and three-body rate coefficients, respectively, and n is the density of a third-body molecule. The magnitude of observed rate constant values is explicable in terms of the collisional de-excitation of two-body formed complex or intermediate species.

3.3. Experimental Methods to Measure Absolute Rate Constants for Recombination

However, in dense media such as high-pressure gases near the critical points, liquids, and solids, experimental studies were relatively few because of difficulties in experimental methods. But recently there has been great progress due to the development of a new powerful method, called the decay-curve analysis method [96,97,99,100]. The following are a brief comparative survey of the experimental methods [99].

1. In a pulse-radiolysis d.c. conductivity method, k_r/μ_e is measured and then k_r is determined from the measured k_r/μ_e and a separately known μ_e.

2. In a pulse-radiolysis charge-clearing method, a residual amount of charge that is left unrecombined is collected and measured. Because the collected charge is expressed as a function of k_r, the value of $k_r/\beta\mu_e$, where β is a correction factor due to a space charge is obtained. The kr value is thus determined by using the value of μ_e known separately and the value of β which is approximately assumed.

3. In a pulse-radiolysis optical-absorption method, the value of k_r/ε, where ε is a molar absorption coefficient, is measured by the time-resolved measurement of the optical absorption of solvated electrons, and then the k_r value is determined by the observed value of k_r/ε and the value of ε known separately.

4. In a pulse-radiolysis d.c. conductivity method with a long pulse width, the resulting steady state current during the long pulse is considered by the assumption that the rate of excess electron generation becomes equal to the rate of electron loss because of the reaction with positive ions and impurities.

5. In a pulse-radiolysis microwave-conductivity method, the k_r value is determined by the time-resolved measurement of microwave conductivity.

6. In a method using a radioisotope, e.g., ^{207}Bi, as an internal radiation source, the difference in a fluorescence intensity with and without a d.c. electric field is assumed to be due to electron–ion recombination, and the k_r value is determined by the time-resolved measurement of the fluorescence with the known number density of ionization, which is separately determined.

7. In the decay-curve analysis method, which is one of the pulse radiolysis d.c. conductivity methods, both k_r and μ_e values are simultaneously determined by analyzing observed transient current decay curves by fitting them to a general formula of conductivity both at a high pulse dose of x-rays where the recombination is predominant and at a low pulse dose where the recombination is negligible. This is a useful method also because of the k_r determination without knowing the pulse dose of x-rays or the free-ion yield. Recently, this method has been improved by taking account of a space-charge effect to obtain more accurate values of k_r.

3.4. Experimental Results: The Breakdown of the Debye Theory

In the liquid phase, observed electron–ion recombination rate constants k_r in a variety of nonpolar media were, as shown in Fig. 15, in good agreement with the values of k_D calculated from the reduced Debye equation,

$$k_D = \frac{4\pi e}{\varepsilon}\mu_e \tag{50}$$

viz., Eq. (36b), in the range of μ_e below about 150 cm^2/V sec, and thus the recombination in these media was concluded to be diffusion controlled [99,100,112].

However, in liquid and high-pressure gaseous methane in which most μ_e values are larger than 150 cm^2/V sec, it has been found [49] that the observed k_r values are much lower than k_D. This experiment has stimulated a great deal of theoretical investigations [24,25,30,38–42,48,113–119] of the recombination process in dense media. Because the

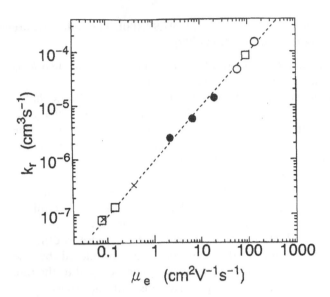

Figure 15 Variation of k_r with μ_e in the μ_e region below about 150 cm^2/V sec in condensed nonpolar media. n-Pentane, n-hexane, TMS ($\square$); n-hexane, cyclohexane ($\times$); neopentane–n-hexane mixtures ($\bullet$); and liquid and solid neopentane ($\bigcirc$). The straight line corresponds to k_D (From Refs. 99,100,112.)

magnitudes of the electron mobilities in dense Ar, Kr, and Xe media are as high as in dense methane media, the values of k_r and μ_e in these media have been also measured [120–123] and compared further with theoretical results. The observed k_r values in these rare-gas media in both liquid and dense gas phases are again found to be much smaller than those calculated by the reduced Debye equation. However, in the solid phase, the observed k_r values are almost in agreement with those calculated by Eq. (50). These results are summarized in Fig. 16. The effect of an external d.c. electric field strength on both k_r and μ_e values has also been measured. An example of the field effect is shown in Fig. 17 for liquid Ar. Recently, the effect of the addition of molecular impurities such as N$_2$ and CH$_4$ on k_r and μ_e values has been further examined to differentiate between the energy and spatial criteria [39,42,114] in expectation that electrons heated up by applied external electric fields may be cooled down by the added molecular impurities due to inelastic collisions between electrons and molecules. Detailed comparisons have been made between these experimental results and the theoretical ones, from which it has been concluded that the recombination in liquid and dense gaseous Ar, Kr, and Xe as well as CH$_4$ is clearly different from a usual diffusion-controlled reaction corresponding to Eq. (50).

3.5. Comparison with Theories

Theoretical investigations trying to explain the k_r deviation from k_D, i.e., the breakdown of the Debye theory, are summarized as follows:

1. A semiempirical treatment taking both the diffusion controlled and electron-energy exchange controlled recombination processes into account [38].
2. A Monte Carlo simulation with the parameter Λ/a, where Λ is the mean free path of electrons and a is the reaction radius [39,42,114].

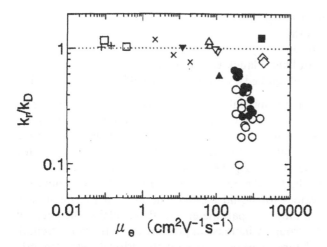

Figure 16 Variation of k_r/k_D with μ_e in condensed nonpolar media [99,100,112]. Neopentane, liquid ($\triangle$) and solid ($\blacktriangle$); neohexane ($\blacktriangledown$); TMS ($\triangledown$); n-pentane, n-hexane ($+$); n-hexane, cyclohexane ($\square$); neopentane–n-hexane mixtures ($\times$); methane, liquid ($\bullet$) and solid ($\blacksquare$); and argon, liquid ($\bigcirc$) and solid ($\diamondsuit$).

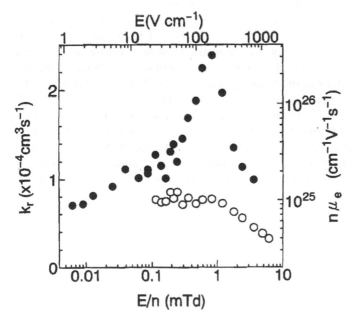

Figure 17 Electric field dependence of k_r ($\bullet$) and $n\mu_e$ ($\bigcirc$) in liquid argon ($T = 87$ K, $n = 2.1 \times 10^{22}/$ cm^3) (From Refs. 99,100,120.)

3. A molecular dynamics simulation [115].
4. A fractal treatment [24,117,118].
5. A gas kinetic approach [116].
6. An approach based on the Fokker–Planck equation [40].
7. Theories to explain the effect of applied external electric fields on k_r [30,48,119].

Warman [38] first proposed a semiempirical formula of the recombination rate constant to explain the new experimental result of $k_r \ll k_D$ [49]. In spite of many assumptions to derive the formula, the calculated value for liquid argon is in rather good agreement with the experimental one. However, for dense gaseous argon, the calculated value is far from the experimental one. Tachiya [39,42,114] pointed out using a Monte Carlo method that if Λ/a is not negligible the rate constant for bulk recombination deviates from k_D. Experimental k_r values for liquid methane lie around the theoretical curve of his simulation which uses Λ/a only as a parameter. He also explains the effect of the field strength on k_r. The values of k_r roughly increase in proportion to the field strength in the lower field strength region and form a peak at around the critical field strength. These results are explained in terms of the effect of the field strength on the trajectory of electrons to be recombined. In the lower field region, the disentangling of random electron trajectories caused by the presence of the external electric field should increase the k_r value. In the field region higher than the critical field strength, the internal energy of an electron–ion pair decreases the k_r value.

Recent experiments [123] on the effects of electric field strength and the addition of molecular impurities on k_r values in dense gaseous Ar and Kr at the density range near the critical points, in which a new experimental method to measure k_r values are developed by including the effect of space charges on k_r and μ_e values in the analysis of observed conductivity signals, have clearly shown again a large deviation of k_r values from the Debye theory. The obtained experimental results are discussed in terms of the electron mean free paths and electron energies based on recent theories.

It is concluded that electron–ion recombination processes in dense high-mobility media such as liquid and dense gaseous Ar, Kr, and Xe as well as CH_4 are clearly different from those expressed by the Debye theory, i.e., a well-established diffusion-controlled theory. Newly developed theories, instead of the Debye theory, as stimulated by the experimental finding of the breakdown of the Debye theory, have been explicable, at least qualitatively, in terms of the experimental results of the breakdown, i.e., $k_r \ll k_D$, but still quantitatively inexplicable not only in the magnitude of the breakdown itself but also in the effects of external electric fields and the addition of molecular impurities on it. Further new investigations are strongly needed both theoretically and experimentally. It is also noted that these investigations may provide a new insight into the fundamental processes of phenomena of ionization and excitation in condensed matter, e.g., those in liquid xenon detectors [124].

REFERENCES

1. Smoluchowski, M. Z. Phys. Chem. 1917, *92*, 129.
2. Onsager, L. Phys. Rev. 1938, *54*, 554.
3. Sano, H.; Tachiya, M. J. Chem. Phys. 1979, *71*, 1276.
4. Tachiya, M. J. Chem. Phys. 1978, *69*, 2375.
5. Collins, F.C.; Kimball, G.E. J. Colloid Sci. 1949, *4*, 425.

6. Wilemski, G.; Fixman, M. J. Chem. Phys. 1973, *58*, 4009.
7. Murata, S.; Tachiya, M. J. Chim. Phys. 1996, *93*, 1577.
8. Que, W.; Rowlands, J.A. Phys. Rev. B 1995, *51*, 10500.
9. Abell, G.C.; Funabashi, K. J. Chem. Phys. 1973, *58*, 1079.
10. Mozumder, A. J. Chem. Phys. 1974, *60*, 4300.
11. Hong, K.M.; Noolandi, J. J. Chem. Phys. 1978, *69*, 5026.
12. Wojcik, M.; Tachiya, M. to be published.
13. Hong, K.M.; Noolandi, J. J. Chem. Phys. 1978, *68*, 5163.
14. Noolandi, J. *Kinetics of Nonhomogeneous Processes*; Freeman, G.R., Ed. Wiley: New York, 1987; 465 pp.
15. Hummel, A.; Infelta, P.P. Chem. Phys. Lett. 1974, *24*, 559.
16. Hummel, A. J. Chem. Phys. 1968, *49*, 4840; M. Tachiya, M.S. Thesis (Univ. of Tokyo, 1969).
17. Tachiya, M. Radiat. Phys. Chem. 1988, *32*, 37; Green, N.J.B.; Pimblott, S.M.; Tachiya, M. J. Phys. Chem. 1993, *97*, 196.
18. Bartczak, W.M.; Hummel, A. J. Chem. Phys. 1987, *87*, 5222.
19. Clifford, P.; Green, N.J.B.; Pilling, M.J.; Pimblott, S.M. J. Phys. Chem. 1987, *91*, 4417; Green, N.J.B.; Pilling, M.J.; Pimblott, S.M.; Clifford, P. J. Phys. Chem. 1989, *93*, 8025.
20. Tachiya, M.; Hummel, A. Chem. Phys. Lett. 1989, *154*, 497.
21. Wojcik, M.; Bartczak, W.M.; Hummel, A. J. Chem. Phys. 1992, *97*, 3688.
22. Tachiya, M. J. Chem. Phys. 1988, *89*, 6929.
23. Tachiya, M.; Schmidt, W.F. J. Chem. Phys. 1989, *90*, 2471.
24. López-Quintela, M.A.; Buján-Núñez, M.C.; Pérez-Moure, J.C. J. Chem. Phys. 1988, *88*, 7478.
25. Mozumder, A. J. Chem. Phys. 1990, *92*, 1015.
26. Rice, S.A.; Butler, P.R.; Pilling, M.J.; Baird, J.K. J. Chem. Phys. 1979, *70*, 4001.
27. Debye, P. J. Electrochem. Soc. 1942, *82*, 265.
28. Hummel, A. Adv. Radiat. Chem. 1974, *4*, 1.
29. Tachiya, M. Radiat. Phys. Chem. 1983, *21*, 167.
30. Isoda, K.; Kouchi, N.; Hatano, Y.; Tachiya, M. J. Chem. Phys. 1994, *100*, 5874.
31. See, for example; Rice, S.A. Diffusion-Limited Reactions. Bamford, C.H. Tipper, C.F.H., Compton, G. eds.; *Comprehensive Chemical Kinetics*; Elsevier: Amsterdam, 1985; Vol. 25.
32. Traytak, S. D.; Tachiya, M. to be published.
33. Scher, H.; Montroll, E.W. Phys. Rev. B 1975, *12*, 2455.
34. Blumen, A.; Klafter, J.; Zumofen, G. *"Optical Spectroscopy of Glasses"*; In Zschokke. I.. Ed. Reidel: Dordrecht, 1986; 199 pp.
35. Metzler, R.; Klafter, J. Phys. Rep. 2000, *339*, 1.
36. Helman, W.P.; Funabashi, K. J. Chem. Phys. 1977, *66*, 5790; 1979, *71*, 2458.
37. Sung, J.; Barkai, E.; Silbey, R.J.; Lee, S. J. Chem. Phys. 2002, *116*, 2338.
38. Warman, J.M. J. Phys. Chem. 1983, *87*, 4353.
39. Tachiya, M. J. Chem. Phys. 1987, *87*, 4108.
40. Sceats, M.G. J. Chem. Phys. 1989, *90*, 2666.
41. Mozumder, A. J. Chem. Phys. 1994, *101*, 10388.
42. Tachiya, M. J. Chem. Phys. 1987, *87*, 4622.
43. Pitaevskii, L.P. Sov. Phys. JETP 1962, *15*, 919.
44. Wojcik, M.; Tachiya, M. J. Chem. Phys. 2000, *112*, 3845.
45. Morgan, W.L. J. Chem. Phys. 1986, *84*, 2298.
46. Wojcik, M.; Tachiya, M. Chem. Phys. Lett. 2002, *363*, 381.
47. Cohen, M.H.; Lekner, J. Phys. Rev. 1967, *158*, 305.
48. Wojcik, M.; Tachiya, M. J. Chem. Phys. 1998, *109*, 3999.
49. Nakamura, Y.; Shinsaka, K.; Hatano, Y. J. Chem. Phys. 1983, *78*, 5820.
50. Bronskill, M.J.; Taylor, W.B.; Wolff, R.K.; Hunt, J.W. Rev. Sci. Instrum. 1970, *41*, 333.
51. Sumiyoshi, T.; Sawamura, S.; Koshikawa, Y.; Katayama, M. Bull. Chem. Soc. Jpn. 1982, *4*, 2341.
52. Beck, G.; Thomas, J.K. J. Phys. Chem. 1972, *76*, 3856.

53. Beck, G. Rev. Sci. Instrum. 1979, *50*, 1147.
54. Jonah, C.D. Rev. Sci. Instrum. 1975, *46*, 62.
55. Jonah, C.D.; Sauer, M.C. Chem. Phys. Lett. 1979, *63*, 535.
56. Tagawa, S.; Katsumura, Y.; Tabata, Y. Chem. Phys. Lett. 1979, *64*, 258.
57. Tagawa, S.; Katsumura, Y.; Tabata, Y. Radiat. Phys. Chem. 1982, *19*, 125.
58. Tagawa, S.; Washio, M.; Tabata, Y. Chem. Phys. Lett. 1979, *68*, 276.
59. Katsumura, Y.; Tagawa, S.; Tabata, Y. Radiat. Phys. Chem. 1982, *19*, 267.
60a. Tagawa, S.; Tabata, Y.; Kobayashi, H.; Washio, M. Radiat. Phys. Chem. 1982, *19*, 193.
60b. Tagawa, S.; Washio, M.; Tabata, Y.; Kobayashi, H. Radiat. Phys. Chem. 1982, *19*, 277.
61. Tagawa, S.; Washio, M.; Kobayashi, H.; Katsumura, Y.; Tabata, Y. Radiat. Phys. Chem. 1983, *21*, 45.
62. Yoshida, Y.; Tagawa, S.; Tabata, Y. Radiat. Phys. Chem. 1984, *23*, 279.
63. Yoshida, Y.; Tagawa, S.; Tabata, Y. Radiat. Phys. Chem. 1986, *28*, 201.
64. Yoshida, Y.; Tagawa, S.; Kobayashi, H.; Tabata, Y. Radiat. Phys. Chem. 1987, *30*, 83.
65. Yoshida, Y.; Tagawa, S.; Washio, M.; Kobayashi, H.; Tabata, Y. Radiat. Phys. Chem. 1989, *34*, 493.
66. Tagawa, S.; Hayashi, N.; Yoshida, Y.; Washio, M.; Tabata, Y. Radiat. Phys. Chem. 1989, *34*, 503.
67. Kobayashi, H.; Ueda, T.; Kobayashi, T.; Tagawa, S.; Yoshida, Y.; Tabata, Y. Radiat. Phys. Chem. 1984, *23*, 393.
68. Tachiya, M. Radiat. Phys. Chem. 1987, *30*, 75.
69. Kobayashi, H.; Tabata, Y. Nucl. Instrum. Methods 1985, *B10/11*, 1004.
70. Kobayashi, H.; Tabata, Y.; Ueda, T.; Kobayashi, T. Nucl. Instrum. Methods 1987, *B24/25*, 1073.
71. Yoshida, Y.; Ueda, T.; Kobayashi, T.; Tagawa, S. J. Photopolym. Sci. Tech. 1991, *4*, 171.
72. Jonah, C.D.; Hart, E.J.; Matheson, M.S. J. Phys. Chem. 1973, *77*, 1838.
73a. Yoshida, Y.; Tagawa, S. Proc. International Workshop Femtosecond Tech., Tsukuba, 1995; 63 pp.
73b. Tagawa, S.; Yoshida, Y.; Miki, M.; Yamamoto, T.; Ushida, K.; Izumi, Y. Proc. International Workshop Femtosecond Tech., Tsukuba p. 31.
73c. Yoshida, Y.; Mizutani, Y.; Kozawa, T.; Saeki, A.; Seki, S.; Tagawa, S.; Ushida, K. Radiat. Phys. Chem. 2001, *60*, 313.
74. Kozawa, T.; Mizutani, Y.; Miki, M.; Yamamoto, T.; Suemine, S.; Yoshida, Y.; Tagawa, S. Nucl. Instrum. Methods 2000, *A440*, 251.
75. Unpublished data measured by T. Kozawa, Y. Yoshida and S. Tagawa
76. Kozawa, T.; Mizutani, Y.; Yokoyama, K.; Okuda, S.; Yoshida, Y.; Tagawa, S. Nucl. Instrum. Methods 1999, *A429*, 471.
77. Kozawa, T.; Yoshida, Y.; Tagawa, S. Jpn. J. Appl. Phys. 2002, *41*, 4208.
78a. In *"Kinetics of Nonhomogeneous Processes"*; Freeman, G.R., Wiley-Interscience Pub., John Wiley and Sons, 1987; 215 pp.
78b. Mozumder, A. *"Fundamentals of Radiation Chemistry"*; Academic Press: San Diego 1999.
79a. Jonah, C.D. Radiat. Phys. Chem. 1983, *21*, 53.
79b. Sauer, M.C., Jr; Werst, D.M.; Jonah, C.D.; Trifunac, A.D. Radiat. Phys. Chem. 1991, *37*, 461.
80. Saeki, A.; Kozawa, T.; Saeki, A.; Yoshida, Y.; Tagawa, S. Radiat. Phys. Chem. 2001, *60*, 319.
81. Saeki, A.; Kozawa, T.; Saeki, A.; Yoshida, Y.; Tagawa, S. Jpn. J. Appl. Phys. 2002, *41*, 4213.
82. Hong, Y.M.; Noolandi, J. J. Chem. Phys. 1978, *68*, 5172.
83. Wojcik, M.; Hummel, A. J. Chem. Phys. 1987, *87*, 5222.
84. Swallen, S.F.; Fayer, M.D. J. Chem. Phys. 1995, *103*, 8864.
85. Kim, H.; Shin, S.; Lee, S.; Shin, K.J. J. Chem. Phys. 1996, *105*, 7705.
86. Yoshida, Y.; Ueda, T.; Kobayashi, T.; Tagawa, S. Nucl. Instrum. Methods 1993, *A327*, 41.
87. Hatano, Y.; Shimamori, H. *Electron and Ion Swarms*; Christophorou, L.G., Ed. Pergamon Press: New York; 1981; 103 pp.

88. Warman, J.M. *The Study of Fast Processes and Transient Species by Electron Pulse Radiolysis*; Baxendale, J.H. Busi, F., Eds. NATO-ASI Series, C86, Dordrecht: D. Reidel Publ. Co, 1982; 433 pp.

89. Christophorou, L.G.; Siomos, K. *Electron-Molecule Interactions and Their Applications, Vol. 2*; Christophorou, L.G., Ed.; Academic Press: New York, 1984, Vol. 2, 221 pp.

90. Johnsen, R.; Lee, H.S. *Swarm Studies and Inelastic Electron-Molecule Collisions*; Pitchford, L.C. Mckoy, B.V., Chutjian, A., Trajmar, S., Eds.; Springer-Verlag: New York, 1985; 23 pp.

91. Morgan, W. L. ibid. (1985) p. 43.

92. Hatano, Y. *Electronic and Atomic Collisions*; Lorents, D.C. Meyerhof, W.F., Peterson, J.R., Eds.; Elsevier: Amsterdam, 1986; 153 pp.

93. Freeman, G.R. *Kinetics of Nonhomogeneous Processes*; Freeman, G.R., Ed; John Wiley and Sons Inc: New York, 1987; 19 pp.

94. Holroyd, R.A. *Radiation Chemistry*; Farhataziz, Rodgers, M.A.J., Ed; VCH Publ. Inc: New York, 1987; 201 pp.

95. Holroyd, R.A.; Schmidt, W.F. Annu. Rev. Phys. Chem. 1989, *40*,439.

96. Shinsaka, K.; Hatano, Y. *Non-Equilibrium Effects in Ion and Electron Transport*; Gallagher, J.W., Hudson, D.F., Kunhardt, E.E., Van Brunt, R.J., Eds.; Plenum Press: New York, 1990; 275 pp.

97. Shinsaka, K.; Hatano, Y. Nucl. Instrum. Methods 1993, *A327*, 7.

98. Sanche, L. *Excess Electrons in Dielectric Media*; Ferradini, C. Jay-Gerin, J.-P., Eds.; CRC Press: Boca Raton, 1991; 1 pp.

99. Hatano, Y. *Linking the Gaseous and Condensed Phases of Matter—The Behavior of Slow Electrons*; Christophorou, L.G. Illenberger, E., Schmidt, W.F., Eds. NATO-ASI Series, Plenum Press: New York, 1994; 467 pp.

100. Hatano, Y. Aust. J. Phys. 1997, *50, 615*.

101. Mitchell, J.B.A.; Yousif, F.B. *Dissociative Recombination: Theory, Experiment, and Applications*; Mitchell, J.B.A., Guberman, S.L., Eds.; World Scientific: Singapore, 1989; 109 pp.

102. Shiu, Y.-J.; Biondi, M.A.; Sipler, D.P. Phys. Rev. 1977, *A 15*, 494.

103. Smith, D.; Adams, N.G. *The Physics of Electronic and Atomic Collisions*; Dalgarno, A., Freund, R.S., Koch, P.M., Lubell, M.S., Lucatorto, T.B., Eds.; AIP Press: New York, 1990; 325 pp.

104. Tanabe, T.; Katayama, I.; Kamegaya, H.; Chiba, K.; Watanabe, T.; Arakaki, Y.; Yoshizawa, M.; Saito, M.; Haruyama, Y.; Hosono, K.; Hatanaka, K.; Honma, T.; Noda, K.; Ohtani, S.; Takagi, H. *The Physics of Electronic and Atomic Collisions*; Dube, L.G. Mitchell, J.B.A., McConkey, J.W., Brion, C.E., Eds.; AIP Press: New York, 1995; 329 pp.

105. A. Muller, ibid., (1995) p. 317.

106. Hatano, Y. Phys. Rep. 1999, *313*, 109.

107. Hatano, Y. J. Electron Spectrosc. Relat. Phenom. 2001, *119*, 107.

108. Warman, J.M.; Senhauser, E.S.; Armstrong, D.A. J. Chem. Phys. 1979, *70*, 995.

109. Senhauser, E.S.; Armstrong, D.A.; Warman, J.M. Radiat. Phys. Chem. 1980, *15*, 479.

110. Van Sonsbeek, R.J.; Cooper, R.; Bhave, R.N. J. Chem. Phys. 1992, *97*, 1800.

111. Cooper, R.; Burgers, M.; Bhave, R.; Van Sonsbeek, R.; Caulfield, K.; Lowke, J. *Gaseous Electronics and Its Applications*; Crompton, R.W. Hayashi, M., Boyd, D.E., Makabe, T., Eds.; KTK Sci. Publ.: Tokyo, 1991; 73 pp.

112. Tezuka, T.; Namba, H.; Nakamura, Y.; Chiba, M.; Shinsaka, K.; Hatano, Y. Radiat. Phys. Chem. 1983, *21*, 197.

113. Morgan, W.L.; Bardsley, J.N. Chem. Phys. Lett. 1983, *96*, 93.

114. Tachiya, M. J. Chem. Phys. 1986, *84*, 6178.

115. Morgan, W.L. J. Chem. Phys. 1990, *92*, 1015.

116. Kaneko, K.; Usami, Y.; Kitahara, K. J. Chem. Phys. 1988, *89*, 6420; Kaneko, K.; Tanimoto, J.; Usami, Y.; Kitahara, K. J. Phys. Soc. Jpn. 1990, *59*, 56.

117. Lopez-Quintela, M.A.; Bujan-Nunes, M.C. Chem. Phys. 1991, *157*, 307.

118. Mozumder, A. J. Chem. Phys. 1993, *98*, 8347; Chem. Phys. Lett. 1993, *245*, 359.

119. Fedorenko, S.G.; Burshtein, A.I. J. Chem. Phys. 1997, *107*, 6659.
120. Shinsaka, K.; Codama, M.; Srithanratana, T.; Yamamoto, M.; Hatano, Y. J. Chem. Phys.
      1988, *88*, 7529; Shinsaka, K.; Codama, M.; Nakamura, Y.; Serizawa, K.; Hatano, Y. Radiat.
      Phys. Chem. 1989, *34*, 519.
121. Honda, K.; Endou, K.; Yamada, H.; Shinsaka, K.; Ukai, M.; Kouchi, N.; Hatano, Y. J.
      Chem. Phys. 1992, *97*, 2386; Honda, K.; Endou, K.; Yamada, H.; Isoda, K.; Shinsaka, K.;
      Ukai, M.; Kouchi, N.; Hatano, Y. ibid, 1992, *98*, 8348.
122. Ukai, M.; Odaka, T.; Yamada, H.; Isoda, K.; Shinsaka, K.; Kouchi, N.; Hatano, Y. Int. J.
      Mass Spectrom. Ion Process. 1995, *149/150*, 451.
123. Takeda, K.; Kato, R.; Hayashida, M.; Odaka, T.; Shinsaka, K.; Kameta, K.; Odagiri, T.;
      Kouchi, N.; Hatano, Y. J. Chem. Phys. 2001, *114*, 3554.
124. Hatano, Y. *Proceedings of Xe-01 Symposium, December 3–4, 2001, Kashiwa*; Suzuki, Y., Ed.;
      World Scientific: Singapore, 2000. In press.

11

Radical Ions in Liquids

Ilya A. Shkrob and Myran C. Sauer, Jr.
Argonne National Laboratory, Argonne, Illinois, U.S.A.

1. INTRODUCTION

The interaction of ionizing radiation—fast electrons, α-particles, x- and γ-rays, and ultraviolet (UV) and vacuum ultraviolet (VUV) photons—with molecular solids and liquids causes the formation of short-lived electron-hole pairs that, in such media, thermalize and, eventually, localize yielding radical ions and/or trapped (solvated) electrons and holes (see Chaps. 1–5). The distinction between the radical anions and the solvated electrons is arbitrary. For the time being, it will be assumed that "radical ions" have an excess electron or electron deficiency in the valence orbitals of a single solvent molecule ("molecular ions" or "monomer ions") or a small group of such molecules ("dimer ions" or "multimer ions") that do not share charge with neutral solvent molecules that "solvate" them. Naturally, the excess electron in a radical anion is indistinguishable from other valence electrons in this anion. By contrast, in the "solvated electron" (also known as "cavity electron," see Chap. 7), the electron density resides mainly in interstitial sites between the solvent molecules ("solvation cavity") that are polarized by the negative charge at its center (thereby forming the outer shell of a "negative polaron"). The underlying assumption of this visualization is that the solvated electron is a single-electron state whose properties can be given by a band model in which the valence electrons in the solvent and the excess electron in the cavity are wholly separately treated [1]—in the exact opposite way to how the electronic structure of the solvent radical ion is viewed. An additional assumption is usually made that the excess electron interacts with (rigid, flexible, or polarizable) solvent molecules by means of an empirical *classical* potential. Both of these simplifying assumptions find little support in structural studies of "trapped electrons" in vitreous molecular solids using magnetic resonance spectroscopy [2].

The "primary" species—solvated/trapped electrons/holes and solvent radical ions—are efficient donors and acceptors of electrons and protons; they readily react with the solvent, with the solute and dopant molecules, with each other, and with the short-lived species (radicals, molecular fragments, and excited states) generated in the ionization and excitation events along with these primary charges. In most radiation chemistry studies, the species of interest are the resulting "secondary" ions, radicals, and excited neutrals. For example, radiolysis of liquid solutions and doped solid matrices is widely used for producing high yields of secondary radical ions (via ion–molecule reactions of the primary ions) for detailed structural studies, e.g., by using magnetic resonance and UV–Vis and/or infrared (IR)-Raman spectroscopies [3]. The general approach in these structural studies is to generate a primary ion that rapidly reacts with a dopant (or solute) of interest yielding a secondary ion that is relatively stable in the matrix (on the time scale of observation). Such stabilization usually requires reduced mobility (low sample temperature), low reactivity, and high ionization potential (IP) or electron affinity (EA) of the solvent/matrix relative to the solute. A classical example of the matrix stabilization is radiolytic generation of hydrocarbon radical cations in frozen halocarbon (freon) solutions [3,4]. Rapid dissociative electron attachment to the halocarbon solvent yields relatively immobile halide anions and C-centered radicals that recombine slowly (if at all) in these low-temperature matrices [3]. The solute radical cations are rapidly formed (on the time scale of the observations) because of the oxidation of the dopant by a mobile solvent hole [3]. As a result, the ionization energy is efficiently converted into the product (radical cation), enabling many types of structural studies. A similar approach is used for stabilization of radical cations in rare gas matrices doped by SF_6 (a matrix favored by IR-Raman spectroscopists [5]), zeolites, and microporous solids (favored by electron paramagnetic resonance (EPR) spectroscopists [6]), etc. A review of the literature shows that almost a third of radiolytic studies on organic radical ions are done in this fashion.

Another popular approach is to react the intermediates of liquid water radiolysis—hydroxyl radicals and hydrated electrons—with the parent molecules of interest (see several representative reviews in Ref. 7). The resulting solute radical ions are observed in a time-resolved fashion, using pulse radiolysis–transient absorption spectroscopy with nanosecond-to-millisecond time resolution; other detection techniques may involve fluorescence, time-resolved EPR, and Raman spectroscopy. The emphasis in such studies is to observe structural rearrangements and reaction patterns of short-lived solute radical ions. The reaction behavior is usually complex (ring opening, protonation/deprotonation, rearrangements, etc.), especially for radical ions whose parent molecules are multifunctional organic compounds, such as biomolecules and biomimetic systems (see, e.g., Chap. 15). Whether the solute is simple (inorganic ions) or complex (biomolecules), the general approach stays the same. A variation of this common technique is pulse radiolysis of solutes in aprotic, high-IP liquids, such as acetone, CCl_4, CH_2Cl_2, 1,2-dichloroethane, and, most often, n-butyl chloride [8]. Unlike the OH radicals in water, solvent holes in such liquids oxidize almost any organic solute; the resulting secondary cations are usually more stable than the same species in water, for lack of efficient donors and acceptors of the proton. There are many examples of using this approach for studies of radical cations of hydrocarbons, heterocyclic compounds, phenols, etc.

Finally, solute radical ions can be generated by light-induced, one-photon or multiphoton ionization of their parent compounds (Chaps. 5 and 16). This approach is particularly useful in the ultrafast studies of short-lived, unstable radical ions that aim to unravel their solvation, recombination, reaction, and vibrational relaxation dynamics of the primary charges (see, e.g., Chap. 10). Whereas the time scale of radiolytic production of secondary ions is always limited by the rate with which the primary species reacts with the dispersed parent molecules, light-induced charge separation can occur in <100 fsec. There are many studies on photoionization of solute molecules in liquid solutions; we do not intend to review these works.

There are also many photoreactions, in which rapid, efficient charge transfer between electron donor and acceptor moieties occurs, yielding a radical ion pair that subsequently dissociates; because the solvent is not oxidized or reduced in these light-induced reactions, these reactions belong to photo- rather than radiation chemistry. Of particular interest, in the context of radiation chemistry studies, are photoreactions in which the *solvent* is oxidized and the resulting hole reacts with the solutes yielding secondary radical ions, just like in radiolysis. This occurs, for example, in one-photon excitation of certain salts (such as N-methyl quinolinium pentafluoride) in low-IP nonpolar fluids (benzene and toluene) [9,10,11]. Upon photoexcitation, the cation moiety accepts a valence electron from the solvent, and the resulting dimer radical cation of the solvent reacts with solute molecules. This reaction is so efficient that it has been used for preparative organic synthesis [9]. Similar photooxidation reactions of aliphatic alcohols, by excited transition metal (e.g., Fe^{III}) [10] or stable organic cations (e.g., methylviologen^{2+}) [11], have been studied. Another class of photooxidation reactions mimicking radiolytic reactions involves electronically excited radical cations [12,13]. *Radical* cations are more efficient oxidizers than stable cations because they are closer in energy to the valence band of the solvent. Photoexcited aromatic radical cations (IR to UV) readily oxidize most solvents, including saturated hydrocarbons [12], water, and alcohols [13]. These charge-transfer-from-solvent (CTFS) photoreactions mirror the better-known charge-transfer-to-solvent (CTTS) reactions in which photoexcited anion (e.g., halide) injects an electron into a polar liquid [14]. The photoelectrons released in the CTTS or photoionization reactions can be used to generate secondary radical anions, although this approach is seldom used in practice.

Understandably, most workers who use radiolysis, photoionization, CTFS, or CTTS as the means for generation of (secondary) radical ions pay little attention to the nature of short-lived precursors of these ions. After all, the subject of interest is a secondary rather than a primary ion. This ad hoc approach is justifiable because radiolytic production is just another means of obtaining a sufficient yield of the radical ion. Quite often in such studies, the radiolysis is complemented by other techniques for radical ion generation, such as plasma oxidation, electron bombardment–matrix deposition, and chemical and electrochemical reduction or oxidation. While the data obtained in these studies are useful, there is little radiation chemistry in such—nominally, radiation chemistry—studies.

There are many excellent books and reviews on the structure and reactions of secondary radical ions generated in radiolytic and photolytic reactions. Common topics include the means and kinetics of radical ion production, techniques for matrix stabilization, electronic and atomic structure, ion–molecule reactions, structural rearrangements, etc. On the other hand, the studies of primary radical ions, viz. solvent radical ions, have not been reviewed in a systematic fashion. In this chapter, we attempt to close this gap. To this end, we will concentrate on a few better-characterized systems. (There have been many scattered pulse radiolysis studies of organic solvents; most of these studies are inconclusive as to the nature of the primary species.)

Before we review specific systems, note that the primary species should be considered on a different footing than the secondary radical ions. The latter ions are well isolated from their parent molecules by the matrix or the solvent. By contrast, in the primary species, the charge is residing on a molecule(s) that is surrounded by like solvent molecules. This often results in unusual properties because the barriers for charge hopping and charge delocalization are lowered. We will examine several examples of the reaction and migration dynamics of such primary radical ions; the multiplicity of examples and commonality of the observed behavior suggest a general pattern.

As demonstrated below, a primary charge viewed as a solvated electron or the molecular ion residing in an inert liquid does not account for experimental observations in many, if not in most, of the systems. While we cannot offer a specific, general model of these "exceptional" ions, we provide a general introduction to the known properties of such species. Furthermore, we argue that these species comprise the rule rather than the exception. The reader is invited to reach his or her own conclusions.

2. ELECTRONS AND SOLVENT ANIONS IN SUPERCRITICAL CO₂

The first "exceptional" system that we review is carbon dioxide [15–21]. Supercritical (sc) CO_2 finds numerous industrial applications as a "green" solvent, and this practical consideration stimulates interest in its radiation chemistry. Though the studies of sc CO_2 are recent, this system is particularly interesting because of the simplicity of the solvent molecule and extensive gas phase and matrix isolation studies of the corresponding ions (see below).

Like other sc fluids, sc CO_2 is a collection of nanoscale molecular clusters that rapidly associate, dissociate, and exchange molecules among each other [22]. Prethermalized electrons in sc CO_2 attach to these $(CO_2)_n$ clusters. In the gas phase, the attachment can be dissociative [reaction (2)] and or nondissociative [reaction (1)],

$$e^- + (CO_2)_n \rightarrow (CO_2)_n^- \tag{1}$$

$$\rightarrow CO + O^-(CO_2)_{n-1} \rightarrow CO + CO_3^-(CO_2)_{n-2} \tag{2}$$

depending on the electron energy [for $n = 2$, reaction (2) is 3.6 eV more endothermic] [23]. Although in radiolysis, a large fraction of electrons have initial excess energies >10 eV, the yield of CO_3^- in 20 MeV electron radiolysis of dense sc CO_2 does not exceed 5% of the total ionization yield [20].

In the gas phase, anions formed in reaction (1) have been extensively studied [24]. A linear CO_2 molecule has negative gas-phase electron affinity (EA_g) of -0.6 ± 0.2 eV. The metastable C_{2v} monomer anion, CO_2^- (with OCO angle of 135° and autodetachment time of <100 μsec) exhibits vertical detachment energy (VDE) of 1.33–1.4 eV. This energy increases to 2.6–2.79 eV in the stable, D_{2d} symmetric C–C bound dimer anion, $C_2O_4^-$, shown in Fig. 1a (with lifetime > 2 msec), and further increases to 3.4 eV for $n = 6$ clusters. In larger clusters, the VDE first decreases to 2.49 eV ($n = 7$), then monotonically increases to 3.14 eV ($n = 13$); then, for $n \geq 14$, the VDE suddenly increases to 4.55 eV. The onset of the photoelectron spectrum increases from 1.5 eV for $n = 2$–7 clusters to 1.8–2 eV for $n = $ 8–13 clusters to 3 eV for $n = $ 15–16 clusters. Tsukuda et al. [24] argue that while the core of small ($n \leq 6$) and large ($n \geq 14$) clusters is a D_{2d} dimer anion, the core of $6 < n < 14$ clusters is a monomer anion weakly coupled to several CO_2 molecules (with binding energy ca. 0.22 eV per molecule). Both of these forms coexist in $n = 6$ and $n = 13$ clusters. It appears that the two isomers are close in energy, and core switching readily occurs when the cluster geometry and size change. Because in sc fluid there is a wide distribution of the cluster sizes and shapes [22], the solvent radical anion does not have a well-defined geometry.

In CO_2 gas, the density-normalized electron mobility $\mu_e \tau_e$ is independent of temperature (2×10^{22} molecule/cm V sec [25]), although the apparent mobility steadily decreases with the pressure: free electrons are trapped by neutral $(CO_2)_n$ clusters ($n \approx 6$) with nearly collisional rates, and the electron affinity of these clusters > 0.9 eV. When extrapolated to solvent densities of $(2–15) \times 10^{21}$ cm^{-3} typical for sc CO_2, these estimates suggest that the free electron mobility μ_e is ca. 1 cm²/V sec and its collision-limited lifetime $\tau_e < 30$ fsec [18]. If the lifetime were this short, the electrons would negligibly contribute either to the conductivity or the product formation. However, this extrapolation is not supported by experiment [18,20].

In low-temperature solid matrices (e.g., Ne and Ar at 5–11 K), the CO_2^- monomer, $(CO_2^-)(CO_2)_n$ multimer anion ($n = 1, 2$), and the $C_2O_4^-$ dimer were observed using IR spectroscopy; the latter species was the only one stable at 25–31 K [26]. In nonpolar

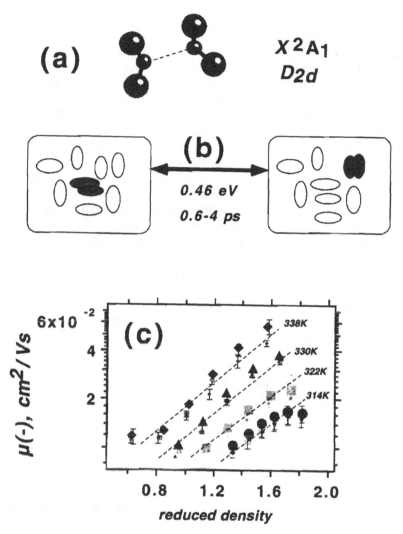

Figure 1 (a) The structure of D_{2d} symmetric $(CO_2)_2^-$ dimer radical anion. (b) Visualization of rapid resonant charge hopping in sc CO_2. The hopping barrier is 0.46 eV; the residence time of the charge on a given (dimer) molecule is 0.6 to 4 psec, depending on the solvent temperature [18]. (c) Reduced density (ρ_r) dependence of solvent anion mobility $\mu(-)$ for four temperatures. The activation energy does not depend on density; the mobility exponentially increases with the solvent density.

liquids, monomer CO_2 is an efficient electron scavenger. Still, the electron can be detached from the monomer anion both thermally and photolytically [27]. For instance, in *iso*-octane, reaction (1) is exothermic by 1.08 eV, the VDE peak is at 3 eV, and the photo-detachment threshold is at 1 eV [27].

As seen from above, the mode of electron trapping in sc CO_2 cannot be deduced from the results obtained in the gas phase or matrix isolation studies. It is not obvious whether the solvent radical anion should be similar to multimer cluster anions found in the gas phase, dimer cation(s) in solid matrices, or monomer CO_2^- anions in inert liquids. Such a situation is typical for other molecular liquids.

Time-resolved laser d.c. photoconductivity and pulse radiolysis–transient (electro)-absorbance studies of sc CO_2 showed that ionization of the solvent (or UV-light absorbing solute) yields two negatively charged species: a metastable (quasifree) conduction band (CB) electron and a rapidly hopping multimer anion (a self-trapped electron) [18,20]. Both of these species exhibit unusual properties that account for many oddities observed in radiolysis of sc CO_2. Quasifree electrons in sc CO_2 have lifetime $\tau_e < 200$ psec and mobility $\mu_e > 100$ cm^2/V sec [18]. For reduced solvent density $\rho_r > 1.2$ (defined as the ratio of the solvent density ρ and the critical density $\rho_c = 0.468$ g/cm^3), the product $\rho_r\mu_e\tau_e$ exponentially increases with ρ_r. The onset for the formation of rapidly migrating quasifree electrons coincides with the emergence of the CB in the solvent [18]. Exactly the same behavior was observed by Itoh et al. [28] for sc saturated hydrocarbons (see Chap. 8). Their studies suggest that the electron mobility increases exponentially with density between $\rho_r = 1$ and 2 and then stabilizes and/or slightly decreases at greater density. In sc CO_2, the product $\mu_e\tau_e$ shows signs of saturation at the reduced density of 1.8 (reaching ca. 2.510^{-9} cm^2/V) [18]. Both this behavior and the high mobility indicate that the quasifree electron is not attached by CO_2 clusters, even temporarily, before it is finally trapped. This is surprising, given the extremely rapid rate of electron attachment to $(CO_2)_n$ clusters in the gas phase. Apparently, once the CB is formed in a dense liquid, the electron dynamics dramatically changes.

In sc CO_2, both the solvent viscosity and the mobility of solute ions (e.g., halide anions and aromatic and alkylamine cations) are a function of the solvent density rather than the solvent temperature and pressure separately [18,21]. In other words, if the density is constant, the ion mobilities do not change with the solvent temperature. At a given temperature, the ion mobility decreases rapidly with ρ_r for $0.2 < \rho_r < 1$ and then decreases slowly for $1 < \rho_r < 2$ [18,21]. In contrast to this behavior, the mobility of the solvent anion exponentially increases with ρ_r, being 2–10 times greater than the mobilities of all other ions in sc CO_2 [18]. The activation energy of the solvent anion migration is 0.46 eV (for constant ρ_r) whereas for the solute ions (Fig. 1c), this energy is less than 20 meV [18,21]. Careful analysis of the data on the solvent anion and electron dynamics and thermochemistry indicate that an equilibrium between the quasifree and trapped electrons, similar to that observed in saturated hydrocarbons (Chap. 8), cannot account for the observed dynamics. This is reasonable, given that the trapping energy of the electron is almost an order of magnitude larger in sc CO_2 than in these hydrocarbons.

Anomalously high mobility and large activation energy for migration of the solvent anion suggests that this anion migrates by rapid charge hopping between the solvent clusters (Fig. 1b); this hopping easily outruns Brownian diffusion of the core anion [18]. The hopping mechanism is also suggested by the fact that the mobility exponentially increases with ρ_r, at any temperature: As the solvent density increases, the cluster-to-cluster distance decreases, and the coupling integral becomes greater; the hopping rate increases accordingly.

Electron photodetachment upon laser excitation of the solvent anion above 1.76 eV was observed (Fig. 2a,c) [18]. The cross section of photodetachment linearly increases between 1.76 and 3 eV (Fig. 2b). Under the same physical conditions, the photodetachment and absorption spectra of the solvent anion are identical (Fig. 2b) [20], suggesting a bound-to-CB transition; the quantum yield of the photodetachment is close to unity. The photodetachment spectrum is similar to the photoelectron spectra of $(CO_2)_{6-9}^-$ clusters observed by Tsukuda et al. [24] in the gas phase; it is distinctly different from the electron photodetachment spectra of CO_2^- in hydrocarbon liquids [27]. This suggests that a C–C bound, D_{2d} symmetric dimer anion constitutes the core of the solvent radical anion [18,19].

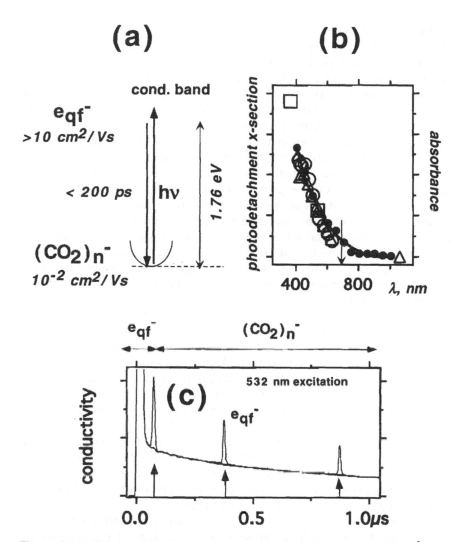

Figure 2 (a) When metastable quasifree electron, e_{qf}^{-} (with mobility > 10 cm^2/V sec) is trapped by dense sc CO$_2$ solvent, a high-mobility radical anion, $(CO_2)_n^{-}$, with $\mu(-)$ of 10^{-2} cm^2/V sec is formed. The binding energy of the electron is 1.6–1.8 eV; the electron can be detrapped by absorption of a photon with energy >1.76 eV. (b) Electron photodetachment (empty symbols) and photoabsorption (filled circles) spectra of solvent radical anion in sc CO$_2$. The arrow points to the onset of the photodetachment band. (c) Photoinduced electron detachment in sc CO$_2$ observed by dc photoconductivity. The initial (clipped) narrow peak is the prompt conductivity signal from free electrons; the time profile of this signal follows the shape of the 248-nm excitation laser pulse. The arrows indicate the delay times at which a second (532 nm) laser pulse was fired. The 532-nm photoexcitation detraps the electrons that subsequently trap within the duration of the 532-nm laser pulse (sharp "spikes").

Both the electrons and the solvent anions react with nonpolar electron acceptor solutes with $EA_g > 0.4$ eV [18]. For oxygen (EA_g of 0.4 eV), the electron transfer from the solvent anion to the solute (which yields a stable CO_4^- anion) is reversible (Fig. 3); the free energy of the corresponding reaction is −0.42 eV. The rates of the electron attachment and solvent anion scavenging correlate with each other and the EA_g. On the other hand, the correlation of these rate constants with the free energy ($\Delta G°$) of the overall reaction (expected from Marcus' theory of electron transfer) is very poor [18]. The same pattern is observed for high-mobility solvent *cations* in nonpolar liquids (Sec. 5). The reason for this behavior is that while the overall $\Delta G°$ depends on what happens to the products (e.g., solute ions) *after* the electron transfer (structural relaxation, solvation, fragmentation, and bonding to the solvent), the activation energy of the reaction depends only on the vertical electron affinity (ionization potential) of the solvent anion (cation) and the solute [18]. This can be rationalized assuming that the scavenging reaction occurs by *direct* electron transfer to (or from) the solute: If the solute needs "reorganization" to accept (donate) the electron, the reaction does not occur and the solvent ion migrates away from the solute molecule. In other words, because of the extremely fast charge hopping, the lifetime of the collision complex is always shorter than the time needed for stabilization of the solute ion by solvation and/or structural relaxation. No such anomalies are observed for (relatively slow) electron transfer reactions that involve *regular ions* in the same nonpolar liquids, and those do conform to the Marcus theory (although, to our knowledge, the inverted region has never been observed in such reactions). For example, rate constants for charge-transfer reactions of solvent *holes* in sc CO_2 with electron donors (CO, O_2, N_2O, and dimethylaniline) systematically increase with the ionization potential of the solute [15,16]. These solvent holes are C_{2h} symmetrical O–O bound $(CO_2)_2^+$ cations that exhibit a prominent charge resonance band in the visible, which suggests strong coupling in the dimer [15,16,19,20]. These solvent cations exhibit normal diffusion properties and rate constants of diffusion-controlled electron transfer reactions [15,16,20]; no ultrafast charge hopping is suggested by the data [20]. Such a situation is common: No known liquid or solid yields both high-mobility anions *and* high-mobility cations. The likely reason is that at least one of these species has a tendency to form strongly bound dimer radical ions that cannot rapidly migrate by the hopping mechanism (see below).

In sc CO_2, only solutes with $EA_g > 2$ eV exhibit diffusion-controlled kinetics for reaction with the solvent anion (which is consistent with the electron trapping energy between 1.6 and 1.8 eV that was estimated from the photodetachment spectrum) [18]. For $\rho_r > 0.85$, the scavenging radii of these diffusion-controlled reactions systematically decrease with the solvent density, ρ_r [18]. Interestingly, in addition to the electron transfer reactions, solvent anions in sc CO_2 form 1:1 and 1:2 complexes with polar molecules that have large dipole moments, such as water, aliphatic alcohols, alkyl halides, and alkyl nitriles [29]. None of these polar low-EA_g solutes directly reacts with quasifree electrons in sc CO_2. The complexation rate is 10–50% of the diffusion-controlled rate, the equilibrium constants of the 1:1 complexation range from 10 to 350 M^{-1} depending on the solute, the reaction heat is −15 to −21 kJ/mol, and the reaction entropy is negative [29]. The stability of these complexes increases with the dipole moment of the polar group and decreases with substitution at the α-carbon. It appears that these complexes are bound by weak electrostatic interaction of the negative charge and the molecular dipole. Previously, such dipole-bound complexes, with monomers of acetonitrile [30] and dimers and higher multimers of aliphatic alcohols [31], were observed for "solvated electrons" in saturated hydrocarbons. For these electrons, the complexation manifests itself through a precipitous decrease in the electron mobility upon the addition of the solute: Thermal emission of trapped electrons

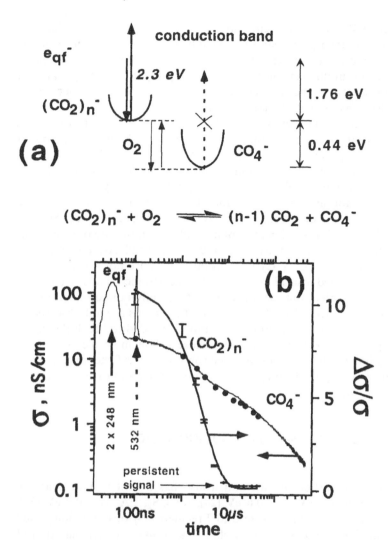

Figure 3 Reversible electron transfer reaction of the solvent radical anion in sc CO_2 with dioxygen. The resulting O_2^- anion rapidly forms O–O bound complex with a solvent molecule, yielding a stable radical anion, CO_4^-. (a) The 532-nm photoexcitation can detach an electron from the solvent anion (Fig. 1a). CO_4^- is a 0.44 eV deeper trap and the electron cannot be detached from it. By determining the magnitude of the "spike" from quasifree electrons (generated by 532-nm laser excitation of the pholysate) as a function of time, one can obtain the decay kinetics of the solvent anion. (b) Experimental realization of this concept for a sc CO_2 solution containing 120 μM of O_2. The conductivity signal (σ) shown on the double-logarithmic scale shows a gradual transformation of the solvent anion to CO_4^-. The ratio $\Delta\sigma/\sigma$ of the 532-nm laser induced conductivity signal to the conductivity tracks the concentration of the solvent anion in a reaction mixture that contains both $(CO_2)_n^-$ and CO_4^-. After the first 10 μsec, this fraction persists at a small value; this represents the settling of the equilibrium between the solvent anion and CO_4^-.

from the complexes to the CB is much less efficient than that of free solvated electrons (Chap. 8). No changes in the absorption band of the solvated electron in the NIR are observed upon the complexation of these electrons with the monomers and dimers of polar molecules [32]. Higher alcohol multimers, such as tetramers, provide very deep traps for these electrons; absorption spectra of such cluster-trapped electrons are almost identical to the spectra of solvated electrons in neat aliphatic alcohols [33].

That the solvent anion in sc CO_2 demonstrates behavior similar to that of solvated electrons in saturated hydrocarbons suggests that the mechanism of binding to the polar molecules must be similar. Apparently, increasing the trapping energy by ca. 0.15 eV because of the electrostatic binding to a monomer or dimer molecule with a dipole moment of (2–4) Debye is sufficient to halt the charge hopping completely [29]. The same effect of polar solutes on the hopping rate of high-mobility solvent radical *cations* (with fairly similar thermochemistry) was observed in neat *cis*- and *trans*-decalin (Sec. 5).

sc CO_2 is not the only liquid for which high-mobility solvent anions are observed (see below), but it is the simplest one. In monoatomic liquids (liquefied Ar, Kr, and Xe), ammonia, and simple hydrocarbons (CH_4, C_2H_6), solvent anions are not formed (quasi-free or solvated electrons are formed instead; see Chap. 8). In diatomic liquids, such as N_2 and O_2, solvent anions are generated, but the formation of strongly bound dimer anions precludes rapid electron hopping. Even for CO_2, high-mobility anions are not observed in the low-temperature liquid: Considerable thermal activation is needed to break the dimer anion prior to every hop of the electron. In liquids whose solute molecules show less tendency to form strongly bound dimer anions, the electron hopping is faster and requires much less activation energy. As argued below, such a situation occurs in carbon disulfide, hexafluorobenzene, benzene, and toluene.

3. SOLVENT ANIONS IN LIQUID CS_2, C_6F_6, AND AROMATIC HYDROCARBONS

Carbon disulfide is isovalent to carbon dioxide and it also has a bent monomer anion. While gas-phase CO_2 has negative EA_g of -0.6 eV [24], for CS_2, EA_g is $+0.8$ eV [34]. Despite this very different electron affinity, Gee and Freeman [34] observed long-lived "electrons" in CS_2 (with lifetime > 500 μsec) with mobility ca. 8 times greater than that of solvent cations. Over time, these "electrons" converted to secondary anions whose mobility was within 30% of the cation mobility. Between 163 and 500 K, the two ion mobilities scaled linearly with the solvent viscosity, as would be expected for regular ions. Of course, Gee and Freeman's identification of the long-lived high-mobility solvent anions as "electron" is just a manner of speech: Obviously, quasifree or solvated electrons cannot survive for over a millisecond in a positive-EA_g liquid.

To the best of our knowledge, pulse radiolysis–transient absorption studies of neat CS_2 have not been reported. "CS_2^- anion" in 0.1 M cyclohexane and 0.1 M THF solutions appears as a single 275-nm peak [35]; there is no charge-resonance band that can be attributed to the dimer anion, at early (< 10 nsec) or later times.

The studies carried out in the gas phase and low-temperature matrices suggest that $(CS_2)_n^-$ anions have somewhat different structure from the $(CO_2)_n^-$ anions discussed in the previous section [24]. Similar to the $(CO_2)_n^-$ anions, the CS_2^- monomer and C–C bound $C_2S_4^-$ dimer anions switch as the core of the $(CS_2)_n^-$ anion [36]. However, unlike the two core anions in the $(CO_2)_n^-$ anions, CS_2^- and $C_2S_4^-$ anions coexist in the clusters of all sizes ($n=2$–6) with the monomer core being statistically prevalent [36]. The dimer core is responsible for the 1.5–1.8 eV peak observed in the photodestruction spectra of $n = 2$–4 anions; no matching peak is observed in the liquid [37]. The dimer anion has either D_{2h} (C–C bound) or C_{2v} symmetry (C–C and S–S bound) rather than the D_{2d} symmetry of the C–C bound $C_2O_4^-$ anion [36–38]. It is almost certain that the S–S bound structure is energetically preferable, because the UV–Vis photoexcitation of $C_2S_4^-(CS_2)_{n-2}$ anions results in the fragmentation of the dimer core to $C_2S_2^-$ and S_2 [37]. The dimer anion was also observed in the EPR spectra of γ-irradiated CS_2-doped frozen alcohols; the g-tensor parameters suggested S–S bonding [39]. Interestingly, while $C_2O_4^-$ is readily observed using IR spectroscopy when electrons are injected in frozen CO_2/Ne or CO_2/Ar mixtures at 4 K [26], only CS_2^- is observed in similar mixtures doped with CS_2 [40]. It seems that dimerization of the molecular anion in CS_2 is less favored than in CO_2, both in the gas and solid phase—perhaps because of the positive electron affinity of the monomer molecule.

Despite these structural differences, both liquid CS_2 and sc CO_2 yield high-mobility solvent anions whose mobilities are similar. The most striking difference is the activation energy for the anion migration. For CS_2, this activation energy is only 5 kJ/mol [34] whereas for CO_2 it is 46 kJ/mol [18]. Such a large difference is surprising because similar transport mechanisms were suggested for both of these anions. This result becomes more understandable when other examples of high-mobility anions are examined.

A relatively long-lived (> 100 nsec) high-mobility solvent anion has been observed by microwave conductivity in room-temperature β-irradiated liquid hexafluorobenzene (C_6F_6) [41]. Like CS_2, hexafluorobenzene has positive EA_g estimated to be between 1 and 2 eV [41]. The anion mobility is 40 times greater than that of the solute ions and the activation energy for the solvent anion migration is 11 kJ/mol [41]. The electron in the solvent anion is strongly bound: The anion does not react with such efficient electron acceptors as SF_6, although it reacts with CBr_4, CCl_4, and $(NC)_2C=C(CN)_2$ (with rate constant as large as 1.5×10^{11} M^{-1} sec^{-1}). Addition of small amounts of inert solvents (benzene, saturated hydrocarbons, C_6F_{12}) results in the exponential decrease in the anion mobility with the molar fraction x of the inert solvent [41,42]. For example, addition of 5 mol% of cyclohexane drops the mobility by 50% relative to neat C_6F_6 [41]. This decrease can be approximated by $(1-x)^n$, where the exponent $n = 15$–20. A "percolation" model of charge migration [42] suggests that the negative charge in C_6F_6 is spread over ca. 12 solvent molecules; this is why even slight dilution has strong effect. The multimerization is also consistent with the emergence of a 675 nm anion band (with molar extinction coefficient of 5000 M^{-1} cm^{-1}) in pulse radiolysis of neat C_6F_6 [41]. This band is different from the $C_6F_6^-$ band observed in dilute solutions of C_6F_6 in inert liquids; the latter is centered at 480 nm [41]. Thermochemistry considerations suggest that the 675-nm band cannot be from a sandwich dimer anion similar to the dimer cations of benzene and other planar aromatic hydrocarbons [43]. This conclusion is consistent with optically detected magnetic resonance (ODMR) and magnetic level-crossing data that indicate that in cold hydrocarbon solutions, the encounter of $C_6F_6^-$ with C_6F_6 results in diffusion-controlled degenerate electron transfer rather than anion dimerization [44]. Thus, while the data clearly point to charge sharing and charge hopping, there seems to be no evidence that a metastable dimer is formed. Once more, one needs to postulate a flexible-structure multimer anion in which fractional negative charge freely exchanges between the solvent molecules.

Importantly, both for liquid C_6F_6 and CS_2, there is a relatively high yield of *ortho*-positronium (*o*-Ps) observed in e$^+$ irradiation of these fluids [34]. The *o*-Ps is formed in the e$^+$e$^-$ recombinations that occur in the end-of-track spurs. The higher the negative charge mobility, the higher the probability that these e$^+$e$^-$ recombinations occur before the e$^+$

undergoes pick-off annihilation with an electron bound in a solvent molecule. Following Gee and Freeman [34], we suggest that high o-Ps yield is a general indicator for the presence of high-mobility solvent anions in molecular fluids with positive EA_g.

Yet another example of a high-mobility anion is given by benzene and toluene [45], whose molecules have *negative* EA_g of -1 eV (see also Chap. 8, Sec. 4.6) In dilute solutions of benzene and toluene in saturated hydrocarbons and tetramethylsilane, there is an equilibrium between the solvated electrons and benzene/toluene anions [45]. This equilibrium is shifted toward the anion ($\Delta G°$ of -9 kJ/mol), and this shift becomes greater at higher pressure. At low pressure, the data for neat liquid benzene and toluene can be interpreted the same way. However, at high pressure (1–2 kbar), the negative charge mobility becomes independent of pressure, which indicates that no volume change occurs during the charge migration [45]. Any mechanism that requires thermal emission of the electron from the bound state back into the CB would require such a change; only resonant charge transfer can account for the zero reaction volume. That the electron in pressurized benzene is not quasifree also follows from the extrapolation of medium-pressure equilibrium constants to the high-pressure range. These estimates give an equilibrium constant of 10–100 for the bound electron. Simple calculation shows that the estimated fraction of free electrons is too small to account for the observed negative charge mobility. Interestingly, the activation energy for anion hopping in liquid benzene and toluene (0.12 and 0.13 eV, respectively [45]) is very close to that in hexafluorobenzene (0.11 eV [41]), despite a large difference in their EA_g; the corresponding anion mobilities are also comparable. Liquid benzene and toluene provide the only known examples of a pressure-induced switch from the thermally activated electron detrapping to charge hopping.

As in the case of hexafluorobenzene solvent anion, EPR and ODMR spectroscopies suggests that no dimerization of monomer radical anions of benzene and toluene occur in liquid benzene and/or in alkane solutions of benzene (whereas the radical cation of benzene is known to dimerize rapidly). The conductivity studies also indicate that there is no volume change associated with the dimerization [45].

We conclude this section with the following observations:

First, high-mobility anions occur both in liquids whose molecules have negative and positive EA_g. The gas-phase electron affinity has no effect on the rate and the activation barrier of electron hopping in neat liquid solvents.

Second, the formation of strongly bound dimer anions is detrimental for rapid charge hopping. Indeed, the dimer must dissociate every time the negative charge moves; this requires thermal activation. As discussed in Sec. 2, high-mobility anions in sc CO_2 have dimer radical anions as their chromophore core; this only results in a higher activation barrier for hopping and a moderate increase in the anion mobility relative to that of solute ions (a factor of 2 at the critical temperature). By contrast, high-mobility solvent anions in liquid CS_2, hexafluorobenzene, benzene, and toluene (for which the tendency for anion dimerization is weak), have 3–5 times lower activation barriers for the charge hopping and substantially higher migration rates (up to 15 times) than the solute ions. We speculate that the only reason why high-mobility solvent anions are observed in sc CO_2 at all is the fact that the core rapidly switches between the monomer and dimer anion with change in the cluster size.

Third, charge delocalization over many solvent molecules, perhaps as many as 10–15, seems to be the only way to explain the observations (such as the effect of the dilution on the conductivity and the emergence of new absorption bands in the

UV–Vis spectra). Classifying these solvent anions as molecular ions solvated by their parent liquids or solvated electrons does not explain these properties.

It may appear from the above that only *nonpolar* liquids yield "nonmolecular" solvent anions upon the ionization. Perhaps this is misleading: Most polar liquids studied by radiation chemists are aliphatic alcohols and water, and these liquids yield solvated electrons rather than radical ions. Although there has been sporadic interest in other polar liquids (e.g., neat acetone), the current state of knowledge of such systems does not allow one to reach *any* conclusion as to the nature of the reducing species observed therein (although, see Sec. 4).

It may also appear that the few nonpolar liquids considered above do not comprise the rule. Again, we stress that the most extensively studied nonpolar polyatomic liquids are saturated hydrocarbons; it so happens that these fluids also yield solvated electrons (that are in a dynamic equilibrium with quasifree CB electrons). Actually, very few nonpolar liquids other than alkanes and liquefied rare gases have been studied by pulse radiolysis or photoconductivity. In almost all such systems, either the "hole" or the "electron" exhibits unusual migration or reaction dynamics that are suggestive of rapid charge hopping. Given that in many liquids, the primary solvent ions are short-lived, we hazard a conjecture that most liquids that do not yield solvated electrons yield high-mobility, multimer solvent anions or (as shown in Sec. 5) solvent holes. The true scope for the occurrence of such ions is not known, but the number of examples steadily increases.

4. SOLVENT ANIONS AND ELECTRON LOCALIZATION IN LIQUID ACETONITRILE

The previous examples of high-mobility solvent radical anions were all in nonpolar liquids. What happens in polar liquids? In some polar liquids (whose molecules have negative EA_g) such as water, mono- and poly-atomic alcohols, and ethers, solvated electron is observed (Chaps. 7 and 8). In this species, the excess electron density is mainly outside the solvent molecules. The electron resides at the cavity center and interacts as a point-like negative charge with the surrounding solvent dipoles. The greater the dipole, the more stable is the solvated electron. A link between these energetics and the solvent polarity is observed in the bell-like visible or near-infrared (NIR) spectra of the solvated electrons [1]. The maximum of the band systematically shifts toward the blue as the polarity of the solvent increases. For obvious reasons, these solvated electrons are not observed in liquids whose molecules have positive EA_g. However, there is no a priori reason to believe that a given negative-EA_g polar liquid will yield solvated electrons upon ionization. Even if the monomer has negative EA_g, the dimer may have positive electron affinity, especially in a liquid solution where the electrostatic field of the solvent (considered as a polarizable dielectric continuum) can stabilize the corresponding dimer anion. For hydroxylated molecules, such as water and aliphatic alcohols, stabilization via the formation of a dimer (or multimer) anion is lacking, because no low-lying unoccupied molecular orbitals (MOs) are readily accessible for the excess electron. By contrast, many organic molecules have readily accessible C and N $2p$ orbitals in their lowest unoccupied molecular orbitals (LUMOs); these are the atomic orbitals involved in the formation of C–C bound dimer anions in CO_2 and CS_2 considered in Secs. 2 and 3, respectively.

Below, we consider a polar, negative-EA_g liquid—acetonitrile—in which the dimer anion formation (because of the electron accessing low-lying π^* orbitals) competes with electron stabilization because of the formation of a polarized solvent cavity [30,46]. The outcome of this competition depends on the solvent temperature, as the dimer anion is in a dynamic equilibrium with a more energetic cavity electron. The latter cannot be regarded as solvated electron because the partition between the valence electrons in the solvent molecules and the excess electron at the cavity center is incomplete [30]. We argue that the properties of these two electron states can only be understood when the traditional one-electron approximation is abandoned in favor of many-electron model.

Like water and aliphatic alcohols, gas-phase CH_3CN monomer has a large dipole moment (4.3 D) and negative vertical EA_g of -2.84 eV (adiabatic EA_g is $+17$ meV) [47]. CH_3CN^- is a classical example of a dipole-bound anion, with the electron in a diffuse orbital (> 3 nm) [47]. While neutral dimers in which the CH_3CN dipoles are coupled in an antiparallel fashion readily form in vapor and in liquid [48], the dimer anion, $\{CH_3CN\}_2^-$, has not been observed in the gas phase. In the neutral trimer, one of.the monomers couples sideways to the antiparallel pair; this molecule binds the electron in the same way as the monomer; the adiabatic electron affinity of this trimer (14–20 meV) is higher than that of the monomer [47]. Higher multimer anions, $\{CH_3CN\}_n^-$, were found only for $n > 12$ [49]. The stabilization of excess electron in solid and liquid acetonitrile is a concerted effect of many solvent molecules. One would expect that the electron in solid and liquid acetonitrile localizes in the same way as the "solvated/trapped electron" in water and alcohols. This expectation is not borne out.

Solid acetonitrile exists in two crystalline forms, a high-temperature phase, α, and a low-temperature phase, β [50]. When ionized at 77 K, α-acetonitrile yields a dimer radical anion, while β-acetonitrile yields a monomer radical anion [51]. The observed dichotomy follows from the crystal structure: In α-acetonitrile, the dimer anion retains the same reflection plane and inversion center as the symmetric antiparallel pair of CH_3CN molecules [52]. β-acetonitrile consists of infinite chains of parallel dipoles (no antiparallel pairs are present) and a monomer anion is formed instead [51,52]. Electron paramagnetic resonance experiments and ab initio calculations of Williams et al. [51] indicate that the dimer radical anion is C_{2h} symmetrical and has the staggered, side-by-side structure shown in Fig. 4a [46,52]. The mechanism for orbital stabilization of bent acetonitrile molecules in the dimer is illustrated in Fig. 4b. The CCN angle is 130° and the distance between cyanide carbons is 0.165 nm. The negative charge and spin are mainly on carbonyl N and methyl C atoms. This structure accounts for the observed EPR parameters and vibronic progressions observed in the charge-resonance band of the dimer radical anion [51,52]. The monomer radical anion in β-acetonitrile is also bent; the CCN angle is close to 131° [52]. In both of these anions, the C–C bond is stretched to 0.153 nm (vs. 0.1443 nm in neutral CH_3CN). Photoexcitation of these radical anions (<650 nm) causes further elongation of the $NC–CH_3$ bonds (because of the promotion of electron into the corresponding C–C antibonding orbital), which leads to their fragmentation to CH_3 and CN^- (see Fig. 4b). Except for the vibronic progressions, both radical anions exhibit similar absorption spectra in the visible (see Fig. 3 in Ref. 51). For the dimer radical anion, the absorption band is centered at 530 nm, for the monomer radical anion at 420 nm [51]. The positions of these bands are in good agreement with ab initio calculations [52]. These calculations indicate that no bound-to-bound transitions in the IR are possible, either for the monomer or the dimer radical anion [46,52].

In liquid acetonitrile, there are two radical anions present shortly after the ionization event: anion-1 that absorbs in the NIR (whose band is centered at 1.45 μm) and anion-2 that absorbs in the 400–800 nm region (whose band is centered at 500 nm); see Fig. 5a [46,53]. These two anions are in a rapid dynamic equilibrium (Fig. 5b): As the liquid is

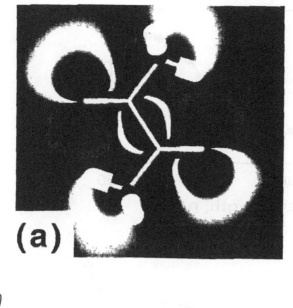

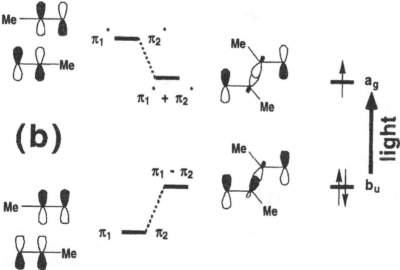

Figure 4 (a) Singly occupied molecular orbital (SOMO) of dimer radical anion of acetonitrile (from a density functional calculation); (b) a scheme for the formation of the SOMO and the doubly occupied subjacent orbital from π and π^* orbitals of neutral acetonitrile molecules.

cooled, the 1.45-μm band becomes more prominent and the 500-nm band less prominent [53]. From the temperature dependencies of the transient absorption spectra. it was estimated that anion-2 is 0.36 eV more stable than anion-1 [53]. The transformation of anion-1 to anion-2 is rapid at room temperature [40,46,53] but fairly slow at the lower temperature; at −30°C, it takes 20–50 nsec [53].

While it was initially suggested that anion-1 and anion-2 are the monomer and the dimer radical anions of acetonitrile [53], respectively, more recent work suggests that anion-1 cannot be a monomer anion (which in any case has a different absorption

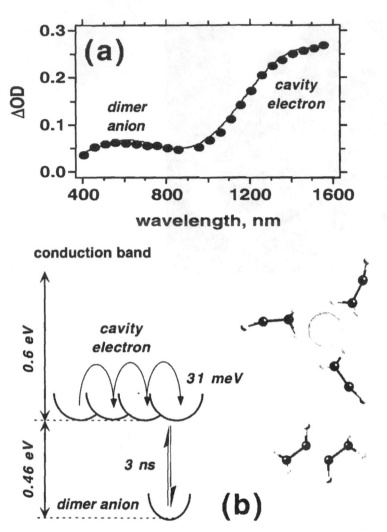

Figure 5 (a) Typical end-of-pulse absorption spectra obtained in pulse radiolysis of room temperature liquid acetonitrile (7-nsec fwhm pulse of 20 MeV electrons). The 500-nm peak is from anion-2 (dimer radical anion); the 1450-nm peak is from anion-1 (cavity electron). (b) Energy diagram and sketches of anion-1 and anion-2 (see the text).

spectrum from anion-1, as explained above) [30,46]. Actually, the absorption spectrum of anion-1 is very similar to that of solvated electron in saturated hydrocarbons. This is understandable because the CN dipole has negative charge on the nitrogen. Consequently, if a cavity electron were formed in acetonitrile, this cavity would be lined by *methyl* rather than CN groups [30]; that is, the first solvation shell of the s electron would resemble that of the solvated/trapped electron in liquid and vitreous alkanes.

The NIR location of the absorption band for a hypothetical cavity electron in acetonitrile makes even more sense if one recalls that there is a linear correlation between the position of the band maximum of a cavity electron in a given polar liquid and the position of the CTTS band maximum for a given halide anion in the same liquid [46,54].

The absorption band of anion-1 perfectly fits on this correlation plot (predicted 1.48 μm [54] vs. the observed 1.40–1.45 μm [46]).

Not only does anion-1 differ from the molecular anions of acetonitrile in its absorption properties, but its dynamic properties are also anomalous. While anion-2 has normal mobility, anion-1 is a high-mobility anion whose room temperature diffusion coefficient is more than three times higher than that of solute ions and anion-2 [30]. The activation energy for this migration is just 3.2 kJ/mol while the value for normal ions (including anion-2) is 7.6 kJ/mol [30]. Electron-transfer reactions that involve anion-1 proceed with rate constants approaching 10^{11} M^{-1} sec^{-1} [30,46]. These reactions can be directly observed on a subnanosecond time scale (before the equilibration of the two anions) using ultrafast pump-probe laser spectroscopy. To this end, Kohler et al. injected the electron into room temperature liquid acetonitrile using one-photon CTTS excitation of iodide [46]. Both anion-1 and anion-2 were observed within 300 fsec after the excitation with a 200-fsec, 260-nm pulse, and rapid decay of anion-1 in the presence of $CHCl_3$ was observed [46]. The same experiment gave an estimate of 0.26 nsec for the settling of the equilibrium between the two anions.

In the time-resolved photoconductivity experiments carried out at Argonne [30], the anion equilibrium was observed via non-Arrhenius temperature dependencies of anion mobility and rate constants of scavenging by electron acceptors, such as CCl_4 [30]. These conductivity experiments clearly demonstrate that the high-mobility anion is anion-1 rather than anion-2, contrary to previous suggestions [55]. To distinguish between the two anions, anion-1 and anion-2 were photoexcited in their respective absorption bands using 1064 and 532 nm, 6-nsec laser pulses; this photoexcitation causes anion fragmentation to CH_3 and CN (with quantum yields of 0.01 and 0.32, respectively) and a decrease in the d.c. conductivity. Using 532 nm photobleaching of anion-2, the equilibrium fraction of this anion between −20°C and 50°C was determined; knowing this fraction, the mobility and reaction constant for each anion were determined and the equilibrium constant (1.3 at 25°C) and the heat of anion conversion (which is ca. −0.46 eV) were estimated. The photon fluence dependencies of the photobleaching efficiency gave estimates for the anion conversion rate. These measurements suggested a longer time constant of 3 nsec (vs. 0.26 nsec obtained in Ref. 46) for settling the equilibrium between the two anions at 25°C.

The formation of CH_3CN^- in solid β-acetonitrile is a result of its favorable crystal structure [50]. According to x-ray diffraction and nuclear magnetic resonance (NMR) data, the short-range structure of liquid acetonitrile is similar to that of crystalline α-acetonitrile, with a pentamer as the basic unit [56]. The prevalent orientation of the acetonitrile molecules in the liquid is the antiparallel pair of the type found in α-acetonitrile. Given that dimerization strongly reduces the energy of the anion [46,51,52], it seems likely that the monomer anion cannot form in liquid acetonitrile, where the "special arrangement" of neighboring molecules needed for the formation of the monomer anion is not possible.

While anion-2 is clearly the dimer radical anion of acetonitrile, identification of anion-1 as a cavity electron requires caution. First, we stress that anion-1 cannot be the monomer anion of acetonitrile. The monomer anion does not absorb in the NIR [30,46,52]. For the monomer anion to occur at all, the neighboring acetonitrile molecules should all be oriented in the same direction, as in β-acetonitrile; otherwise, coupling to a neighboring (antiparallel) molecule reduces the overall energy and causes instant dimer formation. It is difficult to see how such a fortuitous orientation could persist for 0.3–3 nsec in a room temperature liquid. Also, it is not clear why a monomer anion would rapidly migrate. The only migration mechanism possible for this anion would be charge hopping. Assuming that this hopping is between neighboring molecules (separated by 0.4 nm) and the diffusion coefficient is 8.3×10^{-4} cm^2/sec (estimated from the room-temperature mobility of 3.3×10^{-4} cm^2/V sec [30]), the residence time for the charge on a given molecule is 2 psec. This implies that 10^2–10^3 hops occur prior to the transformation of the monomer anion-1 to anion-2. The lowest bending modes of the CCN fragment of acetonitrile molecules and anions are 300–330 cm^{-1}, which is equivalent to 0.1 psec in time units. Thus although the diffusion is fast, the lifetime of a given "monomer anion" is sufficiently long for the structural relaxation; in other words, this monomer anion must be a bent species like CH_3CN^- in β-acetonitrile. Thus the low-barrier resonant charge transfer needed to explain the high mobility would have to be between a strongly bent anion and a linear neutral molecule. Such a process cannot proceed with a low activation energy because bending of the neutral molecule and solvation of the resulting anion require much energy. Furthermore, never once in a series of these 10^2–10^3 hops could the two molecules involved in the resonant charge transfer be in the antiparallel orientation, because then anion-1 would couple to the neighboring molecule yielding anion-2. It appears that the monomer anion cannot account for any property of anion-1.

It is more likely that the high-mobility anion-1 is a multimer anion in which the charge is spread over several acetonitrile molecules, like the analogous species in nonpolar liquids that were examined in Sec. 3. Because of the reduction in the charge on the individual molecules, their bending is less strenuous and the barrier for the migration of the multimer anion is low. Such a multimer anion is actually no different from the solvated electron in alkanes (see below), which accounts for the striking similarity between the absorption spectra of anion-1 and solvated/trapped electrons in saturated hydrocarbons. We suggest that acetonitrile provides a rare example of a liquid in which the solvated electron (multimer anion) coexists with a molecular—dimer—radical anion.

To investigate possible structures of the multimer anion, a $\{CH_3CN\}_3^-$ cluster was modeled using a density functional (B3LYP) method [30]. A $6-31+G^{**}$ basis set that included polarized (d,f) and diffuse functions was used and the C_{3h} symmetry was imposed. A "ghost" hydrogen atom with zero charge was placed at the center of the cluster to provide s-functions for the solvated electron. The polarizable (overlapping spheres) continuum model was implemented in the integral equation formalism. The lowest energy state was a "propeller-like" 2A' state shown in Fig. 6. The CCN angle in the acetonitrile subunits is 178° in vacuum and 168° in solution (vs. 180° in the neutral molecule). This bending is considerably smaller than in the monomer and dimer anions (ca. 130°). The solvated cluster anion is compact: The closest methyl hydrogens are 0.171 nm away from the symmetry center. The C–C bond in the acetonitrile subunits is elongated from 0.144 to 0.148 nm, while the C–N bond is very slightly changed. In this structure, the singly occupied molecular orbital (SOMO) envelopes the whole cluster anion. The main negative nodes are on methyl carbons, while the main positive nodes are at the center of symmetry, on the in-plane hydrogens, and on carbonyl carbons. This structure may be viewed both as a multimer anion and a solvated electron: The SOMO has a noticeable s-character at the symmetry center (ca. 0.34), although the main spin density is on the methyl carbons. The latter atoms exhibit large hyperfine coupling constants (hfcc) for ^{13}C: the isotropic hfcc is 6.9 mT; the anisotropy is negligible. Isotropic hfcc for methyl protons are relatively small: 0.19 mT for in-plain hydrogens (the principal values of the dipole tensor are -0.29, -0.15, and 0.44 mT) and -0.086 mT for out-of-plane hydrogens (-0.29, -0.16, and $+0.45$ mT, respectively). The isotropic hfcc for cyanide ^{13}C and ^{14}N nuclei are 0.4 and 0.36 mT, respectively.

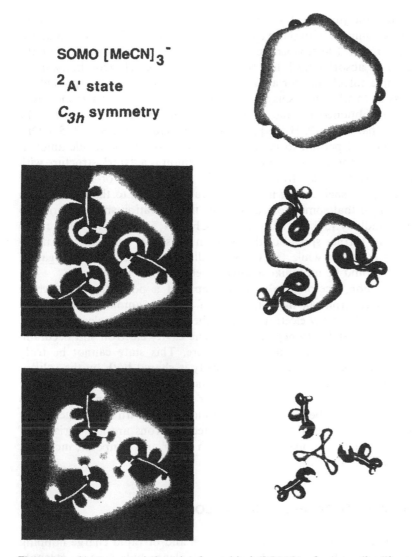

SOMO [MeCN]$_3^-$

^{2}A' state

C_{3h} symmetry

Figure 6 Singly occupied molecular orbital (SOMO) of a propeller-like trimer radical anion of acetonitrile obtained using density functional theory. The structure was "immersed" in a polarizable dielectric continuum with the properties of liquid acetonitrile. Several surfaces (on the right) and midplane cuts (on the left) are shown. The SOMO has a diffuse halo that envelops the whole cluster; within this halo, there is a more compact kernel that has nodes at the cavity center and on the molecules.

The structure bears strong resemblance to the trapped electron in saturated hydrocarbons studied by Kevan et al. [2]. The "electron" is "solvated" by methyl groups; the positive charge on these groups is increased because of the considerable elongation of C–C bonds. This elongation, as demonstrated by our discrete Fourier transform (DFT) calculations, is the consequence of large electron density on the skeletal carbon atoms. In the semicontinuum model of Kevan et al. [2], this (multielectron) interaction is treated in terms of a "polarizable" C–C bond; our calculation justifies their ad hoc approach. The size of the solvation cage, the juxtaposition of methyl groups, and the hfcc tensors for methyl protons favorably compare with those experimentally obtained by Kevan et al. for the trapped electron in frozen 3-methylpentane [2]. Therefore it is reasonable that the multimer $\{CH_3CN\}_n^-$ anion absorbs much like the solvated/trapped electron in alkanes.

While a first-principle calculation for a larger cluster is impractical, it is possible to make an educated guess as to what happens to the anion when the cluster size increases. The "propeller" structure obtained for the $\{CH_3CN\}_3^-$ anion is similar (save for the elongated C–C bonds) to that of the $\{CH_3CN\}_nX^-$ (X = I, Br) cluster for $n = 3$ [57]. One may expect that this trend will pertain to larger size clusters. When the halide anion is solvated by less than seven acetonitrile molecules, the core anion is a "star" structure with radial CH_3CN dipoles looking away from the halide anion [57]. For $n > 9$–12, the molecules in the first solvation shell couple in an antiparallel fashion to the molecules in the second solvation shell, so that some molecules in the first solvation shell are oriented tangentially rather than radially [57]. Perhaps, small $\{CH_3CN\}_n^-$ anions ($n \leq 6$) are also star-shaped. Because of the further spread of the electron density in such clusters, the s-character of the SOMO increases while the CCN bending and C–C bond elongation decreases: Such an anion would be more like a solvated electron.

To conclude this section, acetonitrile is an example of a polar liquid in which stabilization of the excess electron via the formation of a dimer anion is energetically favored over the formation of a cavity electron, despite the fact that the molecule has one of the largest dipole moments and very negative EA_g. The cavity electron still occurs in this liquid as a metastable state at the high temperature. This state cannot be truly regarded as solvated electron because the electron density is shared both by the solvent molecules and interstitial sites; the excess electron is not separable from the valence electrons of the solvent. A similar situation exists for trapped electrons in vitreous hydrocarbons. These species should be regarded as multimer anions with flexible geometry and extensive delocalization of the charge. These solvated electrons are just variants of multimer radical anions that occur in many liquids, both polar and nonpolar, including the several examples examined above.

5. SOLVENT RADICAL CATIONS IN LIQUID CYCLOALKANES

At first glance, it may appear that extensive delocalization and/or rapid charge hopping should not occur for solvent radical *cations* because the valence hole is more strongly associated with the molecule than the excess electron. We have already seen that such expectations are not supported for solvent anions, were the delocalization and degenerate electron exchange occur for liquid solvents whose molecules differ by more than 2 eV in their electron affinity. The same applies to the solvent holes: The fact that a given molecule forms a well-defined radical cation when this molecule is isolated in an inert matrix does not mean that the same species is formed in a liquid where all molecules are alike. The last few examples discussed in this chapter are high-mobility solvent holes in cycloalkanes: cyclohexane, methylcyclohexane, and decahydronaphthalenes (decalins) [58,59,60].

We forewarn the reader that the formation of high-mobility holes is not peculiar to these four cycloalkanes: For instance, cyclooctane [61], squalane [62,63,64], and CCl_4 [65] also yield such holes. However, in these other liquids, the holes are unstable and, consequently, more difficult to study (the lifetimes are 5–20 nsec vs. 1–3 μsec). This explains why convincing demonstrations for the occurrence of high-mobility holes are slow to come. For example, squalane (by virtue of its high viscosity) has been frequently used in the studies on fluorescence and magnetic and spin effects in pulse radiolysis. Despite these many studies, it has only recently been recognized that its short-lived hole (with lifetime < 20 nsec) has abnormally high diffusion and reaction rates [62]. Shortly after this fact was established using transient absorption spectroscopy, subsequent studies confirmed the hopping mechanism, as fast diffusion with degenerate electron exchange and high-rate scavenging reactions of the squalane holes were observed using time-resolved ODMR [62], magnetic level-crossing, and quantum beat spectroscopies [63]. Rapid scavenging reactions of the squalane hole were also found to account for the anomalies in the magnetic field effect observed for delayed fluorescence in the VUV excitation of squalane [64]. Basically, in such systems, one needs to know where to look; once the property is established, it can be demonstrated in several ways, using different techniques.

In cyclooctane, high-mobility solvent holes were observed using time-dependent electric-field-modulated delayed fluorescence [61] and by observation of rapid scavenging of cyclooctane holes by aromatic solutes in the initial stage of radiolysis. Recently, there has been a suggestion of the presence of such holes in cyclopentane and cycloheptane [61]; their natural lifetimes must be < 5 nsec. Faster-than-normal scavenging of short-lived isooctane holes by diphenylsulfide and biphenyl was observed using quantum beat and transient absorption spectroscopies [66]. A controversy exists as to the presence of high-mobility holes in liquid CCl_4 [65].

These disparate findings hint that there may be many examples of rapidly migrating (delocalized) solvent holes in molecular fluids: The known systems are few because it is difficult to establish these properties for short-lived species. As the time resolution improves, more examples might follow. In most saturated hydrocarbons, fragmentation and proton transfer limit the lifetime of the solvent hole to several nanoseconds (or less) [58] and, therefore, little is known about their dynamics. On the other hand, the most studied alkane liquids, paraffins, do not seem to yield high-mobility solvent cations [67]. This is because many conformers coexist in these liquids, some of which have higher ionization potential than others. Variations in the binding energy of the hole stall its rapid hopping because thermal activation is needed to detrap the hole from the low-IP conformers. That conformation dynamics and isomerism play an important role in the charge hopping is supported by many observations (note, e.g., that high-mobility solvent anions are known to occur only in liquids whose molecules are rigid). As for the paraffins, while no rapid hole hopping is observed in liquid alkanes, in low-temperature crystals (where all molecules have the same extended conformation) this exchange is very fast and can be readily observed by means of time-resolved and/or cw ODMR [68]. Ironically, in these n-alkane crystals, the hole migrates much faster than the hole in frozen cycloalkanes, because the latter solids exhibit more structural disorder (because of the formation of plastic crystal and glass phases) detrimental to hole hopping; thus the situation is exactly opposite to that in a liquid. The recent magnetic level-crossing spectroscopy study of Borovkov et al. places an upper estimate of just 10^8 M^{-1} sec^{-1} for degenerate electron exchange between n-nonane$^+$ and the parent alkane molecule in room temperature solution [68].

The reader may notice that only saturated hydrocarbons (with a possible exception of CCl_4) have been observed to yield rapidly migrating solvent holes. As mentioned above, part of this bias is explained by the fact that the holes are usually short-lived, so their dynamic properties are difficult to study. However, in many liquids (such as aromatic hydrocarbons and sc CO_2), the solvent holes are relatively stable, yet no rapid hole hopping is observed. In such liquids, the solvent hole has a well-defined dimer cation core with strong binding between the two halves (in the first place, it is this dimerization that causes the hole stability). For example, solvent holes in aromatic liquids are sandwich dimer cations with overlapping π systems [43]; in sc CO_2, the solvent cation is an O–O bound molecular dimer [19], etc. This strong dimerization is detrimental to charge delocalization and rapid hopping. High temperature is needed to overcome this hindrance; perhaps high-mobility holes more readily occur in hot (e.g., supercritical) liquids. In many room temperature liquids, a catch-22 situation occurs: For the solvent radical cations to be stable toward fragmentation and proton transfer, these holes must dimerize. The dimer radical cations are long lived and can readily be studied; however, they have ordinary dynamic properties. The holes that do not dimerize might have interesting dynamic properties but they are unstable and, therefore, difficult to study. As a result, one is limited to the studies of the few solvent holes that do not dimerize and yet are long-lived.

In cycloalkanes, proton transfer is weakly endothermic, conformational dynamics is slow, dimerization is not favored, and the high-mobility solvent holes can be readily observed [60]. Ionization of cyclohexane, methylcyclohexane, *trans*-decalin, and *cis*-decalin produces cations whose mobilities are 5 to 25 times greater than the mobilities of normally diffusing molecular ions and (in some cases) thermalized electrons in these liquids [58,59,60]. Long lifetime and high mobility makes it possible to study the reactions of these holes using time-resolved microwave and d.c. conductivity, an option that does not exist for other saturated hydrocarbons. The activation energies for the hole mobility range from $-(3\pm1)$ kJ/mol for *trans*-decalin and cyclohexane to $+(7$–$8)$ kJ/mol for methylcyclohexane and *cis*-decalin [58]. Methylcyclohexane has the largest temperature interval where it is liquid at atmospheric pressure and exhibits a single activation energy of hopping (7.8 kJ/mol) between 133 and 360 K [69]. The activation energies for the highest-rate scavenging reactions of the cycloalkane holes range from 4 to 9 kJ/mol [58]. All these activation energies are small, suggesting low barrier for resonant charge transfer.

Dynamic and chemical properties of the cycloalkane holes have been reviewed [58,59], and we refer the reader to these publications for more detail. Below, we briefly summarize the main findings. Although the cycloalkane holes are paramagnetic species, these holes cannot be observed by magnetic resonance techniques, whether in neat cycloalkanes or in dilute solutions in high-IP liquids. Only recently has it been understood that rapid spin-lattice (T_1) relaxation in the high-symmetry cycloalkane radical cations precludes their detection using ODMR [70]. This relaxation is caused by dynamic averaging between the nearly degenerate ground and excited states of the radical cations; this degeneracy results from the Jahn–Teller distortion. For example, *trans*-decalin cation isolated in room-temperature cyclohexane has $T_1 < 7$ nsec [70]. Because it takes several tens of nanoseconds to flip the electron spin for detection, radical cations of these cycloalkanes cannot be detected by ODMR.

In the early studies, the cycloalkane holes were viewed as molecular radical cations that undergo rapid resonant charge transfer. At any given time, the positive charge was assumed to reside on a single solvent molecule and, once in 0.5–2 psec, to hop to a neighboring molecule. The low activation energy was explained by the similarity between the shapes of cycloalkane molecules and their radical cations [60].

This model is consistent with many observations. Dilution of cycloalkanes with high-IP alkanes (or higher-IP cycloalkanes) results in a decrease in the hole mobility that correlates with the mole fraction of the cycloalkane in the mixture: The hopping rate decreases when the density of the like molecules decreases. The occurrence of resonant charge hopping is firmly established experimentally. Charge transfer between c-$C_6D_{12}+$ and c-C_6H_{12} was observed in the gas phase, where it proceeds at 1/3 of the collision rate [71]. The hopping was also observed for radical cations and molecules of *cis*- and *trans*-

decalins in dilute cyclohexane solutions (where it proceeds with a diffusion-controlled rate) [72]. In low-temperature solid hydrocarbons (4–30 K), hole hopping was observed by ODMR [68]. At higher temperatures, the spectral diffusion caused by the hopping causes the ODMR spectrum to collapse to a single narrow line observed using magnetic level-crossing and quantum beat spectroscopies [64,65].

On the other hand, matrix-isolation EPR and ab initio calculations suggest that neutral cycloalkanes and their cations have rather different geometries. In *cis*- and *trans*-decalins, the bridging bond elongates from 0.153–0.156 nm in the neutral molecule to 0.19–0.21 nm in the radical cation [73]. Upon charging, the molecules undergo considerable structural relaxation, losing 0.5–0.7 eV [73]. If the electron transfer were a single-step process, it would require an activation energy of 1–2 eV. What then makes the resonant charge transfer possible? In the gas phase, the electron exchange proceeds through the formation of a collision complex in which the charge is shared by both of the cycloalkane moieties [71]. This sharing considerably reduces the barriers for the structural relaxation. It may be assumed that in liquid cycloalkanes the charge is shared between several solvent molecules (analogous to the situation for solvent anions) and this sharing further reduces the hopping barrier.

The sharing of charge causes delocalization of the hole. The best evidence for the delocalization of cycloalkane holes was provided by large scavenging radii (>2 nm) in cold methylcyclohexane [69] and by hole dynamics in cyclohexane-methylcyclohexane mixtures [74]. While the addition of less than 5–10 vol.% of methylcyclohexane to cyclohexane reduces both the d.c. conductivity signal and its decay rate, further addition of methylcyclohexane yields little change in the conductivity signal and kinetics. The initial reduction is accounted for by rapid reversible trapping of cyclohexane holes by methylcyclohexane [74]. At higher concentration of methylcyclohexane, the equilibrium fraction of the cyclohexane holes becomes very low and the conductivity should decrease. Experimentally, the migration of methylcyclohexane hole in 5 vol.% methylcyclohexane solution is as rapid as that of the solvent holes in neat methylcyclohexane. When the methylcyclohexane is diluted by *n*-hexane instead of the cyclohexane, the conductivity signal decreases proportionally to the fraction of *n*-hexane. These results suggest that the methylcyclohexane holes are coupled to the cyclohexane solvent (the difference in the liquid IPs is <0.11 eV [74]). This coupling makes the charge migration of methylcyclohexane holes in cyclohexane as efficient as in neat methylcyclohexane. From the critical concentration of methylcyclohexane, the delocalization radius was estimated as 1 nm, or 4 to 5 molecular diameters [74]. Thus the degree of charge delocalization in cyclohexane (for the hole) and hexafluorobenzene (for the electron) are comparable.

We turn to the chemical behavior of cycloalkane holes. Several classes of reactions were observed for these holes: (1) fast irreversible electron-transfer reactions with solutes that have low adiabatic IPs (ionization potentials) and vertical IPs (such as polycyclic aromatic molecules); (2) slow reversible electron-transfer reactions with solutes that have low adiabatic and high vertical IPs; (3) fast proton-transfer reactions; (4) slow proton-transfer reactions that occur through the formation of metastable complexes; and (5) very slow reactions with high-IP, low-PA (proton affinity) solutes.

Class (1) reactions were observed in all four cycloalkanes. The highest rate constants were observed for reactions of cyclohexane hole with low-IP aromatic solutes, (3–4.5) × 10^{11} M^{-1} sec^{-1} at 25°C [75]. In these irreversible reactions, a solute radical cation is generated. Class (2) reactions were observed for reactants 1,1-dimethylcyclo-pentane, *trans*-1, 2-dimethylcyclopentane, and 2,3-dimethyl-pentane in cyclohexane [74], *trans*-decalin, bicyclohexyl, and *iso*-propylcyclohexane in methylcyclohexane [69], and benzene in *cis*-

and *trans*-decalins [76] (Fig. 7). In these class (2) reactions, biexponential scavenging kinetics of the solvent hole results because of the dynamic equilibrium between the solvent hole and the corresponding solute cation (in the latter case, the kinetics are complicated by the subsequent dimerization of the benzene cation, Fig. 7). For methylcyclohexane in cyclohexane, the equilibrium is so rapidly reached that the decay kinetics are single exponential at any temperature. Similar equilibria exist for high-mobility holes in mixtures of *cis*- and *trans*-decalins.

The rate constants of the forward class (2) reactions are much slower than those of the class (1) reactions, although some of electron donors have comparably low adiabatic IP_gs. These rate constants do not correlate with the free energies of hole scavenging

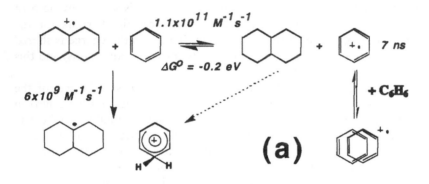

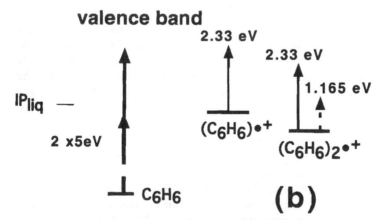

Figure 7 A typical reversible reaction of a cycloalkane hole. (a) *Trans*-decalin hole rapidly reacts with benzene transferring positive charge to the solute. The reverse charge transfer reaction is relatively slow (the free energy decreases by 200 meV), and the lifetime of benzene monomer is ca. 7 nsec. This lifetime is further shortened by dimerization of the monomer; this dimerization shifts the equilibrium to the right side. The charge transfer competes with slow, with the formation of decalyl radical and benzonium carbocation by an irreversible proton transfer. (b) The populations of solute monomer and dimer cations (e.g., benzene cations) can be tracked using a CTFS technique analogous to the electron photodetachment technique discussed in the caption to Fig. 3c. Photoexcited solute radical cations oxidize the solvent and their dynamics can be followed through the observation of increase in conductivity because of generation of the high-mobility solvent hole; this increase is analyzed as a function of the delay time of the excitation pulse (From Ref. 76.)

reactions obtained from the temperature dependencies of equilibria parameters [74]. An explanation proposed was that the rate constants are controlled by the height of the activation barrier determined by the difference in the *vertical* IP of the solute and the adiabatic IP of the solvent (Sec. 2) [74]. A similar mechanism accounts for the chemical behavior of the high-mobility solvent radical anion in sc CO_2 (vide supra) [18].

Class (3) reactions include proton-transfer reactions of solvent holes in cyclohexane and methylcyclohexane [71,74,75]. The corresponding rate constants are 10–30% of the fastest class (1) reactions. Class (4) reactions include proton-transfer reactions in *trans*-decalin and *cis–trans* decalin mixtures [77]. Proton transfer from the decalin hole to aliphatic alcohol results in the formation of a C-centered decalyl radical. The proton affinity of this radical is comparable to that of a single alcohol molecule. However, it is less than the proton affinity of an alcohol dimer. Consequently, a complex of the radical cation and alcohol monomer is relatively stable toward proton transfer; when such a complex encounters a second alcohol molecule, the radical cation rapidly deprotonates. Metastable complexes with natural lifetimes between 24 nsec (2-propanol) and 90 nsec (*tert*-butanol) were observed in liquid *cis*- and *trans*-decalins at 25°C [77]. The rate of the complexation is one-half of that for class (1) reactions; the overall decay rate is limited by slow proton transfer in the 1:1 complex. The rate constant of unimolecular decay is $(5–10) \times 10^6 \text{ sec}^{-1}$; for primary alcohols, bimolecular decay via proton transfer to the alcohol dimer prevails. Only for secondary and ternary alcohols is the equilibrium reached sufficiently slowly that it can be observed at 25°C on a time scale of > 10 nsec. There is a striking similarity between the formation of alcohol complexes with the solvent holes (in decalins) and solvent anions (in sc CO_2).

A detailed analysis of the thermodynamics and energetics of the complexation reactions is given in Ref. 77. The forward reaction has near-zero activation energy, whereas the proton transfer within the complex is thermally activated (20–25 kJ/mol). The stability of the complex increases with the carbon number of the alcohol; the standard heat of the complexation decreases in the opposite direction (from -39 kJ/mol for ethanol to -25 kJ/mol for *tert*-butanol). Complexes of *cis*-decalin$^+$ are more stable than complexes of *trans*-decalin$^+$ because for the former, the standard reaction entropy is 35 J mol^{-1} K^{-1} more positive. The decrease in the entropy is small for both decalins (> -80 J mol^{-1} K^{-1}) and approaches zero for higher alcohols. Similarly small changes in the standard entropy were observed for class (2) reactions of the methylcyclohexane hole [69]. Because the molecular complex formation can only reduce the degrees of freedom, to account for the small change in the entropy, there must be an increase in the solvent disorder. This is consistent with a hole ordering solvent molecules around itself. When the positive charge is compensated, the solvent becomes disordered, and the reaction entropy increases. The same effect is expected to occur for all solvent radical ions considered in this chapter.

Extremely slow class (4) reactions were observed for scavenging of (a) cyclohexane hole by cyclopropane [60] and (b) cyclohexane and decalins holes by O_2 [75]. H atom transfer from the hole to O_2 and H_2^- transfer from cyclopropane to the hole were suggested as the possible reaction mechanisms.

In conclusion, the behavior of high-mobility solvent anions and cations is similar. Both occur only in liquids whose molecules have rigid bodies and exhibit little or no conformational dynamics. Neither occurs in liquids where solvent radical ions have a strong tendency to form dimers with neutral solvent molecules. Both migrate by rapid hopping—sometimes over the entire liquid range of the solvent—and involve charge delocalization over several molecules. The activation energies and the degree of delocalization are roughly the same. Delocalization is required for the hopping to be rapid because it reduces geometric adjustment to charge placement and thereby decreases the activation barrier for charge transfer. Both species rapidly react with electron donors/acceptors with rate constants that are determined only by vertical IP_g or EA_g of the solute. In nonpolar liquids, both species display a strong tendency to form metastable complexes with polar molecules, such as alcohols and nitriles, in which the charge is electrostatically bound to the solute dipole. With respect to this propensity, the high-mobility ions are similar to solvated electrons in saturated hydrocarbons. Even in polar solvents, solvent anions (e.g., the dimer anion in acetonitrile) are protonated only after formation of a complex with the alcohol monomer; the transfer occurs when a second alcohol molecule encounters the complex [30].

6. CONCLUDING REMARKS

The important lesson of this chapter is that there are many ways in which a charge can be localized in a molecular system, and quite a few liquids localize electrons and holes in ways that defy easy classification. One does not need to look far for such "exotic" systems; ordinary solvents will do. In liquid acetonitrile [30,46], a high-energy electron state, a cavity electron, coexists in a dynamic equilibrium with a low-energy state, a dimer radical anion. In liquid benzene [45], the negative charge can migrate both by thermal emission into the CB and by degenerate electron hopping, depending on the pressure. Actually, most liquids seem to exhibit unique charge dynamic properties; there are few general rules.

In the previous four sections, several solvent radical ions that cannot be classified as molecular ions ("a charge on a solvent molecule") were examined. These delocalized, multimer radical ions are intermediate between the molecular ions and "cavity electrons," thereby bridging the two extremes of electron (or hole) localization in a molecular liquid. While solvated electrons appear only in negative-EA_g liquids, delocalized solvent anions appear both in positive and negative-EA_g liquids. Actually, from the structural standpoint, trapped electrons in low-temperature alkane and ether glasses [2] are closer to the multimer anions because their stabilization requires a degree of polarization in the molecules that is incompatible with the premises of one-electron models.

How general is the formation of multimer solvent ions? We reiterate the argument made in Sec. 3 that very few systems apart from water, alcohols, saturated hydrocarbons, and ethers have been studied by pulse radiolysis and laser photolysis, and for most of these liquids the ionic species observed were not primary ions. The incidence of high-mobility primary ions among neat organic liquids is actually high. It should also be kept in mind that if a given liquid does not yield high-mobility solvent ions under normal conditions, this does not necessarily hold for other conditions. High-mobility solvent anions in sc CO_2 occur only in the supercritical phase [18,20]; in the cold liquid, the binding of the dimer anion core is too strong for the rapid charge hopping to occur. High-mobility solvent anions in benzene [45] are observed only under high-pressure conditions, etc. Furthermore, as discussed above, many organic liquids yield solvent ions that are short-lived (< 10 nsec), and their dynamic properties cannot be studied using existing pulse radiolysis techniques. Finally, only in a small subset of liquids (whose molecules have rigid bodies and whose ions do not dimerize) can the delocalization of the excess charge be observed through faster-than-Brownian-diffusion hopping.

The authors believe that the formation of "peculiar" solvent ions is common; however, only in a handful of cases can one clearly demonstrate that such ions are formed. Far from being exotic species, these ions may constitute the rule, whereas the textbook species, solvated electrons and molecular ions, could be rare exceptions. That these exceptions loom large in the collective mind of chemical physicists is because aqueous solutions surround us in everyday life, and most radiation chemistry studies have been carried out on aqueous solutions at 25°C. If human beings were made of sc CO_2, hydrated electrons would be considered marginal and exotic. An alien living in an acetonitrile ocean would be correct in pointing out that terrestrial radiation chemists pay too much attention to water, whose properties are unique. An intelligent bacterium living in a hydrothermal vent would have difficulty understanding our interest in the "normal" conditions. In short, radiation chemistry, as we know it, is biased. Arguably, overcoming this bias, by expanding the range of physical conditions and the number of systems studied, will rejuvenate the twenty-first century radiation chemistry.

ACKNOWLEDGMENT

The preparation of this review was supported under contract No. W-31-109-ENG-38 with US-DOE Office of Basic Energy Sciences, Division of Chemical Sciences.

REFERENCES

1. Ferradini, C., Jay-Gerin, J.-P., Eds. *Excess Electrons in Dielectric Media*. Ann Arbor: CRC Press, 1991; Pikaev, A.K. *The Solvated Electron in Radiation Chemistry*. Israel Program for Scientific Translations: Jerusalem, 1971.
2. Kevan, L. J. Phys. Chem. 1978, *82*, 1144; Kevan, L.; Ichikawa, T. J. Phys. Chem. 1980, *84*, 3260; Kimura, T.; Fueki, K.; Narayana, P.A.; Kevan, L. Can. J. Chem. 1977, *55*, 1940; Feng, D.-F.; Kevan, L.; Yoshida, H. J. Chem. Phys. 1974, *61*, 4440; Nishida, M. J. Chem. Phys. 1977, *65*, 5242; Noda, S.; Kevan, L.; Fueki, K. J. Phys. Chem. 1975, *79*, 2866.
3. Shida, T.; Haselbach, E.; Bally, T. Acc. Chem. Res. 1984, *17*, 180 [Note that Hal radicals that are formed by fragmentation of halocarbon radical anions readily oxidize low-IP solutes: see, for example, Ushida, K. et al. J. Phys. Chem., A 1999, *103*, 4680].
4. Bally, T.; Bernhard, S.; Matzinger, S.; Truttmann, L.; Zhu, Z.; Roulin, J.-L.; Marcinek, A.; Gebicki, J.; Williams, F.; Chen, G.-F.; Roth, H.D.; Torsten, T. Chem. A Eur. J. 2000, *6*, 849; this recent paper illustrates the possibilities and synergetic methods used in the studies of organic radical cations and their rearrangement patterns.
5. See, for example, Bally, T. Chimia 1994, *48*, 378.
6. Werst, D.W.; Trifunac, A.D. In *Radiation Chemistry: Present Status and Future Prospects*; Jonah, C.D., Rao, B.S.M., Eds.; Amsterdam: Elsevier, 2001; 395 pp.
7. See, for example, reviews by Asmus, K.-D.; von Sonntag, C.; Schuchmann, H.-P.; Houee-Levin, C.; Sicard-Roselli, C. In *Radiation Chemistry: Present Status and Future Prospects*; Jonah, C.D.; Rao, B.S.M., Eds.; Elsevier: Amsterdam 2001; pp. 341, 513, and 553, respectively.
8. See, for example, Brede, O. Res. Chem. Intermed. 2001, *27*, 709 and references therein for pulse radiolysis of phenols liquid halocarbons and alkyl halides and Lomoth, R.; Brede, O. Chem. Phys. Lett. 1998, *288*, 47 for pulse radiolysis of DNA bases in liquid acetone.
9. See, for example, Todd, W.P.; Dinnocenzo, J.P.; Farid, S.; Goodman, J.L.; Gould, I.R. J. Am. Chem. Soc. 1991, *113*, 3601 and other publications from the Rochester group.
10. Greatorex, D.; Kemp, T.J. Trans. Faraday Soc. 1971, *67*, 56; Cox, A.; Kemp, T.J. J. Chem. Soc., Faraday Trans. 1 1975, *71*, 2490; Langford, C.H.; Carey, J.H. Can. J. Chem. 1975, *53*, 2430 and 2436; Ferraudi, G.; Endicott, J.F.; Barber, J. J. Am. Chem. Soc. 1975, *97*, 6406; Kiseleva, O.B.; Plyushin, V.F.; Bazhin, N.M. High Energy Chem. 1976, *12*, 77; Dzuba, S.A.; Raitsimring, A.M.; Tsvetkov, Yu.D. Chem. Phys. 1979, *44*, 357; Fadnis, A.G.; Kemp, T.J. J.

Chem. Soc. Dalton Trans. 1989, 1237; Cunningham, J.; Srijaranai, S. J. Photochem. Photobiol. A 1990, *55*, 219.

11. See, for example, Peon, J.; Tan, X.; Hoerner, J.D.; Xia, C.; Luk, Y.F.; Kohler, B. J. Phys. Chem. A 2001, *105*, 5768.

12. Shkrob, I.A.; Sauer, M.C., Jr.; Trifunac, A.D. J. Phys. Chem. B 1999, *103*, 4773.

13. Shkrob, I.A.; Sauer, M.C., Jr.; Liu, A.D.; Crowell, R.A.; Trifunac, A.D. J. Phys. Chem. A 1998, *102*, 4976.

14. For recent femto- and pico- second optical spectroscopy studies of CTTS in aqueous and alcohol solutions, see Refs. 1–11 in Ref. 46; for recent theoretical studies of CTTS dynamics, see Refs. 10 and 11 in Ref. 30.

15. Dimitrijevic, N.M.; Bartels, D.M.; Jonah, C.D.; Takahashi, K. Chem. Phys. Lett. 1999, *309*, 61.

16. Dimitrijevic, N.M.; Takahashi, K.; Bartels, D.M.; Jonah, C.D.; Trifunac, A.D. J. Phys. Chem. A 2000, *104*, 568.

17. Takahashi, K.; Sawamura, S.; Dimitrijevic, N.M.; Bartels, D.M.; Jonah, C.D. J. Phys. Chem. A 2002, *106*, 108.

18. Shkrob, I.A.; Sauer, M.C., Jr. J. Phys. Chem. B 2001, *105*, 4520.

19. Shkrob, I.A. J. Phys. Chem. A 2002, *106*, 11871.

20. Shkrob, I.A.; Sauer, M.C., Jr.; Jonah, C.D.; Takahashi, K. J. Phys. Chem. A 2003, *106*, 11855.

21. Itoh, K.; Holroyd, R.A.; Nishikawa, M. J. Phys. Chem. A 2001, *105*, 703.

22. Tucker, S.C. Chem. Rev. 1999, *99*, 391.

23. Klots, C.E.; Compton, R.N. J. Chem. Phys. 1978, *69*, 1636.

24. Bowen, K.H.; Eaton, J.G. *The Structure of Small Molecules and Ions*; Naamna, R., Vagar, Z., Eds.; Plenum: New York, 1987; 147 pp.; DeLuca, M.J.; Niu, B.; Johnson, M. J. Chem. Phys. 1988, *88*, 5857; Tsukuda, T.; Johnson, M.; Nagata, T. Chem. Phys. Lett. 1997, *268*, 429; Saeki, M.; Tsukuda, T.; Iwata, S.; Nagata, T. J. Chem. Phys. 1999, *11*, 6333.

25. Jacobsen, F.M.; Freeman, G.R. J. Chem. Phys. 1986, *84*, 3396.

26. Zhou, M.; Andrews, L. J. Chem. Phys. 1999, *110*, 2414 and 6820; Thompson, W.E.; Jacox, M.E. J. Chem. Phys. 1999, *111*, 4487; J. Chem. Phys. 1989, *91*, 1410.

27. Holroyd, R.A.; Gangwer, T.E.; Allen, A.O. Chem. Phys. Lett. 1975, *31*, 520; Lukin, L.V.; Yakovlev, B.S. High Energy Chem. 1978, *11*, 440.

28. Itoh, K.; Nakagawa, K.; Nishikawa, M. Radiat. Phys. Chem. 1988, *32*, 221; J. Chem. Phys. 1986, *84*, 391; Nishikawa, M.; Holroyd, R.A.; Sowada, U. Chem. Phys. 1980, *72*, 3081.

29. Shkrob, I.A.; Sauer, M.C., Jr. J. Phys. Chem. B 2001, *105*, 7027.

30. Shkrob, I.A.; Sauer, M.C., Jr. J. Phys. Chem. A 2002, *106*, 9120.

31. Baxendale, J.H.; Sharpe, P.H.G. Chem. Phys. Lett. 1976, *41*, 440; Ito, M.; Kimura, T.; Fueki, K. Can. J. Chem. 1981, *59*, 2803.

32. Baxendale, J.H.; Rasburn, E.J. J. Chem. Soc. Faraday Trans. I 1974, *70*, 705.

33. Kenney-Wallace, G.A.; Jonah, C.D. J. Phys. Chem. 1982, *86*, 2572; Gangwer, T.E.; Allen, A.O.; Holroyd, R.A. J. Phys. Chem. 1977, *81*, 1469.

34. Gee, N.; Freeman, G.R. J. Chem. Phys. 1989, *90*, 5399; Mogensen, O.-E. J. Chem. Phys. 1974, *60*, 998.

35. Azuma, T.; Washio, M.; Tabata, Y.; Ito, Y.; Nishiyama, K.; Miyake, Y.; Nagamine, K. Radiat. Phys. Chem. 1989, *34*, 659.

36. Tsukuda, T.; Hirose, T.; Nagata, T. Chem. Phys. Lett. 1997, *279*, 179.

37. Maeyama, T.; Oikawa, T.; Tsumura, T.; Mikami, N. J. Chem. Phys. 1998, *108*, 1368.

38. Zhang, S.W.; Zhang, C.G.; Yu, Y.T.; Mao, B.Z.; He, F.C. Chem. Phys. Lett. 1998, *304*, 265; Hiraoka, K.; Fujimaki, S.; Aruga, K.; Yamabe, S. J. Phys. Chem. 1994, *98*, 1802; Sanov, A.; Lineberger, W.C.; Jordan, K.D. J. Phys. Chem. A 1998, *102*, 2509.

39. Lea, J.S.; Symons, M.C.R. J. Chem. Soc. Faraday Trans. I 1988, *84*, 1181.

40. Zhou, M.; Andrews, L. J. Chem. Phys. 2000, *112*, 6576.

41. van den Ende, C.A.M.; Nyikos, L.; Sowada, U.; Warman, J.M.; Hummel, A. J. Electrost. 1982, *12*, 97; Radiat. Phys. Chem. 1982, *19*, 297; J. Phys. Chem. 1980, *84*, 1155.

42. Schiller, R.; Nyikos, L. J. Chem. Phys. 1980, *72*, 2245.

43. Badger, B.; Brocklehurst, B. Trans. Faraday Soc. 1969, 65, 2582; Miller, J.H.; Andrews, L.; Lund, P.A.; Schatz, P.N. J. Chem. Phys. 1980, 73, 4932; Mehnert, R. In *Radical Ionic Systems*. Kluwer: Amsterdam, 1991; 231 p.; Mehnert, R.; Brede, O. Radiat. Phys. Chem. 1985, 26, 353; Mehnert, R.; Brede, O.; Naumann, W. Ber. Bunsenges. Phys. Chem. 1984, 88, 71.

44. Anisimov, O.A.; Grigoryants, V.M.; Molin, Yu. N. Chem. Phys. Lett. 1980, 74, 15; Molin, Yu. N.; Anisimov, O.A. Radiat. Phys. Chem. 1983, 21, 77; Grygoryants, V.M.; McGrane, S.D.; Lipsky, S. J. Chem. Phys. 1998, 109, 7355; Werst, D.W. Chem. Phys. Lett. 1993, 202, 101; see also Ref. [72].

45. Itoh, K.; Holroyd, R.A. J. Phys. Chem. 1990, 94, 8850.

46. Xia, C.; Peon, J.; Kohler, B. J. Chem. Phys. 2002, 117, 8855.

47. Illenberger, E. Chem. Rev. 1992, 92, 1589; Abdoul-Carime, H.; Bouteiller, Y.; Desfrançois, C.; Philippe, L.; Schermann, J.P. Acta. Chem. Scand. 1997, 51, 145; Desfrançois, C.; Abdoul-Carmie, H.; Adjouri, C.; Khelifa, N.; Schermann, J.P. Europhys. Lett. 1994, 26, 25.

48. Renner, T.A.; Blander, M. J. Phys. Chem. 1977, 81, 857; Siebers, J.G.; Buck, U.; Beu, T.A. Chem. Phys. 1998, 129, 549.

49. Mitsuke, K.; Kondow, T.; Kuchitsu, K. J. Phys. Chem. 1986, 90, 1505.

50. Barrow, M.J. Acta Crystallogr. B 1981, 37, 2239; Torrie, B.H.; Powell, B.M. Mol. Phys. 1982, 75, 613.

51. Williams, F.; Sprague, E.D. Acc. Chem. Res. 1982, 15, 408.

52. Shkrob, I.A.; Takeda, K.; Williams, F. J. Phys. Chem. A 2002, 106, 9132.

53. Bell, I.P.; Rodgers, M.A.J.; Burrows, H.D. J. Chem. Soc. 1976, 315.

54. Blandamer, M.J.; Catterall, R.; Shiddo, L.; Symons, M.C.R. J. Chem. Soc. 1964, 4357; Fox, M.F.; Hayon, E. Chem. Phys. Lett. 1974, 25 511; J. Chem. Soc. Faraday Trans. 1 1976, 72, 1990.

55. Hirata, Y.; Mataga, N.; Sakata, Y.; Misumi, S. J. Phys. Chem. 1983, 87, 1493; J. Phys. Chem. 1982, 86, 1508; Hirata, Y.; Mataga, N. J. Phys. Chem. 1983, 87, 1680.

56. Michel, H.; Lippert, E. In *Organic Liquids*; Buckingham, A.D., Lippert, E., Bratos, S., Eds.; Wiley: New York, 1978; 293 pp.; see also p. 13; Whittenburg, S.L.; Wang, C.H. J. Chem. Phys. 1977, 66, 4255; Knozinger, K.; Leutloff, D.; Wittenbeck, R. J. Mol. Struct. 1980, 60, 115; Kovacs, H.; Kowalewski, J.; Maliniak, A.; Stilbs, P. J. Phys. Chem. 1989, 93, 962.

57. Markovich, G.; Perera, L.; Berkowitz, M.L.; Cheshnovsky, O. J. Chem. Phys. 1996, 105, 2675; Ayala, R.; Martinez, J.M.; Pappalardo, R.R.; Marcos, E.S. J. Phys. Chem. A 2000, 104, 2799.

58. Shkrob, I.A.; Sauer, M.C., Jr.; Trifunac, A.D. In *Radiation Chemistry: Present Status and Future Prospects*; Jonah, C.D., Rao, B.S.M., Eds.; Amsterdam: Elsevier, 2001; 175 pp.

59. Trifunac, A.D.; Sauer, M.C., Jr.; Shkrob, I.A.; Werst, D.W. Acta Chem. Scand. 1997, 51, 158.

60. Warman, J.M. In *The Study of Fast Processes and Transient species by Electron-Pulse Radiolysis*; Baxendale, J.H., Busi, F., Eds.; Reidel: Dordrecht, The Netherlands, 1982; 433 pp.

61. Borovkov, V.I.; Usov, O.M.; Kobzeva, T.V.; Bagryanskii, V.A.; Molin, Yu. N. Dokl. Phys. Chem. (Engl.) 2002, 384, 97.

62. Shkrob, I.A.; Sauer, M.C., Jr.; Trifunac, A.D. J. Phys. Chem. 1996, 100, 5993; Shkrob, I.A.; Trifunac, A.D. J. Phys. Chem. 1996, 100, 14681.

63. Tadjikov, B.M.; Stass, D.V.; Usov, O.M.; Molin, Yu. N. Chem. Phys. Lett. 1997, 273, 25; Usov, O.M.; Stass, D.V.; Tadjikov, B.M.; Molin, Yu. N. J. Phys. Chem. A 1997, 101, 7711.

64. Brocklehurst, B. Radiat. Phys. Chem. 1997, 50, 213; J. Chem. Soc. Faraday Trans. 1997, 93, 1079.

65. The main problem with CCl_4 is that the absorption band, the yield, and the fate of solvent radical cation are poorly known because cation fragmentation (with the formation of CCl_3^{++} Cl) and contact ion pair formation (with Cl^-) are a serious concern. See Washio, M.; Yoshido, Y.; Hayashi, N.; Kobayashi, H.; Tagawa, S.; Tabata, Y. Radiat. Phys. Chem. 1989, 34, 115; Buehler, R.E. J. Phys. Chem. 1986, 90, 6293; Miyasaka, H.; Masuhara, H.; Mataga, N. Chem. Phys. Lett. 1985, 118, 459; Sumiyoshi, T.; Sawamura, S.; Koshikawa, Y.; Katayama, M. Bull. Chem. Soc. Jpn. 1982, 55, 2341; Brede, O.; Boes, J.; Mehnert, R. Ber. Bunsenges. Phys. Chem. 1980, 84, 63; Radiochem. Radioanal. Lett. 1982, 51, 47; Cooper, R.; Thomas, J.K. Adv. Chem. Ser. 1968, 82, 351.

66. Grigoryants, V.M.; Tadjikov, B.M.; Usov, O.M.; Molin, Yu. N. Chem. Phys. Lett. 1995, *246*, 392.

67. Mehnert, R. In *Radical Ionic Systems, Properties in Condensed Phase*; Lund, A., Shiotani, M., Eds; Dordercht, The Netherlands: Kluver, 1991; 231 pp.; Sviridenko, F.B.; Stass, D.V.; Molin, Yu. N. Chem. Phys. Lett. 1998, *297*, 343.

68. Shkrob, I.A.; Werst, D.W.; Trifunac, A.D. J. Phys. Chem. 1994, *98*, 13262; Shkrob, I.A.; Trifunac, A.D. J. Phys. Chem. 1994, *98*, 13262; Tadjikov, B.M.; Lukzen, N.N.; Anisimov, O.A.; Molin, Yu. N. Chem. Phys. Lett. 1990, *171*, 413; Tadjikov, B.M.; Melekhov, V.I.; Anisimov, O.A.; Molin, Yu. N. Radiat. Phys. Chem. 1989, *34*, 353; Melekhov, V.I.; Anisimov, O.A.; Veselov, V.A.; Molin, Yu. N. Chem. Phys. Lett. 1986, *127*, 97; Borovkov, V.I.; Bagryansky, V.A.; Yeletskikh, I.V.; Molin, Yu. N. Mol. Phys. 2002, *100*, 1379.

69. Shkrob, I.A.; Liu, A.D.; Sauer, M.C., Jr.; Trifunac, A.D. J. Phys. Chem. A 2001, *105*, 7211; Liu, A.D.; Shkrob, I.A.; Sauer, M.C., Jr; Trifunac, A.D. J. Phys. Chem. 1998, *51*, 273; Buehler, R.E. Res. Chem. Intermed. 1999, *25*, 259; Buehler, R.E.; Katsumura, Y. J. Phys. Chem. A 1998, *102*, 111; Katsumura, Y.; Azuma, T.; Quadir, M.A.; Domazou, A.S.; Buehler, R.E. J. Phys. Chem. 1995, *99*, 12814.

70. Tadjikov, B.M.; Stass, D.V.; Molin, Yu. N. J. Phys. Chem. A 1997, *101*, 377.

71. Lias, S.G.; Ausloos, P.; Horvath, Z. Int. J. Chem. Kinet. 1976, *8*, 725.

72. Stass, D.V.; Lukzen, N.N.; Tadjikov, B.M.; Grigoryants, V.M.; Molin, Yu. N. Chem. Phys. Lett. 1995, *243*, 533.

73. Shkrob, I.A.; Liu, A.D.; Sauer, M.C., Jr.; Schmidt, K.H.; Trifunac, A.D. J. Phys. Chem. 1998, *102*, 3363 and references therein.

74. Shkrob, I.A.; Liu, A.D.; Sauer, M.C., Jr.; Schmidt, K.H.; Trifunac, A.D. J. Phys. Chem. 1998, *102*, 3371.

75. Shkrob, I.A.; Sauer, M.C., Jr.; Trifunac, A.D. J. Phys. Chem. 1996, *100*, 7237; Sauer, M.C., Jr; Shkrob, I.A.; Yan, J.; Schmidt, K.H.; Trifunac, A.D. J. Phys. Chem. 1996, *100*, 11325.

76. Shkrob, I.A.; Sauer, M.C., Jr.; Trifunac, A.D. J. Phys. Chem. B 2000, *104*, 3760.

77. Shkrob, I.A.; Sauer, M.C., Jr.; Trifunac, A.D. J. Phys. Chem. B 1999, *103*, 4773; J. Phys. Chem. B 2000, *104*, 3752.

12

The Radiation Chemistry of Liquid Water: Principles and Applications

G. V. Buxton
University of Leeds, Leeds, England

SUMMARY

An overview of the radiation chemistry of water is given from the start of its systematic study some 60 years ago to the present day. Attention is confined to the effects of low linear energy transfer (LET) radiation at ambient temperature and pressure because these are the conditions most commonly used in the application of water radiolysis as a tool in general chemistry. After a brief historical perspective, a scheme for the radiation chemistry of water is presented together with a review of recent results of stochastic and deterministic modelling of the spur reactions that occur between 10^{-12} and 10^{-7} sec. This is followed by a detailed discussion of the evolution of the yields (G-values) of e_{aq}^-, H$^\bullet$, $^\bullet$OH, H$_2$, and H$_2$O$_2$ resulting from the spur reactions and how they are measured using scavengers, or by direct observation using pulse radiolysis. Next, the physical and chemical properties of e_{aq}^-, H$^\bullet$, and $^\bullet$OH are presented, followed by methods of converting these primary radicals to secondary ones of a single kind, either oxidizing or reducing, and covering a wide range of reduction potentials. From this information, one can determine what G-values to use for the radicals and can tune the reduction potential for the chemical system of interest. Finally, reference is made to a recently published textbook [*Radiation Chemistry. Present Status and Future Prospects*, Jonah, C. D., Rao, B. S. M., Eds.; Elsevier: Amsterdam, 2001] that gives details of the wide range of chemical studies that have been made using water radiolysis as a tool.

1. HISTORICAL PERSPECTIVE

Systematic studies of the effects of ionizing radiation on liquid water began in earnest in the 1940s, at the University of Chicago [1], as part of the work on the atomic bomb project. This project involved building a nuclear reactor to produce plutonium and, because of its neutron absorption cross-section, fast-flowing water was deemed to be the best method for cooling the reactor. Fast neutrons knock out protons from water molecules and these energetic charged particles ionize and excite further molecules to produce a multitude of free electrons

331

and positively charged species. The ejected electrons themselves can also have sufficient energy to cause ionization and excitation of water molecules. Qualitatively, whether the energetic particle is a fast neutron, charged particle, or photon (i.e., x-ray or γ-ray), the result is that a single interaction of such a particle causes several water molecules to be ionized or excited. These initial events are classed as radiation physics. The radiation chemistry comprises the sequence of events that lead to the final products of water decomposition resulting from the ionization and excitation processes. The yields of these final products do, however, depend on the nature of the original energetic particle. As will be seen later, this sequence of events is divided into a physicochemical stage, during which the initial products reach thermal equilibrium with the liquid, followed by a kinetically nonhomogeneous reaction stage.

The idea that ionizing radiation decomposes water into hydrogen atoms and hydroxyl radicals according to reaction (1) was suggested first by Risse [2] in 1929 and later by Weiss [3] in 1944:

$$H_2O \rightsquigarrow H^{\bullet} + {}^{\bullet}OH \tag{1}$$

In 1938, Fricke et al. [4] had shown that although x-rays seemed not to decompose pure water, they caused oxidation and reduction to take place in aqueous solutions. On the other hand, it was also known from the earliest days of radiochemistry that α-rays decompose liquid water to H_2, O_2, and H_2O_2. This dependence of the decomposition of water on the linear energy transfer (LET) of the ionizing radiation was first explained by Allen [5], who put forward the concept that the molecular products are formed in tiny regions of high radical concentrations, so-called "spurs," along the track of the ionizing particle. Essentially, there is a competition between the "back" reactions of the radicals in the spurs to form the molecules H_2, H_2O_2, and H_2O, and their escape by diffusion into the bulk liquid. This led to the development of the spur–diffusion model (see Section 2 below) and was a major contribution to the understanding of the radiation chemistry of water, which at that time was represented by reaction (2):

$$H_2O \rightsquigarrow H^{\bullet}, {}^{\bullet}OH, H_2, H_2O_2 \tag{2}$$

The yields of these so-called "primary" species, present at the time when radical combination in, and diffusive escape from, the spurs is complete, were obtained by adding solutes to the water to capture the radicals and by measuring the stable identifying products. It was from a number of these studies that it became clear that the reducing radical must exist in two forms, which turned out to be the hydrogen atom and the hydrated electron (e_{aq}^-). For example, Hayon and Weiss [6] found that the yields of H_2 and Cl^- produced by irradiating solutions of chloroacetic acid varied with pH in a manner that was consistent with the following reactions:

$$e_{aq}^- + H^+ \rightarrow H^{\bullet} \tag{3}$$

$$e_{aq}^- + ClCH_2CO_2H \rightarrow Cl^- + {}^{\bullet}CH_2CO_2H \tag{4}$$

$$H^{\bullet} + ClCH_2CO_2H \rightarrow H_2 + {}^{\bullet}CH(Cl)CO_2H \tag{5}$$

It is now generally agreed that for low LET radiation, such as x-rays, γ-rays, and fast electrons, the radiolysis of water can be represented to a good approximation by reaction (6):

$$4.36H_2O \rightsquigarrow 2.8e_{aq}^-, 0.62H, 2.8OH, 0.47H_2, 0.73H_2O_2, 2.8H^+ \tag{6}$$

The numbers in reaction (6) are the so-called "primary" yields (expressed here in units of 10^{-7} mol J^{-1}) of the radical and molecular products that are present when the spur processes are complete at ca. 10^{-7} sec. The original units for these yields, which are still in use, are molecules $(100 \text{ eV})^{-1}$, and the relationship between the two sets is 1 mol $J^{-1} \equiv 9.65 \times 10^6$ molecules $(100 \text{ eV})^{-1}$. Thus, reaction (6) may also be written as reaction (7):

$$4.2H_2O \rightsquigarrow 2.7e_{aq}^-, 0.6H, 2.7OH, 0.45H_2, 0.7H_2O_2, 2.7H^+ \tag{7}$$

In this chapter, the various radiation chemical yields, known as G-values, are defined as follows: $g(X)$ is the yield of the species in reactions (6) and (7); $G^o(X)$ is the yield of the initial products of water radiolysis at the end of the physicochemical stage; and $G(X)$ is an experimentally measured yield. In some publications G_X, rather than $g(X)$, is used to represent the primary yield of the species X. The yields in reactions (6) and (7) are numerically very similar, but one must be sure of the units that are being used to express them. In the older literature, it was common practice to quote G-values without units when the units were in fact molecules $(100 \text{ eV})^{-1}$.

Shortly after the discovery of the hydrated electron, Hart and Boag [7] developed the method of pulse radiolysis, which enabled them to make the first direct observation of this species by optical spectroscopy. In the 1960s, pulse radiolysis facilities became quite widely available and attention was focussed on the measurement of the rate constants of reactions that were expected to take place in the spurs. Armed with this information, Schwarz [8] reported in 1969 the first detailed spur–diffusion model for water to make the link between the yields of the products in reaction (7) at ca. 10^{-7} sec and those present initially in the spurs at ca. 10^{-12} sec. This time scale was then only partially accessible experimentally, down to ca. 10^{-10} sec, by using high concentrations of scavengers (up to ca. 1 mol dm^{-3}) to capture the radicals in the spurs. From then on, advancements were made in the time resolution of pulse radiolysis equipment from microseconds $(10^{-6}$ sec) to picoseconds $(10^{-12}$ sec), which permitted spur processes to be measured by direct observation. Simultaneously, the increase in computational power has enabled more sophisticated models of the radiation chemistry of water to be developed and tested against the experimental data.

More generally, water radiolysis provides a very versatile method of generating other free radicals and unstable intermediates such as hyperreduced metal ions. This, when combined with the technique of pulse radiolysis, has, without doubt, made an enormous contribution to our knowledge of one-electron redox chemistry in aqueous inorganic, organic, and biochemical systems that cannot be studied readily by thermal or photochemical methods. Most commonly, this means using radiation of low LET such as ^{60}Co γ-rays or beams of high-energy electrons. The effects of increasing LET and/or temperature on the chemical stage of water radiolysis are covered in Chaps. 14 and 23, respectively.

In this chapter, an outline of the radiation chemistry of water is given in Section 2. Section 3 describes how the yields of the radical and molecular products evolve with time, and indicates the bounds of useful experimental conditions for the use of water radiolysis as a tool in general chemistry. In these two sections, emphasis is placed on recent developments in modelling experimental data that shed new light on our understanding of the fundamental aspects of water radiolysis. Section 4 contains a summary of the chemical and physical properties of the primary radicals. In Section 5, examples are given to show how these radicals can be converted into secondary radicals, often of a single kind. This is an important consideration for the study of the chemistry of free radicals and metal ions in unusual oxidation states. Finally, reference is given in Section 6 to examples of applications that have resulted in significant advances being made in general chemistry.

2. SCHEME FOR THE RADIATION CHEMISTRY OF WATER

The approximate time scale of events initiated by the absorption of energy by water from the incident ionizing radiation is shown in the following scheme:

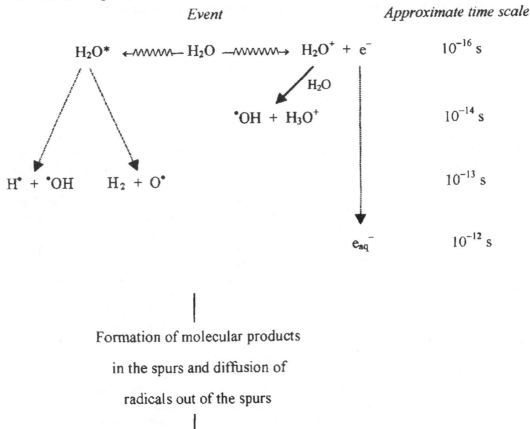

<table>
<tr><td align="center">*Event*</td><td align="right">*Approximate time scale*</td></tr>
</table>

Initially, a water molecule is ionized or electronically excited in about 10^{-16} sec or less, depending on the energy of the particle causing the ionization. The ejected electron generally has sufficient energy to ionize or excite further water molecules. This process continues until the energy of the ejected electron is no longer sufficient to raise a water molecule above its ground electronic state. This so-called subexcitation electron loses the rest of its excess energy by exciting vibrational and rotational modes of the solvent molecules.

At the end of the physical stage, which is within about 10^{-16} sec of the passage of the ionizing particle through the liquid, the track made by the particle contains H_2O^+, subexcitation electrons e_{se}^-, and electronically excited water molecules H_2O^* in small clusters called spurs. From about 10^{-16} to 10^{-12} sec, the following processes are thought to occur and comprise the physicochemical stage [9,10]:

$$H_2O^+ + e_{se}^- \rightarrow H_2O_{vib}^* \tag{I}$$

$$H^{\cdot} + {}^{\cdot}OH \tag{Ia}$$

$$H_2O^*_{vib} \longrightarrow H_2 + O^{\cdot}({}^1D) \quad \rightarrow \quad H_2 + H_2O_2(\text{or } 2\,{}^{\cdot}OH) \tag{Ib}$$

$$2H^{\cdot} + O^{\cdot}({}^3P) \tag{Ic}$$

$$H_2O^+ + H_2O \rightarrow H^+_{aq} + {}^{\cdot}OH \tag{II}$$

$$e^-_{se} + H_2O \rightarrow {}^{\cdot}OH + H^- \rightarrow {}^{\cdot}OH + H_2 + OH^- \tag{III}$$

$$e^-_{se} + H_2O \rightarrow e^-_{aq} \tag{IV}$$

$${}^{\cdot}OH + H^{\cdot} \tag{Va}$$

$$H_2O^*_{elec} \longrightarrow H_2 + O^{\cdot}({}^1D) \quad \rightarrow \quad H_2 + H_2O_2(\text{or } 2\,{}^{\cdot}OH) \tag{Vb}$$

$$2H + O^{\cdot}({}^3P) \tag{Vc}$$

Processes (I) and (II) account for H_2O^+, whereas processes (I), (III), and (IV) describe the fate of e^-_{se}. According to Kaplan et al. [11], process (I) produces water molecules in high vibrational levels of their electronic ground state. The remaining H_2O^+ reacts with water to form H^+_{aq} and ${}^{\cdot}OH$ in process (II). This ion–molecule reaction is known to occur in the gas phase with a rate constant of 8×10^{12} dm^3 mol^{-1} sec^{-1} [12], which, when extrapolated to liquid water, sets the lifetime of H_2O^+ in this medium at less than 10^{-14} sec. However, Hamill [13] pointed out that H_2O^+ initially has the structure of a neutral water molecule so that it may migrate rapidly over distances of a few molecular diameters by resonant electron transfer with a succession of neighboring water molecules.

The subexcitation electrons that escape process (I) either undergo a dissociative attachment to H_2O as in process (III), or thermalize and become solvated (process (IV)). Process (III), which is important in the gas phase, was suggested by Platzman [14] and Faraggi and Desalos [15] to be responsible for an apparently unscavengeable yield of molecular hydrogen in the radiolysis of liquid water. That a precursor of the hydrated electron is the source of at least part of $G(H_2)$ has been established by the finding that scavengers such as selenate ion, dichromate ion, and nitrate ion, which are known to react efficiently with a precursor of the hydrated electron (see Section 3.1.1), lower $G(H_2)$ to near zero at sufficiently high concentrations, for both low [16] and high [17] LET radiation. It has also been shown that when water is photoionized using a subpicosecond laser pulse [18], process (IV) occurs in less than 10^{-12} sec.

Very little is known about the decay channels for excited water molecules in the liquid phase. Consequently, they are generally assumed [9,10] to be essentially the same as those in the gas phase.

2.1. Modelling

In recent years, two different approaches, deterministic [9,19] and stochastic [10,20], have been used with a good level of success to model the radiation chemistry of water. Each approach leads to reasonable agreement between calculated results and experimental data obtained for a wide range of LET from room temperature up to ca. 300°C [9,10]. There are, however, fundamental differences between the two models. The deterministic model is based on the concept of an average spur [8,9,19,23] at the end of the physicochemical stage (ca. 10^{-12} sec), which contains the products of processes (I), (II), (III), (IV), and (V) in certain yields and spatial distributions, and in thermal equilibrium with the liquid. For low LET radiation, the spurs are treated as isolated spheres with usually a Gaussian distribution of each species, and for high LET as continuous cylindrical tracks with a Gaussian distribution having pseudo-two-dimensional symmetry. The initial yields (G^o) of the species and the width of their spatial distributions are chosen to fit with the experiment, but the two sets of parameters are not independent [24]. The nonhomogeneous kinetics of the spur processes of reaction and diffusion is treated as a set of coupled differential equations, which are solved numerically.

The stochastic model [10,20] uses Monte Carlo simulations of the events that take place during the physical stage up to ca. 10^{-15} sec, which, in principle, lead naturally to the spatial distributions at the end of the physicochemical stage (ca. 10^{-12} sec). However, assumptions have to be made about the energy loss processes of the incident ionizing radiation and the secondary electrons that are thereby generated because these processes are not well known quantitatively for liquid water. As in the deterministic approach, the parameters are adjusted to give a fit with experiment. Thereafter, the nonhomogeneous reaction kinetics between ca. 10^{-12} and 10^{-6} sec is treated by the method of Independent Reaction Times (IRT) [25]. This method assumes that the reaction time of each pair of reactants is independent of the presence of other reactants in the system. The important spur reactions are listed in Table 1.

The major difference between the deterministic and stochastic methods lies in the spatial distributions, and their dependence on temperature, of the species at the end of the physicochemical stage (ca. 10^{-12} sec). As mentioned above, the spatial distribution is assumed to be Gaussian in the deterministic approach. For $G^o(e_{aq}^-) = 4.78$ molecules $(100 \text{ eV})^{-1}$, which is the value derived by Schwarz [8], the best fit to the experimental data was obtained with the width parameter r^o of the Gaussian distributions at 25°C ca. equal to 2.3 nm [9] or 2.23 nm [19] for e_{aq}^- and the products of processes (Ia)–(Ic), and 0.85 nm for the other species [9,19]. It was later shown [24] that for $G^o(e_{aq}^-) = 4.1$ molecules $(100 \text{ eV})^{-1}$ (see Section 3.2.1), a good fit could be obtained by increasing r^o for e_{aq}^- to 3.8 nm. In contrast to this, a recent version of the stochastic model [10] generates a very asymmetrical distribution of hydrated electrons with most probable and average thermalization distances (equivalent to r^o) of ~2.5 and 9.2 nm, respectively at 25°C, and $G^o(e_{aq}^-) = 5.3$

Table 1 Reaction Scheme of the Chemical Stage in Water Radiolysis at 298 K for Low LET Radiation

Symbol	Reaction	$k/10^{10} \text{ dm}^3 \text{ mol}^{-1} \text{ sec}^{-1}$ [a]	$\Delta G/\text{molecule } (100 \text{ eV})^{-1}$ [b]
R1	$e_{aq}^- + H^+ \rightarrow H^{\bullet}$	2.3	0.55
R2	$e_{aq}^- + {}^{\bullet}OH \rightarrow OH^-$	3.0	1.15
R3	$H^+ + OH^- \rightarrow H_2O$	14.0[c]	0.65
R4	$e_{aq}^- + e_{aq}^- \rightarrow H_2 + 2OH^-$	$2k = 1.1$	0.27
R5	$e_{aq}^- + H^{\bullet} \rightarrow H_2 + OH^-$	2.5	0.10
R6	$H^{\bullet} + H^{\bullet} \rightarrow H_2$	$2k = 1.5$	0.007
R7	$H^{\bullet} + {}^{\bullet}OH \rightarrow H_2O$	1.5[d]	0.16
R8	${}^{\bullet}OH + {}^{\bullet}OH \rightarrow H_2O_2$	$2k = 1.1$	1.53
R9	$e_{aq}^- + H_2O_2 \rightarrow {}^{\bullet}OH + OH^-$	1.1	0.07

[a] *Source*: Ref. 59.
[b] ΔG is the change in yield of e_{aq}^-, $H^{\bullet}$, or ${}^{\bullet}OH$ for each reaction between 10^{-12} and 10^{-6} sec predicted by the deterministic model [9].
[c] *Source*: Laidler, K.J. *Chemical Kinetics*; McGraw-Hill, New York, 1965.
[d] *Source*: Ref. 63.

molecules $(100 \, eV)^{-1}$ at 10^{-12} sec. The model predicts that $G(e^-_{aq})$ has decreased by 11.6% to 4.7 molecules $(100 \, eV)^{-1}$ at 10^{-10} sec, which is about the shortest time at which experimental measurements have been made. In an earlier version [20], it was recognized that the initial distribution would be asymmetrical but would rapidly become Gaussian through Brownian motion. The initial distribution was therefore assumed to be Gaussian for convenience. The best fit with experiment was obtained with $r^o = 4$ nm and $G^o(e^-_{aq}) = 4.93$ molecules $(100 \, eV)^{-1}$. This version [20] also predicts $G(e^-_{aq}) = 4.9$ and 4.7 molecules $(100 \, eV)^{-1}$ at 10^{-12} and 10^{-10} sec, respectively, a fall of only 2.1%.

In their deterministic modelling, Swiatla-Wojcik and Buxton [9] scaled the change in the initial spur radii r^o with temperature t using Eq. (8):

$$r^o(t) = r^o(25^\circ C)[\rho(25^\circ C)/\rho(t)]^{1/3} \qquad (8)$$

where ρ is the density of water. The use of Eq. (8) results in r^o increasing by no more than 11.5% up to 300°C. In the stochastic model [10], the temperature dependence of the electron thermalization distance r_{th} was deduced from considerations of $g(H_2)$, which was found to be the yield most sensitive to r_{th}. The best fit with the experiment was obtained using Eq. (9):

$$r_{th} = (300^\circ C) = r_{th}(25^\circ C)/2 \qquad (r_{th}(t) \text{ decreasing linearly with } t) \qquad (9)$$

Hervé du Penboat et al. [10c] suggested that g-values at 300°C calculated using the deterministic approach [9] would not be sensitive to a 50% decrease in r_{th}, whereas when this condition is imposed, it is found [26] that $g(e^-_{aq})$ decreases from 3.4×10^{-7} to 2.8×10^{-7} mol J^{-1}, which is a significant change. On the other hand, the g-values of the other products are changed only slightly, the change being within the error of the measured values.

As would be expected, these results indicate that the thermalization distances and spatial distribution of the hydrated electron are key parameters in modelling the radiation chemistry of water. Although the stochastic approach is the more logical one to adopt, its present status does not appear to outweigh the advantages of using the simpler deterministic model to represent the essential features of water radiolysis over a wide range of conditions.

Before leaving the topic of modelling, it is pertinent to note that, at room temperature, the deterministic model [9] and the stochastic model [10b] both predict that ca. 50% of $g(H_2)$ comes from the sum of reactions (R4) and (R5), and negligible amounts from reaction (R6). The values for the deterministic model are given in Table 1. The corresponding values for the stochastic model are 0.24 and 0.15 molecules $(100 \, eV)^{-1}$, respectively [10,b,], and 0.28 and 0.14 molecules $(100 \, eV)^{-1}$, respectively [20]. On the other hand, $g(H_2O_2)$ is predicted to be almost entirely accounted for by reactions (R8) and (R9) in both types of model.

3. YIELDS FOR LOW LET RADIATION

Water radiolysis has proven to be a very useful method of generating free radicals whose reactions can be studied in a wide range of aqueous solutions. Indeed, a wealth of kinetic [27] and mechanistic data has been obtained in this way, almost invariably using low LET radiation, such as ^{60}Co γ-rays for continuous radiolysis and high-energy electron beams for pulse radiolysis. An important quantity in this kind of work is the yield of the radical or molecule under investigation because it is dependent on the experimental conditions. The g-values in Eqs. (6) and (7) have been derived from numerous scavenging studies where the scavenging power, which is defined as $k_{(X+S)}[S]$, is typically ca. $10^7 \, sec^{-1}$, is generally appropriate for studies of the radiolysis of dilute aqueous solutions on the order of

10^{-3} mol dm^{-3}. In more concentrated solutions, one needs to consider scavenging by the solute of e_{aq}^-, H$^{\cdot}$, and $^{\cdot}$OH within the spurs, whereas in very dilute solutions, possible depletion of the solute has to be taken into account. As will become clear in the following section, scavenging studies have played an important role in our current understanding of the chemical stage of water radiolysis.

A generalized reaction scheme for intraspur scavenging is represented by the following reactions, using the principal radicals e_{aq}^- and $^{\cdot}$OH in neutral water as examples:

$$e_{aq}^- + S_1 \rightarrow P_1 \tag{10}$$

$$^{\cdot}OH + S_2 \rightarrow P_2 \tag{11}$$

$$e_{aq}^- + P_2 \rightarrow S_2 \tag{12}$$

$$^{\cdot}OH + P_1 \rightarrow S_1 + OH^- \tag{13}$$

where S and P represent scavenger and product, respectively. As noted above, the effectiveness of a given scavenging solute S is measured by its scavenging power, which, for reactions (10) and (11), is $k_{10}[S_1]$ and $k_{11}[S_2]$ sec^{-1}, respectively. In principle, therefore, data for different S can be normalized in terms of their scavenging power; in practice, however, this procedure is not always valid. One reason for this is that when a single scavenger is present, the product of the scavenging reaction (e.g., reaction (10) for e_{aq}^-) may react to some extent with the sibling radical $^{\cdot}$OH in reaction (13). In such a case, $G(P_1)$ will be smaller than the scavenged yield of e_{aq}^-. A second reason is that some scavengers may react with a precursor of e_{aq}^-, but at relative rates different from those for their reactions with e_{aq}^- itself. Examples of such behavior are described in the following sections.

Although the time dependence of $G(e_{aq}^-)$ and $G(^{\cdot}OH)$ can be obtained by direct observation using pulse radiolysis, as well as by adding scavengers, the evolution of the yields of H$^{\cdot}$, H$_2$, and H$_2$O$_2$ can only be derived from scavenger studies.

3.1. The Molecular Yields $G(H_2)$ and $G(H_2O_2)$

$G(H_2)$

In dilute aqueous solution, $g(H_2) = 0.45$ molecule $(100 \text{ eV})^{-1}$ is well established. In more concentrated solutions of certain solutes, $G(H_2)$ decreases as the concentration of solute increases, showing that precursors of H$_2$ are being scavenged. The question to be answered is: Which are the relevant precursors? According to the scheme presented in Section 2, the possibilities are e_{sec}^-, H$_2$O*, H$^-$, e_{aq}^-, and H$^{\cdot}$. Several studies have been made in attempts to answer this question; the main ideas are summarized below.

Faraggi and Desalos [15] showed that $G(H_2)$ is independent of $[S_1]$ for $S_1 = $ Ni^{2+}, Co^{2+}, and Zn^{2+} although these cations all react efficiently with e_{aq}^-. They concluded, therefore, that H$^-$, rather than e_{aq}^- or H$^{\cdot}$, is the main precursor of H$_2$. On the other hand, Peled and Czapski [28] investigated the effects on $G(H_2)$ of high concentrations of reactive anions and neutral molecules as well as cations, and concluded that e_{aq}^- is the main precursor through reactions (R4) and (R5) in Table 1. The apparent conflict with the results of Faraggi and Desalos [15] for $S_1 = $ Ni^{2+}, Co^{2+}, and Zn^{2+} was effectively resolved by adding 10^{-3} mol dm^{-3} NO$_3^-$ to the solutions and allowing for the effect of ionic strength on k_{10}. The role of NO$_3^-$ is to oxidize the M$^+$ formed in reaction (10) back to M^{2+} to prevent the possibility of H$_2$ being produced in further reactions of M$^+$. Peled and Czapski [28] also showed that the results for anions, neutral molecules, and cations as scavengers for e_{aq}^- could be brought into near coincidence on a single curve of $G(H_2)$ vs. $\log(k_{10}[S_1])$ when ionic strength effects were properly taken into account.

The idea that e_{aq}^- is the main precursor of H_2 was supported by results obtained by Draganic and Draganic [29]. These are exemplified in Fig. 1, which shows that measurements made of the ratios $G(H_2)/g(H_2)$ and $G(H_2O_2)/g(H_2O_2)$ as a function of scavenging power for a number of different scavengers fall very close to a single curve. Draganic and Draganic [29] also found that when S_1 and S_2 are both present, raising $[S_1]$ at fixed $[S_2]$ causes $G(H_2O_2)$ to increase; the corresponding effect is also observed for $G(H_2)$. These results seemed to provide clear evidence that scavenging in the spurs of one of the primary radicals promotes the self-reaction of the other one.

Recent studies by Pastina et al. [16] of solutes having a wider range of reactivity toward the hydrated electron demonstrate that the single scavenging curve shown in Fig. 1 is not always obtained. Some of the results reported [16], which illustrate this point, are shown in Fig. 2. Here, the left-hand set of points is plotted against the scavenging power of S_1 for e_{aq}^-, whereas the right-hand set is plotted against the scavenging power of S_1 for the precursor(s) of e_{aq}^-, designated here as e_{pre}^-. There is much less scatter in the right-hand set of data, which led Pastina et al. [16] to conclude that "the yield of molecular hydrogen is better parameterized by the scavenging capacity for the precursors to the hydrated electron than by the scavenging capacity for the hydrated electron." In determining the rate coefficients for the reactions of the precursor of e_{aq}^-, Pastina et al. [16] used a procedure outlined previously by Pimblott and LaVerne [30]. However, for the same scavengers, Pastina et al. [16] quote values for the rate coefficients that are quite different from those reported by Pimblott and LaVerne [30]. The difference is greatest for H_2O_2, with values of zero [30] and 6.33×10^{13} $dm^3 \, mol^{-1} \, sec^{-1}$ [16], but there are significant differences between the values for other solutes. Pastina et al. [16] concluded that process (III) could make a significant contribution to $g(H_2)$ in the radiolysis of water (reaction (7)). Similarly, a good agreement among SeO_4^{2-}, $Cr_2O_7^{2-}$, and H_2O_2 has been obtained with high LET radiation [17] where $g(H_2)$, as is well

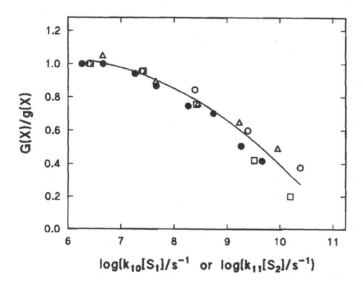

Figure 1 Decrease of $G(H_2)/g(H_2)$ or $G(H_2O_2)/g(H_2O_2)$ with increasing reactivity toward e_{aq}^- or ·OH. For H_2: $S_1 = H_2O_2 + 10^{-3} \, mol \, dm^{-3} \, Br^-$(O); $NO_3^- + 5 \times 10^{-4} \, mol \, dm^{-3} \, I^-$(□); $Cu^{2+} + 10^{-3} \, mol$ $dm^{-3} \, Br^-$(△). For H_2O_2: $S_2 = C_2H_5OH + 2.5 \times 10^{-3} \, mol \, dm^{-3} \, NO_3^-$(●). (From Ref. 29.)

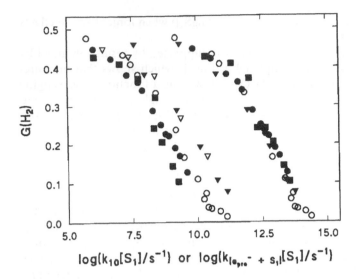

Figure 2 Decrease of $G(H_2)$ with increasing reactivity toward e_{aq}^- (left-hand data set) and its precursor, e^- (right-hand data set). $S_1 = H_2O_2$ ($\blacktriangledown$); SeO_4^{2-} ($\blacksquare$); MoO_4^{2-} ($\bullet$); $Cr_2O_7^{2-}$ (O). (From Ref. 16.) The left-hand set includes data for $S_1 = H_2O_2$ from Figure 1 for comparison ($\triangledown$).

known, is larger. However, process (III) is kinetically first-order, whereas the increase in $g(H_2)$ with increasing LET is expected to involve second-order processes because the local concentration of radiolysis products also increases with LET. On the basis that the lifetime of H_2O^+ is on the order of 10^{-13} sec, rather than 10^{-14} sec (see Section 2), LaVerne and Pimblott [17] concluded that process (I), followed by process (Ib), is even more significant than process (III) as a source of the yield of molecular hydrogen. These results imply that the precursor of e_{aq}^- is scavenged while it still has excess energy [17].

The sources of $g(H_2)$ according to deterministic [9] and stochastic [10] modelling are processes (I), (III), and (V) and also the spur reactions (R4) (R5) (R6). As shown in Table 1, the deterministic model indicates that approximately 50% of $g(H_2)$ is produced in reactions (R4) and (R5), and similar results are predicted by the stochastic model [10b,20]. The models also predict that the extent of reactions (R4) and (R5) increases with LET. For an increase in LET from 0.2 to 60 eV nm^{-1}, $g(H_2)$ increases to 0.27 and 0.42 molecules (100 eV)$^{-1}$ due to reactions (R4) and (R5), respectively [21]. These data, together with numerous experimental results, are not consistent with the conclusion [17] that H_2 is generated mainly through reaction of e_{se}^- with H_2O^+. There will, of course, be no e_{aq}^- produced under conditions where e_{se}^- is scavenged efficiently in sufficiently concentrated solution. However, it does not follow that reactions (R4) and (R5) do not contribute to $g(H_2)$ in dilute solution or pure water.

Reactions of e_{pre}^-. It is convenient to consider these reactions here. That the electron can react with solutes before it becomes solvated is clearly demonstrated by the data in Fig. 3, which were obtained by picosecond pulse radiolysis [31]. The data show that there is no good correlation between the C_{37} value of a solute and its rate constant k for reaction with e_{aq}^- measured at the same high concentrations of solute used to determine C_{37}. The parameter C_{37} is the concentration of solute that lowers the earliest measurable value of $G(e_{aq}^-)$ to 37% of the value obtained in the absence of solute. A linear correlation between C_{37} and k would be expected if the solute merely reacted with e_{aq}^- before this earliest time of

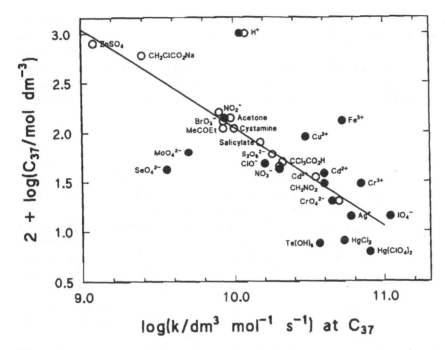

Figure 3 Dependence of C_{37} on $k(e_{aq}^- + S_1)$ obtained at $[S_1] = C_{37}$ mol dm^{-3}. Ref. 32: open points; Ref. 31: solid points.

measurement (50 psec). Such a correlation was reported originally by Lam and Hunt [32]; the straight line of slope -1 in Fig. 3 fits quite closely their data for several solutes. The notable exception is H$^+$.

It is now known that there are a number of different states in which the precursors e_{pre}^- of e_{aq}^- exist [33]. Recent studies [33] show that, using a static scavenging model, the efficiencies with which these precursors are scavenged are very similar for NO$_3^-$, Cd^{2+}, and SeO$_4^{2-}$, whether the energized state is reached by multiphoton ionization of water, photoionization of Fe(CN)$_6^{4-}$, optical excitation of e_{aq}^-, or ionization of H$_2$O by radiolysis. These results are considered to be consistent with there being no energy barrier to the transfer of the electron due to the high exoergicity of these reactions and the availability of open electronic and vibrational channels involving excited states of the scavenger [33]. Kee et al. [33] showed that the rate constant for reaction of the precursors of e_{aq}^- with NO$_3^-$, Cd^{2+}, and SeO$_4^{2-}$ is inversely proportional to the volume of e_{pre}^-. On this basis, they obtained a best-fit value of 9.02×10^{13} sec^{-1} for the static scavenging rate constant for e_{aq}^- (i.e., the rate of the scavenging reaction when e_{aq}^- is within the reaction distance of a scavenger molecule). An anomaly in the radiolysis results was noted by Kee et al. [33]. This disappears when rate coefficients for e_{pre}^- from Ref. 30, which were used by Kee et al. [33], are replaced by those from Ref. 16.

$G(H_2O_2)$

There seems to be little doubt from modelling [9,10,19,20] and experimental results (see Fig. 1) that the major precursor of the molecular yield of H$_2$O$_2$ is ·OH through reaction (R8) in Table 1. That reaction (R8) is a more important spur reaction than reactions (R4) and

(R5) is consistent with ˙OH having a narrower spatial distribution than e_{aq}^-. The unified curve for $G(H_2)/g(H_2)$ and $G(H_2O_2)/g(H_2O_2)$ in Fig. 1 may therefore be a somewhat fortuitous result of the spatial distributions of the radicals and the rate constants of their reactions with one another. If, as proposed by LaVerne and Pimblott [17], process I(b) is an important contributor to $g(H_2)$, then there should also be a significant initial yield of H_2O_2 resulting from the reaction of $O˙(^1D)$ with H_2O, but there is no experimental evidence for such a yield.

3.2. $G(e_{aq}^-)$

Numerous measurements have been made of $g(e_{aq}^-)$ for low LET radiation and most of these have been carried out under steady-state conditions using dilute (ca. 10^{-3} mol dm^{-3}) solutions of a suitable scavenger S_1 to form an identifiable stable product P_1, as in reaction (10). As the scavenging power $k_{10}[S_1]$ increases, $G(P_1)$ also increases because reaction (10) competes with other reactions of e_{aq}^- in the spur (see Table 1). From their extensive results on spur scavenging, Draganic and Draganic [34] obtained unifying curves that describe the dependence of the fractional changes in the radical and molecular yields $G(X)/g(X)$, on scavenging power. These curves are reproduced in Figs. 4 and 5. Draganic and Draganic [34] concluded on the basis of these results that secondary spur reactions such as reactions (12) and (13) are not significant for the scavengers they studied. Moreover, their data showed [34] that efficient scavenging of e_{aq}^- not only increases $G(P_1)$ in reaction (10) but also increases $G(˙OH)$ by reducing the extent of water reformation (see Table 1) and releasing more ˙OH for other reactions such as reaction (11) when the solution contains S_2 as well as S_1. This condition is known as cooperative scavenging. Similarly, the scavenging of ˙OH in reaction (11) leads not only to an increase in $G(P_2)$ but also in $G(e_{aq}^-)$. The results obtained by Draganic and Draganic [34] are presented in Figs. 4 and 5.

Other data for a number of different scavengers showing the effect on $G(P_1)$ of increasing the concentration of the scavenger S_1 are reproduced in Fig. 6 for nitrous oxide

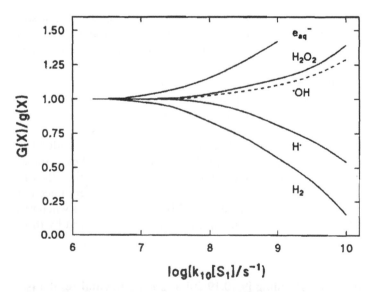

Figure 4 Dependence of the fractional yields $G(X)/g(X)$ of the primary products of water radiolysis (see reaction (7)) on the scavenging power of S_1. (From Ref. 34.)

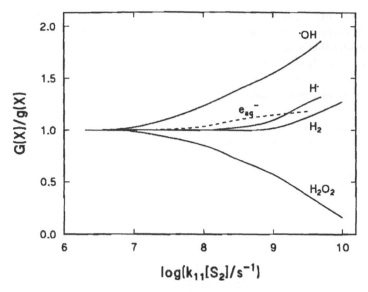

Figure 5 Dependence of the fractional yields $G(X)/g(X)$ of the primary products of water radiolysis (see reaction (7)) on the scavenging power of S_2. (From Ref. 34.)

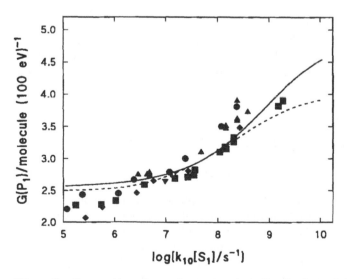

Figure 6 Comparison of experimental and predicted values of $G(P_1)$ from electron scavenging as a function of the scavenger power $k_{10}[S_1]$. $S_1 = N_2O$ (■); CH_3Cl (▲); $C(NO_2)_4$ (♦); glycylglycine (●); NO_3^- (O). The broken line is predicted from direct observation of the time dependence of $G(e_{aq}^-)$ in pulse radiolysis experiments. (From Ref. 49.) The solid line is the fit obtained by Pimblott and LaVerne [43] with the restriction that $G^o(e_{aq}^-) = 4.80$ molecules $(100 \text{ eV})^{-1}$.

[35–37], methyl chloride [38,39], nitrate ion [29,40], glycylglycine [41], tetranitromethane [37], and methyl viologen [40]. Contrary to the unifying curve obtained [34] for e_{aq}^- in Fig. 4, the data in Fig. 6 show considerable scatter, which is probably at least partially due to scavenging of electrons in their presolvated state as discussed in Section 3.1. Note that $G(P_1)$ continues to fall as $k_{10}[S_1]$ is decreased below 10^6 sec^{-1}.

The first attempt at providing a mathematical formulation of the effect of scavengers on G-values in aqueous solution was made by Balkas et al. [38] based on an equation developed by Warman et al. [42] for scavenging of ion pairs in nonpolar liquids. Balkas et al. [38] showed that the yield of chloride ion from reaction (14) can be represented by Eq. (15):

$$e_{aq}^- + CH_3Cl \rightarrow {}^{\bullet}CH_3 + Cl^-$$ (14)

$$G(Cl^-) = 2.55 + \frac{2.23(k[S]/\lambda)^{1/2}}{1 + (k[S]\lambda)^{1/2}}$$ (15)

In Eq. (15), $g(e_{aq}^-) = 2.55$ molecules $(100 \text{ eV})^{-1}$, $G^{\circ}(e_{aq}^-) = 4.78$ molecules $(100 \text{ eV})^{-1}$, and λ is a constant estimated [38] to be $8 \times 10^8 \text{ sec}^{-1}$. The term $(k[S]/\lambda)^{1/2}/(1 + (k[S]/\lambda)^{1/2})$ is referred to as the scavenger function $F(S)$. The value of $G^{\circ}(e_{aq}^-)$ was chosen to agree with the value of Schwarz [8]. The value of $g(e_{aq}^-) = 2.55$ molecules $(100 \text{ eV})^{-1}$ is a fit parameter that refers to very low values of scavenger power $k[S]$ (i.e., $F(S) \rightarrow 0$). This g-value is a little lower than the generally accepted value in Eq. (7), which is appropriate to dilute solutions where, typically, $k[S] = 10^7 \text{ sec}^{-1}$ and $F(S) \approx 0.1$.

The form of Eq. (15) assumes that $G(e_{aq}^-) = 2.55$ molecules $(100 \text{ eV})^{-1}$ is the yield of e_{aq}^- that escapes from the track into the bulk solution at infinite time. This is a reasonable assumption under most experimental conditions where $k_{10}[S_1] \geq 10^6 \text{ sec}^{-1}$. On the other hand, the data in Fig. 6 for lower scavenger power do suggest that the true yield that would escape from a low LET track in pure water is less than 2.55 molecules $(100 \text{ eV})^{-1}$.

La Verne and Pimblott [19] and Pimblott and La Verne [43] refined Eq. (15) and developed an analytical description of these effects of scavengers using the deterministic diffusion kinetic model outlined in Section 2. For a single scavenger, they showed that the dependence of the amount of scavenging reaction on the concentration of S could be better described by [22]:

$$G(X)_s = g(X) + (G^{\circ}(X) - g(X))F(S)$$ (16)

where X is the species of interest (e.g., e_{aq}^-, H_2, etc.) and $F(S)$ is given by:

$$F(S) = ((a[S])^{1/2} + a[S]/2)/(1 + (a[S])^{1/2} + a[S]/2)$$ (17)

Here $a = \alpha k_s$ where α, which is equal to $1/\lambda$ in Eq. (15), is equivalent to a phenomenological decay or formation time of the species X in the absence of scavenger and k_s is the rate constant of the scavenging reaction. Thus, the effects of different scavengers for a given species X can be normalized using the relative values of k_s. Applying this model to scavenging data for e_{aq}^-, LaVerne and Pimblott [19] and Pimblott and LaVerne [43] obtained a good fit to experimental data for the scavengers N_2O, NO_3^-, and CH_3Cl with $\alpha = 0.905$ nsec, $g(e_{aq}^-) = 2.55$ molecules $(100 \text{ eV})^{-1}$, and choosing $G^{\circ}(e_{aq}^-) = 4.80$ molecules $(100 \text{ eV})^{-1}$. The result is shown as the solid line in Fig. 6. They noted [19] that when no restriction was placed on the value of $G^{\circ}(e_{aq}^-)$, a best-fit value of 4.3 molecules $(100 \text{ eV})^{-1}$ was obtained for this parameter.

Pimblott and LaVerne [43] then extended their model to cooperative scavenging and obtained the following expression for cooperative scavenging of e_{aq}^-:

$$G(e_{aq}^-)(S_1, S_2) = 2.55 + 2.25F[(9.1 \times 10^{-10})s_1 + 1.2(F[5.5 \times 10^{-10})\{s_1 + s_2\}]$$

$$- F[(5.5 \times 10^{-10})s_1] \tag{18}$$

where s_1 and s_2 are the scavenging powers of S_1 and S_2, respectively, and the value of α for reaction (R2) in Table 1 is 5.5×10^{-10} sec. The first two terms on the right-hand side of Eq. (18) represent the scavenging of e_{aq}^- by the primary scavenger S_1; the last two terms represent the effect of the secondary scavenger S_2 for $^\cdot OH$ through its interference with the spur reaction $e_{aq}^- + {}^\cdot OH$. The maximum yield of this reaction was estimated [43] to be 1.2 molecules $(100 \, eV)^{-1}$; a very similar value was obtained by Swiatla-Wojcik and Buxton [9] in their deterministic modelling (see Table 1).

Time Dependence of $G(e_{aq}^-)$

The time dependence of the yield of hydrated electrons during the chemical stage of water radiolysis (ca. 10^{-12} to 10^{-6} sec) can be obtained from scavenging studies and by direct observation using pulse radiolysis. In the former case, the scavenger dependence $G(S)$ is converted into the time dependence using the inverse Laplace transform of $F(S)$ to $F(t)$, which is given by Eqs. (19) and (20):

$$F(t) = \exp(\lambda t)\text{erfc}((\lambda t)^{1/2}) \text{ for Eq. (15)} \tag{19}$$

$$F(t) = 2F_f(4.0\lambda t/\pi)^{1/2} \text{ for Eq. (17)} \tag{20}$$

In Eq. (20), F_f is the auxiliary function for the Fresnel integrals [44]. In practice, it is usual to choose empirical scavenging functions $F(S)$ that have analytical inverse Laplace transforms.

The first pulse radiolysis experiments to measure $G^\circ(e_{aq}^-)$ directly were made in the 1970s, with reported values of 4.0 ± 0.2 molecules $(100 \, eV)^{-1}$ at 30 psec [45] and 4.1 ± 0.1 molecules $(100 \, eV)^{-1}$ at > 200 psec [46]. The latter value was subsequently revised to 4.6 ± 0.2 molecules $(100 \, eV)^{-1}$ at 100 psec [47], and later a yield of 4.8 ± 0.3 molecules $(100 \, eV)^{-1}$ at 30 psec was reported by Sumiyoshi et al. [48]. The method of evaluating $G(e_{aq}^-)$ at these short times is either to use dosimetry [45,46] and the molar absorption coefficient of e_{aq}^-, or to compare the optical absorbance at short times with that observed at 10^{-7}–10^{-6} sec and take $G(e_{aq}^-) = 2.7$ molecules $(100 \, eV)^{-1}$ at this time [48]. The causes of the discrepancies between these pulse radiolysis values have been reviewed recently by Bartels et al. [49], who have also made new measurements of the spur decay of e_{aq}^-.

In their pulse radiolysis experiments, Bartels et al. [49] measured directly the decay of e_{aq}^- from 100 psec out to microseconds. By using the four exponential sum in Eq. (21) to fit the observed decay, they obtained a ratio $G^\circ(0)/G_{inf}$ of 1.59, effectively equivalent to $G^\circ(e_{aq}^-)/g(e_{aq}^-)$:

$$G^\circ(t)/G_{inf} = 1 + 0.090\exp(-t/139\text{nsec}) + 0.128\exp(-t/24.4\text{nsec})$$

$$+ 0.255\exp(-t/3.51\text{nsec}) + 0.118\exp(-t/0.480\text{nsec}) \tag{21}$$

From the time dependence of the optical absorbance due to e_{aq}^-, Bartels et al. [49] found that if they assumed $G_{inf} = 2.5$ molecules $(100 \, eV)^{-1}$, then they calculated $G(e_{aq}^-) = 2.7$ molecules $(100 \, eV)^{-1}$ for a scavenging power of 10^7 sec^{-1} (see Fig. 6). Thus, these data are in accord with $G(e_{aq}^-)$ in reaction (7) and with the first term on the right-hand side of Eq. (15). On the other hand, Bartels et al. [49] obtained $G^\circ(0)$ (i.e., $G^\circ(e_{aq}^-) = 4.0 \pm 0.2$ molecules $(100 \, eV)^{-1}$), some 20% lower than the value of Schwarz [8]. Their predicted scavenging curve is compared with that of Pimblott and LaVerne [43] in Fig. 6. Because of the scatter in the experimental data, the fit parameters of the calculated curves are not tightly defined. This is especially true for $G°(e_{aq}^-)$ because the maximum scavenging power that can be used experimentally is limited to ca. 10^{10} sec^{-1}, corresponding to ~100 psec in real time. A reason put forward by Bartels et al. [49] to account for $G(P_1) > G(e_{aq}^-)$ is that the electron may react with the scavenger in its presolvation state e_{pre}^-, as discussed in Section 3.1.

The experimentally determined value of $G°(e_{aq}^-) = 4.0 \pm 0.2$ molecules $(100 \text{ eV})^{-1}$ [49] has implications for the stochastic and deterministic models of the radiolysis of water because, as noted in Section 2, $G°(e_{aq}^-)$ at 1 psec is calculated to be 5.3 molecules $(100 \text{ eV})^{-1}$ [10] and 4.9 molecules $(100 \text{ eV})^{-1}$ [20]. On the other hand, for the deterministic model [9], it has been shown [24] that $G°(e_{aq}^-) = 4.1$ molecules $(100 \text{ eV})^{-1}$ is accommodated by increasing the initial width of the assumed Gaussian distribution of e_{aq}^- from 2.3 to 3.8 nm. Another factor to be considered is the extent to which e_{aq}^- decays in the time interval 1–100 psec, which at present is largely inaccessible experimentally. The ratio calculated from Eq. (21) at 100 psec (i.e., $G(100 \text{ psec})/G_{inf}$) is 1.56, indicating that e_{aq}^- decays by only 1.8% between 10^{-12} and 10^{-10} sec. This is quite close to the decay of 2.1% predicted by the stochastic simulations of Pimblott and LaVerne, assuming a Gaussian distribution of thermalized electrons with $r° = 4$ nm [20], but much smaller than the 11.6% decay obtained by Frongillo et al. [10b] (see Section 2). The deterministic model calculations predict decays of 8.0% and 3.1% for Gaussian distributions with $r° = 2.3$ nm [9] and 3.8 nm [24], respectively. In the deterministic model, $G°(e_{aq}^-)$ and $r°$ are adjustable parameters, whereas the stochastic treatment automatically generates the value of $G°(e_{aq}^-)$ and the distribution of thermalized electrons. If the results obtained by Bartels et al. [49] are correct, then the stochastic calculations [10,20] need to be revisited.

To make a complete interpretation of the scavenging data in Fig. 6, one has to consider the reactivity of the electron in the various states in which it can exist before it reaches its fully solvated state [33] as discussed in Section 3.1. Static scavenging must also be taken into account when e_{aq}^- is formed within the reaction distance of the solute so that it reacts before it can be observed [50]. Hence, it is not surprising that the data in Fig. 6 do not conform to the unified curve for $G(e_{aq}^-)$ in Fig. 4. Therefore, in chemical applications involving concentrated solutions, one should measure $G(P_1)$ in the system of interest. If this is not feasible, then it should be recognized that the value calculated from scavenging functions based only on the reactivity of e_{aq}^- could be in error by as much as 10%.

3.3. $G(^\bullet OH)$

By comparison with $G(e_{aq}^-)$, relatively few independent measurements of $G(^\bullet OH)$ have been made. In contrast to e_{aq}^-, only the relative change in $G(^\bullet OH)$ with time has been reliably measured by pulse radiolysis [51]. In practice, absolute values of $G(^\bullet OH)$ have been obtained from scavenger studies or by material balance (reaction (7)). Fig. 7 shows data for aerated solutions of formate ion [52] and hexacyanoferrate(II) [53] taken from Fig. 1 of Ref. 54. The data for formic acid, which were included by LaVerne and Pimblott [54], have been omitted here because they were obtained at low pH where the primary yields are different (see Section 3.4). The solid line shows the best fit obtained using Eqs. (16) and (17) and the broken line is the best fit when the term $a[S]/2$ is omitted from Eq. (17). The respective sets of parameters are $\alpha = 1.64$ and 1.69 nsec, $g(^\bullet OH) = 2.53$ and 2.50 molecules $(100 \text{ eV})^{-1}$, and $G°(^\bullet OH) = 4.48$ and 4.86 molecules $(100 \text{ eV})^{-1}$. These values differ significantly from those obtained by LaVerne and Pimblott [54], which were $\alpha = 0.258$ nsec, $g(^\bullet OH) = 2.66$ molecules $(100 \text{ eV})^{-1}$, and $G°(^\bullet OH) = 5.50$ molecules $(100 \text{ eV})^{-1}$. The reason for the difference is that LaVerne and Pimblott [54] chose $G°(^\bullet OH) = 5.50$ molecules $(100 \text{ eV})^{-1}$.

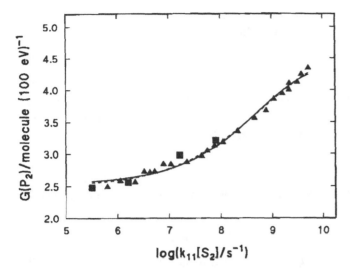

Figure 7 Comparison of experimental and predicted values of $G(P_2)$ from hydroxyl radical scavenging as a function of the scavenger power $k_{11}[S_2]$. $S_2 = HCO_2^-$ (■); $Fe(CN)_6^{4-}$ (▲). The full and broken lines are the best fits of Eqs. (16) and (17) without any restriction on the parameters (see text). (From Ref. 54.)

When no constraints were placed on the fitting parameters, $G^\circ(\cdot OH)$ was found to be 4.4 molecules $(100\ eV)^{-1}$ [54], which is closer to those given above for the data in Fig. 7. Subsequently, Pimblott and LaVerne [20] compared the results of their stochastic simulations with the scavenging data. However, their simulations predicted $G^\circ(e_{aq}^-) \approx 4.9$ molecules $(100\ eV)^{-1}$ and $G^\circ(\cdot OH) \approx 5.5$ molecules $(100\ eV)^{-1}$, respectively, which now seem to be too large when compared with the results and conclusions of Bartels et al. [49] (i.e., $G^\circ(e_{aq}^-) \approx 4.0$ molecules $(100\ eV)^{-1}$ and $G^\circ(\cdot OH) \approx 5.1$ molecules $(100\ eV)^{-1}$).

The question of the magnitude of $G^\circ(\cdot OH)$ has been addressed recently by Jay-Gerin and Ferradini [55] who concluded that a value of 4.6 ± 0.25 molecules $(100\ eV)^{-1}$ at 100 psec is consistent with the scavenger data in Fig. 7 and with stochastic simulation results, which give $G^\circ(\cdot OH)$ equal to 5.8 molecules $(100\ eV)^{-1}$ at 1 psec and $G(\cdot OH) = 4.8$ molecules $(100\ eV)^{-1}$ at 100 psec [10]. This represents a 17.2% decrease in $G(\cdot OH)$ over this period. The effect of using $G^\circ(e_{aq}^-) = 4.1$ instead of 4.78 molecules $(100\ eV)^{-1}$ in the deterministic model [9,24] is to make $G^\circ(\cdot OH)$ 4.9 molecules $(100\ eV)^{-1}$ instead of 5.5 molecules $(100\ eV)^{-1}$. The corresponding decreases in $G(\cdot OH)$ between 10^{-12} and 10^{-10} sec are 8.3% and 14.1%, respectively.

In contrast to the data in Fig. 7, Sutton et al. [56], using pulse radiolysis, measured $G(Br_2^{\cdot-})\varepsilon(Br_2^{\cdot-})$, where ε is the molar absorption coefficient, and found it to be independent of the concentration of Br^- over the range 2×10^{-3} to $1.0\ mol\ dm^{-3}$ in O_2-saturated solution at neutral pH. It was necessary to use a low dose per pulse (26 Gy) to exclude interradical combination during the growth of $Br_2^{\cdot-}$. Under these conditions, the formation of $Br_2^{\cdot-}$ occurs via reactions (22)–(24):

$$\cdot OH + Br^- \rightarrow HOBr^{\cdot-} \tag{22}$$

$$HOBr^{\cdot-} \rightleftharpoons OH^- + Br^\cdot \tag{23}$$

$$Br^\cdot + Br^- \rightleftharpoons Br_2^{\cdot-} \tag{24}$$

To explain their results, Sutton et al. [56] suggested that spur reactions (R2) and (R8) in Table 1 be replaced by reactions (25) and (26), which are equally efficient:

$$e_{aq}^- + Br^\bullet/Br_2^{\bullet-} \rightarrow Br^-/2Br^- \tag{25}$$

$$Br^\bullet + Br^\bullet \rightarrow Br_2 \tag{26}$$

Similarly, it was found that $G(P_2)$ is independent of $[S_2]$ for the $^\bullet OH$ scavengers thiocyanate ion [57] and iodide ion [58]. Clearly, there is good evidence that back reactions such as reaction (25) can occur in the spur and thus modify $G(P_2)$.

$G(^\bullet OH)$ in N_2O-Saturated Solution

In applications of water radiolysis to free radical chemistry, it is particularly convenient to use aqueous solutions saturated with N_2O, where e_{aq}^- is converted to $^\bullet OH$ via reactions (27) and (28) [59]:

$$e_{aq}^- + N_2O \rightarrow O^{\bullet-} + N_2; \qquad k_{27} = 9.1 \times 10^9\,dm^3\,mol^{-1}\,sec^{-1} \tag{27}$$

$$O^{\bullet-} + H_2O \rightleftharpoons {}^\bullet OH + OH^-; \qquad k_{28} = 1.7 \times 10^6\,dm^3\,mol^{-1}\,sec^{-1} \tag{28}$$

$$k_{-28} = 1.2 \times 10^{10}\,dm^3\,mol^{-1}\,sec^{-1}$$

Under normal conditions of room temperature and atmospherical pressure, the saturation concentration of N_2O is ca. $2.5 \times 10^{-2}\,mol\,dm^{-3}$ and its scavenging power is ca. 2.3×10^8 sec^{-1}. Thus, one has to consider that reactions (27) and (28) generate extra $^\bullet OH$ in the spurs because $k_{28}[H_2O] = 9.4 \times 10^7\,sec^{-1}$. This problem was addressed by Schuler et al. [60], who showed that the scavengeable yield of $^\bullet OH$ is described by the empirical equation, Eq. (29), which is analogous to Eq. (15):

$$G(^\bullet OH) = 5.2 + \frac{3.0(k[S]/\lambda)^{1/2}}{1 + (k[S]\lambda)^{1/2}}\ molecule\ (100eV)^{-1} \tag{29}$$

where $\lambda = 4.7 \times 10^8\,sec^{-1}$ and $k[S]$ is the scavenging power of the solute reacting with $^\bullet OH$ (i.e., $k_{11}[S_2]$). Thus, the $^\bullet OH$ that replaces e_{aq}^- in the spurs through reactions (27) and (28) also participates in spur reactions. Values of $G(^\bullet OH)$ calculated from Eq. (29) are plotted vs. scavenging power in Fig. 8.

3.4. pH Dependence of the Primary Yields $g(X)$

This is illustrated in Fig. 9 for low LET radiation. At pH > 3, the g-values are effectively constant; at pH < 3, the extent of reaction (R1) increases through scavenging of e_{aq}^- in the spurs. The observed increase in $g(e_{aq}^-) + g(H^\bullet)$ is probably due to the diffusion coefficient of $H^\bullet$ ($D_H = (7.0 \pm 1.5) \times 10^{-9}\,m^2\,sec^{-1}$ [61]) being larger than that of e_{aq}^- ($D_e = (4.90 \pm 0.02) \times 10^{-9}\,m^2\,sec^{-1}$ [62]), and reaction (R7) ($k_{R7} = 1.5 \times 10^{10}\,dm^3\,mol^{-1}\,sec^{-1}$ [63]) being slower than reaction (R2) ($k_{R2} = 3.0 \times 10^{10}\,dm^3\,mol^{-1}\,sec^{-1}$ [59]), so that conversion of e_{aq}^- to $H^\bullet$ in the spurs favors diffusive escape into the bulk liquid. The change in the individual values of $g(e_{aq}^-)$ and $g(H^\bullet)$ with pH has been measured [6] with chloroacetic as the scavenger, as described in Section 1. The data are shown in Fig. 10:

$$e_{aq}^- + ClCH_2CO_2H \rightarrow Cl^- + {}^\bullet CH_2CO_2H \tag{4}$$

$$H^\bullet + ClCH_2CO_2H \rightarrow H_2 + {}^\bullet CH(Cl)CO_2H \tag{5}$$

The values of $g(^\bullet OH)$ and $g(H_2O_2)$ also increase at pH < 3, as shown in Fig. 9. An increase in the yield of oxidizing equivalents is expected, of course, to maintain the balance

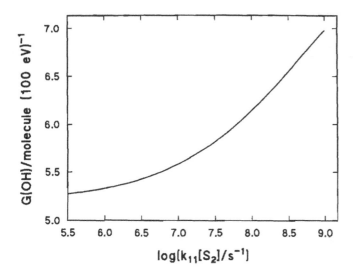

Figure 8 Values calculated from Eq. (29) of $G(\cdot OH)$ vs. scavenging power $k_{11}[S_2]$ in N_2O-saturated solution.

with the increase in $g(e_{aq}^-) + g(H^\cdot)$, but the increase in $g(H_2O_2)$ does suggest that reaction (R8) becomes more important when reaction (R7) replaces reaction (R2) at low pH.

The conversion of $H^\cdot$ to e_{aq}^- takes place at high pH through reaction (30):

$$H^\cdot + OH^- \rightarrow e_{aq}^-; \qquad k_{30} = 2.2 \times 10^7 \, dm^3 \, mol^{-1} \, sec^{-1} \, [59] \tag{30}$$

Although reaction (30) is too slow to interfere significantly with spur processes up to pH 14, it may compete with the reaction of $H^\cdot$ with other solutes in the bulk solution under these conditions and thus increase $G(e_{aq}^-)$.

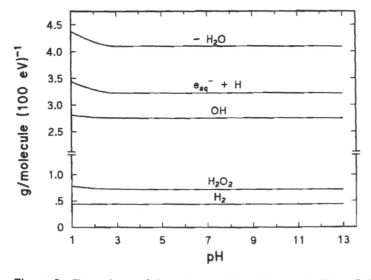

Figure 9 Dependence of the primary yields $g(X)$ on pH. (From Ref. 52.)

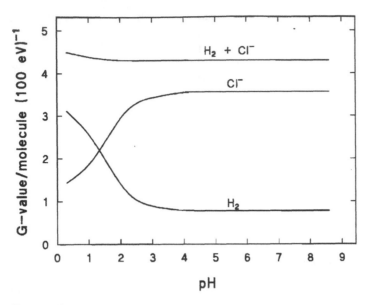

Figure 10 Dependence of $g(e_{aq}^-)$ and $g(H^\cdot)$ on pH. (From Ref. 6.)

4. PROPERTIES OF THE PRIMARY RADICALS

The properties of the primary radicals produced by the radiolysis of water are collected in Table 2.

4.1. The Hydrated Electron

Although its precise structure has not yet been settled, the hydrated electron may be visualized as an excess electron surrounded by a small number of oriented water molecules and behaving in some ways like a singly charged anion of about the same size as the iodide ion. Its intense absorption band in the visible region of the spectrum makes it a simple matter to measure its reaction rate constants using pulse radiolysis combined with kinetic spectrophotometry. Rate constants for several hundred different reactions have been obtained in this way, making e_{aq}^- kinetically one of the most studied chemical entities.

As expected from its standard reduction potential of -2.9 V (Table 2), e_{aq}^- reacts rapidly with many species having more positive reduction potentials. Its mode of reaction can be represented as the one-electron transfer process (reaction (31)):

$$e_{aq}^- + S^n \rightarrow S^{n-1} \tag{31}$$

where n is the charge on the solute. In some cases, the electron adduct S^{n-1} immediately dissociates (e.g., as in reactions (4), (14), and (27) shown above). Values of k_{31} range from ca. 10^1 dm^3 mol^{-1} sec^{-1} up to the diffusion-controlled limit of ca. 10^{10} dm^3 mol^{-1} sec^{-1}. However, the activation energies are invariably small and generally fall in the range 6–30 kJ mol^{-1}, with the majority around 15 kJ mol^{-1}. The latter observation led Hart and Anbar [67] to suggest that reaction (31) has an activation energy associated with reorientation of the solvent shell to facilitate transfer of the electron, and that this reorientation energy is the same as that required for e_{aq}^- to diffuse in water. The corollary to this argument is that the

Table 2 Selected Properties of the Radicals e_{aq}^-, H·, ·OH, and O·⁻ in aqueous solution

Property	e_{aq}^-	H·	·OH	O·⁻
Absorption maximum (nm)	720	<200	~225	240
Molar absorption coefficient ($m^2\ mol^{-1}$)	200 (720 nm) [65]	95 (200 nm) [66]	50 (250 nm) [63]	24 (240 nm) [64]
Diffusion coefficient ($10^{-9}\ m^2\ sec^{-1}$)	4.9 [62]	7 [61]	2.2	
Equivalent conductivity ($S\ cm^2\ mol^{-1}$)	190			
Mobility ($10^{-3}\ cm^2\ V^{-1}\ sec^{-1}$)	1.9			
Reduction potential (V)	-2.9^a	-2.3^b	2.7^c, 1.9^d	1.7^e
pK_a		9.6^f	11.9^g	

The data are taken from Table 1 in Ref. 59 unless stated otherwise.
^{a} ($H_2O + e^- \rightarrow e_{aq}^-$).
^{b} ($H_2O + e^- + H^+ \rightarrow \cdot H_{aq}$).
^{c} ($\cdot OH + e^- + H^+ \rightarrow H_2O$).
^{d} ($\cdot OH + e^- \rightarrow OH^-$).
^{e} ($O\cdot^- + e^- + H^+ \rightarrow OH^-$).
^{f} ($H\cdot + H_2O \rightarrow H_3O^+ + e_{aq}^-$).
^{g} ($\cdot OH + H_2O \rightarrow H_3O^+ + O\cdot^-$).

entropy of activation energy is the dominant kinetic parameter, and this can be understood in terms of the availability of a suitable vacant orbital on the acceptor molecule S^n into which the electron can transfer from its solvent shell. Molecules such as water, simple alcohols, ethers, and amines have no low-lying vacant orbitals, and this explains why solvated electrons can be observed in these liquids.

The kinetics of the reactions of e_{aq}^- can be rationalized as follows. The general expression for a time-independent rate constant in solution is [68]:

$$1/k_{obs} = 1/k_{diff} + 1/k_{read} \tag{32}$$

where k_{obs}, k_{diff}, and k_{react} are the observed, diffusion-controlled, and activation-controlled rate constants, respectively. When the rate of a reaction is diffusion-controlled ($k_{react} \gg k_{diff}$), it is impossible to obtain details of the chemical step in the overall process of encounter and reaction, but such information is revealed by raising the temperature so that the diffusion step becomes much faster than the chemical step. In this way, it has been shown [69] that some reactions of e_{aq}^- (e.g., with NO_2^-, NO_3^-, SeO_4^{2-}, and phenol) are reversible in the chemical step, which accounts for the abnormally low activation energy that was reported for NO_2^- [70]. In fact, the activation energy becomes negative as the temperature is raised above 100°C, as shown in Fig. 11.

These reactions can be described by the following general scheme for transfer of an electron from a donor D to an acceptor A [71]:

$$D + A \rightleftharpoons [D/A] \rightleftharpoons [D/A]^* \rightleftharpoons [D^+/A^-]^* \rightarrow [D^+/A^-] \rightarrow product \tag{33}$$

where $[D/A]^*$ and $[D^+/A^-]^*$ represent the reorganized precursor and successor complexes involved in the electron transfer step. This scheme predicts that the observed activation energy will switch from a positive to a negative value if the relaxation of $[D/A]^*$ back to [D/A] has a larger temperature dependence than the reorganization of [D/A] to $[D/A]^*$. In the

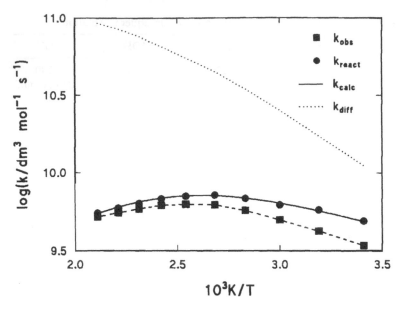

Figure 11   Arrhenius plot of $k(e_{aq}^- + NO_2^-)$. The solid circles show the values of k_{react} obtained from Eq. (32). The solid line shows the fit of reaction (33). Note the influence of k_{diff} on k_{obs}. (From Ref. 69.)

case of NO_2^-, the calculated line in Fig. 11 was obtained with values of E_{act} = 24.9 and 7.2 kJ mol^{-1}, respectively, for these two steps [69].

The hydrated electron behaves as a nucleophile in its reactions with organic molecules and its reactivity is greatly enhanced by electron-withdrawing substituents attached to aromatic rings or adjacent to double bonds. Some of the features of the reactivity of e_{aq}^- are illustrated by the data in Table 3.

4.2. The Hydrogen Atom

The hydrogen atom is the conjugate acid of e_{aq}^- and can be thought of as a weak acid with a formal pK_a of 9.7, obtained from the rate constants of reactions (30) and (34) [59]:

$$e_{aq}^- + H_2O \rightarrow H^\bullet + OH^-; \qquad k_{34} = 19\,dm^3\,mol^{-1}\,sec^{-1} \qquad (34)$$

It is not an important species in the radiation chemistry of neutral and alkaline solution (reaction (7)), but it is the major reducing species in acidic solution (see Fig. 10). Having a reduction potential of -2.3 V (Table 2), $H^\bullet$ is a slightly less powerful reducing agent than e_{aq}^-, but its chemistry is often quite different. A good example of this is provided by reactions (4) and (5). $H^\bullet$ readily reduces inorganic ions having more positive reduction potentials than itself, but often at slower rates than e_{aq}^-. In some cases, in strongly acidic solution, it effectively acts as an oxidant, forming a hydride intermediate, which reacts with H^+ to eliminate H_2 and leave the oxidized form of the solute as in reactions (35) and (36):

$$H^\bullet + Fe^{2+} \rightarrow Fe^{3+}H^- \xrightarrow{H^+} Fe^{3+} + H_2 \qquad (35)$$

$$H^\bullet + I^- \rightarrow HI^- \xrightarrow{I^-} HI_2^{2-} \xrightarrow{H^+} I_2^- + H_2 \qquad (36)$$

Table 3 Rate Constants for Selected Reactions of e_{aq}^- at Room Temperature

Solute (inorganic)	k (dm^3 mol^{-1} sec^{-1})	Solute (organic)	k (dm^3 mol^{-1} sec^{-1})
Ag^+	3.9×10^{10}	C_6H_6	1.1×10^7
H_3O^+	2.3×10^{10}	C_6H_5Cl	6.4×10^8
NH_4^+	1.5×10^6	C_6H_5I	1.2×10^{10}
Cd^{2+}	5.3×10^{10}	$C_6H_5NO_2$	3.7×10^{10}
$Co(NH_3)^{3+}$	8.7×10^{10}	$C_6H_5NH_2$	2.8×10^7
Fe^{3+}	6.0×10^{10}	C_5H_5N (pyridine)	1.0×10^9
ClO_4^-	$<1 \times 10^6$	MV^{2+} [a]	7.2×10^{10}
MnO_4^-	2.6×10^{10}	CH_4	$<1 \times 10^7$
CrO_4^{2-}	1.8×10^{10}	CH_3CN	3.7×10^7
$Fe(CN)_6^{3-}$	3.1×10^9	CH_3I	1.6×10^{10}
$Fe(CN)_6^{4-}$	$<7 \times 10^4$	CH_3OH	$<1 \times 10^4$
CO_2	7.7×10^9	CH_2I_2	3.4×10^{10}
H_2O	19	$C(NO_2)_4$	5.3×10^{10}
H_2O_2	1.1×10^{10}	C_2H_2	2.0×10^7
N_2O	9.1×10^9	$CH_2=CH_2$	$<3 \times 10^5$
O_2	1.9×10^{10}	$CH_2=CCl_2$	2.3×10^{10}

[a] $MV^{2+} = 1,1'$-dimethyl 4,4' bipyridinium ion.
Source: Ref. 59.

H$^{\cdot}$ generally reacts with organic compounds by abstracting a hydrogen atom from saturated molecules and by adding to centers of unsaturation, for example:

$$H^{\cdot} + CH_3OH \rightarrow {}^{\cdot}CH_2OH + H_2 \tag{37}$$

$$H^{\cdot} + CH_2 = CH_2 \rightarrow {}^{\cdot}CH_2CH_3 \tag{38}$$

Because H$^{\cdot}$ absorbs only weakly in the ultraviolet region of the spectrum (Table 2) and $g({}^{\cdot}H)$ is relatively small in neutral solution, its optical signal cannot generally be used for making direct measurements of its rate of reaction with solutes, but in suitable cases, the required data can be obtained by observing the rate of growth of an absorbing product. Rate constants have also been determined directly by observation in pulse radiolysis experiments of the ESR signal of the hydrogen atom, which is possible because of its enhancement by spin polarization effects [72,73].

More generally, the method of competition kinetics is used to determine H-atom rate constants. The hexacyanoferrate(III) ion is a suitable solute because reaction (39) can be followed from the decrease in absorbance at 420 nm due to $Fe(CN)_6^{3-}$ ($\varepsilon_{420} = 104 \, m^2 \, mol^{-1}$). When a second solute is present so that reaction (40) competes with reaction (39), $G(-Fe(CN)_6^{3-})$ is given by Eq. (41):

$$H^{\cdot} + Fe(CN)_6^{3-} \rightarrow Fe(CN)_6^{4-} + H^+ \tag{39}$$

$$H^{\cdot} + S \rightarrow P \tag{40}$$

$$\frac{1}{G(-Fe(CN)_6^{3-})} = \frac{1}{G(H^{\cdot})} \left(1 + \frac{k_{40}[S]}{k_{39}[Fe(CN)_6^{3-}]} \right) \tag{41}$$

4.3. The Hydroxyl Radical

The hydroxyl radical is a powerful oxidant, having a reduction potential of 2.7 V in acidic solution. In neutral solution, where the free energy of neutralization of OH^- by H_3O^+ is not available, the reduction potential decreases to 1.9 V (Table 2). Several inorganic anions and low-valency transition metal ions readily undergo one-electron oxidation by reaction with $^\bullet OH$, which is often represented as a simple electron transfer:

$$^\bullet OH + S^n \rightarrow S^{n+1} + OH^- \tag{42}$$

where n is the ionic charge. However, although the stoichiometry is described by reaction (42), there are many examples which show that the reaction actually proceeds via the formation of an adduct, followed by an inner-sphere electron transfer. In the case of halide and pseudo-halide ions, X^- ($X = Cl$, Br, I, and SCN), the intermediate species $HOX^{\bullet -}$ have been observed by pulse radiolysis (cf. reaction (22)). Similarly, the oxidation of metal ions has been shown to occur as in reaction (43):

$$^\bullet OH + M^{n+} \rightarrow [M^{n+}OH] \rightarrow M^{(n+1)+} + OH^- \tag{43}$$

for $M^{n+} = Tl^+$, Ag^+, Cu^{2+}, Sn^{2+}, Fe^{2+}, Mn^{2+}, and Cr^{3+}. It has been proposed [74] that $^\bullet OH$ is generally unlikely to react by accepting an electron from the donor by a simple outer-sphere transfer because of the large solvent reorganization energy involved in the formation of the hydroxide ion. This idea is certainly supported by the experimental results relating to reactions (42) and (43).

Although $^\bullet OH$ reacts at near-diffusion-controlled rates with inorganic anions [59], there seems to be an upper limit of ca. 3×10^8 dm^3 mol^{-1} sec^{-1} in the case of simple hydrated metal ions, irrespective of the reduction potential of M^{n+}. Also, there is no correlation between the measured values of k_{43} and the rates of exchange of water molecules in the first hydration shell of M^{n+}, which rules out direct substitution of $^\bullet OH$ for H_2O as a general mechanism. Other mechanisms that have been proposed are (i) abstraction of H from a coordinated H_2O [75,76], and (ii) $^\bullet OH$ entering the first hydration shell to increase the coordination number by one, followed by inner-sphere electron transfer [77,78]. Data reported [78] for $M^{n+} = Cr^{3+}$, for which the half-life for water exchange is of the order of days, are consistent with mechanism (ii):

$$^\bullet OH + Cr^{III} \rightleftharpoons [Cr^{III}(^\bullet OH)] \rightarrow Cr^{IV}OH^- \tag{44}$$

In alkaline solution at pH > 10, $^\bullet OH$ reacts rapidly with hydroxide ion to form $O^{\bullet -}$, with $k_{45} = 1.2 \times 10^{10}$ dm^3 mol^{-1} sec^{-1} and $k_{-45} = 9.3 \times 10^7$ sec^{-1} [59] so that equilibrium (reaction (45)) is established when $k_{11}[S_2] < 10^7$ sec^{-1} for $^\bullet OH$ reactions:

$$^\bullet OH + OH^- \rightleftharpoons O^{\bullet -} + H_2O \tag{45}$$

It is important to note that, as expressed here, $K_{45} = 130$ dm^3 mol^{-1} is a concentration equilibrium constant. Thus, when equilibrium is established, $^\bullet OH$ makes up 8% of the total concentration of hydroxyl radicals in 0.1 mol dm^{-3} OH^- solution, and 0.8% in 1 mol dm^{-3} OH^- solution. Consequently, corrections for the presence of $^\bullet OH$ will generally be necessary when measuring rates of $O^{\bullet -}$ reactions when the latter are significantly slower than those of $^\bullet OH$. For the reaction scheme (reactions (45) (46) (47)):

$$^\bullet OH + S \rightarrow P \tag{46}$$

$$O^{\bullet -} + S \rightarrow P \tag{47}$$

the observed rate constant is given by Eq. (48):

$$k_{obs} = \frac{k_{46} + k_{47}\dfrac{K_{45}}{[OH^-]}}{1 + \dfrac{K_{45}}{[OH^-]}} \tag{48}$$

so that by varying $[OH^-]$, one can evaluate k_{46} and k_{47} unambiguously, provided that the reactivity of the solute S does not change in the pH range of interest.

$O^{\cdot-}$ is generally less reactive than $^{\cdot}OH$ with inorganic anions, and the rate is immeasurably slow with Br^-, CO_3^{2-}, and $Fe(CN)_6^{4-}$, although these ions all react rapidly with $^{\cdot}OH$. This difference in reactivity was used to establish the pK_a of $^{\cdot}OH$ as 11.9 (Table 2). $O^{\cdot-}$ reacts rapidly with O_2 as in reaction (49), whereas $^{\cdot}OH$ is unreactive:

$$O^{\cdot-} + O_2 \rightleftharpoons O_3^{\cdot-} \tag{49}$$

For this reaction, $k_{49} = 3.5 \times 10^9$ dm^3 mol^{-1} sec^{-1} and $K_{49} = 1.8 \times 10^6$ dm^3 mol^{-1} at $20\,°C$ [79]. These kinetic properties and the characteristic absorption spectrum of $O_3^{\cdot-}$ ($\lambda_{max} = 430$ nm, $\varepsilon_{430} = 190$ m^2 mol^{-1} [64]) make O_2 a suitable reference solute for the measurement of rate constants of $O^{\cdot-}$ reactions by the method of competition kinetics [79].

In their reactions with organic compounds, $^{\cdot}OH$ is electrophilic and $O^{\cdot-}$ is nucleophilic. Thus, $^{\cdot}OH$ behaves like $H^{\cdot}$, readily adding to double bonds and aromatic rings, whereas $O^{\cdot-}$ does not. Both forms of the radical abstract an H atom from C–H bonds and this can result in different products in neutral and alkaline solutions. This is exemplified by toluene, which reacts with $^{\cdot}OH$ and $O^{\cdot-}$ as in reactions (50) and (51), respectively:

$$^{\cdot}OH + C_6H_5CH_3 \rightarrow [HOC_6H_5CH_3]^{\cdot} \tag{50}$$

$$O^{\cdot-} + C_6H_5CH_3 \rightarrow {}^{\cdot}CH_2C_6H_5 + OH^- \tag{51}$$

with $k_{50} = 5.1 \times 10^9$ dm^3 mol^{-1} sec^{-1} [80] and $k_{51} = 2.1 \times 10^9$ dm^3 mol^{-1} sec^{-1} [81].

Although $^{\cdot}OH$ behaves mechanistically like $H^{\cdot}$ in its reactions with organic molecules, it is less selective and more reactive than $H^{\cdot}$ in H-abstraction reactions because the formation of the H–OH bond is 57 kJ mol^{-1} more exothermic than that of H–H. Examples of the reactivities of $H^{\cdot}$, $^{\cdot}OH$, and $O^{\cdot-}$ are listed in Table 4.

Measurements of the rates of reactions such as reaction (50) over an extended temperature range reveal that k_{obs} increases by less than threefold up to $150\,°C$ and then de-

Table 4 Rate Constants for Selected Reactions of $^{\cdot}OH$, $H^{\cdot}$, and $O^{\cdot-}$ at Room Temperature

| Solute | Reaction type | k (10^7 dm^3 mol^{-1} sec^{-1}) | | |
		$^{\cdot}OH$	$H^{\cdot}$	$O^{\cdot-}$ [a]
$CH_2=CHCONH_2$	Addition	590	3100	65
$C_6H_5CH_3$	Addition	510	260	210
$C_6H_5CO_2^-$	Addition	590	85	~4
CH_3OH	Abstraction	97	0.26	75
$CH_3CH_2CH_2OH$	Abstraction	280	2.4	150
$(CH_3)_2CHOH$	Abstraction	190	7.4	120
HCO_2^-	Abstraction	320	21	140

[a] Abstraction reactions.
Source: Ref. 59.

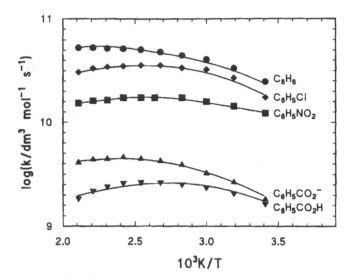

Figure 12 Arrhenius plots for the reaction of ˙OH with some aromatic compounds. (From Ref. 82.)

creases slightly up to 200°C as shown in Fig. 12 for C_6H_6, C_6H_5Cl, $C_6H_5NO_2$, $C_6H_5CO_2H$, and $C_6H_5CO_2^-$ [82].

The variations of k_{obs} with temperature are all very similar, but the absolute values of k_{obs} depend on the substituent on the aromatic ring. To explain these results, Ashton et al. [82] proposed the following mechanism whereby ˙OH adds reversibly to form a π-complex as a precursor of the final σ-bonded hydroxycyclohexadienyl radical:

$$\text{˙OH} + C_6H_5X \rightleftharpoons [HOC_6H_5]^{\bullet}_{\pi} \tag{52}$$

$$[HOC_6H_5]^{\bullet}_{\pi} \rightarrow [HOC_6H_5]^{\bullet}_{\sigma} \tag{53}$$

Application of the steady-state approximation to $[HOC_6H_5]^{\bullet}_{\sigma}$ in this reaction scheme leads to:

$$k_{obs} = k_{52}k_{53}/(k_{-52} + k_{53}) \tag{54}$$

On the assumption that reaction (52) is diffusion-controlled, Ashton et al. [82] found that rearrangement of $[HOC_6H_5]^{\bullet}_{\pi}$ to $[HOC_6H_5]^{\bullet}_{\sigma}$ requires virtually no activation energy, whereas dissociation of $[HOC_6H_5]^{\bullet}_{\sigma}$ in reaction (−52) has an activation energy of ca. 20 kJ mol^{-1}. The differences in the absolute values of k_{obs} are reflected largely in the values of k_{53}, suggesting that the electronic effects of the substituents govern the rate of rearrangement. These parameters are summarized in Table 5.

4.4. The Perhydroxyl Radical $HO_2^{\bullet}$

Although $HO_2^{\bullet}$ and its conjugate base $O_2^{\bullet-}$ are only significant primary radicals for high LET radiation, they are important secondary radicals in oxygenated solution where they are formed in reactions (55)–(58):

$$e^-_{aq} + O_2 \rightarrow O_2^{\bullet-} \tag{55}$$

$$H^{\bullet} + O_2 \rightarrow HO_2^{\bullet} \tag{56}$$

Table 5 Rate Constants and Activation Energies for Reactions (-52) and (53)

X	k (10^9 dm^3 mol^{-1} sec^{-1})		E_{act} (kJ mol^{-1})	
	Reaction (-52)	Reaction (53)	Reaction (-52)	Reaction (53)
H	6.5	14	18	0.5
Cl	4.0	12	22	0.5
NO_2	4.0	4	20	1.3
CO_2^-	3.8	13	21	1.0
CO_2H	4.0	8	23	1.0

Source: Ref. 82.

$$HO_2^\bullet \rightleftharpoons O_2^{\bullet-} + H^+ \tag{57}$$

The pK_a of $HO_2^\bullet$ is 4.8 [83], and both $O_2^{\bullet-}$ and $HO_2^\bullet$ have reducing and oxidizing properties. The reduction potentials are -0.33 V for the $O_2/O_2^{\bullet-}$ couple with respect to a standard state of 1 atm pressure of O_2 [83] and -0.05 V for the $O_2, H^+/HO_2^\bullet$ couple [84] so that $O_2^{\bullet-}$ is the better reductant. Conversely, $HO_2^\bullet$ is the better oxidant, the reduction potentials being 1.45 V for the $(H^+, HO_2^\bullet/H_2O_2)$ couple and 1.03 V for the $(O_2^{\bullet-}, H^+/HO_2^-)$ couple [84].

$HO_2^\bullet$ and $O_2^{\bullet-}$ have characteristic absorption spectra with $\varepsilon_{max} = 140$ m^2 mol^{-1} at 225 nm [83] and $\varepsilon_{max} = 189$ m^2 mol^{-1} at 245 nm [85], respectively, which are sufficiently intense to permit their reactions to be followed by direct observation in pulse radiolysis experiments. Both radicals are relatively unreactive with organic molecules [83], abstracting only weakly bonded hydrogen atoms in, for example, ascorbic acid, cysteine, and hydroquinone. $O_2^{\bullet-}$ undergoes reversible electron transfer in its reaction with quinones (Q), which was used to establish its reduction potential [86]:

$$O_2^{\bullet-} + Q \rightleftharpoons O_2 + Q^{\bullet-} \tag{58}$$

A careful study of the bimolecular decay of $HO_2^\bullet/O_2^{\bullet-}$ in the absence of metal ions has shown [84] that the reactions involved are reactions (59) and (60):

$$HO_2^\bullet + HO_2^\bullet \rightarrow H_2O_2 + O_2 \tag{59}$$

$$HO_2^\bullet + O_2^{\bullet-} \xrightarrow{H^+} H_2O_2 + O_2 \tag{60}$$

There is no evidence that reaction (61) takes place at a measurable rate under ambient conditions, but it is rapid in the presence of aquo-complexes and some organo-complexes of copper ions via reactions (62) and (63):

$$O_2^{\bullet-} + O_2^{\bullet-} + 2H^+ \rightarrow H_2O_2 + O_2 \tag{61}$$

$$O_2^{\bullet-} + Cu^{II} \rightarrow Cu^{I} + O_2 \tag{62}$$

$$O_2^{\bullet-} + Cu^{I} + 2H^+ \rightarrow H_2O_2 + Cu^{II} \tag{63}$$

Reaction (63) is an example of $O_2^{\bullet-}$ acting as an oxidant and it probably proceeds via an inner-sphere electron transfer mechanism in which incompletely coordinated Cu^{I} binds $O_2^{\bullet-}$ prior to electron transfer [87]. $HO_2^{\bullet-}$ and $O_2^{\bullet-}$ also react readily with a number of other transition metal ions, either by electron transfer or through the formation of a complex [83], for example:

$$O_2^{\bullet-} + Mn^{2+} \rightarrow MnO_2^+ \tag{64}$$

$$HO_2^{\bullet-} + Fe^{2+} \rightarrow Fe^{3+} \cdot HO_2^- (\xrightarrow{H^+} Fe^{3+} + H_2O_2) \tag{65}$$

Reactions (66) and (67) provide a very convenient way of converting $^\bullet OH$ to $O_2^{\bullet-}$:

$$^\bullet OH + HCO_2^- \rightarrow CO_2^{\bullet-} + H_2O \tag{66}$$

$$CO_2^{\bullet-} + O_2 \rightarrow O_2^{\bullet-} + CO_2 \tag{67}$$

Thus, all the primary water radicals can be converted into $HO_2^\bullet/O_2^{\bullet-}$ in reactions (55), (56), (66), and (67).

5. GENERATION OF SECONDARY RADICALS

As we have seen in Sections 1 and 4, the principal primary products of the radiolysis of water are powerful oxidizing and reducing radicals in approximately equal yields. For water radiolysis to be a useful tool in general chemistry, it is desirable to convert the primary radicals to a single kind of secondary radical to achieve either totally oxidizing or reducing conditions. Moreover, there is the possibility of designing the system to have the required redox properties by suitable selection of the secondary radicals. Some useful systems that meet these requirements are described below.

5.1. Oxidizing Conditions

As noted in Section 3.3, reactions (27) and (28) provide a very convenient way of converting e_{aq}^- to $^\bullet OH$. Because reaction (68) is relatively slow, approximately 10% of the radicals in N_2O-saturated neutral solution remain as reducing radicals in the form of $H^\bullet$, but this small fraction can be allowed for or even be neglected in chemical applications:

$$H^\bullet + N_2O \rightarrow {}^\bullet OH + N_2; \qquad k_{68} = 2.1 \times 10^6 \, dm^3 \, mol^{-1} \, sec^{-1} \tag{68}$$

At pH > 11, where reaction (30) becomes increasingly important, $G(^\bullet OH)$ can increase by up to 0.6 molecule $(100 \, eV)^{-1}$ (i.e., $g(H^\bullet)$), whereas at pH < 3, $G(^\bullet OH)$ becomes smaller because reaction (3) competes with reaction (27). One should not forget that the product of reaction (27) is $O^{\bullet-}$ rather than $^\bullet OH$ [88,89]. Thus, if a solute that reacts with $O^{\bullet-}$ in competition with its protonation in reaction (28) is present, then the reaction products may not be the same in N_2O-saturated solutions containing hydroxyl radical scavengers in high and low concentrations because $O^{\bullet-}$ can react differently from $^\bullet OH$ (see Section 4.3).

More selective oxidizing conditions can be achieved by converting $^\bullet OH$ into another inorganic radical (e.g., $Br_2^{\bullet-}$, $I_2^{\bullet-}$, $(SCN)_2^{\bullet-}$, $CO_3^{\bullet-}$, $N_3^\bullet$, etc.), thereby tuning the reduction potential (see Table 6). One of the most powerful oxidizing radicals is $SO_4^{\bullet-}$, which is produced in reaction (69):

$$e_{aq}^- + S_2O_8^{2-} \rightarrow SO_4^{\bullet-} + SO_4^{2-} \tag{69}$$

Because of its high reduction potential of 2.43 V (see Table 6), it can be used to generate other strongly oxidizing radicals such as $Cl^\bullet$ and $NO_3^\bullet$ through electron transfer reactions with Cl^- and NO_3^-, respectively [27].

Secondary inorganic radicals are particularly useful for studying redox changes in metalloproteins and organometallic complexes because these radicals are more likely to react at the metal center by electron transfer, whereas $^\bullet OH$ will also attack the organic moiety and, by abstracting H, create a reducing radical there.

Table 6 Reduction Potentials of Some Inorganic Radicals

Redox couple	$E°$ (V)	Redox couple	$E°$ (V)
$Cl^•/Cl^-$	2.41	$CO_3^{•-}/CO_3^{2-}$	1.5
$Cl_2^{•-}/2\,Cl^-$	2.09	$N_3^•/N_3^-$	1.33
$Br^•/Br^-$	1.92	$NO_2^•/NO_2^-$	1.04
$Br_2^{•-}/2\,Br^-$	1.62	$NO_3^•/NO_3^-$	2.5
$I^•/I^-$	1.33	$SO_3^{•-}/SO_3^{2-}$	0.63
$I_2^{•-}/2\,I^-$	1.03	$SO_4^{•-}/SO_4^{2-}$	2.43
$SCN^•/SCN^-$	1.63	$CN^•/CN^-$	2.59
$(SCN)_2^{•-}/2SCN^-$	1.32	$OCN^•/OCN^-$	2.66

Source: Ref. 90.

5.2. Reducing Conditions

Although it is possible to convert $^•OH$ to e_{aq}^- through reactions (70) and (30), the slowness of these reactions means that conditions of 100 atm of hydrogen and high pH are required. Nevertheless, these have been realized in experiments crucial to the measurement of spur reactions (R5) and (R6) in Table 1 [91,92]:

$$^•OH + H_2 \rightarrow H^• + H_2O; \qquad k_{70} = 4.2 \times 10^7\,dm^3\,mol^{-1}\,sec^{-1}\,[59] \tag{70}$$

A much more convenient method of obtaining totally reducing conditions is to convert all the primary radicals to the same secondary reducing radical by adding the appropriate organic solute to N_2O-saturated water. A prime example of this is the generation of $CO_2^{•-}$ through reactions (66) and (71):

$$H^• + HCO_2^- \rightarrow CO_2^{•-} + H_2; \qquad k_{71} = 2.1 \times 10^8\,dm^3\,mol^{-1}\,sec^{-1}\,[59] \tag{71}$$

Alternatively, N_2O can be replaced by CO_2 so that $CO_2^{•-}$ is produced in reaction (72):

$$e_{aq}^- + CO_2 \rightarrow CO_2^{•-}; \qquad k_{72} = 7.7 \times 10^9\,dm^3\,mol^{-1}\,sec^{-1}\,[59] \tag{72}$$

Simple alcohols are often used as the source of the secondary radicals because they also react rapidly with both $^•OH$ and $H^•$ (e.g., 2-propanol in reactions (73) and (74)):

$$^•OH + (CH_3)_2CHOH \rightarrow (CH_3)_2{}^•COH + H_2O \tag{73}$$

$$H^• + (CH_3)_2CHOH \rightarrow (CH_3)_2{}^•COH + H_2 \tag{74}$$

It is the hydrogen atom on the α-carbon that is most readily abstracted, as indicated in reactions (73) and (74). However, $^•OH$ does abstract H from other positions so that a mixture of radicals is produced where only one is desired. The distributions of radicals generated from a number of simple alcohols are given in Table 7.

Radicals derived from simple alcohols have a dissociable proton on the hydroxyl group, for example:

$$(CH_3)_2{}^•COH \rightleftharpoons (CH_3)_2{}^•CO^-; \qquad pK_a = 12.03\,[94] \tag{75}$$

Thus, one should bear in mind that the basic form of the radical will be generated when $pH > pK_a$ and it is usually more strongly reducing than the acidic form. The values of pK_a

Table 7 Percentage Abstraction of H-Atom by $^{\bullet}$OH from Various Positions in Simple Alcohols

Alcohol	α-C—H	Other C—H	—OH
CH_3OH	93.0	—	7.0
CH_3CH_2OH	84.3	13.2	2.5
$CH_3(CH_2)_2OH$	53.4	46.0	<0.5
$(CH_3)_2CHOH$	85.5	13.3	1.2
$CH_3(CH_2)_3OH$	41.0	58.5	<0.5
$(CH_3)_3COH$	—	95.7	4.3
$(CH_2OH)_2$	100	—	<0.1
$CH_3CH(OH)CH_2OH$	79.2	20.7	<0.1
$CH_3CH(OH)CH(OH)CH_3$	71.0	29.0	<0.1

Source: Ref. 93.

and reduction potential for the more commonly used organic radicals are summarized in Table 8.

$CO_2^{\bullet-}$ is a useful reducing radical because of its low pK_a, which means that it can be used in acidic solutions where e_{aq}^- would react predominantly with H^+. However, where it is required to use e_{aq}^- as the reductant, it is common practice to add 2-methyl-2-propanol (*tert*-butanol) as the scavenger for $^{\bullet}$OH to convert it into the relatively unreactive radical $^{\bullet}CH_2(CH_3)_2COH$:

$$^{\bullet}OH + (CH_3)_3COH \rightarrow {}^{\bullet}CH_2(CH_3)_2COH + H_2;$$

$$k_{76} = 6.0 \times 10^8 \, dm^3 \, mol^{-1} \, sec^{-1} \, [59]$$

(76)

In this way, the powerful reducing properties of e_{aq}^- can be exploited to generate hyper-reduced states of metal ions, for example:

$$e_{aq}^- + Cd^{2+} \rightarrow Cd^+$$

(77)

The monovalent ions Cd^+, Co^+, Ni^+, and Zn^+ are themselves quite strong reductants and, being positively charged, can be used to reduce metalloproteins at sites where there is a local negative charge that makes them less accessible to e_{aq}^- [98].

Table 8 Values of pK_a and E° for Commonly Used Reducing Radicals Obtained by Reaction of $^{\bullet}$OH with Simple Alcohols (see Table 7) and Formate Ion

Radical	pK_a [94]	E° (RO, H^+/ROH$^{\bullet}$) (V) [95]	E° (RO/RO$^{\bullet-}$) (V) [95]
$^{\bullet}CH_2OH$	10.71	−1.18	−1.81
$CH_3{}^{\bullet}CHOH$	11.51	−1.25	−1.93
$(CH_3)_2{}^{\bullet}COH$	12.03	−1.39	−2.10
$^{\bullet}CH_2C(CH_3)_2OH$		−0.1 [96]	
$CO_2^{\bullet-}$	−0.2 [97]		−1.90

6. APPLICATIONS IN GENERAL CHEMISTRY

The great advantage of radiolysis over other methods of producing free radicals is that only the solvent absorbs the ionizing radiation in most chemical applications. Moreover, it is clear from the foregoing information that the radiolysis of water provides a powerful method of generating one-electron redox agents that can be finely tuned in terms of reduction potential and electric charge. The use of pulse radiolysis in particular has resulted in a great wealth of information being obtained on the kinetics and mechanisms of the reactions of free radicals, and of metal ions in unusual oxidation states, in aqueous solution. The chapters, each written by experts, in "Radiation Chemistry—Present Status and Future Trends" [99] provide excellent descriptions of the diverse nature and applications of this field of study.

REFERENCES

1. Allen, A.O. The story of the radiation chemistry of water. In *Early Developments in Radiation Chemistry*; Kroh, J., Ed.; Royal Society of Chemistry: London, 1989; p 1.
2. Risse, O. Strahlenchemie 1929, *34*, 578.
3. Weiss, J. Nature 1944, *153*, 7481.
4. Fricke, H.; Hart, E.J.; Smith, H.P. J. Chem. Phys. 1938, *6*, 229.
5. Allen, A.O. J. Phys. Colloid Chem. 1948, *52*, 479.
6. Hayon, E.; Weiss, J.J. Proceedings of the 2nd International Conference on Peaceful Uses Of Atomic Energy; Geneva, 1958, *29*, 80.
7. Hart, E.J.; Boag, J.W. J. Am. Chem. Soc. 1962, *84*, 4090.
8. Schwarz, H.A. J. Phys. Chem. 1969, *73*, 1928.
9. Swiatla-Wojcik, D.; Buxton, G.V. J. Phys. Chem. 1995, *99*, 11464.
10a. Cobut, V.; Frongillo, Y.; Patau, J.P.; Goulet, T.; Fraser, M.-J.; Jay-Gerin, J.-P. Radiat. Phys. Chem. 1998, *51*, 229.
10b. Frongillo, Y.; Goulet, T.; Fraser, M.-J.; Cobut, V.; Patau, J.P.; Jay-Gerin, J.-P. Radiat. Phys. Chem. 1998, *51*, 245.
10c. Hervé du Penboat, M.-A.; Goulet, T.; Frongillo, Y.; Fraser, M.-J.; Bernat, P.; Jay-Gerin, J.-P. J. Phys. Chem. A 2000, *104*, 11757.
11. Kaplan, I.G.; Miterev, A.M.; Sukhonosov, A.M. Radiat. Phys. Chem. 1990, *36*, 493.
12. Lampe, F.W.; Field, F.H.; Franklin, J.L. J. Am. Chem. Soc. 1957, *79*, 6132.
13. Hamill, W.H. J. Phys. Chem. 1969, *73*, 1341.
14. Platzman, R.L. In *Abstracts of Papers, Second International Congress of Radiation Research, Harrogate, August, 5–11, 1962*; Ebert, M., Howard, A., Eds.; North-Holland Publishing: Amsterdam, 1963; p 128.
15. Faraggi, M.; Desalos, J. Int. J. Radiat. Phys. Chem. 1969, *1*, 335.
16. Pastina, B.; LaVerne, J.A.; Pimblott, S.M. J. Phys. Chem. A 1999, *103*, 5841.
17. LaVerne, J.A.; Pimblott, S.M. J. Phys. Chem. A 2000, *104*, 9820.
18. Wiesenfeld, J.M.; Ippen, E.P. Chem. Phys. Lett. 1980, *73*, 47.
19. LaVerne, J.A.; Pimblott, S.M. J. Phys. Chem. 1991, *95*, 3196.
20. Pimblott, S.M.; LaVerne, J.A. J. Phys. Chem. A 1997, *101*, 5828.
21. Swiatla-Wojcik, D.; Buxton, G.V. J. Chem. Soc. Faraday Trans. 1998, *94*, 2135.
22. Hervé du Penboat, M.-A.; Meesungnoen, J.; Goulet, T.; Filali-Mouhim, A.; Manketkorn, S.; Jay-Gerin, J.-P. Chem. Phys. Lett. 2001, *341*, 135.
23. Magee, J.L.; Chatterjee, A. J. Phys. Chem. 1980, *84*, 1937.
24. Swiatla-Wojcik, D.; Buxton, G.V. Phys. Chem. Chem. Phys. 2000, *2*, 5113.
25. Clifford, P.; Green, N.J.B.; Oldfield, M.J.; Pilling, M.J.; Pimblott, S.M. J. Chem. Soc., Faraday Trans. I 1986, *82*, 2673.

26. Swiatla-Wojcik, D. personal communication.
27. www.rcdc.nd.edu.
28. Peled, E.; Czapski, G. J. Phys. Chem. 1970, 74, 2903.
29. Draganic, Z.D.; Draganic, I.G. J. Phys. Chem. 1971, 75, 3950.
30. Pimblott, S.M.; LaVerne, J.A. J. Phys. Chem. A 1998, 102, 2967.
31. Jonah, C.D.; Miller, J.R.; Matheson, M.S. J. Phys. Chem. 1977, 81, 1618.
32. Lam, K.Y.; Hunt, J.W. Int. J. Radiat. Phys. Chem. 1975, 7, 317.
33. Kee, T.W.; Son, D.H.; Kambhampati, P.; Barbara, P.F. J. Phys. Chem. A 2001, 105, 8434 (and references therein).
34. Draganic, Z.D.; Draganic, I.G. J. Phys. Chem. 1973, 77, 765.
35. Head, D.A.; Walker, D.C. Nature 1965, 207, 517.
36. Dainton, F.S.; Logan, S.R. Trans. Faraday Soc. 1965, 61, 715.
37. Buxton, G.V.; Lynch, D.A.; Stuart, C.R. J. Chem. Soc. Faraday Trans. 1998, 94, 2379.
38. Balkas, T.I.; Fendler, J.H.; Schuler, R.H. J. Phys. Chem. 1970, 74, 4497.
39. Schmidt, K.H.; Han, P.; Bartels, D.M. J. Phys. Chem. 1995, 99, 10530.
40. Elliot, A.J.; Chenier, M.P.; Oullette, D.C. J. Chem. Soc. Faraday Trans. 1993, 94, 1193.
41. Yoshida, H. Radiat. Res. 1994, 137, 145.
42. Warman, J.M.; Asmus, K.-D.; Schuler, R.H. J. Phys. Chem. 1969, 73, 931.
43. Pimblott, S.M.; LaVerne, J.A. J. Phys. Chem. 1992, 96, 8904.
44. Ambramwitz, M.; Stegun, I.A. Handbook of Mathematical Functions; Dover: New York, 1970.
45. Wolff, R.K.; Bronskill, M.J.; Aldrich, J.E.; Hunt, J.W. J. Phys. Chem. 1973, 77, 1350.
46. Jonah, C.D.; Hart, E.J.; Matheson, M.S. J. Phys. Chem. 1973, 77, 1838.
47. Jonah, C.D.; Matheson, M.S.; Miller, J.R.; Hart, E.J. J. Phys. Chem. 1976, 80, 1276.
48. Sumiyoshi, T.; Tsugaru, K.; Yamada, T.; Katayama, M. Bull. Chem. Soc. Jpn. 1985, 58, 3073.
49. Bartels, D.M.; Cook, A.R.; Mudaliar, M.; Jonah, C.D. J. Phys. Chem. A 2000, 104, 1686.
50. Czapski, G.; Peled, E. J. Phys. Chem. 1973, 77, 893.
51. Chernovitz, A.C.; Jonah, C.D. J. Phys. Chem. 1988, 92, 5946.
52. Draganic, I.G.; Nenadovic, M.T.; Draganic, Z.D. J. Phys. Chem. 1969, 73, 2564.
53. Schuler, R.H.; Behar, B. In Proceedings of the 5th Tihany Symposium on Radiation Chemistry; Dobo, J. Hedvig, P., Schiller, R., Eds.; Akademai Kiado: Budapest, 1983; Vol. 1, p 183.
54. LaVerne, J.A.; Pimblott, S.M. J. Chem. Soc. Faraday Trans. 1993, 89, 3527.
55. Jay-Gerin, J.-P.; Ferradini, C. Chem. Phys. Lett. 2000, 317, 388.
56. Sutton, H.C.; Adams, G.E.; Boag, J.W.; Michael, B.D. In Pulse Radiolysis; Ebert, M., Keene, J.P., Swallow, A.J., Baxendale, J.H., Eds.; Academic Press: London, 1965; p 61.
57. Adams, G.E.; Boag, J.W.; Currant, J.; Michael, B.D. In Pulse Rdaiolysis; Ebert, M., Keene, J.P., Swallow, A.J., Baxendale, J.H., Eds.; Academic Press: London; 1965; p 117.
58. Buxton, G.V.; Dainton, F.S. Proc. R. Soc. A 1965, 287, 427.
59. Buxton, G.V.; Greenstock, C.L.; Helman, W.P.; Ross, A.B. J. Phys. Chem. Ref. Data 1988, 17, 513.
60. Schuler, R.H.; Hartzell, A.L.; Behar, B. J. Phys. Chem. 1981, 85, 192.
61. Benderskii, V.A.; Krivenko, A.G.; Rukin, A.N. High Energy Chem. 1980, 14, 303.
62. Schmidt, K.H.; Han, P.; Bartels, D.M. J. Phys. Chem. 1992, 96, 199.
63. Buxton, G.V.; Elliot, A.J. J. Chem. Soc., Faraday Trans. 1993, 89, 485.
64. Hug, G.L. Optical Spectra of Nonmetallic Inorganic Transient Species in Aqueous Solution, Nat. Stand. Ref. Data Ser. Nat. Bur. Stand. (U.S.A.). Washington, 1981; p 69.
65. Elliot, A.J.; Oullette, D.C. J. Chem. Soc., Faraday Trans. 1994, 90, 837.
66. Sehested, K.; Christensen, H. Radiat. Phys. Chem. 1990, 36, 499.
67. Hart, E.J.; Anbar, M. The Hydrated Electron, Wiley-Interscience: New York, 1970.
68. Noyes, R.M. In Progress in Reaction Kinetics; Porter, G., Ed.; Pergamon: London, 1961; Vol. 1, p 129.
69. Buxton, G.V.; Mackenzie, S.R. J. Chem. Soc. Faraday Trans. 1992, 88, 2833.
70. Cercek, B. Nature 1969, 223, 491.
71. Newton, M.D.; Sutin, N. Annu. Rev. Phys. Chem. 1984, 35, 437.

72. Fessenden, R.W.; Verma, N.C. Faraday Discuss. Chem. Soc. 1977, *63*, 104.
73. Han, P.; Bartels, D.M. Chem. Phys. Lett. 1989, *159*, 538.
74. Meyerstein, D. Faraday Discuss. Chem. Soc 1977, *63*, 203.
75. Collinson, E.; Dainton, F.S.; Tazuke, S.; Smith, D.R. Nature 1963, *206*, 198.
76. Berdnikov, V.M. Russ. J. Phys. Chem. 1973, *47*, 1547.
77. Meyerstein, D. Acc. Chem. Res. 1978, *11*, 43.
78. Buxton, G.V.; Djouider, F.; Lynch, D.A.; Malone, T.N. J. Chem. Soc. Faraday Trans. 1997, *93*, 4265.
79. Elliot, A.J.; McCracken, D.R. Radiat. Phys. Chem. 1989, *33*, 69.
80. Boder, M.; Wojnarovits, L.; Foldiak, G. Radiat. Phys. Chem 1990, *36*, 175.
81. Christensen, H.C.; Sehested, K.; Hart, E.J. J. Phys. Chem. 1973, *77*, 983.
82. Ashton, L.; Buxton, G.V.; Stuart, C.R. J. Chem. Soc., Faraday Trans. 1995, *91*, 1631.
83. Bielski, B.H.J.; Cabelli, D.E.; Arudi, R.L.; Ross, A.B. J. Phys. Chem. Ref. Data 1985, *14*, 1041.
84. Allen, A.O.; Bielski, B.H.J. In *Superoxide Dismutase*; Oberley, L.W., Ed.; CRC Press: Boca Raton, 1982; Vol. 1, p 125.
85. Elliot, A.J.; Buxton, G.V. J. Chem. Soc., Faraday Trans. 1992, *88*, 2465.
86. Meisel, D.; Czapski, G. J. Phys. Chem. 1975, *79*, 1503.
87. Goldstein, S.; Czapski, G. J. Am. Chem. Soc. 1983, *105*, 7276.
88. Buxton, G.V. Trans. Faraday Soc. 1970, *66*, 1656.
89. Zehavi, D.; Rabani, J. J. Phys. Chem. 1971, *75*, 1738.
90. Stanbury, D.M. Adv. Inorg. Chem. 1989, *33*, 69.
91. Sehested, K.; Christensen, H. Radiat. Phys. Chem. 1990, *36*, 499.
92. Chrisrensen, H.; Sehested, K.; Logager, T. Radiat. Phys. Chem. 1994, *43*, 527.
93. Asmus, K.-D.; Möckel, H.; Henglein, A. J. Phys. Chem. 1973, *77*, 1218.
94. Laroff, G.P.; Fessenden, R.W. J. Phys. Chem. 1973, *77*, 1283.
95. Schwarz, H.A.; Dodson, R.W. J. Phys. Chem. 1989, *93*, 409.
96. Endicott, J.P. In *Concepts of Inorganic Photochemistry*; Adamson, A.W., Fleishauer, P.D., Eds.; Wiley: New York, 1975; p 88.
97. Jeevarajan, A.S.; Carmichael, I.; Fessenden, R.W. J. Phys. Chem. 1990, *94*, 1372.
98. Govindaraju, K.; Christensen, H.E.M.; Lloyd, E.; Olsen, M.; Salmon, G.A.; Tomkinson, N.P.; Sykes, A.G. Inorg. Chem. 1993, *32*, 40.
99. *Radiation Chemistry. Present Status and Future Prospects*; Jonah, C.D., Rao, B.S.M., Eds.; Elsevier: Amsterdam, 2001.

13

Photochemistry and Radiation Chemistry of Liquid Alkanes: Formation and Decay of Low-Energy Excited States

L. Wojnarovits
Hungarian Academy of Sciences, Budapest, Hungary

1. INTRODUCTION

The radiation chemistry of alkanes was reviewed several times in the past. The books of Topchiev [1] and Foldiak (editor) [2] deal mainly with the yield and distribution of final products. In the book edited by Gaumann and Hoigne [3] the yields of radical intermediates, LET effects (linear energy transfer, energy lost when the particle passes unit length) reaction mechanisms, etc. are discussed. "Hydrocarbon chemistry" dominates the book edited by Ausloos [4]. Freeman [5] and, more recently, Hummel [6] explained ionizing-radiation-induced chemical changes in alkanes based on the radiation chemical theories available at the time of writing of their review papers. The most recent review published by Shkrob et al. [7] concentrates on the liquid phase ion chemistry detailing the different views on the high-mobility cations observed in several liquid cycloalkanes; they also discuss some aspects of the excited-state chemistry. The formation and decay of excited alkane molecules was reviewed in a recent book chapter of Mayer and Szadkowska-Nicze [8]. In addition to these special books, book chapters or review articles on hydrocarbon radiolysis, or more specifically on alkane radiolysis, general radiation chemistry books also provide great attention to the radiolysis of alkanes [9–13].

This chapter is concerned mainly with the chemistry of excited molecules in photon and high-energy ionizing-radiation-induced processes of alkanes. Our knowledge on the higher-energy excited states is rather limited, so the results on the reactions of lower energy excited molecules are reviewed. We will compare the processes taking place in radiolysis and photolysis and try to give more chemistry than reported in the previous reviews. Most of the work done in this field was published in the 1980s; however, as can be seen from the reference list, more recent papers also appeared, which are incorporated in the review.

2. PHOTOCHEMISTRY OF LIQUID ALKANES

The electron excitation of molecules in nearly all classes of organic compounds can be performed by visible or ultraviolet light ($\lambda_{ex} > 190$ nm, < 6.5 eV). There are a few exceptions, e.g., the alkanes, where excitation requires higher energies in the so-called vacuum UV (VUV) region. The photon irradiation of alkanes is generally done in the 100- to 180-nm wavelength range; this range corresponds to a photon energy range of 7–12 eV. From the energy dependence and by comparing the data obtained in radiolysis and photolysis, valuable information can be obtained concerning the activated intermediates leading to the formation of stable end products in radiolysis.

Most publications dealing with the photodecomposition of alkanes discuss the processes in the gas phase; several comprehensive works have already been published in this field [14–17]. In the present work, we summarize the results of liquid phase photolytic studies and compare them with those obtained in radiolysis. An early review on liquid alkane photochemistry was published in Ref. 18, a brief overview of the field was given in Ref. 19.

2.1. Absorption and Emission Spectra of Alkanes, Ionization in the Liquid Phase

The VUV absorption spectra of alkanes are rather featureless, usually without marked peaks (as shown in Fig. 1 for n-hexane) in contrast to the spectra of aromatics or to the spectra of compounds containing heteroatoms. Earlier works usually assumed that at low energies in alkanes the excitation results in $\sigma \rightarrow \sigma^*$-type valence-electron transitions, but later works concluded that the excitations, at least partly, result in Rydberg states [20–23].

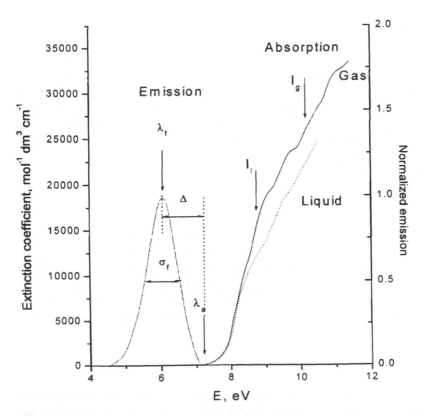

Figure 1   Absorption [20,21] and emission [25] spectra of n-hexane.

Rydberg-type orbitals are extended in space and resemble atomic orbitals, whereas the valence excitations do not lead to significant spatial extension. Consequently, the Rydberg states are, in contrast to the valence-electron excitations, sensitive to external perturbations coming from changes of pressure or phase (liquid–gas). The probability of formation of Rydberg orbitals in condensed phases is less than in gases, although many experimental results indicate their existence, e.g., energy transfer from higher excited states (Sec. 3.2).

During irradiation of many alkanes with VUV photons, with visible or infrared photons causing double or multiphoton excitation, or ionizing radiation, weak fluorescence was first observed at the end of the sixties with quantum yields of $\Phi_f \approx 10^{-5}$ to 10^{-2} [24–26] (Table 1). The fluorescence spectrum is a broad, structureless band around 170–250 nm; the fluorescence intensity for a given alkane has approximately a Gaussian dependence on the fluorescence photon energy. The fluorescence spectrum has been characterized by the wavelength of maximum λ_f and the full width at the half height σ_f of the emission spectrum. λ_f lies by ~1 eV below the adsorption onset. This large energy gap between the absorbed and emitted energy in alkanes (Δ) is presumably a consequence of the large difference between the nuclear distances in the ground and excited states. Thus, Δ has a similar meaning as the Stokes' shift [25]. For aromatics, where a conjugated electron system is excited, the Stokes' shift is only 0.1 eV. For most compounds studied, the fluorescence quantum yields decline monotonically with the energy of photons used for excitation. Φ_f flattens out or reaches a minimum a few tenths of an electron volt above the liquid phase ionization threshold. The minimum is sometimes followed by a slow increase [26,27]. The minimum and the slow increase are attributed to the contribution of recombination fluorescence as will be discussed later in this section. Fluorescence was not detected ($\Phi_f \leq 10^{-5}$) in the irradiation of geminally substituted branched alkanes, such as isooctane, and the C_5, C_7–C_{10} cycloalkanes.

The energy of the relaxed S_1 state of alkanes in the liquid phase is certainly somewhere between the energy of the absorption onset and the energy of the fluorescence maximum. In energy-transfer experiments, the energy of absorption threshold is usually

Table 1 Photon Absorption and Fluorescence Properties of Some Selected Alkanes

Alkanes	λ_a, nm[a]	λ_f, nm[b]	Δ, eV[c]	σ_f, eV[d]	$\Phi_{f,165\ nm}$[e]	I_g, eV[f]	I_l, eV[g]
n-Hexane	171	206	1.24	1.02	0.0006	10.18	8.7
n-Decane	173	207	1.19	0.95	0.0042	9.65	8.4
3-Methylpentane	173	231	1.8	1.24	0.0002	10.08	8.5
3-Methylhexane	174	225	1.6	1.14	0.002	10.0	8.65
2,3-Dimethylbutane	175	242	1.95	1.08	0.0061	10.02	8.6
2,3,4-Trimethylpentane	179	244		1.10	0.0016	9.6	8.4
Cyclohexane	177	201	0.83	0.95	0.0088	9.86	8.4
Methylcyclohexane	179	213	1.08	1.1	0.011	9.64	8.3
trans-Decalin	186	217	0.94	1.1	0.023	9.24	8.0

[a] Wavelength at which the decadic extinction coefficient of the neat liquids equals 5 mol^{-1} dm^3 cm^{-1}.
[b] Wavelength of fluorescence maximum.
[c] The difference between the energies of the absorption onset and the fluorescence maximum.
[d] Full width at the half height of the fluorescence spectrum expressed in energy units.
[e] Quantum yield of fluorescence measured at 165-nm excitation wavelength.
[f] Energy of the gas phase ionization potential.
[g] Energy of the liquid phase ionization potential.
Source: Refs. 25–27, and 38.

regarded as the energy of the fluorescing excited state (S_1). On the contrary, Kimura and Hormes, when they constructed term schemes for excited cyclohexane and n-hexane molecules, used the energy of λ_f as the energy of the lowest lying singlet excited state [28]. Because of the very low yield of fluorescence, other processes, as will be discussed later, chemical decomposition dominates the decay of alkane excited molecules.

As it is obvious from this review, during the last three decades much information has accumulated on the properties of singlet excited alkane molecules. At the same time, little has been known about the properties of triplet excited molecules because there are no direct methods for studying them [29–31]. Our knowledge originates mainly from indirect sources. Because the $S_0 \rightarrow T_n$ optical transitions are spin forbidden, in contrast to $S_0 \rightarrow S_n$ transitions, the optical absorption spectra give little information on the triplet energies. The $S_0 \rightarrow T_n$ transitions do appear in the electron-energy-loss spectra, but the resolution of most of these spectra is poor. The energy of the lowest triplet state of the liquid alkanes can also be estimated from biphotonic sensitization experiments performed with aromatic hydrocarbons in alkanes usually in the solid state [32–34]. In this process, by consecutive absorption of two photons, high-energy triplet states of the aromatic molecules are produced, which by energy transfer may produce the triplet state of the alkane molecule. The transfer is recognized by the fluorescence and phosphorescence properties of the aromatic molecule [33] or by the decomposition of the alkane molecule [32,34]. The T_1 energies estimated from these experiments, e.g., for the cyclohexanes, are all around 5.8–6.0 eV [29,33,34]. Using another technique, i.e., bombardment of thin films of alkanes by low-energy electrons, Leclerc et al. [35] estimated a value of 5.8 eV for the T_1 of n-hexane.

Therefore, the energy separation between S_1 and T_1 is ~1 eV. For Rydberg-type excitations, the separation is, as a rule, ~0.1 eV; for valence-type excitations, it is 1–2 eV [22]. The separation of ~1 eV is in favor of valence excitations in agreement with the results of those quenching experiments in which the "kinetic" distance was found to be comparable with the molecule diameter (Sec. 3.1).

Based on energy-transfer experiments in some works, low-energy (and possibly long-living) alkane triplet state is also indicated; however, there is no direct proof for its existence [7,36,37].

According to precise photoconductivity measurements, the ionization onset, which is usually taken as the ionization potential, is ca. 1.5 eV lower in liquid alkanes than in the gas phase [38]. The ionization potentials in liquid and gas phases (I_l and I_g, respectively) are related by the equation:

$$I_1 = I_g + P_+ + V_o \tag{1}$$

where V_o is the lowest energy of the electron-conducting band and P_+ denotes the polarization energy of the positive ion. P_+ can be represented by Born's equation:

$$P_+ \approx \frac{-e^2}{2R_o}\left(1 - \frac{1}{\varepsilon}\right) \tag{2}$$

where e is the elementary charge, R_o is the ion radius, and ε denotes the dielectric constant of the liquid. For liquid alkanes, P_+ has a value between -1.6 and -1.3 eV and V_o is between -0.6 and $+0.2$ eV, so that Eq. (1) predicts a difference between I_1 and I_g of ~1.5 eV in agreement with the experimental values (Table 1).

Passing through the threshold energy for ionization of liquid alkanes, the quantum yield of ionization, Φ_i, increases very slowly with the photon energy [23,26,38–40].

According to conductivity and fluorescence measurements 1.5 eV above the threshold, Φ_i ≈ 0.1–0.4. Ausloos et al. used chemical decomposition measurements to estimate the ionization quantum yield of liquid cyclopentane: they reported $\Phi_i = 0.26 \pm 0.05$ and 0.4 ± 0.1 at energies of 1.3 and 2.9 eV, respectively, above the liquid phase ionization onset [40]. Thus, 1–3 eV above the ionization threshold the photon absorption mostly produces neutral excited molecules, so-called superexcited molecules. The photoionization is nearly complete at 16- to 20-eV photon energies. Ostafin and Lipsky suggested that the initially produced $e^- + RH^+$ geminate ion pair first rapidly converts internally to an ion-pair state ca. 1.6 eV below the liquid phase ionization threshold, then collapses to generate the fluorescent state of RH with an energy close to its absorption threshold value [26]. Therefore, in (singlet) recombination fluorescing excited molecules form with a high probability.

2.2. Excitation Sources

As mentioned earlier, the light absorption of alkanes causing electron excitation is in the far-UV range, where the emission of the so-called "continuous light sources" is little, e.g., that of high-pressure mercury or xenon lamps. For photodecomposition measurement, mostly line-emitting sources are applied. They give a light intensity of 10^{15} to 10^{16} photon sec^{-1}. The photon energies of a nitrogen lamp emitting at 174 nm (7.1 eV) [41], a bromine lamp emitting at 163 nm (7.6 eV) [42], and a xenon lamp emitting at 147 nm (8.4 eV) [43,44] are just slightly above the VUV absorption onset of alkanes, but they are lower by 2–3 eV than the gas phase ionization potential and by 0–1.5 eV than the liquid phase ionization onset. The energies of the krypton and argon lamps emitting at 124 nm and at 105–107 nm (10 eV and 11.6–11.8 eV) are the ranges of gas phase ionization potentials [43]. In liquids, the absorption takes place in an extremely thin layer (~1 μm); thus, to avoid high local conversion and rise of temperature, vigorous stirring is necessary. Some types of lasers were also utilized for exciting alkanes, e.g., following ArF* excimer laser excitation with 193 nm (6.34 eV) the final products were analyzed [45]. Other lasers (e.g., N_2 laser, Nd^+ YAG laser, and some dye lasers) in double or multiphoton absorption mode because of the weak intensities were used only for fluorescence measurements or transient absorption measurements [46–50]. In fluorescence lifetime measurements applying the single-photon counting technique, the VUV component of the radiation induced by electric discharges in medium-pressure (1–10 bar) gases (H_2, N_2, Xe, or air) was also used [51,52]. In photon absorption and emission studies, synchrotron radiation is often applied [53–56].

Finally, ionizing radiations should be mentioned. A disadvantage of applying ionizing radiation is that the excited molecule formation is a delayed process; most excited molecules form probably in electron–cation neutralization [7]. Another disadvantage is that in the irradiated liquid differently excited molecules and ions are present, including fragments, which may cause complications, e.g., in energy-transfer measurements.

2.3. Lifetime of the Excited Molecules

For the determination of the excited-state lifetime, τ, energy-transfer measurements in steady-state experiments, florescence decay measurements, or transient photon absorption measurements were applied.

In steady-state quenching experiments [57,58], usually the ratio of quantum yields of products or fluorescence photons, respectively, measured in the absence (Φ^0) and presence (Φ) of the quencher showed a linear dependence on the quencher concentration c. From the linear dependencies, $\Phi^0/\Phi = 1 + Kc$, the K quenching parameters were calculated: $K = k_d\tau$. When energy-transfer experiments were used to deduce the τ values, the quenching was assumed to proceed with a diffusion-controlled rate, and the transfer was supposed to take place when the acceptor and donor molecules were in contact (no static quenching). The k_d transfer rate coefficients were calculated using the usual diffusional equations [24,25,59] (see Sec. 3.1).

In Table 2, we show the room temperature τ's for some selected alkanes, and among them there are lifetime data that were determined in quenching experiments. (Data for larger groups of alkanes are discussed in Refs. 48, 49, 51, 59–61, 72, and 73.) For the

Table 2 S_1 Lifetimes (nsec) of Selected Liquid Alkanes at Room Temperature Determined in Steady-State Excited State Quenching or in Transient Fluorescence and Absorption Studies

Alkanes	Quenching	Fluorescence	Absorption
n-Hexane		0.3 [77]; ~0.32 [48]; ≤0.7 [59]; 0.7 [73]	0.25 [47]; 0.28 [49]
n-Heptane		0.73 [48,60]; 0.9 [51]; 1.05 [59]; 1.2 [73]	0.65 [49]
n-Octane		1.08 [48]; 1.35 [81]; 1.47 [51]; 1.5 [59]; 1.6 [73]	>1.0 [49]
n-Decane		1.76 [48]; 2.4 [73,81]; 2.5 [51,59]; 2.6 [60]; 3.0 [54]	>0.6 [49]
n-Dodecane		2.5 [48]; 3.1 [60]; 3.7 [81]; 3.9 [51,59]; 4.0 [73]; 4.2 [54]	>2.8 [49]
n-Tetradecane		3.01 [48]; 4.3 [50,59]; 4.4 [73]	
n-Hexadecane		3.7 [48]; 5.2 [59]	>3.0 [49]
2-Methylbutane		0.75 [59]	
2,3-Dimethylbutane		1.0 [81]; 1.4 [59]	
Cyclohexane(CH)	1.2 [59]	0.3 [60]; 0.68 [51]; 0.9 [46]; 1.1 [59,69,78]; 1.2[72]	>0.3 [49]; 1.0 [47]
MethylCH		0.65 [51]; 0.69 [79] 0.81 [48]; 1.2 [72]; 1.25 [59]	
EthylCH		1.25 [59]	
cis-1.2-DimethylCH		1.02 [79]; 1.15 [59]	
trans-1,2-Dime.CH		0.9 [59]; 1.11 [79]	
cis-Decalin		2.1 [61,63]; 2.18 [51]; 2.45 [54]	
trans-Decalin		2.24 [48]; 2.7 [59]; 2.82 [51]; 2.86 [79]; 3.0 [54]	>1.4 [49]; 3.0 [47]
Bicyclohexyl		1.58 [51]; 1.6 [60]; 1.62 [48]; 1.8 [54]; 1.9 [72]	
n-Dodecane-d$_{26}$		6.5 [83]	
Cyclohexane-d$_{12}$		2.1 [83]; 2.5 [78]	2.7 [47]
Cyclopentane	≤0.1 [59]		0.1 [49]
Cycloheptane	0.1 [59]		
Cyclooctane	0.3 [59]		
2,2,4-Trimethylpentane	≤0.2 [59]		<0.01 [47]; 0.04 [49]

fluorescence lifetime of cyclohexane, the quenching experiments gave 1.2 nsec, a value that agrees with the results of the newer fluorescence and absorbance decay measurements.

The lifetimes of the alkane excited states were generally measured by detecting the decay of fluorescence. The τ values obtained by different excitations, by x-ray pulses [60,62], high-energy electron pulses [61,63–74], pulses of photons [51,52,54,55,75–77], laser pulses causing double or multiphoton excitations [46,48,50,78–83], are generally in good agreement indicating that fluorescence originates in each case from the same excited state. These fluorescence lifetimes also agree with the lifetimes that were measured by detecting the absorbance decay of excited molecules. As regards τ, cyclohexane is a special case. For its excited-state lifetime, Henry and Helman in 1972 suggested 0.3 nsec, based on the very first direct measurements [60]. Two years later, Ware and Lyke found 0.68 nsec [51]. However, over 10 later measurements yielded lifetimes in the 0.8- to 1.2-nsec range: the most probable room-temperature value is 1.0 nsec.

As shown by Tagawa et al. [74], the alkane excited molecules have a broad absorption band in the visible region with maxima increasing from ~430 to ~680 nm between C_5 and C_{20} for the n-alkanes. This spectrum strongly overlaps with the absorption spectrum of the radical cations: with low carbon atom number alkanes the two maxima practically coincide. With increasing carbon atom number, the red shift in the radical cation absorbance is stronger than in the S_1 molecule absorbance [47,49,84–86]. The decay of the excited molecule absorbance was composed of two components with 0.1- and 1.0-nsec decay times in cyclohexane, and 0.17 and 2.7 nsec in perdeuterocyclohexane [47]. The nature of the faster-decaying component is as yet unclear.

As mentioned before, the excited-state lifetimes determined by photon and (low-LET) ionizing radiation excitations practically coincide. However, when 5- to 14-keV soft synchrotron x-rays were used for the irradiation in the experiments of Holroyd et al. [87], instead of the usual fast electrons with energies in the MeV range, shorter τ's were measured than in VUV photon irradiation. The lifetimes and singlet yields decreased with decreasing x-ray energy, thus decreased with increasing LET value. This decrease between 14 and 5 keV for *cis*- and *trans*-decalin and *n*-dodecane is ca. 15%. The decrease was attributed to the quenching of alkane excited molecules by alkyl radicals whose concentrations may be as high as 10 mmol dm^{-3} in the high ionizing density tracks. Shibata et al. [88] explained similarly the decrease of n-dodecane fluorescence lifetime with increasing LET in heavy-ion radiolysis.

In contrast to the liquid phase lifetimes, those in the gas phase showed marked dependence on the energy of excitation [52,55,75]. They decreased with increasing excitation energy, and for some alkanes the τ's at the red edge in the vapor phase were longer than the liquid phase lifetimes. For instance, in the 170-nm photoirradiation of cyclohexane at 1 Torr pressure, Ware and O'Connor found $\tau = 1.2$–1.3 nsec [52], whereas Wickramaaratchi et al. measured 1.5 ± 0.2 nsec [75]. The liquid phase value is around 1 nsec. In the literature, there is no satisfactory explanation for this phase effect. In the lower pressure range, the increase of the vapor pressure, or added inert gases increased the lifetime because of vibrational cooling of excited alkane molecules. Most of the energy dependence in the gas phase is probably because of a competition between relaxation and the chemical decomposition.

In photoirradiated solid cyclohexane (freezing point 6.5°C), much higher fluorescence quantum yields and longer fluorescence lifetimes were observed than in the liquid phase [89]. In solid Ar matrices, the fluorescence characteristics, energy dependence of the lifetime and intensity, were found to be very similar to these characteristics in the gas phase. This points to the importance of cyclohexane–cyclohexane interactions to determine the excited-state characteristics in the liquid phase [76].

The lifetime of the triplet excited molecule with energy ~1 eV below the S_1 energy is probably extremely short: the triplet excitations are suggested to take place to dissociative states [90]. Such conclusion can be obtained from the biphotonic sensitization experiments mentioned in connection with the energy of the triplet states (Sec. 2.1): if the decomposition had not been fast enough ($\tau \leq 10^{-12}$ sec) due to the back transfer of energy, the sensitization could not have been observed [29].

2.4. The Radiative Rate Coefficient at ~25°C

By using the techniques mentioned before, room-temperature τ's for about 50 alkanes were determined. In Fig. 2, we show the fluorescence quantum yields as a function of lifetimes. The Φ_f values were generally taken from the work of Rothman et al. [25]; most of the fluorescence quantum yields were measured using 165-nm photons for excitation. This wavelength is close to the absorption onset of most alkanes and (with the exception of the smaller molecules) the measured quantum yield is close to the fluorescence quantum yield of the relaxed S_1 molecules [26]. The plot in Fig. 2 is similar to the plot we published in Ref. 59 using only our measurements. Here we use practically all the data that are available in the literature. For most of the alkanes, several lifetime measurements were published. When

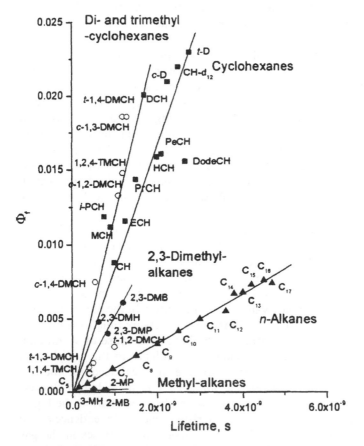

Figure 2 Relationship between lifetime (τ) and fluorescence intensity (Φ_f) for a large group of alkanes.

these measurements do not differ considerably we use the average. When there are larger differences between the results of different laboratories we try to select the most probable value. In Fig. 2, the points representing compounds with similar chemical structure lie around the same straight line. The slope of the lines in the figure is the ratio of the fluorescence intensity and lifetime (or in other words the product of the fluorescence intensity and the decay rate coefficient). Therefore, the slope gives the so-called radiative rate coefficient, R, which is one of the most important photophysical characteristics of excited states. R is a rate coefficient with which the excited molecules would disappear from the solution if there was no chemical decomposition and only fluorescence existed. Five groups can be distinguished; as the sixth group, we mention the nonfluorescing alkanes whose excited-state lifetimes are extremely short (Table 2):

n-Alkanes, $R \approx 1.6 \times 10^6 \text{ sec}^{-1}$
Methylalkanes, $R \approx 2 \times 10^5 \text{ sec}^{-1}$
Vicinally branched dimethylalkanes, $R \approx 5.3 \times 10^6 \text{ sec}^{-1}$
Cyclohexane and alkylcyclohexanes including deuterated cyclohexane and decalins,
$\quad R \approx 8 \times 10^6 \text{ sec}^{-1}$
Di- and trimethylcyclohexanes, $R \approx 1.2 \times 10^7 \text{ sec}^{-1}$
Nonfluorescing alkanes, $R < 3 \times 10^3 \text{ sec}^{-1}$

The similarity of R within a class of alkanes (the fact that within a class the points are on the same straight line) indicates that the properties of excited states are similar. This is also supported by Lipsky's measurements on the fluorescence spectra. Within a class the value of the Stokes' shift (Δ) and the half width of the fluorescence peak (σ_f) are nearly constant (see also Table 1).

For n-alkanes the lifetime increases with the number of carbon atoms in the molecule together with the increase of the fluorescence intensity, whereas the radiative rate coefficient remains constant with a value of $\sim 1.6 \times 10^6 \text{ sec}^{-1}$; that is, in the absence of other competing reactions the excited molecules would decay with fluorescence exhibiting a lifetime of about 600 nsec. Addition of a branch to the carbon skeleton greatly influences the photophysical and photochemical properties. In comparison with n-alkanes, σ_f is larger, the absorption onset is slightly red-shifted, and the emission maximum is strongly red-shifted (Table 1). These properties point to large nuclear distortions between ground state and the first excited state (large Stokes' shift, Δ). The large nuclear distortions lead to rapid decay, low Φ_f, and low R. The high Δ values of the vicinally substituted 2,3-dimethylalkanes and other optical properties indicate larger nuclear distortions than those of the singly substituted ones. These compounds exhibit large fluorescence quantum yields and moderate lifetimes. Their radiative rate coefficients are the highest among the aliphatic alkanes: $R \approx 5.3 \times 10^6 \text{ sec}^{-1}$.

The S_1 excited-state lifetimes of cyclohexane and alkyl cyclohexanes are relatively short, all around 1–2 nsec; however, their Φ_f values are relatively high with the result of relatively high radiative rate coefficient: $\sim 8 \times 10^6 \text{ sec}^{-1}$. R is the largest for the di- and trimethylcyclohexane group, $\sim 1.2 \times 10^7 \text{ sec}^{-1}$.

In the group of nonfluorescing alkanes the lifetimes are very short, $\tau \le 0.3$ nsec; an upper limit of $R < 3 \times 10^3 \text{ sec}^{-1}$ is estimated. The absence of fluorescence for these compounds may have two causes: the low R, i.e., low rate coefficient of the $S_1 \rightarrow S_0$ radiative transition and the short lifetime, i.e., the very fast chemical decomposition. In the C_5 and C_7–C_{10} cycloalkanes the ring strain, which is mainly caused by the repulsive interaction of their unfavorably displaced H atoms, may enhance the rate coefficient of the chemical decay by C–H decompositions. In the excited sates of the geminally branched alkanes the excess energy strongly localized to the branch points can cause rapid local decomposition.

2.5. Temperature Dependence of the S_1 Lifetimes, Depopulation of S_1 Excited Molecules

In the early eighties, several papers [79–81] reported that the fluorescence lifetime exhibits characteristic temperature dependence: it increases with the decreasing temperature and tends to level off at low temperature. The Arrhenius plots, log τ^{-1} versus T^{-1}, did not give straight lines. In photophysics, such behavior is usually interpreted in terms of several competing decay channels, and the description is usually made by the sum of several Arrhenius-type decays. Because at higher temperatures the k decay rate coefficient $(= \tau^{-1})$ can be well described by a single Arrhenius-type equation, the following description was introduced by Orlandi et al.:

$$\tau^{-1} = k = k_0 + A \exp(-E_a/RT) \tag{3}$$

According to this equation the decay is composed of a temperature-independent and a thermally activated part [55,67,75–77,79–83]. For cis- and trans-decalins, however, the temperature dependence was found to be very small [62,67,79,80]. Because of the small temperature dependence, the activated and nonactivated processes were not separated.

The k_0, A, and E_a parameters obtained for a few alkanes are collected in Table 3. k_0 is around 10^8 sec^{-1}, $A \approx 10^{11}$ to 10^{12} sec^{-1}, and $E_a \approx 10$ to 20 kJ mol^{-1}. In principle, the decay of excited states may involve $S_1 \rightarrow S_x$-type internal conversion transitions [IC, where S_x is some singlet state that gives the product(s) of chemical decomposition] and $S_1 \rightarrow T_n$-type intersystem crossing processes (ISC). The temperature-independent decay was attributed, on the basis of the size of the rate parameter ($k_0 \approx 10^8$ sec^{-1}), to $S_1 \rightarrow T_n$-type intersystem crossing. At the same time the temperature-activated decay with a frequency factor of $A \approx 10^{11}$ to 10^{12} sec^{-1} was attributed to an internal conversion process that takes place by overcoming a barrier of $E_a \approx 10$–20 kJ mol^{-1} and leads finally to some

Table 3 Temperature-Independent Term (k_0), Preexponential Factor (A), and Activation Energy (E_a) Obtained by Using Eq. (3) for Selected Groups of Alkanes

Alkanes	k_0, sec^{-1}	A, sec^{-1}	E_a, kJ mol^{-1}	$\Phi_{nonact.}$	$\Phi_{act.}$	$\Phi_{radical}$	$\Phi_{molec.}$
n-Octane	2.9×10^8	1.1×10^{11}	14	0.4	0.6	0.2	0.8
n-Decane	1.5×10^8	5.3×10^{11}	18	0.3	0.7	0.2	0.8
2,3-Dimethylbutane	5.2×10^8	2.0×10^{11}	15	0.5	0.5	0.5	0.5
Cyclohexane (CH)	3.6×10^8	1.6×10^{11}	13	0.3	0.7	0.2	0.8
MethylCH	3.4×10^8	7.5×10^{11}	16	0.2	0.8	0.3	0.7
trans-1,4-DimethylCH	2.9×10^8	6.8×10^{11}	18	0.4	0.6	0.4	0.6
cis-1,3-DimethylCH	3.6×10^8	2.1×10^{12}	20	0.4	0.6	0.5	0.5
IsopropylCH	7.1×10^8	7.5×10^{12}	18	0.6	0.4	0.6	0.4

The table also shows the quantum yields of nonactivated and activated processes and the yields of the radical-forming and unimolecular reactions.
Source: Refs. 29, 67, 79, and 81.

singlet state of the products. We summarized the reaction possibilities and their identification in Fig. 3. (The chemical reactions shown in the figure will be discussed later.)

The identification of the temperature-dependent decay with $S_1 \rightarrow S_x$ IC and the temperature-independent decay with $S_1 \rightarrow T_n$ ISC was confirmed by investigations on xenon heavy-atom effect and on deuterium effect as well. Heavy atoms are known to increase the rates of spin-forbidden processes, whereas the rates of spin-allowed processes remain essentially unchanged. In agreement with the expectations, the A preexponential factor and the E_a activation energy were found to be unaffected by the presence of Xe added to the samples at a concentration of 0.1–0.4 mol dm^{-3} [67,82]. At the same time k_0 was found to increase. For deuteration, the effect was just the reverse. The IC process is connected with chemical decomposition; for the investigated compounds the chemical decomposition was mainly H$_2$ elimination, and therefore its rate should be influenced by deuteration. At the same time, the ISC process is connected with multiplicity changes only; therefore the effect of deuteration on that process should be negligible, in agreement with the experimental findings [83].

2.6. Characteristic Decomposition Modes of the Low-Energy Excited Alkane Molecules

The photodecomposition of n-alkanes at excitation energies slightly above the absorption onset involves both C–H and C–C bond decompositions [18]. The dominant process is the C–H scission, $\Phi(H_2) = 0.8$–0.9, and the contribution of C–C decomposition is small. In the photolysis of cyclohexane, cycloheptane, cyclooctane, and cyclodecane, however, only hydrogen evolution was observed [$\Phi(H_2) \approx 1$]. In the photolysis of cyclopentane, C–C decomposition also takes place with low yield. In the reactions of cyclopropane and

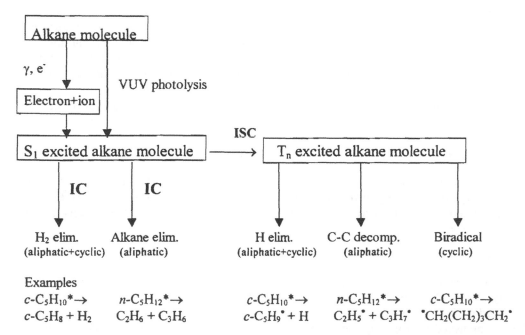

Figure 3 Schematic representation of formation and decay of alkane S_1 excited molecules.

cyclobutane, the ring decomposition is the decisive process; the yield of C–H decomposition is very low. The branchings on the carbon skeleton in isoalkanes and alkyl cycloalkanes increase the yield of C–C decompositions: the highly branched alkanes such as 2,2-dimethylbutane or 2,2,4-trimethylpentane decompose mainly by C–C bond scissions at the branch points.

H_2 and H Elimination

The mechanism of H_2 formation involves both unimolecular H_2 elimination, when both of the H atoms of a hydrogen molecule originate from the same alkane molecule, and H atom elimination with subsequent H atom abstraction reaction. We show the mechanism on the example of cyclohexane photolysis ($hv = 7.6$ eV) [91]:

$$c - C_6H_{12} \xrightarrow{\ hv\ } c - C_6H_{12}* \tag{4}$$

$$c - C_6H_{12}^* \longrightarrow c - C_6H_{10} + H_2 \qquad \Phi(H_2)_u = 0.85 \tag{5}$$

$$\longrightarrow c - C_6H_{11}^\bullet + H \qquad \Phi(H) = 0.14 \tag{6}$$

$$\longrightarrow c - C_6H_{12} + hv' \qquad \Phi_f = 0.01 \tag{7}$$

$$H + c - C_6H_{12} \longrightarrow c - C_6H_{11}^\bullet + H_2 \tag{8}$$

$$2c - C_6H_{11}^\bullet \xrightarrow{\ k_d\ } c - C_6H_{12} + c - C_6H_{10} \tag{9a}$$

$$\xrightarrow{\ k_c\ } (c - C_6H_{11})_2 \tag{9b}$$

The H atom detachment in Eq. (6) is followed by H atom abstraction from another cyclohexane molecule in Eq. (8) and the cyclohexyl radicals disappear in disproportionation and combination reactions giving in nearly equal amounts cyclohexene and dicyclohexyl end products. In the liquid phase, the secondary decomposition of the primarily formed energy-rich reaction products is of minor importance because of the effective collisional deactivation. This is in contrast to the gas phase reaction, where the primarily formed products readily undergo decomposition in the absence of deactivation (at low pressures), e.g.:

$$c - C_6H_{12}* \xrightarrow{\ -H_2\ } [c - C_6H_{10}]^+ \longrightarrow C_2H_4^+ C_4H_6 \tag{5a}$$

The yield of unimolecularly formed cyclohexene, and by that the yield of unimolecular H_2 elimination (quantum yield in photolysis or G value in radiolysis), $X(H_2)$, may be calculated from the yields of alkene and dimer dehydrogenation products by the relation [91,92]:

$$X(H_2)_u = X(\text{alkene}) - (k_d/k_c)X(\text{dimer}) \tag{10}$$

where k_d/k_c stands for the disproportionation-to-combination ratio of the relevant radicals. Using a value of 1.1 for this ratio and the yields of cyclohexene and bicyclohexyl $\Phi(H_2)_u = 0.85$ and $G(H_2)_u = 1.3$–1.5 has been calculated. The molecular and atomic channels can also be distinguished by applying the isotope dilution technique, introduced by Dyne [93] and Dyne and Denhartog [94].

The quantum yields and G values of H_2 elimination for a larger group of alkanes are collected in Table 4; the values were mostly determined by using Eq. (10). Because in the photolysis and radiolysis of the n-alkanes shown in the table various kinds of radicals are produced simultaneously (e.g., n-propyl and sec-propyl from propane), the weighted averages of several k_d/k_c values were used in Eq. (10). The ratios can be determined by suppressing the radiolytic alkene and dimer yields in the presence of radical scavenger (e.g., I_2):

$$\frac{k_d}{k_c} = \frac{\Delta G(\text{alkene})}{\Delta G(\text{dimer})} \tag{11}$$

As shown in the table, during photolysis of n-alkanes and C_5–C_{10} cycloalkanes, $\Phi(H_2)_u$ is 0.7–0.9, slightly exceeding the values of 0.6–0.7 measured at low photon energies in the gas phase [18,19]. Cyclopropane and cyclobutane undergo H_2 elimination to a negligible extent: $\Phi(H_2)_u \approx 0.02$. The value of $\Phi(H_2)_u$ in branched alkanes is much smaller than in normal and cyclic alkanes. The results obtained in the gas phase show that the yields decrease with the increasing branching [18]: for instance, in the series of C_5H_{12} isomers, the isomer with maximum branching, neopentane, essentially does not eliminate hydrogen [16].

In connection with the gas-phase studies, it was first suggested at the beginning of the 1960s [95,96] that the H_2 elimination from the lower-molecular-mass alkanes takes place mainly from a single carbon atom (1,1 elimination) and not from two neighboring ones

Table 4 Quantum Yields and G Values of Unimolecular H_2 Elimination and the G Values of S_1 Molecule Yields Estimated from the Yields of Unimolecular Hydrogen Elimination

Alkanes	$\Phi(H_2)_u$	Energy, eV	$G(H_2)_u$	$G(S_1)$
Propane	0.76	8.4	1.4	1.8
n-Pentane	0.82	8.4	1.4	1.7
n-Hexane	0.9	7.6	1.48	1.6
n-Heptane	0.9	7.6	1.31	1.5
n-Octane	0.92	7.6	1.4	1.5
Cyclopentane	0.88	7.6	1.6	1.8
Cyclohexane (CH)	0.85	7.6	1.3	1.5
Cycloheptane	0.89	7.6	1.3	1.6
Cyclooctane	0.87	7.6	1.9	2.2
Cyclodecane	0.91	7.6	2.0	2.3
2,2,4-Trimethylpentane	0.08	8.4		
Methylcyclopentane	0.46	7.6		
Methylcyclohexane	0.7	7.6		
Ethylcyclohexane	0.63	7.6		
cis-1,2-DimethylCH	0.28	7.6		
$trans$-1,2-DimethylCH	0.21	7.6		
cis-1,3-DimethylCH	0.5	7.6	1–1.5	2–3
$trans$-1,3-DimethylCH	0.53	7.6	0.5–1	1–2
cis-1,4-DimethylCH	0.45	7.6		
$trans$-1,4-DimethylCH	0.45	7.6		

Source: Refs. 18, 29, and 92.

(1,2 elimination). In liquid-phase radiolysis experiments, 1,1 elimination was also demonstrated by deuterium-labeling experiments [97,98]:

$$-CH_2-CH_2-* \longrightarrow \underset{\substack{| \quad\quad | \\ -CH\cdots CH-}}{\overset{H\cdots H}{}} \xrightarrow[\text{1,2 elimination}]{-H_2} -CH=CH- \qquad (12)$$

$$-CH_2-CH_2-* \xrightarrow{-H_2} -\ddot{C}-CH_2- \xrightarrow[\text{1,1 elimination}]{} -CH=CH- \qquad (13)$$

As a result of 1,1 elimination, a carbene intermediate remains that quickly rearranges to alkene. During the photolysis and radiolysis of cycloalkanes with 7–10 carbon atoms, apart from the formation of the usual cycloalkene products, dehydrogenation into cycloalkanes with interesting bridged rings was also observed [99–102]. Similar products were also found in catalytic reactions [103]. The cross-bridged products are formed by stabilization of the carbene in transannular insertion. During photolysis of $C_7 - C_{10}$ cycloalkanes, the distribution of products remaining after H_2 elimination in several solvents [99] agreed with the distribution of products formed during stabilization of carbenes produced in an independent way, by the UV photolysis of cycloalkanone *p*-tosylhydrazones under aprotic conditions, e.g., for the C_8 ring:

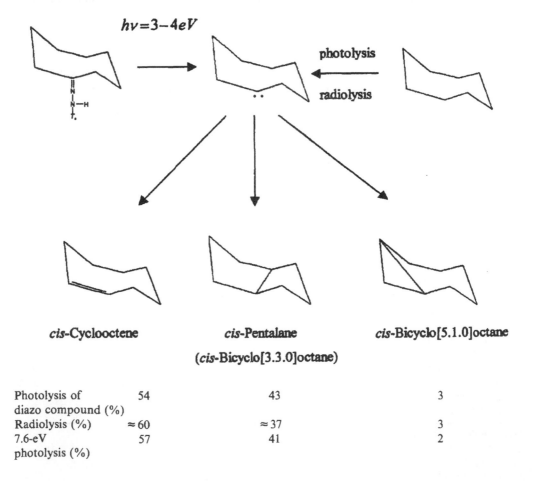

cis-Cyclooctene   *cis*-Pentalane   *cis*-Bicyclo[5.1.0]octane
                                (*cis*-Bicyclo[3.3.0]octane)

Photolysis of diazo compound (%)	54	43	3
Radiolysis (%)	≈ 60	≈ 37	3
7.6-eV photolysis (%)	57	41	2

It means that the H_2 elimination from these compounds takes place exclusively through 1,1 elimination [99,100].

Alkane and Alkyl Radical Elimination

In the alkane elimination, a smaller alkane and an alkene molecule are formed in the decay of excited molecule without intermediate radicals. A few examples of this reaction are shown in Table 5. Homolytic split of alkane molecules to two alkyl radicals and the subsequent reactions of radicals may give the same products. Radical-scavenging techniques or deuterium-labeling studies may be used to distinguish the two processes. Although liquid phase elimination has rarely been investigated, a great amount of data are available in the literature concerning the reaction in the gas phase. For example, Ausloos et al. [16] using partially deuterated compounds, showed that in the photolysis of isopentane during methane elimination (from position 1) the H atom is taken off with comparable probability from positions 2 and 3:

$$\begin{array}{c} 5\ CH_3 \\ | \\ CH_3 \text{---} CH - CH_2 - CH_3 \\ 1\quad 2\quad 3\quad\ 4 \end{array}$$

Alkane elimination has a low yield during the photolysis of liquid n-alkanes (e.g., n-pentane [104,106]). This reaction takes place with high yield only for branched alkanes where it is likely to be a main primary-decomposition step [105,107].

Isomerization of Cycloalkanes into Aliphatic Alkenes

Alkylcyclopentanes and cyclohexanes rearrange to aliphatic alkenes during both photolysis and radiolysis [108–113]. The double bond in the product aliphatic alkenes can be found connected to one of the carbon atoms taking part in ring opening. The derivation of

Table 5 Quantum Yield of Alkane Elimination in the Photolysis of Some Liquid Alkanes

	Φ
Propane, 8.4 eV [105]	
$C_3H_8^* \rightarrow C_2H_4 + CH_4$	0.1
n-Pentane, 8.4 eV [104]	
$C_5H_{12}^* \rightarrow C_2H_4 + C_3H_8$	0.06
$\rightarrow C_3H_6 + C_2H_6$	0.05
$\rightarrow C_4H_8 + CH_4$	0.1
Isobutane, 8.4 eV [105]	
$C_4H_{10}^* \rightarrow C_4H_8 + CH_4$	~0.5
2,2-Dimethylpropane, 7.6 eV [107]	
$C_5H_{12}^* \rightarrow iso\text{-}C_4H_8 + CH_4$	0.57 ± 0.14
2,2,4-Trimethylpentane, 7.6 eV, 8.4 eV [104,106]	
$C_8H_{18}^* \rightarrow iso\text{-}C_4H_8 + iso\text{-}C_4H_{10}$	0.36
$\rightarrow C_3H_6 + neo\text{-}C_5H_{12}$	0.05
$\rightarrow C_7H_{14} + CH_4$	0.17

the alkenes formed is shown schematically on the example of ethylcylohexane:

$CH_3-CH_2-CH=CH-CH_2-CH_2-CH_2-CH_3$ (14a)

$CH_3-CH_2-CH_2-CH_2-CH_2-CH_2-CH=CH_2$ (14b)

$CH_3-CH=CH-CH_2-CH_2-CH_2-CH_2-CH_3$ (14c)

fragmentation to alkenes and cyclopropanes (14d)

reclosure

From this compound as well as from other alkylcyclohexanes the yield of ring-opening products is relatively small, about $\Phi = 0.1$–0.4, and $G = 0.3$–1.6 [108,110] (Table 6), while usually the main decomposition process is the hydrogen formation, which leaves the cyclic structure intact. Here, and with the other alkylcyclohexanes and alkylcyclopentanes, the scission of the ring to smaller molecular mass alkenes and cyclopropane derivates was detected with very low yield.

It was an important observation from the point of view of the opening mechanism that in dimethylcyclohexane isomers the yield of transformation from *cis* into *trans* and from *trans* into *cis* also takes place. In the 7.6-eV photolysis of 1,2-dimethylcyclohexanes the *cis* form isomerizes to *trans* with a yield of $\Phi(c\rightarrow t) = 0.12$, while the reversed reaction has a yield of 0.07. In the photolysis of 1,3-dimethylcyclohexanes $\Phi(c\rightarrow t) = 0.07$ and $\Phi(t\rightarrow c) = 0.15$,

Table 6   Quantum Yields and G Values of C_nH_{2n} Aliphatic Alkenes in the 7.6-eV Photolysis and Radiolysis of C_nH_{2n} Cycloalkanes

C_nH_{2n} cycloalkane	Φ (aliphatic alkenes)	G (aliphatic alkenes)
Cyclopentane	0.04	0.88
Cyclohexane	No product	0.34
Methylcyclohexane	0.10	0.62
cis-1,2-Dimethylcyclohexane	0.37	1.50
trans-1,2-Dimethylcyclohexane	0.39	1.60
cis-1,3-Dimethylcyclohexane	0.10	0.71
trans-1,3-Dimethylcylohexane	0.10	0.60
cis-1,4-Dimethylcyclohexane	0.16	0.68
trans-1,4-Dimethylcyclohexane	0.17	0.73

In radiolysis a fraction of C_nH_{2n} alkenes forms in bimolecular radical reactions.
Source: Refs. 18, 29, 108, 110, and 111.

while for the 1,4-dimethylcyclohexanes $\Phi(c{\rightarrow}t) = 0.13$ and $\Phi(t{\rightarrow}c) = 0.06$ [108]. Similar phenomena were observed during the radiolytic processes as well [110,112]. In dimethylcyclohexanes the methyl groups as substituents can be of equatorial position, falling into the "plane" of the ring, or of axial position, roughly perpendicular to this. If the chair form of cyclohexane rearranges into another chair form, the axial and equatorial substituents change their positions, i.e., the axial position turns into the equatorial one and vice versa in the dimethyl-cyclohexanes. Both (e,e)↦(e,a) and (a,a)↦(e,a) transformations can take place through the previous dissociation of at least one bond of the molecule.

In all cases when photolytic isomerization into aliphatic alkanes was observed it was found that the quantum yield of primary decomposition calculated from the measured end products is slightly smaller than unity.

On the basis of these results it is assumed that isomerization to aliphatic alkenes proceeds through alkyl biradicals which—apart from alkene formation—can reclose again into the initial cycloalkanes giving in this way "products" not measured by the usual techniques. In 1,2-, 1,3-, and 1,4-dimethylcycloalkanes, reclosing can also result in the opposite isomer, thereby providing a possibility for the observation of the process. The biradical may also fragment yielding ethylene and a smaller biradical. The cyclopropanes observed among the products may form in stabilization of the 1,3-biradicals.

Correlation Between Photophysical and Photochemical Data

Because fluorescence plays a very small role in the depopulation of alkane excited molecules the sum of the quantum yields of the chemical decompositions in the thermally activated and nonactivated channels is practically unity: $\Phi(S_1{\rightarrow}S_x) + \Phi(S_1{\rightarrow}T_n)$ is 1. Using Eq. (3), the temperature dependencies of the product yields formed in the activated and nonactivated channels have the following forms:

$$\Phi(S_1 \rightarrow S_x) = \frac{A \exp(-E_a/RT)}{k_0 + A \exp(-E_a/RT)} \tag{15}$$

$$\Phi(S_1 \rightarrow T_n) = 1 - \Phi(S_1 \rightarrow S_x) \tag{16}$$

According to these equations, the product formation in the $S_1{\rightarrow}S_x$ activated channel is increasing with the increasing temperature, whereas the product formation in the $S_1{\rightarrow}T_n$ nonactivated channel is decreasing. Xenon is expected to have an opposite effect on the yields. The temperature and Xe effect studies showed that the H_2 elimination is related to the activated channel, and the H atom elimination to the nonactivated [67,79,81,113]. In the 7.6-eV photolysis of n-decane between -20 and $45°C$, $\Phi(H_2)_u$ increased from 0.62 to 0.82, whereas $\Phi(H)$ decreased from 0.4 to 0.08. Xe in a concentration of 0.2 mol dm^{-3} decreased $\Phi(H_2)_u$ and increased $\Phi(H)$ by $\sim$20% [67]. In cycloalkane photolysis, the yields of ring-opening products showed a definite nonactivated character: in the temperature range mentioned, the yield of linear octenes from ethylcyclohexane decreased from 0.32 to 0.13; Xe increased the yield by $\sim$25% [113]. From the point of view of activated and nonactivated character the C–C decompositions in aliphatic alkane photolysis were not investigated. However, based on the values of the yields, the alkane elimination was tentatively identified as a singlet decomposition, and the homolytic split to two alkyl radicals as a triplet decomposition [81]. In Table 3 we show for some selected alkanes the quantum yields of the decompositions in the activated and nonactivated channels at 25°C calculated using Eqs. (15) and (16). The table also contains the yields of unimolecular (hydrogen and alkane) eliminations and the yields of radical-forming (H atom elimination, C–C scission to radicals or biradicals) reactions. The correlation between the yields of activated decay channel and the unimolecular decompositions and between the yield of the nonactivated decay and the yield of radical/biradical forming reactions supports the identifications of the individual decomposition types.

The very small temperature dependence in the decay of excited cis- and trans-decalin molecules may be interpreted in terms of absence of thermally activated decay. The chemical decomposition of excited decalin molecules yields a large number of products making the identification of the individual decomposition channels and the determination of their yields rather complicated [81,114]. However, from the distribution of products it is certain that the yield of H_2 elimination is low, and the majority of H_2 formation, $\Phi(H_2) = 0.42$ and 0.50 for cis- and trans-decalin, respectively, originates via H atom formation and H abstraction. Another important reaction is the C–C bond decomposition yielding probably a biradical. The formation of C_4–cyclo-C_6 products, cyclodecenes, and partly also cis→trans/trans→cis isomerizations observed in 7.6-eV photolysis is probably due to the stabilization of these biradicals. Both the H-atom- and the biradical-forming reactions occur with a high probability at the tertiary carbon atoms: previously these reactions were attributed to nonactivated (triplet) decompositions. Therefore, the absence of temperature dependence in the excited molecule decay is probably because of a strong localization of the excitation energy between the two tertiary carbons where there is no possibility for molecular elimination reaction.

2.7. Energy Dependence of the Primary Decomposition

Similarly to the fluorescence quantum yields, the yields of individual primary decomposition steps generally show considerable excitation energy dependence: the yields of the unimolecular H_2 and alkane eliminations and also those of the radical-type decompositions show a continuous variation with photon energy [27,39,42,107,115]. In cyclohexane photolysis the sum of the quantum yields of the two primary decompositions described by Reactions (5) and (6) is practically unity between photon energies 7.6 and 11.6 eV: $\Phi(H_2)_u + \Phi(H) \approx 1$. Close to the absorption threshold the H_2 elimination predominates: at 7.6 eV $\Phi(H_2)_u = 0.85$. The yield decreases with the energy, $\Phi(H_2)_u = 0.78$ and 0.64 at 8.4 and 10.0 eV, and it is only 0.4 at 11.6 eV. Thus, at the highest energy the yields of H_2 and H elimination are comparable.

In the photolysis of cyclopentane [39,116] the energy dependencies of the H_2 elimination and H atom detachment are very similar to the energy dependencies of these processes in cyclohexane photolysis. However, in cyclopentane photolysis ring decomposition is also observed producing 1-pentene and ethylene + $C_3H_6 \Phi(1\text{-pentene}) = 0.05$, $\Phi(C_2 + C_3) = 0.05$ at 7.6 eV]. The yields of both ring decomposition reactions were found to increase with the increasing photon energy at the expense of hydrogen formation.

The product yields in the photolysis of n-pentane, n-hexane, n-octane, and n-decane also show decreasing importance of H_2 elimination and increasing importance of H atom detachment with the increasing photon energy [27,117].

In contrast to the results obtained with n-alkanes and cycloalkanes, H_2 elimination is found to be an unimportant process in the photolysis of neopentane [107]. At 7.6 eV the methane elimination and direct C–C bond cleavage to radicals are the predominant processes:

$$neo\text{-}C_5H_{12}{}^* \longrightarrow CH_4 + iso\text{-}C_4H_8 \qquad \Phi = 0.57 \pm 0.14 \tag{17}$$

$$neo\text{-}C_5H_{12}{}^* \longrightarrow CH_3^\bullet + t\text{-}C_4H_9^\bullet \qquad \Phi = 0.38 \pm 0.14 \tag{18}$$

With an increase in photon energy, the importance of the radical-forming reaction increases at the expense of the methane elimination process.

The energy dependence of the photodecomposition is attributed to an energy-dependent relaxation of the higher excited molecules to the relaxed S_1 state. When the photon energy is higher than the energy of the ionization onset ionic processes may also contribute to the energy dependence. Although there are not enough measured data to make a definite conclusion, from the published data it seems that when the energy is increased above the ionization threshold the yield of the molecular-elimination-type reactions continues to decrease and that of the radical-type reactions continues to increase. Thus the chemical decomposition does not parallel the fluorescence where sometimes an increase in the fluorescence yield was observed (Sec. 2.1). The energy dependence is higher for the smaller molecules than for the larger ones.

The ionization, electron–cation recombination, and the time scale of formation, relaxation, and decay of alkane excited molecules were studied several times in pump-and-probe-type subpicosecond laser experiments [49,118–120]. Sander et al. [49] used 0.4-, psec 248.5-nm laser excitation in the double-photon mode. At their probe wavelength of 497 nm, both the excited molecules and the radical cations contribute to the light absorption. These experiments may supply direct data on the time scale of relaxation. After the fast rise of absorbance during the laser pulse there is a 0.4- to 1.4-psec ultrafast component of absorption decay that they attribute to fragmentation from higher singlet excited states, which competes with internal conversion to the S_1 state of the parent molecules. The fast component is followed by a 5- to 15-psec slower component in which for some alkanes there is a re-rise of absorbance. They attribute the second component to geminate recombination. A third component is due to the S_1 molecule decay. It should be noted, however, that other pump-and-probe experiments, e.g., the *f*sec transient measurements of Long et al. [119], suggest much faster geminate recombination than those proposed by Sander et al. Thereby, these experiments contradict the view of Sander et al.

It would be elegant to finish the part on photophysics and photochemistry of liquid alkanes by giving a picture that unifies the temperature- and energy-dependence results obtained in fluorescence and photodecomposition studies. However, the spectroscopic information available for alkane molecules is not sufficient to identify the exact excited states involved in the radiative and nonradiative processes [55]. Because of the lack of information, there are different views on the positions and identities of excited states involved [52,55,83,121,122].

In Fig. 4 we try to explain and, where it is possible, to unify the different views using the usual simplified two-dimensional potential energy diagram for C–H bonds. The potential energy curve of S_1 along the length of a C–H bond is displaced to higher R_{C-H} nuclear distances (reaction coordinate) to explain the strong nuclear distortions and large Stokes' shift reflected by fluorescence. This is in contrast to the views of Plotnikov [122] and Flamigni and Orlandi [83]: they assumed that the potential energy surfaces of the S_0 and S_1 states are essentially identical, except that S_1 is shifted to higher values by ~7 eV, because, in their opinion, the excitation energy of S_1 is delocalized over the entire molecule. Other authors, however, suggest strongly localized alkane excited states. From the S_1 potential energy curve, by passing through a small E_a energy barrier S_x can be reached, which, according to the suggestion of Orlandi's group, may give the carbene + H_2 decomposition products. Wichramaaratchi et al. suggest a dissociative state denoted by S_2 (this is shown in the figure), which crosses S_1, and the crossing point lies near a very low vibrational level of S_1, thereby promoting $S_1 \rightarrow S_2$ IC. S_2 is assumed to be responsible for two reaction channels: one leading to molecular H_2 and other leading to atomic H [55]. If

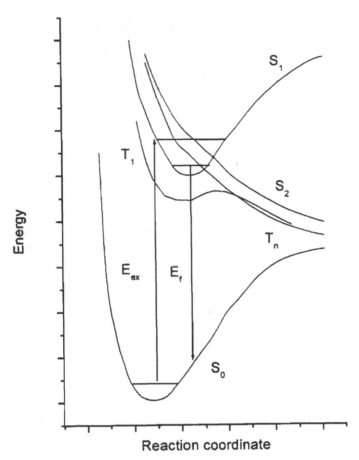

Figure 4   Two-dimensional potential energy curve for C–H bond decomposition.

H atom elimination is of lower importance from this state, $S_1 \rightarrow S_2$ IC in the mechanism suggested by Wichramaaratchi et al. is practically identical with $S_1 \rightarrow S_x$ used in the terminology of Orlandi's group. The authors of Refs. 55, 83, 121, and 122 all agree that there is a dissociative triplet state, T_n, which can easily be populated by ISC from S_1. We show the potential energy curve of this state (T_n) as it has been suggested by Orlandi et al. [83,121]. This state gives only radical decomposition products. In order to explain the results of the biphotonic sensitization experiments (Sec. 2.1) another triplet state ~1 eV below S_1 (T_1) is also needed. Most probably, this triplet is also dissociative.

The proposed mechanism may explain such excited-state characteristics as the temperature dependence of lifetime or λ_{ex} dependence of the fluorescence intensity at low excitation energies. However, in order to explain the energy dependence of the photodecomposition at high energies at least one more dissociative state should be included in the mechanism, which decompose to radicals.

3. ENERGY TRANSFER

Energy transfer from an electronic excited state of an alkane molecule to a solute molecule was first suggested 50 years ago [123]. Since then over 100 papers discussed the various aspects of the transfer, e.g., the properties of excited states involved in the transfer (reaction diameter, lifetime) or the nature of the transfer reaction (diffusional type with short-range contact interaction, Forster type). An important practical aspect of the subject is the protecting effect exerted by different additives on radiation induced decomposition of saturated hydrocarbon systems (lubricants, insulators, polyethylene). Most frequently, the transfer in liquid cyclohexane was studied, but a few papers discussed the transfer reactions also in decalins, 2,2,4-trimethylpentane, and n-alkanes. Mostly the low solute concentration region was studied; however, by photoexcitation or radiolysis in some alkane–alkane mixtures the product formation, and by that the energy transfer, was investigated in the entire concentration range [124–126]. The techniques used to obtain transfer data include stationary photolysis or radiolysis with detection of final products or fluorescence intensity, pulsed photon or electron irradiation, and detection of transient fluorescence or absorbance.

3.1. Energy Transfer from the S_1 Excited State

Based on the magnitude of the transfer rate coefficients and other transfer characteristics it is obvious that in the energy transfer from alkane excited states to solutes long-range interactions play a minor role (Forster-type energy transfer) [70,127]. In the Forster-type energy transfer, the overlap of the donor emission and the acceptor absorption spectra determines the transfer rate coefficients. However, in the quenching of alkane excited states no correlation was found between the overlap integrals and the transfer rate coefficients: the transfer is suggested to take place because of short-range contact interactions. Therefore, in the transfer mutual diffusion of the reactants to each other's proximity has a determining role. The diffusional character is also supported by the temperature dependence of the transfer rate coefficient; this dependence practically agrees with the temperature dependence of the diffusion [42,62,64,128]. However, the simple Stokes–Einstein–Smoluchowski equation (SES), $k_{d,SES} = 8RT/3000\eta$, where η is the solvent viscosity at temperature T, fails to describe properly the values of transfer rate coefficients [42,129–131]. In alkane excited state quenching, where the lifetimes are short ($\tau \approx 1$ nsec), and the quencher concentrations are high, $c = 0.01–0.2$ mol dm^{-3}, perhaps static quenching effects and also transient quenching effects play a role in the transfer in addition to the usual diffusional quenching [131].

When the reaction radius is larger than the sum of the molecular radii of the reacting molecules, there is an instantaneous quenching by the quencher molecules that reside inside the reaction radius of the excited molecule, σ' (also called "kinetic" distance), in the moment of excitation. The term that takes into account instantaneous quenching is called *static quenching* term and usually has the form of $\exp(-NV_1c)$, where c is the quencher concentration and N is Avogadro's number. V_1 is defined as the volume around an excited molecule, where instantaneous quenching occurs:

$$V_1 = \frac{4\pi}{3}(\sigma'^3 - \sigma^3) \tag{19}$$

$\sigma (= r + r_s)$ is the collision distance, i.e., the sum of the molecular radii of the solute and solvent molecules.

The quencher molecules that are near the excited molecule in the moment of excitation have a much higher probability for depleting the excited molecules than the other quencher molecules. For that reason the rate coefficients used to describe the decay of excited molecules often have a so-called *transient term*, $k^0\sigma'(\pi Dt)^{-1/2}$: the transient term has some importance only at short times (t), much shorter than the natural lifetime of the excited molecule.

The third term k^0 is the *usual diffusion controlled rate coefficient*: $k^0 = 4\pi N\sigma' D$, where $D = D_S + D_s$ is the sum of the diffusion coefficients of the solute and solvent molecules.

In the literature there are several, mostly just slightly different, equations that describe the rate coefficient of the diffusion controlled reactions: these equations are usually based on the solutions of Fick II diffusion law assuming that the reaction probability at contact distance is 1. Andre et al. [131] used the following equation to describe the time dependence of excited molecule concentration [RH*] produced by an infinite excitation pulse:

$$[RH^*] = [RH^*]_0 \exp(-NV_1c)\exp[-(k + k^0c(1 + 2\sigma'(\pi Dt)^{-1/2}))t] \tag{20}$$

In Eq. (20), k is the rate coefficient of the excited molecule decay without quencher ($k = \tau^{-1}$). Using this equation in Fig. 5, we show the time dependence of the relative excited cyclohexane molecule concentration in a solution containing 0.05 mol dm^{-3} CCl$_4$. In order to show the effect of static quenching we chose a large reaction radius of $\sigma' = 1.3$ nm. It is

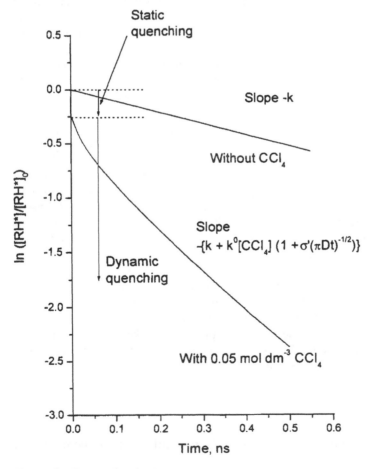

Figure 5   Decay of excited cyclohexane molecules in the pure liquid and in solution containing 0.05 mol dm^{-3} CCl$_4$. Parameters used for the calculations: $k = 1.05 \times 10^9$ sec^{-1} [67], $\sigma' = 1.3$ nm [64], $\sigma = 0.7$ nm, and $D = 3 \times 10^{-9}$ m^2 sec^{-1} [59].

obvious from Fig. 5 that static quenching is important in stationary experiments; however, when the rate coefficient is calculated from decay measurements, the static quenching, because of its instantaneous nature, is unobserved. In pulsed experiments, the quenching rate coefficient has the form of Eq. (21):

$$k_d = k^0[1 + \sigma'(\pi Dt)^{-1/2}] \tag{21}$$

For stationary experiments, Andre et al. [131] suggest the form:

$$k_d = NV_1\tau^{-1} + k^0[1 + \sigma'(D\tau)^{-1/2}] \tag{22}$$

In the usual treatments there is only one parameter to be fitted, i.e., the reaction radius. In practice three cases should be distinguished:

1. The "kinetic" distance exceeds the collision distance, $\sigma' > \sigma$, which is typical for Forster-type energy transfer, but as mentioned before there is evidence against this type of transfer in alkane systems. The reaction radius may also be high if the transfer takes place from a large-orbit Rydberg state. In early studies using steady-state quenching or fluorescence decay experiments, high "kinetic" distances in the order of 1.5 nm were suggested for cyclohexane and the decalins [42,64,129]. Most of the later studies agree, however, that the largest "kinetic" distance values practically agree with the contact distances ($\sim$0.7–0.8 nm) or they are just slightly larger. This result suggests that the relaxed excited states of these alkanes do not have large orbit character and that static quenching has a minor role in the transfer.

2. When $\sigma' \approx \sigma (= 0.8 \pm 0.2 \text{ nm})$ the reaction probability is 1 at the contact distance. This condition is fulfilled for good quenchers such as CCl_4, SF_6, fluorinated hydrocarbons, and in some experiments alkenes or benzene [70,128,130,132]. When $\sigma' \leq \sigma$, the term $NV_1 = 0$.

3. When $\sigma' < \sigma$ not all of the encounters lead to quenching. The p reaction probability may be calculated as the ratio of the measured and calculated rate coefficient using $\sigma' = \sigma$: $p = k_q/k_d$. Reaction probability values determined for larger groups of quenchers in cyclohexane or n-hexadecane have been reported in Refs. 70 and 130.

For the description of partially diffusion-controlled energy-transfer reactions often the reaction scheme and theory introduced by Rehm and Weller [133] as well as by Balzani et al. [134] are applied:

$$\text{RH}^* + Q \overset{k_d}{\underset{k_{-d}}{\leftrightarrow}} \text{RH}^* \cdots Q \overset{k_e}{\underset{k_{-e}}{\leftrightarrow}} \text{RH} \cdots Q^* \overset{k_{-d}}{\underset{k_d}{\leftrightarrow}} \text{RH} + Q^* \tag{23}$$

$$\downarrow 1/\tau \qquad\qquad\qquad\qquad\qquad\qquad \downarrow 1/\tau_q$$

where k_d and k_{-d} are the rate coefficients of the formation (diffusion-controlled rate coefficient) and dissociation of the encounter pair, respectively, and k_e and k_{-e} represent the rate coefficients of the energy transfer step and its reverse, respectively. For the energy-transfer step one can write:

$$\frac{k_{-e}}{k_e} = \exp(\Delta G/RT) \tag{24}$$

$$k_e = k_e^0 \exp(-\Delta G^+/RT) \tag{25}$$

ΔG is the free energy change of the forward energy transfer and k_e^0 and ΔG^+ are the preexponential factor and the free energy of activation of its rate coefficient, respectively.

ΔG can be approximated by the energy differences of excited states of donor and acceptor molecules:

$$-\Delta G \approx \Delta E = E[RH^*(S_1), RH(S_0)] - E[Q^*(S_1), Q(S_0)] \tag{26}$$

ΔG^+ is given by the Rehm–Weller theory:

$$\Delta G^+ = \frac{\Delta G}{2} + \left[\left(\frac{\Delta G}{2}\right)^2 + \left(\frac{\lambda}{4}\right)^2 \right]^{1/2} \tag{27}$$

λ is often an adjustable parameter that is connected with reorganization of the inner coordinates of the reactants and reorganization of solvent molecules. For a given donor molecule and chemically similar acceptor molecules, λ is assumed to be constant. When λ is small and in Eq. (28), which is developed to describe the transfer rate coefficient assuming $1/\tau_q \gg k_d c$, the second term in brackets, $(k_{-d}/k_e^0) \exp(\Delta G^+/RT)$, can be neglected, we get the so-called Sandros equation (29):

$$k_q = k_d \left[\left(1 + \exp\left(\frac{\Delta G}{RT}\right) \right) + \frac{k_{-d}}{k_e^0} \exp\left(\frac{\Delta G^+}{RT}\right) \right]^{-1} \tag{28}$$

$$k_q = \frac{k_d}{1 + \exp(\Delta G/RT)} \tag{29}$$

For excited alkane donors λ is assumed to be high. Eq. (28) was often used to evaluate the molecular structure dependence of the transfer coefficients [69,70,127,135,136]. In Fig. 6, we show the rate coefficients of the energy transfer from excited cyclohexane molecules to alkanes, alkenes, and benzene as a function of the ΔE energy difference based on the experiments of Hermann et al. [135]. The solid lines were calculated with $k_d = 2 \times 10^{10}$ mol^{-1} dm^3 sec^{-1} and $\lambda = 0.5$ or 1 eV. Using a similar treatment to describe the energy transfer from excited cyclohexane molecules, Busi and Casalbore [127] obtained the best fit with $k_d = 3.5 \times 10^{10}$ mol^{-1} dm^3 sec^{-1} and $\lambda = 0.55$ eV. For n-hexadecane solutions a lower value of k_d was applied, $\sim 1 \times 10^{10}$ mol^{-1} dm^3 sec^{-1} [70]. The calculations again strongly support that the transfer occurs with a collision mechanism and the free energy difference determines the rate coefficient of the energy-transfer step.

Seemingly, a different mechanism determines the rate of energy transfer to such inorganic molecules as N_2O, CCl_4, SF_6, and organic molecules with halogen atoms. For these additives, k_q may be high while ΔE is low or negative [130,135]. Those molecules were found to be good quenchers that were good electron scavengers in radiolysis [42]. This correlation suggests that electron transfer and energy transfer may have similar mechanism, e.g.:

$$RH^* + B \rightarrow RH + B^* \rightarrow \text{neutral products} \tag{30}$$

$$RH^* + B \rightarrow [RH^+ + B^-] \rightarrow \text{neutral products} \tag{31}$$

Reactions (30) and (31) may give the same products. In (31) the polarization energy decreases the energy demand for temporal charge separation and it can be exothermic when B has a considerable electron affinity. For aromatic hydrocarbon quenchers (e.g., anthracene) such mechanism leads to dissipation of the excitation energy on the vibrational levels. When the quencher molecules contain Cl or Br atom in the intermediate step, Cl$^-$ or Br$^-$ elimination is expected, e.g., with benzyl chloride additive:

$$C_6H_5CH_2Cl^- \rightarrow C_6H_5CH_2^\bullet + Cl^- \tag{32}$$

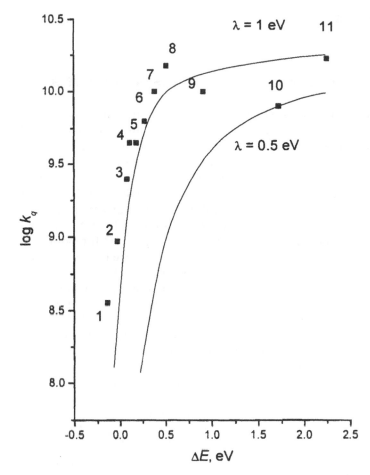

Figure 6 Energy transfer rate coefficients from excited cyclohexane molecules to solutes as a function of the ΔE energy difference: (1) isooctane, (2) n-heptadecane, (3) cyclooctane, (4) *trans*-1,2-dimethylcyclohexane, (5) isopropylcyclohexane, (6) *cis*-1,2-dimethylcyclohexane, (7) decalin, (8) 1-decene, (9) 2-hexene, (10) tetramethylethylene, (11) benzene. (From Ref. 135.)

As final products of the transfer HCl, $C_6H_5CH_3$, $C_6H_5CH_2$-c-C_6H_{11} and $(C_6H_5CH_2)_2$ were found. The same products are expected in the electron-scavenging reaction [130].

3.2. Energy Transfer from Higher-Energy Excited States

The energy transfer from excited alkane molecules to 2,5-diphenyloxazole (PPO) was frequently investigated, often with unusual results. For example, Kimura and Hormes investigated the transfer using for excitation 3.3–10.5 eV photons from a synchrotron source. They reported extremely effective energy transfer at energies 7.81 and 9.67 eV in cyclohexane. The transfer at 7.81 eV was tentatively ascribed to energy transfer from the eighth peak of the Rydberg series with a diameter of ~ 3 nm (static quenching). They did not observe similar efficient transfer in n-heptane [28]. Lipsky et al. also observed efficient transfer in several saturated hydrocarbons at photon energy 7.7 eV. According to Wang et al. [137] and Krishna and Lipsky [138], the transfer takes place from some unidentified ion-pair state of alkanes.

4. FORMATION AND YIELD OF ALKANE EXCITED STATES IN RADIOLYSIS

The yield determined in a certain type of experiment usually strongly depends on the assumptions made about the formation mechanism. In the older literature, the excited molecules were often assumed to be produced solely in neutral excitations [127,139–143] and energy-transfer experiments with Stern–Volmer-type extrapolation (linear concentration dependence) were used to derive $G(S_1)$. For instance, by sensitization of benzene fluorescence, Baxendale and Mayer established $G(S_1) = 0.3$ for cyclohexane [141]. Later Busi [140] corrected this value to $G(S_1) = 0.51$ on the basis that in the transfer, in addition to the fluorescing benzene state S_1, the S_2 and S_3 states also form and the $S_2 \rightarrow S_1$ and $S_3 \rightarrow S_1$ conversion efficiencies are smaller than 1. Johnson and Lipsky [144] reported an efficiency factor of 0.26 ± 0.02 per encounter for sensitization of benzene fluorescence via energy transfer from cyclohexane. Using this efficiency factor the corrected yield is $G(S_1) = 1.15$. Based on energy-transfer measurements Beck and Thomas estimated $G(S_1) = 1$ for cyclohexane [145]. Relatively small $G(S_1)$ values were determined in energy-transfer experiments for some other alkanes as well: n-hexane 1.4, n-heptane 1.1 [140], cyclopentane 0.07 [142] and 0.12 [140], cyclooctane 0.07 [142] and 1.46 [140], methylcyclohexane 0.95, cis-decalin 0.26 [140], and cis/trans-decalin mixture 0.15 [142].

A basic problem of the energy-transfer experiments is that the quenchers used may also react with electron or cationic species, in addition to the excited molecules. If the excited molecules also form after charge recombination (which is now unambiguously established), the quenchers may considerably hinder the formation of excited molecules [8]. Comparison of the results obtained in this manner with those obtained via other techniques shows that the solute technique strongly underestimates the $G(S_1)$ value. It should be mentioned that in most of the sensitization experiments unrealistically high K Stern–Volmer constants were obtained [141,142].

If the excited molecule formation is mainly because of electron–cation recombination the concentration dependence of quenching of the precursor charged species may follow some form of the modified Warman–Asmus–Schuler (WAS) semiempirical equation [8], e.g.:

$$G(\text{Scavenged } S_1 \text{ precursor}) = f_s \left[G_{fi} + G_{gi} \frac{(\alpha c)^n}{1 + (\alpha c)^n} \right] \qquad (33)$$

where n is usually taken as $\frac{1}{2}$ (square-root concentration dependence). In the equation G_{fi} and G_{gi} are the yields of free and geminate ions, respectively, and α is an adjustable parameter characteristic to the scavenger. The fraction of S_1 excited molecules formed in recombination, f_s, can be smaller than 1 mainly because of three reasons:

1. Some of the cations may fragment or undergo ion–molecule reactions before neutralization. As shown in the older literature, the fragmentation has higher yield in the radiolysis of the smaller and/or highly branched alkanes such as neopentane or isooctane. Ion–molecule reactions, such as the H transfer, may also reduce the S_1 yield:

$$c\text{-}C_6H_{12}^{+\bullet} + c\text{-}C_6H_{12} \longrightarrow c\text{-}C_6H_{13}^{+} + c\text{-}C_6H_{11}^{\bullet} \qquad (34)$$

The ion–molecule reactions and the spin effects (see below) were detailed in the recent review of Shkrob et al. [7]; therefore, we only briefly mention these reactions.

2. As discussed earlier, the recombination is nearly completed within a few picoseconds; during this short time the spin correlation is maintained and from the originally singlet geminate ion pair singlet excited molecule forms. However, if several ion pairs are in each other's vicinity (in the spur) there is a high probability for cross-recombination. In cross-recombination in 3:1 ratio triplet and singlet molecules form. In a very large spur in the limit statistically $\frac{1}{4}$ of the recombinations give singlet product. The singlet fraction was calculated several times by using experimentally derived spur-size distribution [8,146–148]. Numerous computer simulations were also made on the singlet/triplet formation probability: the theoretical calculations predict singlet formation probabilities in the 0.5–0.7 range.

3. The singlet excited molecules formed in charge recombination may not convert with 100% efficiency to S_1 as discussed in Sec. 2.1. This effect may be important for smaller molecules; for larger molecules such as the decalins or n-dodecane the efficiency is probably very close to unity.

The experimentally obtained f_s values show larger divergence: based on alkane fluorescence measurement. Walter et al. [148] and Luthjens et al. [65,128] published 0.8–0.9. Later the latter authors modified their value to 0.65 [149]. In solutions of isooctane, cyclohexane, or n-hexane with naphthalene, Sauer and Jonah established this value as 0.5 ± 0.1; this singlet formation probability is approximately constant during the decay of charged species up to 70 nsec [150]. In final product experiments $f_s = 0.34$ [151], 0.53, and 0.47 [84] was estimated for cyclohexane, cis-decalin, and trans-decalin, respectively. The low value for cyclohexane is due to the ionic reactions before recombination.

Most papers published during the last 20 years on the formation mechanism of S_1 alkane molecules agree that they form mostly or nearly exclusively in charge recombination [65,72,73,128,148–153]. In pulse radiolysis done with ~50-psec pulses, definite afterformation of fluorescence was observed for the larger alkanes such as the decalins. In the radiolysis of cyclohexane, Sauer et al. used electron scavenger in subnanosecond fluorescence measurements to determine the formation route. Because the charge scavenging occurs on the time scale of geminate recombination (a few picoseconds), while the excited molecule quenching takes place during the excited molecule decay (~1 nsec), they separated these effects [153]. Based on the effective "initial" inhibition of excited molecule formation they concluded that less than 10% of the S_1 excited cyclohexane molecules were formed in direct excitation. Low direct excitation contributions were suggested also for other alkanes [7,128,148].

The uncertainty about the formation mechanism is eliminated when such method is used for the determination of the $G(S_1)$ value, which is invariant to the formation route. Such method is the comparison of the yields of those products, which are formed only in one way in the reaction of the S_1 molecules. Such products come from the singlet decay channel: fluorescence photons, products of H_2 or alkane elimination. The triplet channel due to the large triplet yield originating from other processes than the $S_1 \rightarrow T_n$ ISC (e.g., direct triplet excitation, or charge recombination with triplet product) cannot be used for the calculation of the S_1 yield. The calculations are based on the equation:

$$G(X) = \Phi(X) \times G(S_1) \tag{35}$$

$G(X)$ and $\Phi(X)$ are the G value and the quantum yield of the product used for the calculation, respectively. Two possible sources of errors should be mentioned. One is the energy dependence of the photodecomposition. In order to obtain "intrinsic" yield $\Phi(X)$ should be measured at photon energies as close as possible to the absorption onset, or appropriate corrections should be used [154,155]. Both photon emission and photodecomposition studies show that the energy dependence is more severe for the smaller molecules than for the larger ones. Therefore, $G(S_1)$ for the larger molecules can be estimated with higher accuracy. The second source of error is that X may not solely form in S_1 molecule transformation.

The fluorescence of liquid alkanes is supposed to originate entirely from the relaxed S_1 state. Walter and Lipsky [154], by measuring the *fluorescence* yields of alkane solutions irradiated with 165 nm photons or ^{65}Kr beta particles (E_{max} = 0.67 MeV) relative to benzene fluorescence, determined the following yields: 2.3-dimethylbutane $G(S_1)$ < 1.3, cyclohexane 1.4–1.7, methylcyclohexane 1.9–2.2, dodecane 3.3–3.9, hexadecane 3.3–3.9, *cis*-decalin 3.4, and bicyclohexyl 3.5. After reinvestigating the "intrinsic" quantum yield of cyclohexane fluorescence, Choi et al. published $G(S_1)$ = 1.45 for this alkane in Ref. 155. For *trans*-decalin a $G(S_1)$ value of 2.8–3.1 has been accepted [65,128,132]. The uncertainties in the values reflect the uncertainties in the intrinsic fluorescence quantum yields.

As discussed in Sec. 2.6, the characteristic singlet photodecomposition mode of *n*-alkanes and cycloalkanes is the H_2-molecule elimination (Table 4) with $\Phi(H_2)_u$ = 0.7–0.9. This process has a high yield also in radiolysis [91–93,111]. Based on experiments with deuterated hydrocarbons, Fujisaki et al. [156] and Shida et al. [157] reported values of $G(H_2)_u$ = 1.4 and 1.6 for the yield of unimolecular H_2 detachment occurring in the radiolysis of propane and *n*-butane liquids. In cyclohexane experiments with c-C_6H_{12}/c-C_6D_{12} mixtures, Dyne and Jenkinson suggested that 10–20% of the hydrogen yield of $G(H_2)$ = 5.6 forms in unimolecular elimination [158]. Using the same technique Nevitt and Remsberg [159] published $G(H_2)_u \leq 1.4$ for cyclohexane. The unimolecular H_2 elimination yields for a large group of alkanes were determined based on the alkene and dimer yields using Eq. (10) [91,92,111,160]. Although this method of $G(H_2)_u$ determination, because of the comparable alkene and dimer yields, is less accurate than the calculation of $\Phi(H_2)_u$, we think that the precision of the $G(H_2)_u$ values in Table 4 are better than ± 15%. The accuracy of the $G(S_1)$ values calculated by using the $\Phi(H_2)_u$ and $G(H_2)_u$ yields and Eq. (35) is around ± 20%. In some works H_2 elimination from the alkane radical cations were also suggested [7,161]:

$$c\text{-}C_6H_{12}{}^{+\bullet} \longrightarrow c\text{-}C_6H_{10}{}^{+\bullet} + H_2 \tag{36}$$

However, the yields are probably very low, and for most compounds this reaction does not change considerably the determination of the $G(S_1)$ values by the H_2-elimination method [7]. When interpreting the $G(S_1)$ values obtained by this method the possibility should also be considered that besides S_1 states, to some extent, higher-energy excited states might also contribute to the unimolecular H_2 elimination. This would cause the $G(S_1)$ data calculated using this method to be higher than the real $G(S_1)$ values. Such an assumption would, however, contradict the general observation that with increasing excitation energy and, consequently, decreasing lifetime of the excited state, molecular processes lose their importance and atomic and molecular processes become predominant [160]. The yield determined by this method for cyclohexane, $G(S_1)$ = 1.4–1.7 [91,92,111,160], agrees with the yield obtained in fluorescence experiments.

In the case of alkanes with tertiary or quaternary carbon atoms the characteristic decomposition mode is usually the fragment alkane elimination (Sec. 2.6) and its yield can be used to estimate $G(S_1)$ values. Based on the *fragment alkane elimination* yield Pitchuozhkin et al. [162] calculated a singlet G value of 3.3 ± 1 for 2,2,4-trimethylpentane.

We collected the singlet yields determined by comparing the fluorescence or product yields measured in photolytic and radiolytic experiments in Fig. 7. The G value of low-energy excited states is increasing with the size of the molecules. The trend is by no means unexpected because the density of excited states is known to increase with the size of molecules, giving rise to a more efficient internal conversion from the higher excited states to the first excited level, S_1. The f_s singlet fraction of the charge recombination may have size dependence also due to the decreasing yield of ion decomposition with the molecular weight of ions [160].

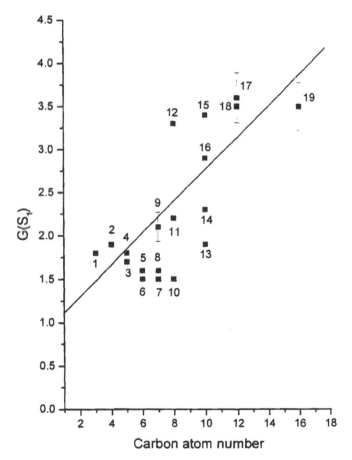

Figure 7 $G(S_1)$ value as a function of the carbon atom numbers in the molecules. When more than one measured value was published, we tried to select the most probable value. Alkanes: (1) propane, (2) *n*-butane, (3) *n*-pentane, (4) cyclopentane, (5) *n*-hexane, (6) cyclohexane, (7) *n*-heptane, (8) cycloheptane, (9) methylcyclohexane, (10) *n*-octane, (11) cyclooctane, (12) isooctane, (13) *n*-decane, (14) cyclodecane, (15) *cis*-decalin, (16) *trans*-decalin, (17) *n*-dodecane, (18) dicyclohexyl, (19) *n*-hexadecane. (From Refs. 18, 29, 65, 92, 148, and 155.)

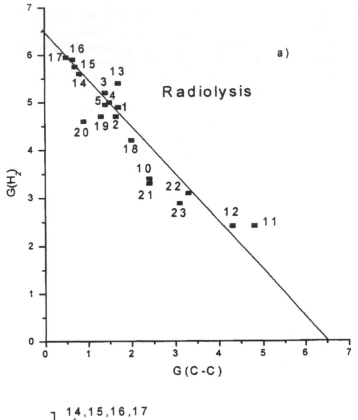

a)

Radiolysis

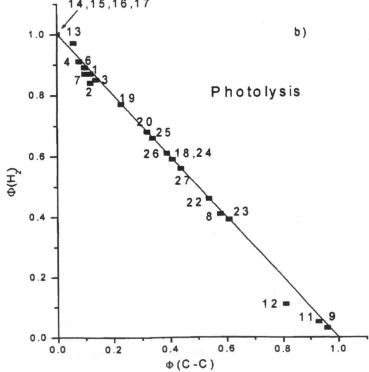

b)

Photolysis

Under high-LET irradiation conditions by using fluorescence detection [87,88] or final product measurements [102,163–165], a continuous decrease of the S_1 alkane excited molecule yields was detected with increasing LET. Holroyd et al. [87], in their soft x-ray experiments, attributed this decrease to the fast quenching of excited molecules by radicals in the spur. LaVerne and coworkers [102,163–165] investigated the distribution of final product of cyclopentane, cyclohexane, and cyclooctane radiolysis in the presence and absence of I_2 radical scavenger in a very wide LET range. They found that the proportion of H-atom-forming reaction is increasing with increasing LET at the expense of other hydrogen-forming reactions, e.g., H_2 molecule elimination. The shift was attributed to changes in charge recombination from basically geminate recombination in low-LET radiolysis to predominantly cross-recombination in high-LET heavy-ion radiolysis. In the high-LET limit the S_1 yields were estimated as $G \approx 1.2$, which is about one fourth of the $G = 4$–5 ion pair yield.

5. THE ORIGIN OF STRUCTURE EFFECT IN ALKANE RADIOLYSIS

The products that form in photolysis and radiolysis of liquid alkanes at room temperature can be divided more or less arbitrarily into two groups: products reflecting primary C–H or primary C–C ruptures. On this ground, one is able to calculate the Φ value or the G value of molecules activated by the photon or ionizing radiation absorption and undergoing primary C–H or C–C bond scission [2,160]. There are several examples for the calculation of the yields in Ref. 2. By investigating the low-LET radiolysis of about 30 alkanes the overall yield of these two processes was found to be approximately constant: $G = 6.5 \pm 1$. The effect of chemical structure can be observed particularly in the ratio of the competing primary C–H and C–C bond rupture reactions. This is obvious from Fig. 8a, where the $G(H_2)$ values are displayed as a function of the G values of the primary C–C bond rupture reactions. In contrast to the decomposition of n-alkanes and cycloalkanes, which yield mostly hydrogen and dehydrogenation products, some branched alkanes, e.g., neopentane and 2,2,4-trimethylpentane, decompose mainly by C–C bond rupture.

In photolysis of liquid alkanes (Fig. 8b) the competition and structure effect are more apparent than during radiolysis: the highly branched alkanes decompose nearly exclusively by C–C bond rupture. Since, during radiolysis of many alkanes, the decompositions originating from low-energy excited states can make a significant contribution (25–55%) to the final product formation, and the reactions from this state show higher selectivity, a considerable part of the structure effect and selectivity in radiolysis is due to the decomposition of low-energy excited molecules. Another important source of the structure

Figure 8 Connection between primary C–H and C–C bond ruptures during radiolysis and photolysis. Alkanes: (1) propane, (2) n-butane, (3) n-pentane, (4) n-hexane, (5) n-heptane, (6) n-octane, (7) n-decane, (8) isobutane, (9) neopentane, (10) 3-methylpentane, (11) 2,2-dimethylbutane, (12) isooctane, (13) cyclopentane, (14) cyclohexane, (15) cycloheptane, (16) cyclooctane, (17) cyclodecane, (18) methylcyclopentane, (19) methylcyclohexane, (20) ethylcyclohexane, (21) 1,1-dimethylcyclohexane, (22) cis-1,2-dimethylcyclohexane, (23) trans-1,2-dimethylcyclohexane, (24) cis-1,3-dimethylcyclohexane, (25) trans-1,3-dimethylcyclohexane, (26) cis-1,4-dimethylcyclohexane, (27) trans-1,4-dimethylcyclohexane. (From Refs. 18, 29, 91, 92, 99, 100, 108, 110, 111, 113, 114, and 160.)

effect is the ionic decomposition, which is effective for highly branched and especially smaller molecular mass alkane radical cations. With most molecules these two types of structure effects can act in the same direction: e.g., in the reaction of highly branched alkanes both the ionic decomposition and S_1 decomposition result in high yield of C–C bond ruptures at the branched carbons. There is no agreement regarding the yields of ionic decompositions, the yields for some molecules may be as high as $G = 1$–3.

It is obvious from this analysis that while the decompositions from the S_1 excited molecules or the radical cations tend to take place selectively, the bond ruptures from other species, e.g., from higher-energy excited molecules are closer to a random distribution.

6. CONCLUDING REMARKS

Because of such difficulties as the featureless absorption and emission spectra in the vacuum ultraviolet region, very weak and energy-dependent fluorescence intensity, short excited-state lifetime, etc. the photophysics and photochemistry of alkanes is much less known than those of other organic molecules, for instance, aromatic hydrocarbons. In this chapter, the present status was reviewed.

At present there are different views about the nature of the alkane excited states. The excited states are probably admixtures of Rydberg- and valence-type states. There are some views that the initial energy absorption takes place in a basically Rydberg state and than it relaxes to a valence-type excitation. The Rydberg-type excitations of alkanes have valence type counterparts with the same symmetry [22]. According to some works the lower-energy states are delocalized all over the molecules, while other works mention strongly localized excited orbitals. Some excited-state properties may point to the localized nature of the excited states. These include the characteristics of fluorescence, the large disparity between the absorption onset and the fluorescence maximum (Stokes' shift, $\Delta \approx$ 0.8–1.6 eV, width of the fluorescence peak ($\sigma_f \approx 1$ eV), and also the molecular structure dependence of these characteristics, e.g., the very strong and characteristic influence of the alkane chain branching and the high probability of decompositions at the tertiary and quaternary carbons. In the photoexcitation of many alkanes the fluorescence quantum yields decrease with increasing photon energy; this dependence is attributed to an energy-dependent relaxation of the higher excited states to the relaxed S_1 state from where the fluorescence occurs.

The excited-state lifetimes ($\tau \approx 1$ nsec) determined by fluorescence decay, absorbance decay, or steady-state quenching experiments after single-photon, multiphoton, or ionizing radiation excitation generally agree indicating that the same excited state (the relaxed S_1) is observed in all experiments. Within a homologous series, e.g., for the n-alkanes, τ is increasing with the size of the molecules, similarly to the fluorescence intensity. The ratio of the two quantities, $R = \Phi_f/\tau$, the so-called radiative rate coefficient, within a group of compounds remains constant: $R \approx 1.6 \times 10^6$ sec^{-1} for n-alkanes, ~2 × 10^5 sec^{-1} for linear methylalkanes, ~5.3 × 10^6 sec^{-1} for vicinally branched dimethylalkanes, ~8 × 10^6 sec^{-1} for cyclohexanes, and ~1.2 × 10^7 sec^{-1} for dimethylcyclohexanes. The lifetimes in the group of the so-called nonfluorescing alkanes (geminally substituted alkanes, C_5, C_7–C_{10} cycloalkanes) are short ($\tau \leq 0.3$ nsec): the absence of fluorescence may have two causes, low radiative rate coefficient and fast chemical decay.

The radiationless $S_1 \rightarrow S_0$ transition has a negligible role in the depopulation of the excited molecules; the process, which competes with the weak fluorescence (~1%) is the chemical decomposition (~99%). Two basic chemical decomposition types are distinguished: hydrogen molecule and alkane molecule elimination (in these reactions no intermediate radicals are observed) and C–H or C–C scission yielding two radicals or C–C scission yielding a biradical (in the reaction of cycloalkanes).

The decomposition of a C_nH_{2n+2} alkane molecule to H_2 molecule and C_nH_{2n} alkene molecule belongs to the first class of reactions. This is the typical decomposition mode of the S_1 excited n-alkanes and C_5–C_{10} cycloalkanes at room temperature. The quantum yield of the reaction decreases with decreasing temperature. Thus, it needs some thermal activation. The reaction is attributed to an $S_1 \rightarrow S_x$-type IC process; S_x denotes some singlet state that gives the products. The elimination takes place mostly or probably entirely from a single carbon atom yielding an H_2 molecule and a carbene intermediate. The carbene, by intramolecular 1,2-H atom shift, rearranges to an olefin. In the cases of C_7–C_{10} cycloalkanes the carbene may also stabilize in the form of cross-bridged bicycloalkanes by 1,3-, 1,5-, or 1,6-H atom shifts.

Alkane elimination is a basic photodecomposition mode of highly branched alkanes, e.g., most of the excited neopentane molecules split directly to methane and isobutene.

The radical-forming reactions are suggested to take place mostly after an $S_1 \rightarrow T_n$ type ISC: the reactions have nonactivated character. The homolytic split to H atom and alkyl radical has a considerable yield in the photolysis of n-alkanes and cycloalkanes, while the scission to two radicals is characteristic of the decay of excited branched alkane molecules.

With the increase of the photon energy used for excitation, the yield of the radical-forming reactions increases, whereas that of the H_2 elimination decreases.

By now it is unambiguously established that the energy transfer from the S_1 excited alkane molecules takes place with a diffusional mechanism at the contact distance: the reaction radii deduced from such experiments (0.8 ± 0.2 nm) practically agree with the sum of the radii of the reacting molecules. In some special cases, however, very fast energy transfer was also observed (e.g., to PPO molecules): the fast transfer was attributed to energy transfer from high-orbit Rydberg excited states or from some unspecified ion pair state.

The properties of the alkane triplet states are much less known than the properties of the singlet states. It seems there is a dissociative, and thereby short-living triplet state ~ 1 eV below the lowest singlet excited state [29]. The other triplet states are probably also dissociative. The low-lying and long-living triplet states, supposed in some works based on unusual energy-transfer results, still needs real experimental verification [7].

In radiolysis most of the S_1 excited molecules (up to 90%) form in charge recombination. The singlet fraction, f_s, based on the newer experimental and theoretical studies is around 0.5–0.7. Both theoretical and experimental work indicate that f_s decreases with increasing LET value of radiation, because of preferred cross-recombination and by preferred triplet formation in the high ionizing density tracks.

The frequency of C–H and C–C bond ruptures in the radiolysis of liquid alkanes is far from a statistical distribution, and there is a correlation between the dissociation reactions and the chemical structure. In contrast to the decomposition of n-alkanes and C_5–C_{10} cycloalkanes, which yield mostly hydrogen and dehydrogenation products, some branched alkanes, e.g., neopentane and 2,2,4-trimethylpentane, decompose mainly by C–C bond rupture. The considerable structure effect can to some extent be traced back to the fact that with many alkanes 25% to 55% of the total primary decomposition of $G = 6.5 \pm 1$ occurs from low-energy singlet excited states. Chemical reactions from these states tend to take place selectively, whereas the bond ruptures from higher-energy states are closer to a random distribution. Fragmentation of the highly branched and especially smaller molecular mass alkane radical cations can also contribute significantly to observed selectivity of bond ruptures in radiolysis.

REFERENCES

1. Topchiev, A.V., In: *Radiation Chemistry of Hydrocarbons*, English Ed.; Holroyd, R.A.; Elsevier; Amsterdam, 1964.
2. Foldiak, G., Ed.; Radiation Chemistry of Hydrocarbons. In *Studies in Physical and Theoretical Chemistry 14*; Elsevier: Amsterdam, 1981.
3. Gaumann, T., Hoigne, J., Eds.; In *Aspects of Hydrocarbon Radiolysis*. Academic Press: London, 1968.
4. Ausloos, P., Ed; In: *Fundamental Process in Radiation Chemistry*. Interscience: New York, 1968.
5. Freeman, G.R. Radiat. Res. Rev. 1968, *1*, 1.
6. Hummel, A. Radiation Chemistry of Alkanes and Cycloalkanes. In *The Chemistry of Alkanes and Cycloalkanes*; Patai, S.; Rappoport, Z., Eds.; John Wiley and Sons: Chichster, 1992; 743 pp.
7. Shkrob, I.A., Sauer, M.C., Jr., Trifunac, A.D. Radiation chemistry of organic liquids: saturated hydrocarbons. In *Radiation Chemistry: Present Status and Future Trends*, Studies in Physical and Theoretical Chemistry 87; Jonah, C.D., Rao, B.S.M., Eds.; Elsevier: Amsterdam, 2001; 175 pp.
8. Mayer, J., Szadkowska-Nicze, M. Excited states in liquid alkanes and related polymers. In *Properties and Reactions of Radiation Induced Transient Species*, Selected Topics; Mayer, J., Ed.; Polish Scientific Publishers: Warsawa, 1999; 77 pp.
9. Spinks, J.W.T.; Woods, R.J. *An Introduction to Radiation Chemistry*; Wiley-Interscience: New York, First ed. (1964), Second ed. (1976), Third ed. (1990).
10. Swallow, A.J. Radiation Chemistry. An Introduction. Longman: London, 1973.
11. Farhataziz, Rodgers, M.A.J., Eds.; In *Radiation Chemistry, Principles and Applications*. VCH Publishers: New York, 1987.
12. *CRC Handbook of Radiation Chemistry*, Tabata, Y., Ed.; CRC Press: Boca Raton, 1991; *Pulse Radiolysis*, Tabata, Y., Ed.; CRC Press: Boca Raton, 1991.
13. Mozumder, A. *Fundamentals of Radiation Chemistry*; Academic Press: San Diego, 1999.
14. Ausloos, P.; Lias, S.G. Radiat. Res. Rev. 1968, *1*, 75; Rev. Phys. Chem. 1971, *22*, 85.
15. Ausloos, P. Mol. Photochem. 1972, *4*, 39.
16. Ausloos, P.; Lias, S.G. Far ultraviolet photochemistry of organic compunds. In *Chemical Spectroscopy and Photochemistry in the Vacuum Ultraviolet*; Sandorfy, C.; Ausloos, P.; Robin, M.B., Eds.; Chemical Spectrospy and Photochemistry in Vacuum Ultraviolet; Reidel: Dordecht, 1974; Vol. 8, Ser. 6, 465 pp.
17. a) Dorofeev, Y.I.; Skurat, V.E. Russ. Chem. Rev. 1982, *51*, 527.
    b) Dorofeev, Y.I.; Skurat, V.E. Radiation Chemistry, Photochemistry, Vol 3. Photochemical Processes in the Presence of Far Ultraviolet Radiation (Alkanes, Alkenes, Halohydrocarbons, Organic Polymers). VINITI, Moscow (in Russian), 1983.
18. Foldiak, G.; Wojnarovits, L. Photochemistry of liquid alkanes, in *Proceedingd of Baxendale memorial symposium*, Centro Stampa"Lo Scarabeo": Bologna, 1983; 21 pp.
19. Oppenlander, T.; Zang, G.; Adam, W. Rearrangements and Photochemical Reactions Involving Alkanes and Cycloalkanes. In *The Chemistry of Alkanes and Cycloalkanes*; Patai, S., Rappoport, Z., Eds.; John Wiley and Sons, Chichester, 1992, 681 pp.
20. Raymonda, J.W.; Simoson, W.T. J. Chem. Phys. 1967, *47*, 430.
21. Sowers, B.L.; Arakawa, M.W.; Hamm, R.N.; Arakawa, E.T. J. Chem. Phys. 1972, *57*, 167.
22. Robin, M.B. *Higher Excited States of Polyatomic Molecules*; Academic Press: New York; Vol. I (1974), Vol. II (1975), Vol. III (1985).
23. Hatano, Y. Radiat. Phys. Chem. 2003, *67*, 187.

24. Hirayama, F.; Lipsky, S. J. Chem. Phys. 1969, *51*, 3616.
25. Rothman, W.; Hirayama, F.; Lipsky, S. J. Chem. Phys. 1973, *58*, 1300.
26. Ostafin, A.E.; Lipsky, S. J. Chem. Phys. 1993, *98*, 5408.
27. Schwarz, F.P.; Smith, D.; Lias, S.G.; Ausloos, P. J. Chem. Phys. 1981, *75*, 3800.
28. Kimura, K.; Hormes, J. J. Chem. Phys. 1983, *79*, 2756.
29. Wojnarovits, L., Foldiak, G. Triplets states of alkanes. In *Fifth Working Meet. Radiat. Interact*; Mai, H., Brede, O., Mehnert, R., Eds.; ZFI: Leipzig, 1991; 64 pp.
30. Klein, G.; Voltz, R. Int. J. Radiat. Phys. Chem. 1975, 7, 155.
31. Tramer, A.; Voltz, R. Time-resolved studies of excited molecules. Excited States. Lim, E.C., Ed.; Academic Press: New York, 1979, *4*, 281 pp.
32. Lamotte, M.; Pereyre, J.; Joussot-Dubien, J.; Lapouyade, R. J. Photochem. 1987, *38*, 177.
33. Smirnov, V.A.; Nazarov, V.B.; Gerko, V.I.; Alfimov, M.V. Chem. Phys. Lett. 1975, *34*, 500.
34. Lamotte, M. J. Phys. Chem. 1981, *85*, 2632.
35. Leclerc, G.; Cui, Z.; Sanche, L. J. Phys. Chem. 1987, *91*, 6461.
36. Hirayama, F.; Lipsky, S. Saturated Hydrocarbons as Donors in Electronic Energy Transfer Processes New York. In *Organic Scintillators and Liquid Scintillation Counting*; Horrocks, D.L., Peng, C.-T., Eds.; Academic Press, 1971; 205 pp.
37. Lipsky, S., Excited states in hydrocarbons. Proceedings 8th International Congress of Radiation Research; Fielden, E.M.J., Hendry, H., Scott, D., Eds.; Taylor and Francis: London, 1987; Vol 2., 72 pp.
38a. Casanovas, J., Grob, R., Delacroix, D., Guelfucci, J.P., Blanc, D. J. Chem. Phys. 1981, 75, 4661
38b. Casanovas, J., Guelfucci, J.P., Terrissol, M. Radiat. Phys. Chem. 1988, *32*, 361.
38c. Casanovas, J., Guelfucci, J.P., Caselles, O. IEEE Trans. Electr. Insul. 1991, *26*, 603.
38d. Guelfucci, J.P., Casanovas, J., Huertas, M.L., Salon, J. J. Phys. Chem. 1993, *97*, 10352.
39. Ausloos, P.; Lias, S.G.; Rebbert, R.E. J. Phys. Chem. 1981, *85*, 2322.
40. Ausloos, P.; Rebbert, R.E.; Schwarz, F.P.; Lias, S.G. Radiat. Phys. Chem. 1983, *21*, 27.
41. Wieckowski, A.; Collin, G.J. J. Phys. Chem. 1977, *81*, 2592.
42. Nafisi-Movaghar, J., Hatano, Y. J. Phys. Chem. 1974, *78*, 1899; Wada, T., Hatano, Y. J. Phys. Chem. 1975, *79*, 2210; Wada, T., Hatano, Y. J. Phys. Chem. 1977, *81*, 1057.
43. Gorden, R., Jr.; Rebbert, R.E.; Ausloos, P. Rare Gas Resonance Lamps. NBS Technical Note 496, 1969.
44. McNesby, J.R.; Braun, W.; Ball, J. Vacuum ultraviolet techniques in photochemistry. In *Creation and Detection of Excited State*; Lamola, A.A., Ed.; Marcel Dekker Inc.: New York; 1971; 503 pp.
45. Ouchi, A.; Yabe, A.; Inoue, Y.; Daino, Y.; Hakushi, T. J. Chem. Soc. Chem. Commun. 1989, 1669.
46. Dellonte, S.; Gardini, E.; Barigelletti, F.; Orlandi, G. Chem. Phys. Lett. 1977, *49*, 596.
47. Miyasaka, H.; Mataga, N. Chem. Phys. Lett. 1986, *126*, 219.
48. Volkmer, A.; Wynne, K.; Birch, D.J.S. Chem. Phys. Lett. 1999, *299*, 395.
49. Sander, M.U.; Brummund, U.; Luther, K.; Troe, J. J. Phys. Chem. 1993, *97*, 8378.
50. Lakowicz, J.R.; Gryczynski, I. Biospectroscopy 1993, *1*, 3.
51. Ware, W.R.; Lyke, R.L. Chem. Phys. Lett. 1974, *24*, 195.
52. Ware, W.R.; O'Connor, D.V. Chem. Phys. Lett. 1979, *62*, 595.
53. Hatano, Y. Phys. Rep. 1999, *313*, 110.
54. Shinsaka, K.; Koizumi, A.; Yoshimi, T.; Kouchi, N.; Nakamura, Y.; Toriumi, M.; Morita, M.; Hatano, Y.; Asaoka, S.; Nishimura, H. J. Chem. Phys. 1985, *83*, 4405.
55. Wickramaaratchi, M.A.; Preses, J.M.; Holroyd, R.A.; Weston, R.E., Jr. J. Chem. Phys. 1985, *82*, 4745.
56. Preses, J.M.; Grover, J.R.; Kvick, A.; While, M.G. Am. Sci. 1990, *78*, 424.
57. Orlandi, G.; Dellonte, S.; Flamigni, L.; Barigelletti, F. J. Chem. Soc. Faraday Trans. I 1982, *78*, 1465.
58. Choi, H.T.; Hirayama, F.; Lipsky, S. J. Phys. Chem. 1984, *88*, 4246.
59. Hermann, R.; Mehnert, R.; Wojnárovits, L. J. Lumin. 1985, *33*, 69.

60. Henry, M.S.; Helman, W.P. J. Chem. Phys. 1972, 56, 5734.
61. Codee, H.D.K. Excited States of a Number of Saturated and Cyclic Hydrocarbons due to Irradiation. INIS-mf-5335, 7.Jun.1979.
62. Helman, W.P. Chem. Phys. Lett. 1972, 17, 306.
63. Luthjens, L.H.; de Haas, M.P.; de Leng, H.C.; Hummel, A.; Beck, G. Radiat. Phys. Chem. 1982, 19, 121.
64. Luthjens, L.H.; Codee, H.D.K.; de Leng, H.C.; Hummel, A. Chem. Phys. Lett. 1981, 79, 444.
65. Luthjens, L.H.; Codee, H.D.K.; de Leng, H.C.; Hummel, A.; Beck, G. Radiat. Phys. Chem. 1983, 21, 21.
66. Luthjens, L.H.; de Leng, H.C.; Wojnarovits, L.; Hummel, A. Radiat. Phys. Chem. 1985, 26, 509.
67. Wojnarovits, L.; Luthjens, L.H.; de Leng, H.C.; Hummel, A. J. Radioanal. Nucl. Chem. Articles, 1986, 101, 509.
68. Grigoryants, V.M.; Lazovoi, V.V. High Energy Chem. 1996, 30, 38.
69. Mehnert, R.; Brede, O.; Naumann, W.; Hermann, R. Ber. Bunsenges. Phys. Chem. 1983, 87, 387.
70. Hermann, R.; Mehnert, R.; Brede, O.; Neumann, W. Radiat. Phys. Chem. 1985, 26, 513.
71. Mehnert, R. ZFI-Mitt. (Leipzig) 1984, 87, 1–101.
72. Katsumura, Y.; Tabata, Y.; Tagawa, S. Radiat. Phys. Chem. 1982, 19, 267.
73. Katsumura, Y.; Yoshida, Y.; Tagawa, S.; Tabata, Y. Radiat. Phys. Chem. 1983, 21, 103.
74. Tagawa, S.; Washio, M.; Kobayashi, H.; Katsumura, Y.; Tabata, Y. Radiat. Phys. Chem. 1983, 21, 45.
75. Wichramaaratchi, M.A.; Preses, J.M.; Weston, R.E., Jr. J. Chem. Phys. 1986, 85, 2445.
76. Preses, J.M. J. Chem. Phys. 1988, 89, 1251.
77. Preses, J.M.; Holroyd, R.A. J. Chem. Phys. 1990, 92, 2938.
78. Barigelletti, F.; Dellonte, S.; Mancini, G.; Orlandi, G. Chem. Phys. Lett. 1979, 65, 176.
79. Flamigni, L.; Barigelletti, F.; Dellonte, S.; Orlandi, G. Chem. Phys. Lett. 1982, 89, 13.
80. Orlandi, G.; Flamigni, L.; Barigelletti, F.; Dellonte, S. Radiat. Phys. Chem. 1983, 21, 113.
81. Dellonte, S.; Flamigni, L.; Barigelletti, F.; Wojnarovits, L.; Orlandi, G. J. Phys. Chem. 1984, 88, 58.
82. Orlandi, G.; Barigelleti, F.; Flamigni, L.; Dellonte, S. J. Photochem. 1985, 31, 49.
83. Flamigni, L.; Orlandi, G. J. Photochem. Photobiol. A 1988, 42, 241.
84. Mehnert, R. Radicals cations in pulse radiolysis. In Radical Ionic Systems. Properties in Condensed Phases; Lund, A., Shiotani, M., Eds.; Kluwer: Dordrecht, 1991, 231 pp.
85. Tagawa, S.; Hayashi, N.; Yoshida, Y.; Washio, M.; Tabata, Y. Radiat. Phys. Chem. 1989, 34, 503.
86. Mehnert, R.; Brede, O.; Naumann, W. J. Radioanal. Nucl. Chem., Articles 1986, 101, 307.
87. Holroyd, R.A.; Preses, J.M.; Hanson, J.C. Radiat. Res. 1993, 135, 312; J. Phys. Chem. A. 1997, 101, 6931.
88. Shibata, H.; Yoshida, Y.; Tagawa, S.; Aoki, Y.; Namba, H. Nucl. Instrum. Methods Phys. Res. A 1993, 327, 53.
89. Andre, J.C.; Lyke, R.; Ware, W.R. React. Kinet. Catal. Lett. 1984, 26, 3.
90. Marconi, G.C.; Orlandi, G.; Poggi, G. Chem. Phys. Lett. 1976, 40, 88.
91. Wojnarovits, L.; Foldiak, G. Acta Chim. Hung. (Budapest) 1980, 105, 27.
92. Wojnarovits, L.; Foldiak, G. ZFI-Mitt. (Leipzig) 1981, 43a, 243.
93. Dyne, P.J. Can. J. Chem. 1965, 43, 1080.
94. Dyne, P.J.; Denhartog, J. Can. J. Chem. 1962, 40, 1616.
95. Okabe, H.; McNesby, J.R. J. Chem. Phys. 1961, 34, 668.
96. McNesby, J.R.; Okabe, H. Adv. Photochem. 1964, 3, 157.
97. Gaumann, T. Liquid hexanes. In Aspects of Hydrocarbon Radiolysis; Gaumann, T.; Hoigne, J. Academic Press: London, 1968; 213 pp.
98. Ballenegger, M.; Ruf, A.; Gaumann, T. Helv. Chim. Acta 1971, 54, 1373.
99. Wojnarovits, L.; Szondy, T.; Szekeres-Bursics, E.; Foldiak, G. J. Photochem. 1982, 18, 273.

100. Wojnarovits, L. J. Chem. Soc. Perkin Trans. 2, 1984, 1449.
101. Foldiak, G.; Wojnarovits, L. Acta Chim. Hung. (Budapest) 1974, 82, 269.
102. Wojnarovits, L.; LaVerne, J.A. J. Phys. Chem. 1994, 98, 8014.
103. Paal, Z. Metal—catalysed cyclization reactions of hydrocarbons. In *Advances in Catalysis*; Eley, D.D.; Pines, H.; Weisz, P.B., Eds.; Academic Press: New York, 1980; Vol. 29, 273 pp.
104. Holroyd, R.A. J. Am. Chem. Soc. 1969, 91, 2208.
105. Horvath, Zs.; Ausloos, P; Foldiak, G. Comparison between liquid phase photolysis and radiolysis of C_3-C_4 hydrocarbons mixtures. In Proceedings Fouth Tihany Symposium on Radiation Chemistry; Hedvig, P.; Schiller, R., Eds.; Akademiai Kiado: Budapest, 1977; 57 pp.
106. Antonova, E.A.; Pichuzhkin, V.I. High Energy Chem. 1977, 11, 201.
107. Rebbert, R.E.; Lias, S.G.; Ausloos, P. Picosecond ultraviloet multiphoton laser photolysis related to radiolysis. J . Photochem. 1975, 4, 121.
108. Wojnarovits, L.; Kozari, L.; Keszei, Cs.; Foldiak, G. J. Photochem. 1982, 19, 79.
109. Wojnarovits, L.; Foldiak, G. Acta Chim. Hung. (Budapest) 1977, 93, 1.
110. Kozari, L.; Wojnarovits, L.; Foldiak, G. Acta Chim. Hung. (Budapest) 1982, 109, 249.
111. Wojnarovits, L.; LaVerne, J.A. J. Phys. Chem. 1995, 99, 3168.
112. Eberhardt, M.K. J. Phys. Chem. 1968, 72, 4509.
113. Foldiak, G.; Wojnarovits, L. Radiat. Phys. Chem. 1988, 32, 335.
114. Hummel, A.; de Leng, H.C.; Luthjens, L.H.; Wojnarovits, L. J. Chem. Soc. Faraday Trans. 1994, 90, 2459.
115. Yang, J.Y.; Servedio, F.M.; Holroyd, R.A. J. Chem. Phys. 1968, 48, 1331.
116. Wojnarovits, L. Radiochem. Radioanal. Lett. 1978, 32, 267.
117. Pitchoozhkin, V.I.; Yamazaki, H.; Shida, S. Bull. Chem. Soc. Jpn. 1973, 46, 67.
118. Miyasaka, H.; Mataga, N. Picosecond Ultraviolet Multiphoton Laser Phtotlysis Related to Radiolysis. In *Pulse Radiolysis*; Tabata, Y., Ed,; CRC Press: Boca Raton, 1991; 173 pp.
119. Long, F.H.; Lu, H.; Eisenthal, K.B. J. Phys. Chem. 1995, 99, 7436.
120. Siebbeles, L.D.A.; Emmerichs, U.; Hummel, A.; Bakker, H.J. J. Chem. Phys. 1997, 107, 9339.
121. Dellonte, S.; Barigelletti, F.; Orlandi, G.; Flamigni, L. Intramolecular Deactivation Processes and Energy Transfer Mecahanism in Liquid Alkanes Studied by N_2 Laser Two Photon Excitation. Proceedings 5th Tihany Symposium on Radiation Chemistry. Dobo, J.; Hedvig, P.; Schiller, R., Eds.; Akademiai Kiado: Budapest, 1983; 437 pp.
122. Plotnikov, V.G. Radiat. Phys. Chem. 1985, 26, 519.
123. Manion, J.P.; Burton, M. J. Phys. Chem. 1952, 56, 560.
124. Szondy, T.; Wojnarovits, L.; Foldiak, G. "Anomalous" Energy Transfer in Radiolysis of Alkanes. A possible Theoretical Explanation. Proceedings 5th Tihany Symposium on Radiation Chemistry. Dobo, J.; Hedvig, P.; Shciller, R., Eds.; Akademiai Kiado, Budapest, 1983; 119 pp.
125. Wojnarovits, L.; Szondy, T.; Foldiak, G. Radiochem. Radioanal. Lett. 1981, 48, 175.
126. Kudo, T.; Shida, S. J. Phys. Chem. 1967, 71, 1971.
127. Busi, F.; Casalbore, G. Gazz. Chim. Ital. 1981, 111, 443.
128. Luthjens, L.H.; de Leng, H.C.; Appleton, W.R.S.; Hummel, A. Radiat. Phys. Chem. 1990, 36, 213.
129. Choi, H.T.; Lipsky, S. J. Phys. Chem. 1981, 85, 4089.
130. Wojnarovits, L. J. Photochem. 1984, 24, 341.
131. Andre, J.C.; Niclause, M.; Ware, W.R. Chem. Phys. 1978, 28, 371.
132. Johnston, D.B.; Wang, Y-M.; Lipsky, S. J. Phys. Chem. 1991, 95, 5524.
133. Rehm, D.; Weller, A. Ber. Bunsenges. Phys. Chem. 1969, 73, 834.
134. Balzani, V.; Bolletta, F.; Scandola, F. J. Am. Chem. Soc. 1980, 102, 2152.
135. Hermann, R.; Brede, O.; Mehnert, R. Energy Transfer from Alkane Singlets and Quenching of AlkaneFluorescences by Alkyl Chlorides. A. Pulse Radiolysis Study. Proceedings 5th Tihany Symposium on Radiation Chemistry; Dobo, J.; Hedvig, P.; Schiller, R., Eds.; Akademiai Kiado: Budapest, 1983; 503 pp.
136. Mehnert, R. .Pulse Radiolysis Investigation of Charge and Energy Transfer Reactions in

Alkanes and Alkyl Chlorides. Proceedings 5th Tihany Symposium on Radiation Chemistry; Dobo, J.; Hedvig, P.; Schiller, R., Eds.; Akademiai Kiado: Budapest, 1983; 489 pp.

137. Wang, Y.-M.; Johnston, D.B.; Lipsky, S. J. Phys. Chem. 1993, *97*, 403.
138. Krishna, T.S.R.; Lipsky, S. J. Phys. Chem. A 1998, *102*, 496.
139. Busi, F.; Casalbone, G. Gazz. Chim. Ital. 1983, *113*, 83.
140. Busi, F. Labile Species and Fast Processes in Liquid Alkanes. In *The Study of Fast Processes and Transient Species by Electron Pulse Radiolysis*; Baxendale, J.F.; Busi, F., Eds.; Reidel; Dordrecht, 1982; 417 pp.
141. Baxendale, J.H.; Mayer, J. Chem. Phys. Lett. 1972, *17*, 458.
142. O'Neill, P.; Salmon, G.A.; May, R. Proc. R. Soc. Lond. A 1975, *347*, 61.
143. Busi, F.; Flamigni, L.; Orlandi, G. Radiat. Phys. Chem. 1979, *13*, 165.
144. Johnston, D.B.; Lipsky, S. J. Phys. Chem. 1991, *95*, 1896.
145. Beck, G.; Thomas, J.K. J. Phys. Chem. 1972, *76*, 3856.
146. Brocklehurst, B. J. Chem. Soc., Faraday Trans. 1992, *88*, 167.
147. Bartczak, W.M. Computer Simulation of Early Physico-Chemical Processes in Irradiated Media. In *Properties and Reactions of Radiation Induced Transient Species, Selected Topics*; Mayer, J., Ed.; Polish Scientific Publishers: Warsawa, 1999; 101 pp.
148. Walter, L.; Hirayama, F.; Lipsky, S. Int. J. Radiat. Phys. Chem. 1976, *8*, 237.
149. Luthjens, L.H.; Dorenbos, P.; De Haas, J.T.M.; Hummel, A.; van Eijk, C.W.E. Radiat. Phys. Chem. 1999, *55*, 255.
150. Sauer, M.C., Jr.; Jonah, C.D. Radiat. Phys. Chem. 1994, *44*, 281.
151. LaVerne, J.A.; Pimblott, S.M.; Wojnarovits, L. J. Phys. Chem. A 1997, *101*, 1628.
152. Yoshida, Y.; Tagawa, S.; Tabata, Y. Radiat. Phys. Chem. 1984, *23*, 279.
153. Sauer, M.C., Jr.; Jonah, C.D.; LeMotais, B.C.; Chernovitz, A.C. J. Phys. Chem. 1988, *92*, 4099.
154. Walter, L.; Lipsky, S. Int. J. Radiat. Phys. Chem. 1975, *7*, 175.
155. Choi, H.T.; Askew, D.; Lipsky, S. Radiat. Phys. Chem. 1982, *19*, 373.
156. Fujisaki, N.; Shida, S.; Hatano, Y. J. Chem. Phys. 1970, *52*, 556.
157. Shida, S.; Fujisaki, N.; Hatano, Y. J. Chem. Phys. 1968, *49*, 4571.
158. Dyne, P.J.; Jenkinson, W.M. Can. J. Chem. 1960, *38*, 539.
159. Nevitt, T.D.; Remsberg, L.P. J. Phys. Chem. 1960, *64*, 969.
160. Wojnarovits, L.; Foldiak, G. Radiat. Res. 1982, *91*, 638.
161. Werst, D.W.; Trifunac, A.D. J. Phys. Chem. 1991, *95*, 3466.
162. Pichuzhkin, V.I.; Antonova, E.A.; Chudakov, V.M.; Bakh, N.A. High Energy Chem. 1978, *12*, 275.
163. LaVerne, J.A.; Schuler, R.H.; Foldiak, G. J. Phys. Chem. 1992, *96*, 2588.
164. Wojnarovits, L.; LaVerne, J.A. J. Phys. Chem. 1995, *99*, 11292.
165. Wojnarovits, L.; LaVerne, J.A. J. Radioanal. Nucl. Chem. 1998, *232*, 19.

14

Radiation Chemical Effects of Heavy Ions

Jay A. LaVerne
University of Notre Dame, Notre Dame, Indiana, U.S.A.

1. INTRODUCTION

The chemical effects induced by the passage of heavy ions in matter have been actively examined for more than 100 years. Curie and Debierne in 1901 performed the first radiation chemistry study on water by any type of radiation using α-particles from solutions of radium salts [1]. All of the early studies examined the production of gases, mainly hydrogen, from radium solutions [2–9]. Radium was used because of its availability, but equally important is the reality that products could not be observed using other types of radiation because of impurities and the crude techniques of the time. In contrast to the observable transitions in radium solutions, Fricke could detect no decomposition of pure, air-free water by x-rays as late as 1933 [10]. Many of the transient species discussed in the preceding chapters of this book were just being discovered at about this time and a complete description of the observed chemical effects was understandably difficult. However, experimental techniques quickly advanced and many early monographs on radiation effects note the different effects observed with heavy ions compared to that found with fast electrons, x-rays or γ-rays [11–16].

A wide variety of early experimental [17] and theoretical studies [18–22] have formed much of the basis for understanding the chemical effects of heavy ions. The general concepts of heavy ion radiolysis have not changed since these studies. However, more information on specific systems is constantly being accumulated, especially in complicated mixtures and in heterogeneous systems. The details of heavy ion radiolysis will not be discussed in depth, but rather an overview of the major effects and the reason for their occurrence will be given here. An examination of rare events specific to heavy ion radiolysis can be found in a recent review [23]. Descriptions of typical experimental techniques used in heavy ion radiolysis are published elsewhere [24,25]. The discussion in this chapter will be limited to the radiolysis of water because most of the work has been performed in this medium and because of the importance of water in a variety of applications. Many of the fundamental ideas discussed here for liquid water are applicable to other liquids. However, the nature of the processes occurring in many solids and almost all gases may lead to very different effects than discussed here.

In addition to the fundamental scientific aspects, many studies on the chemical effects of heavy ion radiolysis have significant practical applications. These applications range from the nuclear power industry [26,27], space radiation effects [28], medical therapy [29], and the environmental management of radioactive waste materials [30]. Much of the future effort on the examination of heavy ion radiolysis will driven by these need-based demands. The complexity of such systems can be overwhelming and good fundamental research on simple systems is required.

The term "heavy ions" usually refers to any ionizing radiation other than fast electrons. In this chapter, the term will apply to charged atomic nuclei of all the elements from protons to uranium ions. The concepts presented may apply to other particles such as muons; however, a number of collision parameters are different between these ions and atomic nuclei and one should refer to specific monographs for details. High velocity molecular ions fragment in the first few collision processes and one gets atomic nuclei of the same velocity, which can then be treated as individual heavy ions. Neutrons, being neutral, are not classified as heavy ions either. They produce recoil ions in passing through matter. The recoil ions, oxygen or hydrogen in water, are usually formed with a wide energy distribution and each ion can be examined independently.

2. ENERGY LOSS PROPERTIES OF HEAVY IONS

Most radiation chemical studies involve ions with initial energy of a few MeV or more. At these energies, the main interaction responsible for energy loss is the Coulombic force between the charge of the incident heavy ion and that of the electrons of the medium. Energy transferred from one of these collisions leads to excitation or ionization of the medium molecule and the rate at which this energy is lost per unit path length, $-dE/dx$, is called the electronic stopping power of the medium. Rutherford was the first to successfully explain the scattering of charged particles using Coulombic forces and classical mechanics [31]. Bohr applied this cross section using an impact parameter approach to determine the stopping power and ranges of ions in matter [32]. A quantum mechanical theory of stopping powers was later developed by Bethe based on energy and momentum transfer [33]. Both the Bohr and Bethe formalisms are equally valid depending on the charge and velocity of the incident heavy ion [34]. Almost all other methods for determining the energy loss by charged particles through electronic processes are modifications of these two early theories and a complete discussion is given in Chapter 2 of this book.

The details of the stopping power equations have been very well summarized elsewhere [22,34–38]. Most applications involve ions of low charge and high velocity in which the Bethe formalism is valid. The nonrelativistic stopping power equation of Bethe for heavy ions is given approximately by,

$$-dE/dx = \frac{4\pi e^4 Z^2 N}{mV^2} \ln\left(\frac{2mV^2}{I}\right) \tag{1}$$

where e and m are the charge and mass of the electron, Z and V are the charge and velocity of the incident particle, and N and I are the electron density and mean excitation potential of the medium. Relativistic terms, shell effects, the Fermi density correction, and other modifications to the stopping power are usually added to the logarithmic term as needed. Sometimes stopping powers are further subdivided into hard and soft [32] or glancing and knock-on [22] collisions. One can also encounter restricted stopping powers in which an upper limit is placed on the energy loss in a given collision [39]. These distinctions are made according to the energy loss in the collision and aided in the early development of track structure as described below. However, such distinctions are rarely necessary for understanding basic radiation chemistry. They can also lead to erroneous assumptions because the energy cutoff is often arbitrary.

Equation (1) is rarely used in its given form to determine stopping powers or ranges of ions. There are too many difficulties in estimating the effective charge of the incident ion, screening effects, effective mean excitation potentials, and other factors that must be taken into account for accurate stopping powers. Most investigators use standard compilations that employ semi-empirical fits to experimental data [40–42]. The compilations typically compare the stopping of different ions in the same target medium and scale to other media or standardize the stopping of protons in different media and scale to other ions. The accuracy of the compilations is thereby increased considerably. However, the predictions of these compilations should be verified for the specific radiolytic condition to be examined as some compilations are more accurate than others for a particular ion in a given energy range. Ref. 42 is developing into the standard compilation for most heavy ion studies, while Ref. 41 is especially suitable for very high Z particles.

Low-energy, high Z particles may also have a sizable nuclear stopping power due to elastic nuclear–nuclear interactions [40–42]. These processes can be thought of as billiard-ball type collisions between the nucleus of the incident ion and the nuclei of the atoms making up the medium molecules. These collisions occur at very low energies and can usually be ignored, e.g., nuclear stopping equals electronic stopping for carbon ions in water at energy of about 20 keV. On the other hand, the nuclear component can be considerable with high Z particles and should be included in range or stopping power determinations, e.g., nuclear stopping equals electronic stopping for uranium ions in water at energy of about 3.5 MeV. However, no specific chemistry of water radiolysis has been identified unequivocally to the nuclear stopping, so it will not be treated separately. The nuclear stopping power is not to be confused with nuclear fragmentation. The latter process occurs at very high energies and leads to fragmentation of the incident heavy ion.

General trends of the stopping power can be easily deduced from Eq. (1). The velocity term in the numerator of the logarithmic term and in the denominator of the pre-logarithmic term gives rise to the familiar Bragg peak, i.e., with decreasing velocity of the incident particle the stopping power increases to a maximum and then decreases at lower velocities. Specifically, the Bragg peak refers to the increase in ionization toward the end of the particle track since he first noticed this effect in the ionization of gases [43]. However, the expression is also commonly used in reference to the stopping power. The distinction is negligible in water because ionization is the dominant outcome due to energy loss by the incident heavy ion. The stopping powers of water for five different ions are shown in Fig. 1 as a function of incident energy [42]. Also shown in this figure is the stopping power for fast electrons [44]. The important aspect of the stopping powers is the huge difference between heavy ions and electrons. Heavy ion stopping powers are typically several orders of magnitude higher than that for fast electrons. There are two major consequences due to the high rate of energy loss for heavy ions that affects the radiation chemistry: the very short range and the track structure.

The dotted lines in Fig. 1 show the stopping powers for the different ions at a constant velocity in units of MeV/amu. This unit of energy is very often used in heavy ion radiolysis and it is based on the classical formula for kinetic energy, $E = \frac{1}{2}MV^2$, where M is the heavy ion mass. As seen in Eq. (1), the ion velocity is a dominant parameter in energy loss processes and the MeV/amu energy unit is more convenient to use than converting to absolute velocity units. Remember that MeV/amu is actually proportional to the square of the velocity.

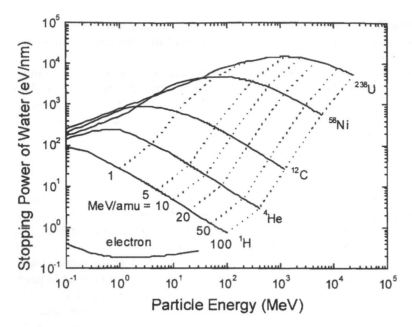

Figure 1 Stopping power, $-dE/dx$, of some heavy ions [42] and electrons [44] in water as a function of energy. The dotted lines show the stopping power for heavy ions of equal velocity.

There are several methods used to describe the range of an energetic heavy ion. The penetration, projected range, or depth refers to the mean displacement in the initial direction of motion and is the most commonly used parameter in radiation chemistry. Scattering will always add a radial component to the trajectory, but the magnitude of this component is usually small for ions of energies typically used in radiation chemistry experiments. Therefore the range of heavy ions often can be determined from the continuous slowing down approximation, csda, by the following equation.

$$R_{csda} = \int_0^{E_o} dE/(dE/dx) \tag{2}$$

where E_o is the initial energy, and the stopping power, dE/dx, is obtained from one of the compilations. Fig. 2 shows the ranges corresponding to the ions in Fig. 1. The extremely short ranges make experiments with heavy ions difficult. A 1-MeV electron has a range of about 4 mm in water and experiments are relatively easy to conduct. On the other hand, a 1-MeV proton has a range of only 24 μm. Window materials and sample thickness become major concerns at these ranges. A proton would need about 20 MeV of energy to have the same range as a 1-MeV electron, and other heavier ions would need considerably higher energy. The ranges given in Fig. 2 are csda ranges and somewhat over estimate the true penetration. Applications requiring greater accuracy should use Monte Carlo simulations of heavy ion trajectories. However, these simulations require detailed knowledge of the elastic cross sections, which are not known for heavy ions in many materials. In practice, the csda ranges are usually sufficient for the determination of penetration, but the true range should always be checked experimentally.

The electronic stopping power involves a collision between the heavy ion nucleus and an electron of the medium. The reduced mass of the encounter can always be assumed to

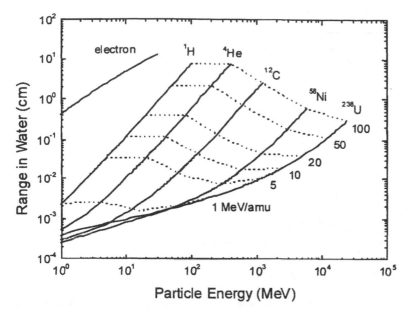

Figure 2   Range, csda, of some heavy ions [42] and electrons [44] in water as a function of energy. The dotted lines show the range for heavy ions of equal velocity.

be equivalent to that of the electron, even with incident protons. Therefore energy loss depends on the velocity and not the mass of the incident ion. In other words, there are really no isotope effects for radiolysis with ions of the same element, provided they have the same velocity. For example, protons and deuterons of equal velocity give similar radiation chemical yields [45]. Subtle mass effects on the stopping power of heavy ions are predicted in more detailed theories [35,46], but their effects are small and have not yet been observed in normal water radiolysis.

The charge and velocity of the incident particle are important parameters in the characterization of energy loss processes as predicted by Eq. (1). The velocity can usually be obtained by energy or time of flight analysis, but the charge is more difficult to determine [47–49]. It is obvious from Eq. (1) that for two ions of the same velocity but with different charges, e.g., He^{+1} and He^{+2}, the ion with twice the charge will have four times the stopping power. Experiments with gas targets show the expected dependence of the cross section on the charge of the incident particle [50]. However, in gases one is dealing with a single isolated collision. The rates of capture and loss of electrons in condensed media are very rapid and the heavy ion attains an equilibrium charge state depending on its velocity [51–53]. Strictly, each individual heavy ion has an integral charge, but the probability of finding that charge state depends on the velocity. However, an average charge per heavy ion is often used in place of Z in stopping power formulas such as Eq. (1). Experimental studies have led to several empirical equations for the average effective charge, Z_{eff}, of heavy ions. The following formula is accurate over a wide energy range of greater than 0.3 MeV/amu [48,49].

$$Z_{eff}/Z_o = 1 - \exp\left[-6.35\left(\chi/Z_o^{4/3}\right)^{1/2}\right] \qquad (3)$$

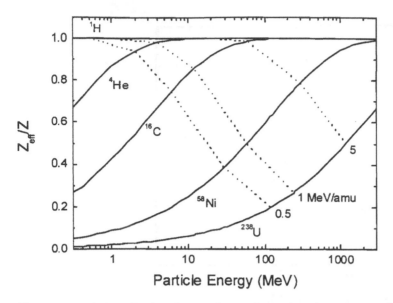

Figure 3 Relative effective charge of some heavy ions in water as a function of incident energy (Eq. (3)). The dotted lines show the relative effective charge for heavy ions of equal velocity.

The energy, χ, is in units of MeV/amu and Z_o is the heavy ion atomic number. This formula assumes that a proton at the same energy is completely stripped and should not be used at energy less than 0.3 MeV/amu. Corrections can be applied to extend Eq. (3) to lower energies [48]. Fig. 3 shows the relative effective charge for several heavy ions as a function of their energy. It can be seen that the lighter ions are generally stripped at energies above 1 MeV/amu. The core electrons of heavy nuclei are tightly bound and considerably more energy is required to strip them in a collision process.

Energy loss due to charge exchange processes and the equilibrium charge are included in the total stopping power found in compilations. This energy does not have to be considered explicitly. There have been several references to "track end" effects that are presumably due to charge exchange or other processes occurring at the end of the heavy ion track. Schuler [54] once suggested that the excess production of molecular hydrogen in benzene is due to charge exchange processes. However, later calculations using more reliable cross sections indicate that charge exchange processes are insufficient to explain all of the hydrogen produced [55]. No other chemistry has been shown to be due exclusively to charge exchange processes.

3. FUNDAMENTALS OF TRACK STRUCTURE

Many of the observed radiation chemical effects in water are due to the geometry of the physical energy deposition events, commonly referred to as the track structure. The observation that the energy absorbed in the medium was more important for determining the resultant chemistry than the energy loss by the incident particle was first proposed by Zirkle et al. who coined the term linear energy transfer, LET, to indicate this significance [56]. Of course, the track structure is largely determined by the energy loss processes. Linear energy transfer is usually assumed to be equivalent to the stopping power of the medium, which is almost always valid for bulk liquid water. However, one should use caution in heterogeneous systems or in isolated nanometer sites. Track structure effects are commonly called LET effects because most of the early studies used this parameter to characterize the different yields for the various heavy ions. Linear energy transfer has continued to be a dominant parameter in the radiation chemistry of heavy ions, but even in the earliest studies it was apparent that LET was not the only factor that determined product yields [57,58]. The variation in product yields for two ions at the same LET was correctly ascribed by early investigators to subtle differences in the track structures [16, 57,58].

The medium absorbing the energy loss by the ionizing radiation is almost always structured. Water is made up of finite dimensioned molecules. Stopping power formulas such as Eq. (1) give average rates of energy loss to a continuum of bound electrons with no regard to the molecularity of the medium. In normal systems, energy is lost by the incident radiation in stochastic processes leading to isolated regions or clusters of excited states and ion pairs of medium molecules. The early cloud chamber pictures of Wilson in the 1920s [59,60] show that β-particle tracks are made up of well-separated ionization events, except near the end of the tracks. The cloud chamber tracks of α-particles appear as solid, continuous strings of ionization events. Theories developed in the 1940s and 1950s used these visualizations as aids to describe radiation chemical events due to the nonhomogeneous distributions of reactive species [12,61,62].

The track structure is determined to a large extent by the distribution of energy absorption by the medium, which is well characterized by the dipole oscillator strength distribution, $f(\varepsilon)$. Classically, the dipole oscillator strength distribution gives the number of electrons of the medium capable of absorbing a photon of energy ε. The complete integral of the dipole oscillator strength distribution over all energies is equal to the total number of electrons in a molecule, Z'. Phase effects and other characteristics of the medium are included in the dipole oscillator strength distribution. A large number of molecules, even in the different phases, have distributions comparable to that for liquid water as shown in Fig. 4 [63]. Therefore the overall track structures of ionizing radiation in most liquid media are similar. Subtle differences will be found because the oscillator strength distributions are not identical for all media and other mechanisms, such as channeling in crystals, may have a significant contribution. The density of the medium has a direct effect on energy loss processes, which is the reason why there is little meaning of track structure in the gas phase.

The dipole oscillator strength distribution does not directly correlate to energy loss by charged particles. However, Monte Carlo simulations of ionizing radiation using cross sections incorporating the dipole oscillator strength distribution can give this information [63,64]. Fig. 4 shows the frequency of a given energy loss for 1-MeV electrons in liquid water. The similarity between the energy loss distribution and the dipole oscillator strength distribution is obvious with both showing peaks at about 20 eV. There is a wide range of possible collisions due to the passage of a fast electron in water, but it can be seen from Fig. 4 that the most probable events involve energy losses of less than 100 eV.

The mean energy loss by a fast electron in liquid water is about 60 eV and somewhat independent of the phase and the initial electron energy [63,64]. Collisions involving this magnitude of energy loss will produce secondary electrons that further lead to an average of one or two ionizations [65]. Low-energy electrons have relatively large total elastic cross sections with a substantial backscattering component [66–68]. In other words, electrons of a few tens of eV do not go very far in liquid water and show a nearly isotropic angular distribution as they thermalize [69]. The electrons will thermalize and eventually become

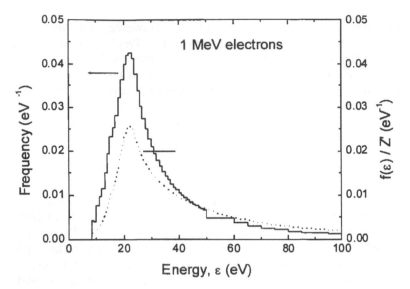

Figure 4 Relative dipole oscillator strength distribution, $f(\varepsilon)$, for liquid water [63] and frequency of a given energy loss by 1-MeV electrons in liquid water [64].

hydrated, adding another spatial delocalization from the initial energy loss event. The net result is that the initial energy loss event leads to a cluster, or spur, of two or three ionizations spatially localized. The concept of a spur in the liquid phase of water is very old [61,62]. The existence of spurs in water manifests itself in the characteristic time dependence of the decay of reactive species produced by water decomposition [70]. The spur is still the main entity used by radiation chemists to explain kinetic effects with low LET radiation. Solvation of the electron does not occur in the gas phase and the delocalization of the electron during this process makes up a substantial fraction of the size of the spur. The spur has no counterpart in the gas phase, which causes much confusion when applying gas phase theory and cross sections to the liquid phase.

Following the initial ionization events, water decomposes within a few picoseconds to give a spur composed of hydrated electrons, OH radicals, H atoms, H_2, and H_2O_2 [16]. The nonhomogeneous distribution of water decomposition products making up the spur relaxes by diffusion and reaction of the sibling radicals and molecules. A combination of direct observations of the hydrated electron decay with model calculations can be used to estimate the average geometry of the spur [71]. Obviously, each spur is slightly different. However, even a single 1-MeV electron produces greater than 10^4 spurs so that its chemistry appears as an average over all spurs. Time decay measurements using low-energy multiphoton ionization just above the ionization potential of water have also been used to estimate the distribution of low-energy electrons in water [72]. These latter experiments agree well with the predicted results for the "typical" spur produced by fast electrons. The observed time decay of the hydrated electron coupled with Monte Carlo calculations suggests that the "typical" spur in electron radiolysis is well approximated by a Gaussian distribution with a characteristic radius of 4–5 nm [73].

The average energy loss per unit track length of a 1-MeV electron is about 0.2 eV/nm [64]. With an average energy loss per collision event of 60 eV the mean separation of spurs is 300 nm, which is much too far apart for interspur reactions. (It is assumed throughout this article that incident particles are isolated from each other, an assumption not necessarily correct at very high dose rates or with very short pulses of intense beams.) On the other hand, a 5-keV electron has an average LET of 3 eV/nm, which corresponds to an initial average separation of only a few spur diameters. The spurs produced by an electron of this energy are not initially overlapping, but they will overlap soon thereafter as they develop in time. The resulting geometry leads to a short track as originally defined by Mozumder and Magee [20]. These authors suggested that even lower energy electrons form blobs, which are elliptical or stretched spurs. All of these geometric shapes are meant to be aids in predicting the resultant chemistry and should not be considered as fixed entities of finite dimensions.

The tracks of heavy ions can be explained in much the same fashion as that for fast electrons, e.g., considering the primary energy loss followed by the delocalization due to the multiple secondary electron thermalization and solvation. If the LET is greater than 3 eV/nm the spurs will be formed overlapping or will overlap shortly thereafter. At LET much less than 3 eV/nm the spurs will exist independently of each other and the observed chemistry of a heavy ion should be much like that of a fast electron. Monte Carlo simulations suggest that the energy loss distributions for high velocity heavy ions are very similar to that for fast electrons, i.e., the frequency of a given energy loss for a heavy ion is nearly the same as shown in Fig. 4 for fast electrons (S.M. Pimblott, unpublished results). Collisions by heavy ions have slightly different kinematic parameters than fast electrons. For instance, an electron can lose up to half its energy in a collision, whereas the maximum energy loss by a heavy ion is $4mE/M$, where m is the electron mass, and E and M are the heavy ion energy and mass, respectively. For both electrons and heavy ions, the energy loss is largely determined by the dipole oscillator strength distribution, which is significantly peaked at low energy. The result is that almost all types of ionizing radiation produce large amounts of low-energy electrons with distributions in water similar to that shown in Fig. 4. However, the initial collision process strongly determines the track structure and resultant chemistry because it determines the spacing of the clusters of low-energy electrons.

Recent studies have applied Monte Carlo techniques well established for electrons to heavy ions [74]. The predicted density normalized mean free path, Λ, of several heavy ions is given in Fig. 5. It can be seen that the mean free path mirrors the stopping power by decreasing with decreasing energy and passing through a minimum at low energies. Spur sizes can vary widely, but from the discussion above a typical spur can be assumed to have a diameter of about 10 nm. Fig. 5 shows the relationship between a spur diameter and the mean free path of several ions. Low-energy heavy ions will produce primary ionization events well within the diameter of a spur. Spatial delocalization due to thermalization and solvation of the secondary electrons will result in one continuous track with a columnar-like structure. Experimentally it has been observed that radical and molecular yields with protons above about 20 MeV, 3 eV/nm, are virtually the same as with fast electrons [75,76]. This observation agrees well with predictions based on the mean free path shown in Fig. 5. Even higher energies are required for a helium or carbon ion track to resemble that of a fast electron. These expected trends are observed experimentally, as will be shown below. It has been speculated that the charge on an ion may slightly alter the distribution of species within the spur [77], but no such process has been confirmed experimentally.

For a given medium such as water, its density and the velocity of the incident heavy ion contribute to the formation of the columnar track structure. The spacing of the primary energy loss events must be within the delocalization of the secondary electrons for effects attributed to high LET to occur. For this reason, typical LET effects will not be

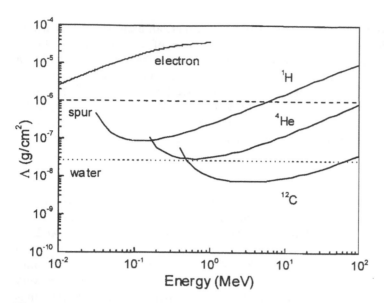

Figure 5   Energy dependence of the density normalized mean free path for e⁻, ¹H, ⁴He, and ¹²C ions in liquid water. The dotted line shows the water diameter and the dashed line shows the typical spur diameter. (From Ref. 74.)

observed in the gas phase. Simple density scaling of experimental or theoretical data between the gas and condensed phases should be performed with caution.

Fig. 5 shows that the mean free path of low-energy carbon ions is less than that of a water molecular diameter. Calculations based on such short mean free paths may predict two energy loss events within the same molecule. However, as discussed elsewhere in this book, the double ionization cross section is much lower than predicted by these results. The problem is that cross sections are normally based on a single isolated collision at gaseous density. Extrapolation to the condensed phase can lead to unrealistic predictions.

The column of species defined by the overlapping spurs along the path of a high LET particle makes up what is commonly referred to as the track core [21]. Its physical parameters are difficult to quantify and it has no corresponding entity in the gas phase. Obviously, some energy loss collisions will be violent enough to form true δ-rays, i.e., secondary electrons with enough energy to make tracks of their own. Various attempts have been made to separate the track core from the region of radiation effects due to secondary electrons (a region sometimes called the penumbra) [21,22,78,79]. There are at least two problems with such an attempt. First, there is no clear demarcation between the two regions, either by energy or chemically. The result has been confusion and different core definitions for various chemical systems. A second difficulty in defining a track core, or any track parameter, is that the track is dynamic. The track is constantly expanding in time due to diffusion of the transient species and reactions initially associated with the track core may envelop those due to the secondary electrons. Despite the inherent problems, some radiation chemical problems have been successfully addressed by treating each part of the track separately [79–83]. This approach to heavy ion track modeling is advantageous because of its relative simplicity. However, Monte Carlo calculations are required to completely elucidate the track structure and resulting chemistry. The few published Monte Carlo studies on heavy ions are quite promising [74,84,85].

4. CHEMICAL EFFECTS OF TRACK STRUCTURE

This section will discuss general effects due to track structure that apply to most condensed phases. Water radiolysis will be used in specific examples because more data exist for this chemical system than any other. As discussed above, the earliest experiments showed that there were differences in product yields for various ionizing radiations in liquid water. With increasing particle LET there is an increase in the yields of molecular products such as H_2 and H_2O_2 and a corresponding decrease in the yields of radicals such as e_{aq}^-, H, and OH [16,86]. The nonhomogeneous distribution of reactive species initially produced by the energy deposition relaxes in time toward a homogeneous distribution. In pure water, a competition is quickly established between radical–radical combination reactions among the sibling radicals and radical diffusion into the bulk medium. Increasing the particle LET increases the probability of radical–radical combination reactions with a corresponding increase in molecular products.

It has always been assumed that the initial decomposition of water is independent of particle type [87–89]. There are experimental results that seem to support this conclusion. Pulsed experiments with protons find that the short time yield of the hydrated electron is similar to that found with fast electrons [90,91]. In addition, W-values, the average energy required per ion pair, for high-energy particles in a given medium are nearly independent of the particle type [92]. Ion pairs are the main initial species formed in the radiolytic decomposition of water so one would expect similar yields for the products due to these species. In special circumstances the initial product yield may be dependent on a particle type, e.g., low-energy heavy ions with large contributions from nuclear scattering processes or radiation with a high probability of causing multiple ionization of the medium molecules. For the most part, observed variations in product yields of a given medium with radiation type or energy are due to effects of track structure on the radiation chemistry.

There are a few important details that must be addressed in discussing product yields with heavy ions because of their extremely short range. For example, a 10-MeV proton (LET = 4.7 eV/nm) has a range in water of 1.2 mm while a 10-MeV carbon ion (LET = 700 eV/nm) has a range of 13 μm. Both particles would be completely stopped in the solution in most practical applications. The chemistry measured in such a configuration is an average over all particle energies from the incident particle energy, E_o, to zero. Radiation chemical yields, G-values, are given as the number of molecules formed or lost per 100 eV deposited, and track segment yields, G_i, are the radiation chemical yields in a discrete segment of path length. Within this small path segment the energy, LET, and other parameters of the heavy ion are well defined and constant. Model calculations normally predict track segment yields while experiments usually, but not always, give track average yields, G_o. The difference between track average and track segment yields can sometimes be significant. The definition of the track segment yield is $G_i = d(G_o E_o)/dE_o = G_o + E_o dG_o/dE_o$. It can be seen that track segment yields approach track average yields as the initial particle energy approaches zero or if the track average yield is independent of energy. Track segment yields are often equivalent to track average yields, but such equality should not be assumed.

Fig. 6 shows a plot of $G_o E_o$ for HO_2 production as a function of the initial energy of protons [75]. The proton was completely stopped in all the experiments and the track average yield of HO_2 measured. The traditional integral or track averaged G-value at 10 MeV is given by the slope of the solid line to be 0.030 molecules/100 eV. The differential or track segment yield at 10 MeV is given by the slope of the dotted line to be 0.023 molecules/100 eV. This 30% difference can be significant in some applications. As

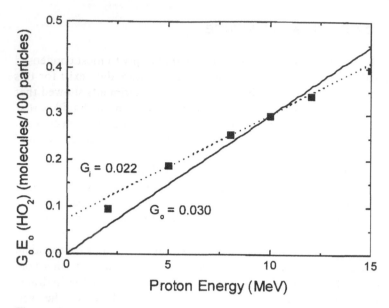

Figure 6 The production of HO_2 in the proton radiolysis of water as a function of initial ion energy [75]. The slope of the solid line (0.030) is the track average yield while the slope of the dotted line (0.023) is the track segment yield for 10-MeV protons.

the proton energy approaches zero, the two types of yields approach each other. However, one must remember that the Bragg peak occurs at the lower energies and yields are not expected to be the same on both sides of it. There is as yet no satisfactory way to deal with this problem. The best technique for obtaining differential yields is to fit the data as presented in Fig. 6 and determine the derivative [75]. However, one needs accurate yield measurements at appropriate energy intervals, data which are seldom available.

There is a considerable amount of effort required to experimentally measure the energy dependence of product yields. One solution is to use high energies where the long ranges allow the direct measurement of track segment yields. However, nuclear fragmentation also occurs at high energies, which complicates interpretation of the data. The track average yields are often the only measured values, especially in the early literature. Track segment yields are often presented as a function of LET, especially when comparing the results for different types of ions. The correctness of using LET will be discussed later, but it is definitely incorrect to couple track average G-values with track segment LET. The track average LET is defined as $(1/E_o)\int_0^{E_o}(dE/dx)dE$, which correctly weighs the LET over the entire track. This method of representation of the data is by far the most common encountered in the literature.

Product yields with heavy ions are generally presented as a function of the differential or integral LET of the irradiating particle depending on the type of G-value. However, LET is not the sole parameter that describes the track structure and thereby the observed yields [16,57,58]. Many experimental observations have shown that for two incident particles at the same LET radical yields are greater and molecular yields are lower for the ion with a higher charge [57,58,77,93]. Fig. 7 shows an example of the LET effect on the production of HO_2 and O_2 in the radiolysis of water [75,94–96]. The experiments actually convert HO_2 to O_2 so both are measured, but HO_2 is the dominant species. Although HO_2 is a radical, it behaves like a molecular species as its yield increases with increasing LET.

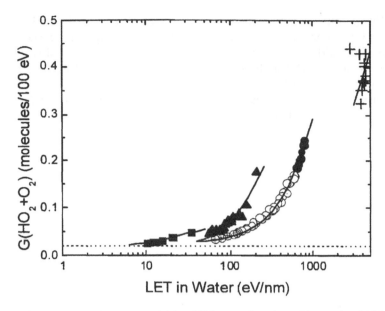

Figure 7   Track average yields of HO_2 as a function of heavy ion LET: (■,▲,●) ¹H, ⁴He, and ¹²C, Ref. 75; (○) ¹²C, Ref. 94; (◎) ¹²C, Ref. 96; (+) ⁵⁸Ni, Ref. 95. The dotted line is the limiting fast electron yield of 0.02 [123].

Many of the expected track effects discussed above are observable with this system. For instance, high-energy protons give about the same HO_2 yield as fast electrons because they both deposit energy in isolated spurs. One can readily observe that LET is not a unique parameter for describing yields.

Equation (1) predicts to a first approximation that for two different types of particles to have the same LET the one with the higher charge will have a higher velocity. A higher incident particle velocity produces a larger radial distribution of energy deposition in part because the distribution of secondary electrons increases with an increase in the heavy ion velocity. Several theoretical studies have examined the average radial distribution of energy about a heavy ion track and the results agree with the observations [21,22,79,97,98]. Although two heavy ions may deposit the same amount of energy per unit path length the ion with higher charge will lose that energy in a larger volume, which leads to a less dense track of reactive species. Different media and even different products in a given media can exhibit a variety of dependencies on particle Z at a given LET.

Early studies on biological media suggested that there must be a better parameter than LET for characterizing the response to heavy ions [99]. These observations led to the proposal that Z^2/β^2 would be a better parameter, where β is the ratio of the ion velocity to the speed of light [100]. The rationale for such a parameter is that the radial energy distribution should be related to the maximum energy of the secondary electron, which is proportional to β^2 or E/M. Equation (1) suggests that the rate of energy loss along the particle path is proportional to Z^2 times the corresponding value for a proton of the same velocity. Combining the estimates for radial and axial energy loss gives a first approximation to the energy density for nonrelativistic ions as Z^2/β^2. At very low energies, the parameter Z should be replaced by the effective charge of the heavy ion, Z_{eff}, as discussed above and given by Eq. (3). Several sets of radiation chemical data have been examined with some success using this parameter [101–104]. However, it was also pointed out that this parameter does not work in some instances [102]. Monte Carlo techniques have been applied to heavy ion tracks in liquid water in order to examine energy loss in detail [74]. Fig. 8 shows the results from that work for the radial energy density or the fraction of energy loss as a function of the radial distance. The radial distributions of energy about protons and helium ions are found to be similar for ions of equal velocity, but not for equal LET [74]. The results also show that the response of the Fricke dosimeter more

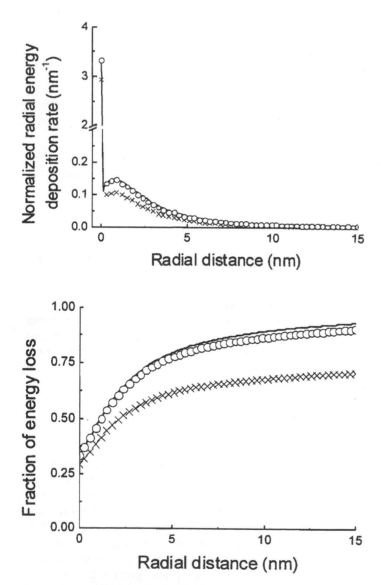

Figure 8   Radial energy deposition rate and fraction of total energy loss in a track segment as a function of radial distance for (solid line) 1 MeV ^{1}H and for ^{4}He ions of (OOOO) equal velocity and (××××) equal LET. (From Ref. 74.)

closely follows the parameter MZ^2/E than LET. This parameter is equivalent to Z^2/β^2 but easier to visualize from common units used to describe heavy ions. The next section will show that the MZ^2/E parameter gives nearly universal agreement for all track average product yields of water radiolysis at long times.

The effects of track structure on product yields are often discussed in terms of one parameter or another that is related to the energy loss of the heavy ion. This approach implies a static track, which is not true. There is a tremendous amount of diffusion and reaction of reactive species following the energy loss processes. One of the biggest challenges for experimentalists and theoreticians is determining the temporal history of products in water radiolysis with heavy ions. Not only do the time dependencies put severe limitations on track models but also they can be vital for predicting radiation damage in concentrated systems that cannot be probed directly. Several pulsed heavy ion experiments have directly measured water product yields at specific times [90,91,105–113]. Except for the work of Burns and coworkers who used water jets, these experiments are limited to determining the product yields on the microsecond time domain. The major difficulty with pulsed experiments is the need for relatively large doses in narrow pulses and the very short heavy ion range. These limitations restrict the number of systems that can be examined and the time scales. Competition kinetics can be used to determine the time scale of chemical reactions at high LET where direct observations are not possible. Solutes that are radical scavengers will stop the competition between radical reaction and diffusion at a time characteristic of the particular scavenging reaction. By changing the solute concentration one can effectively probe the chemistry in the particle track at different times. Even most of the pulsed experiments rely on scavengers to stop the chemistry at short times so product yields can be observed with pulses at longer times. Several experiments using scavenger techniques have been able to determine the temporal variation of radicals and molecular products from nanoseconds to microseconds using a variety of heavy ions [114–118]. Considerably more work is required to obtain a complete set of data for all products of water radiolysis.

5. HEAVY ION RADIOLYSIS OF WATER

The only true method to characterize a product yield in the radiolysis of water is to have the complete temporal dependence for a particular heavy ion with a given energy. The track is constantly evolving in time due to diffusion and reaction of the reactive species and in many instances the yields are never really constant. This result is very different than fast electron or γ-radiolysis where the yields of transients can be considered to be invariant with time from a few microseconds to milliseconds or longer. Track processes are occurring at times less than a microsecond, while homogeneous reactions begin to dominate after a few minutes and products approach steady-state concentrations. The transition from track processes to homogeneous chemistry occurs quickly with heavy ions. It is sometimes difficult to consider the yield of a product as a single valued entity.

Product yields in the radiolysis of water are required for a number of practical and fundamental reasons. Model calculations require consistent sets of data to use as benchmarks in their accuracy. These models essentially trace the chemistry from the passage of the incident heavy ion to a specified point in time. Engineering and other applications often need product yields to predict radiation damage at long times. Consistent sets of both the oxidizing and reducing species produced in water are especially important to have in order to maintain material balance. Finally, it is impossible to measure the yields of all water products for every heavy ion and energy. Therefore one must to be able to interpolate to other ions or energy for which no data exist.

The next few subsections will present data for the products in the radiolysis of water with a wide variety of heavy ions. Product yields with heavy ions are rarely constant in time, but most of them vary slowly in the microsecond region. Furthermore, most chemical systems used to probe yields with heavy ions were developed for examining fast electron or γ-radiolysis in the microsecond region. The following discussion on product yields can be assumed to apply to the microsecond time regime. Because a wide range of systems are used to determine product yields, consistency between the different experiments can be obtained by examination of the material balance. The net decomposition of water can be obtained by setting the number of H atoms in each product equal to the number of H atoms in the net water decomposed to form that product and similarly for O atoms [16].

$$2G(-H_2O) = 2G(H_2) + G(e_{aq}^-) + G(H) + G(HO_2) + 2G(H_2O_2) + G(OH) \qquad (4)$$

$$G(-H_2O) = 2G(H_2O_2) + G(OH) + 2G(HO_2) + 2G(O_2) \qquad (5)$$

Combining these two equations gives the material balance between the oxidizing and reducing species formed in the decomposition of water.

$$2G(H_2) + G(e_{aq}^-) + G(H) = 2G(H_2O_2) + G(OH) + 3G(HO_2) + 4G(O_2) \qquad (6)$$

Molecular oxygen formation is probably negligible at all but the highest LET and HO_2 can be normally ignored with fast electrons and γ-rays.

The discussion in the previous section suggests that the track of a heavy ion becomes more like that of a fast electron with increasing velocity. Therefore one expects that in the high velocity limit the yields of water products with heavy ions are the same as with fast electrons or γ-rays. The yields for the major products of water radiolysis in fast electron or γ-radiolysis are given in Table 1. These values were taken from a number of different sources in conjunction with the results predicted by model calculations [73,116,119–123]. Material balance shows that almost four molecules of water are decomposed for every 100 eV of energy absorbed by fast electrons or γ-rays. Because only about six water molecules are initially decomposed, most of the water products escape intraspur reactions in fast electron or γ-radiolysis.

Linear energy transfer has often been shown to be a poor parameter for characterizing product yields. Therefore in order to compare the results for the different ions the MZ^2/E parameter has been used. This parameter has been successfully used in the past and seems to work well for the data presented here. Note that an increase in MZ^2/E corresponds to an increase in LET so product trends as a function of MZ^2/E presented here are comparable to similar discussions in the literature using LET. Unfortunately, there is no good formalism to fit product yields as a function of MZ^2/E. The solid lines in Figs. 9–14

Table 1 Product Yields in the Radiolysis of Water at about 1 μsec

Particle	e_{aq}^-	H_2	H	OH	H_2O_2	HO_2	$-H_2O$
Fast electron/γ	2.5	0.45	0.56	2.5	0.70	0.02	3.94
(reference)	(73, 119)	(120)	(119, 121)	(73, 122)	(116)	(123)	
1 MeV ^{1}H	0.48	0.94	0.31	0.67	0.91	0.06	2.61
5 MeV ^{4}He	0.26	1.12	0.12	0.38	0.95	0.10	2.48
∞	0.12	2.02	0.0	0.0	0.97	0.56	3.06

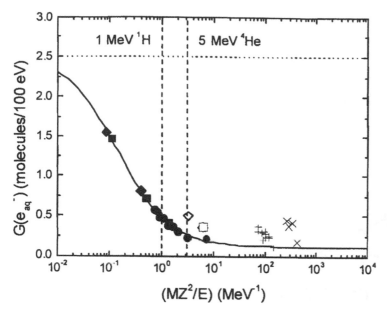

Figure 9 Track average yields of e_{aq}^- as a function of MZ^2/E: (■) ^{1}H, ^{4}He, Ref. 121; (◆) ^{4}He, ^{7}Li, Ref. 26; (●) ^{4}He, Ref. 115; (△) ^{10}B$(n,\alpha)^7$Li, Ref. 124; (□) ^{10}B$(n,\alpha)^7$Li, Ref. 125; (+,×) ^{58}Ni, ^{238}U, Ref. 93. The dotted line is the limiting fast electron yield of 2.5 [73,119]. The dashed lines are for 1 MeV ^{1}H and 5 MeV ^{4}He ions.

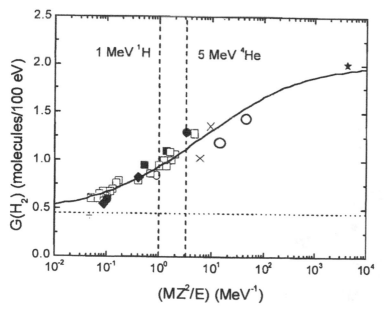

Figure 10 Track average yields of H$_2$ as a function of MZ^2/E: (■) ^{1}H, ^{4}He, Ref. 121; (◆) ^{4}He, ^{7}Li, Ref. 26; (●) ^{1}H, ^{4}He, Ref. 131; (O) ^{1}H, ^{4}He, ^{14}N, ^{20}Ne, Ref. 126; (□) ^{1}H, ^{4}He, Ref. 45; (+) ^{1}H, Ref. 127;(×) ^{14}N, Ref. 128; (★) ff, Ref. 129. The dotted line is the limiting fast electron yield of 0.45 [120]. The dashed lines are for 1 MeV ^{1}H and 5 MeV ^{4}He ions.

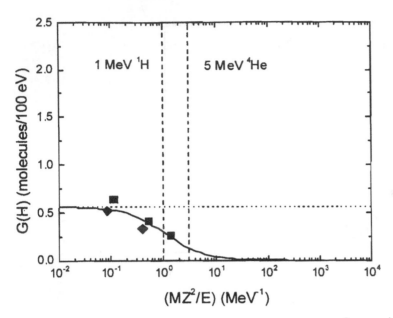

Figure 11 Track average yields of H atom as a function of MZ^2/E: (■) ^{1}H, ^{4}He, Ref. 121; (♦) ^{4}He, ^{7}Li, Ref. 26. The dotted line is the limiting fast electron yield of 0.56 [119,121]. The dashed lines are for 1 MeV ^{1}H and 5 MeV ^{4}He ions.

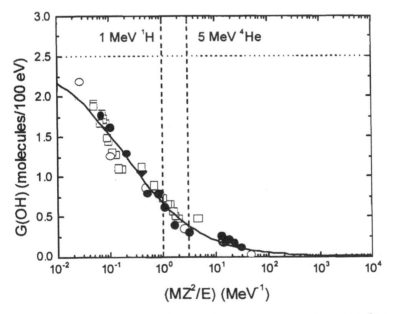

Figure 12 Track average yields of OH radical as a function of MZ^2/E: (♦) ^{4}He, ^{7}Li, Ref. 26; (●) ^{4}He, Ref. 114; (●) ^{1}H, ^{12}C, LaVerne, unpublished results; (○) ^{1}H, ^{4}He, ^{14}N, ^{20}Ne, Ref. 126; (□) ^{1}H, ^{4}He, Ref. 45. The dotted line is the limiting fast electron yield of 2.5 [73,122]. The dashed lines are for 1 MeV ^{1}H and 5 MeV ^{4}He ions.

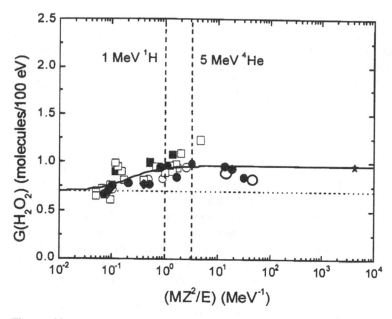

Figure 13 Track average yields of H_2O_2 as a function of MZ^2/E: (■) ^1H, ^4He, Ref. 132; (◆) ^4He, ^7Li, Ref. 26; (●) ^1H, ^4He, ^{12}C, Ref. 116; (O) ^1H, ^4He, ^{14}N, ^{20}Ne, Ref. 126; (□) ^1H, ^4He, Ref. 45; (★) ff, Ref. 133. The dotted line is the limiting fast electron yield of 0.7 [116]. The dashed lines are for 1 MeV ^1H and 5 MeV ^4He ions.

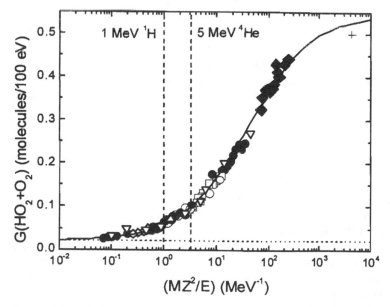

Figure 14 Track average yields of HO_2 and O_2 as a function of MZ^2/E: (●) ^1H, ^4He, ^{12}C, Ref. 75; (O) ^{12}C, Ref. 94; (◎) ^{12}C, Ref. 96; (+) ^{58}Ni, Ref. 95; (∇) ^1H, ^4He, ^{14}N, ^{20}Ne, Ref. 126; (△) ^1H, ^4He, Ref. 121; (+) ff, Ref. 133. The dotted line is the limiting fast electron yield of 0.02 [123]. The dashed lines are for 1 MeV ^1H and 5 MeV ^4He ions.

were obtained by fitting the data for each product Y to the function $G(Y) = G_e(Y) + (G_\infty(Y) - G_e(Y))(ax)^n/(1 + (ax)^n)$, where x is MZ^2/E, a and n are fitting parameters, and G_e and G_∞ are fast electron and infinite MZ^2/E (LET) yields, respectively. This function works well in the inflection region where most of the data are found, but its approach to the limiting values is not good. Therefore the values of the parameters a and n will not be given, but the value of n generally varied from 0.4 to 1.4.

5.1. Hydrated Electron

Hydrated electrons have been widely studied in fast electron radiolysis using pulsed techniques. Only a few pulse radiolysis studies have examined hydrated electron formation with heavy ions [90,81]. Burns and coworkers used 3-MeV protons and reported yields out to 70 nsec, which is too short to compare to the results with solutes [90]. The work of Naleway et al. measured track segment yields in the microsecond region using time resolved techniques [81]. Track segment yields were also measured by Appleby and coworkers [103]. All of the yields for the hydrated electron presented in Fig. 9 are track average yields obtained in scavenging experiments [26,115,121,124,125]. Nitrous oxide and nitrate are most often used as scavengers of the hydrated electron. Fig. 9 shows that there is good agreement for hydrated electron yields over a wide range of MZ^2/E values. The major deviation occurs for the $^{10}B(n,\alpha)^7Li$ recoil experiments [124,125]. An average value of MZ^2/E for the two recoils is used, which is probably giving too high a value. Dosimetry is sometimes a problem in the use of recoil ions.

Hydrated electron yields decrease with increasing MZ^2/E, but they do not seem to decrease to zero. Experiments have been performed on aerated and deaerated Fricke dosimeter solutions using ^{58}Ni and ^{238}U ions [93]. One half of the difference in the ferric ion yields of these two systems is equal to the H atom yield. The Fricke dosimeter is highly acidic so the electrons are converted to H atoms and to a first approximation the initial H atom yield can be assumed to be zero (see below). There is considerable scatter in the data of the very heavy ions, but they seem to indicate that hydrated electron yields decrease to a lower limit of about 0.1 electron/100 eV. The hydrated electron distribution is wider than that of the other water products because of the delocalization due to solvation. This dispersion probably allows some hydrated electrons to escape the heavy ion track at even the highest value of MZ^2/E.

5.2. Molecular Hydrogen

The yields of H_2 are usually straightforward to measure using gas detection techniques. A small amount of scavenger is added to stop the OH radical back reaction from occurring, which would decrease the yields. Fig. 10 shows the results for H_2 as a function of MZ^2/E for the various ions [26,45,121,126–130]. A track segment yield was determined at very high energies, but not included in the figure [131]. There is some scatter in the data, but the overall agreement is good. Molecular hydrogen yields are extremely sensitive to the scavenging capacity of system because this species evolves over a long period of time [130]. The scavenging capacity is equal to the product of the scavenger concentration and the scavenging reaction rate coefficient and reflects the time dependence of H_2 formation. Further detailed experiments on the H_2 yield at different scavenging capacity will probably help clarify the results.

As expected, the H_2 yield increases with increasing MZ^2/E values. Measurements with fission fragments found a limiting yield of 2.0 H_2 molecules/100 eV [129]. The maximum value of MZ^2/E with fission fragments was determined from the average velocity and effective charge of the two main fragments. Although the fission fragment data point may be in some error because of dosimetry, a fit of data suggests that the maximum H_2 yield is about 2 molecules/100 eV. These results show that H_2 is the major product of water radiolysis at high MZ^2/E values.

5.3. Hydrogen Atom

Only two studies have determined H atom yields in neutral water with heavy ions [26,121]. H atom yields are extremely difficult to determine directly because of competition with hydrated electrons or OH radicals. Yields of H atoms are usually estimated from differences in molecular hydrogen yields using various scavengers for the H atom. Fig. 11 gives the H atom yields as a function of MZ^2/E. There are not many data points, but they seem to agree with each other and show a decrease with increasing MZ^2/E values.

The measured H atom G-value is about 0.25 at $MZ^2/E = 1$, while the equivalent yield of hydrated electrons is found at $MZ^2/E = 10$. The persistence of the hydrated electron to higher MZ^2/E values suggests that it does not decrease to zero at an infinite value of MZ^2/E. Most H atoms are produced in conjunction with OH radicals in the core of the heavy ion track. The recombination rate constant is high so there is a small probability that H atoms will escape the track at high LET (MZ^2/E). H atoms can be formed by hydrated electron reactions and their yield cannot decrease to zero if hydrated electron yields do not. However, hydrated electron yields are low at high MZ^2/E values so the H atom yield can be considered negligible in this region.

5.4. Hydroxyl Radical

OH radical yields have been measured over a wide range of energy and particle type and the results are shown in Fig. 12 [26,45,102,114,126]. There is generally good agreement with the data considering that the OH radical is difficult to measure. Most studies use formate or formic acid to give CO_2, while others rely on material balance. There are always experimental problems with H atom reactions contributing to the observed product yields.

Intratrack combination reactions cause the OH radical yields to decrease with increasing MZ^2/E values from a G-value of 2.5 with fast electrons and γ-rays. The lower limit of OH radical yield appears to be zero for high MZ^2/E values. OH radicals are highly reactive and they are formed in the center of the heavy ion track so it is not surprising that there is little probability for them to escape the track. The results also suggest that OH radical production reactions, e.g., H atom reaction with H_2O_2, are negligible at high MZ^2/E values.

5.5. Hydrogen Peroxide

Fig. 13 gives the results of several studies on the measurement of H_2O_2 with heavy ions [26,45,116,126,132,133]. The yield of H_2O_2 never varies significantly from its value with fast electrons and γ-rays. There is a significant amount of scatter in the data so predicting trends is difficult. Large scatter in the data is expected for yields determined by material balance or other manipulation of two or more sets of measurements. However, the production of H_2O_2 is usually determined directly using spectroscopic techniques. Hydrogen peroxide is fairly reactive with all the radical species produced in water radiolysis. The production of H_2O_2 is due to OH radical combination reactions, but the yields are sensitive to added scavengers for all radicals. Even variations in the concentration of hydrated electron scavengers have been shown to affect H_2O_2 yields [134]. This sensitivity to the particular radiolysis conditions makes comparison of different sets of data difficult.

Some studies have suggested that H_2O_2 yields reach a maximum corresponding to the LET of 5-MeV helium ions and then decrease with increasing LET [116]. The decrease was thought to be due to OH radical reactions with H_2O_2 in the dense tracks at high LET. Experiments with fission fragments suggest that the H_2O_2 yield does not decrease to zero at very high MZ^2/E values [133]. The latter result was obtained in acidic solutions, but the outcome should be somewhat similar for neutral solutions. If this assumption is valid, then the limiting value of H_2O_2 at high MZ^2/E values is about 1 molecule/100 eV, which is not significantly different from the value of 0.7 found with fast electrons and γ-rays. It should be stressed that the small variation in H_2O_2 yields with the MZ^2/E parameter is due to multiple compensating reactions and not due to its inertness.

5.6. Perhydroxyl Radical

The mechanism for the production of HO_2 is probably the most uncertain in water radiolysis. Its yield is extremely small with fast electrons and gamma rays, but HO_2 is the most abundant radical produced at high LET. A wide variety of studies have determined HO_2 yields over a huge range of MZ^2/E values as shown in Fig. 14 [75,94–96,121,126,133]. The data agree very well, which is expected because all the sources used ferrous sulfate–cuprous sulfate aqueous solutions. Cuprous sulfate is reduced in this system by the HO_2 to give O_2, which is then measured by a variety of gaseous techniques. However, O_2 produced directly or by track reactions is also determined using this technique. Therefore the yields generally quoted for HO_2 also include O_2. Molecular oxygen yield is expected to be low at all but the highest MZ^2/E values.

The yield of HO_2 seems to increase to a value of over 0.5 molecule/100 eV at high MZ^2/E values. Extrapolation of the present data suggests a maximum value of 0.56, which agrees well with the measured value of 0.5 with fission fragments [133]. A previous work estimated 0.57 molecules/100 eV for the track segment yield of HO_2 at infinite LET. The limiting values agree well with each other. If HO_2 has a simple precursor, such as the O atom, then the limiting value of HO_2 at high MZ^2/E values represents the yield of this precursor in the initial decomposition of water. Further experimental and model studies are necessary to separate the relative yields of HO_2 and O_2 and the mechanisms for their formation.

5.7. Protons and Helium Ions

Many applications of heavy ion radiolysis involve α-particles or neutrons. The former are well represented by the results for 5-MeV helium ions while the latter mainly give protons with a distribution of energy. (Neutrons will also produce low-energy oxygen ion recoils in water.) For the sake of comparison in this work, 1-MeV protons are assumed to represent neutron yields. Individual G-values at other proton energies can be determined from Figs. 9–14 and used in specific applications involving well described energy distributions.

Each of the figures (Figs. 9–14) has dashed lines showing the corresponding MZ^2/E values for 1-MeV protons and 5-MeV helium ions.

Table 1 contains the tabulated G-values for water products estimated for 1-MeV protons and 5-MeV helium ions. The data are consistent with the curves in Figs. 9–14 and give material balance between the oxidizing and reducing species. The major uncertainties are in the yields for H_2O_2 because of the large amount of scatter in the data for this product. Net water decomposition yields determined from the data of Table 1 and Eq. (5) show a net decrease from fast electrons to 1-MeV protons to 5-MeV helium ions. The increasing density of reactive species in the tracks of these particles leads to a net recombination of water. However, the net water decomposition seems to increase at the highest MZ^2/E values. The decreased yields of radicals at high MZ^2/E values allow the molecular products to survive the intratrack reactions and diffuse into the bulk. About half of the initial water decomposed by heavy ions is not recovered on the microsecond time scale.

6. FUTURE

Time resolved and track segment yields are important for fundamental studies on water radiolysis with heavy ions. Advances in accelerators and laser probe systems will make direct time resolved measurements easier. Scavenger concentration studies for obtaining time dependencies will remain the most common experimental heavy ion technique. Scavenger systems are amendable to track segment yield determination and the use of older nonpulsed accelerators. Some products such as molecular hydrogen may never be determined using time resolved techniques. Second-order reactions in the high-density regions of heavy ion tracks may be a problem with some solutes. However, there is considerable confidence in the use of some of the systems that have been examined thoroughly. New systems are constantly being evaluated and developed. All of the studies probing the fundamental water products will be invaluable in predicting radiation damage in unexamined systems and for the development of track models. Detailed Monte Carlo track codes for the radiation chemistry in heavy ion tracks must be advanced beyond the present level.

Much of the future work on heavy ion radiolysis will be largely driven by practical applications. Corrosion processes associated with nuclear reactors and waste storage containers will be a popular area of research because of the impact to society and the availability of funds [11,132]. Scavenger and temperature effects on the main oxidizing species involved in corrosion, e.g., H_2O_2, will be examined for application to the cooling water in nuclear reactors. Cosmic studies, especially those related to planetary evolution, will require extensive examination of the radical reactions in ices. Health effects ranging from space travel, evaluation of the hazards of radon, and medical therapy will dominate much of the work on water with biological endpoints. A large amount of work will be performed on water in heterogeneous situations, such as absorbed water or bulk water near solid surfaces. The interfacial effects are especially challenging because of the differences in initial energy distribution and in the type of radiolysis products formed.

ACKNOWLEDGMENTS

The research described herein was supported by the Office of Basic Energy Sciences of the U.S. Department of Energy. This document is NDRL-4424 from the Notre Dame Radiation Laboratory.

REFERENCES

1. Curie, P.; Debierne, A. Comptes Rendus 1901, *132*, 768.
2. Geisel, F. Ber. Deut. Chem. Ges. 1902, *35*, 3608.
3. Geisel, F. Ber. Deut. Chem. Ges. 1903, *36*, 342.
4. Ramsay, W.; Soddy, F. Proc. R. Soc. 1903, *72*, 204.
5. Ramsay, W. J. Chem. Soc. (London) 1907, *91*, 931.
6. Cameron, A.T.; Ramsay, W. J. Chem. Soc. (London) 1908, *93*, 966.
7. Kernbaum, M. Radium 1910, *7*, 242.
8. Duane, W.; Scheuer, O. Radium 1913, *10*, 33.
9. Duane, W.; Scheuer, O. Comptes Rendus 1913, *156*, 466.
10. Fricke, H.; Brownscombe, E.R. Phys. Rev. 1933, *44*, 240.
11. Lind, S.C. *"The Chemical Effects of Alpha Particles and Electrons"*, Second edition; The Chemical Catalog Company: New York, 1928.
12. Lea, D.E. *"Actions of Radiations on Living Cells"*; Cambridge University Press: New York, 1946.
13. Burton, M. Chem. Eng. News 1948, *26*, 1764.
14. Bacq, Z.M.; Alexander, P. *"Fundamentals of Radiobiology"*; Butterworths Scientific Publications: London, 1955.
15. Hart, E.J. J. Chem. Educ. 1959, *36*, 266.
16. Allen, A.O. *"The Radiation Chemistry of Water and Aqueous Solutions"*; Van Nostrand: New York, 1961.
17. LaVerne, J.A.; Schuler, R.H.; Ross, A.B.; Helman, W.P. Radiat. Phys. Chem. 1981, *17*, 5.
18. Platzman, R.L. In *"Symposium on Radiobiology, the Basic Aspects of Radiation Effects on Living Systems"*; Nickson, J.J., Ed.; John Wiley & Sons: New York, 1952; 97 pp.
19. Platzman, R.L. In *"Radiation Biology and Medicine"*; Claus, W.D., Ed.; Addison-Wesley Publishing Company: Reading, Mass, 1958; 15 pp.
20. Mozumder, A.; Magee, J.L. Radiat. Res. 1966, *28*, 203.
21. Mozumder, A.; Chatterjee, A.; Magee, J.L. In *"Advances in Chemistry Series 81"*; Gould, R.F., Ed.; American Chemical society: Washington D.C., 1968; 27 pp.
22. Mozumder, A. In *"Advances in Radiation Chemistry"*; Burton, M. Magee, J.L., Eds.; Wiley-Interscience: New York, 1969; 1 pp. Mozumder, A. *"Fundamentals of Radiation Chemistry"*. Academic Press: San Diego, 1999.
23. LaVerne, J.A. Radiat. Res. 2000, *153*, 487.
24. LaVerne, J.A. Nucl. Instrum. Meth. Phys. Res. B 1996, *107*, 302.
25. Kudoh, H.; Katsumura, Y. In *"Ion-Beam Radiation Chemistry in Radiation Chemistry: Present Status and Future Trends"*; Jonah, C.D. Rao, B.S.M., Eds.; Elsevier: London, 2002; 37 pp.
26. Elliot, A.J.; Chenier, M.P.; Ouellette, D.C.; Koslowsky, V.T. J. Phys. Chem. 1996, *100*, 9014.
27. McCracken, D.R.; Tsang, K.T.; Laughton, P.J. *"Aspects of the Physics and Chemistry of Water Radiolysis by Fast Neutrons and Fast Electrons in Nuclear Reactors"*. AECL Publication 11895, Atomic Energy of Canada Limited: Chalk River, Ontario, 1998.
28. Schimmerling, W., Budinger, T.F., Eds.; *"Manned Exploration of Deep Space, Cosmic Particle Biology and Health"*; Lawrence Berkeley Laboratory: Berkeley, California, 1989.
29. Raju, M.R. *"Heavy Particle Radiotherapy"*; Academic Press: New York, 1980.
30. Choppin, G. In *"Research Needs and Opportunities in Radiation Chemistry Workshop"*; U.S. Department of Energy: New York, 1998; 25 pp.
31. Rutherford, E. Philos. Mag. 1911, *21*, 669.
32. Bohr, N. Philos. Mag. 1913, *25*, 10.
33. Bethe, H. Ann. Phys. 1930, *5*, 325.
34. Bohr, N. Det. klg. Danske Videnskabernes Selskab. 1948, *18*, 1.
35. Bethe, H.A.; Ashkin, J. In *"Experimental Nuclear Physics"*; Segrè, E., Ed.; John Wiley & Sons: New York, 1953; 166 pp.
36. Fano, U. Ann. Rev. Nucl. Sci. 1963, *13*, 1.

37. Northcliffe, L.C. Ann. Rev. Nucl. Sci. 1963, *13*, 67.
38. Inokuti, M. Rev. Mod. Phys. 1971, *43*, 297.
39. *"Linear Energy Transport, ICRU Report 16"*; International Commission on Radiation Units and Measurements: Washington D.C., 1970.
40. Northcliffe, L.C.; Schilling, R.F. Nucl. Data. Sect. A 1970, *7*, 233.
41. Hubert, F.; Fleury, A.; Bimbot, R.; Gardes, D. Ann. Phys. Suppl. 1980, *5*, 1.
42. Ziegler, J.F.; Biersack, J.P.; Littmark, U. *"The Stopping and Range of Ions in Solids"*; Pergamon: New York, 1985.
43. Bragg, W.H. Philos. Mag. 1905, *10*, 318.
44. *"Stopping Powers for Electrons and Positrons, ICRU Report 37"*; International Commission on Radiation Units and Measurements: Bethesda, MD, 1984.
45. Anderson, A.R.; Hart, E.J. Radiat. Res. 1961, *14*, 689.
46. Bichsel, H.; Inokuti, M. Nucl. Inst. Meth. Phys. Res. B 1998, *134*, 161.
47. Northcliffe, L.C. Phys. Rev. 1960, *120*, 1744.
48. Pierce, T.E.; Blann, M. Phys. Rev. 1968, *173*, 390.
49. Bloom, S.D.; Sauter, G.D. Phys. Rev. Lett. 1971, *26*, 607.
50. Toburen, L.H. In *"High-Energy Ion-Atom Collisions"*; Berényi, D. Hock, G., Eds.; Elsevier: New York, 1982; 53 pp.
51. Lo, H.H.; Fite, W.L. Atomic Data 1970, *1*, 305.
52. Wittkower, A.B.; Betz, H.D. Atomic Data 1973, *5*, 113.
53. Allison, S.K.; Garcia-Munoz, M. In *"Atomic and Molecular Processes"*; Bates, D.R., Ed.; Academic Press: New York, 1962; 721 pp.
54. Schuler, R.H. Trans. Faraday Soc. 1965, *61*, 100.
55. LaVerne, J.A.; Schuler, R.H. J. Phys. Chem. 1982, *86*, 2282.
56. Zirkle, R.E.; Marchbank, D.F.; Kuck, K.D. J. Cell. Comp. Phys. 1952, *39*, 75.
57. Miller, N. Radiat. Res. 1958, *9*, 633.
58. Schuler, R.H.; Allen, A.O. J. Am. Chem. Soc. 1957, *79*, 1565.
59. Wilson, C.T.R. Proc. R. Soc. A 1923, *104*, 192.
60. Wilson, C.T.R. Proc. Camb. Philos. Soc. 1923, *21*, 405.
61. Magee, J.L. J. Am. Chem. Soc. 1951, *73*, 3270.
62. Samuel, A.H.; Magee, J.L. J. Chem. Phys. 1953, *21*, 1080.
63. LaVerne, J.A.; Pimblott, S.M. Radiat. Res. 1995, *141*, 208.
64. Pimblott, S.M.; LaVerne, J.A.; Mozumder, A. J. Phys. Chem. 1996, *100*, 8595.
65. Pimblott, S.M.; Mozumder, A. J. Phys. Chem. 1991, *95*, 7291.
66. Hayashi, M. In *"Atomic and Molecular Data for Radiotherapy"*; International Atomic Energy Agency: Vienna, 1989; 193 pp.
67. Michaud, M.; Sanche, L. Phys. Rev. A 1987, *36*, 4672.
68. Michaud, M.; Sanche, L. Phys. Rev. A 1987, *36*, 4684.
69. LaVerne, J.A.; Pimblott, S.M. J. Phys. Chem. A 1997, *101*, 4504.
70. Schwarz, H.A. J. Phys. Chem. 1969, *73*, 1928.
71. Pimblott, S.M.; LaVerne, J.A.; Bartels, D.M.; Jonah, C.D. J. Phys. Chem. 1996, *100*, 9412.
72. Crowell, R.A.; Bartels, D.M. J. Phys. Chem. 1996, *100*, 17940.
73. Pimblott, S.M.; LaVerne, J.A. J. Phys. Chem. A 1997, *101*, 5828.
74. Pimblott, S.M.; LaVerne, J.A. J. Phys. Chem. A 2002, *106*, 9420.
75. LaVerne, J.A.; Schuler, R.H.; Burns, W.G. J. Phys. Chem. 1986, *90*, 3238.
76. LaVerne, J.A.; Schuler, R.H. J. Phys. Chem. 1987, *91*, 5770.
77. LaVerne, J.A.; Schuler, R.H. J. Phys. Chem. 1986, *90*, 5995.
78. Chatterjee, A.; Schaefer, H.J. Radiat. Environ. Biophys. 1976, *13*, 215.
79. Magee, J.L.; Chatterjee, A. J. Phys. Chem. 1980, *84*, 3529.
80. Chatterjee, A.; Magee, J.L. J. Phys. Chem. 1980, *84*, 3537.
81. Naleway, C.A.; Sauer, M.C., Jr.; Jonah, C.D.; Schmidt, K.H. Radiat. Res. 1979, *77*, 47.
82. LaVerne, J.A.; Schuler, R.H. J. Phys. Chem. 1984, *88*, 1200.
83. Burns, W.G.; Marsh, W.R. Trans. Faraday Soc 1968, *64*, 2375.

84. Cobut, V.; Frongillo, Y.; Patau, J.P.; Goulet, T.; Fraser, M.-J.; Jay-Gerin, J.-P. Radiat. Phys. Chem. 1998, *51*, 229.
85. Frongillo, Y.; Goulet, T.; Fraser, M.-J.; Cobut, V.; Patau, J.P.; Jay-Gerin, J.-P. Radiat. Phys. Chem. 1998, *51*, 245.
86. Draganic, I.G.; Draganic, Z.D. *"The Radiation Chemistry of Water"*; Academic Press: New York, 1971.
87. Allen, A.O. J. Phys. Colloid Chem. 1948, *52*, 479.
88. Hart, E.J. Radiat. Res. 1954, *1*, 53.
89. Allen, A.O. Radiat. Res. 1954, *1*, 85.
90. Burns, W.G.; May, R.; Buxton, G.V.; Tough, G.S. Faraday Disc. Chem. Soc 1977, *63*, 47.
91. Burns, W.G.; May, R.; Buxton, G.V.; Wilkinson-Tough, G.S. J. Chem. Soc. Faraday Trans. 1981, *77*, 1543.
92. *"Average Energy Required to Produce an Ion Pair, ICRU Report 31"*; International Commission on Radiation Units and Measurements: Washington, D. C., 1979.
93. LaVerne, J.A.; Schuler, R.H. J. Phys. Chem. 1996, *100*, 16034.
94. LaVerne, J.A.; Schuler, R.H. J. Phys. Chem. 1985, *89*, 4171.
95. LaVerne, J.A.; Schuler, R.H. J. Phys. Chem. 1987, *91*, 6560.
96. LaVerne, J.A.; Schuler, R.H. J. Phys. Chem. 1992, *96*, 7376.
97. Chatterjee, A.; Maccabee, H.D.; Tobias, C.A. Radiat. Res. 1973, *54*, 479.
98. Fain, J.; Monnin, M.; Montret, M. Radiat. Res. 1974, *57*, 379.
99. Turner, J.E.; Hollister, H. Health Phys. 1969, *17*, 356.
100. Katz, R. Health Phys. 1970, *18*, 175.
101. Christman, E.A.; Appleby, A.; Jayko, M. Radiat. Res. 1981, *85*, 443.
102. Appleby, A.; Christman, E.A.; Jayko, M. Radiat. Res. 1985, *104*, 263.
103. Appleby, A.; Christman, E.A.; Jayko, M. Radiat. Res. 1986, *106*, 300.
104. Sauer, M.C., Jr.; Schmidt, K.H.; Hart, E.J.; Naleway, C.A.; Jonah, C.D. Radiat. Res. 1977, *70*, 91.
105. Sauer, M.C., Jr.; Schmidt, K.H.; Hart, E.J.; Naleway, C.A.; Jonah, C.D. Radiat. Res. 1977, *77*, 47.
106. Sauer, M.C., Jr.; Schmidt, K.H.; Jonah, C.D.; Naleway, C.A.; Hart, E.J. Radiat. Res. 1978, *75*, 519.
107. Sauer, M.C., Jr.; Jonah, C.D.; Schmidt, K.H.; Naleway, C.A. Radiat. Res. 1983, *93*, 40.
108. Chitose, N.; Katsumura, Y.; Zuo, Z.; Domae, M.; Ishigure, K.; Murakami, T. J. Chem. Soc. Faraday Trans. 1997, *93*, 3939.
109. Chitose, N.; Katsumura, Y.; Domae, M.; Zuo, Z.; Murakami, T. Radiat. Phys. Chem. 1999, *54*, 385.
110. Chitose, N.; Katsumura, Y.; Domae, M.; Zuo, Z.; Murakami, T.; LaVerne, J.A. J. Phys. Chem. A 1999, *103*, 4769.
111. Chitose, N.; Katsumura, Y.; Domae, M.; Cai, Z.; Muroya, Y.; Murakami, T.; LaVerne, J.A. J. Phys. Chem. A 2001, *105*, 4902.
112. Baldacchino, G.; Bouffard, S.; Balanzat, E.; Gardès-Albert, M.; Abedinzadeh, Z.; Jore, D.; Deycard, S.; Hickel, B. Nucl. Inst. Meth. Phys. Res. B 1998, *146*, 528.
113. Baldacchino, G.; LeParc, D.; Hickel, B.; Gardès-Albert, M.; Abedinzadeh, Z.; Jore, D.; Deycard, S.; Bouffard, S.; Mouton, V.; Balanzat, E. Radiat. Res. 1998, *149*, 128.
114. LaVerne, J.A. Radiat. Res. 1989, *118*, 201.
115. LaVerne, J.A.; Yoshida, H. J. Phys. Chem. 1993, *97*, 10720.
116. Pastina, B.; LaVerne, J.A. J. Phys. Chem. A 1999, *103*, 1592.
117. Pimblott, S.M.; LaVerne, J.A. J. Phys. Chem. 1992, *96*, 746.
118. Pimblott, S.M.; LaVerne, J.A. J. Phys. Chem. 1994, *98*, 6136.
119. LaVerne, J.A.; Pimblott, S.M. J. Phys. Chem. 1991, *95*, 3196.
120. Pastina, B.; LaVerne, J.A.; Pimblott, S.M. J. Phys. Chem. A 1999, *103*, 5841.
121. Appleby, A.; Schwarz, H.A. J. Phys. Chem. 1969, *73*, 1937.
122. LaVerne, J.A. Radiat. Res. 2000, *153*, 196.

123. Bjergbakke, E.; Hart, E.J. Radiat. Res. 1971, *45*, 261.
124. Henglein, A.; Asmus, K.-D.; Scholes, G.; Simic, M. Z. Physik. Chem. 1965, *45*, 39.
125. Yokohata, A.; Tsuda, S. Bul. Chem. Soc. Jpn. 1974, *47*, 2869.
126. Burns, W.G.; Sims, H.E. J. Chem. Soc. Faraday Trans. 1981, *1, 77*, 2803.
127. Nichiporov, F.G. High Energy Chem. 1986, *20*, 233.
128. Baverstock, K.F.; Cundall, R.B.; Burns, W.G. In "Proc. 3rd. Tihany Symp. On Radiation Chemistry"; Dobo, J. Hedvig, P., Eds.; Akademiai Kiado: Budapest, 1972; 1133 pp.
129. Sowden, R.G. Trans. Faraday Soc. 1959, *55*, 2084.
130. LaVerne, J.A.; Pimblott, S.M. J. Phys. Chem. A 2000, *104*, 9820.
131. Appleby, A.; Christman, E.A.; Jayko, M. Radiat. Res. 1989, *118*, 401.
132. Schwarz, H.A.; Caffrey, J.M., Jr.; Scholes, G. J. Am. Chem. Soc. 1959, *81*, 1801.
133. Bibler, N. J. Phys. Chem. 1975, *79*, 1991.
134. Hiroki, A.; Pimblott, S.M.; LaVerne, J.A. J. Phys. Chem. A 2002, *106*, 9352.

15

DNA Damage Dictates the Biological Consequences of Ionizing Irradiation: The Chemical Pathways

William A. Bernhard
University of Rochester, Rochester, New York, U.S.A.

David M. Close
East Tennessee State University, Johnson City, Tennessee, U.S.A.

1. BIOLOGICAL CONSEQUENCES OF EXPOSURE TO IONIZING RADIATION

1.1. Cellular Endpoint

The effects of ionizing radiation on DNA have been studied for many years because of the central role the DNA plays as the major cellular "target." Cell killing, mutagenesis, and transformation are all caused by damage to DNA.

The ability of external penetrating ionizing radiation to cause biological damage was recognized soon after the discovery of x-rays in 1896. Radiation protection standards for workers were introduced in the 1920s. These first standards were based on the ability of ionizing radiation to produce acute effects such as erythema and could be observed immediately after exposure. At the same time, the appearance of cancers among radiation workers showed that radiation could produce neoplastic effects. This led to the recognition of long-term epidemiological studies for understanding radiation risks and setting standards.

1.2. DNA is the Critical Target

When ionizing radiation affects just one nucleotide in a sequence, this may produce a point mutation. Most point mutations are of little consequence because the same protein or a functional variation is made anyway. However, some point mutations result in a nonsense message from which a nonfunctional protein is constructed, while other point mutations give a meaningful but changed message leading to a protein with altered properties.

Radiation can also cause larger-scale errors involving many nucleotides. In some cases, the change in the chromosome is visible with a microscope (called chromosome aberrations). Some important types of chromosome aberrations are deletions, amplification, and translocation. Deletions are simply loss of a segment of DNA with the consequence that a number of different proteins, coded by the deleted region, are not made at all. This kind of error is thought to occur in the childhood cancer retinoblastoma. The amplification type of error, with repetition of a whole sequence of DNA bases, also occurs in some childhood cancers. Translocation is what happens when one piece of DNA is physically moved and joined onto some unrelated piece of DNA. Some kinds of human leukemia may result from this type of error.

One effect of complex DNA damage is a cell unable to produce copies of itself. This is called reproductive death; the cell can no longer divide. In tissue, a very large number of cells will be present, so death of a few cells makes little difference to the function of the tissue. If the radiation dose is increased, some significant effects may occur quickly. In humans, such effects include radiation sickness, skin burns, or failure of blood cell production. For each of these effects to be large enough to be noticeable, a minimum radiation dose, or threshold, has to be exceeded. For all doses above this, the severity of the effect becomes greater with increasing dose. These are nonstochastic or deterministic effects.

Some radiation effects result from nonlethal damage to a single cell. These effects are called stochastic. They have the property that there is no threshold for these effects to occur. It is the probability of occurrence rather than its severity which increases with dose. The causation of some cancers may be rooted in a stochastic effect.

The dose from low-LET (linear energy transfer) ionizing radiation is delivered by high-speed electrons (Compton and photoelectrons) traveling through the cell and creating primary ionization tracks. One track of ionizations is the minimum disturbance at the cellular level. Paretzke [1] has shown how one can convert the energy of an x-ray or a γ-ray into electrons and their distribution. One can then determine how many photons of a given energy are required to deposit a given dose. Tables are available which present tissue dose in centi-Gray (cGy) when the average track rate per cell nucleus is 1. For example, there would be, on average, one track per cell nucleus for a dose of 1.0 Gy from a medical x-ray. The same would be true for a dose of 0.33 Gy from ^{137}Cs.

Some radiobiologists believe that cancer can be initiated as a result of a single radiation track passing through a single cell nucleus. That would mean cancer induction can be a unicellular process following the rules of chance. This could be taken to imply that there is no safe dose or no safe dose rate. Of course, most DNA damage is corrected by repair enzymes. So a key question then becomes this: does repair of carcinogenic injuries operate flawlessly when dose is sufficiently low? If every carcinogenic lesion to DNA was faithfully repaired, then the net effect of a particular dose to cancer induction would be zero. It follows then that many small doses, with corresponding times in between, could be absorbed without increasing the cancer rate. Human epidemiological studies show that repair fails to prevent radiation-induced cancer, even at doses where the repair system has ample time to deal with only a few tracks at a time. Therefore, a question of considerable current interest is whether or not there is an inherent inability of the repair system to fix certain types of complex damage to DNA.

1.3. Unique Characteristic of Radiation Damage

DNA damage is caused routinely, and frequently, by normal endogenous biochemistry. This type of damage is restricted to single sites and therefore to a single strand. Damage to a single strand is relatively easy to faithfully repair since the complementary strand is available as a template. Radiation damage to DNA is very different since much of the damage is produced in clusters; one example of which are the double-strand breaks (dsb). Double-strand breaks pose a greater risk of misrepair. A dsb can lead to loss of base sequence information or can permit the two ends of a break to separate and rejoin with the wrong partner. Therefore the repair of dsb is essential in preventing chromosomal fragmentation, translocations, and deletions.

1.4. Genomic Instability

The Low Dose/Low-Dose Rate Conundrum

The current health risks associated with exposure to low-dose radiation are extrapolated from high-dose data taken from the Life Span Study of the Japanese atomic bomb survivors. Currently, a linear no threshold extrapolation is recommended. The numerous technical reports and scientific papers about the Japanese A-bomb survivors were widely interpreted as showing that the effects of occupational exposures to radiation would be too small to detect in epidemiological studies. However, questions about the reliability of the A-bomb results were presented by Stewart and Kneale [2]. Their Oxford Childhood Study observed that children whose in utero exposures were as little as 10 to 20 mSv had 40% more childhood leukemias than those who were not exposed. No similar effects are reported in the A-bomb data. Of course, the finding of "no effect" is not a compelling argument for or against a safe dose.

During World War II, a different exposure situation was created in the United States by the Manhattan Project. Several hundreds of thousands of workers were recruited to a new industry in remote and secret locations. Since some of the hazards of working with radiation were recognized in advance, workers at many sites were monitored for radiation exposure. A new discipline, Health Physics, was created to provide radiological protection. The systematic collection of dosimetry records created an opportunity to investigate relationships between repeated exposure to small doses of radiation and disease.

Studies of nuclear workers have suggested that radiation risk estimates based on A-bomb survivors could be underestimating the cancer risks from extended low-level exposure to radiation [3]. Due to such studies, and advances in our understanding of basic mechanisms in radiation biology, the importance of determining the consequences of low dose/low dose rate exposure has become more apparent. The BEIR (Biological Effects of Ionizing Radiation) VII Committee is currently evaluating new DOE Low-Dose Radiation studies (http://lowdose.tricity.wsu.edu/). The current protection standard used by the International Commission of Radiation Protection is 20 mSv whole body annual dose equivalent; in the United States, 50 mSv per year is used.

Bystander Effect

A central tenet in radiation studies has been that energy must be deposited in a cell nucleus to elicit a biological effect. Now, a number of nontargeted effects of radiation exposure have been described which suggest that one must reevaluate this central tenet. One of these nontargeted effects is the bystander effect, which describes the ability of the cell affected by radiation to induce instability in neighboring cells that have not absorbed any radiation energy. Convincing demonstrations of the bystander effect have come from studies using charged particle microbeams.

Using the Columbia microbeam, Zhou et al. [4] passed one alpha particle through the nuclei of mammalian cells. When 10% of the cells were irradiated with a single alpha particle, the mutation yield was similar to that observed when 100% of the cells were irradiated. Clearly, a single alpha particle can induce genomic instability in cells that were not irradiated.

Hormesis

Working in the opposite direction of the bystander effect is hormesis, which is the stimulation of a protective effect by small doses of radiation. It has been proposed that a low priming dose may increase the number or quality of enzymes available to repair subsequent radiation damage. But to date the underlying biochemical basis of hormesis is not known. In addition, it is not clear how to extrapolate the information gained from studies of lower organisms to humans. UNSCEAR (United Nation Scientific Committee on the Effects of Atomic Radiation) [5] examined the notion of "adaptive response" in 1993 and concluded that while it was an interesting phenomenon that occurs in some cell systems at various stages of development, it has little relevance in radiation protection. While it seems likely that the biochemical trigger for hormesis is damage to the DNA, this is not yet proven.

2. ENERGY DEPOSITION

2.1. Ionization and Excitation

Most of the energy associated with an incident x-ray or γ-ray is absorbed by ejected electrons. These secondary electrons are ejected with sufficient energy to cause further ionizations or excitations. The consequences of excitations may not represent permanent change, as the molecule may just return to the ground state by emission or may dissipate the excess energy by radiationless decay. In the gas phase, excitations often lead to molecular dissociations. In condensed matter, new relaxation pathways combined with the cage effect greatly curtail permanent dissociation. Specifically in DNA, it is known that the quantum yields for fluorescence are very small and relaxation is very fast [6]. For these reasons, the present emphasis will be on the effects of ionizations.

The initial chemical events involving the deposition of energy in DNA are conveniently divided into two parts: (1) energy deposited in water and (2) energy deposited in the DNA itself. These are often called indirect and direct effects. Since some of the water in a cell is intimately associated with the DNA, these terms must be used with caution. The presence of DNA close to an energy deposition event in the water will effect the fate of the species produced, and, likewise, water molecules closely surrounding the energy deposition event in the DNA will modify the subsequent fate of the initial species. So the presence of each component modifies the behavior of the other.

2.2. Indirect Effects

The initial ionization of a water molecule produces an electron and the water radical cation. The water radical cation is a strong acid and rapidly loses a proton to the nearest available water molecule to produce an $HO^{\bullet}$ radical and H_3O^+. The electron will lose energy by causing further ionizations and excitations until it solvates (to produce the solvated electron e_{aq}^-). In addition to the two radical species $HO^{\bullet}$ and e_{aq}^-, a smaller quantity of H-atoms, H_2O_2, and H_2 are also produced.

Of the two radical species $HO^{\bullet}$ and e_{aq}^-, the hydroxyl radical is more important in the radiation chemistry of DNA. The e_{aq}^- adds selectively to the DNA bases. The radiation chemistry of the DNA base radical anions is discussed in Sec. 3. The one-electron reduced bases are viewed as less important in the overall scheme since they do not lead to strand breaks [7].

About 20% of the HO$^{\bullet}$ radicals interact with the sugar phosphate by H-atom abstraction and about 80% react by addition to the nucleobases. In model sugar compounds, the H abstraction would occur evenly between the hydrogens on C1', C2', C3', C4', and C5'. In DNA, H abstraction occurs mainly at C4', since the C4'–H is in the minor groove and to some extent with the C5'–H$_2$.

The HO$^{\bullet}$ radical is electrophilic and can interact by addition with the unsaturated bonds of the nucleobases. For the pyrimidines, this would be the C5=C6 double bond [8,9]. For the purines, this would include predominately C4 and C8 addition, with a minor amount of C5 addition [10,11].

2.3. Direct Effect

Since there is such an imprecise division between direct and indirect effects in the literature, some experimental results are presented to clarify this classification. Basically, one cannot detect HO$^{\bullet}$ radicals at low DNA hydrations (ca. 10 water molecules per nucleotide) [12]. This means that in the first step of ionization, the hole produced in the DNA hydration shell transfers to the DNA. It is impossible to distinguish the products from the hole or electron initially formed in the water from the direct-effect damage products. For this discussion, direct-type damage will be considered to arise from direct ionization of DNA or from the transfer of electrons and holes from the DNA solvation shell to the DNA itself.

2.4. Direct-Type vs. Indirect-Type Damage

The DNA solvation shell consists of about 20–22 water molecules per nucleotide; of these, ~15–17 waters associate with the nucleoside and ~5 waters associate with the phosphate group [13,14]. Water outside the solvation layer is termed "bulk water." Upon freezing, the DNA solvation water forms two primary phases: the ice phase, consisting of one or more of the crystalline forms of ice, and a DNA-associated phase, consisting of ordered water which comes in direct contact with the DNA (primary layer) and disordered water in the secondary layer. DNA hydration is expressed in terms of Γ, the number of water molecules per nucleotide.

Sevilla et al. have shown that HO$^{\bullet}$ is not detected in relatively dry DNA ($\Gamma < 8$), but is detected in the $\Gamma > 8$ waters per nucleotide, suggesting that all holes do not transfer to DNA in the regime ($8 > \Gamma > 22$). The sites where these holes are initially produced are not particularly good hole traps. The holes move about until they encounter deep traps such as guanine.

The aqueous electron will react rapidly with all four of the nucleobases. Reaction of the aqueous electron with deoxyribose or ribose phosphate is 2 orders of magnitude lower, so the dominant interaction is with the bases.

2.5. Focus of this Chapter is on Direct-Type Damage

Von Sonntag [7] has estimated that the direct effects contribute about 40% to cellular DNA damage, while the effects of water radicals amount to about 60%. A paper by Krisch et al. [15] on the production of strand breaks in DNA initiated by HO$^{\bullet}$ radical attack has the direct effects contribution at 50%.

Indirect-type damage is much better characterized, both quantitatively and mechanistically, than its direct-type counterpart. Since indirect-type damage has been thoroughly reviewed by von Sonntag [7] and by O'Neill [16], the emphasis of the present chapter will be direct-type damage.

3. DISTRIBUTION OF INITIAL DIRECT-TYPE DAMAGE IN DNA

3.1. Predictions Based on Physics of Ionization and Electron Trapping

Ionizing radiation produces nonspecific ionizations; it ionizes DNA components approximately in direct proportion to the number of electrons. The sugar-phosphate backbone contains 52% of the electrons; the "average base" contains 48%. Electron paramagnetic resonance (EPR) experiments have shown, however, that the final damage to DNA is not a random distribution among these three components. Rather, the majority of the radicals are on the DNA bases. Furthermore, the radiation damage is not randomly distributed among the bases. At low temperatures (4–10 K), one finds initial trapping of the excess electron primarily by cytosine and secondarily by thymine and, upon warming to 180k, primarily by thymine. The radical cations are localized almost exclusively on guanine (which has the lowest gas-phase ionization potential).

The low-temperature EPR experiments used to determine the DNA ion radical distribution make it very clear that electron and hole transfer occurs after the initial random ionization. What then determines the final trapping sites of the initial ionization events? To determine the final trapping sites, one must determine the protonation states of the radicals. This cannot be done in an ordinary EPR experiment since the small hyperfine couplings of the radicals only contribute to the EPR linewidth. However, detailed low-temperature EPR/ENDOR (electron nuclear double resonance) experiments can be used to determine the protonation states of the low-temperature products [17]. These protonation/deprotonation reactions are readily observed in irradiated single crystals of the DNA base constituents. The results of these experiments are that the positively charged radical cations tend to deprotonate and the negatively charged radical anions tend to protonate.

To predict which of the initial ionization events will recombine, and which ones will lead to a stably trapped radical, one must consider the molecular environment. For example, after irradiation, 1-methylcytosine (1-meCyt) is known to have a very low free radical yield, so it is argued that a large percentage of the initial radicals formed by the ionizing radiation must recombine. The hydrogen-bonding network of 1-meCyt does not favor long-range proton displacements [18]. Consequently, there are no energetically favorable paths which would promote the separation of unpaired spin and charge, leaving the initial sites prone to recombination. On the other hand, in many of the systems considered here, there are efficient pathways for returning ionization sites to their original charge states, thereby effectively inhibiting recombination. As a consequence, many of the radiation-induced defects reported are not the primary radiation-induced events, i.e., radical cations or anions, but rather neutral products (deprotonated cations or protonated anions), which are less susceptible to recombination.

These ideas have been illustrated in a recent study of the co-crystalline complex of 1-meCyt:5-FUra [19]. Using model calculations, it was shown how the hydrogen-bonding network of the crystal is able to sustain a proton shuttle which leads to the selective formation of certain radicals. Calculations predict that the site of reduction would be the cytosine base, yielding the N3 protonated cytosine anion, $Cyt(N3+H)^{\bullet}$, while the uracil base would be the site of oxidation, yielding the N1 deprotonated uracil cation, $Ura(N1-H)^{\bullet}$

(see Fig. 1 for radical structures and notation). These are indeed the primary radiation-induced species observed experimentally [19].

There seems, however, to be a problem with the above reasoning. If all the free radical damage ends up on the DNA bases, then one has problems explaining significant lesions such as single-strand breaks. How then does one explain strand breaks? The question could be rephrased by asking if there could be damage to the ribose-phosphate which is difficult to detect.

A single ionization event creates a hole and an ejected electron. To maintain charge neutrality, if a hole is trapped, then an electron is also trapped. Therefore the radical yield must consist of equal oxidation (e⁻ removal) and reduction (e⁻ trapping) events. The electron addition half occurs exclusively at the bases, while the electron loss half occurs according to the number of electrons per component. That means 52% of the holes are initially generated on the sugar phosphate, corresponding to 26% of the initial radicals. A

a. Sugar centered radicals

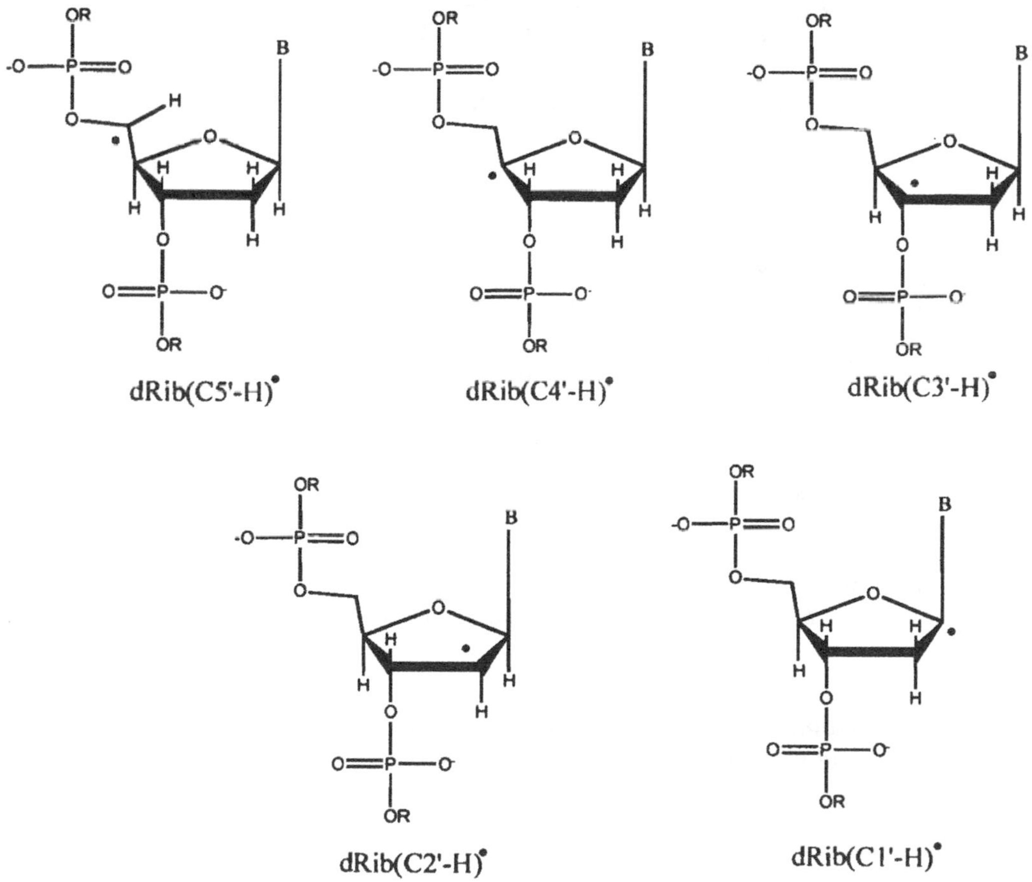

dRib(C5'-H)˙   dRib(C4'-H)˙   dRib(C3'-H)˙

dRib(C2'-H)˙   dRib(C1'-H)˙

Figure 1 Free radical structures, parent compounds, and stable end products for the various components of DNA: (a) deoxyribose, (b) guanine, (c) adenine, (d) thymine, and (e) cytosine. Panel (f) shows trapping of the electron and hole by proton transfer in the GC base pair in duplex DNA.

b. guanine radicals and products

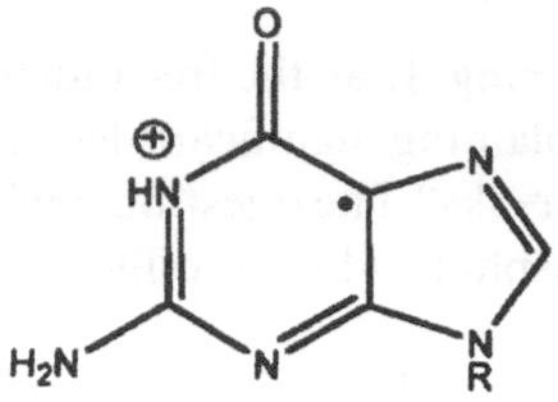

Gua·+ 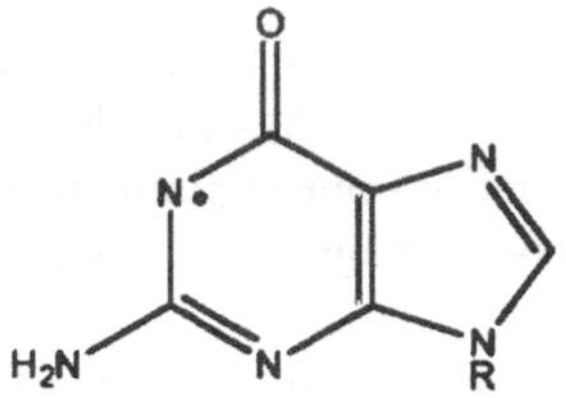   Gua(N1-H)· 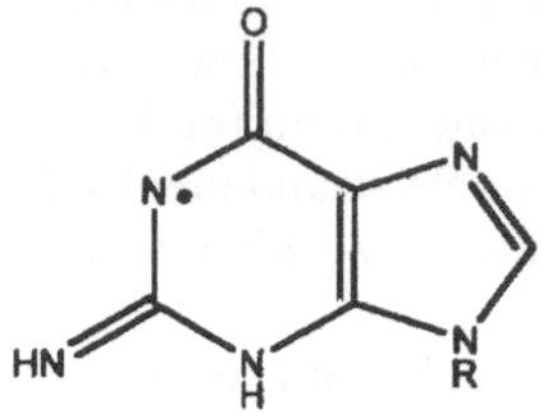   Gua(N2-H)·

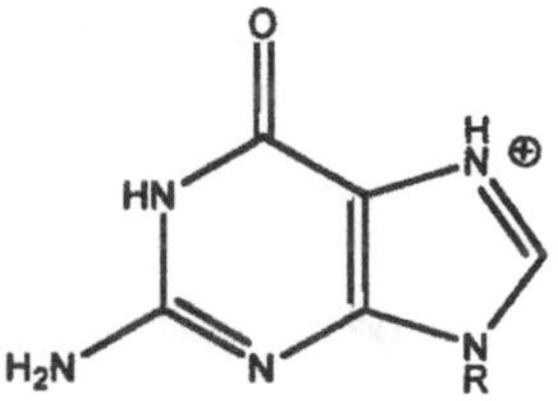

Gua(N7+H⁺)⁺ 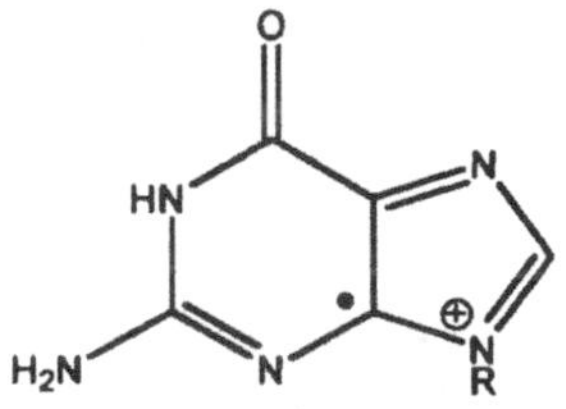   Gua(N7+H⁺, N7-H)·+

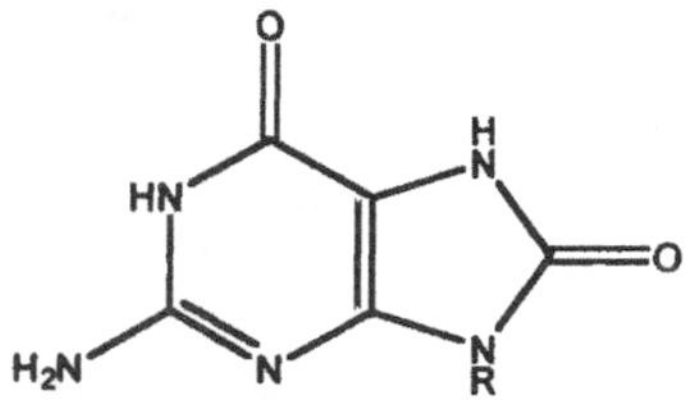

8OxyGua 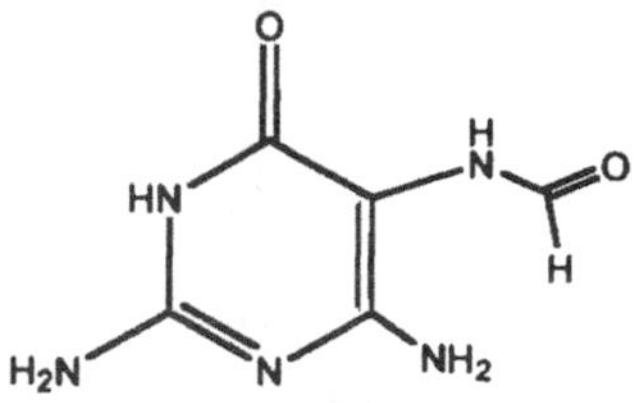   FapyGua

Figure 1   Continued.

recent paper has demonstrated that small percentages of the common H-abstraction sugar radicals could be present in irradiated DNA but could not be detectable in powder samples [20].

3.2.   Initial Trapping Sites of Holes and Excess Electrons Observed by Low-Temperature EPR/ENDOR

Model Systems

Before discussing the types of radicals actually observed in DNA, it is important to first summarize the work on model systems. These include studies of irradiated nucleosides and nucleotides from which one can usually determine the structures of the free radical products. The emphasis here will be to summarize the results on EPR/ENDOR studies of irradiated DNA bases at low temperatures in efforts to study the primary radiation-induced defects.

c. adenine radicals and protonated parent

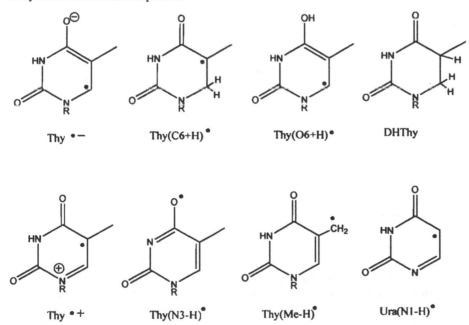

d. thymine/uracil radicals and products

Figure 1   Continued.

e. cytosive radicals and products

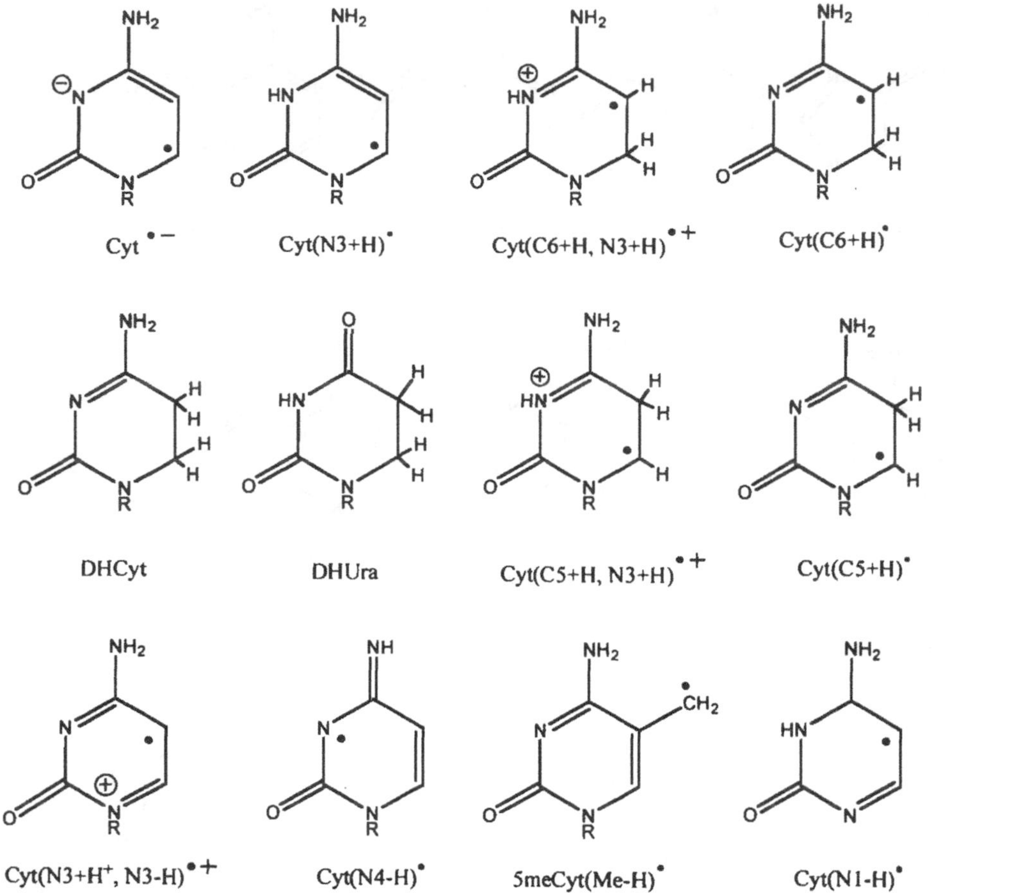

| Cyt$^{\bullet -}$ | Cyt(N3+H)$^{\bullet}$ | Cyt(C6+H, N3+H)$^{\bullet +}$ | Cyt(C6+H)$^{\bullet}$ |

| DHCyt | DHUra | Cyt(C5+H, N3+H)$^{\bullet +}$ | Cyt(C5+H)$^{\bullet}$ |

| Cyt(N3+H$^+$, N3-H)$^{\bullet +}$ | Cyt(N4-H)$^{\bullet}$ | 5meCyt(Me-H)$^{\bullet}$ | Cyt(N1-H)$^{\bullet}$ |

Figure 1 Continued.

Reduction of cytosine produces a radical with sites of unpaired spin density at C2, C4, and C6. The hyperfine coupling of the unpaired spin with the C6–H$_\alpha$ produces a −1.4-mT doublet, which is the main EPR feature of one-electron reduced cytosine observed in various cytosine derivatives [21].

The cytosine anion, Cyt.$^{\bullet -}$, is a strong base (p$K_a > 13$) [22] and is therefore expected to rapidly protonate in solution. Hissung and von Sonntag [23] have shown by con-ductance techniques that the radical anion of cytosine is protonated by water (most likely at N3 or O2) in less than 4 nsec. From studies of cytosine monohydrate single crystals irradiated at 10 K, Sagstuen et al. [24] concluded that the primary reduction product is the N3 protonated anion, Cyt(N3 + H)$^{\bullet}$. ENDOR experiments detected the C6–H$_\alpha$ hyperfine coupling, the N3–H$_\alpha$ hyperfine coupling, and one of the small couplings to the N4–H$_2$ protons.

Oxidation of cytosine produces a radical with sites of unpaired spin density at N1, N3, and C5. The cytosine cation has a p$K_a < 4$ and in solution deprotonates at NH$_2$ [22]. In the solid state, Sagstuen et al. [24] assigned the primary oxidation radical observed in cytosine monohydrate as the N1 deprotonated cation, Cyt(N1 −H)$^{\bullet}$. It is known from the ENDOR

f. proton transfer in GC base repair

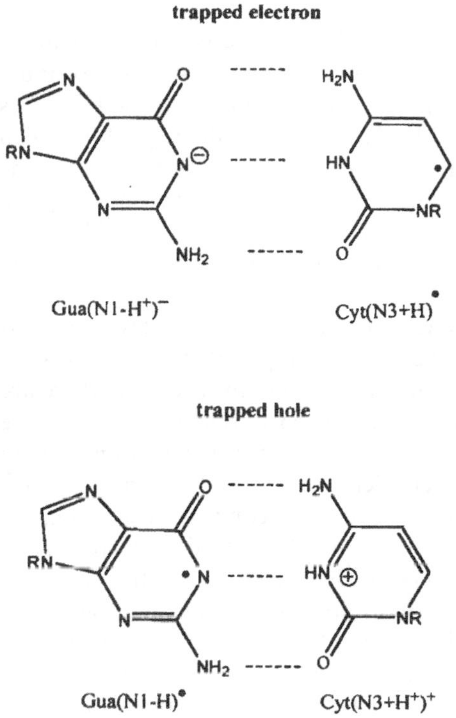

trapped electron

Gua(N1-H$^+$)$^-$ Cyt(N3+H)$^{\bullet}$

trapped hole

Gua(N1-H)$^{\bullet}$ Cyt(N3+H$^+$)$^+$

Figure 1 Continued.

experiment that the unpaired electron densities, ρ, are $\rho(N1) = 0.30$ and $\rho(C5) = 0.57$. Furthermore, there are two small exchangeable N–H couplings whose angular variations correlate well with the exocyclic N4H$_2$ protons.

Since an oxidation product on cytosine in DNA could not deprotonate at N1, it may be more relevant to look at oxidation in a nucleotide. In 5'-dCMP (where N3 is protonated in the parent molecule), oxidation produces the N3 deprotonated cation, Cyt(N3 + H$^+$, N3–H)$^{\bullet}$, with $\rho(N1) = 0.30$ and $\rho(C5) = 0.60$ [25].

Some time ago, an allyl-like radical was observed in irradiated crystals of 5'-dCMP [26]. This radical was thought to be a sugar radical, although no likely scheme was proposed for its formation. It now appears that this radical is formed on 5-methyl cytosine impurities in these crystals [27]. This radical forms by deprotonation at the methyl group of the cytosine cation, 5meCyt(Me–H)$^{\bullet}$, and may have important consequences in the radiation chemistry of DNA since the ionization potential of 5-methyl cytosine is lower than that of either cytosine or thymine.

The thymine anion radical has been identified in duplex DNA when it is exposed to ionizing radiation at temperatures of 77 K or below [28]. Evidence for this radical assignment comes from measurements that demonstrate the conversion of an EPR doublet associated with the thymine anion into a readily identifiable eight-line spectrum of the 5,6-dihydrothymine-5-yl radical [29,30]. This conversion is due to the protonation of the thymine anion at C6. While this process seems rather straightforward, questions remain about the fraction of the free radical population assignable to the thymine anion (reduction will also occur at cytosine) and whether or not the thymine anion protonates at C4═O prior to protonation at C6.

The thymine anion is only a weak base ($pK_a = 6.9$) [22]. This means that protonation of the anion may depend on the specific environment. The primary reduction product observed in the solid state in thymine derivatives is the C4–OH protonated anion [17]. This species exhibits significant spin density at C6 and O4. Here one must distinguish between two different situations. In single crystals of thymidine, the C4–OH$_\beta$ proton is out of the molecular plane which gives rise to an additional 33.1-MHz isotropic hyperfine coupling [31]. A similar situation is observed in single crystals of anhydrous thymine [32]. In 1-meThy, however, the C4–OH$_\beta$ proton is in the molecular plane. Consequently, the proton coupling is very small.

There is not much discussion of thymine oxidation products since they are viewed as unimportant in the radiation chemistry of DNA, the feeling being that in DNA, most of the holes will be trapped by the purines. However, when model systems are used, there are several known pathways that involve oxidation of the thymine base. When a thymine base is ionized, the resulting thymine cation is an acid with $pK_a = 3.6$ for deprotonation in aqueous solution [22]. The thymine cation will likely deprotonate at N3, Thy(N3−H)$^{\bullet}$, although one must look for alternative routes for the cation to eliminate excess charge if N3 is not hydrogen-bonded to a good proton acceptor. One could have reversible deprotonation of the thymine cation at N3 or irreversible deprotonation at the C5–CH$_3$.

In all thymine derivatives studied so far in the solid state, there is always a significant concentration of a radical formed by net H abstraction from the >C5–CH$_3$ group, Thy(Me−H)$^{\bullet}$ [17]. This allyl-like radical is present at helium temperatures. From studies of frozen thymine solutions and poly(T), it can be shown that the precursor of the allyl-like radical is the thymine cation [33,34].

There has been continued interest in the radiation chemistry of the purines since early reports on oriented DNA by Gräslund et al. [35] which suggest that the main trapping site of one-electron oxidation in DNA is the guanine base. It is remarkable that in aqueous solution, the electron adducts of the purine nucleosides and nucleotides undergo irreversible protonation at carbon with a rate constant 2 orders of magnitude higher than that for carbon protonation of the electron adduct in thymidine [36]. It is therefore important to know the properties of the various purine reduction products and to ask why they have not been observed in irradiated DNA.

In single crystals of 5′-dGMP, the native molecule is not protonated at N7. EPR/ENDOR experiments detected a narrow doublet whose hyperfine coupling correlates with a C8–H$_\alpha$ interaction. The computed spin density was $\rho(C8) = 0.11$. This radical was unstable on warming above 10 K, and therefore it was proposed that the radical responsible was the pristine radical anion [37]. However, it is possible that the guanine C6–OH protonated anion could explain these data. One-electron reduced guanine deprotonates at 10 K, but then the protonation reverses upon warming, leaving the original anion, which is then subject to recombination.

The guanine radical cation is a weak acid ($pK_a = 3.9$) [22]. Therefore deprotonation will depend on the environment. Bachler and Hildenbrand [38] have studied the guanine oxidation product in aqueous solution of 5′-dGMP. The best fit to their EPR spectra seems to be from the radical cation (Gua$^{\bullet +}$).

In the solid state, oxidation of 5′-dGMP at 10 K leads to deprotonation at the exocyclic nitrogen, Gua(N2−H)$^{\bullet}$, which is characterized by $\rho(C8) = 0.175$ and $\rho(N2) = 0.33$ [37]. The same radical was detected in crystals of 3′,5′-cyclic guanosine 5′-monohydrate. In this second study, the N3 spin density was determined to be 0.31 [39].

Some experiments have been performed on guanine molecules that were originally protonated at N7. Subsequent electron loss by this molecule leads to deprotonation at N7 yielding Gua(N7 + H$^+$, N7−H)$^\bullet$, which is equivalent to the guanine radical cation. The experimental results from this guanine cation have ρ(C8) = 0.18, ρ(N2) = 0.17, and ρ(N3) = 0.28 [40].

It is not clear what the structure of one-electron oxidized guanine is in DNA. The amino-deprotonated product observed in 5-dGMP does not seem to fit parameters of the oxidation species observed in DNA. Recently, Reynisson and Steenken have proposed that the one-electron oxidized species found in ds DNA is the radical cation [41].

The adenine anion has a pK_a = 3.5 [22]. After electron capture, the negative charge of the adenine radical anion resides mainly on N1, N3, and N7 and therefore protonation likely occurs at one of these nitrogens. The results show that in single crystals examined at 10 K, that reduction of adenine leads to the N3 protonated adenine radical anion, Ade(N3 + H)$^\bullet$, with spin densities of ca. ρ(C2) = 0.41, ρ(C8) = 0.14, and ρ(N3) = 0.12 [42].

The adenine radical cation was observed in a single crystal of adenine hydrochloride hemihydrate [43]. In this crystal, the adenine is protonated at N1. After electron loss, the molecule deprotonates at N1, giving Ade(N1 + H, N1−H)$^{\bullet+}$. This produces a radical that is structurally equivalent to the cation of the neutral adenine molecule with spin density on C8 and N6 [ρ(C8) = 0.17 and ρ(N6) = 0.25]. The adenine radical cation is strongly acidic (pK_a < 1) [22]. This strong driving force makes the reaction independent of environmental conditions. In single crystals of adenosine [42] and anhydrous deoxyadenosine [44], the N6 deprotonated cation [Ade(N6−H)$^\bullet$] is observed which is characterized by ρ(C8) = 0.16 and ρ(N6) = 0.42. The experimental isotropic hyperfine couplings are N6−H$_\alpha$ = 33.9 MHz and C8−H$_\alpha$ = 12.4 MHz.

In single crystals of deoxyadenosine [45], the site of oxidation seems to be the deoxyribose moiety. This brings up an interesting point. In studies of the radiation-induced defects in nucleosides and nucleotides, one often sees evidence of damage to the ribose or deoxyribose moiety. These radicals have not been discussed here because much less is known about sugar-centered radicals in irradiated DNA.

DNA

The previous section outlined the typical e$^-$ loss and e$^-$ gain products observed in the nucleic acid bases in the solid state. These studies can be applied to the study of the radiation chemistry of DNA. The relevance of the study of model systems is shown by considering the following remarkable observations. Years ago, Ehrenberg et al. showed the EPR spectra of the 5,6-dihydrothymine-5-yl radical observed in thymine, thymidine, and DNA. The spectra are nearly identical [46]. The reduction product observed in cytosine monohydrate is the N3 protonated anion. In solution, this reduction product gives rise to a 1.4-mT EPR doublet. The same feature is present in irradiated DNA at 77 K. Likewise, the result of e$^-$ loss in guanine bases is characterized by a broad EPR singlet. The same feature is also evident in the EPR spectrum of DNA irradiated and observed at 77 K.

The EPR spectrum of irradiated DNA is not very well resolved. Attempts to improve the spectral resolution have used deuterated DNA and samples of oriented DNA. Using oriented DNA, Gräslund et al. [35] suggested the thymine anion Thy$^{\cdot-}$ and the cytosine cation Cyt$^{\cdot+}$ were also considered possibilities. Later work reported that only Gua$^{\cdot+}$ and Thy$^{\cdot-}$ were present [47].

Another method is to produce EPR basis spectra by irradiating various nucleic acid bases. The EPR spectrum of DNA is simulated by taking various combinations of the basis spectra. For example, Bernhard looked at one-electron reduction of d(pApGpCpT) at 4 K. The spectrum of the reduced tetramer was simulated by using basis spectra of Thy, Cyt, Ade, and Gua radical anions [48]. The conclusions were that 80% of the trapped electrons resided on cytosine and the remaining 20% on thymine. It should be pointed out that it is difficult to know these percentages precisely since the EPR spectra of reduced Thy and Cyt are quite similar. (At X-band microwave frequencies, this is more so than at Q-band.) Therefore small changes in the Thy/Cyt ratio do not affect the outcomes of the simulations very much.

Sevilla et al. [49] have simulated the EPR spectrum of whole DNA equilibrated with D_2O irradiated and observed at 77 K. The results were 77% $Cyt^{\bullet-}$ and 23% $Thy^{\bullet-}$ for the anions and $>90\%$ $Gua^{\bullet+}$ for the cations. The analysis produced a small imbalance in the cations (44%) and anions (56%). It was suggested that some holes remain trapped in the solvation shell. It is also possible that this reflects small errors in the treatment of the basis spectra, or that some DNA radicals are not accounted for because their EPR signal is too broad and poorly resolved.

Another approach by Hüttermann et al. [50] has been to use pulsed EPR techniques to study the radicals present in DNA fibers equilibrated in D_2O and then irradiated and observed at 77 K. This work supports the conclusions that the primary radiation-induced defects are $Cyt^{\bullet-}$ and $Gua^{\bullet+}$. Also reported are contributions from $Thy^{\bullet-}$ and an allyl radical found on thymine [Thy(Me−H)$^{\bullet}$]. Also discussed are three components tentatively assigned as adenine and guanine anions and a species whose dominant hyperfine interaction involves the N1 of cytosine.

More recently, this same group has studied DNA with high-field EPR (245 GHz) [51]. This study shows nice spectra of the Thy(Me−H)$^{\bullet}$ radical. Also, the authors discuss the effects of hydration levels on the production of the various base radicals. A more recent paper by Weiland and Hüttermann [52] considers the same base radicals in DNA at 77 K and then looks at the transformation of these radicals into the more stable room-temperature products. Among these are C1$'$ and C3$'$ sugar radicals, dRib(C1$'$−H)$^{\bullet}$, and dRib(C3$'$−H)$^{\bullet}$, respectively.

3.3. Predictions Based on Computational Chemistry

Several recent papers have reported Density Functional Theory (DFT) calculations on the primary oxidation and reduction products observed in irradiated single crystals of the common nucleobases: thymine [53], cytosine [54], guanine [55], and adenine [56]. The theoretical calculations include estimates of spin densities and isotropic and anisotropic hyperfine couplings which can be compared with experimental results (obtained from detailed EPR/ENDOR experiments).

It should also be pointed out that the calculation of accurate hyperfine coupling constants is rather difficult. Two factors are involved: the isotropic component (A_{iso}) and the anisotropic component (T_{xx}, T_{yy}, T_{zz}). One must have a good description of electron correlation and a well-defined basis set in order to calculate accurate isotropic hyperfine couplings. This is not easy to do with molecules the size of the DNA bases. Even when the computational demands are met, the theoretical calculations may deviate more than 20% from the experimental results. Recently, it has been shown that the calculation of anisotropic hyperfine couplings for hydrogens is often within 5–10% of the experimental values [54]. The goal is to make comparisons of calculated and experimental isotropic and anisotropic hyperfine couplings a useful guide in identifying radiation-induced free radicals.

The present level of theoretical work can be seen in a recent paper by Sevilla's group. They have employed DFT to study the details of proton transfer reactions in Gua:Cyt base pairs [57]. Using the DFT functional B3LYP with the 6–31 + G(d) basis set on the entire Gua:Cyt molecule (19 heavy atoms), the results are presented which show that it is energetically favorable for the N1 proton of guanine to transfer to the N3 of cytosine after reduction of cytosine or after oxidation of guanine.

Several additional problems must also be considered. The calculations discussed here are computationally challenging. They involve a single-point calculation on the optimized structure using triple-zeta plus polarization functions [B3LYP/6–311G(2df,p)] in order to compute spin densities. Also, the detailed DFT calculations were performed on isolated molecules, whereas some of the experimental results reported involve free radical formation in the solid state, mainly in single crystals. Therefore the theoretical calculations are ignoring the electrostatic environment of the radicals discussed, in particular, the intricate hydrogen-bonding structure that the free radicals are imbedded in. This often leads to nonplanar radicals which may or may not represent what is believed to be observed experimentally in the solid state.

In many cases, the theoretical and experimental results agree rather well. In other cases, there are discrepancies between the theoretical and experimental results. A review of the successes and failures of using DFT to calculate spin densities and hyperfine couplings of the primary radiation-induced free radicals observed in the nucleobases has recently appeared [58].

3.4. DNA Stopping Power

Invaluable to understanding the spatial distribution of damage in DNA are calculations that employ Monte-Carlo track-structure simulations [59–62]. Since most of the energy loss is through inelastic interactions between fast electrons and the DNA, the cross sections for energy loss by electrons are of critical importance. While most theoretical calculations are based on the stopping power of gaseous water, it has been pointed out by La Verne and Pimblott [63,64] that the most probable energy loss and the mean energy loss for electrons in DNA do not compare favorably with gaseous water but are reasonably close to liquid water. Their calculation was based on inelastic cross sections that are obtained from the dipole oscillator strength distribution of DNA. They calculate a mean energy loss for a 1-MeV electron in DNA of 57.9 eV, close to the value of 56.8 eV for liquid water but differing substantially from 50.9 eV for gaseous water. In order to compare experimentally determined yields and degree of clustering with the calculated track structures, it would help considerably to have more concrete information on the stopping power of DNA.

One important application of track-structure calculations is in understanding experimental measurements of free radical yields in DNA, particularly crystalline DNA. Currently, the yield of initial ionizations in DNA is unknown. As a proxy, one can use the values for water (low LET): G(initial ion) ~G(OH) + G(eaq) + G(H) = 0.59 + 0.50 + 0.09 = 1.18 μmol/J [65]. If the stopping power of DNA is the same as liquid water, then samples such as crystalline DNA should yield 1.18 μmol/J initial ionizations at low LET. At 4 K, the observed yields of 0.55–0.78 μmol/J [66,67] would require that 50–70% of the initial ionizations are stably trapped. To the degree that the DNA stopping power might be greater than that of water, this percentage goes down and the range of electron or hole transfer must be increased. If the data on cross sections of low-energy electrons with DNA [68] could be integrated into the Monte-Carlo track-structure calculations, it should help close this gap between experiment and theory.

4. TRACK EXPANSION AND CHARGE MOBILITY IN DNA

4.1. Track Expansion Affects the Spatial Distribution of Products

The nonhomogeneous distribution of energy absorption is a defining characteristic of ionizing radiation. If this characteristic had been absent, the unique impact of ionizing radiation in biology and medicine would almost certainly not exist. But the fact that energy deposition is spatially nonhomogeneous is not, in itself, sufficient to explain the high toxicity of ionizing radiation. The high toxicity (lethality, mutagenicity, and carcinogenicity) is a consequence of the fact that the *final* damage produced in DNA is spatially nonhomogeneous. The severity of the biological consequences depends on the degree to which stable products are formed in clusters [69] or, in different words, occur as multiply damaged sites [70]. An inevitable conclusion is therefore that the rate of product formation must be competitive with the rate of track expansion. Clustered products evolve from clustered ionizations.

Fig. 2 depicts, in schematic, the two extremes plus an intermediate case of track expansion. In Fig. 2A, the initial distribution on ions is shown at zero time, where $t = 0$ is after thermalization but before charge transfer. The initial cluster density is a function of LET, and even at very low LET (e.g., ^{60}Co γ-rays), a large percentage of energy is deposited in clusters [71]. In Fig. 2B, ion mobility is high, which would be the case if DNA were a good conductor. After a very short time, combination reactions fully annihilate the charges in DNA. In Fig. 2C, the holes and excess electrons have low mobility, moving a short distance before trapping. Closely spaced, opposite charges combine. In Fig. 2D, the mobility is zero; the track is etched in place. We consider case B first, the case where holes and electrons freely conduct through the base stacks, consequently leaving no damage on the base stack.

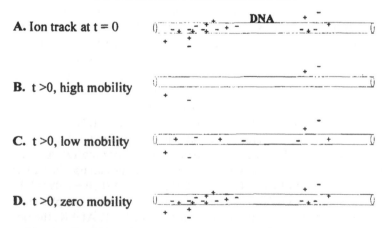

Figure 2   A schematic representation of track expansion is shown assuming different mobilities. (**A**) Spatial distribution of radical ions at the sites of initial thermalization. (**B**) When there is high mobility, that of a conductor is assumed, no radical ions are trapped on DNA. (**C**) If there is low mobility, some radical ions are trapped while others recombine. (**D**) At zero mobility, the spatial distribution of trapped radicals is the same as in the initial track, i.e., the same as in (**A**).

If DNA has the properties of a molecular wire, as proposed in 1993 [72] and subsequent work (e.g., Refs. 73–76), then the spatial distribution in DNA would look something like that in Fig. 2B. If this was the case, it is very difficult to reconcile a large body of knowledge in radiation biology because not only does it dramatically reduce the degree of clustering, it would also make DNA remarkably radiation-resistant. Indeed, base damage produced via the direct effect would not exist. Publications by radiation chemists and theoreticians, before and after the 1993 proposal, provide rather strong evidence that DNA is not a molecular wire [77–79]. Joining that body of work is a recent series of photochemistry investigations that shows that charge mobility through DNA is that expected of an insulator (a poor semiconductor) [80–82]. Therefore the situation under Fig. 2B does not occur.

The other extreme shown in Fig. 2D also does not occur. Holes are mobile. We know this because, as discussed above, the probability of finding a hole on a particular component does not correlate with the probability of ionization. Holes are trapped preferentially at guanine [83], to a lesser extent at the sugar [84], and rarely at the other bases. This redistribution requires hole transfer. Trapping of the excess electron is also selective, occurring primarily at cytosine [49,85,86]. This implies that after thermalization, the electron transfers between bases until finding a stable trapping site. Therefore the situation is as depicted in Fig. 2C. A key question is "What is the range of both the excess electron and the hole within DNA and through its local environment?"

4.2. Hole and Electron Transfer

Studies of the direct effect have been largely confined to DNA samples in the solid state. This is done in order to maximize direct-type damage and minimize indirect-type damage. In addition, low temperatures are often employed both as a means of sequestering the DNA from the bulk water and as a means of stabilizing free radical intermediates. In frozen DNA samples, the mobility of holes and excess electrons differs for the different sample components: ice, solvation shell, DNA backbone, and base stacks. We start with the ice phase.

Ice

In an aqueous solution of DNA, the water outside of the solvation shell is referred to as bulk water. When DNA solutions are frozen, the bulk water crystallizes as a separate phase—ice. Ice does not form if the concentration of DNA is brought to a level where only the solvation shell remains, about 20–22 waters/nucleotide. If brought to this concentration slowly, a film is formed. Freezing a film does not create ice. Another type of sample is prepared by first lyophilizing DNA and then letting it sit at a preselected humidity that determines the level of hydration, typically $2.5 < \Gamma < 22$. Subsequent freezing of these cotton-like samples does not yield ice.

During irradiation, the holes produced in the ice phase are sequestered by the ice. This is not strictly true for the excess electrons; some escape the ice phase and are selectively scavenged by DNA. The degree to which this increases the yield of electrons trapped by DNA is considered to be relatively small and therefore when computing yields of direct damage in DNA, the mass of the ice phase is usually excluded from the target mass. While ice is a relatively passive component of the system, it does provide an important reference point with respect to product yields. Electrons and holes are quite mobile in ice [87,88]; it is a decent conductor. Consequently, even at low temperatures, the free radical yields are quite low, e.g., the yield of $HO^{\bullet}$ in ice at 77 K is 0.037 µmol/J [83]. This is just one example of a general principle; high charge mobility promotes recombination and therefore results in low yields of trapped free radicals.

DNA Solvation Shell

Ionization of DNA's solvation shell produces water radical cations ($H_2O^{\bullet+}$) and fast electrons. The fate of the hole is dictated by two competing reactions: hole transfer to DNA and formation of $HO^{\bullet}$ via proton transfer. If the ionized water is in direct contact with the DNA ($\Gamma < 10$), hole transfer dominates. If the ionized water is in the next layer out ($9 < \Gamma < 22$), $HO^{\bullet}$ formation dominates [67,89,90]. The thermalized excess electrons attach preferentially to bases, regardless of their origin. Thus the yield of one-electron reduced bases per DNA mass increases in lockstep with increasing Γ, up to an Γ of 20–25. This means that when Γ exceeds 9, there will be an imbalance between holes and electrons trapped on DNA, the balance of the holes being trapped as $HO^{\bullet}$. At $\Gamma = 17$, an example where the water and DNA masses are about equal, the solvation shell doubles the number of electron adducts, increasing the DNA-centered holes by a bit over 50% [91–93].

Samples that are prepared by freezing DNA solutions result in a DNA aggregate phase in which ice formation draws water from the solvation shell. The net result is that the DNA aggregates have an average Γ of ~15 [89,90]. For DNA prepared in this manner, the target mass is DNA + 15 waters [91]. Since the solvation shell is not pure water, containing about one cation per nucleotide, it is more accurate if one replaces water mass by the equivalent cation mass. Taking Na^+ as an example, 15 waters would be replaced by the sum of 1 Na + 13.5 waters. A rule of thumb is that if one assumes that only the DNA is the target mass, the solvation layer of DNA doubles the direct-type damage.

DNA Backbone

Forward reactions of the one-electron oxidized sugar-phosphate backbone are of two types: hole transfer to the base stack or formation of a neutral sugar radical by irreversible deprotonation. In much of the older literature, it was assumed that the former eclipsed the latter. This was largely because free radicals centered on the sugar moiety were not detected by EPR. As has been pointed out, this does not prove that sugar radicals are not formed [94,95], and indeed, computed simulations have shown that such radicals would give EPR spectra that would be difficult to detect in the presence of the intense signal from the base-centered radicals [20,96]. Recent results leave little doubt that a substantial fraction (~25%) of the holes are irreversibly trapped on the sugar [84,97–99].

Evidence that direct-type damage occurs at the sugar consists of the following. Becker et al. [100] observed sugar radicals in DNA exposed to high LET radiation. Swarts et al. [101] observed free base release in DNA and Razskazovskiy et al. [98] observed free base release in crystalline oligodeoxynucleotides. Free base release correlates with sugar damage. The yields of strand-break products have been measured by Debije et al. [97] and Razskazovskiy et al. [99] in crystalline oligomers x-irradiated at 4 K. Debije has identified a 3′-centered sugar radical in crystals of d(CTCTCGAGAG)$_2$, a B-form DNA [84]. Broadly speaking, 10–20% of the radicals initially stabilized on DNA at 4 K are sugar radicals. This corresponds to about half of the backbone-centered holes (formed directly and indirectly by transfer from water) leading to neutral sugar radicals via deprotonation of the one-electron oxidized sugar-phosphate backbone. Also, it correlates well with the observed yields for direct strand breaks, 0.05–0.15 μmol/J, in high molecular weight DNA (Sec. 5).

DNA Bases

Radiation Chemistry. It has been known for decades that charge migrates through the stacked bases of DNA [102,103]. Compelling evidence comes from studies employing EPR [30,77,104–106], pulse radiolysis [107,108], and product analysis [101,109,110]. The range of migration for electron and holes at 4–77 K is at least 2 bp and the mean is between 3 and 11 bp [106,111]. Warming samples above 150 K extends thermally activated migration to 30 bp or more [106]. Clearly, the DNA bases are very effective traps for both holes and excess electrons. Because the activation energy needed to detrap these species is very disperse [104], charge migration by hopping can be observed over a wide temperature rage, < 20 to > 150 K. The dielectric relaxation that creates traps most assuredly occurs even at room temperature, competing with other processes such as tunneling. While it may be difficult to observe these trapping events at room temperature, one should anticipate that they moderate the speed and range of charge transport, even at high temperatures and in dilute aqueous solution.

Photochemistry. The proposal by Murphy et al. [72] that DNA is a molecular wire stimulated new interest in the possibility that DNA conducts charge through its base stacks, particularly among photochemists. Based on an observed absence of fluorescence, they concluded that electron transfer occurs through a length of stacked bases ("π-ways") > 4.0 nm in $< 10^{-9}$ sec. They calculated $\beta_e \sim 0.2$ Å^{-1} for the equation $k_{ET} = k_0 exp\{-\beta_e(R-R_0)\}$, where, for the case of tunneling, the electron transfer rate is expected to drop off exponentially as a function of distance $(R-R_0)$. Conductors do have very low values of β_e, such as Murphy et al.'s group observed. The high conductivity of DNA was eventually promoted as a paradigm shift [112,113]. Others, using the tools of photochemistry, have measured the values of β, making sure of high quantum yields and confirming transfer with corroborating evidence. For electron transfer and for hole transfer, $\beta_e \sim 1.0$ Å^{-1} and $\beta_h \sim 0.6$ Å^{-1} are obtained, respectively, values typical of insulators [114–116]. If experiments are designed that are permissive for hopping, then low values of β_h, < 0.2 Å^{-1}, can be observed, but under these circumstances, electron transfer is not fast [117]. In our opinion, the evidence is conclusive; DNA is an insulator. The current questions of interest focus on how excess electrons and holes move through DNA and what reactions compete with charge transfer. Indeed, electron transfer in DNA proceeds independently of the "π-ways," the main function of which is to promote base stacking and thereby stabilize the linearity of DNA.

Interstrand and Interduplex

DNA–DNA. While it is difficult to design an experiment that proves that electron/ hole transfer occurs between the strands of a single duplex (interstrand), it is rather simple to prove that transfer occurs from one duplex to another (interduplex). This was done by monitoring the concentration of holes and excess electrons trapped in crystalline DNA as a function of temperature. Given that the annealing characteristics are relatively insensitive to DNA conformation, sequence, or base stacking continuity, it is readily shown that electron and/or hole transfer must be intermolecular [104]. In more elegant experiments (see below), it is shown that interduplex tunneling occurs for both electrons and holes [118–121]. These new results make it clear that, with regard to charge transfer via tunneling, the key variable is distance, not the intervening covalent or π-bonds. Thus in a crystalline lattice, where the tunneling distance between helices is comparable to that within a strand, the hole and electron move in three dimensions.

If hole and electron transfer occurs interduplex, it follows that it will also occur interstrand. The tunneling distances are comparable. This is elaborated in Sec. 4.3.

DNA–Protein. A large number of proteins in nature perform their function by expediting electron transfer, and there is an extensive literature on electron transfer through proteins (see Refs.122–124 for reviews). Relevant here are the observations that the excess electron has a large range (not readily trapped) [125] while the hole is relatively immobile (trapped by deprotonation at the peptide bond giving amido radicals) [126]. This raises the expectation that electrons but not holes could be transferred from protein to DNA. This has been observed by a number of groups [127–131].

In one of the more detailed studies to date, Weiland and Hüttermann [131] compare the trapped radical distribution of DNA alone with that of dry chromatin. Using high-field EPR (285 GHz), these experiments significantly reduce the spectral overlap between one-electron oxidized guanine (Gua_{ox}) and the other DNA radicals thereby improving the precision over EPR measurements made at lower fields (9 GHz). Following x-irradiation at 77 K, the DNA samples trap out 55% of the radicals as one-electron reduced pyrimidines (Pry_{re}) and 22% as one-electron oxidized guanine. Under the same conditions, chromatin gives 26% and 5%, respectively. The radical concentration at a specific dose increases by $1.4\times$ in chromatin vs. DNA alone. Given that ~47% of the radicals are preferentially on DNA and that the chromatin samples consisted of ~30% DNA and ~70% histone proteins, the authors find that the concentration of radicals trapped by DNA in chromatin is ~2× that of DNA alone. The data can be used to estimate the influx of electrons (x) and presumed efflux of holes (y). Solving the two equations, $100 + x - y = 200$ and $26/5 = (55 + x)/(22 - y)$, one obtains $x = 94$ and $y = -6$. Thus these results indicate that when DNA is packaged in chromatin (dry), it traps twice as many electrons while hole trapping is relatively unchanged. Warming the 77 K irradiated chromatin to 300 K reduces the total radical concentration by at least 95% while not causing any marked changes in the DNA–protein radical ratio. This means that electrons or holes are thermally mobilized in both components, but there is no net exchange between the components.

The amount of information is still insufficient to predict how packaging of DNA in chromatin modifies the final product distribution in DNA. If the stable end products reflect a doubling in reductive damage while oxidative damage remains the same, then one would expect an increase in frequency and complexity of clustered lesions. If that proves true, the histone proteins and attending nucleosomes structure would act as radiation sensitizers.

4.3. Tunneling and Hopping

Charge transfer in DNA occurs under two distinctly different conditions: one which is temperature-independent and the other which is not. The mechanism underpinning the former is tunneling, while the latter is controlled by an activation barrier that must be "hopped" over. Of these two, tunneling is the one better characterized.

Tunneling

Tunneling as a mechanism of charge transfer in DNA has received widespread attention (see, for example, Refs. 117 and 132). Recently, definitive proof that tunneling in DNA occurs has been reported by Sevilla et al. [118–121,133]. Key to this proof is the demonstration that the observed rates of hole and electron transfer are independent of temperature from 4 to 130 K. We summarize their findings here.

DNA, laced with an intercalator characterized by a high electron affinity, is γ-irradiated and observed by EPR. The one-electron reduced intercalator presents an EPR spectrum that is readily distinguishable from that of the DNA-trapped radicals. A key example is mitoxantrone (MX), with an electron affinity of 6.25 eV and a radical anion spectrum that is a sharp singlet. Charges are injected into the DNA by γ-irradiation at a preselected temperature (4–130 K). Holding the temperature constant, the EPR spectrum changes as a function of time (0.5–30 h). Thereby, a direct measure of the rate of electron transfer from one-electron reduced pyrimidines (Pry_{re}) to the intercalator, e.g., MX, is measured. The tunneling rate is observed to depend on the electron affinity (EA) of the acceptor (intercalator): for MX, $\beta_e = 0.9\pm0.1$ Å^{-1} and $k_0 \sim 10^{11}$ sec^{-1}, and for ethidium bromide (EA = 4.32 eV), $\beta_e = 1.2 \pm 0.1$ Å^{-1} and $k_0 \sim 10^5$ sec^{-1}. It also depends on the EA of the donor (Pry_{re}). The calculated EA of the Ade:Thy and Gua:Cyt base pairs are 0.30 and 0.49 eV, respectively [57,134]. The values of β_e are 0.7 ± 0.1 and 1.4 ± 0.1 Å^{-1} for Thy_{re} and Cyt_{re}, respectively. In addition, the rate depends on DNA–DNA distance, proving that tunneling is intermolecular as well as intramolecular. In addition, the values of β are not strongly dependent on the medium lying between the donor and acceptor. If intermolecular transfer is not taken into account, calculation of the transfer distance gives a deceptively large number, e.g., 31 base pairs reported by Pezeshk et al. [135]. Taking intermolecular transfer into account, the tunneling distance at long times (1 hr) is remarkably reproducible for MX-DNA over a wide range of DNA packing configurations, 35–38 Å (10–11 base pairs). For hole transfer in MX-DNA, β_h equals 1.1 Å^{-1} assuming a value of k_0 of 1×10^{11} sec^{-1}.

Two important insights into the spatial separation of end products grow out of the recent findings on tunneling in DNA. (1) In order for the distance between the stable products derived from oxidized guanine and reduced pyrimidines to be less than 30–40 Å (<10 bp), a competing irreversible reaction must trap the radical at a rate competitive with that of tunneling. While this is highly improbable at low temperatures, <120 K, it is likely to be significant at RT. More needs to be learned about the rates of irreversible trapping. (2) The rate of irreversible trapping of holes on the sugar must be fast relative to hole tunneling to the bases. This means that the distance between a strand break (derived from the sugar-centered hole) and base damage may be substantially less that 30 Å. The neutral sugar radical will not be a strong getter for base radicals. One predicts therefore that a clustering of strand breaks with base damages will be more prevalent than clustering of base damage with base damage when DNA is irradiated at low temperatures. How this shifts, when irradiation is at RT instead of <120 K, depends on the rates of base damage fixation. These rates are governed by an activation energy, which is our next topic.

Hopping

The trapping rate, for both holes and excess electrons, is competitive with the tunneling rate. This must be the case because DNA is a very effective trap of both holes and electrons; there are no other materials that are better traps that we are aware of. Thus the way to maximize trapped radical density is to maximize DNA density, and this is why crystalline DNA tops the scale in trapping capacity, with trapped radical yields of 0.5–0.7 µmol/J at 4 K [66,67]. Remarkably, only 30–50% of all initial ionizations are lost to recombination. This is remarkable because tunneling distances, either Gua–Gua or Pyr–Pyr, are typically short, < ~7 Å, in all directions [104]. This means that the trapping rate must be fast $[k > 10^{11}$ sec^{-1} $\exp(-1.0$ $\text{Å}^{-1} \times 7$ $\text{Å}) \sim 10^9$ $\text{sec}^{-1}]$, and the trap depth must be low enough to shut down tunneling. In order to mobilize trapped charges, an activation energy is required. Charge transfer governed by an activation energy is referred to as hopping.

Trapping entails dielectric relaxation about the charge altered site. This is analogous to the Marcus reorganization energy in the liquid state [136]. But in polar solids at 4 K, large displacements such as molecular translation and rotation cannot occur as a part of the reorganization. It is constrained to relatively small displacements of the atoms surrounding the newly charged site. The displacements optimize dipole–dipole, dipole–monopole, and monopole–monopole interactions. The most pliable particles in the DNA matrix (at <120 K) are the protons participating in highly polar bonds ($N-H$ and $O-H$). Nearly every such bond in a crystal lattice participates in a hydrogen bond, most notably the $N-H\cdots O$ and the $N-H\cdots N$ Watson–Crick bonds that zip together the two strands. When a charge attaches to a particular site, the surrounding hydrogen-bonded protons are relatively "free" to shift in response to the new charge distribution. In the case of hole trapping, this proton shift is so pervasive that we do not know of any case where an organic radical cation has been trapped within a hydrogen-bonded lattice; the radical cations are observed to consistently deprotonate. Furthermore, if lattice structure does not provide an opportunity for proton displacements across hydrogen bonds, the probability of stabilizing the hole decreases and, if stabilized, it is through carbon deprotonation [137–139]. Insight into what makes DNA such an efficient electron and hole trap has been obtained by studying the annealing properties of traps.

Annealing DNA irradiated at 4 K leads to a steady decline in EPR signal intensity; the radical concentration is approximately a monotonic function of temperature. If the sample is held at any given annealing temperature, the concentration initially decreases and then plateaus. Detrapping of the radicals is therefore characterized by a large dispersion in activation energies. This is, in contrast to the annealing properties of a number of crystalline DNA components, where discrete activation energies can be measured [140–142]. In our working model, the DNA's annealing behavior is explained by three classes of traps: (a) very shallow, (b) shallow, and (c) deep. The first two are reversible states and the third is irreversible.

Very Shallow Traps. It has been proposed that the neutral $\text{Gua(N1}-\text{H)}^{\bullet}$ radical, formed by proton transfer from the $\text{Gua}^{\bullet+}$ radical by proton transfer from N1 of Gua to N3 of Cyt, is a shallow trap [143,144]. This proposal is based on projections from pK_a made on monomers in dilute aqueous solution, which predict that proton transfer is favored by 2.3 kJ/mol [22,145]. Ab initio calculations are in excellent agreement with this value [146,147]. So one expects that an energy of at least 0.025 eV is needed to activate the return of the proton to N1 Gua, reforming $\text{Gua}^{\bullet+}$. Once $\text{Gua}^{\bullet+}$ is reformed, tunneling to nearby guanines is reestablished as a competitive pathway. Proton transfer therefore is a gate for hole transfer. Proton-coupled hole transfer describes the thermally driven transfer of holes from one Gua:Cyt base pair to another.

Charge mobilization is initiated at very low temperatures, <20 K [104]. It has been proposed that the mobile species at these low temperatures is the guanine-centered hole [144]. But if activation of hole detrapping entailed only reverse proton transfer, one would anticipate a relatively discrete activation energy (~0.025 eV) and instead a large dispersion is observed. Detrapping, presumably of holes, occurs from ~10 to ~120 K. The energetics of reorganization (dielectric relaxation) therefore must vary significantly from site to site. This observation means that the $N1-H\cdots N3$ proton transfer is just one of a large number of proton shifts. The hydration layer of DNA, even in crystals of DNA oligomers, is disordered. It is therefore not difficult to envisage a wide range of trapping depths that depend on not only the energy of proton transfer within the Watson–Crick hydrogen bond, but also on the proton shifts within the immediate environment.

Shallow Traps. The depth of the traps for the excess electron is expected to be deeper than the holes. Once more, using the findings of Steenken [22,145], the change in free energy for proton transfer can be estimated. In this case, one-electron reduced Cyt (Cyt$^{\bullet-}$) acquires a proton at N3 of guanine via the N3$\cdots$H$-$N1 hydrogen bond, forming the neutral Cyt(N3 + H)$^{\bullet}$ radical. Steenken estimates that this transfer is favorable by 20 kJ/mol and this value is also found by ab initio calculations [146,147]. In order to reverse the proton transfer, at least 0.17 eV is needed. Reversal recreates Cyt$^-$ and permits tunneling to resume. Just like hole transfer, electron transfer is gated by proton transfer. Proton-coupled electron transfer has been observed in aqueous DNA systems [148].

There is good evidence that the excess electron is mobilized at temperatures between 140 and 180 K. At these temperatures, the Cyt(N3 + H)$^{\bullet}$ concentration decreases, a decrease that correlates with the formation of a neutral radical, Thy(C6 + H)$^{\bullet}$, produced by protonation of one-electron reduced thymine (Thy$^-$)$^{\bullet}$ at C6. In order for this to occur, electrons trapped at cytosine must be mobilized [30,49,106]. This increases the probability that Thy$^{\bullet-}$ will be formed and, if the temperature is high enough to activate proton addition to C6, formation of the Thy(C6 + H)$^{\bullet}$ radical becomes competitive with tunneling. While the activation energy for electron detrapping shows dispersion, it is less so than for the hole. This makes sense in that the energy of proton transfer is an order of magnitude larger than that of the hole. This relatively discrete component would therefore be more dominant relative to the more disperse contributions to reorganization by the local environment.

Deep Traps. A hole produced by ionization of the sugar-phosphate backbone is permanently trapped by deprotonation from any of the five ribose carbons, producing a neutral sugar radical. The probability that any of the sugar radicals [dRib(C1'$-$H)$^{\bullet}$, etc., in Fig. 1a] will return to the native structure is very small. This is in part a consequence of these radicals being relatively stable [142,149–151] and in part due to the deprotonation step being irreversible. The lost proton may attach to water in the hydration shell or to the DNA bases, e.g., cytosine [30]. Either way, the increased distance between the unpaired electron and site of altered charge helps protect the radical from recombination events [137]. The sugar radicals are therefore relatively radiation-resistant. While annealing may activate unimolecular reactions [150–153], the reactions remain centered at the initial site of hole formation. In this case, the damage is fixed at the site of ionization.

The hole formed by ionizations of the backbone, if not trapped by deprotonation, will transfer to the bases. If the temperature is low enough ($<$ ~120 K), the hole will be trapped by guanine in a very shallow trap. But if the temperature is high enough, another reaction pathway is activated, HO$^-$ addition to the guanine radical cation [154]. The resulting neutral radical may be one-electron oxidized to yield 8-oxoguanine or, in the absence of an oxidant, lead to formamido-pyrimidine (8oxoGua and fapyGau in Fig. 1b). While it is not yet known what temperature is required to activate HO$^-$ addition, this reaction is known to occur at room temperature [154].

Thus we find that the hole is deeply trapped by two distinctly different mechanisms. One that requires no activation energy, sugar radical formation occurs even at 4 K, and the other that does, HO$^-$ addition to the guanine radical cation. Similarly, the excess electron is irreversibly trapped by two distinct mechanisms, one that is activated and the other which is not.

A deep trap for the excess electron is the Thy(C6 + H) radical. Irradiation of DNA at temperatures $<$ 80 K results in a population of shallowly trapped electrons that consists of about 80% Cyt(N3 + H)$^{\bullet}$ and 20% Thy$^{\bullet-}$ [85,86,155–157]. As mentioned above, the electron is mobilized upon warming to 140–200 K. In this same temperature range, irreversible protonation of Thy$^{\bullet-}$ at C6 is activated and the excess electron is locked in place as the Thy(C6 + H) radical [30,49]. Under anaerobic conditions, this radical will lead to the formation of 5,6-dihydrothymine as a stable end product [158].

An electron trapping reaction analogous to that in thymine might be expected for cytosine: Cyt$^{\bullet-}$ + H$^+$ → Cyt(C6 + H)$^{\bullet}$ (see Fig. 1e for structures). But this reaction is not observed when DNA, irradiated at < 80 K, is warmed [159]. It is presumably quenched by the faster reaction of Cyt$^{\bullet-}$ + H$^+$ → Cyt(N3 + H)$^{\bullet}$. But a reaction exists by which the excess electron is deeply trapped by cytosine, and it is not thermally activated. The Cyt(C6 + H, N3 + H$^+$)$^{\bullet}$ radical is formed in DNA irradiated at 4 K [144,159]. In addition, there is evidence of the closely related radical formed by the net gain of hydrogen at C5 of cytosine: Cyt(C5 + H)$^{\bullet}$ or Cyt(C5 + H, N3 + H$^+$)$^{\bullet}$ (the protonation state at N3 is undetermined). The proposed mechanism for formation of Cyt(C6 + H)$^{\bullet}$ invokes spatially correlated events: electron capture by cytosine accompanied by proton transfer to N3 correlated with the release of a proton by a nearby sugar-phosphate radical cation. Cytosine, thereby, gains one electron and two protons. This radical would be, thereby, a signature of a clustered lesion. If correct, one would expect to find two end products, dihydrouracil and a single-strand break, in close proximity to one another.

General Scheme

Charge transfer in DNA is summarized by the scheme shown in Fig. 3. Holes on the sugar-phosphate (SP) tunnel to the nearest base at a rate k_t in competition with the deprotonation rate, k_{irr}, forming sugar products (spr$_n$). Holes on the bases (B$_n$) tunnel between

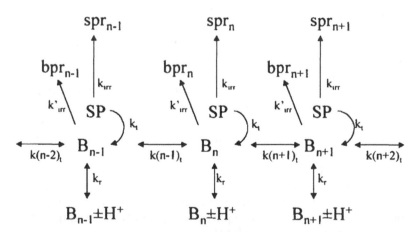

Figure 3 A working model for charge transfer and trapping in DNA is shown for three deoxynucleotides (base B$_n$ attached to sugar phosphate SP) juxtaposed in space. These deoxynucleotides may or may not be in the same strand or the same duplex. Hole tunneling occurs from SP to B$_n$ at a rate k_t in competition with the hole trapping rate k_{irr}, the rate of deprotonation at one of the sugar carbons resulting in stable sugar product radicals (spr$_n$). Holes transferred to B$_n$ or holes (excess electrons) initially formed on B$_n$ tunnel to other bases at a rate $k(n \pm 1)_t$ governed primarily the through-space distance and the ionization energy (electron affinity) of B($n \pm 1$). Competing with hole (excess electron) transfer is reversible proton transfer, forming thermally labile trap sites, B($n \pm$ H$^+$), at a rate k_r. Also competing with hole (excess electron) transfer is irreversible hydroxide (proton) addition, forming stably trapped base product radicals, bpr$_n$, at a rate k'_{irr}. Not shown in this scheme is the fact that the k values for holes are different than the k values for excess electrons.

guanines at rate $k(n)_t$ that depends on the distance to the nearest guanine; otherwise, the rate is relatively independent of whether the transfer is intrastrand, interstrand, or interduplex. Competing with hole tunneling between the bases is reversible proton transfer at a rate k_r. At low temperatures, the hole is locked in place by this reaction but, if the temperature is sufficiently high, back transfer of the proton reenables tunneling. At even higher temperatures, irreversible reactions (k'_{irr}) lead to base products (bp_n).

The excess electron follows the scheme in a similar way, except that there is no attachment or reaction involving the sugar-phosphate backbone. It attaches to the pyrimidines where its tunneling rate $[k(n)_t]$ is controlled by the distance to neighboring pyrimidines and by the competing reversible proton transfer from guanine to the cytosine radical anion. At low temperatures, proton transfer shuts down tunneling. At higher temperatures, back transfer permits tunneling. If the temperature is sufficiently high, irreversible protonation terminates transfer, forming base products (bp_n).

Ultimately, if the end products are formed in clusters, the rates k_{irr} and k'_{irr} must be fast relative to $k(n)_t$. It will be important to determine these rates.

5. DISTRIBUTION OF STABLE END PRODUCTS IN MODEL SYSTEMS

The ultimate goal is to use our understanding of DNA radiation physics and chemistry to predict what types of damage will occur in DNA in vivo, including yields and spatial distribution. A step toward achieving this goal is measuring nonradical (diamagnetic) damage produced by the direct effect. Until recently, there has been a paucity of information on direct-type end products because of the difficulties inherent to isolating and analyzing products produced in very small amounts, on the order of 1 part in a thousand. In this section, we review some of the relevant findings on stable products produced in directly ionized DNA, then outline a model by which end product distribution can be linked to our current knowledge of free radical intermediates and conclude with a few remarks on how the reaction pathways will be influenced by a chromatin environment.

5.1. High Molecular Weight DNA

Much of the early work on measuring direct damage in DNA focused on cleavage of the sugar-phosphate backbone, called strand breaks. While strand breaks represent an important class of products, the specific chemical structure is generally left undetermined. The importance of this type of lesion is primarily due to the fact that it can be detected with very high sensitivity and secondarily due to observed correlations with biological endpoints. High detection sensitivity is a consequence of the fact that strand breaks lead to large changes in DNA conformation.

Among the earliest work, properties such as viscosity and molecular weight were used to demonstrate that single- and double-strand breaks are produced by direct ionization and that these products are influenced by water content [160]. Progress on the measurement of strand-break yields accelerated with the discovery by W. D. Taylor that closed circular DNA (φX174 viral DNA) provided an exquisitely sensitive system for detection of single-strand (ssb) and double-strand breaks (dsb) [161]. Closed circular DNA adopts a tight supercoil conformation, which relaxes into an open circle when one strand is broken and into a linear form when two strands are broken within close proximity ($\sim$10 base pairs) of each other. Separation and quantification of these three forms were first performed by ultracentrifugation and subsequently by gel electrophoresis.

Products Characterized by a Single-Strand Break

The yield of single-strand breaks [G(ssb)] and double-strand breaks [G(dsb)] measured under nitrogen using low LET radiation are summarized in Table 1. The reported G(ssb) is remarkably consistent considering that these results come from five different laboratories. The top nine entries report prompt strand breaks. The first four are for relatively dry DNA, irradiated at different temperatures: 77 K, 6°C, and 25°C. The values cluster tightly around G(ssb) = 0.05 µmol/J. The next five values are for DNA that is approximately fully hydrated; in each case, there is a large uncertainty as to the actual hydration level. It is important to note that these yields were calculated assuming that only DNA comprises the target mass; for example, if the sample contains an average of 20 waters (MW = 18)/nucleotide (average nucleotide MW = 308), then G(dry DNA) = G(hydrated DNA) × (308 + 20 × 18)/308 = 2.2 × G(hydrated DNA). Inclusion of the counter ion in the hydration layer increases this factor. Thus if the energy absorbed by the DNA hydration layer is converted to strand breaks with the same efficiency as the energy absorbed by DNA itself, G(ssb) should more than double in going from dry to fully hydrated. For hydrated DNA, the observed values of G(ssb) give an average of 0.07 ± 0.02 µmol/J. Using this average, one obtains G(dry DNA) = 1.4×G(hydrated DNA), which is consistent with transfer of holes from the solvation shell to the DNA increasing strand breakage by 40%. If all the holes from the tightly bound water ($\Gamma < 11$) transfer to DNA, as is observed at low temperatures, and 50% of those holes localize on the sugar so as to give strand breaks, then one would expect an enhancement of roughly (10/20)×0.5, i.e., about a 25% increase due to energy transfer from the hydration layer. Within this framework of logic, a 40% increase can be explained by either hole transfer occurring from an even larger sphere of water ($\Gamma \sim 16$) or by 80% of holes localizing on the backbone or a combination of these.

In addition to prompt strand breaks, there are other types of damage that, under certain conditions, will develop into strand breaks. One such class is alkali labile sites; the existence of this class of damage is revealed by the early work shown in lines 10 and 11 of Table 1. Another class involves specific types of base damage that are substrates for enzymatic repair and, consequently, treatment with the appropriate enzyme results in an ssb. Entries 12–15 compared to 1–4 show that the yield of this class of damage is 1 to 4 that of prompt strand breaks. Yokoya et al. [162] employed endonuclease III (Nth) and formamidopyrimidine-DNA glycosylase (Fpg); these enzymes act on products stemming from the one-electron reduction of cytosine and thymine and one-electron oxidation of guanine, respectively. These results are discussed further below in the section on base damage.

Products Characterized by Double-Strand Break

In Table 1, we see that the spread in values of G(dsb) is larger than in values of G(ssb). This is not surprising given the inherent experimental difficulties that revolve around measuring the formation of relatively small amounts of linear DNA in the presence of large amounts of open circular DNA. These difficulties aside, the reported ratios for ssb/dsb fall primarily in the 10–20 range. It has been noted [163,164] that this ratio is much too small to be accounted for by two ssb occurring independently on opposite strands within close proximity (about 10 base pairs) of each other. While Boon et al. [163] suggest a mechanism where base-centered radicals give rise to strand breaks, Ito et al. [164] suggest a mechanism where localized multiple ionizations result in spatially correlated ssb, resulting in a dsb. There is now considerable evidence that the base radicals, predominantly oxidized guanine and reduced pyrimidines, are relatively ineffective in causing strand breaks [98,154], making the later explanation more plausible. Indeed, evidence continues

Table 1 Single- and Double-Strand Break Yields for Low LET Radiation in the Absence of O_2

	DNA	Conditions[a]	G(ssb)[b] (μmol/J)	G(dsb)[b] (μmol/J)	ssb/dsb	Reference
1	pBR322	Dry/77 K	0.047	0.0014	35	[164]
2	pBR322	Dry/25°C	0.056	0.0031	18	[164]
3	pUC18	rh < 0.1 (Γ ~4)/6°C	0.054	0.0053	10	[162]
4	pUC18	rh < 0.1 (Γ ~4)/6°C + 37°C	0.047	0.0043	11	[162]
5	qpBR322	Humid (Γ ~20)/77 K	0.034	0.001	33	[164]
6	pBR322	humid (Γ ~20)/298 K	0.120	0.011	11	[164]
7	pUC18	rh = 0.97 (Γ ~34)/6°C	0.067	0.0072	10	[162]
8	pUC18	(Γ ~34)/6°C + 37°C	0.067	0.0072	9	[162]
9	pBR322	Frozen aqueous/77 K	0.04–0.07		10–20	[163]
10	φX174	Alkaline sucrose assay	0.070	0.013	5	[188]
11	SV40	Alkaline sucrose assay	0.110	0.011	10	[189]
12	pUC18	Γ ~4/6°C + 37°C + Nth	0.10	0.0069	15	[162]
13	pUC18	Γ ~34/6°C + 37°C + Nth	0.18	0.0171	11	[162]
14	pUC18	Γ ~4/6°C + 37°C + Fpg	0.094	0.0103	9	[162]
15	pUC18	Γ ~34/6°C + 37°C + Fpg	0.16	0.0193	8	[162]
16	pUC18	Γ ~34/6°C + 37°C + Nth + Fpg	0.24	0.0237	10	[162]

[a] For the first two entries, alkaline sucrose gradients were used; all others employed gel electrophoresis. Degree of hydration, deduced from known relative humidity, is given as Γ, the molar ratio of water to mononucleotide. Sample temperature during radiation.
[b] Yields are based on dry mass of DNA.

to build in favor of dsb arising from two radicals, spatially correlated, on opposing strands [165].

Complimenting the strand-break studies on DNA irradiated in the solid phase are studies that use high concentrations of radical scavengers to divide out the indirect from the direct effect in DNA irradiated in aqueous solution (see Refs. 15, 166, and 167). The work of Krisch et al. was one of the early studies that provided persuasive evidence that the direct effect is responsible for about half of the dsb produced under in vivo conditions. The absolute yields obtained from these cell mimetic conditions, however, are difficult to compare with the solid-state results.

Production of Strand Breaks by Monoenergetic Electrons and Photons

An incident beam, whether comprised of high-energy photons or fast charged particles, deposits most of its energy via fast electrons. Because the energy of fast electrons is dissipated by producing secondary electrons, a large fraction of the radiation chemistry is due to electrons with very low energies, <20 eV [168]. It is of considerable value therefore to investigate the efficiency by which low-energy electrons create strand breaks.

Pioneering the investigation of strand breaks using monoenergetic electrons, Folkard et al. [169] discovered that energies <25 eV are sufficient to cause ssb, and between 25 and 50 eV, the onset of dsb occurs. They employed pBR322 DNA, under vacuum, spread approximately as a monolayer. At 50 eV, the efficiency for ssb/dsb is 2.0; it climbs to 5 for 100–500 eV, and it plateaus around 10 for 2000–4000 eV. More recently, the energy of electron beams has been extended down to 3 eV by the Sanche laboratory, measuring ssb and dsb in plasmid DNA under ultrahigh vacuum [68,170,171]. Surprisingly, strand breaks are observed in substantial yields even at energies below ionization thresholds, 1.5–10 eV. At energies below 20 eV, the authors provide convincing evidence that dissociative electron attachment (DEA) is the major pathway for producing a wide range of damage, including ssb and dsb. Just what fraction of strand breaks, under in vivo condition, is due to DEA vs. one-electron oxidation is an important question that remains to be answered.

Hieda and Ito [172] measured strand-break yields as a function of monochromatic x-ray energy, using 2.146, 2.153, and 2.160 keV, corresponding to below, on, and above the K-edge for phosphorus. While the efficiency of single-strand breaks did not vary significantly, the efficiency of double-strand breaks did increase by about 30% at the K resonance peak. These authors raise the possibility that dsb efficiency is enhanced by an Auger cascade in phosphorus. The enhancement of dsb/ssb by resonant excitation of phosphorus provides a method for assessing the relative contribution of the direct effect to biological endpoints such as cell death and mutation [173].

Products Derived from the DNA Bases

The formation and identification of the free bases, and subsequently other types of base damage, have been reported by Swarts et al. [101,174–176]. These papers present the first measurements of specific products produced by the direct effect in DNA. Equally important, the sample type and preparation correspond closely to that employed in a wide range of EPR studies that identify free radical intermediates trapped on DNA [14,30,83].

The free bases (thymine, cytosine, adenine, and guanine) are the simplest form of product in this class. But the simplicity of these chemical products belies the complexity of their origin. The yield of each free base was measured using HPLC on salmon sperm

DNA, variably hydrated ($2 < \Gamma < 33$), subjected to γ-irradiation at RT under N_2 or O_2, and subsequently dissolved in water [101]. Focusing on the N_2 results, the yield of all four bases combined holds relatively constant in the hydration range $2 < \Gamma < 33$ (0.05–0.07 μmol/J,) where the G values are based on the weight of DNA plus its hydration shell. In order to compare these values with those above (Table 1), one needs to recalculate these based on the dry weight of DNA, which gives a plateau of 0.06–0.07 μmol/J for $2 < \Gamma < 13$ and then climbs to 0.17 μmol/J at $\Gamma = 33$. The yields for free base release therefore correlate well with the ssb yields shown in Table 1, entries 1–9. Since it is well known that in dilute aqueous solutions, free base is released as a consequence of radical formation on the sugar-phosphate backbone [177], it is likely that some (or perhaps all) of the released bases are derived from sugar-centered radicals. When Swarts et al. analyzed the yields of each of the free bases separately, they found Gua to be about 60% that of Thy, Cyt, and Ade, with the yields of the latter three being the same. It is important to reconcile this with the observation that in crystalline deoxynucleotides [98], the yield of guanine is not reduced relative to the other three bases, but when DNA is irradiated in dilute aqueous solution [178], the yield of guanine is reduced. One possibility is that in high molecular weight DNA, both solid and in dilute solution, the radical formed by hydrogen abstraction from C4′is the predominant sugar-centered radical. The C4′radical undergoes fast β-elimination of the 3′ phosphate, forming an enol–ether radical cation [179] which is capable of oxidizing an adjacent guanine via electron transfer [180]. Such an electron transfer would prevent release of guanine as a free base. Additionally, one must propose that in crystalline oligodeoxynucleotides, either hydrogen abstraction from C4′ is not predominant or 3′ phosphate β-elimination is inhibited.

Swarts et al. [175] measured the yields of 14 base end products in addition to the free bases. For $2 < \Gamma < 13$, the yields (based on total weight) are relatively constant and are largest for 8-oxoGua (0.084–0.101 μmol/J), FapyGua (0.012–0.021 μmol/J), and 5,6-diHThy (0.026–0.067 μmol/J). This result is consistent with holes and electrons produced in the tightly bound water transferring to the DNA and creating the same spectrum of products, at approximately the same yields, as for holes/electrons formed by direct ionization of the DNA molecule. The ratio of yields for products stemming from one-electron oxidation vs. one-electron loss is unexpectedly large: 0.201/0.028, 0.158/0.068, and 0.150/0.052, for Γ of 2.5, 10.3, and 13.2, respectively. Since this ratio is expected to be close to 1.0, the authors conclude that a large fraction of the reduction products were not detected. The authors present a working model for oxidation product formation by direct effects in DNA that combines their findings with the preexisting knowledge on free radical intermediates. This model, representing a major advance in the field, proposes that (1) hole transfer from the sugar-phosphate backbone to the bases competes with damage localization on the sugar, (2) damage localized on the sugar will give rise to a free base, (3) holes transferred to the base stack are trapped preferentially by guanine, and (4) irreversible hole trapping occurs primarily by OH^- addition. We build on this model in Sec. 5.3 below.

Clustered Damage

One type of clustered damage is double-strand breaks, consisting of two single-strand breaks in close proximity to one another. Since techniques for measuring dsb have been available for some time, a fair amount of data is available, as summarized above. But until recently, there has been very little experimental information on other types of clustered damage produced by direct effects. The work of Yokoya et al. extends our knowledge to clustered damage consisting of one base damage in close proximity to either a strand break or another base damage on the opposing strand. Yields of dsb are compared with the yields of clusters containing base damage. They employ the Nth enzyme to create a strand break at sites containing 5,6-dihydrothymine, thymine glycol, and abasic sites and Npg to create breaks at 2,6-diamino-4-hydroxy-5-N-methylformamidopyrimidine, 7,8-dihydro-8-oxo-2'deoxyguanine, and apurinic sites. Comparing lines 8 and 16 in Table 1, it is seen that treatment with both enzymes raises the ssb and dsb yields by about 3. The value of 0.024 μmol/J (based on dry DNA weight at $\gamma = 34$) represents one of the best measures to date for the yield of clustered damage, 1/3 due to the dsb type and 2/3 due to base-damage-opposed-to-ssb-or-base-damage type. Since the yield of total clustered damage is almost certainly significantly larger, we see that the probability of producing clustered damage by direct effects is comparable to that of producing ssb.

5.2. Oligodeoxynucleotides

DNA samples consisting of oligodeoxynucleotides provide better-defined samples for investigating the influence of variables such as base sequence, conformation, degree of hydration, chemical environment, and DNA packing density. This is particularly true if the oligomers are in crystalline form where these variables are defined with atomic precision. When high molecular weight DNA is condensed, the packing density tends to be highly heterogeneous and difficult to reproduce from sample to sample [91,92]. Crystalline DNA reduces this variability dramatically, making it possible to measure free radical yields with a typical precession of ±10% [66]. This kind of reproducibility, which is a direct consequence of the sample homogeneity, when coupled with atomic resolution knowledge of sample structure, provides a powerful system for investigations of DNA radiation physics and chemistry.

Free Base Release

The yields of free base from crystalline oligodeoxynucleotides were measured by Razskazovskiy et al. [98]. Three types of crystals, d(GCACGCGTGC)$_2$, d(CACGCG):d (CGCGTG), and d(CGATCG)$_2$:2D(D = daunomycin or adriamycin), were x-irradiated (70 keV) at 4 K, warmed to RT, dissolved in water, and released base analyzed using reverse phase HPLC. Free radical yields were measured on the same samples, prior to dissolution, using EPR at 4 K. The yield (based on total weight) of the combined four bases ranges from 0.06 to 0.11 μmol/J, 10–20% of the yield for total free radicals trapped at 4 K (prior to annealing). The fact that the intercalated anthracycline moiety of daunomycin effectively scavenges all the holes and electrons from the bases [105] eliminates base-centered radicals as precursors to base release. The observation that hydroxyl radicals are not trapped in the crystalline matrix rules out hydroxyl radical reactions. It is concluded, thereby, that radicals fixed on the sugar are precursors to base release. This strongly suggests that direct ionization of the sugar-phosphate backbone is followed by deprotonation of the primary radical cation and that deprotonation competes with hole transfer from the radical cation to the bases. From this result, the yield of DNA strand breaks for DNA hydrated in the $4 < \Gamma < 14$ range is predicted to be ~0.1 μmol/J.

Strand-Break Products

By employing crystalline DNA, the first measurements of specific strand-break products were possible [97,99]. Crystalline d(CGCG)$_2$, d(CGCGCG)$_2$, d(CACGCG):d(CGCGTG), and d(CGCACG:CGTGCG) were irradiated at 4 and 293 K, dissolved in water, and analyzed with anion exchange HPLC. The major products were free bases and oligodeoxy-nucleotides with 3'- and 5'-phosphate end groups. Based on just the later products, the yield of immediate strand breaks varies from 0.06 to 0.16 μmol/J. There is increased damage to the terminal sugars relative to the internal sugars, perhaps reflecting the presence of a free OH group in the terminal sugars (none of the oligomers are phosphorylated at the termini). Accumulation of strand-break products is remarkably resistant to dose saturation, remaining linear out to doses of ~500 kGy. This is consistent with a neutral sugar radical as the precursor to strand breaks. In addition, strand-break yields are not affected greatly by sample temperature during radiation exposure. Most striking is that there is no obvious preference for the cleavage at Ade, Thy, Cyt, or Gua. The lack of influence of the base ionization potentials can be explained by the high exothermicity of the process, making hole transfer from the sugar-phosphate radical cation independent of the base itself. Prediction of the DNA strand-break yields ($4 < \Gamma < 8$ for these crystals) is also ~0.1 μmol/J.

In general, it appears difficult to correlate the results obtained using vacuum-UV with those obtained using high-energy ionizing radiation. We should note, however, that when vacuum-UV is used to directly ionize deoxyoligonucleotides in solid form, a dependence on base sequence is observed [181]. In order to explain the observed products, these workers proposed that an initial site of damage is the deoxyribose moiety.

Sugar Radical Intermediate

If sugar-centered radicals are indeed the primary precursors to strand breaks, then one should be able to observe these with EPR. But EPR detection is difficult because the diversity in radical structure and conformation tends to make the spectrum broad, with poorly resolved hyperfine structure and much of the signal masked by the more intense signal from the base radicals [20,96]. To some degree, these problems can be overcome by applying EPR to study a single crystal of DNA. This has been done for crystals of d(CTCTCGAGAG)$_2$ where x-irradiation at 4 K is shown to produce a radical due to the net loss of hydrogen from C3'. The yield of the C3' radical is 0.03 ± 0.01 μmol/J. It does not dose-saturate up to 100 kGy, and it makes up about 4.5% of the total radical population trapped at 4 K. Since this study does not exclude formation of sugar radicals at the other four carbon sites, this observation fits nicely with projected strand-break yields of ~0.1 μmol/J.

Other Precursors to Strand Breaks

Although a reasonably good correlation exists between the yield of trapped sugar radicals and the yield of strand breaks, this does *not* prove that sugar radicals produced by direct ionization are the only, perhaps not even the predominant, precursor to strand breaks. As discussed above, Sanche et al. have proven that low-energy electrons (<20 eV) produce strand breaks in plasmid DNA by dissociative electron attachment (DEA). This work has been extended to studies of low molecular weight fragments produced by monoenergetic electrons with energies from 1 to 30 eV [182–185]. A series of pyrimidine oligodeoxy-nucleotides {d(T$_6$), d(T$_{12}$), d(C$_6$), d(C$_{12}$), d(T$_9$), d(C$_9$), d(T$_6$C$_3$), d(C$_3$T$_6$), d(C$_6$T$_3$)} were chemisorbed onto a gold substrate via a thiol modification at the 3'-end and exposed to the electron beam under ultrahigh vacuum. The major fragments are CN, OCN, and/or H$_2$NCN, and when Thy is present, there is a small contribution tentatively assigned to CH$_3$CCO. One of the main conclusions is that, for incident electrons with <20 eV, fragmentation occurs at the bases through DEA, in contrast to >20 eV where fragmentation is controlled by nonresonance processes. Another is that base fragmentation in single-stranded DNA depends on base type and sequence. A challenge for future work is to determine the degree to which these processes play a role in solid-state DNA when ionization is initiated at high energies in the absence of the vacuum and metal interfaces. Currently, it is difficult to reconcile the base ring fragmentation products and properties such as sequence dependence with the lack of evidence for such events from studies using EPR and HPLC. One possibility is that the resonance states that yield these fragments are quenched in bulk samples and electron capture by the pyrimidines dominates. This remains to be tested.

5.3. A Working Model

Oligodeoxynucleotides, Crystalline Samples

A model that is consistent with the observed free radical intermediates trapped in crystalline deoxyoligonucleotides x-irradiated at 4 K and warmed to RT has been published [144]. Here we review the salient points of that model and then extend it to other types of DNA samples.

The crystal-based model assumes that half of the radicals trapped at 4 K are derived from one-electron loss and the other half from one-electron gain. Following thermalization, electrons are captured exclusively by the bases and the holes are distributed according to electron densities of each component: bases, sugar-phosphate backbone, and solvation shell. Holes initially formed in the tightly bound solvation shell of DNA rapidly transfer to the sugar and bases [67,93]. Because all of the end products are assumed to come from free radical precursors [143], possible damage from excitations and dissociative electron attachment [170,183] are excluded. There are predominately five radical types produced by electron loss or electron gain: one-electron oxidized sugar phosphate (SP$^{\bullet}$), guanine radical cation (Gua$^{\bullet+}$), cytosine radical anion (Cyt$^{\bullet-}$), thymine radical anion (Thy$^{\bullet-}$), and, perhaps, adenine radical anion (Ade$^{\bullet-}$). These radicals, with the possible exception of Thy$^{\bullet-}$, are unstable at 4 K. Reversible proton transfer, across the Watson–Crick hydrogen bond, stabilizes the hole at Gua and the excess electron at Cyt and Ade [22]. Irreversible deprotonation from the sugar-phosphate backbone traps out five different radicals due to the net loss of hydrogen from each of the five deoxyribose carbons (Fig. 1a). The protons released by the one-electron oxidized sugar phosphate directly (or indirectly) promote proton addition to carbon sites of the base radical anions, yielding base radicals that are irreversibly protonated [30,159]. At 4 K, the ratio of hole trapping at guanine vs. SP is ~76:24 and the ratios for electrons trapping at Cyt/Thy/Ade is ~86:8:6. The base-centered radicals are thermally more labile than the sugar-centered radicals.

The model proposes that the deprotonated guanine radical cation [Gua(N1–H)$^{\bullet}$] is the shallowest trap, possessing an activation energy with large dispersion. Thus hole detrapping occurs between 4 and ~140 K. Back transfer of the proton is thermally activated, creating Gua$^{\bullet+}$ and opening the gate for tunneling to other guanines in close proximity [186]. In a crystal lattice, where the guanine–guanine intermolecular distances are comparable to the interstrand and intrastrand distances [104], the mobilized hole undergoes a random walk in three dimensions. Given sufficient thermal energy, the hole continues its walk until it recombines with another radical or it is irreversibly trapped by OH^{-} addition [154]. Under gradual annealing conditions, recombination predominates. After annealing to 120 K, the radical population consists primarily of H-addition to C5–C6 of cytosine, H-addition to C5 of thymine, H-addition to C5/C8 of adenine, and H abstraction from sugar. The relative concentration of these is approximately 0.3:0.1:0.1:0.5, respectively. Evidence for the first two radical types comes from single-crystal EPR [94]; evidence for the third is based on work on reduction of adenine in monomeric and oligomeric systems [85,95,98,187], and evidence for the fourth comes from the yields of end products [97,98] related to sugar damage. Continued annealing from 140 to 240 K detraps the excess electron by reforming $Cyt^{\bullet-}$; this mobile state of the excess electron is formed by back transfer of the proton, from N3 of cytosine to N1 of guanine. The electron walk is terminated by combination reactions with base-trapped radicals or by irreversible protonation at the C5–C6 bond of cytosine or thymine. These later reactions are relatively minor because once permissive temperatures (> 140 K) are reached, the concentration of $Cyt^{\bullet-}$ has been reduced significantly by combination reactions [144].

The proposal that holes are detrapped at lower temperatures than the excess electrons is based on the observations discussed above in *Very Shallow Traps* and *Shallow Traps* (Sec. 4.3.2). One expects that the activation energy needed to detrap the hole from Gua in duplex DNA is relatively small, an order of magnitude less than that needed to detrap the electron. This fits well with the observation that upon warming 4 K irradiated crystalline DNA to 77 K, 10–30% of the radicals anneal out, i.e., at least one of the trapping sites [fide infra $Gua(N3–H)^{\bullet}$] is very shallow.

From the yields of 5,6-dihydropyrimidine radicals, we predict reduction product yields of 0.04–0.06 μmol/J for 5,6-dihydrouracil and 0.03–0.05 μmol/J for 5,6-dihydrothymine for B-form DNA hydrated to 9 waters per nucleotide. With respect to oxidation products, we predict strand-break yields ~0.10 μmol/J. A very surprising prediction of this model is that the yield of damaged guanine is nil. Half the damage is oxidized sugar products and the other half is reduced pyrimidines.

High Molecular Weight DNA, Amorphous Samples

Samples of high molecular weight DNA are generally amorphous; the local chemical environment is heterogeneous and difficult to characterize. The heterogeneity and lower DNA packing density of the amorphous samples are believed to be important factors with regard to understanding the differences between direct effects in amorphous vs. crystalline DNA samples. Packing density reflects the global average of the interhelix distances. When this distance is short, the DNA scavenges a larger fraction of the radicals created in the solvent compartments. Higher scavenging reduces radical recombination and results in a higher yield of DNA-trapped radicals. Variable packing density has been evoked to explain the threefold variation in free radical yields for DNA films having the same average value for Γ [91,92]. It was also used to explain why the yield of trapped radicals for DNA in the amorphous state is about 1/2 that of the crystalline state [66]. Note that across this range of DNA samples, the yield of radicals that are initially trapped varies by a factor of 3 to 6 even if variables such as Γ, base sequence, and conformation are held constant.

The events that occur after initial trapping should also be governed by the interhelix distance. Given that the tunneling rates are strongly dependent on distance and relatively independent of what occupies the intervening space [118–120], the average rate of intermolecular hole/electron transfer should be much slower in amorphous DNA compared with crystalline DNA. Therefore the random walk in amorphous samples tends toward one-dimensional, in contrast to that in crystals where it is known to be three-dimensional [104]. The dimensionality of the walk will influence the probability of radical recombination. Since tunneling is temperature-independent, this will influence the yield of trapped radicals even at 4 K. This effect will also influence the rate of radical recombination when the DNA samples are annealed.

Another consequence of increased interhelix distance is related to differing DNA scavenging capacities for electrons vs. holes. In the $\Gamma < 10$ range, electrons and holes produced in the solvent layer are scavenged by DNA with equal probability. For $9 < \Gamma < 30$, electron scavenging capacity changes very little while hole scavenging drops rapidly to zero [92]. As a result, we expect that the stoichiometry of trapped holes to trapped electron be ~50:50 in crystalline DNA [144] and ~40:60 in amorphous DNA [14,30]. When coupled with the dimensionality of the walk, this should have significant consequences on the distribution of end products derived from base radical intermediates.

In crystalline DNA irradiated at 4 K and annealed to RT, the model predicts that at low temperatures (40–140 K), the guanine radical cation undergoes a three-dimensional walk. If the range of holes is large, termination by combination with Cyt(N3 + H)$^{\bullet}$ radicals has a high probability. The net result is that damage to guanine is quenched and the remaining reductive damage to the pyrimidines will be about the same as the yield of sugar damage (strand breaks). In contrast, amorphous DNA irradiated at low temperatures contains a larger imbalance between holes and electrons on the base stacks. If 85% of the holes are trapped by guanine, it means that 34% (0.85×40%) of all radicals are due to one-electron oxidized guanine while 60% are due to one-electron reduced pyrimidines. If the range of the mobilized hole (or electron) is large during an anneal to RT, there should be very little oxidized guanine products and the yields of reduced pyrimidine products should be about twice the strand-break yield.

Impact of Sample Temperature During Irradiation

Much of the information on direct damage to DNA comes from EPR studies of free radical intermediates that are, out of necessity, trapped by irradiating samples at low temperatures. In order to predict the biological consequences of direct effects, that information must be extrapolated to sample irradiation at room temperature. In doing so, there are at least 10 competing reactions to consider: (1) hole transfer from the solvent shell to the DNA, (2) hole transfer from the sugar-phosphate backbone to the bases, (3) hole transfer between bases, (4) electron transfer between bases, (5) rate of deprotonation of the solvent hole (HO$^{\bullet}$ formation), (6) reversible proton transfer between the base pair of the trapped hole, (7) reversible proton transfer between the base pair of the trapped electron, (8) irreversible proton transfer from the sugar phosphate-centered hole, (9) irreversible proton transfer to pyrimidine radical anions, and (10) irreversible hydroxide addition to the guanine radical cation. Extrapolation to room temperature comes down to what is known about the rates of these reactions at room temperature. Although some of the parameters are known, such as the two values for hole and electron tunneling, most are unknown, such as the frequency factors for the tunneling reactions and activation energies for the thermally activated reactions. However, based on temperature stabilities of various intermediates and a small amount of data on samples irradiated at variable temperatures, some educated guesses can be made.

The yield of strand breaks appears to be relatively independent of sample irradiation temperature, as discussed above. This implies that competing processes do not have much impact on reactions 1, 2, 5, and 8. That is, the competitions between holes tunneling from the solvent to DNA and deprotonation of $H_3O^{\bullet+}$ and between hole tunneling form the sugar phosphate to the bases and deprotonation of the sugar are fairly temperature-insensitive (from 4 to 300 K). In contrast, the mobility of the holes and excess electrons centered on the bases is very temperature-sensitive, zero at 4 K, onset at ~40 K, and highly mobile at 180 K. By our model, the mobility is controlled by the proton transfer rate across the base-pair hydrogen bond. Based on the temperatures for mobility onset, the activation energy is relatively low and therefore at RT, the hoping rate for both holes and excess electron must be appreciable (as observed). Competing with the base-to-base hole/electron transfer reactions (3, 4, 6, and 7), there are irreversible reactions (9 and 10) that lock out the damage site. The activation energy for reaction 9 is almost certainly less than that of reaction 10. We expect therefore that the range of the excess electron will be shorter than that of hole. The shorter the range, the less likely recombination will occur, favoring higher yields. The yield of base damage therefore should increase with increasing sample irradiation temperature.

Based on the above, we make the following predictions for direct damage in DNA irradiated at 297 K compared to 4 K (in the absence of oxygen). (1) Prompt strand-break yields will be about the same, ~0.1 μmol/J. (2) Oxidized guanine products increase substantially; yields become comparable to, or higher than, that of prompt strand breaks. (3) Reduced pyrimidine products increase substantially with yields greater than or equal to the sum of the yields for prompt strand breaks and oxidized guanine products. (4) The ratio of base damage to prompt strand breaks will be between 2:1 and 4:1 (compared to 1:1 for 4 K).

5.4. Extrapolation to Chromatin DNA

A full understanding of the radiation chemistry of DNA in vivo requires an understanding of how the cellular environments of DNA modify the events discussed above. While the most common environment is that of DNA "at rest" in tightly packaged chromatin, it is important to consider the environment when DNA is being translated and transcribed. Also, the environment surrounding mitochondrial DNA differs from that of nuclear DNA. The influence of these in vivo environments is likely to be considerable. While this area has been researched by a number of laboratories in the past, this area is poised for rapid progress due to the new array of molecular biological tools that improve our ability to dissect the relevant parts of the living cell. While we have not covered such topics as DNA–DNA and DNA–protein cross-links, this is not because we view them as unimportant. Rather, we believe it is premature to try to integrate this information with that on systems containing DNA in isolation.

6. SUMMARY

This review has looked at the direct effects of ionizing radiation on nucleic acids. The first step was to review detailed EPR/ENDOR experiments on irradiated model compounds at low temperatures in order to study the primary radiation-induced defects. Next, it was shown how these EPR spectra are used to simulate the EPR spectra of the DNA polymer.

A section on track expansion and charge mobility covered the processes by which excess electrons and holes move through DNA, and what reactions compete with charge transfer. Current work on determining the conductivity of DNA is discussed. Based on the high yields of radiation damage products in DNA, it is concluded that DNA is an insulator. Next, the protonation/deprotonation schemes are discussed, which can account for the types of free radicals that are trapped in DNA. A general scheme is presented to summarize charge transfer in DNA.

A discussion of the distribution of stable end products produced by direct ionization in DNA is presented. Single- and double-strand break yields for low LET radiation are discussed. Then work on end products derived from the DNA bases is reviewed. Recent work on oligodeoxynucleotides, including free base release, strand-break products, and sugar radical intermediates, is summarized. Finally, a model for end product formation by direct effects in DNA, based on the knowledge of free radical intermediates, is presented.

REFERENCES

1. Paretzke, H.G. In *Kinetics of Nonhomogeneous Processes*; Freeman, G.R., Ed.; Wiley: New York, 1989; 89 pp.
2. Stewart, A.; Kneale, G.W. Lancet 1971, *1*, 42.
3. Wing, S.; Shy, C.M.; Wood, J.L.; Wolf, S.; Cragle, D.L.; Frome, E.L. J. Am. Med. Assoc. 1988, *265*, 1397.
4. Zhou, H.; Suzuki, M.; Randers-Pehrson, G.; Vannais, D.; Chen, G.; Trosko, J.E.; Waldren, C.A.; Hei, T.K. Proc. Natl. Acad. Sci. 2001, *98*, 14410.
5. *UNSCEAR Report, Sources and Effects of Ionizing Radiation*; United Nations: New York, 1993.
6. Georghiou, S.; Nordlund, T.M.; Saim, A.M. Photochemistry and Photobiology 1985, *41*, 209.
7. von Sonntag, C. *The Chemical Basis of Radiation Biology*; Taylor and Francis: New York, 1987.
8. Hazra, D.K.; Steenken, S. J. Am. Chem. Soc. 1983, *105*, 4380.
9. Fujita, S.; Steenken, S. J. Am. Chem. Soc. 1981, *103*, 2540.
10. O'Neill, P. Radiat. Res. 1983, *96*, 198.
11. Vieira, A.J.S.C.; Steenken, S. J. Am. Chem. Soc. 1990, *112*, 6986.
12. La Vere, T.; Becker, D.; Sevilla, M.D. Radiat. Res. 1996, *145*, 673.
13. Tao, N.J.; Lindsay, S.M.; Rupprecht, A. Biopolymers 1989, *28*, 1019.
14. Wang, W.; Yan, M.; Becker, D.; Sevilla, M.D. Radiat. Res. 1994, *137*, 2.
15. Krisch, R.E.; Flick, M.B.; Trumbore, C.N. Radiat. Res. 1991, *126*, 251.
16. O'Neill, P. In: *Radiation Chemistry: Present Status and Future Trends*; Jonah, C.D., Rao, B.S.M., Eds.; Elsevier: Amsterdam, 2001; 585 pp.
17. Close, D.M. Radiat. Res. 1993, *135*, 1.
18. Bernhard, W.A.; Barnes, J.; Mercer, K.R.; Mroczka, N. Radiat. Res. 1994, *140*, 199.
19. Close, D.M.; Eriksson, L.A.; Hole, E.O.; Sagstuen, E.; Nelson, W.H. J. Phys. Chem. B 2000, *104*, 9343.
20. Close, D.M. Radiat. Res. 1997, *147*, 663.
21. Barnes, J.P.; Bernhard, W.A. J. Phys. Chem. 1994, *98*, 887.
22. Steenken, S. Free Radical Res. Commun. 1992, *16*, 349.
23. Hissung, A.; von Sonntag, C. Int. J. Radiat. Biol. Relat. Stud. Phys. Chem. Med. 1979, *35*, 449.
24. Sagstuen, E.; Hole, E.O.; Nelson, W.H.; Close, D.M. J. Phys. Chem. 1992, *96*, 8269.
25. Close, D.M.; Hole, E.O.; Sagstuen, E.; Nelson, W.H. J. Phys. Chem. A 1998, *102*, 6737.
26. Close, D.M.; Fouse, G.W.; Bernhard, W.A. J. Chem. Phys. 1977, *66*, 4689.
27. Close, D.M. J. Phys. Chem. B 2003, *107*, 864.
28. Bernhard, W.A.; Patrzalek, A.Z. Radiat. Res. 1989, *117*, 379.
29. Ormerod, M.G. Int. J. Radiat. Biol. 1965, *9*, 291.
30. Wang, W.; Sevilla, M.D. Radiat. Res. 1994, *138*, 9.
31. Hole, E.O.; Sagstuen, E.; Nelson, W.H.; Close, D.M. J. Phys. Chem. 1991, *95*, 1494.
32. Sagstuen, E.; Hole, E.O.; Nelson, W.H.; Close, D.M. J. Phys. Chem. 1992, *96*, 1121.
33. Sevilla, M.D. J. Phys. Chem. 1971, *75*, 626.
34. Spalletta, R.A.; Bernhard, W.A. Radiat. Res. 1993, *133*, 143.
35. Gräslund, A.; Ehrenberg, A.; Rupprecht, A. Internat. Conf. on Electron Spin Resonance and Nuclear Magnetic Resonance in Biology and Medicine and Fifth Internat. Conf. on Magnetic Resonance in Biological Systems, 1972.
36. Candeias, L.P.; O'Neill, P.; Jones, G.D.D.; Steenken, S. Int. J. Radiat. Biol. 1992, *61*, 15.
37. Hole, E.O.; Nelson, W.H.; Sagstuen, E.; Close, D.M. Radiat. Res. 1992, *129*, 119.
38. Bachler, V.; Hildenbrand, K. Radiat. Phys. Chem. 1992, *40*, 59.

39. Hole, E.O.; Sagstuen, E.; Nelson, W.H.; Close, D.M. Radiat. Res. 1992, *129*, 1.
40. Close, D.M.; Sagstuen, E.; Nelson, W.H. J. Chem. Phys. 1985, *82*, 4386.
41. Reynisson, J.; Steenken, S. Phys. Chem. Chem. Phys. 2002, *4*, 527.
42. Close, D.M.; Nelson, W.H. Radiat. Res. 1989, *117*, 367.
43. Nelson, W.H.; Sagstuen, E.; Hole, E.O.; Close, D.M. Radiat. Res. 1992, *131*, 272.
44. Nelson, W.H.; Sagstuen, E.; Hole, E.O.; Close, D.M. Radiat. Res. 1998, *149*, 75.
45. Close, D.M.; Nelson, W.H.; Sagstuen, E.; Hole, E.O. Radiat. Res. 1994, *137*, 300.
46. Ehrenberg, A.; Ehrenberg, L.; Loefroth, G. Nature 1963, *200*, 376.
47. Gräslund, A.; Ehrenberg, A.; Rupprecht, A.; Stroem, G.; Crespi, H. Int. J. Radiat. Biol. Relat. Stud. Phys. Chem. Med. 1975, *28*, 313.
48. Bernhard, W.A. J. Phys. Chem. 1989, *93*, 2187.
49. Sevilla, M.D.; Becker, D.; Yan, M.; Summerfield, S. J. Phys. Chem. 1991, *95*, 3409.
50. Hüttermann, J.; Voit, K.; Oloff, H.; Köhnlein, W. Faraday Discuss. Chem. Soc. 1984, *78*, 135.
51. Weiland, B.; Hüttermann, J.; Van Tol, J. Acta Chem. Scand. 1997, *51*, 585.
52. Weiland, B.; Hüttermann, J. Int. J. Radiat. Biol. 1998, *74*, 341.
53. Wetmore, S.D.; Boyd, R.J.; Eriksson, L.A. J. Phys. Chem. B 1998, *102*, 5369.
54. Wetmore, S.D.; Himo, F.; Boyd, R.J.; Eriksson, L.A. J. Phys. Chem. B 1998, *102*, 7484.
55. Wetmore, S.D.; Boyd, R.J.; Eriksson, L.A. J. Phys. Chem. B 1998, *102*, 9332.
56. Wetmore, S.D.; Boyd, R.J.; Eriksson, L.A. J. Phys. Chem. B 1998, *102*, 10602.
57. Li, X.; Cai, Z.; Sevilla, M.D. J. Phys. Chem. B 2001, *105*, 10115.
58. Close, D.M. In: *Computational Chemistry, Review of Current Trends*; Leszczynski, J., Ed.; World Scientific: Singapore, 2002; p. *in press*.
59. Holley, W.R.; Chatterjee, A.; Mian, I.S.; Rydberg, B. Schr. Forsch.zent. Jul., Bilater. Semin. Int. Bur. 1998, *29*, 12.
60. Ottolenghi, A.; Merzagora, M.; Tallone, L.; Durante, M.; Paretzke, H. Radiat. Environ. Biophys. 1995, *34*, 239.
61. Nikjoo, H.; Goodhead, D.T.; Charlton, D.E.; Paretzke, H.G. Int. J. Radiat. Biol. 1991, *60*, 739.
62. Goodhead, D.T.; Leenhouts, H.P.; Paretzke, H.G.; Terrissol, M.; Nikjoo, H.; Blaauboer, R. Radiat. Prot. Dosim. 1994, *52*, 217.
63. La Verne, J.A.; Pimblott, S.M. Radiat. Res. 1995, *141*, 208.
64. Pimblott, S.M.; LaVerne, J.A. In: *Radiation Damage in DNA: Structure/Function Relationship at Early Times*; Ficiarelli, A. F., Zimbrick, J.D., Eds.; Batelle Press: Columbus, OH, 1995; 3 pp.
65. Chatterjee, A.; Koehl, P.; Magee, J.L. Adv. Space Res. 1986, *6*, 97.
66. Debije, M.G.; Bernhard, W.A. Radiat. Res. 1999, *152*, 583.
67. Debije, M.G.; Strickler, M.D.; Bernhard, W.A. Radiat. Res. 2000, *154*, 163.
68. Boudaïffa, B.; Hunting, D.; Cloutier, P.; Huels, M.A.; Sanche, L. Radiat. Res. 2002, *157*, 227.
69. Goodhead, D.T. Can. J. Phys. 1990, *68*, 872.
70. Ward, J.F. Radiat. Res. 1981, *86*, 185.
71. Pimblott, S.M.; La Verne, J.A.; Mozumder, A.; Green, N.J.B. J. Phys. Chem. 1990, *94*, 488.
72. Murphy, C.J.; Arkin, M.R.; Jenkins, Y.; Ghatlia, N.D.; Bossmann, S.H.; Turro, N.J.; Barton, J.K. Science 1993, *262*, 1025.
73. Dandliker, P.J.; Holmlin, R.E.; Barton, J.K. Science 1997, *275*, 1465.
74. Kelley, S.O.; Barton, J.K. Met. Ions Biol. Syst. 1999, *36*, 211.
75. Wan, C.; Fiebig, T.; Schiemann, O.; Barton, J. Proc. Natl. Acad. Sci. U. S. A. 2000, *97*, 14052.
76. Williams, T.T.; Odom, D.T.; Barton, J.K. J. Am. Chem. Soc. 2000, *122*, 9048.
77. Debije, M.G.; Milano, M.T.; Bernhard, W.A. Angew. Chem. Int. Ed. 1999, *38*, 2752.
78. Warman, J.M.; de Haas, M.P.; Rupprecht, A. Chem. Phys. Lett. 1996, *249*, 319.
79. Beratan, D.N.; Priyadarshy, S.; Risser, S.M. Chem. Biol. 1997, *4*, 3.
80. Meade, T.J.; Kayyem, J.F. Angew. Chem. 1995, *34*, 352.
81. Lewis, F.D.; Wu, W.; Zhang, Y.; Letsinger, R.L.; Greenfield, S.R.; Wasielewski, M.R. Science 1997, *277*, 673.
82. Meade, T. In: *Metal Ions in Biological System*; Sigel, A., Sigel, H., Eds.; Marcel Dekker, Inc.: New York, 1996; Vol. 32, 453 pp.

83. Wang, W.; Becker, D.; Sevilla, M.D. Radiat. Res. 1993, *135*, 146.
84. Debije, M.G.; Bernhard, W.A. Radiat. Res. 2001, *155*, 687.
85. Bernhard, W.A. In: *The Early Effects of Radiation on DNA, Vol. Ser. H 54*; Fielden, E.M., O'Neill, P., Eds.; Springer-Verlag: Berlin Heidelberg, 1991; 141 pp.
86. Bernhard, W.A. Free Radical Res. Commun. 1989, *6*, 93.
87. Warman, J.M.; Jonah, C.D. Chem. Phys. Lett. 1981, *79*, 43.
88. Warman, J.M.; de Haas, M.P.; Verberne, J.B. J. Phys. Chem. 1980, *84*, 1240.
89. Becker, D.; La Vere, T.; Sevilla, M.D. Radiat. Res. 1994, *140*, 123.
90. LaVere, T.; Becker, D.; Sevilla, M.D. Radiat. Res. 1996, *145*, 673.
91. Milano, M.T.; Bernhard, W.A. Radiat. Res. 1999, *152*, 196.
92. Milano, M.T.; Bernhard, W.A. Radiat. Res. 1999, *151*, 39.
93. Becker, D.; Sevilla, M.D.; Wang, W.; LaVere, T. Radiat. Res. 1997, *148*, 508.
94. Bernhard, W.A. In: *Advances in Radiation Biology*; Lett, J.T., Adler, H., Eds.; Academic Press: New York, 1981; 199 pp.
95. Kar, L.; Bernhard, W.A. Radiat. Res. 1983, *93*, 232.
96. Close, D.M. Radiat. Res. 1997, *148*, 512.
97. Debije, M.G.; Razskazovskiy, Y.; Bernhard, W.A. J. Am. Chem. Soc. 2001, *123*, 2917.
98. Razskazovskiy, Y.; Debije, M.G.; Bernhard, W.A. Radiat. Res. 2000, *153*, 436.
99. Razskazovskiy, Y.; Debije, M.G.; Bernhard, W.A. Radiat. Res. 2003, *159*, 663.
100. Becker, D.; Razskazovskii, Y.; Sevilla, M.D. Radiat. Res. 1996, *146*, 361.
101. Swarts, S.G.; Sevilla, M.D.; Becker, D.; Tokar, C.J.; Wheeler, K.T. Radiat. Res. 1992, *129*, 333.
102. Sevilla, M.D. Jerus. Symp. Quantum Chem. Biochem. 1977, *10*, 15.
103. Gregoli, S.; Olast, M.; Bertinchamps, A. Radiat. Res. 1982, *89*, 238.
104. Debije, M.G.; Bernhard, W.A. J. Phys. Chem. B 2000, *104*, 7845.
105. Milano, M.T.; Hu, G.G.; Williams, L.D.; Bernhard, W.A. Radiat. Res. 1998, *150*, 101.
106. Razskazovskii, Y.; Swarts, S.G.; Falcone, J.M.; Taylor, C.; Sevilla, M.D. J. Phys. Chem. B 1997, *101*, 1460.
107. Anderson, R.F.; Patel, K.B.; Wilson, W.R. J. Chem. Soc. Faraday Trans. 1991, *87*, 3739.
108. Anderson, R.F.; Wright, G.A. Phys. Chem. Chem. Phys. 1999, *1*, 4827.
109. Fuciarelli, A.F.; Sisk, E.C.; Miller, J.H.; Zimbrick, J.D. Int. J. Radiat. Biol. 1994, *66*, 505.
110. Fuciarelli, A.F.; Sisk, E.C.; Zimbrick, J.D. Int. J. Radiat. Biol. 1994, *65*, 409.
111. Spalletta, R.A.; Bernhard, W.A. Radiat. Res. 1992, *130*, 7.
112. Turro, N.J.; Barton, J.K. J. Biol. Inorg. Chem. 1998, *3*, 201.
113. Kelley, S.O.; Barton, J.K. Science 1999, *283*, 375.
114. Netzel, T.L., In: *Organic and Inorganic Photochemistry*; Ramamurthy, V., Eds.; Marcel Dekker, Inc.: New York, 1998; 1 pp.
115. Meade, T.J. Met. Ions Biol. Syst. 1996, *32*, 453.
116. Krider, E.S.; Meade, T.J. J. Biol. Inorg. Chem. 1998, *3*, 222.
117. Grozema, F.C.; Berlin, Y.A.; Siebbeles, L.D.A. J. Am. Chem. Soc. 2000, *122*, 10903.
118. Cai, Z.; Gu, Z.; Sevilla, M.D. J. Phys. Chem. B 2000, *104*, 10406.
119. Cai, Z.; Sevilla, M.D. J. Phys. Chem. B 2000, *104*, 6942.
120. Messer, A.; Carpenter, K.; Frozley, K.; Buchanan, J.; Yang, S.; Razskazovskii, Y.; Cai, Z.; Sevilla, M. J. Phys. Chem. B 2000, *104*, 1128.
121. Cai, Z.; Gu, Z.; Sevilla, M.D. J. Phys. Chem. B 2001, *105*, 6031.
122. Gray, H.; Winkler, J. Ann. Rev. Biochem. 1996, *65*, 537.
123. McLendon, G. Acc. Chem. Res. 1988, *21*, 160.
124. Peterson-Kennedy, S.F.; McGourty, J.L.; Ho, P.S.; Lian, N.; Zemel, H.; Blough, N.V.; Margoliash, F.; Hoffman, B.M. Coord. Chem. Rev. 1985, *64*, 125.
125. Symons, M.C.R.; Peterson, R.L. Biochim. Biophys. Acta 1978, *535*, 241.
126. Symons, M.C.R.; Taiwo, F.A.; Suistunenko, D.A. J. Chem. Soc. Faraday Trans. 1993, *89*, 3071.
127. Alexander, P.; Lett, J.T.; Ormerod, M.G. Biochim. Biophys. Acta 1961, *51*, 209.

128. Kuwabara, M.; Yoshii, G. Biochim. Biophys. Acta 1976, *432*, 292.
129. Cullis, P.M.; Jones, G.D.D.; Symons, M. Nature 1987, *330*, 773.
130. Faucitano, A.; Buttafava, A.; Marinotti, F.; Pedraly-Noy, G. Internat. J. Radiat. Biol. 1992, *40*, 357.
131. Weiland, B.; Hüttermann, J. Int. J. Radiat. Biol. 2000, *76*, 1075.
132. Harriman, A. Angew. Chem. Int. Ed. 1999, *38*, 945.
133. Razskazovskiy, Y.; Roginskaya, M.; Jacobs, A.; Sevilla, M.D. Radiat. Res. 2000, *154*, 319.
134. Li, X.; Cai, Z.; Sevilla, M.D. J. Phys. Chem. A 2002, *106*, 1596.
135. Pezeshk, A.; Symons, M.C.R.; McClymont, J.D. J. Phys. Chem. 1996, *100*, 18562.
136. Marcus, R.; Sutin, N. Biochim. Biophys. Acta 1985, *811*, 265.
137. Bernhard, W.A.; Barnes, J.; Mercer, K.R.; Mroczka, N. Radiat. Res. 1994, *140*, 199.
138. Hole, E.O.; Sagstuen, E.; Nelson, W.H.; Close, D.M. Radiat. Res. 1991, *125*, 119.
139. Hole, E.O.; Sagstuen, E.; Close, D.M.; Nelson, W.H. In: *Radiation Damage in DNA: Structure/Function Relationships at Early times*; Fucianelli, A.F.; Zimbrick, J.D., Eds.; Battelle Press: Columbus, Ohio, 1995; 105 pp.
140. Bernhard, W.; Snipes, W. J. Chem. Phys. 1967, *46*, 2848.
141. Bernhard, W.A.; Hüttermann, J.; Müller, A.; Close, D.M.; Fouse, G.W. Radiat. Res. 1976, *68*, 390.
142. Hüttermann, J.; Bernhard, W.A.; Haindl, E.; Schmidt, G. J. Phys. Chem. 1977, *81*, 228.
143. Bernhard, W.A.; Mroczka, N.; Barnes, J. Int. J. Radiat. Biol. 1994, *66*, 491.
144. Debije, M.G.; Bernhard, W.A. J. Phys. Chem. A 2002, *106*, 4608.
145. Steenken, S. Biol. Chem. 1997, *378*, 1293.
146. Colson, A.O.; Besler, B.; Sevilla, M.D. J. Phys. Chem. 1992, *96*, 9787.
147. Colson, A.O.; Besler, B.; Close, D.M.; Sevilla, M.D. J. Phys. Chem. 1992, *96*, 661.
148. Shafirovich, V.; Dourandin, A.; Luneva, N.P.; Geacintov, N.E. J. Phys. Chem. B 2000, *104*, 137.
149. Bernhard, W.; Snipes, W. Proc. Natl. Acad. Sci. U. S. A. 1968, *59*, 1038.
150. Madden, K.P.; Bernhard, W.A. J. Phys. Chem. 1980, *84*, 1712.
151. Madden, K.P.; Bernhard, W.A. J. Chem. Phys. 1979, *70*, 2431.
152. Madden, K.P.; Bernhard, W.A. J. Phys. Chem. 1979, *83*, 2643.
153. Madden, K.P.; Bernhard, W.A. J. Phys. Chem. 1982, *86*, 4033.
154. Cullis, P.M.; Malone, M.E.; Merson-Davies, L.A. J. Am. Chem. Soc. 1996, *118*, 2775.
155. Yan, M.; Becker, D.; Summerfield, S.; Renke, P.; Sevilla, M.D. J. Phys. Chem. 1992, *96*, 1983.
156. Cullis, P.M.; McClymont, J.D.; Malone, M. J. Chem. Soc. Perkin Trans. 1992, *2*, 1695.
157. Cullis, P.M.; Evans, P.; Malone, M.E. Chem. Commun. 1996, *985*.
158. Swarts, S.G.; Sevilla, M.D.; Wheeler, K.T. *The Ninth International Congress of Radiation Research*; Academic Press, New York, 1991; 1 pp.
159. Debije, M.G.; Close, D.M.; Bernhard, W.A. Radiat. Res. 2002, *157*, 235.
160. Lett, J.T.; Alexander, P. Radiat. Res. 1961, *15*, 159.
161. Taylor, W.D.; Ginoza, W. Proc. Natl. Acad. Sci. U.S.A. 1967, *58*, 1753.
162. Yokoya, A.; Cunniffe, S.M.T.; O'Neill, P. J. Am. Chem. Soc. 2002, *124*, 8859.
163. Boon, P.J.; Cullis, P.M.; Symons, M.C.R.; Wren, B.W. J. Chem. Soc. Perkin Trans. 1984, *2*, 1393.
164. Ito, T.; Baker, S.C.; Stickley, C.D.; Peak, J.G.; Peak, M.J. Int. J. Radiat. Biol. 1993, *63*, 289.
165. Prise, K.M.; Gillies, N.E.; Michael, B.D. Radiat. Res. 1999, *151*, 635.
166. Milligan, J.R.; Aguilera, J.A.; Ward, J.F. Radiat. Res. 1993, *133*, 158.
167. Milligan, J.R.; Aguilera, J.A.; Ward, J.F. Radiat. Res. 1993, *133*, 151.
168. Cobut, V.; Frongillo, Y.; Patau, J.P.; Goulet, T.; Fraser, M.-J.; Jay-Gernin, J.-P. Radiat. Phys. Chem. 1998, *51*, 229.
169. Folkard, M.; Prise, K.M.; Vojnovic, B.; Davies, S.; Roper, M.J.; Michael, B.D. Int. J. Radiat. Biol. 1993, *64*, 651.
170. Boudaïffa, B.; Cloutier, P.; Hunting, D.; Huels, M.A; Sanche, L. Science 2000, *287*, 1658.
171. Boudaïffa, B.; Hunting, D.; Cloutier, P.; Huels, M.A.; Sanche, L. Int. J. Radiat. Biol. 2000, *76*, 1209.

172. Hieda, K.; Ito, T. In: Handbook of Synchrotron Radiation; Ebashi, S., Koch, M., Rubenstein, E., Eds.; Elsevier Science B.V., North-Holland, 1991; Vol 4, 431 pp.
173. Le Sech, C.; Frohlich, H.; Saint-Marc, C.; Charlier, M. Radiat. Res. 1996, *145*, 632.
174. Swarts, S.G.; Miao, L.; Wheelen, K.T.; Serilla, M.D.; Becker, D. In: *Radiation Damage in DNA; Structure/Function Relationships at Early Times*; Zimbrick, J.D. a. Fuciarelli, A.F., Eds.; Battelle Press: Columbus, OH, 1995; 131 pp.
175. Swarts, S.G.; Becker, D.; Sevilla, M.; Wheeler, K.T. Radiat. Res. 1996, *145*, 304.
176. Swarts, S.G.; Smith, G.S.; Miao, L.; Wheeler, K. Radiat. Environ. Biophys. 1996, *35*, 41.
177. Pogozelski, W.K.; Tullius, T.D. Chem. Rev. 1998, *98*, 1089.
178. Ward, J.F.; Kuo, I. Radiat. Res. 1976, *66*, 485.
179. Gugger, A.; Batra, R.; Rzadek, P.; Rist, G.; Giese, B. J. Am. Chem. Soc. 1997, *119*, 8740.
180. Meggers, E.; Kusch, D.; Spichty, M.; Wille, U.; Giese, B. Angew. Chem. Int. Ed. 1998, *37*, 460.
181. Ito, T.; Saito, M. Photochem. Photobiol. 1988, *48*, 567.
182. Abdoul-Carime, H.; Dugal, P.-C.; Sanche, L. Radiat. Res. 2000, *153*, 23.
183. Abdoul-Carime, H.; Sanche, L. Radiat. Res. 2001, *156*, 151.
184. Abdoul-Carime, H.; Dugal, P.C.; Sanche, L. Surf. Sci. 2000, *451*, 102.
185. Dugal, P.-C.; Huels, M.A.; Sanche, L. Radiat. Res. 1999, *151*, 325.
186. Bernhard, W.A.; Debije, M.G.; Milano, M.T.; Razskazovskiy, Y. *Eleventh International Congress of Radiation Research* (Dublin, Ireland) Allen Press, Inc., Lowrence, KS, 2000; 321 pp.
187. Lee, J.Y.; Bernhard, W.A. Radiat. Res. 1981, *86*, 287.
188. Lücke-Huhle, C.; Braun, A.; Hagen, U. *Z. Naturforsch.* 1970, *25b*, 1264.
189. Kessler, B.; Bopp, A.; Hagen, U. Int. J. Radiat. Biol. 1971, *20*, 75.

16

Photon-Induced Biological Consequences

Katsumi Kobayashi
High Energy Accelerator Research Organization
Tsukuba, Japan

1. INTRODUCTION

In this chapter, recent research on photo- or radiobiological phenomena is reviewed in which the underlying processes are considered to be initiated from specified photoabsorption events in biologically relevant atoms or molecules. The energy range of photons covered in this chapter is above 5 eV. Therefore synchrotron radiation (SR) is a key word in the chapter. Molecules or samples reviewed in this chapter are mainly restricted to biologically relevant molecules or living cells. Because of the complexity of these samples, entire mechanisms from photoabsorption to biological consequences have not been elucidated in most cases, and much remains to be studied in the future. Because earlier studies published before 1990 have already been reviewed elsewhere [1–3], most of the referred papers in this chapter are those after 1990. Also, many vacuum ultraviolet (VUV) studies appeared in *Photochem. Photobiol.* Vol. 44, No. 3 (1986), which is a special issue dedicated to VUV work. What characterize this chapter are samples, which include biomolecules, water, and living cells.

Photons induce various types of changes in molecules. Photon-induced molecular processes, and hence molecular products, depend on the energy of the illuminating photons, reflecting the amount of absorbed energy. When a molecule is a constituent of a biological system, such as living cells, photoabsorption can cause various types of biological effects. One of the most well-known examples of photon-induced biological effects might be the sterilizing action of ultraviolet light (4.88 eV or 253.7 nm) from low-pressure mercury lamps. This mechanism has been well studied for many years, and has been well understood. For further details concerning these phenomena, see review articles, e.g., Ref. 4. The purpose of this chapter is to review the action of photons, the energy of which is higher than the UV light from a germicidal lamp, namely, from 5 eV to the x-ray region. X-rays or γ-rays are also known to induce various biological effects. However, these actions of radiation are understood to be initialized mainly through energy transfer from secondary electrons to the biological molecules including water. The amount of energy transferred to molecules from electrons cannot be specified in most studies. These radiobiological studies are not described in this chapter. Studies in which the amount transferred to molecules was specified by the method of illuminating monochromatic photon beams to the biological system, and studies in which the photon energy dependence was observed as the result of specific photoabsorption events, are described in this chapter. To study the photon energy dependence, monochromatic photon beams are needed over a wide energy range, including ultraviolet (UV), vacuum UV, and a higher-energy region up to x-rays. Only a light source having a continuous spectrum in such a wide spectral range with practical intensity would be synchrotron radiation (SR). Recent studies with biologically relevant atoms or molecules and with living cells using SR as light source are introduced.

When we consider that probability of energy transfer from electrons to molecules (inelastic scattering) is closely related to the absorption spectrum of the molecule, we would soon realize that photobiology in the vacuum UV and soft x-ray regions would greatly contribute to our understanding radiobiological processes. Photoabsorption spectroscopy is very important because photoabsorption is the first step in the whole process leading to biological consequences.

2. IRRADIATION SYSTEM WITH SYNCHROTRON RADIATION

Presently, more than 20 synchrotron facilities are working or under construction in the world. Synchrotron radiation generated from electron storage rings is introduced into the experimental area situated around the ring through beamlines. Most of the beamlines are equipped with monochromators to monochomatize the white radiation. These beamlines are usually optimized for high-resolution spectroscopy. When we wish to study photochemical reactions, a sufficiently high flux is required to produce enough product for detection. A high flux is usually not consistent with a high resolution. Beamlines with a sufficient flux should be constructed with a reasonable compromise concerning the resolution. It should also be noticed that a wide beam is often necessary to uniformly illuminate a sample, because it sometimes happens that only a small fraction on the surface can absorb photons in a thick sample because of the large absorption cross section in the vacuum UV region. One method to fulfill both requirements is to accommodate two sample positions in the beamline, one at the focal point for a high flux density, the other at an off-focal point for a wider and more uniform beam [5].

There were only a very few beamlines where these requirements were taken into consideration. The first vacuum UV beamline dedicated to photobiological research was constructed in 1980 at Synchrotron Radiation Laboratory, University of Tokyo at Tanashi, Tokyo [6]. X-ray beamlines for radiobiology were constructed at the Photon Factory, Tsukuba, Japan [7], after some experience concerning monochromators and irradiation apparatus. These beamlines were optimized for irradiation. At the Photon Factory, supporting facilities for radiobiological experiments were also constructed near the beamline, which is essential for visiting radiobiologists to perform experiments in good condition.

In the UV and vacuum UV regions, the efficiency of the biological effects is usually evaluated based on the incident photon. The efficiency is expressed in units of m^2, and is usually called the action cross section. Another definition used to evaluate the efficiency is based on the absorbed energy, and is expressed in Gy^{-1} ($= kg/J$). This is commonly used in radiobiology. In this chapter, the word efficiency is used based on the incident photon, except when particularly described, because the interactions of photons with molecules are neglected when discussed using Gy.

3. EFFECTS ON DNA AND ITS CONSTITUENTS IN THE DRY STATE

3.1. Vacuum Ultraviolet Region

DNA molecules have now become very familiar, not only to scientists, but also to nonscientific people. However, it is surprising that the photoabsorption cross section of this molecule has not yet been measured over the whole energy region. A direct measurement of absorption by the transmission method is difficult because of the large absorption coefficient in the vacuum UV region. The absorption spectrum calculated from experimentally obtained optical constants is so far available only below 82 eV by Inagaki et al. [8]. In 1986, using synchrotron radiation, the absorption spectrum of DNA was obtained using a photoacoustic technique [9].

The energy of 253.7 nm (4.9 eV) photons is not large enough to ionize a DNA molecule, because the ionizing potential of organic molecules lies around 8 eV. In the vacuum UV region, we can see the transition from excitation to ionization in the photon energy dependence of biological phenomena. Below 8 eV, the production cross sections of cis–syn type and (6–4) photoproduct type thymine dimers were measured in the energy region between 4.3 eV (290 nm) and 8.3 eV (150 nm) using sublimed film of thymine as a sample [10]. The authors reported that both action spectra of these products are similar, and that the quantum yields stay almost constant above 200 nm, 0.008 and 0.013, respectively, but decrease by about half at 150 nm. This suggests that the mechanism to form dimers is independent of the energy of the photon illuminated, and that another reaction channel may open when excited with higher-energy photons. Similar results were also obtained with a DNA molecule using an enzyme-linked immunosorbent assay to detect thymine photodimers [11].

When higher-energy photons are illuminated to DNA, another type of damage begins to emerge, which is breakage of the DNA main chain consisting of deoxyribose and phosphate; in other words, strand breaks of DNA. Because of the double-helix structure of DNA, these are classified into two types, single strand breaks (ssb) and double strand breaks (dsb). Hieda et al. [12] extended their earlier work [13] on strand break induction in plasmid DNA up to 20.7 eV (60 nm). They demonstrated that photons of 8.3 eV (150 nm) can induce double strand breaks in DNA, and that the induction cross section of dsb increases with an increase of the photon energy more rapidly than that of ssb induction. Although the strand breaks of DNA have been commonly observed, there have been very few studies on the chemical nature of the cleaved edge. Ito and Saito [14] studied the edge structure of strand breaks induced by vacuum UV, using oligonucleotides as model compounds of DNA. It was demonstrated that, on illumination by VUV photons, strand breakage of DNA is a result of the destruction of deoxypentose, releasing a free base not of the simple breakage of a phosphoester bond. They also discussed the base dependence of the breakage efficiency [15]. Saito and Hieda [16] succeeded to demonstrate this transition from UV type, base damage, to strand-break-type damage using thymidylyl-(3′→5′)-thymidine (dTpdT), with an increase of the photon energy from 4.1 eV (300 nm) to 8.3 eV (150 nm). The action cross sections of UV-type damage was larger than that of strand-break-type damage when the photon energy was less than 5.9 eV (210 nm); above this energy, UV-type damage became the minor fraction.

Recently, a group in the United Kingdom started to study the effect of vacuum UV photons using the Daresbury synchrotron facility. They measured the yields of ssb and dsb in plasmid DNA, and proposed that the mechanism or a precursor to produce both types of strand breaks are common, because the photon energy dependence is similar for both types, although the absolute yield differed by 50-fold [17]. They confirmed that double strand breaks could be induced by low-energy photons of 7 eV, as had been reported by Hieda et al. [12]. From an energetic point of view, this observation would be a great challenge to theoretical radiobiologists, who usually assume that about 20 eV is the minimum energy to induce double strand break.

3.2. Soft X-Ray Region

Soon after commissioning of the Photon Factory, KEK Japan, the K-shell resonance photoabsorption by a phosphate group in a DNA molecule was found, and the effect of phosphorus photoabsorption has been studied using various samples [18,19]. Inner-shell photoabsorption by phosphorus has attracted much attention from two reasons: (1) inner-shell photoabsorption is followed by the Auger effect, which emits Auger electrons and, consequently, a multiply charged atom or molecule is produced, and (2) phosphate constitutes the main chain of a DNA molecule together with deoxyribose. Also practically important is that the absorption cross section at this resonance is about fivefold of the off-resonance K-shell absorption cross section. Therefore we can expect efficient photo-absorption by phosphorus in DNA, and hence, an efficient production of strand breaks by illuminating on-resonance x-rays. The true effect of phosphorus photoabsorption can be deduced by comparing the efficiency by monochromatic soft x-rays at the resonance absorption (2153 eV) with that of off-resonance x-rays (2147 eV), because the resonance is so sharp that 2147-eV x-rays cannot be absorbed by the K-shell of phosphorus.

Ito et al. [20] studied the mechanism to produce strand breakage by phosphorus photoabsorption using a dinucleotide as a sample, and revealed that even with photo-absorption by phosphorus strand breakage occurs through the destruction of deoxyribose of the 5′ end, leaving adenine and 5′-AMP as products. Yamada et al. [21] used penta-deoxythymidylic acid (d(pT)$_5$) as a sample, and analyzed the products quantitatively with on-resonance (2153 eV) and off-resonance (2147 eV) x-rays. They found that the distribution of the products was independent of the x-ray energy, and that the yields of the products were proportional to the absorption cross section of the sample. Unexpectedly, they could not find any evidence of Auger-specific products, neither qualitatively nor quantitatively.

The yields of single and double strand breaks were studied with plasmid DNA at five specific photon energies around the K-shell absorption edge of phosphorus, including the on-resonance energy. The yields of ssb and dsb were parallel with the absorption spectrum of DNA. It was found that the yield of double strand breaks per absorbed energy in the sample was higher at the on-resonance photon energy than at off-resonance, while the yield of single strand breaks was independent of the photoabsorption site [22]. Because the prepared plasmid samples were in the solid state, such as a pellet, plasmid molecules are closely packed, neighboring with each other. Under this circumstance, photo- or Auger electrons generated at one molecule can hit other molecules nearby, which may cause an increase in the yields of strand breaks. In a following study, they prepared a plasmid sample buried in an overwhelming amount (100-fold in weight) of cytidine, to minimize the probability of electrons hitting other DNA molecules. The yields of strand breaks per one K-shell photoabsorption in phosphorus were carefully calculated using the measured yields of ssb and dsb with photons of on-resonance and off-resonance. The results were unexpectedly small, 0.2 for ssb and 0.009 for dsb [23]. These values could be compared with the quantum yield in other photochemical reactions.

In the energy region between 20 eV and 2 keV, the yields of strand breaks in plasmid DNA were measured at 388, 435, and 573 eV, and it was found that the efficiency of strand breaks per absorbed energy was much less than that in the 2 keV region [24].

It is not easy to compare the yield of strand breaks in the vacuum UV region with those obtained with x-rays or γ-rays, because of the difference in the dosimetry systems commonly used. However, when we took the ratio of ssb per dsb, the value is independent of the dosimetry, and the difference in the value could be considered as being the result of different mechanisms contributing to the production of dsb. By compiling the data presented by Hieda's group [12,13,22,25], we could plot the ratio over the whole energy range from UV to x-ray regions, as shown in Fig. 1. From this figure, they discussed the following: (1) In the photon energy region around 8 eV, a strand break would be mainly produced via a superexcited state and dsb could be produced with the least probability by some mechanism, such as a radical swing over hypothesis (1 dsb over nearly 100 ssb). (2) Around 20 eV, photoelectrons are emitted from the molecules, which may contribute to

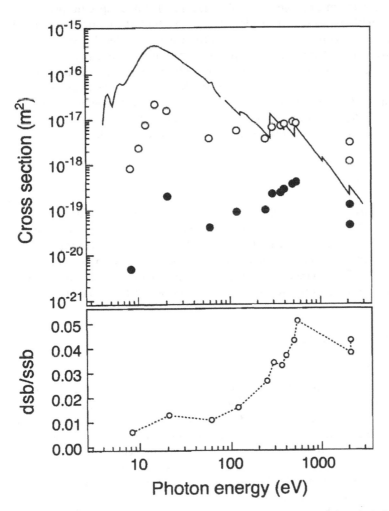

Figure 1 Action spectra for the induction of single strand break (ssb, open circle) and double strand break (dsb, closed circle) in dry plasmid DNA. Ratio of the cross section of ssb against dsb is also shown in lower panel. Solid line is reconstructed absorption cross section of the plasmid. (From Refs. 12, 13, 22, and 25.)

the production of dsb; the contribution of these electrons increases up to the keV region (1 dsb over 50 ssb). (3) Above 1 keV, the contribution of secondary electrons in the production of dsb would become dominant, and the ratio would not increase any further (1 dsb over 20 ssb). We could say that at least two mechanisms would contribute to producing dsb, the ratio of the contribution being dependent on the illuminated photon energy.

Biological effects of photoabsorption by exogenous atoms are also interesting because of a potential application to the radiotherapy of cancer (Photon Activation Therapy). Photoabsorption by the inner shell of a heavy atom has been focused on because of the large penetration depth of x-rays, and because of the expected large Auger cascade. Le Sech et al. [26,27] demonstrated that a Pt-containing intercalator, chloroterpyridine platinum, would act as a "photosensitizer" when photons corresponding to the absorption edge of inner shell of Pt are illuminated. They also reported that the ratio of dsb to ssb increased on on-resonance photoabsorption.

To elucidate the mechanism and the molecular nature of Auger-specific products, the development of new methods to detect or identify the product at the molecular level is badly needed. One of the trials to develop a new method was reported by Svensson et al. [28]. They successfully detected fragments from ATP irradiated with x-rays, or bombarded by electrons, the energy of which was tuned to the binding energy of the K-shell electron of phosphorus, and fragments from cystine bombarded with electron tuned to the binding energy of sulfur K-shell, using a time-of-flight mass spectrometer. Although this method is presently applicable only to small molecules of high vaporing pressure, it would be a powerful method for studying the mechanism to produce molecular damage.

4. EFFECT ON AMINO ACIDS

Using the method mentioned above, the effect of photoabsorption by sulfur atoms in amino acids was studied by Yokoya et al. [29]. They chose cystathionine as a model molecule because of the structural advantage that most of the decomposed fragments are easily detectable amino acids. A comparison of the distribution pattern of fragments between on-resonance and off-resonance lead to the conclusion that cystathionine was more frequently cleaved around a sulfur atom than at a bond apart from the sulfur.

5. EFFECT ON WATER MOLECULE

5.1. Vacuum Ultraviolet Region

About 70% in weight of living cells is water, and chemical reactions necessary to keep the cell alive proceed under aqueous circumstances. For this reason, we need to consider the photochemistry of water when a cell is illuminated with photons of various energies. Very few data are available so far on the absorption cross section of liquid water. The cross section starts to increase at around 6.5 eV (190 nm), and grows very rapidly; and the penetration depth of a photon at 8.3 eV (150 nm) becomes as short as 0.1 μm. A water molecule, on the absorption of vacuum UV photons, decomposes into H and OH radicals. For example, these reactive radicals can attack DNA to form lethal or mutagenic damage, as is well known in radiation biochemistry. Therefore determining the yield of radicals is very essential to understand the action mechanism of vacuum UV photons on living cells. Watanabe et al. [30] measured the yield by a Fricke dosimeter, which is commonly used in γ-rays dosimetry. Because of the cutoff of the window material (MgF_2), which can transmit vacuum UV photons, but is not soluble in water, the obtained values are limited to less than 8.5 eV. The calculated quantum yield at 8.3 eV (150 nm) was 1.7. For determining the quantum yield between 6.5 and 7.1 eV, more reliable data on the absorption cross section are needed. Very recently, Hayashi et al. [31] reported the complete optical spectrum of liquid water, measured by inelastic x-ray scattering. This method is greatly expected to solve the above problem.

5.2. Soft X-Ray Region

The Fricke dosimeter is one of the standard systems for the dosimetry of ionizing radiation. Hoshi et al. [32] applied this method for dosimetry of monochromatized synchrotron x-rays at 8.86 and 13.55 keV. In the soft x-ray region, the Fricke yield was measured with a photon energy down to 1.8 keV. The yield was found to decrease with a decrease in the photon energy [33]. The data that they presented was the first experimental data in this energy region, which should be compared with the values recommended by ICRP report No.17 (1970). In this energy region, energetic photoelectrons are produced in water, most of which are ejected from the K-shell of oxygen because of the largest cross section. The energy of the electrons is the photon energy minus 0.54 keV (binding energy of K shell electron of oxygen). Therefore the observed photon energy dependence could be considered to be a consequence of the difference in the structure, or distribution, of the energy deposition events along the electron track, not the nature of a photoabsorbing atom or molecule. To characterize the structure of energy deposition by a charged particle, linear energy transfer, or LET, is commonly used. However, for electrons, this parameter might not be appropriate, because the track of an electron is not as straight as those of heavy particles. To study the characteristics of low-energy electrons, simulation method seems to be the only way.

The Fricke solution contains iron ion in its constituent, and one might expect the effect of inner-shell photoabsorption of iron around the energy region of the Fe K-shell absorption edge. No photon-energy dependence of the Fricke yield around the K-shell absorption edge of the ion was found, which was explained by the small abundance of iron in the sample.

6. EFFECT ON DNA IN SOLUTION

6.1. Vacuum Ultraviolet Region

When DNA is illuminated with vacuum UV photons in an aqueous solution, the induced reactions become more complicated. In the energy region less than 6.5 eV (190 nm) where water is transparent, the photoabsorbing molecule is DNA itself; but with 7.7 eV (150 nm) photons, the water molecules absorb almost all photons illuminated onto the sample. In this transient region, the sensitivity of single strand breaks were measured using plasmid DNA under different buffering conditions [34,35]. They found that the peak sensitivity moves depending on the scavenging condition in the sample, although they confirmed that the main attacking agents are water radicals produced by the absorption of VUV photons. They discussed this dependence from the viewpoint of the absorption cross section of the constituents of the sample solution and from the difference in the initial distribution of short-lived OH radicals, depending on the photon energy [36]. The complicated behavior of the photon energy dependence was recently discussed again by Nikjoo et al. [37].

6.2. Soft X-Ray Region

When aqueous samples are irradiated with soft x-rays, fewer photons are absorbed by the target atom in the molecule because of the less abundant atoms in the sample. Also, a greater fraction of the photon energy is taken away by photoelectrons and/or Auger electrons from the atom in which the photoabsorption event has occurred than in the case of low-energy VUV photons. For this reason, the effect of specific photoabsorption events, such as inner-shell photoabsorption followed by Auger cascade, is generally believed to become less important in the final consequences. However, studies on DNA or nucleotides in aqueous solution are absolutely necessary for understanding the effect on living cells, because living cells contain about 70% of water in total weight.

The first work was performed by Takakura [38], who irradiated a concentrated solution (200 mg/mL) of bromo-deoxyuridine-monophosphate (Br-dUMP) with monochromatic x-rays below (13.43 keV) and above (13.49 keV) the absorption edge of bromine. Br-dUMP can be incorporated into a living cell instead of dTMP. Radiolytic products, such as deoxyuridine-monophosphate (dUMP), uracil, and bromo-uracil (Br-uracil), were quantitatively analyzed using high-performance liquid chromatography (HPLC) as a function of the absorbed dose in the sample solution. The photon energy dependence was studied on the yield per absorbed energy. The ratios of the yields of products between x-rays of 13.49 and 13.43 keV were 2.2 for dUMP, 1.02 for Br-uracil, and 1.23 for uracil. On the other hand, only a small difference between x-rays of 13.49 and 13.43 keV was observed for the yield of uracil released from dUMP irradiated in aqueous solutions. The energy-dependent enhancement of debromination was considered to be caused by the inner-shell photoabsorption of bromine, followed by Auger effects. Also, interesting results were that the yield of debromination per absorbed energy of 100 eV (G-value) was 10.5 at 13.49 keV and 4.7 at 13.43 keV. This means one photon induces $13.49/0.1 = 134.9$ debromination events at 13.49 keV. This suggests that a sort of chain reaction participates in debromination, and that this reaction is enhanced by the inner-shell photoabsorption of bromine.

Further work was performed by the same group using an adenosine $5'$-triphosphate (ATP) solution as the sample. They irradiate a concentrated ATP solution (155 mg/mL) with three energies (2.146, 2.153, and 2.160 keV) of monochromatic soft x-rays around the absorption edge of phosphorus; the yields of adenine, adenosine $5'$-monophosphate (AMP) and adenosine $5'$-diphosphate (ADP) were analyzed with HPLC. The yields of these three products clearly showed an energy dependence, indicating that the decomposition of ATP is enhanced by the photoabsorption of phosphorus [39]. The yield of adenine at the resonance photoabsorption of phosphorus (2.153 keV) was 1.6 per absorbed energy of 100 eV, indicating that one photoabsorption produces $2.153/0.1 = 21.5$ adenine molecules. It should be noted that the number of Auger electrons emitted from phosphorus is smaller than that from the bromine K-shell vacancy. Using the same sample system, Kobayashi et al. [40] studied the yield of adenine from an ATP solution with different concentrations. The enhancement ratio of the yields, yield at 2.153 keV/yield at 2.146 keV, was found to be concentration dependent. It was considered that one of the reasons why the effect of inner-shell photoabsorption could be observed in aqueous solution is the high concentration of the target molecules in the solution, which increase the atomic abundance in the sample. However, the concentration dependence of the enhancement cannot be explained. One hypothesis to explain the concentration-dependent enhancement might be that the influence of the inner-shell photoabsorption followed by Auger effect is restricted within a close vicinity of the photoabsorption site.

Strand breaks in plasmid DNA in aqueous solution were also studied by several groups. Takakura et al. [41] irradiated a plasmid DNA solution (1 mg/mL) with monochromatic soft x-rays around the K-shell absorption edge of phosphorus. Single and double strand breaks were analyzed with gel electrophoresis, the same method as that in the case of a dry sample. They succeeded to observe an enhancement in the yields of ssb and dsb in plasmid by the resonance photoabsorption of phosphorus, although the photon energy dependence was not the same. It was surprising that the enhancement was observed at low concentrations of DNA solution (1 mg/mL), much lower than in the case of Br-dUMP or ATP. The enhanced induction of ssb and dsb might be explained as follows: The microscopic density of nucleotides could be considered to be high in the local volume including DNA and, hence, the concentration-dependent enhancement of the inner-shell photoabsorption might be observable, as was observed in a concentrated ATP solution.

To investigate the possibility of photon activation therapy, the effect of the inner-shell photoionization of exogenous atoms on plasmid was studied. Takakura [42] studied the effect of the inner-shell photoabsorption of calcium, and observed that the mono-chromatic x-rays tuned to the absorption edge of calcium. Recently, the effect of Pt-containing chemicals has been studied. An intercalating agent, chloroterpyridine platinum (PtTC), was mixed with plasmid DNA with a ratio of 1 Pt atom per 10 nucleotides, and the solution was irradiated with monochromatic X rays tuned to the resonant photo-absorption energy of the L(III) shell of the platinum atom. Although the atomic abundance of Pt was very small in the sample, the yield of dsb was higher with an energy above the absorption edge than below the edge [43]. They discussed the possibility that energetic electrons generated in the sample lose their energy more densely around the platinum atoms because of the high electron density of platinum, and hence more water radicals are produced around platinum that bind to DNA.

The photon energy dependence of ssb and dsb induction over a wide energy range, between 2.14 and 10 keV, was investigated by Tomita et al. [44]. They found that the yield of ssb increased with an increase in the photon energy, similar to the Fricke yield, but the yield of dsb showed an opposite dependence as shown in Fig. 2. These phenomena do not have any relation with absorption spectrum of the sample, but are related to the energy of photoelectrons generated from oxygen in water. In other words, from the observed results, lower-energy electrons produce dsb more efficiently, but ssb less efficiently. This also demonstrates again that the induction mechanism of dsb is different, at least in part, from that of ssb. The difference in the distribution and/or density of energy deposition events along the electron track is considered to be important, which may cause a local difference in the concentration of water radicals.

It is suggested from the results with aqueous samples that local distribution of energy deposition events would be very important in determining the efficiency of a biological effect. The initial process of radiobiological phenomena could be said to be a dissipation process of the radiation energy. Soon after being deposited from photons or electrons, the local energy density is very high, and every kind of "instability," such as radicals, which can cause changes in biomolecules, are produced around the site. The reactive species then diffuse away from the deposition site to reach a homogeneous distribution. During early times of these processes, the distribution of radicals retains the original site where the energy

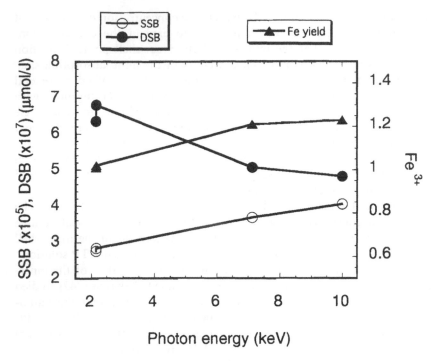

Figure 2 Photon energy dependence of the yield of ssb (open circle) and dsb (closed circle) per absorbed energy in plasmid DNA irradiated in aqueous solution [44]. The Fricke yield is also shown for comparison (closed triangle).

deposition event took place. For example, the dsb of DNA seems to be more likely produced at such an early stage of the process, while ssb can be produced even after homogeneity is attained. Radiation chemistry in the early stage should be promoted more. When we move on to the circumstances such as inside the cell, where biomolecules are densely and inhomogeneously dissolved, a more complex reaction system is waiting to be resolved.

7.   EFFECT ON BIOLOGICAL CELLS

7.1.   Vacuum Ultraviolet Region

In this section, studies using living cells and phages in the energy region below 140 nm are described. The energy dependence observed in most of these studies could be explained by the transition in the mode of action on DNA by photon, or by the penetration depth of vacuum UV photons in living systems, especially in living cells containing a large amount of water. The water molecules work as a shielding layer as well as attacking agents when they absorb photons. The simplest biological system including DNA would be viruses or phages. These can survive in a vacuum because they do not contain any liquid water. A coating layer of protein is so thin that vacuum UV photons can reach DNA and produce damage, as observed in the case of dry plasmid DNA. Maezawa et al. [45] reported that the energy dependence of the inactivation of the T1 phage was similar to the spectrum of DNA. The transition of DNA damage from dimer type to another one was substantiated by the difference in the susceptibility to the repair ability of a host cell or to the photo-reactivation treatment. Similar results were obtained using a spore of bacterium, another small and dry system, by Munakata et al. [46].

When larger systems, such as yeast cells, were used, the penetration depth of vacuum UV photons into cell nuclei becomes important. Ito et al. [47] found that yeast cells were killed by vacuum UV as well as UV, but that genetic changes were not induced by vacuum UV. Vacuum UV was suspected not to produce any DNA damage in yeast cells, because genetic changes can be considered to be a result of DNA damage. Instead, they found that damage in the cell membrane was produced. These findings suggest that vacuum UV cannot penetrate into cell nuclei and, hence, cannot produce DNA damage, and that cell death by vacuum UV is attributable to damage in the cell membrane. They pointed out that the action spectra of inactivation and of membrane damage were similar to the absorption spectrum of liquid water in vacuum UV region, suggesting that membrane damage was produced by an attack of the water radicals, the dissociative fragments of water molecules. These works and related articles have been reviewed by Ito [1].

7.2. Soft X-Ray Region

Living cells consist of various components, such as nucleic acids, proteins and amino acids, lipids, and water. Because of a large abundance of water, most photoabsorption events occur at oxygen of water, and nearly monoenergetic electrons (kinetic energy 0.5 keV less than the photon energy). The biological effects of monochromatic photons have been studied from two points of view: (1) An enhancement of biological effects by inner-shell photoabsorption, followed by the Auger effect has been investigated in relation to the possibility of the photon activation therapy of cancers. (2) Rather basic studies on the electron energy dependence of the radiobiological effect were investigated to understand the biological effects of high-energy γ-rays.

In 1985, preliminary data on the enhanced killing of HeLa cells prelabeled with 5-bromodeoxyuridine were reported by Shinohara et al. [48]. The enhancement by bromine inner-shell photoabsorption was intensively studied by Maezawa et al. [49] using E. coli cells. They clearly observed the enhancement by comparing the effect of monochromatic x-rays above the absorption edge with those of x-rays below the edge. They also found that the enhancement was more clearly observed in the presence of dimethylsulfoxide, a radical scavenger, suggesting less importance of the radical-mediated processes. Similar work was performed using a mammalian cell line sensitized with 5-bromodeoxyuridine [50] or with 5-iododeoxyuridine [51]. Usami et al. [52] discussed the enhancement by bromine incorporation into DNA from the view point of the absorbed energy, or dose, using the substituted fraction of thymine with bromo-uracil, and concluded that the enhancement could not be solely explained by the increased absorbed energy, or dose, and that some damage specific to the Auger cascade might be produced. Jonsson et al. [53] tried to find the enhancement by inner-shell ionized indium atoms incorporated into the cell. No significant difference in survival between irradiations above and below the K-edge could be observed, because of the small amount of indium incorporated into the cell. This demonstrates the importance of selecting the chemicals or methods to deliver the target atoms into the cell, when photon activation therapy becomes effective.

The effect of phosphorus photoabsorption in DNA has been intensively inves-tigated because the discovery of the resonance absorption at the K-shell absorption edge (Fig. 3). Soon after the discovery of large resonance photoabsorption in the K-shell edge

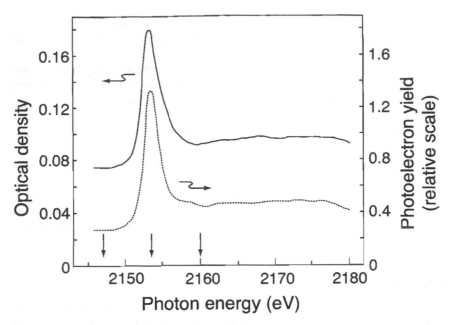

Figure 3 Absorption spectrum of the DNA film around the K-shell absorption edge of phosphorus [18]. Dotted line is the total photoelectron yield of KH_2PO_4. Three vertical arrows indicate the X-ray energies used for the irradiation experiments shown in Fig. 4.

(2.15 keV) of phosphorus in DNA, Kobayashi et al. [18] studied the biological effect of phosphorus photoabsorption in yeast cells. They found that the photon energy dependence of the killing effect and of the induction of genetic changes were very similar to the resonance absorption spectrum of DNA (Fig. 4). On the other hand, they could not detect any impairment of the cell membrane consisting of a bilayer of phospholipids. These results clearly indicate that soft x-ray photoabsorption by phosphorus produces DNA damage, which causes biological effects, such as cell killing and the induction of genetic changes. For the case of induction of genetic changes, they calculated the quantum efficiency of induction to be 0.13, assuming the target size to be the size of the gene concerned.

The enhancement of biological effects by inner-shell photoabsorption in DNA could be considered to be caused either by the increase in the amount of energy deposited locally around the photoabsorption site, namely DNA, or by the production of DNA damage specific to inner-shell photoabsorption, which could be repaired less by the cellular repair system. The former possibility can only be discussed from a theoretical nanodosimetric point of view. Experimentally, the latter hypothesis has been challenged by radiobiologists using repair-deficient strains. The most probable candidate of nonrepairable damage has been considered to be a complex damage composed of more than one dsb, or dsb and other types of damages, produced very near each other. This type of damage is called "clustered damage," and is now being intensively studied by many radiobiologists and biochemists.

Watanabe at al. [54] compared the induction efficiencies of various biological endpoints using Syrian golden hamster embryo (SHE) cells. They found that monochromatic soft x-rays tuned to the resonance absorption of phosphorus (2.153 keV) were more

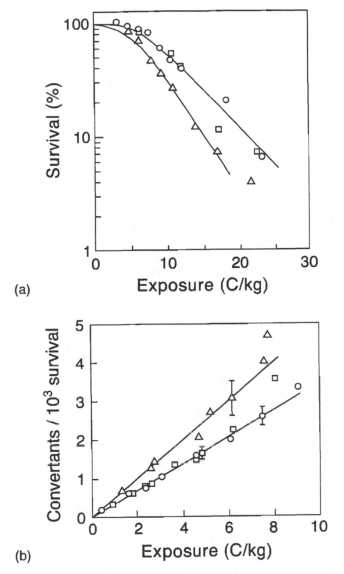

Figure 4 (a) Lethal effect and (b) induction of genetic changes in yeast cells irradiated with x-rays tuned to resonance absorption peak [2.153 keV (triangle)] and both sides of the peak [2.147 eV (square) and 2.160 keV (circle)]. (From Ref. 18.)

efficient in the induction of mutations and chromatid aberrations than off-peak x-rays (2.147 and 2.159 keV). When compared at equal survival levels, there was no difference in the frequencies of mutations and chromatid aberrations in cells irradiated with soft x-rays. On the other hand, a morphological transformation was more efficiently induced with 2.147-keV x-rays than with 2.153- and 2.159-keV x-rays. To see the nature of DNA damage produced by inner-shell photoabsorption, studies using repair-deficient stains were made. Maezawa et al. [55] compared the enhancement by phosphorus photoabsorption between wild type and several repair-deficient strains of *E. coli*. The enhancement ratios, defined as the ratio of the killing sensitivity of 2.153 keV to that at 2.146 keV, were 1.32 for the wild type, repair-proficient strain, and 1.42 to 1.54 for repair-deficient strains. They also calculated the number of lethal hits per Auger cascade for the strains examined. The value for the recombination-deficient strain was highest (0.65), while the value for the wild strain was 0.11. This suggests that most of the DNA damage produced by the Auger cascade would be double strand breaks, which can be repaired by the recombinational repair pathway, and that some part of the damage produced by Auger cascades was repaired in wild-type strains. Usami et al. [56] investigated the enhancement by phosphorus photoabsorption using a repair-deficient strain of yeast, which cannot repair double strand breaks of DNA. Double strand breaks were expected to be more efficiently induced by phosphorus photoabsorption. They found that the enhancement disappeared in the repair-deficient strain. On the other hand, when wild-type cells were given more time for repair, the enhancement became larger than in the reference condition. These results were explained as meaning that the fraction of nonrepairable dsb produced by inner-shell photoabsorption was too small to be detected in the cell-killing effect where the damage was not repaired at all, while the fraction of nonrepairable damage became larger after dsb repair had fully operated compared to before. Seemingly, nonrepairable damage is concentrated by the repair system of the cell. The difference in the enhancement in repair-deficient strains between *E. coli* and yeast might be because of the difference in the DNA damage repair system of each species.

The repair of damage produced by phosphorus photoabsorption was more directly investigated by Maezawa et al. [57]. They measured chromatin breaks of cells irradiated with x-rays of on- and off-peak the resonance absorption of phosphorus using the premature chromosome condensation (PCC) method. It was found that chromosomal damage produced with 2.147-keV x-rays was repaired to one-third by the action of the cellular repair system, while the damage by on-peak x-rays (2.153 keV) was not repaired at all. The elucidation of the mechanism by which DNA damage leads to chromosomal damage might help us understand the chemical nature of Auger-specific damage.

These lines of results using repair-deficient strains suggest that inner-shell photoabsorption, followed by the Auger effect, more efficiently produces a nonrepairable type of damage than x-rays with other energies. However, studies to find or identify these types of damages at the molecular level have not yet been successful so far.

Another type of research using monochromatic x-rays is to investigate the electron energy dependence of biological effects. In living cells containing a large amount of water, most of the photoabsorption events of low-energy x-rays occur at the oxygen atom of water, which generate monoenergetic photoelectrons in the sample. The induction of chromosome aberrations was studied in human peripheral blood lymphocytes irradiated by monochromatic soft x-rays of quantum energy in a range between 4.8 and 14.6 keV [58]. These x-rays were found to be more effective in producing chromosome aberrations (dicentrics and rings) than ^{60}Co γ-rays. The observed dependence of the efficiency on the LET of the photoelectrons and their associated Auger electrons was discussed in relation to the dual nature of chromosome aberration formation. The lethal effect of ultrasoft x-rays was investigated using Chinese hamster V79 and mouse C3H 10T1/2 cells [59]. They chose 273- and 860-eV ultrasoft x-rays, both of which have the same attenuation depth into the cells. Such an isoattenuating energy pair allows a direct examination of biological effectiveness of ultrasoft x-rays. They could not find any significant difference in the lethal action of these x-rays. These results should be discussed from the viewpoint of the energy distribution of the generated photoelectrons and Auger electrons in the sample.

The photon energy dependence of inactivation and mutagenesis in the whole energy region between 0.1 and 300 nm was investigated by Munakata et al. [46,60,61] using *Bacillus subtilis* spores. Spores can survive in a high vacuum, in which vacuum UV and ultrasoft x-ray photons can be exposed on samples. Also, its size is small enough for vacuum UV photons to penetrate into and reach DNA. They constructed the photon energy dependence of inactivation and mutagenesis [62]. Some of the characteristics that they found are:

(1) Shift in the nature of DNA damage that occurred between 5.6 eV (220 nm) and 8.3 eV (150 nm). This shift is considered to be a shift from the base dimer-type damage to the strand-break type, which could be expected with work using DNA molecules.

(2) A decrease in the biological efficiency was observed in the vacuum UV region because of the shielding effect of the outer materials in a spore.

(3) In the soft x-ray region, an enhancement by the photoabsorption of phosphorus and of calcium was observed. The condensation of calcium near DNA in spores, as a form of dipicolinic acid, has already been reported [63]. The calcium atom has acted in the same manner as exogenous sensitizers.

8. FUTURE PERSPECTIVE

Data on the photon energy dependence of the biological effect has been accumulated fairly well with strand-break induction in plasmid DNA, and with an inactivation of the spore of a bacterium. These studies gave us many clues to understand the induction mechanism of biological effects in cells. However, for a complete understanding, research in the following areas should be highly promoted.

(1) The spectroscopy of biomolecules, themselves alone, and in complexes with water molecules over the whole energy region from vacuum UV to x-rays.

(2) Photochemistry and radiation chemistry of biomolecules in vacuo and in water. Radiolytic products from Auger cascade are interesting from the viewpoint of radiotherapy. The development of irradiation systems is required in which liquid samples can be irradiated with vacuum UV or ultrasoft x-rays of high intensity.

(3) From the viewpoint of the electron energy dependence of biological effects, the efficiency of low-energy x-rays on living cells should be studied. Because the electron track is composed of a track segment of different energy, data over a wide energy range are necessary.

(4) The development of new methods for studying the photon-induced decomposition, or damages, of biomolecules and for studying radical reactions in the early stage in water.

ACKNOWLEDGMENTS

The author expresses sincere gratitude to Prof. K. Hieda for the critical reading of this manuscript. Thanks are also due to Dr. N. Usami for her help in preparing the manuscript.

REFERENCES

1. Ito, T. Vacuum ultraviolet photobiology with synchrotron radiation. In: *Sweet, R.M. Woodhead, A.D.;* Synchrotron Radiation in Structural Biology. Plenum Publishing Corporation, 1989; 221–241 pp.

2. Hieda, K.; Ito, T. Radiobiological experiments in the X-ray region with synchrotron radiation. Elsevier Science Publishers B. V.: Amsterdam, 1991; Vol. 4.

3. Ito, T. The effects of vacuum-UV radiation (50–190 nm) on microorganisms and DNA. Adv. Space Res. 1992, *12*,(4), 249–253.

4. Smith, K.C. Science of Photobiology. Plenum, 1989.

5. Kobayashi, K. Radiobiology using Synchrotron Radiation. Bruttini, E. Balerna, A., Eds.; Proceedings of the International School of Physics 'Enrico Fermi.' IOS Press:Amsteradm, 1996; Vol. Course CXXVII, 333–352 pp.

6. Ito, T.; Kada, T.; Okada, S.; Hieda, K.; Kobayashi, K.; Maezawa, H.; Ito, A. Synchrotron system for monochromatic uv irradiation (greater than 140 nm) of biological material. Radiat. Res. 1984, *98* (1), 65–73.

7. Konishi, H.; Yokoya, A.; Shiwaku, H.; Motohashi, H.; Makita, T.; Kashihara, Y.; Hashimoto, S.; Harami, T.; Sasaki, T.S.; Maeta, H.; Ohno, H.; Maezawa, H.; Asaoka, S.; Kanaya, N.; Ito, K.; Usami, N.; Kobayashi, K. Synchrotron radiation beamline to study radioactive materials at the photon factory. Nucl. Instrum. Methods 1996, *A372*, 322–332.

8. Inagaki, T.; Hamm, R.N.; Arakawa, E.T.; Painter, L.R. Optical and dielectric properties of DNA in the extreme ultraviolet. J. Chem. Phys. 1974, *61*, 4246.

9. Inagaki, T.; Ito, A.; Hieda, K.; Ito, T. Photoacoustic spectra of some biological molecules between 300 and 130 nm. Photochem. Photobiol. 1986, *44* (3), 303–306.

10. Yamada, H.; Hieda, K. Wavelength dependence (150–290 nm) of the formation of the cyclobutane dimer and the (6–4) photoproduct of thymine. Photochem. Photobiol. 1992, *55* (4), 541–548.

11. Matsunaga, T.; Hieda, K.; Nikaido, O. Wavelength dependent formation of thymine dimers and (6–4) photoproducts in DNA by monochromatic ultraviolet light ranging from 150 to 365 nm. Photochem. Photobiol. 1991, *54* (3), 403–410.

12. Hieda, K.; Suzuki, K.; Hirono, T.; Suzuki, M.; Furusawa, Y. Single- and double-strand breaks in pBR322 DNA by vacuum-UV from 8.3 to 20.7 eV. J. Radiat. Res. (Tokyo) 1994, *35* (2), 104–111.

13. Hieda, K.; Hayakawa, Y.; Ito, A.; Kobayashi, K.; Ito, T. Wavelength dependence of the formation of single-strand breaks and base changes in DNA by the ultraviolet radiation above 150 nm. Photochem. Photobiol. 1986, *44* (3), 379–383.

14. Ito, T.; Saito, M. Degradation of oligonucleotides by vacuum-UV radiation in solid: roles of the phosphate group and bases. ([published erratum appears in Photochem. Photobiol. 1989 Jun; 49(6):845]) Photochem. Photobiol. 1988, *48* (5), 567–572.

15. Ito, T.; Saito, M. Effects of vacuum ultraviolet radiation on deoxyoligonucleotides in solids in the wavelength region around and above ionization potential—With special reference to the chain scission. Radiat. Phys. Chem. 1991, *37* (5/6), 681.

16. Saitou, M.; Hieda, K. Dithymine photodimers and photodecomposition products of thy-midylyl-thymidine induced by ultraviolet radiation from 150 to 300 nm. Radiat. Res. 1994, *140* (2), 215–220.

17. Prise, K.M.; Folkard, M.; Michael, B.D.; Vojnovic, B.; Brocklehurst, B.; Hopkirk, A.; Munro, I.H. Critical energies for SSB and DSB induction in plasmid DNA by low-energy photons: action spectra for strand-break induction in plasmid DNA irradiated in vacuum. Int. J. Radiat. Biol. 200, *76* (7), 881–890.

18. Kobayashi, K.; Hieda, K.; Maezawa, H.; Furusawa, Y.; Suzuki, M.; Ito, T. Effects of K-shell X-ray absorption of intracellular phosphorus on yeast cells. Int. J. Radiat. Biol. 1991, *59* (3), 643–650.

19. Furusawa, Y.; Maezawa, H.; Suzuki, K. Enhanced killing effect on 5-bromodeoxyuridine

labelled bacteriophage T1 by monoenergetic synchrotron X-ray at the energy of bromine K-shell absorption edge. J. Radiat. Res. (Tokyo) 1991, 32 (1), 1–12.

20. Ito, T.; Saito, M.; Kobayashi, K. Dissociation of a model DNA compound dApdA by monochromatic soft X-rays in solids and comments on the high selectivity for 3' breakage in the phosphoester bond. Int. J. Radiat. Biol. 1992, 62 (2), 129–136.

21. Yamada, H.; Kobayashi, K.; Hieda, K. Effects of the K-shell X-ray absorption of phosphorus on the scission of the pentadeoxythymidylic acid. Int. J. Radiat. Biol. 1993, 63 (2), 151–159.

22. Hieda, K.; Hirono, T.; Azami, A.; Suzuki, M.; Furusawa, Y.; Maezawa, H.; Usami, N.; Yokoya, A.; Kobayashi, K. Single- and double-strand breaks in pBR322 plasmid DNA by monochromatic X-rays on and off the K-absorption peak of phosphorus. Int. J. Radiat. Biol. 1996, 70 (4), 437–445.

23. Hieda, K.; Tomita, M.; Takemura, T.; Kobayashi, K. Do Auger events of phosphorus by monochromatic X-rays induce DNA double strand breaks efficiently? Moriarty, M., Mothersill, C., Seymour, C., Edington, M., Ward, J.F., Fry, R.J.M., Eds.; Proceedings of 11th International Congress of Radiation Research, Allen Press, 1999; 142–145.

24. Yokoya, A.; Watanabe, R.; Hara, T. Single- and double-strand breaks in solid pBR322 DNA induced by ultrasoft X-rays at photon energies of 388, 435 and 573 eV. J. Radiat. Res. (Tokyo) 1999, 40 (2), 145–158.

25. Hieda, K. DNA damage induced by vacuum and soft X-ray photons from synchrotron radiation. Int. J. Radiat. Biol. 1994, 66 (5), 561–567.

26. Le Sech, C.; Takakura, K.; Saint-Marc, C.; Frohlich, H.; Charlier, M.; Usami, N.; Kobayashi, K. Strand break induction by photoabsorption in DNA-bound molecules. Radiat. Res. 2000, 153 (4), 454–458.

27. Le Sech, C.; Takakura, K.; Saint-Marc, C.; Frohlich, H.; Charlier, M.; Usami, N.; Kobayashi, K. Enhanced strand break induction of DNA by resonant metal-innershell photoabsorption. Can. J. Physiol. Pharm. 2001, 79 (2), 196–200.

28. Svensson, A.; Bordas, J.; Hughes, E.A.; Mant, G. Radiation damage induced in free nucleotides and sulfur-containing amino-acids by monoenergetic X-rays. Chance, B., Deisenhofer, J., Ebashi, S., Goodhead, D.T., Helliwell, J.R., Huxley, H.E., Iizuka, T., Kirz, J., Mitsui, T., Rubenstein, E., Sakabe, N., Schmahl, G., Stuhrmann, H.B., Wuthrich, K., Zaccai, G., Eds. Synchrotron Radiation in the Biosciences, 1994; 721–729.

29. Yokoya, A.; Kobayashi, K.; Usami, N.; Ishizaka, S. Radiolytic degradation of cystathionine irradiated with monochromatic soft X-rays at the K-shell resonance absorption of sulfur. J. Radiat. Res. (Tokyo) 1991, 32 (2), 215–223.

30. Watanabe, R.; Usami, N.; Takakura, K.; Hieda, K.; Kobayashi, K. Water radical yields by low energy vacuum ultraviolet photons as measured with Fricke dosimeter. Radiat. Res. 1997, 148, 489–490.

31. Hayashi, H.; Watanabe, N.; Udagawa, Y.; Kao, C. The complete optical spectrum of liquid water measured by inelastic x-ray scattering. Proc. Natl. Acad. Sci. U. S. A. 2000, (12), 6264–6266.

32. Hoshi, M.; Uehara, S.; Yamamoto, O.; Sawada, S.; Asao, T.; Kobayashi, K.; Maezawa, H.; Furusawa, Y.; Hieda, K.; Yamada, T. Iron(II) sulphate (Fricke solution) oxidation yields for 8.9 and 13.6 keV X-rays from synchrotron radiation. Int. J. Radiat. Biol. 1992, 61 (1), 21–27.

33. Watanabe, R.; Usami, N.; Kobayashi, K. Oxidation yield of the ferrous ion in a Fricke solution irradiated with monochromatic synchrotron soft X-rays in the 1.8–10 keV region. Int. J. Radiat. Biol. 1995, 68 (2), 113–120.

34. Takakura, K.; Ishikawa, M.; Hieda, K.; Kobayashi, K.; Ito, A.; Ito, T. Single-strand breaks in supercoiled DNA induced by vacuum-UV radiation in aqueous solution. Photochem. Photobiol. 1986, 44 (3), 397–400.

35. Takakura, K.; Ishikawa, M.; Ito, T. Action spectrum for the induction of single-strand breaks in DNA in buffered aqueous solution in the wavelength range from 150 to 272 nm: dual mechanism. Int. J. Radiat. Biol. 1987, 52 (5), 667–675.

36. Ito, T.; Takakura, K. Interpretation of DNA–SSB-sensitivity spectra in aqueous solution from the roles of active species. J. Radiat. Res. (Tokyo) 1996, 37, 316.

37. Nikjoo, H.; Goorley, T.; Fulford, J.; Takakura, K.; Ito, T. Quantitative analysis of the energetics of DNA damage. Radiat. Prot. Dosim. 2002, 99, 91–98.

38. Takakura, K. Auger effects on bromo-deoxyuridine-monophosphate irradiated with monochromatic X-rays around bromine K-absorption edge. Radiat. Environ. Biophys. 1989, 28 (3), 177–184.

39. Watanabe, R.; Ishikawa, M.; Kobayashi, K.; Takakura, K. Damage to adenine-triphosphate induced by monochromatic X-rays around the K-shell absorption edge of phosphorus. Howell, R.W., Narra, V.R., Sastry, K.S.R., Rao, D.V.; Biophysical Aspects of Auger Processes. American Association of Physicists in Medicine, 1992; 24–36 pp.

40. Kobayashi, K.; Usami, N.; Watanabe, R.; Takakura, K. Production yield of adenine form ATP irradiated with monochromatic X-rays in aqueous solution of different concentrations. Goodhead, D.T., O'Neill, P., Menzel, H.G., Eds.; Microdosimetry—An Interdisciplinary Approach, 1997; 65–69 pp.

41. Takakura, K.; Maezawa, H.; Kobayashi, K.; Hieda, K. Strand breaks in DNA in buffered solution induced by monochromatic X-rays around the K-shell absorption edge of phosphorus. In: Synchrotron Radiation in Biosciences. Chance, B., Deisenhofer, J., Ebashi, S., Goodhead, D.T., Helliwell, J.R., Huxley, H.E., Iizuka, T., Kirz, J., Mitsui, T., Rubenstein, E., Sakabe, N., Schmahl, G., Stuhrmann, H.B., Wuthrich, K., Zaccai, G., Eds.; Oxford University Press: Oxford, 1994; 756–764 pp.

42. Takakura, K. Double-strand breaks in DNA induced by the K-shell ionization of calcium atoms. Acta Oncol. 1996, 35 (7), 883–888.

43. Kobayashi, K.; Frohlich, H.; Usami, N.; Takakura, K.; Le Sech, C. Enhancement of X-ray-induced breaks in DNA bound to molecules containing platinum: a possible application to hadrontherapy. Radiat. Res. 2002, 157 (1), 32–37.

44. Tomita, M.; Hieda, K.; Watanabe, R.; Takakura, K.; Usami, N.; Kobayashi, K.; Hieda, M. Comparison between the yields of DNA strand breaks and ferrous ion oxidation in a Fricke solution induced by monochromatic photons, 2.147–10 keV. Radiat. Res. 1997, 148, 490–491.

45. Maezawa, H.; Ito, T.; Hieda, K.; Kobayashi, K.; Ito, A.; Mori, T.; Suzuki, K. Action spectra for inactivation of dry phage T1 after monochromatic (150–254 nm) synchrotron irradiation in the presence and absence of photoreactivation and dark repair. Radiat. Res. 1984, 98 (2), 227–233.

46. Munakata, N.; Hieda, K.; Kobayashi, K.; Ito, A.; Ito, T. Action spectra in ultraviolet wavelengths (150–250 nm) for inactivation and mutagenesis of Bacillus subtilis spores obtained with synchrotron radiation. Photochem. Photobiol. 1986, 44 (3), 385–390.

47. Ito, T.; Ito, A.; Hieda, K.; Kobayashi, K. Wavelength dependence of inactivation and membrane damage to Saccharomyces cerevisiae cells by monochromatic synchrotron vacuum-uv radiation (145–190 nm). Radiat Res. 1983, 96 (3), 532–548.

48. Shinohara, K.; Ohara, H.; Kobayashi, K.; Maezawa, H.; Hieda, K.; Okada, S.; Ito, T. Enhanced killing of HeLa cells pre-labeled with 5-bromodeoxyuridine by monochromatic synchrotron radiation at 0.9 A: an evidence for Auger enhancement in mammalian cells. J. Radiat. Res. (Tokyo) 1985, 26 (3), 334–338.

49. Maezawa, H.; Hieda, K.; Kobayashi, K.; Furusawa, Y.; Mori, T.; Suzuki, K.; Ito, T. Effects of monoenergetic X-rays with resonance energy of bromine K-absorption edge on bromouracil-labelled E. coli cells. Int. J. Radiat. Biol. Relat. Stud. Phys. Chem. Med. 1988, 53 (2), 301–308.

50. Larson, D.; Bodell, W.J.; Ling, C.; Phillips, T.L.; Schell, M.; Shrieve, D.; Troxel, T. Auger electron contribution to bromodeoxyuridine cellular radiosensitization. Int. J. Radiat. Oncol. Biol. Phys. 1989, 16 (1), 171–176.

51. Laster, B.H.; Thomlinson, W.C.; Fairchild, R.G. Photon activation of iododeoxyuridine: biological efficacy of Auger electrons. Radiat. Res. 1993, 133 (2), 219–224.

52. Usami, N.; Kobayashi, K.; Maezawa, H.; Hieda, K.; Ishizaka, S. Biological effects of Auger

processes of bromine on yeast cells induced by monochromatic synchrotron X-rays. Int. J. Radiat. Biol. 1991, 60 (5), 757–768.

53. Jonsson, A.C.; Jonsson, B.A.; Strand, Se.; Grafstrom, G.; Spanne, P. Cell survival after Auger electron emission from stable intracellular indium exposed to monochromatic synchrotron radiation. Acta Oncol. 1996, 35 (7), 947–952.

54. Watanabe, M.; Suzuki, M.; Watanabe, K.; Suzuki, K.; Usami, N.; Yokoya, A.; Kobayashi, K. Mutagenic and transforming effects of soft-X-rays with resonance energy of phosphorus K-absorption edge. Int. J. Radiat. Biol. 1992, 61 (2), 161–168.

55. Maezawa, H.; Furusawa, Y.; Kobayashi, K.; Hieda, K.; Suzuki, M.; Usami, N.; Yokoya, A.; Mori, T. Lethal effect of K-shell absorption of intracellular phosphorus on wild- type and radiation sensitive mutants of *Escherichia coli* [published erratum appears in Acta Oncol. 1997;36(2):238] Acta Oncol. 1996, 35 (7), 889–894.

56. Usami, N.; Yokoya, A.; Ishizaka, S.; Kobayashi, K. Reparability of lethal lesions produced by phosphorus photoabsorption in yeast cells. J. Radiat. Res. (Tokyo) 2001, 42 (3), 317–331.

57. Maezawa, H.; Suzuki, M.; Yokoya, A.; Usami, N.; Kobayashi, K. Repair of chromatin breaks produced by monochromatic X-rays with energies of around K-shell absorption edge of phosphorus in V79 Cells: Photon Factory Activity Report 1997, Vol. 14, 420.

58. Sasaki, M.S.; Kobayashi, K.; Hieda, K.; Yamada, T.; Ejima, Y.; Maezawa, H.; Furusawa, Y.; Ito, T.; Okada, S. Induction of chromosome aberrations in human lymphocytes by monochromatic X-rays of quantum energy between 4.8 and 14.6 keV. Int. J. Radiat. Biol. 1989, 56 (6), 975–988.

59. Hill, C.K.; Nelms, B.E.; MacKay, J.F.; Pearson, D.W.; Kennan, W.S.; Mackie, T.R.; DeLuca, P.M., Jr., Lindstrom, M.J.; Gould, M.N. Synchroton-produced ultrasoft X rays: equivalent cell survival at the isoattenuating energies 273 eV and 860 eV. Radiat. Res. 1998, 150 (5), 513–520.

60. Munakata, N.; Saito, M.; Hieda, K. Inactivation action spectra of *Bacillus subtilis* spores in extended ultraviolet wavelengths (50–300 nm) obtained with synchrotron radiation. Photochem. Photobiol. 1991, 54 (5), 761–768.

61. Munakata, N.; Hieda, K.; Usami, N.; Yokoya, A.; Kobayashi, K. Inactivation action spectra of *Bacillus subtilis* spores with monochromatic soft X rays (0.1–0.6 nm) of synchrotron radiation. Radiat. Res. 1992, 131 (1), 72–80.

62. Munakata, N. Action spectra for inactivation and mutagenesis of *Bacillus subtilis* spores in wavelength ranges between 0.1 and 300 nm. In: *Synchrotorn Radiation in the Biosciences*. Chance, B., Deisenhofer, J., Ebashi, S., Goodhead, D.T., Helliwell, J.R., Huxley, H.E., Iizuka, T., Kirz, J., Mitsui, T., Rubenstein, E., Sakabe, N., Schmahl, G., Stuhrmann, H.B., Wuthrich, K., Zaccai, G., Eds.; Oxford University Press, 1994; 765–774.

63. Nishihara, T.; Kondo, M.; Nonaka, T.; Higashi, Y. Location of calcium and phosphorus in ashed spores of *Bacillus megaterium* determined electron probe X-ray microanalysis. Microbiol. Immunol. 1982, 26, 167–172.

17

Track Structure Studies of Biological Systems

Hooshang Nikjoo
Medical Research Council, Harwell, Oxfordshire, England

Shuzo Uehara
Kyushu University, Fukuoka, Japan

1. INTRODUCTION

Although radiation biology over the past decade has made striking progress in understanding the components of radiation damage in the cell, the future direction is in understanding the biological systems, using theory, computation, and simulations as an integral part of experimental method. Useful biophysical models give predictions that can be related to other biological phenomena and are not just simulations of experimental data and observation. An immediate question comes to mind, whether there is a role for mathematics in biology and cancer research. Unlike in more pure sciences such as physics, most biological work is highly descriptive and entities are complex and not easily expressed as a set of invariant quantities. Despite such difficulties, over the past half century, there has been a rapid surge in mathematical theories along with the advances in laboratory and clinical cancer research to provide generalized description, e.g., observations on population and dynamics in tumor growth, diffusion processes, protein regulations, and the role of kinase pathways in human diseases and others in genetics and ecology. To obtain an insight into the mechanism of radiation oncogenesis especially at low doses relevant to radiation protection, it is necessary to formulate a radiation risk model that attempts to relate the early molecular damage to the observed effects. Such an attempt requires a quantitative description of the general features of radiation insult in terms of DNA damage as a starting platform to investigate the role of repair and cell survival using more exact mathematical expressions. Although such a goal is yet a long way off, this chapter shares with the reader some aspects of knowledge and experience gained in description and quantification of the spectrum of DNA damage and deeper understanding of the nature of radiation insult.

A current concern in radiation biology is understanding the mechanism of damage to cells and tissue and quantification of the health hazards of ionizing radiation of different qualities, and in particular quantification of human cancer risks at low doses and dose rates to the general population [1], and specific group of workers such as space-radiation exposures to astronauts [2]. This concern is evident from major scientific programs currently in progress in Europe [3] and the United States [4–8].

In the recent past, our perceived understanding of DNA as the target for ionizing radiation has become more complex with discoveries of new phenomena such as the by-stander effect [9], the adaptive response [10], and the genetic instability [11,12]. These new phenomena have shed doubt on the estimation of cancer risk solely by extrapolation of dose response from high to low dose and dose rates [13,14]. With recent accelerated progress in molecular biology techniques and advances in theoretical methods, attention has become more focused on mechanistic studies of effects of ionizing radiation. To this end, track structure has provided a theoretical/simulation framework to investigate those parameters of ionizing radiation that predominantly determine the nature and magnitude of the final effect [15,16] (see also Chap. 4). In general, "microdosimetry" and "track structure" have tried to scrutinize and understand aspects of radiation damage in cellular structures from a theoretical approach based on fundamental physical and chemical principles and provide hypotheses that are experimentally testable. In particular, "track structure" provides a basis for understanding the underlying mechanism that shapes the dose–effect relationship. There is a wealth of information and data accumulated on harmful effects of radiation track that needs to be placed in the framework of a general descriptive theory. For example, there are considerable data: on the early effects of radiation damage on DNA [17]; on chromosome aberration [18–21], mutational events [22,23], and genetic instability [11]; on early clonal expansion of the cell to neoplasia and on the final expression of malignancy [24]. This chapter includes a brief description of application of track structure in model calculations of DNA damage, application of DNA damage in radioprobing of novel DNA structures, bystander effect, and latest progress in track simulation. The model calculations presented here are a testimony to the depth and breadth of track structure calculations that have been applied in a broad range of subjects.

2. TIME SCALE

Biological responses to ionizing radiation, broadly speaking, depending on the end point, can be divided into early and late effects. Early effects are those that become apparent within milliseconds to days following irradiation. Late effects manifest themselves within weeks to years following exposure. Radiation injury begins with the physical processes of energy deposition in the medium, at times less than femtoseconds, leading to ionizations and exci-tations. The initial physical processes lead to molecular damage by direct interaction of the radiation with the target and through the generation of free radicals, oxidizing agents and other molecular species, and the bystander signals. Molecular damage is usually expressed through alteration of biochemical processes and is amplified by biological mediators. In addition, damage may persist, which is not generally visible or clinically detectable, leading to genetic instability. Damage can also indirectly arise from signals received from the neigh-boring hit cells. In both direct and indirect modes, we assume DNA is the primary target leading to observed cellular effects. Fig. 1 shows a flowchart of such ideas including the DNA repair pathways and other processes leading to cell inactivation and cancer.

The sequence of events starting with the transfer of energy from the primary particle to the molecules of the medium initiates a chain of events depending on the magnitude of energy being transferred and electronic structure of the material of the medium. These interactions, in the form of clusters of ionizations and excitations, set in motion one or more electrons in the surrounding molecules. These events have a random distribution and normally no two particles produce the same distribution. The energy degradation of the

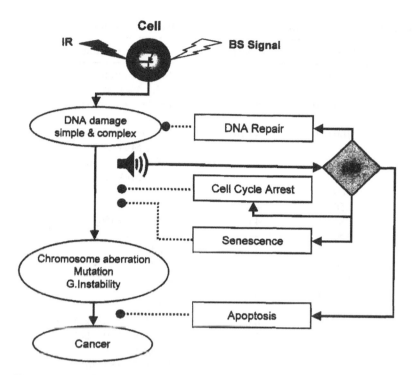

Figure 1 Schematic sequence of events following DNA damage.

primary particle continues by successive interactions with the neighboring molecules and the ejection of secondary electrons until its excess energy is completely lost and the electron becomes trapped by the electrostatic charges. The trapped electron is usually referred to as the hydrated electron (e_{aq}^-), which behaves as a free radical, diffuses, and interacts with other atoms until captured. The creation of hydrated electrons is regarded as the start of the chemical stage at which time the system is in thermal equilibrium (10^{-12} sec and < 0.025 eV). A simple consequence of free radical diffusion and reactions with other molecules is the modification and amplification of existing molecular damage from the physical stage as the consequence of direct interactions in the target site.

Chemical alterations due to direct ionization of the molecules and free radical reactions lead to degradation of the biomolecule and induction of cross-linking, and intercrosslinking of the DNA and proteins [25]. These and other changes to the conformation of the biomolecules may change the enzymatic activity of the cell [26]. Table 1 provides a quantitative summary of the number of events produced by 1 Gy of radiation of different quality in a mammalian cell. The differences observed in biological effects of radiations of different quality may arise not only from differences in their track structures but also from differences in physical and physiological conditions. It is noted that although energy deposited in the cell by 1 Gy of radiation produces a large number of events, only a fraction of this leads to induction of molecular damage. In mammalian cells, the majority of single-strand breaks (SSB), in the form of single nucleotide gaps, and damage to DNA bases, in the form of base products, are readily repaired [27]. Although the frequencies of observed changes in a single cell induced by ionizing radiations are very low, it is widely believed that most cancers arise

Table 1 Average Yield of Damage in a Single Mammalian Cell after 1 Gy of Radiation

Radiation	Low-LET	High-LET	Comments
Tracks in nucleus	1,000	4	100 keV electrons and 3.2 MeV α[a]
Ionizations in nucleus	100,000	8×10^5	100 keV electrons and 3.2 MeV α[b]
Excitations in nucleus	100,000	8×10^5	100 keV electrons and 3.2 MeV α[b]
Base damage (BD): SSB	3.3	3.4	100 keV electrons and 3.2 MeV α[c]
8-Hydroxyadenine	700	—	[222]
Thymine damage	250	—	[222]
DSB: initial	38	46	3.4 MeV α [56]
DNA protein cross link	150	—	[222]
Chromosome aberration	1	3	[220]
Dicentric per cell	0.1	0.4	[220]
HPRT mutation	6.9×10^{-6}	8.3×10^{-5}	[219]
HPRT mutation (BG)	3.2×10^{-6}	3×10^{-5}	[219]
Lethal lesions	0.5	2.6	[221]
Lethal lesions per DSB	0.01	0.045	250 KV X, 100 keV/μm α [221]
DSB per lethal lesion	87	22	250 KV X, 100 keV/μm α [221]
Cell inactivation	30%	85%	[223,224]
Complex SSB	4%	30%	100 keV electrons, 3.2 MeV α [79–81]
Complex SSB_B	40%	70%	100 keV electrons, 3.2 MeV α [79–81]
Complex DSB	20%	70%	100 keV electrons, 3.2 MeV α [79–81]
Complex DAB_B	60%	90%	100 keV electrons, 3.2 MeV α [79–81]
Fract hybrid DSB	0.38	0.25	100 keV electrons, 3.2 MeV α [79–81][d]

[a] Data from authors' Monte Carlo track structure calculations.
[b] For a nucleus 10-μm diameter.
[c] See Table 5.
[d] Hybrid DSB: when a SSB induced by direct energy deposition converted to a DSB by OH.

from multiple changes in a single cell, in which at least one of the initial events could have been induced by radiation in the damaged cell or its progeny [28].

3. SIZE SCALE

3.1. Model of DNA

Dimensions of the DNA and its higher-order molecular structures, the nucleosome and the basic fiber, are given in Table 2. It is assumed that the genome of the mammalian cell con-

Table 2

Structure	Diameter (nm)	Segment Length (nm)	Number per Genome	Comment
DNA segment	2.3	2.3	9.3×10^8	Assuming 6 pg of DNA in nucleus, MW of 660, DNA duplex 1.86-m length
Nucleosome	10	5 single nucleosome	2.9×10^7	190 bp containing 50 nm of DNA including the linker
Chromatin segment	30	30	1.9×10^6	Solenoid of nucleosomes

tains 6 pg of DNA in a nucleus, a physical size of 6349 Mbp [29], a molecular weight of an average base pair of 660, and a length of a base pair of about 0.34 nm.

Models of DNA of various degrees of sophistication have been used in the calculations of radiation-induced DNA damage. These can be classified into three groups. The simplest of DNA models is a linear segment in the form of a cylinder [30]. These segments can be generated along random chords cutting a convex body (spherical nucleus) using the method of μ-randomness [31]. This model has mainly been used to obtain frequencies of energy depositions in macromolecular structures without a priori assumption of the role of the atomic structure in determining the biological responses. A more realistic model used in our early studies of DNA damage is the volume model of DNA [32]. In the volume model, DNA is in its native B-form with a diameter of 2.3 nm and divided into 0.34-nm slices. In turn, each slice is divided into three volumes comprising the central core representing the volumes of the complementary paired nucleobases and the two arches, each representing the volumes occupied by the deoxyribosophosphate backbone of each strand. This model does not take into account the detailed atomic structure of DNA. The third category is the sophisticated atomic models of DNA that have been available for a number of years, and can now readily be generated using commercially available programs such as Newhelix [33], Curves, and MidasPlus [34,35]. Further refinements of the latter models include the distribution of solvent molecules around the sugar–phosphate backbone and the nucleobases [36 37]. Modeling of higher-order structures of DNA, nucleosome, chromatin, and other forms such as triplexes, have been made by several authors [38–41] and also available from the PDB library.

3.2. A Model of the First Hydration Shell of DNA

The following describes the development of an atomistic model of DNA including the first hydration shell of a B-DNA [37]. Water plays an important role in the three-dimensional structure of DNA, the conformation it adopts, and its response to radiation. Influence on the conformation of DNA by the solvent has been known since the earliest fiber diffraction studies [42]. The native conformation of the macromolecule to a large degree is influenced by inter- and intramolecular forces, such as hydrogen bonding forces, repulsive and dispersive forces, and interaction with the solvent. At high relative humidity or water activity, corresponding to low salt or low atomic organic solvent content, DNA adopts the B-form, which is considered to be the most biologically relevant form of DNA. If the relative humidity is lowered, DNA transforms from B-form to A-form.

The basic building block of DNA is the nucleotide. The backbones run in opposite directions with each base pair having a purine base (A/G) hydrogen bonded to a pyrimidine base (T/C). The purine and pyrimidine bases are on the inside of the helix while the phosphate and sugar are on the outside. The two chains are held together by hydrogen bonds between the base pairs. Adenine is paired to thymine by means of two hydrogen bonds while guanine is paired to cytosine by three hydrogen bonds. A single nucleotide consists of three chemical parts; a simple sugar molecule, an inorganic phosphate and a nitrogen-containing base. Successive nucleotides are linked together via a phosphodiester bond between the sugar and phosphate of adjacent nucleotides. The nitrogen-containing bases are not involved in any covalent linkages other than their attachment to the sugar–phosphate backbone. It is the sequence of these nitrogen-containing bases along the sugar–phosphate backbone that constitutes the unique structural and functional individuality of the DNA molecule. The nitrogen bases are linked at the C_1 position while the C_2 and C_4 hydroxyl groups participate in phosphodiester bonds. The nitrogen-containing bases are either the double-ringed purines or the single-ring pyrimidines. The phosphate group is present as part of a diester linkage. The DNA fibers can exist in three major forms, as B-DNA under conditions of high humidity, as A-DNA, and a left-handed family of Z-DNA. The differences between the families occur in the way the sugar–phosphate backbone is wrapped around the helix axis, the way in which the base pairs are stacked and in the pucker of the furanose ring [43]. The A-DNA form with 11 base pairs per pitch of 28 Å is much more compressed than the B-DNA form with 10 base pairs per pitch of 34 Å. Another difference between the A and B forms is that the sugar in the B-DNA is puckered in the C(2')-endo family region, extending from O(4')-endo to C(1')-exo, resulting in a wide interphosphate separation of 6.6 Å along the chain. For the A-DNA, the sugar puckering is confined to the C(3')-endo resulting in a shorter interphosphate separation of approximately 6 Å. Thus phosphate groups in A-DNA can be bridged by water molecules as the interphosphate distances is not too large while all phosphate groups are individually hydrated because the interphosphate separation is too large [44]. These differences results in different hydration patterns around the phosphate groups, which may offer a scheme for the transition from A- to B-DNA. At lower humidity, the hydration becomes more economical and the sugar puckering changes, effectively reducing the interphosphate separation and so allowing water molecules to bridge the free phosphate oxygen atom [45].

DNA hydration studies with various techniques have shown that there are approximately 10 and 20 water molecules in the first hydration of A- and B-DNA, respectively. These water molecules bind in decreasing order of strength to the anionic oxygens of the phosphate group, to the ester oxygens of the phosphodiester linkage, to the O(4') oxygen of the furanose ring, and to the electronegative atoms of the base pairs [46].

Distribution of water molecules around the polar atoms of a decamer canonical B-DNA structure were obtained from quantitative analysis of the solvent interactions within hydrogen bonding distances of polar atoms of oligonucleotides using 12 B-DNA oligonucleotide crystal structures [37]. Table 3 shows the distribution of water molecules around the bases, the phosphate group, and the sugar. The data are presented in terms of polar coordinates around a polar atom in the plane of abc as indicated in the table. Data shows that water interactions around the bases mainly occur with the polar atoms of the major and minor grooves. In this manner, the data in Table 3 have been used to generate variable DNA segments. The first atom in the table given in bold type is the atom at the origin of the r, ϑ, Φ coordinate system.*

---

*Atomic coordinates of the hydrated B-DNA can be obtained from the author.

Table 3 Distribution of Water Sites for B-DNA

	Atom a–b–c	$r(A)$	$\theta°$	$\phi°$
Phosphate	O1P–P–O2P	2.91	53	145
		2.63	32	103
	O2P–P–O1P	2.95	48	179
		2.76	46	−74
Sugar	O4′–C1′–C2′	2.84	53	−174
Cytosine	O2–C2–N3	2.94	22	−160
	N4–C4–C5	3.07	63	9
Thymine	O2–C2–N3	2.77	22	−123
		2.84	47	104
	O4–C4–C5	2.68	45	15
Adenine	N3–C4–C9	2.84	43	23
	N6–C6–N1	3.18	41	176
	N7–C5–C6	2.70	43	−7
Guanine	N2–C2–N3	3.36	24	−35
	O6–C6–N1	2.70	43	−176
	N3–C4–N9	2.91	45	7
	N7–C5–C6	2.77	46	−9

4. APPLICATION OF TRACK-STRUCTURE CALCULATIONS IN UNDERSTANDING RADIOBIOLOGICAL MECHANISMS

Motivation for understanding mechanism of cellular damage by ionizing radiation tracks dates back to the pioneering work of Lea [47]. Investigators since then have tried to relate the initial events produced when a track of ionizing radiation traverses the cell nucleus to the observed biological lesions. In most cases, such a description starts with knowledge of spatial distribution of energy loss in subcellular structures such as DNA. For example, for practical reasons, assessment of risk of radiation exposure requires allowances to be made for the differences in biological effectiveness of different types of radiation. Because of the stochastic nature of radiation, such assessments depend on the absorbed dose in the volume of the target, the quality of radiation, and the size of the target. We also know that relevant biological effects of radiation are mostly because of damage in individual cells. To this end, it has also been established that the spatial and temporal distribution of radiation interactions within the cell or its nucleus has an important influence on its biological effectiveness. Therefore these distributions must be considered if we are seeking either practical quantities for comparison of different radiations in risk assessment and therapy, or seeking a greater understanding of the fundamental problems in the mechanism of radiation action. In pursuit of a suitable description of the mechanism(s) of radiation effects, a number of biophysical models have been put forward including the "exchange" theory of Revel, the "breakage first" theory of Lea and the "lesion–nonlesion" interaction by Chadwick and Leenhouts [20,48,49,82].

4.1. Energy Depositions in Molecular Targets

Although local energy depositions can be experimentally measured over subcellular distances of nearly 1-μm diameter, biophysical theories require knowledge of energy deposition in small targets of the dimensions of DNA molecules. To this end, Monte Carlo track structure methods allow calculations of energy depositions and DNA damage at resolutions down to subnanometer dimensions (see, e.g., Chap. 4). To obtain the distribution of energy deposition in molecular targets, the conceptual framework of the method is to place the track in a virtual volume big enough to contain the entire track and then randomly place the target structure in the volume. The method ensures the establishment of an electronic equilibrium and accurate calculation of absorbed dose in the target. In this way, the energy deposited in the target can be grouped together in slices of 0.1-nm length to give the total energy in the target. Fig. 2 shows examples of such distributions in three volumes pertaining to a segment of DNA, a nucleosome, and a chromatin fiber segment for different radiation qualities. The left ordinate shows comparison of distributions of absolute frequency of energy depositions in the target, randomly positioned and oriented in water and irradiated with 1 Gy of the given radiation. The right ordinate gives the corresponding average number of events in the target volume in one typical mammalian cell. Subsequently, more complex forms of DNA structures could be used to obtain more detailed physical and chemical information on the nature of damage from the distributions of energy depositions.

4.2. Summary

Frequency of energy deposition in volumes of dimensions similar to DNA and other biological molecules provides a method of interpretation of mechanism of radiation effects. In this method, the assumption is that radiation quality effects are mostly because of the spatial properties of the radiation on a microscopic scale. The database of absolute frequencies of energy depositions in cylindrical targets of dimensions 1–100 nm include:

* Electrons—10 eV to 100 keV [50].
* Ultrasoft x-rays (C, Al, Ti, and Cu) [51].
* Alpha particles and protons 0.3–4 MeV/u (track segments) [52].
* Protons 1-26 keV (full-slowing-down-tracks) [53].
* Alpha particles 2 keV-1 MeV (full-slowing-down-tracks) [54].

Table 4 provides a summary of selected particles, energies, and microdosimetric parameters available from the monographs. All data are taken from Refs. 50–55. In Table 4, $f(>0)$ is the probability of hit of any size = $\bar{z}_F^{-1}$, $\bar{z}_F$, and $\bar{z}_D$ are the frequency- and dose-averaged mean-specific energies.

Figure 2 Frequency distributions of energy deposition in a target corresponding approximately in size to a segment of DNA (2.3-nm diameter, 4.6-nm length), a nucleosome (10-nm diameter × 5-nm width), and a segment of chromatin fiber (30 × 30 nm), in water irradiated with soft x-ray, protons, and alpha particles. The left ordinate gives the absolute frequency $f(>E)$ of deposition events greater than the energy E (eV) in the target volume when randomly placed in water uniformly irradiated with 1 Gy of the given radiation. The right ordinate is the corresponding average number of events in such targets in a typical mammalian cell, using the factors in Table 2. The frequency of hits of any size is given by $f(>0) = 1/\bar{z}_F$ and the number of events, corresponding to the frequency $f(>E)$, is obtained from $N = [f(>E)/f(>0)]M$, where M is the total number of hits.

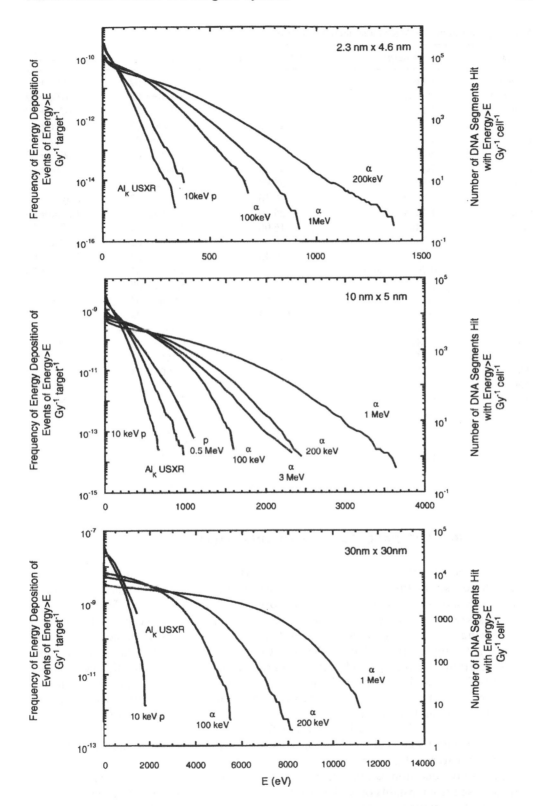

Table 4 Summary of Microdosimetry Parameters

Particle	Energy	LET (keV/μm)	$\bar{z}_F$ (kGy)	$\bar{z}_D$ (kGy)
Electron	1 keV	—	8.300e + 02	1.500e + 03
Electron	100 keV	—	6.300e + 02	1.310e + 03
Proton	1 keV	8.97	3.255e + 02	8.071e + 02
Proton	4 keV	25.75	4.309e + 02	1.007e + 03
Proton	10 keV	53.19	5.934e + 02	1.322e + 03
Proton	20 keV	79.56	7.496e + 02	1.605e + 03
Proton	26 keV	91.20	8.016e + 02	1.714e + 03
Proton	300 keV	58.90	1.014e + 03	cGy
Proton	500 keV	40.10	8.596e + 02	cGy
Proton	1 MeV	25.80	7.518e + 02	1.580e + 03
Proton	2 MeV	16.00	7.063e + 02	1.390e + 03
Proton	4 MeV	8.8	6.528e + 02	1.300e + 03
Proton	200 MeV	0.45	1.100e + 02	1.100e + 03
Deutron	0.79 MeV/amu	30.656	8.107e + 02	1.476e + 03
^{4}He	2 keV	—	5.615e + 02	1.191e + 03
^{4}He	10 keV	39.52	1.064e + 03	1.981e + 03
^{4}He	100 keV	125.94	1.575e + 03	3.478e + 03
^{4}He	200 keV	176.57	1.655e + 03	3.983e + 03
^{4}He	1 MeV	—	1.690e + 03	4.719e + 03
^{4}He	2 MeV	167.2	1.356e + 03	—
^{4}He	4 MeV	103	1.090e + 03	2.260e + 03
^{4}He	8 MeV	61.4	9.300e + 02	1.800e + 03
^{4}He	16 MeV	40.3	7.800e + 02	1.550e + 03
^{4}He	20 MeV	32.2	7.200e + 02	1.480e + 03
^{4}He	8.89 MeV/amu	21.5	6.925e + 02	1.223e + 03
C_x	278 eV	—	1.053e + 03	—
Al_x	1487 eV	—	9.067e + 02	—
Ti_x	4509 eV	—	7.168e + 02	—

5. MODELING AND CALCULATIONS OF DNA DAMAGE

Two recent papers provide comprehensive summaries and references to the experimental data of double strand breaks (DSB) [56,57]. Experimental data of DSB in mammalian cells using PFGE shows that for a given radiation dose, the relative biological effectiveness of all radiations is nearly unity [58,59,152]. Such observation begs to ask some questions including the following: Is cell sensitivity to radiation damage due to biological processing of initial damage? Is the initial critical damage the same for both high and low linear energy transfer (LET) radiations? Do biological effectiveness of radiations correlate with differences in track structure of high- and low-LET radiations? Do biological differences correlate with temporal differences of radiations? Is there a correlation between the cell survival and the numzber of double strand breaks? Is there a correlation between the cell survival (or other end points such as aberration, mutation, and transformation) and the quality of double strand breaks? Is the effectiveness of low doses of radiation predominantly determined by the local microscopic features of individual tracks?

Not all the question can be categorically answered. However, our understanding has been increased from insights offered by modeling and simulation work. In general, it is believed that there exists a general correlation between the nature of the initial physical features of the track and the possible final biological lesions [60]; it is believed that damage of DNA is a key step in the sequence of events that leads to, e.g., radiation-induced cell death [61–65]; it is now widely believed that the initiating events are mostly because of DNA double strand breaks [66]; it is believed that critical initial damages are mainly produced by localized clusters of DNA damage [67,28]; and it is believed, in general, that a linear relationship exists between dose and number of DSB, indicating double strand breaks are formed as the result of a single track interaction [68].

Two approaches to modeling and calculation of strand breaks in general and yield of double strand breaks in particular have been used. One method is fitting various phenomenological parameters for particle tracks or energy depositions to measured double strand breaks data [69–71], and the other is the application of mechanistic models to cellular DNA damage. Application of mechanistic models has become a reality through the advances in particle track simulation [60,72–74], availability of fast computers, and realistic models of DNA in simple and sophisticated forms. These mechanistic biophysical descriptions have progressed from considerations of direct effects of radiation alone [30], and contribution of radical species [75–79] in the environment of the cell nucleus to cause DNA damage, in the form of single strand breaks, double strand breaks, base damage, and complex combinations within the cluster of damage [80 81], to include cellular processes such as DNA repair [82] and its consequences [83,84].

5.1. Classification of DNA Damage

Two alternative classification schemes have been used for DNA breaks (Fig. 3) [79]. The scheme considers damage in DNA as: (A) patterns of breaks according to the complexity of damage (single, double, and their major combinations) on one or both strands and in (B) the damage has been classified according to the origin of the breaks either from direct energy deposition or reactions of OH radicals. In these calculations, it is assumed that the yield of damage from other radical species is negligible. The top and bottom lines represent a two-dimensional projection of the double-helix sugar–phosphate backbone and the two middle dashed lines represent the individual bases. Vertically, the two center dashes and the corresponding sugar–phosphates on the top and bottom strands form a two-dimensional projection of a nucleotide base pair. The maximum distance between two single strand breaks on complementary strands that are classed as producing a double strand break was set at 10 base pairs [85].

5.2. Parameters of DNA Damage Modeling

In "jump-the-detail" method, two principle parameters need to be considered. The first one is the quantity of energy, or the threshold energy, required to induce a DNA strand break. A major problem in quantification of energy required for induction of a single strand break arises from variability of the total yield of strand breaks in different cell lines under different conditions, with different types of radiations, and measured with different techniques [56, 57]. However, measurements of single strand breaks that are more relevant to the energetic of DNA damage are less available. Experimentally, such measurements are much more demanding and where these exist, such measurements have been carried out under dissimilar conditions unsuitable for comparisons [56,86]. In general, although experimental yield of strand breaks provide a reliable test for modeling, they do not provide a means to derive accurate and quantitative values of parameters needed in biophysical modeling. For the purposes of modeling, we seek information from experiments under controlled conditions with a resolution of a single base pair. Experiments with the latter criteria can mainly be

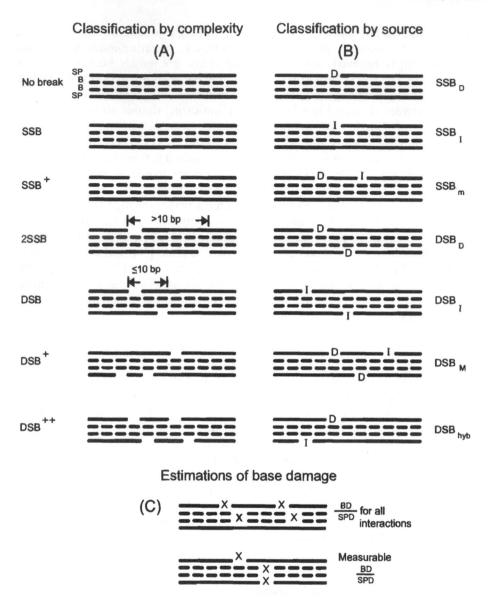

Figure 3 Various possible types of damage in the sugar–phosphate of the DNA induced by direct energy deposition or by diffusing hydroxyl radicals. Only one type of damage is assigned to each DNA segment. On the left, DNA damage is shown by an ×. Combination of one, two, three, or greater than three single strand breaks on the same strand or on opposite strands have been assigned as SSB, SSB+, 2SSB, DSB, DSB+, and DSB+. The seven modes of strand breaks can also be combined with base damages on one or both base moieties (diagrams not shown). The model on the right-hand side explicitly show the origin of the strand break × whether arising from a direct energy deposition "D" or from an OH radical "I."

found among those using cells labeled with a suitable radionuclide, such as Auger emitting ^{125}I, rather than external radiations [88,89].

In modeling of DNA damage, various authors have adopted a number of simplifying and convenient assumptions. These assumptions allow jump-the-details of a full-scale molecular analysis of the formation of strand breaks by a particle track. Such detailed analysis at the level of genome is beyond the capability of present theoretical chemistry and molecular dynamic systems [90]. The simplifying assumptions in jump-the-detail include using probabilities associated with the induction of single and double strand breaks by direct energy depositions and the reaction of hydroxyl radicals with the nucleobases. Estimation of energy required for induction of single strand breaks, by direct energy depositions, could be divided into three categories: those calculations dealing with the stochastic of energy deposition in the volume of DNA [68]; calculations using physical parameters such as oscillator strength of the target (e.g., DNA) and collective excitation [91]; and threshold energy derived from simulation of specific experiments for induction of single strand breaks with a single base pair resolution information [92,93]. Although all three categories listed above suggest a range of values for induction of SSB and/or DSB, intrinsically, they do not provide elucidation of the biochemical mechanism leading to DNA damage/strand breaks. Of the three categories, the first one is primarily based on microdosimetry and track structure calculations of energy depositions in target volumes pertaining to DNA or higher-order structures. In such calculations, one usually seeks quantities of energy depositions and spatial dimensions per unit dose of ionizing radiation as a function of energy deposited in the volume under consideration by a single particle track that do, or do not, correlate with the observed biological effectiveness. Correlation of local energy depositions, in terms of their clustering properties, have revealed that typical energy depositions of the order of tens of electron volts, in DNA-size targets, predominantly lead to single strand type breaks, and damage from energy depositions of the order of 100 eV is predominantly of double strand break type [68].

An alternative way of using jump-the-detail provided by the Monte Carlo track structure approach is to examine the way in which energy is deposited in the volume of reaction of sugar–phosphate, purine, and pyrimidine bases [97]. To convert these energy depositions into damage requires an understanding of the relationship between track events (ionizations, excitations, and radical products) and their actions on DNA. To establish such a coupling, modeling of an experiment in which single strand breaks induced by the decay of ^{125}I incorporated into the plasmid DNA was originally carried out by Charlton and Humm [92]. It was found, assuming an energy deposition greater than 17.5 eV in the sugar–phosphate moiety induced a single strand break (SSB), that the experimental distribution of SSB could be reproduced [89,92,93]. The distribution of energy depositions along the DNA could now be converted into distributions of SSBs and DSBs. Use of such a technique has provided reasonable agreement with experimental data. Similar jump-the-detail modeling has also been carried out by other workers using different values for the threshold energy needed for induction of a SSB. These include a single ionization [94], energy deposition greater than ionization threshold 10.5 eV [39], 10–13 eV energy deposition [77], collective excitation energy 21.6 eV [95], and DNA dipole oscillator strength 30 eV [91]. It should not be surprising that all the values listed above, used in the respective modeling work, produce agreement with the experimental values of total yield of strand breaks. However, such overall agreements may not necessarily provide a unique description of the spectrum and complexity of DNA damage produced in the hit region of the genome. The spectrum of DNA damage produced by radiation track becomes crucially important when one is interested in kinetic modeling of DNA damage and repair and relating results of track structure calculations for different radiations and doses to corresponding observed biological effects.

The second parameter in jump-the-detail method is the efficiency of the reaction of hydroxyl radical with DNA leading to strand breakage. Hydroxyl radicals are important oxygen species, which upon reaction with DNA produce strand break and base damage, and their contribution to the total yield of damage is a function of the hydration level surrounding the DNA. The more dilute the solution, the longer the lifetime of the radicals. The environment of the cell contains high levels of scavengers that reduce the number of radicals available for reaction with DNA. In a cell mimetic condition, a diffusion distance of 4–6 nm has been used [79]. Only a few quantitative experimental reports are available that provide values for the probability of induction of DNA strand breaks by OH radicals, as not all radicals lead to strand breaks [96,98]. Based on the findings of experimental work the probability of SSB formation per OH radical has been considered to be 0.12 ± 0.01 upon reaction with DNA [79,99]. It is usually considered that the efficiencies of other radical species to produce a strand break are negligible. In a number of studies, probability of OH radical for induction of SSB has been assumed as unity [39,100,101].

An alternative method to investigate DNA strand breakage by OH radicals considers the surface accessibility of hydrogen atoms of the DNA backbone [102]. The solvent accessibility is ~ 80% for the sugar–phosphates and ~20% for the bases. This method allows a more direct determination of reaction of OH radicals with the individual deoxyribose hydrogens [103,104]. Recent studies show trends in reactivity of OH radicals closely follow the accessibility of the solvent to various deoxyribose hydrogens [105,106].

Production of strand breaks by very low energy electrons (5–25 eV) in thin solid DNA films using ultrahigh vacuum systems have been reported in a number of studies [107–109]. Such studies have demonstrated the efficiencies of low energy electrons and photons to induce DNA damage. In the vacuum ultraviolet (UV) region, examination of experimental data [86,110,111] shows that the induction of strand breaks depends on the absorption spectrum of the components in the medium and the sensitivity spectrum of DNA [112]. Introduction of a variable with the wavelength for the induction of SSB by OH radicals, in conjunction with a fixed value for the quantum efficiency for the production of OH radical ($\varphi_{OH} = 0.04$), allows an interpretation of the vacuum UV sensitivity spectrum for induction of SSB in aqueous system [112].

5.3. Base Damage

The measurement and quantification of base damage depends on environmental conditions, cell type, and the technique used. For the purpose of modeling, an experimentally related quantity is the ratio of the yield of base radicals to sugar–phosphate damage. A preliminary estimate of the ratio of base to sugar–phosphate damage can be calculated either by independently counting all hits that lead to damage in bases or sugar–phosphates, or by counting those segments of DNA that contain at least one base damage and two strand breaks, one on each strand within ≤10 bp (Fig. 3C). The latter ratio should be comparable with the experimentally measurable quantity. In the absence of rigorous experimental data on energetic of base damage, the preliminary calculations of the yield of base damage have been obtained by considering a similar kinetic as those used for the production of single strand breaks. A probability of 0.8 was assumed for induction of base damage from the reaction of OH radicals with the nucleobases. For direct damage, recent experimental data by Abdoul Carime et al. [113] and modeling by Watanabe and Nikjoo [76] show a threshold energy of ~10 eV for the activation of a base moiety for direct energy deposition to produce an electron adduct or the protonated form of the base.

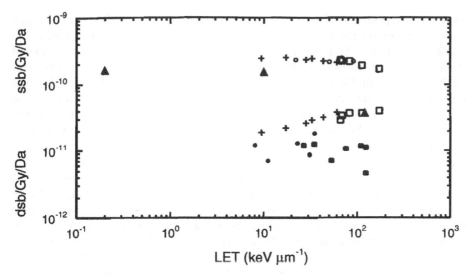

Figure 4 Yield of single and double strand breaks as a function of LET for ions.

5.4. Summary

Current jump-the-detail technique of modeling DNA damage requires at least two essential parameters: (1) energy for direct induction of SSB, which has been determined from modeling a well-controlled experiment with internal Auger emitters, and (2) probability for induction of a SSB by reaction of an OH radical with DNA—a value of 0.12±0.01 was adopted, based on experimental findings in a cell mimetic condition. Our data suggests a threshold value of 17.5 ± 2 eV as the most probable value for the energy deposited in the volume of a sugar–phosphate by a single track for the induction of SSB by direct mechanism. Based on these assumptions, Fig. 4 shows the yields of SSB and DSB as the function of LET and in comparison with experimental observations. A wide range of energies, from 10 to 30 eV, have been used in various modeling works. Such a broad range on the choice of the threshold energy for induction of a SSB emanates from the type of experimental data used for normalization of calculated data and differences in track structure codes. For estimation of base damage, similar kinetics as those for induction of SSB have been used, viz, 17.5 eV for induction of base damage by direct energy deposition and a probability of 0.8 from reaction of OH radicals, but in the light of more recent reports [76,113], these values may have to be revised.

6. SPECTRUM OF DNA DAMAGE

Earlier in this chapter, a number of questions were considered on correlation between biological effectiveness of ionizing radiation and severity or the quality of damage. In this respect, the hypothesis put forward in the 1980s by two of its protagonists [15,16] articulated that the quality of damage by ionizing radiation is a major factor in determining the survival and mutagenecity of the cell. Evidence in support of the hypothesis has come mainly from modeling and calculations [32,68,114,115].

In essence, the hypotheses "Clustered Damage" and the "Locally Multiply Damaged Sites" attempt to correlate the frequencies of various types of DNA damage with the observed biological lesions. The approaches were originally based on the classification of damage either by energy deposition or by number of ionizations and reactions of hydroxyl radicals in DNA. Brenner and Ward [116], using a track structure approach, concluded that yields of clusters of multiple ionizations within 2–3 nm sites correlate well with observed yields of double strand breaks. Goodhead [28,114], using the database of frequencies of energy depositions in small volumes of dimensions similar to important biological structures for a range of radiations, deduced a classification based on the size of energy deposition in the target, which could correlate with particular biological effects. The classification includes four classes of initial physical damage in terms of sparse energy depositions of a few tens of electron volts in DNA producing simple damage predominantly of single strand type, moderate energy depositions of nearly 100 eV in DNA producing damage predominantly of double strand break type, large energy depositions of the order of 400 eV in nucleosome DNA producing multiple DSB and protein cross-link, and very large clusters of ~ 800 eV pertaining to chromatin size targets producing large deletion and gross damage. With the emergence of more sophisticated techniques in modeling and quantification of damage in DNA, it is now possible to estimate the spectrum of DNA damage in terms of complexity of damage for low and high-LET radiations [32,75,79,81,87,91,117–122].

Simple experimental approaches to this problem recently started postulate that the repair of clustered DNA damage leads to conversion of nonlethal lesions, e.g., dihydrothymine, or mutagenic lesions, such as 8-OxoGuanine, into lethal double strand breaks. These early experiments have studied kinetics and influence of excision of base lesion within clustered DNA damage by *E. coli* and nuclear extracts [27,123–129].

For a graphical and quantitative demonstration of some of the ideas expressed above, data are presented for DNA damage induced by 3.2 MeV alpha particles. Patterns of energy depositions in individual DNA segments were analyzed for induction of strand breaks and base damage using an atomistic model of B-DNA. Fig. 5 shows a few examples of the hit region of DNA by tracks of 3.2 MeV alpha particles. The "×" indicates strand-break induction by direct energy deposition. The numerals/symbols indicate the damage by direct energy depositions that did not lead to strand break. "H"s are the sites of reaction of OH radical with the DNA leading to strand break. Track structure calculations have shown that a substantial proportion of the dose, ~ 30% from low-LET irradiation, is deposited by low-energy secondary electrons [115] that are efficient in producing clustered damage at nanometer scale [68,115]. Fig. 6 shows the frequency of complex double strand breaks. The data shows with increasing LET the proportion of complex double strand breaks increases reaching to more than 70%. The data presented in Fig. 6 do not include the contribution of base damage. On the assumption that base damage more frequently occurs than strand breaks (Table 1), it is likely that a high proportion of double strand breaks contain additional base damage.

Table 5 provides calculation of strand breakage with the additional information on the number of base damages for 3.2 MeV alpha particles. The left-hand side of Table 5 presents frequencies of strand breaks containing none or at least one or more base damage on purine or pyrimidine bases. The right-hand side of the table shows a more detailed analysis of the frequencies of breaks in italic-bold. For example, 12.5% of 58% DNA segments with "no strand breaks" contains two base damages. Of these, 3.6% (1.7 + 1.9) are located within 3 bp of each other, while 8.9% (4.3 + 4.3) are located at distances >3 bp from each other.

SCORING IN 23 Å DNA FROM 3.2MeV α-PARTICLES

```
.......-.......................................     (base damage only)
H.....H....H...H.H.............................
............x..-....H.........................
..............................................
..............................................     (2ssb)
................................x.............
............H3................................
.........HH.........H..........x....H.........
H.............................................
.........-.......x1-.........-x...............     (dsb++)
.........-......H.H-........H-................
H.........H......1.......H........1...........
...............x.........H....................
.......H......................................     (dsb+)
......H....H..................................
....H..H......................................
1.......x......H..............................
...............................1......1.x..-.......     (dsb)
H..............H.-......H....H....-...........
..................1...........................
...............-............x1........H.......
.......1........1.............................     (no strand break)
.......H......HH..............................
H..H..............H...........................
..............................................
..............................................     (ssb+)
..............H..............................
..HH....HH...................................
H....x........................................
.............x................................     (ssb)
H.............................................
...........H..................................
..............................................
```

Figure 5 Individual examples of simulated sites of damage induced by 3.2 MeV alpha particles in DNA. In each example, the outer and inner rows represent the sugar–phosphate moieties and the pairs of bases, respectively, with single base pair resolution (dots). An × or "H" represent energy deposition or reaction of hydroxyl radical leading to induction of a single strand break or base damage. A "–" indicates hit sites that did not lead to strand breaks (SB) or base damage (BD). Nomenclature: no strand break (No SB); single strand break (SSB), (SSB$^+$), (2SSB); double strand break (DSB), (DSB$^+$), (DSB$^+$).

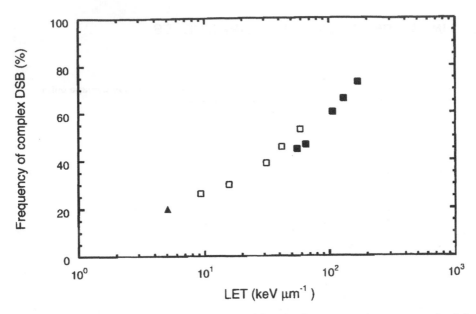

Figure 6 Fraction of complex double strand breaks for protons (open squares), alpha particles (solid squares) as a function of LET, and electrons (triangle). The contribution of base damage is not included in these calculations.

6.1. Summary

Complex damage presents an increased challenge to the repair processes operating in the cell and is postulated to be largely responsible for the greater biological effectiveness of high-LET radiations.

- Experiments have shown slower repair of double strand breaks by high-LET radiations [66,152,226].
- Modeling and calculations show that with increasing LET, there is a shift toward a greater complexity of clustered damage in DNA [30,81,87,119].

Table 5 Frequency Distribution of a Number of Base Damage

		3.2 MeV alpha particles										
		No. of Base						BD≤3 bp		BD>3 bp		
Damage		0	1	2	3	4	≥5	adj	opp	adj	opp	Total Hits
Model 1	No break	—	31.4	12.5	6.9	3.7	3.5	1.7	1.9	4.3	4.6	58.0
Model 2	SSB	5.1	4.8	4.0	4.8	2.6	2.6	1.1	0.9	1.3	1.6	23.9
Model 3	SSB+	0.1	0.5	1.3	1.0	1.0	2.4	0.1	0.1	0.0	0.3	6.3
Model 4	2SSB	0.2	0.1	0.1	0.2	0.3	0.8	0.0	0.1	0.0	0.0	1.7
Model 5	DSB	0.2	0.6	0.7	1.1	1.0	1.7					5.3
Model 6	DSB+	0.1	0.1	0.2	0.6	1.0	1.7					3.6
Model 7	DSB++	0.0	0.1	0.1	0.0	0.2	0.8					1.2

Ratio of base to sugar/phosphate damage = 3.4.

- Inclusion of base damage substantially increase the proportion complex damages (both SSB and DSB) [80,81,87].
- Experiments as well as calculations show the yield of strand breaks per unit dose is nearly constant over a wide range of LETs [75,80,81].
- The majority of strand breaks is of simple type containing a single strand break; however, if base damage is taken into account, the majority of single strand breaks appears to be complex [81].
- For low-LET radiations, nearly 30% of dose is deposited by low-energy secondary electrons [115].
- Low-energy electrons are efficient in producing clustered damage at DNA level [115].
- These clusters occur with high frequencies per unit dose [68].
- A high proportion of double strand breaks ($\sim$ 20%) produced by low-LET radiation are of complex origin [79–81].
- A significant proportion of non-double strand breaks are of complex type (Table 5).
- Including the base damage, $\sim$ 50% of all double strand breaks contain additional breaks [79–81].
- For high-LET radiations, per unit dose fewer regions of genomes are hit, i.e., smaller number of tracks per unit dose (Table 1).
- For high-LET radiations, reaching $\sim$70% of all double strand breaks contain additional strand breaks, if base damage is included, this proportion increases to $\sim$90% [79,81].

7. RADIOPROBING OF NOVEL DNA STRUCTURES

Since the determination of the double-helical structure based on x-ray fiber data and chemical modeling [130,131], considerable effort has been made to obtain an ever more detailed picture of the conformation(s) of DNA alone and in complexes with drugs or protein. The x-ray crystallographic analysis of mono- and dinucleotides leads to the atomistic determination of the structure of the bases, sugar, and base pairs, while x-ray fiber diffraction data have been used to determine the overall shape of the various helical families [132,133]. Other techniques for probing DNA structures include enzymes and chemicals [134]. Crystallographic studies need production of large and well-defined crystals that are not always possible. Enzymes and chemical methods have been used to study the local structural perturbation in DNA from local reactivity to enzyme cleavage and chemical modification. Enzymatic probes are mostly nucleases that probe accessibility of the enzyme at each phosphodiester bond. There is a wide range of chemical probes available for the study of DNA structures in the neighborhood of the bound drug molecule, proteins, and base mismatches. Rydberg et al. [40] reported a new method for probing the conformation of chromatin in living cells. They used x-ray and accelerated heavy ions, and from the characteristics of patterns of radiation-induced strand breaks the authors proposed a zigzag model of chromatin structure in eukaryotic cells. In addition to stable structures of DNA, there is an important class of DNAs that cannot easily be obtained in a single crystal such as three- and four-stranded structures. This section provides a short description of a new method for the study of DNA structures such as triplex and quadruplex DNAs using Auger electrons as a probing tool [41].

Auger electrons have extensively been used as a tool in molecular and biophysical aspects of radiation action at DNA and cell level. Historically, in fact, although Auger electrons was discovered in 1925 by Pierre Auger, it was not long after that two French biologists [135] suggested its use as a magic bullet in radiation therapy.

The ability of nucleic acids to form triple helical structures may play an important role in mechanisms of DNA recombination and gene transcription, and also represents an important new approach to sequence-specific drug targeting [136,137]. Although triple helices have been known for over 40 years [138], details of their molecular structure still remain elusive. Early models of triple helices, based on fiber diffraction data, suggested an A-form conformation [139], while more recent nuclear magnetic resonance (NMR) and modeling studies support the view that these structures have more in common with the B-form of duplex DNA [136,140]. In this study, we used a computational method to simulate the damage to a triplex DNA resulting from Auger electrons emitted through the decay of an incorporated ^{125}I. In comparing theoretical frequency distributions of single strand breaks with the experimental data [141], the results are very sensitive to the conformation of the triplex model used. We find that the best fit to the experimental data results from using a hybrid triplex model, in which the base-step geometry is A-like, while the sugar puckers adopt the B-like C'_2-endo conformation.

Radionuclide such as ^{125}I decay by emitting a dense cascade of electrons (Auger electrons), most of which are short range with energies less than 1 keV [142]. The release of these electrons results in a highly charged residual atom and the deposition of a large amount of energy near the site of decay. In recent years, it has become possible to position Auger emitting radionuclide at specific locations along a defined DNA target molecule [143–146]. This has been carried out either by direct incorporation of, e.g., iodine-labeled deoxynucleotide, or by binding of a ^{125}I-labeled sequence-selective DNA ligand. After allowing the accumulation of decay in the target, resulting in DNA strand breaks and base damage, DNA sequencing techniques have been used to locate the positions of strand breaks relative to the site of decay with a precision of a single nucleotide base. Previously, we and others have developed Monte Carlo track structure methods to simulate the process of DNA damage [92,93,100,147] and obtained frequency distributions of strand breaks induced in plasmid DNA labeled with an Auger emitting radionuclide that is in good agreement with experimental data. These calculations have shown that most damage occurs directly because of the emitted electrons within about 10 base pairs either side of the decay site, while the damage caused by hydroxyl radicals generated in the water surrounding the DNA is mainly long range, only making a minor contribution to DNA damage near the site of decay [93, 148]. These results prompted us to consider the approach as a method for structure determination, or at least evaluation, in the neighborhood of the site of decay. Panyutin et al. [141,149–151] have published the results of experiments measuring the profile of DNA damage induced in the polypurine–polypyrimidine region of the *nef* gene of the human immunodeficiency virus through interaction with a triplex forming oligonucleotide labeled with ^{125}I. The experiment [141] was the first attempt to use the antigene approach to deliver radio nucleotides to a target DNA and so produce sequence-specific DNA breaks. Nikjoo et al. [41] have reported on the modeling of this experiment and the sensitivity of the results to structural parameters. As a result of comparing calculated frequency distributions of fragment lengths with those obtained from the experiment, it was concluded that currently accepted models for triplex structure are not optimal, and propose a modified structure that better fits the radioprobing results, while maintaining agreement with the fiber diffraction and NMR data.

8. RADIATION BYSTANDER EFFECT

Radiation-induced genomic instability and bystander effects are now well-established consequences of exposure of living cells to ionizing radiation. Cells not directly traversed by radiation may still exhibit radiation effects. This phenomenon, known as bystander effect, has become a major activity in radiation biology and in some cases has challenged the conventional wisdom. An example is the currently accepted models used for low-dose extrapolation of radiation risks. The currently used models assume that cells in an irradiated population respond individually rather than collectively. If bystander effects have implications for health risks estimates from exposure to ionizing radiation, then the question of whether this is a general phenomenon or solely a characteristic of a particular type of cell and the radiation under test becomes an important issue.

Dose–response curve for all solid cancers based on atomic bomb survivors data indicate a linear form down to doses as low as 50 mSv. However, risk estimates for radiation-induced oncogenesis at very low doses (< 50 mSv), at a region where direct experimental observations are not available, are usually extrapolated from high doses of ionizing radiation using a linear nonthreshold model [14,153]. The validity of this approach is still a subject of investigation and discussion. There are a number of factors affecting the shape of dose–response curve at low doses of irradiation including low-dose hyper-radiosensitivity [154], adaptive dose response [155], genomic instability [156,157], and more recently bystander effect [9]. The latter is considered as one phenomenon that may be of significant importance in influencing the shape of dose–response curves.

The bystander effect has been observed for a variety of biological end points such as cell survival [158 159], mutation [160–162], sister chromatid exchanges [163], cell transformation [164,165], micronucleated and apoptosis [166], gene expression [167], and radiation genomic instability [168–170].

Theoretical and modeling works have also been recently published [171–173]. The model of Brenner et al. [171] is based on a binary phenomenon in a small sensitive subpopulation of cells. Their model suggested that the bystander effect is important only at small doses. The papers published by Nikjoo and Khvostunov [172,173] investigated whether the bystander effect depends on the mode of irradiation (broad beam or microbeam systems), hypothesizing a particular mechanism of cell-to-cell communication assuming the signal has a protein-like nature, and is communicated by diffusion in the medium. The authors investigated the bystander phenomenon for the broad beam and microbeam irradiation systems for cell inactivation and oncogenic cell transformation in C3H10T1/2 cells. The general conditions for the bystander effect to dominate were formulated based on the fraction of hit and nonhit cells. The model shows the bystander phenomenon is the dominant factor at low doses in the case of the broad beam irradiation, while for the microbeam irradiation both at low and high dose.

9. TRACK STRUCTURE SIMULATION

This section provides a brief description of theoretical bases of the Monte Carlo track simulation codes we have developed for electrons and ions. Our database of Monte Carlo track simulation codes include: electrons (code *kurbuc* −10 eV to 10 MeV) [174], protons (*lephist* −1 keV to 1 MeV) [175], alpha particles (*leahist* −1 keV to 8 MeV) [176], all ions (pits 0.3 MeV/u to GeV) [74] and the code *chemkurbuc* for description of prechemical and chemical stages of electron track in liquid water (unpublished).

A complete model of track structure consists of descriptions for inelastic and elastic interactions. As charged particles pass through matter, they lose energy primarily through collisions with bound electrons. Ionization cross sections for all projectile and secondary electron energies are needed to follow the history of an incident particle and its products, covering all range of energies transferred in individual collisions. Elastic interactions take place when the incoming particle interacts with the atomic field and is deviated from its path by a small angle without the loss of energy.

Not all cross sections are available for materials of interest in radiation biology. Cross-section data for liquid water are scarce, as measurements are either impractical or very difficult. For practical reasons, we use water vapor cross sections for total and partial ionization and excitation cross sections. In general, a Monte Carlo track structure code requires at least eight sets of cross sections: (1) total ionization, (2) total excitation, (3) total elastic scattering, (4) partial ionization cross sections for five electron orbitals of water molecule, (5) partial excitation cross sections for various excited states, (6) secondary electron energy spectrum, (7) angular distribution of elastic scattering, and (8) angular distribution of secondary electrons.

9.1. Electrons

Ionization Cross Sections

Inelastic cross sections for ionizations and excitations were compiled for low- and high-energy electrons. The experimental ionization cross sections for water vapor in the energy range of 10 eV to 10 keV [177–181] were least-squares fitted as shown by a solid line in Fig. 7 using a model function:

$$\sigma_{exp} = c + at^r \, e^{-(b_1 t + b_2 t^2)} \tag{1}$$

$$t = \ln\frac{T}{15}$$

in which the unit of σ is 10^{-16} cm^2 and that of T, the particle kinetic energy, is eV. The fitting parameters are given by

$$a = 2.07201 \quad b_1 = 0.271302 \quad b_2 = 0.119638 \quad r = 1.46521 \quad c = 0.074$$

At energies higher than 10 keV, ionization cross sections were calculated using Seltzer's formula [183]. Seltzer's formula gives the partial ionization cross sections for five molecular orbitals of water. The total ionization cross sections in the energy range of 10 keV to 10 MeV were obtained by summing up all the partial cross sections (Fig. 8).

Excitation Cross Sections

There are various modes of excitation for $T > 7.4$ eV. As reported experimental excitation cross sections are fragmentary, we used the compiled data of Paretzke [184] that includes all the major excitation modes. Paretzke fitted the experimental cross sections for the 10 major individual states using a model function:

$$\sigma_{exc} = \Phi_3(T)\frac{4\pi a_0^2 R}{T} M_a^2 \ln\frac{4c_s T}{R} \tag{2}$$

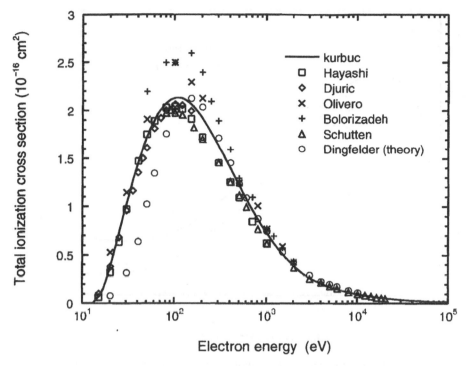

Figure 7   Experimental total ionization cross section as a function of electron energy. The solid line is the least-squares fit to all data [177–181]. The calculated data of Dingfelder et al. [182] for liquid water is plotted for comparison.

in which $\Phi_3(T)$ is an empirical correction factor of the form

$$\Phi_3(T) = 1 - e^{-0.25(T/E_a - 1)} \tag{3}$$

where a_0 is the Bohr radius $= 0.5293 \times 10^{-8}$ cm, and R is the Rydberg constant $= 13.6$ eV. Parameters M_a^2, E_a, and c_s for each level are given by Paretzke. An additional three levels were fitted by the analytical model given by Gren and Stolarski [185]. Table 6 lists eight major excitations and excitation energies considered in this work. The total excitation cross section in the energy range of 10 eV to 10 keV was obtained by summing up all the individual cross sections. The total excitation cross section for the high-energy region up to 10 MeV is given by an empirical formula derived from the Fano plot, with fitting parameters determined by Berger and Wang [186]:

$$\sigma_{exc} = 4\pi \left(\frac{a_0}{137\beta}\right)^2 \left[12.30 + 1.26\left\{\ln\frac{\beta^2}{1 - \beta^2} - \beta^2\right\}\right] \tag{4}$$

in which β is the ratio of electron velocity to the velocity of light. Total excitation cross section in the energy range of 10 eV to 10 MeV is shown in Fig. 8.

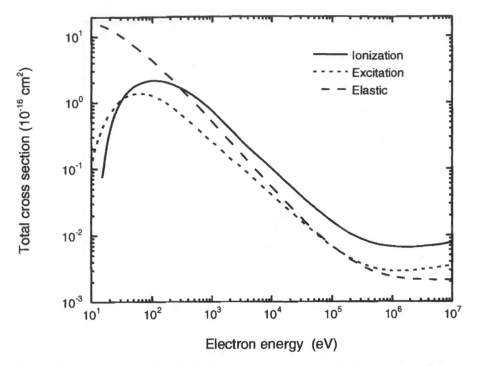

Figure 8 Total cross sections for ionization, excitation, and elastic scattering of electrons in water vapor in the energy range of 10 eV to 10 MeV.

Elastic Scattering Cross Sections

Elastic scattering cross sections were calculated using the Rutherford formula, taking into account the screening parameter given by Moliere [187]. The differential cross section $(d\sigma/d\Omega)_{el}$ and total cross sections σ_{el} for each molecule are represented by

$$\left(\frac{d\sigma}{d\Omega}\right)_{el} = Z(Z+1)r_e^2 \frac{1-\beta^2}{\beta^4} \frac{1}{(1-\cos\theta+2\eta)^2} \tag{5}$$

$$\sigma_{el} = \pi Z(Z+1)r_e^2 \frac{1-\beta^2}{\beta^4} \frac{1}{\eta(\eta+1)} \tag{6}$$

Table 6 Excitations and Transition Energy Considered in the *Kurbuc* Code

Excitation state	Energy (eV)
A^1B_1	7.4
B^1A_1	9.7
Diffuse band	13.3
Rydberg $(A+B)$	10.0
Rydberg $(C+D)$	11.0
H* Lyman α	21.0
H* Balmer α	21.0
OH*	9.0

The screening parameter η is given by

$$\eta = \eta_c \times 1.7 \times 10^{-5} Z^{2/3} \frac{1}{\tau(\tau+2)} \tag{7}$$

where the effective atomic number of the water molecule, Z, was assumed to be 7.42, r_e is the classical electron radius, $\tau = T/m_0 c^2$ is the kinetic energy in units of the electron rest mass.

$$\eta_c = \begin{cases} = 1.198 & T < 50\,\text{keV} \tag{8a} \\ = 1.13 + 3.76\left(\frac{Z}{137\beta}\right)^2 & T \geq 50\,\text{keV} \tag{8b} \end{cases}$$

The parameter η_c was determined by the least-square fitting to the experimental data [188–191]. Total elastic scattering cross section in the energy range of 10 eV to 10 MeV is shown in Fig. 8. Angular distributions for elastic scattering below 1 keV were obtained by direct sampling of various experimental data [189,190,192].

Secondary Electrons

The calculation of the secondary electron spectrum was carried out using the method of Seltzer [183]. For the jth orbital of a molecule, the cross-section differential in kinetic energy w of the ejected electron is written as the sum of close and distant collisions.

$$\frac{d\sigma^{(j)}}{dw} = \frac{d\sigma_c^{(j)}}{dw} + \frac{d\sigma_d^{(j)}}{dw} \tag{9}$$

The first term describes collision between two electrons.

$$\frac{d\sigma_c^{(j)}}{dw} = \frac{2\pi r_e^2 m_0 c^2 n_j}{\beta^2} \frac{T}{T + B_j + U_j}$$
$$\times \left\{ \frac{1}{E^2} + \frac{1}{(T-w)^2} + \frac{1}{T^2}\left(\frac{\tau}{\tau+1}\right)^2 - \frac{1}{E(T-w)}\frac{2\tau+1}{(\tau+1)^2} + G_j \right\} \tag{10}$$

$$G_j = \frac{8U_j}{3\pi}\left[\frac{1}{E^3} + \frac{1}{(T-w)^3}\right]\left[\tan^{-1}\sqrt{y} + \frac{\sqrt{y}(y-1)}{(y+1)^2}\right] \tag{11}$$

in which n_j is the number of electrons in the orbital, B_j is the orbital binding energy, U_j is the mean kinetic energy of the target electron in the orbital, E is the energy transfer ($= w + B_j$) and $y = w/U_j$. Table 7 shows electron number, binding energy [193], and kinetic energy [194]

Table 7 Electron Number n_j, Binding Energy B_j, and Kinetic Energy U_j for the jth Orbital of a Water Molecule

j	Orbital	n_j	B_j (eV)	U_j (eV)
1	$1b_1$	2	12.62	61.91
2	$3a_1$	2	14.75	59.52
3	$1b_2$	2	18.51	48.36
4	$2a_1$	2	32.4	70.71
5	$1a_1$	2	539.7	1589.5

for the jth orbital of a water molecule. The second term is described in terms of the interaction of the equivalent radiation field with the orbital electrons.

$$\frac{d\sigma_d^{(j)}}{dw} = n_j I(E) \sigma_{PE}^{(j)}(E) \tag{12}$$

where $\sigma_{PE}^{(j)}$ is the photoelectric cross section for the jth orbital (per orbital electron), for an incident photon of energy E, $I(E)$ is the virtual-photon spectrum. Fig. 9 shows the distribution of secondary electrons for 1 and 10 keV primary electrons calculated by Eqs. (10)–(12) in comparison with the calculated data of Paretzke [184] and the experimental data of Vroom and Palmer [195] and Bolorizadeh and Rudd [180].

Angular distribution of secondary electrons for the ejected energy $w \geq 200$ eV was calculated using the kinematical relationships

$$\cos\theta = \sqrt{\frac{w(T + 2m_0c^2)}{T(w + 2m_0c^2)}} \tag{13}$$

For $w < 200$ eV, regardless of the primary electron energy, angular distributions were randomly sampled using the experimental data of Opal et al. [196].

Monte Carlo Electron Code

The Monte Carlo track structure code *kurbuc* simulates electron tracks in water vapor for initial electron energies 10 eV–10 MeV [174]. The code *kurbuc* provides all coordinates of

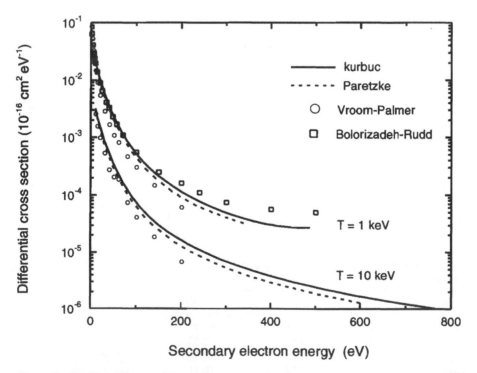

Figure 9 Energy spectra of secondary electrons for 1 and 10 keV electrons produced by the Seltzer's theory in comparison with the calculations by Paretzke [184] and the experimental data [180,195].

molecular interaction in water vapor, amount of energy deposited at each event, and the type of interaction at each event. Using tracks generated by *kurbuc*, various physical parameters such as penetration distances, W values, radial distributions of interactions, point kernel, and frequencies of energy deposition in small cylindrical targets can be obtained. As an example of track structure analysis, Fig. 10 shows results of W values of low-energy electrons in water vapor. W values are derived from scoring the number of electron–H_2O^+ pairs because of ionization process. Calculated W value at the high-energy limit agrees with the experimentally determined asymptotic limit of 30.4 eV.

9.2. Ions

Ionization Cross Sections

Here we use the term "ions" limiting to protons H^+ and alpha particles He^{2+}. For fast ions, the majority of energy is transferred in ionizing collisions, resulting in energetic free electrons and the potential energy of residual ions. Ionizing collisions involving bare ion projectiles, unless they are very slow, can be treated by using a classical description for the trajectory of the projectile with a quantal description of the target. When fast ions slow down around the Bragg peak (0.3 MeV/u), interactions involving electron capture and loss by the moving ions become an increasingly important component of the energy loss process. Detailed descriptions for charge transfer processes will be given in Sec. 9.3.

Total ionization cross sections for bare ions (H^+ and He^{2+}) were obtained by fitting polynomial functions to the experimental data given by Rudd et al. [198] for protons, and

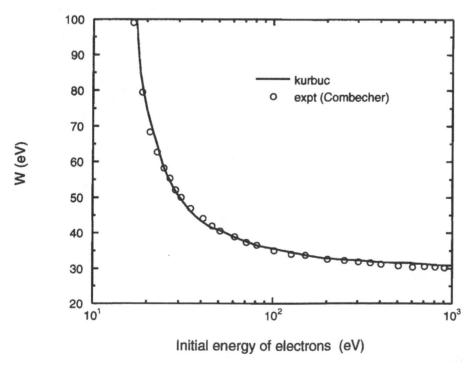

Figure 10 Calculated W values for electrons in water vapor as a function of the initial energy of electrons in comparison with the data of Combecher [197].

Rudd et al. [199] and Toburen et al. [200] for alpha particles.

$$\sigma_{ion} = \sum_j c_j (\log_{10} T)^{j-l} \tag{14}$$

in which T is in keV/u. Where data were lacking, extrapolation was made by taking into account the reproducibility of stopping powers. Fig. 11 shows total ionization cross sections of water vapor by H^+ (left) and He^{2+} (right).

Excitation Cross Sections

There are no experimental data of excitation cross sections for proton and alpha particle impact for water. The proton cross sections were obtained by scaling of the electron excitation cross sections for high-energy protons >500 keV [201]. For the lower-energy regions, the semiempirical model developed by Miller and Green [202] was adopted, which is based on the electron impact excitation. They assumed an analytical function for each excited level of the form

$$\sigma_e = \frac{\sigma_0 (Za)^{\Omega} (T - W)^{\nu}}{J^{\Omega + \nu} + T^{\Omega + \nu}} \tag{15}$$

where $\sigma_0 = 10^{-16} \, cm^2$, $Z = 10$, and the parameters W, a, J, Ω, and ν for 28 excitation states are given in Table 3 of Ref. 202. We assumed cross sections for protons can be applied to the impact excitation cross sections for alpha particles as shown in Fig. 11.

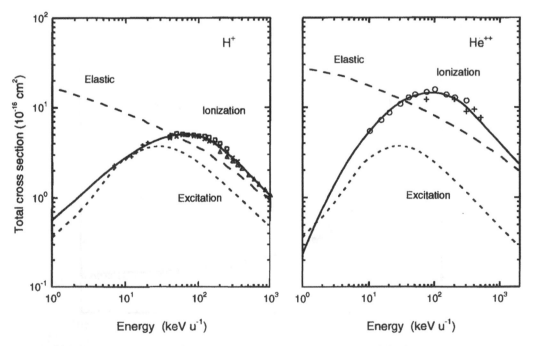

Figure 11 Total cross sections due to proton (left) and alpha particle (right) impact on water vapor. Total ionization cross sections were obtained by fitting polynomial functions to the experimental data [198–200]. The curve for excitation was assumed to be the same between protons and alpha particles. Elastic scattering was evaluated by the classical mechanics trajectory calculations [Eqs. (16) and (17)].

Elastic Scattering Cross Sections

High-energy ions proceed along a straight-line trajectory. However, the effects of elastic nuclear collisions will become more prominent in the low-energy region. Elastic collisions transfer little or no energy but can have a significant effect on the spatial character of the track structure at very low energies. Also, the nuclear energy loss cannot be neglected <10 keV/u. Nuclear elastic scattering was evaluated using the classical mechanics trajectory calculations taking into account the screening by atomic electrons [203]. For a particle scattered in a central potential $V(r)$, the deflection angle θ is obtained as a function of the impact parameter p [204] such that

$$\theta = \pi - 2 \int_{r_{\min}}^{\infty} \frac{p}{r^2\sqrt{1 - V(r)/T_{\mathrm{cm}} - p^2/r^2}}\, dr \tag{16}$$

where $r_{\min}$ is the distance of closest approach, T_{cm} is the particle energy in the center-of-mass system. Equation (16) was numerically solved using procedures given by Everhart et al. [205]. The differential elastic scattering cross section can be obtained by the classical relationship [206],

$$\frac{d\sigma_{\mathrm{el}}}{d\Omega} = -\frac{p}{\sin\theta}\frac{dp}{d\theta} \tag{17}$$

To reduce the divergence of the total cross section, which is obtained by integral of Eq. (17), the cutoff angle was set so as to limit the increase in the scattering probability at low scattering angle. Fig. 11 shows calculated total cross section for H^+ (left) and He^{2+} (right).

Secondary Electrons Distribution

Energy spectra of secondary electrons ejected by ion impact were calculated using the empirical model given by Rudd [207]

$$\frac{d\sigma_j(\varepsilon)}{d\varepsilon} = \frac{S}{B}\frac{F_1 + F_2 w}{(1 + w)^3\{1 + \exp[\alpha(w - w_{\mathrm{c}})/v]\}} \tag{18}$$

where $w = \varepsilon/B$, $v = (T/\lambda B)^{1/2}$, $w_{\mathrm{c}} = 4v^2 - 2v - R/4B$, $S = 4\pi a_0^2 N z^2 (R/B)^2$, ε = electron energy. B = binding energy for each orbital, $a_0 = 0.529 \times 10^{-8}$ cm, N = electron number in orbital. z = charge of the projectile nucleus, and $R = 13.6$ eV,

$$F_1(v) = \begin{cases} L_1 + H_1 \\ L_1 = C_1 v^{D_1}/(1 + E_1 v^{D_1+4}) \\ H_1 = A_1 \ln(1 + v^2)/(v^2 + B_1/v^2) \end{cases}$$

$$\tag{19}$$

$$F_2(v) = \begin{cases} L_2 H_2/(L_2 + H_2) \\ L_2 = C_2 v^{D_2} \\ H_2 = A_2/v^2 + B_2/v^4 \end{cases}$$

The fitting parameters for water vapor are given by Rudd et al. [208]

$A_1 = 0.97$	$B_1 = 82$	$C_1 = 0.40$	$D_1 = -0.30$	$E_1 = 0.38$
$A_2 = 1.04$	$B_2 = 17.3$	$C_2 = 0.76$	$D_2 = 0.04$	$\alpha = 0.64$

This model gives the single differential cross sections (SDCS) for secondary electron production by ion impact, and also the partial ionization cross sections for each orbital. The boundary energy for energetic secondary electrons generated by ionization was determined as 1 eV. Fig. 12 shows the randomly sampled energy spectra of ejected electrons for various energies of H^+ (left) and He^{2+} (right) in comparison with experimental data [200,209].

Experimental angular distributions of electrons ejected from water by H^+ impact are given by Bolorizadeh and Rudd [209], and by Toburen and Wilson [210] in the energy range between 15 keV and 1 MeV. Available experimental angular distributions for He^{2+} are limited. To cover the broad ion energy range, available H^+ data were used assuming that differences of angular distributions between different ions are not very large. Interpolation was made for the intermediate energies of electrons and ions.

9.3. Low-Energy Ions

Charge Transfer Cross Sections

Charge transfer can produce residual ions without the release of free electrons, and free electrons can be ejected from the moving ion (or neutral) with no residual ions being formed. Charge transfer cross sections are generally designated as σ_{if} where "i" is the initial and "f" is the final charge state of the moving particle. Table 8 lists the interactions for low-energy proton (H^+) and neutral hydrogen (H^0) with water. Table 9 lists the interactions for low-energy He^{2+}, He^+, and He^0. To develop track structure models, cross sections are needed for both electron capture and electron loss. For incident ions with speeds comparable to or

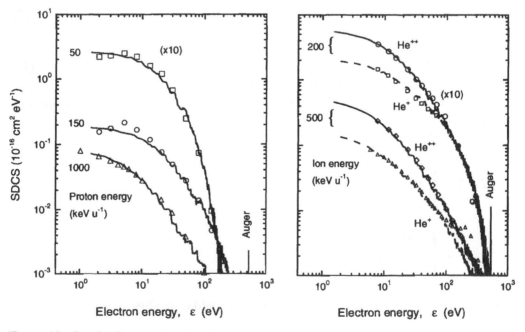

Figure 12 Randomly sampled energy spectra for secondary electrons ejected by proton (left) and alpha particle (right) impact in comparison with experimental data [200,217]. The lines are the results obtained from sampling 5×10^5 times using calculated data from the Rudd model [Eqs. (18) and (19)].

Table 8 Interactions for Low-Energy H^+ and H^0 with Water

Interactions		$H^+ + H_2O \rightarrow$	$H^0 + H_2O \rightarrow$
(i)	Elastic scattering	$H^+ + H_2O$	$H^0 + H_2O$
(ii)	Target ionization	$H^+ + H_2O^+ + e$	$H^0 + H_2O^+ + e$
	Target excitation	$H^+ + H_2O^*$	$H^0 + H_2O^*$
(iii)	Electron capture	$H^0 + H_2O^+$ (σ_{10})	
	Electron loss		$H^+ + e + H_2O$ (σ_{01})
(iv)	Electron capture and target ionization	$H^0 + H_2O^{2+} + e$	
	Electron loss and target ionization		$H^+ + e + H_2O^+ + e$

lower than the orbital speeds of bound electrons in the target, the capture of electrons from the target is a significant mechanism for the ionization of the target without production of secondary electrons. If the projectile is a dressed ion (i.e., an ion bearing one or more electrons), a collision can eject electrons from either the projectile or the target. For a fast projectile, one can consider the projectile as a simple collection of a nucleus and electrons traveling together at the same speed. After the charge transfer, the dressed ion or the neutral atom becomes either stripped or ionizes a water molecule. In the former case, stripped ion, an electron is ejected in a forward direction with nearly the same velocity as the projectile. In the latter case, ionization of water molecule without changing the ionic charge state, an electron is ejected with a different distribution of electron energy and with a similar ejection angle as that of the bare ions.

Fig. 13 shows charge exchange cross sections based on available experimental data for H^+ and H^0 (upper left) [211,212], He^{2+} (lower left) [199], He^+ (lower right) [213,214], and He^0 (upper right) [214,215]. There is a probability of two-electron transfer for helium atom, such as σ_{20} and σ_{02}. Total cross sections for electron capture σ_{10} for H^+ and electron loss σ_{01} for H^0 were evaluated using the analytical functions developed by Miller and Green [202]. Cross sections for He ions were least-square fitted by a simple polynomial function similar to Eq. (14). Smooth extrapolation was carried out where the experimental

Table 9 Interactions of Low-Energy He^{2+}, He^+, and He^0 with Water

Interactions		$He^{2+} + H_2O \rightarrow$	$He^+ + H_2O \rightarrow$	$He^0 + H_2O \rightarrow$
(i)	Elastic scattering	$He^{2+} + H_2O$	$He^+ + H_2O$	$He^0 + H_2O$
(ii)	Target ionization	$He^{2+} + H_2O^+ + e$	$He^+ + H_2O^+ + e$	$He^0 + H_2O^+ + e$
	Target excitation	$He^{2+} + H_2O^*$	$He^+ + H_2O^*$	$He^0 + H_2O^*$
(iii)	Two-electron capture	$He^0 + H_2O^{2+}$ (σ_{20})		
	One-electron capture	$He^+ + H_2O^+$ (σ_{21})	$He^0 + H_2O^+$ (σ_{10})	
	One-electron loss		$He^{2+} + e + H_2O$ (σ_{12})	$He^+ + e + H_2O$ (σ_{01})
	Two-electron loss			$He^{2+} + e + e + H_2O$ (σ_{02})
(iv)	One-electron capture and target ionization	$He^+ + H_2O^{2+} + e$	$He^0 + H_2O^{2+} + e$	
	One-electron loss and target ionization		$He^{2+} + e + H_2O^+ + e$	$He^+ + e + H_2O^+ + e$

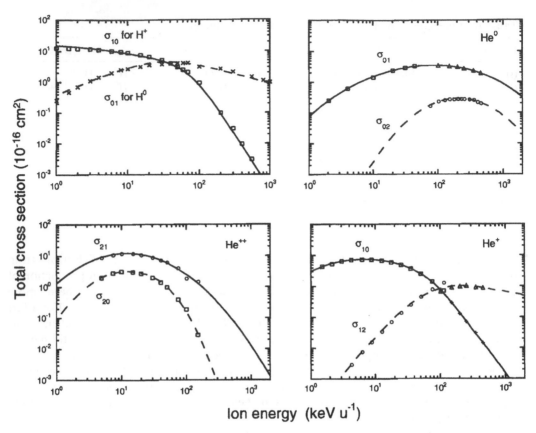

Figure 13   Charge exchange cross sections due to H$^+$ and H^0 (upper left panel), He^{2+} (lower left panel), He$^+$ (lower right panel), and He^0 (upper right panel) impact on water vapor. The curves are fitted to experimental data [199,211–215] by various model functions and extrapolated where data are lacking.

data were lacking. Fig. 14 shows calculated equilibrium charge fractions as a function of ion energy [215]. These fractions are useful for the analytical calculations of electronic stopping powers [216].

Cross Sections for Dressed Ions

The total ionization cross section for neutral hydrogen impact was obtained by a least-square fitting the experimental data of Bolorizadeh and Rudd [217]. Assuming similarity with proton impact ionization, cross sections were approximated by scaling proton ionization to give smooth transitions for the energy regions <20 or >150 keV/u, taking into account the reproducibility of electronic stopping powers. Total ionization cross sections for He$^+$ were fitted to the experimental data of Rudd et al. [213] and Toburen et al. [200]. Total ionization cross sections for He^0 at energies lower than 100 keV/u were adjusted to fit the electronic stopping powers tabulated in ICRU Report 49 [203]. In the region below 30 keV/u, the stopping powers for He^0 is the main contributor. At energies above 100 keV/u, the total cross sections were assumed to be the same as those of He$^+$. Fig. 15 shows the total ionization cross sections for H^0, He$^+$, and He^0.

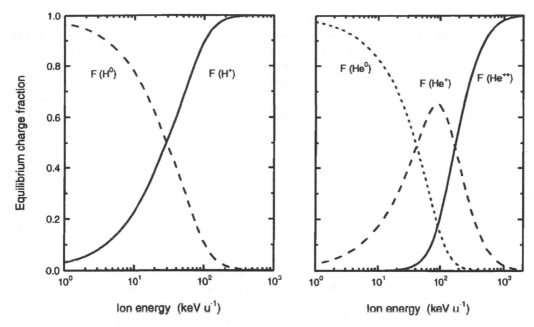

Figure 14 Equilibrium charge fractions of hydrogen beam (left panel) and helium beam (right panel) in H_2O calculated by the theory of Allison [215].

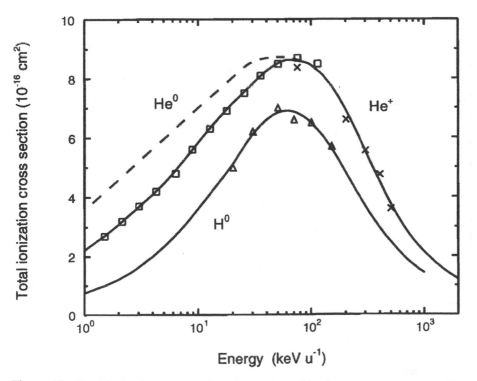

Figure 15 Total ionization cross sections due to dressed ion impact on water vapor. Cross sections for H^0 and He^0 were adjusted to reproduce stopping powers for the lower energy protons and alpha particles. (Experimental data from Refs. 200, 213, and 217.)

It was assumed that the spectra and angular distributions of electrons ejected from the target molecule by neutral hydrogen impact were equal to that for protons. Toburen et al. [200] have provided the ratios of SDCSs for ionization of water vapor by He^+ to those for He^{2+} as a function of the ejected electron energy. This ratio is distributed linearly from 0.4 to 1 for $0 < \varepsilon < 90$ eV for various projectile energies between 200 and 500 keV/u. The energy spectra for secondary electrons because of He^+ impact were obtained by modifying the Rudd's model [Eqs. (18) and (19)] using a simple function $g(\varepsilon)$ such that:

$$\frac{d\sigma_i^{He^+}(\varepsilon)}{d\varepsilon} = g(\varepsilon)\frac{d\sigma_i(\varepsilon)}{d\varepsilon} \tag{20}$$

in which

$$g(\varepsilon) = 0.4 + \varepsilon/150 \quad for \, \varepsilon < 90 \text{ eV}$$
$$= 1 \qquad\qquad for \, \varepsilon \geq 90 \text{ eV} \tag{21}$$

Fig. 12 (right) shows a good agreement between the calculated spectra and the experimental data for He^+ energies of 200 and 500 keV/u. Available experimental angular distributions of electrons ejected from water by He^+ are limited. To cover the broad ion energy range, available proton data were used assuming that differences of angular distributions between different ions are not very large. Again, we assumed the spectra and angular distributions of electrons ejected from the target molecule by He^0 impact to be equal to that for He^+. We assumed both excitation cross sections for protons [Eq. (15)] and elastic scattering cross

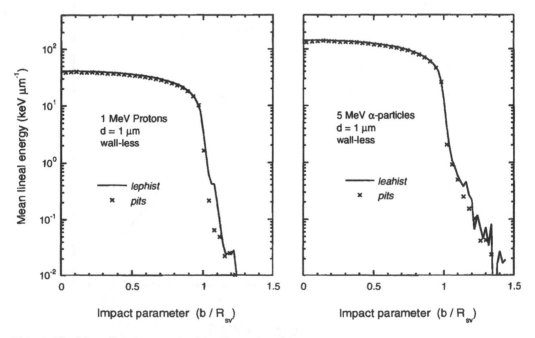

Figure 16 Mean lineal energy for 1 MeV proton tracks (left panel) and 5 MeV alpha particle tracks (right panel) passing through and near a 1-μm diameter site. (———), present calculations; (×) pits. (From Ref. 74.)

sections for bare ions [Eqs. (16) and (17)] can be applied to respective cross sections for three dressed ions (H^0, He^+, and He^0).

Monte Carlo Codes for Low-Energy Ions

Based on the evaluated cross sections, a new generation of Monte Carlo track structure codes were developed for simulation of tracks of low energy protons (the code *lephist*) [175] and alpha-particles (the code *leahist*) [218] in water. The effective energy ranges are 1 keV–1 MeV for protons and 1 keV–8 MeV for alpha particles. The tracks of secondary electrons were generated using the electron track code *kurbuc*. The codes can be operated to generate tracks both in the track segment and the full-slowing-down modes. Tracks were analyzed to provide confirmation on the reliability of the code and information on physical quantities, such as range, W values, restricted stopping power, and radial dose profiles.

Model Calculation

Tissue equivalent proportional counter (TEPC) is the instrument for measuring the distributions of energy deposited in a specific volume. The instrument is commonly designed as a spherical or a cylindrical volume with varying sensitive volume. Fig. 16 shows the distribution of mean lineal energy as a function of impact parameter b normalized by the radius of the sensitive volume. The parameter b (the eccentricity) is the distance from the center of the cavity to the chord of the ion cutting the sphere and R_{sv} is the radius of the counter. Distribution of the mean lineal energy were simulated for 1 MeV protons and 5 MeV alpha particles through a wall-less counter of 1-μm diameter.

REFERENCES

1. UNSCEAR, Sources and effects of ionizing radiation. UNSCEAR 2000 Report. Supplement N-46(A/55/46) volume 2. United Nations Publications, 2000.
2. UNSCEAR. Sources and effects of ionizing radiation. United Nations Scientific Committee on the Effects of Atomic Radiation. 1993 Report to the General Assembly. Annex F: Influence of dose and dose rate on stochastic effects of radiation, 1993.
3. The Low Dose Radiation Risks is part of The Sixth Framework Program of the European Union. http://www.cordis.lu/fp6
4. The Low Dose Radiation Research Program of the U.S. Department of Energy http://low-dose.org.
5. National Council on Radiation Protection and Measurements, NCRP Report 132. Bethesda MD, 2000. http://www.ncrp.com/
6. Strategic program plan for space radiation health research, Life Sciences Division, Office of Life and Microgravity Sciences and Application, NASA, Washington DC 20546.
7. National Academy of Sciences Space Science Board, Report of the Task Group on the Biological Effects of Space Radiation. Radiation Hazards to Crews on Interplanetary Mission National Academy of Sciences, Washington, DC, 1997.
8. Modeling Human Risk: Cell and Molecular Biology in Context. Report No. LBNL-40278. DOE, National Technical Information Service. US Department of Commerce, 5285 Port Royal Rd, Springfield, VA 22161, 1997.
9. Mothersill, C.; Seymour, C. Radiat. Res. 2001, *155*, 759.
10. Sasaki, M.S.; Ejima, Y.; Tachibana, A., et al. DNA damage response pathway in radioadaptive

response. Mutat. Res. Fundam. Mol. Mech. Mutagen. JUL 25 2002, *504* (1–2), 101–118. Special Issue.

11. Lorimore, S.A.; Wright, E.G. Int. J. Radiat. Biol. 2002. *in press.*
12. Baverstock, K. Mutat. Res. Fundam. Mol. Mech. Mutagen. 2000, *454* (1–2), 89.
13. Health Effects of Exposure to Low Levels of Ionizing Radiation: BEIR V Committee on the Biological Effects of Ionizing Radiation (BEIR V). National Research Council, 1990.
14. ICRP, 1990, Recommendations of the International Commission on Radiological Protection. Publication 60, Annals of the ICRP, Pergamon Press: Oxford, 1990; Vol. 21, 1–3.
15. Goodhead, D.T. Int. J. Radiat. Biol. 1994, *65*, 7.
16. Ward, J.F. Int. J. Radiat. Biol. 1994, *66*, 427.
17. Prise, K.M.; Pinto, M.; Newman, H.C.; Michael, B.D. Radiat. Res. 2001, *156*, 572.
18. Savage, J.R.K. BI Radiol. 62, 507.
19. Sachs, R.K.; Chen, A.M.; Brenner, D.J. Int. J. Radiat. Biol. 1989, *71*, 1.
20. Ottolenghi, A.; Ballarini, F.; Merzagora, M. Radiat. Environ. Biophys. 1999, *38*, 1.
21. Edwards, A.A. Int. J. Radiat. Biol. 2002, *78*, 551.
22. Bridges, B.A. Radiat. Res. 2001, *156*, 631.
23. Sankaranarayanan, K.; Chakraborty, R. Radiat. Res. 2001, *156*, 648.
24. Cox, R. Int. J. Radiat. Biol. 1994, *65*, 57.
25. Sharer, O.D.; Jiricny, J. BioEssays 2001, *23*, 270.
26. Ward, J.F. In *DNA Damage and Repair*; Nickoloff, J.A. Hoekstra, M.F., Eds.; Human Press, 1998; 65 pp.
27. Harrison, L.; Hatahet, Z.; Wallace, S.S. J. Mol. Biol. 1999, *290*, 667.
28. Goodhead, D.T. Int. J. Radiat. Biol. 1989, *56*, 623.
29. Newton, E.; Morton, E. Proc. Natl. Acad. Sci. U. S. A. 1991, *88*, 7474.
30. Nikjoo, H.; Goodhead, D.T.; Charlton, D.E.; Paretzke, H.G. Phys. Med. Biol. 1989, *34*, 691.
31. Fitzsimons, C.J.; Nikjoo, H.; Bolton, C.E.; Goodhead, D.T. Math Biosci. 1998, *154*, 103.
32. Charlton, D.E.; Nikjoo, H.; Humm, J.L. Int. J. Radiat. Biol. 1989, *56*, 1.
33. Dickerson, R.E.; Drew, H.E.; Conner, B.N.; Wing, R.M.; Fratini, A.V.; Kopla, M.L. Science 1982, *216*, 475.
34. Lavery, R.; Sklenar, H. J. Biomol. Struct. Dyn. 1988, *6*, 63.
35. Ferrin, T.E.; Huang, C.C.; Jarvis, L.E.; Langridge, R. J. Mol. Graph. Model, 6, 13–27, 36–37, 1988. http://www.cgl.ucsf.edu/Outreach/midasplus/
36. Berman, H.M.; Olson, W.K.; Beveridge, D.L.; Westbrook, J.; Gelbin, A.; Demeny, T.; Hsieh, S.-H.; Srinivasan, A.R.; Schneider, B. Biophys. J. 1992, *63*, 751.
37. Umrania, Y.; Nikjoo, H.; Goodfellow, J.M. Int. J. Radiat. Biol. 1995, *67*, 145.
38. Joshua-Tor, L.; Sussman, J. Curr. Opin. Struct. Biol. 1993, *3*, 323.
39. Friedland, W.; Jacob, P.; Paretzke, H.G.; Stork, T. Radiat. Res. 1998, *150*, 170.
40. Rydberg, B.; Holley, W.R.; Mian, S.M.; Chatterjee, A. J. Mol. Biol. 1988, *284*, 71.
41. Nikjoo, H.; Laughton, C.A.; Terrissol, M.; Panyutin, I.G.; Goodhead, D.T. Int. J. Radiat. Biol. 2000, *76*, 1607.
42. Franklin, R.E.; Gosling, R.G. Nature 1953, *171*, 740.
43. Dickerson, R.E.; Drew, H.R.; Conner, B.N.; Wing, R.M.; Fratini, A.V.; Kopka, M.L. Science 1982, *216*, 475.
44. Saenger, W. Annu. Rev. Biophys. Biophys. Chem. 1987, *16*, 93.
45. Saenger, W.; Hunter, W.N.; Kennard, O. Nature 1986, *324*, 385.
46. Saenger, W. Principles of Nucleic Acid Structure. Springer: New York, 1983.
47. Lea, D.E. Action of Radiation on Living Cells. Cambridge University Press: London, 1955.
48. Savage, J.R.K. Mutat. Res. 1998, *404*, 139.
49. Hlatky, H.; Sachs, R.K.; Vazquez, M.; Cornforth, M.N. BioEssays 2002, *24*, 714.
50. Nikjoo, H.; Goodhead, D.T.; Charlton, D.E.; Paretzke, H.G. Energy deposition by mono-energetic electrons in cylindrical targets. MRC Monograph, 1994.
51. Nikjoo, H.; Goodhead, D.T.; Charlton, D.E.; Paretzke, H.G. Energy deposition by C, Al, Ti

and Cu x-rays in cylindrical volumes within mammalian cells. MRC Radiobiology Unit Monograph, 1988.

52. Charlton, D.E.; Goodhead, D.T.; Wilson, W.E.; Paretzke, H.G. Energy deposition in cylindrical volumes: (a) Protons, energy 0.3 MeV to 4 MeV, (b) Alpha particles, energy 1.2 MeV to 20 MeV. MRC Radiobiology Unit Monograph, 1985.

53. Nikjoo, H.; Boyden-Pratt, N.; Uehara, S.; Goodhead, D.T. Energy deposition by monoenergetic protons in cylindrical targets. MRC Monograph, 2001.

54. Nikjoo, H.; Boyden-Pratt, N.; Girard, P.; Uehara, S.; Goodhead D.T. Energy deposition by alpha-particles in cylindrical targets. MRC Monograph, 2002.

55. Cucinotta, F.A.; Nikjoo, H.; Goodhead, D.T. Radiat. Res. 2000, *153*, 459.

56. Frankenberg, D.; Brede, H.J.; Schrewe, U.J.; Steinmetz, C.; Frankenberg-Schwager, M.; Kasten, G.; Pralle, E. Radiat. Res. 1999, *151*, 540.

57. Prise, K.M.; Ahnstrom, G.; Belli, M.; Carlsson, J.; Frankenberg, D.; Keifer, J.; Lobrich, M.; Michael, B.D.; Nygren, J.; Simone, G.; Stenerlow, B. Int. J. Radiat. Biol. 1998, *74*, 173.

58. Belli, M.; Cherubini, R.; Dalla Vecchia, M.; Dini, V.; Moschini, G.; Signoretti, C.; Simone, G.; Tabocchini, M.A.; Tiveron, P. Int. J. Radiat. Biol. 2000, *76*, 1095.

59. Prise, K.M.; Folkard, M.; Davies, S.; Michael, B.D. Int. J. Radiat. Biol. 1990, *58*, 261.

60. Nikjoo, H.; Uehara, S.; Wilson, W.E.; Hoshi, M.; Goodhead, D.T. Int. J. Radiat. Biol. 1998, *73*, 355.

61. Frankenberg-Schwager, M.; Frankenberg, D. Int. J. Radiat. Biol. 1990, *58*, 569.

62. Radford, I.R. Int. J. Radiat. Biol. 1985, *48*, 45.

63. Blocher, D. Int. J. Radiat. Biol. 1988, *54*, 761.

64. Foray, N.; Arlett, C.F.; Malaise, E.P. Biochemistry 1997, *79*, 567.

65. Olive, P.L. Radiat. Res. 1998, *150*, S42.

66. Jenner, T.J.; DeLara, C.M.; O'Neill, P. Int. J. Radiat. Biol. 1993, *64*, 265.

67. Ward, J.F. Radiat. Res. 1995, *142*, 362.

68. Goodhead, D.T.; Nikjoo, H. Int. J. Radiat. Biol. 1989, *55*, 513.

69. Leenhouts, H.P.; Chadwick, K.H. Radiat. Prot. Dosim. 1985, *13*, 267.

70. Booz, J.; Finnendegen, L.E. Int. J. Radiat. Biol. 1988, *53*, 13.

71. Cucinotta, F.A.; Nikjoo, H.; Goodhead, D.T. Radiat. Environ. Biophys. 1999, *38*, 81.

72. Ballarini, F.; Biaggi, M.; Merzagora, M.; Ottolenghi, A.; Dingfelder, M.; Friedland, W.; Jacob, P.; Paretzke, H.G. Radiat. Environ. Biophys. 2000, *39*, 179.

73. Cobut, V.; Frongillo, Y.; Patau, J.P.; Goulet, T.; Fraser, M.-J.; Jay-Gerin, J.-P. Radiat. Phys. Chem. 1998, *51*, 229.

74. Wilson, W.E.; Nikjoo, H. Radiat. Environ. Biophys. 1999, *38*, 97.

75. Rydberg, B.; Heilbronn, L.; Holley, W.R.; Lobrich, M.; Zeitlin, C.; Chatterjee, A.; Cooper, P.K. Radiat. Res. 2002, *158*, 32.

76. Watanabe, R.; Nikjoo, H. Int. J. Radiat. Biol. 2002, *78*, 953.

77. Moiseenko, V.V.; Hamm, R.N.; Waker, A.J.; Prestwich, W.V. Int. J. Radiat. Biol. 1998, *74*, 533.

78. Begusova, M.; Sy, D.; Charlier, M.; Spotheim-Maurizot, M. Int. J. Radiat. Biol. 2000, *76*, 1063.

79. Nikjoo, H.; O'Neill, P.; Goodhead, D.T.; Terrissol, M. Int. J. Radiat. Biol. 1997, *71*, 467.

80. Nikjoo, H.; Bolton, C.E.; Watanabe, R.; Terrissol, M.; O'Neill, P.; Goodhead, D.T. Radiat. Prot. Dosim. 2002, *99*, 77.

81. Nikjoo, H.; O'Neill, P.; Wilson, W.E.; Goodhead, D.T. Radiat. Res. 2001, *156*, 577.

82. Cucinotta, F.A.; Nikjoo, H.; O'Neill, P.; Goodhead, D.T. Int. J. Radiat. Biol. 2000, *76*, 1463.

83. Ottolenghi, A.; Ballarini, F.; Merzagora, M. Radiat. Environ. Biophys. 1999, *38*, 1.

84. Friedland, W.; Li, W.B.; Jacob, P.; Paretzke, H.G. Radiat. Res. 2001, *155*, 703.

85. van Rijn, K.; Mayer, T.; Blok, J.; Bverberne, J.; Loman, H. Int. J. Radiat. Biol. 1985, *47*, 309.

86. Ito, T.In *Radiation Damage*; Fuciarelli, A.F., Zimbrick, J.D., Eds.; Battelle Press: Columbus, 1995; 259 pp.

87. Nikjoo, H.; O'Neill, P.; Terrissol, M.; Goodhead, D.T. Radiat. Environ. Biophys. 1999, *38*, 31.

88. Kandaiya, S.; Lobachevsky, P.; Martin, F. Acta Oncol. 1996, *35*, 803.
89. Martin, R.F.; Haseltine, W.A. Science 1981, *213*, 896.
90. Ninbar, A.; Goodfellow, J.M. Radiat. Environ. Biophys. 1999, *38*, 23.
91. Holley, W.R.; Chatterjee, A.; Magee, J.L. Radiat. Res. 1990, *121*, 161.
92. Charlton, D.E.; Humm, J.L. Int. J. Radiat. Biol. 1988, *53*, 353.
93. Nikjoo, H.; Martin, R.F.; Charlton, D.E.; Terrissol, M.; Kandaiya, S.; Lobachevsky, P. Acta Oncol. 1996, *35*, 849.
94. Terrissol, M.; Demonchy, M.; Pomplun, E. In *Microdosimetry—An Interdisciplinary Approach*; Goodhead, D.T., O'Neill, P., Menzel, H.G., Eds.; Royal Society of Chemistry: Cambridge, 1997; 15 pp.
95. Michalik, V. Int. J. Radiat. Biol. 1992, *62*, 9.
96. Scholes, G.; Willson, R.L.; Ebert, M. Pulse radiolysis of aqueous solutions of deoxyribonucleotides and of DNA: reaction with hydroxyl radicals. Chem. Commun. 1969, *17*.
97. Buxton, G.V.; Greenstock, C.L.; Helman, W.P.; Ross, A.B. J. Phys. Chem. 1988, *17*, 513.
98. Milligan, J.R.; Aguilera, J.A.; Ward, J.F. Radiat. Res. 1993, *133*, 151.
99. Chatterjee, A.; Magee, J.L. Radiat. Prot. Dosim. 1985, *13*, 137.
100. Terrissol, M. Int. J. Radiat. Biol. 1994, *66*, 447.
101. Tomita, H.; Kai, M.; Kusama, T.; Aoki, Y.; Ito, A. Int. J. Radiat. Biol. 1994, *66*, 669.
102. Alden, C.J.; Kim, S.H. J. Mol. Biol. 1979, *132*, 411.
103. Makrigiorgos, G.M.; Bump, E.; Huang, C.; Baranowska-Kortylewicz, J.; Kassis, A.I. Int. J. Radiat. Biol. 1994, *66*, 247.
104. Michalik, V.; Begusova, M. Int. J. Radiat. Biol. 1994, *66*, 267.
105. Balasubramanian, B.; Pogozelski, W.K.; Tullius, T.D. Proc. Nat. Acad. Sci. U. S. A. 1998, *95*, 9738.
106. Aydogan, B.; Marshall, D.T.; Swarts, S.G.; Turner, J.E.; Boone, A.J.; Richards, N.G.; Bolch, W.E. Radiat. Res. 2002, *157*, 38.
107. Folkard, M.; Prise, K.M.; Vojnovic, B.; Davies, S.; Roper, M.J.; Michael, B.D. Int. J. Radiat. Biol. 1993, *64*, 651.
108. Boudaiffa, B.; Cloutier, P.; Hunting, D.; Huels, M.A.; Sanche, L. Science 2000, *287*, 1658.
109. Sanche, L. Radiat. Prot. Dosim. 2002, *99*, 57.
110. Hieda, K. Int. J. Radiat. Biol. 1994, *66*, 561.
111. Takakura, K.; Ishikawa, M.; Ito, T. Int. J. Radiat. Biol. 1987, *52*, 667.
112. Nikjoo, H.; Goorley, T.; Fulford, J.; Takakura, K.; Ito, T. Radiat. Prot. Dosim. 2002, *99*, 91.
113. Abdoul-Carime, H.; Dugal, P.C.; Sanche, L. Radiat. Res. 2000, *153*, 23.
114. Goodhead, D.T. Can. J. Phys. 1990, *68*, 872.
115. Nikjoo, H.; Goodhead, D.T. Phys. Med. Biol. 1991, *36*, 229.
116. Brenner, D.J.; Ward, J.F. Int. J. Radiat. Biol. 1992, *61*, 737.
117. Michalik, V.; Frankenberg, D. Radiat. Environ. Biophys. 1996, *35*, 163.
118. Holley, W.R.; Chatterjee, A. Radiat. Res. 1996, *145*, 188.
119. Ottolenghi, A.; Merzagora, M.; Tallone, L.; Durante, M.; Paretzke, H.G.; Wilson, W.E. Radiat. Environ. Biophys. 1995, *34*, 239.
120. Ottolenghi, A.; Merzagora, M.; Paretzke, H.G. Radiat. Environ. Biophys 1997, *36*, 97.
121. Moiseenko, V.V.; Hamm, R.N.; Waker, A.J.; Prestwich, W.V. Radiat. Environ. Biophys. 2001, *40*, 23.
122. Goodhead, D.T.; Nikjoo, H. Radiat. Res. 1997, *148*, 485.
123. Weinfeld, M.; Rasouli-Nia, A.; Chaudhry, M.A.; Britten, R.A. Radiat. Res. 2001, *156*, 584.
124. Sutherland, B.M.; Bennett, P.V.; Sutherland, J.C.; Laval, J. Radiat. Res. 2002, *157*, 611.
125. Sutherland, B.M.; Bennett, P.V.; Sidorkina, O.; Laval, J. Biochemistry 2000, *39*, 8026.
126. David-Cordonnier, M.H.; Laval, J.; O'Neill, P. J. Biol. Chem. 2000, *275*, 11865.
127. David-Cordonnier, M.H.; Boiteux, S.; O'Neill, P. Nucleic Acids Res. 2001, *29*, 1107.
128. Jenner, T.J.; Fulford, J.; O'Neill, P. Radiat. Res. 2001, *156*, 590.
129. Cunniffe, S.; O'Neill, P. Radiat. Res. 1999, *152*, 421.

130. Watson, J.D.; Crick, F.H.C. Nature 1953, *171*, 737.
131. Franklin, R.E.; Gosling, R.G. Nature 1953, *171*, 740.
132. Saenger, W. *Principles of Nucleic Acid Structure*. Springer Verlag, 1986.
133. Neidle, S. DNA Structure and Recognition. Focus Series. Oxford: IRL Press, 1994.
134. Lilley, D.M.J. *Methods in Enymology*; Lilley, D.M., Dahlberg, J.E., Eds.; 1992; Vol. 212B, 133.
135. Regaud, C.; Lacassagne, A. Radiophysiol. Radither. 1927, *1*, 95.
136. Frank-Kamenetskii, M.D.; Mirkin, S.M. Annu. Rev. Biochem. 1995, *64*, 65.
137. Praseuth, D.; Guieysse, A.L.; Hélène, C. Biochim. Biophys. Acta 1999, *1489*, 181.
138. Felsenfeld, G.; Davis, D.R.; Rich, A. J. Am. Chem. Soc. 1957, *79*, 2023.
139. Arnott, S.; Bond, P.J.; Selsing, E.; Smith, P.J. Nucleic Acids Res. 1976, *11*, 4141.
140. Raghunathan, G.; Miles, H.T.; Sasisekharan, V. Biochemistry 1993, *32*, 455.
141. Panyutin, I.G.; Neumann, R.D. Nucleic Acids Res. 1994, *22*, 4979.
142. Charlton, D.E. Radiat. Res. 1986, *107*, 163.
143. Hofer, K.G. Acta Oncol. 1996, *35*, 789.
144. Adelstein, S.J.; Kassis, A. Acta Oncol. 1996, *35*, 797.
145. Walicka, M.A.; Ding, Y.; Roy, A.M.; Harpanhalli, R.S.; Adelstein, S.J.; Kassis, A.I. Int. J. Radiat. Biol. 1999, *75*, 1579.
146. Lobachevsky, P.N.; Martin, R.F. Radiat. Res. 2000, *153*, 263.
147. Pomplun, E.; Terrissol, M. Radiat. Environ. Biophys. 1994, *33*, 279.
148. Karamychev, V.N.; Zhurkin, V.B.; Garges, S.; Neumann, R.D.; Panyutin, I.G. Nat. Struct. Biol. 1999, *6*, 747.
149. Panyutin, I.G.; Neumann, R.D. Nucleic Acids Res. 1997, *25*, 883.
150. Panyutin, I.G.; Neumann, R.D. Acta Oncol. 1996, *35*, 817.
151. Panyutin, I.V.; Luu, A.N.; Panyutin, I.G.; Neumann, R.D. Radiat. Res. 2001, *156*, 158.
152. de Lara, C.M.; Hill, M.A.; Jenner, T.J.; Papworth, D.; O'Neill, P. Radiat. Res. 2001, *155*, 440.
153. NCRP, Report No. 136. Evaluation of the linear-nonthreshold dose–response model for ionizing radiation. National Council on radiation Protection and Measurements. Bethesda, Maryland, 30814–5916.
154. Joiner, M.C. Int. J. Radiat. Biol. 1994, *65*, 79.
155. Azzam, E.I.; Toledo, S.M.; Gooding, T.; Little, J.B. Radiat. Res. 1998, *150*, 497.
156. Hei, T.K.; Wu, L.J.; Liu, S.X.; Vannia, D.; Waldren, A.; Randers-Pheterson, G. Proc. Natl. Acad. Sci. U. S. A. 1997, *94*, 3765.
157. Kadhim, M.A.; Macdonald, D.A.; Goodhead, D.T.; Lorimore, S.A.; Marsden, S.J.; Wright, E.G. Nature 1992, *355*, 738.
158. Mothersill, C.; Seymour, C. Int. J. Radiat. Biol. 1997, *71*, 421.
159. Mothersill, C.; Seymour, C. Radiat. Res. 1998, *149*, 256.
160. Nagasawa, H.; Little, J. Radiat. Res. 1999, *152*, 552.
161. Wu, L.-J.; Randers-Pherson, G.; Xu, A.; Waldern, CA.; Geard, C.R.; Yu, Z.-L.; Heui, T.K. Proc. Natl. Acad. Sci. U. S. A. 1999, *96*, 4959.
162. Zhou, H.; Randers-Pherson, G.; Waldern, C.A.; Vannias, D.; Hall, E.J.; Hei, T.K. 2000, *97*, 2099.
163. Deshpande, A.; Goodwin, E.H.; Bailey, S.M.; Marrone, B.L.; Lehnert, B.E. Radiat. Res. 1996, *145*, 260.
164. Sawant, S.G.; Randers-Pehrso, G.; Geard, C.R.; Brenner, D.J.; Hall, E.J. Radiat. Res. 2001, *155*, 397.
165. Miller, R.C.; Randers-Pehrson, G.; Geard, C.R.; Hall, E.J.; Brenner, D.J. Proc. Natl. Acad. Sci. U. S. A. 1999, *96*, 19.
166. Prise, K.M.; Belyakov, O.V.; Folkard, M.; Michael, B.D. Int. J. Radiat. Biol. 1998, *74*, 793.
167. Azzam, E.I.; Raaphorst, G.P.; Mitchel, R.E.J. Radiat. Res. 1994, *138*, S28.
168. Lorimore, S.A.; Kadhim, M.A.; Pocock, D.A.; Papworth, D.; Stevens, D.L.; Goodhead, D.T.; Wright, E.G. Proc. Natl. Acad. Sci. U. S. A. 1998, *95*, 5730.
169. Watson, G.E.; Lorimore, S.A.; Macdonald, D.A.; Wright, E.G. Cancer Res. 2000, *60*, 5608.

170. Seymour, C.B.; Mothersill, C. Radiat. Res. 2000, *153*, 508.

171. Brenner, D.J.; Little, J.B.; Sachs, R.K. Radiat. Res. 2001, *155*, 402.

172. Nikjoo, H.; Khvostunov, I.K. Int. J. Radiat. Biol. 2002. *in press.*

173. Khvostnov, I.K.; Nikjoo, H. J. Radiol. Prot. 2002, *22*, A33.

174. Uehara, S.; Nikjoo, H.; Goodhead, D.T. Phys. Med. Biol. 1993, *38*, 1841.

175. Uehara, S.; Toburen, L.H.; Nikjoo, H. Int. J. Radiat. Biol. 2001, *77*, 139.

176. Uehara, S.; Nikjoo, H. J. Phys. Chem. 2002, *106*, 11051.

177. Hayashi, M. TECDOC-506. Vienna: IAEA, 1989; 193 pp.

178. Djuric, N.Lj.; Cadez, I.M.; Kurepa, M.V. Int. J. Mass Spectrom. Ion Process. 1988, *83*, R7.

179. Olivero, J.J.; Stagat, R.W.; Green, A.E.S. J. Geophys. Res. 1972, *77*, 4797.

180. Bolorizadeh, M.A.; Rudd, M.E. Phys. Rev. A 1986, *33*, 882.

181. Schutten, J.; de Heer, F.J.; Moustafa, H.R.; Boerboom, A.J.H.; Kistemaker, J. J. Chem. Phys. 1966, *44*, 3924.

182. Dingfelder, M.; Hantke, D.; Inokuti, M.; Paretzke, H.G. Radiat. Phys. Chem. 1998, *53*, 1.

183. Seltzer, S.M. In *Monte Carlo Transport of Electrons and Photons*; Jenkins, T.M., Nelson, W.R., Rindi, A., Eds.; Plenum Press: New York, 1988; 81 pp.

184. Paretzke, H.G. GSF-Bericht 24. Institut fur Strahlenschutz der Gesellschaft fur Strahlenund Umweltforschung: Munich, 1988; 1 pp.

185. Green, A.E.S.; Stolarski, R.S. J. Atmos. Terr. Phys. 1972, *34*, 1703.

186. Berger, M.; Wang, R. In *Monte Carlo Transport of Electrons and Photons*; Jenkins, T.M., Nelson, W.R., Rindi, A., Eds.; New York: Plenum Press, 1988; 21 pp.

187. Moliere, G. Naturforsch Z. 1948, *3a*, 78.

188. Shyn, T.W.; Cho, S.Y. Phys. Rev. A 1987, *36*, 5138.

189. Katase, A.; Ishibashi, K.; Matsumoto, Y.; Sakae, T.; Maezono, S.; Murakami, E.; Watanabe, K.; Maki, H. J. Phys., B 1986, *19*, 2715.

190. Nishimura, H. In *Electronic and Atomic Collisions*; Oda, N., Takayanagi, K., Eds.; Amsterdam: North-Holland, 1979; 314 pp.

191. Danjo, A.; Nishimura, H. J. Phys. Soc. Jpn. 1985, *54*, 1224.

192. Trajmar, S.; Williams, W.; Kuppermann, A. J. Chem. Phys. 1973, *58*, 2521.

193. Zaider, M.; Brenner, D.J.; Wilson, W.E. Radiat. Res. 1983, *95*, 231.

194. ICRU, Secondary Electron Spectra from Charged Particle Interations, Report 55. ICRU, Bethesda, 1996.

195. Vroom, D.A.; Palmer, R.L. J. Chem. Phys. 1977, *66*, 3720.

196. Opal, C.B.; Beaty, E.C.; Peterson, W.K. At. Data 1972, *4*, 209.

197. Combecher, D. Radiat. Res. 1980, *84*, 189.

198. Rudd, M.E.; Goffe, T.V.; DuBois, R.D.; Toburen, L.H. Phys. Rev. A 1985, *31*, 492.

199. Rudd, M.E.; Goffe, T.V.; Itoh, A. Phys. Rev. A 1985, *32*, 2128.

200. Toburen, L.H.; Wilson, W.E.; Popowich, R.J. Radiat. Res. 1980, *82*, 27.

201. Dingfelder, M.; Inokuti, M.; Paretzke, H.G. Radiat. Phys. Chem 2000, *59*, 255.

202. Miller, J.H.; Green, A.E.S. Radiat. Res. 1973, *54*, 343.

203. ICRU, Stopping Powers and Ranges for Protons and Alpha Particles, Report 49. ICRU, Bethesda, 1993.

204. Mott, N.F.; Massey, H.S.W. *The Theory of Atomic Collisions,* 3rd ed; Oxford University Press, London, 1965.

205. Everhart, E.; Stone, G.; Carbone, R.J. Phys. Rev 1955, *99*, 1287.

206. Goldstein, H. *Classical Mechanics.* Addison-Wesley: Reading, MA, 1950.

207. Rudd, M.E. Phys. Rev. A 1988; *38*, 6129.

208. Rudd, M.E.; Kim, Y.K.; Madison, D.H.; Gay, T.J. Rev. Mod. Phys. 1992, *64*, 441.

209. Bolorizadeh, M.A.; Rudd, M.E. Phys. Rev. A 1986, *33*, 888.

210. Toburen, L.H.; Wilson, W.E. J. Chem. Phys 1977, *66*, 5202.

211. Toburen, L.H.; Nakai, M.Y.; Langley, R.A. Phys. Rev. 1968, *171*, 114.

212. Dagnac, R.; Blanc, D.; Molina, D. J. Phys. B 1970, *3*, 1239.

213. Rudd, M.E.; Itoh, A.; Goffe, T.V. Phys. Rev. A 1985, *32*, 2499.

214. Sataka, M.; Yagishita, A.; Nakai, Y. J. Phys. B 1990, *23*, 1225.
215. Allison, S.K. Rev. Mod. Phys. 1958, *30*, 1137.
216. Uehara, S.; Toburen, L.H.; Wilson, W.E.; Goodhead, D.T.; Nikjoo, H. Radiat. Phys. Chem 2000, *59*, 1.
217. Blorizadeh, M.A.; Rudd, M.E. Phys. Rev. A 1986, *33*, 893.
218. Uehara, S.; Nikjoo, H. Radiat. Prot. Dosim. 2002, *99*, 53.
219. Ward, J.F. The radiation-induced lesions which trigger the bystander effect. Mutat. Res. Fundam. Mol. Mech. Mutagen 2002, *499*, 151.
220. Moiseenko, V.V.; Edwards, A.A.; Nikjoo, H.; Prestwich, W.V. Radiat. Res 1997, *147*, 208.
221. Prise, K.M. Int. J. Radiat. Biol 1994, *65*, 43.
222. Ward, J.F. Prog. Nucleic Acids Res. Mol. Biol. 1988, *35*, 95.
223. Botchway, S.W.; Stevens, D.L.; Hill, M.A.; Jenner, T.J.; O'Neill, P. Radiat. Res. 1997, *148*, 317.
224. Cox, R.; Thacker, J.; Goodhead, D.T. Int. J. Radiat. Biol. 1977, *31*, 561.

18

Microdosimetry and Its Medical Applications

Marco Zaider
Memorial Sloan-Kettering Cancer Center, New York, New York, U.S.A.

John F. Dicello
Johns Hopkins University School of Medicine, Baltimore, Maryland, U.S.A.

1. INTRODUCTION

The quantity *absorbed dose* is used in medical physics on the assumption that when its value is the same, irradiation of equal objects results in equal biological effects. In reality, most of the biological effects of radiation depend on the microscopical pattern of energy deposition and the absorbed dose is only the expected value (average) of the energy deposited per unit mass. That the same dose of radiation may result in different effects, when delivered by different types of radiation, is well documented. For instance, at low doses, the *relative biological effectiveness* (RBE) of neutrons relative to γ-rays to induce a cell to become malignant is on the order of 100, which means that the ratio of absorbed doses of γ-rays and neutrons causing equal effects is about 100 [1–13].

The "invention" of microdosimetry some 50 years ago is a direct result of the empirical observation that radiations have $RBE \neq 1$. This was unexpected because, on first principles, one would anticipate that biological effects would be proportional to the (average) number of biomolecular disruptions produced by radiation in the sensitive sites of the cell. On the other hand, all radiations act on biomatter primarily through the agency of ionizations and excitations (presumably, the events that initiate biological damage). Over a fairly large range of energies, the average energy expended to produce an ion pair (~34 eV) is the same for all radiations. Thus, one expects proportionality between average effect (E) and the average number of, say, ionizations in the cell:

$$Effect \propto z \tag{1}$$

In this expression, z, termed *specific energy*, is the energy imparted per unit mass to a volume V; the expected value of z (see below) is the absorbed dose D. It follows that, on average:

$$\overline{Effect} \propto \bar{z} = D \tag{2}$$

which means that RBE = 1. Conversely, RBE ≠ 1 entails a *nonlinear* relationship between effect and specific energy. For instance, if:

$$Effect \propto z^2 \tag{3}$$

then:

$$\overline{Effect} \propto \overline{z^2} \neq (\bar{z})^2 = D^2 \tag{4}$$

This means that in order to obtain a physical predictor of biological effect, one needs information on the distribution $f(z)$ of specific energy z. It also means—and this is the essential point—that the spatial pattern of energy deposition, which will change from one radiation to another and thus account for differences in $f(z)$ among radiations, is an important predictor of biological effect.

The application of microdosimetry to medical physics derives from biological models of radiation action (primarily, the Theory of Dual Radiation Action [14]) that explicitly utilize for their predictions a microdosimetric description of the radiation field.

Specifically, they concern the following two problems:

(a) The RBE of a radiation field with respect to conventional treatment modalities (photon or high-energy electrons) for which extensive empirical data already exist. This includes, as particular cases, RBE variations *within* the treatment field as well as brachytherapy, where low-energy photons (with RBE ≠ 1) are frequently used.
(b) Predicting effect modifications due to the temporal pattern of dose delivery.

We shall start with definitions of the principal microdosimetric quantities.

2. MICRODOSIMETRIC QUANTITIES

The basic concept of microdosimetry is the *microdosimetric event* (or, simply, *event*), which is energy deposition by a charged particle and its statistically correlated particles (e.g., secondary electrons from the same particle track) [15–17]. Events are important because they are statistically independent entities. The principal microdosimetric quantity, specific energy z, is defined as follows:

$$z = \frac{\varepsilon}{m} \tag{5}$$

where ε is the energy imparted by ionizing radiation to matter of mass m. The *absorbed dose D* is the average energy imparted, per unit mass, by ionizing radiation. Formally:

$$D(\vec{r}) = \lim_{V \to 0,\, \vec{r} \subset V} \bar{z} \tag{6}$$

where $\bar{z}$ is the average specific energy in a domain of volume V centered at $\vec{r}$, and the limit is taken at constant density ρ.

A second quantity (physically equivalent to specific energy) is the *lineal energy*, which is the quotient of ε by $\bar{l}$, where $\bar{l}$ is the mean chord length in that volume:

$$y = \frac{\varepsilon}{\bar{l}} \tag{7}$$

In this definition, ε refers to energy deposition imparted in a *single event*, a restriction that does not apply to z.

In the matter of mass density ρ, the relationship between z (in a single event) and y is:

$$z = \frac{4y}{\rho S} \tag{8}$$

This follows from a theorem by Cauchy [18] according to which the mean chord length $\bar{l}$ in a convex site is given by $4V/S$, where V and S are, respectively, the volume and surface area of the site. For a spherical site of diameter d:

$$\bar{l} = \frac{2}{3}d \qquad (9)$$

and thus:

$$z = \frac{4y}{\pi\, d^2 \rho} \qquad (10)$$

which, for $\rho = 1$ g/cm^3 and with units as indicated, becomes:

$$\frac{z}{\text{Gy}} = \frac{0.204}{\left(\dfrac{d}{\mu\text{m}}\right)^2} \frac{y}{\text{keV}/\mu\text{m}} \qquad (11)$$

The probability that the lineal energy delivered in *single* events is in the interval $[y, y+dy]$ is denoted as $f(y)dy$. The corresponding distribution in z is $f_1(z)dz$, where the subscript makes explicit the fact that this distribution refers to single events only. The distributions $f(z)$, $f_1(z)$, and $f(y)$ depend on the geometry of the site where specific energy is determined. Two moments of these distributions that are often invoked are the *frequency average* y_F (or z_F) and the *dose average* y_D (or z_D). They are defined as follows:

$$y_F = \int_0^\infty y f(y)\,dy; \qquad z_F = \int_0^\infty z f_1 dz$$

$$z_D = \frac{1}{y_F}\int_0^\infty y^2 f(y)\,dy; \qquad z_D = \frac{1}{z_F}\int_0^\infty z^2 f_1(z)\,dz \qquad (12)$$

The first and second moments of the multievent distribution $f(z;n)$ are denoted $\bar{z}$ and $\overline{z^2}$, respectively. Because, by definition, $D = \bar{z}$, the average number of events n at dose D is:

$$n = \frac{D}{z_F} \qquad (13)$$

At a given dose D, the distribution $f(z;n)$ [or, equivalently, $f(z;D)$] is determined uniquely by $f_1(z)$. This can be understood from the following argument: By definition, the probability of exactly v events is given by the Poisson distribution:

$$p(v, n) = e^{-n}\frac{n^v}{v!} \qquad (14)$$

If we denote by $f_v(z)$ the distribution of z in *exactly* v events, then $f_v(z)$ can be evaluated by successive convolutions because, by definition, events are statistically independent:

$$f_0(z) = \delta(z)$$

$$f_1(z) \equiv f_1(z)$$

$$f_2(z) = \int_0^\infty f_1(z')f_1(z - z')\,dz' \qquad (15)$$

$$\vdots$$

$$f_v(z) = \int_0^\infty f_1(z')f_{v-1}(z - z')\,dz'$$

The distribution $f(z;n)$ becomes:

$$f(z;n) = \sum_{v=0}^{\infty} p(v;n) f_v(z) \tag{16}$$

which proves that if $f_1(z)$ is known, one can calculate $f(z;n)$ for any value of n (i.e., D). One can also demonstrate a general relationship between the moments of $f(z;n)$ and $f_1(z)$ [17], namely:

$$m_{k+1} = \frac{M_{k+1}}{n} - \sum_{j=0}^{k-1} C_k^j M_{k-j} m_{j+1}, \qquad k = 0, 1, 2, \ldots \tag{17}$$

Here, m_k is the kth moment of $f_1(z)$ and M_k is the kth moment of $f(z;n)$ and:

$$C_k^j = \frac{k(k-1)\ldots(k-j+1)}{j!} \tag{18}$$

With the notations used above (Eq. (12)):

$$z_F = m_1, \qquad z_D = m_2/m_1 \tag{19}$$

In particular:

$$M_1 = \bar{z} = n z_F \tag{20}$$

$$M_2 = \bar{z}(\bar{z} + z_D) \tag{21}$$

or, equivalently:

$$M_2 = z_D D + D^2 \tag{22}$$

an expression that provides the mathematical justification for the much used (and abused) linear–quadratic expression (see below).

The dose-averaged specific energy z_D plays a fundamental role in microdosimetric-based models of radiation action. For instance, it will be shown below that its magnitude relative to D determines the extent to which dose–rate effects are important. If the micro-dosimetric spectrum $f_1(z)$ is known, z_D can be obtained by straightforward integration. There is, however, a second more powerful procedure for obtaining this quantity and this will be now explained.

z_D depends on two elements: the structure of the charged particles tracks in the radiation field, and the geometry of the site where z is determined. One may adopt the point of view that $f_1(z)$ obtains from the random overlap of two "objects": the particle track (represented as a collection of local energy depositions, ionizations, and excitations) and the site of interest. Consider then two objects (three-dimensional domains) of volumes V_1 and V_2 and let v represent the volume of their overlap when they are placed randomly in space. One can show [19] that the first two moments of v, $E(v)$ and $E(v^2)$ satisfy:

$$\frac{E(v^2)}{E(v)} = V_1 V_2 \int \frac{p_1(r) p_2(r)}{4\pi r^2} \, dr \tag{23}$$

where p_1 is the *point-pair distribution of distances* (ppdd) between pairs of points randomly chosen in V_1 (and similarly for p_2). This result is remarkable because it shows that the ratio of the second to the first moments of v can be factorized in functions (ppdd) that characterize each object individually. This result can be immediately translated to microdosimetry [20]. Assume that in a charged particle track, the "volume" occupied by each transfer point (ionization or excitation) equals the energy locally deposited. The overlap volume becomes the energy e deposited in the site (V_2), V_1 becomes the total energy T in the track, and V_2 is the site volume. With this:

$$\frac{E(e^2)}{E(e)} = TV_2 \int \frac{p_{\text{track}}(r) p_{\text{site}}(r)}{4\pi r^2} \, dr \tag{24}$$

The product between the ppdd and the volume of the site is termed the proximity function of that domain. Thus, we have a proximity function of energy deposition:

$$t(r) = (\text{track energy}) \, p_{\text{track}}(r) \tag{25}$$

and the proximity function of the site:

$$s(r) = (\text{site volume}) \, p_{\text{site}}(r) \tag{26}$$

It is immediately recognized that:

$$z_D = \frac{1}{m} \int \frac{t(r)s(r)}{4\pi r^2} \, dr \tag{27}$$

where m is the mass contained in the site. A few examples will illustrate the concept of proximity function. For an object that occupies the entire space:

$$t(r) - 4\pi r^2 \tag{28}$$

For a track represented by a straight line of infinite length:

$$t(r) = 2 \tag{29}$$

For a track of finite length L:

$$t(r) = 2\left(1 - \frac{r}{L}\right) \qquad r \in [0, L] \tag{30}$$

For a spherical site of diameter d:

$$t(r) = 4\pi r^2 \left(1 - \frac{3r}{2d} + \frac{r^3}{2d^3}\right) \qquad r \in [0, d] \tag{31}$$

For other regular volumes, see Ref. 21.

In summary, microdosimetry is the study and quantification of the *spatial* and *temporal* distributions of absorbed energy in irradiated matter [15,17,22,23]. One makes a distinction between *regional microdosimetry* [the object of which is the study of microdosimetric distributions $f(z)$] and *structural microdosimetry* (a mathematically more advanced approach, which is concerned with characterizing the spatial distribution of individual energy deposition events, i.e., ionizations and/or excitations). Regional microdosimetry asserts that the effect is entirely determined by the amount of specific energy deposited in the relevant site (typically, a cell nucleus). The two "kinds" of microdosimetry, regional and structural, were shown to be in fact mathematically equivalent—once the sensitive site is judiciously determined [16].

3. RADIOBIOLOGICAL CONSIDERATIONS

3.1. The Theory of Dual Radiation Action

In order to obtain a quantity that relates to biological effect, one must postulate a functional dependence of effect probability on specific energy. One such approach, the Theory of Dual

Radiation Action, takes as starting point the empirical observation that for numerous effects, the yield of *lesions* (alterations responsible for the endpoint observed) depends quadratically on specific energy [14]:

$$E(z) = \beta z^2 \tag{32}$$

where β is a positive constant found to be only weakly dependent on radiation quality. The form of this equation obtains by postulating that a lesion is the result of two molecular *sublesions*, produced each at a rate proportional to z. On average:

$$\overline{E(D)} = \beta \bar{z}^2 = \beta(z_D D + D^2) \tag{33}$$

The last step follows from Eq. (22). The expression, Eq. (33), is the familiar *linear–quadratic equation*. In a more familiar notation:

$$\overline{E(D)} = \alpha D + \beta D^2 \tag{34}$$

Thus:

$$z_D = \frac{\alpha}{\beta} \tag{35}$$

Although not strictly valid, it is possible to extend this formulation to the probability of cellular survival [reproductive inactivation $S(D)$] by making the assumptions that (1) one lesion inactivates the cell, and (2) the number of lesions produced is Poisson-distributed. When valid (mostly for low-LET radiation), it leads to:

$$S(D) = e^{-\alpha D - \beta D^2} \tag{36}$$

Examples of sublesions are single chromosome breaks that combine to produce a dicentric aberration or single-strand DNA breaks, which, when in close proximity, result in double-stranded DNA breaks. The expression, Eq. (35), provides the link between the micro-dosimetric quantity z_D, and the dose–effect coefficients α and β.

3.2. RBE

Within the linear–quadratic formulation, the RBE of radiation H (α_H, β_H) relative to radiation L (α_L, β_L) is given by [9,24,25]:

$$RBE(D_H) = \frac{\alpha_L}{2\beta_L D_H}\left[\sqrt{1 + \frac{4\beta_L(\alpha_H D_H + \beta_H D_H^2)}{\alpha_L^2}} - 1\right] \tag{37}$$

Note that the RBE is a decreasing function of D_H. If $\beta_H \equiv \beta_L$, and taking Eq. (35) into account, one obtains the simplified expression:

$$RBE = \frac{1}{2D_H}\left[\sqrt{z_{D,L}^2 + 4D_H(z_{D,H}^2 + D_H)} - z_{D,L}\right] \tag{38}$$

If, furthermore, $\beta_H = \beta_L \approx 0$, the RBE is independent of dose:

$$RBE = \frac{\alpha_H}{\alpha_L} \approx \frac{z_{D,H}}{z_{D,L}} = \frac{y_{D,H}}{y_{D,L}} \tag{39}$$

The efficacy of radiotherapy is generally considered to depend on the probability of cell killing, $1 - S(D)$. Because RBE depends nonlinearly on absorbed dose, both the absorbed dose and (through z_D) the microdosimetric properties of the radiation field must be known at locations in the tumor and in the surrounding healthy tissues.

3.3. Temporal Effects

The temporal pattern of dose delivery, an important parameter in radiotherapy, further changes the expressions derived thus far. To reinforce these ideas, consider the archetypical lesion, a chromosomal dicentric aberration that obtains when two single chromosome breaks (sublesions) misjoin. A pair of sublesions can be produced by a single track (single event) or by two independent tracks. The yield of lesions that result via the first mechanism (termed *intratrack* action) is proportional to dose because the number of events is proportional to dose. In contrast, and for analogous reasons, the second mechanism (*intertrack* action) generates lesions at a rate proportional to D^2. The two mechanisms are represented in Eq. (34) by the linear (αD) and quadratic (βD^2) terms, respectively.

An important consequence of this interpretation is the fact that the magnitude of the quadratic term (but not the linear one) depends on dose rate. Indeed, in most cases, sublesions undergo repair and thus the lower the dose rate [i.e., the larger the (average) time interval that separates the formation of the two sublesions, the lower the yield of intertrack lesions]. With this, the linear–quadratic equation becomes:

$$\overline{E(D)} = \alpha D + \beta q(t)D^2 \tag{40}$$

where $q(t)$ is a function—termed *dose–rate factor*—that modifies the quadratic term to account for the temporal distribution of dose.

Mathematical expressions for the function $q(t)$ can be obtained for practically any temporal pattern of dose delivery. It is commonly assumed that during the time interval, t repair eliminates sublesion damage according to $\exp(-t/t_0)$, where t_0 is a characteristic time required for repair (typically on the order of 1 hr). For instance [17]:
(a) For irradiations at a constant dose rate:

$$q(r) = \frac{2}{r} - \frac{2}{r^2}(1e^{-r}) \tag{41}$$

where $r = t/t_0$. In particular, when $t \gg t_0$:

$$q \cong \frac{2t_0}{t} \tag{42}$$

(b) For f well-separated fractions (complete sublesion repair in between fractions):

$$q = \frac{1}{f} \tag{43}$$

In a typical *low-dose rate* (LDR) treatment in brachytherapy, the total irradiation time is on the order of several days and therefore $q \approx 0$. It follows that the probability of cell survival $S(D)$, is quasi-exponential and the RBE is determined by the linear (α) coefficient, or, in microdosimetric terms, by z_D.
(c) If the dose rate is decreasing exponentially (e.g., a temporary or permanent implant with radioactive seeds), then [26,27]:

$$q(t) = \frac{2(\lambda t)^2}{(\mu t)^2(1 - \lambda^2/\mu^2)(1 - e^{-\lambda t})^2} \left[e^{-(\lambda+\mu)t} + \mu t\left(\frac{1 - e^{-2\lambda t}}{2\lambda t}\right) - \frac{1 + e^{-2\lambda t}}{2} \right] \tag{44}$$

Here $\mu = \log(2)/t_0$. At any given time t after the implantation, the absorbed dose $D(t)$ is:

$$D(t) = \frac{R_0}{\lambda}\left[1 - e^{-\lambda t}\right] \tag{45}$$

where R_0 is the initial dose rate and λ is the radioactive decay constant of the radioisotope used.

(d) In the most general case [28]:

$$q = \frac{\sum\limits_{i=1}^{n} a_i^2 t_0\left[c_i + t_0\left(e^{c_i/t_0} - 1\right)\right] + \sum\limits_{i=1}^{n}\sum\limits_{j=i+1}^{n} a_i a_j t_0^2 e^{-(b_j - b_i)/t_0}\left(e^{-c_j/2t_0} - e^{c_j/2t_0}\right)\left(e^{-c_i/2t_0} - e^{c_i/2t_0}\right)}{\frac{1}{2}\sum\limits_{i=1}^{n} a_i^2 c_i^2 + \sum\limits_{i=1}^{n}\sum\limits_{j=i+1}^{n} a_i a_j c_i c_j} \tag{46}$$

This is the dose rate factor for any arbitrary dose–rate distribution that consists of n contiguous segments, and where a_i, b_i, and c_i are, respectively, the dose rate, the center, and the time required to deliver the dose–rate segment i.

3.4. Microdosimetric Constraints on Models of Radiation Action

An Upper Boundary for Effect as a Function of Dose

If each cell responds to radiation autonomously (i.e., independently of any other cell), an upper limit of effect probability $E(D)$ is the probability that at least one energy deposition event actually occurred in it. This latter is given [by Poisson statistics, see Eqs. (13) and (14)] by $1 - \exp(-D/z_F)$ and thus, for any dose, one must have:

$$E(D) \leq 1 - e^{-D/z_F}$$
$$S(D) \equiv 1 - E(D) \geq e^{-D/z_F} \tag{47}$$

where $S(D)$ is the survival probability—meaning here "absence of effect". It is obvious that at sufficiently large doses the linear–quadratic expression, $S(D) = \exp(-\alpha D - \beta D^2)$ does *not* satisfy this condition and thus—as a matter of principle and independent of any biological assumption—can not be valid.

Definition of Low Dose and Low-Dose Rate

Microdosimetry provides an unambiguous definition of *physical low dose*, namely a dose such that $n = D/z_F 1$. Indeed, under these conditions, each cell is essentially traversed by at most one particle. Furthermore:

$$f(z; D) \approx e^{-n}\delta(z) + e^{-n}n f_1(z) \tag{48}$$

For any arbitrary expression $\varepsilon(z)$ describing the probability of effect as a function of z, one has:

$$E(D) = \int\varepsilon(z)f(z; D)dz \approx D\int\frac{\varepsilon(z)}{z_F}f_1(z)dz \tag{49}$$

which shows that at physical low doses, the effect probability $E(D)$ must be a linear function of dose irrespective of any specific biological mechanism. The expression, Eq. (49), is often used in the form [29]:

$$E(D) = D \int Q(z) f_1(z) dz \tag{50}$$

where $Q(z)$ is taken as an empirical response function, the shape of which is obtained by solving the system of integral equations:

$$E_i(D) = D \int Q(z) f_{1,i}(z) dz \qquad i = 1, 2, \ldots \tag{51}$$

where $E_i(D)$ is empirical (measured) probability of effect produced by radiation i with single-event microdosimetric spectrum $f_{1,i}(z)$ [30,31].

A further consequence is that—with no more than one event involved—*effects must be independent of dose rate* if the cells respond autonomously.

In most situations, only a small fraction of events traversing the cell is causative. For instance, only one 1 of 6000 electrons traversing a cell nucleus will produce a lethal lesion [32]. A possibly more meaningful definition of low dose ("biological" low dose) can be made in terms of these causative events (hits) by asking that the contribution of the quadratic term in dose (βD^2) be less than a certain fraction, say 10%, of the total effect. Let Δ be this limiting dose. By definition:

$$\beta \Delta^2 = 0.1(\alpha \Delta + \beta q \Delta^2) \tag{52}$$

When cells are irradiated at a constant dose rate D over a time t, q is given by Eq. (45) and this expression becomes:

$$1 - \frac{1}{18(Dt_0)} \left(\frac{\alpha}{\beta} \right) = \frac{(Dt_0)}{\Delta} \left[1 - e^{-\Delta/(Dt_0)} \right] \tag{53}$$

Note that this is an equation for Δ as a function of α/β, both in units of Dt_0. For tumor cells ($\alpha/\beta = 10$ Gy, $t_0 = 0.5$ hr, $D = 2$ Gy/min), $q \approx 1$ and $\Delta = 1.1$ Gy, which is significantly larger than the physical low-dose limit of 2 mGy. Note that Δ is an upper limit for the doses where Eq. (53) is valid.

The Kellerer–Hug Theorem

This theorem, given here without demonstration, sets up a lower limit for the mean number of inactivating events m. Given a probability of effect $E(D)$ or its complement $S(D) = 1 - E(D)$, define the mean inactivation dose [33]:

$$\delta = \int S(D) dD \tag{54}$$

and its variance:

$$\sigma^2 = 2 \int D S(D) dD - \delta^2 \tag{55}$$

The mathematical statement of this theorem is:

$$m \geq \frac{\delta^2}{\sigma^2} \tag{56}$$

The mean number of inactivating events is δ/z_F.

4. MICRODOSIMETRIC DISTRIBUTIONS: EXPERIMENTAL TECHNIQUES

The experimental techniques used in microdosimetry are quite complex, and a complete description is beyond the scope of this chapter (see Ref. 17). Conventional experimental techniques do not provide methods for measuring frequency distributions of energy depositions in sites with dimensions in the micrometer range or smaller, and a number of alternate methods that are capable of providing equivalent information for many biomedical applications have been developed. The underlying concept for most of these methods is as follows: The mean energy imparted by a particle traversing a thickness x of specified material of density ρ depends only on x and ρ. One can obtain equal energy depositions by manipulating x and ρ such that their product $x\rho$ remains constant. Thus, a microscopical volume ($x \sim 1\,\mu m$, $\rho = 1\,g/cm^3$) can be simulated with a larger cavity (x a few centimeters) by correspondingly reducing the density. For instance, a 2.5-cm diameter sphere filled with a propane-based tissue-equivalent (TE) gas at 17 Torr is equivalent to a 1-μm diameter of unit density tissue. In practice, it is usually not possible to exactly match the atomic compositions of the gas and the solid. A typical counter consists of a spherical or cylindrical cavity (called sensitive volume) filled with gas and delimited by a solid wall material, and a central wire placed along the counter diameter and isolated electrically from the counter wall. The traversal of the counter by charged particles results in the production of ion pairs. If a voltage is applied between the (conductive) wall and the central wire, the ions will drift along the electrical field and induce an electrical signal at the collecting electrode. The amount of charge collected is a measure of the energy deposited in the counter.

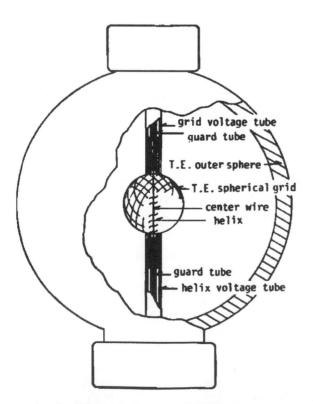

Figure 1   Schematic diagram of the Rossi proportional counter (see text for details).

For x-rays or γ-rays, the average signal thus obtained is very weak relative to background, and typically one runs the detector in the so-called proportional mode. This is achieved by setting a sufficiently large voltage for the drifting ions to acquire enough energy to produce second-generation and higher-generation ions through collisions with gas molecules. A schematic diagram of a typical detector, usually called a Rossi proportional counter in reference to H. H. Rossi whose group pioneered this field, is shown in Fig. 1.

5. MICRODOSIMETRIC DISTRIBUTIONS: THEORETICAL METHODS

For various technical reasons [17], the empirical determination of microdosimetric spectra is not always possible, particularly for site sizes significantly smaller than 1 μm or for volumes of irregular shape. In these situations, calculations may be used instead. A key element facilitating this approach is the possibility of regarding the spectrum of energy deposition as resulting from the random overlapping of two "objects": the charged-particle track (essentially, a collection of energy transfer points), and the sensitive site [14,15, 22,23,34]. Theoretical microdosimetric distributions are generated by randomly placing a site on a Monte Carlo-generated track. In a sense, this is the opposite of what happens in an experiment where the detector (a fixed object) is traversed by random tracks. To the extent that accurate cross sections for the interaction of the ionizing particles with the medium are known and the code can adequately simulate the spatial and structural geometries, theoretical calculations potentially represent the best procedure for obtaining microdosimetric quantities.

A full description of computational techniques in microdosimetry is beyond the charter of this chapter and the interested reader may consult the textbook by Rossi and Zaider [17]. Briefly, the calculation proceeds along the following lines:

For a given source, one generates at each point in space the electron distribution in energy (slowing down spectrum) $n(E)$; commercially available codes are available for this purpose.

For each electron energy E, the interactions with tissue (frequently represented by liquid water) are simulated event by event, which means that one obtains a detailed description of the geometrical coordinates, energy locally deposited, and type of energy transfer (ionization or excitation) for each interaction point, primary or secondary.

Microdosimetric spectra $f(y;E)$ for events transferring energy E are generated by randomly placing the volume of interest (commonly, spheres or cylinders) on each track and then evaluating, at each instance, the total energy deposited in it.

The lineal energy spectrum of the brachytherapy source is evaluated from:

$$f(y) = \frac{\displaystyle\int_0^\infty \frac{n(E)E}{y_F(E)} f(y;E)\,dE}{\displaystyle\int_0^\infty \frac{n(E)E}{y_F(E)}\,dE} \tag{57}$$

6. MICRODOSIMETRIC DISTRIBUTIONS FOR RADIATIONS USED IN RADIOTHERAPY

6.1. Brachytherapy

Brachytherapy physics is concerned with electrons, whether primary or secondary, as in the case of photon sources (Table 1).

Table 1 Physical Characteristics of Radioactive Sources Currently Used in Brachytherapy

	$\overline{E_\gamma}$ [MeV]	E_β(max) [MeV]	$T_{1/2}$
^{137}Cs	0.662	1.17	30 years
^{198}Au	0.416	0.96	2.7 days
^{192}Ir	0.380	0.67	74.2 days
^{125}I	0.028	—	60.2 days
^{103}Pd	0.021	—	17 days
^{131}I	0.364	0.61	8.06 days
^{90}Sr/^{90}Y	—	2.27	28.9 years
^{32}P	—	1.71	14.3 days
^{106}Ru/^{106}Rh	—	3.55	367 days

At low doses or *low-dose rates*, the RBE of 100 keV photons relative to, for example, 1-MeV photons may exceed 2 [35,36]. For instance, the RBE for ^{125}I has been extensively studied and results of 1.2–2 have been reported for the dose–rate range of 0.03–9 Gy/hr [37]. Similarly, for ^{103}Pd, a study performed at 0.07–0.8 Gy/hr reported RBE values of 1.9±0.7 [37–40]. While considerably lower RBE values apply to the higher doses or dose rates usually employed in external radiotherapy, differences in biological effectiveness on the order of 10–15% remain.

Microdosimetric distributions have been determined for a wide range of photon energies, including those of the photon sources quoted in Table 1. Kliauga and Dvorak [41] measured lineal energy spectra for photon energies in the range 12–1250 keV and for site diameters of 0.24–7.7 µm. Fig. 2 shows these spectra as a function of photon energy.

Microdosimetric distributions for ^{103}Pd-, ^{125}I-, ^{241}Am-, and ^{192}Ir-encapsulated sources were published by Wuu et al. [36] and are shown in Fig. 3. Table 2 gives y_D and RBE values for some of the distributions shown in Fig. 3.

These spectra have notable common characteristics. All the spectra have a modal energy deposition that is related to the average energy lost per unit path length by the primary particles. Because the lower energy electrons have a higher mean dE/dx, the spectra for the lower energy photons and electrons tend to peak at the higher lineal energies. Conversely, the higher energy particles have peaks at lower lineal energies. There is a decreasing probability of events at lower lineal energies characteristic of the frequency distribution of ionizations associated with the production of a single electron with energy close to zero (i.e., an energy deposition just sufficient to produce an ion pair). Irrespective of the mean energy deposited, there is a kinematic upper limit, which represents secondary electrons with precisely enough energy to make it across the diameter, thus depositing their total energies in the sensitive volume. Electrons of lower kinetic energy must deposit less total energy, and higher energy electrons will deposit only a fraction of their energy with a lower dE/dx.

6.2. High-Energy Electrons, X-Rays, and γ-Rays External Beam Therapy

Particles emitted from radioactive isotopes are generally too low in energy to provide the penetration required for conventional treatments of tumors with external radiation beams. Most external beam radiation therapies are performed with high-energy x-rays or electrons produced with compact linear accelerators with accelerating potentials between about 4 and 20 MeV. One notable exception is certain devices designed for stereotactical radiosurgery or radiation therapy of superficial tumors that use cobalt-60 γ-rays. The

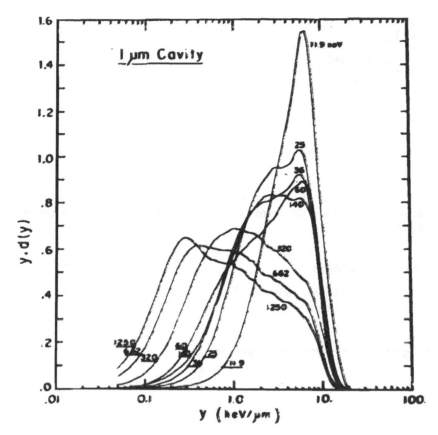

Figure 2 Microdosimetric distributions for photons of different energies. (From Ref. 41.) The spectra were measured in a 1-μm diameter spherical cavity.

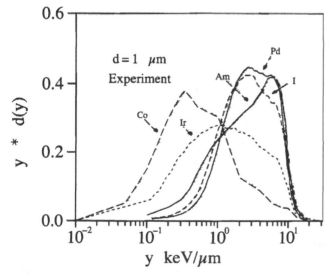

Figure 3 Microdosimetric distributions for several photon radiations commonly used in brachytherapy. (From Ref. 36.)

Table 2 Dose-Averaged Lineal Energy for the
Distributions Shown in Fig. 3

Radionuclide	y_D [keV/μm]	RBE
^{103}Pd	3.8	2.3
^{125}I	3.5	2.1
^{241}Am	3.5	2.1
^{192}Ir	2.0	1.3
^{60}Co	1.6	1.0

higher average energy of these therapy beams provides reduced attenuation and, therefore, more dose at depth. At the same time, this results in events of lower average lineal energy, as shown in Fig. 4 [42].

6.3. Heavy Particles in Medicine

It was quickly recognized after the development of high-energy particle accelerators that accelerator-produced particles provided better penetration and more dose to the treatment

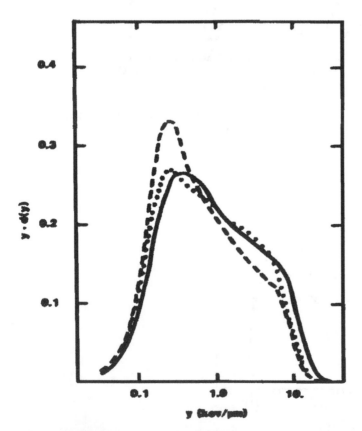

Figure 4 Microdosimetric distributions as a function of lineal energy in a 2-μm equivalent spherical detector. The solid line is a spectrum for cobalt-60 γ-rays, and the dashed and dotted lines are the spectra for 10- and 15-MeV bremsstrahlung x-rays.

volume while sparing the surrounding normal tissues. In addition, different particle types produced different biological responses in both the malignant and normal tissues. This led research groups to speculate that particles other than photons or electrons might have physical and/or biological advantages in comparison. Clinical trials have taken place using proton beams, neutrons, energetic heavy ions, and even negative pions. The protons, pions, and heavy ions, in contrast to the exponential attenuation associated with electrons and photons, have a relatively constant dE/dx with depth until near the end of their range, where there is a rapid increase in energy deposited, reaching a maximum at what is called the Bragg peak and followed by a rapid decrease. This type of depth–dose curve is better suited to deposit more dose in the treatment volume while sparing the surrounding normal tissues. Negative pions have the added characteristic that, at the end of their range, they disintegrate, transforming most of their rest mass (about 140 MeV equivalent) into secondary radiations. This is literally a localized submicroscopical nuclear explosion. Today, protons and, to a lesser extent, energetic heavy ions are being used routinely to treat cancer patients in a limited number of facilities. Although such particles showed that dose localization can be a major advantage for specific diseases, their higher costs in relation to electron accelerators [43] have restricted their use to a few centers, instead stimulating the development of improved technologies with photons to achieve increasingly better dose distributions at, unfortunately, increasing costs.

Therapeutic beams of energetic protons and heavy ions result not only in a different depth dose distribution but also in different microdosimetric spectra as well. Energies for protons used for therapy are nearly minimum ionizing and their microdosimetric spectra are similar in the range of lineal energy to that of higher-energy electrons and photons, as seen in Fig. 5 [44]. For the same energy per nucleon, the average dE/dx of the particles increases with increasing atomic mass, with a corresponding increase in the linear energies, as seen in Fig. 6 [45].

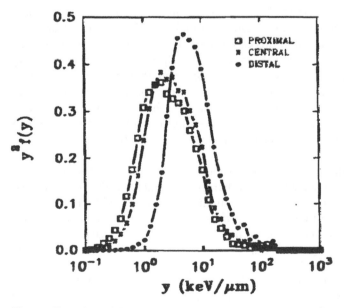

Figure 5 Microdosimetric spectra at the proximal, central, and distal positions of a therapeutic proton beam with an initial energy of 155 MeV.

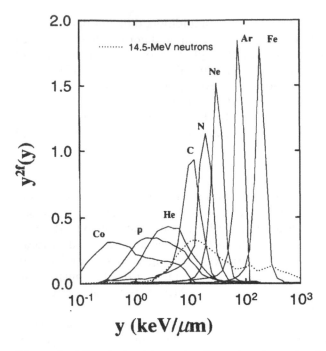

Figure 6 Microdosimetric spectra for various heavy ion beams as noted with nominal energies of approximately 600 MeV per nucleon. Corresponding spectra for cobalt-60 γ-rays, protons, and 14.5-MeV neutrons are included for comparison.

Moreover, for monoenergetic heavy ions of the same energy per nucleon, the average number of ion pairs increases with a corresponding decrease in the deviation about the mean, so we have a higher modal value with a more sharply defined Bragg peak.

The previous theoretical discussion tells us that increasing ionization density, such as that observed with increasing ion mass, could result in a higher biological effectiveness, at least until saturation sets in. Saturation in this case refers to large energy depositions that have such a high probability of producing the biological endpoint that increased energy deposited has little, if any, increased effect. We might expect, then, that changing the particle type and energy might change the relative biological effectiveness of the particle compared with photons and, therefore, the clinical outcome. This is indeed the case, rendering the RBE a clinically useful parameter to vary, with the expectation that particles with higher RBEs might be more effective in eradicating tumor cells. This is in part the motivation for the use of beams of energetic heavy ions.

Unfortunately, energetic heavy ions require expensive, large accelerators to produce. If increased biological effectiveness were the main therapeutic benefit, with improved dose localization not a major issue, then neutron beams might be more desirable. Fig. 7 shows microdosimetric spectra for neutron beams as a function of the nominal neutron energy.

Experiments generally show that energetic neutrons are at least as effective as most heavy ions for producing most biological effects. Although they are generally no better than photons in terms of dose localization, they are much less expensive to produce. Therefore, if biological effectiveness is the main clinical requirement and dose localization is not important, then neutron beams may be the modality of choice. Neutrons have been the focus of clinical investigations for at least six decades, and today they are used for conventional

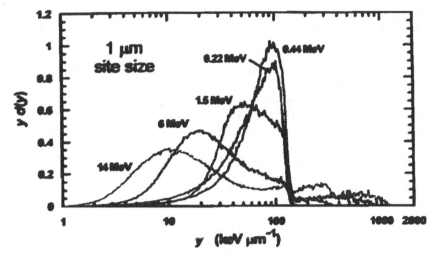

Figure 7 Microdosimetric spectra in a 2-μm equivalent sphere as a function of neutron energy as noted.

external beam therapy in a few therapy centers. Clinical investigations of boron neutron capture therapy are underway to take advantage of the increased RBEs with neutrons and, and the same time, to confine the dose primarily to malignant cells. Lower-energy neutrons have an exceptionally high interaction cross section with boron, so the boron is preferentially loaded into cancerous cells via appropriately selected antibodies.

We had mentioned that negative pions have been used in clinical trials for cancer therapy. They approach the dose localization of protons with an enhanced RBE in comparison with protons or photons. Because the pion mass is lighter than that of protons and heavier ions, the scattering is greater and edges are less sharp. Moreover, although the effectiveness of pions can be greater than that of protons or photons, it is generally lower than that of neutrons or heavy ions. Finally, pions are produced as secondary particles at higher energies and, therefore, are more expensive to produce at intensities suitable for therapy. Although a few institutions have used pions in clinical trials with results comparable to those of neutrons and heavy ions, there are no such studies underway at this time.

In summary, the variety of radiation types available for both external beam radiotherapy and brachytherapy provides multiple means of localizing the dose for maximum effectiveness in specific applications. The variation in the microdosimetric spectra results in corresponding variety in radiation quality and biological effectiveness, providing insight into the mechanistic processes of energy transfer and different mechanisms for attacking persistent diseases. Building upon this cadre of radiations, after a hundred years in the clinic, radiation therapy remains one of the primary oncological methods for controlling cancer.

REFERENCES

1. Rossi, H.H.; Bateman, J.L.; Biavati, B.J. NYO-2740-4. NYO Reports 1967; 121–127.
2. Bateman, J.L.; Rossi, H.H.; Kellerer, A.M.; Robinson, C.V.; Bond, V.P.Radiat. Res. 1972, *51* (2), 381–390.
3. Bateman, J.L.; Johnson, H.A.; Bond, V.P.; Rossi, H.H. Radiat. Res. 1968, *35* (1), 86–101.

4. Bateman, J.; Rossi, H.H. NYO-2740-6. NYO Reports 1969; 235–237.
5. Hall, E.J.; Kellerer, A.M.; Rossi, H.H.; Lam, Y.M. Int. J. Radiat. Oncol. Biol. Phys. 1978, 4 (11–12), 1009–1013.
6. Hall, E.J.; Kellerer, A.M. Radiat. Res. 1973, 55 (3), 422–430.
7. Wolf, C.; Lafuma, J.; Masse, R.; Morin, M.; Kellerer, A.M. Radiat. Res. 2000, 154 (4), 412–420.
8. Kellerer, A.M.; Rossi, H.H. Br. J. Radiol. 1972, 45 (536), 626.
9. Kellerer, A.M.; Rossi, H.H. Radiat. Res. 1971, 47 (1), 15–34.
10. Bond, V.P.; Meinhold, C.B.; Rossi, H.H. Health Phys. 1978, 34 (5), 433–438.
11. Hall, E.J.; Novak, J.K.; Kellerer, A.M.; Rossi, H.H.; Marino, S.; Goodman, L.J. Radiat. Res. 1975, 64 (2), 245–255.
12. Kellerer, A.M.; Hall, E.J.; Rossi, H.H.; Teedla, P. Radiat. Res. 1976, 65 (1), 172–186.
13. Wolf, C.; Lafuma, J.; Masse, R.; Morin, M.; Kellerer, A.M. Radiat. Res. 2000, 154 (4), 412–420.
14. Kellerer, A.M.; Rossi, H.H. Radiat. Res. 1978, 75 (3), 471–488.
15. International Commission on Radiation Units and Measurements. Microdosimetry 32; ICRU: Bethesda, MD, 1983.
16. Zaider, M.; Rossi, H.H. Radiat. Res. 1988, 113 (1), 15–24.
17. Rossi, H.H.; Zaider, M. Microdosimetry and Its Applications; Springer-Verlag: Berlin, 1996.
18. Cauchy, A. Memoire sur la Rectification des Courbes et la Quadrature des Sourface Courbes. Qevres Completes; Gauthier Villard: Paris, 1908.
19. Robbins, H.E. Ann. Math. Stat. 1944, 16, 342–347.
20. Kellerer, A.M.; Rossi, H.H. Radiat. Res. 1978, 75 (3), 471–488.
21. National Council on Radiation Protection and Measurements; Conceptual Basis for Calculations of Absorbed-Dose Distributions Recommendations 168; National Council on Radiation Protection and Measurements: Bethesda, MD, 1991.
22. Kellerer, A.M.; Chmelevsky, D. Radiat. Environ. Biophys. 1975, 12 (3), 205–216.
23. Kellerer, A.M.; Chmelevsky, D. Radiat. Environ. Biophys. 1975, 12 (4), 321–335.
24. Rossi, H.H. Radiat. Environ. Biophys. 1979, 17 (1), 29–40.
25. Rossi, H.H. Ann. N.Y. Acad. Sci. 1969, 161 (1), 260–271.
26. Dale, R.G. Br. J. Radiol. 1985, 58 (690), 515–528.
27. Zaider, M.; Wuu, C.S. Br. J. Radiol. 1995, 68, 58–63.
28. Zaider, M.; Dicello, J. RBEOER: a Fortran program for the computation of RBEs, OERs, survival ratios and the effects of fractionation using the Theory of Dual Radiation Action. Los Alamos Scientific Laboratory Report LA-7196-MS, 1978; 1–19.
29. Zaider, M.; Brenner, D.J. Radiat. Res. 1985, 103 (3), 302–316.
30. Bond, V.P.; Varma, M.; Feinendegen, L.E.; Wuu, C.S.; Zaider, M. Health Phys. 1995, 68 (5), 627–631.
31. Zaider, M. Health Phys. 1996, 70 (6), 845–851.
32. Zaider, M.; Rossi, H.H. Int. J. Radiat. Biol. 1998, 74 , 633–637.
33. Kellerer, A.M.; Hug, O. Random factors in the survival curve. Adv. Biol. Med. Phys. 1968, 12, 353–366.
34. Kellerer, A.M.; Chmelevsky, D. Radiat. Environ. Biophys. 1975, 12 (1), 61–69.
35. Wuu, C.S.; Zaider, M. Med. Phys. 1998, 25 (11), 2186–2189.
36. Wuu, C.S.; Kliauga, P.; Zaider, M.; Amols, H.I. Int. J. Radiat. Oncol. Biol. Phys. 1996, 36 (3), 689–697.
37. Ling, C.C.; Li, W.X.; Anderson, L.L. Int. J. Radiat. Oncol. Biol. Phys. 1995, 32 (2), 373–378.
38. Zellmer, D.L.; Shadley, J.D.; Gillin, M.T. Radiat. Prot. Dosim. 1994, 52 (1–4), 395–403.
39. Nath, R.; Meigooni, A.S.; Melillo, A. Int. J. Radiat. Oncol. Biol. Phys. 1992, 22 (5), 1131–1138.
40. Zellmer, D.L.; Gillin, M.T.; Wilson, J.F. Int. J. Radiat. Oncol. Biol. Phys. 1992, 23 (3), 627–632.
41. Kliauga, P.; Dvorak, R. 1199. Radiat. Res. 1978, 73 (1), 1–20.
42. Amols, H.A.; Zellmer, D.L. Med. Phys. 1984, 11, 247–253.
43. Dicello, J.; Slater, J.M. Trans. Am. Nucl. Soc. 1993, 68B, 37–40.
44. Dicello, J. Chapman and Hall: New York, 1995; 35–44.
45. Dicello, J.; Wasiolek, M.; Zaider, M. IEEE Trans. Nucl. Sci. 1991, 38, 1203–1209.

19

Charged Particle and Photon-Induced Reactions in Polymers

S. Tagawa, S. Seki, and T. Kozawa
Osaka University, Osaka, Japan

SUMMARY

Basic and applied researches on charged particle and photon-induced reactions of polymers are surveyed. The basic parts are fundamentals of radiation effects on polymers and pulse radiolysis studies on polymers. The intermediate parts are a great diversity of radiation effects on polymers and reaction mechanisms of electron beam (EB) and x-ray resists. The applied parts are economic scale of utilization of radiation and industrial application of radiation to polymers.

1. INTRODUCTION

Since the first studies of radiation effects on polymeric materials, fundamental researches on radiation-induced reactions in polymers have been widespread, and the field of their applications has been growing until now. Seeds of typical application of different types of radiation were introduced by Chapiro and by Charlesby during the 1950s, and great deal of work has been published in this field concerning various polymeric materials and radiation sources. This is due to quick or sensitive responses of macromolecules to radiation, because few reactions can dramatically change such macroscopic properties of polymeric materials as solubility, mechanical properties, and thermal stability. Cross-linking and chain-scission reactions caused by irradiation have been treated as the key reactions controlling the properties of macromolecules. Decision of the predominant reaction in a polymer for a radiation source has been very important to predict the possibility of modification of the polymer for engineering applications. This approach to the combination of radiation and polymers has been practically useful because polymeric materials have been the leading structural materials, so that polymers are now used in every stage of our daily life.

However, recent progress in both radiation and polymer researches indicates a dramatic change in the circumstance of this research field, and the trend has been rapidly expanding, especially in the last decade. Polymers will be used not only as structural materials but also as electronic, optical, medical, and pharmaceutical materials, and radiation will be a powerful candidate to manipulate the structure and property of polymers. In this section, the potentials of radiation and polymers are reviewed for the forecast of technology in the twenty-first century.

2. ECONOMIC SCALE OF UTILIZATION OF RADIATION

The economic scale of utilization of radiation has been surveyed several times by many organizations. Mainly from the viewpoint of public acceptance (PA), Management Information Services, Inc., in the United States conducted a survey to clarify the size of those economic scale [1] and the results were presented by Alan E. Walter, the former president of the American Nuclear Society, at conferences [2]. Similar studies were done in Japan, but the data had poor consistency for comparison. To overcome this shortcoming, in 1999, the Special Committee for Evaluation of Economy on Utilization of Radiation examined the Japanese economic scale of utilization of radiation, consistently and systematically, for the first time, based on reliable published data and reasonable assumption both under the sponsorship of the Ministry of Education, Culture, Sports, Science and Technology (MEXT) and with the support of the Japan Atomic Energy Research Institute (JAERI). The evaluation was based on economic indices indicating the magnitude of the market created by products manufactured through use of radiation.

We performed a detailed comparison of the economic scale of radiation use in the fields of industry [4], agriculture [5], and medicine [6] in Japan and the United States. The comparison between Japan and the United States was also done for an economic index regarding market creation of products obtained from use of both radiation and nuclear energy [7]. The results of our study are compared for the year 1997.

The economic scale is compared between the United States and Japan with selected industrial parameters such as sterilization, semiconductors, radiographic testing, and radial tire production, because the very large industrial markets make a whole comparison difficult. The economic scale revealed in selected industrial fields was about $56 billion for the United States and $39 billion for Japan. The former is larger by a factor of ~ 1.4 [4].

The economic scale of the application of radiation in the field of agriculture in Japan was estimated from public documents to be about $964 million in 1997. The economic scale survey in food irradiation and mutation breeding was extended to the United States for a direct comparison to the situation in Japan. The maximum estimation amounted to $3.2 billion for food irradiation and $11.2 billion for mutation breeding. The economic scale for products in selected agricultural fields was $14.5 billion for the United States and about $0.8 billion for Japan, implying that the former is larger in magnitude by a factor of about 18 [5].

The economic scale of the use of radiological technology in medicine in Japan in 1997 was about $10 billion. The economic scale in 1997 was compared between Japan and the United States [6]. Within the same limited parameters, the economic scale of the use of radiological technology in medicine was approximately $49 billion for the United States and $12 billion for Japan. The former is larger in magnitude by a factor of 4, implying that practitioners in the United States do not hesitate to use radiological technologies for medical applications. This tendency was clearly observed by the expenditure on prostate cancer, FDG-PET, and so on.

In 1997, the economic scale of radiation usage was estimated to be $119 billion (1.4% of GDP) in the United States and $52 billion ($1 = 121 yen) (1.2% of GDP) in Japan. In the United States, the use of radiation technology not only in industry but also in medicine is significantly advanced. The usage in agriculture is relatively small compared to those in industry and medicine both in the United States and Japan. Usage in the agricultural sector in Japan is much smaller than in the United States [7].

3. RADIATION-INDUCED REACTIONS IN POLYMERS

3.1. Fundamentals of Radiation Effects on Polymers

Early processes caused by high-energy charged particles and photons have already been described in other chapters. Some characteristic processes occurring in the polymeric systems are focused on in the present chapter. It is well known that in some radiation chemical processes, the energy consumption in the process is far smaller than the energy deposited by incident radiation, indicating that most of the energy is consumed by radiative and/or nonradiative decay of excited states via energy transfer processes instead of producing reactive intermediates. One very important example of this is the so-called protection effects of polyolefin by addition of aromatic molecules [8]. Energy and charge transfer processes are in relation to each other, and these are systematized theoretically in the condensed phases including polymeric systems. It is because of the high mobility, hence, of the large diffusion constant of charged species even in the polymeric system. It provides a validity of the formulation developed in the liquid or solid phases for extension to polymeric systems. In contrast, diffusion processes of neutral reactive intermediates strongly depend on the structure, molecular mobility, dielectric constants, etc. of solid materials. Although the neutral species including (free) radicals bear strict limitation in their migration they are very important promoters of radiation-induced reaction in the polymeric systems.

Energy Transfer Process

Several different processes have been accounted for transfer of electronic excitation energy between molecules or functional groups in macromolecules: (1) electron exchange interaction, (2) radiative excitation transfer, (3) excimer migration, (4) exciplex excitation transfer, and (5) long-range excitation transfer. Electron exchange interaction occurs over a very short distance via some overlap of molecular orbitals of the excitation donor and acceptor, called a Dexter-type energy transfer process [9]. Only the limitation of this transfer process is the conservation of the total spin of the system. Radiative transfer is due to the reabsorption of emission from a donor molecule by an acceptor via a photon. It needs the overlap between an acceptor's optical absorption and emission spectrum of a donor. There is no total spin conservation in the system; however, the triplet state of the acceptor is not formed because direct transition is usually spin forbidden. Excimers and exciplexes are the pairs of an excited molecule and an unexcited molecule [10], and the excimer is the pair formed between two consistent molecules [10]. Excimer formation has been reported in many systems of solids, liquids, and gases. One example of the excimer formation in radiation chemical processes is the excimer of phenyl groups in polystyrene of which the yield, absorption, kinetics, and emission have been precisely reported to date [11,12]. Long-range excitation transfer is mainly caused by a Coulomb interaction between an excited and an unexcited molecule that are making simultaneous coupled electronic transitions of almost equal energy. This is the so-called Förster-type energy transfer (inductive resonance or vibrational relaxation transfer) (for a review, see Ref. 13). The transfer occurs with a relatively slow rate constant because the motive force of the transfer, namely, the dipole–dipole interaction (coupling) of molecules, is very small compared to the width of the vibrational band of the excited state of the donor molecule. This leads to the transfer from the lowest vibrational excited state of donor molecules after vibrational relaxation, but enables long-range excitation transfer over ~ 10 nm in a single step.

In the real polymeric system, a large number of modeling efforts have been made for energy transfer processes caused by radiation as well as low-energy photons, and direct or indirect observation of excited-state dynamics has also been performed for a variety of polymeric systems. The earliest modeling work of excitation dynamics in polymeric systems was given by Simpson for polyenes by treating them as the chained analogs of simple molecules having π conjugation [14]. The developed modeling and experimental evidence is noted later. To date, the formation of the excited states, the energy transfer, and decomposition from excited states in the radiation chemical processes have been studied over a wide range of polymeric systems: hydrocarbon polymers without double bonds (polyethylene, polypropylene, etc. [15–19]), saturated polymers with aromatics (e.g., polystyrene, poly-α-methylstyrene, polyvinylcarbazole [11,12,20–22]), π-conjugated polymers (e.g., polyphenylenevinylene, polythiophene [22–25]), and σ-conjugated polymers (e.g., polysilane, polygermane [26,27]). LET and temperature effects on excited states of solid-state polystyrene [28a,b] and solid-state polysilanes [28c] were studied by ion beam pulse radiolysis.

Radical Processes

When a polymeric system is exposed to high-energy radiation, the system undergoes main-chain scission and the creation of cross-links, end-links, double bonds, free radicals, etc. The neutral radicals are the key promoters of the reactions above, and the structure, reactivity (stability), migration, etc., have been extensively investigated by many techniques including ESR. Details of the techniques and their results are described in other chapters or reviews (for a book, see Ref. 29), and the effects of the above reactions on polymeric systems are mainly discussed in the present section.

On the basis of no difference in the probability of main-chain scission in any part of the polymer backbone (no chain end effect), an approach to the statistical treatment of main-chain scission was given by Charlesby in 1954 [30]. Although main-chain scission is directly related to the degradation of a polymeric system, cross-linking and end-linking reactions cause more dynamic changes in its macroscopic behavior such as gelation of the polymers, which have produced many industrial applications: radiation curing, stiffening, grafting, etc. It is also a reasonable assumption that cross-linking and end-linking reactions are produced randomly by low-LET radiation and the number of links is proportional to the radiation dose, so long as the number of cross-links is sufficiently small in comparison to the total number of structural units. Classical statistical theory of main-chain scission and cross-linking were given by Kuhn in 1930 for chain scission [31], by Flory for cross-linking in 1941 [32], and modified by Charlesby in 1953 [33]. The initial formulation of main-chain scission and cross-linking probability is given as Charlesby–Pinner relationship by

$$1/M_n = 1/(M_n)_0 + (p_0 - 0.5q_0)D/m \qquad (1)$$

$$1/M_w = 1/(M_w)_0 + (0.5p_0 - q_0)D/m \qquad (2)$$

where p_0 is the probability of scission and q_0 is the probability of cross-linking. M_n and M_w denote the number and weight average molecular weight with their initial values of $(M_n)_0$ and $(M_w)_0$, respectively, m is the molecular weight of a unit monomer, and D (MGy) is absorbed dose. G values of main-chain scission and cross-linking are related to the values of p_0 and q_0 as follows,

$$G(s) = 9.6 \times 10^3 \times p_0 \qquad (3)$$

$$G(x) = 4.8 \times 10^3 \times q_0 \qquad (4)$$

where $G(x)$ is the G value of cross-linking and $G(s)$ is the G value of main-chain scission. The above statistical equations were superior for the ease of calculation to other classical theories; however, they were not sufficient to give an entire solution of main-chain scission and cross-linking of a polymeric system because of the assumption of the theory of an initial arbitrary molecular size. This gave some difficulties regarding average molecular weight, solubility, viscosity, etc. Problems with the molecular weight distribution and its change with the irradiation were solved by Saito [34] and by Inokuti [35]. In their theory, molecular weight distribution was expanded by the Poisson and/or Schulz–Zimm distributions, and changes in the distribution by the simultaneous reactions of main-chain scission and cross-linking were analytically traced. All of these initial works were done in parallel, and the following work developed on the starting points so provided.

O'Donnell et al. used a binominal expansion of the initial molecular weight distribution and discarded higher terms than a cubic factor [36]. This leads to a computer simulation to trace the changes in the molecular weight of polymers without using a mathematical table. In the 1990s, modified formulations of chain scission and cross-linking were proposed by Zhang et al. [37] and Olejniczak et al. [38]. Sun introduced a parameter β, which was related to the glass transition temperature of the polymer, and took the effects of molecular stiffness of the polymers into account. Rosiak proposed a deductive distribution function of molecular weight on the basis of an arbitrary distribution and extended the validity of the Charlesby–Pinner relationship. Major difficulties have been steadily solved by many attempts to improve the formulation; however, some new problems have also arisen in recent progress in the investigation of radiation-induced reactions in some polymeric systems. Nowadays, approach to main-chain scission and cross-linking reactions still contains some assumptions: (1) $G(s)$ and $G(x)$ are constant with radiation dose, (2) occurrence of scission and cross-linking may be treated separately and independently, (3) random scission and cross-linking occur in the system, (4) not more than one cross-link connects any two polymer molecules, etc. Some of these assumptions have been revealed to be incorrect in some polymeric systems and for high-LET radiation. The problems will be discussed in the following sections.

3.2. Pulse Radiolysis Studies on Polymers

Recent pulse radiolysis studies on polymers [39,40] and monomers [41] are reviewed from 1990. Very high quality of pulse radiolysis systems is required for pulse radiolysis studies on polymers and polymeric systems, especially polymer and polymeric intermediates, because of the complexity of reactions and small absorption coefficients of reactive intermediates of both polymers and polymeric systems compared with the reactive intermediates of aromatic molecules such as pyrene and biphenyl. Therefore, most of pulse radiolysis studies on polymers are studies on reactive intermediates of solute molecules, such as pyrene in polymer and polymeric systems. Recent progress in pulse radiolysis studies on polymers and polymeric systems has been enhanced mainly by the development of both pulse radiolysis techniques and advanced technology such as lithography. The development of pulse radiolysis techniques has been done by improvement of electron beam intensity and stability, sensitivity and dynamic range of photodetectors, time resolution of pulse radiolysis, wavelength region of monitor light, and experimental temperature. Especially, mention may be made of the development of laser synchronized picosecond [42] and subpicosecond [43] pulse radiolysis systems and the application of low-temperature nanosecond pulse radiolysis technique over a wide range of temperature [44a] with the monitoring wavelength region from 300 to 1600 nm [44b] to polymers and polymeric systems. Details of the progress in laser synchronized picosecond and subpicosecond pulse radiolysis are described in another chapter.

Polystyrene and Related Compounds

The radiation chemistry of polystyrene (PS) in solution has been studied by many groups. Absorption peaks of the intramolecular excimers of PS at 530 nm and poly-α-methylstyrene at 520 nm in p-dioxane solutions [45a], and absorption peaks of a charge transfer (CT) complex between a polystyrene phenyl ring and Cl atom at 320 and about 500 nm in chloroform solutions [45b], were observed for the first time by using laser flash photolysis in 1980. Later, the CT complex between poly-α-methylstyrene and Cl atom [45c] was observed. Intermediate species of PS in solutions, such as the CT complex [21a] and the intramolecular excimer [46], were studied by using pulse radiolysis. The lifetime of the excimer absorption and fluorescence of polystyrene is 20 nsec [21b,45a,46]. The intramolecular excimer formation and energy migration of oligostyrenes [21c,21d] were studied by using pulse radiolysis. The CT complex between a PS phenyl ring and a Cl atom was formed through a geminate ion recombination [21a] whose kinetics was different from that between the CT complex and the excimer. Lifetimes of the CT complexes in CCl_4 and $CHCl_3$ were about 200 and 420 nsec, respectively, dominated by hydrogen abstraction from PS by Cl atom [21a,46]. In the pulse radiolysis study on polystyrene solution, only the tail of absorption band due to polystyrene intramolecular dimer cation radical was observed in less than 1000 nm by previous laser flash photolysis and pulse radiolysis of PS in both solution and solid film [21b].

Recently, by using improved nanosecond pulse radiolysis with the monitoring wavelength region from 300 to 1600 nm [44], absorption spectra due to main reactive intermediates such as the intramolecular dimer cation radical in the near-IR wavelength region were clearly observed in the pulse radiolysis of polystyrene in various solutions [47]. For example, Fig. 1 shows the absorption spectra observed in the pulse radiolysis of polystyrene solutions in CH_2Cl_2.

Saturated Hydrocarbon Polymers

Absorption due to main intermediates such as polymer cation radicals and excited states, electrons, and alkyl radicals of saturated hydrocarbon polymers had not been observed for a long time by pulse radiolysis [39]. In 1989, absorption due to the main intermediates was observed clearly in pulse radiolysis of saturated hydrocarbon polymer model compounds except for electrons [39,48]. In 1989, the broad absorption bands due to polymer excited states in the visible region and the tail parts of radical cation and electrons were observed in pulse radiolysis of ethylene–propylene copolymers and the decay of the polymer radical cations were clearly observed [49]. Recently, absorption band due to electrons in saturated hydrocarbon polymer model compounds was observed clearly by pulse radiolysis [49] as shown in Fig. 2. In addition, very broad absorption bands in the infrared region were observed clearly in the pulse radiolysis of ethylene–propylene copolymers [50] as shown in Fig. 3. Radiation protection effects [51] and detailed geminate ion recombination processes [52] of model compounds were studied by nano-, pico-, and subpicosecond pulse radiolyses.

Polysilanes

Very intense and sharp near-UV absorption bands due to radical ions of polysilanes [53a,b] and polygermanes [53c] were observed by nanosecond pulse radiolysis. Broad visible and IR absorption spectra due to the radical ions of polysilyne [54] and polygermyne [54] were also observed. Very systematic pulse radiolysis studies on many different kinds of polysilanes [55] have been made by our improved nanosecond pulse radiolysis system over a wide range of

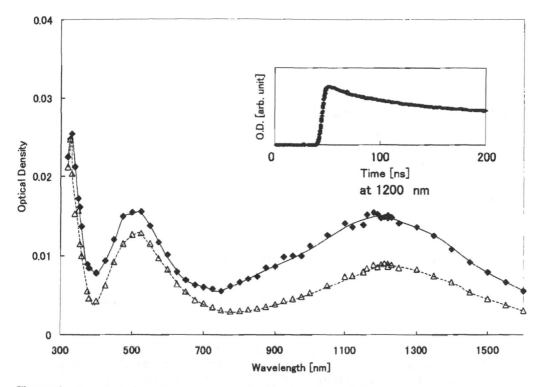

Figure 1 Transient absorption spectra obtained in the pulse radiolysis in 200 mM (base mM unit) polystyrene solutions in CH_2Cl_2 at the pulse end (♦) and 100 nsec after the pulse (Δ). Inset: time-dependent behavior observed at 1200 nm.

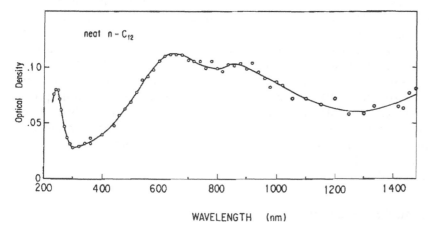

Figure 2 Transient absorption spectra obtained in the pulse radiolysis of neat n-dodecane.

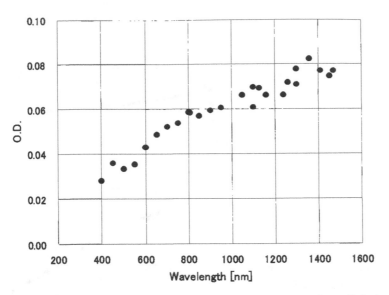

Figure 3   Transient absorption spectra obtained in the pulse radiolysis of neat n-polyolefin.

temperature and with the monitoring wavelength region from 300 to 1600 nm [44]. Both radical cations and anions of polysilanes have very intense sharp near-UV and very intense and broad absorption bands as shown in Fig. 4 [55a]. Transient absorption spectra due to excited states [26,27a,b] and radical cations [27c] of polysilanes were also observed by laser flash photolysis. Detailed studies on radical ions of oligogermanes were performed by nanosecond pulse radiolysis [56].

Poly(Methyl Methacrylate)

A broad, structureless absorption in the 350- to 500-nm region was observed in pulse radiolysis of additive-free poly(methyl methacrylate) (PMMA) in 1969, but the assignment of the absorption was not possible at that time [57]. Then absorption peaks at 725 and 440 nm were ascribed to the radical anion and the radical cation of PMMA, respectively, in 1983 [58]. However, these assignments were not compatible with the results obtained in pulse radiolysis of solutions of PMMA and its substituted analogues [59]. No distinct absorption peak at 440 nm was observed and a short-lived absorption component around 440 nm was attributed to PMMA radical anions [59]. For pulse-irradiated solid PMMA, no distinct absorption peak around 440 nm was observed by the pulse radiolysis of solid PMMA with 0.1-μsec time resolution, from 320 to 620 nm, and the short-lived absorption component around 440 nm was attributed to PMMA radical anions [60]. Transient absorption spectra of irradiated additive-free solid PMMA containing MMA monomers were clearly observed by our improved nanosecond pulse radiolysis system over a wide range of temperature and with the monitoring wavelength region from 300 to 1600 nm [44]. Absorption bands due to PMMA radical anions and MMA dimer anions [61] were observed in the UV and near-IR regions, respectively, as shown in Fig. 5.

3.3.   Diversity of Radiation Effects on Polymers

Classical investigations of radiation effects on polymers have been summarized in some books or review articles in the 1970s to the 1990s [29,62–64]. Many of the works were focused

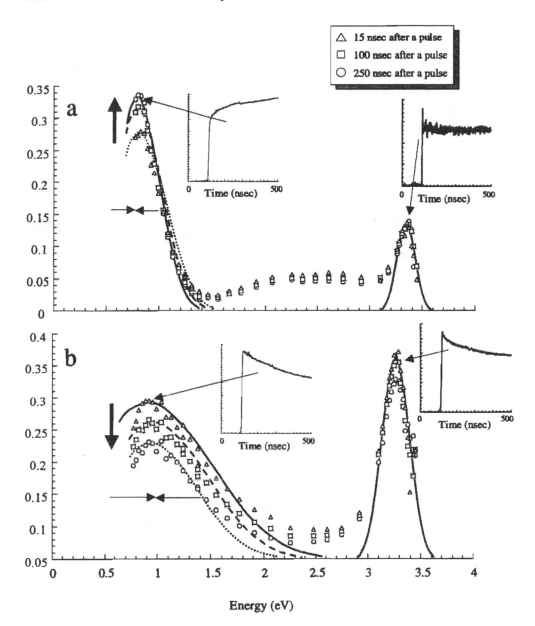

Figure 4 Transient absorption spectra of radical anions (a) and radical cations (b) of polymethylphenylsilane at 15, 100, and 250 nsec after a pulse. Superimposed figures indicate the kinetics traces of transient absorption.

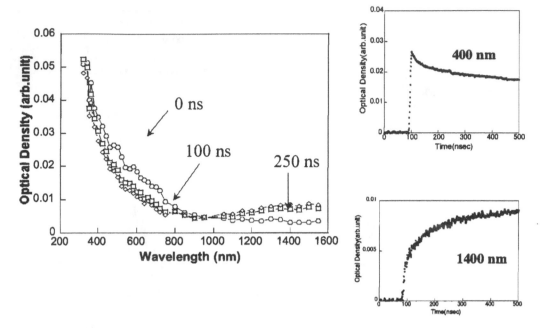

Figure 5 Transient absorption spectra of irradiated additive-free solid PMMA containing MMA monomers observed by the authors' improved nanosecond pulse radiolysis system over a wide range of temperatures and with the monitoring wavelength region from 300 to 1600 nm.

on the reaction mechanisms including main-chain scission and cross-linking induced by the irradiation of low-LET radiation such as ^{60}Co γ-rays, high-energy electron beams, etc. Throughout these studies, the main and general objective was to develop an understanding of the different chemical reactions that occur when a polymer is subjected to radiation and to relate these to observed changes in physical properties. Pioneering studies of the simplest polymeric material, polyethylene, were started by Dole [65] and Charlesby [66] in 1948–1952, demonstrating the drastic improvement in the thermal stability and insolubility for solvents. Since this pioneering study, radiation-induced reaction mechanisms in polyethylene have been vigorously and extensively investigated in detail, which brought up many questions in the fundamental aspects of radiation chemical processes in polymeric materials.

Dose Rate. The dose rate factor is closely related to the atmosphere, temperature, shape of irradiated specimen, etc.; thus the effect of dose rate is hardly discussed separately from the others. An accumulation of reactive intermediates such as radicals basically plays an important role in the dose rate effects in the ratio of main-chain scission and cross-linking reactions. The accumulation, namely, the density of reactive intermediates, is a slow process competing with the consumption by reaction with the molecules in the atmosphere, radical recombination, etc., as well as the consumption by other reactions such as double bond formation. In many cases of irradiation with relatively high dose rate by high-energy EB, these effects may often be neglected because the consumption by atmospheric molecules occurs via diffusion in a polymeric material. However, the effects become crucial for cross-linking in reactive atmosphere such as hydrogel formation or grafting [67], radiation-induced conductivity measurement [68], or sterilization [69] by ionizing radiation. These effects also play an important role in the ultra-high-density energy deposition by charged particles, and will be discussed later.

Temperature. Temperature determines the chain motion of irradiated polymer molecules and is easily predicted to be a crucial factor controlling the efficiency of the reactions induced by radiation. Cross-linking reactions in polyethylene showed a large dependence of the yield on its crystallinity, suggesting the motion of the chain (ends) directly controls the efficiency of cross-linking and main-chain scission. The yield of cross-linking tends to be higher with an increase in the irradiation temperature. One of the recent typical examples was reported by Oshima et al. for the radiation-induced reactions in poly(tetrafluoroethylene) (PTFE) [70]. PTFE was so sensitive to ionizing radiation that its mechanical properties degrade with either a very low dose in air or even under vacuum irradiation, and this had been considered a typical main-chain scission type polymeric system. Oshima et al. reported that the cross-linking reactions of PTFE became predominant by the irradiation at its molten state just above the melting temperature (~608 K) under oxygen-free atmosphere, and antichemical solvents, high frictional properties, and cross-linked PTFE shows remarkable improvement in radiation resistance, mechanical properties, frictional properties, visible light transparency, and so on, compared with non-cross-linked PTFE. There is no other way to attain the cross-linked PTFE than by the above technique. Similar temperature effects were also observed for polycaprolactone, which showed a remarkable increase in the yield of cross-linking at the temperature above its melting point (333 K), keeping the high yield even at RT in the quenched specimen [71]. The inverse effects of temperature on the radiation-induced cross-linking were observed in poly(di-*n*-hexylsilane) of which liquid crystalline transition temperature was 313 K. The polymer indicated efficient formation of cross-linking for EB and ion beam irradiation at RT; however, the predominant reaction abruptly changed into main-chain scission at above that temperature for the same radiation sources. This was mainly attributed to the reduction of very efficient main-chain scission reaction by radical recombination within the tightly packed crystalline phase of the polymers, providing an exceptional example of radiation-induced cross-linking reactions [72].

Structures. Stereotacticity is also an important factor that determines the efficiency of main-chain scission and cross-linking. Polypropylene is one of the sensitive polymers in which radiation-induced reaction depends on the stereoregularity of the polymer backbone. Wei et al. reported that G values of H_2 and CH_4 became smaller with increasing isotacticity of polypropylene, suggesting the higher yield of cross-linking reaction in polypropylene with higher isotacticity [73]. Reduced yield of main-chain scission reaction was also reported in syndiotactic polystyrene in comparison with that in static polystyrene by Takashika et al. [74]. Remarkable differences in the G values of H_2 were also recorded in atactic and syndiotactic polystyrene. The tacticity itself was also affected by the irradiation. High-temperature irradiation with γ-rays caused the reduction of isotacticity of isotactic PMMA [75]. The changes were mainly caused by temporal main-chain scission followed by recombination. The trace of the tacticity change enabled the evaluation of the efficiency of temporal main-chain scission, $G(\mathrm{TCS})$, giving $G(\mathrm{TCS}) = 18.6$, which was much higher than the G value of permanent main-chain scission. The high value of $G(\mathrm{TCS})$ is also very important to reveal the amount of the potential reactive intermediates, hence for the fate of energy deposited by radiation in the polymeric system.

3.4. Reaction Mechanisms of Electron Beam and X-Ray Resists

Electron beam resist has been a key material for mask fabrication in the semiconductor industry. EB and x-ray lithography have recently attracted much attention as not only a next-generation lithography in the semiconductor industry but also a nanofabrication tool in the field of nanotechnology and science. Under these circumstances, the research and development of resist materials for EB and x-ray lithography have become more important than ever. Many types of EB and x-ray resists have been developed. Details of these resists will be discussed in Sec. 19.4.1. In this section, the radiation-induced reaction mechanisms of resists are described.

Conventional resists such as PMMA and chlorinated poly(methylstyrene) (CMS) utilize main-chain scission or cross-linking reactions induced by radiation. As the reaction mechanisms of conventional resists have been reviewed in many books [76], this section will be focused on the reaction mechanisms of chemically amplified EB and x-ray resists expected as a next-generation resist for mass production. For chemically amplified resists, the latent images are at first formed with the acids generated by an exposure. The acids catalyze the reactions such as a deblocking reaction to form real pattern in the following lithographic process, typically during postexposure bake (PEB). In this process, the chain reactions by acid catalysis enhance resist sensitivity. Since the role of radiation is to generate acids in polymer matrix, acid-generation processes will be discussed.

Onium salts have been widely used as an acid generator for photo-, EB, and x-ray resist. In addition, aromatic polymers such as novolak and polyhydroxystyrene have been often used as a base polymer for EB and x-ray resist. The reaction mechanisms in a typical resist system have been investigated by pulse radiolysis [43,52,77–88], SR exposure [79,80,83–85], and product analysis [88]. Figure 6 shows the acid-generation mechanisms induced by ionizing radiation in triphenylsulfonium triflate solution in acetonitrile. The yields of products from electron beam and KrF excimer laser irradiation of 10 mM triphenylsulfonium triflate solution in acetonitrile are shown in Fig. 7 to clarify the

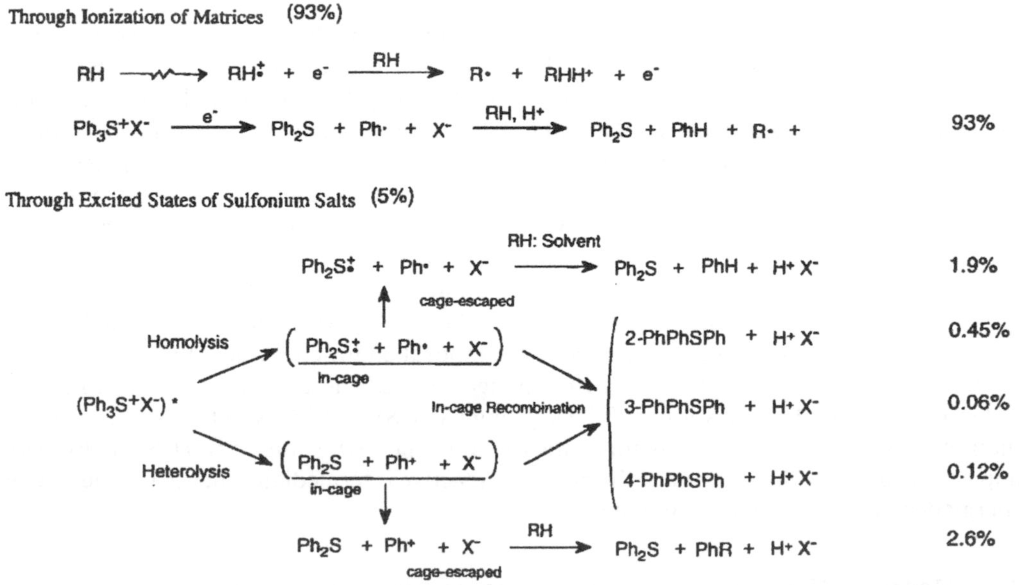

Figure 6 Acid-generation reaction mechanisms induced by ionizing radiation triphenylsulfonium triflate solution. Percentages in the scheme are contributions of each reaction to total acid generation for the 10 mM triphenylsulfonium triflate in acetonitrile.

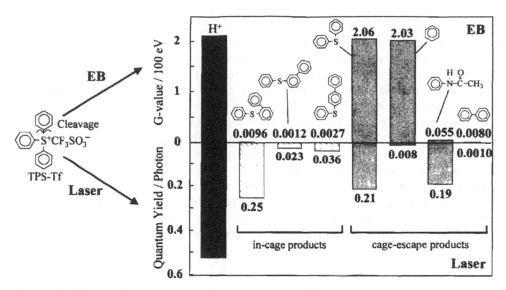

Figure 7 Yields of products from electron beam and KrF excimer laser irradiation of 10 mM triphenylsulfonium triflate solution in acetonitrile.

difference between photoresist and radiation resist. In the case of KrF excimer resist, the acid-generation processes mostly start from the direct excitation of onium salts. The main products from the decomposition of excited sulfonium salts are in-cage recombination products. This means the main proton source is hydrogen in the phenyl moiety in sulfonium salts. Contrary to the photoinduced reaction, the reaction path from the excitation of sulfonium salts accounts little for the yields of acids by EB and x-ray exposure [88].

In EB and x-ray lithography, the energy of radiation is deposited on resist materials mostly via an ionization process. Cation radicals of base resin and electrons are generated by the ionization. The electrons generated by ionization lose their energy through the interaction with surrounding molecules and eventually thermalized. The initial separation distance on average is thought to be approximately several nanometers, depending on the type of materials. Cation radicals of base resin produce proton adducts via ion molecular reaction with the base resin. In the case of novolak, the protons originate in the hydroxyl groups of novolak. Therefore, the yield of acids is thought to be influenced by the protection of hydroxyl groups, which are often partially protected in the polarity change type resists. For chemically amplified EB and x-ray resists, not only the dissolution characteristics of the resists, but also the efficiency of acid generation should be considered when hydroxyl groups are protected in the design of the chemically amplified resists.

Onium salts scavenge electrons generated by ionization and release counteranions of acids. As for various kinds of acid generators including nonionic acid generators, the rate constant of their reactions with solvated electrons in methanol was evaluated [85]. All of these are greater than 1.0×10^{10} M^{-1} sec^{-1} and are considered to be diffusion-controlled reactions. The acids are thought to be formed through the recombination reaction of the proton adducts with decomposed anions of onium salts. The electron- scavenging effect of onium salts delays the recombination of cationic intermediates with electrons and prolongs the lifetime of cationic intermediates. Yields of proton adducts increase with this effect.

In the chemically amplified resists, both the cation radicals of base resin and the electrons generated by the ionization play an important role in the formation process of acids [77]. Therefore, the geminate recombination of the cation radical of base resin with the thermalized electrons is predicted to cause the decrease of the sensitivity. In addition, the migration of thermalized electrons is predicted to cause the severe degradation of space resolution of resists. The high efficiency of reaction of onium salts with thermalized electrons blocks both the geminate recombination and the migration of thermalized electrons in the resist matrix. This prevents the degradation of sensitivity and spatial resolution of the resist caused by the behavior of the thermalized electrons. This mechanism works well in the current chemically amplified resists with spatial resolution around 100 nm. As the dimension of resist pattern becomes closer to the initial separation distance of thermalized electrons, the thermalization process and the subsequent reaction become more important. The initial separation between cation radicals and electrons should be considered in the resist design for nanolithography, as well as the scattering of secondary electrons and the diffusion of acids.

4. INDUSTRIAL APPLICATION OF RADIATION TO POLYMERS

4.1. Electron Beam and X-Ray Resists

With the advent of microelectronics, the human society has progressed to a highly sophisticated information-oriented state. Microelectronic circuits have been fabricated with a high resolution printing technique called lithography. The advance of micro-electronics has highly depended on the technological innovation of lithography. Light sources have changed toward shorter wavelength from the g-line (436 nm) of a Hg lamp to the i-line (365 nm) of a Hg lamp to a KrF excimer laser (248 nm). An ArF excimer laser (193 nm) is ready to be deployed as an exposure tool for mass production. An F_2 excimer laser (157 nm) is being researched as the next-generation light source following the ArF excimer laser. However, the resolving power of light sources will fall short of the market demand in the near future. Radiation sources such as electron beam and x-ray are expected to take the place of the light sources. Furthermore, nanotechnology has recently attracted enthusiastic attention in many fields such as materials science, bioscience, and information technology. EB and x-ray lithography are also promising tools for the mass production of nanostructures. Research and development of each element such as a resist, optics, and a mask for EB and x-ray lithography have been pursued. Especially, the resist is a key technology for mass production and has been investigated energetically.

Mass production poses strict requirements on resist materials, most important of which are sensitivity, spatial resolution, contrast, and etch resistance. The sensitivity of the next-generation resists is required to be less than $10\,\mu C/cm^2$ (EB), $100\,mJ/cm^2$ (x-ray), and 25 mJ/cm^2 (EUV) [89]. It should be noted that the resist sensitivity is traditionally expressed not by absorbed dose but exposure charge or energy per unit area. As for the spatial resolution, 45 nm is needed for the production of dynamic random access memory (DRAM) in 2010 [89]. Although resist patterns below 10 nm are presently fabricated by some kinds of resists, they do not have enough sensitivity required for the mass production [90,91].

Electron beam lithography has been widely used in mask fabrication for photolithography. X-ray lithography has been also used for the fabrication of micromachines such as the LIGA (Lithographie, Galvanoformung, Abformung) process [92]. Many types of resists have been developed for these purposes. Some positive and negative resists for EB lithography are listed in Tables 1 and 2, respectively.

Table 1   Typical Conventional Positive EB Resists

Poly(methyl methacrylate)
Poly(butenesulfone)
Poly(hexafluorobutylmethacrylate)
Poly(trifluoroethyl-α-chloroacrylate)
Poly(methacrylic acid-*co*-phenylmethacrylate)
Poly(2-methylpentene-1-sulfone)/novolac
Poly(dimethyltetrafluoropropylmethacrylate)
Poly(glycidylmethacrylate-*co*-hexafluoropropylmethacrylate)
Poly(α-chloroacrylate)
Poly(isobutylene-*co*-methylmethacrylate)
Poly(methyl-isopropenylketone)
Poly(acrylonitrile-*co*-methylmetacrylate)

From the necessity of higher sensitivity resists than conventional ones, a chemically amplified resist was developed originally for photolithography by Ito of IBM [93]. The concept of "chemically amplified" made a great impact on lithography. Chemically amplified resists have been used on a full scale in the mass-production line of semiconductors since the KrF excimer laser was deployed as an exposure source. The use of chemically amplified resists is also one of the most promising technologies in EB and x-ray lithography. Many types of chemically amplified resists have been developed and many kinds of acid generators, cross-linkers, dissolution inhibitors, and resins have been reported as shown in Table 3. Although chemically amplified resists based on acid catalytic chain reaction mechanisms show high performance such as high sensitivity and high contrast, they have very narrow process latitudes compared with conventional resists. The sensitivities of chemically amplified resists tend to be degraded by impurities as these depend on prebake temperature, prebake time, PEB temperature, PEB time, delay time from exposure to development, and shelf life, among others. These problems are common among photo, EB, and x-ray resists. Early researches focused on these problems to make the chemically amplified resists applicable to the process lines, while the development of new chemically amplified resists have been pursued [94–100]. The feature required for the next generation is so small that line-edge roughness and pattern collapse during development are also problems and have been investigated [101,102]. Line-edge roughness should be less than 5.5 nm (3 sigma) in the production of DRAM in 2010 [89].

Although single-layer resists are preferred from the viewpoint of simplicity of the process, two- or three-layer resists are also promising [103]. These resists have an advantage

Table 2   Typical Conventional Negative
EB Resists

Poly(glycidylmethacrylate)
Epoxy poly(butadiene)
Poly(ethylacrylate-*co*-glycidylmethacrylate)
Chlorinated poly(methylstyrene)
Poly(iodostyrene)
Chlorinated poly(methylnaphthylmethacrylate)

Table 3 Typical Chemically Amplified EB Resists

Resist	Component	Type
No name	Poly(4-*tert*-butoxycarbonyloxystyrene)	Positive
[IBM]	Acid generator	
SAL series	Novolak resin	Negative
[Shipley]	Cross-linker	
	Acid generator	
RAY-PN (AZPN100)	Novolak resin	Negative
[Hoechst AG]	Melamine cross-linker	
EXP	Novolak resin	Positive
[NTT]	2,2-*bis*(*t*-Butoxycarbonyloxyphenyl)-propane	
	bis(*p-t*-Butylphenyl)iodonium trifluoromethanesulfonate	
PTBSS	Poly(4-*tert*-butoxycarbonyloxystyrene-*co*-sulfur dioxide)	Positive
[AT & T]		

over the single-layer resists in that multilayer resists can allot different roles to each layer. Typically, a top layer bears the role of patterning and lower layers have the responsibilities for etch resistance and planation of the substrate surface. Surface imaging technique is also one of the candidates of the next-generation processes [104]. Imaging by silylation of commercially available resists was reported. These techniques are expected to be useful especially for EUV and a low-energy (a few kilovolts) EB lithography such as low energy e-beam proximity projection lithography (LEEPLE).

In recent progress, calixarene resists have been prominent [105]. Hexaacetate *p*-methylcalix[6]arene was demonstrated to work as a high-resolution negative resist. This resist also shows high etch resistance. Calixarene resist has an advantage in its molecular size (about 1 nm). Liquid crystal resists and inorganic resists show high resolution [106]. These resists are suitable for the fabrication of nanostructures. However, the resist sensitivities are lower than those of chemically amplified resists, even PMMA.

It has been reported that the incorporation of nanoparticles such as fullerene and oligosilsesquioxane (POSS) to resist matrix improves the resist performance such as etch resistance and mechanical properties [107]. These resists are called nanocomposite resists. A single-component chemically amplified resist that incorporates not only POSS but also an acid generator into its main chain has been reported [108].

Although many types of excellent resists have been developed based on novel ideas, the requirements for the next-generation resists have not yet been achieved. The quest for the ultimate resist materials will continue.

4.2. Other Applications of Radiation to Polymers

Preparation, characterization, and fabrication of nanostructured materials have elicited great interest in view of their possibility of innovation in science, industry, environment, and in our daily lives in the twenty-first century. The research field called "nanoscience" or "nanotechnology" has been expanded and widespread since the end of the last century, and the trend is expected to be wider and accelerated in this and in the next decade. Current interest in the nanostructured materials predominantly comes from their unique physico-chemical properties on account of their finite small sizes giving peculiar effects such as quantum size effect and quantum entrapment. Recent progress on the nature of nano-structured materials has revealed unique electronic, magnetic, optical, and mechanical properties, and the application of these properties has already started in industry. As introduced in the previous section, nanoscale materials, to say nothing of their importance in the semiconductor industry, are also expected to play a crucial role in the integrated electronic circuit in the near future, which is obvious in the well-known Moore's law and its leading road map of semiconductor manufacturing.

In view of the preparation and fabrication of nanostructured materials, there are apparently two kinds of trends. One is the assembling of molecules or atoms by using their nature such as self-organization, self-reproduction, selective reaction, and condensation, which is called "bottom-up" technology, to prepare the nanostructured materials. The other approach, called the "top-down" technique, producing very small structures by using the reaction caused by charged particles and photons, has been the major trend of micro- and nanofabrication to date, and will be a key technology even in the future, because it is necessary to prepare the interface of nanostructured materials for their real application even if the materials themselves are produced by the bottom-up technique.

It has been considered that the most important advantage of radiation-induced re-actions in the fundamental research is their nonselectivity in the media or spatial homoge-neity of deposited energy, hence the distribution of reactive intermediates. It has been the basic feature of radiation chemistry, not only in polymeric systems but also in organic and inorganic systems, which has given much benefit to us as an accurate quantitativeness typically represented by G values. However, this nonselective and homogeneous concept is no longer valid for the reaction induced by high-LET radiation including ion beams, which has been suggested by the theoretical aspects on track structure in the condensed media. The ultrahigh density of excitation and ionization within a charged particle track produce extremely nonhomogeneous distribution of reactive intermediates along a particle trajec-tory. The density of the excitation, ionization, and intermediates cannot be realized by any other physicochemical techniques, and has potentials to cause "brand-new" chemical reactions in the matter. If the yield of a chemical reaction can be controlled by the density of the intermediates, we can freely control the size of the nonhomogeneous field of chemical reaction, which is called a "chemical core," because the density of deposited energy is basically presented as the function of radial distance from a particle trajectory.

Several experimental studies on the particle track suggested that the size of the chemical core is controllable by selecting the target media and/or chemical treatment after irradiation within the range from a few to a few tens of nanometers, which shows the best fit to produce the nanosized materials. An incident charged particle promotes a top-down-style nonhomogeneous field of chemical reaction, and its high controllability is provided by the bottom-up nature of the target material. This is the concept of the complex of top-down and bottom-up styles, which is one of the practical examples of radiation chemistry contributing to nanoscience and nanotechnology.

Basics. It has been suggested that spatial distribution of deposited energy by charged ions has played a significant role in chemical reactions occurring in the target materials. Models of the energy distribution were proposed experimentally and theoretically such as "Track Core" and "Penumbra" models by Magee and Chattarjee (for a review, see Ref. 109), Katz et al. [110], Waligorski et al. [111], Varma et al. [112,113]. Wingate and Baum [114], Wilson [115], and other groups. In spite of the theoretical modeling effort, there are still unknown factors in the relationship between the ion track structure and the values of track radii that were experimentally obtained by the analysis of irradiation products [72,116–118].

As one of the basic formulations accounting for the radial dose distribution in an ion track, the following formulas were given [92]:

$$\rho_c(r) = \frac{LET}{2}\left[\pi r_c^2\right]^{-1} + \frac{LET}{2}\left[2\pi r_c^2 \ln\left(\frac{e^{1/2}r_p}{r_c}\right)\right]^{-1} \qquad r \leq r_c$$

$$\rho_p(r) = \frac{LET}{2}\left[2\pi r^2 \ln\left(\frac{e^{1/2}r_p}{r_c}\right)\right]^{-1} \qquad r_c < r \leq r_p$$

(5)

where ρ_c and ρ_p are the deposited energy density at the core and the penumbra region, respectively, r_c and r_p are the radii of the core and the penumbra, respectively, and e is the base of natural logarithm.

Experimental approaches to measure the radial dose distribution are also in progress [119], and it was found that the distribution follows r^{-2} law in the inner region of a critical distance and obeys r^{-3} law outside of that region. LaVerne and Schuler reported the considerable decrease in the radiation chemical yield for ferric production in the Fricke dosimeter, suggesting a model of a deposited energy density in an ion track, which depends on the LET and the atomic number of an irradiation particle [120,121].

Puglisi and Licciardello [122], Licciardello and Puglisi [123], and Calcagno et al. [124] also reported the effects of ion beam bombardment to polystyrene leading to the aggregation of molecules and cross-linking reactions. The abnormal change in molecular weight distribution was ascribed to the intratrack reaction; however, the estimated size of ion tracks was also larger than that of the track core. In many cases, the effects of the intratrack nonhomogeneous reaction have been presented as the LET effects on the chemical yield of radiation-induced reaction in the polymeric system. The LET effects on the yield of cross-linking reaction were reported in polystyrene [125], polyethylene [126], polysulfone [126,127], and polysilane [72,116,128], showing remarkable increase in the yield with an increase in the value of LET of incident particles. Especially, polysulfone and polysilane showed a drastic conversion from scission type (for low LET) to cross-linking (for high LET) with the threshold value of LET [72,126]. Decrease in the yield of main-chain scission was also observed in PMMA with an increase in the LET. This experimental evidence can be explained in part by the model of the nonhomogeneous reaction in an ion track in which high-density reactive intermediates cause an increase in the yield of radical recombination, cross-linking reaction, excited-state relaxation, cross-recombination of charges, etc. Ion-beam-induced chemical reaction in other media was reviewed in other books [129] or proceedings, and the present section introduces the present status of application of charged particles in relation with nanoscience and nanotechnology.

Charged Particles for Nanofabrication. The basic approaches to nanofabrication by using the nonhomogeneous reaction for nanofabrication can be divided into the next two concepts: (1) making pores in polymer film along a particle track and filling with metals, semiconductors, ceramics, polymers, etc. and (2) directly making a wire along an ion track based on the target polymer materials. The schematic view of these ideas is displayed in Fig. 8. The first approach surpasses the second in the choice of the materials to make, and the second can prepare more precise structures of sizes smaller than 10 nm. The first approach began from the preparation of membranes having several microsized pores produced by etching of charged particle tracks in nuclear materials (for reviews, see Refs. 130 and 131). The concept of etched ion track and its filling was first presented by Vetter and Spohr in 1993 [132] and at almost the same time by Ferain and Legras [133]. Their investigations were done in parallel and overlapped with each other.

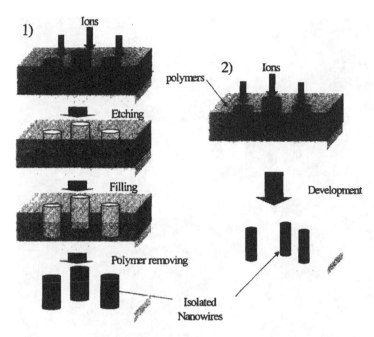

Figure 8 Schematic view of the two different concepts to prepare nanostructured materials using chemical reactions in a particle track (chemical core). Bar length: 10 μm.

The attempt to use the membrane as the template of metallic replicas was successful for preparation of copper whiskers on the substrate [132]. Vetter and Spohr used polycarbonate membranes with a thickness of 10–150 μm, and the films were irradiated by heavy-ion beam (Xe and heavier). The irradiated films were etched by alkaline solution, and the etched pore sizes were controlled by the time of the treatment. The copper metal was deposited by electrolysis in a galvanic cell, and after the deposition, the membrane was removed mechanically or by a suitable solvent. Figure 9 shows the SEM image of copper microwhiskers on the substrate prepared by the above technique. The size of the whiskers was a few micrometers as for the diameter of the cross section in this image. Fundamental studies on the mechanism of etching processes have shown remarkable progress in the controllability of the size of the etched pores [134], and are going on still now with the collaboration of other groups to produce a drug-releasing system [135], thermoresponsive membrane [136], etc.

Similar approach has also been taken by Ferain and Legras [133,137,138] and De Pra et al. [139] to produce nanostructured materials based on the template of the membrane with etched pores. Polycarbonate film was also of use as the base membrane of the template, and micro- and nanopores were formed by precise control of the etching procedure. Their most resent report showed the successful formation of ultrasmall pores and electrodeposited materials of which sizes were as much as 20 nm [139]. Another attractive point of these studies is the deposited materials in the etched pores. Electrochemical polymerization of conjugated polymer materials was demonstrated in these studies, and the nanowires based on polypyrrole or polyaniline were formed with a fairly cylindrical shape reflecting the side wall structure of the etched pores. Figure 10 indicates the shape of the polypyrrole microwires with their dimension changes by the limitation of the thickness of the template.

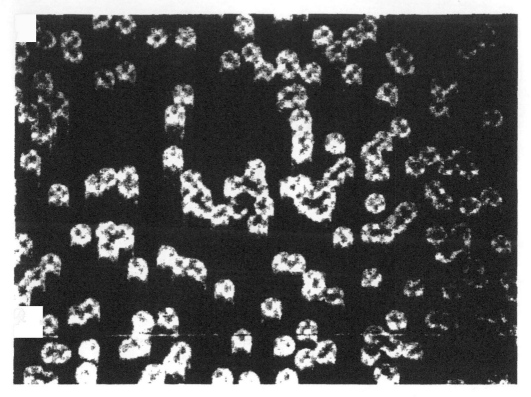

Figure 9 Overview of copper whisker array prepared by electrodeposition in the membrane with micropores, and dissolution of the polymer matrix by a solvent.

The direct formation of nanostructured materials was first reported by Seki et al. as the nonhomogeneous cross-linking reactions of polysilane giving "nanogels" or nanowires along a particle trajectory in the polymer [72,116,140]. The schematic procedure to obtain isolated nanowires is presented in Fig. 8. An incident ion deposits its energy densely within a limited area, and a cross-linked, cylinder-like structure is formed along the projectile. The uncross-linked part of a film is washed by solvent, leading to the isolated nanowires of cross-linked polysilanes. Figure 11 shows a series of atomic force microscope (AFM) images for nanowires formed on a Si substrate, together with the enlarged view of a wire and the schematic of a cross section. The section of a wire indicates a complete cylindrical shape. The number of observed nanowires in a unit area shows good agreement with that of incident particles. In spite of a development procedure by solvents, the wires remain adhered to the Si substrate and the nanowires remain isolated from each other. This indicates that one end of the wire is tightly connected to the substrate by chemical bonds. Another interesting feature of the wires is their length. The film with 170-nm initial thickness is used to obtain the images in Fig. 11. The mean length of the nanowires in Fig. 11 is 180 ± 9 nm, suggesting that the length of a wire is fairly controlled by simply changing the thickness of a target film. Figure 12 shows the AFM images of the nanowires produced in polysilanes with a variety of molecular weights and by the different ion beams. The structure of the nanowire apparently changes from a thick rodlike to a fine wormlike wire with a decrease in the molecular weight, and reflects the film thickness as the length of nanowires. The thickness of the nanowire obviously depends also on the LET of ion beams.

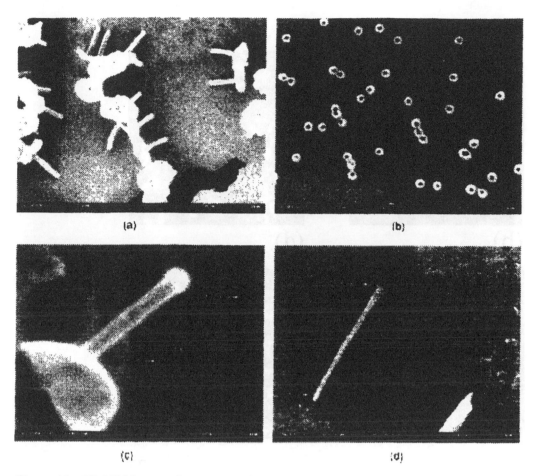

Figure 10 FE-SEM images of polypyrrole/ClO4 nanotubes obtained by electropolymerization in the pores of supported nanoporous template with thickness of 350 nm (a, b, c) and 1.3 μm (d).

The formation of nanowire by the nonhomogeneous reaction brought up the need for a new model dealing with the cross-linking reaction and gelation caused by high-LET radiation on the basis of the nonrandom, nonhomogeneous spatial distribution of cross-links. A preliminary expression of form for gel and sol fractions is given by [72]

$$s = \{1 - (r + \delta r)\}^f$$
$$g = 1 - \{1 - (r + \delta r)\}^f \tag{6}$$

where s denotes the fraction of a soluble part of a film, g is the fraction of insoluble parts reflecting the total volume of the wires, r is the radius of a chemical track determined by the energy deposition rate of an incident ion, δr is the differential radius of the track, which depends on the shape and size of an individual polymer molecule, and f is the fluence of the incident ion beams. The trace of the total volume of nanowires (gels) give the value of r in the polymeric system for the incident charged particle based on the Eq. (6). The obtained values of r show good agreement with those measured from AFM images of nanowires for polysilane and heavy ions with high LET [140].

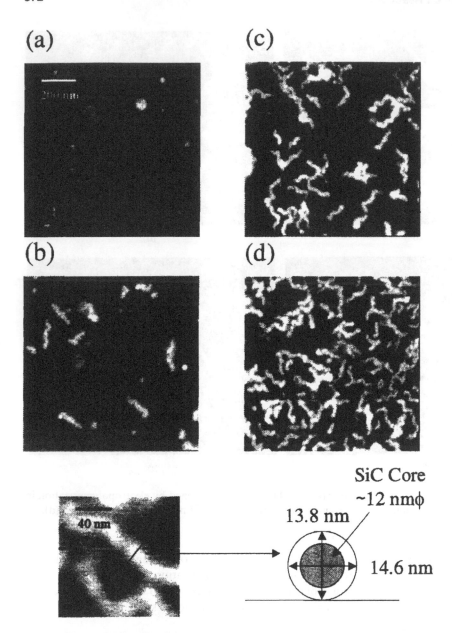

Figure 11 AFM images of the nanowires. (a) Top view observed after development of nonirradiated films. Images (b), (c), and (d) indicate the surface morphology of the developed films of poly(methylphenylsilane) after ion irradiation by 450-MeV $^{129}Xe^{23+}$ at 2.2×10^9, 7.1×10^9, and 1.1×10^{10} ions/cm^2, respectively. The tone changing from dark to bright in this figure implies the height as much as 24 nm. The enlarged view of a nanowire is observed for the same specimen as in (c).

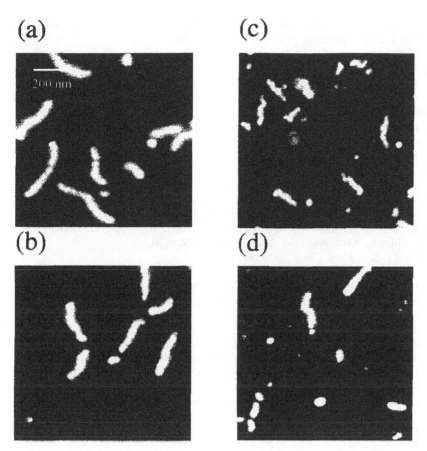

Figure 12 Changes in the sizes of the nanowires. Images (a), (b), and (c) are the top view of nanowires formed by the irradiation of 450-MeV $^{129}\text{Xe}^{23+}$ to a poly(methylphenylsilane) (PS) (Mn = 4.2×10^5) film at 700-nm thickness, a PS (Mn = 1.5×10^5) film at 400-nm thickness, and a PS (Mn = 5.0×10^3) film at 180-nm thickness, respectively. (d) Observed for a PS (Mn = 4.2×10^5) film at 310-nm thickness irradiated by 520-MeV $^{84}\text{Kr}^{20+}$.

The nonhomogeneous reaction caused by high-LET charged particles is a unique technique to produce nanosized cylindrical structures. It not only produces isolated nanowires on substrate but also controls the size and length of wires. It is needless to say that homogeneous processes by radiation have played a very important role in the radiation chemistry; however, nonhomogeneous processes might take part in future technologies as a powerful tool for the preparation of nanostructured materials, although the process is a rather extraordinary case in the field of conventional radiation chemistry.

REFERENCES

1a. Management Information Services, In *The Untold Story: Economic and Employment Benefits of the Use of Radioactive Materials*; Organizations United for Responsible Low-Level Radioactive Waste Solutions, 1994.

1b. *Economic and Employment Benefits of the Use of Nuclear Energy to Produce Electricity*; The U.S. Council for Energy Awareness: Washington, DC, 1994.

1c. In *The Untold Story: Economic and Employment Benefits of Nuclear Technologies*; Organizations United for Responsible Low-Level Radioactive Waste Solutions, 1997.

2. Walter, A.E. The visible but not imaginary numbers. In *ANS International Conference on Mathematics and Computations*, Reactor Physics and Environmental Analyses: Portland, Oregon, 1995.

3. Non-reactor nuclear technology applications in the 21st Century, In ANS Meeting, Albuquerque: New Mexico, 1997.

4. Tagawa, S.; Kashiwagi, M.; Kamada, T.; Sekiguchi, M.; Hosobuchi, K.; Tominaga, H.; Makuuchi, K. J. Nucl. Sci. Technol. 2002, *39*, 1002.

5. Kume, T.; Amano, E.; Nakanishi, T.M.; Chino, M. J. Nucl. Sci. Technol. 2002, *39*, 1106.

6. Inoue, T.; Hayakawa, K.; Shiotari, H.; Takada, E.; Torikoshi, M.; Nagasawa, K.; Hagiwara, K.; Yanagisawa, K. J. Nucl. Sci. Technol. 2002, *39*, 1114.

7. Yanagisawa, K.; Kume, T.; Makuuchi, K.; Tagawa, S.; Chino, M.; Inoue, T.; Takehisa, M.; Hagiwara, M.; Shimizu, M. J. Nucl. Sci. Technol. 2002, *39*, 1120.

8. Burton, M.; Lipsky, S. J. Phys. Chem. 1957, *61*, 1461.

9a. Dextor, D.L. J. Chem. Phys. 1953, *21*, 836.

9b. Inokuti, M.; Hirayama, F. J. Chem. Phys. 1965, *43*, 1978.

10a. Birks, J.B.; Dyson, D.J.; Munro, I.H. Proc. Roy. Soc. A 1963, *275*, 575.

10b. Birks, J.B. Nature 1967, *214*, 1187.

11a. Basile, L.J. Trans. Faraday Soc. 1964, *52*, 4987.

11b. Kistiakowsky, G.B.; Parmenter, C.S. J. Chem. Phys. 1965, *42*, 2942.

12. Vala, M.T.; Haebig, J.; Rice, S.A. J. Chem. Phys. 1965, *43*, 886.

13a. Förster, T. In *Modern Quantum Chemistry*; Sinanoglu, O. Ed.; Academic Press: New York, 1966; Vol. 3, 93 pp.

13b. Förster, T. In *Energetics and Mechanisms in Radiation Biology*; Phillips, G.O. Ed.; Academic Press: New York, 1968; 183 pp.

14a. Simpson, W.T. J. Am. Chem. Soc. 1951, *73*, 5363.

14b. Simpson, W.T. J. Am. Chem. Soc. 1955, *77*, 6164.

15. Dole, M.; Fallgatter, M.B.; Katsuura, K. J. Phys. Chem. 1966, *70*, 628.

16a. Holroyd, R.A.; Yang, J.Y.; Servedio, F.M. J. Chem. Phys. 1967, *46*, 4540.

16b. Holroyd, R.A.; Yang, J.Y.; Servedio, F.M. J. Chem. Phys. 1968, *48*, 1331.

17a. Partridge, R.F. J. Chem. Phys. 1970, *52*, 1277.

17b. Partridge, R.F. J. Chem. Phys. 1970, *52*, 2501.

18. Sauer, M.C.; Mani, I. J. Phys. Chem. 1968, *72*, 3856.

19. Zeman, A.; Heusinger, H. Radiochim. Acta. 1967, *8*, 149.

20a. Harrah, L.A. Mol. Cryst. Liq. Cryst. 1969, *9*, 197.

20b. Wilske, J.; Heusinger, H. J. Polym. Sci. Part A 1969, *7*, 995.

21a. Tagawa, S.; Schnabel, W.; Washio, M.; Tabata, Y. Radiat. Phys. Chem. 1981, *18*, 1087.

21b. Tagawa, S. Radiat. Phys. Chem. 1986, *27*, 455.

21c. Itagaki, H.; Horie, K.; Mita, I.; Washio, M.; Tagawa, S.; Tabata, Y. J. Chem. Phys. 1983, *79*, 3996.

21d. Itagaki, H.; Horie, K.; Mita, I.; Washio, M.; Tagawa, S.; Tabata, Y.; Sato, H.; Tanaka, Y. Macromolecules 1987, *20*, 2774.

22. Tagawa, S.; Washio, M.; Tabata, Y. Chem. Phys. Lett. 1979, *68*, 276.

23. Rauscher, H.; Bässler, H.; Bradley, D.D.C.; Hennecke, M. Phys. Rev. B 1990, *42*, 9830.

24. Hennecke, M.; Damerau, T.; Müllen, K. Macromolecules 1993, *26*, 3411.

25. Heun, S.; Mahrt, R.F.; Greiner, A.; Lemmer, U.; Bässler, H.; Halliday, D.A.; Bradley, D.D.C.; Burn, P.L.; Holmes, A.B. J. Phys. Condens. Matter 1993, *5*, 247.

26a. Thorne, J.R.G.; Hochstrasser, R.M.; Zeigler, J.M. J. Phys. Chem. 1988, *92*, 4275.

26b. Kim, Y.R.; Lee, M.; Thorne, J.R.G.; Hochstrasser, R.M. Chem. Phys. Lett. 1988, *145*, 75.

27a. Matsui, Y.; Seki, S.; Iwamoto, T.; Tagawa, S. Chem. Lett. 1998, *861*.

27b. Matsui, Y.; Seki, S.; Tagawa, S. Chem. Phys. Lett. 2002, *357*, 346.

27c. Seki, S.; Tsuji, S.; Nishida, K.; Matsui, Y.; Yoshida, Y.; Tagawa, S. Chem. Lett. 2002, *1187*.

28a. Kouchi, N.; Tagawa, S.; Kobayashi, H.; Tabata, Y. Radiat. Phys. Chem. 1989, *34*, 453.

28b. Kouchi, N.; Aoki, Y.; Shibata, H.; Tagawa, S.; Kobayashi, H.; Tabata, Y. Radiat. Phys. Chem. 1989, *34*, 759.

28c. Shibata, H.; Seki, S.; Tagawa, S.; Yoshida, Y.; Ishigure, K. Nucl. Instrum. Methods Phys. Res. B 1995, *105*, 42.

29. Dole, M. The Radiation Chemistry of Macromolecules. Academic Press: New York, 1972.

30a. Charlesby, A. Proc. R. Soc. Lond. A 1954, *222*, 60.

30b. Charlesby, A. Proc. R. Soc. Lond., A 1954, *224*, 120.

31. Kuhn, W. Berichte der Deutschen Chemischen Gesellschaft 1930, *63*, 1503.

32. Flory, P.J. J. Am. Chem. Soc. 1941, *63*, 3083, 3091, 3096.

33. Charlesby, A. J. Polym. Sci. 1953, *11*, 513.

34a. Saito, O. J. Phys. Soc. Jpn. 1958, *13*, 198, 1451, 1465.

34b. Saito, O.; Kang, H.Y.; Dole, M. J. Chem. Phys. 1967, *46*, 3607.

35a. Inokuti, M. J. Chem. Phys. 1960, *33*, 1607.

35b. Inokuti, M. J. Chem. Phys. 1963, *38*, 2999.

36. O'Donnell, J.H.; Winzor, C.L.; Winzor, D.J. Macromolecules 1990, *23*, 167.

37. Zhang, Y.F.; Ge, X.W.; Sun, J.Z. Radiat. Phys. Chem. 1990, *35*, 163.

38. Olejniczak, J.; Rosiak, J.; Charlesby, A. Radiat. Phys. Chem. 1991, *37*, 499.

39. Tagawa, S. In *Radiation Effects on Polymers: ACS Symposium Series 475*; Clough, R.L., Shalaby, S.W., Eds.; American Chemical Society: Washington, DC, 1991; 2 pp.

40. Tagawa, S. In *CRC Handbook of Radiation Chemistry*; Tabata, Y., Ito, Y., Tagawa, S., Eds.; CRC Press, Inc.: Boston, 1991; Chap. XIV, 739 pp.

41. Tagawa, S. In *CRC Handbook of Radiation Chemistry*; Tabata, Y., Ito, Y., Tagawa, S., Eds.; CRC Press, Inc.: Boston, 1991.

42. Yoshida, Y.; Mizutani, Y.; Kozawa, T.; Saeki, A.; Seki, S.; Tagawa, S.; Ushida, K. Radiat. Phys. Chem. 2001, *60*, 313.

43a. Kozawa, T.; Mizutani, Y.; Miki, M.; Yamamoto, T.; Suemine, S.; Yoshida, Y.; Tagawa, S. Nucl. Instrum. Methods, A 2000, *440*, 251.

43b. Kosawa, T.; Saeki, A.; Yoshida, Y.; Tagawa, S. Jpn. J. Appl. Phys. 2002, *41*, 4208.

44a. Seki, S.; Kunimi, Y.; Nishida, K.; Yoshida, Y.; Tagawa, S. J. Phys. Chem. B 2001, *105*, 900.

44b. Seki, S.; Yoshida, Y.; Tagawa, S.; Asai, K. Macromolecules 1999, *32*, 1080.

45a. Tagawa, S.; Schnabel, W. Chem. Phys. Lett. 1980, *75*, 120.

45b. Tagawa, S.; Schnabel, W. Makromol. Chem. Rapid Commun. 1980, *1*, 345.

45c. Tagawa, S.; Schnabel, W. Polym. Photochem. 1983, *3*, 203.

46. Washio, M.; Tagawa, S.; Tabata, Y. Radiat. Phys. Chem. 1983, *21*, 239.

47. Okamoto, K.; Kozawa, T.; Yoshida, Y.; Tagawa, S. Radiat. Phys. Chem. 2001, *60*, 417.

48. Tagawa, S.; Hayashi, N.; Yoshida, Y.; Washio, M.; Tabata, Y. Radiat. Phys. Chem. 1989, *34*, 503.

49. Yoshida, Y.; Ueda, T.; Kobayashi, T.; Tagawa, S. J. Photopolym. Sci. Tech. 1991, *4*, 171.

50. Tagawa, S.; Yoshida, Y. Unpublished data.

51. Tabuse, S.; Izumi, Y.; Kojima, T.; Yoshida, Y.; Kozawa, T.; Miki, M.; Tagawa, S. Radiat. Phys. Chem. 2001, *62*, 179.

52a. Saeki, A.; Kozawa, T.; Yoshida, Y.; Tagawa, S. Radiat. Phys. Chem. 2001, *60*, 319.

52b. Saeki, A.; Kozawa, T.; Yoshida, Y.; Tagawa, S. Jpn. J. Appl. Phys. 2002, *41*, 4213.

53a. Ban, H.; Sukegawa, K.; Tagawa, S. Macromolecules 1987, *20*, 177.

53b. Ban, H.; Sukegawa, K.; Tagawa, S. Macromolecules 1988, *21*, 45.

53c. Ban, H.; Tanaka, A.; Hayashi, N.; Tagawa, S.; Tabata, Y. Radiat. Phys. Chem. 1989, *34*, 587.

54. Watanabe, A.; Komatsubara, T.; Matsuda, M.; Yoshida, Y.; Tagawa, S. Macromol. Chem. Phys. 1995, *196*, 1229.

55a. Seki, S.; Yoshida, Y.; Tagawa, S. Macromolecules 1999, *32*, 1080.

55b. Seki, S.; Yoshida, Y.; Tagawa, S. Radiat. Phys. Chem. 2001, *60*, 411.

55c. Seki, S.; Kunimi, Y.; Nishida, K.; Yoshida, Y.; Tagawa, S. J. Phys. Chem. B 2001, *105*, 900.

55d. Seki, S.; Yoshida, Y.; Tagawa, S. J. Phys. Chem. B 2002, *106*, 6849.

56a. Mochida, K.; Hata, R.; Chiba, H.; Seki, S.; Yoshida, Y.; Tagawa, S. Chem. Lett. 1998, *263*.

56b. Mochida, K.; Kuwano, N.; Nagao, H.; Seki, S.; Yoshida, Y.; Tagawa, S. Chem. Lett. 1999, *3*.

56c. Mochida, K.; Kuwano, N.; Nagao, H.; Seki, S.; Yoshida, Y.; Tagawa, S. Inorg. Chem. Commun. 1999, *2*, 238.

57. Ho, S.K.; Siegel, S. J. Chem. Phys. 1969, *50*, 1142.

58. Tabata, M.; Nilsson, G.; Lund, A.; Shoma, J. J. Polym. Sci. Polym. Chem. Ed. 1983, *21*, 3257.

59. Ogasawara, M.; Tanaka, M.; Kobayashi, H. J. Phys. Chem. 1987, *91*, 937.

60. Szadkowska-Nicze, M.; Kiszka, M.; Mayer, J. J. Polym. Sci. Polym. Chem. Ed. 1983, *21*, 3257.

61. Nakano, A.; Okamoto, K.; Kozawa, T.; Tagawa, S. Unpublished data.

62a. Clough, R.L.; Shalaby, S.W. In *Radiation Effects on Polymer: ACS Symposium Series 475*. American Chemical Society: Washington, DC, 1991.

62b. Clough, R.L.; Shalaby, S.W. In *Irradiation of Polymers: ACS Symposium Series 620*. American Chemical Society: Washington, DC, 1996.

63. Tabata, Y.; Ito, Y.; Tagawa, S. In *CRC Handbook of Radiation Chemistry*. CRC Press: Boston.

64. Jonah, C.D.; Rao, B.S.M. In *Radiation Chemistry Present Status and Future Trends*. Elsevier: Amsterdam, 2001.

65. Dole, M. In *Early Developments in Radiation Chemistry*; Kroh, J. Ed.; Royal Society of Chemistry: London, 1989; Chap. 2. 81 pp.

66. Charlesby, A. In *Early Developments in Radiation Chemistry*; Kroh, J. Ed.; Royal Society of Chemistry: London, 1989; Chap. 3. 29 pp.

67. Wach, R.A.; Mitomo, H.; Toshii, F.; Kume, T. J. Appl. Polym. Sci. 2001, *81*, 3030.

68. Zhutayeva, Y.A.; Khatipov, S.A. Nucl. Instrum. Methods Phys. Res. B 1999, *151*, 372.

69. Woo, L.; Ling, M.T.K.; Ding, S.Y.; Westphal, S.P. Thermochim. Acta 1998, *324*, 179.

70a. Oshima, A.; Tabata, Y.; Kudoh, H.; Seguchi, T. Radiat. Phys. Chem. 1995, *45*, 269.

70b. Oshima, A.; Ikeda, S.; Seguchi, T.; Tabata, Y. Radiat. Phys. Chem. 1997, *50*, 519.

70c. Oshima, A.; Seguchi, T.; Kudoh, H.; Tabata, Y. Radiat. Phys. Chem. 1997, *50*, 611.

71. Darwis, D.; Mitomo, H.; Enjoji, T.; Yoshii, F.; Makuuchi, K. J. Appl. Polym. Sci. 1998, *68*, 581.

72a. Seki, S.; Kanzaki, K.; Yoshida, Y.; Tagawa, S.; Shibata, H.; Asai, K.; Ishigure, K. Jpn. J. Appl. Phys. 1997, *36*, 5361.

72b. Seki, S.; Maeda, K.; Kunimi, Y.; Yoshida, Y.; Tagawa, S.; Kudoh, H.; Sugimoto, M.; Sasuga, T.; Seguchi, T.; Iwai, T.; Shibata, H.; Asai, K.; Ishigure, K. J. Phys. Chem. B 1999, *103*, 3043.

73. Wei, G.; Qiao, J.; Xuan, H.; Zhang, F.; Wu, J. Radiat. Phys. Chem. 1998, *52*, 237.

74. Takashika, K.; Oshima, A.; Kuramoto, M.; Seguchi, T.; Tabata, Y. Radiat. Phys. Chem. 1999, *55*, 399.

75. Dong, L.; Hill, D.J.T.; O'Donnell, J.H.; Carswell-Pomerantz, T.G.; Pomery, P.J.; Whittaker, A.K.; Hatada, K. Macromolecules 1995, *28*, 3681.

76. For example, Polymeric Materials for Microelectronic Applications, eds. Ito, H.; Tagawa, S.; Horie, K. American Chemical Society: Washington, DC, 1994.

77. Kozawa, T.; Yoshida, Y.; Uesaka, M.; Tagawa, S. Jpn. J. Appl. Phys. 1992, *31*, 4301.

78. Kozawa, T.; Uesaka, M.; Yoshida, Y.; Tagawa, S. Jpn. J. Appl. Phys. 1993, *32*, 6049.

79. Watanabe, T.; Yamashita, Y.; Kozawa, T.; Yoshida, Y.; Tagawa, S. *Polymeric Materials for Microelectronic Applications*; Ito, H., Tagawa, S., Horie, K., Eds.; American Chemical Society: Washington, DC, 1994; Chap. 8 110–120.

80. Watanabe, T.; Yamashita, Y.; Kozawa, T.; Yoshida, Y.; Tagawa, S. J. Vac. Sci. Technol. B 1994, *12*, 3879.

81. Kozawa, T.; Uesaka, M.; Watanabe, T.; Yamashita, Y.; Shibata, H.; Yoshida, Y.; Tagawa, S. *Polymeric Materials for Microelectronic Applications*; Ito, H., Tagawa, S., Horie, K., Eds.; American Chemical Society: Washington, DC, 1994; Chap. 9 121–129.

82. Kozawa, T.; Uesaka, M.; Watanabe, T.; Yamashita, Y.; Yoshida, Y.; Tagawa, S. J. Photopolym. Sci. Technol. 1995, *8*, 37.

83. Nagahara, S.; Yamashita, Y.; Taguchi, T.; Kozawa, T.; Yoshida, Y.; Tagawa, S. Jpn. J. Appl. Phys. 1996, *35*, 6491.
84. Nagahara, S.; Yamashita, Y.; Taguchi, T.; Kozawa, T.; Tagawa, S. J. Photopolym. Sci. Technol. 1996, *9*, 619.
85. Kozawa, T.; Nagahara, S.; Yoshida, Y.; Tagawa, S.; Watanabe, T.; Yamashita, Y. J. Vac. Sci. Technol. B 1997, *15*, 2582.
86. Nagahara, S.; Kozawa, T.; Yamamoto, Y.; Tagawa, S. J. Photopolym. Sci. Technol. 1998, *11*, 577.
87. Tsuji, S.; Kozawa, T.; Yamamoto, Y.; Tagawa, S. J. Photopolym. Sci. Technol. 2000, *13*, 733.
88. Tagawa, S.; Nagahara, S.; Iwamoto, T.; Wakita, M.; Kozawa, T.; Yamamoto, Y.; Werst, D.; Trifunac, A.D. Proc. SPIE 3999: California, 2000; 204 pp.
89. *International Technology Roadmap for Semiconductors*; 2001 Edition; Semiconductor Industry Association (SIA) et al.
90. Emoto, F.; Gamo, K.; Namba, S.; Samoto, N.; Shimizu, R. Jpn. J. Appl. Phys. 1985, *24*, L809.
91. Fujita, J.; Watanabe, H.; Ochiai, Y.; Manako, S.; Tsai, J.S.; Matsui, S. Appl. Phys. Lett. 1995, *66*, 3065.
92. Beckera, E.W.; Ehrfeldb, W.; Hagmannc, P.; Manerd, A.; Münchmeyer, D. Microelectron. Eng. 1986, *4*, 35.
93. Ito, H.; Willson, C.G. Polym. Eng. Sci. 1983, *23*, 1012.
94. Ito, H. Jpn. J. Appl. Phys. 1992, *31*, 4273.
95. Blum, L.; Perkins, M.E.; Liu, H.-Y. J. Vac. Sci. Technol. B 1988, *6*, 2280.
96. Fedynyshyn, T.H.; Cronin, M.F.; Szmanda, C.R. J. Vac. Sci. Technol. B 1991, *9*, 3380.
97. Fedynyshyn, T.H.; Cronin, M.F.; Poli, L.C.; Kondek, C. J. Vac. Sci. Technol. B 1990, *8*, 1454.
98. Novembre, A.E.; Tai, W.W.; Kometani, J.M.; Hanson, J.E.; Nalamasu, O.; Taylor, G.N.; Reichmanis, E.; Thompson, L.F.; Tomes, D.N. J. Vac. Sci. Technol. B 1991, *9*, 3338.
99. Pan, S.-W.; Reilly, M.T.; Taylor, J.W.; Cerrina, F. J. Vac. Sci. Technol. B 1993, *11*, 2845.
100. Ban, H.; Nakamura, J.; Deguchi, K.; Tanaka, A. J. Vac. Sci. Technol. B 1991, *9*, 3387.
101. Ocola, L.E.; Orphanos, P.A.; Li, W.F.; Waskiewicz, W.; Novembre, A.E.; Sato, M. J. Vac. Sci. Technol. B 2000, *18*, 3435.
102. Nakasugi, T.; Ando, A.; Inanami, R.; Sasaki, N.; Sugihara, K.; Miyoshi, M.; Fujioka, H. Jpn. J. Appl. Phys. 2002, *41*, 4157.
103. Lee, K.Y.; Hsu, Y.; Le, P.; Tan, Z.C.H.; Chang, T.H.P.; Elian, K. J. Vac. Sci. Technol. B 2000, *18*, 3408.
104. Rao, V.; Hutchinson, J.; Holl, S.; Langston, J.; Henderson, C.; Wheeler, D.R.; Cardinale, G.; O'Connell, D.; Bohland, J.; Taylor, G.; Sinta, R. J. Vac. Sci. Technol. B 1998, *16*, 3722.
105. Fujita, J.; Onishi, Y.; Ochiai, Y.; Matsui, S. Appl. Phys. Lett. 1996, *68*, 1297.
106. Robinson, A.P.G.; Palmer, R.E.; Tada, T.; Kanayama, T.; Allen, M.T.; Preece, J.A.; Harris, K.D.M. J. Phys. D 1999, *32*, L75.
107. Ishii, T.; Nozawa, H.; Tamamura, T.; Ozawa, A. J. Vac. Sci. Technol. B 1997, *15*, 2570.
108. Wu, H.; Gonsalves, K.E. Adv. Mater 2001, *13*, 670.
109. Magee, J.L.; Chattarjee, A. In *Kinetics of Nonhomogeneous Processes*; Freeman, G.R., Ed.; John Wiley and Sons: New York, 1987; Chap. 4 171 pp.
110. Katz, R.; Sinclair, G.L.; Waligorski, M.P.R. Radiat. Meas. 1986, *11*, 301.
111. Waligorski, M.P.R.; Hamm, R.N.; Katz, R. Radiat. Meas. 1986, *11*, 309.
112. Varma, M.N.; Baum, J.W.; Kuehner, A.J. Radiat. Res. 1975, *62*, 1.
113. Varma, M.N.; Baum, J.W.; Kuehner, A.J. Radiat. Res. 1977, *70*, 511.
114. Wingate, C.L.; Baum, J.W. Radiat. Res. 1976, *65*, 1.
115. Wilson, W.E. Radiat. Res. 1994, *140*, 375.
116. Seki, S.; Kanzaki, K.; Kunimi, Y.; Tagawa, S.; Yoshida, Y.; Kudoh, H.; Sugimoto, M.; Sasuga, T.; Seguchi, T.; Shibata, H. Radiat. Phys. Chem. 1997, *50*, 423.
117. Koizumi, H.; Ichikawa, T.; Yoshida, H.; Shibata, H.; Tagawa, S.; Yoshida, Y. Nucl. Instrum. Methods Phys. Res. B 1996, *117*, 269.

118. Taguchi, M.; Matsumoto, Y.; Namba, H.; Aoki, Y.; Hiratsuka, H. Nucl. Instrum. Methods Phys. Res. B 1998, *134*, 427.

119. Furukawa, K.; Ohno, S.; Namba, H.; Taguchi, M.; Watanabe, R. Radiat. Phys. Chem. 1997, *49*, 641.

120. LaVerne, J.A.; Schuler, R.H. J. Phys. Chem. 1987, *91*, 5770.

121. LaVerne, J.A.; Schuler, R.H. J. Phys. Chem. 1996, *100*, 16034.

122. Puglisi, O.; Licciardello, A. Nucl. Instrum. Methods Phys. Res. B 1994, *91*, 431.

123. Licciardello, A.; Puglisi, O. Nucl. Instrum. Methods Phys. Res. B 1994, *91*, 436.

124. Calcagno, L.; Percolla, R.; Foti, G. Nucl. Instrum. Methods Phys. Res. B 1994, *91*, 426.

125. Aoki, Y.; Kouchi, N.; Shibata, H.; Tagawa, S.; Tabata, Y. Nucl. Instrum. Methods Phys. Res. B 1988, *B33*, 799.

126. Sasuga, T.; Kudoh, H.; Seguchi, T. Polymer 1999, *40*, 5095.

127. Sasuga, T.; Kawanishi, S.; Seguchi, T.; Kohno, I. Polymer 1989, *30*, 2054.

128. Seki, S.; Shibata, H.; Yoshida, Y.; Ishigure, K.; Tagawa, S. Radiat. Phys. Chem. 1996, *48*, 539.

129. Kudoh, H.; Katsumura, Y. Ion-beam radiation chemistry. In *Radiation Chemistry Present Status and Future Trends*; Jonah, C.D., Rao, B.S.M., Eds.; Elsevier: Amsterdam, 2001, 37–66.

130. Fleisher, R.L.; Price, P.B.; Walker, R.M. *Nuclear Tracks in Solids: Principles and Applications*; Fleisher, R.L. Ed.; University of California Press: Berkeley, 1975; 1 pp.

131. Spohr, R. *Ion Tracks and Microtechnology*; Vieweg: Anfahrt, 1990.

132. Vetter, J.; Spohr, R. Nucl. Instrum. Methods Phys. Res. B 1993, *79*, 691.

133. Ferain, F.; Legras, R. Nucl. Instrum. Methods Phys. Res. B 1993, *82*, 539.

134. Molarres, M.E.T.; Buschmann, V.; Dobrev, D.; Neumann, R.; Scholz, R.; Schuchert, I.U.; Vetter, J. Adv. Mater. 2001, *13*, 62.

135. Spohr, R.; Reber, N.; Wolf, A.; Alder, G.M.; Ang, V.; Bashford, C.L.; Pasternak, C.A.; Omichi, H.; Yoshida, M. J. Controlled Release 1998, *50*, 1.

136. Reber, N.; Omichi, H.; Spohr, R.; Wolf, A.; Yoshida, M. Nucl. Instrum. Methods Phys. Res. B 1999, *151*, 146.

137. Ferain, F.; Legras, R. Nucl. Instrum. Methods Phys. Res. B 2001, *174*, 116.

138. Ferain, F.; Legras, R. Radiat. Meas. 2001, *34*, 585.

139. De Pra, L.D.; Ferain, F.; Legras, R.; Champagne, S.D. Nucl. Instrum. Methods Phys. Res. B 2002, *196*, 81.

140. Seki, S.; Maeda, K.; Tagawa, S.; Kudoh, H.; Sugimoto, M.; Morita, Y.; Shibata, H. Adv. Mater. 2001, *13*, 1663.

20

Charged Particle and Photon Interactions in Metal Clusters and Photographic Systems Studies

Jacqueline Belloni and Mehran Mostafavi
UMR CNRS–UPS, Université Paris-Sud, Orsay, France

1. INTRODUCTION

A century ago, the reduction of silver ions in photographic plates helped W. Roentgen [1], then H. Becquerel [2], to discover x-rays and radiation of radioactive elements, respectively. Various metal ions were subsequently used widely in aqueous solutions as radical scavengers and redox indicators of the short-lived primary radiolytic species, allowing their identification and the calibration of their yield of formation [3–5]. Some underwent reduction by γ_- [6] or pulse radiolysis [7] to the zero-valence metal, to form colloids and then precipitates [7,8].

Due to the accurate knowledge of the dose used, a control of the progressive extent of the radiolytic reduction was achieved. However, quite often, puzzling data were reported when the zero-valent metal was formed, such as an induction time for precipitation, a sensitivity of the radiolytic yields to the initial presence of added particles, and an unusually weak reproducibility [9,10]. Moreover, oxidation of silver atoms by molecular oxygen was observed [7], although the process was thermodynamically improbable for a noble metal such as silver.

It was also observed, in 1973, that the fast reduction of Cu$^-$ ions by solvated electrons in liquid ammonia did not yield the metal and that, instead, molecular hydrogen was evolved [11]. These results were explained by assigning to the "quasi-atomic state" of the nascent metal, specific thermodynamical properties distinct from those of the bulk metal, which is stable under the same conditions. This concept implied that, as soon as formed, atoms and small clusters of a metal, even a noble metal, may exhibit much stronger reducing properties than the bulk metal, and may be spontaneously corroded by the solvent with simultaneous hydrogen evolution. It also implied that for a given metal the thermodynamics depended on the particle nuclearity (number of atoms reduced per particle), and it therefore provided a rationalized interpretation of other previous data [7,9,10]. Furthermore, experiments on the photoionization of silver atoms in solution demonstrated that their ionization potential was much lower than that of the bulk metal [12]. Moreover, it was shown that the redox potential of isolated silver atoms in water must be lower than that of the silver electrode, $E°(Ag^+/Ag_{met}) = 0.79 \, V_{NHE}$, by the sublimation energy of the metal equal to 2.6 V and $E°(Ag^+/Ag^0) = -1.8 \, V_{NHE}$ [13]. In the early 1980s, an increasing number of experimental work emphasized, for metal or semiconductor particles prepared by various ways in the gaseous and condensed phases, the nuclearity-dependent properties of clusters of atoms or molecules [14–17], theoretically predicted earlier by Kubo [18]. For two decades, the importance of these fundamental aspects and the consequences for various applications, from electronics to material science, catalysis, photography, nonlinear optics, etc. have given rise to a new research domain focused on the synthesis and properties of these mesoscopic phases.

The radiation-induced method, in the γ- or pulse regime [19,20] and, to a certain extent, the photo-induced method [21], provide a particularly powerful means to produce in condensed media, metal, and semiconductor clusters from monomers as precursors, to study their properties and to understand the exotic phenomena which occur whenever a new phase of oligomeric particles is formed in the bulk of a homogeneous mother phase [22–26], phenomena which are therefore rather frequent in physics and chemistry. Unlike a recent review concerning semiconductor clusters [19], the present chapter is specifically focused on metal clusters, induced by ionizing radiation or ultraviolet (UV)–visible photons.

2. METAL CLUSTER NUCLEATION AND GROWTH

2.1. Principles of Metal Atom Formation

The specificity of the radiation-induced reduction of metal ion precursors into metal atoms, which then coalesce into clusters, is attributable to strong reducing agents such as the

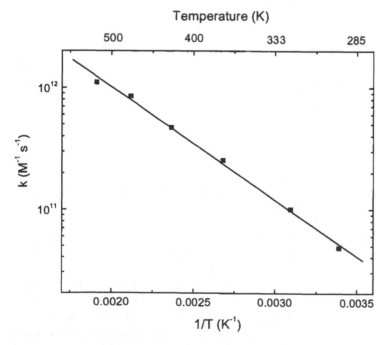

Figure 1 Arrhenius plot of the rate constant of the hydrated electron with silver cation Ag^+. (From Ref. 50.)

species solvated electrons e_{solv}^-, and H atoms arising from the radiolysis of liquids [27]. Thus, no reducing additive is necessary and no other product is formed. Because the ionizing radiation penetrates throughout the solution, these species are produced in the bulk everywhere in the surroundings of the ions. Due to their very negative redox potential ($E°(H_2O/e_{aq}^-) = -2.87 \ V_{NHE}$ and $E°(H^+/H) = -2.3 \ V_{NHE}$ in water) [27], they easily reduce metal ions up to the zero-valent state (Fig. 1). The reduction is achieved at room temperature, so that the inclusion of clusters into thermo-sensitive systems is possible.

Generally, the reduction is achieved under deaerated conditions to avoid a competitive scavenging of e_{solv}^- and H atoms by oxygen. These atoms are as homogeneously distributed as the ions and the reducing species, and they are therefore produced at first as isolated entities. Similarly, multivalent ions are reduced by multistep reactions, including disproportionation of intermediate valencies. Such reduction reactions have been observed directly by pulse radiolysis for a variety of metal ions (Fig. 2), mostly in water [28], but also in other solvents where the ionic precursors are soluble. Most of their rate constants are known and the reactions are often diffusion controlled.

In contrast, OH radicals, which are formed concomitantly in water radiolysis [27], are able to oxidize the ions or the atoms into a higher oxidation state and thus to counterbalance the previous reductions. For that reason, the solution is generally added with a scavenger of OH radicals. Among various possible molecules, the preferred choice is for solutes whose oxidation by OH yields radicals, which are unable to oxidize the metal

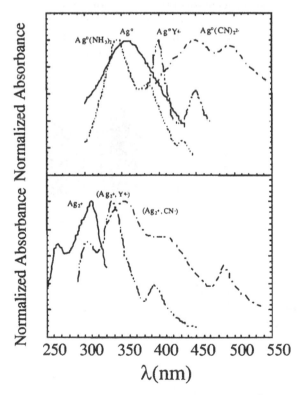

Figure 2 Absorption spectra of silver atom Ag^0 (top) and charged dimer Ag_2^+ (bottom) complexed by CN^- [37], EDTA [37], and NH_3 [36,47] in solution. The spectra of uncomplexed Ag^0 and Ag_2^+ are shown for comparison. (From Ref. 35.)

ions, but in contrast, exhibit strong reducing power. Such solutes are alcohols (such as 2-propanol) or formate anions whose monoelectronic oxidized forms ($(CH_3)_2C^{\cdot}OH$, $COO^{\cdot -}$) are strong reducing free radicals ($E°((CH_3)_2COH/(CH_3)_2C^{\cdot}OH) = -1.8$ V_{NHE} [29] at pH 7 and $E°(CO_2/COO^{\cdot -}) = -1.9$ V_{NHE}) [30]. $H^{\cdot}$ radicals ($E°(H^{\cdot}/H_2) = +2.3$ V_{NHE}) oxidize these molecules as well by H abstraction. Alcohol and formyl radicals can reduce, directly, or via complexation with the ion, the metal ions into lower valencies or into atoms for monovalent cations.

Some low-valency cations produced in early steps and at partial reduction are immediately neutralized by the counter anion if the solubility product is quite low, and they behave indeed as transient monomers able to coalesce into semiconductor clusters [19].

2.2. Cluster Nucleation

As the early species produced after reduction such as metal atoms, dimers and oligomers are short-lived; time-resolved observations of the reactions of these transients are carried out by the pulse radiolysis method, coupled with optical absorption or conductivity. Generally, kinetics are studied in the absence of oxygen or a stabilizer, unless their specific interaction has to be known. The earliest [7,8,20] and most complete data [31–39] on the nucleation mechanism were obtained on silver clusters. Indeed, silver may be considered as a model system owing to the one-step reduction of the monovalent silver ions, hydrated or complexed with various ligands, and to the intense absorption bands of the transient oligomers (Fig. 2) and final clusters (Fig. 3). The wavelength of the absorption band maxima of the atom Ag^0 and of the charged dimer Ag_2^+ in aqueous solutions and the rate constants of their formation are given in Table 1. Atoms and charged dimers of other metals, such as Tl [40,41], In [42], Au [43,44], and Cu [42], are formed by homolog reactions (Table 1).

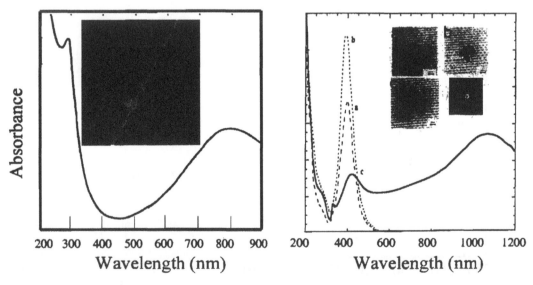

Figure 3 Left: Optical absorption spectrum and STM image of radiation-induced oligomers Ag_7^{3+} at partial reduction and stabilized by PA [85,86]. Right: Absorption spectrum of radiation-induced silver clusters Ag_n in the presence of EDTA. (a) After partial reduction (10 krad), (b) after 4 days, (c) after 8 days. Inset: TEM micrographs of (a), (b), (c), and electron diffraction pattern of sample c. (From Ref. 145.)

Table 1 Formation Rate Constants (mol l^{-1} sec^{-1}) and Optical Absorption Maxima of Metal Atoms, Hydrated or Complexed, and of the Corresponding Charged Dimers in Water [19]

Metal Ions	$k_{M+e_{aq}^-}$	λ (nm) of M^0	k_{M+M^0}	λ (nm) of M_2^+
Ag^+	3.6×10^{10}	360 [13,35]	8×10^9	290, 315 [13,35]
Tl^+	3×10^{10}	450, 260 [40,41]	1.4×10^9	700, 420, 245 [40,41]
In^+	–	500 [42]	1.5×10^9	310, 460 [42]
$Ag(CN)_2^-$	5×10^9	450, 500 [37]	2×10^{10}	350, 410, 490 [37]
$Ag(EDTA)^{3-}$	1.7×10^9	400, 450 [37]	1.6×10^9	310, 340, 400, 475 [37]
$Ag(NH_3)_2^+$	3.2×10^{10}	350 [8], 385 [47]	–	315, 340 [48]
$Au(CN)_2^-$	1.1×10^{10}	420 [43,44]	–	–
Ag^I, PA	3.6×10^9	360 [39]	8.9×10^9	310, 450 [39]
Ag^I, gelatin	1.1×10^{10}	360 [34]	1.1×10^{10}	290, 315, 308, 325 [34]
$Cu(Cl)_3^{2-}$	2.7×10^{10}	380 [42]	4.9×10^7	360 [42]

The band maxima of metal species are different in the gaseous and condensed phases. The optical absorption bands of Ag^0 and Ag_2^+ are highly dependent on the environment [45]. As shown in Table 1 and Fig. 2, the influence of the interaction of ligands CN^-, NH_3, or EDTA with the atom or the dimer is important [46]. The transient product of the reduction in water of complexed silver ions is not the isolated atom but a complexed silver atom, $Ag^0(CN)_2^{2-}$ [37], $Ag^0(NH_3)_2$ [47], or $Ag^0(EDTA)$ [37,48], respectively (Fig. 2), as well as for $Au^0(CN)_2^{2-}$ [44] or $CuCl_3^{2-}$ [42].

The optical absorption spectra depend on the solvent. They are red-shifted with the decreasing polarity of the solvents as in EDA, liquid NH_3, where they appear at a longer wavelength than in water [49], and blue-shifted in methanol [38]. Moreover, the maximum in NH_3 is red-shifted with the increase of temperature. In water, the Ag^0 band is almost unchanged in the range 20–200 °C, while that of Ag_2^+ is markedly shifted to the red [50]. Electron spin echo modulation analysis of Ag^0 in ice or methanol glasses has concluded to a charge transfer character to solvent (CTTS) of the absorption band [51].

The multistep reduction mechanism of multivalent cations is known partially from pulse radiolysis studies [19]. For example, the reduction of $Au^{III}Cl_4^-$ into $Au^{II}Cl_3^-$ and the disproportionation of Au^{II} into Au^I and Au^{III} have been directly observed and the rate constants determined [52]. However, the last step of reduction of Au^I complexed by Cl^- into Au^0 is not observed by pulse radiolysis because the e_{aq}^- scavenging by the precursors Au^{III} is more efficient. Moreover, the disproportionation of Au^I or of other monovalent cations is thermodynamically hindered by the quite negative value of $E°(M^I/M^0)(E°(M^I/M^0) < E°(M^{II}/M^I))$ (Section 20.4.4).

2.3. Cluster Growth

The binding energy between two metal atoms is stronger than the atom–solvent or atom–ligand bond energy. Therefore the atoms spontaneously dimerize when encountering or associate with excess ions (Fig. 2). Then, by a cascade of coalescence process, these species progressively coalesce into larger clusters M_{n+x}^{y+}, where n represents the nuclearity, i.e., the number of reduced atoms they contain, x is the number of associated ions, and y ($y \geq x$) is the charge number of adsorbed ions. When these ions are complexed by negatively charged ligands, the cluster charge may be negative.

The radius of M_n increases as $n^{1/3}$. The fast reactions of ion association with atoms or clusters play an important role in the cluster growth mechanism. First, the homolog charge of clusters slow down their coalescence. Second, the subsequent reduction in situ of the ions fixed on the clusters favors their growth rather than the generation of new isolated atoms. The competition between reduction of free ions and of adsorbed ions is controlled by the formation rate of the reducing radicals, that is, by the dose rate. Coalescence reactions obey second-order kinetics. Therefore the cluster formation by reduction into isolated atoms followed by coalescence is predominant at high irradiation dose rate, particularly in the pulsed regime compared to continuous irradiation at much lower dose rate.

Moreover, almost in all the early steps, the redox potential of the clusters, which decreases with the nuclearity, is quite negative. Therefore the growth process undergoes another competition with a spontaneous corrosion by the solvent and the radiolytic protons, corrosion which may even prevent the formation of clusters, as mostly in the case of nonnoble metals. Monomeric atoms and oligomers of these elements are so fragile to reverse oxidation by the medium that H_2 is evolved and the zerovalent metal is not formed [11]. For that reason, it is preferable in these systems to scavenge the protons by adding a base to the solution and to favor the coalescence by a reduction faster than the oxidation [53].

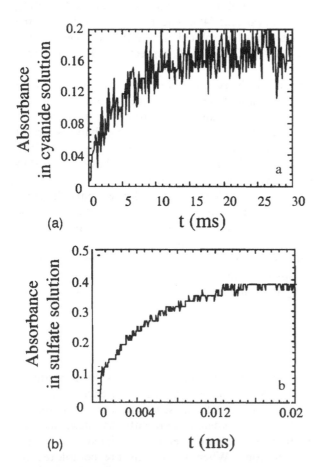

Figure 4 Growth kinetics of silver clusters observed through their absorbance at 400 nm in the presence of (a) cyanide or (b) sulfate. (From Ref. 54.)

After the formation of Ag^0 and Ag_2^+, followed by the reaction of Ag_2^+ with Ag^+ yielding Ag_3^{2+} ($\lambda_{max} = 315$ and 260 nm) [35], or its dimerization into Ag_4^{2+} ($\lambda_{max} = 265$ nm), the multistep coalescence of oligomers result in clusters of increasing nuclearity Ag_n. The absorption spectrum is shifted to the surface plasmon band of Ag_n at 380–400 nm, the coalescence rate depending on the ligand [54]. At the same initial concentration of atoms, the plateau is reached at an almost 10^3 longer time for $Ag_n(CN^-)$ [54] than $Ag_n(SO_4^{2-})$ [31] (Fig. 4). In methanol, the coalescence rate constants are also much lower [38]. Indeed, it is known, according to the Mie theory [55] and its extension [56], that the interaction of light with the electrons of small metal particles results in an absorption band whose shape and intensity depend on the cluster size, the complex dielectric constants of the metal and the environment (solvent, ligand, molecules adsorbed). During the coalescence, the total amount of silver atoms formed by the pulse is constant, but they are aggregated into clusters of increasing nuclearity n with a decreasing concentration. Thus the absorbance increase observed at 400 nm (Fig. 4) is assigned to the increase with the nuclearity of the extinction coefficient *per* silver atom. It was shown from the kinetics analysis that the plasmon band is totally developed with the constant value per atom $\varepsilon = 1.5 \times 10^4$ l mol^{-1} cm^{-1} beyond $n = 13$. [57] Then, the coalescence into larger clusters is still continuing, although the spectrum is unchanged.

3. REACTIVITY OF TRANSIENT METAL CLUSTERS

A number of rate constants for reactions of transients derived from the reduction of metal ions and metal complexes were determined by pulse radiolysis [58]. Because of the short-lived character of atoms and oligomers, the determination of their redox potential is possible only by kinetic methods using pulse radiolysis. In the couple M_n^+/M_n, the reducing properties of M_n as electron donor as well as oxidizing properties of M_n^+ as electron acceptor are deduced from the occurrence of an electron transfer reaction with a reference reactant of known potential. These reactions obviously occur in competition with the cascade of coalescence processes. The unknown potential $E°(M_n^-/M_n)$ is derived by comparing the action of several reference systems of different potentials.

3.1. Redox Potentials of Metal Atoms and Dimers

The oxidation reactions of Ag^0 and Ag_2^+ [7,20], of $Au^0(CN)_2^{2-}$ [44], or of Tl^0 and Tl_2^- [41,59], by even *mild oxidizing* molecules (O_2, CCl_4, CH_3NO_2, N_2O, Fe^{III}, for example) were observed by pulse radiolysis. The high rate constants found indicate the strong electron donating character of atomic silver, gold, or thallium. This confirms the evaluation of $E°(Ag^+/Ag^0) = -1.8$ V_{NHE} [13] and $E°(Tl^+/Tl^0) = -1.9$ V_{NHE} [41], which were derived from the difference between the metal electrode potential, $E°(M^+/M_{met})$, and the metal sublimation energy (Table 2). More generally, because of the high value of the sublimation energy, the potential of any metal $E°(M^+/M^0)$ is expected to be quite negative.

Table 2 Redox Potentials (V_{NHE}) of the Hydrated and Complexed M^I, M^0 Couple in Water

Couple M^I/M^0	$Au^+/$ Au^0	$Cu^+/$ Cu^0	$Tl^+/$ Tl^0	$Ag^+/$ Ag^0	$Ag(CN)_2^-/$ $Ag^0(CN)_2^{2-}$	$Ag(NH_3)_2^+/$ $Ag^0(NH_3)_2$	$Ag(EDTA)^{3-}/$ $Ag^0(EDTA)^{4-}$
$E°$	-1.4 [44]	-2.7 [13]	-1.9 [41]	-1.8 [13]	-2.6 [36,61]	-2.4 [47]	-2.2 [37]

The redox potential values of all metal atoms, except alkaline and earth-alkaline metals [60], are higher than that of $E°(H_2O/e_{aq}^-) = -2.87\ V_{NHE}$. However, some complexed ions are *not reducible* by alcohol radicals under basic conditions and thus $E°(M^+L/M^0L) < -2.1\ V_{NHE}$ (Table 2). The results were confirmed by SCF calculations of Ag^+L and Ag^0L structures associated with the solvation effect given by the cavity model for $L = CN^-$ [61] or NH_3 [47], respectively.

3.2. Redox Potentials of Oligomeric Metal Clusters

For clusters of higher nuclearity too, the kinetic method for determining the redox potential $E°(M_n^+/M_n)$ is based on electron transfer, for example, from *mild reductants* of known potential which are used as reference systems, towards charged clusters M_n^+. [31] Note that the redox potential differs from the microelectrode potential $E°(M^+, M_n/M_n)$ by the adsorption energy of M^+ on M_n (except for $n = 1$). The principle [31] is to observe at which step n of the cascade of coalescence reactions, a reaction of electron transfer occurring between a donor S^- and the cluster M_n^+ could compete with the coalescence. Indeed, n is known from the coalescence rate constant value, measured in the absence of S, and from the time elapsed from the atom appearance to the start of coalescence. The donor S^- is produced by the same pulse as the atoms M^0, the radiolytic radicals being shared between M^+ and S. One form at least in the couple S/S^- should possess intense optical absorption properties to permit a detailed kinetics study.

The transfer requires that $E°(M_n^+/M_n)$, which increases when n becomes higher than the reference $E°(S/S^-)$, thus fixing a critical nuclearity n_c. As far as $n < n_c$, the coalescence occurs as in the absence of S^-, whose concentration remains constant during an induction time. As soon as the nuclearity increases, such as $n \geq n_c$, the electrons are transferred from S^- to M_n^+, which is thus indirectly reduced, and the S^- concentration decreases. The donor S^- behaves as an electron relay. After the adsorption of another cation, the potential of the new reduced cluster $E°(M_{n+1}^+/M_{n+1})$ is more positive and another electron is transferred from S^- repeatedly. The clusters grow now mostly by alternate reactions of electron transfer and adsorption of surrounding metal ions. It has been shown that once formed, a critical cluster, of silver for example, behaves indeed as a nucleus of an autocatalytic growth [31]. The branching ratio between the direct reduction of metal ions by radiolytic species and the indirect reduction of M^+ by the donor is fixed by the initial M^+ and S concentration ratio. The distribution of final sizes is totally governed by this competition because, at the increasing S/M^+ concentration ratio, less nuclei are formed and they develop to larger sizes [31].

The value of the critical nuclearity n_c allowing the transfer from the monitor depends on the redox potential of this selected donor S^-. The induction time and the donor decay rate both depend on the initial concentrations of metal atoms and of the donor [31,62]. The critical nuclearity n_c corresponding to the potential threshold imposed by the donor and the transfer rate constant value, which is supposed to be independent of n, are derived from the fitting between the kinetics of the experimental donor decay rates under various conditions and numerical simulations through adjusted parameters (Fig. 5) [54]. By changing the reference potential in a series of redox monitors, the dependence of the silver cluster potential on the nuclearity was obtained (Fig. 6 and Table 5) [26,63].

Clusters M_n^+ may also behave as electron donors when formed in the presence of *oxidizing* reactants. For example, as far as $n < n_c$, a cascade of electron transfers from M_n to S leading to corrosion of M_n is possible because $E°(M_n^+/M_n) < E°(S/S^-)$, and it should be taken into account in the simulation. This corrosion process is generally negligible because

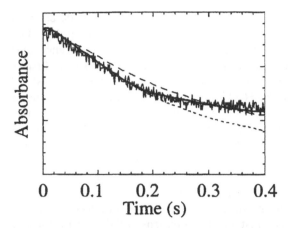

Figure 5 Comparison between simulated and experimental decay of MV$^{+\cdot}$ in the presence of growing clusters Ag$_{n,CN^-}$ including oxidation of subcritical oligomers by MV^{2+} and development of supercritical clusters by MV$^{+\cdot}$. The best fit yields $n_c = 5$. (From Ref. 54.)

the step of M$^+$ release is not fast enough. However, it is observed when the coalescence is also slow, for example in the case of ions included in cavities of a Nafion membrane, where the coalescence of Ag$_n$ is controlled by the very slow diffusion between the cavities (10^4 l mol^{-1} sec^{-1}) [64,65]. It has been found that the smallest clusters could be oxidized by the protons H$_3$O$^+$, which are highly concentrated at the cavity surface. In contrast, when clusters reach by coalescence the critical nuclearity for which their potential is higher than $E°(H_3O^+/H_2) = 0$ V$_{NHE}$, they escape corrosion and are observed by optical absorption. The

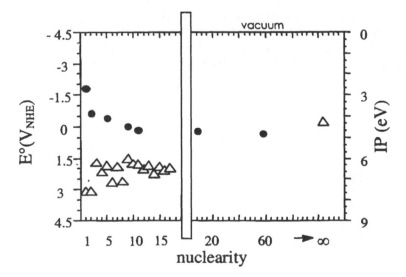

Figure 6 Size-dependence of the redox potential of silver clusters in water ($\bullet$) [63] and of the corresponding ionization potential in the gas phase ($\triangle$) [67,68]. The redox potentials refer to the normal hydrogen electrode whose Fermi potential is at 4.5 eV.

corrosion is particularly important when the dose per pulse is low and the coalescence is slower (the final amount of Ag_n is 20 times less for a four times lower dose per pulse). The numerical simulation of the kinetic signal including the cascades of coalescence and oxidation reactions yields the value $n = 8$ for the upper limit of nuclearity of silver clusters oxidized by H_3O^+ in the Nafion® cavities [65]. Therefore, $E°(Ag_8^+/Ag_8) > E°(H_3O^+/H_2) = 0\ V_{NHE}$. Contrarily, such a corrosion by H_3O^+ is not observed under conditions of free diffusion of the clusters as in Ag^+ solutions [49], when the coalescence is much faster.

When silver or gold atoms are generated from $Ag^I(CN)_2^-$ or $Au^I(CN)_2^-$ in the presence of the methylviologen redox couple $MV^{2+}/MV^{+\bullet}$, oxidation of the smallest clusters is also observed, because coalescence in cyanide solutions is slow (Fig. 4) [54,66]. While supercritical silver clusters ($n > 6 \pm 1$) (Table 3) accept electrons from $MV^{+\bullet}$ with a progressive increase of their nuclearity, the subcritical clusters undergo a progressive oxidation by MV^{2+} (Fig. 5). Actually, the reduced ions $MV^{+\bullet}$ so produced act as an electron relay favoring the growth of large clusters at the expense of the small ones. If cyano-silver clusters require a higher nuclearity than aquo clusters to react with $MV^{+\bullet}$ (Table 3), it means that the ligand CN^- lowers the redox potential at a given n, as in the case of the bulk metal. In the case of gold clusters complexed by cyanide, the coalescence is still slower and oxidation of Au_{n,CN^-} by MV^{2+} is alone observed [66].

Fig. 6 compares the nuclearity effect on the redox potentials [19,31,63] of hydrated Ag_n^+ clusters $E°(Ag_n^+/Ag_n)_{aq}$ together with the effect on ionization potentials IP_g (Ag_n) of bare silver clusters in the gas phase [67,68]. The asymptotic value of the redox potential is reached at the nuclearity around $n = 500$ (diameter $\approx$ 2 nm), which thus represents, for the system, the transition between the mesoscopic and the macroscopic phase of the bulk metal. The density of values available so far is not sufficient to prove the existence of odd–even oscillations as for IP_g. However, it is obvious from this figure that the variation of $E°$ and IP_g do exhibit opposite trends vs. n, for the solution (Table 5) and the gas phase, respectively. The difference between ionization potentials of bare and solvated clusters decreases with increasing n as $n^{-1/3}$ which corresponds fairly well to the solvation free energy of the cation M_n^+ deduced from the Born solvation model [45] (for the single atom, the difference of 5 eV represents the solvation energy of the silver cation) [31].

A variation of $E°(Cu_n^+/Cu_n)$, similar to that of silver, is expected for copper because the atomic potential is $E°(Cu^+/Cu^0) = -2.7\ V_{NHE}$ [13] (Table 2), that of Cu, is $E°(Cu_7^+/Cu_7) = -0.4\ V_{NHE}$ [69] (Table 5), and the bulk electrode potential is at $0.52\ V_{NHE}$.

Table 3 Nuclearity-Dependence of $E°(M_n^+/M_n)$ (in V_{NHE}) [63]

Reference System (Electron Donor)	$E°(S/S^-)$ (Ref. Couple)	Metal Cluster (Electron Acceptor)	n_c (Reduced Atoms/Cluster)
Ni^+	−1.9	Ag_n	1
$(CH_3)_2C^\bullet OH$ (pH 5)	−1.8	Ag_n	1
$SPV^{-\bullet}$	−0.41	Ag_n	4
Cu^+	0.16	Ag_n	11
$Q^{-\bullet}$ (pH = 4.8)[a]	0.22	Ag_n	85 ± 5
$Q^{-\bullet}$ (pH = 3.9)	0.33	Ag_n	500 ± 30
$MV^{+\bullet}$	−0.41	$Ag_{n,CN-}$	5–6
$MV^{+\bullet}$	−0.41	$Ag_{n,PA}$	4
$MV^{+\bullet}$	−0.41	$Cu_{n,Cl}^-$	6 ± 1

[a] $Q^{-\bullet}$ is the semiquinone of naphtazarin.

Table 4 Radiation-Induced Metal Clusters in Water

Metal	λ_{max} (nm)	Ref.	Metal	λ_{max} (nm)	Ref.	Metal	λ_{max} (nm)	Ref.
Co	UV	[104,114]	Pd	205	[84,110,112]	Ir	UV	[88,105]
Ni*	UV	[53,104,115,116]	Ag*	380	[8,31,35,57]	Pt*	215	[52,92,93,106]
Cu*	570	[42,84,93,112,113]	Cd*	260	[84,117,118]	Au*	520	[43,44,52,107–109]
Zn	UV	[104]	In*	270	[42]	Hg*	500	[42,104]
Mo	UV	[104]	Sn	200	[104,119]	Tl*	300	[40–42,59,120]
Ru	UV	[104]	Sb	UV	[104]	Pb*	220	[42,104,121,122]
Rh	UV	[104]	Os	UV	[104]	Bi	253	[42,104,123]

* indicates that the earliest steps of the cluster formation have been studied by pulse radiolysis.

Some calculations [70] were made to derive the microelectrode potential $E°(M^-, M_{n-1}/M_n)$ for silver and copper from the data in the gas phase [nuclearity-dependent M–M_n bond energy and $IP_g(M_n)$]. The potential $E°(M^+,M_{n-1}/M_n)$ presents odd–even oscillations with n (more stable for n even) as for IP_g, but again, the general trends are opposite, and an increase is found in solution because to the solvation energy.

3.3. Nucleation, Growth, and Reactivity of Transient Bimetallic Clusters

The spectra of silver and gold nanoclusters are intense and distinct (Table 4). They are thus particularly suitable to detect the evolution of a cluster composition during the construction of a bimetallic cluster in mixed solution. The system studied by pulse radiolysis was the radiolytic reduction of a mixed solution of two monovalent ions, the cyano-silver and the cyano-gold ions $Ag(CN)_2^-$ and $Au(CN)_2^-$ (Fig. 7) [66]. Actually, the time-resolved observation demonstrated a two-step process. First, the atoms Ag^0 and Au^0 are readily formed after the pulse and coalesce into an alloyed oligomer. However, due to

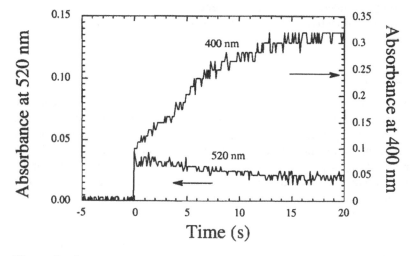

Figure 7 Correlated signals at 400 and 520 nm with a single pulse in equimolar mixed solution of gold and silver cyanide, $Ag(CN)_2^-$ and $Au(CN)_2^-$, in the presence of 2-propanol. (From Ref. 66.)

a slow electron transfer (within seconds) from Au^0 atoms of the alloy to the unreduced ions Ag^I, a supplementary formation of Ag^0 correlates with the complete dissolution of Au^0 as in an intermetallic displacement.

Another way to favor the statistical reduction leading to alloyed particles of silver and gold is to allow the clusters grow more rapidly, before any observable intermetal transfer, and consequently to let their redox potential increase in order to lower the risk of the displacement. This can be achieved through a fast chemical reduction by a donor S^- at an alloyed nucleus surface as for monometallic clusters (Section 20.4.4). The reaction competing with the intermetal exchange is now the chemical reduction, which can be very fast provided nuclei are present. In a pulse radiolysis study, the transient reduced viologen formed by an electron pulse has been observed to react with the small alloyed $(AuAg)_{n,CN^-}$ clusters and the developed alloyed clusters are stable toward segregation [66]. Their final size is quite large because the ions reduced by the donor are concentrated on a small number of nuclei. Moreover, it was found [66] that the redox potential of such critically alloyed AgAu clusters is almost the same (-0.4 V_{NHE} at $n_c = 6$) as that of the more noble metal (silver in cyanide environment) [54].

4. SYNTHESIS OF STABLE MONOMETALLIC NANOCLUSTERS

Stability means that clusters do not undergo coalescence nor corrosion by the medium, at least in the absence of oxygen. The quite negative value of $E°(M^1/M^0)$ and the dependence of the cluster redox potential on the nuclearity have crucial consequences in the formation of early nuclei, their possible corrosion or their growth. As an example, the faster the coalescence, the lower is the probability of corrosion of the small clusters by the medium. The property of stability offers the means to apply to these clusters a larger amount of suitable characterization techniques than to transient oligomers.

Various applications (Sections 20.5.3 and 20.7.3) require that the synthesis of stable clusters be as small as possible, and the coalescence should be prevented somewhat by a stabilizer or a support. Moreover, stable metal clusters can be used as precursors to synthesize, by functionalizing their surface, highly elaborated materials, acquiring specific properties [21], For example, fullerene- [71,72] or pyrene-functionalized gold clusters [73] are used as self-assembled photoactive antenna units which could serve as building blocks in the design of light-harvesting systems.

4.1. Cluster Size Stabilization

Metal atoms formed by irradiation or any other method tend to coalesce into oligomers, which themselves progressively grow into larger clusters and eventually into precipitates, as found in early radiolytic experiments. However, for studying stable clusters or for applications, the coalescence must be limited as in other chemical methods of cluster synthesis [74], either by adding a polymeric molecule acting as a cluster stabilizer or by generating the atoms in a confined structure where diffusion is prevented. Some ligands such as CN^- under deaerated conditions, or ethylenediaminetetraacetic acid (EDTA) [37] under complete radiolytic reduction conditions, are able by themselves to stabilize small-sized particles (Fig. 3).

Functional groups of a polymer with high affinity for the metal ensure the anchoring of the molecule at the cluster surface, while the polymeric chain protects the cluster from coalescing with the next one and thus inhibits, at an early stage, further coalescence through electrostatic repulsion or steric hindrance. When metal clusters are to be prepared by irradiation, the stabilizing polymers must be selected for their inability to directly reduce the ions before irradiation. Various stabilizers have been studied [19]: poly(vinyl alcohol) (PVA) [75], sodium dodecyl sulfate (SDS) [75,76], sodium poly(vinylsulfate) (PVS) [77], poly(acrylamide) (PAM) [78,79] or poly(N-methylacrylamide) (PNPAM) [78], carbo-wax [80], poly(ethyleneimine) (PEI) [81,82], polyphosphate (PP) [83], and gelatin [84]. Some functional groups such as alcohol are OH⁻ scavengers and these may contribute to the reduction under irradiation. The final size of metal clusters stabilized by these polymers lies in the nanometer range but it increases with the ratio metal atoms/polymer. Sodium polyacrylate (PA) is a much stronger stabilizer which allows the formation of long-lived metal oligomers (Fig. 3) [85,86].

The coalescence of atoms into clusters may also be restricted by generating the atoms inside confined volumes of microorganized systems [87] or in porous materials [88]. The ionic precursors are included *prior* to irradiation. The penetration in depth of ionizing radiation permits the ion reduction in situ, even for opaque materials. The surface of solid supports, adsorbing metal ions, is a strong limit to the diffusion of the nascent atoms formed by irradiation at room temperature, so that quite small clusters can survive.

4.2. Long-Lived Metal Oligomers

Metal oligomers are stabilized at quite small nuclearity when formed at low dose in the presence of polyacrylate PA [85]. From pulse radiolysis of Ag^+–PA solutions, it appears that the very slow (10^5 l mol^{-1} sec^{-1}) dimerization of Ag_4^{2+} *silver* oligomers [39,89] (275 and 350 nm) results into a "blue-silver" clusters ($n = 4$) absorbing at 292 and 800 nm [90], and stable in air for years (Fig. 3). The 800-nm absorption band is assigned to a cluster–ligand PA interaction [86,91]. Clear images by STM show flat clusters of 0.7 nm with atoms spaced by 0.25 nm (Fig. 3) [85]. Each cluster contains seven atoms (possibly with an eighth atom in the central position). Because only four atoms were reduced, they correspond to the stoichiometry Ag_7^{3+} (or Ag_8^{4+}).

Similarly, when $PtCl_6^{4-}$ ions are irradiated in the presence of PA, which prevents coalescence, most of the Pt atoms are found via STM imaging in the form of very small *platinum* oligomers of 3–7 atoms only [92]. In the presence of PP, $PtCl_6^{4-}$ irradiated solutions present a UV band with a maximum at 215 nm [93]. About 20% of the initial signal is lost by oxidation when the sample is exposed to air, the rest being stable. This result again confirms that the redox potential of the smallest clusters is markedly shifted to the negative values.

When $CuSO_4$ solutions are γ-irradiated with PA at pH 10, new absorption bands at 292, 350, and 455 nm are observed, provided the reduction of copper ions is partial [94]. At an increasing dose, the intensity of the peaks assigned to small, stabilized copper oligomers decreases correlatively with the plasmon band increase which develops at 570 nm. The UV species are very sensitive to oxygen. Under acidic conditions, large clusters are observed but not the UV oligomers. A pulse radiolysis study of monovalent Cu^+ ions, complexed in the presence of a high concentration of Cl^- without polymeric stabilizer, has shown that short-lived states of reduced copper corresponding to the early steps of growth also absorb in the range 355–410 nm [95].

Among the nonnoble metals that can be synthesized by radiation-induced reduction in solution, nickel raises some difficulties because the atom formation and aggregation processes undergo the competition of oxidation reactions of highly reactive transients, such as monovalent Ni^+ ion, Ni atom, and the very first *nickel* oligomers. Nevertheless, in the presence of PA, a new absorption band develops at 540 nm with increasing doses [96]. The formation of the same absorption band was observed by pulse radiolysis. It increases simultaneously with the formation of the very first Ni atoms and it is assigned to a cluster–ligand interaction. Because of the high reactivity of nickel oligomers, the band disappears within 24 hr, even when solutions are preserved from oxygen, through a spontaneous reaction with the solvent.

When the irradiation of metal ion solutions is performed in the presence of the ligands CO or PPh_3, metal reduction, ligandation, and aggregation reactions compete, leading to reduced metal complexes and then to stable molecular clusters, such as Chini clusters $[Pt_3(CO)_6]_m^{2-}$ with $m = 3$–10 (i.e., 9–30 Pt atoms) [97], or other metal clusters [98]. The synthesis is selective and m is controlled by adjusting the dose (m decreases at high doses). The mechanism of the reduction has been determined recently by pulse radiolysis [99]. Molecular clusters $[Pt_3(CO)_6]_5^{2-}$ have been observed by STM [100].

Colloidal supports, such as small colloidal SiO_2 particles, restrict the interparticle diffusion of silver atoms when formed by radiolysis of the ions at their surface. The silver oligomers absorbing at 290 and 330 nm are observed by pulse radiolysis. They are stable with respect to coalescence but they are oxidized by MV^{2+}, O_2, Cu^{2+}, and $Ru(NH_3)_6^{3+}$ [101].

4.3. Nanometric Metal Clusters

Most of mono- or multivalent metals, except alkaline and earth-alkaline metals, have been prepared by γ-radiolysis in the form of nanometric clusters, stabilized with respect to further coalescence by polymers or simply by ligands. The final size depends on the type of polymer or ligand, on the ratio metal/polymer but also on the dose rate (Fig. 8) [19]. Indeed, sudden generation of a high concentration of isolated atoms at a high dose rate yields, by coalescence, more numerous and smaller nuclei clusters (Fig. 8a) [102]. Instead, the in situ reduction of adsorbed ions at the surface of clusters, which is predominant at decreasing dose rate and which is not prevented by the polymer, contributes to the development of less numerous growth centers and results therefore in larger clusters (Fig. 8b). In both cases, however, the nuclei generation being strictly reproducible, the distribution of long-term sizes presents an exceptional homodispersity. Table 4 presents the maximum wavelengths of the surface plasmon absorption band of clusters as prepared by radiolysis in water. They are in agreement with the Mie model [55,56]. Most of the clusters absorbed in the UV and the nanocolloids are brown. The intense colors of some ultradivided metals, red for gold ($\lambda_{max} = 520$ nm), yellow for silver ($\lambda_{max} = 380$ nm), and pink for copper ($\lambda_{max} = 570$ nm) have been well known for a long time [56,74].

In solution, the particles are most often spherical. Nevertheless, some of these clusters, which are generated in solution and then deposited on a grid for HRTEM and electron diffraction observations, exhibit clearly pentagonal shapes with an fcc structure, which suggests the formation of twin fcc crystallites growing from the faces of an icosahedral nucleus. Phase imaging in tapping-mode AFM is a powerful tool for the characterization of clusters stabilized by polymer matrices as shown recently in observing silver and gold samples [103], and helps to discriminate between amorphous polymer and metal clusters.

Noble metal clusters have been most often studied (Table 4) [104], such as Ag_n [8,31,35,57], Ir_n [88,105], Pt_n [52,92,93,106], Au_n [43,44,52,107–109], Pd_n [84,110–112], Ru_n [104], Rh_n [104], and Os_n [104]. However, silver and gold clusters, when produced in the presence of the ligand CN^- [44,54,66], or in methanol [38], even at nanometric size, are spontaneously corroded as soon as in contact with oxygen.

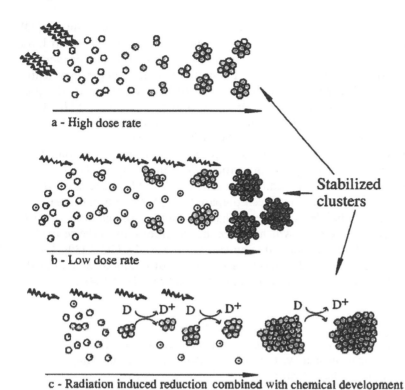

a - High dose rate

Stabilized
clusters

b - Low dose rate

c - Radiation induced reduction combined with chemical development

Figure 8 Nucleation and growth of clusters generated by radiolytic radicals at various dose rates, with or without electron donor D. The stabilizing effect of the polymer prevents exclusively coalescence beyond a certain limit of nuclearity, but does not prevent successive ion and electron transfers at low dose rate from the radicals. The donor allows the cluster to become much larger at any dose rate. (From Ref. 19.)

Nonnoble metal clusters are considerably more fragile to corrosion by the solvent [11] compared to noble metal clusters. Therefore the production of stable small particles results from a compromise between the smallest size and the longest stability [104].

Long-lived clusters, Cu_n [42,84,93,112,113], Co_n [104,114], Ni_n [53,104,115,116], Cd_n [84,117,118], Sn_n [104,119], Tl_n [40–42,59,120], Pb_n [42,104,121,122], Bi_n [42,104,123], Hg_n [42,104], In_n [42], Mo_n [104], and Sb_n [104] (Table 4), may be formed in deaerated basic medium, but Zn_n [104] clusters are oxidized into zinc hydroxide within a few weeks even in the absence of O_2. Clusters Cd_n are better stabilized by gelatin [84]. Ni_n or Co_n clusters display ferromagnetic properties and at high concentration, and the sols behave as ferrofluids [124]. Whereas all of these nonprecious metal clusters are easily oxidized in solution by O_2, they may be stabilized in air after drying under inert atmosphere and observed by microscopy [53].

Adsorption of ions or molecules on metal clusters markedly affects their optical properties. It was shown that the intensity and the shape of the surface plasmon absorption band of silver nanometric particles, which is close to 380 nm, change upon adsorption of various substances [125]. The important damping of the band generally observed is assigned to the change of the electron density of the thin surface layer of the particle where electrons are, for instance, injected by adsorbed electrophilic anions [24]. This is followed by a red shift of the damped absorption band corresponding to loose agglomeration of the clusters [126]. Silver particles covered by I^- become particularly sensitive to oxidation in air, because the Fermi level is shifted to a more negative potential by the electron donation [24]. Likewise, binding of electronegative groups to the gold surface leads to damping of the surface plasmon band of gold nanoparticles [127]. It was shown that $(SCN)_2^-$ radicals generated by pulse radiolysis bind strongly to the gold nanoparticles, oxidizing the gold clusters to form $[Au(SCN)2]^-$ [128].

Although the mechanism of the photo-induced generation of mono- and bimetallic metal clusters, except for the photographic application (Section 20.6), has been studied with considerably less detail than for the radiolytic route, some stable clusters, mostly of noble metals (Ag, Au, Pt, Pd, Rh), have also been prepared by UV excitation of metal ion solutions [129–141]. Generally, halides and pseudo-halides counter anions are known to release, when excited, solvated electrons, which reduce the metal ions up to the zerovalent state. Oxalate excitation yields the strong reducing carbonyl radical $COO^{-\cdot}$ [30]. Photosensitizers are likewise often added [142]. Metal clusters are photo-induced as well at the surface of photo-excited semiconductors in contact with metal ions [143,144].

4.4. Developed Clusters

Actually, the kinetics study of the redox potential of transient clusters (Section 20.3.2) has shown that beyond the critical nuclearity, they receive electrons without delay from an electron donor already present. The critical nuclearity depends on the donor potential and then the autocatalytic growth does not stop until the metal ions or the electron donor are not exhausted (Fig. 8c). An extreme case of the size development occurs, despite the presence of the polymer, when the nucleation induced by radiolytic reduction is followed by a chemical reduction. The donor D does not create new nuclei but allows the supercritical clusters to develop. This process may be used to select the cluster final size by the choice of the radiolytic/chemical reduction ratio. But it also occurs spontaneously any time when even a mild reducing agent is present during the radiolytic synthesis. The specificity of this method is to combine the ion reduction successively:

(i)   By radiolytic reduction. The atoms independently formed in the bulk coalesce into oligomers.

(ii) By chemical reduction beyond a certain oligomer nuclearity. This chemical agent is unable thermodynamically to reduce the ions directly into atoms (Section 20.3.1), but it achieves the reduction of the rest of the ions after they have been adsorbed on the radiation-induced clusters acting as nuclei of catalytic reduction and growth. Note that the polymer restricts the coalescence of the nuclei or of the final particles but does not prevent at all the development by diffusion and adsorption of excess ions and electron transfer from the donor as shown for some of the following systems.

The final *silver* cluster diameter increases, at given initial Ag^+ and PVA concentrations, for example from 15 to 50 nm (*n* 50 times larger), when the part of reduction is increasingly achieved by the donor SPV^-, rather than by radiolytic radicals [31]. A red shift correlates with the growth in size in the final optical surface plasmon band. Nonirradiated solutions of EDTA silver complex are stable because EDTA does not reduce the ions directly. However, after the appearance of the 400-nm spectrum of silver clusters formed in a partially radiation-reduced solution (spherical particles of 10–15 nm diameter), the band intensity increases for days as a posteffect and the silver ions are totally reduced (Fig. 3) [37,145]. The postirradiation reduction is assigned to the lone pair N atom electrons of the EDTA ligand, provided the silver ions are fixed on clusters formed by irradiation acting as nuclei. Once the development is over, the Mie band at 400 nm is less intense and a new band appears around 1000 nm as the particle shape changed from spherical to triangular pellets (large particles of 100–150 nm). It is clear that this unusual orientated growth favors the 111 surface.

Nickel oligomers prepared in the presence of PA ($\lambda_{max} = 540$ nm) (Section 20.4.2) may also act as catalysts for the reduction of Ni^{2+} by hypophosphite ions. This requires, as shown by pulse radiolysis, a critical nuclearity, while free Ni^{2+} cannot be reduced directly by $H_2PO_2^-$. Very low radiation dose conditions, just initiating the formation of a few supercritical nuclei, will lead to large particles of nickel [96].

When trivalent chloro *gold* ions $Au^{III}Cl_4^-$ are γ-irradiated at increasing doses, the reduction occurs by successive steps. However, Au^I ions are not reduced as far as Au^{III} ions are more concentrated than Au^I, nor do they disproportionate because the redox potential involving the single atom $E°(Au^I/Au^0)$ is quite negative (Table 2). Thus, $Au^I Cl_2^-$ ions accumulate and the cluster appearance is delayed by an induction time [108]. With higher doses, Au^I ions become more abundant than Au^{III} and are also reduced into atoms and clusters. In the presence of alcohols or of preformed clusters, Au^I ions do not accumulate because the potential order of (Au^I/Au^0) and (Au^{II}/Au^I) couples is now inverted as a result of complexation or adsorption, respectively. Au^I ions are thus allowed to disproportionate, giving rise to atoms and growing clusters. In addition, a very slow chemical reduction by the alcohol is found, which occurs exclusively at the surface of the clusters formed by irradiation so that all gold ions may be reduced, whatever the initial dose [108]. As seen by AFM observations, the cluster size remains constant and their concentration increases with dose during the γ-irradiation, while during the postirradiation reduction by alcohol the size increases at constant density of particles. The lower the dose, the larger the final size of the homodisperse cubic crystallites, which may range from 10 to 500 nm, depending on the respective parts of radiolytic and chemical types of reduction.

Gold ions $Au^I(CN)_2^-$ are not reducible in solution by alcohol radicals, owing to their much lower redox potential in the complexed form [44], than in the hydrated form (Section 20.4.1 and Table 2). This was recently confirmed [109] by the observation of a very long induction time before the appearance of the plasmon band of gold clusters generated by a low amount of solvated electrons only. However, the complexed ions are reduced when adsorbed at the particle surface and the gold clusters are developed to an extent depending on added gold ion amount [146].

A stabilization of low valencies of metals and an induction time before cluster formation have been observed as well in the case of *iridium* [105], *platinum* [53], *palladium* [147], *copper* [94], or *nickel* [115].

The time-resolved studies of the cluster formation achieved by pulse radiolysis techniques allow one to better understand the main kinetic factors which affect the final cluster size found, not only in the radiolytic method but also in other reduction (chemical or photochemical) techniques. Generally, reducing chemical agents are thermodynamically unable to reduce directly metal ions into atoms (Section 20.4) unless they are complexed or adsorbed on walls or dust particles. Therefore, we explain the higher sizes and the broad dispersity obtained in this case by in situ reduction on fewer sites. A classic example is given by the reduction of silver ions by reducing additives. In the bulk of the solution, the redox potential of the latter is not negative enough to reduce the ions into free atoms. In contrast, due to the much more positive redox potential of the ions adsorbed on walls of the reaction vessel, their reduction is allowed and the result is a silver mirror.

4.5. Supported and Confined Nanoclusters

The radiolytic synthesis consists of: (1) either preparing first nanometric metal clusters in solution, which are then put in contact with the support (possibly by filtration), or (2) irradiating in situ the ionic precursors after their adsorption onto the supporting material.

Positively charged metal ions easily diffuse into the cavities of a Nafion® *polymeric membrane* by ion exchange of the counter cations of the constitutive sulfonic groups (Section 20.4.2). The size of channels and cavities is controlled by the proportion of alcohol in aqueous solution, which also governs the final size of silver or nickel clusters after irradiation [64]. During irradiation of a solution of metal ions (silver, palladium or nickel) containing PVA without OH scavenger, the ion reduction occurs simultaneously with the cross-linking of the polymer [25]. Finally, after drying, the clusters formed are trapped in a polymeric film. In the case of nickel, the thin film is ferromagnetic. Irradiation of reverse *micelles* containing gold ions in the aqueous phase generates small gold clusters [87]. Silver oligomers ($n < 10$) in AOT micelles absorb in the UV [148].

Iridium clusters supported on Al_2O_3 or TiO_2 *oxides* were prepared by irradiation of ions adsorbed on the oxides [149,150]. Radiation-induced grafting of Ir, Pt, or Ru clusters was achieved on SnO_2 [151,152]. Optically clear *alumina–silica* xerogels with glassy aspect containing silver clusters of 2.5–4.5 nm were also obtained [153,154]. Irradiation by MeV ion beams of copper oxide containing *glasses* followed by a moderate thermal annealing at 673 K, allows the control of atom coalescence and to produce nanometric copper clusters which are protected from further reaction by the glass matrix [155]. Homogeneously dispersed Pt nanoparticles were obtained from radiolytic reduction of molecular carbonyl clusters on *carbon fibers* or *powders* [156,157]. It was shown by EXAFS that the prismatic structure of radiation-induced Chini molecular clusters $[Pt_3(CO)_6]_{4-6}^{2-}$, when supported on *carbon* under inert atmosphere, was transformed into a planar layer with conservation of trimeric units, and partially into small fcc Pt clusters on contact with the air. On α-Al_2O_3, larger particles are formed [158].

Irradiation by ionizing radiation, capable of penetrating into a *zeolite* material exchanged by metal ions, allows the generation of metal atoms and clusters in situ in the cavities [159–161]. The observation of irradiated faujasite (Na–Y zeolite) by optical absorption spectroscopy at increasing doses and, at low silver content, demonstrates the formation of two bands, at 265 and 305 nm, which have been assigned to the charged trimer Ag_3^{2+}. Then, the ESR observation, after a dehydration step, indicates the reduction of this species into Ag_3^0 [162]. The photoreduction of H_2PtCl_6 solution included in mesoporous channels of molecular sieves induces nanowires of platinum [163]. Chini molecular clusters of platinum carbonyl can also be synthezised in situ in these channels ("ship-in-bottle" synthesis); then, they are photoreduced by UV–vis light to nanowires of 2.5 nm diameter and 50–300 nm length (Fig. 9) [137]. The temperature dependence of the unique magnetic properties of these Pt nanowires has been studied.

Pulse radiolysis studies of the reactivity of 25-nm AgI particles with e_{aq}^- in the presence of alcohol have shown that, first, the semiconductor spectrum at 360 nm is bleached, then silver atoms and clusters at 450–600 nm are formed [164,165]. The electron

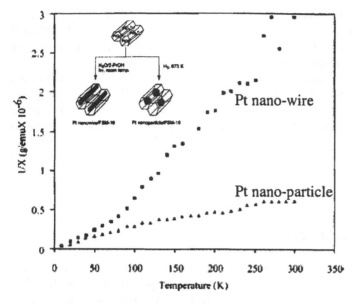

Figure 9 Left: Schematic representation of chemical or UV-induced synthesis of Pt nanowires and nanoparticles in zeolites FSM-16. Right: Comparison of temperature dependence of magnetization for Pt nanowires and nanoparticles in FSM-16. (From Ref. 137.)

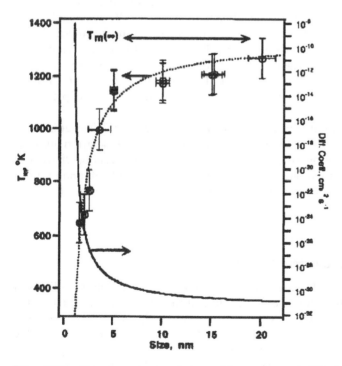

Figure 10 Size-dependence of the melting point and diffusion coefficient of silica-encapsulated gold particles. The dotted curve is calculated by the equation of Buffat and Borel. The bulk melting temperature of Au is indicated by the double arrow as $T_m(\infty)$. The solid curve (right-hand side axis) is a calculated Au self-diffusion coefficient. (From Ref. 146.)

transfer between the couple $MV^{2+}/MV^{-\bullet}$ and Ag_n clusters formed on an AgCl crystallite was studied by pulse radiolysis [166]. The coalescence of Ag atoms at the AgCl surface is slow so that, as in the presence of CN^-, an electron transfer from subcritical clusters to MV^{2+} precedes the electron transfer from $MV^{+\bullet}$ to supercritical Ag_n^+.

Recently, gold clusters of increasing size from 1.5 to 20 nm, after being chemically prepared, radiolytically or by radiolytic enlargement, were coated with a silica shell [146]. Because the gold/silica concentration ratio is constant, the silica thickness increases upon increasing the core size and is 5 nm at the maximum. For all particles, the melting point is noticeably lower than that of the bulk. It decreases markedly with the cluster size, implying an increase of the self-diffusion of gold atoms, from 2×10^{-32} in the bulk to $10^{-24}\, cm^2\, sec^{-1}$ for 2-nm clusters (Fig. 10).

5. SYNTHESIS OF BIMETALLIC CLUSTERS

Composite clusters, alloyed or bilayered, are of great interest because they enlarge the number of the possible types of clusters and also of their applications. It is therefore important to be able to select the conditions of the synthesis of composite metal clusters containing M^0 and M'^0 in variable proportions, with either an alloyed or a core/shell structure. The latter can be obtained by a two-step process [19,82,167–171]. The method of generating first gold clusters radiolytically, which are then coated with a silver shell by irradiating them in the second step in the presence of silver ions, was recently applied to produce core/shell Au/Ag nanoparticles of 2.5–20 nm diameter [172]. The smaller the initial gold core diameter (≤ 4.6 nm), the more important is the spontaneous interdiffusion after a week at room temperature of both metals at the core–shell interface, leading to alloying, which is detected by XAFS. The self-diffusion of atoms increases upon decreasing the cluster size [146], but the diffusion coefficient alone is too low to account for the observations. However, the migration of atoms is catalyzed by vacancy defects present at the interface [172].

Actually, when a mixed solution of two ionic precursors M^+ and M'^+ is irradiated or chemically reduced, both situations of alloyed or bilayered cluster formation may be encountered without clear prediction [102,173]. Moreover, an unambiguous characterization of the intimate structure of nanometric mixed clusters is quite difficult and requires appropriate methods, applied at different steps (or different doses) of the mixed cluster construction.

5.1. Core–Shell and Alloyed Clusters Formation

In many cases, although M^+ and M'^+ are both readily reduced by radiolytic radicals, a further electron transfer from the more electronegative atoms (for example, M') to the more noble ions M^+ ($E°(M'^+/M'^0) < E°(M^+/M^0)$) systematically favors the reduction into M^0. If the ionic precursors are plurivalent, an electron transfer is also possible between the low valencies of both metals, so increasing the probability of segregation [174]. The intermetal electron transfer has been observe directly by pulse techniques for some systems [66,175,176], and the transient cluster $(MM')^+$ sometimes identified such as $(AgTl)^+$ or $(AgCo)^{2+}$ [176]. The less noble metal ions act as an electron relay toward the precious metal ions, so long as all M^+ are not reduced. Thus, monometallic clusters M_n are formed first and M'^+ ions are reduced afterward in situ when adsorbed at the surface of M_n as in the cases of Ag/Cd [177], Ag/Cu [178], and Ag/Tl [120]. The final result is a core–shell cluster M_m/M'_n, where the more noble metal M is coated by the second M'.

In some other cases, the intermetal electron transfer does not occur even during hour-long irradiations. The initial simultaneous reduction reactions of M^+ and M'^+ are followed by mixed coalescence and association of atoms and clusters with ions, homolog or not. Besides the dimerization of atoms of the same metal into M_2 and M'_2, coalescence of both types of atoms occurs twice more frequently. Subsequently, mixed coalescence and reduction reactions progressively build bimetallic alloyed clusters according to the statistics of encounters, therefore to the relative initial ion abundance [53].

5.2. Dose Rate Effects

The possible formation of an alloyed or a core–shell cluster depends on the kinetic competition between, on one hand, the irreversible release of the metal ions displaced by the excess ions of the more noble metal after electron transfer and, on the other hand, the radiation-induced reduction of both metal ions, which depends on the dose rate (Table 5). The pulse radiolysis study of a mixed system [66] (Fig. 7) suggested that a very fast and total reduction by the means of a powerful and sudden irradiation delivered for instance by an electron beam (EB) should prevent the intermetal electron transfer and produce alloyed clusters. Indeed, such a decisive effect of the dose rate has been demonstrated [102]. However, the competition imposed by the metal displacement is more or less serious, because, depending on the couple of metals, the process may not occur [53], or, on the contrary, may last only hours, minutes, or even seconds [102].

The alloyed or layered character of a small bimetallic cluster structure is generally quite difficult to conclude experimentally [102,179]. Even if the surface plasmon transitions of both pure metals are specific (with one possibly in the UV), the unknown spectra of alloyed or bilayered clusters are both expected in the same intermediate region. The

Table 5 Multimetallic Clusters

	Synthesis and Irradiation Conditions	
Mixed Salts	Dose Rate (kGy hr^{-1}) Irradiation Source	Particle Structure (Ref.)
$Ag(CN)_2^-$, $Au(CN)_2^-$	γ, 35	Ag/Au, Bilayer [102]
$Ag(CN)_2^-$, $Au(CN)_2^-$	EB, 7.9×10^3	AgAu, Alloy [102]
Ag^+, $AuCl_4^-$	γ, 3.8	Au/Ag, Bilayer [102]
Ag^+, $AuCl_4^-$	γ, 35; EB, 7.9×10^3	AgAu, Alloy [102]
Ag^+, Cd_2^+	γ, 0.87	Ag/Cd, Bilayer [177]
Ag^+, Pd_2^+	γ, 35	AgPd, Alloy [180]
Ag^+, $PtCl_6^{2-}$	γ, 30	AgPt, Alloy [183]
Ag^+, Tl^+	γ, 0.6	Ag/Tl, Bilayer [119]
$AuCl_4^-$, $PtCl_6^{2-}$	EB, 7.9×10^3; γ, 0.5–40	Au/Pt, Bilayer [184]
$Au(CN)_2^-$, $PtCl_6^{2-}$	EB, 7.9×10^3; γ, 0.5–40	Pt/Au, Bilayer [184]
Cu^{2+}, Cd^{2+}	γ, 0.42, 0.48	CuCd, Alloy [83,182]
Cu^{2+}, Pd^{2+}	γ, 30	CuPd, Alloy [53]
$AuCl_4^-$, Pd^{2+}	γ 35	Au/Pd, AuPd, Bilayer [181]
$AuCl_4^-$, Pd^{2+}	EB, 7.9×10^3	Au/Pd, AuPd, Alloy [181]
Ag^+, Rh^{2+}	γ, 35	Rh, Ag, Segregation [185]
Ag^+, Rh^{2+}	EB, 7.9×10^3	Rh, Ag, Segregation [185]

structure of the composite cluster should be derived indeed from the evolution observation of the absorption spectrum at increasing dose during the cluster building [102,180,181]. Because the spectrum is attributable to surface phenomena, it changes with dose if one metal is preferentially reduced first. The surface composition of a bilayered cluster is shifted at increasing dose from the spectrum of the more noble metal to that of a cluster coated by the second. In contrast, the spectrum shape of alloyed clusters is unchanged at increasing dose while the intensity increases. X-ray analysis of clusters at partial reduction (increasing dose) clearly indicates at each step whether atoms of only one or of both types of metals are present in the particles [102]. In the case of an alloyed cluster, electron and X-diffraction methods may reveal superlattices that correspond to a perfectly ordered atom arrangement [4,99,140]. Provided that the lattice constants of the metals are somewhat different, the distances between atomic plans in HRTEM images may indicate the possible alloyed character of clusters.

For example, when the mixed solution of $Ag(CN)_2^-$ and $Au(CN)_2^-$ is irradiated by γ-radiolysis at increasing dose, the spectrum of pure silver clusters is observed first at 400 nm, because Ag is more noble than Au due to the CN^- ligand. Then, the spectrum is red-shifted to 500 nm when gold is reduced at the surface of silver clusters in a bilayered structure [102], as when the cluster is formed in a two-step operation [168] (Table 5). However, when the same system is irradiated at a high dose rate with an electron beam, allowing the sudden (out of redox thermodynamics equilibrium) and complete reduction of all the ions *prior* to the metal displacement, the band maximum of the alloyed clusters is at 420 nm [102].

Similarly, at moderate dose rate for the couple $Au^{III}Cl_4^-$, Ag^+, gold initially appears at 520 nm. Therefore, Ag^+ ions essentially act as an electron scavenger, and as an electron relay toward more noble gold ions as far as gold ions are not totally reduced. Then silver-coated gold clusters are formed and the maximum is shifted to 400 nm, which is that of silver (Fig. 11) [102]. But at higher γ- or EB dose rate (irradiation time of a few seconds), the electron transfer is too slow to compete with coalescence and the spectrum of alloyed clusters

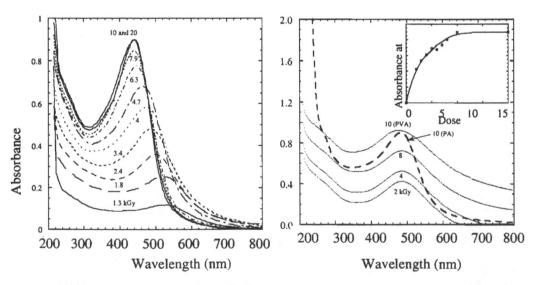

Figure 11 Evolution with increasing irradiation dose of the absorption spectra of a mixed solution $AuCl_4^-/Ag^+$ (50/50). Left: with PVA at a dose rate of 3.8 kGy hr^{-1}. Right: 35 kGy hr^{-1}. Dotted line: spectrum with PA. Inset: Evolution with increasing dose of the absorbance at $\lambda = 480$ nm. (From Ref. 102.)

surrounded by Cl⁻ develops at 480 nm without any shift from the lowest doses (Fig. 11). Once this quenching radiolytic reduction has consumed all the ions, the intermetal displacement becomes in fact excluded at room temperature. In both cases, low- and high-dose rates, a transverse analysis of a single cluster in a scanning transmission electron microscope equipped with a field emission gun and x-ray detector confirmed the cluster composition in depth, bilayered and alloyed, respectively [179].

The same evolution of the absorption spectrum with the dose has been found in a high dose rate for various values of the Ag and Au ion fraction in the initial solution. Clusters $Ag_{1-x}Au_x$ are alloyed with the same composition. The maximum wavelength λ_{max} and the extinction coefficient ε_{max} of the alloy depend on x. The experimental spectra are in good agreement with the surface plasmon spectra calculated from the Mie model at x values for which optical data are available (Fig. 12) [102]. Similar calculations were carried out for the alloy Ag_xPd_{1-x} obtained at a moderate dose rate [180].

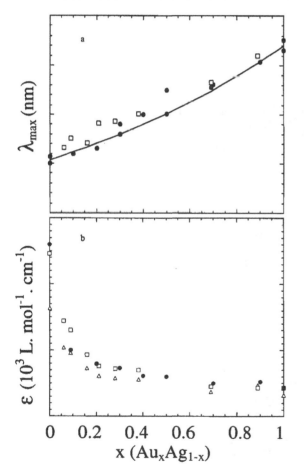

Figure 12 Top: Maximum wavelength of the plasmon band of alloyed gold–silver clusters as a function of the mole fraction x of gold in alloyed gold–silver clusters, produced at the dose rate 7.9 MGy hr⁻¹ and the dose 20 kGy. ● Experiments, □ calculated values by Mie model. Bottom: Extinction coefficient at the maximum of the plasmon band as a function of the mole fraction x of gold in alloyed gold-silver clusters. ● Experiments, □ calculated values from Kreibig equation [74] with $r = 5$ nm; △ with $r = 3$ nm. (From Ref. 102.)

The same contrast between bilayered and alloyed clusters of Au–Pd has been observed at low- and high-dose rate, respectively [180], by using spectrophotometry, EDAX, and XPS analysis at partial and complete reduction.

Other examples of bimetallic clusters are presented in Table 5. Depending on the rate of the intermetal electron transfer competing with the dose rate-dependent coalescence (Fig. 13 and Table 5):

1. Alloying may occur spontaneously, even at low reduction rate (γ-irradiation with low dose rate) [53], as for alloyed Cu_3Pd [53], CuPd [53], Ni–Pt [53], CuCd [84,182], AgCd [177], Ag_xPd_{1-x} [180], and Ag–Pt clusters [183].
2. Alloying may require higher γ-dose rates as for Ag_xAu_{1-x} [102].
3. Alloying is only obtained through a short and intense irradiation provided by an electron accelerator as for Ag_xAu_{1-x} [102], and Au–Pd [180].
4. Alloying is even not obtained. In extreme cases, the electron transfer is achieved within times shorter than the irradiation time required for complete reduction

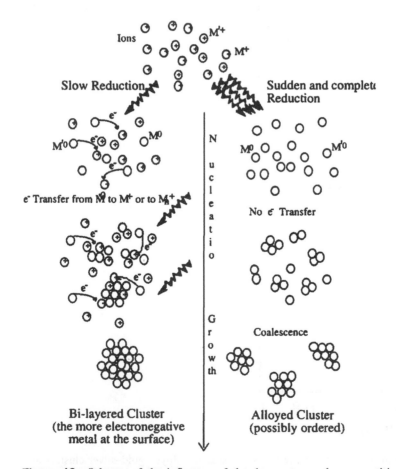

Figure 13 Scheme of the influence of the dose rate on the competition between the inter-metal electron transfer and the coalescence processes during the radiolytic reduction of mixed metal ion solutions. Sudden irradiation at high dose rates favor alloying, whereas low dose rates favor core–shell segregation of the metals because of metal displacement in the clusters.

(a few seconds with EB) and, in spite of the very fast reduction, metals are segregated into core–shell structures (Au/Pt, [184]) or into distinct clusters (Ag_n and Rh_n) [185]. For multivalent ion precursors, intermetal electron transfer reactions may also occur between the transient lower valency ions.

5.3. Catalytic Properties of Metal Clusters

One of the important applications of mono- and multimetallic clusters is to be used as catalysts [186]. Their catalytic properties depend on the nature of metal atoms accessible to the reactants at the surface. The possible control through the radiolytic synthesis of the alloying of various metals, all present at the surface, is therefore particularly important for the catalysis of multistep reactions. The role of the size is twofold. It governs the kinetics by the number of active sites, which increase with the specific area. However, the most crucial role is played by the cluster potential, which depends on the nuclearity and controls the thermodynamics, possibly with a threshold. For example, in the catalysis of electron transfer (Fig. 14), the cluster is able to efficiently relay electrons from a donor to an acceptor, provided the potential value is intermediate between those of the reactants [49]. Below or above these two thresholds, the transfer to or from the cluster, respectively, is thermodynamically inhibited and the cluster is unable to act as a relay. The optimum range is adjustable by the size [63].

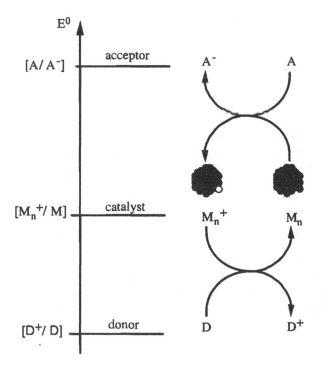

Figure 14 Mechanism of catalytic electron transfer involving metal clusters as relay. The thermodynamic conditions to be fulfilled are that the cluster redox potential be higher than the donor D and lower than the acceptor A potential, which implies that the cluster itself is a size range that offers the efficient redox potential. (From Ref. 63.)

Clusters, as possible catalytic reactors, are perfectly dispersed in solutions. They are thus suitable systems for observing, under quasi-homogeneous conditions by time-resolved techniques, the kinetics of catalyzed electron transfer, which would be inaccessible on a solid catalyst. It was demonstrated that the reaction of radiation-induced free radicals $CO_2^{\cdot-}$ and $(CH_3)_2C^{\cdot}OH$ catalyzed by metal clusters started by the storage of electrons on clusters as charge pools and that electrons were then transferred pairwise to water-producing molecular hydrogen [22,75].

The mechanism of the electron transfer from electron donor such as MV^+ to H^+ producing molecular hydrogen and catalyzed by Au [187] or Pt clusters [188], and that of the disproportionation reaction of the superoxide radical anion O_2^- by Pt clusters in solution or supported on colloidal TiO_2 particles [189], have been studied by pulse radiolysis.

In some examples, the reaction catalyzed by radiation-induced clusters was studied during long periods under the conditions of a pilot system to compare their properties with catalysts prepared by chemical reduction. They easily fulfill the conditions of controlled, nanometric and homodisperse size, of alloying for multimetallic clusters and of strong grafting on various supports. They were found efficient catalysts, for example, in the case of alumina-supported iridium clusters for hydrazine decomposition in thrusters developed for spacecraft orbit and attitude control systems [88]. Radiation-induced mono- and multimetallic clusters imbedded in Nafion® [64] or grafted on metal [190], on SnO_2 transparent counterelectrodes [152,152], or on carbon electrodes [25,100] were found to be quite active in electrochemical, photovoltaic, and fuel cells (methanol or hydrogen oxidation), respectively. Multimetallic clusters deposited on gelatine silver bromide emulsions are excellent catalysts for silver reduction in the offset process [191] The catalytic performances of photo-induced Pt nanowires in zeolites were tested by using the water–gas shift reaction [137].

6. PHOTOGRAPHY

Silver photography [192–195], and later, radiography [1,2], are the oldest applications of the unique properties of oligomers, long before the mechanisms were understood. The time-resolved techniques were particularly helpful to improve our understanding of the intimate reaction process. Since its invention, innumerable improvements to photography have been introduced to capture an image as close as possible to our vision of the light intensities and the colors of a scene, and to enhance the photosensitivity of the photographic layer. However, the basic principle of silver photography lies on the photosensitivity to visible light of minute silver halide crystals, which are fixed in a gelatine layer as a mosaic. Each crystal will behave as a picture element (pixel). During *exposure*, the light generates clusters of a few silver atoms on the crystals. Their ensemble constitutes the latent image which is of an extremely weak intensity and therefore invisible. Then the *development* consists of converting chemically into metal particles the crystals containing a cluster with a super-critical number of photo-induced silver atoms and of catalytically transforming the latent image into a visible picture. In the last step of *fixing*, undeveloped crystals which are still photosensitive should be eliminated and are dissolved by complexation. The result is the negative image, which is then used as a mask to produce the positive image.

6.1. Photographic Development

It is well known from photographic experience that critical nuclearity, which is required for the development, depends on the redox potential of the developer (it is an electron donor in excess in the development bath). Subsequently, the autocatalytic growth does not stop until all the metal ions of the exposed crystals containing supercritical clusters are not exhausted and the crystals transformed into silver metal grains. For over a century, numerous theoretical models have been proposed, without success [196,195], to explain how the supercritical clusters created by the light act as nuclei to catalyze development. Trautweiler [197] had the intuition in his speculative model that the ionization potential of supercritical sizes should lie above that of the developer. However, the only data then available did not confirm this view.

Actually, the kinetic study of the cluster redox potential by pulse radiolysis [31] (Section 20.3.2) somewhat mimics the process of the black-and-white photographic development, except that clusters are free in the solution (not fixed on AgBr crystals), and that they are produced by ionizing radiation (as in radiography and not by visible photons; but the last choice had been incompatible with the time-resolved optical detection in the visible. Beyond the critical nuclearity, they receive electrons without delay from the developer already present (actually, the photographic development is achieved in a delayed step).

However, the similarity of the features presented by the processes in both cases, reduction catalyzed by supercritical clusters in solution and photographic development on AgBr crystals, allows us to extend to photographic development the same mechanism demonstrated by pulse radiolysis [31,198]. Consequently, the mechanism results from the increase of the cluster redox potential with the nuclearity when the developer solution is in contact with silver bromide crystals. It explains the existence of a critical size in the photographic development, and the threshold of the discrimination between underexposed and exposed crystals for complete reduction is imposed by the first monoelectronic redox potential of the developer. It also explains the sensitizing effect [66] of gold ions more noble than Ag^+, and the fragility to oxidation of the smallest clusters of the latent image [5]. With standard developers and sulfide-gold sensitization, the amplification of the cluster size by the development is from the critical number [194] $n_c = 3$ to $n = 10^{8-9}$, that is the total number of silver ions in the AgBr crystal (about 1 μm).

Apart from the development in photography, most of nucleation and growth mechanisms based on a chemical reduction (Section 20.4.4) behave as development processes, and are likewise controlled by the nuclearity dependence of the cluster redox potential and by the potential of the electron donor.

6.2. Enhancement of Photographic Sensitivity

According to the Gurney–Mott model [199], the primary light effect on a AgBr crystal is to produce as many electron–hole pairs (e^-–h^+) as photons absorbed. The hole corresponds to the electron vacancy created by the electron photoejection. The electron reduces a silver cation into an atom, generally close to a sensitizer adsorbed at specific surface sites of AgBr (sulfide or/and gold centers). Then a cation adjacent to the atom and constituting a charged dimer is reduced by another electron, and so on, the result being a cluster of a few atoms. But an important part of electrons are lost by direct recombination with the parent holes, before these diffuse to the surface where they are irreversibly scavenged by gelatine and additives (Fig. 15, top, left). Holes are also able to oxidize the newly formed silver atoms, so counterbalancing somewhat the reduction by electrons (indirect recombination). If R is the fraction of initial pairs lost by both types of recombination, and $\Phi_{theor} = 1$ pair/photon the theoretical quantum yield, the effective quantum yield Φ_{eff} is $\Phi_{eff} = \Phi_{theor}(1-R)$. In sulfide/gold sensitized crystals, the yield is about $\Phi_{eff} = 0.10$–0.30

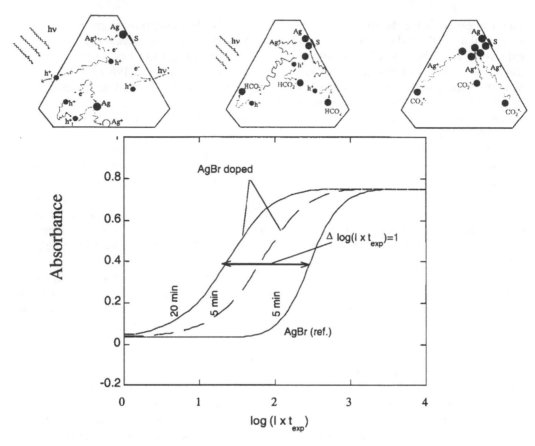

Figure 15 Top: Photographic latent image formation in undoped (left) and formate-doped and gold-sulfide sensitized AgBr crystals with the hole-scavenging step (center). Secondary reduction step by formyl radical (right). Bottom: Sensitometry curves for gold-sulfide sensitized emulsions, undoped or formate-doped, and developed after 5 or 20 min ($t_{exp} = 10^{-2}$ sec, development with aminophenol and ascorbic acid). The same absorbance is observed for a number of photons absorbed 5 or 10 times less, respectively, than in the undoped emulsion. (From Ref. 200.)

(the fraction R is thus 0.90–0.70). Thus, apart from the absorption properties and the area of the crystal, the photosensitivity of a film depends directly on the quantum yield of the cluster formation.

A recent approach to enhance the sensitivity was to specifically scavenge the holes, faster than their possible recombination with the electrons or the atoms, with the help of a dopant [200]. In addition to a small size and an ionic character to allow its inclusion in the AgBr crystal, the dopant should obey other strict criteria concerning the redox potential. The dopant should be a bielectronic donor with a very weak first redox potential, in order to let the dopant scavenge the holes, without spontaneously reducing free Ag^+ and producing fogging in the dark, and with a very negative second redox potential, to avoid a possible hole-like behavior of the dopant oxidized form, so blocking any reversible oxidation (Fig. 15, top, center). In addition, a second silver ion can be reduced (Fig. 15, top, right). From radiation chemical knowledge, the anion formate HCO_2^- fulfills all the conditions, at least in solution ($E°(CO_2^{-•}/HCO_2^-) = 1.07$ V_{NHE} and $E°(CO_2/CO_2^{-•}) = -1.9$ V_{NHE} in water) [30].

Sensitometric tests at variable light intensity I under conditions of photography, achieved on emulsions doped at the relative concentration of 10^{-6} mol HCO_2^- per mol Ag^+, confirmed the photo-induced bielectronic transfer (Fig. 15, bottom) [200]. The emulsion is completely stable in the dark. The number of photons required to induce development of the same grain population fraction is 5 times less (after immediate development) or 10 times less (development delayed by 20 min after exposure) in doped than in undoped emulsions where $\Phi_{eff} = 0.20$. The quantum yield is thus close to the theoretical limit $\Phi_{eff} = 1$ atom/photon in immediately developed doped emulsion ($R = 0$), and $\Phi_{eff} = 2$ atom/photon if the development is delayed for 20 min, and the slow additional reduction of a silver ion by $CO_2^{-•}$ can occur, thus doubling the gain. An important sensitivity enhancement is similarly observed for dye-sensitized crystals [200]. These doping studies [200,201] not only provide a better understanding of the mechanism of the latent image formation, but also offer a promising route for improving the performance of all kinds of silver emulsions, for black-and-white and color photography, radiography, holography, etc.

7. CLUSTER FORMATION AND CLUSTER EXCITATION BY POWERFUL LASERS

Metal clusters exhibit intense optical absorption due to the surface plasmon band, some in the visible spectrum (Table 4). The response to light excitation is linear as far as the source is of weak intensity. However, other phenomena appear when pulses of extremely high intensity are used. A burst activity has been seen in recent years in the area of interaction of light from powerful laser pulses with metal in order to produce nanoparticles by laser ablation, or with nanoparticles in order to let them absorb suddenly a high amount of energy and to observe the transient and the irreversible effects on the clusters. Eventually, a change in the morphological shape induced by the laser or a fragmentation of nanoclusters may occur (Fig. 16).

7.1. Nanoparticles Formation by Laser Ablation

The method of laser ablation from a metal has been used to prepare nanoparticles dispersed in a solution [202–207]. Recently, colloidal gold nanoparticles in water having an average diameter of 5.5 nm were prepared by laser ablation at 1064 nm (800 mJ/cm^2) from a gold metal plate [208]. The final nanoparticle size depends on the laser fluence and the stabilizer concentration (Fig. 17).

7.2. Nanoparticles Melting and Fragmentation

Gold clusters can be tailored by subjecting them to melting or to fragmentation and bimetallic nanoparticles can be alloyed after melting under irradiation by a pulsed laser at 532 nm in the vicinity of the wavelength of the surface plasmon band [209–212]. In fact, the electrons of the plasmon band are highly excited. The hot electrons transfer the energy to the lattice in less than a few picoseconds. The transfer can occur before or after the electrons

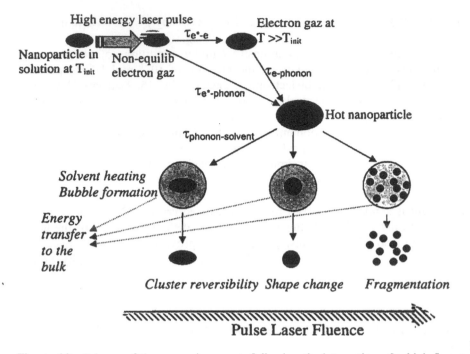

Figure 16 Scheme of the successive events following the interaction of a high fluence laser pulse with metal clusters. First, the energy is confined in a hot cluster, then the effects depend on the fluence and on the cluster surface to volume ratio. Low fluence: the energy is transferred to the solvent, which expands into bubbles scattering the light; meantime, the cluster is reversibly cooled without change of the shape. Medium fluence: the cluster temperature is high enough to obtain cluster melting and changing to spherical shape. High fluence: the energy confined within the cluster creates a fast step of expansion–vaporization of the metal particle which acts as a light scattering center. After the energy has been transferred to the solvent, the solvent bubbles become secondary light scattering centers and the metal vapor condenses into fragmented clusters. Eventually, the solvent bubbles cool down by energy dissipation to the bulk.

have reached thermal equilibrium (Fig. 16). Eventually, the energy of the phonons is transferred to the surrounding solvent.

If the laser pulse is very short and its energy high enough, the temperature of the lattice can reach several thousands degrees. Depending on the energy of the pulse, melting, fragmentation, and even vaporization of the particles is thus possible. It was shown that silver and gold clusters undergo significant morphological change under laser irradiation. Gold nanorods can be transformed into spherical particles by exciting the metal clusters with femtosecond laser pulses (Fig. 18) [211]. The amount of energy needed to transform a gold nanorod into a sphere is more than 60 fJ. The time corresponding to the melting is around 35 psec, which is shorter than the period required for the energy transfer from the lattice to the solvent (100 psec). Therefore, sufficient energy is accumulated by the lattice arising from photon absorption and the temperature increases enough to melt and change the morphology of the gold nanorods. Therefore, if the laser energy increases above the threshold of melting, the fragmentation channel becomes open. Transmission electron microscopy (TEM) measurements of nanorods irradiated by a high-energy nanosecond laser pulse clearly showed that the nanorods are fragmented into small spheres.

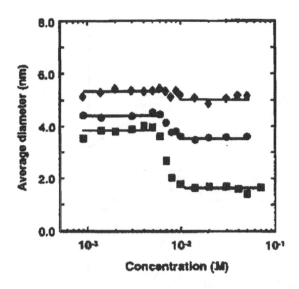

Figure 17 Average diameter of gold nanoparticles after laser ablation at 532 nm and a laser fluence of (■) 320, (●) 480, and (◆) 1200 mJ pulse^{-1} cm^{-2} as a function of the SDS concentration in the solution. (From Ref. 208.)

7.3. Optical Limitation Induced by Nanoparticles

The nonlinear optical properties of nanoclusters were also studied as potential light limitation systems [213]. The nonlinear optical response of gold particles (of 5-, 10-, or 30-nm diameter, the latter being enlarged by development) prepared by γ-radiolysis in water solution and stabilized by PVA is size-dependent. The threshold and the amplitude of limitation depend on the particle size (Fig. 19). At a given total atom concentration, the light transmission at 532 nm of nanosecond laser pulses at very high fluence is weakly limited by 5-nm clusters, but strongly limited by the 30-nm clusters. The optical limiting effect induced by gold clusters, measured as a function of the excitation wavelength, reaches the highest efficiency below 530 nm, and the wavelength dependence of the limiting effect amplitude induced at a fixed excitation wavelength increases monotonically to the UV as expected from photo-induced scattering centers.

The time delay of the amplitude maximum and the fluence threshold, where nonlinear effects are observed, differ for the time-resolved signals after nanosecond or picosecond pulses [213]. Therefore, two types of scattering centers are successively responsible for the optical limitation. The fast mechanism which reaches an amplitude maximum in less than 1 nsec appears at relatively high fluence (>0.94 J cm^{-2}, Fig. 20) and with picosecond pulses, that is for high-energy confinement. It is assigned to light scattering by gold clusters that have expanded and vaporized after having absorbed an important amount of light energy (Fig. 16). The slow limitation mechanism occurs in a second step when the energy is transferred subsequently from the cluster to the solvent, and is a result of light scattering by solvent bubbles. The second mechanism is observed alone for 5-nm clusters even at high fluences, or for 10- and 30-nm clusters at moderate fluences because the energy dissipation from small clusters into the solvent producing bubbles is then more efficient and precludes metal particle vaporization.

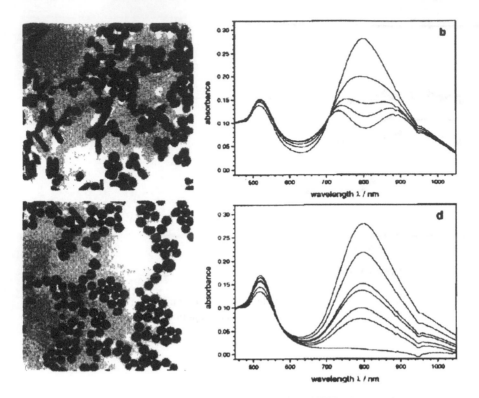

Figure 18   Comparison of optical absorption data and TEM images (at the same magnification) for two gold samples irradiated by laser pulses having the same fluence (0.25 J cm^{-2}) and different pulse width. Top: with a 7-nsec laser pulse at 800 nm. (a) Partial melting of the gold nanorods and (b) selective optical hole burning in the near-IR band corresponding to the nanorods. Bottom: with laser pulses of 100 fsec. (c) Complete melting of the gold nanorods into nanodots and (d) complete depletion of the nanorod band. (From Ref. 211.)

The tests for limitation efficiency as a function of the number of pulses (8 nsec, 5 J cm^{-2}) have shown that the 10-nm clusters are irreversibly modified and lose their activity after 20 pulses [213]. It was concluded that the limitation decrease is due to the decrease after each pulse of the size of part of the clusters below the size threshold for optical limitation. For 30-nm clusters, the condensation following the vaporization leads to smaller clusters, but the size after relaxation is still large enough to cause optical limitation, and the cluster is able to absorb again enough energy to vaporize. After 20 pulses, an equilibrium size, with a two times lower efficiency, seems to be reached.

8.   CONCLUSION

Radiation chemical methods have been proven to have a potential to induce and to study the dynamics of nucleation and growth and of the reactivity of metal clusters from the monomer to the stable nanoparticle.

Pulse radiolysis permits the study, via the time-resolved technique, of the detailed mechanisms of metal ion reduction in different environments (ligand, solvent, support, . . .),

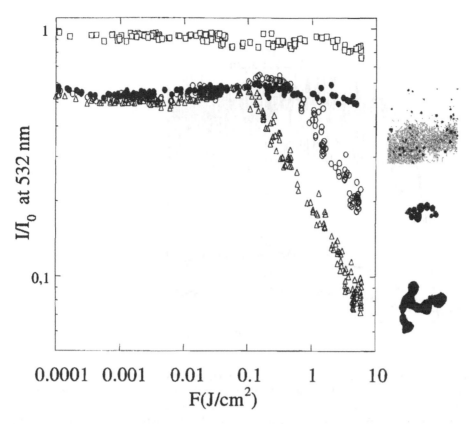

Figure 19 Size dependence of the optical transmission obtained at 532 nm vs. the laser fluence for γ-induced gold clusters. (●) 5-nm diameter (PVA 0.1 mol L^{-1}). (○): 10 nm diameter (PVA 10^{-3} mol L^{-1}). (△): 30 nm diameter (PVA 0.1 mol L^{-1}, MV^{2+} 10^{-3} mol L^{-1}). (□): Reference solution containing PVA and MV^{2+} without clusters. TEM micrographs of the samples I, II, and III are on the right. (From Ref. 213.)

of nucleation of atoms into oligomers, of their growth into stable metal clusters, and of the reactivity of the transient species at each step, also in mixed solutions of two different metal ions. Thus during this growth process, the nuclearity-dependent properties of oligomers and clusters were determined, such as the redox potential, which is of crucial importance in certain catalytic reactions and for the growth itself which can be autocatalyzed, and for the stability of the clusters at the final size.

Pulse radiolysis and pulsed laser methods are powerful tools to study the reactions at the interface between clusters and solution, such as fast heterogeneous catalytic processes, or transfer of the charges generated inside the particle.

The generation of clusters through irradiation of a homogeneous ionic solution has numerous advantages: (1) an ideal instantaneous mixing of the radiolytic species, acting as the reductant, with the ions; (2) the strong reducing properties of these species even at room temperature and a formation of atoms without energetic barrier; (3) a strict control of the reduction equivalents by the choice of the dose; (4) a control of the competition between the creation of isolated atoms and the growth of nuclei by the choice of the dose rate from slow to quasi-sudden regimes. The chances to create isolated atoms in the first step are thus

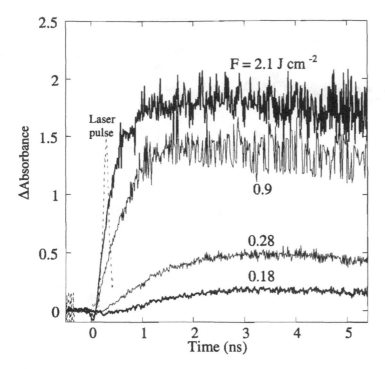

Figure 20 Time profiles of induced absorbance at 580 nm after focusing a 30-psec, 532-nm laser pulse at various fluences F onto gold clusters, 30 nm diameter. Dotted line: time profile of the pulse. The fast amplitude increase, up to a maximum in less than 1 nsec for fluences ≥ 0.9 J cm^{-2}, is assigned to light scattering by expanded metal clusters. The delayed increase with a maximum at 3 nsec occurs at fluences ≥ 0.18 J cm^{-2} already, and is assigned to light scattering by solvent bubbles. (From Ref. 213.)

higher than in any other method. (5) The possible synthesis in situ of clusters inside cavities of opaque porous material. These conditions explain why the general concept of size effects on cluster properties, which are the most drastic at low nuclearity, was indeed pointed out first in radiolytic experiments. They also explain that the growth can be restricted to very small final sizes and that their distribution is exceptionally narrow.

The knowledge of the kinetics mechanism, explaining the properties of the final particles from the conditions of the radiolytic synthesis, has been exploited for guiding the procedure of obtaining clusters with definite size and composition according to the needs of various studies and applications. Due to this mastering, an increasing number of data have been obtained on the cluster size- and shape-dependent properties, such as optical and magnetic properties, fusion temperature, cluster-to-solvent thermal energy transfer, optical limitation, etc. The understanding of the detailed mechanism of bimetallic cluster synthesis has also permitted the control of conditions favoring alloying, possibly with ordered atom arrangement, or segregation in a core–shell structure. Moreover, except for the reduction step, the mechanism is not specific to the radiation-induced method and will be helpfully extended to other ways of chemical or photochemical synthesis. In photography, this understanding resulted in a new interpretation of the critical nuclearity and of the selective amplification occurring during the development. Besides, conditions were found of favoring the latent image formation and markedly enhancing the photographic sensitivity by suppressing the electron–hole recombination.

However, the variety of composite materials to be elaborated by the method is still barely explored. For example, bi- and multimetallic nanoparticles, included in different matrices (polymeric membranes, porous supports, ...) or functionalized, have promising applications. New methods of cluster characterization at this extremely low size scale are developed and will improve their study.

Moreover, the interpretation of experimental data on clusters in solution requires more elaborate theoretical models to include the solvation effects around the structure of a small metal cluster. New kinetic models must be developed to describe nucleation, which governs the phase transition from a solute to a small solid phase.

REFERENCES

1. Roentgen, W.C. Sitz.-Ber. Phys.-Med. Ges. Wuerzburg 1896, 11 pp.
2. Becquerel, H. C. R. Acad. Sci. 1896, *122*, 420.
3. Fricke, H. J. Chem. Phys. 1934, *2*, 556.
4. Haïssinsky, M.; Pujo, A.M. C. R. Acad. Sci. 1955, *240*, 2530.
5. Haïssinsky, M. La Chimie Nucléaire et ses Applications. Masson et Cie: Paris, 1957.
6. Koulkes-Pujo, A.M.; Rashkov, S. J. Chim. Phys. 1967, *64*, 534.
7. Baxendale, J.H.; Fielden, E.M.; Keene, J.P.; Ebert, M. In *Pulse Radiolysis*; Keene, J.P., Swallow, A., Baxendale, J.H., Eds.; Acad. Press: London, 1965; 207 pp.
8. Von Pukies, J.; Roebke, W.; Henglein, A. Ber. Bunsenges. Phys. Chem. 1968, *72*, 842.
9. Haïssinsky, M. In *Radiation Chemistry*; Dobo, J., Hedwig, P., Eds.; Akad. Kiado: Budapest, 1972; *2*, 1353. And following discussion.
10. Belloni, J. In *Actions Chimiques et Biologiques des Radiations*; Haïssinsky, M., Ed.; Masson et Cie: Paris, 1971; 47 pp.
11. Delcourt, M.O.; Belloni, J. Radiochem. Radioanal. Lett. 1973, *13*, 329.
12. Basco, N.; Vidyarthi, S.K.; Walker, D.C. Can. J. Chem. 1973, *51*, 2497.
13. Henglein, A. Ber. Bunsenges. Phys. Chem. 1977, *81*, 556.
14. Halperin, W.H. Rev. Mod. Phys. 1986, *58*, 533.
15. Morse, M.D. Chem. Rev. 1986, *86*, 1046.
16. Schumacher, E. Chimia 1988, *42*, 357.
17. Haberland, H., Ed.; In *Clusters of atoms and Molecules*; Springer, 1994.
18. Kubo, R. J. Phys. Soc. Jpn. 1962, *17*, 975.
19. Belloni, J.; Mostafavi, M. In *Studies in physical and theoretical Chemistry 87. Radiation Chemistry: Present Status and Future Trends*; Jonah, C.D., Rao, B.M., Eds.; Elsevier, 2001; 411 pp.
20. Henglein, A. Ber. Bunsenges. Phys. Chem. 1995, *99*, 903.
21. Kamat, P.V. J. Phys. Chem. B 2002, *106*, 7729.
22. Henglein, A. Chem. Rev. 1989, *89*, 1861.
23. Belloni, J.; Amblard, J.; Marignier, J.L.; Mostafavi, M. In *Clusters of Atoms and Molecules*; Haberland, H. Ed.; Springer-Verlag, 1994; *2*, 290.
24. Tausch-Treml, R.; Henglein, A.; Lilie, J. Ber. Bunsenges. Phys. Chem. 1978, *82*, 1335.
25. Belloni, J.; Mostafavi, M.; Remita, H.; Marignier, J.L.; Delcourt, M.O. New J. Chem. 1998, *22*, 1239.
26. Belloni, J.; Mostafavi, M. In *Metal Clusters in Chemistry*; Braunstein, P., Oro, R., Raithby, J., Eds.; Wiley, 1999; 1213 pp.
27. Baxendale, J.H.; Busi, F. The Study of Fast Processes and Transient Species by Electron Pulse Radiolysis. NATO ASI Ser. 86. D. Reidel, 1982.

28. Buxton, G.V.; Greenstock, C.L.; Helman, W.P.; Ross, A.B. J. Phys. Chem. Ref. Data 1988, *17*, °2n.
29. Elliott, J.; Simon, A.S. Radiat. Phys. Chem. 1984, *24*, 229.
30. Schwarz, H.A.; Dodson, R.W. J. Phys. Chem. 1989, *93*, 409.
31. Mostafavi, M.; Marignier, J.L.; Amblard, J.; Belloni, J. Radiat. Phys. Chem. 1989, *34*, 605.
32. Janata, E.; Lilie, J.; Martin, M. Radiat. Phys. Chem. 1994, *43*, 353.
33. Janata, E. Radiat. Phys. Chem. 1994, *44*, 449.
34. Kapoor, S.; Lawless, D.; Kennepohl, P.; Meisel, D.; Serpone, N. Langmuir 1994, *10*, 3018.
35. Janata, E.; Henglein, A.; Ershov, B.G. J. Phys. Chem. 1994, *98*, 10888.
36. Texier, I.; Mostafavi, M. Radiat. Phys. Chem. 1997, *49*, 459.
37. Remita, S.; Mostafavi, M.; Delcourt, M.O. J. Phys. Chem. 1997, *100*, 10187.
38. Mostafavi, M.; Dey, G.R.; François, L.; Belloni, J. J. Phys. Chem. 2002, *106*, 10184.
39. Mostafavi, M.; Delcourt, M.O.; Keghouche, N.; Picq, G. Radiat. Phys. Chem. 1992, *40*, 445.
40. Cercek, B.; Ebert, M.; Swallow, A.J. J. Chem. Soc. A, 1966, 612.
41. Butler, J.; Henglein, A. Radiat. Phys. Chem. 1980, *15*, 603.
42. Ershov, B.G.; Sukhov, N.L. Radiat. Phys. Chem. 1990, *36*, 93.
43. Ghosh-Mazumdar, A.S.; Hart, E.J. Adv. Chem. Ser. 1968, *81*, 193.
44. Mosseri, S.; Henglein, A.; Janata, E. J. Phys. Chem. 1989, *93*, 6791.
45. Belloni, J.; Khatouri, J.; Mostafavi, M.; Amblard, J. In *Ultrafast Reaction Dynamics and Solvent Effects*; Rossky, P.J., Gauduel, Y., Eds. Am. Inst. Phys. 1993; 541 pp.
46. Mostafavi, M.; Belloni, J. Recent Res. Dev. Phys Chem. 1997, *1*, 459.
47. Texier, I.; Remita, S.; Archirel, P.; Mostafavi, M. J. Phys. Chem. 1996, *100*, 12472.
48. Mostafavi, M.; Remita, S.; Delcourt, M.O.; Belloni, J. J. Chem. Phys. 1996, *93*, 1828.
49. Belloni, J.; Delcourt, M.O.; Marignier, J.L.; Amblard, J. In *Radiation Chemistry*; Hedwig, P., Schiller, L.R., Eds.; Akad. Kiado: Budapest, 1987; 89 pp.
50. Mostafavi, M.; Lin, M.; Wu, G.; Katsumura, Y.; Muroya, Y. J. Phys. Chem. A 2002, *106*, 3123.
51. Kevan, L. J. Phys. Chem. 1981, *85*, 1828.
52. Ghosh-Mazumdar, A.S.; Hart, E.J. Int. J. Radiat. Phys. Chem. 1969, *1*, 165.
53. Marignier, J.L.; Belloni, J.; Delcourt, M.O.; Chevalier, J.P. Nature 1985, *317*, 344.
54. de Cointet, C.; Mostafavi, M.; Khatouri, J.; Belloni, J. J. Phys. Chem. 1997, *101*, 3512.
55. Mie, G. Ann. Phys. 1908, *25*, 377.
56. Creighton, J.A.; Eadon, D.J. J. Chem. Soc. Faraday Trans. 1991, *87*, 3881.
57. Henglein, A.; Tausch-Treml, R. J. Colloid Interface Sci. 1981, *80*, 84.
58. Buxton, G.V.; Mulazzani, Q.G.; Ross, A.B. J. Phys. Chem. Ref. Data 1995, *24*, 3.
59. Rao, P.S.; Hayon, E. J. Phys. Chem 1975, *79*, 865.
60. Renou, F.; Mostafavi, M. Chem. Phys. Lett. 2001, *335*, 363.
61. Remita, S.; Archirel, P.; Mostafavi, M. J. Phys. Chem. 1995, *99*, 13198.
62. Khatouri, J.; Ridard, J.; Mostafavi, M.; Amblard, J.; Belloni, J. Z. Phys. D 1995, *34*, 57.
63. Khatouri, J.; Mostafavi, M.; Belloni, J. In *Photochemistry and Radiation Chemistry: Complementary Method for the Study of Electron Transfer*; Wishart, J., Nocera, D., Eds.; ACS, 1998; 293 pp.
64. Platzer, O.; Amblard, J.; Marignier, J.L.; Belloni, J. J. Phys. Chem. 1992, *96*, 2334.
65. Amblard, J.; Platzer, O.; Ridard, J.; Belloni, J. J. Phys. Chem. 1992, *96*, 2340.
66. de Cointet, C.; Khatouri, J.; Mostafavi, M.; Belloni, J. J. Phys. Chem. 1997, *101*, 3517.
67. Jackschath, C.; Rabin, I.; Schulze, W. Z. Phys. D 1992, *22*, 517.
68. Alameddin, G.; Hunter, J.; Cameron, D.; Kappes, M.M. Chem. Phys. Lett. 1992, *192*, 122.
69. Khatouri, J.; Mostafavi, M.; Amblard, J.; Belloni, J. Z. Phys. D 1993, *26*, 82.
70. Henglein, A. J. Phys. Chem. 1993, *97*, 5457.
71. George Thomas, K.; Biju, V.; George, M.V.; Guldi, D.M.; Kamat, P.V. J. Phys. Chem. B 1999, *103*, 8864.
72. Sudeep, P.K.; Ipe, B.I.; George Thomas, K.; George, M.V.; Barazzouk, S.; Hotchandani, S.; Kamat, P.V. Nano Lett. 2002, *2*, 29.

73. Ipe, B.I.; George Thomas, K.; Barazzouk, S.; Hotchandani, S.; Kamat, P.V. J. Phys. Chem. B 2002, *106*, 18.
74. Kreibig, U.; Vollmer, M. Optical properties of metal clusters, Springer, Berlin, 1995. Sinzig, J., Radtke, U., Quinten, M., Kreibig, U., Eds. Z. Phys. D 1993, *26*, 242.
75. Henglein, A. J. Phys. Chem. 1979, *83*, 2209.
76. Zhu, Y.; Qian, Y.; Zhang, M.; Chen, Z.; Lu, B.; Wang, C. Mater. Lett. 1993, *17*, 314.
77. Jonah, C.D.; Matheson, M.S.; Meisel, D. J. Phys. Chem. 1977, *81*, 1805.
78. Rafaeloff, R.; Haruvy, Y.; Binenboym, J.; Baruch, G.; Rajbenbach, L.A. J. Mol. Catal. 1983, *22*, 219.
79. Zhu, Y.; Qian, Y.; Li, X.J.; Zhang, M. Chem. Commun. 1997, *12*, 1081.
80. Graetzel, M. Acc. Chem. Res. 1981, *14*, 376.
81. Ershov, B.G. Russ. Chem. Bull. 1994, *43*, 16.
82. Sosebee, T.; Giersig, M.; Holzwarth, A.; Mulvaney, P. Ber. Bunsenges. Phys. Chem. 1995, *99*, 40.
83. Henglein, A. Chem. Phys. Lett. 1989, *154*, 473.
84. Kumar, M.; Kapoor, S.; Gopinathan, C. Radiat. Phys. Chem. 1997, *50*, 465.
85. Remita, S.; Orts, J.M.; Feliu, J.M.; Mostafavi, M.; Delcourt, M.O. Chem. Phys. Lett. 1994, *218*, 115.
86. Mostafavi, M.; Keghouche, N.; Delcourt, M.O.; Belloni, J. Chem. Phys. Lett. 1990, *167*, 193.
87. Kurihara, K.; Kizling, J.; Stenius, P.; Fendler, J.H. J. Am. Chem. Soc. 1983, *105*, 2574.
88. Belloni, J.; Delcourt, M.O.; Leclere, C. Nouv. J. Chim. 1982, *6*, 507.
89. Mostafavi, M.; Delcourt, M.O.; Picq, G. Radiat. Phys. Chem. 1992, *41*, 453.
90. Mostafavi, M.; Keghouche, N.; Delcourt, M.O. Chem. Phys. Lett. 1990, *169*, 81.
91. Ershov, B.G.; Henglein, A. J. Phys. Chem. 1998, *102*, 667 and 10663.
92. Keita, B.; Nadjo, L.; de Cointet, C.; Amblard, J.; Belloni, J. Chem. Phys. Lett. 1996, *249*, 297.
93. Henglein, A.; Ershov, B.G.; Malow, M. J. Phys. Chem. 1995, *99*, 14129.
94. Khatouri, J.; Mostafavi, M.; Amblard, J.; Belloni, J. Chem. Phys. Lett. 1992, *191*, 351.
95. Sukhov, N.L.; Akinshin, M.A.; Ershov, B.G. Khim. Vys. Energ. 1986, *20*, 292.
96. Lin, M.Z. Thesis, University of Paris-Sud (1996).
97. Remita, H.; Derai, R.; Delcourt, M.O. Radiat. Phys. Chem. 1991, *37*, 221.
98. Belloni, J.; Mostafavi, M.; Remita, H.; Marignier, J.L.; Delcourt, M.O. In *Synthesis Chemistry and Some Properties of Metal Nanoparticles*; Bradley, J., Chaudret, B. Eds; New J. Chem. 1998, *22*, 1239.
99. Treguer, M.; Remita, H.; Pernot, P.; Khatouri, J.; Belloni, J. J. Phys. Chem. A 2001, *105*, 6102.
100. Le Gratiet, B.; Remita, H.; Picq, G.; Delcourt, M.O. J. Catal. 1996, *164*, 36.
101. Lawless, D.; Kapoor, S.; Kennepohl, P.; Meisel, D.; Serpone, N. J. Phys. Chem. 1994, *98*, 9616.
102. Treguer, M.; De Cointet, Ch.; Remita, H.; Khatouri, J.; Mostafavi, M.; Amblard, J.; Belloni, J.; De Keyzer, R. J. Phys. Chem. 1998, *102*, 4310.
103. Keita, B.; Nadjo, L.; Gachard, E.; Remita, H.; Khatouri, J.; Belloni, J. New J. Chem. 1997, *21*, 851.
104. Belloni, J.; Marignier, J.L.; Delcourt, M.O.; Minana, M. US Patent No 4 629 709 (1986). CIP No 4 745 094 (1987).
105. Mills, G.; Henglein, A. Radiat. Phys. Chem. 1985, *26*, 391.
106. Delcourt, M.O.; Belloni, J.; Marignier, J.L.; Mory, C.; Colliex, C. Radiat. Phys. Chem. 1984, *23*, 485.
107. Westerhausen, J.; Henglein, A.; Lilie, J. Ber. Bunsenges. Phys. Chem. 1981, *85*, 182.
108. Gachard, E.; Remita, H.; Khatouri, J.; Belloni, J.; Keita, B.; Nadjo, L. New J. Chem. 1998, *22*, 1257.
109. Henglein, A.; Meisel, D. Langmuir 1998, *14*, 7392.
110. Ershov, B.G.; Sukhov, N.L.; Troistskii, D.I. Russ. J. Phys. Chem. 1994, *68*, 734.
111. Michaelis, M.; Henglein, A. J. Phys. Chem. 1992, *96*, 4719.

112. Ershov, B.G.; Sukhov, N.L.; Troistskii, D.I. Russ. J. Phys. Chem. 1994, 68, 734.
113. Ershov, B.G.; Janata, E.; Michaelis, M.; Henglein, A. J. Phys. Chem. 1991, 95, 8996.
114. Liu, Y.; Zhu, Y.; Zhang, Y.H.; Qian, Y.; Zhang, M.; Yang, L.; Wang, C. J. Mater. Chem. 1997, 7, 787.
115. Marignier, J.L.; Belloni, J. J. Chem. Phys. 1988, 85, 21.
116. Fujita, N.; Matsuura, C.; Hiroishi, D.; Saigo, K. Radiat. Phys. Chem. 1998, 53, 603.
117. Henglein, A.; Lilie, J. J. Phys. Chem. 1981, 85, 1246.
118. Henglein, A.; Gutierrez, M.; Janata, E.; Ershov, B.G. J. Phys. Chem. 1992, 96, 4598.
119. Henglein, A.; Giersig, M. J. Phys. Chem. 1994, 98, 6931.
120. Buxton, G.V.; Rhodes, T.; Seller, R. J. Chem. Soc. Faraday Trans. 1 1982, 78, 3341.
121. Breitenkamp, M.; Henglein, A.; Lilie, J. Ber. Bunsenges. Phys. Chem. 1976, 80, 973.
122. Henglein, A.; Janata, E.; Fojtik, A. J. Phys. Chem. 1992, 96, 4734.
123. Gutierrez, M.; Henglein, A. J. Phys. Chem. 1996, 100, 7656.
124. Lian R. Thesis, USTC, Hefeï (2001).
125. Henglein, A.; Mulvaney, P.; Linnert, T. Faraday Discuss. 1991, 92, 31.
126. Strelow, F.; Henglein, A. J. Phys. Chem. 1995, 99, 11834.
127. Linnert, T.; Mulvaney, P.; Henglein, A. J. Phys. Chem. 1993, 97, 679.
128. Dawson, A.; Kamat, P.V. J. Phys. Chem. 2000, 104, 1184.
129. Teo, B.K.; Keating, K.; Kao, Y.-H. J. Am. Chem. Soc. 1987, 109, 3494.
130. Noshima, N. J. Macromol. Sci. Chem. A 1990, 27, 1225.
131. Sato, T.; Kuroda, S.; Takami, A.; Yonezawa, Y.; Hada, H. Appl. Organomet. Chem. 1991, 5, 261.
132. Kobayashi, M.; Sato, H. Chem. Lett. 1993, 1659.
133. Torigoe, K.; Esumi, K. Langmuir 1992, 8, 59 and 1993, 9, 1964.
134. Esumi, K.; Suzuki, A.; Aihara, N.; Usui, K.; Torigoe, K. Langmuir 1998, 14, 3157.
135. Sasaki, M.; Osada, M.; Higashimoto, N.; Inagaki, S.; Fukushima, Y.; Fukuoka, A.; Ishikawa, M. Microporous Mesoporous Mater. 1998, 21, 597; J. Mol. Catal. A 1999, 141, 223.
136. Pal, T. Talanta 1998, 46, 583.
137. Fukuoka, A.; Higashimoto, N.; Sakamoto, Y.; Sasaki, M.; Sugimoto, N.; Inagaki, S.; Fukushima, Y.; Ishikawa, M. Catal. Today 2001, 66, 23.
138. sau, T.K.; Pal, A.; Jana, N.R.; Wang, Z.L.; Pal, T. J. Nanopart. Res. 2001, 3, 257.
139. Kim, F.; Song, J.H.; Yang, P. J. Am. Chem. Soc. 2002, 124, 14316.
140. Mallik, K.; Mandal, M.; Pradhan, N.; Pal, T. Nano Lett. 2001, 1, 319.
141. Abid, J.P.; Wark, A.W.; Brevet, P.F.; Girault, H.H. Chem. Commun. 2002; 792.
142. Itakura, T.; Torigoe, K.; Esumi, K. Langmuir 1995, 11, 4129.
143. Fernandez, A.; Gonzalez-Elipe, A.R.; Real, C.; Caballero, A.; Munuera, G. Langmuir 1993, 9, 121.
144. Stathatos, E.; Lianos, P.; Falaras, P.; Siokou, A. Langmuir 2000, 16, 2398.
145. Remita, S.; Mostafavi, M.; Delcourt, M.O. New J. Chem. 1994, 18, 581.
146. Dick, K.; Dhanasekaran, T.; Zhang, Z.Y.; Meisel, D. J. Am. Chem. Soc. 2002, 124, 2312.
147. Michaelis, M.; Henglein, A. J. Phys. Chem. 1992, 96, 4719.
148. Marignier, J.L.; Dokuchaev, A.; Hautecloque, S.; Grand, D. Proc. 7th Int. Symp. Small Part. Inorg. Clusters, 1994; 189 pp.
149. Belloni, J.; Delcourt, M.O.; Leclere, C. Nouv. J. Chim. 1982, 6, 507.
150. Belloni, J.; Lecheheb, M. Radiat. Phys. Chem. 1987, 29, 89.
151. Bruneaux, J.; Cachet, H.; Froment, M.; Amblard, J.; Belloni, J.; Mostafavi, M. Electrochim. Acta 1987, 32, 1533.
152. Bruneaux, J.; Cachet, H.; Froment, M.; Amblard, J.; Mostafavi, M. J. Electroanal. Chem. 1989, 269, 375.
153. Gacoin, T.; Chaput, F.; Boilot, J.P.; Mostafavi, M.; Delcourt, M.O. In Eurogel 91; Vilminot, S., Nass, R., Schmidt, H., Eds.; North Holland: E-MRS, 1991; 159 pp.
154. Zhu, X.; Qian, Y.; Zhang, M.; Chen, Z.; Zhou, G. J. Mater. Chem. 1994, 4, 1619.
155. Valentin, E.; Bernas, H.; Ricolleau, C.; Creuzet, F. Phys. Rev. Lett. 2001, 86, 99.

156. Keghouche N. Thesis, University Constantine, 1993.
157. Le Gratiet B. Thesis, University Paris-Sud, Orsay, 1996.
158. Torigoe, K.; Remita, H.; Picq, G.; Belloni, J.; Bazin, D. J. Phys. Chem. B 2000, *104*, 7050.
159. Zhang, G.; Liu, X.; Thomas, J.K. Radiat. Phys. Chem. 1998, *51*, 135.
160. Michalik, J.; Azuma, N.; Sadlo, J.; Kevan, L. J. Phys. Chem. 1995, *99*, 4679.
161. Michalik, J.; Wasowicz, T.; Sadlo, J.; Reijerse, E.J.; Kevan, L. Radiat. Phys. Chem. 1996, *47*, 75.
162. Gachard, E.; Belloni, J.; Subramanian, M.A. J. Mater. Chem. 1996, *6*, 867.
163. Sasaki, M.; Osada, M.; Higashimoto, N.; Yamamoto, T.; Fukuoka, A.; Ishikawa, M. J. Mol. Catal. A 1999, *141*, 223.
164. Vucemilovic, M.I.; Micic, O.I. Radiat. Phys. Chem. 1988, *32*, 79.
165. Micic, O.I.; Meglic, M.; Lawless, D.; Sharma, D.K.; Serpone, N. Langmuir 1990, *6*, 487.
166. Marignier, J.L.; Ashokkumar; Mostafavi, M. Proc. 50th Annual IS&T Conf. on Silver Halides, 1997; 67 pp.
167. Michaelis, M.; Henglein, A.; Mulvaney, P. J. Phys. Chem. 1994, *98*, 6212.
168. Henglein, A.; Mulvaney, P.; Holzwarth, A.; Sosebee, T.E.; Fojtik, A. Ber. Bunsenges. Phys. Chem. 1992, *96*, 754.
169. Mulvaney, P.; Giersig, M.; Henglein, A. J. Phys. Chem. 1992, *96*, 10419.
170. Henglein, F.; Henglein, A.; Mulvaney, P. Ber. Bunsenges. Phys. Chem. 1994, *98*, 180.
171. Henglein, A.; Giersig, M. J. Phys. Chem. 1994, *98*, 6931.
172. Shibata, T.; Bunker, B.A.; Zhang, Z.; Meisel, D.; Vardeman, C.F. Jr.; Gezelter, D. Am. Chem. Soc. 2002, *124*, 11989.
173. Belloni, J. Curr. Opin. Colloid Interface Sci. 1996, *1*, 184.
174. Malkov, A.; Belloni, J. J. Chem. Phys. 1992, *89*, 885.
175. Ershov, B.G.; Janata, E.; Henglein, A. J. Phys. Chem. 1994, *98*, 7619 and 10891.
176. Ershov, B.G.; Janata, E.; Henglein, A. Radiat. Phys. Chem. 1996, *47*, 59.
177. Henglein, A. J. Phys. Chem. 1992, *96*, 2411.
178. Khatouri, J.; Mostafavi, M.; Amblard, J.; Belloni, J. Z. Phys. D 1993, *26*, 82.
179. De Vyt, A.; Gijbels, R.; Davock, H.; Van Roost, C.; Geuens, I. J. Anal. At. Spectrom. 1999, *14*, 499.
180. Remita, H.; Khatouri, J.; Treguer, M.; Amblard, J.; Belloni, J. Z. Phys. D 1997, *40*, 127.
181. Remita, H.; Etcheberry, A.; Belloni, J. J. Phys. Chem. 2002, *106*. in press.
182. Kumar, M.; Kapoor, S.; Gopinathan, C. Radiat. Phys. Chem. 1999, *54*, 39.
183. Remita, S.; Mostafavi, M.; Delcourt, M.O. Radiat. Phys. Chem. 1996, *47*, 275.
184. Remit, S.; Picq, G.; Khatouri, J.; Mostafavi, M. Radiat. Phys. Chem. 1999, *54*, 463.
185. Torigoe, K.; Remita, H.; Beaunier, P.; Belloni, J. Radiat. Phys. Chem. 2002, *64*, 215.
186. Braunstein, P.; Oro, R.; Raithby, J. Metal Clusters in Chemistry, Eds; Wiley, NY, 1999; 1213 pp.
187. Meisel, D.; Mulac, W.A.; Matheson, M.S. J. Phys. Chem. 1981, *85*, 179.
188. Delcourt, M.O.; Keghouche, N.; Belloni, J. Nouv. J. Chim. 1983, *7*, 131.
189. Belloni, J.; Lecheheb, A. Radiat. Phys. Chem. 1987, *29*, 89.
190. Amblard, J.; Belloni, J.; Platzer, O. J. Chim. Phys. 1991, *88*, 835.
191. Agfa GV. Eur. Patent 95203 1706 (1995).
192. Hamilton, F. In "*Theory of the Photographic Process,*" 4th Ed.: James, T.H., Ed.; MacMillan: NY, 1977.
193. Glafkides, P. Chimie et Physique Photographiques, 5th Edition, Editions de l'Usine Nouvelle, 1987.
194. Tani, T. Photographic Sensivity. Theory and Mechanisms. NY: Oxford University Press, 1995.
195. Belloni, J. In *C. R. Acad. Sci. Phys.*; Bréchignac, C., Cahuzac, P., Eds.; 2002, *3*, 381.
196. Hamilton, J.F. In *Growth and Properties of Metal Clusters*; Bourdon, J., Ed., Elsevier Publ., 1980; 289 pp.
197. Trautweiler, F. Photogr. Sci. Eng. 1968, *12*, 138.

198. Mostafavi, M.; Marignier, J.L.; Amblard, J.; Belloni, J. J. Imaging Sci. 1991, *35*, 68.
199. Gurney, R.W.; Mott, N.F. Proc. R. Soc. Lond. A 1938, *164*, 485.
200. Belloni, J.; Treguer, M.; Remita, H.; de Keyzer, R. Nature 1999, *402*, 865.
201. Treguer, M.; Remita, H.; Belloni, J.; De Keyzer, R. J. Imaging Sci. 2002, *46*, 193.
202. Fojtik, A.; Henglein, A. Ber. Bunsenges. Phys. Chem. 1993, *97*, 252.
203. Neddersen, J.; Chumanov, G.; Cotton, T.M. Appl. Spectrosc. 1993, *47*, 1959.
204. Sibbald, M.S.; Chumanov, G.; Cotton, T.M. J. Phys. Chem. 1996, *100*, 4672.
205. Yeh, M.S.; Yang, Y.S.; Lee, Y.P.; Lee, H.F.; Yeh, Y.H.; Yeh, C.S. J. Phys. Chem. 1999, *103*, 6851.
206. Chen, Y.H.; Yeh, C.S. Colloids Surf. 2002, *197*, 133.
207. Mafuné, F.; Kohno, J.-Y.; Takeda, Y.; Kondow, T.; Sawabe, H. J. Phys. Chem. 2001, *105*, 5114.
208. Mafuné, F.; Kohno, J.-Y.; Takeda, Y.; Kondow, T. J. Phys. Chem. 2002, *106*, 85555.
209. Kamat, P.V.; Flumiani, M.; Hartland, G.V. J. Phys. Chem. B 1998, *102*, 3123.
210. Hodak, J.H.; Henglein, A.; Giersig, M.; Hartland, G.V. J. Phys. Chem. B 2000, *104*, 11708.
211. El-Sayed, M.A. Acc. Chem. Res. 2001, *34*, 257.
212. Kamat, P.V. J. Phys. Chem. B 2002, *106*, 7729.
213. Mostafavi, M.; François, L.; Belloni, J.; Delaire, J. Phys. Chem. Chem. Phys. 2001, *3*, 4965.

21

Application of Radiation Chemical Reactions to the Molecular Design of Functional Organic Materials

Tsuneki Ichikawa
Hokkaido University, Sapporo, Japan

Although radiation chemical reactions have been applied to many industrial fields, especially to polymer industries, they are mainly aimed at improving the mechanical, thermal, and surface properties of materials [1]. Applications of radiation chemical reactions to the construction of modern functional organic materials such as electronic and optical materials are relatively few. However, radiation chemical reactions have the following characteristics that are advantageous for studying and constructing functional organic materials:

1. The reactions take place even in a solid without structural deformation, which is advantageous for modifying the chemical and physical properties of precisely shaped solid substances.
2. The reactions take place within the region where the radiation energy is absorbed, so that only a part of a solid substance can be modified with precise spatial resolution. Moreover, the region is controllable three dimensionally by changing the field of exposure and the depth of penetration that depends on the energy of ionizing radiation.
3. The reactions can be induced without additives, which is advantageous for maintaining the purity of materials.

In the following section, we will show how radiation chemical reactions are useful for the molecular design of functional organic materials by taking our recent studies on next-generation resist polymers and organic semiconductors as examples.

1. APPLICATION OF DISSOCIATIVE ELECTRON ATTACHMENT TO THE CONSTRUCTION OF RADIATION RESIST

Because of their short wavelength, ionizing radiations such as ultra-short-wave UV light, x-rays, and electron and ion beams have been considered to be the light sources of next-generation lithography for mass production of very large scale integrated circuits [2–5]. The lithography process is based on radiation-induced change in the dissolution rate of resist polymers. Ionizing radiation induces either degradation or cross-linking of resist polymers, which increases or decreases the dissolution rate, respectively.

Poly(methyl methacrylate) (PMMA) has been widely used as a radiation-degradable resist polymer [6]. Although PMMA has high spatial resolution of less than 10 nm, the sensitivity to ionizing radiation is poor. Several attempts to increase the sensitivity have been made by modifying the molecular structure of polymers [7–10]. However, the sensitivity is still not enough for mass production. Another shortcoming is the limited spatial resolution. The main chains of PMMA are randomly destroyed by homolytic scission [11–13], so that the molecular weight of the decomposed polymer has a wide distribution, which limits the resolution if a spatial resolution of less than the original molecular size is expected.

A novel approach to increase the sensitivity and the spatial resolution is to use a polymer that can be cut just in half by ionizing radiation. Because the selective scission of the polymer skeleton does not leave a longer fragment that is difficult to be dissolved, the sensitivity and the resolution are expected to be improved [14,15] by using such a polymer as a radiation resist (see Fig. 1).

Dissociative electron attachment is a radiation chemical reaction suitable for cutting a polymer just in half. If two equivalent polymer skeletons R are connected with a functional group XY that has a large cross section of dissociative electron attachment, the polymer captures an ejected electron at the center of the polymer skeleton and is broken into two fragments with similar molecular weight, as $R-XY-R + e^- \rightarrow R-X^- + Y-R$. The key to construct such a polymer is to find a functional group that is possible to connect two polymer chains, to capture an electron efficiently, and to dissociate into two fragments after the capture.

1.1. Molecular Design for Position-Selective Scission of Polymer Skeleton

The dissociation of a functional group connecting two polymer skeletons must be fast enough for competing the charge recombination due to the migration of positive charges. Assuming the linear free energy relationship, a functional group with larger heat of dissociative electron attachment is expected to have lower activation energy of dissociation. Esters of carboxylic acids are strong candidates for the dissociative electron attachment,

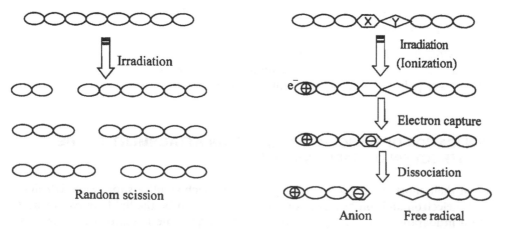

Figure 1 Schematic representation of the formation of polymer fragments by random scission (left) and selective scission by dissociative electron attachment (right).

since the dissociation of ester anions is expected to be easy due to high stability of resultant carboxylate anions [16].

Table 1 shows molecular products generated by electron attachment to several esters at 77 K in 2-methyltetrahydrofuran (MTHF) matrices. The concentration of the esters is ca. 0.05 mol/dm^3. Table 1 also shows the changes in enthalpies for the dissociation of anion radicals into carboxylate anions and neutral free radicals that are calculated from the experimental values of electron affinities of carboxylates and the dissociation energies of ester bonds [17]. Within the solutes examined, only benzyl esters dissociate into carboxylate anions and benzyl radicals that show intense absorption and fluorescence bands with maxima at 320 and 475 nm [18], respectively (see Fig. 2). Dissociative electron attachment to alkyl esters is endothermic, so that the electron attachment does not result in the dissociation but simply gives their molecular radical anions with broad absorption bands at < 600 nm, which is attributable to ketyl-type anions with the unpaired electron on the carbonyl carbon and the excess charge on the carbonyl oxygen [19]. Although the heat of dissociative electron attachment is higher for $C_2H_5COOCH_2C_{16}H_9$ and $CF_3COOC_2H_5$ than for $C_2H_5COOCH_2C_6H_5$, as shown in Fig. 3, electron attachment to $C_2H_5COOCH_2C_{16}H_9$ and $CF_3COOC_2H_5$ give the radical anions showing the absorption spectrum very similar to the pyrenyl radical anion [20] and the ketyl-type anion, respectively.

These results indicate that the dissociation of an ester radical anion into a carboxylate anion and a neutral free radical is controlled by two factors. One is the exothermicity of the reaction and the other is the distribution of the excess electron on the radical anion. Dissociation of an ester radical anion takes place if the change in the free energy is negative. Because the change in entropy is positive for the dissociation, the change in the free energy, $\Delta G = \Delta H - T\Delta S$, is always negative if the change in the enthalpy is negative. However, the negative change in the free energy does not certify the dissociation of the radical anion. The radical anion does not dissociate if the activation energy for the dissociation is too high. The activation energy is lowered if the radical anion of the ester R_1COO-R_2 shares the excess electron with R_1COO- and $-R_2$ groups, because the sharing causes the formation of

Table 1 Reaction of Free Electron with Esters in MTHF

Solute	Expected reaction	ΔH/kJ·mol^{-1}	Observed radical[a]	Yield of acid[b]
$C_6H_5COOCH_2C_6H_5$	$C_6H_5COO^-$ + $CH_2C_6H_5$	-48	$CH_2C_6H_5$	~2.5
$C_2H_5COOCH_2C_6H_5$	$C_2H_5COO^-$ + $CH_2C_6H_5$	-17	$CH_2C_6H_5$	~2.5
$C_{11}H_{23}COO-$ $CH(C_6H_5)C_8H_{17}$	$C_{11}H_{23}COO^-$ + $CH(C_6H_5)C_8H_{17}$	-17	$CH (C_6H_5)C_8H_{17}$	~2.3
$C_2H_5COOCH_2C_{16}H_9$	$C_2H_5COO^-$ + $CH_2C_{16}H_9$	-52	$[C_2H_5COOCH_2C_{16}H_9]^-$	<0.15
$CF_3COOC_2H_5$	CF_3COO^- + C_2H_5	-50	$[CF_3COOC_2H_5]^-$	<0.15
$C_2H_5COOC_2H_5$	$C_2H_5COO^-$ + C_2H_5	57	$[C_2H_5COOC_2H_5]^-$	<0.15
$C_2H_5COOC(CH_3)_3$	$C_2H_5COO^-$ + $C(CH_3)_3$	40	$[C_2H_5COOC(CH_3)_3]^-$	<0.15

[a] In solid matrices at 77 K.
[b] Number of molecules per 100 eV radiation energy absorbed to MTHF solution at 293 K.

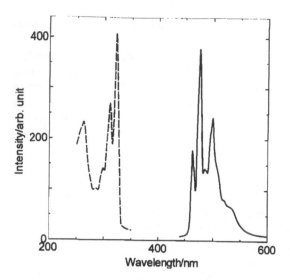

Figure 2 Excitation (broken line, observed wavelength = 475 nm) and fluorescence (solid line, excitation wavelength = 320 nm) spectra of the benzyl radical generated in γ-irradiated MTHF at 77 K by dissociative electron attachment to $C_2H_5COOCH_2C_6H_5$.

an antibonding orbital on the ester bond and therefore the reduction of the activation energy for the dissociation. However, because the electron affinity of $–C_{16}H_9$ or $CF_3CO–$ is much higher than those of their counterparts, the radical anion of $C_2H_5COO–CH_2C_{16}H_9$ or $CF_3COO–C_2H_5$ does not share the excess electron but is localized on $–C_{16}H_9$ or $CF_3CO–$, respectively. The radical anion of $C_2H_5COOCH_2C_{16}H_9$ or $CF_3COOC_2H_5$ is therefore stable. The excess electron on benzyl esters may occupy the antibonding orbital of the ester

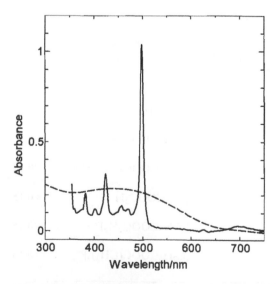

Figure 3 Absorption spectra for γ-irradiated MTHF at 77 K containing $C_2H_5COOCH_2C_{16}H_9$ (solid line) and $CF_3COOC_2H_5$ (broken line).

bond, so that the activation energy is low enough to dissociate into the carboxylate anion and the benzyl radical even at 77 K (see Fig. 4).

Figure 5 shows the relationship between the concentration of benzyl esters and the yield of benzyl radicals that is determined from the absorption band at 320 nm. The yield is saturated at 0.03 mol/dm^3, which indicates that 0.03 mol/dm^3 of benzyl esters capture all electrons generated in MTHF and toluene matrices at 77 K. It turns out that one benzyl ester molecule can capture an electron that is generated within the radius of 2.4 nm. Assuming the molecular weight of one monomer unit to be 100, one benzyl ester in 333 monomer units is enough for collecting all electrons generated by ionizing radiation from a polymer with the specific density of unity. The yield of benzyl radicals in toluene is about one fifth of that in MTHF, which is due to the protection effect of the aromatic solvent; the yield of ejected electrons available for the dissociative electron is much lower in toluene. Assuming that the G value (the number of events per 100 eV radiation energy absorbed) of free-electron formation is 2.5, the radiation dose for reducing the initial molecular weight of 33,300 to 16,650 is 120 kGy.

1.2. Synthesis of Polymer by Living Radical Polymerization

Since the benzyl esters are found to be a suitable functional group for connecting two polymer skeletons, the next problem is to connect two equivalent polymer skeletons with benzyl esters. This can be achieved by applying recently developed living radical polymerization [21–23].

Synthesis of polymer molecules with the same molecular weight necessitates living polymerization. Although living polymerization in homogeneous systems had been limited to ionic polymerization, recent development of polymer science makes it possible to synthesize polymers with narrow polydispersity by radical polymerization. Living polymerization necessitates propagating polymer ends that do not deactivate during the polymerization. Regular radical polymerization is not living because the propagating radicals quickly combine with each other. Living radical polymerization protects terminal radicals from the combination by using a radical capping agent such as stable nitroxyl radicals. Heating a mixture of monomer M and a initiator R–ONX with a capping agent ONX starts the dissociation of the initiator into reactive and stable free radicals, and therefore starts the polymerization at about the same time, as

$$R - ONX \ \rightleftharpoons \ R + ONX, \quad R + M \rightarrow RM.$$

The capping agent protects the deactivation of propagating radicals during the polymerization, because the steady-state concentration of reactive propagating radicals is lowered by the reaction with the capping agent.

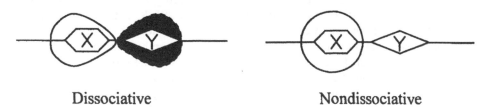

Dissociative   Nondissociative

Figure 4 Schematic representation of the molecular orbitals of an excess electron on dissociative (left) and nondissociative (right) functional groups.

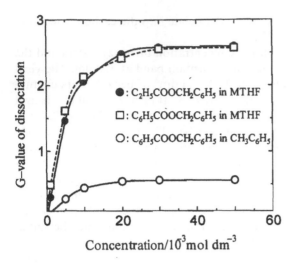

Figure 5   Relation between the radiation chemical yield of the benzyl radical and the concentration of benzyl esters in 2-methyltetrahydrofuran and toluene at 77 K.

$$RM_n + ONX \xrightarrow{\longleftarrow} \quad RM_n - ONX$$

There are two strategies for constructing a polymer with benzyl ester in the middle of the skeleton. One is to make polymer skeletons with the same molecular weight, and then combine two skeletons with benzyl ester. The other is to synthesize a chemical with propagation sites of polymerization at both sides of benzyl ester, and use the chemical as an initiator of living polymerization. Since the former strategy did not work well, probably due to low reactivity of polymer molecules with benzyl esters, the latter approach will be mentioned.

We synthesized polystyrene with benzyl ester in the middle of the skeleton by using the following initiator that dissociates into stable tetramethylpyperidinoxyl radical and reactive benzyl-type radical.

Figure 6 shows the results of the living radical polymerization of styrene at 400 K. The linear dependence of the molecular weight on the conversion ratio and the narrow distribution of the molecular weight are the evidence of living radical polymerization.

Figure 7 shows the profiles of the size exclusion chromatography (SEC) of coupled polystyrene that was dissolved in MTHF and irradiated at 295 K with a Co 60 γ-ray source. The irradiation cuts the main chains just in half, which indicates that the scission of the skeleton exclusively takes place in the middle of the skeleton by dissociative electron attachment to the benzyl ester. Figure 8 shows the distribution of the molecular weight of polystyrene after γ-irradiation at 295 K of MTHF solution containing 10 wt.% of coupled polystyrene. The G value of the dissociation into two polymer fragments is about 5, which is twice as much higher as that of corresponding $C_6H_5COOCH_2C_6H_5$. The efficiency of the scission in toluene solution is about one fifth of that in MTHF, which is due to the protection effect of the aromatic solvent.

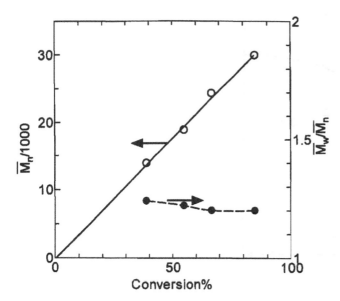

Figure 6 Relation between the conversion ratio and the molecular weight of coupled polystyrene generated by living radical polymerization at 300 K.

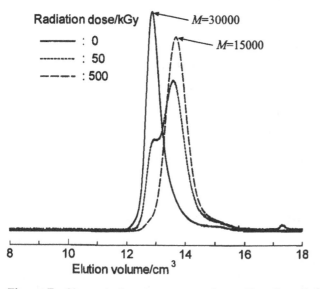

Figure 7 Size exclusion chromatography profiles of coupled polystyrene before and after γ-irradiation in MTHF at 295 K.

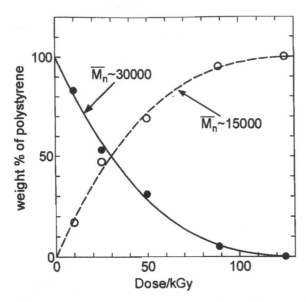

Figure 8 Distribution of the molecular weight of polystyrene after γ-irradiation at 295 K of MTHF solution containing 10 wt.% of coupled polystyrene.

As shown in Fig. 9, irradiation of neat polystyrene does not result in the efficient scission but increases the molecular weight due mainly to cross-linking, though a small bump at the low molecular weight side of the SEC pattern indicates the dissociative electron attachment. The cross-linking is suppressed by addition of hydroquinone as a radical scavenger, which suggests that combination reaction of polymer radicals takes place in the neat polymer, as

$$-CH_2CH(C_6H_5)- \rightarrow [CH_2CH(C_6H_5)]^+ - + e^-, \quad -CH_2C(C_6H_5) - + H$$

$$-[CH_2CH(C_6H_5)]^+ - \rightarrow -CH_2C(C_6H_5) - + H^+$$

$$-CH_2CH(C_6H_5) - + H \rightarrow -CH_2C(C_6H_5) - + H_2$$

$$e^- + -COOCH(CH_3)(C_6H_5)- \rightarrow -COO^- + CH(CH_3)(C_6H_5)-$$

$$-CH_2C(C_6H_5) - + - CH_2C(C_6H_5)- \rightarrow [-CH_2C(C_6H_5)-]_2 \quad \text{(cross-link)}$$

$$- CH_2C(C_6H_5) - + CH(CH_3)(C_6H_5)- \rightarrow$$
$$- CH_2C(C_6H_5)[CH(CH_3)(C_6H_5)-] - \quad \text{(graft)}$$

The cross-linking and the grafting of the polymer radicals are suppressed by the addition of hydroquinone, as

$$-CH_2C(C_6H_5) - + HOC_6H_4OH \rightarrow -CH_2C (CH_6H_5) - + OC_6H_4OH$$

$$CH(CH_3)(C_6H_5) - +HOC_6H_4OH \rightarrow CH_2(CH_3)(C_6H_5) - +OC_6H_4OH$$

Although the irradiation of 200 kGy decomposes about 80% of polystyrene in toluene by the dissociative electron attachment, the yield of the decomposition is only 20% for solid toluene. Because of its low efficiency of scission, the coupled polystyrene may not be a polymer suitable as a radiation resist. However, the present study has shown that a polymer that can be decomposed into two equivalent skeletons by ionizing radiation is possible to be

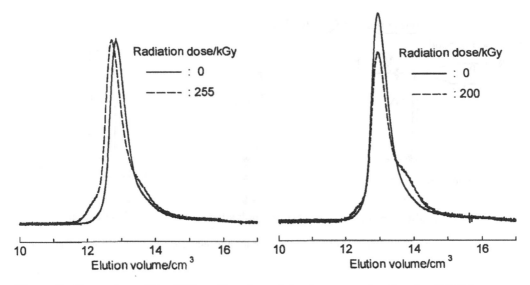

Figure 9 Comparison of the SEC profiles of coupled polystyrene γ-irradiated at 295 K in vacuum without additive (left) and with 0.07 mol/kg of hydroquinone (right).

constructed by living radical polymerization. Our preliminary study on polymethylacrylate containing benzyl ester in a part of the main chain showed that the efficiency of the scission for the neat polymer was comparable to that of the coupled polystyrene in MTHF, though the scission does not take place without benzyl ester.

2. MOLECULAR DESIGN OF POLYSILANE AS A CONDUCTIVE POLYMER

Polysilanes are σ-conjugated polymers composed of Si–Si skeletons and organic pendant groups. They are insulators with filled intramolecular valence bands and empty intramolecular conduction bands. However, because of strong σ conjugation, they have rather narrow band gaps of less than 4 eV [24,25] and are converted to conductors by photoexcitation or by doping electron donors or acceptors. Recently they have attracted much attention because of their potential utility as one-dimensional conductors, nonlinear optical materials, and electroluminescent materials [26–28].

The optical and electronic functions of polysilanes owe to their delocalized highest occupied molecular orbital (HOMO) and lowest unoccupied molecular orbital (LUMO) that are occupied by holes and conduction electrons, respectively. The polymer does not show high conductivity or optical nonlinearity if the electrons or holes are localized on a small part of the polymer chain. To elucidate the structure of HOMO and LUMO is therefore important for the molecular design of polysilanes as functional materials.

Analyses of the electronic and electron spin resonance (ESR) spectra of the radical cation and anion of polysilanes make it possible to elucidate the structure of HOMO and LUMO, because an unpaired electron in the radical anion or cation occupies HOMO or LUMO, respectively. As schematically depicted in Fig. 10, the radical ions of polysilanes show absorption bands in UV and near-IR regions [29–31]. The former band corresponds to intraband transitions between valence and conduction bands. The latter band corresponds to transitions within the valence or the conduction band [32,33]. Because the near-

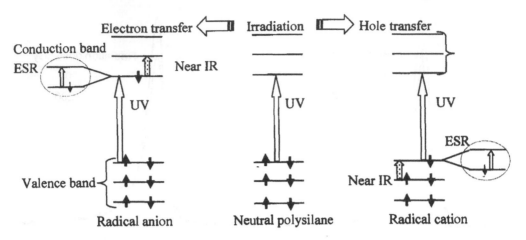

Figure 10 Schematic representation of the electronic and electron spin transitions of polysilane and its radical ions.

IR band is characteristic to the transition of a hole in the radical cation or a conduction electron in the radical anion, analysis of the near-IR band gives useful information on the structure of these charge carriers. Analysis of the ESR spectra of radical ions is also useful for elucidating the electronic structure of the carriers. The ESR spectra give direct information on the structure of charge carriers in HOMO and LUMO, because ESR transitions take place between the spin sublevels of ground-state unpaired electrons.

The radiation chemical matrix isolation is a useful method for the spectroscopic study of radical ions. In this method, a solution containing a small amount of solute molecules is rapidly frozen to make a rigid solid matrix and is then irradiated at cryogenic temperature. The radical ions of solute molecules generated by ionizing radiation are immobilized in the matrix, so that the lifetime of otherwise unstable radical ions is long enough for precise spectroscopic measurements. The radical anion of solute molecules is selectively obtained by using ether or alcohol, RH, as a solvent

RH + ionizing radiation $\rightarrow RH^+ + e^-$ (ionization of solvent molecule)

$RH^+ + RH \rightarrow RH_2^+ + R$ (formation of solvent radical by proton transfer)

S (solute) $+ e^- \rightarrow S^-$ (electron capture by solute molecule)

The radical cation is similarly obtained by using a halogenated molecule as a solvent.

RX (solvent molecule, $X = Cl$ or Br) + ionizing radiation $\rightarrow RX^+ + e^-$

$e^- + RX \rightarrow R + X^-$ (dissociative electron attachment)

$RX^+ + RX \rightarrow RX + RX^+$ (hole transfer)

$RX^+ + S \rightarrow RX + S^+$ (hole transfer from solvent to solute)

The advantage of this radiation chemical method is that the by-products of radical ions are scarcely generated, because the direct decomposition of solute molecules by ionizing radiation is negligible due to low solute concentration. Another advantage is that the molecular orbitals of radical ions are scarcely disturbed by their counterpart ions, because electrons or holes are captured by solute molecules after traveling more than 5 nm from the point of ionization [34].

The electronic structures of polysilane radical ions have also been studied by pulse radiolysis of the liquid solution [35–40]. However, due to short lifetime of the radical ions, the measurement is limited to electronic absorption spectroscopy.

The bulk conduction in solid polysilane can be divided into intramolecular and intermolecular conductions. Intermolecular conduction is the migration of a charge in the polymer skeleton, whereas intermolecular conduction is the charge migration between polymer molecules. Both the fast intramolecular and intermolecular conductions are desirable as a good conductive polymer. In the following section, we will show how the radiation chemical matrix isolation method is useful for obtaining information crucial for the molecular design of conductive polysilanes.

2.1. Intramolecular Conduction

Intramolecular conduction of polysilanes arises from σ conjugation between adjacent Si–Si bonds. Because σ orbitals are axially symmetric with respect to their Si–Si bonds, HOMO and LUMO are considered to be delocalized over a significant part of the polymer skeleton no matter whether the polymer skeleton is distorted or not. However, this is not necessarily the case. Figure 11 shows the ESR spectra of the radical cations of permethyloligosilanes [33]. The hexamethyldisilane radical cation shows 19 hyperfine lines with the hyperfine coupling constant of 0.566 mT and the binominal intensity ratio. The ESR line width of oligosilane radical cations decrease with increasing length of the Si–Si main chain, which is due to the reduction of hyperfine interaction with methyl protons by the decrease of the spin density on each monomer unit. The observed spectra are in good agreement with the simulated ones that are obtained by using hyperfine parameters estimated from the ESR spec-

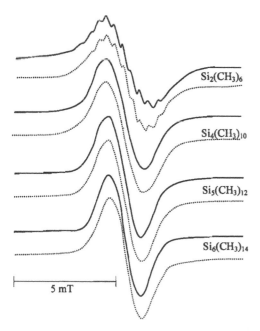

Figure 11 Observed (—) and simulated (---) ESR spectra of permethyloligosilane radical cations at 77 K in γ-irradiated freon matrices. The simulations were carried out under the assumption that an unpaired electron is delocalized over the entire Si–Si chain.

trum of the hexamethyldisilane radical anion under the assumption that the unpaired electrons are delocalized all over the main chains. The degree of delocalization on polysilanes is therefore possible to be estimated from the ESR line width.

Figure 12 compares the ESR spectra of the radical cations of silane dimers and their polymeric analogues. The observed spectra of polysilane radical cations are reproducible by using hyperfine parameters of the corresponding dimer radical cations and by assuming that the partial delocalization of the unpaired electron on only six Si atoms.

The localization of an unpaired electron on a part of the polymer skeleton is also supported by the analysis of the near-IR absorption spectrum. Figure 13 shows the absorption spectra of permethyloligosilanes and poly(cyclohexylmethylsilane) radical cations at 77 K. Decrease of the energy of the near-IR band with increasing the main-chain length of oligosilanes is due to the increase of σ conjugation length. The energy of the near-IR band is expected to be zero for polysilanes with electrons delocalized all over the polymer chain, because the energy levels composed of Si–Si σ orbitals become continuous. Observation of the near-IR band around 0.8 eV therefore implies that the unpaired electron is not completely delocalized but is confined on a part of the polymer chain. The other polysilanes such as poly(methylphenylsilane), poly(dibutylsilane), and poly(dihexylsilane) also show the near-IR bands around 0.8 eV, which indicates that the HOMO of polysilanes is generally localized on a part of the polymer skeleton.

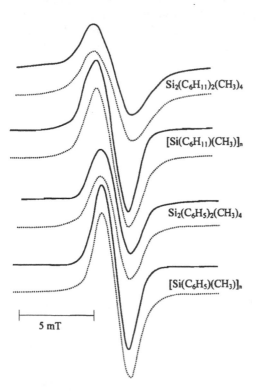

$Si_2(C_6H_{11})_2(CH_3)_4$

$[Si(C_6H_{11})(CH_3)]_n$

$Si_2(C_6H_5)_2(CH_3)_4$

$[Si(C_6H_5)(CH_3)]_n$

5 mT

Figure 12   Observed (—) and simulated (- - -) ESR spectra of disilane and polysilane radical cations at 77 K in γ-irradiated freon matrices. The simulation for polymers was carried out under the assumption that an unpaired electron is localized on a part of polymer skeleton composed of about six Si atoms.

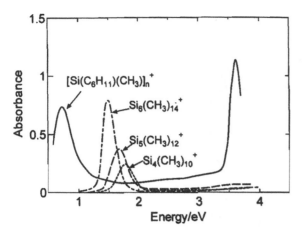

Figure 13 Absorption spectra of permethylorigosilane and poly(cyclohexylmethylsilane) radical cations at 77 K in γ-irradiated freon matrices.

As shown in Fig. 14, the energy of the near-IR band of the radical anions of permethyloligosilanes also decreases with increasing main-chain length [32]. The near-IR absorption band of the radical anion of poly(cyclohexylmethylsilane) at around 0.8 eV indicates that the unpaired electron is also localized on a part of the main chain composed of probably six Si atoms. The degree of delocalization of an unpaired electron in the radical anion of polysilanes cannot be estimated from the ESR line width, because the spectra scarcely show the hyperfine interaction with protons in the side chains. The solvatochromatic shift of poly(cyclohexylmethylsilane) radical anion also supports the localization of the unpaired electron [41]. Figure 15 shows the absorption spectra of the radical ions of poly(cyclohexylmethylsilane) in the mixed solvent of MTHF and methylcyclohexane at 77 K. Increase of the concentration of polar MTHF in nonpolar methylcyclohexane causes the blue shift of the near-IR band. Solvatochromic shift by polar solvents arises from the difference of polarization energies between ground and excited electronic states and is smaller for molecules

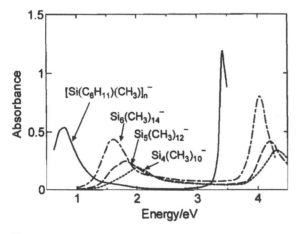

Figure 14 Absorption spectra of permethylorigosilane and poly(cyclohexylmethylsilane) radical anions at 77 K in γ-irradiated MTHF matrices.

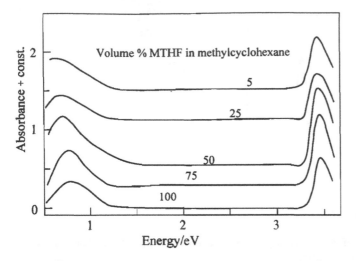

Figure 15 Solvatochromic shift for poly(cyclohexylmethylsilane) radical anion in the mixed solvent of MTHF and methylcyclohexane at 77 K.

with diffusive charge distribution. The observed solvatochromic shift proves that a negative charge or an unpaired excess electron is confined on a part of the polymer skeleton, because the solvatochromic shift is not expected for a polymer ion with the charge distribution over the entire main chain. Analysis of the solvation structure for poly(cyclohexylmethylsilane) radical anion by electron spin–echo envelope modulation also indicates that the unpaired electron is confined in a segment composed of six to eight Si atoms [42].

Understanding of the localization of unpaired electrons on a part of polymer skeletons necessitates detailed knowledge about the structure of HOMO and LUMO. As shown in Fig. 16, the intramolecular valence and conduction bands are composed of bonding and antibonding Si–Si σ orbitals, respectively. The HOMO locating at the top of the valence band is given by a linear combination of the bonding orbitals with alternating phase. The atomic orbitals at Si atoms composing the HOMO are then expressed by

$$\mathrm{HOMO} = \frac{1}{\sqrt{2}} \left[\sigma_{\mathrm{Si-Si}}(sp^3) - \sigma_{\mathrm{Si-Si}}(sp^3) \right]$$

$$= \frac{1}{\sqrt{2}} \left[\left(\frac{1}{\sqrt{2}} 3p_x + \frac{1}{2} 3p_x - \frac{1}{2} 3s \right) + \left(\frac{1}{\sqrt{2}} 3p_x - \frac{1}{2} 3p_y + \frac{1}{2} 3s \right) \right] = 3p_x \tag{1}$$

where $\sigma_{\mathrm{Si-Si}}(sp^{-3})$ and $\sigma_{\mathrm{Si-Si'}}(sp^{-3})$ denote hybrid orbitals composed of two adjacent Si–Si σ bonds. The HOMO is composed of only Si $3p_x$ atomic orbitals, so that this is a pseudo-π orbital. The LUMO locating at the bottom of the conduction band is given by a linear combination of the antibonding orbitals with the same phase, so that the atomic orbitals for the LUMO are expressed by

$$\mathrm{LUMO} = \frac{1}{\sqrt{2}} \left[\sigma^*_{\mathrm{Si-Si}}(sp^3) - \sigma^*_{\mathrm{Si-Si}}(sp^3) \right] = \frac{1}{\sqrt{2}} \left[\left(\frac{1}{\sqrt{2}} 3p_x + \frac{1}{2} 3p_y - \frac{1}{2} 3s \right) \right.$$

$$\left. + \left(-\frac{1}{\sqrt{2}} 3p_x + \frac{1}{2} 3p_y - \frac{1}{2} 3s \right) \right] = \frac{1}{\sqrt{2}} (3p_y - 3s) \tag{2}$$

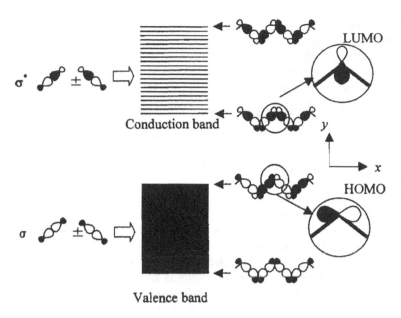

Figure 16 Schematic representation of the HOMO and LUMO of polysilanes.

where $\sigma^*_{Si-Si}(sp^3)$ and $\sigma^*_{Si-Si'}(sp^3)$ denote hybrid orbitals composing two adjacent anti-bonding Si–Si σ^* bonds. The LUMO is further coupled with Si–C antibonding orbitals to eliminate Si 3s orbital, since the ESR spectra of permethylorigosilane and polysilane radical anions do not show ^{29}Si hyperfine splitting arising from Fermi contact interaction by an unpaired electron in the Si 3s orbital [32]. The actual atomic orbital for the LUMO is therefore $3p_y$. The reason for mixing with Si–C antibonding orbitals will be given in the next section.

The delocalization of the HOMO and LUMO arises from pseudo-π orbital interactions between adjacent $3p_x$ and $3p_y$ atomic orbitals of Si atoms, respectively. Since the 3p orbital is on the Si–Si–Si plane, the degree of the interaction depends on the dihedral angle between adjacent Si–Si–Si planes. Fluctuation of the dihedral angles along the polymer skeleton then causes the so-called Anderson localization of the HOMO and LUMO [41].

Anderson localization is the localization of electrons on low-dimensional materials, which is induced by the irregularity of the periodic potential field [43]. Figure 17 gives a schematic representation of Anderson localization of a particle in one-dimensional box. The same is true for an electron on a polymer skeleton. A localized state in a completely periodic

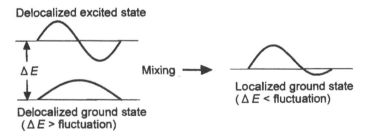

Figure 17 Schematic representation of Anderson localization of a particle in one-dimensional box.

field is expressed by the superposition of steady-state delocalized wave functions of ground and excited states, so that the energy of a localized state is always higher than that of the delocalized ground state. Increase of the irregularity of the periodic field causes the increase of the ambiguity of the energies of delocalized states, which reduces the energy differences between the delocalized ground and excited states. The energy for generating a localized state therefore decreases with increasing irregularity, and a localized state becomes a ground state if the irregularity exceeds the energy gaps between adjacent delocalized states. In other words, the irregularity of the potential energy field allows the mixing of delocalized ground and excited states to generate a localized ground state. The Anderson localization is easier for a longer polymer skeleton, because the energy gaps between adjacent energy levels decrease with increasing length of the skeleton.

Anderson localization is difficult to take place for conformationally constrained polymers with narrow distributions of dihedral angles. Introduction of bulky side chains is therefore expected to enhance the delocalization of the HOMO and LUMO of polysilanes [44]. Figure 18 shows the absorption spectra of the radical ions of poly(n-decyl-(S)-2-methylbutylsilane) that is a rigid rodlike polysilane with narrow distribution of the torsional angles [45,46]. The energy of the near-IR bands is much lower than that of poly(cyclohexylmethylsilane), which indicates that the Anderson localization is suppressed in poly(n-decyl-(S)-2-methylbutylsilane). The absorption spectrum of the radical anion of poly(n-decyl-(S)-2-methylbutylsilane) in nonpolar methylcyclohexane matrix is the same as that in polar MTHF. Observation of no solvatochromic shift also indicates the suppression of the Anderson localization, because the solvatochromic shift is not expected for a polymer ion with extended charge distribution.

The sharpness of the UV absorption and emission bands corresponding to HOMO↔LUMO transition is also a measure of delocalization. The transition energy depends on the location of a polymer skeleton where a photon is absorbed or emitted, if electron orbitals are localized on parts of the skeleton. A sharp UV band of poly(n-decyl-(S)-2-methylbutylsilane) indicates the delocalization of HOMO and LUMO. Table 2 summarizes the location

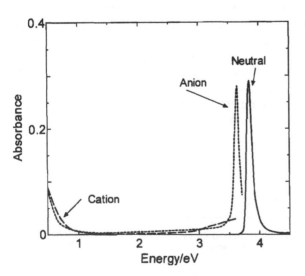

Figure 18 Absorption spectra of poly(n-decyl-(S)-2-methylbutylsilane) at 295 K in MTHF and its radical cation and anion at 77 K in γ-irradiated freon and MTHF matrices, respectively.

Table 2 Absorption and Emission Bands of Polysilanes in MTHF at 296 K

Pendant groups	$M_n/10^4$	$M_w/10^4$	Absorption band[a]		Emission band[a]	
			λ_{max}	Width	Stokes shift	Width
1-Methylpropyl, decyl	0.39	0.52	no peak			
(S)-2-methylbutyl, decyl	67.2	94.1	3.84	0.10	0.05	0.12
2-Methylpropyl, decyl	1.61	3.15	3.88	0.10	0.05	0.15
2-Ethylbutyl, decyl	1.25	2.66	3.83	0.1	0.04	0.10
(2-Methylpropyl)₂	0.22	0.24	3.97	0.22	0.25	0.21
3-Methylpentyl, decyl	29.5	41.3	3.87	0.22	0.13	0.28
n-Butyl, decyl	1.74	7.53	3.91	0.45	0.29	0.24
2-Methylpropyl, methyl	8.93	18.8	4.21	0.62	0.5	0.28
2-Methylpropyl, ethyl	3.55	5.23	3.97	0.28	0.27	0.32
2-Methylpropyl, propyl	2.38	2.93	4.05	0.28	0.20	0.27
2-Methylpropyl, butyl	3.60	5.30	3.98	0.18	0.12	0.19
2-Methylpropyl, pentyl	0.76	4.72	3.92	0.14	0.07	0.17
2-Methylpropyl, hexyl	1.94	3.55	3.88	0.10	0.04	0.15

[a] In electron volts.

and the width of the UV absorption and emission bands for several polysilanes. Polysilanes with pendant groups composed of $\geq C_6$ linear alkanes and branched alkanes with the branch at the β position show sharp UV bands and low Stokes shifts, which indicates that these pendant groups are effective for stretching the polymer skeleton [47] and therefore for preventing the Anderson localization.

The structures of polysilanes with flexible and rigid skeletons are schematically shown in Fig. 19 together with the distribution of charge carriers on the skeletons. In conclusion, to replace the pendant groups of polysilanes with bulky ones is an effective way of suppressing the undesired Anderson localization of charge carriers on a part of the polymer skeleton. An ideal polysilane quantum wire with high intramolecular charge mobility can be synthesized by using bulky pendant groups.

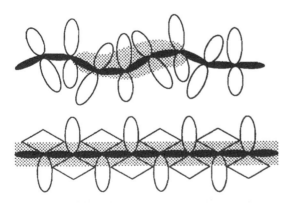

Figure 19 Schematic representation of the molecular structures of flexible polysilane with Anderson-localized charge carrier (upper) and rigid polysilane with delocalized charge carrier.

2.2. Intermolecular Migration of Charge Carriers

It has been established from conductivity measurements that thermally activated and field-assisted hole hopping is responsible for the charge transport in solid polysilanes [48,49]. The mobility of the hole is as high as 10^{-8} m²/V sec, while the mobility of the electron is a few orders of magnitude lower. In this section, we will show the reason why only the hole is mobile in polysilanes and how we can construct electron-conductive polysilanes.

It has been shown in the previous section that the structure of conduction electrons on polysilane skeletons is essentially the same as that of holes. The suppression of electron conduction is therefore attributable to low electron conductivity of side chains, because interchain charge hoppings through the side chains are necessary for the migration of conduction electrons between polymer molecules [50]. Figure 20 compares the ESR spectra of the radical anion and cation of poly(cyclohexyl methylsilane) and poly(methylphenylsilane) in MTHF and freon mixture, respectively. The line width for the radical anions primarily arises from the anisotropy of the ESR g factors. The two radical anions show the same g factors, which indicates that the helical structures of the polymer skeletons are the same [51]. The spectrum of the radical cation is more than five times broader than that of the radical anion. This is caused by stronger hyperfine interactions between the unpaired electron and the protons of the side-chain alkyl groups. The stronger hyperfine interaction arises from the delocalization of the hole onto the alkyl substituents. The lack of a detectable hyperfine interaction in the polysilane radical anion suggests that the excess electron in the LUMO or the intramolecular conduction band is not delocalized onto the side chains so that the side chains act as good intermolecular insulators for the conduction electron. On the other hand, since the hole is delocalized onto the side chains, it can migrate to an adjacent main chain via the side chains.

The optical absorption spectra of polysilane radical ions also support the validity of the above-mentioned mechanism of charge migration. Figure 21 compares the absorption

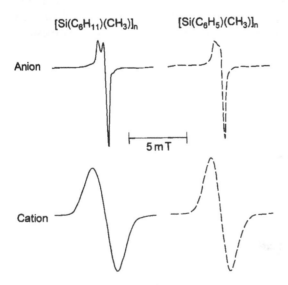

Figure 20 Comparison of the ESR spectra of radical cations and anions of poly(cyclohexylmethylsilane) and poly(methylphenylsilane) at 77 K in γ-irradiated freon and MTHF matrices, respectively.

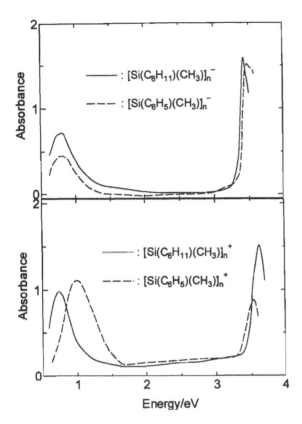

Figure 21 Comparison of the electronic absorption spectra of poly(cyclohexylmethylsilane) and poly(methylphenylsilane) radical anions and cations at 77 K in γ-irradiated MTHF and freon matrices, respectively.

spectra of the radical anion and cation of both poly(cyclohexylmethylsilane) and poly-(methylphenylsilane). The near-IR band of the radical anion is scarcely influenced by the substitution of the side chains from the cyclohexyl group to the phenyl group. On the other hand, the near-IR band of the radical cation is blue-shifted by the substitution. This indicates that the unpaired electron of the radical anion is confined within the main chain, whereas that of the radical cation is extended onto the side chains.

The delocalization of the conduction electron onto the side chains would be expected if the pendant groups were replaced with more electrophilic substituents than the phenyl group. However, this is not the case. Figure 22 shows the absorption spectrum of poly-(methylnaphthylsilane) radical anion. The absorption spectrum is very similar to that of the naphthalene radical anion, which implies that the unpaired electron is localized on the pendant group. Increase of the electron affinity of pendant groups does not necessarily cause the delocalization.

Replacing all the pendant groups with aryl groups attains the delocalization of the unpaired electron. Figure 23 compares the absorption spectra of the radical anions of poly-(4-ethylphenyl-phenylsilane) and poly(dicyclohexylsilane) in MTHF at 77 K. Although the molecular structures of these polysilanes are similar, the absorption spectra are different. The absorption spectrum of the poly(dicyclohexylsilane) radical anions is similar to that of

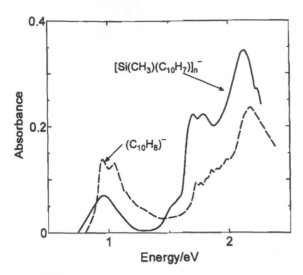

Figure 22 Comparison of the electronic absorption spectra of poly(methylnaphthylsilane) and naphthalene radical anions at 77 K in γ-irradiated MTHF matrices.

the poly(cyclohexylmethylsilane) radical anion; the unpaired electron is confined within the main chain. The poly(4-ethylphenylphenylsilane) radical anion shows a blue-shifted near-IR band and several peaks in visible and near-IR regions. The blue shift and the peaks strongly suggest that the unpaired electron is extended onto the pendant groups. Recent studies on the electroluminescence of polysilanes revealed that poly(di-4-*t*-butylphenylsilane) is electron conductive [52], which also supports the delocalization of conduction electron onto the pendant groups. Table 3 summarizes the experimental results on the near-

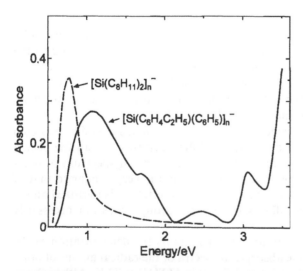

Figure 23 Comparison of the electronic absorption spectra of poly(4-ethylphenyl-phenylsilane) and poly(dicyclohexylsilane) radical anions at 77 K in γ-irradiated MTHF matrices.

Table 3 Near-IR Spectra and the Location of Conduction Electron on Polysilanes

Pendant 1	Pendant 2	E_a (pendant 2)/eV	Near-IR band/eV	Electron location
Methyl	Phenyl	−1.15	0.8	Main chain
Methyl	p-Fluorophenyl	−0.89	0.8	Main chain
Methyl	Trifluorophenyl		Trifluorophenyl anion	Pendant 2
Methyl	1-Naphthyl	−0.19	Naphthalene anion	Pendant 2
Phenyl	4-Ethylphenyl	−1.17	1.1	Main and side chains
Phenyl	4-Propylphenyl	Approx. −1.1	1.1	Main and side chains
Phenyl	4-tert-Butylphenyl	−1.06	1.1	Main and side chains
Cyclohexyl	Cyclohexyl		0.8	Main chain
Cyclohexyl	Methyl		0.8	Main chain

IR bands of polysilane radical anions and the location of conduction electrons. Except for poly(diarylsilanes), the conduction electron is confined either in the skeleton or in one of the pendant groups.

The difference of the spatial distribution of the conduction electron and hole on polysilanes can be explained by taking the shapes of the molecular orbitals on the main and side chains into account. Figure 24 shows HOMO and LUMO of the model compounds of polysilanes obtained by semiempirical PM3 calculations. The HOMO of all the model compounds is extended to both the Si–Si–Si skeletons and pendant groups. The LUMO of $SiHC_2H_5(SiH_3)_2$, the model compound of poly(dialkylsilane), and $SiHC_6H_5(SiH_3)_2$, the model compound of poly(methylphenylsilane), is not extended onto the pendant group. The LUMO of $SiHC_{10}F_7(SiH_3)_2$, the model compound of poly(methylnaphthylsilane), is on the pendant group but not on the skeleton. Although the LUMO + 1 of $SiHC_{10}F_7(SiH_3)_2$ has approximately the same energy as that of LUMO, it does not mix with LUMO but is localized on the pendant group. These results indicate that the mixing of the LUMO of the skeleton and the pendant group is very difficult.

The difficulty of the mixing arises from the symmetry of the atomic orbitals of LUMO on the skeleton with respect to molecular orbitals on the pendant group. Let us assume orbital interactions between polymer skeletons and pendant groups are much weaker than those within the skeletons and within the pendant groups. The molecular orbitals of polysilanes are then divided into three parts, a delocalized orbital on a polymer skeleton, a delocalized orbital on a pendant group, and a localized orbital on an Si–C bond. The entire orbitals are reconstructed from these partial orbitals by treating orbital interactions between the skeletons and the pendant groups as perturbations. Figure 25 illustrates the shape of the partial orbitals before and after the reconstruction. As shown in Fig. 15, the atomic orbitals of the delocalized HOMO and LUMO on the Si skeleton are given by $3p_x$ and $(3p_y - 3s)/\sqrt{2}$, respectively. The atomic orbital on the Si–C carbon of the delocalized HOMO of alkyl or aryl groups is $2p_x$, so that the delocalized HOMO on the Si skeleton and pendant groups are mixed with each other through π-type interactions with $3p_x$ and $2p_x$

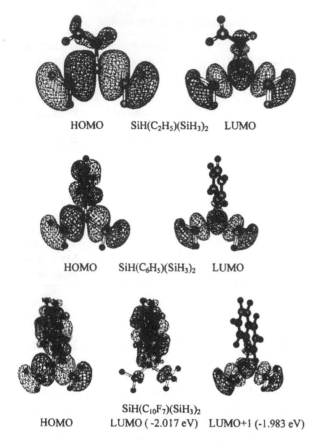

HOMO   SiH(C₂H₅)(SiH₃)₂   LUMO

HOMO   SiH(C₆H₅)(SiH₃)₂   LUMO

SiH(C₁₀F₇)(SiH₃)₂
HOMO   LUMO (-2.017 eV)   LUMO+1 (-1.983 eV)

Figure 24 Isosurfaces at ± 0.02 of molecular orbitals of the model compounds of polysilanes obtained by semiempirical PM3 calculation.

atomic orbitals. The delocalized HOMO on the Si skeleton does not mix with the Si–C σ orbitals, because $3p_x$ is orthogonal to the σ orbitals.

The LUMO on the Si skeleton interacts not with the LUMO on pendant groups but with Si–C antibonding σ orbitals. The atomic orbital of the LUMO, $(3p_y - 3s)/\sqrt{2}$, is divided into two sp^3 σ orbitals with the symmetry axes pointed toward two Si–C bonds, as

$$\frac{1}{\sqrt{2}}(3p_y - 3s) = \frac{1}{\sqrt{2}}\left[\left(\frac{1}{2}3p_y + \frac{1}{\sqrt{2}}3p_z - \frac{1}{2}3s\right) + \left(\frac{1}{2}3p_y - \frac{1}{\sqrt{2}}3p_z - \frac{1}{2}3s\right)\right] \qquad (3)$$

These orbitals completely overlap with two Si–C antibonding orbitals, since the antibonding orbitals are expressed by

$$\frac{1}{\sqrt{2}}\left(\frac{1}{2}3p_y \pm \frac{1}{\sqrt{2}}3p_z - \frac{1}{2}3s\right) - \frac{1}{\sqrt{2}}(\text{carbon sp}^3) \qquad (4)$$

The LUMO on the Si skeleton therefore mixes up with the antibonding orbitals to generate $3p_y$ atomic orbitals, as

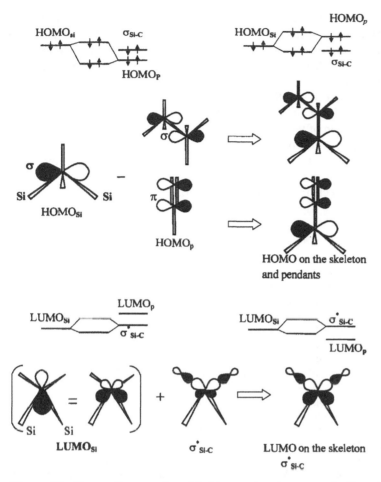

Figure 25 Energy diagrams for orbital interactions between polysilane skeleton and pendant group. HOMO$_{Si}$, HOMO$_p$ and LUMO$_{Si}$, LUMO$_p$ denote the HOMO and LUMO delocalized on the skeleton and a pendant group, respectively. $\sigma_{Si\text{-}C}$ and $\sigma^*_{Si\text{-}C}$ are bonding and antibonding Si–C σ orbitals, respectively.

$$\frac{1}{\sqrt{2}}\left[\frac{1}{\sqrt{2}}(3p_y - 3s) + \frac{1}{\sqrt{2}}\left(\frac{1}{2}3p_y + \frac{1}{\sqrt{2}}3p_z + \frac{1}{2}3s\right) - \frac{1}{\sqrt{2}}\left(\text{carbon sp}^3\right)_1\right.$$
$$\left. + \frac{1}{\sqrt{2}}\left(\frac{1}{2}3p_y - \frac{1}{\sqrt{2}}3p_z + \frac{1}{2}3s\right) - \frac{1}{\sqrt{2}}\left(\text{carbon sp}^3\right)_2\right] \tag{5}$$
$$= 3p_y - \frac{1}{2}\left[\left(\text{carbon sp}^3\right)_1 + \left(\text{carbon sp}^3\right)_2\right]$$

The LUMO on the Si skeleton is impossible to be extended onto the pendant group because the Si–C antibonding orbital is orthogonal to the delocalized LUMO on the pendant group. Observation of no Si 3s component on the ESR spectra of oligosilane and polysilane radical anions [32] supports the conclusion that the LUMO on the Si skeleton is mixed with the antibonding Si–C σ orbitals. The energy diagram for the orbital interactions is also shown in Fig. 25.

The mechanism of orbital interactions mentioned above is not applicable if each monomer unit of polysilanes has sufficient electron affinity and two pendant groups of the monomer unit have approximately the same electron affinity. The LUMO of polysilanes is then regarded as a series of the LUMOs of monomer units that is perturbed by Si–Si bonds. The LUMO of the monomer unit has some electron density on the Si atom, since the unpaired electron in the LUMO is delocalized over the two pendant groups with approximately the same electron density. Connection of the monomer units with Si–Si bonds therefore causes the delocalization of the LUMO on adjacent monomer units and therefore on both the Si skeleton and pendant groups.

In conclusion, radiation chemical matrix isolation technique gives precise information on the electronic structure of charge carriers, which is useful for the molecular design of organic semiconductors as electronic and optical materials.

REFERENCES

1. Clough, R.L. Nucl. Instrum. Methods B 2001, *185*, 8.
2. Tolfree, D.W.L. Rep. Prog. Phys. 1998, *61*, 313.
3. Melngaillis, J.; Mondelli, A.A.; Berry, I.L. III; Mohondro, R. J. Vac. Sci. Technol. B 1998, *16*, 927.
4. Wallraff, G.M.; Hinsberg, W.D. Chem. Rev. 1999, *99*, 1801.
5. Cerrina, F. J. Phys. D: Appl. Phys. 2000, *33*, R103.
6. Hatzakis, M. J. Electrochem. Soc. 1969, *116*, 1033.
7. Bowden, M.J. Materials for microlithography. ACS Symp. Ser. 1984, *266*, 39.
8. Reichmanis, E.; Thompson, L.F. Chem. Rev. 1989, *89*, 1273.
9. Iwayanagi, T.; Ueno, T.; Nonogaki, S.; Itoh, H.; Willson, C.G. Electronic and photonic applications of polymers. Adv. Chem. Ser. 1988, *218* (Chapter 3).
10. Bouden, M.J.; Thompson, L.F. J. Appl. Polym. Sci. 1973, *17*, 3211.
11. Ichikawa, T.; Yoshida, H. J. Polym. Sci. Part A: Polym. Chem. 1990, *28*, 1185.
12. Yates, B.M.; Shiozaki, D.M. J. Polym. Sci. Part B: Polym. Phys. 1991, *31*, 1779.
13. Ichikawa, T.; Oyama, K.; Kondoh, T.; Yoshida, H. J. Polym. Sci. Part A: Polym. Chem. 1994, *32*, 2487.
14. Bignozzi, M.C.; Ober, C.K.; Nobvembre, A.J.; Knurek, C. Polym. Bull. 1999, *43*, 93.
15. Barclay, G.G.; King, M.; Orellana, A.; Malenfant, P.R.L.; Sinta, R.; Malmstrom, E.; Ito, H.; Hawker, C.J. Organic thin films. ACS Symp. Ser. 1998, *695*, 360.
16. Ichikawa, T.; Ueda, H.; Koizumi, H. Chem. Phys. Lett. 2002, *363*, 13.
17. Lias, S.G.; Bartmess, J.E.; Liebman, J.F.; Holmes, J.L.; Levin, R.D.; Mallard, W.G. J. Phys. Chem. Ref. Data 1988, *17*(Suppl. 1).
18. Izumida, T.; Ichikawa, T.; Yoshida, H. J. Phys. Chem. 1980, *84*, 60.
19. Ogasawara, M.; Tanaka, M.; Yoshida, H. J. Phys. Chem. 1987, *91*, 937.
20. Shida, T. Electron Absorption Spectra of Radical Ions. Elsevier: Amsterdam, 1998; 85 pp.
21. George, M.K.; Veregin, R.P.N.; Kazmainer, P.M.; Hamer, G.K. Trends Polym. Sci. 1994, *2*, 66.
22. Hawker, C.J. Acc. Chem. Res. 1997, *30*, 373.
23. Sawamoto, M.; Kamigaito, M. Trends Polym. Sci. 1996, *4*, 371.
24. Takeda, K.; Shiraishi, K. Phys. Rev. B 1989, *39*, 11028.
25. Yokoyama, K.; Yokoyama, M. Chem. Lett. 1989, *1989*, 1005.
26. Miller, R.D.; Michl, J. Chem. Rev. 1989, *89*, 1359.
27. Michl, J. Synth. Met. 1992, *49–50*, 367.
28. Suzuki, H.; Hoshino, S.; Furukawa, K.; Ebata, K.; Yuan, C.H.; Bleyl, I. Polym. Adv. Technol. 2000, *11*, 460.

29. Irie, S.; Oka, K.; Nakao, R.; Irie, M. J. Organomet. Chem. 1990, *388*, 253.

30. Irie, S.; Irie, S.M. Macromolecules 1992, *25*, 1766.

31. Ushida, K.; Kira, A.; Tagawa, S.; Yoshida, Y.; Shibata, H. Proc. Am. Chem. Soc. Div. Polym. Mater. 1992, *66*, 299.

32. Kumagai, J.; Yoshida, H.; Koizumi, H.; Ichikawa, T. J. Phys. Chem. 1994, *98*, 13117.

33. Kumagai, J.; Yoshida, H.; Ichikawa, T. J. Phys. Chem. 1995, *99*, 7965.

34. Ichikawa, T. J. Phys. Chem. B 2002, *106*, 10684.

35. Ban, H.; Sukegawa, K.; Tagawa, S. Macromolecules 1987, *20*, 1775; 1988, *21*, 45.

36. Seki, S.; Matsui, Y.; Yoshida, Y.; Tagawa, S.; Koe, JR.; Fujiki, M. J. Phys. Chem. B 2002, *106*, 6849.

37. Seki, S.; Yoshida, Y.; Tagawa, S. Radiat. Phys. Chem. 2001, *60*, 411.

38. Seki, S.; Kunimi, Y.; Nishida, K.; Yoshida, Y.; Tagawa, S. J. Phys. Chem. B 2001, *105*, 900.

39. Seki, S.; Kunimi, Y.; Nishida, K.; Aramaki, K.; Tagawa, S. J. Organomet. Chem. 2000, *611*, 64.

40. Seki, S.; Yoshida, Y.; Tagawa, S.; Asai, K. Macromolecules 1999, *32*, 1080.

41. Ichikawa, T.; Kumagai, J.; Koizumi, H. J. Phys. Chem. B 1999, *103*, 3812.

42. Ichikawa, T.; Koizumi, H.; Kumagai, J. J. Phys. Chem. B 1997, *101*, 10698.

43. Skinner, J.L. J. Phys. Chem. 1994, *98*, 2503.

44. Ichikawa, T.; Yamada, Y.; Kumagai, J.; Fujiki, M. Chem. Phys. Lett. 1999, *306*, 275.

45. Fujiki, M. J. Am. Chem. Soc. 1994, *116*, 6017.

46. Fujiki, M. Appl. Phys. Lett. 1994, *65*, 3251.

47. Terao, K.; Terao, Y.; Teramoto, A.; Nakamura, N.; Terakawa, I.; Sato, T. Macromolecules 2001, *34*, 2682.

48. Samuel, L.M.; Sanda, P.N.; Miller, R.D. Chem. Phys. Lett. 1989, *159*, 227.

49. Vanderlaan, G.P.; Dehaas, M.P.; Hummel, A.; Frey, H.; Moller, M. J. Phys. Chem. 1996, *100*, 5470.

50. Kumagai, J.; Tachikawa, H.; Yoshida, H.; Ichikawa, T. J. Phys. Chem. 1996, *100*, 16777.

51. Ichikawa, T.; Sumita, M.; Kumagai, J. Chem. Phys. Lett. 1999, *307*, 81.

22
Applications to Reaction Mechanism Studies of Organic Systems

Tetsuro Majima
Osaka University, Osaka, Japan

1. INTRODUCTION

1.1. Radical Ions and the Generation Methods

Radical cation and anion ($M^{\cdot+/\cdot-}$) generated from one-electron oxidation and reduction are short-lived species, as important intermediates, in chemistry, physics, and biology. The formation and the reactivity of $M^{\cdot+/\cdot-}$ have been extensively studied with various research aspects [1–19]. $M^{\cdot+/\cdot-}$ has essentially radical- and ion-type reactivities which are different from free radicals and ions. Excitation of $M^{\cdot+/\cdot-}$ can generate $M^{\cdot+/\cdot-}$ in the excited state ($M^{\cdot+}*/\cdot\ *$), and the chemistry involving $M^{\cdot+}*/\cdot-*$ is an interesting research subject [20–35].

Many organic $M^{\cdot+/\cdot-}$ can be generated at high concentration of 10^{-6}–10^{-3} M from electrochemical reactions and chemical oxidation by using one-electron oxidants. However, heterogeneous formation on the electrode and subsequent two-electron oxidation are involved in the former oxidation, and the kinetic measurements is essentially not possible in the latter oxidation. Therefore, both methods can be used for the generation of $M^{\cdot+/\cdot-}$, but not for the kinetic study involving $M^{\cdot+/\cdot-}$. Photochemical reactions such as photoinduced electron transfer (PET) and resonant two-photon ionization (RTPI) are used to generate $M^{\cdot+/\cdot-}$ and $M^{\cdot+}$, respectively. Laser flash photolysis (LFP) is a useful technique for the kinetic study involving $M^{\cdot+/\cdot-}$. However, generation of $M^{\cdot+\cdot-}$ in solution by PET must be accompanied by ion pair formation, e.g., $M^{\cdot+\cdot-}$, $M^{\cdot+}$/anion, etc. Because solvated electron (e_s^-) is also generated together with $M^{\cdot+}$ in RTPI, recombination of e_s^- and $M^{\cdot+}$ decreases the yield of $M^{\cdot+}$. Such accompanied species influence the transient behavior of $M^{\cdot+/\cdot-}$. Moreover, M is limited for PET and RTPI, in which M should have appropriate character such as lower oxidation potential (E_{ox}) and strong absorption at the laser wavelength (λ).

On the other hand, $M^{\cdot+}$ or $M^{\cdot-}$ can be selectively generated in radiation chemical reactions of any M in solution via pulse radiolysis (PR) and γ-radiolysis (γ-R) techniques [1,36–41], which are used in the present study. Pulse radiolysis has been widely used for the kinetic study involving $M^{\cdot+/\cdot-}$. Various processes, such as ionization, excitation, electron transfer, solvation, relaxation, decomposition, etc., occur initially in the radiation chemical reaction in solutions. Chemical species generated from the initial processes react with M as a solute molecule to generate effectively and selectively $M^{\cdot+}$, $M^{\cdot-}$, or M in the triplet excited state ($^3M^*$) at nsec–100 μsec after an electron pulse (e$^-$) depending on the solvent. M$^{\cdot+}$ is generated during PR of M in alkylhalides such as 1,2-dichloreethane (DCE) or butyl chloride (BC), M$^{\cdot-}$, in basic solvents such as *N*,*N*-dimethylformamide (DMF), hexamethylphosphoric triamide (HMPA), tetrahydrofuran (THF), and 2-methyltetrahydrofuran (MTHF), and $^3M^*$, in benzene. The established reaction mechanisms for generation of M$^{\cdot+/\cdot-}$ are shown in Eqs. (1–6). Because electron and hole are trapped by DCE and DMF, respectively, as stable species, the reactivities of M$^{\cdot+}$ and M$^{\cdot-}$ can be studied in the time scale of nsec–100 μsec.

$$DCE \rightsquigarrow DCE^{\cdot+} + e_s^- \tag{1}$$

$$DCE^{\cdot+} + M \rightarrow M^{\cdot+} + DCE \tag{2}$$

$$e_s^- + DCE \rightarrow ClCH_2CH_2^{\cdot} + Cl^- \tag{3}$$

$$DMF \rightsquigarrow DMF^{\cdot+} + e_s^- \tag{4}$$

$$e_s^- + M \rightarrow M^{\cdot-} \tag{5}$$

$$DMF^{\cdot+} + DMF \rightarrow DMF(-H^+)^{\cdot} + DMF(+H^+)^+ \tag{6}$$

Either M$^{\cdot+}$ or M$^{\cdot-}$ at 10^{-5} M is generated during PR as the only species absorbing light in the visible and near-infrared (NIR) regions [42–54]. Because free M$^{\cdot+}$ or M$^{\cdot-}$ can be generated during PR, effects of counter ionic species and charge recombination are not involved in the time scale of nsec–100 μsec. Thus PR has been extensively used for the kinetic study involving various M$^{\cdot+/\cdot-}$.

The γ-R of glassy rigid matrix of degassed BC or MTHF solutions containing M at 77 K can generate M$^{\cdot+}$ or M$^{\cdot-}$, respectively, as stable species [40,41]. This technique is useful to measure the absorption spectra of M$^{\cdot+}$ or M$^{\cdot-}$ at 77 K, although the kinetic study involving M$^{\cdot+}$ or M$^{\cdot-}$, particularly bimolecular reactions M$^{\cdot+}$ or M$^{\cdot-}$ with other molecules, cannot be carried out because of the rigid matrix.

Because M$^{\cdot+/\cdot-}$ has strong absorption in the visible region, photoirradiation of M$^{\cdot+/\cdot-}$ with $h\nu$ at λ tuned to the absorption generates M$^{\cdot+*/\cdot-*}$. The photochemical transformation of M$^{\cdot+/\cdot-}$ generated in glassy rigid matrix of BC or MTHF at 77 K can be easily studied with the conventional absorption measurement [55–63], while the transient phenomena of M$^{\cdot+*/\cdot-*}$, including bimolecular reactions, can be carried out using the PR–LFP combined method.

1.2. Pulse Radiolysis and Pulse Radiolysis-Laser Flash Photolysis

Pulse radiolysis was performed using e$^-$ from a linear accelerator at Osaka University [42–48]. The e$^-$ has an energy of 28 MeV, single-pulse width of 8 nsec, dose of 0.7 kGy, and a diameter of 0.4 cm. The probe beam for the transient absorption measurement was obtained from a 450-W Xe lamp, sent into the sample solution with a perpendicular intersection of the electron beam, and focused to a monochromator. The output of the monochromator was monitored by a photomultiplier tube (PMT). The signal from the PMT was recorded on a transient digitizer. The temperature of the sample solution was controlled by circulating thermostated aqueous ethanol around the quartz sample cell. Sample solution of M (5×10^{-3}–10^{-1} M) was prepared in a 1×1 cm rectangular Suprasil cell.

The PR–LFP combined method was constructed by PR as the first irradiation and LFP as the second irradiation with delay times of -10 nsec to 2 μsec, controlled by a delay generator. A laser flash at 532 nm (hv_{532}) was obtained by using the second-harmonic oscillation from a Nd:YAG laser (Quantel Model Brilliant), which was operated by a large current pulsed-power supply, which was synchronized with e^-. The single hv_{532} has a 0.5-cm diameter, 5-nsec duration, 100 mJ, and an incident photon number of 1.4×10^{18} pulse^{-1} cm^{-2}. The laser beam was directed in the opposite direction of the electron beam through the sample solution. The laser beam intersected the electron beam at an angle of 20°. An almost collinear arrangement was used, such that the narrow laser beam passed through the central part of the irradiated volume. Transient absorption was measured by using the probe beam mentioned above. The PR–LFP combined method at low temperatures over the range of 140–300 K was performed in the cell that was mounted in a variable-temperature liquid-nitrogen cryostat.

Transient fluorescence was measured without the probe beam. Fluorescence from the sample solution was collected with a 50-mm lens with focal length of 70 mm and was focused into a quartz optical fiber with diameter 250 mm and length of 10 m, which transported the fluorescence to a gated-multichannel/monochromatic spectrometer. The gate time of the image intensifier in the fluorescence was set to be longer than 20 nsec, typically 50 nsec, to eliminate the effect of the mode dispersion of the optical fiber. The emission from the sample was measured for e/hv_{532}, e^-, and hv_{532} irradiations. Fluorescence spectra were obtained by subtraction of the spectra for e^- and hv_{532} irradiations from that for e^-/hv_{532} irradiation to eliminate the Rayleigh and Raman scatterings and other emissions induced by hv_{532} irradiation. Because e^- irradiation gave no emission except the Cherenkov radiation, whose duration was the same as that of e^-, emission from the Cherenkov radiation was omitted with a proper delay time longer than 10 nsec. Fluorescence quantum yields (φ_f) were determined by the comparison of the integrated fluorescence intensity normalized by $(1-10^{\Delta O.D.532/d})$, where $\Delta O.D._{532}$ is the optical density at 532 nm, and d is the optical path length equal to the beam diameter of e^- (7 mm), with that of a dilute cyclohexane solution of zinc(II) tetraphenylporphyrin with $\varphi_f = 0.03$ as a standard.

1.3. γ-Radiolysis

γ-Radiolysis (γ-R) of DCE or BC solutions, at a concentration of 1.0×10^{-2} M, was carried out in a Pyrex tube with an inner diameter of 1.0 cm or in 1.5-mm-thick Suprasil cells cooled in liquid nitrogen at 77 K for UV–vis absorption measurements, respectively, using a ^{60}Co γ source (dose, 2.6×10^2 Gy) [42–48]. Optical absorption spectra were taken by a spectrophotometer and a multichannel photodetector. Sample solutions were degassed by freeze–pump–thaw cycles. The irradiated sample at 77 K was placed in a precooled quartz Dewar vessel, and the absorption spectra were measured at 77 K. Warming of the irradiated sample was performed as follows. Immediately after the liquid nitrogen was removed from the vessel, the sample was again placed in the vessel and the absorption spectra were measured with the MCPD every 3 sec for 3 min during an increase in temperature up to 90 K, measured using a thermocouple. Because it took only 10 msec to measure each spectrum with the MPCD, the temperature was kept constant while measuring the spectrum. Photoirradiation of solutions after γ-R was carried out in the vessel with a 150-W Xe lamp through a glass filter (UV-39) at 77 K.

Product analysis of γ-R was carried out in a Pyrex tube with an inner diameter of 1.0 cm at room temperature (r.t.). After γ-R, the reaction mixtures were directly analyzed by GC and GC-MS. The G values were calculated from the product yields, and the absorbed dose measured by the ferrous sulfate dosimeter (Fricke dosimeter).

2. REACTIONS OF RADICAL CATIONS

2.1. Isomerization, Oxidation, and Dimerization of Radical Cations of Stilbene Derivatives

Numerous studies on the reactions of radical cations of aromatic olefins $(ArO^{\cdot +})$ have been reported from the synthetic and mechanistic points of view [1,2,4,8,42,64–70]. It is well known that $cis(c)$–$trans(t)$ isomerization, dimerization, and addition of a nucleophile occur in $ArO^{\cdot +}$ as typical reactions [1,2,4,8,42,64–70]. The reactivities depend on the structure or substituents of $ArO^{\cdot +}$. No unimolecular c–t isomerization occurs in stilbene$^{\cdot +}$ $(St^{\cdot +})$ [64–66], although it does so in c-4,4'-dibromostilbene$^{\cdot +}$ and c-4,4'-dimethylstilbene$^{\cdot +}$ [66]. Dimerization of $St^{\cdot +}$ with St occurs at $k_d = (3.5–3.9) \times 10^8 \ M^{-1} \ sec^{-1}$ [42,67] to yield π-dimer$^{\cdot +}$ $(\pi\text{-}St_2^{\cdot +})$ with overlapping π-electrons between the two benzene rings, which converts to σ-dimer$^{\cdot +}$ $(\sigma\text{-}St_2^{\cdot +})$ with an acyclic linear structure having both a radical and a cation on the 1- and 4-positions of the C_4 linkage, decomposing to t-$St^{\cdot +}$ and St as final products [42]. Although t-$St^{\cdot +}$ has little reactivity toward O_2, a high reactivity of $k_{O_2} = 2.6 \times 10^8 \ M^{-1} \ sec^{-1}$ has been reported for (E)-2,3-diphenyl-2-butene$^{\cdot +}$ [70].

Three reactions of eight stilbene derivatives$^{\cdot +}$ $(S^{\cdot +} = RCH = CR'R'' = 1^{\cdot +} - 8^{\cdot +})$ (Scheme 1), unimolecular isomerization from c-$S^{\cdot +}$ to t-$S^{\cdot +}$ [Eq. (7)], and bimolecular reactions with O_2 [oxidation, Eq. (8)] and a neutral S [dimerization, Eq. (9)] during PR and γ-R in DCE or BC will be described here [44,45,71].

$$c - S^{\cdot +} \rightarrow t - S^{\cdot +}; \ k_i \text{ for } c - t \text{ unimolecular isomerization} \tag{7}$$

$$S^{\cdot +} + O_2 \rightarrow RC^+HC(OO\cdot)CR'R''; k_{O_2} \text{ for oxidation} \tag{8}$$

$$S^{\cdot +} + S \rightarrow S_2^{\cdot +}; k_d \text{ for dimerization} \tag{9}$$

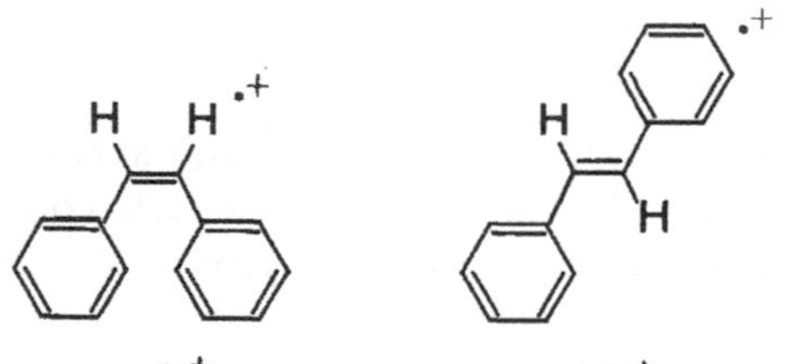

$$c\text{-}S^{\cdot +} \qquad\qquad t\text{-}S^{\cdot +}$$

RCH=CR'R''(S)

	R	R'	R''
1	C_6H_5	H	C_6H_5
2	$4\text{-}CH_3C_6H_4$	H	C_6H_5
3	$4\text{-}CH_3OC_6H_4$	H	C_6H_5
4	$2,4\text{-}(CH_3O)_2C_6H_3$	H	C_6H_5
5	$3,4\text{-}(CH_3O)_2C_6H_3$	H	C_6H_5
6	$3,5\text{-}(CH_3O)_2C_6H_3$	H	C_6H_5
7	$4\text{-}CH_3OC_6H_4$	CH_3	C_6H_5
8	$4\text{-}CH_3OC_6H_4$	H	$4\text{-}CH_3OC_6H_4$

Scheme 1 Eight stilbene derivatives (S).

Formation and Reactions of c-$S^{\cdot+}$ and t-$S^{\cdot+}$

It is well established that $1^{\cdot+}$ is formed during PR and γ-R in alkyl halide such as DCE and BC [40,72–74]. The absorption spectra of c-$S^{\cdot+}$ and t-$S^{\cdot+}$ ($S^{\cdot+} = 1^{\cdot+}$–$8^{\cdot+}$) generated by γ-R of c-S and t-S in BC at 77 K show bands similar to those of c-$1^{\cdot+}$ and t-$1^{\cdot+}$. The transient absorption spectra immediately after e^- were assigned to c-$1^{\cdot+}$–$7^{\cdot+}$ and t-$1^{\cdot+}$ – $8^{\cdot+}$ [40,72,73], formed from capture of the hole generated in the initiation step during PR of S in DCE (Figs. 1 and 2, Table 1). It is well known that t-$1^{\cdot+}$ shows two bands at 480 and 750 nm assigned to $D_2 \leftarrow D_0$ and $D_1 \leftarrow D_0$ transitions, respectively, while c-$1^{\cdot+}$ also shows two bands at 515 and 780 nm, respectively [40,72,73]. The absorption bands of c-$1^{\cdot+}$ are broader and weaker than those of t-$1^{\cdot+}$. Delocalization of π-electrons is less in c-$1^{\cdot+}$ than in t-$1^{\cdot+}$.

The transient absorptions of t-$S^{\cdot+}$ decayed monotonously with first-order kinetics and did not show any shifts or spectral changes. The transient absorptions of c-$1^{\cdot+}$, c-$2^{\cdot+}$, and c-$6^{\cdot+}$ without p-methoxyl group (p-MeO) decayed with k_{obs}, in a similar way to those of t-$1^{\cdot+}$–$8^{\cdot+}$. On the other hand, formation of the band of t-$3^{\cdot+}$–$5^{\cdot+}$ and t-$8^{\cdot+}$ after e^- was observed with decay of the band of c-$3^{\cdot+}$–$5^{\cdot+}$ and c-$8^{\cdot+}$ with p-MeO, respectively. $k_i = 4.5 \times 10^6$ to 1.4×10^7 sec^{-1} for the unimolecular was isomerization of c-$S^{\cdot+}$ to t-$S^{\cdot+}$ was calculated from the rise of the peaks of $t^{\cdot+}$, which did not depend on [c-S].

Dimerization was observed in $1^{\cdot+}$–$3^{\cdot+}$ and t-$6^{\cdot+}$, because the decay of the absorption of $1^{\cdot+}$–$3^{\cdot+}$ and t-$6^{\cdot+}$ increased with increasing [S], respectively. The transient absorption of c-$1^{\cdot+}$ shifted to that of t-$1^{\cdot+}$ during the decay because of the bimolecular isomerization of c-$1^{\cdot+}$ to t-$1^{\cdot+}$ via σ-$1_2^{\cdot+}$. Similarly, the transient absorption of c-$2^{\cdot+}$ shifted to that of t-$2^{\cdot+}$ through bimolecular isomerization. $k_d = (2.0$–$4.3) \times 10^8$ M^{-1} sec^{-1} were calculated from the dependence of k_{obs} on [S] (Table 2). No dimerization of c-$6^{\cdot+}$ was observed, because k_{obs} did not depend on [c-6]. The transient absorption of c-$7^{\cdot+}$ with p-MeO and methyl group (Me) on the olefinic carbon decayed with k_{obs} according to first-order kinetics without any shifts or dependence of [c-7]. In other words, neither unimolecular isomerization nor dimerization was observed in c-$7^{\cdot+}$.

The k_{obs} of t-$S^{\cdot+}$ (S = $3^{\cdot+}$–$5^{\cdot+}$ and $7^{\cdot+}$) with p-McO was accelerated in the presence of O_2, and $k_{O_2} = (1.2$–$4.5) \times 10^7$ M^{-1} sec^{-1} for the oxidation was calculated (Table 2). On the other hand, no oxidation was observed on a similar time scale in t-$1^{\cdot+}$, t-$2^{\cdot+}$, and t-$6^{\cdot+}$ without p-MeO nor in t-$8^{\cdot+}$ with two p-MeOs. Oxidation of c-$7^{\cdot+}$ was also observed with k_{O_2} similar to that for t-$7^{\cdot+}$, while no oxidation was observed in c-$1^{\cdot+}$, c-$2^{\cdot+}$, and c-$6^{\cdot+}$ without p-MeO and in c-$8^{\cdot+}$ with two p-MeOs. In the case of c-$3^{\cdot+}$–$5^{\cdot+}$ with p-MeO, k_{O_2} could not be estimated because of the fast unimolecular isomerization of c-$3^{\cdot+}$–$5^{\cdot+}$ to t-$3^{\cdot+}$–$5^{\cdot+}$ at $k_i = 4.5 \times 10^6$–1.3×10^7 M^{-1} sec^{-1}.

Dimerization Mechanism of c-$S^{\cdot+}$ and t-$S^{\cdot+}$

Decay of c-$1^{\cdot+}$ has been analyzed by neutralization with Cl$^-$ formed from dissociative electron attachment to DCE [Eq. (3)] at $k_n = 1.6 \times 10^{11}$ M^{-1} sec^{-1} [Eq. (10)] when [c-1] is low (10^{-4} M) [42,74]. It is reported that the dimerization of c-$1^{\cdot+}$ with c-1 occurs at $k_d^c = 3.5 \times 10^8$ M^{-1} sec^{-1} at r.t. to form the π-type dimer$^{\cdot+}$ of c-1 [π-(c-1)$_2^{\cdot+}$, Eq. (11)], which changes to σ-type dimer$^{\cdot+}$ of 1 [σ-$1_2^{\cdot+}$, Eq. (12)] [42]. Such fast formation of σ-$1_2^{\cdot+}$ was confirmed by absorption measurement during γ-R of c-1 in BC rigid matrix at 77 K (see Section 2.2). The unimolecular isomerization of c-$1^{\cdot+}$ to t-$1^{\cdot+}$ does not occur, although the bimolecular isomerization of c-$1^{\cdot+}$ to t-$1^{\cdot+}$ is proposed to occur through σ-$1_2^{\cdot+}$ on the time scale of 100–300 nsec at $(5$–$10) \times 10^{-3}$ M at r.t. [42,67]. Thermally unstable σ-$1_2^{\cdot+}$ decomposes

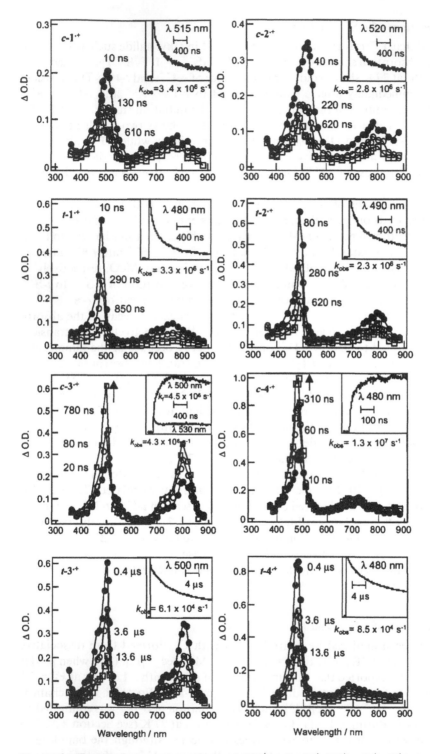

Figure 1 Transient absorption spectra of c-$S^{\cdot+}$ and t-$S^{\cdot+}$ ($S^{\cdot+} = 1^{\cdot+}$–$4^{\cdot+}$) recorded at time t after e^- during PR of c-S and t-S (S = 1–4) with [S] = 5.0×10^{-3} M in DCE at r.t. Insets: kinetic traces illustrating the time profiles of the D_2 band at λ as a function of t. Both t and the observed decay rate constant (k_{obs}) and k_i are mentioned in the figure.

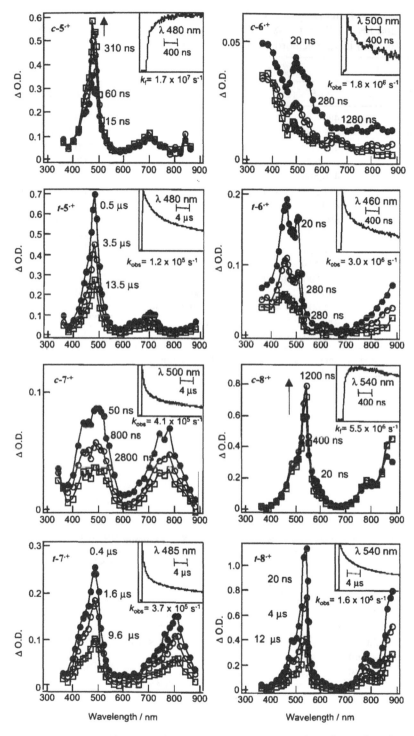

Figure 2 Transient absorption spectra of c-S$^{\cdot+}$ and t-S$^{\cdot+}$ (S$^{\cdot+}$ = **5**$^{\cdot+}$–**8**$^{\cdot+}$) recorded at time t after e$^-$ during PR of c-S and t-S (S = **5–8**) with [S] = 5.0×10^{-3} M in DCE at r.t. Insets: kinetic traces illustrating time profiles of the D$_2$ band at λ as a function of t. Both t and the observed decay rate constant (k_{obs}) and k_i are mentioned in the figure.

Table 1 Transient Absorption Peaks (λ_{max}) and the Molar Absorption Coefficients at the Peak (ε_λ) of D_1 and D_2 of c-$S^{\cdot+}$ and t-$S^{\cdot+}$ ($S^{\cdot+} = \mathbf{1}^{\cdot+}$–$\mathbf{8}^{\cdot+}$) Obtained by Pulse Radiolysis (PR) of c-S and t-S ($S = \mathbf{1}$–$\mathbf{8}$) (5.0×10^{-3} M) in 1,2-Dichloroethane (DCE) at r.t.[a]

$S^{\cdot+}$	c-$S^{\cdot+}$ (D_2)	t-$S^{\cdot+}$ (D_2)	c-$S^{\cdot+}$ (D_1)	t-$S^{\cdot+}$ (D_1)
	\multicolumn{4}{c}{λ_{max} [nm] (ε_λ [M^{-1} cm^{-1}])}			
$\mathbf{1}^{\cdot+}$	515 (180,000)	480 (65,000)	780 (6,180)	750 (11,400)
$\mathbf{2}^{\cdot+}$	520 (30,900)	490 (80,900)	780 (11,500)	790 (19,600)
$\mathbf{3}^{\cdot+}$	530 (13,200)	470,500 (39,200; 74,800)	nd	720,800 (14,700; 41,700)
$\mathbf{4}^{\cdot+}$	530 (nd)	480 (105,000)	nd	680 (14,700)
$\mathbf{5}^{\cdot+}$	540 (nd)	480 (84,600)	nd	700 (14,700)
$\mathbf{6}^{\cdot+}$	520 (3,270)	460,510 (24,500; 208,000)	700 (1,240)	>900 (>11,000)
$\mathbf{7}^{\cdot+}$	450,500 (8,820; 9,710)	440,490 (18,400; 29,400)	760,780 (7,940; 7,940)	740,800 (9,810; 184,000)
$\mathbf{8}^{\cdot+}$	nd	480,540 (33,100; 76,000)	nd	780,880 (24,500; 61,300)

[a] ε_λ was estimated from ΔO.D.$_\lambda$ and the optical length of 1 cm with the assumption of $[S^{\cdot+}] = 8 \times 10^{-6}$ M which was calculated from ΔO.D.$_{480}$, the optical length of 1 cm, and $\varepsilon_{480} = 6.5 \times 10^4$ M^{-1} cm^{-1} for $\mathbf{1}^{\cdot+}$ in DCE.

Table 2 Rate Constants for the Reactions of $S^{\cdot+}$ ($S^{\cdot+} = \mathbf{1}^{\cdot+}$–$\mathbf{8}^{\cdot+}$) During PR in DCE at r.t.[a]

$S^{\cdot+}$	\multicolumn{3}{c}{c-$S^{\cdot+}$}			\multicolumn{2}{c}{t-$S^{\cdot+}$}	
	k_i[sec^{-1}]	k_d[M^{-1} sec^{-1}]	k_{O_2}[M^{-1} sec^{-1}]	k_d[M^{-1} sec^{-1}]	k_{O_2}[M^{-1} sec^{-1}]
$\mathbf{1}^{\cdot+}$	<10^6	3.9×10^8	<10^6	3.9×10^8	<10^6
$\mathbf{2}^{\cdot+}$	<10^8	3.4×10^8	<10^6	4.3×10^8	<10^6
$\mathbf{3}^{\cdot+}$	4.5×10^6	2.0×10^8	nd	3.0×10^8	4.5×10^7
$\mathbf{4}^{\cdot+}$	1.3×10^7	nd	nd	<10^6	4.0×10^7
$\mathbf{5}^{\cdot+}$	1.4×10^7	nd	nd	<10^6	1.2×10^7
$\mathbf{6}^{\cdot+}$	<10^6	<10^6	<10^6	3.4×10^8	<10^6
$\mathbf{7}^{\cdot+}$	<10^6	<10^6	2.5×10^7	<10^6	2.8×10^7
$\mathbf{8}^{\cdot+}$	nd	nd	<10^6	<10^6	<10^6

[a] k_i, calculated from the formation of t-$S^{\cdot+}$ at 5.0×10^{-3} M of c-S; k_d, measured from the dependence of the decay of $S^{\cdot+}$ on [S]; k_{O_2}, obtained from the dependence of the decay of c-$S^{\cdot+}$ or t-$S^{\cdot+}$ on [O$_2$]. $k_i < 10^6$ sec^{-1} denotes that the formation of t-$S^{\cdot+}$ was not observed. $k_d < 10^6$ M^{-1} sec^{-1} and $k_{O_2} < 10^6$ M^{-1} sec^{-1} denote that there was no dependence of the decay of c-$S^{\cdot+}$ or t-$S^{\cdot+}$ on [c-S], [t-S], or [O$_2$].

rapidly into the thermodynamically stable t-$1^{\cdot +}$ and t-1 [Eq. (13)], as proposed [42,67]. Therefore the bimolecular isomerization is enhanced with increasing [c-1].

$$S^{\cdot +} + Cl^- \rightarrow \text{neutral products; } k_n \text{ for neutralization} \tag{10}$$

$$c - S^{\cdot +} + c - S \rightarrow \pi - (c - S)_2^{\cdot +} \ (S = 1 - 3); \ k_d^c \tag{11}$$

$$\pi - (c - S)_2^{\cdot +} \rightarrow \sigma - S_2^{\cdot +} \tag{12}$$

$$\sigma - S_2^{\cdot +} \rightarrow t - S^{\cdot +} + t - S \tag{13}$$

Concentration effects suggest that the bimolecular isomerization of c-$2^{\cdot +}$ and c-$3^{\cdot +}$ to t-$2^{\cdot +}$ and t-$3^{\cdot +}$ involve the dimerization of c-$2^{\cdot +}$ and c-$3^{\cdot +}$ to form π-$(c$-$2)_2^{\cdot +}$ and π-$(c$-$3)_2^{\cdot +}$ and then σ-$2_2^{\cdot +}$ and σ-$3_2^{\cdot +}$ [Eqs. (11) and (12), respectively]. Similar k_d^c for c-$1^{\cdot +}$–c-$3^{\cdot +}$ indicates that p-Me and p-MeO substitutions have little effect on the dimerization of c-$S^{\cdot +}$. Decay of the absorption of other c-$S^{\cdot +}$ at higher [c-S] could not be detected, because the decay was too fast to be detected within the experimental limits, or [c-4–8] could not be increased because of low solubilities in DCE. k_d^c for c-$8^{\cdot +}$ with p-MeO is expected to be 10^8 M^{-1} sec^{-1} from analogy with c-$3^{\cdot +}$, while the steric hindrance resulting from the twisted structure with the 3,5-dimethoxyphenyl ring rotating relative to the C=C double bond and to Me on the olefinic carbon probably inhibits the dimerization of c-$6^{\cdot +}$ and $7^{\cdot +}$, respectively.

Similarly to c-$1^{\cdot +}$, t-$1^{\cdot +}$ decays via neutralization with Cl^- [Eq. (10)] and dimerization with t-1 at $k_d^l = 3.9 \times 10^8$ M^{-1} sec^{-1} to form the π-type dimer$^{\cdot +}$ of t-1 [π-$(t$-$1)_2^{\cdot +}$, Eq. (14)] with overlapping of π-electrons between the two benzene rings of t-$1^{\cdot +}$ and t-1.[3d,11] Neither the unimolecular nor the bimolecular isomerization of t-$1^{\cdot +}$ to c-$1^{\cdot +}$ occurs. It is found that π-$(t$-$1)_2^{\cdot +}$ generated at 77 K converts to σ-$1_2^{\cdot +}$ upon warming [Eq. (15)], and finally decomposes into the thermodynamically stable t-$1^{\cdot +}$ and t-1 [Eq. (13)] as discussed above for c-$1^{\cdot +}$ [42].

$$t - S^{\cdot +} + c - S \rightarrow \pi - (t - S)_2^{\cdot +} \ (S = 1 - 3); \ k_d^l \tag{14}$$

$$\pi - (t - S)_2^{\cdot +} \rightarrow \sigma - S_2^{\cdot +} \tag{15}$$

t-$2^{\cdot +}$, t-$3^{\cdot +}$, and t-$6^{\cdot +}$ show behavior similar to that of t-$1^{\cdot +}$ (Table 2). Similar k_d^l for t-$1^{\cdot +}$–t-$3^{\cdot +}$ and t-$6^{\cdot +}$ suggests that p-Me and m-MeO substitutions have little effect on dimerization of t-$S^{\cdot +}$ and the similar dimerization mechanism [Eqs. (11–13)]. On the other hand, k_{obs} of t-$4^{\cdot +}$, t-$5^{\cdot +}$, t-$7^{\cdot +}$, and t-$8^{\cdot +}$ were much slower than those of t-$1^{\cdot +}$– $3^{\cdot +}$ and t-$6^{\cdot +}$ because k_{obs} did not change with varying concentration (5×10^{-3}–10^{-1} M). This indicates that no dimerization of t-$4^{\cdot +}$, t-$5^{\cdot +}$, t-$7^{\cdot +}$, and t-$8^{\cdot +}$ occurs. It is suggested that the dimerization involves the initial formation of π-complex with overlapping of the two benzene rings and that the π-complex formation is inhibited by steric hindrance of the substituents on the benzene rings and olefinic carbons.

Enhancements of Unimolecular Isomerization and Oxidation by Charge–Spin Separation

In contrast to c-$1^{\cdot +}$ and $2^{\cdot +}$, the unimolecular isomerization of c-$3^{\cdot +}$–$5^{\cdot +}$ and c-$8^{\cdot +}$ having p-MeO occurs to yield t-$3^{\cdot +}$–$5^{\cdot +}$ and t-$8^{\cdot +}$, respectively, k_i for c-$4^{\cdot +}$ and c-$5^{\cdot +}$ are larger than those for c-$3^{\cdot +}$ and c-$8^{\cdot +}$. Neither the unimolecular nor bimolecular isomerization occurs in c-$6^{\cdot +}$ and c-$7^{\cdot +}$. The oxidation of t-$3^{\cdot +}$–$5^{\cdot +}$ and $7^{\cdot +}$ having p-MeO with O_2 occurs at $k_{O_2} = (1.2$–$4.5) \times 10^7$ M^{-1} sec^{-1}, while no oxidation occurs in c-$1^{\cdot +}$, c-$2^{\cdot +}$, or c-$6^{\cdot +}$ without p-MeO and in t-$8^{\cdot +}$ with two p-MeOs. The unimolecular isomerization and oxidation with O_2 are remarkably enhanced in $3^{\cdot+}$–$5^{\cdot+}$ and $7^{\cdot+}$ with p-MeO as an electron-donating substituent on the benzene ring.

It is reported that t-$1^{\cdot+}$ has a planar structure with two benzene rings and a C=C double bond as the most stable conformation, while c-$1^{\cdot+}$ has a twist angle of 26° between the two benzene rings on the basis of MO calculations [75]. Although the positive charge and unpaired electron are delocalized on the SOMO involving π-orbital of sp^2 C atoms and n-orbital of O atom of p-MeO in $1^{\cdot+}$–$8^{\cdot+}$, as shown in structure (A), a quinoid-type structure (B) is considered to contribute in $3^{\cdot+}$–$5^{\cdot+}$, $7^{\cdot+}$, and $8^{\cdot+}$ with p-MeO but not in $1^{\cdot+}$, $2^{\cdot+}$, and $6^{\cdot+}$ without p-MeO (Scheme 2). Alternatively, the n-orbital of O atom of the p-MeO and the π-orbital of the olefinic β-carbon participate considerably in the SOMO of $3^{\cdot+}$–$5^{\cdot+}$, $7^{\cdot+}$, and $8^{\cdot+}$. Therefore, it is suggested that p-MeO in $3^{\cdot+}$–$5^{\cdot+}$, $7^{\cdot+}$, and $8^{\cdot+}$ causes separation and localization of the positive charge on O atom of the p-MeO and an unpaired electron on the olefinic β-carbon (charge–spin separation), which induces a twisted conformation with a large twist angle and C–C single-bond character for the central C=C double bond in (B). Such charge–spin separation is less important in $1^{\cdot+}$, $2^{\cdot+}$, and $6^{\cdot+}$ than in $3^{\cdot+}$–$5^{\cdot+}$, $7^{\cdot+}$, and $8^{\cdot+}$. It is suggested that distonic radical cations known in the gas phase [76] can be used to explain the reactivities of radical cations even in solution [77].

Neither the unimolecular isomerization of c-$7^{\cdot+}$ nor the oxidation of t-$8^{\cdot+}$ occurred, although $7^{\cdot+}$ and $8^{\cdot+}$ have p-MeO. These are possibly explained in terms of a barrier to the twisting of the C=C double bond and the spin density on the olefinic carbon, respectively. The contribution of (B) is decreased by the electron-donating Me on the olefinic carbon; therefore, it is suggested that the single-bond character of the C=C double bond is lower in c-$7^{\cdot+}$ than in c-$3^{\cdot+}$ and that the barrier to the twisting of the C=C double bond is higher in c-$7^{\cdot+}$ than in c-$3^{\cdot+}$. An unpaired electron appears on the olefinic carbon on the side of the p-MeO in $3^{\cdot+}$ because of contribution of (B), while it appears on both olefinic carbons in

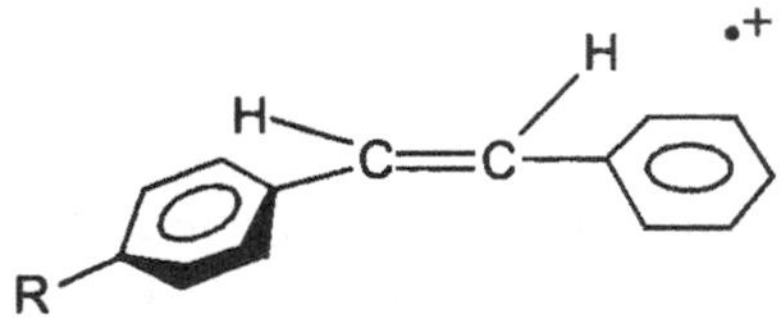

(A)

charge-spin delocalization

$1^{\cdot+}$ - $8^{\cdot+}$

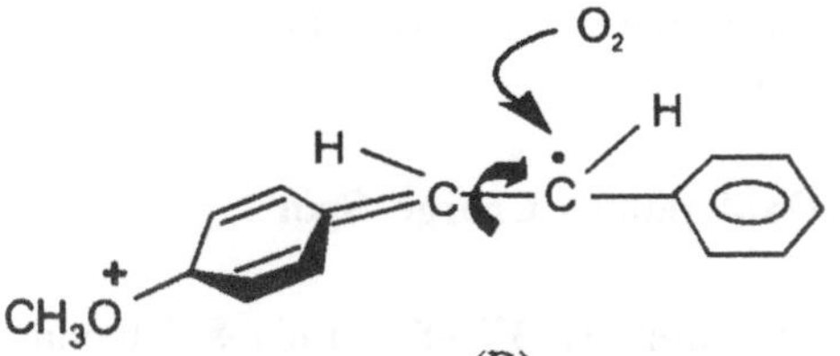

(B)

charge-spin separation

$3^{\cdot+}$, $4^{\cdot+}$, $5^{\cdot+}$, $7^{\cdot+}$, $8^{\cdot+}$

Scheme 2 Charge–spin delocalization (A) and separation (B) in $S^{\cdot+}$.

$8^{\cdot+}$ with two symmetrical p-MeOs. Therefore, the density of an unpaired electron on the olefinic carbon in t-$8^{\cdot+}$ is lower than that in t-$3^{\cdot+}$. Neither the unimolecular c–t isomerization nor the oxidation of c-$6^{\cdot+}$ with an extremely small molar absorption coefficient (ε) at λ_{max}, but similar to that of 1,3-dimethoxybenzene$^{\cdot+}$ [78] may be interpreted by a twisted structure with the 3,5-dimethoxyphenyl ring rotating relative to the C=C double bond, because (B), having m-MeO, is not possible in c-$6^{\cdot+}$.

The k_i for c-$8^{\cdot+}$ was measured at various temperature. From the Arrhenius plot of ln k_i and the reciprocal of the temperature (T^{-1}), the activation energy (E_a) and the pre-exponential factor (A) were determined to be 1.3 kcal mol^{-1} and $10^{7.5}$ sec^{-1}, respectively, which are smaller than E_a = 7.7 kcal mol^{-1} and A = $10^{11\pm2}$ sec^{-1} for c-4,4'-dibromostilbene$^{\cdot+}$ and E_a = 3.3 kcal mol^{-1} and A = $10^{9\pm2}$ sec^{-1} for c-4,4'-dimethylstilbene$^{\cdot+}$ [66]. This result is consistent with the larger k_i for c-$8^{\cdot+}$ than for c-4,4'-dibromo- and c-4,4'-dimethylstilbene$^{\cdot+}$, and suggests that E_a predominantly controls the isomerization. Based on the differences in E_{ox} between c-S and t-S (Table 3), E_a is too high to be overcome in the isomerization of t-$S^{\cdot+}$ to c-$S^{\cdot+}$ (11–14 kcal mol^{-1}). This renders the isomerization one-way from c-$S^{\cdot+}$ to t-$S^{\cdot+}$ for S=1–6.

The k_{O_2} for free radical-O_2 reactions is found to be nearly equal to the diffusion rate constant (k_{diff}). For example, k_{O_2} = 2.4 × 10^9 M^{-1} sec^{-1} for the benzyl radical-O_2 reaction is similar to k_{diff} = 6.7 × 10^9 M^{-1} sec^{-1} in cyclohexane. On the other hand, k_{O_2} = (1.2–4.5) × 10^7 M^{-1} sec^{-1} for the oxidation of t-$3^{\cdot+}$–$5^{\cdot+}$ and t-$7^{\cdot+}$ are found to be two orders of magnitude smaller than k_{O_2} for free radicals and k_{diff} = 7.8 × 10^9 M^{-1} sec^{-1} in DCE. This suggests that an unpaired electron is not completely localized on the olefinic carbon in t-$3^{\cdot+}$–$5^{\cdot+}$ and t-$7^{\cdot+}$, and that the positive charge interferes with the reactivity of t-$3^{\cdot+}$–$5^{\cdot+}$ and t-$7^{\cdot+}$ as a radical toward O_2 because of the electrophilic character of O_2.

Chain Isomerization and Oxidation of S$^+$

In order to confirm the enhancement of c–t isomerization and oxidation by charge–spin separation in $S^{\cdot+}$ with p-MeO, product analyses were carried out. Isomerization from c-S to t-S was found to occur stoichiometrically with large G values (5–121) and G_{lim} (20–10^4) in γ-R of c-S in DCE at r.t. (Table 4). The larger G values for the yield of t-3–5 and t-8 during γ-R of c-3–5 and c-8, respectively, with p-MeO than for c-1, 2, and 6 without p-MeO suggest a smaller barrier to c–t isomerization for c-$3^{\cdot+}$–$5^{\cdot+}$ and c-$8^{\cdot+}$ than for c-$1^{\cdot+}$, $2^{\cdot+}$, and $6^{\cdot+}$ because of the greater single bond character of the former central C=C double bond for c-$3^{\cdot+}$–$5^{\cdot+}$ and $8^{\cdot+}$ relative to c-$1^{\cdot+}$, $2^{\cdot+}$, and $6^{\cdot+}$ because of the con-

Table 3 Half-Peak Oxidation Potentials ($E_{1/2}^{ox}$) of c-S and t-S (S=1–6) and Difference Between $E_{1/2}^{ox}(c$-S$)$ and $E_{1/2}^{ox}(t$-S$)^a$

S	R of RCH=CHPh	$E_{1/2}^{ox}(c$-S$)$[V]	$E_{1/2}^{ox}(t$-S$)$[V]	$\Delta E_{1/2}^{ox}(c$-$t)$[V]
1	Ph	1.25	1.14	0.11
2	4-CH$_3$C$_6$H$_4$	1.16	1.05	0.11
3	4-CH$_3$OC$_6$H$_4$	0.94	0.82	0.12
4	2,4-(CH$_3$O)$_2$C$_6$H$_3$	0.80	0.64	0.16
5	3,4-(CH$_3$O)$_2$C$_6$H$_3$	0.87	0.72	0.15
6	3,5-(CH$_3$O)$_2$C$_6$H$_3$	1.23	1.03	0.20

^{a} $E_{1/2}^{ox}(c$-S$)$ and $E_{1/2}^{ox}(t$-S$)$ of c-S and t-S at 1.0 × 10^{-2} M, reference electrode of Ag/AgNO$_3$, supporting electrolyte of 10^{-1} M tetraethylammonium tetrafluoroborate in acetonitrile; $\Delta E_{1/2}^{ox}(c$-$t) = E_{1/2}^{ox}(c$-S$)$-$E_{1/2}^{ox}(t$-S$)$.

Table 4 Products From γ-Radiolysis (γ-R) of c-S (S = 1–8)[a] and k_i of c-S$^{\cdot +}$ [b]

	Product from γ-R				
S	Conv. of c-S [%]	Yield of t-S [%]	G-value of t-S	G_{lim}	k_i [sec^{-1}]
1	6.0	94	18.9	30	$< 10^6$
2	5.4	100	18.2	200	$< 10^6$
3	32.5	76	81.5	$> 10^3$	4.5×10^6
4	26.0	100	105.6	$> 10^4$	1.3×10^7
5	22.7	97	94.7	$> 4 \times 10^3$	1.4×10^7
6	2.9	100	10.2	140	$< 10^6$
7	2.0	85	5.4	20	$< 10^6$
8	39.1	100	120.6	$> 10^4$	5.5×10^6

[a] Dose, 2.6×10^2 Gy; [S] = 1.0×10^{-2} M in DCE; r.t.; the yield of t-S based on conversion of c-S; G value of t-S, the yield of t-S; G_{lim}, the G value of t-S at $[c\text{-S}]_\infty$ in the plots of G^{-1} of t-S vs. $[c\text{-S}]^{-1}$.
[b] k_i, calculated from the formation of t-S$^{\cdot +}$ during PR of c-S at 5.0×10^{-3} M in DCE.

tribution of (B) with separation and localization of the positive charge on O atom of p-MeO and an unpaired electron on the β-olefinic carbon (Scheme 2). The isomerization of c-**3**–**5** and c-**8** proceeds via a chain reaction mechanism involving c–t unimolecular isomerization and endergonic hole transfer (HT) from t-S$^{\cdot +}$ to c-S. On the other hand, the isomerization of c-**1** and -**2** proceeds via a chain reaction mechanism involving dimerization and decomposition into t-S$^{\cdot +}$ and t-S.

The formation of oxidation products **a**–**c** in a range of G values (0.7–3.8) during the γ-R of S in O$_2$-saturated DCE suggests that **a**–**c** would be produced from complicated reactions of peroxy radicals with S (Table 5). On the other hand, the regioselective formation of **3d** with large G values (2.6–3.0) in oxidation of **3**$^{\cdot +}$ with O$_2$ is explained by spin localization on the β-olefinic carbon because of the contribution of (B) in **3**$^{\cdot +}$. The results of products analyses are essentially identical with prediction based on k_i and k_{O_2} for S$^{\cdot +}$ measured with PR. It should be emphasized that the reactivities of c–t unimolecular isomerization and reaction of S$^{\cdot +}$ with O$_2$ can be understood in terms of charge–spin separation induced by p-MeO.

In conclusion, c-**3**$^{\cdot +}$–**5**$^{\cdot +}$ and c-**8**$^{\cdot +}$ with p-MeO generated during PR or γ-R isomerize unimolecularly to the corresponding t-S$^{\cdot +}$ at k_i = 4.5×10^6–1.4×10^7 sec^{-1}, while t-**3**$^{\cdot +}$–**5**$^{\cdot +}$ and **7**$^{\cdot +}$ with p-MeO are oxidized with O$_2$ at k_{O_2} = $(1.2–4.5) \times 10^7$ M^{-1} sec^{-1}. On the other hand, neither the unimolecular isomerization from t-S$^{\cdot +}$ to c-S$^{\cdot +}$ nor the oxidation with O$_2$ occurs in t-S$^{\cdot +}$ without p-MeO. It should be noted that an unpaired electron is not completely localized on the olefinic carbon in t-**3**$^{\cdot +}$–**5**$^{\cdot +}$ and **7**$^{\cdot +}$ and that a positive charge interferes the reactivity of t-**3**$^{\cdot +}$–**5**$^{\cdot +}$ and **7**$^{\cdot +}$ as a radical toward O$_2$ because of the electrophilic character of O$_2$.

2.2. Cycloreversion of Tetraphenylcyclobutane Radical Cations (TPCB$^{\cdot +}$)

Cleavage of a C—C bond gives a distonic radical cation as an intermediate, while concerted cleavage of two C—C bonds yields the corresponding ArO$^{\cdot +}$ and ArO in cycloreversion of aryl-substituted cyclobutane$^{\cdot +}$. Therefore, the cycloreversion mechanism is related to dimerization of ArO$^{\cdot +}$, where π- and σ-dimers$^{\cdot +}$ are detected during PR of ArO such as

Table 5 Products From γ-R of S (S = 1–8) in the Presence of O_2[a] and k_{O_2} of $S^{\cdot+}$[b]

| S | RCH=CR'Ph | | G value or yield | | | | k_{O_2} [M^{-1} sec^{-1}] |
	R	R'	PhCHO a	RCHO b	RCH[O]CHPh[c] c	RCH$_2$C(=O)Ph[d] d	
c-1	Ph	H	1.7	—	0.9	0.1	$<10^6$
t-1	Ph	H	1.6	—	1.1	0.3	$<10^6$
t-2	4-CH$_3$C$_6$H$_4$	H	1.1	1.3	0.7	0.05	$<10^6$
c-3	4-CH$_3$OC$_6$H$_4$	H	1.0	1.6	1.2	3.0	nd
t-3	4-CH$_3$OC$_6$H$_4$	H	0.9	1.4	1.4	2.6	4.5×10^7
t-4	2,4-(CH$_3$O)$_2$C$_6$H$_3$	H	1.3	1.5	nd	nd	4.0×10^7
t-5	3,4-(CH$_3$O)$_2$C$_6$H$_3$	H	1.3	2.3	nd	nd	1.2×10^7
t-6	3,5-(CH$_3$O)$_2$C$_6$H$_3$	H	1.0	0.9	nd	nd	$<10^6$
t-7	4-CH$_3$OC$_6$H$_4$	CH$_3$	3.8	1.7	nd	0	2.8×10^7
c-7	4-CH$_3$OC$_6$H$_4$	CH$_3$	3.5	1.7	nd	0	2.5×10^7
t-8	(4-CH$_3$OC$_6$H$_4$)$_2$	—	—	1.0	0.8	0.2	$<10^6$

[a] Dose, 4.9×10^4 Gy; [S] = 1.0×10^{-2} M in O_2-saturated DCE; G value, the yield of oxidation products. Isomerization products were also observed with small G values at higher conversions in prolonged γ-R. nd denotes that the values were note measured.

[b] k_{O_2}, obtained from the dependence of the decay of $S^{\cdot+}$ on the concentration of O_2 during pulse radiolysis of S (5.0×10^{-3} M) in the presence of O_2 in DCE.

[c] 1-Aryl-2-phenyl-1,2-epoxyethane (aryl = R).

[d] Arylmethyl phenyl ketone (aryl = R).

styrene derivatives, 1,1-diphenylethylene, and St at higher concentrations [32,67,79–83]. Cycloreversion of r-1,c-2,t-3,t-4- and r-1,t-2,c-3,t-4-tetraphenylcyclobutanes$^{\cdot\,+}$ (t,c,t-TPCB$^{\cdot\,+}$ and t,t,t-TPCB$^{\cdot\,+}$, respectively) during γ-R in the rigid matrix of glassy BC of TPCB (5×10^{-3} M) at 77 K will be described here [42].

Cycloreversion of t,c,t-TPCB$^{\cdot\,+}$

Tetraphenylcyclobutane radical cation is expected to decompose into π-type t-St$_2^+$, σ-type St$_2^+$, or t-St$^{\cdot\,+}$ plus t-St. A broad absorption spectrum of t,c,t-TPCB$^{\cdot\,+}$ was detected at 77 K with weak peaks at 370, 390, 485, 550, and 780 nm, which changed to have clear peaks at 370, 485, 550, and 780 nm upon warming up to 90 K (Fig. 3a). The 48-nm peak is assigned to t-St$^{\cdot\,+}$. Photoirradiation of t,c,t-TPCB$^{\cdot\,+}$ at 77 K at λ over 390 nm gave bands at 400–470, 470–500, and 680–800 nm (Fig. 3b). Warming this sample caused a large increase of the peak at 350, 550, and 780 nm. The difference spectrum was obtained by subtracting the spectra after photoirradiation and warming and indicated the collapse of bands at 400–470 and 470–500 nm and the formation of peaks at 350, 550, and 780 nm (Fig. 3c).

Observed spectral changes can be explained by four species: t,c,t-TPCB$^{\cdot\,+}$, t-St$^{\cdot\,+}$, π-(t-St)$_2^{\cdot\,+}$ with peaks at 400–470 and 680–800 nm [84], and σ-St$_2^+$ with peaks at 350, 550, and 780 nm. Thus t,c,t-TPCB$^{\cdot\,+}$ converts to σ-St$_2^+$ upon warming [Eq. (16)], while photoirradiation of t,c,t-TPCB$^{\cdot\,+}$ at 77 K causes cycloreversion into π-(t-St)$_2^{\cdot\,+}$ [Eq. (17)], which converts to σ-St$_2^+$ upon warming [Eq. (18)]. Finally, σ-St$_2^+$ decomposes into the thermodynamically stable t-St$^{\cdot\,+}$ and t-St [Eq. (19)].

$$t,c,t - \text{TPCB}^{\cdot+} + \text{heat} \rightarrow \sigma - \text{St}_2^{\cdot+} \tag{16}$$

$$t,c,t - \text{TPCB}^{\cdot+} + h\nu \rightarrow \pi - (t - \text{St})_2^{\cdot+} \tag{17}$$

$$\pi - (t - \text{St})_2^{\cdot+} + \text{heat} \rightarrow \sigma - \text{St}_2^{\cdot+} \tag{18}$$

$$\sigma - \text{St}_2^{\cdot+} + \text{heat} \rightarrow t - \text{St}^{\cdot+} + t - \text{St} \tag{19}$$

Cycloreversion of t,t,t-TPCB$^{\cdot\,+}$

A broad absorption spectrum of t,t,t-TPCB$^{\cdot\,+}$ was observed at 77 K, at 400–670 nm with a peak at 480 nm, and collapsed without any particular spectral change upon warming up to 90 K (Fig. 4). However, the spectrum changed to have clear peaks at 485 and 780 nm upon photoirradiation at 77 K (Fig. 4a). The absorption spectrum then changed to have a peak at 550 nm upon warming. The difference spectrum, obtained by subtracting the spectra both before and after warming, clearly indicates the collapse of the peak at 485 nm and the formation of bands at 370, 550, and 770 nm (Fig. 4b).

These results can be interpreted by four species: t,t,t-TPCB$^{\cdot\,+}$, t-St$^{\cdot\,+}$, t-St$^{\cdot\,+}$/t-St pair with peaks at 485 and 780 nm, and σ-St$_2^+$. As there was no overlapping of π-electrons between two benzene rings in the t-St$^{\cdot\,+}$/t-St pair, it is anticipated that the pair has the same absorption as t-St$^{\cdot\,+}$. Thus t,t,t-TPCB$^{\cdot\,+}$ is rather stable up to 90 K and does not convert to π-(t-St)$_2^+$ nor σ-St$_2^+$. Photoirradiation of t,t,t-TPCB$^{\cdot\,+}$ at 77 K causes cycloreversion into t-St$^{\cdot\,+}$/t-St pair [Eq. (20)], which converts to σ-St$_2^+$ upon warming [Eq. (21)].

$$t,t,t - \text{TPCB}^{\cdot+} + h\nu \rightarrow t - \text{St}^{\cdot+}/t - \text{St} \tag{20}$$

$$t - \text{St}^{\cdot+}/t - \text{St} + \text{heat} \rightarrow \sigma - \text{St}_2^{\cdot+} \tag{21}$$

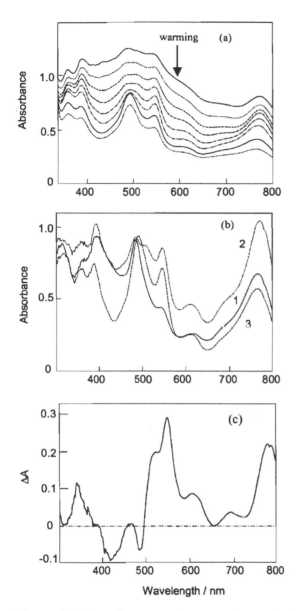

Figure 3 Absorption spectral changes of t,c,t-TPCB$^{\cdot+}$ recorded immediately after γ-R of t,c,t-TPCB (5×10^{-3} M) in glassy BC matrices at 77 K and after warming (a), or after γ-R and subsequent photoirradiation at λ longer than 390 nm (1) and after warning (in the order of 2 and 3) (b), and difference spectra between spectra after photoirradiation and after warming ($\Delta A = A_{No.\,1} - A_{No.\,2}$) (c).

Intermediacy of Stilbene Dimer Radical Cations (St$_2^{\cdot+}$)

Cycloreversion of TPCB$^{\cdot+}$ and dimerization of St$^{\cdot+}$ are summarized in Scheme 3, in which π-(t-St)$_2^{\cdot+}$, σ-St$_2^{\cdot+}$, and the t-St$^{\cdot+}$/t-St pair are involved as the key intermediates. π-(t-St)$_2^{\cdot+}$ takes various overlapping arrangements of the π-electrons between two benzene rings, and exhibits a large charge resonance (CR) band at shorter λ of 680–800 nm than those of other dimer$^{\cdot+}$ reported [84]. On the other hand, σ-St$_2^{\cdot+}$ has an acyclic linear structure,

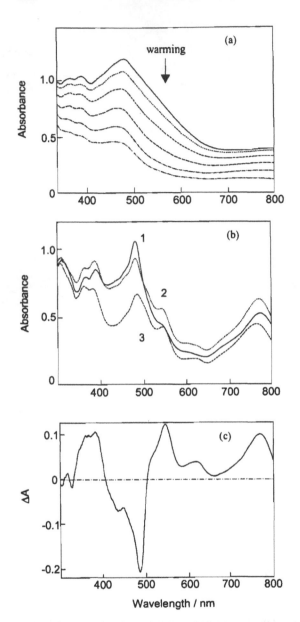

Figure 4 Absorption spectral changes of t,t,t-TPCB$^{\cdot+}$ recorded immediately after γ-R of t,t,t-TPCB $(5 \times 10^{-3}$ M) in glassy BC matrices at 77 K and after warming (a), or after γ-R and subsequent photoirradiation at λ longer than 390 nm (1) and after warming (in the order of 2 and 3) (b), and difference spectra between spectra after photoirradiation and after warming ($\Delta A = A_{\text{No. 1}} - A_{\text{No. 2}}$) (c).

having both a radical and a cation on the 1- and 4-positions of the C$_4$ linkage. Although the c-St$_2^{\cdot+}$ was not detected, the bimolecular reaction of c-St$^{\cdot+}$ and c-St yields σ-St$_2^{\cdot+}$ or c-St$^{\cdot+}/c$-St pair, which decomposes into thermodynamically stable t-St$^{\cdot+}$ and t-St.

It should be noted that t,c,t-TPCB$^{\cdot+}$ decomposes into σ-St$_2^{\cdot+}$, but t,t,t-TPCB$^{\cdot+}$ does not upon warming up to 90 K. On the other hand, photochemical cycloreversion of both TPCB$^{\cdot+}$ occurs through cleavage of two C–C bonds to give π-(t-St)$_2^{\cdot+}$ or t-St$^{\cdot+}/t$-St pair. This may account for the different interaction between two phenyl (Ph) groups through the

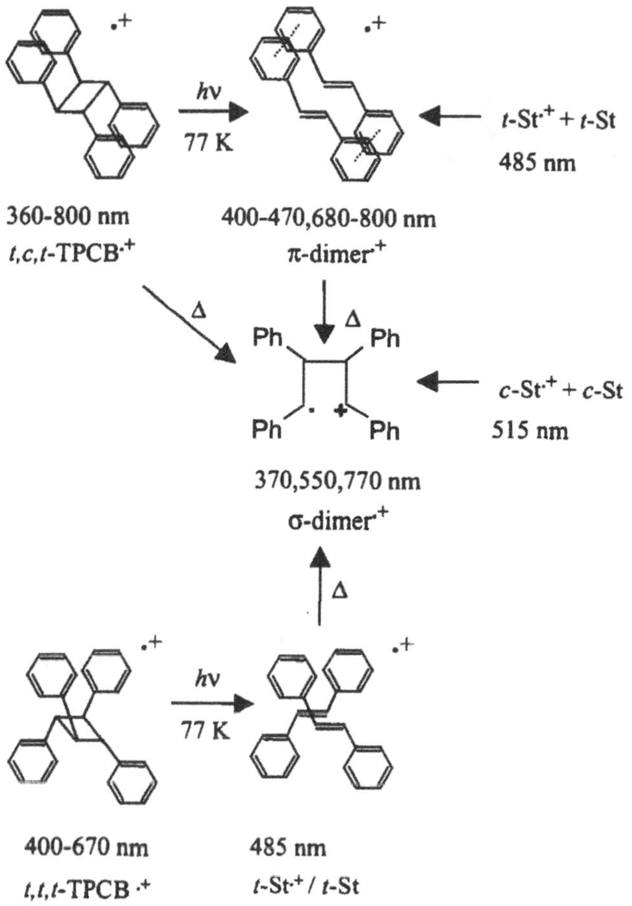

Scheme 3 Cycloreversion of TPCB$^{\cdot+}$ and dimerization of St$^{\cdot+}$.

C—C bond of cyclobutane ring of t,c,t-TPCB$^{\cdot+}$ and t,t,t-TPCB$^{\cdot+}$ in the ground and excited states.

2.3. Reactions of Radical Cations of Organophosphorus Compounds

Little is known about the reactions of radical cations of organophosphorus compounds (P$^{\cdot+}$) such as aryl phosphates and phosphonates as caged compounds and reagents in organic syntheses [85–91]. To elucidate PET reactions including P$^{\cdot+}$ as the key intermediate, PR and γ-R are used here.

Intramolecular Reaction of Radical Cations of Trinaphthyl Phosphates and Dinaphthyl Methylphosphonates

From product analysis, fluorescence quenching, and LFP, initially formed trinaphthyl phosphates (**9$^{\cdot+}$**) and dinaphthyl methylphosphonates (**10$^{\cdot+}$**) decompose into 1,1′-binaphthyl (Np$_2$) and 1-naphthyl phosphate or methyl phosphate, respectively. Although it is suggested that **9$^{\cdot+}$** and **10$^{\cdot+}$** change to the intramolecular π-dimer$^{\cdot+}$ between two Np

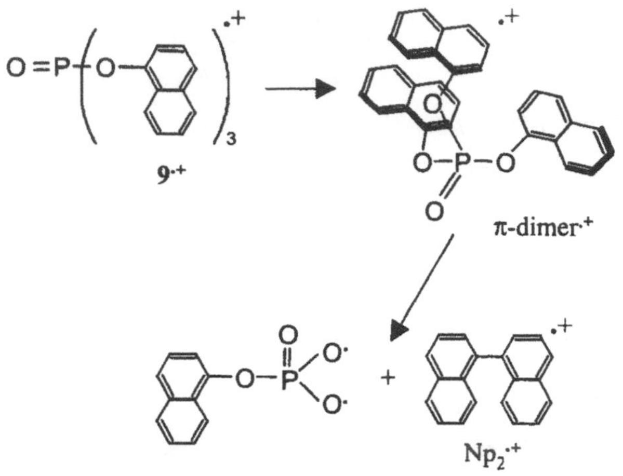

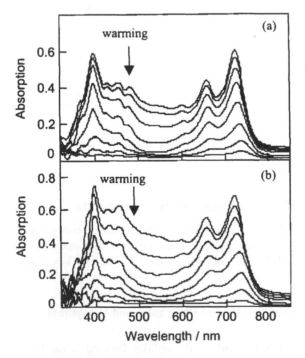

Scheme 4   Elimination of $Np_2^{\cdot+}$ from $9^{\cdot+}$ through the intramolecular π-dimer$^{\cdot+}$.

groups and $Np_2^{\cdot+}$ is subsequently eliminated (Scheme 4), no direct evidence of the transformation [90,91].

The assignments of $9^{\cdot+}$, $10^{\cdot+}$, and $Np_2^{\cdot+}$ can be confirmed by their absorption spectra observed during γ-R in BC rigid matrices at 77 K. Three peaks of $9^{\cdot+}$ at 400, 680, and 720 nm and two broad bands in the region of 500–600 and 800–850 nm were observed after the γ-R of **9** in BC rigid matrix at 77 K (Fig. 5). The absorption spectrum

Figure 5   Absorption spectral changes of $9^{\cdot+}$ and $10^{\cdot+}$ recorded immediately after γ-R of **9** at 3 × 10^{-2} M (a) and **10** at 1 × 10^{-2} M (b) in BC rigid matrices at 77 K and after warming. Arrows show decrease in the peak upon warming.

disappeared with no formation of a peak upon warming with independence on [9]. The broad band in the region of 500–600 nm, assigned to an intramolecular π-dimer$^{\cdot+}$ of $\mathbf{9}^{\cdot+}$ with face-to-face interaction between the two Np groups, disappeared faster than the peaks of $\mathbf{9}^{\cdot+}$ at 400, 680, and 720 nm (Fig. 5a). Similar spectral changes of $\mathbf{10}^{\cdot+}$ were observed during γ-R of **10** in BC rigid matrix at 77 K (Fig. 5b).

The transient absorption spectrum assigned to $\mathbf{9}^{\cdot+}$ and intramolecular π-dimer$^{\cdot+}$ of $\mathbf{9}^{\cdot+}$ was observed immediately after 8-nsec e$^-$, and decayed with the formation of new peaks at 480 and 520 nm assigned to $Np_2^{\cdot+}$ (Fig. 6). This indicates that intramolecular dimerization in $\mathbf{9}^{\cdot+}$ occurs to reversibly form π-dimer$^{\cdot+}$ at $k_d^{intra} = 1.0 \times 10^8 - 1.0 \times 10^9$ sec^{-1}. In DCE, $\mathbf{9}^{\cdot+}$ and π-dimer$^{\cdot+}$ disappeared by neutralization with Cl$^-$ generated by the initial radiolytic processes at $k_n = 4.7 \times 10^{10}$ M^{-1} sec^{-1}. From the time profile of the transient absorption at 520 nm involving the formation of $Np_2^{\cdot+}$ and decay of π-dimer$^{\cdot+}$ of $\mathbf{9}^{\cdot+}$, $k_r = 5.3 \times 10^5$ sec^{-1} is estimated for the reaction.

Warming up to 90 K is not enough for the activation barrier for the formation of $Np_2^{\cdot+}$ from π-dimer$^{\cdot+}$ of $\mathbf{9}^{\cdot+}$, while the reaction is completely attained even for e$^-$ duration of 8 nsec at r.t. The peaks of π-dimer$^{\cdot+}$ of $\mathbf{9}^{\cdot+}$ were much broader than that of π-dimer$^{\cdot+}$ of 1,3-di-1-naphthylpropane$^{\cdot+}$ (720 nm) because of the different spacers, O−P(O)−O and (CH$_2$)$_3$, between the two Np groups in $\mathbf{9}^{\cdot+}$ and 1,3-di-1-naphthylpropane$^{\cdot+}$, respectively. No elimination of $Np_2^{\cdot+}$ occurs in 1,3-di-1-naphthylpropane$^{\cdot+}$ and bis(1-naphthyloxy) methane$^{\cdot+}$. The electron withdrawing character of the P(O) group in the O−P(O)−O spacer may be responsible for the elimination of $Np_2^{\cdot+}$.

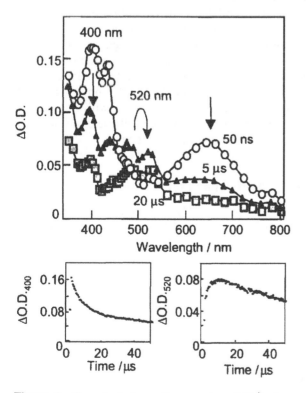

Figure 6 Transient absorption spectra of $\mathbf{9}^{\cdot+}$ observed immediately (open circle), 5 μsec (solid triangle), and 20 μsec after e$^-$ (gray square) during PR of 1.0×10^{-2} M of **9** in DCE at r.t. Attachments are time profiles of ΔO.D.$_{400}$ and ΔO.D.$_{520}$.

Bimolecular Reaction of *p*-Substituted Triphenylphosphines Radical Cation with Water

From the product analyses of PET reactions of trivalent phosphorus compounds such as phosphines and phosphates, it is found that $P^{\cdot +}$ reacts with nucleophiles such as water and alcohol to yield phosphine oxides and phosphates [92–95]. During a reaction of triphenylphosphines $(P(C_6H_4\text{-}X\text{-}p)_3{}^{\cdot +})$ with water to give phosphine oxides $O{=}P(C_6H_4\text{-}X\text{-}p)_3$ (Scheme 5), electron-donating character of the *p*-substituent shifted the bands of $P(C_6H_4\text{-}X\text{-}p)_3{}^{\cdot +}$ to longer λ (520–610 nm) and decreased the yield of $O{=}P(C_6H_4\text{-}X\text{-}p)_3$ (0.26–0.02) on the basis of LFP, PR, γ-R, and product analysis [96].

The positive charge on the P atom of $P(C_6H_4OMe\text{-}p)_3{}^{\cdot +}$ is less than $P(C_6H_5)_3{}^{\cdot +}$ because of conjugation with the π-electrons of the benzene ring and the *n*-electrons of O atom of *p*-MeO. Thus, the k_r of $P(C_6H_4OMe\text{-}p)_3{}^{\cdot +}$ with H_2O is smaller than that of $P(C_6H_5)_3{}^{\cdot +}$. The delocalization of the positive charge of $3^{\cdot +}$ may be related to the longer λ shift of the bands with an increase in the electron-donating character of *p*-substituent. The difference of reactivities of $P(C_6H_4\text{-}X\text{-}p)_3{}^{\cdot +}$ from those of $9^{\cdot +}$ and $10^{\cdot +}$ is a consequence of the respective HOMO, which are the phosphorus nonbonding orbital and Np π-orbital, respectively.

3. DIMERIZATION OF RADICAL ANIONS

Whether or not the radical anion of aromatic hydrocarbons $(ArH^{\cdot -})$ is stabilized by CR π-interaction with ArH, giving the dimer radical anion, $(ArH)_2^{\cdot -}$, is an interesting subject. Although much effort has been concentrated on detecting $(ArH)_2^{\cdot -}$ with a sandwich structure between two aromatic groups of $ArH^{\cdot -}$ (and ArH, the formation of $(ArH)_2^{\cdot -}$ (has not been observed [46,47,73,97–111]. As an exceptional case, $(\text{anthracene})_2^{\cdot -}$ has been formed from the cleavage of a $4\pi + 4\pi$ anthracene photodimer$^{\cdot -}$ in a rigid matrix at 77 K [102]. Radical anions of several olefins with electron-withdrawing substituents such as fumaronitrile, maleic anhydride, and acrylonitrile have been observed to dimerize to give the dimer$^{\cdot -}$ which are assumed to have a broad CR band in the 800–2000 nm range [101]. In contrast to $St^{\cdot +}$ forming π- and/or σ-$(St)_2^{\cdot +}$, no dimerization of $St^{\cdot -}$ with St has been observed. It has been reported that $St^{\cdot -}$ reacts with $St^{\cdot -}$ to give the stilbene dianion (St^{2-}) and St via disproportionation [106,107] at higher concentrations of $St^{\cdot -}$.

3.1. Dimer Radical Anions of Aromatic Acetylenes [48]

Diphenylacetylene$^{\cdot -}$ ($11a^{\cdot -}$) reacts with $11a^{\cdot -}$ to give the $11a$ dimer dianion, $(11a)_2^{2-}$ with a peak at 470 nm ($\varepsilon_{470} = 9.0 \times 10^3$ and $1.1 \times 10^4\,M^{-1}\,cm^{-1}$ in THF [107] and HMPA [108],

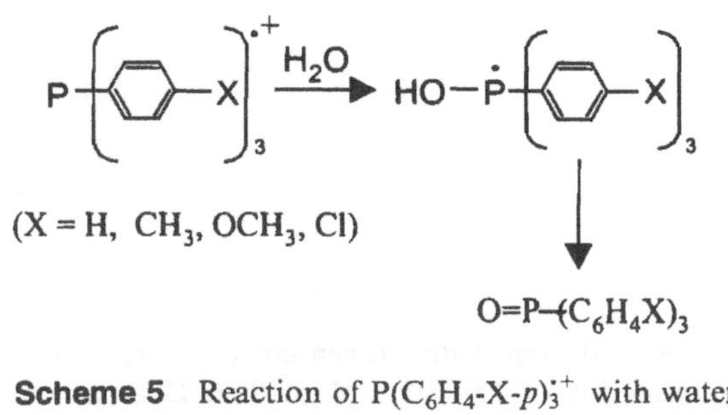

$$P\left(\!\!-\!\!\bigcirc\!\!-\!\!X\right)_3^{\cdot +} \xrightarrow{H_2O} HO-\overset{\cdot}{P}\left(\!\!-\!\!\bigcirc\!\!-\!\!X\right)_3$$

$(X = H,\ CH_3,\ OCH_3,\ Cl)$

$$O{=}P{-}(C_6H_4X)_3$$

Scheme 5 Reaction of $P(C_6H_4\text{-}X\text{-}p)_3^{\cdot +}$ with water to give $O{=}P(C_6H_4\text{-}X\text{-}p)_3$.

respectively) at higher $[\mathbf{11a}^{\cdot-}] = (2.6\text{--}7.3) \times 10^{-2}$ and 3.5×10^{-3} M in THF[107] and HMPA [108], respectively. The structure of $(\mathbf{11a})_2^{2-}$ is assumed to be the 1,2,3,4-tetraphenylbut-1,3-ene-1,4-diyl dianion ($PhC^- = CPh\text{--}CPh = C^-Ph$) with the formation of one C—C bond between two sp carbons [107,108]. On the other hand, it was found that aromatic acetylene$^{\cdot-}$ ($\mathbf{A}^{\cdot-}$) (Scheme 6) such as $\mathbf{11a}^{\cdot-}\text{--}\mathbf{11g}^{\cdot-}$ and 1,4-diphenyl-1,3-butadiyne$^{\cdot-}$ ($\mathbf{12}^{\cdot-}$) dimerizes with $\mathbf{A}$ to form $\mathbf{A}_2^{\cdot-}$. The intramolecular dimerization occurs in 1,ω-di-(4-(diphenylacetylenyl))-ethane$^{\cdot-}$ ($\mathbf{13}^{\cdot-}(n = 2)$), -propane$^{\cdot-}$ ($\mathbf{13}^{\cdot-}$ ($n = 3$)), -butane$^{\cdot-}$ ($\mathbf{13}^{\cdot-}$ ($n = 4$)), -pentane$^{\cdot-}$ ($\mathbf{13}^{\cdot-}$ ($n = 5$)), and -hexane$^{\cdot-}$ ($\mathbf{13}^{\cdot-}$ ($n = 6$)) to give the intramolecular dimer$^{\cdot-}$ between two $\mathbf{11a}$-chromophores during γ-R in MTHF rigid matrices at 77 K and PR in HMPA and DMF solutions at r.t.

Formation of Intermolecular Dimer Radical Anions

Transient absorption spectrum with peaks at 450 and > 840 nm assigned to $\mathbf{11a}^{\cdot-}$ was observed immediately after e^- during PR of $\mathbf{11a}$ at $5.0 \times 10^{-3}\text{--}4.0 \times 10^{-2}$ M in HMPA at r.t. (Fig. 7) and γ-R in an MTHF rigid matrix at 77 K (Fig. 8) [73], and during the reduction by Na in THF and HMPA [107,108]. The transient absorption of $\mathbf{11a}^{\cdot-}$ decayed on a time scale of a few μsec with formation of a band at 500 nm with isosbestic points at 470 and 630 nm at r.t. The pseudo-first-order rate constants of the decay of 450-nm band and formation of 500-nm band were calculated to be $k_d^{450} = 1.0 \times 10^6$ and $k_f^{500} = 1.3 \times 10^6$ sec^{-1}, respectively, at $[\mathbf{11a}] = 10^{-2}$ M. It is found that k_{decay}^{450} and k_{form}^{500} increase with increasing $[\mathbf{11a}]$. This suggests that $\mathbf{11a}^{\cdot-}$ dimerizes with $\mathbf{11a}$ to give $(\mathbf{11a})_2^{\cdot-}$ with the formation of 500-nm band. A similar absorption spectral change of $\mathbf{11a}^{\cdot-}$ was observed in the MTHF rigid matrix at 77 K during γ-R of $\mathbf{11a}$ (Fig. 8). Formation of $(\mathbf{11a})_2^{\cdot-}$ is completely contrary to previous results of electrolysis, the reduction by Na, photolysis, and radiolysis of $\mathbf{11a}$ where no dimerization was observed. $k_d = (3.8\text{--}6.6) \times 10^7$ M^{-1} sec^{-1} for the dimerization was calculated from the linear plots of k_{decay}^{450} and k_{form}^{500} vs. $[\mathbf{11a}]$. σ-Type or diene-type stucture is most likely for $(\mathbf{11a})_2^{\cdot-}$ with 500-nm band (Scheme 7), because $(\mathbf{11a})_2^{2-}$, $PhC^- = CPhCPh = C^-Ph$, has similar band at 470 nm [107,108].

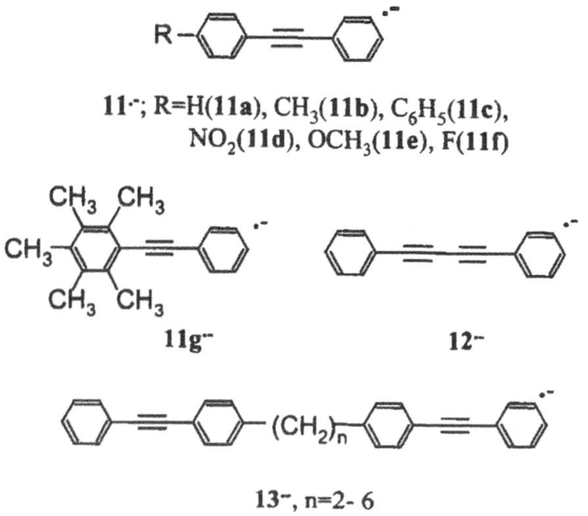

$\mathbf{11}^{\cdot-}$; R=H($\mathbf{11a}$), CH$_3$($\mathbf{11b}$), C$_6$H$_5$($\mathbf{11c}$),
NO$_2$($\mathbf{11d}$), OCH$_3$($\mathbf{11e}$), F($\mathbf{11f}$)

$\mathbf{11g}^{\cdot-}$ $\mathbf{12}^{\cdot-}$

$\mathbf{13}^{\cdot-}$, n=2- 6

Scheme 6 Aromatic acetylene$^{\cdot-}$ ($\mathbf{11a}^{\cdot-}\text{--}\mathbf{11g}^{\cdot-}$, $\mathbf{12}^{\cdot-}$, and $\mathbf{13}^{\cdot-}$ ($n = 2\text{--}6$)).

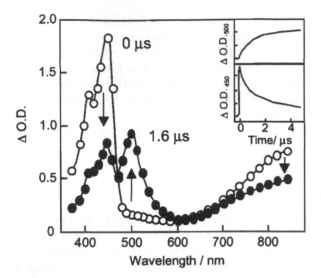

Figure 7 Transient absorption spectral changes of **11a·⁻** during PR of **11a** at 1.0×10^{-2} M in HMPA at r.t. Insets: time profiles of ΔO.D.$_{450}$ and ΔO.D.$_{500}$.

k_d for dimerization of other **11·⁻** and **12·⁻** were determined (Table 6). $k_d = (4.1–6.6) \times 10^7$ M^{-1} sec^{-1} for **11a·⁻–11c·⁻** with weak electronic substituents such as 4-Me and 4-Ph is larger than $k_d = 7.3 \times 10^6–1.7 \times 10^7$ M^{-1} sec^{-1} for **11d·⁻–11g·⁻** with strong electronic substituents such as 4-nitro, 4-MeO, 4-fluoro, and 2,3,4,5,6-pentamethyl groups. The dimerization of **11·⁻** and **11** is inhibited by steric and electronic effects of the substituents causing a higher barrier to a C−C bond formation between two sp carbons.

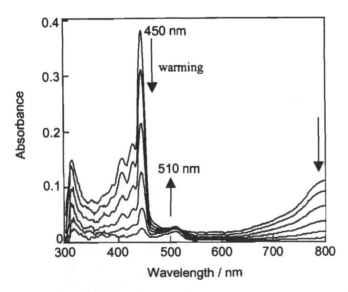

Figure 8 Absorption spectral changes of **11a·⁻** upon warming up to 90 K after γ-R of **11a** at 5.0×10^{-3} M in MTHF rigid matrix at 77 K.

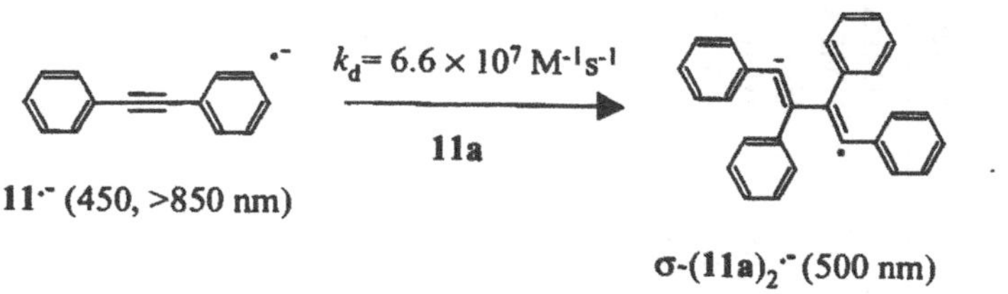

$$k_d = 6.6 \times 10^7 \, M^{-1}s^{-1}$$

11a

$11^{\cdot-}$ (450, >850 nm)

$\sigma\text{-}(11a)_2^{\cdot-}$ (500 nm)

Scheme 7 σ-Type or diene-type dimer$^{\cdot-}$ $(11a)_2^{\cdot-}$ from dimerization of $11a^{\cdot-}$ and $11a$.

Almost equivalent results were observed during PR and γ-R of **12**. Therefore, $12^{\cdot-}$ dimerizes with **12** to $12_2^{\cdot-}$ with a peak at 560 nm (Scheme 8). k_d for $12^{\cdot-}$ is almost equivalent with that for $11a^{\cdot-}$.

Formation of Intramolecular Dimer Radical Anions

The formation of intramolecular dimer$^{\cdot-}$, $\sigma\text{-}(11a)_2^{\cdot-}$ between $11a^{\cdot-}$ and $11a$ was found to occur in $13^{\cdot-}$ ($n = 2$–6) having two **11a**-chromophores linked by a di-to-hexa methylene (Scheme 9). $k_d^{intra} = (3.4–6.3) \times 10^5 \, sec^{-1}$ for the intramolecular dimerization was almost constant for $13^{\cdot-}$ ($n = 2$–6). The intermolecular dimerization of $13^{\cdot-}$ and **13** at [13] $= 1.0 \times 10^{-3}$ to 2.5×10^{-2} M was not observed on a time scale of a few μsec. This seems to be attributed to the 4-C_6H_4CCPh substituent of $13^{\cdot-}$ and **13** at the 4-position because of the electronic and steric effects.

The $\Delta O.D._{510}$ of $\sigma\text{-}(11a)_2^{\cdot-}$ in $13^{\cdot-}$ ($n = 2, 3, 5,$ and 6) was much smaller in comparison, while that of $\sigma\text{-}(11a)_2^{\cdot-}$ in $13^{\cdot-}$ ($n = 4$) was comparable to that of intermolecular $\sigma\text{-}(11a)_2^{\cdot-}$. The yield of $\sigma\text{-}(11a)_2^{\cdot-}$ in $13^{\cdot-}$ ($n=4$) was significantly high among $13^{\cdot-}$ ($n = 2$–6) and comparable with that of intermolecular $(11a)_2^{\cdot-}$. Although the $n=3$ rule is effective for the intramolecular π-interaction of Ar$-(CH_2)_n-$Ar in the excited state [112], $n=4$ is effective for that of intramolecular $\sigma\text{-}(11a)_2^{\cdot-}$ in $13^{\cdot-}$.

No direct evidence for the formation of $\pi\text{-}A_2^{\cdot-}$ suggests that $A^{\cdot-}$ is stabilized by the formation of $\sigma\text{-}A_2^{\cdot-}$ with a C$-$C bond between two sp carbons of $A^{\cdot-}$ and A and that the formation of $\sigma\text{-}A_2^{\cdot-}$ is responsible for the C$-$C triple bonds of $A^{\cdot-}$ and A.

3.2. Intramolecular Dimer Radical Anions in Diarylmethanols

The formation of an intramolecular dimer$^{\cdot-}$ of diaryllmethanol (Ar$_2$CHOH, **14a–e**) was found as the intermediates in the attachment of e_s^- (Scheme 10) during PR and γ-R in DMF and 77 K MTHF matrix, respectively [113].

Transient absorption spectrum showed a peak at 330 nm assigned to diphenylmethyl radical Ph$_2$CH$^{\cdot}$ immediately after e$^-$ (8 nsec) during PR of **14a** (5.0×10^{-3} M) in DMF. Transient absorption of $14a^{\cdot-}$ was not observed. On the other hand, the transient absorption spectrum with a peak at 328 nm assigned to $14b^{\cdot-}$ was observed immediately after e$^-$, and decayed with formation of peaks at 362 and 530 nm assigned to Ar$_2$CH$^{\cdot}$ and a byproduct from $14b^{\cdot-}$, respectively, during PR of **14b** (Fig. 9). This indicates that introduction of electron-withdrawing groups such as CN and CO$_2$CH$_3$ on the benzene ring retarded the C$-$O bond dissociation of $14^{\cdot-}$. **14b–e**$^{\cdot-}$ decayed slowly to give the corresponding Ar$_2$CH$^{\cdot}$ obeying a first-order kinetics ($k_r = (2–3) \times 10^5 \, sec^{-1}$) and τ of $14^{\cdot-}$ was independent of the mono- or disubstitution.

Table 6 Intermolecular Dimerizations of $A^{\cdot-}$ and A (A = 11 and 12)[a]

$A^{\cdot-}$	Substituents	λ_{max} [nm]		$\Delta O.D._{max}$		$[A_2^{\cdot-}]_{max}/[A^{\cdot-}]_0$	$\varepsilon_{max}[M^{-1}\,sec^{-1}]$ of $A^{\cdot-}$	$k_d[M^{-1}\,sec^{-1}]$	Solvent
		$A^{\cdot-}$	$A_2^{\cdot-}$	$A^{\cdot-}$	$A_2^{\cdot-}$				
11a$^{\cdot-}$	4-H	450	500	1.8	1.0	0.56	5.4×10^4	6.6×10^7	HMPA
11b$^{\cdot-}$	4-CH$_3$	450	510	1.4	0.9	0.66	4.1×10^4	5.7×10^7	HMPA
11c$^{\cdot-}$	4-C$_6$H$_5$	500	550	1.6	1.2	0.77	4.7×10^4	4.1×10^7	HMPA
11d$^{\cdot-}$	4-NO$_2$	440	490	0.95	0.09	0.10	2.7×10^4	9.0×10^6	HMPA
11e$^{\cdot-}$	4-OCH$_3$	440	500	1.7	0.15	0.09	5.2×10^4	1.0×10^7	HMPA
11f$^{\cdot-}$	4-F	440	500	1.8	0.25	0.15	5.4×10^4	1.7×10^7	HMPA
11g$^{\cdot-}$	(CH$_3$)$_5$	450	540	3.9	0.25	0.06	1.2×10^5	7.3×10^6	HMPA
11a$^{\cdot-}$	4-H	450	510	0.61	0.05	0.09	5.4×10^4	8.0×10^6	DMF
12$^{\cdot-}$	H	480	560	1.0	0.10	0.10	8.9×10^4	8.2×10^6	DMF

[a] $[A_2^{\cdot-}]_{max}/[A^{\cdot-}]_0 = (\Delta O.D._{max}$ at λ_{max} for $A_2^{\cdot-})/(\Delta O.D._{-0}$ at λ_{max} for $A^{\cdot-})$. ε_{max} was calculated from $\Delta O.D._{-0}$ at λ_{max} for $A^{\cdot-}$, $\varepsilon_{max} = 5.4 \times 10^4\,M^{-1}\,cm^{-1}$ for **11a**$^{\cdot-}$, $l = 0.4$ cm, and $[11^{\cdot-}]_0 = 8.3 \times 10^{-5}$ M in HMPA, while ε_{480} for **12**$^{\cdot-}$ was calculated from $\varepsilon_{450} = 5.4 \times 10^4\,M^{-1}\,cm^{-1}$ for **11a**$^{\cdot-}$, $l = 0.4$ cm, and $[11a^{\cdot-}]_0 = 2.8 \times 10^{-5}$ M in DMF. k_d for the dimerization of $A^{\cdot-}$ and A was calculated from the linear plot of k_f for the formation $A_2^{\cdot-}$ vs. [A].

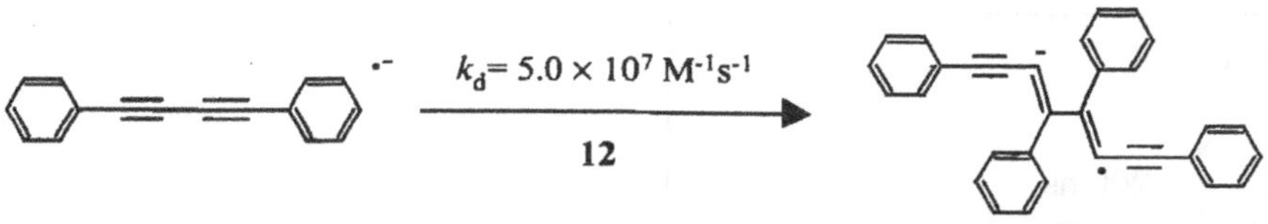
k_d= 5.0 × 10⁷ M⁻¹s⁻¹
12

$12^{•-}$ (480, >850 nm)

σ-$12_2^{•-}$ (560 nm)

Scheme 8 σ-Type or diene-type dimer$^{•-}$ $12_2^{•-}$ from dimerization of $12^{•-}$ and 12.

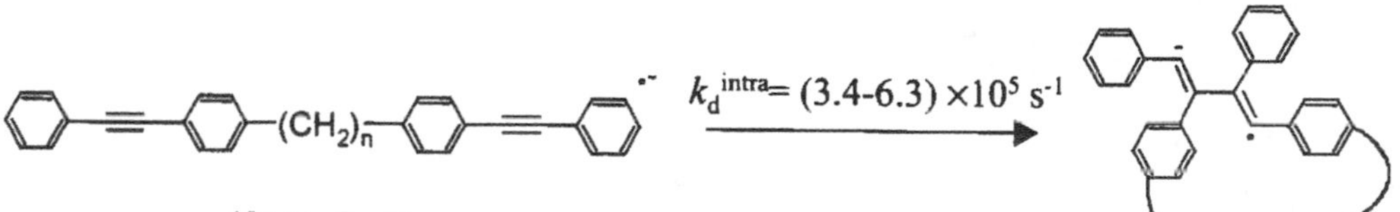
(CH2)n
k_d^{intra}= (3.4-6.3) ×10⁵ s⁻¹
(CH2)n

$13^{•-}$ (n= 2 - 6)
440-450, >850 nm

intramolecular σ-$(11a)_2^{•-}$
510 nm

Scheme 9 Intramolecular dimerization of $11a^{•-}$ and 11a in 13^{-}.

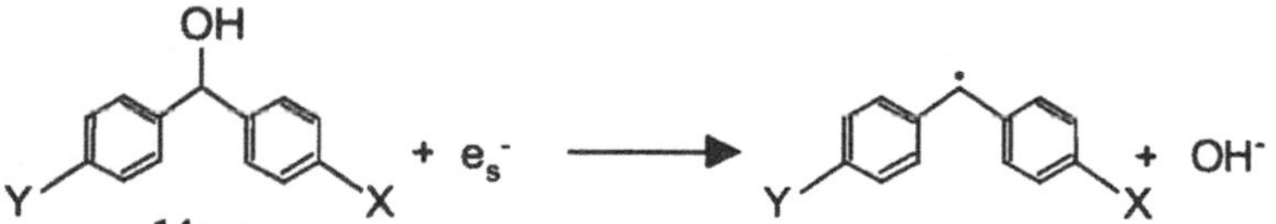
OH
Y
X
+ e_s^-
Y
X
+ OH⁻

14a-e

a: X=Y=H; **b:** X=Y=CN; **c:** X=Y=CO_2CH_3;
d: X=CN, Y=H; **e:** X=CO_2CH_3, Y=H

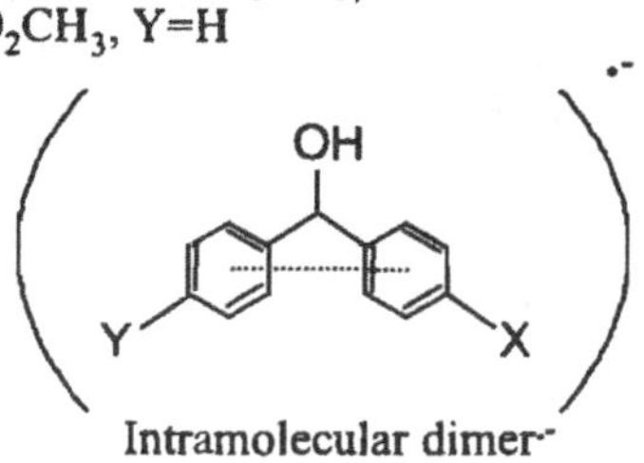
OH
Y
X

Intramolecular dimer$^{•-}$

Scheme 10 Intramolecular dimer^{-} during electron attachment of **14a–e**.

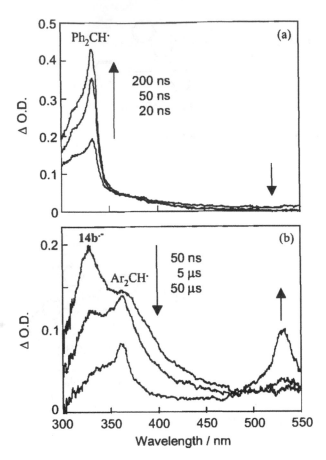

Figure 9 Transient absorption spectra of **14**$^{\cdot-}$ observed during PR of the DMF solutions of **14a** (a) and **14b** (b).

The time-resolved absorption spectra of e_s^- in DMF and **14b**$^{\cdot-}$ in the range of 800–1600 nm were obtained from the decay curve measured at every 100-nm step (Fig. 10). While the peak of e_s^- was at 1400 nm, that of **14b**$^{\cdot-}$ was at >1600 nm. Although the NIR absorption spectra peaked at 1780 and 1930 nm were observed during the γ-R of **14b** and **14c** (5×10^{-3} M), respectively, in MTHF at 77 K, no peak was observed in this region for **14d,e** (Fig. 11). From the structureless feature of the spectra similar to those reported for CR band of dimer$^{\cdot-}$[101,102,114,115], we assigned them to the CR band of the intramolecular dimer$^{\cdot-}$ of **14b,c**$^{\cdot-}$. From the peak λ, CR energies (ΔH_{CR}) were determined to be 8.0 and 7.4 kcal mol^{-1} for **14b**$^{\cdot-}$ and **14c**$^{\cdot-}$, respectively. The absence of CR band for mono-substituted alcohols **14d,e**$^{\cdot-}$ also supported the assignment of the intramolecular dimer$^{\cdot-}$ of **14b,c**$^{\cdot-}$.

Although an intramolecular singlet excimer (^{1}Ar$_2^*$) fluorescence of 1,3-bis(4-cyanophenyl)propan-1-ol was observed in MTHF at r.t., no CR band of 1,3-bis(4-cyanophenyl)-propan-1-ol$^{\cdot-}$ was observed during PR and γ-R. An unsymmetrical structure of 1,3-bis(4-cyanophenyl)propan-1-ol does not inhibit ^{1}Ar$_2^*$ formation [112]. Although two Np-chromophores have been suggested to form intramolecular Np$_2^{\cdot-}$ in 1,3-di-β-naphthylpropanes$^{\cdot-}$, no CR band of 1,3-di-α-naphthylpropane$^{\cdot-}$ and 1,3-di-1-pyrenyl-propane$^{\cdot-}$ was

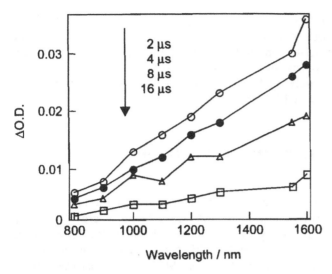

Figure 10 Transient absorption spectral change of **14b**$^{\cdot-}$ in the NIR region (CR band) obtained from decay curves during PR of **14b** in DMF.

observed during γ-R in 77 K MTHF matrix. The PR of 1,3-di-1-pyrenylpropane in THF and 1,3-di-β-naphthylpropan-1-ol in DMF at r.t. showed transient absorptions of pyrene$^{\cdot-}$ and Np$^{\cdot-}$, respectively, without the CR band of $Ar_2^{\cdot-}$, although PR of these compounds in DCE gave the corresponding $Ar_2^{\cdot+}$ of pyrene$^{\cdot+}$ and Np$^{\cdot+}$. $Ar_2^{\cdot-}$ of methylterephthalate-chromophore was assumed from the kinetic study on the anion radical transfer from methylterephthalate to 1,4-dicyanobenzene in diester of 1,n-alkanediols [116]. These suggest that $Ar_2^{\cdot-}$ formation may differ from the $n = 3$ rule for $^1Ar_2^*$ formation, probably due to the repulsion energy higher than ΔH_{CR} [112]. The 1,1-diarylmethyl structure is considered to be

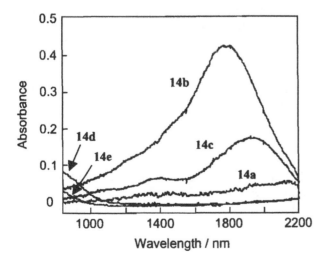

Figure 11 NIR absorption spectra of **14a–e**$^{\cdot-}$ during γ-R of **14a–e** in MTHF at 77 K. The spectra were subtracted by those before photoirradiation.

effective for $Ar_2^{\cdot-}$ formation being different from $^1Ar_2^*$ as suggested for intramolecular $^3Np_2^*$ [117]. While the sandwich-pair is preferred for $^1Ar_2^*$, the L-shaped conformation as in the most symmetric conformation of 1,1-di-α-naphthylmethane and cis-6b,12b-dihydroace-naphth(1,2-a)acenaphthylene (the Agosta dimer) could be favored for $^3Np_2^*$ [117] and the 1,1-diarylmethyl structure would be preferred for intramolecular $Ar_2^{\cdot-}$ of **1b,c**$^{\cdot-}$ rather than displaced sandwich-pair structure favored for $Ar_2^{\cdot+}$.

4. LIFETIMES OF RADICAL CATIONS AND ANIONS IN THE EXCITED STATE

Excitation of $M^{\cdot+/\cdot-}$ increases its electron acceptor and donor characteristics, respectively, because an electron is promoted from a low-energy MO to a higher-energy MO [1,5]. Electron transfer (ET) reactions catalyzed by $M^{\cdot+*/\cdot-*}$ have been studied at electrode surfaces [20,118] and in solution [23,25,26,33,119]. $M^{\cdot+*/\cdot-*}$ is usually nonluminescent in solution, although weak luminescences of thermally stable $M^{\cdot+/\cdot-}$ such as the anthra-quinone$^{\cdot-}$ and N-methylphenothiazine$^{\cdot+}$ have been reported at 470–580 nm [120,121] and at 600 nm [25], respectively. Breslin and Fox [33] have reported the weak luminescence of two substituted triarylamine$^{\cdot+}$ with luminescence peaks at 790 and 803 nm and with $\varphi_f = 10^{-5}$, while the other thermally stable $M^{\cdot-/\cdot-}$ investigated was nonluminescent. They have concluded that the internal conversion from the lowest excited doublet state (D_1) to the ground doublet state (D_0) proceeds as the primary deactivation pathway for $M^{\cdot+/\cdot-}$ (D_1) because of the small energy between D_1 and D_0 states.

4.1. Lifetime of Stilbene Radical Cation in the Excited State ($St^{\cdot+*}$) by the Hole Transfer Quenching

It is assumed that the photochemical c→t one-way isomerization of c-$St^{\cdot+}$ to t-$St^{\cdot+}$ occurs in rigid matrices at 77 K [40,122] and in solution at r.t. on the basis of LPF of c-$St^{\cdot+}$ formed by using PR in DCE [83] or secondary ET in acetonitrile [32]. The photochemical c→t isomerization has been reported to take place in the second excited doublet state (D_2) but not in c-$St^{\cdot+}$ (D_1) [83]. Because $St^{\cdot+}$ (D_2) is nonluminescent, τ of $St^{\cdot+}$ (D_2) must be determined by using other experiments such as the excitation energy (E_{ex}) quenching or hole transfer (HT) quenching. Because $St^{\cdot+}$ (D_2) has a higher E_{ox} than $St^{\cdot+}$ (D_0) [1,5], HT occurs from $St^{\cdot+}$ (D_2) to an appropriate quencher (Q). τ of $St^{\cdot+}$ (D_2), as well as the 1,2-diphenylcyclo-1-butene$^{\cdot+}$ (DPCB$^{\cdot+}$), as a structurally constrained derivative of c-St, was estimated from selective HT quenching of $St^{\cdot+}$ (D_2) using anisole (AN) as Q through the PR-LFP combined method [43,83,123–125].

Lifetime of t-$St^{\cdot+*}$

It is well established that c-$St^{\cdot+}$ and t-$St^{\cdot+}$ are selectively generated during the PR of c-St and t-St in DCE at r.t., and they show bands at 420–580 and 700–850 nm assigned to $D_2\leftarrow D_0$ and $D_1\leftarrow D_0$ transitions, respectively, and peaks at 515 and 780 nm for c-$St^{\cdot+}$ and 480 and 760 nm for t-$St^{\cdot+}$ [42,73,83]. Excitation of $St^{\cdot+}$ with hv_{532} generates $St^{\cdot+}$ (D_2) ($St^{\cdot+*}$) with E_{ex} = 50 and 53 kcal mol^{-1} for c-$St^{\cdot+*}$ and t-$St^{\cdot+*}$, respectively, that were calculated from the red edges of bands at 420–580 nm.

Irradiation of t-$St^{\cdot+}$ with hv_{532} exhibited no change in the transient absorption spectra and time profiles of ΔO.D.$_{480}$, where t-$St^{\cdot+}$ shows a peak (Fig. 12a). Therefore t-

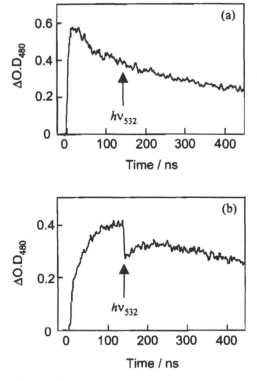

Figure 12 Kinetic traces illustrating time profiles of ΔO.D.$_{480}$ as a function of time during PR–LFP experiment of t-St (5×10^{-3} M) in the absence (a) and presence (b) of AN (1 M) in DCE.

$St^{\cdot+}*$ does not isomerize to c-$St^{\cdot+}$ (D$_0$) with photoirradiation, but decays to t-$St^{\cdot+}$ (D$_0$) through internal conversion at k^t_{ic}(Scheme 11).

The photoirradiation of t-$St^{\cdot+}$ in the presence of AN caused a decrease of ΔO.D.$_{480}$ immediately after $h\nu_{532}$ (Fig. 12b). $|\Delta$O.D.$_{480}|$ increased with increasing [AN]. It is obvious that t-$St^{\cdot+}*$ is quenched by AN via HT quenching to give t-St and AN$^{\cdot+}$ with the bimolecular HT rate constant of k_{ht} (Scheme 11). The recovery of ΔO.D.$_{480}$ after $h\nu_{532}$ corresponds to HT from AN$^{\cdot+}$ to t-St to yield AN and t-$St^{\cdot+}$ [73], which occured at $k_{diff} = 7.8 \times 10^9$ M^{-1} sec^{-1}. The $k_{ht} = k_{diff}$ can be assumed because of the large exothermicity of

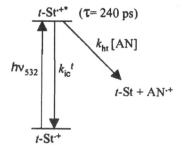

Scheme 11 Hole transfer quenching of t-$St^{\cdot+}*$ by AN.

$\Delta G = -25$ kcal mol^{-1} for HT from St$^{\cdot+}$* to AN. From the linear Stern–Volmer plots of Y_{-t}^{-1} vs. [AN]$^{-1}$, where Y_{-t} is the chemical yield of $[t\text{-St}^{\cdot+}]_{\text{disapp}}$, τ_t, and $k_{\text{ic}}{}'$ for $t\text{-St}^{\cdot+}$ (D$_2$) were calculated to be 240 ± 50 psec and $(4.1 \pm 1.1) \times 10^9$ sec^{-1}, respectively.

Lifetime of c-St$^{\cdot+}$* and DPCB$^{\cdot+}$* (DPCB = 1,2-diphenylcylclobutene)

In contrast to $t\text{-St}^{\cdot+}$*, $c \rightarrow t$ one-way isomerization of $c\text{-St}^{\cdot+}$* to $t\text{-St}^{\cdot+}$ was observed within $h\nu_{532}$ (5 nsec) [32,83]. From the transient spectral changes and time profiles of ΔO.D.$_{515}$ and ΔO.D.$_{480}$ of $c\text{-St}^{\cdot+}$ and $t\text{-St}^{\cdot+}$, respectively (Fig. 13), the photochemical conversion of $c\text{-St}^{\cdot+}$ per $h\nu_{532}$ is $(80 \pm 15)\%$ at $[c\text{-St}^{\cdot+}]_0 = 1.5 \times 10^{-5}$ M before photo-irradiation, and the chemical yield of $t\text{-St}^{\cdot+}$ from $c\text{-St}^{\cdot+}$ (D$_2$) is $(75 \pm 15)\%$. The remaining $(25 \pm 15)\%$ of $c\text{-St}^{\cdot+}$ that disappeared is considered to convert to c-St, t-St, or a radical cation as a product such as dihydrophenanthrene.

The φ of photochemical $c \rightarrow t$ isomerization of $c\text{-St}^{\cdot+}$ to $t\text{-St}^{\cdot+}$ was subsequently determined to be 0.65 ± 0.15 from $c\text{-St}^{\cdot+}$ that disappeared (φ_{-c}) and 0.49 ± 0.12 from $t\text{-St}^{\cdot+}$ that formed (φ_t). Taking into account φ_{-c} and φ_t, as well as the conversion of $(80 \pm 15)\%$ for $c\text{-St}^{\cdot+}$ per $h\nu_{532}$, $c\text{-St}^{\cdot+}$* isomerizes to $t\text{-St}^{\cdot+}$ at $(49 \pm 12)\%$, converts to other products at $(16 \pm 4)\%$, and deactivates to $c\text{-St}^{\cdot+}$ at $(35 \pm 8)\%$ yield.

The transient phenomena of $c\text{-St}^{\cdot+}$* that involved the isomerization and HT quenching of $c\text{-St}^{\cdot+}$* using AN are shown in Scheme 12. According to Scheme 12, the linear Stern–Volmer plots of $\tau_t/\tau_t^{\text{AN}}$ vs. [AN] afforded $\tau_c = 120 \pm 30$ psec being one-half of τ_t at $k_{\text{ht}} = k_{\text{diff}} = 7.8 \times 10^9$ M^{-1} sec^{-1}. k_i, k_p, and $k_{\text{ic}}{}^c$ were determined to be $(4.1 \pm 1.0) \times 10^9$, $(1.3 \pm 0.3) \times 10^9$, and $(2.9 \pm 0.7) \times 10^9$ sec^{-1}, respectively.

Because 1,2-diphenylcyclo-1-butene (DPCB) has a rigid planar structure with the c-St-chromophore structurally constrained by the cyclobutene ring, and because it is stable for geometrical isomerizations, TPCB$^{\cdot+}$ is expected to have the same rigid planar structure. Pulse radiolysis of DPCB in DCE produces DPCB$^{\cdot+}$, which shows bands at

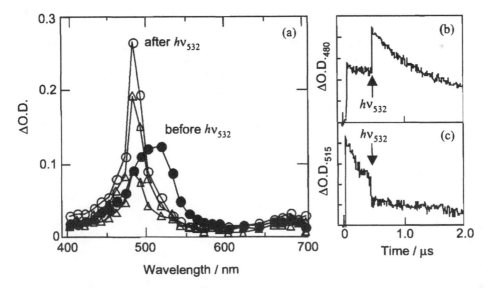

Figure 13 Transient absorption spectra recorded before $h\nu_{532}$ (O) and immediately after (●) and 200 nsec (△) and 1 μs (▲) after $h\nu_{532}$ during PR–LFP of c-St (5×10^{-3} M) in DCE (a). Kinetic traces illustrating time profiles of ΔO.D.$_{480}$ (b) and ΔO.D.$_{515}$ (c) as a function of time after e$^-$.

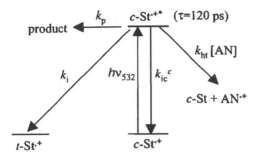

Scheme 12   Reactions of c-St$^{\cdot+}$* involving isomerization, ic, and HT quenching.

420–560 and 700–850 nm that are assigned to $D_2 \leftarrow D_0$ and $D_1 \leftarrow D_0$ transitions, respectively [42,73,83]. Excitation of DPCB$^{\cdot+}$ at 532 nm exhibited no change in the transient absorption. Therefore, DPCB$^{\cdot+}$* is deactivated to DPCB$^{\cdot+}$ (D_0) within $h\nu_{532}$, similarly to t-St$^{\cdot+}$* (Scheme 11). Photoirradiation of DPCB$^{\cdot+}$ in the presence of p-methylanisole (MA) caused a decrease of ΔO.D.$_{480}$ immediately after $h\nu_{532}$ with an increasing [MA]. It is obvious that DPCB$^{\cdot-}$* is quenched by MA via HT quenching to give DPCB and MA$^{\cdot+}$. From the linear Stern–Volmer plots of Y_{-CB}^{-1} vs. $[MA]^{-1}$, τ_{DPCB} and k_{ic}^{DPCB} of DPCB$^{\cdot+}$* were calculated to be 380 ± 30 psec and $(2.6 \pm 0.2) \times 10^9$ sec^{-1}, respectively, at $k_{ht} = k_{diff} = 7.8 \times 10^9$ M^{-1} sec^{-1}. Consequently, it is concluded that τ_{DPCB} is longer than τ_c and τ_t, whereas k_{ic}^{DPCB} is comparable with k_{ic}^c and k_{ic}^t (Table 7).

Comparison of c-St$^{\cdot+}$*, t-St$^{\cdot+}$*, and DPCB$^{\cdot+}$*

Similar to most M$^{\cdot+}$*, c-St$^{\cdot+}$*, t-St$^{\cdot+}$*, and CB$^{\cdot+}$* are nonluminescent. Isomerization of c-St$^{\cdot+}$* to t-St$^{\cdot+}$, which proceeds via twisting about the central C=C double bond as a main process, competes with the internal conversion of c-St$^{\cdot+}$*. However, t-St$^{\cdot+}$* and DPCB$^{\cdot+}$* are deactivated to the D_0 state without any other reactions. Although t-St$^{\cdot+}$* could slightly twist, it does not isomerize to c-St$^{\cdot+}$. The twisting is faster in c-St$^{\cdot+}$* than in t-St$^{\cdot+}$* because of the interactions between the Ph groups of c-St. Because the twisting is severely hindered by structural constraints in DPCB, DPCB is unable to change its initial planar rigid structure. Both DPCB$^{\cdot+}$* and DPCB$^{\cdot+}$ (D_0) probably have the rigid planar structure. The energy levels of c-St$^{\cdot+}$*, t-St$^{\cdot+}$*, and DPCB$^{\cdot+}$* are assumed to be almost equivalent because of the similar $E_{p/2}^{ox}$ for c-St, t-St, and DPCB, as well as the similar E_{ex} of c-St$^{\cdot+}$*, t-St$^{\cdot+}$*, and DPCB$^{\cdot+}$* (Table 7). It is suggested that c-St$^{\cdot+}$*, t-St$^{\cdot-}$*, and DPCB$^{\cdot+}$* exist in potential minima, and that the order $\tau_c < \tau_t < \tau_{DPCB}$ is mainly attributed

Table 7   Properties of t-St$^{\cdot+}$, c-St$^{\cdot+}$, and DPCB$^{\cdot+}$ in the D_2 State: Lifetime (τ), Rate Constants for Internal Conversion (k_{ic}), c–t Isomerization (k_i), and Product Formation (k_p), and Excitation Energy (E_{ex})

Radical Cations in D_2	t-St$^{\cdot+}$*	c-St$^{\cdot+}$*	CB$^{\cdot-}$*
τ[psec]	240 ± 50	120 ± 30	380 ± 30
k_{ic}[sec^{-1}]	$(4.1 \pm 1.1) \times 10^9$	$(2.9 \pm 0.7) \times 10^9$	$(2.6 \pm 0.2) \times 10^9$
k_i[sec^{-1}]	0	$(4.1 \pm 1.0) \times 10^9$	0
k_p[sec^{-1}]	0	$(1.3 \pm 0.3) \times 10^9$	0
E_{ex}[kcal mol^{-1}]	53	50	51

to the isomerization via twisting only in c-St$^{\cdot+}$*. It is clear that the twisting is favorable in c-St$^{\cdot+}$*, slightly favorable in t-St$^{\cdot+}$*, but not at all in DPCB$^{\cdot+}$*.

Considering φ_f, τ, and c–t isomerization for St*, it is clearly suggested that the transient phenomena of St$^{\cdot+}$ (D$_2$) cannot be assumed on the basis of those for St(S$_1$). Judging from $\tau_c \sim 120$ psec and $\tau_t \sim 240$ psec, a diabatic process is proposed for the isomerization of c-St$^{\cdot+}$* to t-St$^{\cdot+}$. It has been proposed the analogous process in which c-St$^{\cdot+}$* avoidably crosses with c-St$^{\cdot+}$ (D$_1$), followed by an allowed crossing to t-St$^{\cdot+}$ (D$_0$) at the 40° twisted geometry against a planar c-St$^{\cdot+}$*, where the D$_2$, D$_1$, and D$_0$ states closely lie, on the basis of MO calculations [32]. The barrier for the twisting of c-St$^{\cdot+}$* is estimated to be 0.32 kcal mol^{-1}, if k_i values are assumed to be 5.3×10^9 sec^{-1} for the largest limit and 2.9×10^9 sec^{-1} for the smallest limit at 300 and 140 K, respectively, on the basis of the error limits ($\pm 30\%$) for φ and k_i. This suggests that c-St$^{\cdot+}$* is stabilized by twisting at a shallow potential minimum, and that conversion to another product can compete with isomerization. Consequently, the diabatic isomerization from c-St$^{\cdot+}$* to t-St$^{\cdot+}$ is strongly expected to take place via an avoidable crossing from c-St$^{\cdot+}$ (D$_2$) to c-St$^{\cdot+}$ (D$_1$), and then to t-St$^{\cdot+}$ (D$_0$) with a small barrier.

In conclusion, selective HT quenching of St$^{\cdot+}$* and CB$^{\cdot+}$* by AN and MA yielded $\tau_c = 120$, $\tau_t = 240$, and $\tau_{DPCB} = 380$ psec, which is attributed to the isomerization via the twisting only in c-St$^{\cdot+}$*. It is also suggested that diabatic isomerization from c-St$^{\cdot+}$* to t-St$^{\cdot+}$ proceeds via an avoidable crossing of the D$_2$, D$_1$, and D$_0$ states. To determine the τ of M$^{\cdot+}$* with $k_{ht} = k_{diff}$ can be essentially applied to any M$^{\cdot+}$*, although it is necessary to find an appropriate molecule as a selective hole quencher. Moreover, the τ of M$^{\cdot+}$* must be longer than 50 psec, because [AN] is 2 M at the maximum to keep the solvent property.

4.2. Lifetime of St$^{\cdot-}$* by the Electron Transfer Quenching

Characterizations of St$^{\cdot-}$ (D$_2$) are necessary to elucidate the isomerization of St$^{\cdot-}$ (D$_2$), which is nonluminescent, similarly to radical anions of aromatic compounds in the excited states [25,33,118,120,121,126–130]. Selective electron transfer (ET) quenching of St$^{\cdot-}$ (D$_2$) was performed using biphenyl (Bp) as an electron acceptor and τ of St$^{\cdot-}$ (D$_2$) was estimated using a PR–LFP combined method [43,46,83,123–125,131,132].

It is well established that c-St$^{\cdot-}$ and t-St$^{\cdot-}$ are selectively generated during PR of c-St and t-St in DMF at r.t., and these show bands at 420–580 and 700–850 nm which are assigned to D$_2 \leftarrow$ D$_0$ and D$_1 \leftarrow$ D$_0$ transitions, respectively, and peaks at 515 and 780 nm for c-St$^{\cdot-}$ and 500 and 760 nm for t-St$^{\cdot-}$ [42,49,83,133]. Excitation of St$^{\cdot-}$ with $h\nu_{532}$ generates St$^{\cdot-}$(D$_2$) (St$^{\cdot-}$*) with $E_{ex} = 45$ and 48 kcal mol^{-1} for c-St$^{\cdot-}$* and t-St$^{\cdot-}$*, respectively. Irradiation of t-St$^{\cdot-}$ with $h\nu_{532}$ caused a decrease of ΔO.D. immediately after $h\nu_{532}$, while no spectral shift was observed.

Photoelectron Ejection from t-St$^{\cdot-}$* and Electron Transfer Quenching of St$^{\cdot-}$*

The differences in ΔO.D. before and after $h\nu_{532}$ ($|\Delta$O.D.$|$) decreased with increasing [t-St]. The spectrum of t-St$^{\cdot-}$ was recovered on a time scale of 10–30 nsec after $h\nu_{532}$, depending on [t-St]. These results show that photoelectron ejection from t-St$^{\cdot-}$ occurs and that the electron is captured by t-St. It is also indicated that t-St$^{\cdot-}$* does not isomerize to c-St$^{\cdot-}$ (D$_0$), but decays to t-St$^{\cdot-}$ (D$_0$) with k_{ic}^t of t-St$^{\cdot-}$*. $\varphi_e^t = 0.06 \pm 0.02$ of photoelectron ejection from t-St$^{\cdot-}$ was measured.

The ET quenching of c-St$^{\cdot-}$ (D$_2$) and t-St$^{\cdot-}$ (D$_2$) (c-St$^{\cdot-}$* and t-St$^{\cdot-}$*) by an ET quencher Bp was examined. Transient absorption spectra composed of bands of St$^{\cdot-}$ and

$Bp^{\cdot-}$ with peaks at 405 and 630 nm at the ratio of $[Bp]/[St] = 2$–100 were observed immediately after e^- during PR of mixture of St and Bp with $[St] = 5 \times 10^{-3}$ M and $[Bp] = 1 \times 10^{-2}$–5×10^{-1} M (Fig. 14). The transient absorption spectrum changed to that of $St^{\cdot-}$ on the time scale of 50–200 nsec after e^- depending on $[St]$, because the exergonic ET from $Bp^{\cdot-}$ to St ($\Delta G = -11$ kcal mol^{-1}) proceeds stoichiometrically at k_{diff} [134,135]. When t-$St^{\cdot-}$ was irradiated with $h\nu_{532}$, $\Delta O.D._{500}$ decreased immediately after $h\nu_{532}$, and $\Delta O.D._{405}$ increased with an increase of $[Bp]$ (Fig. 15). This indicates that t-$St^{\cdot-}*$ is quenched by Bp via ET to give t-St and $Bp^{\cdot-}$ within $h\nu_{532}$. The decay in $\Delta O.D._{405}$ and recovery of $\Delta O.D._{500}$ after $h\nu_{532}$ correspond to ET from $Bp^{\cdot-}$ to t-St to yield Bp and t-$St^{\cdot-}$ at $k_{et} = k_{diff}$.

Electronic transfer quenching of t-$St^{\cdot-}*$ proceeds at $k_{et} = k_{diff} = 7.1 \times 10^9$ M^{-1} sec^{-1} in DMF. Similar to radical ion pair, $(D^{\cdot+}/A^{\cdot-})_{solv}$ formed during ET between donor (D) and acceptor (A) molecules in the excited singlet or triplet state, it is suggested that ET quenching initially gives $(St/Bp^{\cdot-})_{solv}$ with competition of the internal conversion of $St^{\cdot-}*$ to $St^{\cdot-}$. $(St/Bp^{\cdot-})_{solv}$ then undergoes solvent separation into St and $Bp^{\cdot-}$ at k_{sep} or returns to $St^{\cdot-}$ and Bp via back ET at k_{-et}. Therefore, the fraction of free $Bp^{\cdot-}$ formed is represented by $R = k_{sep}/(k_{-et} + k_{sep})$.

Transient Phenomena of t-$St^{\cdot-}*$ and c-$St^{\cdot-}*$

Photoelectron ejection from $ArH^{\cdot-}$ in nonpolar solvents has been interpreted in terms of autoionizing excited states with $\tau = 10^{-14}$ sec, which can also be internally converted to the D_0 states [34,35,136,137]. It is assumed that t-$St^{\cdot-}*$ in the vibrational excited state $((t$-$St^-*)^\ddagger)$ undergoes vibrational relaxation on the time scale of 10^{-14} sec to yield t-$St^{\cdot-}*$, which has a long enough lifetime $\tau_{t*} = (k_d')^{-1}$ to be quenched by Bp via ET involving $(t$-$St/Bp^{\cdot-})_{solv}$ (Scheme 13). $\varphi_e^t = 0.06 \pm 0.02$ suggests that $(t$-$St^{\cdot-}*)^\ddagger$ undergoes photoelectron ejection and vibrational relaxation to t-$St^{\cdot-}*$ in 6% and 94% yields, respectively. The experimental results show that 0–14% of t-$St^{\cdot-}*$ is quenched by Bp depending on $[Bp]$, while the rest of t-$St^{\cdot-}*$ undergoes internal conversion to the D_0 state. From the linear plots of $(\varphi_{-t}^{Bp} - \varphi_e^t)^{-1}$ or

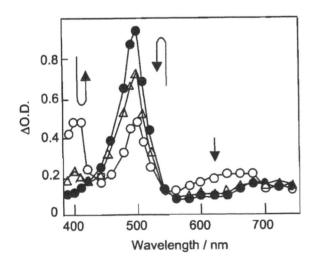

Figure 14 Transient absorption spectra recorded immediately after e^- (O), and before (●) and immediately after (△) $h\nu_{532}$ during PR–LFP of a mixture of t-St (5×10^{-3} M) and Bp (1×10^{-1} M) in DMF. The delay time of $h\nu_{532}$ from e^- was 75 nsec.

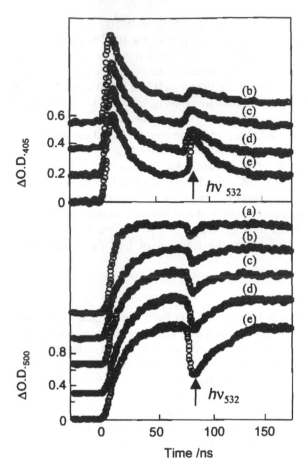

Figure 15 Kinetic traces illustrating the time profiles of $\Delta O.D._{405}$ and $\Delta O.D._{500}$ as a function of time after e^- during PR–LFP of mixtures of t-St and Bp. $[t$-St$] = 5 \times 10^{-3}$, $[Bp] = 0$ (a), 2×10^{-2} (b), 5×10^{-2} (c), 1×10^{-1} (d), and 5×10^{-1} M (e). The delay time of $h\nu_{532}$ from e^- was 75 nsec.

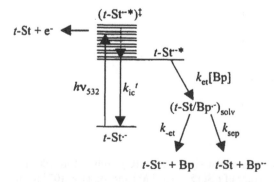

Scheme 13 Electron transfer quenching of t-St^{-*} by Bp.

φ_{Bp}^{-1} vs. $[Bp]^{-1}$, τ_t and k_{ic}^t were estimated to be 2.5 ± 0.7 nsec and $(4.0 \pm 1.2) \times 10^8$ sec^{-1}, respectively. Because $R = k_{sep}/(k_{-et} + k_{sep}) = 0.15 \pm 0.4$ is calculated from the intercept and $\varphi_t = 0.94$, the back ET with $\Delta G = -11$ kcal mol^{-1} occurs six times faster than the solvent separation within $(t\text{-St}/Bp^{\bullet-})_{solv}$.

Isomerization of $c\text{-St}^{\bullet-*}$ to $t\text{-St}^{\bullet-}$ was observed within $h\nu_{532}$. The time profiles of $\Delta O.D._{520}$ and $\Delta O.D._{500}$ assigned to $c\text{-St}^{\bullet-}$ and $t\text{-St}^{\bullet-}$, respectively (Fig. 16), demonstrated that the photochemical conversion of $c\text{-St}^{\bullet-}$ per $h\nu_{532}$ is 49% and that the chemical yield of $[t\text{-St}^{\bullet-}]_{form}$ from $[c\text{-St}^{\bullet-}]_{cons}$ is 67%. The remaining 33% of $c\text{-St}^{\bullet-}$ consumed is considered to give c-St via photoelectron ejection. The φ's of $[c\text{-St}^{\bullet-}]_{cons}$ and $[t\text{-St}^{\bullet-}]_{form}$ were measured to be $\varphi_{-c} = 0.21 \pm 0.05$ and $\varphi_t = 0.14 \pm 0.03$, respectively. Photoelectron ejection of $(c\text{-St}^{\bullet-*})^{\ddagger}$ occurs competitively with vibrational relaxation to $c\text{-St}^{\bullet-*}$, and therefore, φ_e^c was calculated from $\varphi_e^c = \varphi_{-c} - \varphi_t$ to be 0.07 ± 0.02, which is similar to that for $(t\text{-St}^{\bullet-*})^{\ddagger}$, $\varphi_e^t = 0.06 \pm 0.02$. Summarizing, $(c\text{-St}^{\bullet-*})^{\ddagger}$ undergoes photoelectron ejection and vibrational relaxation to $c\text{-St}^{\bullet-*}$ in 7% and 93% yields, and the isomerization and internal conversion of $c\text{-St}^{\bullet-*}$ occur in 15% and 85% yields, respectively.

After the exergonic ET from $Bp^{\bullet-}$ to c-St yielding Bp and $c\text{-St}^{\bullet-}$ stoichiometrically at k_{diff}, $c\text{-St}^{\bullet-}$ was irradiated with $h\nu_{532}$ during the PR–LFP of mixture of c-St and Bp (Fig. 17). $\Delta O.D._{520}$ decreased and $\Delta O.D._{405}$ increased immediately after $h\nu_{532}$, while $\Delta O.D._{500}$ increased on the time scale of 10–100 nsec depending on [Bp] (Fig. 18). Based

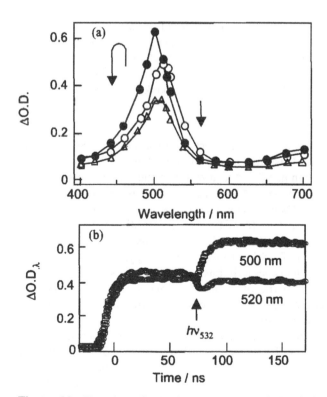

Figure 16 Transient absorption spectra recorded before $h\nu_{532}$ (O) and immediately (●) and 1.4 μsec (△) after $h\nu_{532}$ during PR–LFP of c-St (5×10^{-3} M) in DMF (a). Kinetic traces illustrating time profiles of $\Delta O.D._{500}$ and $\Delta O.D._{520}$ as a function of time after e$^-$ (b). The delay time of $h\nu_{532}$ from e$^-$ was 70 nsec.

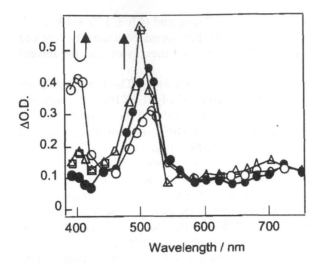

Figure 17 Transient absorption spectra recorded immediately after e⁻ (O), before (●), and immediately after (Δ) $h\nu_{532}$ during PR–LFP of c-St (5×10^{-3} M) and Bp (1×10^{-1} M) mixture in DMF. The delay time of $h\nu_{532}$ from e⁻ was 60 nsec.

on Scheme 14, the linear plot of φ_{Bp}^{-1} vs. $[Bp]^{-1}$ was obtained. τ_c was calculated to be 1.5 ± 0.4 nsec and found to be shorter than τ_t, from $k_{et}\tau_c$, assuming $k_{et} = k_{diff}$. On the basis of $k_i : k_{ic}^c = 1{:}5.7$ and τ_c, k_i and k_{ic}^c were calculated to be $(1.0 \pm 0.3) \times 10^8$ and $(5.6 \pm 1.7) \times 10^8$ sec^{-1}, respectively. $R = k_{sep}/(k_{-et} + k_{sep}) = 0.18 \pm 0.4$ is almost equal to R for $(t\text{-St/Bp}^{\cdot-})_{solv}$ because of the nearly equivalent $\Delta G = -11$ kcal mol^{-1} for the back ET in $(St/Bp^{\cdot-})_{solv}$.

Comparisons Between $St^{\cdot-}$* and $St^{\cdot+}$*

Isomerization of c-St$^{\cdot-}$* to t-St$^{\cdot-}$ competes with the internal conversion of c-St$^{\cdot-}$*, while only internal conversion occurs in t-St$^{\cdot-}$*. Isomerization of c-St$^{\cdot-}$* to t-St$^{\cdot-}$ proceeds via the twisting which is faster in c-St$^{\cdot-}$* than in t-St$^{\cdot-}$* as a result of the interactions between the Ph groups of c-St. It is suggested that c-St$^{\cdot-}$* and t-St$^{\cdot-}$* exist in potential minima, that the internal conversion from D_2 to D_0 proceeds with $k_{ic}^c \sim k_{ic}^t = (4.0{-}5.6) \times 10^8$ sec^{-1}, and that only c-St$^{\cdot-}$* undergoes isomerization to t-St$^{\cdot-}$ via the twisting with $k_i = (1.0 \pm 0.3) \times 10^8$ sec^{-1}. A diabatic process similar to that for the $c{\to}t$ isomerization of c-St$^{\cdot+}$* to t-St$^{\cdot+}$* is proposed for the isomerization of c-St$^{\cdot-}$* to t-St$^{\cdot-}$. It is suggested that c-St$^{\cdot-}$* is twisted at a potential minimum which undergoes a diabatic crossing to t-St$^{\cdot-}(D_0)$ having a twisting form without a large barrier. On the other hand, a planar t-St$^{\cdot-}$* exists in a potential minimum with the relatively large barrier for the twisting and undergoes internal conversion to t-St$^{\cdot-}(D_0)$ having a planar form.

Photoelectron ejection occurs in $(St^{\cdot-}*)^{\ddagger}$, while $c{\to}t$ isomerization occurs in c-St$^{\cdot-}$* and c-St$^{\cdot+}$*. However, the k_i of c-St$^{\cdot-}$* is 41 times smaller than that of c-St$^{\cdot+}$*, and φ_i of c-St$^{\cdot-}$* is 3.5 times smaller than that of c-St$^{\cdot+}$* (Table 8). k_i is 5.6 times smaller than k_{ic} of c-St$^{\cdot-}$*, while k_i is slightly greater than k_{ic} of c-St$^{\cdot+}$*. The k_{ic} of St$^{\cdot-}$* is an order of magnitude smaller than that of St$^{\cdot+}$*. These differences are responsible for the potential surfaces of St$^{\cdot-}(D_2)$ and St$^{\cdot-}(D_0)$ and their relations, which are probably different from those of St$^{\cdot+}$ [138]. It may be considered that the D_2 and D_0 states lie less closely in St$^{\cdot-}$ than in St$^{\cdot+}$. $\tau = 1.0$ and $70{-}83$ psec for c-St and t-St in the S_1 states have been reported

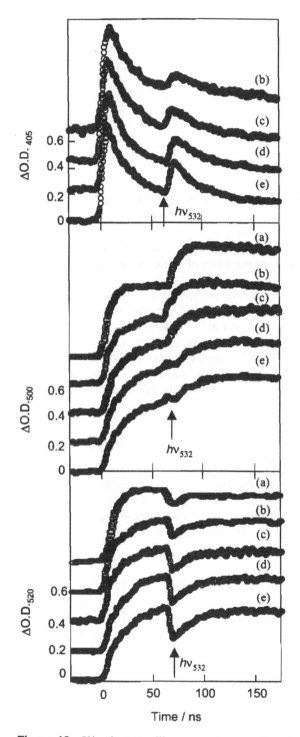

Figure 18 Kinetic traces illustrating time profiles of ΔO.D.$_{405}$, ΔO.D.$_{500}$, and ΔO.D.$_{520}$ as a function of time after e^- during PR–LFP of mixtures of c-St and Bp. [c-St] = 5.0×10^{-3}, [Bp]=0 (a), 1×10^{-2} (b), 2×10^{-2} (c), 5×10^{-2} (d), and 1×10^{-1} M (e). The delay time of hv_{532} from e^- was 60 nsec.

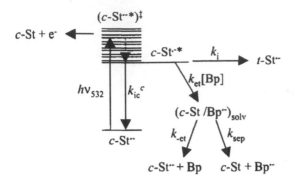

Scheme 14   Reactions of c-St$^{\bullet-}$* involving isomerization, ic, and ET quenching.

from $S_1 \rightarrow S_0$ fluorescence measurement [139–145]. It is clearly suggested that the transient phenomena of $M^{\bullet-}(D_2)$ cannot be assumed on the basis of those of neutral molecules in the S_1 state or $M^{\bullet+}(D_2)$.

In conclusion, Bp acts as a selective electron quencher for St$^{\bullet-}$* via ET quenching. Assuming that $k_{et} = k_{diff}$, τ_c and τ_t are estimated to be 1.5 and 2.5 nsec, respectively, which are longer than $\tau = 120$ and 240 psec for c-St$^{\bullet+}$* and t-St$^{\bullet+}$* (Table 8). The order of $\tau_c < \tau_t$ is attributed to the isomerization via the twisting only in c-St$^{\bullet-}$*, because twisting is favorable in c-St$^{\bullet-}$* compared with t-St$^{\bullet-}$*. It is suggested that diabatic isomerization from c-St$^{\bullet-}$* to t-St$^{\bullet-}$ proceeds via crossing of the D_2 and D_0 states. On the basis of the present results, the method can essentially work to determine the lifetimes of any $M^{\bullet-}$* by the selective ET quenching assuming $k_{et} = k_{diff}$.

4.3.   Lifetime of Radical Anions of Dicyanoanthracene, Phenazine, and Anthraquinone in the Excited States

The lifetimes of $M^{\bullet-}$*, such as 9,10-dicyanoanthracene (DCA), phenazine (PZ), anthraquinone (AQ), and substituted anthraquinones [RAQ = 2-methylanthraquinone (2-MAQ), 2-*tert*-butylanthraquinone (2-BAQ), 1-chloroanthraquinone (1-CAQ), and 2-chloroanthraquinone (2-CAQ)], have been measured by the selective ET quenching of $M^{\bullet-}$* by an electron quencher (Q) during PR–LFP in DMF [47].

Selective Electron Transfer Quenching of Radical Anions in the Excited State

The transient absorption spectra of $M^{\bullet-}$ recorded at $t = 20$ nsec after e$^-$ showed three peaks at 510, 640, and 710 nm for DCA$^{\bullet-}$, and one strong peak at 530–560 nm for PZ$^{\bullet-}$, AQ$^{\bullet-}$, and RAQ$^{\bullet-}$. When $M^{\bullet-}$ was irradiated with $h\nu_{532}$ at $t = 200$ nsec after e$^-$, no

Table 8   Properties of t-St$^{\bullet+}$, c-St$^{\bullet+}$, t-St$^{\bullet+}$, and c-St$^{\bullet+}$ in the D_2 States: τ,[a] k_{ic} k_i, ϕ_i, and E_{ex}

Radical Ions (D_2)	τ [nsec]	k_{ic} [sec^{-1}]	k_i [sec^{-1}]	ϕ_1	E_{ex} [kcal/mol]
t-St$^{\bullet-}$*	2.5 ± 1.0	$(4.0 \pm 1.6) \times 10^8$	0	0	48
c-St$^{\bullet-}$*	1.5 ± 0.4	$(5.6 \pm 1.7) \times 10^8$	$(1.0 \pm 0.3) \times 10^8$	0.14 ± 0.05	45
t-St$^{\bullet+}$*	0.24 ± 0.05	$(4.1 \pm 1.1) \times 10^9$	0	0	53
c-St$^{\bullet+}$*	0.12 ± 0.03	$(2.9 \pm 0.7) \times 10^9$	$(4.1 \pm 1.0) \times 10^9$	0.49 ± 0.12	50

[a] From $k_{et}\tau$ or $k_{ht}\tau$ assuming that $k_{et} = k_{ht} = k_{diff}$ with Q = Bp for St$^{\bullet-}$* and AN for St$^{\bullet+}$*, respectively.

change occurred in the transient absorption spectra and the time profiles. Therefore $M^{\cdot-}*$ converts to $M^{\cdot-}$ within $h\nu_{532}$ duration of 5 nsec.

No transient absorption was observed in the range 450–740 nm immediately after e^-, while the absorption of $DCA^{\cdot-}$ with three peaks appeared and grew until $t = 150$ nsec from exergonic ET from $FN^{\cdot-}$ during PR–LFP of DCA–fumaronitrile (FN) mixture. When $DCA^{\cdot-}$ was irradiated at $t = 190$ nsec with $h\nu_{532}$, the transient absorption of $DCA^{\cdot-}$ decreased within $h\nu_{532}$ (Fig. 19), as a result of the selective ET quenching of $DCA^{\cdot-}*$ by FN proceeds to yield DCA and $FN^{\cdot-}$ at k_{et}. Similar ET quenching of $PZ^{\cdot-}*$ or $AQ^{\cdot-}*$ by FN and of $M^{\cdot-}*$ by p-dicyanobenzene (DCB) was observed. Assuming $k_{et} = k_{diff} = 7.2 \times 10^9$ $M^{-1}\,sec^{-1}$ in DMF as discussed above, τ for $M^{\cdot-}*$ was estimated (Table 9). It should be emphasized that $M^{\cdot-}*$ has a nearly equal τ of 4–5 nsec. The three $M^{\cdot-}*$ have different characteristics from each other in their structure and electronic properties but an equivalent $E_{ex} = 46$–48 kcal mol^{-1}. E_{ex} might be important for τ of $M^{\cdot-}*$. The four $RAQ^{\cdot-}*$ have similar τ of 1.0–1.4 nsec, but they are shorter than τ for $AQ^{\cdot-}*$. It is suggested that the substituents have little effect in size ($2\text{-}MAQ^{\cdot-}*$ vs. $2\text{-}BAQ^{\cdot-}*$), electronic character ($2\text{-}MAQ^{\cdot-}*$ vs. $2\text{-}CAQ^{\cdot-}*$), and position ($1\text{-}CAQ^{\cdot-}*$ vs. $2\text{-}CAQ^{\cdot-}*$) on τ but that the internal conversion from $RAQ^{\cdot-}*$ to the ground state is accelerated by the substituents. The structural character of $M^{\cdot-}*$ influences only τ in a series of $M^{\cdot-}*$. Because the energy gaps between $M^{\cdot-}*$ and $M^{\cdot-}$ are almost equivalent for AQ and RAQ (Table 9), a shorter τ for $RAQ^{\cdot-}*$ than for $AQ^{\cdot-}*$ is probably attributed to rotation of the substituents which accelerate the internal conversion of $RAQ^{\cdot-}*$ to $RAQ^{\cdot-}$.

$\tau = 1.0$–4.4 nsec of $M^{\cdot-}*$ are comparable with $\tau = 1.5$ and 2.5 nsec of $c\text{-}St^{\cdot-}*$ and $t\text{-}St^{\cdot-}*$, respectively. These indicate that internal conversion of $M^{\cdot-}*$ and $St^{\cdot-}*$ occurs with similar k_{ic} as the main process. The structural and electronic characteristics of $M^{\cdot-}*$ are much different from those of $St^{\cdot-}*$, while $E_{ex} = 46$–48 kcal mol^{-1} for $M^{\cdot-}*$ is similar to $E_{ex} = 50$–53 kcal mol^{-1} for $c\text{-}St^{\cdot-}*$ and $t\text{-}St^{\cdot-}*$. Therefore, E_{ex} is the significant factor for the rate of internal conversion of $M^{\cdot-}*$ to $M^{\cdot-}$.

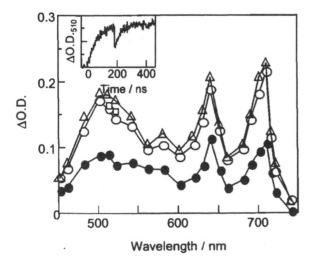

Figure 19 Transient absorption spectra recorded at $t = 190$ nsec after e^- (before $h\nu_{532}$, $\bigcirc$) and immediately after ($\bullet$) and at 100 nsec ($\triangle$) after $h\nu_{532}$ during PR–LFP of DCA (2.5×10^{-3} M)–FN (5.0×10^{-1} M) mixture in DMF. Inset: kinetic trace illustrating time profile of $\Delta O.D._{510}$ as a function of time after e^-.

Table 9 Half-Peak Reduction Potential ($E_{1/2}^{red}$) of M and Q, E_{ex} of M'⁻*, Free Energy Changes (ΔG and ΔG^*) in the ET from Q'⁻ to M and from M'⁻* to Q, τ of M'⁻*, $k_{ic} = \tau^1$ of M'⁻*, ϕ_{-M} of M'⁻ that Disappeared, and Relative Separation factor ($R = k_{sep}/(k_{-et} + k_{sep})$) of Solvated Pairs ((M'/Q'⁻)$_{solv}$) Based on 1.0 for (PZ/FN'⁻)$_{solv}$[a]

M'⁻*	Q	$E_{1/2}^{red}$ [V vs. SCE]		E_{ex} [kcal mol⁻¹]	ΔG [kcal mol⁻¹]	ΔG^* [kcal mol⁻¹]	τ [nsec]	$k_{ic} = \tau^{-1}$ [sec⁻¹]	ϕ_{-M}	R
		M	Q							
DCA'⁻*	FN	-0.89	-1.36	48	-11	-37	4.4	2.3×10^8	0.25	0.71
DCA'⁻*	o-DCB	-0.89	-1.87	48	-23	-26	4.7	2.1×10^8	0.13	0.40
DCA'⁻*	m-DCB	-0.89	-1.92	48	-24	-25	5.4	1.9×10^8	0.086	0.27
DCA'⁻*	p-DCB	-0.89	-1.72	48	-20	-29	5.0	2.0×10^8	0.11	0.34
DCA'⁻*	BN	-0.89	-2.49	48	-37	-11			0	
PZ'⁻*	FN	-1.23	-1.36	46	-3	-43	4.1	2.4×10^8	0.34	1.0
PZ'⁻*	p-DCB	-1.23	-1.72	46	-11	-35	4.8	2.1×10^8	nd	nd
AQ'⁻*	FN	-0.87	-1.36	48	-11	-36	4.0	2.5×10^8	0.095	0.42
2-MAQ'⁻*	FN	-1.0	-1.36	48	-8	-39	1.3	7.7×10^8	0.095	0.36
2-BAQ'⁻*	FN	-1.0	-1.36	48	-8	-39	1.4	7.2×10^8	0.10	0.32
1-CAQ'⁻*	FN	-0.83	-1.36	48	-12	-36	1.1	9.1×10^8	0.063	0.28
2-CAQ'⁻*	FN	-0.78	-1.36	48	-13	-34	1.0	1.0×10^9	0.076	0.27

[a] DCA = 9,10-dicyanoanthracene, PZ = phenazine, AQ = anthraquinone, 2-MAQ = 2-methylanthraquinone, 2-BAQ = 2-tert-butylanthraquinone, 1-CAQ = 1-chloroanthraquinone, 2-CAQ = 2-chloroanthraquinone, FN = fumaronitrile, DCB = dicyanobenzene, and BN = benzonitrile. $E_{1/2}^{red}$ (vs. SCE) was calculated from $E_{1/2}^{red}$ (vs. SCE) = $E_{1/2}^{red}$ (vs. Ag/Ag⁻) + 0.25 V, when $E_{1/2}^{red}$ (vs. Ag/Ag⁻) was given in the reference. $\Delta G = E_{1/2}^{red}$(Q) − $E_{1/2}^{red}$(M) and $\Delta G^* = E_{1/2}^{red}$(M) − $E_{1/2}^{red}$(Q). E_{ex}. ϕ_{-M} was measured under the conditions of [M] = 2.5×10^{-3} [FN] = 5.0×10^{-1}, and [DCB] = 3.0×10^{-1} M. k_{sep} and k_{-et} denote the rate constants of separation to M and Q'⁻ and back electron transfer to M'⁻ and Q from (M'/Q'⁻)$_{solv}$, respectively.

Intermediacy of Radical Anion-Neutral Molecule Pair in the Electron Transfer Quenching

As shown in Table 9, φ_{-M} depended on the combination of $M^{\bullet-}*$ and Q. φ_{-PZ} in $PZ^{\bullet-}*$–FN was the largest among the $M^{\bullet-}$–Q pairs, while φ_{-AQ} in $AQ^{\bullet-}*$–FN was considerably smaller than φ_{-DCA} in $DCA^{\bullet-}*$–FN and φ_{-PZ} in $PZ^{\bullet-}*$–FN. φ_{-RAQ} in $RAQ^{\bullet-}*$–FN was comparable with φ_{-AQ} in $AQ^{\bullet-}*$–FN, whereas φ_{-DCA} in $DCA^{\bullet-}*$–DCB was much smaller than φ_{-DCA} in $DCA^{\bullet-}*$–FN. These results show that the efficiency of the disappearance of $M^{\bullet-}$ and the formation of $Q^{\bullet-}$ depends on the combination of $M^{\bullet-}*$–Q and suggest the intermediacy of $(M/Q^{\bullet-})_{solv}$ in the selective ET quenching of $M^{\bullet-}*$ by Q. In other words, ET proceeds with k_{et} to initially give $(M/A^{\bullet-})_{solv}$, with competition of the internal conversion of $M^{\bullet-}*$ to $M^{\bullet-}$ with k_{ic}. $(M/Q^{\bullet-})_{solv}$ separates into M and $Q^{\bullet-}$ at k_{sep} or returns to $M^{\bullet-}$ and Q via back ET at k_{-et}. Because φ_{-M} is proportional to the separation factor of $(M/Q^{\bullet-})_{solv}$, relative separation factors of $R = k_{sep}/(k_{-et} + k_{sep})$ are calculated on the basis of $R = 1.0$ for $(PZ/FN^{\bullet-})_{solv}$. As k_{sep} of $(D^{\bullet-}/Q^{\bullet-})_{solv}$ is assumed not to depend on the structures of $D^{\bullet-}$ and $Q^{\bullet-}$ in photoinduced ET reactions of D–Q mixtures, the difference in R is responsible for k_{-et} controlled by ΔG in $(M/Q^{\bullet-})_{solv}$. Indeed, R has a correlation with ΔG but not completely (Table 9). Although electrostatic interaction are preferential for R in $(D^{\bullet+}/A^{\bullet-})_{solv}$, structural relations might also be important for R in $(M/A^{\bullet-})_{solv}$.

In conclusion, the selective ET quenching of $M^{\bullet-}*$ (M = DCA, PZ, AQ, and RAQ) by Q (FN and DCB) occurs at $k_{et} = k_{diff}$ because of the large exergonic character. τ of $M^{\bullet-}*$ (M = DCA, PZ, and AQ) is estimated to be 4 nsec, while $\tau = 1.0$–1.3 nsec of $RAQ^{\bullet-}*$ is shorter than that of $M^{\bullet-}*$.

5. FLUORESCENCE OF RADICAL CATIONS IN THE EXCITED STATE

Fluorescence spectroscopy has sensitivity as high as up to the single-molecular detection level. However, fluorescence detection of $M^{\bullet+/\bullet-}$ has been usually unsuccessful by their extremely low fluorescence quantum yields (φ_f) in solution even if their parent molecules are highly fluorescent. This is attributed to the low E_{ex} favorable for internal conversion and the high chemical reactivity of $M^{\bullet+}*/^{\bullet-}*$ [33,146,147]. Only a few tens of examples of fluorescent $M^{\bullet+/\bullet-}$ have been reported but some of them are quite doubtful [148–150].

5.1. Fluorescence of 1,3,5-Trimethoxybenzene Radical Cation (TMB$^{\bullet+}$)

Efficient fluorescence ($\varphi_f = 1.1 \times 10^{-3}$) of 1,3,5-trimethoxybenzene radical cation in the excited state ($TMB^{\bullet-}*$) has been observed during excitation of $TMB^{\bullet+}$ generated from PET of TMB-1,4-dicyanonaphthalene (DCN)–Bp mixture in acetonitrile by two-laser, two-step excitation technique [147–153]. Excitation with a XeCl excimer laser (308 nm, 8 nsec) initiated ET between $^1DCN^*$ and Bp giving $DCN^{\bullet-}$ and $Bp^{\bullet+}$, which led to the secondary ET with O_2 and TMB to give $O_2^{\bullet-}$ and $TMB^{\bullet+}$ with a peak at 590 nm, respectively. Excitation of $TMB^{\bullet+}$ with $h\nu_{532}$ (6 nsec) gave fluorescence spectrum around 620 nm symmetry to the absorption spectrum. $\varphi_f = 1.1 \times 10^{-3}$ allowed an estimation of τ = 210 psec using the Strickler–Berg relationship, giving $k_f = 9.5 \times 10^6 \ sec^{-1}$ for natural radiative rate constant of $TMB^{\bullet+}*$.

Formation Kinetics of [TMB$^{\bullet+}$/Cl$^-$] Ion Pair during PR

Pulse radiolysis of M in chlorinated solvents has been a standard method to generate $M^{\bullet+}$ due to the high ionization potential (IP) of the solvent molecules [146,154–156]. However, the oxidation process is rather complicated because of the generation of Cl⁺, which is also an oxidant together with solvent⁺⁺ during PR [125,157–160]. Cl⁺ forms complexes or an ion pair directly with M [157]. In addition, such species is also formed from M⁺⁺ and Cl⁻ in the course of their neutralization [161,162]. Therefore observed transient absorption spectra would be a sum of spectra of free ions and complexes or ion pairs.

Pulse radiolysis of TMB in chlorinated solvents gave TMB⁺⁺ with a broad spectrum peaked at 500 nm during PR of TMB in highly chlorinated solvent such as CCl_4, CH_2Cl_2, or $CHCl_3$, although the absorption spectrum of TMB⁺⁺ in acetonitrile showed a sharp peak at 590 nm during LFP [125,159,160]. The transient absorption spectrum was time-dependent in DCE (Fig. 20) showing a spectrum similar to that in acetonitrile immediately after e⁻ and similar to that in CCl_4 at >200 nsec. This spectral change was attributed to the ion pair ([TMB⁺⁺/Cl⁻]) formation in DCE. The fast and slow processes for ion pair formation are considered: trapping of Cl⁺ by TMB within a 5-nsec flash and collision of TMB⁺⁺ and Cl⁻ at $k_n = 3.0 \times 10^{11}$ M^{-1} sec⁻¹, respectively.

Fluorescence of Free TMB⁺⁺ and [TMB⁺⁺/Cl⁻] Ion Pair

Generation of M⁺⁺/⁺⁻ in solution includes one-electron transfer and should be accompanied with ion pair formation, e.g., M⁺⁺–M⁺⁻, M⁺⁺–anion, etc. in any processes. Absorption spectroscopy of M⁺⁺/⁺⁻ are not sufficiently sensitive to distinguish ion pair from free ion, while fluorescence spectroscopy is enough sensitive. The φ_f of TMB⁺⁺* can be a measure of the ratio of free TMB⁺⁺ to the ion pair [TMB⁺⁺/Cl⁻], because the excitation of TMB⁺⁺ in the ion pair would cause no fluorescence or weak fluorescence through rapid HT quenching of TMB⁺⁺* by Cl⁻ within the ion pair as compared to free TMB⁺⁺. The φ_f of TMB⁺⁺ as a function of delay time (10–1000 nsec) decreased rapidly during PR–LFP of TMB in DCE

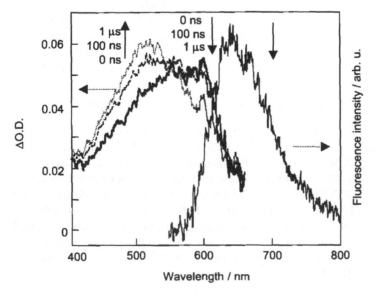

Figure 20 Time-resolved transient absorption spectra observed at 0 nsec (solid line), 100 nsec (dashed line), and 1 μsec (dotted line) after e⁻ and transient fluorescence spectrum of TMB⁺⁺* observed at 300 nsec after e⁻ during PR–LFP of TMB (1.0×10^{-2} M) in DCE. Energy of $h\nu_{532}$, 130 mJ.

(Fig. 21). The addition of 1.2×10^{-2} M of tetrabutylammonium chloride as a source of Cl$^-$ drastically accelerated the decrease of φ_f, as well as the blue shift of the transient absorption. φ_f fell into $(2-3) \times 10^{-5}$ within 20 nsec after e$^-$, which would represent the φ_f of [TMB$^{\cdot+}$*/ Cl$^-$] and the low φ_f was attributed to HT quenching of TMB$^{\cdot+}$* by Cl$^-$ in [TMB$^{\cdot+}$*/Cl$^-$]. τ of TMB$^{\cdot+}$* in the ion pair in DCE and k_{et} for HT quenching of TMB$^{\cdot+}$* by Cl$^-$ within the ion pair were estimated to be 3.2 psec and 3.0×10^{11} sec^{-1}, respectively. On the other hand, free TMB$^{\cdot+}$ was generated from HT during PR of TMB in the presence of Bp in DCE. Free TMB$^{\cdot+}$* emits a fluorescence with $\varphi_f = 1.1 \times 10^{-3}$.

Free ion fraction in the total ion amount can be estimated from $\varphi_f = \varphi_f$ (free ion) $+ \varphi_f$ (ion pair). Pulse radiolysis of TMB (1.0×10^{-2} M) in DCE gave TMB$^{\cdot+}$, whose ion pair fraction is 50% immediately after e$^-$ and 75% at 1 μsec. Possibly fast ion pair formation could be attributed to the reaction of Cl$^{\cdot}$ with TMB. From the φ_f for TMB$^{\cdot+}$* in the ion pair, we estimated free ion fraction ($R_f = $[free TMB$^{\cdot+}$]/([free TMB$^{\cdot+}$]$+$[TMB$^{\cdot+}$/Cl$^-$]), $R_f = $ 0.55 at 0 nsec and $R_f = 0.2$ at 1 μsec after e$^-$ during PR of TMB in DCE. $\varphi_f = k_f/(k_f + k_{ic} + k_{ht}) = k_{fc}$ is represented by k_{ic} and k_{ht} of internal conversion and HT quenching within the ion pair, respectively. Therefore, τ of free TMB$^{\cdot+}$* and TMB$^{\cdot+}$* in the ion pair is estimated to be 116 and 3.2 psec, while the k_{ic} and k_{ht} were also calculated to be 8.5×10^9 and 3.0×10^{11} sec^{-1}, respectively. The transient behaviors of free TMB$^{\cdot+}$* and TMB$^{\cdot+}$* in the ion pair are summarized in Scheme 15.

$\varphi_f = 3.6 \times 10^{-5}$ and 1.9×10^{-4} of TMB$^{\cdot+}$* in CCl$_4$ and CH$_2$Cl$_2$, respectively, at a delay time of 100 nsec were also estimated in a similar manner. These results can be explained by the ion pair formation from TMB and Cl$^{\cdot}$ in these solvents. This indicates two mechanisms for the ion pair formation through ET from TMB to Cl$^{\cdot}$ as a fast process and through the recombination of TMB$^{\cdot+}$ and Cl$^-$ as a slow process.

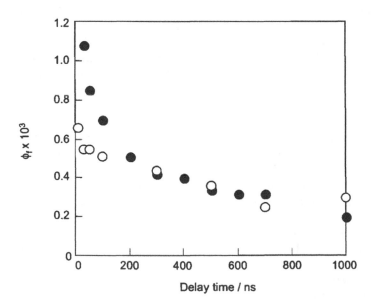

Figure 21 φ_f of TMB$^{\cdot+}$ as a function of the delay time of $h\nu_{532}$ relative to e$^-$ during PR–LFT of TMB (1.0×10^{-2} M) (open circle) and Bp (5.0×10^{-2} M)–TMB (1.0×10^{-3} M) mixture (solid circle) in DCE.

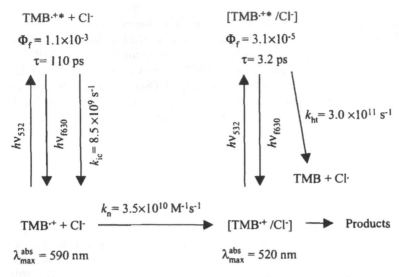

Scheme 15 Transient behaviors of free $TMB^{\cdot+}*$ and $[TMB^{-+}*/Cl^-]$.

5.2. Fluorescence of TMB$^{\cdot+}$ Derivatives and the Quenching

Although the fluorescence of the $TMB^{\cdot+}*$ family was studied [125,159,160], fluorescence was observed only in $TMB^{\cdot+}*$ among mono-, di-, tri-, tetra-, and penta-methoxybenzenes and among 1,2,3-, 1,2,4-, and 1,3,5-trimethoxybenzenes. This indicates that 1,3,5-trioxyl substitution is necessary for the fluorescence. Pseudo-D_{3h} symmetry cause the degenerate of the singly occupied and second highest occupied molecular orbitals (SOMO and SHOMO, respectively) in the $TMB^{\cdot+}$ family bearing 1,3,5-trioxybenzene structure. Therefore, the $TMB^{\cdot+}$ family has low D_0–D_1 and high D_0–D_2 energy separations. This makes the $D_2 \rightarrow D_0$ transition radiative enough to compete with $D_2 \rightarrow D_1$ internal conversion.

Fluorescence was observed for the $TMB^{\cdot+}*$ family such as 3,5-dimethoxyphenol, 1,3-dihydroxy-5-methoxybenzene (5-methoxyresorcinol), 1-acetoxy-3,5-dimethoxybenzene, and 1,3,5-trimethoxy-2-methylbenzene. These results indicated that complete symmetry of the substitution on O atoms is not necessary to observe fluorescence from the $TMB^{\cdot+}*$ family, and that the variation of parent molecules of fluorescent radical cation is possibly performed [153]. Fluorescence was also detected from hexamethxybenzene$^{\cdot+}*$ as an example of pseudo-D_{6h} molecules. The discussion of the symmetry has been described here on the fluorescence from fluorobenzenes$^{\cdot+}*$ in the vapor-phase or noble gas matrices.

The φ_f of $TMB^{\cdot+}$ generated during PR in DCE decreased with the introduction of alkyl or acetyl substituents at O atoms suggesting that they act as quenchers. For example, $\varphi_f = 2 \times 10^{-3}$, 1.1×10^{-3}, and ≈ 0 for 1,3,5-trihydroxybenzene$^{\cdot+}$, $TMB^{\cdot+}$, and 1,3,5-triacetoxybenzene$^{\cdot+}$ ($TAB^{\cdot+}$), respectively. However, the mechanisms involved in the quenching of 1,3,5-trioxybenzenes$^{\cdot+}*$ are not the same. Internal conversion promoted by the C−H vibration would operate in $TMB^{\cdot+}*$ and intramolecular HT would take place in $TAB^{\cdot+}*$.

Excitation of $TMB^{\cdot+}$ generated by PR in benzonitrile or in DCE in the presence of benzonitrile gave weak or zero fluorescence as a result of intermolecular HT quenching [153]. This process has an exothermicity of ≈ -0.57 eV estimated from the ionization potentials (IP) of TMB and benzonitrile (IP = 8.11 and 9.62 eV, respectively) and E_{ex} of $TMB^{\cdot+}$ (2.03 eV). Similarly, fluorescence from $TMB^{\cdot+}*$ was not observed in acetone

(IP = 9.67 eV) and cyclohexane (IP = 10.32 eV). On the other hand, fluorescence of TMB$^{\bullet +}$* was observed in acetonitrile, DCE, 1,1,1,3,3,3-hexafluoro-2-propanole, 2,2,2-trifluoroethanol, and other chlorinated solvents (IP > 10.4 eV). The absence of the fluorescence from TAB$^{\bullet +}$* could be a result of intramolecular HT quenching of the TMB$^{\bullet +}$* family by the acetyl group (IP = 10 eV).

5.3. Fluorescence of 3,5-Dimethoxyphenol Radical Cation (DMP$^{\bullet +}$)

If strong fluorescence of M$^{\bullet +}$ is found, it will be a "fluorescent M$^{\bullet +}$ probe" to monitor ET/HT phenomena in homogeneous and heterogeneous environment with a high sensitivity. DMP$^{\bullet +}$ showed a fluorescence stronger than TMB$^{\bullet +}$ in DCE [150,163].

A transient absorption spectrum with a peak at 575 nm assigned to DMP$^{\bullet +}$ was observed immediately after e$^-$ during PR of DMP (5.0 × 10^{-3} M) in DCE. The time-resolved transient absorption spectra showed a slow growth of a new peak at 500 nm assigned to 3,5-dimethoxyphenoxy radical (DMP$^{\bullet}$) (Fig. 22). The peak at 575 nm due to DMP$^{\bullet +}$ decreased with time, and most of DMP$^{\bullet +}$ disappeared at 2 μsec after e$^-$ to yield DMP$^{\bullet}$ through deprotonation of DMP$^{\bullet +}$. It has been shown that bimolecular deprotonation of DMP$^{\bullet +}$ occurs to yield DMP$^{\bullet}$ at k_{-H^+} = 5.5 × 10^{10} M^{-1} sec^{-1} in the presence of basic reagents. The basic species for the deprotonation is probably Cl$^-$ generated by the radiation chemical primary processes of DCE. The plausible mechanism involved therein is summarized in Scheme 16.

The similarity of the absorption spectrum of DMP$^{\bullet +}$ to that of TMB$^{\bullet +}$ indicates a similar electronic structure. The structure of TMB$^{\bullet +}$ is deviated from pseudo-D$_{3h}$ symmetry because of repulsion of three Me groups, while that of DMP$^{\bullet +}$ is closer to pseudo-D$_{3h}$ symmetry. When D$_{3h}$ symmetry is an important factor for higher efficiency of the fluorescence and Me group enhances internal conversion, DMP$^{\bullet +}$* fluoresces stronger than TMB$^{\bullet +}$*. It is interesting to know whether unimolecular deprotonation from DMP$^{\bullet +}$*

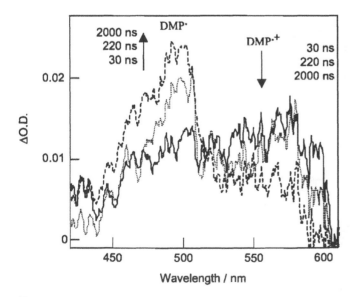

Figure 22 Time-resolved absorption spectra measured at 30, 220, and 2000 nsec (gate time: 30 nsec) after e$^-$ during PR of DMP (5.0 × 10^{-3} M) in DCE.

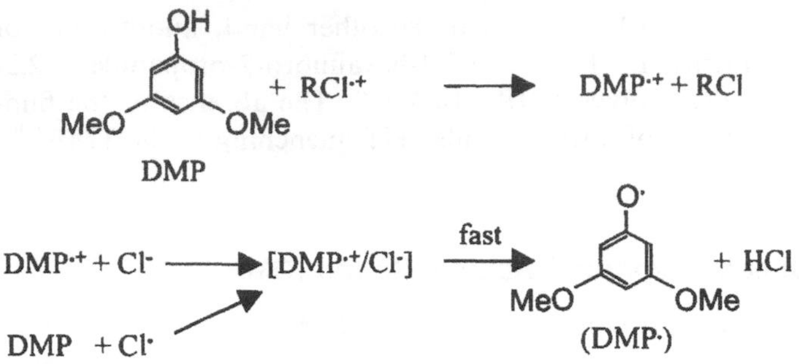

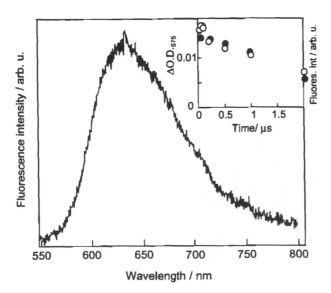

Scheme 16 Deprotonation of DMP$^{\cdot+}$ through [DMP$^{\cdot+}$/Cl$^-$].

competes with the radiative transition or not. If it occurs, the fluorescence efficiency of DMP$^{\cdot+}$ becomes lower than TMB$^{\cdot+}$.

Excitation of DMP$^{\cdot+}$ with hv_{532} gave a broad fluorescence around 600–750 nm (Fig. 23). The fluorescence spectrum showed a mirror image symmetry to the absorption spectrum of DMP$^{\cdot+}$. The integrated fluorescence intensity monitored decreased with the increase of the delay time of hv_{532} relative to e$^-$ in a similar manner to the temporal profile of the transient absorption monitored at 575 nm (Fig. 23b). This also supports the assignment of the fluorescence to that of DMP$^{\cdot+}$*. When DMP$^{\cdot+}$ was excited with hv_{532}, the absorption of DMP$^{\cdot+}$ was unchanged and no increase in that of DMP$^{\cdot}$ was observed. This indicates that DMP$^{\cdot+}$* does not undergo deprotonation, unlike its neutral molecule [164]. Because the acidity of the OH group of DMP$^{\cdot+}$ in the ground state is enhanced by the positive charge of DMP$^{\cdot+}$ and deprotonation takes place even in the presence of weak bases such as Cl$^-$ or H$_2$O, and $E_{ex} = 47$ kcal mol^{-1} and the resonance stabilization of the

Figure 23 Fluorescence spectrum of DMP$^{\cdot+}$ measured at 100 nsec after e$^-$. Energy of hv_{532}, 130 mJ. Inset: a plot of the integrated fluorescence intensity of DMP$^{\cdot+}$ as a function of the delay time of hv_{532} relative to e$^-$ (open circle) is superimposed with a plot of ΔO.D.$_{575}$ (solid circle).

DMP$^{\cdot}$ (12 kcal mol^{-1}) would further enhance the acidity, it is surprising to observe no spontaneous deprotonation of the OH group of DMP$^{\cdot+}$ upon excitation. The absence of the deprotonation upon excitation strongly suggests that MO incorporated in the excitation of DMP$^{\cdot+}$ does not correlate with the σ^* orbital of the OH bond.

The φ_f of DMP$^{\cdot+}$ in DCE was measured from the fluorescence intensity as a function of the delay time of hv_{532} relative to e$^-$ and the time dependence of ΔO.D.$_{532}$. $\varphi_f = (2\pm0.3) \times 10^{-3}$ was independent of the delay time except for $\varphi_f = 1.3 \times 10^{-3}$ at 30 nsec. The low φ_f at 30 nsec may have resulted from HT quenching of DMP$^{\cdot+}$* by Cl$^-$ in [DMP$^{\cdot+}$*/Cl$^-$]. $\varphi_f = (2\pm0.3) \times 10^{-3}$ for DMP$^{\cdot+}$* was higher than $\varphi_f = 1.1 \times 10^{-3}$ for free TMB$^{\cdot+}$* in DCE. The intense fluorescence of DMP$^{\cdot+}$* ruled out the demand of complete pseudo-D$_{3h}$ structure in M$^{\cdot+}$* and suggested a possibility of substitution of H atom of the OH group of DMP with an alkyl group as an attachment group for the introduction of DMP unit to various molecules or surfaces as a fluorescent radical cation probe. $k_f = 5.8 \times 10^6$ sec^{-1} was estimated for the fluorescence decay of DMP$^{\cdot+}$* using the Strickler–Berg relationship. Although k_f was smaller than that for TMB$^{\cdot+}$* (9.5×10^6 sec^{-1}), φ_f was larger by a factor of 1.8. It is suggested that internal conversion in DMP$^{\cdot+}$* is slower by a factor of 3.0 as compared to TMB$^{\cdot+}$*. Because the deprotonation from DMP$^{\cdot+}$* is not an important process for internal conversion, the Me group would play an important role in internal conversion of DMP$^{\cdot+}$* and TMB$^{\cdot+}$* as reported for that of ^{1}alkylbenzenes* [165]. $\tau \approx 350$ psec of DMP$^{\cdot+}$* was estimated from $\varphi_f = k_f /(k_f + k_{ic}) = k_f\tau \approx k_f/k_{ic}$. It is strongly suggested that the introduction of Me groups in 1,3,5-trioxybenzenes causes the deviation from the pseudo-D$_{3h}$ structure by their mutual repulsion, reducing φ_f by the enhancement of internal conversion.

It is found that radical cations of 1,3,5-trixybenzene derivatives in solution fluoresce with moderately high $\varphi_f \approx 10^{-3}$ in the rage of 580–800 nm.

6. CONCLUDING REMARKS

New aspects of radical ion chemistry in solutions have been described in this chapter. These studies were performed by PR, γ-R, and PR–LFP techniques in which M$^{\cdot+}$ or M$^{\cdot-}$ can be selectively generated at 10^{-5} M.

Remarkable enhancements of the unimolecular c–t isomerization of c-S$^{\cdot+}$ with p-MeO and oxidation of S$^{\cdot+}$ with p-MeO are explained by charge–spin separation in such S$^{\cdot+}$. Unimolecular c–t isomerization of such c-S$^{\cdot+}$ proceeds with a chain mechanism, while regioselective oxidation occurs in such S$^{\cdot+}$ because of the spin localization. Cycloreversion of t,c,t-TPCB$^{\cdot+}$ occurs to give a σ-St$_2^{\cdot+}$, while the photochemical cycloreversion of t,c,t-TPCB$^{\cdot+}$ and t,t,t-TPCB$^{\cdot+}$ gives π-St$_2^{\cdot+}$ and t-St$^{\cdot+}$/t-St pair, respectively. Radical cations of phosphorus compounds (**9**$^{\cdot+}$ and **10**$^{\cdot+}$) form intramolecular π-dimer$^{\cdot+}$ between two Nps from which Np$_2^{\cdot+}$ forms. Formation of intermolecular σ-dimer$^{\cdot-}$ of aromatic acetylene (**11**$^{\cdot-}$ and **12**$^{\cdot-}$) and intramolecular dimer$^{\cdot-}$ of **13**$^{\cdot-}$ and diarylmethanol$^{\cdot-}$ was observed, and the $n = 3$ rule is not effective for intramolecular dimer$^{\cdot-}$.

Anisole and Bp act as selective hole and electron quenchers for St$^{\cdot+}$* and St$^{\cdot+}$* via HT and ET quenching, respectively. Assuming that $k_{ht} = k_{et} = k_{diff}$, τ_c and τ_t were estimated to be 120 and 240 psec for c-St$^{\cdot+}$* and t-St$^{\cdot+}$*, respectively, which are shorter than $\tau = 1.5$ and 2.5 nsec for c-St$^{\cdot-}$* and t-St$^{\cdot-}$*. The order of $\tau_c < \tau_t$ is attributed to the isomerization via the twisting only in c-St$^{\cdot+}$* and c-St$^{\cdot-}$*, because twisting is favorable in c-St$^{\cdot+}$*/$^{\cdot-}$* compared with t-St$^{\cdot+}$*/$^{\cdot-}$*. Diabatic isomerization from c-St$^{\cdot-}$*/$^{\cdot-}$* to t-St$^{\cdot+}$/$^{\cdot-}$ is suggested to proceed via crossing of the D$_2$ and D$_0$ states. The selective ET quenching of

$M^{\cdot-}*$ (M = DCA, PZ, AQ, and RAQ) by Q (FN and DCB) occurs to give $\tau = 4$ nsec of $M^{\cdot-}*$ (M = DCA, PZ, and AQ) and $\tau = 1.0$–1.3 nsec of $RAQ^{\cdot-}*$. Intermediacy of the $M/Q^{\cdot-}$ pair is suggested during ET quenching of $M^{\cdot-}*$ by Q.

It is found that radical cations of 1,3,5-trioxybenzene derivatives such as $TMB^{\cdot+}$ and $DMP^{\cdot+}$ in solution fluoresce with moderately high $\varphi_f \approx 10^{-3}$ in the range of 580–800 nm. Large φ_f was obtained for free $TMB^{\cdot+}$, while small φ_f was obtained for the $[TMB^{\cdot+}/Cl^-]$ ion pair because of fast HT quenching of $TMB^{\cdot+}*$ by Cl^- in the ion pair generated from ET from TMB to $Cl^{\cdot}$ and collision between $TMB^{\cdot+}$ by Cl^-. 1,3,5-Trioxyl substitution is necessary for the efficient fluorescence because of the pseudo-D_{3h} symmetry causing low D_0–D_1 and high D_0–D_2 energy separations. Intramolecular and intermolecular HT quenching of $TMB^{\cdot+}*$ occurred by substituents of the O atom and hole acceptor molecules with low IP.

Because $M^{\cdot+}*/^{\cdot-}*$ has spin, charge, and E_{ex}, chemistry of $M^{\cdot+}*/^{\cdot-}*$ is interesting and still a new subject. The PR method can allow us to study various $M^{\cdot+/\cdot-}$ as the starting molecule. Because we used the 532-nm laser flash for excitation of $M^{\cdot+/\cdot-}$ in this study, only $M^{\cdot+/\cdot-}$ with absorption at 532 nm can be used as the starting molecule. When we can use any λ for LFP, various $M^{\cdot+/\cdot-}$ can be excited at λ tuned to the absorption of $M^{\cdot+/\cdot-}$. And because various excited states exist in $M^{\cdot+}*/^{\cdot-}*$, selective excitation of $M^{\cdot+/\cdot-}$ to $M^{\cdot+/\cdot-}$ (D_n) is of particular interest to elucidate the transient phenomena depending on the excited state. Although it has been considered that $M^{\cdot+}*/^{\cdot-}*$ is nonluminescent, we have found that 1,3,5-trioxybenzenes$^{\cdot+}$ fluoresces with moderately high $\varphi_f \approx 10^{-3}$. Considering that fluorescence was readily observed with conventional photomultiplier as a detector, we can expect the use of these type of compounds as ET/HT probe for homogeneous solutions and also for heterogeneous systems such as photocatalysts, and biological oxidation in future.

ACKNOWLEDGMENTS

The author thanks his collaborators, particularly Mrs. Sachiko Tojo, Drs. Akito Ishida, Nobuyuki Ichinose, Mitsunobu Nakamura, and his students, as well as the members of the Radiation Laboratory of ISIR, Osaka University, for running the linear accelerator. This work was partly supported by a Grant-in-Aid from the Ministry of Education, Culture, Sports, Science and Technology (MEXT) of Japanese Government.

REFERENCES

1. Fox, M.A. Chem. Rev. 1979, 79, 253.
2. Fox, M.A., Chanon, M., Eds.; Photoinduced Electron Transfer; Elsevier: Amsterdam, 1988.
3. Eberson, L., Ed.; Electron Transfer Reactions in Organic Chemistry; Springer-Verlag: Berlin, 1987.
4. Mattes, S.L.; Farid, S. In Organic Photochemistry; Padwa, A., Ed.; Marcel Dekker: New York, 1983; 233 pp.
5. Julliard, M.; Chanon, M. Chem. Rev. 1983, 83, 425.
6. Kavarnos, G.J.; Turro, N.J. Chem. Rev. 1986, 86, 401.
7. Roth, H.D. Acc. Chem. Res. 1987, 20, 343.
8. Yoon, U.C.; Mariano, P.S. Acc. Chem. Res. 1992, 25, 233.
9. Peters, K.S.; Cashin, A. J. Phys. Chem. A 2000, 104, 4833.

10. Lewis, F.D.; Letsinger, R.L.; Wasielewski, M.R. Acc. Chem. Res. 2001, *34*, 159.
11. Stevenson, J.P.; Jackson, W.F.; Tanko, J.M. J. Am. Chem. Soc. 2002, *124*, 4271.
12. Pascaly, M.; Yoo, J.; Barton, J.K. J. Am. Chem. Soc. 2002, *124*, 9083.
13. Nakatani, K.; Dohno, C.; Saito, I. J. Am. Chem. Soc. 2002, *124*, 6802.
14. Kanvah, S.; Schuster, G.B. J. Am. Chem. Soc. 2002, *124*, 11286.
15. Sinnecker, S.; Koch, W.; Lubitz, W. J. Phys. Chem. B 2002, *106*, 5281.
16. Fukuzumi, S.; Ohkubo, K.; Chen, Y.H.; Pandey, R.K.; Zhan, R.Q.; Shao, J.G.; Kadish, K.M. J. Phys. Chem. A 2002, *106*, 5105.
17. Horner, J.H.; Taxil, E.; Newcomb, M. J. Am. Chem. Soc. 2002, *124*, 5402.
18. Baciocchi, E.; Bietti, M.; Salamone, M.; Steenken, S. J. Org. Chem. 2002, *67*, 2266.
19. Adam, W.; Librera, C.P. J. Org. Chem. 2002, *67*, 576.
20. Shukla, S.S.; Rusling, J.F. J. Phys. Chem. 1985, *89*, 3353.
21. Galland, B.; Moutet, J.C.; Reverdy, G. Electrochim. Acta 1987, *32*, 175.
22. Moutet, J.C.; Reverdy, G. Nouv. J. Chim. 1983, *7*, 105.
23. Moutet, J.C.; Reverdy, G. Tetrahedron Lett. 1979, *20*, 2389.
24. Labbe, P.; Moutet, J.C.; Paltrier, M.; Reverdy, G. Nouv. J. Chim. 1984, *8*, 627.
25. Shine, H.J.; Zhao, D.C. J. Org. Chem. 1990, *55*, 4086.
26. Moutet, J.C.; Reverdy, G. J. Chem. Soc., Chem. Commun. 1982, 654.
27. Nelleborg, P.; Lund, H.; Eriksen, J. Tetrahedron Lett. 1985, *26*, 1733.
28. Breslin, D.T.; Fox, M.A. J. Org. Chem. 1994, *59*, 7557.
29. Fox, M.A.; Dulay, M.T.; Krosley, K. J. Am. Chem. Soc. 1994, *116*, 10992.
30. Wang, Z.; Mcgimpsey, W.G. J. Phys. Chem. 1993, *97*, 3324.
31. Wang, Z.; Mcgimpsey, W.G. J. Phys. Chem. 1993, *97*, 5054.
32. Kuriyama, Y.; Hashimoto, F.; Tsuchiya, M.; Sakuragi, H.; Tokumaru, K. Chem. Lett. 1994, 1371.
33. Breslin, D.T.; Fox, M.A. J. Phys. Chem. 1994, *98*, 408.
34. Ichikawa, T.; Moriya, T.; Yoshida, H. J. Phys. Chem. 1976, *80*, 1278.
35. Holroyd, R.A. J. Phys. Chem. 1982, *86*, 3541.
36. In *CRC Handbook of Radiation Chemistry*. Tabata, Y., Ed.; CRC Press: Boca Raton, 1991; 63 pp.
37. Hart, E.D.; Boag, J.W. J. Am. Chem. Soc. 1962, *84*, 4090.
38. Hunt, J.W.; Thomas, J.K. Radiat. Res. 1967, *32*, 149.
39. In *CRC Handbook of Radiation Chemistry*; Tabata, Y., Ed.; CRC Press: Boca Raton, 1991; 439 pp.
40. Shida, T.; Hamill, W.H. J. Chem. Phys. 1966, *44*, 2375.
41. Shida, T.; Haselbach, E.; Bally, T. Acc. Chem. Res. 1984, *17*, 180.
42. Tojo, S.; Morishima, K.; Ishida, A.; Majima, T.; Takamuku, S. Bull. Chem. Soc. Jpn. 1995, *68*, 958.
43. Ishida, A.; Fukui, M.; Ogawa, H.; Tojo, S.; Majima, T.; Takamuku, S. J. Phys. Chem. 1995, *99*, 10808.
44. Majima, T.; Tojo, S.; Ishida, A.; Takamuku, S. J. Org. Chem. 1996, *61*, 7793.
45. Majima, T.; Tojo, S.; Ishida, A.; Takamuku, S. J. Phys. Chem. 1996, *100*, 13615.
46. Majima, T.; Fukui, M.; Ishida, A.; Takamuku, S. J. Phys. Chem. 1996, *100*, 8913.
47. Fujita, M.; Ishida, A.; Majima, T.; Takamuku, S. J. Phys. Chem. 1996, *100*, 5382.
48. Majima, T.; Tojo, S.; Takamuku, S. J. Phys. Chem. A 1997, *101*, 1048.
49. Yamamoto, Y. Trends. Org. Chem. 1992, *3*, 93.
50. Tamai, T.; Mizuno, K.; Hashida, I.; Otsuji, Y.; Ishida, A.; Takamuku, S. Chem. Lett. 1994, 149.
51. Kimura, N.; Takamuku, S. J. Am. Chem. Soc. 1994, *116*, 4087.
52. Tsuchida, A.; Ikawa, T.; Yamamoto, M.; Ishida, A.; Takamuku, S. J. Phys. Chem. 1995, *99*, 14793.
53. Kimura, N.; Takamuku, S. J. Am. Chem. Soc. 1995, *117*, 8023.
54. Kojima, M.; Kakehi, A.; Ishida, A.; Takamuku, S. J. Am. Chem. Soc. 1996, *118*, 2612.

55. Shida, T.; Momose, T. J. Mol. Struct. 1985, *126*, 159.
56. Suzuki, H.; Koyano, K.; Shida, T.; Kira, A. Bull. Chem. Soc. Jpn. 1982, *55*, 3690.
57. Shida, T.; Kato, T.; Nosaka, Y. J. Phys. Chem. 1977, *81*, 1095.
58. Shida, T.; Momose, T.; Ono, N. J. Phys. Chem. 1985, *89*, 815.
59. Toriyama, K.; Nunome, K.; Iwasaki, M.; Shida, T.; Ushida, K. Chem. Phys. Lett. 1985, *122*, 118.
60. Dunkin, I.R.; Andrews, L.; Lurito, J.T.; Kelsall, B.J. J. Phys. Chem. 1985, *89*, 1701.
61. Ushida, K.; Shida, T.; Shimokoshi, K. J. Phys. Chem. 1989, *93*, 5388.
62. Momose, T.; Suzuki, T.; Shida, T. Chem. Phys. Lett. 1984, *107*, 568.
63. Shida, T. J. Phys. Chem. 1978, *82*, 991.
64. Lewis, F.D.; Dyksta, R.E.; Gould, I.R.; Farid, S. J. Phys. Chem. 1988, *92*, 7042.
65. Lewis, F.D.; Bedell, A.M.; Dykstra, R.E.; Elbert, J.E.; Gould, I.R.; Farid, S. J. Am. Chem. Soc. 1990, *112*, 8055.
66. Kuriyama, Y.; Arai, T.; Sakuragi, H.; Tokumaru, K. Chem. Phys. Lett. 1990, *173*, 253.
67. Kuriyama, Y.; Sakuragi, H.; Tokumaru, K.; Yoshida, Y.; Tagawa, S. Bull. Chem. Soc. Jpn. 1993, *66*, 1852.
68. Eriksen, J.; Foote, C.S. J. Am. Chem. Soc. 1980, *102*, 6083.
69. Lewis, F.D.; Petisce, J.R.; Oxman, J.D.; Nepras, M.J. J. Am. Chem. Soc. 1985, *107*, 203.
70. Konuma, S.; Aihara, S.; Kuriyama, Y.; Misawa, H.; Akaba, R.; Sakuragi, H.; Tokumaru, K. Chem. Lett. 1991, 1897.
71. Tojo, S.; Morishima, K.; Ishida, A.; Majima, T.; Takamuku, S. J. Org. Chem. 1995, *60*, 4684.
72. Hamill, W.H. In *Radical Ions*. Kaiser, E.T., Kevan, L., Eds.; Interscience: New York, 1968; 405 pp.
73. Shida, T. *Electronic Absorption Spectra of Radical Ions*; Elsevier: Amsterdam, 1988.
74. Yamamoto, Y.; Aoyama, T.; Hayashi, K. J. Chem. Soc., Faraday Trans. I 1988, *84*, 2209.
75. Kikuchi, O.; Oshiyama, T.; Takahashi, O.; Tokumaru, K. Bull. Chem. Soc. Jpn. 1992, *65*, 2267.
76. Hirota, S.; Ogura, T.; Appelman, E.H.; Shinzawaitoh, K.; Yoshikawa, S.; Kitagawa, T. J. Am. Chem. Soc. 1994, *116*, 10564.
77. Tojo, S.; Toki, S.; Takamuku, S. J. Org. Chem. 1991, *56*, 6240.
78. Takamuku, S.; Komitsu, S.; Toki, S. Radiat. Phys. Chem. 1989, *34*, 553.
79. Badger, B.; Brockleh, B. Trans. Faraday Soc. 1969, *65*, 2576.
80. Ichikawa, T.; Ohta, N.; Kajioka, H. J. Phys. Chem. 1979, *83*, 284.
81. Brede, O.; Bos, J.; Helmstreit, W.; Mehnert, R. Radiat. Phys. Chem. 1982, *19*, 1.
82. Yamamoto, Y.; Chikai, Y.; Hayashi, K. Bull. Chem. Soc. Jpn. 1985, *58*, 3369.
83. Ebbesen, T.W. J. Phys. Chem. 1988, *92*, 4581.
84. Tsuchida, A.; Yamamoto, M. J. Photochem. Photobiol. A 1992, *65*, 53.
85. Sluggett, G.W.; McGarry, P.F.; Koptyug, I.V.; Turro, N.J. J. Am. Chem. Soc. 1996, *118*, 7367.
86. Ganapathy, S.; Dockery, K.P.; Sopchik, A.E.; Bentrude, W.G. J. Am. Chem. Soc. 1993, *115*, 8863.
87. Barth, A.; Hauser, K.; Mantele, W.; Corrie, J.E.T.; Trentham, D.R. J. Am. Chem. Soc. 1995, *117*, 10311.
88. Furuta, T.; Torigai, H.; Sugimoto, M.; Iwamura, M. J. Org. Chem. 1995, *60*, 3953.
89. Givens, R.D.; Kueper, L.W. J. Chem. Soc., Chem. Rev. 1993, *93*, 55.
90. Nakamura, M.; Majima, T. Chem. Commun. 1997, 1291.
91. Nakamura, M.; Dohno, R.; Majima, T. J. Org. Chem. 1998, *63*, 6258.
92. Powell, R.L.; Hall, C.D. J. Am. Chem. Soc. 1969, *91*, 5403.
93. Pandey, G.; Hajra, S.; Chorai, M.K.; Kumar, K.R. J. Am. Chem. Soc. 1997, *119*, 8777.
94. Pandey, G.; Pooranchand, D.; Bhalerao, U.T. Tetrahedron Lett. 1991, *47*, 1745.
95. Yasui, S.; Shioji, K.; Ohno, A.; Yoshihara, M. J. Org. Chem. 1995, *60*, 2099.
96. Nakamura, M.; Miki, M.; Majima, T. J. Chem. Soc., Perkin Trans. 2, 2000, 1447.
97. Hamill, W.H. In *Radical Ions*; Kaiser, E.T., Kevan, L., Eds.; Interscience: New York, 1968; 321 pp.
98. In *CRC Handbook of Radiation Chemistry*; Tabata, Y., Ed.; CRC Press: Boca Raton, 1991; 395 pp.

99. Arai, S.; Dorfman, L.M. J. Chem. Phys. 1964, *41*, 2190.
100. Arai, S.; Grev, D.A.; Dorfman, L.M. J. Chem. Phys. 1967, *46*, 2537.
101. Arai, S.; Kira, A.; Imamura, M. J. Phys. Chem. 1977, *81*, 110.
102. Shida, T.; Iwata, S. J. Chem. Phys. 1972, *56*, 2858.
103. Shida, T.; Hamill, W.H. J. Chem. Phys. 1966, *44*, 2369.
104. Shaede, E.A.; Dorfman, L.M.; Flynn, G.F.; Walker, D.C. Can. J. Chem. 1973, *51*, 3905.
105. Cserhegy, A.; Chaudhur, J.; Franta, E.; Jagurgro, J.; Szwarc, M. J. Am. Chem. Soc. 1967, *89*, 7129.
106. Wang, H.C.; Levin, G.; Szwarc, M. J. Am. Chem. Soc. 1977, *99*, 2624.
107. Dadley, D.; Evans, A.G. J. Chem. Soc. B 1967, 418.
108. Levin, G.; Jagurgro, J.; Szwarc, M. J. Am. Chem. Soc. 1970, *92*, 2268.
109. Yamamoto, S.; Yamamoto, Y.; Hayashi, K. Bull. Chem. Soc. Jpn. 1991, *64*, 346.
110. Aoyama, T.; Yamamoto, Y.; Hayashi, K. J. Chem. Soc. Faraday Trans. I 1989, *85*, 3353.
111. Yamamoto, Y.; Nishida, S.; Ma, X.H.; Hayashi, K. J. Phys. Chem. 1986, *90*, 1921.
112. Hirayama, F. J. Chem. Phys. 1965, *42*, 3163.
113. Ichinose, N.; Hobo, J.; Tojo, S.; Majima, T. Chem. Phys. Lett. 2000, *330*, 97.
114. Ishitani, A.; Nagakura, S. Mol. Phys. 1967, *12*, 1.
115. Kira, A.; Ishiwata, M.; Imamura, M.; Tabata, Y. Radiat. Phys. Chem. 1979, *15*, 663.
116. Tsuchida, A.; Masuda, N.; Yamamoto, M.; Nishijima, Y. Macromolecules 1986, *19*, 1299.
117. Lim, B.T.; Lim, E.C. J. Chem. Phys. 1983, *78*, 5262.
118. Lund, H.; Carlsson, H.S. Acta Chem. Scand., Ser. B 1978, *B32*, 505.
119. Nelleborg, P.; Lund, H.; Eriksen, J. Tetrahedron Lett. 1985, *26*, 1773.
120. Eriksen, J.; Jerrgensen, K.A.; Linderberg, J.; Lund, H. J. Am. Chem. Soc. 1984, *106*, 5083.
121. Eriksen, J.; Lund, H.; Nyvad, A.I. Acta Chem. Scand., Ser. B 1983, *B37*, 459.
122. Sazhnikov, V.A.; Rakhmatov, M.; Alfimov, M.V. Chem. Phys. Lett. 1980, *71*, 33.
123. Ebbesen, T.W.; Akaba, R.; Tokumaru, K.; Washio, M.; Tagawa, S.; Tabata, Y. J. Am. Chem. Soc. 1988, *110*, 2147.
124. Sumiyoshi, T.; Sakai, H.; Kawasaki, M.; Katayama, M. Chem. Lett. 1992, 617.
125. Sumiyoshi, T.; Kawasaki, M.; Katayama, M. Bull. Chem. Soc. Jpn. 1993, *66*, 2510.
126. Carlsson, H.S.; Lund, H. Acta Chem. Scand., Ser. B 1980, *B34*, 409.
127. Hiratsuka, H.; Yamazaki, T.; Maekawa, Y.; Kajii, Y.; Hikida, T.; Mori, Y. Chem. Phys. Lett. 1987, *139*, 187.
128. Ramamurthy, V.; Caspar, J.V.; Corbin, D.R. J. Am. Chem. Soc. 1991, *113*, 594.
129. Ikematsu, S.; Hikida, T. J. Photochem. Photobiol. A: Chem. 1990, *52*, 193.
130. Oomori, T.; Hikida, T. Chem. Phys. 1993, *178*, 477.
131. Bromberg, A.; Schmidt, K.H.; Meisel, D. J. Am. Chem. Soc. 1985, *107*, 83.
132. Ishida, A.; Yamamoto, K.; Takamuku, S. Bull. Chem. Soc. Jpn. 1992, *65*, 3186.
133. Arai, S.; Grev, D.A.; Dorfman, L.M. J. Chem. Phys. 1967, *46*, 2572.
134. Kira, A.; Arai, S.; Imamura, M. J. Chem. Phys. 1971, *54*, 4890.
135. Kira, A.; Arai, S.; Imamura, M. J. Phys. Chem. 1972, *76*, 1119.
136. Sawada, U.; Holroyd, R.A. J. Phys. Chem. 1981, *85*, 541.
137. Vanderde, Gm.; Dousma, J.; Speiser, S.; Kommande, J. Chem. Phys. Lett. 1973, *20*, 17.
138. Oshiyama, T.; Takahashi, O.; Morihashi, K.; Kikuchi, O.; Tokumaru, K. Bull. Chem. Soc. Jpn. 1993, *66*, 1622.
139. Saltiel, J.; Waller, A.S.; Sears, D.F. J. Am. Chem. Soc. 1993, *115*, 2453.
140. Bartocci, G.; Mazzucato, U. Chem. Phys. Lett. 1977, *47*, 541.
141. Saltiel, J.; Waller, A.S.; Sears, D.F.; Garrett, C.Z. J. Phys. Chem. 1993, *97*, 2516.
142. Heisel, F.; Miehe, J.A.; Sipp, B. Chem. Phys. Lett. 1979, *61*, 115.
143. Courtney, S.H.; Balk, M.W.; Philips, L.A.; Webb, S.P.; Yang, D.; Levy, D.H.; Fleming, G.R. J. Chem. Phys. 1988, *89*, 6697.
144. Lee, M.; Bain, A.J.; Mccarthy, P.J.; Han, C.H.; Haseltine, J.N.; Smith, A.B.; Hochstrasser, R.M. J. Chem. Phys. 1986, *85*, 4341.
145. Todd, D.C.; Jean, J.M.; Rosenthal, S.J.; Ruggiers, A.J.; Yang, D.; Fleming, G.R. J. Chem. Phys. 1990, *93*, 8658.

146. Burrows, H.D.; Kemp, T.J.; Greatore, D. J. Phys. Chem. 1972, 76, 20.
147. Shkrob, I.A.; Sauer, M.C.; Liu, A.D.; Crowell, R.A.; Trifunac, A.D. J. Phys. Chem. A 1998, 102, 4976.
148. Cook, A.R.; Curtiss, L.A.; Miller, J.R. J. Am. Chem. Soc. 1997, 119, 5729.
149. Zimmer, K.; Hoppmeier, M.; Schweig, A. Chem. Phys. Lett. 1998, 293, 366.
150. Ichinose, N.; Tanaka, T.; Kawanishi, S.; Suzuki, T.; Endo, K. J. Phys. Chem. A 1999, 103, 7923.
151. Ichinose, N.; Majima, T. Chem. Phys. Lett. 2000, 322, 15.
152. Ichinose, N.; Tanaka, T.; Kawanishi, S.; Majima, T. Chem. Phys. Lett. 2000, 326, 293.
153. Ichinose, N.; Majima, T. The Spectrum 2001, 13, 14.
154. Grodkowski, J.; Neta, P. J. Phys. Chem. 1984, 88, 1205.
155. Sujdak, R.J.; Jones, R.L.; Dorfman, L.M. J. Am. Chem. Soc. 1976, 98, 4875.
156. Mehnert, R.; Brede, O.; Bos, J.; Naumann, B. Bunsenges. Phys. Chem. 1979, 83, 992.
157. Alfassi, Z.B.; Mosseri, S.; Neta, P. J. Phys. Chem. 1989, 93, 1380.
158. Emmi, S.S.; Beggiato, G.; Casalbore-Miceli, G. Radiat. Phys. Chem. 1989, 33, 29.
159. Chateauneuf, J.E. J. Am. Chem. Soc. 1990, 112, 442.
160. Shoute, L.C.T.; Neta, P. J. Phys. Chem. 1990, 94, 2447.
161. Mah, S.; Yamamoto, Y.; Hayashi, K. J. Phys. Chem. 1983, 87, 297.
162. Yamamoto, Y.; Nishida, S.; Hayashi, K. J. Chem. Soc., Faraday Trans. 1 1987, 83, 1795.
163. Ichinose, N.; Tojo, S.; Majima, T. Chem. Lett 2000, 1126.
164. Gadosy, T.A.; Shukla, D.; Johnston, L.J. J. Phys. Chem. A 1999, 103, 8834.
165. Schloman, W.W.; Morrison, H. J. Am. Chem. Soc. 1977, 99, 3342.

23

Application of Radiation Chemistry to Nuclear Technology

Yosuke Katsumura
The University of Tokyo, Tokyo, Japan

INTRODUCTION

Nuclear technology is defined as a technology to use the nuclear energy through nuclear fission and fusion reactions as nuclear power and to develop useful processing and application with radiation and radioisotopes. Therefore radiation-induced effect is one of the important issues when nuclear technology is developed and employed. In the present chapter, three important applications of radiation chemistry in nuclear technology will be discussed: coolant water in nuclear power stations, spent nuclear fuel reprocessing, and radioactive high-level waste repository.

1. RADIOLYSIS OF WATER AT ELEVATED TEMPERATURE—RADIATION EFFECT OF COOLANT WATER IN NUCLEAR REACTORS

1.1. Importance for the Understanding of the Radiolysis at Elevated Temperatures

All over the world, 432 nuclear power reactors are under operation and more than 36 GW of electricity could be produced as of December 31, 2001. There are several types of reactors such as boiling water reactor (BWR), pressurized water reactor (PWR), Canada deuterium uranium (CANDU), and others. In these reactors, light water is normally used not only as a coolant, but also as a moderator. On the contrary, in CANDU reactors, heavy water is taken. It is widely known that the quality control of coolant water, the so-called "water chemistry," is inevitably important for keeping the integrity of the plant.

   The coolant water at around 300°C is irradiated by both γ-rays and fast neutrons at the core region with a dose rate of several 10 kGy/sec, which is more than 3 orders of magnitude of the dose higher than those obtained in conventional ^{60}Co γ-ray irradiation facilities. As a result, O_2 and H_2O_2 are formed as radiolysis products. It is well known that the concentrations of O_2 and H_2O_2 strongly affect the stress corrosion cracking (SCC) and the electrochemical corrosion potential of the primary circuit. In order to avoid the SCC, it is necessary to keep the lower level of these concentrations and hydrogen injection (HWC), and, recently, noble metal chemical addition (NMC) has been widely employed in BWRs. In the hydrogen injection, both H_2O_2 and O_2 would be converted into water molecules by radiolytic processes (see below).

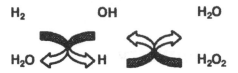

To grasp the chemical condition of water in the pressure vessel, direct measurement is practically impossible because of high pressure, high temperature, and intense radiation. In order to predict the concentrations of water decomposition products, a computer simulation should be applied. This idea was found in 1960s [1–3]. To perform the simulation, both a set of G-values for water decomposition products and a set of reactions for transient species are necessary. For these two decades, much effort has been made in Sweden, Denmark, United Kingdom, Canada, and Japan to evaluate the G-values and rate constants of the reactions at elevated temperatures up to 300°C, and now there are practically enough accumulated data. There are several reviews of water radiolysis at elevated temperatures [4–7] and examples of practical application of the radiolysis in reactors [8,9].

1.2. Temperature Dependence of Water Decomposition Products up to 300°C

The coolant water at around 300°C is irradiated mainly at the core of the reactor. At the initial stage to determine the G-values of water decomposition products at elevated temperatures, the Fricke dosimeter was chosen [10–14] because the mechanism of the reaction has been established. Since the reactions in neutral solution are of practical interest, intensive measurement of the G-values of water decomposition products at elevated temperatures in neutral solutions has been done [15–21].

For the measurement, techniques of product analysis and pulse radiolysis were employed. In product analysis, H_2 yield in degassed 2 mM KBr or 1 mM KNO_2 solutions were taken to evaluate $g(H_2)$ and degassed solutions of 1 mM Cd^{2+} with 10 mM methanol, 1 mM acetone with 10 mM methanol, and 1 mM acetone with 10 mM 2-rpropanol have been selected for $g(H) + g(H_2)$. For the measurement of $g(e_{aq}^-) + g(H) + g(H_2)$, degassed solution of 1 mM $HClO_4$ with methanol was taken. In pulse radiolysis measurement, $Fe(CN)_6^{3-}$ and $CO_3^{\cdot-}$ were used for $g(OH)$ evaluation and 0.25 mM methyl viologen solution with 10 mM $tert$-butanol for $g(e_{aq}^-)$. Many other systems have been used. In the presence of formate, the yield of $G(MV^+)$ would be equal to $g(e_{aq}^-) + g(H) + g(OH)$. Here $g(X)$ indicates the primary yield of water decomposition product X and $G(Y)$ is corresponding to the experimentally determined yield of species Y.

The scavenging capacity is defined as the product of the rate coefficient of the scavenging reaction and the scavenger concentration. The inverse of the scavenging capacity is equal to the lifetime of the reactive species. The scavenging capacity for the transient species at room temperature in each solution is 10^7 sec^{-1}. In other words, these values are corresponding to the yields of the transients in the time range of 100 nsec. With increasing temperature, the reactivity of the scavenger would be accelerated and dependent on the kind of scavenger. However, experimental data using different scavengers do not show significant difference. Furthermore, at elevated temperatures, the time range where the yield was determined is not clearly shown yet because the rate constants of the reactions are not precisely known.

Reported values as a function of temperature are summarized in Fig 1 [17–20,22–24]. In spite of the experimental error, temperature dependence is similar among the evaluations. The G-values of e_{aq}^- and OH are increasing with increasing temperature. On the contrary, the value of H_2O_2 is decreasing. The value for H_2 tends to increase slightly. Elliot et al. [19] proposed the fitting equations as follows.

$$g(e_{aq}^-) = 2.56 + 3.40 \times 10^{-3}t \tag{1}$$

$$g(OH) = 2.64 + 7.17 \times 10^{-3}t \tag{2}$$

$$g(H_2) = 0.43 + 0.69 \times 10^{-3}t \tag{3}$$

$$g(H_2) + G(H) = 0.97 + 1.98 \times 10^{-3}t \tag{4}$$

$$g(H_2O_2) = 0.72 - 1.49 \times 10^{-3}t \tag{5}$$

Here t stands for temperature in °C.

From the difference between Eqs. (4) and (3), the temperature dependence of $g(H)$ can be calculated as

$$g(H) = 0.54 + 1.28 \times 10^{-3}t \tag{6}$$

These lines are also drawn in Fig. 1. It is clear that the agreement is quite nice. Similar fitting equations in heavy water for CANDU reactors were also obtained [19]. Furthermore, temperature-dependence measurements with fast neutrons and high LET ion beams were also investigated [20,23].

Common features of water radiolysis at elevated temperatures up to 300°C are summarized as below.

1. $g(e_{aq}^-)$, $g(OH)$, $g(H_2)$, and $g(H)$ are all increasing almost linearly with increasing temperature up to 300°C. About 40% increase was found for $g(e_{aq}^-)$ and more than 80% for $g(OH)$. This is due to the lower reactivity of OH as compared with that of e_{aq}^- at elevated temperatures.
2. $g(H_2O_2)$ is decreasing with temperature. In spite of the short lifetime of H_2O_2 due to the thermal decomposition, 50 min at 150°C, 5–10 min at 150°C, and 1 min at 200°C, this decrease is obvious.

As pointed out before, time scale of scavenging, corresponding to G-values at elevated temperatures, is not clear yet and further investigation is needed.

1.3. Rate Constants of the Reactions up to 300°C

It is known that more than 30 reactions are needed to reproduce the radiation-induced reactions occurring in pure water. Intensive measurements with a pulse radiolysis method have been done at elevated temperature up to 300°C [25–42], and the temperature dependence of some reactions does not exhibit a straight line but a curved one in Arrhenius plot. These examples are the reactions of the hydrated electron with N_2O, NO_3^-, NO_2^-, phenol, SeO_4^{2-}, $S_2O_3^{2-}$, and Mn^{2+} [33,35], and two examples, $e_{aq}^- + NO_3^-$ and $e_{aq}^- + NO_2^-$, are shown in Fig. 2. The rate constant for the reaction of hydrated electron with NO_3^- is near diffusion-controlled reaction at room temperature and is increasing with increasing temperature. Above 100°C, the rate does not increase and reaches the maximum at 150°C, and then decreases. Therefore the curve is concave upward in Arrhenius plot.

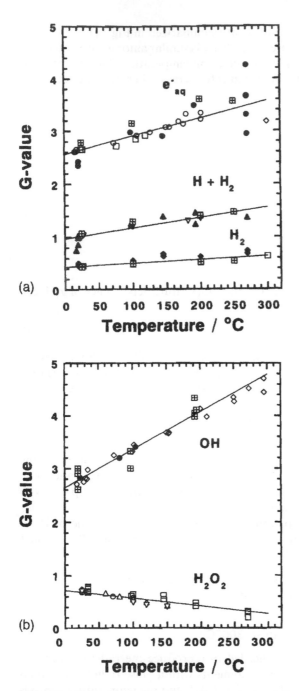

Figure 1 Variations of G-values (in aprt./100 eV) for the radiolysis of light water as a function of temperature; (a) "reducing" species G-values for $G(e_{aq}^-)$, $G(H + H_2)$, and $G(H_2)$ and (b) "oxidizing" species. Data are taken from Ref. 19 (○□◇▽), Ref. 18 (●▲◆), and Refs. 20 and 23 (⊞) for reducing species and from Ref. 19 (◇●○), Ref. 17 (⊞□), and Ref. 24 (△◆▽) for oxidizing species. It is noted that because of the large uncertainties in the G-values determined in Refs. 20 and 23 for both $G(OH)$ and $G(H_2O_2)$ from $HClO_4 + 10^{-2}$ mol kg^{-1} methanol chemical system at above 100°C, these data are not shown in (b). Lines are corresponding to Eqs. (1)–(5).

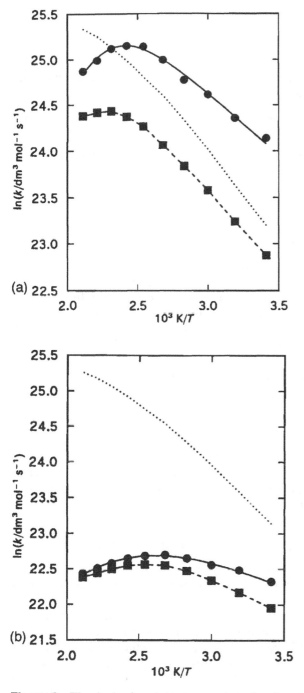

Figure 2 The Arrhenius plot rate constants for the reactions of e_{aq}^- with NO_3^- (a) and NO_2^- (b): k_{obs} (■), k_{react} (●), k_{diff} (·· ·), calculated fit of k_{react} (—). A_1/M^{-1} and A_2 are $(5.7\pm2.1) \times 10^{12}$ and $(4.4 \pm 6.0) \times 10^5$, respectively, for (a). A_1/M^{-1} and A_2 are $(2.0 \pm 0.8) \times 10^{11}$ and $(2.1 \pm 0.7) \times 10^3$, respectively, for (b). (From Ref. 35.)

In general, observed reaction rate constant is written as below with a diffusion term and a reaction term [43].

$$1/k_{obs} = 1/k_{diff} + 1/k_{react} \tag{7}$$

The reaction of the hydrated electron would be considered as a charge transfer reaction and elementary processes can be described as below [44].

$$D + A \overset{K_A}{\rightleftharpoons} D/A \overset{k_1}{\underset{k_2}{\rightleftharpoons}} [D/A]* \overset{v}{\rightleftharpoons} [D^+/A^-]* \overset{k_3}{\rightarrow} [D^+/A^-] \rightarrow product \tag{8}$$

Based on the above scheme, an equation for the overall reaction is derived and setting $k_3 = k_2$ gives a next one [44].

$$k_{react} = K_A k_1/(2 + (k_2/v)) \tag{9}$$

This is rearranged into Eq. (10)

$$k_{react}(T) = A_1 \exp(-E_1/RT)/[2 + A_2 \exp(-E_2/RT)] \tag{10}$$

Here $K_A k_1 = A_1 \exp(-E_1/RT)$ and $k_2/v = A_2 \exp(-E_2/RT)$.

The temperature dependence of k_{diff} can be evaluated by the diffusion coefficients of the reactants as a function of temperature. The term of k_{obs} is experimentally observed. Then, k_{react} can be evaluated by fitting procedures. In Fig. 2, k_{obs}, k_{diff}, and k_{react} were all shown [35]. Similar analysis was made for the experimentally obtained reactions of OH with $Fe(CN)_6^{4-}$, OH, and HCO_2^-, H with MnO_4^- and O_2, e_{aq}^- with H^+, H_2O_2, and Cd^{2+}, and so on [34,37,42].

Recent measurement of the hydrated electron at elevated temperature by Shiraishi et al. [41] shows clearly the existence of an equilibrium between the hydrated electron and the proton.

$$e_{aq}^- + H^+ \rightleftharpoons H, \tag{11}$$

In order to calculate the reactions in pure water at elevated temperatures, more than 30 reactions as a function of temperature are needed, especially for the reactions in nuclear reactors. Much work has been done and a reaction set for calculation at 285°C and 300°C is summarized in Table 1 [4] for light water and in Ref. 5 for heavy water.

1.4. Theoretical Calculations of High-Temperature Radiolysis of Water

In order to reproduce the temporal behavior of water decomposition products, two theoretical approaches based on spur diffusion model and Monte Carlo calculations have been developed.

In the spur diffusion calculation, initial distribution of the water decomposition after thermalization is assumed as a spherical one and is set normally as a Gaussian distribution where hydrated electron has a radius of 2.3 nm, which is larger than the radii of 0.85 nm for OH, H, H_3O^+, H, H_2, and H_2O_2 at ambient temperature. Then, the spur reactions start taking place after 1 psec during the diffusion of each species. These calculations have explained and predicted well the time dependence of the water decomposition products at room temperature. This method was successfully applied to the calculation for the radiolysis with high LET radiation, where a cylindrical initial distribution is taken. Essence of the idea has been clearly summarized by Draganic and Draganic [45]. This calculation has been extended to the water radiolysis at elevated temperatures. Although the initial yields of water decomposition products at 1 psec are fixed as room temperature, the initial dis-

Table 1 Calculated Rate Constants for all Reactions in the Radiolysis of Light Water at 25°C, 285°C, and 300°C

No.	Reaction	k_{25}	k_{285}	k_{300}
(1)	$e_{aq}^- + e_{aq}^- \rightarrow$	6.44E09[a]	2.93E11	3.29E11
(2)	$e_{aq}^- + H \rightarrow$	2.64E10	3.68E11	3.98E11
(3)	$e_{aq}^- + OH \rightarrow$	3.02E10	1.34E11	1.40E11
(4)	$e_{aq}^- + H_2O_2 \rightarrow$	1.41E10	2.53E11	2.76E11
(5)	$e_{aq}^- + O_2 \rightarrow$	1.79E10	2.57E11	2.78E11
(6)	$e_{aq}^- + O_2^- \rightarrow$	1.28E10	1.65E11	1.78E11
(7)	$e_{aq}^- + HO_2^- \rightarrow$	1.28E10	1.65E11	1.78E11
(8)	$H + H \rightarrow$	5.43E09	9.20E10	1.00E11
(9)	$H + OH \rightarrow$	1.53E10	6.58E10	6.87E10
(10)	$H + H_2O_2 \rightarrow$	5.16E07	1.03E09	1.13E09
(11)	$H + O_2^- \rightarrow$	1.32E10	9.72E10	1.03E11
(12)	$H + HO_2 \rightarrow$	9.98E09	7.33E10	7.78E10
(13)	$H + O_2 \rightarrow$	9.98E09	7.33E10	7.78E10
(14)	$OH + OH \rightarrow$	4.74E09	2.00E10	2.09E10
(15)	$OH + H_2 \rightarrow$	4.15E07	1.26E09	1.40E09
(16)	$OH + H_2O_2 \rightarrow$	2.87E07	5.42E08	5.91E08
(17)	$OH + HO_2 \rightarrow$	1.08E10	3.10E10	3.20E10
(18)	$OH + O_2^- \rightarrow$	1.10E10	8.44E10	8.97E10
(19)	$HO_2 + HO_2 \rightarrow$	6.64E05	2.41E07	2.69E07
(20)	$HO_2 + O_2^- \rightarrow$	7.58E07	3.82E08	4.01E08
(21)	$O_2^- + O_2^- \rightarrow$	—	—	3.50E07
Equilibria				
(22)	$H_2O \rightarrow$	1.95E−05	7.57E−02	6.90E−02
(−22)	$H^+ + OH^- \rightarrow$	1.10E11	9.76E11	1.07E12
(23)	$H_2O_2 \rightarrow$	7.86E−02	4.40E01	4.15E01
(−23)	$H^+ + HO_2^- \rightarrow$	4.78E10	5.42E11	6.12E11
(24)	$H_2O_2 + OH^- \rightarrow$	1.27E10	1.37E11	1.47E11
(−24)	$HO_2^- + H_2O \rightarrow$	1.36E06	1.31E08	1.39E08
(25)	$H \rightarrow$	6.32E00	1.36E05	1.40E05
(−25)	$H^+ + e_{aq}^- \rightarrow$	2.25E10	4.31E11	5.15E11
(26)	$e_{aq}^- + H_2O \rightarrow$	1.57E01	8.28E03	9.94E03
(−26)	$e_{aq}^- + OH^- \rightarrow$	2.49E07	3.40E10	4.22E10
(27)	$OH \rightarrow$	7.86E−02	4.40E01	4.15E01
(−27)	$H^+ + O^- \rightarrow$	4.78E10	5.42E11	6.12E11
(28)	$OH + OH^- \rightarrow$	1.27E10	1.37E11	1.47E11
(−28)	$O^- + H_2O \rightarrow$	1.36E6	1.31E08	1.39E08
(29)	$HO_2 \rightarrow$	7.14E05	2.70E05	1.98E05
(−29)	$H^+ + O_2^- \rightarrow$	4.78E10	5.42E11	6.12E11
(30)	$HO_2 + OH^- \rightarrow$	1.27E10	1.37E11	1.47E11
(−30)	$O_2^- + H_2O \rightarrow$	1.36E06	1.31E08	1.39E08
Alkaline solution				
(31)	$O^- + H_2 \rightarrow$	1.21E08	1.62E09	1.75E09
(32)	$O^- + H_2O_2 \rightarrow$	5.53E08	1.04E10	1.13E10
(33)	$OH + HO_2 \rightarrow$	8.29E09	1.56E11	1.70E11
(34)	$O^- + OH \rightarrow$	7.60E09	3.23E10	3.38E10
(35)	$e_{aq}^- + HO_2 \rightarrow$	3.50E09	6.33E10	6.90E10
(36)	$e_{aq}^- + O^- \rightarrow$	2.31E10	1.02E11	1.07E11
(37)	$O^- + O_2 \rightarrow$	3.70E09	3.04E10	3.24E10
(−37)	$O_3^- \rightarrow$	2.68E03	1.45E07	1.88E07

The unit for the second-order reactions is M^{-1} sec^{-1} and, for the first order, they are sec^{-1}. For a reaction between like species, the value of k, not $2k$, is given.
[a] Read as 6.44×10^9.
Source: Ref. 4.

tribution would be varied as a function of temperature. In addition, the rate constants dependent on temperature were also taken into consideration. In the first attempt by Pimblott and LaVerne, the radius for the hydrated electron changes as a function of temperature was assumed to be expressed with a certain activation energy and compared with experimental results. The activation energy of 9.6 kJ/mol was selected to reproduce the experimental data [46]. Swiatla-Wojcik and Buxton [47] considered that the radii are changed according to the density change of water at elevated temperature. These calculations could reproduce the yields of water decomposition products up to 300°C for radiolysis with low LET radiation. Swiatla-Wojcik and Buxton extended the calculation for high LET radiation and for heavy water [48,49].

The Monte Carlo calculation is a stochastic approach, while the spur diffusion calculation has been classified as a deterministic approach. The most important difference is that, contrary to the concentration gradient playing an essential role in spur diffusion calculation, the stochastic nature of the reactions is taken into consideration. In the Monte Carlo calculation, the initial distribution is derived from the cross-section data between electrons and water molecules, thermalization distance, and geminate recombination cross section. The latter two parameters are hardly predicted from theory and are taken as adjustable ones. Once the initial distribution is fixed at 1 psec time, mutual diffusion and reaction processes are calculated with the aid of independent reaction time method [50–55]. This calculation has been developed by several groups and has been applied to radiation biology especially to the radiation damage of DNA [56]. The Monte Carlo calculation has been applied to the high-temperature radiolysis by Jay-Gerin et al. and could reproduce well the experimentally evaluated water decomposition products as a function of temperature up to 300°C and LET dependence [57–61].

Both calculations are able to reproduce well the experimental results for high-temperature radiolysis. While in the spur diffusion model the initial distribution becomes broader with increasing temperature, the thermalization distance becomes smaller in Monte Carlo simulation. Although it seems contradictory, it is difficult to explain the origin of the difference.

1.5. Above 300°C and Supercritical Water

Above 374°C and 22.1 MPa, liquid and gas phase of water are merged into a single phase called supercritical water, as shown in the phase diagram of water (Fig. 3). The property of the supercritical water is very different from that of water at room temperature. Its density is dependent on pressure, for example, at 400°C, 0.166 g/cm^3 under 25 MPa and 0.475 g/cm^3 under 35 MPa. The dielectric constant of water at room temperature is 79, but, in supercritical state, the value is less than 10. Thus the supercritical water is not a highly polar solvent anymore. In fact, in supercritical state, water can be mixed with benzene at any composition. On the contrary, inorganic acid, alkaline, and salt compounds have much lower dissociation constant in higher temperature and in the critical one. At room temperature, hydrogen bonds are expanded all over the water molecules. However, the hydrogen network is broken in supercritical water and there exist monomers and dimers. Furthermore, pK_w of water becomes large and also strongly dependent on temperature and pressure in supercritical condition. By using the above peculiar property, many possible applications of supercritical water have been proposed. Supercritical water oxidation (SCWO) received much attention and intensive development has been done because hazardous organic compounds such as polychlorobiphenyl (PCB) are quickly and completely decomposed into CO_2, H_2O, and Cl^- in supercritical water in the presence of oxidant like O_2 or H_2O_2.

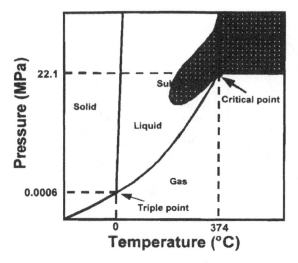

Figure 3 The phase diagram of water. Critical temperature and pressure of light water, H_2O, are 374°C and 22.1 MPa.

This can be used for the decomposition of chemical weapons. Another application is that the supercritical water is a special reaction medium to synthesize the new advanced materials and medicines. There appeared several reviews on the chemistry and application of supercritical water [62–66].

Recently, a new concept of nuclear reactor using supercritical water as a coolant has been proposed by a group of the University of Tokyo [67], and this was selected a Generation IV reactor by DOE in the United States. First pulse radiolysis work was reported by Ferry and Fox in 1998 [68] and 1999 [69]. Then, in 1999, two more groups at the University of Tokyo [70] and Argonne National Laboratory [71] started the pulse radiolysis work on supercritical water. Of course, the main purpose of the study is to evaluate the G-values of water decomposition and to obtain the rate constants for the reactions in forthcoming supercritical water-cooled reactors. However, the study has much interest from the viewpoints of radiation chemistry and practical applications because the radiation-induced reaction would be affected by the change of the dielectric constant of water and the reaction of OH plays a key role for SCWO. Thus the assessment of the OH radical in supercritical water provides basic data to practical application of SCWO. Further, the results obtained in supercritical water would provide good subjects for theoretical consideration.

For pulse radiolysis study of supercritical water, a specially designed experimental system is taken as shown in Fig 4. The sample solution is pumped out by using an HPLC pump and the solution temperature is raised at a preheater and the solution attains the desired temperature in an irradiation cell. The cell has two optical windows made of sapphire in place of quartz, which cannot be used because of insufficient strength and high solubility at supercritical water condition. The solution passing through the cell is cooled down and going to a drain through a back-pressure regulator. The cell is made of Hastelloy, a Ni-based alloy, which is resistant to corrosion of supercritical water.

The hydrated electron is one of the main water decomposition products, and it is known that the absorption band of the hydrated electron shifts to longer wavelength with

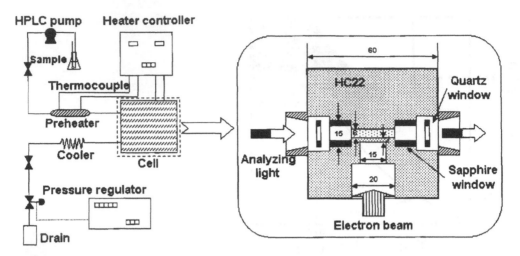

Figure 4 Schematic diagram for the pulse radiolysis of supercritical water and cell structure.

increasing temperature [72–74]. Behavior of the hydrated electron has been investigated above 300°C and in supercritical water [70,71,75]. The spectrum shift continues beyond 300°C, and its peak is placed at around 1200 nm in supercritical water. Under 25 MPa, the change of the absorption spectrum of the hydrated electron as a function of temperature is shown in Fig. 5(a) [76]. In supercritical fluids, even when the temperature is fixed, the density varies with changing pressure. At fixed temperature of 380°C, the change of the absorption at different pressure conditions is shown in Fig. 5(b) [76]. It is clear that the spectrum is strongly dependent on temperature and pressure in supercritical water. This is strongly related to the solvation structure of the hydrated electron and the number of the surrounding water molecules. However, precise analysis would be a future subject. Separated evaluation of G-value or absorption coefficient of the hydrated electrons through this measurement seems impossible because only the product of G and absorption coefficient (ε) can be determined through the optical measurement. Only when the value of G or ε is known separated evaluation becomes possible. It is noted that the above data were obtained mainly in heavy water (D_2O) and not in light water (H_2O). The light water has an absorption band above 1200 nm, and it is not easy to do the absorption measurement above 1200 nm. On the contrary, the D_2O is transparent up to 1600 nm. The critical temperature and pressure for D_2O are 643.9 K and 21.7 MPa, respectively, not quite different from those of H_2O: 647.1 K and 22.1 MPa [77].

Other transient radicals such as $(SCN)_2^{\cdot-}$ [78], carbonate radical $(CO_3^{\cdot-})$ [79], Ag^0 and Ag^{2+} [80], and benzophenone ketyl and anion radicals [81] have been observed from room temperature to 400°C in supercritical water. The $(SCN)_2^{\cdot-}$ radical formation in aqueous solution has been widely taken as a standard and useful dosimeter in pulse radiolysis study [82,83]. The lifetime of the $(SCN)_2^{\cdot-}$ radical is longer than 10 μsec at room temperature and becomes shorter with increasing temperature. This dosimeter is not useful anymore at elevated temperatures. The absorption spectrum of the $(SCN)_2^{\cdot-}$ radical again shows a red shift with increasing temperature, but the degree of the shift is not significant as compared with the case of the hydrated electron. It is known that the $(SCN)_2^{\cdot-}$ radical is equilibrated with $SCN^{\cdot}$, and precise dynamic equilibration as a function of temperature has been analyzed to reproduce the observation [78].

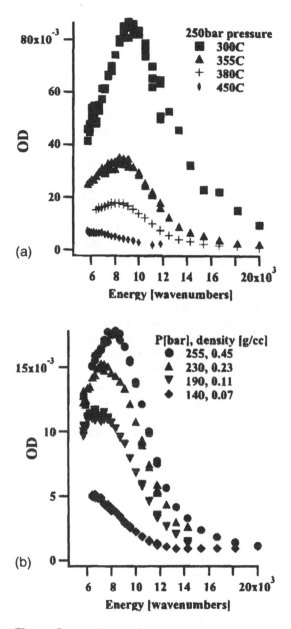

Figure 5 (a) Hydrated electron spectrum in D_2O at 250 bars at increasing temperature and (b) hydrated electron spectrum in D_2O at different pressure at 380°C. (From Ref. 76.)

The carbonate radical has been measured up to 300°C, and the temperature-independent spectrum is known [19]. This independence is kept up to 400°C into supercritical water. Formation and decay behavior were measured precisely [69], and a dimer model for the carbonate radical has been proposed [79].

Benzophenone solution was taken because it is stable in supercritical water at 400°C. It is known that the reaction of hydrated electron with benzophenone produces an anion, and this anion is equilibrated with ketyl radical with pK_a of 9.2 or 9.25 at room temperature [84–86].

$$\phi_2CO + e_{aq}^- \rightarrow \phi_2CO^{\cdot-} \tag{12}$$

$$\phi_2CO^- + H^+ \rightleftharpoons \phi_2C^{\cdot}OH \tag{13}$$

Therefore adjustment of pH can control the chemical form. Both spectra of anion and ketyl radical were measured at different temperatures up to 400°C. The spectrum of the anion shows red shift, but the ketyl radical indicates blue shift.

A pulse radiolysis of Ag^+ solution was studied, and the behavior of formed silver atom (Ag^0) and dimer cation (Ag_2^-) was measured [80]. The absorption band for dimer shows a significant red shift with increasing temperatures, which implies to the CTTS (charge transfer to solvent) character of the band.

It is common that the absorption bands are normally dependent on temperature, and most of them show red shift at elevated temperatures. These behaviors are closely related to the solvation and the formation of hydrogen bonds between transients and water molecules. Precise theoretical consideration is not made yet.

As mentioned before, the spectroscopic behavior of the transients is strongly dependent on temperature and absorption coefficients are not constant with temperature. At present, it is difficult to evaluate the G-values of water decomposition products at elevated temperatures. One of the recent attempts for the evaluation with a use of methylviologen (MV^{2+}) is shown in Fig. 6 [87]. The methylviologen has been taken to evaluate the G-values up to 200°C [16,31]. Similar method was applied to higher temperatures up to 400°C assuming that the absorption coefficients would be estimated by extrapolation of the data known up to 200°C. As shown in Fig. 6(a), the G-value of hydrated electron is increasing up to 300°C, which is consistent with reported as described as summarized in Fig. 1. The value is decreasing above 300°C, and above critical temperature, 374°C, a significant pressure-

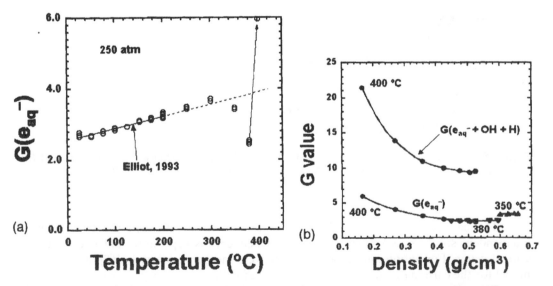

Figure 6 (a) $g(e_{aq}^-)$ as a function of temperature at 250 bars and (b) $g(e_{aq}^-)$ as a function of density at 400°C. (From Ref. 86.)

dependent yield was observed, as shown in Fig. 6(b). When ethanol is taken as a scavenger, the ethanol radical reduce MV^{2+} and the $G(MV^{\cdot+})$ is equal to $g(OH) + g(H) + g(e_{aq}^-)$. It is evident that not only temperature but also pressure, in other words, density of water, are the key parameters.

As for the rate constants above 300°C, little experimental results have been reported. Again, not only temperature, but also pressure dependence was reported. One of the important issues is the decrease of the dielectric constant of water in supercritical water, where it is less than 10 and much lower than the value of 79 at room temperature. It was pointed out that the Coulombic interaction becomes important and the radiolysis of supercritical water resembles that of the organic liquids with low dielectric constant [88]. In addition, it should be noted that the solubility of the solute is quite low and ion pairing would have a significant role [89–91], which reflects the difficulties for the sample preparation in the actual experiment.

Above observations are the special aspects of the peculiarity of water radiolysis in supercritical water and thus much further investigation is needed.

2. RADIATION EFFECTS IN SPENT NUCLEAR FUEL REPROCESSING

2.1. Purex Process

In nuclear power reactors, fission reactions of U-235 are taking place continuously and energy of 200 MeV per fission is released. At the electric power plants, fission energy is converted into thermal energy at first and finally to electricity. In commercial reactors, after the operation for a certain period, new UO_2 fuel is introduced in place of the old one. The used nuclear fuel is called spent fuel and it is composed of 950 kg U, 9 kg Pu, 75 g Np, 140 g Am, 47 g Cm, and 31 kg fission products (FPs) in 1 t of UO_2 after the output of 30 GW day/t [92]. This composition is a typical one and dependent on both the operation history of the reactor and the storage time of the fuel after taking out from the reactor. If Pu is separated from the spent fuel and used as fuel again, high utilization of the fuel is attained. Chemical processing for the separation of fissile material and removal of the FPs from the spent fuel is called reprocessing. Therefore the main purposes of commercial reprocessing are (1) to increase the available energy from fissile and fertile atoms and (2) to reduce hazards and costs for handling the high-level wastes. Two other reasons are sometimes mentioned: (3) to reduce the cost of thermal reactor fuel cycle and (4) to extract valuable by-products from the high active waste. While several reprocessing methods have been proposed, the Purex process is the most popular one and has been widely used all over the world. The Purex is an abbreviation of plutonium uranium refining by extraction, which employs tributyl phosphate, TBP, with $(C_4H_9)_3PO_4$ as an extractant. Since the TBP is rather viscous with a density of 0.973 g/cm^3, in an actual process, a 30-vol.% solution in dodecane (n-$C_{12}H_{26}$) or kerosene as a diluent is employed. The mixture solution makes a quick separation of the organic phase from aqueous phase and exhibits proper viscosity for extraction process. The spent fuel is dissolved into highly concentrated nitric acid, and UO_2^{2+} and Pu^{4+} are the chemical forms for U and Pu, respectively. After the mixing of organic solvent and aqueous phase containing dissolved spent fuel, both UO_2^{2+} and Pu^{4+} in aqueous phase are selectively extracted from aqueous phase to the organic phase and the FPs remain in aqueous phase. This step is called *extraction*. Next step is a separation of Pu from organic phase. After reduction of Pu^{4+} to Pu^{3+} by chemical or electrochemical method, Pu^{3+} goes back to the aqueous phase selectively, while UO_2^{2+} remained in organic phase, which is called *partition*. These processes are precisely explained in textbooks [92,93].

The extractant should satisfy several requirements as follows: (1) high distribution coefficient for U and Pu; (2) low distribution coefficient for FPs; (3) fast attainment to equilibrium; (4) stability to water and acid; (5) radiation resistance; (6) low price; (7) low flammability; (8) big density difference from water; and (9) appropriate viscosity and surface tension. The mixture of TBP in *n*-dodecane satisfies the above requirements and thus has been widely employed as an extractant in the Purex process. Since the processes are operated under strong radiation field not only of γ- and β-rays but also of α-rays, degradation of the solvent takes place and the following detrimental effects appear: (1) loss of U and Pu; (2) low decontamination factor (DF) for U and Pu product; (3) lower extraction efficiency such as precipitation, long separation time, the third phase, and emulsion formation; (4) lower efficiency of the process; and (5) instability and danger of the process due to the criticality. Therefore the understanding of the solvent degradation is inevitable to keep the integrity of the plant and thus much work has been conducted.

In the field of the reprocessing, radiation energy released from radionuclides in the solution is normally expressed in a unit of W hr/l. If all the energy is absorbed in the medium, the absorbed dose can be calculated as:

$$1 \text{ W hr}/l \approx 3600 \text{ J/kg} = 3600 \text{ Gy.} \tag{14}$$

2.2. Radiolysis of TBP and *n*-Dodecane

Radiolysis of TBP has been investigated and reported from the 1950s. The main products are dibutyl phosphate (DBP), monobutyl phosphate (MBP), and gaseous H_2. Therefore CO bond scission occurs easier than CC bond scission. The G-value of DBP is higher than $G(H_2)$. TBP dimer as a polymeric product is also observed. In addition, other minor products such as butane, butene, C_1–C_3 hydrocarbons, phosphoric esters, and butanol are also observed. Reported G-values of TBP radiolysis products are summarized in Table 2 [94–98]. The values are scattered probably because of the difference of the TBP purity in each experiment. From ESR measurement of irradiated phosphades alkyl (R), alkylphosphade $[(RO)(R^{\bullet}O)POO^-]$, phosphoranyl $[((RO)_2P^{\bullet}OO^-)^-]$, and phosphonyl $[(RO)P^{\bullet}OO^-]$ radicals have been observed [99–101]. Pulse radiolysis study reported the observation of the solvated electron in TBP liquid at lower temperatures: at 198 K, the peak position is 1580 nm. The position is shifted shorter wavelength with increasing temperature and the peak position of 1280 nm was observed at 223 K [102]. It was found from the scavenger effect that the solvated electron reacts with TBP leading to alkyl radical and DBP:

$$e_{sol}^- + (C_4H_9O)_3P = O \rightarrow {}^{\bullet}C_4H_9 + (C_4H_9O)_2P(O)O^- \tag{15}$$

Geminate recombination process and reactivity and spectroscopic character of the solvated electron in a variety of phosphates were also investigated [103,104].

It is noted that the decomposition of TBP takes place not only through radiolysis, but also through chemical degradation. In chemical degradation, pyrolysis, oxidative decomposition, and hydrolysis play important roles. Above 150°C, the pyrolysis reaction takes place and the CO bond is broken to lead DBP and MBP formation. The P–O scission is also occurring but minor. The oxidation reaction proceeds quickly in 5 M HNO_3 at 75°C, leading to NO_x. In the presence of uranium salt, this reaction is accelerated and $UO_2(NO_3)_2$ is formed, followed by the formation of butane, butene, N_2, NO_2, CO, CO_2, NO, butyne, and butanol. It is known that the accidents at Savannah River in 1953 and Oak Ridge in 1959 were triggered by this oxidation reaction. The hydrolysis reaction of TBP occurs in alkaline, acid, and organic phases leading to DBP and MBP. The diluent will react with

Table 2 *G*-Values of TBP γ-Radiolysis Products

Product	Burr [94]	Wilkinson and Williams [95]	Wagner et al. [96]	Holland et al. [97]	Burger and McClanahan [98]
H_2	1.73	1.11	1.59	2.02	—
CH_4	0.072	0.05	0.07	0.032	—
C_2	0.18	0.24	0.13	—	—
C_3 and C_4	0.66	0.45	0.75	—	—
DBP	2.44	1.52	2.25	3.71	1.9
MBP	0.14	0.12	0.39	—	0.3
Polymer	2.47	—	—	—	—

HNO_3 leading to the formation of carboxyl acids, nitric acid ester, and nitro and nitroso compounds. Rate parameters for chemical degradation under different conditions have been investigated [106,107]. The reaction of *n*-dodecane with HNO_3 is quite slow.

Radiolysis of hydrocarbons has been investigated and there is a review [108]. *n*-Dodecane is one of the saturated hydrocarbons, and its radiation degradation is rather known as compared with that of TBP. Main degradation products are H_2 and dimers, as summarized in Table 3 [109,110]. Again, the values are dependent on the reports. Recently, systematic study on the radiolysis of n-paraffins has been reported [111–113]. The formation process of dimers and double bonds is assumed as follows.

$$RH \leadsto R\cdot + H\cdot \tag{16}$$

$$\leadsto R(-H) + H_2 \tag{17}$$

$$R_1\cdot + R_2\cdot \rightarrow R_1(-H) + R_2H \tag{18}$$

$$\rightarrow R_1R_2 \tag{19}$$

Here R(–H) is corresponding to alkenes.

2.3. Radiolysis of HNO_3

Radiolysis of aqueous nitrate solution has been investigated not only from practical viewpoints in nuclear technology, but also from scientific interests in order to understand the radiolysis of water because nitrate ion is an efficient scavenger for hydrated electron [114] and even for the precursor of the hydrated electron [115]. In practical process, the concentration range of nitrate covers from millimolars to 10 M and the radiation effect not only in diluted nitrate solutions but also in concentrated ones should be considered.

Table 3 *G*-Values of *n*-Hexadecane γ-Radiolysis Products

Product	Holland et al. [97]	Dewhurst [109]	Rappoport and Gäuman [110]
H_2	6.71	4.9	4.9
CH_4	0.05	0.05	0.01
Dimer	—	—	1.4

In diluted solutions, the reduction of NO_3^- with e_{aq}^- and H generates NO_2 radicals through the transient intermediates of $NO_3^{2-\cdot}$, $HNO_3^-\cdot$, and $H_2NO_3\cdot$ which are in equilibrium with one another. The reactivity of H atoms towards NO_3^- is much lower than e_{aq}^-.

$$NO_3^- + e_{aq}^- \rightarrow NO_3^{2-\cdot} \qquad\qquad k = 9.7 \times 10^9 \ M^{-1}\, sec^{-1} \ [114] \qquad\qquad (20)$$

$$NO_3^- + H \rightarrow HNO_3^-\cdot \qquad\qquad k = 1.0 \times 10^7 \ M^{-1}\, sec^{-1} \ [114] \qquad\qquad (21)$$

$$NO_3^{2-\cdot} \Leftrightarrow HNO_3^-\cdot \Leftrightarrow H_2NO_3\cdot \qquad pK_1 = 4.8, pK_2 = 7.5 \ [114] \qquad\qquad (22)$$

$$NO_3^{2-\cdot} + H_2O \rightarrow NO_2\cdot + 2OH^- \quad k = 1.0 \times 10^3 \ M^{-1}\, sec^{-1} \ [116,117] \qquad (23)$$

$$HNO_3^-\cdot \rightarrow NO_2\cdot + OH^- \qquad\qquad k = 2.0 \times 10^5 \ M^{-1}\, sec^{-1} \ [116,117] \qquad (24)$$

$$H_2NO_3\cdot \rightarrow NO_2\cdot + H_2O \qquad\qquad k = 7.0 \times 10^5 \ M^{-1}\, sec^{-1} \ [116,117] \qquad (25)$$

The $NO_2\cdot$ radicals tend to recombine with each other and, as a result, N_2O_4 appears. The N_2O_4 decays slowly into HNO_2 and HNO_3 through hydrolysis reaction.

$$NO_2\cdot + NO_2\cdot \Leftrightarrow N_2O_4 \qquad\qquad k_{forward} = 4.5 \times 10^8 M^{-1}\, sec^{-1} \ [118] \qquad (26)$$
$$k_{backward} = 6.0 \times 10^3 M^{-1}\, sec^{-1} \ [118]$$

$$N_2O_4 + H_2O \rightarrow HNO_2 + HNO_3 \qquad k = 18 M^{-1}\, sec^{-1} \ [118] \qquad\qquad (27)$$

In the radiolysis of nitric acid, HNO_2 and H_2O_2 are formed. They strongly affect the solvent degradation, oxidation states of metal ions, and corrosion condition of the material. However, HNO_2 and H_2O_2 are not coexisting because the next reaction will take place.

$$HNO_2 + H_2O_2 \rightarrow HNO_3 + H_2O \qquad\qquad\qquad\qquad\qquad\qquad\qquad (28)$$

This reaction is composed of several steps as follows [119,120].

$$HNO_2 + H^+ \Leftrightarrow H_2NO_2^+ \qquad\qquad\qquad\qquad\qquad\qquad\qquad\qquad (29)$$

$$H_2NO_2 + H_2O_2 \rightarrow HOONO + H_2O \qquad\qquad\qquad\qquad\qquad\qquad\qquad (30)$$

$$HOONO \rightarrow NO_3^- + H^+ \qquad k = 4.6 \times 10^3 [H^+] M^{-1} sec^{-1} \qquad\qquad\qquad (31)$$

In concentrated solutions, direct action of nitrate ion is also taken into consideration. In addition, in concentrated nitric acid, nitric acid molecule (HNO_3) coexists. Although it is known that $NO_3\cdot$ plays an important role in concentrated nitric acid, precise formation process was not known. Recently, it was clarified that the formation of the $NO_3\cdot$ radical is formed by two different processes: reaction of OH with molecular HNO_3 and direct action of radiation to HNO_3 and NO_3^- [121]. Experimentally obtained temporal behavior of $NO_3\cdot$ and its yield as a function of nitrate ions are shown in Figs. 7 and 8, respectively.

$$OH + HNO_3 \rightarrow NO_3\cdot + H_2O \qquad k = 1.4 \times 10^8 M^{-1} sec^{-1} \qquad\qquad\qquad (32)$$

Therefore the $NO_3\cdot$ radical is not formed in diluted nitrate solutions because OH does not react with NO_3^-. This finding suggests that the OH reacts with molecular HNO_3 through H atom abstraction. The formation of $NO_3\cdot$ through direct process is in proportion to the electron fraction of nitrate. In nitric acid, $NO_3\cdot$ is formed through two formation processes, but, in concentrated nitrate solutions, only direct process plays a role. The direct action of radiation is commonly observed in concentrated solutions and examples are reviewed in Ref. 122.

For the formation of HNO_2 and O_2, the following reactions due to the direct effect to the solute have been proposed in aqueous solutions [123], molten salts, and crystals [124], having the G-values of g_{s2} and g'_{s2}.

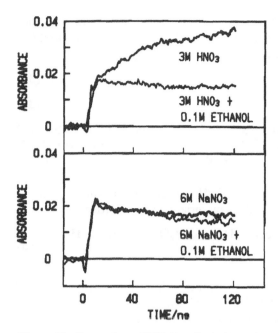

Figure 7   Formation of $NO_3 \cdot$ radical observed at 640 nm in 3 M HNO_3 and 6 M $NaNO_3$ solutions in the presence and absence of 0.1 M ethanol. (From Ref. 121.)

$$NO_3^- \rightsquigarrow O + NO_2^- \tag{33}$$

$$g_{s2}(-NO_3^-) = g_{s2}(NO_2^-) = g_{s2}(O) \tag{34}$$

$$HNO_3 \rightsquigarrow O + HNO_2 \tag{35}$$

$$g'_{s2}(-HNO_3) = g'_{s2}(HNO_2) = g'_{s2}(O) \tag{36}$$

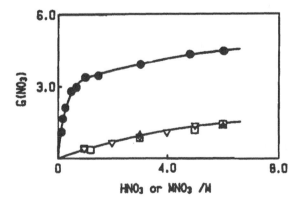

Figure 8   Dependence of $G(NO_3 \cdot)$ as a function of nitrate concentration in HNO_3 and $NaNO_3$, respectively: total ($\bullet$) and fast formation process ($\blacktriangle$) yields in HNO_3 solution; fast formation process yields in $LiNO_3$ (∇) and $NaNO_3$ ($\square$) solutions. (From Ref. 121.)

The O formed is in a singlet (^1D) or triplet (^3P) state. It is known that the O (^1D) will react with water to yield H_2O_2 and this does not contribute to the formation of HNO_2 and O_2. Therefore O (^3P) is needed to explain the formation of O_2 and HNO_2.

$$O(^1D) + H_2O \rightarrow H_2O_2 \tag{37}$$

$$O(^3P) + NO_3^- \rightarrow O_2NO_2^- \tag{38}$$

The rate constant of the second reaction is estimated to be $k[O\,(^3P) + NO_3^-] = 2.2 \times 10^8\ M^{-1}$ sec^{-1} from the values of $[O\,(^3P) + O_2] = 4.0 \times 10^9\ M^{-1}\ sec^{-1}$ [125] and $k[O\,(^3P) + NO_3^-]/k[O\,(^3P) + O_2] = 5.6 \times 10^{-2}\ M^{-1}\ sec^{-1}$ [126]. However, another possible direct path, $NO_3^- \rightsquigarrow O^- + NO_2$, does not contribute to the O_2 formation.

The yield of HNO_2 from water decomposition, $G^W(HNO_2)$, can be calculated by using the yields of water decomposition products and their material balance relation:

$$G^W(HNO_2) = f_W[0.5g_W(e_{aq}^- + H) - 0.5g_W(OH) - g_W(H_2O_2)] = -f_W g_W(H_2)$$
$$= -G(H_2) \tag{39}$$

Here f_W stands for the electron fraction of water in the solution. Thus the electron fraction of the solute is $f_S = 1 - f_W$. In addition, $G(HNO_2)$, $G(H_2)$, and $G(O_2)$ are G-values of HNO_2, H_2, and O_2 formation, respectively, experimentally determined in the solution. Reported $G(H_2)$ and $G(O_2)$ in nitric acid and sodium nitrate solutions are summarized in Fig. 9 [127].

The total yield of HNO_2 in nitric acid, $G(HNO_2)$, coming from the direct and indirect effect, $G^W(HNO_2)$ and $G^S(HNO_2)$, can be expressed as:

$$G(HNO_2) = G^W(HNO_2) + G^S(HNO_2) = -G(H_2) + G^S(HNO_2) \tag{40}$$

The values of $G(HNO_2) + G(H_2)$ and $G(O_2)$ could be expressed by using α, the dissociation constant of molecular nitric acid [128].

$$G(HNO_2) + G(H_2) = 2f_S[\alpha g_{s2}(-NO_3^-) + (1 - \alpha)g'_{s2}(-NO_3^-)] \tag{41}$$

$$G(O_2) = f_S[\alpha g_{s2}(-NO_3^-) + (1 - \alpha)g'_{s2}(-NO_3^-)] \tag{42}$$

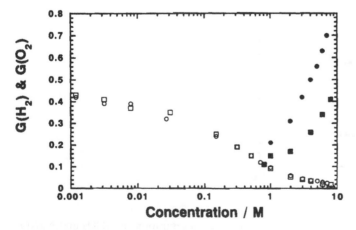

Figure 9 Reported $G(H_2)$ and $G(O_2)$ as a function of nitrate concentration in γ-radiolysis of HNO_3 and $NaNO_3$ solutions. (From Ref. 127.)

From the experimentally determined values of $G(HNO_2)$, $G(H_2)$, and $G(O_2)$, the values of $g_{s2}(-NO_3^-) = 1.7$ and $g'_{s2}(-NO_3^-) = 2.2$ were obtained [129]. The value of $2G(O_2)$ is expected to be equal to $G(HNO_2)$; the relation is almost satisfied, although there are large experimental errors. From this experiment, the possibility of O (^{1}D) formation is not excluded, but the experimental finding that H_2O_2 does not increase at higher concentrations suggests that it is not significant. The NO_2 and NO_3 radicals including other nitrogen oxides in radiolysis of nitrate aqueous solutions and G-values of reported H_2, O_2, NO_2^-, and H_2O_2 have been reviewed [130].

The H_2 evolution is an important safety assessment in the commercial process and much work has been conducted [131]. The $G(H_2)$ is decreasing with increasing of the concentration of nitric acid. Coexisting ions affect the H_2 evolution slightly. Ions such as Rh^{3+}, Sm^{3+}, Pr^{3+}, Ce^{3+}, Cd^{2+}, Ce^{4+}, and Eu^{3+} increase the H_2 evolution in this order and, contrarily, the presence of Fe^{3+} and Cr^{3+} reduces the evolution. At higher temperature, the $G(H_2)$ is higher. It was also found that H_2 evolution is strongly dependent on the depth of the liquid of nitric acid solution: deeper liquid gives smaller evolution than a shallower one [131,132]. The effect was observed in both γ- and α-radiolysis, and the effect is more significant in γ-radiolysis [133]. In Fig. 10 [133–136], the reported $G(H_2)$ and $G(O_2)$ by α-radiolysis in nitric acid are summarized, and it is clear that $G(H_2)$ in α-radiolysis is higher than that in γ-radiolysis in diluted solution, but the difference is not significant in concentrated solutions.

2.4. Radiolysis of the Mixture of TBP, n-Hexadecane, and HNO$_3$

Since, during the actual process, TBP, n-dodecane, and HNO_3 are used in a mixture or contact condition, the degradation is affected by many factors: composition, concentration of nitric acid, concomitant metal ions, and irradiation conditions such as aerated or deaerated, stirring or settling, and high or low dose rate. As a result, a variety of products are formed as summarized in Fig. 11 [137]. Under aerated condition, organic radicals react with O_2 quickly to form peroxy-radical and after successive reactions bring variety of alcohols, ketones, peroxides, and carbonyl compounds. These products are roughly classified into nitration and oxidation products. The ratio of the nitration products to oxidation product, $\eta = $ (nitration/oxidation), is 0.8 under sufficient oxygen supply but is increasing to 8 under insufficient oxygen. Recent advancement of the analytical technique brings more precise data [138].

2.5. Radiation Chemistry of Actinide Ions

Because of the multivalent nature of the actinide ions, understanding the radiation-induced change of the valence-state of the actinide in solutions under self-irradiation or external irradiation is a challenge in radiation chemistry. Some of the ions are strong α-emitters. It is also important from a practical viewpoint that the solution chemistry of actinide ions is closely related to the storage and the repository of the wastes. Much work combined with experiment and simulation has been conducted and reviews were summarized [136,140–144].

2.6. Storage of Radioactive Waste

In 50 years of production of electrical power and weapons from the nuclear fuel, the United States has accumulated millions of cubic meters and tens of billions curies of radioactive

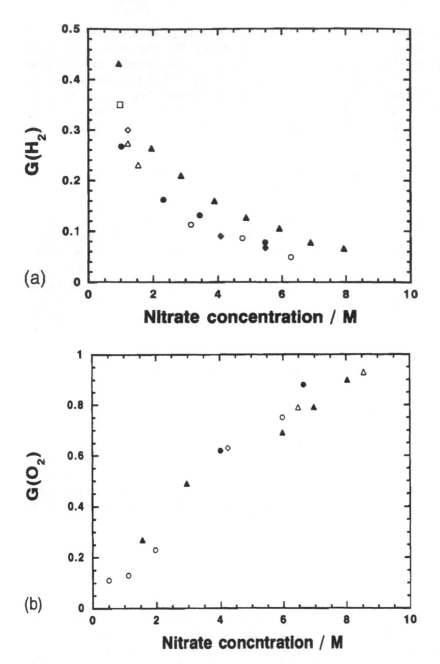

Figure 10   Reported (a) $G(H_2)$ and (b) $G(O_2)$ as a function of nitrate concentration in α-radiolysis of HNO_3 solution. In (a), from Ref. 134 (▲) and Ref. 133 (□: 9.3 g/L, ◇: 13 g/L, Δ: 18.5 g/L, ●: 30 g/L, ■: 37 g/L, ○: 100 g/L, and ◆: 160 g/L). In (b), from Ref. 134 (○), Ref. 135 (▲), and Ref. 133 (Δ: 210 g/L, ◇: 160 g/L, and ●: 100 g/L). (From Ref. 133.)

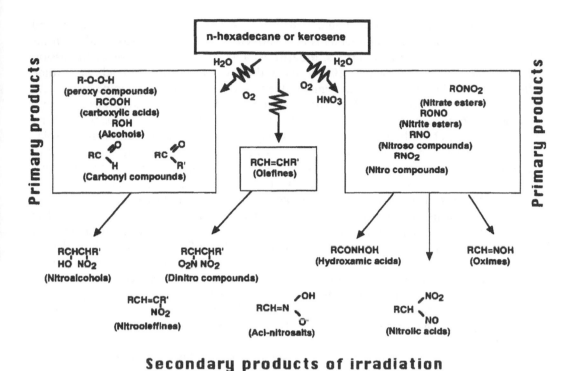

Secondary products of irradiation

Figure 11 Radiolysis products in Purex solvents (TBP, *n*-dodecane or kerosene, and HNO₃). (From Ref. 137.)

wastes [145]. It is widely recognized that safe treatment and storage of these wastes is an urgent problem to be solved. At a radioactive waste storage tank, the so-called SY-101 in Hanford, an intermittent gas evolution was found in 1980s. The released gas, composed of H_2, N_2O, and minor NH_3, in addition to air was emitted every 8–13 weeks. The content of the tank is a mixture of chirate agents and their fragments, carboxylic acids, and alkanes in addition to the sodium nitrate and nitrite salts, which were produced during the processes to separate plutonium from uranium and the FPs in spent metallic uranium fuel. This gas evolution was considered to be a very important issue to keep safe storage of the waste, and much radiolysis study was conducted to understand the mechanism of the gas evolution. Through the work, subjects on (1) direct effect in concentrated solutions, (2) redox reactions of NO_x compounds, (3) radiation degradation of chilators, and (4) reactions in grout, precipitation, and slurry have been pointed out as key issues for understanding. Details have been reported elsewhere [146–148].

3. RADIATION EFFECTS RELEVANT TO RADIOACTIVE HIGH-LEVEL WASTE REPOSITORY

3.1. Once Through, Reprocessing, and High-Level Waste Repository

After the operation of the nuclear power stations for a certain period, the spent nuclear fuel is produced. The spent fuel is reprocessed, and both U and Pu are recovered. At the same time, radioactive high-level liquid waste (HLLW) composed of fission products and trace amount of actinide compounds is also formed. HLLW emits strong radiation including γ-rays, high-energy electrons, α-particles, and fast neurons. Most of the fission products have a half-life less than 1000 years, and activity is dropped 3 orders magnitude after 1000 years but still the activity remains due to the ^{99}Tc, ^{93}Zr, and actinide nuclei with long lifetimes of million years. Therefore isolation of HLW from the activities and environment of human being is the most essential requirement in HLW repository. Many possible methods have been discussed such as disposal into the outer space from the solar system, into the deep sea, under the ice in polar area, and so on. Actually, a disposal to deep underground, geological deposition, has been selected as a reasonable and practical one. In fact, at first, the high-level liquid wastes are solidified with silicate glass into a metal container called canister and are packed into metal container called an overpack made of thick copper, stainless steel, or carbon steel, which depends on the choice. The assembly would be placed in the bentonite clay mantle, which is built in the tunnel with the depth up to more than 1000 m from the surface. Solidification, overpack, and bentonite clay act as the multiple barriers which will prevent the invasion of groundwater to the overpack and reduce the migration of radioactive and hazardous nuclei to biosphere. Another option is that the spent fuel is directly contained without reprocessing into an overpack for disposal, which is called *once through*. The same geological disposal method will be applied. The precise procedures of the depository have been summarized in textbooks [92,93].

In order to assess the integrity of the system, we should know what kind of reactions would take place when the groundwater invades and the overpack is corroded. Consequently, the solidified waste or spent fuel itself will be in contact with groundwater. Since the waste would still be seriously activated, radiolysis of groundwater will take place and change the chemical condition, which might affect the dissolution of the solidified waste or UO_2 of the spent fuel.

3.2. Carbonate Solution

It is known that (bi)carbonate ions are predominant anions in groundwater with a concentration as high as 50 mM. Therefore it is necessary to know the radiolysis effects on groundwater by ionizing radiation from the HLW. A computer set of reaction steps and reliable kinetic data for the radiolysis of (bi)carbonate solutions at ionic strength close to groundwater is essential for the overall safety evaluation of geological disposal. The reactions of water decomposition products toward (bi)carbonate ions have been investigated and rate constants have been accumulated. However, a little experimental evaluation of the formation yield for the products such as formate, oxalate, H_2O_2, and so on has been reported [149–152]. Two sets of the reaction mechanism in carbonate solution were reported [152,153]. In Table 4, recent list is summarized [152].

Here it should be mentioned that the chemical structure of carbonate radical is still under debate. The reaction of the carbonate radical has been intensively investigated by pulse radiolysis [154–156] and laser photolysis [157], and the pK_a of HCO_3 radical was determined to be 7.6 [158] and 9.6 [157,159,160] from the change of the reactivity as a function of pH. However, recent studies of Raman spectroscopy [160] and pulse radiolysis [161] of carbonate solution proposed that the carbonate radical is a strong acid and keeps a chemical form of CO_3^- even at below pH 0. Furthermore, dimer formation mechanism has also been proposed [78].

Table 4 Rate Constants of Radiation-Induced Reactions in Sodium (Bi)carbonate Aqueous Solution

Reactions	Rate constants $(M^{-1} \, sec^{-1})$
Reactions of CO_2, HCO_3^-, and CO_3^{2-}	
(1) $H^+ + CO_3^{2-} = HCO_3^-$	$5E10^a$
(2) $CO_2 + H_2O = H^+ + HCO_3^-$	$2E4$
(3) $H^+ + HCO_3^- = CO_2 + H_2O$	$5E10$
(4) $HCO_3^- = H^+ + CO_3^{2-}$	2
(5) $CO_2 + e_{aq}^- = CO_2^{\cdot -}$	$7.7E9$
(6) $HCO_3^- + {}^{\cdot}OH = CO_3^{\cdot -} + H_2O$	$8.5E6$
(7) $CO_3^{2-} + {}^{\cdot}OH = CO_3^{\cdot -} + OH^-$	$3.9E8$
(8) $HCO_3^- + H = H_2 + CO_3^{\cdot -}$	$4.4E4$
(9) $CO_3^{2-} + e_{aq}^- = CO_2^{\cdot -} + 2OH^- - H_2O$	$3.9E5$
Reactions of $CO_3^{\cdot -}$	
(10) $CO_3^{\cdot -} + CO_3^{\cdot -} = C_2O_6^{2-}$	$1.4E7$
(11) $CO_3^{\cdot -} + H_2O_2 = CO_3^{2-} + O_2^{\cdot -} + 2H^+$	$9.8E5$
(12) $CO_3^{\cdot -} + HO_2^- = CO_3^{2-} + O_2^{\cdot -} + H^+$	$1E7$
(13) $CO_3^{\cdot -} + O_2^{\cdot -} = CO_3^{2-} + O_2$	$4E8$
(14) $CO_3^{\cdot -} + CO^{\cdot -}_{\,7} = CO_3^{2-} + CO_2$	$(3\pm1)E8$
Reactions of CO_2^-	
(15) $CO_2^{\cdot -} + e_{aq}^- = HCOO^- + OH^- - H_2O$	$1E9$
(16) $CO_2^{\cdot -} + CO_2^{\cdot -} = C_2O_4^{2-}$	$6.5E8$
(17) $CO_2^{\cdot -} + O_2 = CO_2 + O_2^{\cdot -}$	$2E9$
(18) $CO_2^{\cdot -} + H_2O_2 = CO_2 + OH^- + {}^{\cdot}OH$	$7.3E5$
(19) $CO_2^{\cdot -} + HCO_3^- = HCOO^- + CO_3^{\cdot -}$	$(1.0\pm0.1)E3$
Reactions of $C_2O_6^{2-}$ and $C_2O_4^{2-}$	
(20) $C_2O_6^{2-} = C_2O_4^{2-} + O_2$	1^b
(21) $C_2O_6^{2-} = HO_2^- + OH^- + 2CO_2 - H_2O$	200 ± 50^b
(22) $CO_3^{\cdot -} + C_2O_4^{2-} = C_2O_4^{\cdot -} + CO_3^{2-}$	$(3\pm1)E3$
(23) $C_2O_4^{2-} + e_{aq}^- = C_2O_4^{\cdot 3-}$	$3.1E7$
(24) $C_2O_4^{2-} + {}^{\cdot}OH = C_2O_4^{\cdot -} + OH^-$	$7.7E6$
Reactions of $HCOO^-$	
(25) $CO_3^{\cdot -} + HCOO^- = HCO_3^- + CO_2^{\cdot -}$	$(1.5\pm0.5)E5$
(26) $HCOO^- + {}^{\cdot}OH = H_2O + CO_2^{\cdot -}$	$3.2E9$
(27) $HCOO^- + H = H_2 + CO_2^{\cdot -}$	$2.1E8$
(28) $HCOO^- + e_{aq}^- = H_2 + CO_2^{\cdot -} - H^+$	$8E3$

^{a} Read as 5×10^{10}.
^{b} There are relative rate constants. They mean 0.5% of $C_2O_6^{2-}$ decomposes via reaction (20) and 99.5% of $C_2O_6^{2-}$ via reaction (21).
Source: Ref. 152.

3.3. Oxychloride and Saline Solution

Seawater is invaded into the groundwater and thus groundwater sometimes contains sodium chloride, which is called the *seawater*-type one. One of the possible sites for the repository is a salt mine. It is assumed that the site does not receive water invasion for so long, and this indicates the stability of the site. If groundwater invade the site, saline solution would be irradiated. Thus understanding of the radiolysis of water containing NaCl is also taken into consideration after the corrosion of the overpack of the HLW. Chloride ion is one species among oxychlorides and redox reactions induce conversion from one species to another, which complicates. Recently, it is recognized that oxychlorides play important roles in atmospheric and environmental chemistry. Oxychlorides are widely used to breach and to sterilize water. Some oxychlorides exhibit chemical oscillation which is discussed from viewpoints of chaos and complex system. Much knowledge has been accumulated. While the reliable reaction sets of oxychloride in aqueous solution are not fully established, a typical example of the literature survey is shown in Table 5 [162–185].

Table 5 Rate Constants of Radiation-Induced Reactions in Aqueous Oxychloride Solution

	Reactions	Rate constants ($M^{-1} sec^{-1}$)	Reference
(1)	$Cl^- + OH \rightarrow ClOH^-$	$4.3E9^a$	[162]
(2)	$ClOH^{\cdot-} \rightarrow OH + Cl^-$	$6.1E9 \ sec^{-1}$	[162]
(3)	$ClOH^{\cdot-} + H^+ \rightarrow \cdot Cl + H_2O$	$2.1E10$	[162]
(4)	$\cdot Cl + H_2O \rightarrow ClOH^{\cdot-} + H^+$	$1.6E5 \ sec^{-1}$	[163]
(5)	$\cdot Cl + OH^- \rightarrow ClOH^{\cdot-}$	$1.8E10$	[163]
(6)	$\cdot Cl + OH \rightarrow HClO$	$1E9$	Assumed
(7)	$\cdot Cl + Cl \rightarrow Cl_2$	$8.8E7$	[164]
(8)	$\cdot Cl + Cl^- \rightarrow Cl_2^{\cdot-}$	$2.1E10$	[162]
(9)	$Cl_2^- \rightarrow \cdot Cl + Cl^-$	$1.1E5 \ sec^{-1}$	[162]
(10)	$\cdot Cl + Cl^- \rightleftharpoons Cl_2^{\cdot-}$	$K = 1.9E5, 4.7E3 \ M^{-1}$	[162,165]
(11)	$ClOH^{\cdot-*} \rightarrow \cdot Cl + OH^-$	$23 \ sec^{-1}$	[163]
(12)	$ClOH^- + Cl^- \rightarrow Cl_2^- + OH^-$	$9E4$	[166]
(13)	$Cl_2^- + OH^- \rightarrow ClOH^{\cdot-} + Cl^-$	$4.5E7$	[167]
(14)	$Cl_2^- + \cdot H \rightarrow 2Cl^- + H^+$	$8E9, (7E9)$	[168,169]
(15)	$Cl_2^- + \cdot OH \rightarrow HClO + Cl^-$	$1E9$	[170]
(16)	$Cl_2^- + H_2O \rightarrow \cdot OH + H^+ + 2Cl^-$	$7.2E3, 1.3 \ E3 \ sec^{-1}$	[170]
(17)	$OH + H^+ + 2Cl^- \rightarrow Cl_2^- + H_2O$	$2.11E11 \ M^{-3} \ sec^{-1}$	[170]
(18)	$Cl_2^- + H_2O_2 \rightarrow HO_2^\cdot + 2Cl^- + H^+$	$1.4E5$	[167]
(19)	$Cl_2^- + HO_2^\cdot \rightarrow 2Cl^- + H^+ + O_2$	$1.0E9$	[169]
(20)	$Cl_2^- + Cl_2^- \rightarrow Cl_3^- + Cl^-$	$4.0E9$	[169] (2k)
(21)	$Cl_2^- + \cdot Cl \rightarrow Cl_3^-$	$6.3E8$	[164]
(22)	$Cl_2 + HO_2^\cdot \rightarrow Cl_2^{\cdot-} + H^+ + O_2$	$1.0E9$	[171]
(23)	$Cl_2 + Cl^- \rightarrow Cl_3^-$	$1.8E5$	[172]
(24)	$Cl_2 + H_2O \rightarrow HClO + H^+ + Cl^-$	$11 \ sec^{-1}$	[173]
(25)	$HClO + H^+ + Cl^- \rightarrow Cl_2 + H_2O$	$1.8E4 \ M^{-2} \ sec^{-1}$	[173]
(26)	$Cl_2 + OH^- \rightarrow HClO + Cl^-$	$3.8E11$	[153]
(27)	$Cl_2 + 2ClO_2^- \rightarrow 2ClO_2^\cdot + 2Cl^-$	$2.1E1 \ [Cl_2(aq)] \ [ClO_2^-]$	[174]
(28)	$HClO + 2ClO_2^- + H^+ \rightarrow 2ClO_2^\cdot + Cl^- + H_2O$	$1.06E6 \ [HOCl] \ [ClO_2^-] \ [H^+]$	[174]
(29)	$2HClO + ClO_2^- \rightarrow ClO_3^- + Cl_2 + H_2O$	$2.1 \ E{-}3 \ M^{-2} \ sec^{-1}$	[175]
(30)	$HClO + \cdot Cl \rightarrow Cl^- + ClO\cdot + H^+$	$3E9$	[163]
(31)	$HClO + H_2O_2 \rightarrow H_2O + O_2 + H^+ + Cl^-$	$3.4E3$	Assumed
(32)	$Cl_2 + H_2O \rightarrow Cl_2OH^- + H^+$	$6.12E2 \ sec^{-1}$	[173]
(33)	$Cl_2OH^- + H^+ - Cl_2 + H_2O$	$2E10$	[173]
(34)	$Cl_2OH^- + H^+ \rightarrow Cl_2 + OH^- + H^+$	$2E5$	[173]
(35)	$Cl_2 + OH^- + H^+ \rightarrow Cl_2OH^- + H^+$	$1E10$	[173]
(36)	$Cl_3^- \rightarrow Cl_2 + Cl^-$	$1.6E6$	[172]
(37)	$Cl_3^- + \cdot H \rightarrow Cl^- + Cl_2^- + H^+$	$3E10$	[176]
(38)	$Cl_3^- + HO_2^\cdot \rightarrow Cl_2^- + Cl^- + H^+ + O_2$	$1.0E9$	[171]
(39)	$HClO + e_{aq}^- \rightarrow Cl^- + \cdot OH$	$6.5E8$	[177]
(40)	$ClO^- + e_{aq}^- \rightarrow Cl^- + O^{\cdot-}$	$5.3E10, 8.3E9, 7.0E9$	[168,177,178]
(41)	$HClO + \cdot H \rightarrow HCl + OH$	$3E9$	Assumed
(42)	$HClO + \cdot OH \rightarrow Cl^- + H_2O$	$1.4E8$	[177]

Table 5 Continued

Reactions	Rate constants (M^{-1} sec^{-1})	Reference
(43) $ClO^- + \cdot OH \rightarrow ClO \cdot + OH^-$	9.0E9, 2.7E9	[168,177]
(44) $ClO^- + O^{\cdot -}(+H_2O) \rightarrow ClO \cdot + 2OH^-$	2.4E8	[168]
(45) $ClO^- + \cdot Cl \rightarrow ClO + Cl^-$	8.2E9	[163]
(46) $ClO^- + O_2^- \rightarrow O^- + O_2 + Cl^-$	7.7E6	[179]
(47) $ClO^- + H_2O_2 \rightarrow H_2O + Cl^- + O_2$	3.4E3	[180]
(48) $ClO^- + HO_2^- \rightarrow OH^- + Cl^- + O_2$	4.4E7	[181]
(49) $HClO \rightarrow ClO^- + H^+$	3.0E2	pKa = 7.53
(50) $ClO^- + H^+ \rightarrow HClO$	1E10	[181]
(51) $HClO + ClO_2^- + H^+ \rightarrow Cl_2O_2 + H_2O$	1.12E6 M^{-2} sec^{-1}	[175]
(52) $Cl_2O_2 + ClO_2^- \rightarrow 2ClO_2 + Cl^-$	5.4E4(~1E5)	[175]
(53) $2Cl_2O_2 \rightarrow 2ClO_2' + Cl_2$	6E10 M^{-1} sec^{-1}	[175]
(54) $ClO_2^- + Cl_2 \rightarrow Cl_2O_2 + Cl^-$	4.0×10^4	[175]
(55) $Cl_2O_2 + H_2O \rightarrow ClO_3^- + Cl^- + 2H^+$	1, 10 sec^{-1}	[175]
(56) $ClO \cdot + ClO \cdot + H_2O \rightarrow ClO^- + ClO_2^- + 2H^+$	2.5E9	[163]
(57) $ClO \cdot + ClO_2^- \rightarrow ClO^- + ClO_2 \cdot$	9.4E8	[182]
(58) $ClO_2^- + e_{aq}^- \rightarrow ClO^- + O^{\cdot -}$	4.5E10, 4.0E9, 2.5E9	[168,177,183]
(59) $ClO_2^- + \cdot OH \rightarrow ClO_2 \cdot + OH^-$	6.3E9, 7.9E9, 7E9	[168,177,183]
(60) $ClO_2^- + O^-(+H_2O) \rightarrow ClO_2 \cdot + 2OH$	2.0E8, 1.9E8	[168,183]
(61) $H^+ + ClO_2^- \rightarrow HClO_2$	1.0E10	pKa = 2.5
(62) $HClO_2 \rightarrow H^+ + ClO_2^-$	3.16E7 sec^{-1}	[174]
(63) $ClO_2 \cdot + HO_2^- \rightarrow ClO_2^- + HO_2$	8E4, 1.3E5	[183,184]
(64) $ClO_2 + O_2^- \rightarrow ClO_2^- + O_2$	3.3E9	[183]
(65) $ClO_2 \cdot + ClO \cdot \rightarrow Cl_2O3$	1E9	Assumed
(66) $Cl_2O3 + H_2O \rightarrow ClO^- + ClO_3^- + 2H^+$	1E4	Assumed
(67) $ClO_2 \cdot + \cdot OH \rightarrow overall$	3.5E9, 4.0E9	[177,185]
(68) $ClO_2 \cdot + \cdot OH \rightarrow HClO + O_2$	1.4E9	[185]
(69) $ClO_2 \cdot + \cdot OH \rightarrow ClO_3^- + H^+$	2.6E9	[185]
(70) $ClO_2 \cdot + O^{\cdot -} \rightarrow overall$	2.7E9	[185]
(71) $ClO_2 \cdot + O^\cdot \rightarrow ClO^- + O_2$	4.9E8	[185]
(72) $ClO_2 \cdot + O^{\cdot -} \rightarrow ClO_3^-$	2.2E9	[185]
(73) $ClO_2 \cdot + e_{aq}^- \rightarrow overall$	$kq = 2.1E10$	[177]
(74) $ClO_2 \cdot + \cdot e_{aq}^- \rightarrow ClO_2^-$	qkq	**
(75) $ClO_2 \cdot + \cdot e_{aq}^- \rightarrow ClO \cdot + \cdot O^{\cdot -}$	$(1-q)kq$	**
(76) $ClO_2 \cdot + \cdot H \rightarrow overall$	$ks = 1 \times 10^9$	Assumed
(77) $ClO_2 \cdot + \cdot H \rightarrow H^+ + ClO_2^-$	sks	**
(78) $ClO_2 \cdot + \cdot H \rightarrow ClO \cdot + \cdot OH$	$(1-s)ks$	**

*The dissociation rate is calculated after the assumption of the recombination rate constant of H^+ with anion to be 10^{10} M^{-1} sec^{-1} when the pK_a value is known.
**Branching reaction. q (s) and $1-q$ $(1-s)$ are the branching fraction.
[a] Read as 4.3×10^9.

In repository of HLW in salt mine, the understanding of radiolysis of concentrated NaCl up to 5 M is important. In addition, α-radiolysis by actinide ions dissolved is also taken into account. There are several reports mentioned above [186–192].

3.4. Radiation-Induced Dissolution of UO$_2$ Fuel

In direct geological depository of the spent fuel, dissolution of UO_2 matrix into groundwater plays a key role for the release of the radionuclides to the biosphere. Therefore mechanism of the dissolution processes has been clarified by using irradiated and unirradiated UO_2 under various conditions. Influence of many factors such as radiolysis, pH, temperature, groundwater composition, and formation of corrosion deposits has been investigated. An excellent review has been published elsewhere [193]. Among many factors, it was revealed that radiolysis accelerates the dissolution of UO_2.

The accelerated dissolution of UO_2 and U_2O_8 in acid or carbonate solutions under the irradiation of γ-rays was found by Gromov in 1981 [194]. Since then, much experimental results have been accumulated as summarized in a review of Cristensen and Sunder [195]. The radiolysis products such as OH and H_2O_2 react with UO_2, and its surface is oxidized to form higher oxidizing phase: UO_{2+x}, $UO_{2.33}$, UO_2^{2+}, and so on. The measurement of the corrosion potential of the specimen was also done and showed the correlated behavior of the oxidizing process. There is a time-dependent generation of oxidants and reductants at fuel/water interface due to α-, β-, and γ-radiolysis. Based on the accumulated experimental data, a calculation code to predict the oxidation and dissolution process was proposed and the calculated results are in good agreement with the experimental ones [195].

Nuclear spent fuel UO_x emits not only gamma and beta, but also alpha particles. The alpha activity lasts over a million years, much longer than the gamma and beta activities which decay within a few hundred years. In long-term assessment, α-radiolysis should be also taken into consideration. Recently, a model experiment using He^{2+} ion beam from an accelerator has been investigated and a significant acceleration of dissolution under beam irradiation was found, and the formation of uranium peroxide hydrate UO_42H_2O (meta-studtite) was detected from x-ray diffraction at UO_2 surface [196,197].

4. CONCLUSION

It is clear that the development of nuclear technology is impossible without the support of radiation chemistry. Thus the radiation chemistry is important not only in basic science, but also in technology.

ACKNOWLEDGMENT

This is supported by the Japan Society for the Promotion of Science under the contract JSPS-RFTE 98P00901. This is also partly supported by the innovative basic research program in the field of high-temperature engineering using HTTR conducted by Japan Atomic Energy Research Institute.

REFERENCES

1. Jenks, G.H. ORNL-3848, UC-80-Reactor Technology, TID-4500, 1965.
2. Jenks, G.H.; Griess, J.C. ORNL-4173, UC-80-Reactor Technology, 1967.
3. Burns, W.G.; Moore, P.B. Radiat. Effects 1976, *30*, 233.
4. Elliot, A.J. AECL-11073, COG-94-167, 1994.
5. Elliot, A.J.; Ouellette, D.C.; Stuart, C.R. AECL 11658, COG-96-390-1, 1996.
6. McCracken, D.R.; Tsang, K.T.; Laughton, P.J. AECL-11895, 1998.
7. Buxton, G.V. Radiation chemistry: present status and future trends. In: Jonah, C.D. Rao, B.S., eds. Studies in Physics and Theoretical Chemistry 87. Elsevier Science, 2001, 195 pp.

8. Ruiz, C.P.; Lin, C.C.; Robinson, R.; Burns, W.G.; Curtis, A.R. Water Chem. Nucl. React. Syst. 1989, *4*, 131. BNES, London.
9. Ruiz, C.P.; Lin, C.C.; Robinson, R.N.; Burns, W.G.; Henshaw, J.; Pathania, R. Water Chem. Nucl. React. Syst. 1992, *6*, 141. BNES, London.
10. Kabakchi, S.A.; Lebedeva, I.E. High Energy Chem. 1984, *18*, 166.
11. Katsumura, Y.; Takeuchi, Y.; Ishigure, K. Radiat. Phys. Chem. 1988, *32*, 259.
12. Katsumura, Y.; Takeuchi, Y.; Hiroishi, D.; Ishigure, K. Radiat. Phys. Chem. 1989, *33*, 299.
13. Elliot, A.J.; Ouellette, D.C.; Reid, D.; McCracken, D.R. Radiat. Phys. Chem. 1989, *34*, 747.
14. Katsumura, Y.; Yamamoto, S.; Hiroishi, D.; Ishigure, K. Radiat. Phys. Chem. 1992, *39*, 383.
15. Burns, W.G.; Marsh, W.R. J. Chem. Soc., Faraday Trans. 1 1981, *77*, 197.
16. Shiraishi, H.; Buxton, G.V.; Wood, N.D. Radiat. Phys. Chem. 1989, *33*, 519.
17. Kent, M.C.; Sims, H.E. AEA-RS-2302, PWR/CTG/(92) 084, 1992.
18. Kent, M.C.; Sims, H.E. Proceedings of the 6th International Conference of Water Chemistry of Nuclear Reactor Systems; British Nuclear Energy Society: London, 1992, 153 pp.
19. Elliot, A.J.; Chenier, M.P.; Ouellette, D.C. J. Chem. Soc., Faraday Trans. 1993, *89*, 1193.
20. Sunaryo, G.R.; Katsumura, Y.; Hiroishi, D.; Ishigure, K. Radiat. Phys. Chem. 1995, *45*, 131.
21. Ishigure, K.; Katsumura, Y.; Sunaryo, G.R.; Hiroishi, D. Radiat. Phys. Chem. 1995, *46*, 557.
22. Elliot, A.J.; Chenier, M.P.; Ouellette, D.C.; Koslowsky, V.T. J. Phys. Chem. 1996, *100*, 9014.
23. Katsumura, Y.; Sunaryo, G.R.; Hiroishi, D.; Ishigure, K. Prog. Nucl. Energy 1998, *32*, 113.
24. Stefanic, I.; LaVerne, J.A. J. Phys. Chem. A 2002, *106*, 447.
25. Christensen, H.; Sehested, K. Radiat. Phys. Chem. 1980, *16*, 183.
26. McCracken, D.R.; Buxton, G.V. Nature 1981, *292*, 439.
27. Christensen, H.; Sehested, K. J. Phys. Chem. 1988, *92*, 3007.
28. Shiraishi, H.; Katsumura, Y.; Hiroishi, D.; Ishigure, K.; Washio, M. J. Phys. Chem. 1988, *92*, 3011.
29. Buxton, G.V.; Wood, N.D.; Dyster, S. J. Chem. Soc., Faraday Trans. 1, 1988, *84*, 1113.
30. Elliot, A.J.; McCracken, D.R. Radiat. Phys. Chem. 1989, *33*, 69.
31. Buxton, G.V.; Wood, N.D. Radiat. Phys. Chem. 1989, *34*, 699.
32. Elliot, A.J. Radiat. Phys. Chem. 1989, *34*, 753.
33. Elliot, A.J.; McCracken, D.R.; Buxton, G.V.; Wood, N.D. J. Chem. Soc., Faraday Trans. 1990, *86*, 1539.
34. Elliot, A.J.; Buxton, G.V. J. Chem. Soc., Faraday Trans. 1992, *88*, 2465.
35. Buxton, G.V.; Mackenzie, S.R. J. Chem. Soc., Faraday Trans. 1992, *88*, 2833.
36. Hickel, B.; Sehested, K. Radiat. Phys. Chem. 1992, *39*, 355.
37. Buxton, G.V.; Elliot, A.J. J. Chem. Soc., Faraday Trans. 1993, *89*, 485.
38. Christensen, H.; Sehested, K.; Løgager, T. Radiat. Phys. Chem. 1993, *41*, 575.
39. Christensen, H.; Sehested, K.; Løgager, T. Radiat. Phys. Chem. 1994, *43*, 527.
40. Elliot, A.J.; Ouellette, D.C. J. Chem. Soc., Faraday Trans. 1994, *90*, 837.
41. Shiraishi, H.; Sunaryo, G.R.; Ishigure, K. J. Phys. Chem. 1994, *98*, 5164.
42. Ashton, L.; Buxton, G.V.; Stuart, C.R. J. Chem. Soc., Faraday Trans. 1995, *91*, 1631.
43. Noyes, R.M. In: Progress in Reaction Kinetics; Porter, G., Ed.; Pergamon: London, 1961: Vol. 1, 129 pp.
44. Newton, M.D.; Sutin, N. Annu. Rev. Phys. Chem. 1984, *35*,437.
45. Draganic, I.; Draganic, Z.D. *Radiation Chemistry*; Academic Press: New York, 1971, 170 pp.
46. LaVerne, J.A.; Pimblott, S.M. J. Phys. Chem. 1993, *97*, 3291.
47. Swiatla-Wojcik, C.; Buxton, G.V. J. Phys. Chem. 1995, *99*, 11464.
48. Swiatla-Wojcik, C.; Buxton, G.V. J. Chem. Soc., Faraday Trans. 1998, *94*, 2135.
49. Swiatla-Wojcik, C.; Buxton, G.V. Chem. Phys. Phys. Chem. 2000, *2*, 5113.
50. Pimblott, S.M.; LaVerne, J.A. J. Phys. Chem. A 1997, *101*, 5828.
51. Frongillo, Y.; Goulet, T.; Fraser, M.-J.; Cobut, V.; Patau, J.P.; Jay-Gerin, J.P. Radiat. Phys. Chem. 1998, *51*, 245.
52. Goulet, T.; Fraser, M.-J.; Frongillo, Y.; Jay-Gerin, J.-P. Radiat. Phys. Chem. 1998, *51*, 85.
53. Clifford, P.; Green, N.J.B.; Pilling, M.J. J. Phys. Chem. 1982, *86*, 1318–1322.

54. Clifford, P.; Green, N.J.B.; Oldfield, M.J.; Pilling, M.J.; Pimblott, S.M. J. Chem. Soc., Faraday Trans. 1 1986, *82*, 2673.
55. Clifford, P.; Green, N.J.B.; Pilling, M.J.; Pimblott, S.M. J. Phys. Chem. 1987, *91*, 4417.
56. Nikjoo, H.; Uehara, S.; Wilson, W.E.; Hoshi, M.; Goodhead, D.T. Int. J. Radiat. Biol. 1998, *73*, 355.
57. Hervé du Penhoat, M.-A.; Goulet, T.; Frongillo, Y.; Fraser, M.-J.; Bernat, Ph.; Jay-Gerin, J.-P. J. Phys. Chem. A 2000, *104*, 11757.
58. Hervé du Penhoat, M.-A.; Meesungnoen, J.; Goulet, T.; Filali-Mouhim, A.; Mankhetkorn, S.; Jay-Gerin, J.-P. Chem. Phys. Lett. 2001, *341*, 135.
59. Meesungnoen, J.; Jay-Gerin, J.-P.; Filali-Mouhim, A.; Mankhetkorn, S. Chem. Phys. Lett. 2001, *335*, 458.
60. Meesungnoen, J.; Jay-Gerin, J.-P.; Filali-Mouhim, A.; Mankhetkorn, S. Can. J. Chem. 2002, *80*, 68.
61. Meesungnoen, J.; Jay-Gerin, J.-P.; Filali-Mouhim, A.; Mankhetkorn, S. Can. J. Chem. 2002, *80*, 767.
62. Savage, P.E. Chem. Rev. 1999, *99*, 603.
63. Tester, J.W.; Cline, J.A. Corrosion 1999, *55*, 1088.
64. Schmieder, H.; Abeln, J. Chem. Eng. Technol. 1999, *22*, 903.
65. Bröll, D.; Kaul, C.; Krämer, A.; Krammer, P.; Richter, T.; Jung, M.; Vogel, H.; Zehner, P. Angew. Chem., Int. Ed. 1999, *38*, 2998.
66. Kajimoto, O. Chem. Rev. 1999, *99*, 55.
67. Oka, Y. Physics of supercritical-pressure light water cooled reactors. *Proc. 1998 Frederic Joliot Summer School in Reactor Physics, Caderache, France*; and references cited herein, 1998; 240–259 pp.
68. Ferry, J.L.; Fox, M.A. J. Phys. Chem. A 1998, *102*, 3705.
69. Ferry, J.L.; Fox, M.A. J. Phys. Chem. A 1999, *103*, 3438.
70. Wu, G.; Katsumura, Y.; Muroya, Y.; Li, X.; Terada, Y. Chem. Phys. Lett. 2000, *325*, 531.
71. Takahashi, K.; Cline, J.A.; Bartels, D.M.; Jonah, C.D. Rev. Sci. Instrum. 2000, *71*, 3345.
72. Michael, B.D.; Hart, E.J.; Schmidt, K.H. J. Phys. Chem. 1971, *75*, 2798.
73. Dixson, R.S.; Lopata, V.J. Radiat. Phys. Chem. 1978, *11*, 135.
74. Christensen, H.; Sehested, K. J. Phys. Chem. 1986, *90*, 186.
75. Wu, G.; Katsumura, Y.; Muroya, Y.; Li, X.; Terada, Y. Radiat. Phys. Chem. 2001, *60*, 395.
76. Cline, J.; Jonah, C.D.; Bartels, D.M.; Takahashi, K. Proc. of the 1st International Symposium on Supercritical Water-cooled Reactors, Design and Technology, Nov. 6–9, 2000, Univ. of Tokyo; Tokyo; Japan; 2000; 194 pp.
77. Levelt Sengers, J.M.H.; Straub, J.; Watanabe, K.; Hill, P.G. J. Phys. Chem. Ref. Data. 1985, *14*, 193.
78. Wu, G.; Katsumura, Y.; Muroya, Y.; Lin, M.; Morioka, T. J. Phys. Chem. A 2002, *106*, 2430.
79. Wu, G.; Katsumura, Y.; Muroya, Y.; Lin, M.; Morioka, T. J. Phys. Chem. A. 2001, *105*, 4933.
80. Mostafavi, M.; Lin, M.; Wu, G.; Katsumura, Y.; Muroya, Y. J. Phys. Chem. A 2002, *106*, 3123.
81. Wu, G.; Katsumura, Y.; Lin, M.; Morioka, T.; Muroya, Y. PCCP 2002, *4*, 3980.
82. Schuler, R.H.; Patterson, L.K.; Janata, E. J. Phys. Chem. 1980, *84*, 2089.
83. Buxton, G.V.; Stuart, C.R. J. Chem. Soc., Faraday Trans. 1995, *91*, 279.
84. Hayon, E.; Ibata, T.; Lichtin, N.N.; Simic, M. J. Phys. Chem. 1972, *76*, 2072.
85. Adams, G.E.; Wilson, R.L. J. Chem. Soc., Faraday Trans. 1 1973, *69*, 719.
86. Brede, O.; Helmstreit, W.; Mehnert, R. Z. Phys. Chem. (Leipzig) 1975, *256*, 513.
87. Lin, M.; Katsumura, Y.; Wu, G.; Muroya, Y.; He, H.; Kudo, H. Proc. of the 45th Annual Meeting on Radiation Chemistry in Japan, Oct. 9–11, 2002, Kyushu University; Fukuoka; Japan; 2002; 193 pp.
88. Takahashi, K.; Bartels, D.M.; Cline, J.A.; Jonah, C.D. Chem. Phys. Lett. 2002, *357*, 358.
89. Ho, P.C.; Palmer, D.A.; Wood, R.H. J. Phys. Chem. B 2000, *104*, 12084.
90. Ho, P.C.; Palmer, D.A.; Gruszkiewicz, M.S. J. Phys. Chem. B 2001, *105*, 1260.

91. Chialvo, A.A.; Ho, P.C.; Palmer, D.A.; Gruszkiewicz, M.S.; Cummings, P.T.; Simonson, J.M. J. Phys. Chem. B 2002, *106*, 2041.

92. Benndict, M.; Pigford, T.H.; Levi, H.W. Nuclear Chemical Engineering. 2nd Ed. McGraw-Hill: New York, 1981.

93. Choppin, G.; Liljenzin, J.O.; Rydberg, J. Radiochemistry and Nuclear Chemistry. 2nd Ed. Butterworth-Heinemann: Oxford, 1995.

94. Burr, J.G. Radiat. Res. 1958, *8*, 214.

95. Willkinson, R.W.; Williams, T.F. J. Chem. Soc., 4098.

96. Wagner, R.M.; Kinderman, E.M.; Towle, L.H. Ind. Eng. Chem. 1959, *51*, 45.

97. Holland, J.P.; Merklin, J.F.; Razvi, J. Nucl. Instr. Methods 1978, *153*, 589.

98. Burger, L.L.; McClanahan, E.D. Abstracts of ACS meeting at Miami, 1957.

99. Nelson, D.; Symons, M.C.R. J. Chem. Soc. Perkin II, 286.

100. Zaitsev, V.D.; Karasev, A.L.; Khaikin, G.I.; Egorov, G.F. High Energy Chem. 1988, *22*, 351.

101. Zaitsev, V.D.; Khaikin, G.I. High Energy Chem. 1994, *28*, 308.

102. Zaitsev, V.D.; Khaikin, G.I. High Energy Chem. 1989, *23*, 79.

103. Zaitsev, V.D.; Protasova, E.L.; Khaikin, G.I. High Energy Chem. 1989, *24*, 58.

104. Zaitse, V.D.; Protasova, E.L.; Khaikin, G.I. High Energy Chem. 1993, *27*, 28.

105. Zaitsev, V.D.; Protasova, E.L.; Khaikin, G.I. High Energy Chem. 1994, *28*, 30.

106. von Stieglitz, L.; Becker, R. Atomkernenerg. Kerntech. 1985, *46*, 76.

107. Tallent, O.K.; Mailen, J.C.; Dodson, K.E. Nucl. Technol. 1985, *71*, 417.

108. Földiak, G. Radiation Chemistry of Hydrocarbons. Elsevier: Amsterdam, 1981.

109. Dewhurst, H.A. J. Phys. Chem. 1957, *61*, 1466.

110. Rappoport, S.; Gäuman, T. Helv. Chim. Acta 1973, *56*, 531. Helv. Chim. Acta 1973, *57*, 2861.

111. Seguchi, T.; Hayakawa, N.; Tamura, N.; Hayashi, N.; Katsumura, Y.; Tabata, Y. Radiat. Phys. Chem. 1988, *32*, 753.

112. Soebianto, Y.S.; Yamaguchi, T.; Katsumura, Y.; Ishigure, K.; Kubo, J.; Koizumi, T. Radiat. Phys. Chem. 1992, *39*, 251.

113. Katsumura, Y. Angew. Makromol. Chem. 1997, *252*, 89.

114. Buxton, G.; Greenstock, G.L.; Hermann, W.P.; Ross, A.B. J. Phys. Chem. Ref. Data 1988, *17*, 513.

115. Wolff, R.K.; Bronskill, M.J.; Hunt, J.W. J. Chem. Phys. 1970, *53*, 4211.

116. Gräzel, M.; Henglein, A.; Lilie, J.; Beck, G. Ber. Bunsenges. Phys. Chem. 1970, *74*, 292.

117. Løgager, T.; Sehested, K. J. Phys. Chem. 1993, *97*, 6664.

118. Gräzel, M.; Henglein, A.; Lilie, J.; Beck, G. Ber. Bunsenges. Phys. Chem. 1969, *73*, 646.

119. Bhattacharyya, P.K.; Veeraraghavan, R. Int. J. Chem. Kinet. 1977, *9*, 629.

120. Damaschen, D.E.; Martin, L.R. Atmos. Environ. 1983, *17*, 2005.

121. Katsumura, Y.; Jiang, P.Y.; Nagaishi, R.; Oishi, T.; Ishigure, K.; Yoshida, Y. J. Phys. Chem. 1991, *95*, 4435.

122. Katsumura, Y. Radiation chemistry: present status and future trends. In: Jonah, C.D. Rao, B.S.M.; Eds.; Studies in physics and theoretical chemistry 87; Elsevier Science: Amsterdam, 2001, p. 163.

123. Daniels, M. Radiation Chemistry I. Gould, F. Ed.; American Chemical Soc., Advances in Chemical Series; American Chemical Soc., Washington; 1968, *81*,153.

124. Pogge, H.B.; Jones, F.T. J. Phys. Chem. 1970, *74*, 1700.

125. Kläning, U.K.; Sehested, K.; Wolff, T. J. Chem. Soc., Faraday Trans. 1 1984, *80*, 2969.

126. Amichai, O.; Treinin, A. Chem. Phys. Lett. 1969, *3*, 611.

127. Mahlman, M.A. J. Chem. Phys. 1961, *35*, 936.

128. Redlich, O.; Duerst, R.D.; Merbach, A. J. Chem. Phys. 1968, *49*, 2986.

129. Nagaishi, R.; Jiang, P.Y.; Katsumura, Y.; Ishigure, K. J. Chem. Soc., Faraday Trans 1994, *90*, 591.

130. Katsumura, Y. In: Alffasi, Z.B. Ed.; N-Centered Radicals; John Wiley & Sons, 1998, 393 pp.

131. Nakagiri, N.; Miyata, T. J. Atom. Energ. Soc. Jpn. 1994; *37*, 1119; 1995; 1996; *38*, 992 pp. in Japanese.

132. Specht, S. In KfK 2940, 1980; 196 pp.
133. Kuno, Y.; Hina, T.; Masui, J. J. Nucl. Sci. Technol. 1993, 30, 919.
134. Savel'ev, Yu., I.; Savel'ev; Ershova, Z.V.; Vladimirova, M.V. Sov. Radiochem. 1967, 9, 221.
135. Becker, R.; Burkhardt, H.G.; Neeb, K.H.; Würtz, R. IAEA-SM-245/13, 1979.
136. Bibler, N.E. J. Phys. Chem. 1974, 78, 211.
137. Huggard, A.J.; Warner, B.F. Nucl. Sci. Eng. 1963, 17, 168.
138. Tripathi, S.C.; Bindu, P.; Ramanujam, A. Sep. Sci. Technol. 2001, 36, 1463; Tripathi, S.C.; Ramanujam, A.; Gupta, K.K.; Bindu, P. Sep. Sci. Technol. 2001, 36, 2863.
139. Miner, F.J.; Seed, J.R. Chem. Rev. 1967, 67, 299.
140. Bibler, N.E. J. Phys. Chem. 1975, 79, 1991.
141. Pikaev, A.K.; Shilov, V.P.; Spitsyn, V.I. Radiolysis of Aqueous Solutions of Lanthanides and Actinides; Nauka: Moscow, 1983. in Russian.
142. Gordon, S.; Sullivan, J.C.; Ross, A.B. J. Phys. Chem. Ref. Data 1986, 15, 1357.
143. Frolov, A.A.; Andreychuk, N.N.; Rotmanov, K.V.; Frolova, L.M.; Vasiliev, V.Ya. J. Radioanal. Nucl. Chem. 1990, 143, 433.
144. Bhattàcharyya, P.K.; Natarajanin, P.R. In Handbook on the Physical Chemistry of Actinides; Freeman, A.J. Keller, C., Eds.; Elsevier Science Publishers B.V., Chapt. 13, 97 pp.
145. Ahearne, J.F. Phys. Today, 24 (June).
146. Tank Waste Science Panel, PNL-7595 DE91 009369, 1991.
147. Reynolds, D.A.; D.D. Siemer, D.M. Strachan, R.W. Wallace, PNL-7520 DE91 009680, 1991.
148. Jonah, C.D.; Kapoor, S.; Matheson, M.S.; Mulac, W.A.; Meisel, D. ANL-94/7, 1994, references cited herein.
149. Draganic, Z.D.; Negron-Mendoza, A.; Navarro-Conzalez, R.; Vujosevic, S.I. Radiat. Phys. Chem. 1987, 30, 229.
150. Eriksen, T.E.; Dalamba, P.N.; Christensen, H.; Bjergbakke, E. J. Radioanal. Nucl. Chem. 1989, 132, 19.
151. Draganic, Z.D.; Negron-Mendoza, A.; Sehested, K.; Vujosevic, S.I.; Navarro-Gonzalez, R.; Albarran-Sanches, M.G.; Draganic, I.G. Radiat. Phys. Chem. 1991, 38, 317.
152. Cai, Z.; Li, X.; Katsumura, Y.; Urabe, O. Nucl. Technol. 2001, 136, 231.
153. Sunder, S.; Christensen, H. Nucl. Technol. 1993, 104, 403.
154. Weeks, J.L.; Rabani, J. J. Phys. Chem. 1966, 70, 2100.
155. Behar, D.; Czapski, G.; Duchovny, I. J. Phys. Chem. 1970, 73, 2206.
156. Chawla, O.P.; Fessenden, R.W. J. Phys. Chem. 1975, 79, 2693.
157. Chen, S.-N.; Cope, V.W.; Hoffman, M.Z. J. Chem. Soc. Chem. Commun., 1970; 991; J. Phys. Chem. 1970, 77, 1111.
158. Eriksen, T.E.; Lind, J.; Merènyi, G. Radiat. Phys. Chem. 1985, 26, 197.
159. Zuo, Z.; Cai, Z.; Katsumura, Y.; Chitose, N.; Muroya, Y. Radiat. Phys. Chem. 1999, 55, 15.
160. Bisby, R.H.; Johnson, S.A.; Parker, A.W.; Tavender, S.M. J. Chem. Soc., Faraday Trans. 1998, 94, 2069.
161. Czapski, G.; Lymar, S.V.; Schwarz, H.A. J. Phys. Chem. A 1999, 103, 3447.
162. Jayson, G.G.; Parsons, B.J.; Swallow, A.J. J. Chem. Soc., Faraday Trans. 1 1973, 69, 1597.
163. Kläning, U.K.; Wolff, T. Ber. Bunsenges. Phys. Chem. 1985, 89, 243.
164. Wu, D.; Wong, D.; DiBartolo, B. J. Photochem. 1980, 14, 303.
165. Adams, D.J.; Barlow, S.; Buxton, G.V.; Malone, T.M.; Salmon, G.A. J. Chem. Soc., Faraday Trans. 1995, 91, 3303.
166. Bjergbakke, E.; Sehested, K.; Rusmussen, L.O.; Christensen, H. Risø M-2430, 1984.
167. Hasegawa, K.; Neta, P. J. Phys. Chem. 1978, 82, 854.
168. Buxton, G.V.; Subhani, M.S. J. Chem. Soc., Faraday Trans. 1 1972, 68, 947.
169. Navaratnam, S.; Parsons, B.J.; Swallow, A.J. Radiat. Phys. Chem. 1980, 15, 159.
170. Wagner, I.; Karthäuser, J.; Strehlow, H. Ber. Bunsenges. Phys. Chem. 1986, 90, 861.
171. Bjergbakke, E.; Navaratnam, S.; Parsons, B.J.; Swallow, A.J. J. Am. Chem. Soc. 1981, 103, 5926.
172. Scott, R.L. J. Am. Chem. Soc. 1953, 75, 1550.

173. Eigen, M.; Kustin, K. J. Am. Chem. Soc. 1962, *84*, 1355.
174. Epstein, I.R.; Kustin, K.; Simonyi, R.H. J. Phys. Chem. 1992, *96*, 5852.
175. Peintler, G.; Nagypál, I.; Epstein, I.R. J. Phys. Chem. 1990, *94*, 2954.
176. Gogolov, A.Y.; Makarov, I.E.; Pikaev, A.K. High Energy Chem. 1984, *18*, 390.
177. Zuo, Z.; Katsumura, Y.; Ueda, K.; Ishigure, K. J. Chem. Soc., Faraday Trans. 1997, *93*, 1885.
178. Amber, M.; Hart, E.J. Adv. Chem. Ser. 1968, *81*, 79.
179. Long, C.A.; Bielski, B.H.J. J. Phys. Chem. 1980, *84*, 555.
180. Held, A.M.; Halko, D.J.; Hurst, J.K. J. Am. Chem. Soc. 1978, *100*, 5732.
181. Morris, J.C. J. Phys. Chem. 1966, *70*, 3798.
182. Alfassi, Z.B.; Huie, R.E.; Mosseri, S.; Neta, P. Radiat. Phys. Chem. 1988, *32*, 85.
183. Eriksen, T.E.; Lind, J.; Merenyi, G. J. Chem. Soc., Faraday Trans. 1 1981, *77*, 2115.
184. Hoigné, J.; Bader, H. Water Res. 1994, *28*, 45.
185. Kläning, U.K.; Sehested, K. J. Phys. Chem. 1991, *95*, 740.
186. Kim, J.I.; Lierse, Ch.; Büppelmann, K.; Magirius, S. Mater. Res. Soc. Symp. Proc. 1986, *84*, 603.
187. Büppelmann, K.; Magirius, S.; Lierse, Ch.; Kim, J.I. J. Less-Common Met. 1986, *122*, 329.
188. Lierse, C.; Sullivan, J.C.; Schmidt, K.H. Inorg. Chem. 1987, *26*, 1408.
189. Grego'rev, A.E.; Makarov, I.E.; Pikaev, A.K. High Energy Chem. 1987, *21*, 99.
190. Büppelmann, K.; Kim, J.I.; Lierse, Ch. Radiochim. Acta 1988, *44/45*, 65.
191. Kelm, M.; Pashalidis, I.; Kim, J.I. Appl. Radiat. Isotopes 1999, *51*, 637.
192. Janata, E.; Kelm, M.; Ershov, B.G. Radiat. Phys. Chem. 2002, *63*, 157.
193. Shoesmith, D.W. J. Nucl. Mater. 2000, *282*, 1.
194. Gromov, V. Radiat. Phys. Chem. 1981, *18*, 135.
195. Cristensen, H.; Sunder, S. Nucl. Technol. 2000, *131*, 102, and references cited herein.
196. Corbel, C.; Sattonnay, G.; Lucchini, J.-F.; Ardois, C.; Barte, M.-F.; Huet, F.; Dehaudt, P.; Hickel, B.; Jegou, C. Nucl. Instr. Methods B 2001, *179*, 255.
197. Sattonnay, G.; Ardois, C.; Corbel, C.; Lucchini, J.-F.; Barte, M.-F.; Garrido, F.; Gosset, D. J. Nucl. Mater. 2001, *288*, 11.

24

Electron Beam Applications to Flue Gas Treatment

Hideki Namba
Japan Atomic Energy Research Institute, Takasaki, Gunma, Japan

1. INTRODUCTION

The emission of toxic gases, sulfur oxides (SO_x), and nitric oxides (NO_x) from industrial plants has become a serious problem in many countries. Fig. 1 shows the sources of SO_x and NO_x emission, which are recognized as the origin of dry and wet deposition known as "acid rain" [1].

These toxic components sometimes travel more than a thousand kilometers and cause problems in other places and, in some cases, in other countries. Therefore the problem of air pollution attracts worldwide attention. Among these, the coal-fired flue gas from thermal power plants has been recognized as one of the main sources of environmental pollution, because of its larger amount of S and N components. Now many countries are introducing stricter emission control regulations to solve the problem.

Fig. 1 also shows the chemical reaction mechanism of SO_x and NO_x in air for producing acid rain [2]. The radicals (O, OH, and HO_2) produced in air play an important role in oxidizing SO_x and NO_x to produce sulfuric acid and nitric acid, which are the main components of acid rain. Some of these reactions also occur in a reaction chamber of electron beam treatment of flue gas.

Electron beam treatment of flue gas was first performed in Japan [3,4] and is now regarded as a promising pollution control method with many advantages [5,6]: simultaneous reduction of SO_x and NO_x, being a dry process, yields a by-product which can be used as an agricultural fertilizer, and so on. Fig. 2 shows a flow diagram of electron beam treatment process for flue gas treatment. In this method, combustion flue gas containing NO_x (main component is NO) and SO_x (main component is SO_2) is introduced in a spray cooler to decrease the temperature. After adding ammonia, the flue gas is irradiated with electron beam in a process vessel. NO and SO_2 are converted to nitric acid and sulfuric acid and are finally changed to ammonium nitrate and ammonium sulfate which may be used as agricultural fertilizers. Cleaned flue gas is emitted from a stack. Radiation chemical reactions induced by electron beam irradiation and related chemical processes will be studied in detail in this chapter.

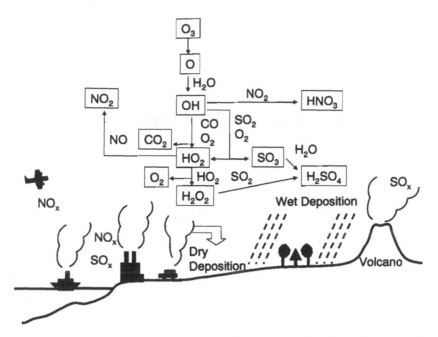

Figure 1   Origin of acid rain and chemical reactions in air. (From Refs. 1 and 2.)

2. INCIDENT ELECTRONS

2.1. Electron–Nuclei and Electron–Electron Interactions

Accelerated, or energized, electrons by an accelerator are used in flue gas treatment. The energy of incident electrons for practical treatment process is usually between 300 and 800 kV. The reason for using this energy region will be discussed in Sec. 2.4. The main energy loss processes of electrons in the region while passing through matter are (1) the interaction with nuclei to emit x-ray, so-called bremsstrahlung; and (2) the interaction with electrons of the materials. The contribution ratio of (1) and (2) is different depending on the atomic number

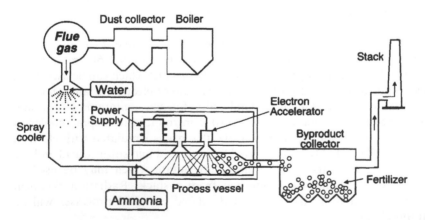

Figure 2   Flow diagram of electron beam treatment process for flue gas treatment.

(Z) of materials. At nearly 1 MeV of incident energy, the ratio is $EZ/800$, where E is the energy represented in MeV unit. As flue-gas components consist of light atoms, as shown in Tables 1 and 2, the contribution of (1) is negligible. That means almost all the energy loss of incident electrons is through inelastic collisions to generate ionization and excitation.

2.2. Stopping Power

In this condition, Bethe formulated the stopping power for electron according to the Born approximation. Stopping power is a property of irradiated materials and gives the amount of energy deposited per unit path length, $-dE/dx$.

$$-\frac{dE}{dx} = \frac{2\pi e^4}{m_0 v^2} NZ \left\{ \ln \frac{m_0 v^2 E}{2\bar{I}^2(1-\beta^2)} - \left(2\sqrt{1-\beta^2} - 1 - +\beta^2\right)\ln 2 + 1 - \beta^2 + \frac{1}{8}\left(1 - \sqrt{1-\beta^2}\right)^2 \right\}$$

(1)

Where, E is the energy, x is the length, m_0 and e are the rest mass and charge of electrons, respectively, v is the velocity of electron, N and Z are the number of atoms in unit volume and atomic number of the irradiated material, respectively, and β is the relative velocity represented by v/c, where c is the velocity of light.

In this equation, $\bar{I}$ is the mean ionization potential, which corresponds to the average energy used to ionize all the electrons in the atom, and is obtained experimentally. The values of mean ionization potential for elements in flue gas are listed in Table 1. The ratio of $\bar{I}/Z$ decreases with increasing Z and becomes almost constant for substances with Z greater than sulfur to be 10 ± 1.

For low energy electron region, namely, below $200 \times Z^2$ eV, this equation will not be valid because of the limitation of Born approximation; in this energy region, the calculations will be made using electron-impact ionization and excitation cross sections for gaseous targets [7].

To get more accurate stopping powers, two corrections are added to the basic stopping power formula. One is shell correction and the other is density effect correction. Including these effects, the formula will be written as follows [8,9]:

$$-\frac{dE}{dx} = \frac{2\pi e^4}{m_0 v^2} NZ \left\{ \ln \frac{m_0 v^2 E}{2\bar{I}^2(1-\beta^{2)}} - \left(2\sqrt{1-\beta^2} - 1 + \beta^2\right)\ln 2 + 1 - \beta^2 + \frac{1}{8}\left(1 - \sqrt{1-\beta^2}\right)^2 - 2\frac{C}{Z} - \delta \right\}$$

(2)

Table 1 Mean Ionization Potentials for Elemental Substances Composing Flue Gas

Z	Element	$\bar{I}$/eV	$\Delta\bar{I}$/eV	$(\bar{I}/Z)$/eV
1	H (gas)	19.2	0.4	19.2
6	C (graphite)	78.0	7	13.0
7	N (gas)	82.0	2	11.7
8	O (gas)	95.0	2	11.9
16	S (solid)	(180)		11.3

Source: From Ref. 7.

Table 2 Typical Concentration of Components in Coal-Fired Flue Gas and Their Electron Fraction

Components	Concentration	Ratio of total number of electrons
N_2	72%	0.685
O_2	6%	0.065
H_2O	10%	0.068
CO_2	12%	0.179
NO	225 ppm	0.0002
SO_2	800 ppm	0.0017

where C is the total shell correction factor, and δ is the density effect correction factor that can be expressed as

$$\delta = \ln\left(\frac{\hbar^2\omega_p^2}{\bar{I}^2(1-\beta^2)}\right) - 1 \tag{3}$$

where $\omega_p = (4\pi Ne^2 Z/m)^{1/2}$ is the so-called plasma frequency.

These corrections, however, are not so significant in flue gas treatment conditions.

2.3. Range

The range is the distance traveled by an electron from its incident point to the point where it has lost its energy completely; therefore it can be given by the integration of $-(dx/dE)$ over dE from the initial energy to 0. Fig. 3 shows the relation between range and energy of electrons in water calculated with continuous slowing-down approximation (csda) [10].

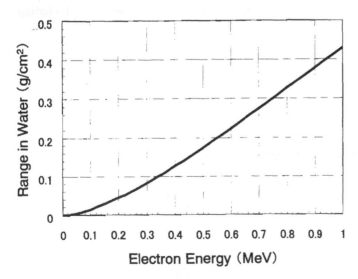

Figure 3 Relation between range and energy of electrons in water calculated with csda. (From Ref. 7.)

For practical use, it is important to know the linear distance in the substances from the incident point, i.e., an accelerator window, to the point at which the energy of electrons has been exhausted. The definitions of range for theoretical consideration and for practical use are somewhat different, however; the range shown in Fig. 3 gives a good indication for estimating the distance of electrons traveling in substances, which is called maximum range. Rough estimation of range is possible for other materials, by simply dividing their density. For example, the density of air is 0.0013 g/cm^3 at normal pressure and temperature; the range may be estimated about 770 times larger than in water, while that calculated by csda is about 860 times in this energy region [11]. As the density of the flue gas is almost the same as that of air, the range in flue gas will be the same as that in air.

2.4. Depth–Dose Curve and Energy Loss in Windows

The dose distribution in the materials is given as a depth–dose curve. An example of the curve is illustrated in Fig. 4 obtained with the irradiation of electron from 0.5 to 1.0 MeV using cellulose triacetate (CTA) film dosimeter [12]. The existence of the maximum dose is an important characteristic of the depth–dose curve. Irradiation from two opposite sides by using two accelerators was proposed in order to give better uniformity in water [13]. The uniform irradiation is also important for flue gas treatment. Better efficiency of NO_x removal was proved with both-side irradiation by using three accelerators for coal-fired flue gas than single-side irradiation at the same dose [14].

Fig. 5 shows a schematic representation of an accelerator and a reaction chamber. Double windows are usually used to separate the accelerator and the chamber. The primary window keeps the accelerator in vacuum; therefore the window material must have enough strength to bear the pressure difference of more than 1 atm. The secondary window prevents the primary window from flue gas in which acidic chemical components are produced during

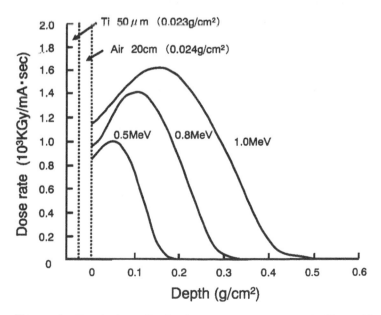

Figure 4 Depth–dose distribution curves in CTA stack films with the irradiation of electron. (From Ref. 11.)

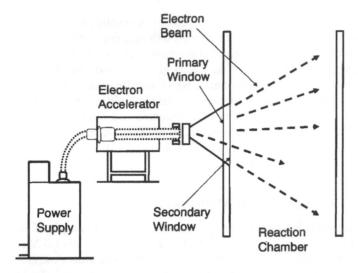

Figure 5 Schematic representation of accelerator and reaction chamber for flue gas treatment.

irradiation. That means the window's material might be anticorrosive. In order to minimize the energy loss in the windows, both windows should be thin. In most cases, the windows are made of metal or plastic films with a thickness of a few micrometers to a few tens of micrometers. Titanium or its alloy foils are commonly used for industrial-scale plants. Table 3 shows the power penetration efficiency of the windows and air-layer, which are the inverse of energy loss and beam current loss [15]. At lower energy of incident electrons, the energy loss is not negligible at the windows region. The efficiency increases with increasing energy of accelerated electron; however, we need a rather large installation to obtain higher energy of electrons. Thus accelerators with an energy region of 300 to 800 keV are commonly used for large-scale practical plants.

Table 3 Power Penetration Efficiency of Electron with Different Incident Energy at Windows and Air Layer

Materials	Ti (30 µm)	Air (5 cm)	Ti (17 µm)	Total
300 keV				0.56
Energy	0.87	0.945	0.89	
Beam	0.86	0.995	0.9	
500 keV				0.79
Energy	0.93	0.974	0.95	
Beam	0.945	1.0	0.97	
300 keV				0.88
Energy	0.96	0.985	0.97	
Beam	0.975	1.0	0.98	
1 MeV				0.90
Energy	0.97	0.99	0.975	
Beam	0.98	1.0	0.98	

Source: From Ref. 15.

3. INITIAL PROCESS

3.1. Energy Absorption to Produce Primary Species

The energy, or power, of electron beam induced in the flue gas is divided and absorbed by their gas components roughly depending on their electron fraction. Therefore almost all the energy is absorbed by the main components of the flue gas, namely, N_2, O_2, CO_2, and H_2O. Table 2 shows a typical concentration of the components in coal-fired flue gas in Japan. The ratio of the total number of electrons in each gas components is also listed in the same table. The energy absorbed directly by the toxic components (SO_2 and NO) is negligibly small. For electron beam treatment of flue gas, ammonia gas is added to the flue gas before the irradiation. The amount of ammonia is usually set as stoichiometrically, i.e., $2\Delta[SO_2] + \Delta[NO]$, where $\Delta[SO_2]$ and $\Delta[NO]$ are the concentrations of SO_2 and NO intended to be treated, respectively. The concentration of ammonia is usually higher than the initial concentration of SO_2 and NO; however, it is still far lower than that of the main components.

The first processes produced by high-energy charged particles are ionization and excitation. Evaluating the amount of energy deposited in molecules is one of the key factors in radiation chemistry, both theoretically and experimentally. The incident electron produces other electrons while losing its energy during ionization. The ionized electrons also produce other electrons till the energy reached is less than ionization potential. These are called "secondary" electrons. As the W-value in air is ~30 eV, a single electron with an energy of 750 kV will produce 25,000 electrons. That means energy deposition to the materials is mainly by the secondary electrons.

In order to estimate the contribution of secondary electrons, degradation spectrum, or in other words slowing down spectrum, is used. Degradation spectrum is defined as the length of the secondary electron trajectory, where the electron with the initial energy E_o dissipates its energy between E and $E+dE$.

Once we can obtain the degradation spectrum and the cross section of each reaction to produce the initial species, the yields of all the initial species are determined with the following equation [16];

$$N_s = N \sum_i n_i \int_{T_z}^{T_{max}} y(T)Q_s(T) \mathrm{d}\,T \tag{4}$$

or

$$N_s = N \sum_i n_i \int_{T_z}^{T_{max}} Ty(T)Q_s(T) \mathrm{d}\ln T \tag{5}$$

where N_s is the number of produced initial species, N is the number density, n_i is the number of electron at ith discreet energy level, T is the energy of electron (eV), $Q_s(T)$ is the cross section (cm^2), and $y(T)$ is the degradation spectrum (cm/eV). Fig. 6 shows the electron degradation spectrum in gaseous nitrogen calculated by binary-encounter collision theory with an irradiation of 100 keV electron [17]. Examples of contribution of the electron degradation spectrum to ionization and excitation are also listed in the same figure. The ionization in the figure shows an ionic process, $N_2 \rightsquigarrow N_2^+ (X^2\Sigma_g^+) + e^-$. The hatched area corresponds to the integral term of Eq. (3) for the process. The difficulty, however, is how to get all the precise cross-section data in order to calculate the accurate amount of energy separated from the discreet energy levels.

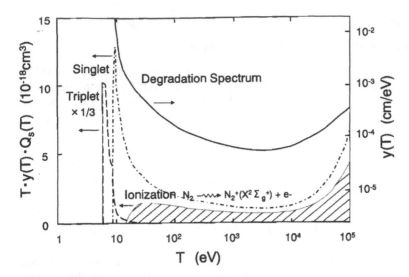

Figure 6 Contribution of the electron degradation spectrum to ionization and excitation in gaseous nitrogen calculated by binary-encounter collision theory with irradiation of 100-keV electron. (From Ref. 15.)

To avoid this problem, Wills and Boyd have applied a semi-empirical method [18]. They used relative cross sections for 100 eV electrons and calculated the yields for the various processes based on the W-values. This method is rather simple and contains many assumptions; nevertheless, the results are a good indication for radiation chemists to provide a specific amount of initial species. The following equations [19] show the amounts of initial species from the main components of flue gas with electron beam irradiation based on the data listed in Ref. 18.

$$4.43N_2 \rightsquigarrow$$
$$2.27N_2^+ + 0.69N^+ + 2.96e^- + 0.885N(^2D) + 0.295N(^2P) + 1.87N(^4S) + 0.29N_2^*$$

$$5.377O_2 \rightsquigarrow$$
$$2.07O_2^+ + 1.23O^+ + 3.3e^- + 2.25O(^1D) + 2.8O(^3P) + 0.18O^* + 0.077O_2^*$$

$$7.33H_2O \rightsquigarrow \tag{6}$$
$$1.99H_2O^+ + 0.01H_2^+ + 0.57OH^+ + 0.67H^+ + 0.06O^+ + 3.3e^- + 0.46O(^3P)$$
$$+ 4.25OH + 4.15H + 0.51H_2^*$$

$$7.54CO_2 \rightsquigarrow$$
$$2.24CO_2^+ + 0.51CO^+ + 0.07C^+ + 0.21O^+ + 3.03e^- + 5.16O(^3P) + 4.72CO$$

where the numerical coefficients are the amounts of decomposed molecules or produced species with 100-eV energy absorption, i.e., the G-values.

Positive ions, electrons, and excited molecules are produced after the irradiation of flue gas as shown in these equations. These ions and radicals play the most important role in radiation chemical reactions in flue gas.

3.2. Ions and Electrons

As the W-values of the major component of flue gas are nearly 30 eV, the total amount of positive ions produced will be ~ 3 in G-value. The most important reaction of positive ions is that with H_2O. For example, N_2^+ reacts with H_2O to produce OH radical, decomposing H_2O through charge transfer reaction.

$$N_2^+ + 2H_2O \rightarrow H_3O^+ + OH + N_2 \tag{7}$$

In a real case, part of the positive ion will be produced through more complex pathways, such as forming N_4^+, forming ion clusters with H_2O, and so on. In any case, however, the final positive ion will be H_3O^+ or its clustered ions, $H_3O^+(H_2O)_n$. Other positive ions, O_2^+, H_2O^+, and CO_2^+, will behave in the same way.

$$O_2^+ + 2H_2O \rightarrow H_3O^+ + OH + O_2$$

$$H_2O^+ + 2H_2O \rightarrow H_3O^+ + OH + H_2O \tag{8}$$

$$CO_2^+ + 2H_2O \rightarrow H_3O^+ + OH + CO_2$$

On the analogy of these reactions, the radical ions, N^+, O^+, and H^+, will produce radicals and positive ions in the following events.

$$N^+ + 2H_2O \rightarrow H_3O^+ + OH + N$$

$$O^+ + 2H_2O \rightarrow H_3O^+ + OH + O \tag{9}$$

$$H^+ + 2H_2O \rightarrow H_3O^+ + OH + H$$

These positive ions are the main source of the OH radical, which is one of the most important radicals for electron beam treatment of flue gas as will be discussed in the next section.

Electrons react mainly with O_2 to produce O_2^- or its clustered ion.

$$e^- + O_2 + M \rightarrow O_2^- + M \tag{10}$$

where M is a third body for the reaction.

These positive and negative ions neutralize to form HO_2.

$$H_3O^+ + O_2^- \rightarrow HO_2 + H_2O \tag{11}$$

Therefore one pair of ions produces one OH and one HO_2 radicals. The total amount of radicals, which are produced in flue gas by electron beam irradiation, is possible to calculate using reported G-values. The main radicals produced initially through direct and ionic decomposition processes are OH, N, HO_2, O, and H.

4. CHEMICAL REACTIONS

4.1. Radical Reactions

The radicals produced by irradiation will react with NO and SO_2, which are well-known good radical scavengers. These radical reactions related to NO_x and SO_x removals are very complicated with more than a few hundreds of reaction pathways [20]. The main radical reactions committed to the removal of NO_x and SO_2 are summarized in Table 4.

The total amount of radicals produced after irradiation is estimated at ~10 molecules/ 100 eV (G-value) from Eq. (6) and from discussions in the previous section. The following equation will help us to compare the amount of radicals with the concentration of NO and SO_2 that we are going to treat:

$$C = 0.104 \times M_w \times D \times G \tag{12}$$

where C is the concentration (ppm), M_w is the molecular weight (g/mol), D is the dose (kGy), and G is G-value (molecules/100 eV). Because M_w is almost the same as air (29 g/mol) in flue gas, the concentration will be 300 ppm at $G = 10$ and $D = 10$ kGy.

Some radicals produce other radicals through the reactions shown in Table 4. Comparing the calculated amount of NO and SO_2 removals through radical reactions to the experimental results [21], it indicates that NO_x removals mainly occur through radical reactions, while SO_2 removals mainly occur through other processes, so-called "thermal" reactions, or heterogeneous reactions. More details will be given in the following sections.

Table 4 Main Radical Reactions Related to the Removal of NO_x and SO_2

$OH + NO + M$	$\rightarrow$	$HNO_2 + M$
$OH + SO_2 + M$	$\rightarrow$	$HSO_3 + M$
$OH + NO_2 + M$	$\rightarrow$	$HNO_3 + M$
$OH + NO_2$	$\rightarrow$	$NO + HO_2$
$OH + HNO_2$	$\rightarrow$	$NO_2 + H_2O$
$OH + NH_3$	$\rightarrow$	$NH_2 + H_2O$
$OH + CO$	$\rightarrow$	$CO_2 + H$
$N(^4S) + NO$	$\rightarrow$	$N_2 + O$
$N(^4S) + NO_2$	$\rightarrow$	$N_2O + O$
$N(^2D) + O_2$	$\rightarrow$	$NO + O$
$N(^2D) + NH_3$	$\rightarrow$	$NH + NH_2$
$N(^2P) + O_2$	$\rightarrow$	$NO + O$
$N(^2P) + NH_3$	$\rightarrow$	$NH + NH_2$
$HO_2 + NO$	$\rightarrow$	$NO_2 + OH$
$HO_2 + SO_2$	$\rightarrow$	$SO_3 + OH$
$O(^1D) + H_2O$	$\rightarrow$	$2OH$
$O(^1D) + M$	$\rightarrow$	$O(^3P) + M$
$O(^3[Pub]) + O_2 + M$	$\rightarrow$	$O_3 + M$
$O(^3[Pub]) + NO + M$	$\rightarrow$	$NO_2 + M$
$O(^3[Pub]) + NO_2$	$\rightarrow$	$NO + O_2$
$H + O_2 + M$	$\rightarrow$	$HO_2 + M$
$NH_2 + NO$	$\rightarrow$	$N_2 + H_2O$
$NH_2 + NO_2$	$\rightarrow$	$N_2O + H_2O$
$NH + NO$	$\rightarrow$	$N_2 + OH$
$NH + NO_2$	$\rightarrow$	$N_2O + OH$
$HSO_3 + O_2$	$\rightarrow$	$SO_3 + HO_2$
$O_3 + NO$	$\rightarrow$	$NO_2 + O_2$
$SO_3 + H_2O$	$\rightarrow$	H_2SO_4

M is a third body molecule.

4.2. NO$_x$ Removal

Fig. 7 shows the main reaction pathways to remove NO$_x$. NO will react with O radical or O$_3$, in which O$_3$ is produced with the reaction of O$_2$.

$$NO + O(^3P) + M \rightarrow NO_2 + M$$

$$O(^3P) + O_2 + M \rightarrow O_3 + M \qquad (13)$$

$$NO + O_3 + M \rightarrow NO_2 + O_2 + M$$

Another important reaction to oxidize NO is that with HO$_2$.

$$NO + HO_2 \rightarrow NO_2 + OH \qquad (14)$$

This NO$_2$ is converted to nitric acid with the reaction of OH and finally changes to ammonium nitrate with the chemical reaction of ammonia.

$$NO_2 + OH + M \rightarrow HNO_3 + M$$

$$HNO_3 + NH_3 \rightarrow NH_4NO_3 \qquad (15)$$

This oxidization is the main process of NO$_x$ removal.

Another important pathway for removing NO$_x$ is reaction of NO with ground-state nitrogen radical N(^{4}S) or NH$_2$ radical, in which NH$_2$ radical is produced with the reaction of NH$_3$ and OH. Nitrogen molecules are produced with these reactions [22].

$$NO + N(^4S) \rightarrow N_2 + O$$

$$NH_3 + OH \rightarrow NH_2 + H_2O \qquad (16)$$

$$NO + NH_2 \rightarrow N_2 + H_2O$$

With the same type of reaction, dinitrogen monoxide is produced [23].

$$NO_2 + N \rightarrow N_2O + O$$

$$NO_2 + NH_2 \rightarrow N_2O + H_2O \qquad (17)$$

Nitrous acid is also produced [24] with the reaction of NO and OH; however, this acid will be oxidized to NO$_2$ or HNO$_3$ finally.

$$NO + OH + M \rightarrow HNO_2 + M \qquad (18)$$

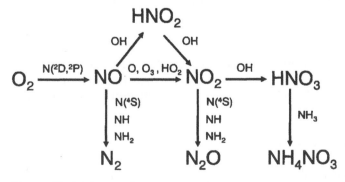

Figure 7 Main reaction pathways for removing NO$_x$. Solid line: radical reaction. Dotted line: "thermal" reaction.

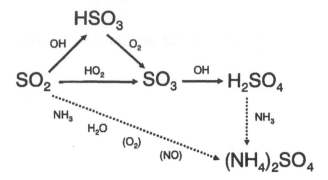

Figure 8 Main reaction pathways for removing SO_2. Solid line: radical reaction. Dotted line: "thermal" reaction.

The ground-state nitrogen radical $N(^4S)$ decreases NO with reducing reaction; however, excited-state nitrogen radicals, $N(^2D)$ and $N(^2P)$, behave perfectly in another way to produce NO.

$$N(^2D,^2P) + O_2 \rightarrow NO + O \tag{19}$$

4.3. SO_2 Removal

The main reaction pathways for SO_2 removal are illustrated in Fig. 8. The most important radical reaction for removing SO_2 is the following reaction to produce HSO_3.

$$SO_2 + OH + M \rightarrow HSO_3 + M \tag{20}$$

This HSO_3 will react with oxygen to form SO_3 and easily react with water to form sulfuric acid.

$$HSO_3 + O_2 \rightarrow SO_3 + HO_2$$
$$SO_3 + H_2O \rightarrow H_2SO_4 \tag{21}$$

The acid will react with ammonia and become ammonium sulfate.

$$H_2SO_4 + 2NH_3 \rightarrow (NH_4)_2SO_4 \tag{22}$$

Sulfur dioxide may react with HO_2 to form SO_3; however, the reaction (Eq. (14)) will be overcome when NO simultaneously exists.

$$SO_2 + HO_2 \rightarrow SO_3 + OH \tag{23}$$

In addition to this SO_x removal process with OH radicals, there exists a so-called "thermal reaction" in which SO_2 reacts with NH_3 [25–27]. This reaction depends on the concentrations of NH_3, SO_2, and H_2O, and it has a very large temperature dependence.

5. INDUSTRIAL-SCALE PLANTS AND OTHER FLUE GAS

Large-scale pilot plants have been installed in Indianapolis, USA, in 1984 [28], in Karlsruhe, Germany, in 1985 [29], in Warsaw, Poland, in 1990 [30], and in Nagoya, Japan, in 1991 [14] for the treatment of coal-fired flue gas with a flow rate of more than 10,000 m^3/hr at normal temperature and pressure. With the success of these pilot-scale tests demonstrating the feasibility of the process, two large industrial-scale plants were installed in Chengdu, China, in 1997 and in Pomorzany, Poland, in 2000. The plant in Chengdu is equipped with two accelerators with a total capacity of 640 kW for treating 300,000 m^3/hr of flue gas with 1800 ppm SO_2 and 400 ppm NO_x. The plant is now successfully operating with 80% removal of SO_2 [31]. The plant in Pomorzany [32] has four accelerators with a total maximum power of 1200 kW to treat 270,000 m^3/hr of flue gas and has started operation.

The technology for treating flue gas from oil burning boilers is essentially the same as that from coal-fired boilers, because of the same components to treat, i.e., NO_x and SO_x. Containing high concentration of sulfur, heavy oil is regarded as an urgent target; therefore intensive investigations are going on in many countries. Among them, an industrial-scale plant for flue gas of 620,000 m^3/hr will start operation in Nagoya, Japan, in 2003 [33].

The flue gas from municipal waste incinerator boilers contains SO_2, NO_x, and HCl. To remove these harmful components simultaneously by dry process, electron beam treatment method was investigated. The pilot-scale test was conducted in Matsudo, Japan, in 1992 with a flue gas of 1000 m^3/hr [34]. Recently, dioxins, namely, poly-chlorinated-di-benzo-paradioxins (PCDDs) and poly-chrorinated-di-benzo-furan (PCDFs), from incinerators have become a very serious problem because of their high toxicity. Pilot-scale tests to decompose dioxins by electron beam irradiation were conducted in Karlsruhe, Germany [35], and in Takahama, Japan [36], using almost the same capacity of flue gas, 1000 m^3/hr. Very promising results were obtained with decomposing more than 90% of dioxins.

REFERENCES

1. Okita, T. In *Sanseiu (Acid Rain)*; Hakuyusya, 1996; 25 pp (in Japanese).
2. Mohnen, V.A. Sci Am Aug. 1988, *259* (2), 14.
3. Kawamura, K.; Aoki, S. J. At Energy Soc. Jpn. 1972, *14*, 597.
4. Machi, S.; Tokunaga, O.; Nishimura, K.; Hashimoto, S.; Kawakami, W.; Washino, M.; Kawamura, K.; Aoki, S.; Adachi, K. Radiat. Phys. Chem. 1977, *9*, 371.
5. "Electron beam processing of combustion flue gases," IAEA-TECDOC-428, 1987.
6. "Environmental Applications of Ionizing Radiation"; Cooper, W.J., Curry, R.D., O'Shea, K.E., Eds.; John Wiley and Sons, Inc.: New York, 1998.
7. Watanabe, T.; Shimamura, I.; Karashima, S. In *CRC Handbook of Radiation Chemistry*; Takahashi, T., Tabata, Y., Itoh, Y., Tagawa, S., Eds.; CRC Press: Boca Raton, 1991; 31 pp.
8. Chatterjee, A. *Radiation Chemistry—Principles and Applications*; Farhatazis, Rodgers, M.A.J., Eds.; VCH Publishers, Inc.: New York, 1987.
9. Mozumder, A. *Fundamentals of Radiation Chemistry*; Academic Press: San Diego, 1999.
10. Watanabe, T.; Shimamura, I.; Karashima, S.; Takahashi, T. In *CRC—Handbook of Radiation Chemistry*; Tabata, Y. Itoh, Y., Tagawa, S., Eds.; CRC Press: Boca Raton, 1991; 47 pp.
11. "Stopping powers for Electrons and Positrons," ICRU Report 37, International Commission on Radiation Units and Measurements, Bethesda, 1984.
12. Kanazawa, T.; Haruyama, Y.; Ueno, S.; Yotsumoto, K.; Tanaka, R.; Washino, M.; Yoshida, K. JAERI-M 86-005, 1986.
13. Woods, J.; Pikaev, K. *Applied Radiation Chemistry: Radiation Processing*; Wiley-Interscience: New York, 1994.
14. Namba, H.; Tokunaga, O.; Hashimoto, S.; Tanaka, T.; Ogura, Y.; Doi, Y.; Aoki, S.; Izutsu, M. Radiat. Phys. Chem. 1995, *46*, 1103.
15. Sakamoto, I.; Mizusawa, K.; Kashiwagi, M. In *Electron beam processing of combustion flue gases*; IAEA-TECDOC-428, 201, 1987.
16. Rau, A.R.P.; Inokuti, M.; Douthat, D.A. Phys. Rev. 1978, *A18*, 971.

17. Okazaki, K.; Sato, S. Bull. Chem. Soc. Jpn 1975, 48, 3523.
18. Willis, C.; Boyd, A.W. Radiat. Phys. Chem. 1976, 8, 71.
19. Mätzing, H. In Adv. Chem. Phys., LXXX; Prigogine, I. Rice, S.A., Eds.; John Wiley and Sons Inc., 1991; 315 pp.
20. Mätzing, H. In Adv. Chem. Phys., LXXX; Prigogine, I. Rice, S.A., Eds.; John Wiley and Sons Inc., 1991; 360 pp.
21. Namba, H.; Tokunaga, O.; Tanaka, T.; Ogura, Y.; Aoki, S.; Suzuki, S. Proceedings of the International Conference on Evolution in Beam Applications, Takasaki, Japan, 1991; 476 pp.
22. Namba, H.; Aoki, Y.; Tokunaga, O.; Suzuki, R.; Aoki, S. Chem. Lett. 1988, 1465.
23. Namba, H.; Tokunaga, O.; Suzuki, R.; Aoki, S. Appl. Radiat. Isotopes 1990, 41, 569.
24. Mätzing, H.; Namba, H.; Tokunaga, O. Radiat. Phys. Chem. 1994, 43, 215.
25. Machi, S.; Namba, H.; Suzuki, N. "Electron beam processing of combustion flue gases," IAEA-TECDOC-428, 13, 1987.
26. Paur, H.-R.; Jordan, S. Radiat. Phys. Chem. 1988, 31, 9.
27. Hirota, K.; Niina, T.; Anwar, E.; Namba, H.; Tokunaga, O.; Tabata, Y. Kankyo Kagaku Kaishi 1993, 6, 143.
28. Frank, N.; Hirano, S.; Kawamura, K. Radiat. Phys. Chem. 1988, 31, 57.
29. Platzer, K.-H.; Willibald, U.; Gottstein, J.; Tremmel, A.; Angele, H.-J.; Zellner, K. Radiat. Phys. Chem. 1990, 35, 427.
30. Cheimielewski, A.G.; Iller, E.; Zimek, Z.; Licki, J. Radiat. Phys. Chem. 1992, 40, 321.
31. Doi, Y.; Nakanishi, I.; Shi, J.; Fujita, K.; Habe, H.; Suzuki, R.; Hayashi, H.; Ikeda, K.; Konno, Y.; Amano, T.; Hagiwara, I.; Satoh, K. Ebara Jiho, 1999, 183, 51.
32. Chmielewski, A.G.; Iller, E.; Zimek, Z.; Romanowski, M.; Koperski, K. Radiat. Phys. Chem. 1995, 46, 1063.
33. Tanaka, T. Genshiryoku Eye 1998, 44, 31 (in Japanese).
34. Hirota, K.; Tokunaga, O.; Miyata, T.; Sato, S.; Osada, Y.; Sudo, M.; Doi, T.; Shibuya, E.; Baba, S.; Hatomi, T.; Komiya, M.; Miyajima, K. Radiat. Phys. Chem. 1995, 46, 1089.
35. Paur, H.-R.; Bauman, W.; Mätzing, H.; Jay, K. Radiat. Phys. Chem. 1998, 52, 355.
36. Hashimoto, S.; Hirota, K.; Hakoda, T.; Taguchi, M. Proceeding of the 23rd KAIF-JAIF Seminar on Nuclear Industry, Seoul, Korea, 2001; 251 pp.

25

Ion-Beam Therapy: Rationale, Achievements, and Expectations

André Wambersie and John Gueulette
Université Catholique de Louvain, Brussels, Belgium

Dan T. L. Jones
iThemba Laboratory for Accelerator Based Sciences, Somerset West, South Africa

Reinhard Gahbauer
Ohio State University, Columbus, Ohio, U.S.A.

ABSTRACT

Radiation therapy is a very important and effective treatment modality for almost all types of malignancies. Its importance in the local control of primary or bulky cancers will even increase in the future if more effective chemotherapeutic or other systemic treatments become available to treat metastatic disease.

Improvements in the physical selectivity, from orthovoltage x-ray to cobalt-60 and high-energy linear accelerators, combined with more powerful diagnostic tools and radiation delivery methods have continuously improved the results of photon therapy (3-D or inverse planning, conformal- and intensity-modulated radiation therapy, and stereotactic methods). The safety and the reliability of photon therapy are well established.

However, new types of radiation continue to be explored with the aim of improving the physical selectivity, the radiobiological differential effect, or both. In this context, the role of fast neutrons with their biological selectivity in selected cancer sites and of protons with their superb physical selectivity will be reviewed.

Fast neutrons were the first nonconventional radiation used in cancer therapy. Fast neutrons (a high-LET radiation) were introduced for the following radiobiological reasons: (1) a reduction of the OER with increasing LET; (2) a reduction in the difference in radiosensitivity related to the position of the cells in the mitotic cycle; (3) and less repair and thus less clinical relevance of the different repair mechanisms. The best and clinically proven indications for fast neutrons are salivary gland tumors, locally advanced prostatic adenocarcinomas, and slowly growing, well-differentiated sarcomas.

Radiobiological issues, mainly related to the RBE of high-LET radiation, are discussed.

Proton beams bring a significant improvement in the physical selectivity of the treatment. The number of proton therapy centers in operation and in the planning stage increases continuously worldwide. The best clinical results, so far, have been reported for uveal melanoma, tumors of the base of skull, and some brain tumors in children.

Heavy ions are very promising by combining the advantage of the excellent physical selectivity of protons with the radiobiological advantages of fast neutrons for some tumor types. Heavy ions were applied at Berkeley from 1975 to 1992 and at NIRS in Chiba since 1994. A pilot study started at GSI-Darmstadt in 1997.

For historical reasons and for their general principles, pion therapy and the still experimental Boron Neutron Capture Therapy are also briefly discussed.

Besides the technological and radiobiological aspects considered here, other promising approaches are now being investigated in the different therapy centers and research laboratories worldwide in order to improve the effectiveness of radiation therapy. They include better, individualized fractionation and time factor, association with drugs to modulate the radiation sensitivity, combination with gene therapy and immunotherapy, and protocols combining radiation therapy with surgery and chemotherapy in a more effective way. These approaches are outside the scope of this chapter.

1. INTRODUCTION: RADIATION THERAPY (AND ION BEAMS)—AN EFFECTIVE APPROACH IN CANCER TREATMENT

1.1. Cancer Incidence and Social Implications

Radiation therapy constitutes, together with surgery and chemotherapy, one of the three "traditional" and recognized methods of cancer treatment. In addition, novel approaches, such as immunotherapy and gene therapy, are developing and appear to be promising.

With increased longevity of the populations resulting from improved control of epidemic and infectious diseases, the frequency of cancer—and the often associated suffering—has raised cancer awareness in the 20th century [1].

In developing countries, cancer cases have risen from 2 million in 1985 to 5 million in 2000 and are projected to number 10 million in 2015. By contrast, in developed countries, there were 5 million cases in 1985 as well as in 2000; no increase is projected to 2015.

For example, in the United States, in 1991, more than 1 million invasive cancers occurred, i.e., an incidence of about 400 per 100,000 per year. In addition, more than 600,000 nonmelanomatous skin cancers occurred; most of them can now be cured. In developed countries, the probability of dying from cancer is 20–25% and is expected to further increase.

1.2. Present Situation in Cancer Cure Rate

The different techniques mentioned above—surgery, radiation therapy, and chemotherapy—are more and more frequently applied in combinations in modern protocols rather than in competition. Today, about 45% of all cancer patients can be cured (Table 1) [2].

Radiation therapy contributes to the cure of 23% of all cancer patients [alone (12%) or in combination with surgery (6%) or chemotherapy/immunotherapy (5%)]. Thus about half of the cancer patients who are cured benefit from radiation therapy at least for part of their treatment: this proportion illustrates the important role of radiation therapy in cancer management.

Sixty-five percent of the cancer patients present themselves for the first consultation with a localized tumor. About 1/3 of these fail, i.e., nearly 25% of the total number of cancer

Table 1 Summary of the Present Situation Concerning Cancer Cure Rate

	Cure rate (%)
Patients appearing with localized tumor (65%):	
Cured by surgery	22
Cured by radiotherapy	12
Cured by combination of surgery and radiotherapy	6
Patients appearing with inoperable or metastatic disease (35%):	
Cured by combined complex treatment including, for example, chemotherapy and immunotherapy	5
Total[a]	45

[a] Excluding nonmelanoma skin cancers.
Source: Ref. 2.

patients. This situation has to be improved mainly by improving the efficiency of "local" treatments, i.e., radiation therapy and/or surgery.

This is the challenge for the coming years for teams involved in development of new radiation therapy modalities.

The importance of radiation therapy in the local control of primary or bulky cancers will even increase in the future if more effective chemotherapeutic or systemic therapy (such as immunology, gene therapy, etc.) becomes available to treat (prevent) metastatic disease.

Besides increasing the "cure rate," it is important to improve the tolerance to the treatment and reduce the long-term sequelae. Also, palliative treatments (in particular, pain relief) deserve attention for those patients for whom there is no hope anymore of a cure.

1.3. Radiation Therapy in Cancer Treatment

At present, in industrialized countries, about 70% of the cancer patients are referred to a radiation therapy department for at least part of the treatment. The majority of them are treated with "conventional" photon beam therapy (i.e., the reference treatment modality as defined in Table 2.1).

Experience accumulated over more than a century indicates that the major improvements in the effectiveness of radiation therapy were always associated with (or made possible by) significant progress in technology [3–5].

The major steps involved were first the move from 200-kV x-rays to cobalt-60 and then to modern electron linear accelerators providing x-ray beams of about 20 MV (and electrons when needed for some patients). Improvements in the beam penetration were, in general, combined with improvements in beam delivery systems, e.g., isocentric gantries, variable collimators, later on multileaf collimators, etc.

In addition, 3-D dosimetry and treatment planning were introduced and, last but not the least, dramatic progress was achieved in imaging and related disciplines.

It is recognized that, besides the technological and radiobiological aspects considered here, other promising approaches are now being investigated in the different therapy centers and research laboratories worldwide in order to improve the effectiveness of radiation therapy. They include better individualized fractionation and time factor, association with drugs to modulate the radiation sensitivity, combination with gene therapy and immunotherapy, and protocols combining radiation therapy with surgery and chemotherapy in a more effective way. These approaches are outside the scope of this chapter.

Before reviewing the results achieved with ion beams in cancer therapy, a short survey of the general situation in radiation therapy and the main trends are presented in the next section.

2. GENERAL TRENDS IN RADIATION THERAPY (IN RELATION WITH TECHNOLOGICAL DEVELOPMENTS)

2.1. "Conformal Radiation Therapy" and Highly Selective and Novel Photon Irradiation Techniques

For several decades, regular and continuous technical improvements were made to the linear accelerators. Similar improvements were achieved in dosimetry and imaging, and, as a consequence, the quality and effectiveness of the treatments progressively improved.*

Recently, dramatic and novel approaches were introduced in radiation therapy planning and delivery: "conformal radiation therapy," "inverse treatment planning," and "intensity-modulated radiation therapy." These improved the optimization of all treatment parameters in order to achieve the desired and optimal dose distribution, i.e., to match ("conform") as closely as possible the treated volume and the planning target volume (PTV) and at the same time reduce the dose to the organs at risk (OAR). These techniques should allow us to bring the conformity index close to unity [6,7]. This goal became possible due to the contribution of (at least) the three following factors.

1. *Patient data.* Complete and accurate anatomical information, including the location of the different target volumes and organs at risk, became possible with the latest developments in modern imaging techniques, including computer tomography (CT), magnetic resonance imaging (MRI), positron emission tomography (PET), etc. In addition, better knowledge of the natural cancer history contributes to the selection of the optimum strategic approach (e.g., selection of CTV, PTV, etc.).

2. *Medical physics and dosimetry.* Powerful and fast 3-D treatment planning, including "inverse planning" with on-line 3-D dose computation, became available. Heterogeneities can be easily identified and appropriate corrections can be applied. In addition, the radiation field can be made to conform to irregular contours and shapes of the organs.

3. *Accelerator engineering.* Full, reliable, and on-line control of the electron accelerator, beam delivery system, and multileaf collimator is the basis of the new conformal treatment and IMRT approaches. Patient-beam positioning became more accurate and fully reliable.

Of course, appropriate QA programs are required not only to check each of the above three factors, but also to ensure that they are adequately combined.

From a technical point of view, intensity-modulated radiation therapy (IMRT) optimizes several parameters: selection of multiple beams and, for each beam, optimization of dose, dose rate, size and shape, etc. Tomotherapy optimizes the dose distribution by IMRT in successive sections with immediate verification; gamma knife, cyberknife, and radiosurgery concentrate high radiation doses in well-defined small volumes.

---

* The brachytherapy applications are not discussed in this chapter.

These new techniques are rapidly expanding and are becoming available in an increasing proportion of hospitals, at least in industrialized countries. They are commercially available. They constitute the most important trend in radiation therapy today.

2.2. Does Effectiveness of Photon Beam Therapy Reach a Plateau?

The energy of photons with their optimal beam penetration, the reliability of the beam delivery and collimation systems, and the mechanical stability of the new generation of accelerators have achieved a nearly optimum level of performance. Actually, with the modern linear electron accelerators, it is now possible to irradiate at the prescribed dose (nearly) any target volume of any shape with reduced irradiation of the surrounding organs at risk (OAR).

However, the impressive development and progress in conformal therapy with photons raise a difficult issue: to what extent has photon beam therapy reached a kind of plateau as far as physical selectivity is concerned? This important question is still controversial and heavily debated. If photon therapy had reached a plateau, the search for improvements would be directed to alternative radiation modalities [8,9].

In particular, it is recognized that when optimizing the treatment parameters in order to improve, e.g., the conformity index in conformal therapy or IMRT, other indices can change adversely, for example, the size of the irradiated volume, the integral dose, and the dose homogeneity throughout the PTV.

2.3. New Types of Radiations

An alternative to further improve or optimize the photon techniques is to replace the conventional photon beams with new types of radiation. Indeed, since the beginning of radiation therapy, the radiation oncologists have always been eager to search new types of beams (different from conventional x-rays/photons) in order to improve the efficacy of radiation therapy. In principle, different approaches can be adopted (Table 2).

Improving the physical selectivity. The most straightforward method to improve the efficacy of radiation therapy is to improve the physical selectivity of the irradiation, i.e., by using beams with better physical characteristics (penetration, collimation, etc.). This is the rationale for the introduction of proton beams: no radiobiological advantage is expected.

Improving the radiobiological differential effect between the cancer and normal cell populations. This second approach is more complex and involves radiobiological considerations. New and different "radiation qualities" have been considered such as high-LET radiations. Some of the radiobiological issues involved in the use of high-LET radiation, in particular the RBE concept, are discussed in Sec. 3.

Table 2   New Types of Beams in Radiation Therapy (Alternative to Conventional Photon Beam Therapy)

(1) Improving the physical selectivity: proton beams
(2) Improving the radiobiological differential effect: fast neutrons and other high-LET radiations
(3) Combination of physical selectivity and radiobiological differential effect: heavy ions

Among high-LET radiations, fast neutrons were first introduced to exploit a radio-biological differential effect; no benefit is expected from the physical selectivity.

However, clinical evidence has demonstrated that, from a pure safety point of view, a physical selectivity at least as good as that obtained with conventional photon beams is absolutely required for neutrons as well.

The logical combination of these two approaches is the rationale for the application of heavy ions. The use of heavy ions in cancer therapy is, in principle, easy to justify by several sets of physical and radiobiological arguments. However, heavy ions are expensive, require sophisticated quality assurance, and raise complex radiobiological and physical issues as long as robust clinical data are not yet available.

The specific techniques of brachytherapy and the administration of unsealed sources of radionuclides, in nuclear medicine [10], are not dealt with in this chapter.

2.4. Reference Treatment Modality

Evaluation of the relative merits of different treatment techniques or strategic approaches—existing or newly introduced—requires the definition of a reference treatment modality.

Fractionated photon beam therapy, as defined in Table 3, is generally accepted as the reference therapy modality. This is due to its broad application, the experience accumulated over several decades, and its recognized effectiveness [4,5,8].

However, when making these comparisons, a difficulty arises from the fact that the efficiency of photon therapy (i.e., the "reference") is improving continuously, as clearly shown above (Sec. 2.1) [9].

3. HIGH-LET RADIATION AND RBE: RADIOBIOLOGICAL CONCEPT AND CLINICAL APPLICATION

3.1. From Absorbed Dose to Biological Effects

Absorbed dose is a scientifically rigorously defined quantity which is used to quantify the exposure of humans, biological systems, and any type of material to ionizing radiation.

The concept and quantity "absorbed dose" were introduced by the ICRU in 1951, with the special unit rad. A new special unit gray (Gy) was introduced in 1972 to be in correspondence with the SI system of quantities and units (Système International des Grandeurs et Unités) [11,12].

Absorbed dose is a fundamental and basic physical quantity which can be used in all fields where ionizing radiations are used. It is directly related to the physical, chemical, and biological effects produced by the irradiation. The concept of absorbed dose thus has broad applications and is indeed widely used. Metrological institutions provide standards and calibration of instruments in terms of absorbed dose.

Table 3 The Reference Radiation Therapy Modality

External photon beam therapy, involving (multiple) beams of conformally adjusted size and shape, adequately orientated to cover the PTV(s) in a homogeneous way, with photon energy ranging from a few megavolts to about 20 MV, using fractionated irradiation with daily fractions of 2 Gy, five times a week, over 4–6–7 weeks depending on the clinical situation.

Although the biological effect is directly related to the quantity absorbed dose, there is no unique relationship between absorbed dose and induced biological/clinical effects, and other factors have to be considered, such as:

1. Radiation "quality"
2. Dose rate, fractionation, and temporal distribution of the irradiation
3. Level of biological effect/endpoint
4. More generally, the technical conditions in which radiological treatment is delivered (resulting in different dose distributions and different doses to normal tissues)

A "weighting factor" or "weighting function," W, thus is applied to correlate and link absorbed dose and biological effect [13,14]. The factors listed above affect the weighting function or weighting factors independently. Different empirical formulations are used in radiation therapy to express this biological weighting function. Some agreement has been reached on the numerical values of the involved quantities, but only for well-established radiotherapy techniques.

3.2. Energy Distribution at the Microscopic Level, Microdosimetry, and Biological Efficiency

Radiation quality is defined by the nature, charge, and energy spectrum of the particles and can be characterized by the linear energy transfer (LET) or, alternatively, by the microdosimetric spectra at the point of interest under the actual irradiation conditions.

With hadrons (i.e., neutrons, protons, and heavy ions), new "radiation qualities" are introduced in therapy. The distributions of the ionizations (and energy deposition events) along the particle tracks are different, and, as a result, different and increased biological effects (at equal absorbed dose) may be expected compared with the "conventional" photon beams. Fig. 1 illustrates the differences in dose necessary to produce a given biological effect as a function of radiation quality [15].

Fig. 2 illustrates schematically the distribution of the ionizations along the particle tracks after photon and neutron irradiation [1,16]. In a more quantitative way, Fig. 3 presents the microdosimetric lineal energy, y, spectra for two clinical fast neutron beams: the highest and lowest energy used in cancer therapy (measurements performed at Louvain-la-Neuve and Essen, respectively) [17].

These large differences in energy distribution, at the microscopic level, between radiation qualities at equal absorbed dose, produce different biological effects: this leads to the concept of RBE.

3.3. The RBE Concept

The concept of relative biological effectiveness (RBE) was introduced by the ICRP and the ICRU [18,19].

The RBE is defined as a dose ratio, between two doses delivered with two radiation qualities, which produces the same effect on a given biological system, under the same conditions.*

---

* If "A" is the *test* radiation quality and "B" is the *reference* radiation quality, and if D_A and D_B are the doses necessary to produce the effect of interest with radiation A and B, respectively, the RBE of radiation A, relative to radiation B, is D_B/D_A.

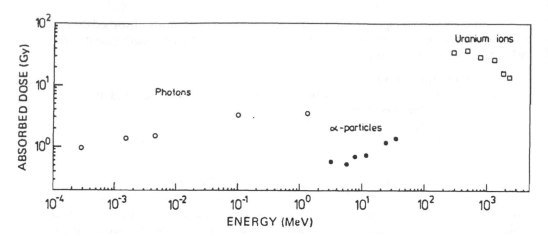

Figure 1 Absorbed dose necessary to produce a given biological effect on a given system, in the present case: 50% inactivation of V79 hamster cells in exponential growth phase. The dose needed depends on the type of the particles (photons, alpha particles, or uranium ions) and on their energy. (From Ref. 15.)

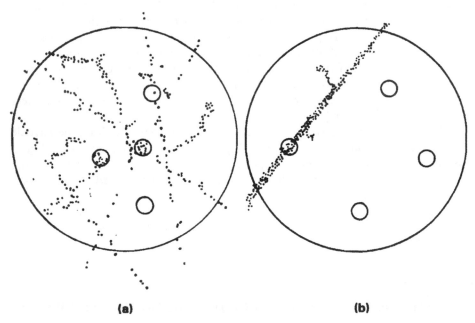

Figure 2 Distribution of ionizations in a medium irradiated by x-rays (a) and neutrons (b). The black dots represent the ionizations produced along the tracks of electrons set in motion by x-rays (a) and of protons, α-particles, or heavy ions set in motion by neutrons (b). The distribution of ionizations is very different. After irradiation by neutrons, when a vital sensitive structure (or target, represented by the circles) is crossed by a track, the deposition of energy (or the damage) is so great that there is a high probability of cell death whatever the cell line, the position in the cell cycle, the degree of oxygenation, etc. After irradiation by x-rays, the energy depositions are smaller and more variable. In some cases, a single particle deposits enough energy to kill the cell; in others, death of the cell requires the accumulation of damage produced by several tracks. (From Ref. 16, adapted in Ref. 3.)

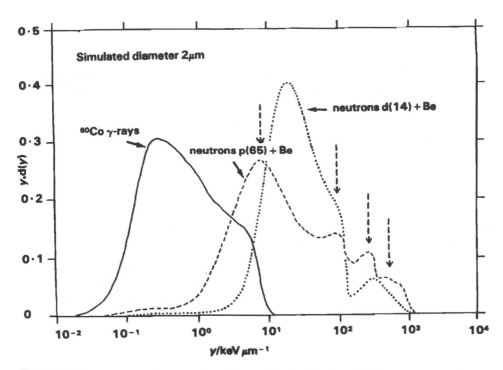

Figure 3 Comparison of energy depositions after irradiation with fast neutrons and x-rays. The curves indicate distributions of individual energy-deposition events in a simulated volume of tissue 2 μm in diameter; the parameter y (lineal energy) represents the energy deposited by a single charged particle traversing the sphere, divided by the mean cord length. The maximum with x-rays is at 0.3 keV μm^{-1} and with d(14) + Be neutrons at 20 keV μm^{-1}. The spectrum for p(65) + Be neutrons shows 4 peaks: the first is at 8 keV μm^{-1} and corresponds to high-energy protons, the second at 100 keV μm^{-1} corresponds to low-energy protons, the third at 300 keV μm^{-1} is due to α-particles, and the last, at 700 keV μm^{-1}, is due to recoil nuclei. (From Ref. 17.)

The RBE is a clear, unambiguous, and well-defined radiobiological concept. A RBE value is the result of an experiment and is thus associated with an experimental uncertainty. The biological system, type, and level of effect, the dose, and the experimental conditions in which a given RBE value has been obtained must be specified.

3.4. RBE Values and Modifying Factors

The RBE of a given radiation quality relative to ^{60}Co gamma rays is not a unique value but varies to a large extent with dose, biological system, and effect.

Fast neutrons are selected here as an example to illustrate the RBE variations because their RBE variations are significant and are well documented. They can be summarized as follows (Figs. 4 and 5) [20,21]:

1. RBE values range within large limits (from less than 2 up to about 10).
2. The RBE increases with decreasing dose.
3. The RBE values for late effects are significantly higher than for early effects (Figs. 4 and 5).
4. The RBE of the fast neutron beams used in therapy varies with energy, and this variation is significant (Fig. 6) [22].

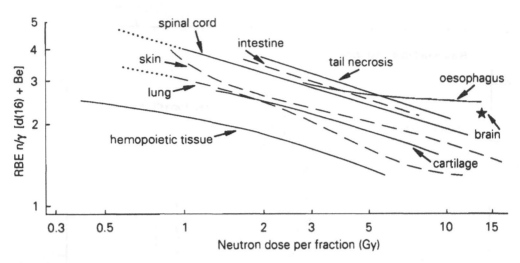

Figure 4 RBE/dose relationships for different normal tissues. Summary of the pretherapeutic radiobiological experiments performed at the Hammersmith cyclotron with d(16) + Be neutrons. The increase of RBE with decreasing dose is obvious for all tissues investigated. (From Ref. 20.)

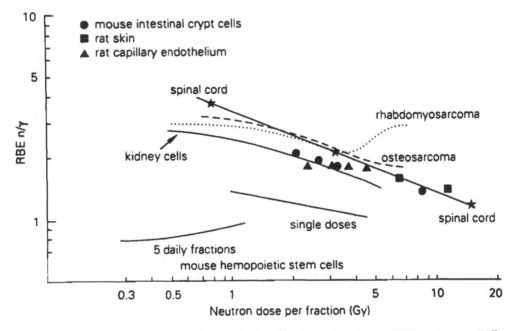

Figure 5 RBE/dose relationships for 15-MeV neutrons produced by a (d,T) generator. Different biological endpoints in normal tissues and tumors are investigated. For late tolerance of spinal cord, the RBE increases from 1.2 to 3.7 when the neutron dose per fraction decreases from 16 to 0.8 Gy. Higher RBE values were found later on for spinal cord at lower doses. (From Ref. 21.)

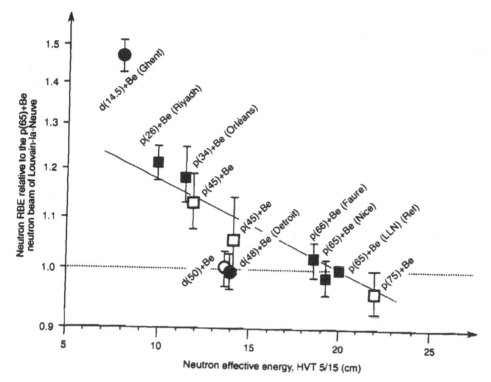

Figure 6 RBE variation of neutron beams as a function of energy. The p(65) + Be beam of Louvain-la-Neuve is taken as reference (RBE = 1). The closed squares and circles correspond to six different visited neutron facilities. The open squares and circles correspond to beams produced at the variable-energy cyclotron of Louvain-la-Neuve. On the abscissa, the "effective energy" of the neutron beams is expressed by their half-value thickness (HVT 5/15) measured in specified conditions. Intestinal crypt regeneration in mice, after a single fraction irradiation, is taken as the biological system. The 95% confidence intervals are shown. A straight line is fitted through the points (squares) corresponding to neutron beams produced by the (p + Be) reaction. For comparison, the neutron beams produced by the (d + Be) reaction are represented by circles. (From Ref. 22.)

A high RBE of a new radiation quality does not per se imply a therapeutic advantage. The therapeutic gain is actually the ratio of the RBE for the effects on the cancer cell population and the RBE for the effects on the normal tissues (under the actual irradiation conditions).

Further information about the RBE and the specific biological effects of the different hadrons used in therapy are found in Sec. 4.

3.5. From Radiobiology to Clinical Application of RBE: The RBE Weighting Factor, W_{RBE}

When prescribing the dose in hadron (neutron) therapy, the radiation oncologist has to take into account the (set of) reported RBE values, and, from them, the radiation oncologist has to *select one* "RBE value" to weight the absorbed dose in order to obtain the desired clinical effect [13,14].

This value is, in a strict sense, no longer a RBE but a weighting factor (selected of course on the basis of RBE data) applied to compensate for RBE differences/variations. It is the "RBE weighting factor," W_{RBE}.

It is current medical practice to select the weighting factors as a function of two main criteria: (1) late tolerance in normal tissues and (2) in relation to a daily fractionation of 2 Gy (photons or ^{60}Co-gamma rays). Therefore among the published RBE values, those obtained for late effects on normal tissues, with 2 Gy of photons per fraction as reference, are the most relevant for the selection of the RBE weighting factor and could be defined as "reference conditions." The reference RBE is then, to some extent, an "average" or "overall" RBE for late tolerance of normal tissues at risk, evaluated vs. 2 Gy per fraction of photon radiation.

The reference RBE provides a first indication to select the RBE weighting factor W_{RBE}. However, other factors, which are not always clearly identified, may also influence the selection of the RBE weighting factor:

1. Past personal clinical experience
2. Clinical experience reported from other centers
3. Some differences in time factors
4. Some volume effects related to the beam characteristics, geometry, or technical conditions

Selection of the RBE weighting factor, W_{RBE}, implies the (informed) judgment and clinical experience and thus the responsibility of the radiation oncologist in charge of the patient.

This "RBE weighting factor" has been called "clinical RBE" in the past; this terminology may be confusing.

3.6. The Biologically Weighted Dose for RBE, D_{RBE}

The product of the neutron absorbed dose D by W_{RBE} is the "biologically weighted dose" for RBE, D_{RBE}:

$$D_{RBE} = DW_{RBE}$$

where D and D_{RBE} are expressed in Gy but a subscript may be useful to avoid confusion between the (physical) absorbed dose D and the biologically weighted dose D_{RBE}.

In collaborative studies involving several neutron therapy centers, the same biologically weighted doses D_{RBE} should be prescribed. If the participating centers use neutron beams with different RBEs, different W_{RBE} values thus have to be introduced and different (physical) neutron doses have to be prescribed.

In addition to D_{RBE}, the delivered absorbed doses and the selected RBE weighting factors W_{RBE} should always be reported, in addition to D_{RBE}, to allow for eventual later reevaluation of the results [23].

A similar approach can be followed with protons and heavy ions. There is an additional difficulty with heavy ions (e.g., carbon ions or neon ions) because the RBE varies significantly with depth in the tissues (see Sec. 4.4). With proton beams, there are only discrete RBE variations (less than 10–20%) (see Sec. 4.2).

3.7. Discussion

Absorbed dose has proven to be a fundamental and very useful quantity in radiation therapy which should always be reported. However, absorbed dose alone cannot predict the biological effects, and different "weighting factors" have to be introduced depending on the radiation therapy modality.

In external photon beam therapy, weighting factors, $W_{\alpha/\beta}$, are currently estimated based on the α/β model to compensate for differences in fractionation.

In brachytherapy, with the modern equipment now available, the clinically applied dose rates vary within large limits. The ICRU and the international brachytherapy community are recommending weighting factor, W_{HDR} and W_{PDR}, to compensate for the differences in dose rate for high dose-rate (HDR) and pulse dose-rate (PDR) brachytherapy, respectively.

In hadron therapy, a similar approach and symbolism are presented to take into account the RBE differences. RBE values for neutrons are well documented, and reliable data are now becoming available for protons and heavy ions, on which safe weighting factors W_{RBE} can be based. Besides radiobiological determinations, microdosimetry brings independently additional information on the radiation quality and improves confidence in both sets of data. Microdosimetry provides an objective description of the radiation quality at the point of interest under the actual irradiation conditions. Correlation between this set of information and the experimental RBE results is of great scientific and clinical value in hadron therapy [24,25].

For the different therapy modalities, the products of the absorbed doses by the biological weighting factors ($W_{\alpha/\beta}$, W_{HDR}, W_{PDR}, and W_{RBE}) are the biologically weighted doses ($D_{\alpha/\beta}$, D_{HDR}, D_{PDR}, and D_{RBE}, respectively). They are expressed in Gy like the absorbed dose, but, to avoid confusion, subscripts ($Gy_{\alpha/\beta}$, Gy_{HDR}, Gy_{PDR}, and Gy_{RBE}) must indicate that the absorbed doses have indeed been weighted.

The biologically weighted dose is a quantity intended to be correlated, as closely as possible, to the relevant biological/clinical effect. However, in all radiotherapy reports, the (physical) absorbed doses D and the biological weighting factors ($W_{\alpha/\beta}$, W_{HDR}, W_{PDR}, W_{RBE}, etc.) should always be indicated separately and not just the biologically weighted doses ($D_{\alpha/\beta}$, D_{HDR}, D_{PDR}, D_{RBE}, etc.).

The weighting factors, W, as introduced above, aim to be a pragmatic approach to harmonize prescription and reporting of radiation treatments. They are based on actual clinical practice, current techniques, and available radiobiological data. Their use should be limited to the technical treatment modality for which they were designed. They are not "universally" applicable factors, and they should not be multiplied or combined between each other.

Finally, it must be stressed again that the selected weighting factors W (and thus the biologically weighted doses) are based on available radiobiological data and clinical observation, but they always imply clinical judgment and experience.

4. NONCONVENTIONAL RADIATION THERAPY MODALITIES (RATIONALE, TECHNICAL ASPECTS, AND SHORT SURVEY OF CLINICAL DATA)

4.1. Fast Neutron Therapy

Rationale: Radiobiological Basis

Fast neutrons were the first "nonconventional" radiations to be used in clinical radiation therapy. The pioneering work in fast neutron therapy was performed by Stone [26] and his associates at Berkeley between 1938 and 1943. In the late 1960s, after extensive radiobiological experiments, a neutron therapy program was initiated at the Hammersmith

Hospital in London and, a few years later, in several centers in Europe and in the United States [27,28].

The radiobiological rationale for introducing high linear energy transfer (LET) radiation in cancer therapy, as proposed in the 1960s, is still valid and has not been contradicted by more recent radiobiological findings [3].

The radiobiological and clinical data obtained with fast neutrons are reviewed in the following section; they deserve a detailed and careful analysis as they constitute the rationale of the clinical application of any high-LET radiation and, in particular, the justification of the modern heavy-ion therapy programs.

The differences in the radiobiological effects produced by low- and high-LET radiations are related to the pattern of energy distribution at the level of the particle tracks. The distributions of energy (or ionizations) at the microscopic level after low- and high-LET radiation exposure are compared in Figs. 2 and 3 [16,17,29]. For fast neutrons, at the energies used in therapy, the energy deposited in a 2-μm sphere by one particle track (proton) is 50 to 100 times higher than after gamma-ray irradiation (secondary electrons). Such differences result in relative biological effectiveness (RBE) of neutrons higher than unity (ranging between 3 and 5) and in different shapes of the dose–effect relationships.

The arguments for using high-LET radiation in cancer therapy are based on these well-established differences in energy distribution at the microscopic level and can be summarized as follows:

A reduction in the oxygen enhancement ratio (OER) with increasing LET
A reduction of the differences in radiosensitivity related to the position of the cell in
    the mitotic cycle
Less repair and thus less clinical relevance of the different repair mechanisms

Historically, the oxygen effect was the main rationale for introducing fast neutrons in radiotherapy. Three sets of data support this rationale:

1. Hypoxic cells are present in the malignant tumors; they result from the fast proliferation of the cancer cells.
2. Hypoxic cells are three times more radio-resistant than well-oxygenated cells (OER = 3) for low-LET radiation. The presence of a small percentage of hypoxic cells (1% or even 0.1%) can thus make the tumor radio-resistant.
3. The OER decreases when LET increases (Fig. 7) [30]. It decreases to about 1.6 for fast neutrons and is close to unity for alpha particles. OER = 1.3 and 1.0 for alpha particles of 4 and 2.5 MeV (i.e., for 110 and 160 keV/μm), respectively.

Fig. 8 illustrates how increasing LET reduces the differences in radiosensitivity related to the position of the cells in the mitotic cycle [31]. Cells in stationary phase and in S phase are significantly more radio-resistant than mitotic cells.

Lastly, because of the large amount of energy deposited in the critical cellular target by a single high-LET particle track, repair of sublethal damage is reduced (or less relevant). High-LET radiation is thus particularly efficient against cells which have a high capability for repair of sublethal damage. Therefore the dose per fraction is less important with the high-LET than with the low-LET radiation, which gives the radiation oncologist more freedom in the selection of the fractionation scheme.

The radiobiological arguments discussed above indicate that the high-LET radiation could bring a benefit in the treatment of some cancer types. However, they also imply the need for the development of "predictive tests," allowing the radiation oncologist to

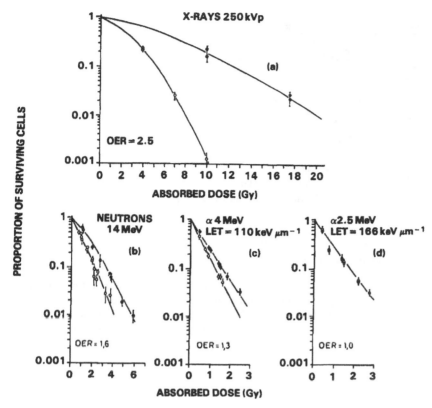

Figure 7 Survival curves of human kidney cells T1 irradiated under hypoxic and aerobic conditions with different qualities of radiation: (a) 250-kV x-rays (LET of about 1.3 keV μm⁻¹); (b) 14-McV neutrons produced by the (d,T) reaction (LET of about 12 keV μm⁻¹); (c) 4-MeV α-particles (LET = 110 keV μm⁻¹); (d) 2.5-MeV α-particles (LET = 166 keV μm⁻¹). (From Ref. 30.)

select the best radiation quality (low or high LET) for a given patient group or individual patient.

The observations of Battermann et al. [32] on lung metastases indicate that slowly growing tumors (well differentiated) could benefit from high-LET treatment (Fig. 9). The clinical results accumulated over more than 25 years confirmed these observations.

Technological Aspects

Many centers, including 15 in Europe (former Soviet Union excluded), have applied fast neutron therapy using different types of generators, which can be schematically distributed into four groups:

Reactors, using fast neutrons in the beam (limited application)
"Low-energy" cyclotrons, using mainly incident deuterons with energies ranging from 13 to 16 MeV (d + Be reaction)
(d + T) generators
"High-energy" cyclotrons or linear accelerators (d/p + Be reaction)

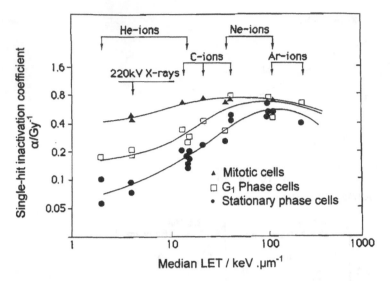

Figure 8 Differences in radiosensitivity with the position of the cells in the mitotic cycle. The differences are reduced with increasing LET. Single-hit inactivation coefficients (α) for homogeneous populations of mitotic, G1-phase, and stationary phase Chinese hamster cells irradiated with 220-kV x-rays and various charged particle beams, as a function of median LET (in keV μm^{-1}). (From Ref. 31.)

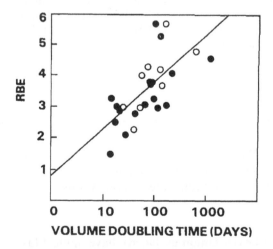

Figure 9 Relation between the RBE of neutrons for regression of lung metastases in patients and their doubling time. The closed circles correspond to measured values of RBE. The open circles correspond to values estimated from irradiation with neutrons only. For the 15-MeV neutrons used in this study [produced by a (d,T) generator], the RBE for the tolerance of the most important normal tissues is about 3. As a consequence, neutrons are a good indication (RBE > 3) for tumors having doubling times greater that $\approx$100 days. In contrast, they should not be used for rapidly growing tumors. (From Ref. 32.)

Only for the fourth group can the physical selectivity and the technical conditions be considered to be sufficient (or nearly sufficient) for adequate treatments, especially in comparison with modern linear accelerators.

In some centers, large clinical programs were completed from which important radiobiological and clinical conclusions could be derived. In some other centers, the facilities were shut down abruptly:

> In most centers, due to technical difficulties; for example, all $(d + T)$ generators are now shut down.
>
> In other centers, due to patient recruitment problems.
>
> In general, due to "suboptimal" physical selectivity.

Today, in the majority of centers still active in neutron therapy, the technical conditions are becoming progressively comparable to those in modern photon beam therapy. In addition, a few new high-energy facilities have been proposed (e.g., in China, Germany, Poland, Slovakia, and South Africa).

Clinical Aspects (Short Survey)

Salivary gland tumors and locally extended prostatic adenocarcinoma were proven to be an indication for fast neutrons.

Salivary Gland Tumors. Neutron beam therapy should be considered as the treatment of choice in patients with unresectable malignant salivary gland tumors or in patients where radical resection would require facial nerve sacrifice (Table 4) [33].

For inoperable primary or recurrent tumors, a randomized cooperative study showed, at 2 years, a significant advantage for neutrons compared to photons for loco-regional control (76% vs. 17%, $P < 0.005$) and a trend towards improved survival (62% vs. 25%). Ten-year analysis continued to show a striking difference in loco-regional control (56% for neutrons vs. 17% for photons, $P = 0.009$), but both groups experienced a high rate of metastatic failure (Fig. 10) [34].

Prostatic Adenocarcinoma. For prostatic adenocarcinomas, their typical slow growth rate and low cycling fraction provide a logical radiobiological rationale for exploring neutrons (high LET) in the treatment of this disease.

Table 4 Pooled European Data of Local Control in Advanced Salivary Gland Tumors

Reference	No. of patients	Local control
Catterall (1987)	65	48 (74%)
Battermann and Mijnheer (1986)	32	21 (66%)
Duncan et al. (1987)	22	12 (55%)
Prott et al.(1996)	64	39 (61%)
Kovács et al. (1987)	15	13 (87%)
Krüll et al.(1995)	74	44 (59%)
Skolyszweski et al. (1982)	3	2
Overall	275	179 (65%)

Source: Ref. 33.

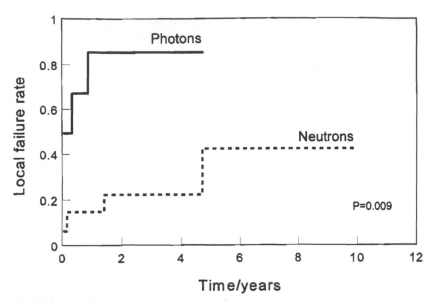

Figure 10 Neutron therapy of salivary gland tumors. Probability of local–regional failure for unresectable salivary gland tumors. Starting values of the curves represent initial local–regional failure rates. (From Ref. 34.)

The Radiation Therapy Oncology Group (RTOG), 1997, compared "mixed beam" (a combination of photons and neutrons) to conventional photons for locally advanced prostatic cancer. Loco-regional control as well as survival were significantly superior after mixed-beam irradiation (Fig.11) [34].

In 1986, the Neutron Therapy Collaborative Working Group (NTCWG) compared neutrons (alone) and conventional photons. A significant difference ($P < 0.01$) was observed in "clinical" loco-regional failure, with actuarial 5-year failure rates of 11% vs. 32% after neutrons and photons, respectively (Fig. 12) [34]. Inclusion of routine posttreatment biopsies resulted in 5-year "histological" local–regional failure rates of 13% and 32%, respectively ($P = 0.01$).

Due to the long natural history of recurrent prostate cancer, longer follow-up is required to assess the ultimate impact of the improved local control on survival. However, the prostate specific antigen (PSA) levels could provide an indication: at 5 years, 17% of the neutron patients showed elevated PSA levels compared to 45% for the photon patients ($P < 0.001$).

Late sequelae (mainly large bowel complications) were worse in the neutron-treated group (11% vs. 3%). However, no colostomy was required in 51 patients treated with a multileaf collimator at the University of Washington, while 6/38 patients, from other centers using movable jaw or fixed cone collimator, required colostomy.

The data from Louvain-la-Neuve suggest that mixed neutron–photon therapy is particularly efficient in patients with unfavorable prognostic factors, such as PSA >20 ng/mL [35].

Other Tumor Sites or Types. For other tumor sites or types, such as slowly growing soft tissue sarcomas, fixed lymph nodes in the cervical area, locally extended antrum tumors, and some bronchus carcinomas, the available clinical results tend to show a benefit with neutrons. However, they need to be confirmed by randomized studies [5,28,36].

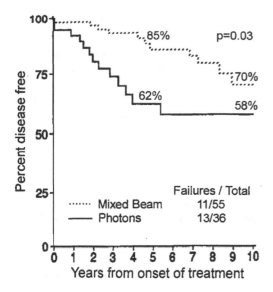

Figure 11 Clinical loco-regional control in patients treated with mixed (neutron/photon) beams or photons only (RTOG randomized trial) for locally extended prostatic adenocarcinoma. (From Ref. 34.)

4.2. Proton Beam Therapy

Proton beam therapy was a significant step towards better physical selectivity, and it is the most straightforward approach to improve the efficacy of therapeutic irradiations.

The first cyclotron was built by Ernest O. Lawrence in 1932, and since 1938, cyclotrons have been used for patient treatment. In Berkeley, in 1954, the first human target irradiated with protons was the pituitary gland with the aim to suppress its function for slowing down the metastatic development of breast cancer.

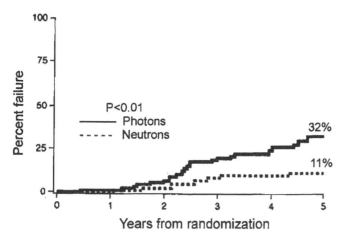

Figure 12 Actuarial clinical loco-regional failure in patients with locally advanced prostate cancer. Results of NTCWG trial on prostate. (From Ref. 34.)

Today, proton beam therapy is applied in more than 20 centers worldwide and new facilities are in preparation (or under discussion). So far, about 30 000 patients have been treated with protons (Tables 5a and 5b). There is a continuously increasing number of proton therapy centers in operation and an impressive number of projects under construction or study.

On the other hand, proton beams have no radiobiological advantages and a constant relative biological effectiveness (RBE) with respect to photons of 1.0–1.1 is generally assumed. The vast clinical experience accumulated with photons can thus be transferred directly to proton therapy.

Rationale: The "Bragg Peak"

The depth–dose curve in a proton beam exhibits a characteristic shape. The absorbed dose is low at the entrance ("initial plateau"), then increases in depth more and more steeply, to

Table 5a Worldwide Charged Particle Patients Totals (HTCOG—Hadron Therapy Cooperative Group; Janet Sisterson, 2001)

Who	Where	What	Date 1st RX	Date last RX	Recent patient total	Date of total
Berkeley 184	California, U.S.A.	p	1954	1957	30	
Berkeley	California, U.S.A.	He	1957	1992	2054	Jun. 1991
Uppsala	Sweden	p	1957	1976	73	
Harvard	Massachusetts, U.S.A.	p	1961		8747	Jan. 2001
Dubna	Russia	p	1967	1974	84	
Moscow	Russia	p	1969		3268	Jun. 2000
St. Petersburg	Russia	p	1975		1029	Jun. 1998
Berkeley	California, U.S.A.	HI	1975	1992	433	Jun. 1991
Chiba	Japan	p	1979		133	Apr. 2000
PMRC, Tsukuba	Japan	p	1983		629	Jul. 1999
PSI (72 MeV)	Switzerland	p	1984		3253	Dec. 2000
Dubna	Russia	p	1987		79	Dec. 2000
Uppsala	Sweden	p	1989		236	Jun. 2000
Clatterbridge	England	p	1989		999	Jun. 2000
Loma Linda	California, U.S.A.	p	1990		5638	Dec. 2000
Louvain-la-Neuve	Belgium	p	1991	1993	21	
Nice	France	p	1991		1590	Jun. 2000
Orsay	France	p	1991		1894	Jan. 2001
N.A.C.	South Africa	p	1993		380	Nov. 2000
MPRI	Indiana, U.S.A.	p	1993		34	Dec. 1999
UCSF-CNL	California, U.S.A.	p	1994		284	Jun. 2000
HIMAC, Chiba	Japan	HI	1994		745	Dec. 1999
TRIUMF	Canada	p	1995		57	Jun. 2000
PSI (200 MeV)	Switzerland	p	1996		41	Dec. 1999
GSI Darmstadt	Germany	HI	1997		72	Jun. 2000
HMI, Berlin	Germany	p	1998		166	Dec. 2000
NCC, Kashiwa	Japan	p	1998		35	Jun. 2000
					3304 Ions	
					28,700 Protons	
				Total	33,104 All particles	

Table 5b Charged Particle New Facilities (HTCOG—Hadron Therapy Cooperative Group; Janet Sisterson, 2001)

Institution	Place	Type	1st
INFN-LNS, Catania	Italy	p	2001
NPTC (Harvard)	Massachusetts, U.S.A.	p	2001
Hyogo	Japan	p, ion	2001
NAC, Faure	South Africa	p	2001
Tsukuba	Japan		2001
Wakasa Bay	Japan		2002
Bratislava	Slovakia	p, ion	2003
IMP, Lanzhou	P.R. China	C–Ar	2003
Shizuoka Cancer Center	Japan		2003
Rinecker, Munich	Germany	p	2003?
CGMH, northern Taiwan	Taiwan	p	2001?
Erlangen	Germany	p	2002?
CNAO, Milan and Pavia	Italy	p, ion	2004?
Heidelberg	Germany	p, ion	2006
AUSTRON	Austria	p, ion	?
Beijing	China	p	?
Central Italy	Italy	p	?
Clatterbridge	England	p	?
TOP project ISS Rome	Italy	p	?
Three projects in Moscow	Russia	p	?
Krakow	Poland	p	?
Proton Development N.A. Inc.	Illinois, U.S.A.	p	?
PTCA, IBA	United States	p	?

reach a maximum at the level of the "Bragg peak" (Fig. 13). The depth of the Bragg peak in the tissues depends on energy. There is no dose beyond the depth of the Bragg peak and thus full sparing of the tissues behind the target volume.

The Bragg peak has to be spread over the depth occupied by the planning target volume [spread-out Bragg peak (SOBP)] (Fig. 14).

Technological Aspects: Beam Delivery

In the past, proton therapy was performed with complex physics machines, adapted to clinical needs, not always available full time, often expensive to maintain, and difficult to tune. Today, several commercial companies offer "turn key equipment" for proton therapy, adapted to the needs (or to the financial limitations) of the center. It is likely that this trend will develop further.

The establishment of the Northeast Proton Therapy Center (NPTC) in Boston was obviously the trigger of the movement, but such a fast proliferation was certainly not expected [37].

In general, a new radiotherapy technique may be considered to have gained its place among the other ones when the equipment can simply be purchased commercially. This is obviously now the case with proton therapy.

From a technical point of view, two methods can be used to deliver proton beam therapy: passive scattering and scanning beam (Fig. 15). The use of a scanning beam is a more complex approach, but it allows close matching of the treated volume with the PTV. In

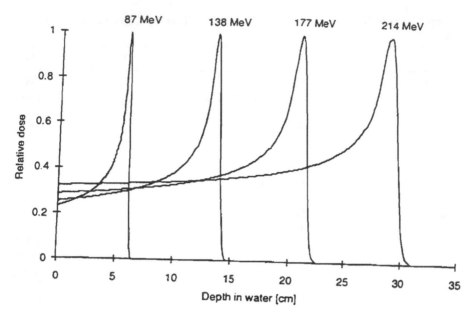

Figure 13 Depth–dose curves for proton beams of different energy. The position of the Bragg peak depends on energy and can thus be adjusted according to the clinical requirements. (From PSI, Villigen, courtesy of Pedroni and Scheib.)

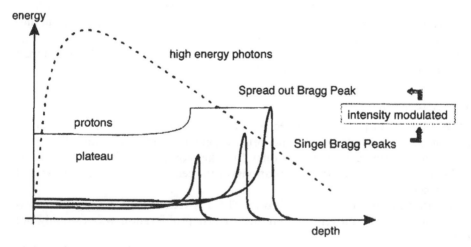

Figure 14 Proton beam irradiation of a deep-seated large tumor. Single Bragg peaks of different energy are combined, in adequate proportions, to obtain a homogeneous dose distribution at the level of the SOBP. The depth–dose curve of a photon beam, shown for comparison, is inferior compared to the proton curve. However, an optimized multifield photon treatment allows to reach better irradiation conditions. (From Ref. 43.)

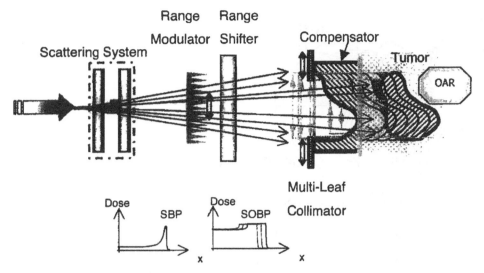

Figure 15 Schematic representation of a purely passive spreading system. A set of scatterers guarantees a flat transverse dose distribution over a large field. Starting with a single Bragg peak, a ripple filter (in combination with a flat energy degrader as range shifter) produces a SOBP with a homogeneous dose at each depth. A special compensator conforms the distal surface of the radiation field with the shape of the planning target volume. (From Ref. 43.)

Europe, the Paul Scherrer Institute (PSI), Villigen, Switzerland has developed (and is planning to further improve) a highly sophisticated beam scanning system in order to exploit all the physical advantages of the proton beams. The scanning beam technique will be dealt with in more detail in the section on heavy ions (Sec. 4.4).

Of course, like with photons, several intersecting proton beams can be used in order to further improve the final dose distribution. Fixed beams or an isocentric gantry can be used.

Clinical Aspects (Short Survey)

The clinical results available so far can be summarized as follows [2,5].

Uveal Melanoma. Charged particle beams and, in particular, protons are ideal for treating intraocular lesions since they can be made to deposit their absorbed dose in the target volume, while significantly limiting the irradiation of the noninvolved ocular and orbital structures (Fig. 16).

Large series of patients with uveal melanoma were treated with protons in several centers worldwide. The Massachusetts General Hospital/Harvard Cyclotron Laboratory (MGH/HCL) in Boston played a pioneering role, and 2568 uveal melanoma patients were treated through September 1998 [38].

Local control of the tumor within the treated eye which was 96% and 95% at 60 and 84 months, respectively, was reported. Eye retention probability after proton therapy depends on tumor size, being 97%, 93%, and 78% for patients with small, intermediate, and large tumors, respectively. Survival at 5 years of the patients treated with protons or enucleation is similar (about 80%).

Tumors of the Base of Skull and Cervical Spine. Proton or other charged particle beams are a treatment of choice for skull base and cervical spine tumors: irradiation can be

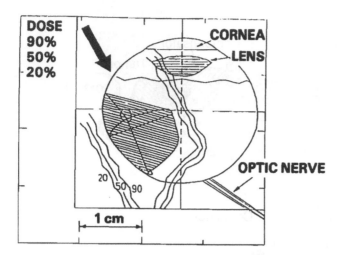

Figure 16 Proton therapy for uveal melanoma. Dose distribution obtained with a beam of 60-MeV protons with an appropriate spread-out Bragg peak (energy modulated from 14–60 MeV). Transverse section through the center of the eye. The position of the tumor [gross tumor volume (GTV)] is indicated by the posterior hatched area. Protons allow to obtain a homogeneous dose over the whole GTV with effective sparing of the normal structures. This implies a great precision in patient-beam positioning. (Courtesy from PSI, cited in Ref. 3.)

focused in the target volume, while achieving significant sparing of the brain, brain stem, cervical cord, optical nerves, and chiasma.

At HCL/MGH in Boston, 621 patients with chordomas and low-grade chondrosarcomas of the skull base and cervical spine were treated with protons between 1975 and January 1998. For skull base tumors, with follow-up ranging from 1 to 254 months (median of 41 months), local recurrence-free survival is significantly better for chondrosarcomas than for chordomas. It is 98% at 5 years and 94% at 10 years for chondrosarcomas and 73% at 5 years and 54% at 10 years for chordomas (Fig. 17) [39]. Overall survival is also significantly better: 91% vs. 80% at 5 years and 88% vs. 54% at 10 years, respectively.

For cervical spine tumors, with follow-up ranging from 1 to 172 months (median of 36 months), local recurrence-free survival was not significantly different for chondrosarcomas and chordomas: 54% vs. 69% at 5 years and 54% vs. 48% at 10 years, respectively. The overall survival at 5 years for chondrosarcomas and chordomas were 48% and 80%, respectively, but at 10 years, 48% and 33%, respectively [38].

CNS Tumors in Children. Pediatric tumors located in the CNS are particularly challenging and deserve highly refined techniques of radiation therapy like proton therapy (Fig. 18) [40–42]. The preliminary results from Loma Linda, MGH/HCL, and Centre de Proton Thérapie d'Orsay are promising and show an excellent immediate and late tolerance.

The reduced integral dose may be of particular importance in the pediatric population.

Other Localizations. Results of proton beam therapy for other tumor types and localizations have been reported, in particular, for retinoblastoma and age-related macular degeneration [2]. In the pioneering centers, patient selection, special care applied to each treatment, as well as expertise of the multidisciplinary clinical teams make it difficult to conclude whether the excellent results were due to the protons themselves or reflect the expertise of the different teams.

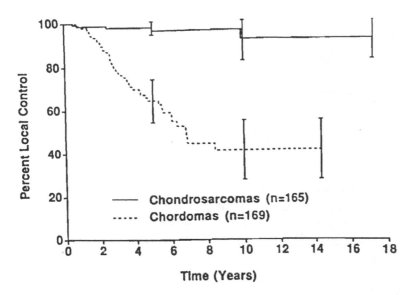

Figure 17 Proton beam therapy of tumors of the base of skull. Probability of local control in patients with chordomas ($n = 169$) and low-grade chondrosarcomas ($n = 165$). (From Ref. 39.)

4.3. Pion Therapy

Between 1974 and 1994, 1100 patients were treated with negative pions at three centers. These particles, which provide a mixture of high- and low-LET components and some physical selectivity advantage, were very expensive to generate and the clinical results were unconvincing. All three facilities have now been closed down [5].

4.4. Heavy-Ion Therapy

Introduction: Terminology

Strictly speaking, hadrons include protons, neutrons, and mesons, but, in the context of hadron therapy, it is common practice to include (charged) ions heavier than protons and often to exclude neutrons. As seen above (Sec. 4.3), π-mesons are no longer used in therapy.

In the radiation oncology (or radiobiology) community, "hadron therapy" currently includes the application of protons, helium, carbon, neon, and argon ions. Carbon, neon, or argon ions are called "heavy ions," while protons and helium ions are called "light ions." In contrast, in the physics community, all ions mentioned above are called "light ions" compared to much heavier ions such as, for example, uranium [43].

In the following section, heavy ions include ions heavier than protons (and helium ions). They combine the advantages of an excellent physical selectivity comparable to that of protons with the radiobiological advantages of high-LET radiations for some types of tumors (as discussed for fast neutrons in Sec. 4.1.1). However, as carbon ions are, at the moment, the only type of heavy ions used in therapy, the next section will deal mainly with carbon ions.

Rationale: The "Bragg Peak" and High-LET

Four advantages can be identified for heavy ions in cancer therapy.

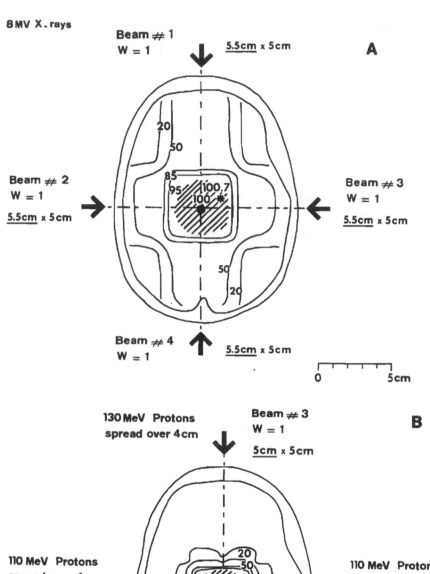

8 MV X . rays

Beam # 1
W = 1   5.5cm x 5cm   A

Beam # 2
W = 1
5.5cm x 5cm

Beam # 3
W = 1
5.5cm x 5cm

Beam # 4
W = 1   5.5cm x 5cm

0   5cm

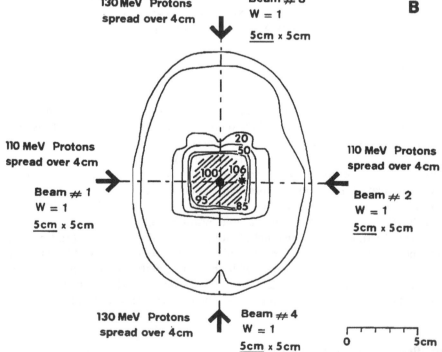

130 MeV Protons
spread over 4cm

Beam # 3
W = 1
5cm x 5cm   B

110 MeV Protons
spread over 4cm

Beam # 1
W = 1
5cm x 5cm

110 MeV Protons
spread over 4cm

Beam # 2
W = 1
5cm x 5cm

130 MeV Protons
spread over 4cm

Beam # 4
W = 1
5cm x 5cm

0   5cm

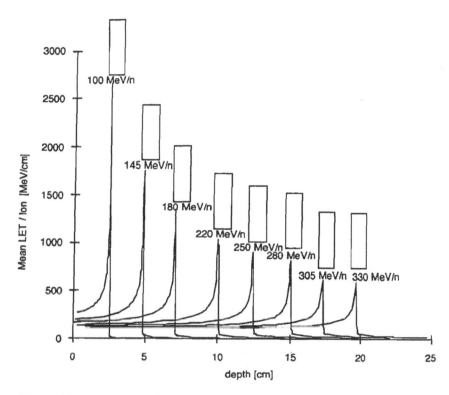

Figure 19 Bragg curves for carbon beams with different energies. When increasing the energy, the range of the particles increases and the Bragg peak occurs at larger depths. Concurrently, the height of the Bragg peak decreases. (From Ref. 44.)

Bragg Peak and Physical Selectivity. The physical selectivity of carbon ions, other heavy ions, and protons is quite similar. Fig. 19 shows the "Bragg peak" for carbon ions as a function of depth [44]. The Bragg peak has to be spread out, as for protons, to fully cover the planning target volume (PTV). The ratio between the dose at the level of spread-out Bragg peak (SOBP) and of the initial plateau is compared for carbon ions and protons and for different ions in Figs. 20 and 21 [44,45].

Nuclear fragmentation observed with carbon and other heavy ions could be a disadvantage as some energy is deposited beyond the primary Bragg peak. This issue is probably not clinically relevant as the involved absorbed dose level is low and the fragments are low-LET particles. The penumbra is somewhat narrower with carbon ions than with protons as the particles are heavier.

High LET and RBE. More important is the fact that the LET—thus also the RBE—increases with depth in the hadron beam (Figs. 22 and 23) [43]. This further increases the

Figure 18 Comparison of photon and proton dose distributions for the treatment of a craniopharyngioma in a child. Typical planning sections for a large suprasellar craniopharyngioma, in a 3-year-old child, treated with photons (A) or protons (B). The planning target volume (PTV) is indicated by the hatched area. For photons and protons, four equally weighted beams are used and the normalization point was chosen at the intersection of the beam axes. For the four proton beams, the Bragg peak was spread over 4 cm. (From Ref. 41.)

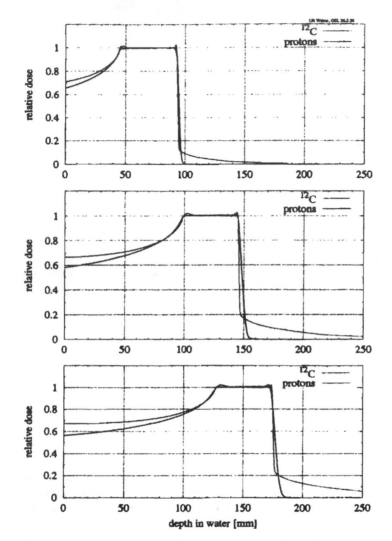

Figure 20 Comparison of calculated extended Bragg peaks of protons and carbon ions (^{12}C) at different penetration depths. (From Ref. 45.)

ratio of the "biological equivalent doses" between the SOBP and the initial plateau (Fig. 24) [44]. The clinical benefit of a higher RBE at the level of the distal part of the SOBP is illustrated in Fig. 25 [45].

High-LET Advantage. Thirdly, high-LET irradiation is delivered at the level of the SOBP where the PTV (thus the cancer cell population) is located. Heavy-ion beams are thus specifically effective against some tumor types as discussed for fast neutrons (see Secs. 4.1.1 and 4.1.3).

Repair Capacity. Finally, when fractionated treatment is applied, the high-LET radiation at the level of the SOBP (i.e., PTV) partly prevents or reduces cell repair. In contrast, the normal tissues irradiated at the level of the initial plateau are exposed to

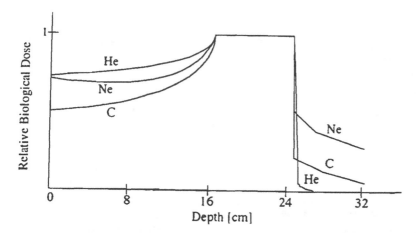

Figure 21 Comparison of the "biological" dose profiles for He, C, and Ne ions. For carbon ions, the ratio of the "biological" dose at the level of the SOBP and in the initial plateau is maximized and the fragmentation tail behind the Bragg peak remains relatively low. These arguments lead to select carbon ions as most appropriate for heavy-ion beam therapy. (From Ref. 44.)

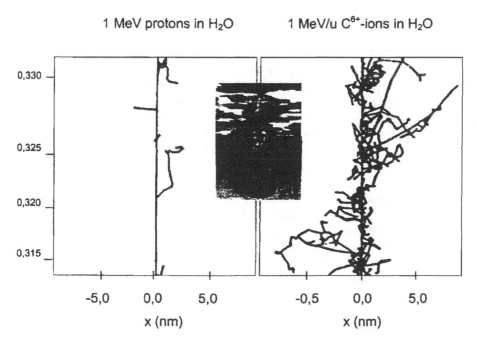

Figure 22 Comparison of interactions of 1-MeV protons and 1 MeV/u C^{6+} ions with water. (From Ref. 43.)

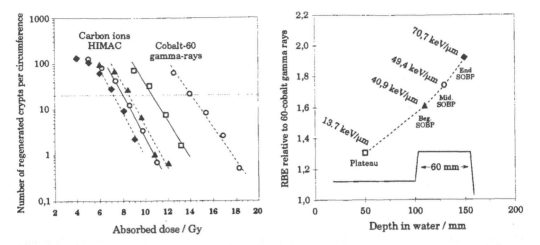

Figure 23 Variation of RBE as a function of depth in the carbon-ion beam used for clinical applications at HIMAC, Chiba, Japan (carbon-12, 290 MeV/u, SOBP 60 mm). The biological system is the well-codified intestinal crypt regeneration in mice. The selected criterion is 20 regenerating crypts per circumference after a single fraction irradiation. RBE determinations were performed at the beginning, middle, and end of the SOBP and at the level of the initial plateau. The dose–effect relationship for cobalt-60 is indicated for comparison. An estimation of the LET is presented for each depth where biological determinations were made. (From Gueulette, unpublished.)

low(er)-LET radiation and can thus fully benefit from repair mechanisms if fractionated irradiation is used.

Technological Aspects—Beam Delivery

The technical problems related to beam delivery are, to some extent, similar to those encountered with protons [46]. They are, however, more complex with carbon (heavier) ions and involve more cumbersome and more expensive equipment.

The scanning beam allows a better shaping of the treated volume. This is especially important when the PTV is close to a critical normal structure. This advantage holds for any shape of the PTV. The principle of the scanning beam technique is presented in Fig. 26 [45].

A rotating isocentric gantry is, or will be, installed in many of the carbon therapy facilities. The possibility of orientating the beam in any suitable direction is obviously an advantage for accurate patient-beam positioning and for allowing the selection of the optimum treatment plan. Isocentric gantries are used in all modern photon linear accelerators. For making any comparison relevant and reliable, it is thus important that carbon-ion therapy could be delivered with the same geometrical possibilities and with the same accuracy as the classical photon treatments. Mounting an isocentric gantry for carbon-ion therapy raises complex mechanical problems and requires huge and expensive magnets (Fig. 27) [43].

Clinical Results (Short Survey)

The Berkeley Program. The first heavy-ion therapy program was initiated at Berkeley, and 433 patients were treated between 1975 and 1992 [47]. The program was limited by the availability of the machine and its complexity (which resulted in many unscheduled down times), and, as a consequence, there was a patient recruitment problem.

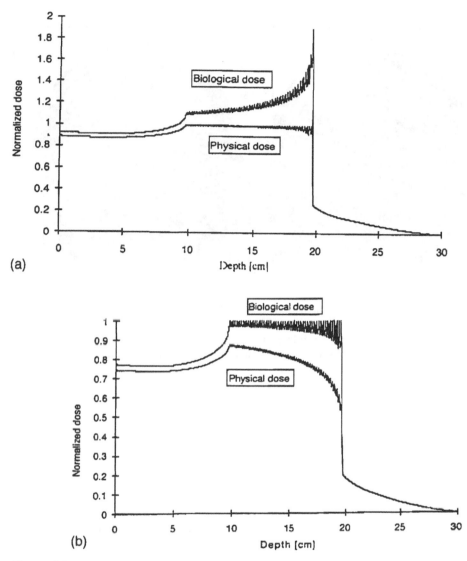

Figure 24 (a) Comparison of the physical and "biological" depth–dose curves for a carbon beam with a 10-cm SOBP. The biological dose is obtained by multiplying, at each depth, the absorbed dose by a weighting factor W_{RBE} which takes into account the RBE increase. The highest RBE is obtained in the deepest part of the PTV where the dose is delivered by a single Bragg peak (thus with high LET and RBE). In contrast, the dose in the proximal part of the PTV is due to many initial-plateau contributions (thus with a lower LET and RBE). (b) In order to cover the whole PTV with a homogeneous biological dose, the shape of the physical dose has to be adapted taking into account the RBE variation over the PTV. (From Ref. 44.)

Nevertheless, a great deal of valuable radiobiological and clinical information was obtained (Table 6).

A summary of the clinical results obtained with neon ions in Berkeley is presented in Table 7 [4,47]. Some fast neutron therapy results are also presented. Although the recruitments are not comparable, it should be pointed out that tumor types or sites for which an advantage was found with neon ions are those for which an advantage was also found with fast neutrons. This suggests a specific "high-LET" effect.

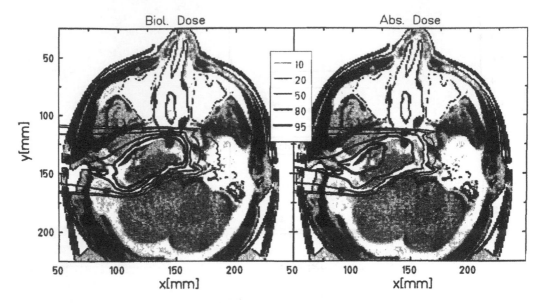

Figure 25 Treatment of a tumor of the base of skull with a carbon-ion beam. Comparison of the (physical) absorbed dose and of the "biological dose" (i.e., the dose weighted for the RBE variation in the carbon beam). The treatment is carefully planned in order to obtain a homogeneous biological dose to the PTV. For the same "biological" dose to the PTV, the normal tissues in the initial plateau receive a lower dose. (From Ref. 45.)

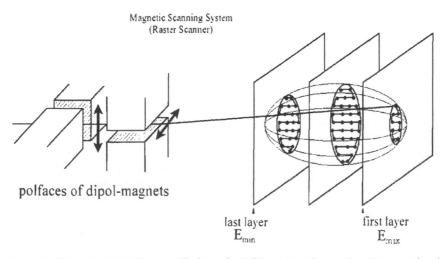

Figure 26 Principle of the irradiation of a PTV of any shape using the scanning-beam technique. The PTV is stratified into a series of layers, perpendicular to the incident beam. The different layers are irradiated successively with narrow "pencil beams" of equal range, scanned over the surface (different shape at each depth) by two deflecting magnets. (From Ref. 45.)

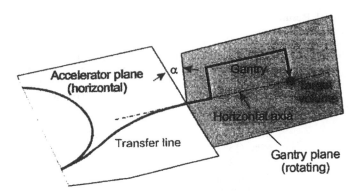

Figure 27 Isocentric gantry for irradiation of the PTV from any direction. The use of an isocentric gantry for carbon-ion therapy requires huge and expensive magnets. (From Ref. 43.)

The HIMAC-NIRS Program. After the shutdown of the therapy facility at Berkeley, the National Institute for Radiological Sciences (NIRS) in Chiba, Japan initiated a therapy program with carbon ions. In 1993, construction and installation of the Heavy Ion Medical Accelerator in Chiba (HIMAC) was completed. This was the world's first carbon-ion accelerator fully dedicated to medical use in a hospital environment. The HIMAC project was part of the Japanese government's 10-year plan against cancer.

There was no limitation on machine time, and it was thus possible to study common tumors on a large scale. A total number of 1297 patients were treated between 1994 and August 2002.

Some of the clinical results are summarized in Table 8. More detailed information on lung cancers [nonsmall cell lung cancer (NSCLC)] is given in Table 9, in particular, about patient selection and treatment conditions [48]. The possibility of hypofractionation has been investigated especially for lung and liver cancers with no increase in toxicity.

The clinical experience previously gained in the same center, during the last decades, with fast neutrons was reported to be useful to identify the patients suitable for high-LET

Table 6 Summary of the Clinical Results Obtained with Helium Ions and Neon Ions at Berkeley

Tumor site	Local control rate with		
	Helium ions	Neon ions	Conventional treatment
Salivary gland		80% (10 patients)	28% (188 patients)[a]
Nasopharynx paranasal sinus	53% (13 patients)	63% (21 patients)	21% (97 patients) (UCSF)
Sarcoma	65% (17 patients)	45% (24 patients)	28%[a]
Prostate		100% (9 patients)	60–70%[a]
Lung		39% (18 patients)	22–40% (UCSF)
Brain/glioblastoma (median survival)		17 months (13 patients)	9–12 months (UCSF, RTOG, NCOG)

[a] Literature review.
Source: Ref. 47.

Table 7 Comparison of Some Clinical Results Obtained with Neon Ions and Fast Neutrons (i.e., External Beam and High-LET Radiation)

	Local control rates after treatment with[a]	
Tumor site or type	Fast neutrons (pooled data)	Neon ions (Berkeley)
Salivary gland tumors	67% (24%)	80% (28%)
Paranasal sinuses	67%	63% (21%)
Fixed cervical lymph nodes	69% (55%)	
Sarcomas	53% (38%)	45% (28%)
Prostatic adenocarcinoma	77% (31%)	100% (60–70%)

[a] For comparison, the best estimates of local control rates currently obtained with conventional photon beam therapy are given in parentheses.
Source: Ref. 2.

therapy and to select the best weighting factor W_{RBE} to take into account the RBE effects (see Secs. 3.5 and 3.6).

A second medical facility dedicated to cancer treatment has been opened at Hyogo, Japan in 2001.

The European Experience. In Europe, the first patients were treated with carbon ions at the Gesellschaft für Schwerionenforschung (GSI), Darmstadt, Germany at the end of

Table 8 Results of Carbon-Ion Beam Therapy at NIRS (Treatments: June 1994–August 2001)

Protocol	Phase	Material	Treatment (fractions/week)	No. of patients	Response rate (%)[a]	Two-year local control (%)[b]	Three-year survival (%)
H&N-1	I/II	Locally advanced	18/6	17	73	80	44
H&N-2	I/II	Locally advanced	16/4	19	68	71	44
H&N-3	II	Locally advanced	16/4	134	52	61	42
Lung-1	I/II	Stage I (peripheral)	18/6	47(+1)	54	62	88
Lung-2	I/II	Stage I (peripheral)	9/3	34	85	86	65
Lung-3	I/II	Stage I (Hillar)	9/3	10	90	100	—
Lung-4	II	Stage I (peripheral)	9/3	50(+1)	65	100	73
Lung-6	I/II	Stage I (peripheral)	4/1	18	67	—	—
Liver-1	I/II	T2~4 MONO	15/5	24(+1)	75	79	50
Liver-2	I/II	T2~4 MONO	4~12/1~3	82(+4)	72	83	45
Liver-3	II	T2~4 MONO	4/1	11	55	—	—
Prostate-1	I/II	B2~C	C ion+hormone	35	—	100	94
Prostate-2	I/II	A2~C	C ion+hormone	61	—	100	97
Prostate-3	II	TIC~C	C ion+hormone	47	—	—	—
Uterus-1	I/II	III = Iva(ACC)	24/6	30	100	50	40
Uterus-2	I/II	Iib-Iva(SCC)	24/6	14	100	67	36
Uterus-3	I/II	Iib-Iva(SCC)	20/5	11	100	—	—
Uterus	I/II	Advanced (Adenoca)	20/5	12	100	38	39
Bone/Soft-1	I/II	unresectable	16/4	57(+7)	36	77	50
Bone/Soft-2	II	unresectable	16/4	30(+1)	57	—	—

[a] Response rate: percent of tumors with >50% reduction in size.
[b] Local control rate: percent of tumors with no evidence of local recurrence or relapse.
Source: Ref. 48.

Table 9 Results of Carbon-Ion Beam Therapy at NIRS: Clinical Studies for Stage I Nonsmall Cell Lung Cancer (NSCLC)

Protocol no.	Lung-1 (9303)	Lung-2 (9701)	Lung-3 (9801)	Lung-4 (9802)	Lung-6 (0001)
Phase	I/II	I/II	I/II	II	I/II
Period of the study	Oct. 1994 to Sep. 1997	Sep. 1997 to Feb. 1999	Apr. 1998 to present	Apr. 1999 to Dec. 2000	Dec. 2000 to present
Tumor type	All type[a]	Peripheral	Central	Peripheral	Peripheral
Total Dose (GyE)	59.4 ~ 95.4	68.4 ~ 79.2	57.6 ~ 64.8	72	54 or 60
Fraction/weeks (fixed)	18f/6w	9f/3w	9f/3w	9f/3w	4f/1w
# Pats (# Tumors)	47(48)	34(34)	15(15)	50(51)	35(35)
Adenoca/SCC/Large cellca	26/22/0	18/15/1	13/2/0	32/19/0	23/11/1

[a] All type includes both peripheral and central type of tumor.
Source: Ref. 48.

1997. Because of the strict limitation in the availability of the machine time, patients with difficult tumor localizations were selected. For these patients, full advantage could be taken from the scanning beam and the energy modulation system [45].

At the end of 2002, 156 patients were treated, a large proportion of them for a tumor of the base of skull. At the end of 2001, for chondrosarcomas (23 pts), no local recurrence was observed, while, for chordomas (44 pts), the local control rate was 87% (Kraft, personal communication). The same difference in prognosis between the two histologies is found as with protons (Sec. 4.2.2).

Taking into account the clinical experience reported from Berkeley and Chiba and the clinical results of the pilot study at the GSI in Darmstadt, both confirming the radio-biological and physical arguments in favor of heavy ions, the German Cancer Research Center decided to build a carbon-ion therapy facility at the University Hospital in Heidelberg. The technical and medical experience gained at the GSI, mainly by Kraft and his team, will help to select the optimum configuration. The possibility to treat with protons at the same facility is also planned: this would allow the medical team to evaluate the benefit of high-LET vs. low-LET radiations delivered with the best possible physical selectivity. It would of course be of great interest to have the possibility to treat either with low- or high-LET radiations depending on the tumor characteristics and patient conditions, in the same center, under the same technical conditions and with the same medical team.

Four other hadron-therapy centers are planned in Europe: MED-AUSTRON in Wiener Neustadt in Austria, Espace de Traitement Oncologique par Ions Légers dans le Cadre Européen (ETOILE), TERA in north of Italy, and the Swedish project in Stockholm.

A European network "ENLIGHT" has been initiated in 2002 by the European Society for Therapeutic Radiology and Oncology (ESTRO) and supported by the European Commission (EC). The goal of the European Network on Light Ion Therapy (ENLIGHT) is to coordinate the five European programs and, in particular, facilitate the mobility of scientists and exchange information. The scientific coordination of ENLIGHT is the responsibility of A. Wambersie (ESTRO-Brussels), J.-P. Gérard (Lyon), and R. Pötter (Vienna).

4.5. Boron Neutron Capture Therapy

Only a short review of BNCT is presented here. The rationale for BNCT is to reach a physical selectivity at the cellular level [5,49,50]. BNCT is a bimodal, binary therapy where an alpha particle is produced at the cellular target by the capture reaction of boron compounds and thermal/epithermal neutrons.

BNCT using thermal neutron beams was started in the United States in 1951 at the Massachusetts Institute of Technology (MIT) and at the Brookhaven National Laboratory (BNL). The clinical results were very poor. The technique was introduced in Japan by Hatanaka in 1968, and some promising results were obtained.

BNCT was restarted in the United States in September 1994 at Brookhaven National Laboratory and shortly thereafter at MIT using epithermal neutron beams (BNL trials ended in 1999 after the treatment of 53 patients but continued at MIT); these programs are supported by the Department of Energy. Forty patients were treated by the end of 1997. In Europe, the European Commission supports a BNCT program in Petten, The Netherlands. The three first patients were treated in 1997. The thermal neutron beam program continues in Japan.

So far, the only available sources of epithermal neutrons, with sufficient output, are nuclear reactors. However, construction of compact proton accelerators, producing epithermal neutrons of adequate energy (e.g., 2.5-MeV protons on a lithium target), has been envisaged.

The main expected advantage is that such an accelerator could be hospital-based, and thus fractionation (of the irradiation and drug administration) could be optimized since the machine would be available 24 hr a day. In addition, patient positioning could be made easier and more accurate.

Lastly, BNCT is used today in combination with fast neutron therapy in some centers such as Seattle, Essen, and Orléans. Boron is incorporated in the tumor cells; it captures thermal neutrons produced in the body by the fast neutron beam. Combination of BNCT and external photon beam therapy has also been suggested [51].

In conclusion, boron neutron capture therapy (BNCT) is still in an experimental phase. Although the rationale is particularly attractive (selectivity at the cellular level), it is difficult to draw conclusions from the available clinical data.

5. SUMMARY AND CONCLUSIONS: PRESENT TRENDS AND EXPECTATIONS IN RADIATION ONCOLOGY

5.1. Radiation Oncology Over the Past Decades

Historically, the progress in radiation therapy has been linked mainly to technological developments. The physical selectivity of the irradiations was significantly increased when 200-kV x-rays were progressively replaced by cobalt-60, betatrons, and linear accelerators. As a consequence, the clinical results were dramatically improved.

More recent technical developments of the electron linear accelerators are impressive. The reliability, mechanical stability, beam delivery, and collimation systems of the new generation of accelerators have reached a high level of quality [52]. These machines now allow irradiation of nearly any target volume with reduced irradiation of the surrounding organs at risk (OAR).

This progress in the treatment machines could be fully exploited because of the dramatic and impressive developments in imaging (such as CT, MRI, and PET). In addition, our better knowledge of the natural history of cancer helps in selecting the optimum treatment strategy.

For the future, (at least) three main approaches can be identified for improving the effectiveness of radiation therapy.

5.2. Optimizing Photon Techniques

Powerful irradiation techniques are now available or under development for photons: intensity-modulated radiation therapy (IMRT), "tomotherapy," "gamma knife," "cyber-knife," radiosurgery, etc. These techniques make it possible to irradiate any volume of any complex shape with minimum irradiation of critical normal structures.

This technical progress has been made possible by:

1. Advances in modern imaging techniques, which have made available complete and accurate anatomical information, including the location of the different target volumes and organs at risk.
2. Developments in medical physics and dosimetry that provide powerful and fast 3-D treatment planning, including "inverse planning," and take into account heterogeneities. Fast computation is absolutely required for IMRT.
3. Engineering improvements that allow full and reliable control of the machine and beam delivery system and patient immobilization.

These new techniques are becoming available in more and more hospitals.

However, challenges remain.

If one matches the "treated volume" (TV) and the planning target volume (PTV) as closely as possible, the conformity index will approach unity [6,7,53]. Different technical approaches are used to reach this goal; among them is IMRT. However, when optimizing the treatment parameters to improve the conformity index in IMRT, other indices can change adversely. For example, the irradiated volume increases (integral dose) and the homogeneity throughout the PTV may become worse. The compromises to be made vary with the treatment modality, and the true therapeutic gain of the newer modalities still needs to be evaluated.

This brings us to an important question: has the efficacy of photon beam therapy now reached a plateau (at least as far as physical selectivity is concerned)? If this is the case, little additional clinical benefit can be expected from further technical improvements with photons, and one has to search for other beam or radiation qualities to further improve the effectiveness of radiation therapy.

This is an important point to consider; however, there is no general consensus as yet.

5.3. Proton Beam Therapy

The introduction of proton beams aims at further improving the physical selectivity of the irradiation. The clinical results obtained by the pioneers in proton therapy, with machines in physics laboratories, were sufficiently convincing to justify building and buying dedicated hospital-based proton machines.

The challenge for the next few years will be to confirm, on a larger scale, the results of the pioneers and to confirm that their excellent results were actually due to the protons themselves and not only to the expertise and the motivation of the teams.

The benefit expected from the better physical selectivity of protons will have to be evaluated in comparison with conventional photon beam therapy, which remains the reference radiation therapy modality.

When making this comparison, one must remain aware of the modern developments in photon therapy such as conformal therapy (3DCRT), IMRT, tomotherapy, gamma knife, stereotactic x-ray techniques, iodine-125 (and palladium-103) seeds and plaques, etc.

There are, however, some (rare) tumor localizations for which nothing can compete with the excellent physical selectivity of protons: for example, tumors adjacent to the spinal cord, adjacent or invading the brain stem, tumors of the base of skull, and certain (especially pediatric) brain tumors (Fig. 18).

Another advantage of protons, relative to the best photon techniques, is a reduction of the integral dose and of the irradiated volumes. This factor could influence the risk of radio-induced cancers, although, so far, epidemiological evidence is still lacking.

Proton therapy equipment is now commercially available from several companies. This is a sign that protons have gained their place in the radiation therapy arsenal. Indeed, the number of proton therapy facilities increases worldwide, both the hospital-based, therapy-dedicated facilities and the facilities in physics laboratories adapted for medical applications.

5.4. High-LET Radiations, Fast Neutrons, and Hadrons

The third approach is the introduction of another type of "radiation quality": high-LET radiation. Clinical experience with neutrons has demonstrated that high-LET radiations are superior to low-LET radiations for some tumor types or sites. Fast neutrons were indeed the first high-LET radiations to be applied clinically (see Sec. 4.1). Although in the first studies they were applied in "suboptimal" conditions from a technical or dose distributions point of view, their advantage for some types of tumors is well established, particularly for slowly growing, well-differentiated tumors. Randomized trials have indeed shown their superiority over conventional photons for salivary gland tumors and prostatic adenocarcinomas.

Unfortunately, several neutron therapy centers were closed for technical reasons or patient recruitment difficulties. However, the centers applying neutron therapy today benefit from the same conditions of safety, reliability, and physical selectivity as photon therapy centers.

Heavy ions combine the benefit of the high physical selectivity of proton beams and the biological advantage of high-LET radiation for some tumor types. They appear today to be one of the most promising radiation therapy modalities when the clinical indication is correctly selected (some tumor types and/or localizations).

When comparing the relative merits of the different types of radiation used in therapy, two criteria have to be considered: the physical selectivity and the radiation quality (actually the LET). They are schematically presented in Fig. 28 [54]. Improving the physical selectivity is per se always an advantage. In contrast, selection between low and high LET depends on the tumor characteristics and should be based on the histology, grade, doubling time, etc. It is a pure radiobiological and medical issue.

Reliable criteria to identify the (individual) patients suitable for high-LET radiation therapy need to be developed. At present, the available criteria are derived (mainly) from the clinical fast neutron experience. It can be expected that novel approaches based on modern techniques involving molecular biology or gene identification may provide appropriate and still missing information. They may also provide information on the susceptibility or risk for secondary radio-induced cancer.

5.5. Reporting the Treatments

It is important for the future that protocols and results of hadron therapy be reported and analyzed using the terminology, definitions, concepts, and approaches currently in use for the other radiation therapy modalities [7,53].

As fractionated photon beam therapy is the reference radiation therapy modality, used for more than 80% of the patients, the terminology in use for photons should be applied for

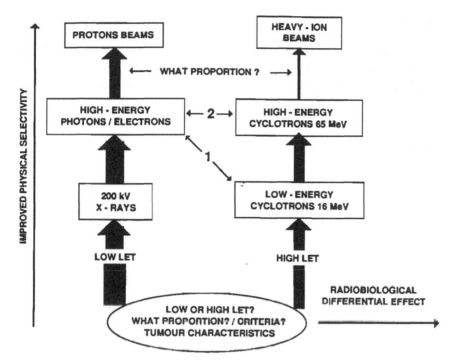

Figure 28 Schematic presentation of the relative situation of the different types of radiations used in therapy. Two criteria are considered: the physical selectivity and the LET (or radiobiological properties). For the low-LET radiations, the physical selectivity was improved from the "historical" 200-kV x-rays to cobalt-60 gamma rays and the modern linacs. Even with the linacs today, significant improvement is continuously achieved (IMRT, etc.). Among the low-LET radiation, the proton beams have the best physical characteristics, but one of the issues is the proportion of patients who will benefit from proton irradiation. A similar scale can be drawn for high-LET radiation: the heavy-ion beams have a physical selectivity similar to protons. Selection between low- and high-LET radiation is a biological/medical problem; it depends on the tumor characteristics, and reliable criteria still need to be established (see text). (From Ref. 54.)

the other techniques. Recommendations for reporting are contained in several ICRU reports [6,7,53]. It is, however, recognized that each technique has its specificity and may sometimes require introduction of specific concepts or definitions.

5.6. Global Approach and Collaboration in Cancer Treatment Strategy

The subject of this chapter is to report the rationale, achievements, and expectation of ions in cancer therapy. To illustrate their potential, the proven advantages of low- and high-LET ions are discussed in detail.

In the field of radiation therapy, it is recognized that there are other promising approaches to improve the efficiency of the treatment such as better individual adaptation of fractionation or association with drugs to modulate the radiation sensitivity. Combination of radiotherapy with the two other "classical" methods of cancer treatment, surgery, and chemotherapy may also be optimized.

Last but not the least, the progress in gene therapy, immunotherapy, and other approaches based on novel molecular biology techniques has to be kept in mind as well.

The recent impressive developments in ion beam therapy should not be regarded as being in competition with the other approaches mentioned above. In contrast, the advantages of combinations should be considered and a potential enhancement of the respective effectiveness may reasonably be expected [5].

ACKNOWLEDGMENT

This work was supported by the European Commission, Directorate General Research-Quality of Life and Management of Living Resources, contract: QLGI-CT-2002-01574.

REFERENCES

1. Levin, V.; Meghzifene, A.; Izewska, J.; Tatsuzaki, H. IAEA, International Atomic Energy Agency, Wagramerstrasse 5, A-1400: Vienna, Austria, 2001; 25 pp.
2. Wambersie, A.; Chauvel, P.; Gademann, G.; Gérard, J.-P.; Sealy, R. EULIMA, Final Report, European Commission, Rue de la Loi, 200, B-1049: Brussels, Belgium, 1992; 2 pp.
3. Tubiana, M.; Dutreix, J.; Wambersie, A. *Introduction to Radiobiology*; Taylor & Francis: London, 1990.
4. R., Pötter, T., Auberger, A., Wambersie., Eds.; Hadrons—A challenge for high-precision radiotherapy. Strahlenther. Onkol. 1999, *175* (Suppl. II),1–128.
5. Wambersie A.; Gahbauer R. Radiochim. Acta 2001, *89*, 245.
6. International Commission on Radiation Units and Measurements (ICRU). Prescribing, Recording and Reporting Photon Beam Therapy, ICRU Report 50; Bethesda, Maryland, 1993.
7. International Commission on Radiation Units and Measurements (ICRU). Prescribing, Recording and Reporting Photon Beam Therapy (Supplement to ICRU Report 50), ICRU Report 62; Bethesda, Maryland, 1999.
8. Wambersie A.; T. Auberger; Gahbauer R.; Jones D.; Pötter R. Strahlenther. Onkol. 1999, *175*, 122.
9. Gahbauer R. Strahlenther. Onkol. 1999, *175*, 121.
10. Wambersie A. International Atomic Energy Agency (IAEA), International Nuclear Data Committee, Nuclear Data for Production of Therapeutic Radioisotopes, INDC (NDS)-432, IAEA: Vienna, 2002.
11. International Commission on Radiation Units and Measurements (ICRU). Quantities and Units in Radiation Protection Dosimetry, ICRU Report 51; Bethesda, Maryland, 1993.
12. International Commission on Radiation Units and Measurements (ICRU). Fundamental Quantities and Units for Ionizing Radiation, ICRU Report 60, Bethesda, Maryland, 1998.
13. Wambersie A.; Menzel H.G.; Gahbauer R.A.; Jones D.T.L.; Michael B.D.; Paretzke H. Radiat. Prot. Dosim. 2002, *99*, 445.
14. Wambersie A.; Menzel H.G.; Gahbauer R.A.; DeLuca P.; Whitmore G. In *Progress in Radio-Oncology VII;* Kogelnik, H.D., Lucas P., Seldmayer, F., Eds.; Monduzzi Editore: Bologna, Italy, 2002, 361 pp.
15. Goodhead D.T. *Nuclear and Atomic Data for Radiotherapy and Related Radiobiology*; IAEA: Vienna, 1987; 37 pp.
16. Barendsen G.W. Radiat. Res. 1994, *139*, 257.
17. Pihet, P. *Etude microdosimétrique de faisceaux de neutrons de haute énergie. Applications dosimétriques et radiobiologiques*, Thesis; Université Catholique de Louvain: Louvain-la-Neuve, Belgium, 1989.

18. International Commission on Radiological Protection (ICRP) Report of the RBE Subcommittee to the International Commission on Radiological Protection and the International Commission on Radiation Units and Measurements. Health Phys. 1963, *9*, 357.

19. International Commission on Radiation Units and Measurements (ICRU). Quantitative Concepts and Dosimetry in Radiobiology, ICRU Report 30, Bethesda, Maryland, 1979.

20. Field S.B.; Hornsey S. In *High-LET Radiations in Clinical Radiotherapy*; Barendsen, G.W., Broerse, J.J., Breur, K. Eds.; Pergamon Press: Oxford, 1979; 181 pp.

21. van der Kogel, A.J. *Late effects of radiation on the spinal cord: dose effect relationships and pathogenesis* (Thesis); University of Amsterdam, Publication of the Radiobiological Institute TNO: Rijswijk, The Netherlands, 1979.

22. Gueulette J.; Menzel H.G.; Pihet P.; Wambersie A. Fast neutrons and high-LET particles in cancer therapy. In *Recent Results in Cancer Research*. Engenhart-Cabillic, R., Wambersie, A. Eds.; Springer-Verlag: Heidelberg, 1998; Vol. 150, 31.

23. Wambersie A.; Menzel H.G. Strahlenther. Onkol. 1993, *169*, 57.

24. Menzel H.G.; Pihet P.; Wambersie A. Int. J. Radiat. Biol. 1990, *57*, 865.

25. Pihet P.; Menzel H.G.; Schmidt R.; Beauduin M.; Wambersie A. Radiat. Prot. Dosim. 1990, *31*, 437.

26. Stone R.S. Am. J. Roentgenol. 1948, *59*, 771.

27. Wambersie A.; Barendsen G.W.; Breteau N. J. Eur. Radiothér. 1984, *5*, 248.

28. Engenhart-Cabillic, R.,Wambersie, A. Eds.; *Fast Neutrons and High-LET Particles in Cancer Therapy, Recent Results in Cancer Research*; Springer: Heidelberg, 1998; 1–209.

29. Pihet P.; Norman C.; Gueulette J.; Menzel H.G., Wambersie A. *Radiophysique*; XXIV Congrès de la Société Française des Physiciens d'Hôpital. Tours, Société Française des Physiciens d'Hôpital, Fondation Curie, 26 Rue d'Ulm, F-75248: Paris, France, 1985; 269 pp.

30. Barendsen G.W. Curr. Top. Radiat. Res. Q. 1968, *4*, 293 (North Holland, Amsterdam).

31. Chapman J.D. In *Radiation Biology in Cancer Research*; Meyn, R.E.,Withers, H.R. Eds.; Raven Press: New York, 1981; 21 pp.

32. Battermann J.J.; Breur K.; Hart G.A.M.; van Peperzeel H.A. Eur. J. Cancer 1981, *17*, 539.

33. Krüll A.; Schwarz R.; Brackrock S.; Engenhart-Cabillic R.; Huber P.; Prott F.J.; Breteau N.; Favre A.; Lessel A.; Koppe H.; Auberger T. In *Fast Neutrons and High-LET Particles in Cancer Therapy;* Engenhart-Cabillic, R.,Wambersie, A. Eds.; Springer: Heidelberg, 1998; 88 pp.

34. Lindsley K.L.; Cho P.; Stelzer K.J.; Koh W.J.; Austin-Seymour M.; Russel K.J.; Laramore T.W.; Griffin T.W. Bull. Cancer, Radiothér. 1996, *83* (Suppl. 1), 78s.

35. Scalliet, P.; Remouchamps, V.; Lhoas, F.; Van Glabbeke, M.; Curran, D.; Ledent, T.; Wambersie, A.; Richard, F.; Van Cangh, P. Proceedings of the XXIX PTCOG Meeting. DKFZ. Heidelberg and GSI, Darmstadt, Germany, 1998.

36. Breteau, N., Le Bourgeois, J.-P., Barendsen, G.W., Stannard, C.E., Rosenwald, J.-C., Wambersie, A. Eds.; *Hadrons in radiation therapy*. Bull. Cancer, Radiothér. 1996, *83* (Suppl. 1), 1s–230s.

37. Suit H.D. Int. J. Radiat. Oncol. Biol. Phys. 2002, *53*, 798.

38. Munzenrider J.E. Strahlenther. Onkol. 1999, *175* (Suppl. II), 68.

39. Munzenrider J.E.; Liebsch N.J. Strahlenther. Onkol. 1999, *175* (Suppl. II), 57.

40. Habrand J.L.; Mammar H.; Ferrand R.; Pontvert D.; Bondiau P.-Y.; Kalifa C.; Zucker J.-M. Strahlenther. Onkol. 1999, *175* (Suppl. II), 91.

41. Wambersie A.; Grégoire A.; Brucher J.-M. Int. J. Radiat. Oncol. Biol. Phys. 1992, *22*, 275.

42. Jones D.T.L.; Schreuder A.N.; Symons J.E.; de Kock E.A.; Vernimmen F.J.A.; Stannard C.E.; Wilson J.; Schmitt G. Strahlenther. Onkol. 1999, *175* (Suppl. I), 30.

43. Regler, M.; Benedikt, M.; Poljanc, K. CERN Accelerator School, Seville, Spain, October 2001, Hephy-PUB-757/023, 2002.

44. Grundinger, U., Ed.; Nachrichten GSI, Gesellschaft für Scherionenforschung mbH, D-64291 Darmstadt, Germany, 1993; 11–93.

45. Kraft G. *The Physics of Highly and Multiple Charged Ions*; Currel, F.J., Ed., Kluwer Academic Publisher, 2002 Chapter 10, pages 145–192.

46. Jones D.T.L. Radiochim. Acta 2001, *89*, 235.

47. Castro J.R.; Chen G.T.Y.; Blakeley E.A. Radiat. Res. 1985, *104* (Suppl. 8), S263.

48. Tsujii H.; Morita S.; Miyamoto T.; Mizoe J.-E.; Kamada T.; Kato H.; Tsuji H.; Yamada S.; Yamamoto N.; Murata H. *Annual Report NIRS*; National Institute of Radiological Sciences: Chiba-shi, Japan, 2002; 61 pp.

49. Gahbauer R.; Gupta N.; Blue T.; Goodman J.; Barth R.; Grecula J.; Soloway A.H.; Sauerwein W.; Wambersie A. *Fast Neutrons and High-LET Particles in Cancer Therapy*; Engenhart-Cabillic, A., Wambersie, A. Eds.; Springer: Heidelberg, 1998; 183 pp.

50. Rorer, R.; Wambersie, A.; Whitmore, G.; Zamenhof, R. Current status of neutron capture therapy, International Atomic Energy Agency, IAEA-TECDOC-12233, 2001.

51. Barth, R.F.; Grecula, J.C.; Yang, W.; Rotaru, J.H.; Nawrocky, M.; Gupta, N.; Albertson, B.J.; Ferketich, A.M.; Moeschberger, M.L.; Coderre, J.; Rofstad, E.K. Int. J. Radiat. Oncol. Biol. Phys. *in press*.

52. Wambersie A. Radiochim. Acta 2001, *89*, 255.

53. International Commission on Radiation Units and Measurements (ICRU). Prescribing, Recording and Reporting Electron Beam Therapy, ICRU Report 71, Bethesda, Maryland, 2004.

54. Wambersie A. *Progress in Radio-Oncology V*; Kogelnik, H.D. Ed.; Monduzzi, Eds: Bologna, Italy, 1995; 685 pp.

26
Food Irradiation

József Farkas
Szent István University, Budapest, Hungary

1. INTRODUCTION

Food irradiation is the process of exposing food, either prepackaged or in bulk, to controlled levels of certain types of ionizing radiation to increase storage life of food, reduce postharvest food losses, and inactivate specific food-borne pathogenic organisms. Food irradiation is one of the most thoroughly and intensively investigated methods of food preservation.

The ionizing radiation applied in food irradiation is limited to high-energy electromagnetic radiation (gamma rays or x-rays) with energies up to 5 MeV or high-energy accelerated electrons with energies up to 10 MeV. These radiations are chosen because

1. They produce the desired effects with respect to the food.
2. They do not induce radioactivity in foods or their packaging materials.
3. They are available in quantities and at costs that allow practical uses of the process.

Other kinds of ionizing radiation, in some respect, do not suit the requirements of food irradiation. Except for difference in penetrability, electromagnetic ionizing radiations and accelerated electrons are equivalent in food irradiation and can be interchangeably used.

Radiation treatment can be considered as one type of "nonthermal processing" of food because even at the largest absorbed radiation dose (see Sec. 4.11) to be applied, which is about 50 kGy, the amount of energy is equivalent to 50 J. At this dose level, if all the absorbed radiation energy would be degraded to heat, the temperature of a high-moisture food, thermodynamically roughly equivalent to water, would rise about 12°C. Because of the negligible heating effect, irradiation treatment is able, e.g., to kill the cells of microbes contaminating frozen foods without thawing them up, or, those in fresh commodities without changing the original physical state of the product.

Irradiation can be applied through any packaging materials including those that cannot withstand heat. This also means that radiation treatment can be performed after packaging, thus avoiding recontamination or reinfestation of the product.

The useful effects of ionizing radiation as a food processing means are summarized in Table 1. In addition to the preservative effects listed in Table 1, in some cases even

Table 1 Useful Effects of Irradiation as a Food Processing Treatment

Effects	Results
Inhibition of sprouting of tubers and bulbs	Increased storability
Decrease of after-ripening and delaying senescence of some fruits and vegetables	Increased shelf life
Killing or sterilizing stored product insects	Insect disinfestation of food
Inactivation of parasites transmissible by food	Prevention of food-borne parasitic diseases
Inactivation of food-borne microorganisms	Microbial decontamination of food: Increased shelf life and/or prevention of food poisoning

improvement of certain functional or sensory quality characteristics of food can be achieved with irradiation.

The basic process is the application of a prescribed amount of ionizing radiation to foods, plus the eventual use of certain other procedures that may be needed to accomplish the purpose of the processing.

Food irradiation is a very complex topic and has an enormous literature. Apart from thousands of journal articles and proceedings of large number of international conferences and panel meetings, its state of the art has been extensively reviewed during the course of decades by a number of noteworthy books [1–5]. The present brief chapter mainly focuses on the principles and some potential applications of food irradiation and refers to some most recent research and developments in these regards.

1.1. Principles of Radiation Sources

Two basic types of radiation sources can satisfy the requirement of industrial use of food irradiation:

1. Machine sources such as electron accelerators and those converting accelerated electron beams to x-ray photons. Accelerated electrons have low penetrability. Thus they cannot meet all the goals of food irradiation. The 10 MeV electrons, the highest energy level of electron irradiation presently recommended by the Codex Alimentarius [6], can penetrate food with typically about 4-cm thickness.
2. Long half-life, man-made radionuclides emitting gamma rays. The most readily available is cobalt-60. Much more limited is the use of cesium-137. Gamma rays and x-rays have highly penetrating characteristics; thus, they can be used to treat food in large containers.

Typical irradiation facilities consist of a process chamber containing the radiation source, some sort of conveyor systems to transport products inside and outside the shielding walls, and sophisticated control and safety systems. Irradiation facilities are built with several layers of redundant protection to detect equipment malfunctions and protect employees from accidental exposure. Technical details depend on the type of irradiation. Typical processing parameters are compared in Table 2 [7].

Table 2 Comparison of Typical Processing Parameters

	Gamma	X-ray	E-beam
Typical source power	3.5 MCi	25 kW	35 kW
Typical processing speed	12 tonnes/ hr at 4 kGy	10 tonnes/ hr at 4 kGy	10 tonnes/ hr at 4 kGy
Source energy	1.33 MeV	5 MeV	5–10 MeV
Penetration depth	80–100 cm	80–100 cm	8–10 cm
Dose homogeneity	High	High	Low
Dose rate	Low	High	Higher
Best application	Bulk processing of large boxes or palletized product in shipping cartons in a warehouse environment	Bulk processing large boxes or palletized product in shipping cartons in a warehouse environment	Sequential processing of primary or secondary packaged product in-line or at-line

Source: Ref. 7.

1:2. Typical Dose Requirements

The technological feasibility of a food irradiation treatment depends on how much irradiation the food withstands without adversely changing its qualities, i.e., how much useful effect can be achieved without significant change to the chemical composition, nutritional value, and sensory properties of the product. Generally, there is a minimum dose requirement. Whether every mass element of a food requires irradiation will depend

Table 3 Dose Requirements of Various Applications of Food Irradiation

Application	Dose Requirement (kGy)
Inhibition of sprouting of potatoes and onions	0.03–0.12
Insect disinfestation of seed products, flours, fresh and dried fruits, etc.	0.2–0.8
Parasite disinfestation of meat and other foods	0.1–3.0
Radurization of perishable food items (fruits, vegetables, meat, poultry, fish)	0.5–10
Radicidation of frozen meat, poultry, eggs and other foods and feeds	3.0–10
Reduction or elimination of microbial population in dry food ingredients (spices, starch, enzyme preparations, etc.)	3.0–10
Radappertization of meat, poultry, and fishery products	25–60

on the purpose of the treatment. In some cases, irradiation of the surface will suffice; in others, the entire food must receive the minimum dose.

Ranges of dose requirement of various applications are listed in Table 3. The details of the requirements for each food must be specifically considered.

2. BIOLOGICAL EFFECTS OF IONIZING RADIATION

Chemical and biological effects of ionizing radiation are thought to occur through two main mechanisms: direct interaction of the radiation with food components and living cells in materials exposed to it, and indirect action from radiolytic products, such as the radicals formed from water molecules (see Chap. 12).

The primary target of biological effects appears to be the DNA (see Chap. 15), although effect on the cytoplasmic membrane may also play a role [8]. Ionizing radiation affects DNA in a number of ways. It causes chemical changes in specific nucleotide bases; it causes single-strand breaks; and it causes double-strand breaks in a dose-dependent manner.

2.1. Radiation Sensitivity of Stored Product Insects and Food-Borne Parasites

The radiation dose required to kill an insect depends on the species and a number of other factors such as age, sex, and stage of development [9]. In general, radiation sensitivity is highest at the egg stage and the lowest at the adult stage of development. Fruit flies are the most radiation-sensitive insect pests while the moths are the most resistant ones.

Radiation effects on parasitic protozoa and helminths are associated with loss of infectivity, loss of pathogenicity, interruption or prevention of completion of life cycle, and death of parasites [10].

2.2. Radiation Resistance of Food-Borne Microorganisms

The actual number of cells or percentage of microbial population that will be killed by an absorbed radiation dose depends also on various factors such as the inherent resistance of the particular organism, the growth stage, as well as environmental factors such as tem-

Table 4 D_{10} Values (kGy) of Some Nonsporeforming Bacteria

Bacteria	Nonfrozen Food	Frozen Food
Vibrio spp.	0.02–0.14	0.04–0.44
Yersinia enterocolitica	0.04–0.21	0.20–0.39
Campylobacter jejuni	0.08–0.20	0.18–0.32
Aeromonas hydrophila	0.11–0.19	0.21–0.34
Shigella spp.	0.22–0.40	0.22–0.41
Escherichia coli (incl. O157:H7)	0.24–0.43	0.30–0.98
Staphylococcus aureus	0.26–0.57	0.29–0.45
Salmonella spp.	0.18–0.92	0.37–1.28
Listeria monocytogenes	0.20–1.0	0.52–1.4

perature during irradiation, oxygen presence, and water content. Vegetative cells of pathogenic bacteria, the main concern in many food-borne infections and intoxication, are relatively radiation sensitive, including both long-time recognized pathogens as well as "emerging" or "new" ones.

Table 4 shows the ranges of decimal reduction doses (D_{10} values) of the most important nonsporeforming pathogens determined in various atmospheres and foods, summarized from publications by a large number of laboratories [11,12]. Because of the effect of irradiation temperature on the radiation resistance of microorganisms, a detailed knowledge of product temperature profile is critical for effective ionizing radiation pasteurization of foods, particularly meat products [12,13]. Some nonpathogenic microorganisms and the bacterial spores in general are more resistant to radiation than those listed in Table 4. Viruses, mycotoxins produced by certain types of molds, and prion particles thought to be responsible for bovine spongiform encephalpathy (BSE) in cattle are highly resistant to irradiation [11].

3. RADIATION-INDUCED CHEMICAL CHANGES IN MAIN CONSTITUENTS OF FOODS

More or less, water is present in almost all foods. Therefore the radiolysis of water is of particular interest in food irradiation. However, this subject is amply dealt with in Chap. 12. The end products of water radiolysis, $^{\bullet}OH$, e_{aq}^-, and $^{\bullet}H$, are very reactive transient species. The hydroxyl radical is a powerful oxidizing agent whereas the hydrated electron is a strong reducing agent. The stable end-products such as hydrogen and hydrogen peroxide are of less significance because they are largely consumed with their respective reactions with e_{aq}^- and $^{\bullet}OH$ [14].

The presence or absence of oxygen during irradiation can have an important influence on the course of radiation-induced changes of food components. When foods are irradiated, oxygen can add to some of the radicals produced to form peroxy radicals, $^{\bullet}RO_2$. Through this reaction, the small amount of oxygen normally present in a food can be quickly consumed during irradiation. Because diffusion of oxygen from the atmosphere is slow, electron irradiation, due to its high dose rate, can create an anoxic condition in the food, whereas gamma sources have much lower dose rates than electron accelerators. Thus anoxic conditions are not necessarily created unless the food is gamma irradiated in an oxygen-free package. This might create a dose rate effect that is actually an oxygen effect. The radiolysis of water is pH-dependent, too. Thus pH influences the result of a radiation treatment.

The temperature during irradiation also influences the chemical changes. Freezing can have a strong protective effect because reactive intermediates of water radiolysis are trapped in the ice of frozen foods and are thus kept from reacting with other food components. During thawing of frozen, irradiated food, the radiolytic products of water apparently react preferentially with each other rather than with other food components. Thus freezing of food has a certain protective effect on some radiation-sensitive vitamins [15] and decreases the associated chemical changes, e.g., with the formation of volatiles producing off-flavor in some irradiated foods [16].

Because in multicomponent systems such as foodstuffs, a mutual protection of different components is exerted, irradiation does not cause much chemical change in foods.

Minerals and trace elements cannot be affected under the process conditions of food irradiation.

3.1. Effects on Main Organic Components

Apart from water, the major constituents of most foods are carbohydrates, proteins, and lipids. These organic components of foods are less sensitive to irradiation than they are in their pure solutions as a single-component system.

Irradiation of *sugars* results in the formation of low levels of radiolytic products mostly derived from the reaction with hydroxyl radicals by hydrogen abstraction from the C–H bonds, resulting in products such as sugar acids, keto sugars, and deoxy compounds [17]. When *polysaccharides* are irradiated, the reactions observed with monosaccharides can occur, and, additionally, the glycosidic bonds that connect the monosaccharide units can be broken. This reduces the degree of polymerization. Especially in case of starch and pectin, this causes changes in the physical properties of the foods that contain them. Changes in properties such as viscosity, mechanical strength, swelling, and solubility are likely to reduce their functionality in food; however, sometimes, the changes improve a particular function.

Major in vitro reactions of irradiated *amino acids* involve decarboxylation and oxidative (if oxygen is present) or reductive (if anoxia exists) deamination [14]. The sulfur amino acids such as cysteine, cystine, and methionine may act as free radical scavengers, thus ameliorating the degradative effects on other components of food. However, their breakdown also generates end products such as hydrogen sulfide, which has undesirable sensory effects. Radiation damage to constituent amino acids in irradiated food is very limited. With some 20 amino acids, the total range of possible reaction products is great, but the quantitative effect on *proteins* in foods during irradiation is small. Consequently, *enzyme* inactivation is also insignificant. Actually, radiation-sterilized foods destined for long-term storage must receive a heat treatment (blanching) in addition to the radiation treatment to prevent enzymatic spoilage (see Sec. 4.11). In some irradiated plant tissues, the in vivo enzyme activities may increase as a result of release, or diffusion, through "leaky" membranes more easily reaching their hitherto unavailable substrates. When meat is irradiated at high doses (10–50 kGy), radiation-induced aggregation of some proteins may occur, which leads to decreased protein solubility [14].

Upon irradiation of *fats*, the formation of a multitude of products is possible after primary ionization and excitation, and deprotonation followed by various dimerization, disproportion reactions, dissociations, or decarboxylation. It is generally assumed that irradiation in the presence of oxygen leads to accelerated autooxidation of lipids, and that the pathways are the same as in light-induced or metal-catalyzed autooxidation.

The irradiation of unsaturated *fatty acids* in foods predominantly results in the formation of a hydroperoxyl radical and then the formation of a hydroperoxide. The hydroperoxides are generally unstable in foods and break down to form mainly carbonyl compounds, many of which have low odor threshold, and contribute to the rancid notes often detected when fat-rich foods are irradiated [18]. In the absence of air, their formation is limited.

In studies on radiation effects on *cholesterol* in meat [19] and egg powder [20], elevated levels of the same oxides were found that are known to result from autooxidation during storage of unirradiated foods. Vacuum packaging or addition of antioxidants can largely prevent such formation of cholesterol oxidation products [21,22].

Radiation effects on *vitamins* will be discussed in Sec. 5.

4. PRACTICAL APPLICATIONS OF FOOD IRRADIATION

4.1. Control of Sprouting and Germination of Vegetable Crops

Inhibition of sprouting of various vegetable crops at low doses was one of the earliest extensively studied application possibility of food irradiation. Most important from these opportunities is the control of sprouting to extend the storability of tubers and bulbs [23]. These studies included the response of varieties/cultivars to radiation treatment, optimal dose, dose-rate effect, time interval between harvest and irradiation treatment for efficacy, susceptibility of irradiated batches to storage rots, biochemical mechanisms underlying sprout inhibition, influence of storage conditions, and the effect of irradiation on technological properties influencing the utilization of irradiated crops for processed products.

The dose required to inhibit sprouting of onions, shallots, and garlics is 0.03 0.12 kGy. For good sprout control of tubers such as potatoes and yams, somewhat higher doses, 0.08–0.14 kGy, are required. Because of decreased wound-healing ability after irradiation, doses in excess of 0.15–0.2 kGy may induce increased microbial rot in storage [24].

An important factor determining the efficacy of radiation treatment of tubers and bulbs is the time delay between harvest and irradiation. The sprout inhibition is most pronounced if the irradiation of tubers and bulbs is applied shortly after harvest, when they are still in their dormancy stage. However, the dormancy period may vary among cultivars and cropping season, and is also dependent on the postharvest storage temperature.

In onion bulbs, if some growth of inner buds takes place already before irradiation, the treatment causes the death and discoloration of the inner buds. The area of this discoloration depends on the size of the inner buds at the time of irradiation. Pilot scale studies in Hungary have shown that the yield of unirradiated onions prepared for drying after storage was 29% of the original compared to 55% in irradiated onions. The dehydrated onions prepared from irradiated bulbs had better quality than did the controls. No discoloration of the inner buds occurred in bulbs of cultivar Alsógödi irradiated at 0.05 kGy and stored up to 8 months [25]. Similar results were reported earlier from Egypt [26] whereas some studies in the United States have indicated that inner-bud discoloration of irradiated onions may lower the quality of dehydrated onion slices or powder prepared from them [27].

In irradiated potatoes, especially in some varieties and as a function of cultivating conditions of the raw material, after-cooking darkening may occur. This discoloration is attributed to formation of ferric-phenolic complexes. This phenomenon depends on the iron content, and is related to increased polyphenol formation and reduced citric acid levels, which are influenced by agronomic and climatic factors. Various technological measures have been developed to prevent this after-cooking darkening [23].

Other types of radiation-induced discoloration in potatoes have also been reported in some cases and measures to avoid them have been studied [28,29].

Because irradiation interferes with the natural wound-healing process, it is an important prerequisite of successful application of irradiation that mature tubers with fully developed periderm are suitable for radiation processing and tubers must be properly cured from harvesting and handling injuries before irradiation.

Because potatoes are good source of vitamin C, it is important to point out that irradiation does not adversely affect the vitamin C levels [23,30]. Although some ascorbic acid is converted into dehydroascorbic acid on irradiation, the latter is also biologically active.

In Japan, where the use of chemical sprout suppressants is not permitted, commercial irradiation of potatoes has been introduced in 1973 in Shihoro, Hokkaido, where an industrial scale irradiator has been processing about 15,000–20,000 tons of potatoes annually [31,32]. The success of this system is due in large measure to careful handling of the product before and after treatment.

Since the late 1990s, commercial irradiation of bulb crops, particularly garlic, has shown a steady increase using ^{60}Co irradiators in the major garlic-producing provinces of China [23].

A commercial demonstration facility for irradiation of onions having a processing capacity of 10 tons/hr was reported in 1999 to be under construction in Nasik District, Maharashtra State, India [23].

Pilot-scale irradiation and consumer acceptance studies with tubers and bulbs have been performed with positive results already in the 1970s in a number of other countries, e.g., Argentina, Bangladesh, Chile, German Democratic Republic, Hungary, Israel, the Philippines, Thailand, and Uruguay, showing the techno-economic feasibility of irradiation of these crops. However, in countries under temperate climates, cool storage by circulating cold outside air, together with the use of chemical sprout inhibitors, provide inexpensive and satisfactory sprout control with reasonably good product quality for both industrial processing and household consumption. Thus industrial interest in commercial implementation in Western countries for sprout inhibition is still very low, basically because it involves high capital investment for relatively low-price crops, perceived consumer opposition to irradiated foods, regulation and trade limitations, and labeling requirements. However, the use of chemicals is coming under increasing scrutiny from the viewpoints of environmental pollution and health risk from residues left in the products. Therefore it is likely that their continued use will be restricted.

In yams, an important crop in tropical climates, neither chemical sprout inhibitors nor cool storage are effective for long-term storage. On the other hand, irradiation would provide an effective alternative for postharvest treatment.

Controlling the germination of malting barley is an interesting potential application of low-dose irradiation. Doses of 0.25–0.5 kGy applied to air-dried barley do not prevent the emergence of shoot tips and tendrils during malting but markedly retard the root growth. In this way, high-quality malt can be obtained while the losses resulting from root growth are reduced [33]. Because this effect of radiation processing persist for at least 7 months, the above radiation treatment applied before the barley is put into storage has the added benefit of destroying insect pests that may be present in the grain.

4.2. Insect Control in Stored Foods

Radiation disinfestation of stored food offers a viable alternative to chemical disinfestation without adverse effects on the product quality. The use of methyl bromide, the only broad-spectrum fumigant used for the disinfestation of stored products, was anticipated to be phased out by around 2001 [34], which highlights the urgency of an alternative treatment.

Disinfestation of stored food by irradiation was extensively studied already in the 1960s and 1970s [9,35], and it has shown that radiation disinfestation can be efficiently applied to almost all dried foods.

The Codex Alimentarius Commission recommends a dose of 1 kGy for killing insects in all food and agricultural products [6]. However, if Good Manufacturing Practices (GMP) and Good Irradiation Practices (GIP) are followed, according to subsequent recommendations published by the International Consultative Group on Food Irradiation (ICGFI), radiation disinfestation of stored product commodities should be achievable at doses up to 0.5 kGy [34]. The radiation doses used to control various stored product insects are summarized in Table 5.

Presently, many commodities have to be fumigated more than once to control insects. Irradiation by a single treatment sterilizes or kill, depending on the dose and the time interval allowed after irradiation, all developmental stages of the common insect pests, including eggs deposited inside grains or even the weevil that may lodge deep inside the seed of the mango.

With the availability of inexpensive, convenient, and easily applied pest control methods based on application of pesticides, irradiation disinfestation has not yet received priority. Nevertheless, radiation disinfestation was performed on an industrial scale in the

Table 5 Radiation Doses Used to Control Stored Product Insects

Species	Stage	Dose (kGy)
Coleoptera		
Sitophilus oryzae	All	0.16
S. granarius	All	0.16
S. zeamais	All	0.16
Tribolium castaneum	All	0.20
T. confusum	All	0.20
T. destructor	All	0.20
T. madeus	All	0.20
Rhyzopertha dominica	Larvae	0.25
Latheticus oryzae	Adults	0.20
Oryzaephilus surinamensis	All	0.20
O. mercator	All	0.20
Callosobruchus analis	All	0.20
C. chinensis	All	0.20
C. maculatus	All	0.20
Bruchus rufimanus	All	0.40
Bruchidius incarnatus	All	0.40
Trogoderma granarium	All	0.25
Dermestes maculatus	All	0.50
Lasioderma serricorne	All	0.50
Nerobia rufipes	All	0.30
Araecerus faciculatus	All	0.75
Lepidoptera		
Anagastus kuehniella	Larvae, pupae	0.60
Plodia interpunctella	Larvae	0.45
Cadra cautella	Larvae, pupae	0.45
Sitotroga cerealella	All	0.60
Nemapogon granellus	All	0.60

Source: Ref. 34.

Soviet Union, where an electron irradiation plant to treat imported grains went into operation in 1980 at Port Odessa and some 400,000 tonnes/year of grain were successfully treated by two electron accelerators. This facility is not currently in use in the Ukraine, after the collapse of the Soviet Union.

The doses applied for the widely used microbial decontamination by irradiation of spices, dried herbs, and dry vegetable seasonings (see Sec. 4.9) are much higher than the disinfestation doses. Thus radiation decontamination of these commodities is more than enough to kill also any insects eventually infesting them.

Dried fruits, vegetables, and nuts, as well as dried fish, an important source of protein in many developing countries, are also good candidates for radiation disinfestation. Application of 0.2–0.7 kGy doses, if they have been suitably packaged to prevent reinfestation, can eliminate the insect problem from these products that cannot be effectively disinfested by either chemical or physical means other than irradiation.

As a residue-free physical treatment, irradiation does not provide any protection from insects that might reinfest the product after postirradiation storage. Therefore it is of paramount importance that proper care is needed to store, transport, and market irradiated products in insect-proof containers or packages [34].

4.3. Irradiation as a Quarantine Treatment

Radiation treatment can significantly contribute to the variety of means for quarantine disinfestation in the international trade to prevent the importation of invasive nonnative insects with food and agricultural commodities that can harbor them. The use of irradiation disinfestation in quarantine treatment has great potential especially against fruit flies. A number of past quarantine treatments have been recently prohibited, e.g., fumigation with ethylene dibromide.

Irradiation is the fastest among quarantine treatments available, although it cannot be used to treat so large loads at once as can be carried out with fumigation or by cold storage. Irradiation treatment is an effective alternative for many types of fresh produce because it can be used on riper fruit and on fruit that cannot tolerate, e.g., heat treatment. However, one characteristic of irradiation quarantine treatment that needs specific consideration by regulatory/inspecting agencies is that irradiation does not provide significant acute mortality (within 48 h) at doses tolerated by fresh agricultural produce; thus identification methods for irradiated commodities and insects, or acceptance of certification as proof of adequate irradiation, are needed [36].

Quarantine treatment doses of several pests (mainly fruit flies) supported by adequate research vary between 0.07 and 0.225 kGy [36]. Such low doses are tolerated by most fruits. However, because of the difficulties in evaluating efficacy by acute mortality and uncertainties of dose measurements, radiation quarantine treatment research shows some inconsistencies, and thus further work is required. Nevertheless, motivated by banning fumigation with ethylene dibromide (EDB), several semicommercial trial shipments of tropical fruits have been irradiated and marketed in the continental United States since 1986, and a commercial linear accelerator e-beam/converted x-ray facility was built in Hawaii to treat fruit and began operating in August 2000 [36].

4.4. Parasite Disinfection

Various animal parasites are associated with certain raw and partially processed food, not only in the developing part of the world but also in developed countries. In the United

States, a recent survey by the Centers for Disease Control and Prevention estimated that there are 2.5 million cases annually due to food- and beverage-borne parasites [37].

There are a number of traditional and new control measures to protect consumers from food-borne parasites. Ionizing radiation is one of the new technologies to control or limit the impact of a number of food-borne parasites on public health [10].

Although relatively high doses (4–6 kGy) are required to kill food-borne parasites, much lower doses are adequate to prevent their reproduction and maturation resulting in loss of infectivity. Table 6 summarizes the effect of ionizing radiation on the most important parasites that may be associated with fishery products and meats [38]. Thus the parasitic roundworm *Trichinella spiralis*, which causes trichinosis, the pork and beef tapeworms, the protozoon in pork responsible for toxoplasmois, and various flukes that infest fish, can be rendered noninfective by low-dose radiation treatment. Doses of

Table 6 The Effect of Irradiation on Parasites

Parasite	Mode of Infection	Dose (kGy)	Effect of Irradiation
Clonorchis spp.	Chinese liver fluke, occurs in raw fish	0.15	In vitro minimum effective dose
Opistorchis viverrini	Liver fluke found in contaminated raw, pickled or smoked fish	0.1	In vitro minimum effective dose
Paragonimus spp.	Parasitic worm found in crabs and crayfish in Asia	0.1	In vitro minimum effective dose
Gnathostoma spinigerum	Parasitic worm found in raw, undercooked or fermented fish	7	Reduces worm recovery rate in mice
Angiostrongylus cantonensis	Parasitic worm found in uncooked molluscs, shellfish	2	Minimum effective dose
Anisakis spp.	Nematode is ingested if fish is eaten raw or slightly salted	2–10	Reduces infectivity of larvae
Trichinella spiralis	Nematode occurs in raw or inadequately cooked pork	0.3	Minimum effective dose
		0–3–1	FDA permitted dose to control trichina in pork
Toxoplasma gondii	Consumption of undercooked meat or poultry; or contact with infected animals	0.7	Minimum effective dose for fresh pork
Cysticercus bovis (*Taenia saginata,* in man)	Tapeworm found in uncooked or undercooked beef; causes taeniasis	0.3	Preliminary minimum effective dose
Cysticercus cellulosae (*Taenia solium,* in man)	Tapeworm found in pork	0.3	Preliminary minimum effective dose

Source: Ref. 38.

irradiation applied to reduce microbial load (to be discussed in further chapters) are sufficiently high to control also many parasites.

4.5. Extension of Shelf Life of Fresh Fruits and Vegetables

Both ripening and reducing fungal decay in many species of harvested fruits and vegetables have been studied in relation to food irradiation alone or in combination with other treatments [39].

Appreciable delay of ripening and consequent enhancement of shelf life have been noted in some *tropical fruits* such as bananas, plantain, and mangoes after low-dose irradiation (0.2–0.7 kGy) [40,41]. Maximum delay of ripening has been observed with fruits of lower maturity. However, these feasible levels of doses are close to those that induce phytotoxicity, and many factors (varietal differences, the fruit's stage of maturity at the time of irradiation, etc.) make the outcome of the treatment uncertain.

More promising is the delay of maturation in *cultivated mushroom* (*Agaricus* sp.) where a dose of 1–2 kGy when applied soon after picking at the closed button stage can extend the shelf life by oppressing cap opening and stalk elongation. At optimal doses, darkening of the gills, cap and stalk, shriveling, and surface mold development are also inhibited [42,43].

The feasibility of irradiation of *temperate fruits*, particularly pome and stone fruits, is generally limited by adverse effects of texture (softening of the tissue) due to the radiation sensitivity of pectin. Among temperate fruits, strawberries seem to be the most amenable to radiation processing, which extends the market life of this perishable fruit by controlling molds that otherwise quickly cause decay, particularly *Botrytis* and *Rhizopus* rots. The dose requirement for this decay control is 1.5–2 kGy, with an upper limit of 3 kGy, without unduly affecting quality [40]. Thereby, strawberries can be harvested when fully ripened and can be transported and displayed for longer periods while maintaining desirable sensory qualities longer than nonirradiated strawberries.

From intact *vegetable crops*, apart from tuber, bulb, and root crops, tomatoes and asparagus show potential for radiation treatment. The development of fungal decay in tomatoes caused by *Alternaria*, *Botrytis*, or *Rhizopus* spp. can be controlled using doses of approx. 3 kGy. However, softening and the loss of characteristic flavor may occur. Doses of 0.1 kGy and above result in a delay of ripening of tomatoes [43].

Irradiation (1–2 kGy) of fresh asparagus in combination with wrapping with polyvinyl chloride (PVC) film can double the shelf life under refrigeration with no effect on color and flavor [44].

4.6. Irradiation of Fresh Meat and Poultry

Increase of shelf life under refrigeration and control of pathogenic nonsporeforming bacteria in fresh meat and poultry can be achieved by a 1–3 kGy dose. Doses for irradiation are selected under the consideration of threshold dose levels for sensory changes (off-odor), which depends on the type of animal meat (Table 7) [45]. Off-odor is due to the generation of volatile compounds from lipids and nitrogenous compounds formed by the reaction of these constituents with the reactive species produced by the radiolysis of water.

The considerable extension of the shelf life by doses as low as 1 kGy under aerobic conditions is due to the radiation sensitivity of *Pseudomonas* spp. [46] and other Gram-negative bacteria mainly associated with spoilage of fresh meat and poultry. Also, a

Table 7 Threshold Doses for
Some Foods of Animal Origin for
an Organoleptically Detectable
Off-Flavor

Food	Threshold Dose[a] (kGy)
Turkey	1.5
Pork	1.75
Beef	2.5
Chicken	2.5
Lobster	2.5
Shrimp	2.5
Rabbit	3.5
Frog	4.0
Trout	4.5
Halibut	5.0
Lamb	6.25
Horse	6.5

[a] Irradiated at 5–10°C.

significant reduction in the number of pathogens including *Salmonella*, *Campylobacter*, *Yersinia*, *Staphylococcus*, *Escherichia coli*, and *Listeria* can be achieved in fresh or frozen meat and poultry by radiation pasteurization [47,48]. The coliforms, including *E. coli* O157:H7, are quite sensitive to ionizing radiation; thus a standard irradiation pasteurization process with 1.5–2.5 kGy causes its 5–6 logs reduction in ground beef, and it would reduce also *Salmonella* and *Listeria monocytogenes* by 2–4 logs [49]. In other recent studies in Hungary, in comparison of gamma irradiation and high hydrostatic pressure-induced effects on minced beef, radiation pasteurization proved to be superior in maintaining quality [50]. Ground meat poses particular food safety concerns because the grinding process can spread pathogens present on the meat's surface throughout the product.

Recent results in the United States [51] showed that irradiation of pork trim with a processing dose of min. 1.25 kGy and max. 1.9 kGy prior to production of dry fermented sausage such as pepperoni can yield a sausage product with quality indicators closely resembling those of the traditional dry sausage while still providing the 5-log reduction of *E. coli* O157:H7 as mandated by the U.S. Department of Agriculture (USDA). A heat treatment of the pepperoni sticks performed on the finished product to an internal temperature of 60°C as decontamination alternative significantly altered the texture and color.

Because freezing confers greater radiation resistance of microorganisms, higher doses are needed under freezing [52] (see also Table 4) but the sensory results can be better.

One of the major concerns in irradiating meats is its effect of meat quality. Minced beef is susceptible to lipid oxidation. However, our own studies in cooperation with the University College Cork, Ireland [53], demonstrated that increased endogenous alpha-tocopherol concentration by dietary vitamin E supplementation prior to slaughter of the animals, and combination of this feeding supplementation with a rosemary extract as natural antioxidant additive to the ground beef, resulted in a color-stabilizing and lipid-protecting effect. Studies by Giroux et al. [54] also showed that incorporation of ascorbic acid into the beef patties before irradiation (0.5–4 kGy) stabilize the color during refrigerated storage, and its use in combination with irradiation increased the shelf life of ground meat without detrimental effects on taste and odor. Dark firm dry (DFD) pork, highly susceptible to microbial spoilage, as a result of a higher pH than normal, was resistant to oxidative changes; thus the microbial safety and the shelf life of DFD meat can be significantly improved by irradiation, thereby improving the utilization of raw DFD pork [55].

Pink color formation in irradiated broiler breast fillets after cooking, and the characteristic irradiation odor, can be significantly reduced through shelf display of raw fillets under aerobic conditions [56].

Radiation decontamination of meat was first commercially implemented in Brittany, France, when e-beam irradiation treatment was established for frozen slabs of mechanically separated chicken meat [57,58].

Serious outbreaks due to contamination with and survival of enterohemorrhagic E. coli in improperly cooked hamburger patties in the west coast of the United States in 1993 prompted a petition to the Food and Drug Administration (FDA) for approval of meat irradiation. This approval was granted in 1997 in view of new outbreaks caused by this pathogen and the subsequent recall of some 10,000 tons of ground beef because of contamination with E. coli O157:H7 [59]. The final rule authorizing irradiation of refrigerated or frozen raw meat products permits the use of a dose of 4.5 kGy for refrigerated meats and a dose of 7 kGy for frozen products [60].

4.7. Irradiation of Fish and Shellfish Products

Fish and many other fishery products are perishable commodities with short shelf lives even under refrigeration. Fish and shellfish can be contaminated by pathogenic microorganisms, either because of polluted water or handling after catch [61]. Irradiation could effectively extend the storage life and reduce the levels of pathogenic organisms. The role of radiation processing in the improvement of the hygienic quality of fishery products has been reviewed, e.g., by Venugopal et al. [62].

In a recent literature, Kamat and Thomas [63] reported that radiosensitivity of four food-borne pathogens, L. monocytogenes, Yersinia enterocolitica, Salmonella typhimurium, and Bacillus cereus was not influenced by the fat content of Indian fish varieties and application of 3 kGy dose at refrigeration temperature would effectively inactivate approx. 10^5 CFU/g of all the organisms tested, except spores of B. cereus. Savvaidis et al. [64] reported recently from Greece that a shelf life of 28 days at 4°C was recorded for salted, vacuum-packaged freshwater trout irradiated at 2 kGy, compared with a shelf life of 7 days for the unirradiated sample based on sensory odor scores. Under the same conditions, the growth of L. monocytogenes inoculated into the samples was suppressed by 2 log cycles after 2 kGy irradiation and storage for up to 18 days at 4°C.

The increased incidence of food-borne diseases from shellfish contaminated with Vibrio species have brought forward renewed interest in irradiation of shellfish and other seafood. Results are very promising for clams and oysters without killing the molluscs [65]. The radiation decimal reduction dose (D_{10}) determined for Vibrio cholerae O1 biotype El Tor inoculated into various molluscs was 0.14 kGy [66]. Similar radiation sensitivity of this organism was found in inoculated fish fillets and shrimp tails [67]. The radiation dose to eliminate as high as 10^7 CFU/g Vibrio spp. in oysters was 1.2 kGy [68]. The radiation D_{10} value for V. cholerae O1 biotype El Tor in marine snails was 0.11 kGy [69].

Petitions by the National Fisheries Institute for approval of irradiation treatment to control food-borne pathogens in raw or processed crustaceans are pending at the regulatory authorities in the United States [70].

In France, most, if not all, frog legs marketed are being treated by irradiation to ensure their hygienic quality [71].

4.8. Irradiation of Minimally Processed or Ready-To-Eat Foods

"Minimally processed" is an equivocal term that is applied to such different types of products as precut, prepackaged fresh produce, and mildly cooked or pasteurized foods (meals or meal components) that can be stored under refrigeration for more than 1 week. Some conventional products such as cured meats can be considered as minimally processed but more frequently a new generation of partially processed, refrigerated foods are described with this term.

There is an increasing consumer trend and interest by the catering industry toward less extensively processed, convenient, or ready-to-eat foods. In response to the demands, minimally processed foods are gaining importance, and consumers perceive these foods to be superior because they are chilled rather than canned, dried, or frozen [72].

Many fresh-cut produce takes the advantage of the internal development of a modified atmosphere packaging (MAP), which improves the retention of quality under delicate interaction between respiratory increase of CO_2, gas transmission properties of packaging materials, and storage temperature.

Various mildly cooked chilled product are also marketed under MAP, packaged with elevated concentrations of CO_2 over the product. CO_2 suppresses the growth of aerobic spoilage bacteria. Special versions of cook-chill foods are the "sous-vides." Sous-vide is a process where the food is cooked under controlled conditions of temperature and time under 100°C inside heat stable, vacuumed plastic pouches [73].

The safety of all these foods largely depends on the temperature of refrigerated storage and eventually on other hurdles to microbial growth (reduced pH, reduced water activity, or antimicrobials). With the extended shelf life, there are opportunities for growth of surviving psychrotrophic microorganisms, and even mesophilic species in case of abusive temperature.

In products prepared without pasteurization, nonsporeforming as well as sporeforming pathogens should be considered as potential hazard. With mildly cooked minimally processed food, the pathogens of greatest concern are the psychrotrophic nonproteolytic types of *Clostridium botulinum* and certain strains of *B. cereus* [74].

Considering the generally quality-friendly character of treatment with ionizing radiation, and by adopting a "hazard analysis and critical control point" (HACCP)-based approach to risk management, irradiation processing offers a physical critical control point for improving microbiological safety of minimally processed foods [75,76]. (A critical control point is a point, step, or procedure at which control can be applied and a food safety hazard can be prevented or reduced to acceptable level.)

Regarding irradiation of minimally processed fresh produce, irradiation treatment at 1 kGy reduced the viable cell count of *L. monocytogenes* inoculated onto precut packaged vegetables such as sliced bell pepper, carrot cubes, shredded white cabbage, and sliced radish [77]. The irradiation also drastically reduced the viable load of spoilage bacteria thereby improving the microbiological shelf life and extending the sensorial keeping quality of the precut vegetables. Loss of vitamin C content as an effect of irradiation did not exceed that in untreated samples throughout their useful life. Similarly, the beneficial effects of irradiation at 1.0 kGy of diced celery were reported by Prakash et al. [78].

Howard et al. [79] reported promising results on the effects of gamma radiation processing (1 kGy) for extending the shelf life of refrigerated "pico de gallo," a Mexican-style cold salad prepared by chopping and mixing fresh tomatoes, onions, and "jalepeno peppers." Microbiological shelf life and microbiological safety of diced tomatoes and cantaloupes could be improved at 0.5–1.0 kGy without significant loss of consumer quality [80,81]. Similarly, low doses of 0.35 and 0.55 kGy can improve the safety and microbiological shelf life of cut romaine lettuce packaged under modified atmosphere and shredded iceberg lettuce, respectively [82,83].

Ionizing radiation effectively inactivates nonsporeforming pathogenic bacteria also from cured or cooked meats [84]. Pathogen reduction and shelf life extension in cooked pork chops and cured hams inoculated with *L. monocytogenes* and *S. typhimurium* were studied by Fu et al. [85]. Low-dose irradiation (0.75–0.9 kGy) reduced *L. monocytogenes* by more than 2 logs and *S. typhimurium* by 1 to 3 logs. Alur et al. [86] described eradication of *Salmonella* and *Staphylococcus* from several processed pork products in India by gamma radiation doses of 2.5–4 kGy. When Sommers and Thayer [87] surface-inoculated several brands or types of commercially available frankfurters with a "cocktail" of four *L. monocytogenes* strains and vacuum packaged, the gamma radiation D values ranged from 0.49 to 0.71 kGy depending on the individual product formulation.

The effect of irradiation (2 and 3 kGy) and storage at 2–3°C on the sensory quality of cook-chill ready meals consisting of roast beef and gravy, and cauliflower together with roast and mashed potatoes were assessed by McAteer et al. [88]. The authors noted that growth of *Pseudomonas* spp. caused obvious spoilage of the nonirradiated meals after 15 days of storage, whereas irradiation reduced the number of microorganisms in the meal to less than 100 CFU/g initially, and significant microbial growth did not occur during the entire storage. However, using sensory profiling techniques, a trained panel found that the organoleptic effects of irradiation and chilled storage were most apparent in the cauliflower and mashed potato components, and mostly occurred in the color, appearance, and textural attributes. In another study, the same research group performed a consumer trial with 107 consumers to assess the acceptability of an irradiated (2 kGy) chilled meal consisting of beef and gravy, Yorkshire pudding, carrot, broccoli, and roast potato. The testing after 4 days after radiation treatment indicated that untrained consumers found the irradiated meal moderately to very acceptable, and not significantly different from the nonirradiated meal [89]. Only for the Yorkshire pudding was there a significant difference between the nonirradiated and irradiated samples.

In a later series of experiments on other types of ready-to-eat meals (roast pork and gravy and mixed vegetables), 2 kGy was selected as the most appropriate irradiation dose [90]. Microbial numbers in control meals reached unacceptable levels by 6- and 4-day at 3°C and 10°C, respectively, while samples irradiated at 2 kGy remained acceptable throughout the 14-day storage at 3°C, but irradiated samples stored at 10°C had unacceptably high numbers by day 6. The 2 kGy radiation dose reduced thiamin content in the pork component of the ready-to-eat meal by 22%. After storage at 3°C for 14 days, a further 26% thiamin loss was measured, and a further 14% decrease occurred when the meals were reheated as per the manufacturer's instruction at 160°C for 25 min. Irradiation caused 51% loss in total vitamin C content of potatoes in the ready meals. The vitamin C content of potatoes greatly decreased in both nonirradiated and irradiated potatoes during storage. Reheating had a highly detrimental effect on the levels of vitamin C.

In a recent study in the United States, irradiation of a prepared meal consisting of Salisbury steak, gravy, and mashed potatoes at 5.7 kGy effectively eliminated the background microbial population and high concentrations of *L. monocytogenes* contamination without causing adverse effects on quality [91].

Sous-vide cooking is thought to be quality friendlier than regular cooking due to the mild heat treatment and vacuum packaging. The combined effect of irradiation and sous-vide cooking of chicken breast meat was investigated with respect to survival and growth of *L. monocycotenes*, shelf life, thiamin content, and sensory quality by Shamsuzzaman et al. [92]. Combining electron irradiation of 2.9 kGy and cooking in vacuum packaging to an internal temperature of 65.6°C, reduction of more than 5.5 logs of CFUs of *L. mono-cytogenes* was achieved and the pathogen remained undetectable in the irradiated product during an 8-week storage at 2°C. Regarding the general microbiological quality, the product that received the combined treatment had at least 8 weeks shelf life under refrigeration. The electron beam treatment had little effect on odor and flavor, and the thiamin content was reduced by only 5%. In a second experiment [93], the same product inoculated with *L. monocytogenes* to approx. 10^6 CFU/g was cooked to 71°C. This sous-vide cooking alone reduced *Listeria* counts by only 1.5 log cycles; the survivors multiplied quickly during the 8°C storage and nonirradiated samples spoiled within 2 weeks. The combination of 3.1 kGy and sous-vide cooking reduced *Listeria* counts to undetectable levels without adversely affecting the sensory quality and prevented microbial spoilage for at least 8 weeks. The loss of thiamin content due to combined treatment varied from 23% to 46%.

In studies of Farkas et al. [94], the survival and growth of spores of a psychrotrophic *B. cereus* strain inoculated in a meal of smoked-cured pork in boiled beans sauce were investigated as affected by combination of 5 kGy and sous-vide cooking and storage time at 10°C. The microbiological analyses demonstrated that this medium dose irradiation prior to sous-vide cooking sensitized surviving spores to the subsequent heat treatment and provided microbiological stability for at least 2 months.

4.9. Radiation Decontamination of Dry Food Ingredients

The microbiological action of ionizing radiation is well proven to decrease the viable cell counts in dry food ingredients, thus improving the microbiological safety of such products and enhancing the storage stabilities of foods prepared with them. Spices, dry vegetable seasonings, herbs, protein preparations, and commercial enzyme preparations used in the food industry can be sufficiently decontaminated with doses of 3–10 kGy without altering their flavor, texture, or other important technological and sensory properties. It is also of practical significance that the surviving microflora of radiation-pasteurized ingredients, consisting mainly of bacterial spores, becomes more sensitive by the radiation treatment to subsequent food processing treatments than the microflora of untreated ingredients [95,96]. Heat processing techniques have a much more limited feasibility of these commodities because of the heat sensitivity of many of these ingredients. The irradiation process is a viable alternative to the formerly used fumigation with ethylene oxide, which was prohibited by a European Union (EU) directive in 1991 and has been banned in a number of other countries, too, for health, environmental, or occupational safety reasons, due to its carcinogenicity [97].

A detailed monograph on irradiation of dry food ingredients has been published in 1988 [98], and an updated shorter summary appeared recently [99]. A Code of Good Irradiation Practice for the Control of Pathogens and Other Microflora in Spices, Herbs, and Other Vegetable Seasonings has been issued by the International Consultative Group on Food Irradiation [100].

Presently, irradiation of spices is the most widely utilized application of food irradiation that is practiced in more than 20 countries, including Argentina, Belgium,

France, Hungary, Mexico, The Netherlands, Norway, South Africa, and the United States, and global production of irradiated spices has increased from about 5000 tons in 1990 to over 60,000 tons in 1997. In the United States alone, over 30,000 tons of spices, herbs, and dry vegetable seasonings were irradiated in 1997, as compared to only 4500 tons in 1993 [101]. In 1999, about 95 million pounds of these products were irradiated accounting for about 10% of their total consumption [102].

In addition to reducing microbial contamination, irradiation is an effective method for increasing of extraction yield of medicinal herbs [103].

4.10. Combination Processes in Food Irradiation

When irradiation is used with other preservative or antimicrobial factors, the global efficiency is reinforced through additive or synergistic action. The combination of irradiation with mild heat treatment has a number of advantageous effects [104]. The possibility of using heat in combination with irradiation was first suggested in the 1950s when synergistic effects were observed with a variety of biological systems including bacteria. More recently, Thayer et al. [105] investigated the effects of heat and ionizing radiation on *S. typhimurium* in mechanically deboned chicken meat and reported that irradiation to a dose of 0.9 kGy caused heat-sensitization of *S. typhimurium*.

The synergy between irradiation and heat, coupled with the persistence of the radiation-induced heat sensitization, could be used to advantage in foods that are cooked before consumption, and cook-chill ready meals that are reheated prior to consumption [106] (see also Sec. 4.8).

The synergistic effect of irradiation plus heat on vegetative organisms may be due to the inability of cells to repair radiation damage because heating might inactivate repair enzymes. In bacterial spores, the synergism [107] may be related to the partial rehydration of the spore core due to the radiation-induced degradation of the cortex peptidoglycan that is providing otherwise an osmoregulative effect maintaining the dehydrated state of the core [108]. A recent paper [109] demonstrated a method to quantify the radiation-induced sensitivity of bacterial spores to heat in conduction heating foods and illustrated the associated savings in the heating cycle of the ultimate thermal process.

Mild heat treatment such as a hot water dip in combination with low-dose irradiation found to be efficient to decrease rot and without adverse effect on nonmicrobial qualities of certain fruits such as mangoes [110,111] and clementines [112].

In the field of muscle foods, the use of marination before irradiation reduced the dose necessary to eliminate *Salmonella* in poultry [113]. Some antimicrobial additives, especially the natural ones (e.g., bacteriocins) [94] and GRAS (generally recognized as safe) preservatives [114] can be usefully combined with irradiation to reduce dose requirements. Some antioxidants have also been used to prevent the undesirable oxidative effects in irradiated foods.

4.11. Radiation Sterilization of Food (Radappertization)

The purpose of this process is the production of foodstuffs that are shelf stable at ambient temperatures and that have better quality characteristics than the corresponding heat-sterilized products. Radappertization dose of low-acid foods must ensure elimination of the spores of the most resistant bacterial pathogen, *C. botulinum*. The dose selected for the purpose is around 50 kGy, 12 times of the D_{10} value. As the product is intended to be shelf-stable at ambient temperature, while autolytic enzymes have high radiation resistance, the radiation treatment is supplemented with a mild heat treatment to inactivate the enzymes. This high-dose treatment must be delivered to a vacuum-packaged and deeply frozen ($-30°C$, or less) product to avoid flavor changes [115]. The whole process thus involves the following steps:

1. Heating to an internal temperature of $65-75°C$.
2. Packaging under vacuum in a sealed container impermeable to moisture, air, light, and microorganisms.
3. Cooling/freezing to irradiation temperature.
4. Irradiation.

The technology has been developed mainly for meats, poultry, and certain seafoods in the United States, and radappertized foods were successfully used in space missions [116]. Similar radiation-sterilized products were manufactured by the Atomic Energy Corporation of South Africa for hikers [117].

Radiation sterilization of deep frozen meals prepared for hospital patients whose immune systems have been suppressed is approved in several countries such as the United Kingdom, The Netherlands, and the Federal Republic of Germany.

5. WHOLESOMENESS OF IRRADIATED FOODS

Wholesomeness (toxicological innocuity, nutritional adequacy, and microbiological safety) of irradiated food has been carefully evaluated by an unprecedented width of research and testing over more than 50 years. All scientifically acceptable evidence resulted from these studies supports the safety of irradiated foods for consumption [14,118–121].

At low and medium doses, it is well established that the nutritional value of proteins, carbohydrates, and fats as macronutrients are not significantly impaired by irradiation, and neither the mineral bioavailability is impacted. Like all other energy depositing process, the application of ionizing radiation treatment can reduce the levels of certain sensitive vitamins. Nutrient loss can be minimized by irradiating food in a cold or frozen state and under reduced levels of oxygen. Thiamin and ascorbic acid are the most radiation sensitive, water-soluble vitamins, whereas the most sensitive, fat-soluble vitamin is vitamin E. In chilled pork cuts at the 3 kGy maximum at $0-10°C$, one may expect about 35–40% loss of thiamin; in frozen, uncooked pork meat irradiated at a 7 kGy maximum at $-20°C$ approx., 35 % loss of it can be expected [122].

The Joint Food and Agriculture Organization (FAO)/International Atomic Energy Agency (IAEA)/World Health Organization (WHO) Expert Committee on Food Irradiation (JECFI) has evaluated the very extensive literature on wholesomeness of irradiated food and concluded already in 1980 that "...the irradiation of any food commodity up to an overall average dose of 10 kGy presents no toxicological hazard, hence, toxicological testing of foods so treated is no longer required," and "...irradiation of foods up to an overall average dose of 10 kGy introduces no special nutritional or microbiological problems" [119]. Independent evaluations by experts, e.g., in Denmark, France, Japan, The Netherlands, the United Kingdom, and the United States found neither toxic effects as a result of consuming irradiated food. The FAO/WHO Food Standards Program, Codex Alimentarius, accepted the JECFI's recommendations and established a "Codex General Standard for Irradiated Food and a Recommended International Code of Practice for the Operation of Radiation Facilities used for the Treatment of Foods" [6].

There is no evidence or reason to expect that irradiation produces more virulent pathogens among those that survive irradiation treatment [123,124].

After publishing an up-to-date review on safety and nutritional adequacy of irradiated food [120], the FAO/IAEA and the WHO jointly conveyed a Study Group on High Dose Irradiation in 1997, which reviewed all relevant data related to the toxicological, microbiological, nutritional, radiation chemical, and physical aspects of food exposed to doses greater than 10 kGy. The Study Group came to the conclusion that foods treated also with doses of radiation sterilization can be considered safe and nutritionally adequate when produced under established Good Manufacturing Practices, and recommended approval of food irradiation without any dose maximum limit [121]. The recommendation has been submitted to the Codex Alimentarius Commission for possible amendment of the 1984 Codex General Standard on Irradiated Foods.

6. LEGISLATION OF FOOD IRRADIATION

The WHO closely collaborates with its Member States and the other international organizations, particularly through the International Consultative Group on Food Irradiation (ICGFI) established in 1984 under the aegis of FAO, IAEA, and WHO. Through the efforts of the ICGFI, which had a membership of 46 governments in 1998, harmonized regulations on food irradiation are being promulgated in developing countries in Asia, the Pacific, Africa, Latin America, and the Middle East [125]. The ICGFI issues numerous publications relating to food irradiation, including codes of good irradiation practice for various classes of foods [126] and compilations of technical data for authorization and control of food irradiation.

All these international documents and activities of the specialized agencies of the UN facilitated legislation of food irradiation in many countries. According to the database developed by the ICGFI's Secretariat at the Food and Environmental Protection Section of the Joint FAO/IAEA Division in Vienna, currently some 50 countries granted national clearances of irradiation of at least one or more food items of food classes. The itemized ICGFI database on these clearances can be visited on the website http://www.iaea.org/cgi-bin/rifa-ste.1. Legislatory authorities require that irradiated food products be labeled. In general, the international food irradiation symbol, the so-called Radura logo:

is required with a statement that the product has been intentionally subjected to radiation. The severity of labeling requirements for multi-ingredient products to identify irradiated components on the list of ingredients is not the same with each authority.

After 10 years of debate, the European Parliament and the Council of the European Union issued on February 22, 1999, a "framework" *Directive 1999/2/EC* on approximation of the laws of the Member States concerning foods and food ingredients treated with ionizing radiation [127], and a *Directive 1999/3/EC* on the establishment of a *Community List* of foods and food ingredients treated with ionizing radiation [128]. The present category of foodstuffs authorized by this Directive for irradiation treatment is "dried aromatic herbs, spices, and vegetable seasonings," and the permitted maximum overall average absorbed dose is 10 kGy. The Directive enforces strict labeling. Unfortunately, all requisite labeling of irradiated foods appears much more in the form of a warning than an item for neutral information.

Working on the subject between 1986 and 1998, the EU Scientific Committee of Food (SCF), an independent expert body advising the European Commission on health matters, concluded that the food irradiation process posed no problem for health over a wide ranged of uses, if used under prescribed conditions and endorsed for authorization of 17 irradiated foods/food classes [129]. The Commission proposal in December 1998 for an EU directive contained eight food categories and three products.

By the published Directive 1999/2/EC, the European commission was charged to develop a final "positive list" of permitted items until the end of 2000, but the contents of this list are not yet published. There was even a "consultation" on this problem on the Internet and industry and consumer organizations responded [130]. As long as the "positive list" is not adopted, Member States can maintain the existing regulations except for spices [131].

Regarding packaging materials for irradiated foods, most commonly used food packaging material are suitable for the purpose. However, irradiation of prepackaged food requires approval of its packaging material [132].

7. FOOD IRRADIATION PROCESS CONTROL AND DETECTION OF IRRADIATED FOODS

Proper control of food irradiation applications should fulfill the requirements for both food technologies and radiation technologies. Application of well-established methods for measurement of absorbed radiation dose and the dose distribution helps to provide assurance that the radiation treatment is both effective and legally correct [133]. Computer tomography (CT) can provide detailed, high-resolution, and accurate dose maps for any arbitrary product and package configurations [134]. Such dose maps are an essential part of process validation.

Significant progress has been made in the field of analytical detection of irradiated food to improve consumers' confidence and to assist international trade of irradiated food [135,136]. Due to national and international programs and activities of the European Committee for Standardization (CEN), five validated and standardized detection methods are now available. The CEN is also considering the adoption of further five detection methods: three will be screening methods (positive results from a screening method must be confirmed using a standardized method) [136].

One of the standardized methods, electron spin resonance (ESR) technique, permits identification of food that contains a hard, dry matrix, e.g., bone. When food containing bone is irradiated, free radicals are produced and trapped in the crystal lattice of the bone, which can be detected by ESR spectroscopy [137]. Thermoluminescence of contaminating minerals for detection of radiation treatment of, e.g., spices and dried fruits can be successfully applied [138, 139]. Another standardized method that has been developed for identification of irradiated fat-containing foods is the mass-spectrometric detection of radiation-induced 2-alkylcyclobutanones after gas-chromatographic separation [140]. The

"DNA comet assay" (analysis of DNA fragmentation by irradiation) can be applied as a fast and simple screening method [141]. For simple screening of irradiation decontamination of dry ingredients of high starch contents, viscosity measurements of their heat-gelified suspensions is suitable because of their reduced viscosity due to the radiation-induced starch damage [142–144].

8. ECONOMIC FEASIBILITY OF FOOD IRRADIATION

The costs of irradiation are influenced by a large number of factors depending on the actual use and local conditions. To optimize processing costs, one must fully understand the processing scenario and then design an industrial facility that best matches it [145]. The many different potential applications of food irradiation create many different requirements in irradiator design [146]. The throughput requirement can range from a few to hundreds of tons per hour. With such a variety of process specifications discussed in the previous chapters, the cost of irradiating food also substantially varies from one application to another. As any other processing technologies, due to varying conditions, cost of the same type of treatment may differ from place to place, and a cost estimate in one country is not necessarily representative of other countries.

The capital costs to build a commercial food irradiation plant is about a few millions of U.S. dollars but within the range of plant costs for other physical methods of food preservation. Total annual operating costs are close to $1 million. The total processing cost is the sum of the total operating costs plus the "cost of money," and the depreciation of capital (amortization) [146]. The unit cost of irradiation equals the total annual processing costs divided by the annual throughput. This unit processing cost in all irradiators rapidly decreases with the initial increase of throughput because the fixed costs are spread over a larger number of units [147]. The economic throughput is a function of many factors and these economics on scale are very pronounced at smaller sizes [148]. The effect of increasing dose on unit processing costs is linear because unit processing costs inversely relates to annual throughput.

The ICGFI [149] estimates that irradiation cost range from $10 to $15 per tonne for a low-dose application (e.g., inhibition of sprouting of potatoes or onions), and $100 to $250 per tonne for a high-dose application (e.g., to ensure hygienic quality of spices). These unit costs are considered to be competitive with alternative treatments.

In addition to economic feasibility, radiation processing is less energy consumptive than other decontamination techniques—an increasingly encouraging feature of radiation technology [150].

9. RECENT DEVELOPMENT OF COMMERCIAL IMPLEMENTATION OF FOOD IRRADIATION

Acceptance of irradiated food was greatly influenced by the perception of food irradiation as a nuclear technology. As Professor J.F. Diehl, one of the world authorities in food irradiation, rightly points out [151]: "Much more than other modern methods of food processing, food irradiation has to overcome barriers created by prejudice, misleading information, restrictive legal and regulatory measures, and a resulting reluctance of food manufacturers and food trade to make use of the new technology." A significant barrier to adoption of the irradiation technology was the "wait and see" attitude of the food industry and trade because they either perceived significant consumer resistance or because they had concerns about being seen as the leader in a technology perceived as controversial [125]. Still much needs to be carried out for the wider commercial application of the technology and public understanding and acceptance of irradiated food. However, in spite of certain anti-irradiation groups that have been effecting slow take-off of food irradiation at work with the news media, the notion of overwhelming consumer resistance to the technology is a "myth" and a changing attitude can be noted on market research studies. Numerous consumer acceptance studies and market tests in several countries during the past decades indicate that majority of the consumers are willing to purchase irradiated food once they understand the safety and benefit of the process [152–155]. A group of rejectors is that segment that generally rejects any new product and no amount of information would convince this group [156].

Besides the pioneering implementations of specific low-dose applications as well as the widely utilized irradiation of spices now, mentioned in Secs. 4.1, 4.2, and 4.9, small-scale commercial application of irradiation to ensure hygienic quality of food, especially those of animal origin, has been carried out in Chile, China, Indonesia, and Thailand in the past two decades. In the recent years, new commercial irradiators including some that are dedicated to food irradiation have been commissioned in Brazil (which plans to add up to 10 facilities in the coming years), China, India, Republic of Korea, Mexico, and Thailand [157].

Very important recent progress is noted in the United States. Following the approval in December 1999 of irradiation of refrigerated or frozen raw meat and meat products to eliminate or significantly reduce *E. coli* O147:H7 and other hazardous bacteria, in May 2000, commercial production and marketing of irradiated frozen beef patties started, electron-beam treated by Sure-Beam Corp., Sioux City, Iowa. This irradiation plant can process yearly more than 100 thousand tons of beef [158]. The other irradiation company, Food Technology Service, Inc., Mulberry, Florida, which has been irradiating poultry since 1996 with its MDS Nordion-designed cobalt-60 plant, also began treating fresh ground beef and frozen beef patties. The Ion Beam Applications (IBA) company was expected to complete in mid-2001 a 135 kW electron accelerator to be used in x-ray mode for food. Because several other irradiation plants are under construction in various states in the United States [159], a nationwide distribution of these irradiated products becomes possible. Due to low conversion rate of production of x-rays, the x-ray technology is probably not as economical in food irradiation as gamma rays or electron beam technology. Nevertheless, an x-ray plant has been recently constructed in Hawaii to disinfest tropical fruit [160]. An important step toward wider future applications is a petition in 1999 to the FDA, for approval of irradiation of many types of ready-to-eat food by the Coalition of Food Irradiation, which involves approx. 25 different associations of food producers.

Due to legislative stumbling blocks (described in Sec. 6), after early pioneering efforts in several countries, e.g., Netherlands, France, and Belgium, where considerable amounts of frozen seafood, frog legs, and dry food ingredients have irradiated already been in the 1980s, progress in Europe is now lagging behind the developments in the United States. However, increasing activities on food irradiation in the United States, which is one of the largest markets in the world, and also similar activities in some countries of Asia and Latin America, will likely influence other countries and regions to implement the use of irradiation for commercial purposes.

REFERENCES

1. Josephson, E.S.; Peterson, M.S. *Preservation of Food by Ionizing Radiation.* Vol. 1–3. CRC Press, Inc.: Boca Raton, FL, 1982–1983.
2. Urbain, W.M. *Food Irradiation*; Academic Press, Inc.: Orlando, 1986.
3. *Food Irradiation. A Technique for Preserving and Improving the Safety of Food*; World Health Organization: Geneva, 1988.
4. Satin, M. *Food Irradiation: A Guidebook*; Technomic Publishing Co., Inc.: Lancaster-Basel, 1996.
5. Food Irradiation: Principles and Applications; Molins, R., Ed.; Wiley-Interscience: New York, 2001.
6. *Codex General Standard for Irradiated Foods and Recommended International Code of Practice for the Operation of Radiation Facilities Used for the Treatment of Food,* CAC/Vol. XV-Ed. 1, Food and Agriculture Organization of the United Nations: Rome, 1984.
7. *Understanding the Key Radiation Processing Parameters—A Guide for Food Producers*, Leaflet published by PURIDEC Irradiation Technologies, 2001.
8. Kim, A.; Thayer, D.W. Appl. Environ. Microbiol. 1996, *62*, 1759.
9. Tilton, E.W.; Brower, J.H. In *Preservation of Food by Ionizing Radiation*; Josephson, E.S., Peterson, M.S., Eds.; CRC Press, Inc.: Boca Raton, FL, 1983; Vol. 2, 269.
10. Farkas, J. Int. J. Food Microbiol. 1998, *44*,189.
11. *High-Dose Irradiation: Wholesomeness of Food Irradiated with Doses Above 10 kGy*, World Health Organization: Geneva, 1999.
12. Thayer, D.W.; Boyd, G. J. Food Prot. 2001, *64*, 1624.
13. Sommers, C.H.; Niemira, B.A.; Tunick, M.; Boyd, G. Meat Sci. 2002, *61*, 323.
14. Diehl, J.F. *Safety of Irradiated Foods,* 2nd Ed.; Marcel Dekker, Inc.: New York, 1995.
15. Proctor, B.E.; O'Meara, J.P. Ind. Eng. Chem. 1951, *43*, 718.
16. Merritt, L.C. Jr.; Angelini, P.; Wierbicki, E.; Shults, G.W. J. Agric. Food Chem. 1975, *23*, 1037.
17. von Sonntag, C. Adv. Carbohydr. Chem. Biochem. 1980, *37*, 7.
18. Hammer, C.T.; Wills, E.D. Int. J. Radiat. Biol. 1979, *35*, 323.
19. Zabielski, J. Radiat. Phys. Chem. 1989, *34*, 1023.
20. Lebovics, V.K.; Gaál, Ö.; Somogyi, L.; Farkas, J. J. Sci. Food Agric. 1992, *60*, 251.
21. Lebovics, V.K.; Gaál, Ö.; Farkas, J.; Somogyi, L. J. Sci. Food Agric. 1994, *66*, 71.
22. Lebovics, V.K.; Farkas, J.; Andrássy, É.; Mészáros, L.; Lugasi, A.; Gaál, Ö. *Reduction of Cholesterol and Lipid Oxidation in Radiation Decontaminated Mechanically Deboned Turkey Meat,* Poster presented at the 48th ICoMST, Rome, 25–30 August 2002.
23. Thomas, P. In *Food Irradiation: Principles and Applications;* Molins, R., Ed.; Wiley-Interscience: New York, 2001; 241 pp.
24. Matsuyama, A.; Umeda, K. In *Preservation of Food by Ionizing Radiation*; Josephson, E.S., Peterson, M.S. Eds.; CRC Press, Inc.: Boca Raton, FL, 1983; 159 pp.
25. Kálmán, B.; Kiss, I.; Farkas, J. In *Food Preservation by Irradiation*; International Atomic Energy Agency: Vienna, 1978; Vol. 1, 113.
26. Salems, A. J. Sci. Food Agric. 1974, *25*,257.
27. Dallyn, S.I.; Sawyer, R.L. Proc. Am. Soc. Hortic. Sci. 1954, *73*, 398.
28. Ogawa, M.; Uritani, I. Agric. Biol. Chem. 1970, *34*, 870.
29. Mondy, N.I.; Gosselin, B. J. Food Sci. 1989, *54*, 982.
30. Thomas, P. CRC Crit. Rev. Food Sci. Nutr. 1984, *21*, 95.
31. Umeda, K. Food Irradiat. Inf. 1978, *8*, 31.
32. Umeda, K. Food Irradiat. Newsl. 1983, 7, 19.
33. Farkas, J.; Kiss, I.; Rázga, Z.; Vas, K. KÉKI-Közlemények 1963, *I–II*, 19.
34. Ahmed, M. In *Food Irradiation: Principles and Applications;* Molins, R.A., Ed.; Wiley-Interscience: New York, 2001; 77 pp.
35. *The Entomology of Radiation Disinfestation of Grain*; Cornwell, P.B., Ed.; Pergamon Press: London, 1966.

36. Hallmann, G.J. In *Food Irradiation: Principles and Applications*; Molins, R.A. Ed.; Wiley-Interscience: New York, 2001; 113 pp.
37. Orlandi, P.A.; Chu, D.-M.T.; Bier, J.W.; Jackson, G.J. Food Technol. 2002, *36*(4), 72.
38. Wilkinson, V.M.; Gould, G.W. *Food Irradiation: A Reference Guide*; Butterworth-Heinemann: Oxford, 1996.
39. Thomas, P. In: *Food Irradiation: Principles and Applications*; Molins, R.A., Eds; Wiley-Interscience: New York, 2001; 213 pp.
40. Thomas, P. CRC Crit. Rev. Food Sci. Nutr. 1986, *23*, 147.
41. Aina, O.J.; Adesiji, O.F.; Ferris, S.R.B. J. Sci. Food Agric. 1999, *79*, 653.
42. Kovács, E.; Vas, K. Acta Aliment. 1974, *3*, 11.
43. Thomas, P. CRC Crit. Rev. Food Sci. Nutr. 1988, *26*, 313.
44. Lescano, G.; Narvaiz, P.; Kairiyama, E. Lebensm. Wiss. Technol. 1993, *26*, 411.
45. Sudarmadji, S.; Urbain, W.M. J. Food Sci. 1972, *37*, 671.
46. Tiwary, N.P.; Maxcy, R.B. J. Food Sci. 1971, *36*, 833.
47. Giddings, G.G.; Mercotte, M. Food Rev. Int. 1991, *7*, 259.
48. Radomyski, T.; Murano, E.A.; Olson, D.G. Dairy Food Environ. Sanit. 1993, *13*, 398.
49. Kiss, I.F.; Mészáros, L.; Kovács-Domján, H. In *Irradiation for Food Safety and Quality*;. Loaharanu, P., Thomas, P. Eds.; Technomic Publishing Co., Inc.: Lancaster, PA, 2001; 81 pp.
50. Hassan, Y.; Mészáros, L.; Simon, A.; Tuboly, E.; Mohácsi-Farkas, Cs.; Farkas, J. Acta Aliment. 2002, *31*(3) (in print).
51. Johnson, S.C.; Sebranek, J.G.; Olson, D.G.; Wiegand, B.R. J. Food Sci. 2000, *65*, 1260.
52. Thayer, D.W.; Boyd, G. J. Food Prot. 2001, *62*, 1136.
53. Formanek, Z.; Kerry, J.P.; Galven, K.; Buckley, D.J.; Farkas, J. In *Meat for the Consumer. Poster Proceedings of the 42nd International Congress of Meat Science and Technology, 1-6 September 1996, Lillehammer, Norway*; MATFORSK, Norwegian Food Research Institute: Oslo, 1996; 90 pp.
54. Giroux, M.; Ouattara, B.; Yefsah, R.; Smoragiewicz, W.; Saucier, L.; Lacroix, M. J. Agric. Food Chem. 2001, *49*, 919.
55. Ahn, D.U.; Nam, K.C.; Du, M.; Jo, C. Meat Sci. 2001, *57*, 419.
56. Du, M.; Hur, S.J.; Ahn, D.U. Meat Sci. 2001, *61*, 49.
57. Sadat, T.; Volle, C. Radiat. Phys. Chem. 2000, *57*, 613.
58. Sadat, T.; Vassenaix, M. Radiat. Phys. Chem. 1990, *36*, 661.
59. Loaharanu, P. *Trends on the Use of Irradiation as a Sanitary and Phytosanitary Treatment for Food and Agricultural Commodities*. Paper presented at ICGFI Workshop on Trade Opportunities for Irradiated Food, Kona, Hawaii, 22–24 May 2000.
60. Irradiation of meat and meat products. Fed. Regist. 1999, *64*, 9089.
61. Garrett, E.S.; Jahncke, M.L.; Tennyson, J.M. J. Food Prot. 1997, *60*, 1409.
62. Venugopal, V.; Doke, S.N.; Thomas, P. CRC Crit. Rev. Food Sci. Nutr. 1999, *39*, 391.
63. Kamat, A.; Thomas, P. J. Food Saf. 1999, *19*, 35.
64. Savvaidis, I.N.; Skandamis, P.; Riganakos, K.A.; Panagiotakis, N.; Kontominas, M.G. J. Food Prot. 2002, *65*, 515.
65. Kilgen, M.B.; Hernard, M.T.; Duet, D.; Rabalais, S. In *Irradiation to Control Vibrio Infection from Consumption of Raw Seafood and Produce*; IAEA-TECDOC-1213, International Atomic Energy Agency: Vienna, 2001; 57 pp.
66. Torres, Z.; Bernuy, B.; Zapata, G.; Vivanco, M.; Kahn, G.; Guzman, E.; Leon, R. In *Irradiation to Control Vibrio Infection from Consumption of Raw Seafood and Fresh Produce*; International Atomic Energy Agency: Vienna, 2001; 37 pp.
67. Torres, Z.; Kahn, G.; Vivanco, M.; Guzman, G.; Bernuy, B. In *Irradiation to Control Vibrio Infection from Consumption of Raw Seafood and Fresh Produce*; International Atomic Energy Agency: Vienna, 2001; 47 pp.
68. Cisneros Despaigne, E.; Leyva Castillo, V.; Castillo Rodriguez, E.; Martinez, L.L.; Lara Ortiz, C. In *Irradiation to Control Vibrio Infection from Consumption of Raw Seafood and Fresh Produce*; International Atomic Energy Agency: Vienna, 2001; 7 pp.

69. Torres, Z.; Arias, F. In *Irradiation to Control Vibrio Infection from Consumption of Raw Seafood and Fresh Produce*; International Atomic Energy Agency: Vienna, 2001, 31 pp.
70. Anon. *Irradiation (Cold Pasteurization) of Molluscan Shellfish*. National Fisheries Institute News Release 99-41. June 25, 1999 (available online at www.nfi.org/hdlines).
71. Anon. *Facts About Food Irradiation*; International Consultative Group on Food Irradiation: Vienna, 1999.
72. Stringer, M. Food Manuf. April 1990, *39*.
73. Schellekens, M. In *Proc. 1st Eur. Sous Vide Cooking Symp., March 25–26, 1993*, Food-Linked Agro-Industrial Research (FLAIR), 1989–1993; 3 pp.
74. Martens, T., Ed.; *Harmonization of Safety Criteria for Minimally Processed Foods. Rational and Harmonization Report*; FAIR Concerted Action, FAIR CT96-1020, European Commission, 1999.
75. Molins, R.A.; Motarjemi, Y.; Käferstein, F. In *Irradiation for Food Safety and Quality*. Loaharanu, P., Thomas, P. Eds.; Technomic Publishing Co., Inc.: Lancaster-Basel, 2001; 55 pp.
76. Farkas, J. In *Food Irradiation: Principles and Applications*. Molins, R.A. Ed.; Wiley-Interscience: New York, 2001; 243 pp.
77. Farkas, J.; Mohácsi-Farkas, Cs.; Andrássy, É.; Mészáros, L.; Sáray, T.; Horti, K. In: *Proceedings of the 4th World Congress on Foodborne Infections and Intoxications, Berlin, June 1998*, 1998, Vol. 1, 593.
78. Prakash, A.; Guner, A.R.; Caporaso, E.; Foley, D.M. J. Food Sci. 2000, *65*, 549.
79. Howard, L.R.; Miller, G.H., Jr; Wagner, A.B. J. Food Sci. 1995, *60*, 461.
80. Mohácsi-Farkas, Cs.; Kiskó, G. In *Report of First FAO/IAEA Research Coordination Meeting on the Coordinated Research Programme on Use of Irradiation to Ensure Hygienic Quality of Fresh, Pre-Cut Fruits and Vegetables and Other Minimally Processed Food of Plant Origin;* D6-RC-844; International Atomic Energy Agency: Vienna, 2001; 11 pp.
81. Prakash, A.; Manley, J.; DeCosta, S.; Caporaso, F.; Foley, D. Radiat. Phys. Chem. 2002, *63*, 387.
82. Prakash, A.; Inthajak, P.; Huibregtse, H.; Caporaso, F.; Foley, D.M. J. Food Sci. 2000, *65*, 1070.
83. Foley, D.M.; Dufour, A.; Rodriguez, L.; Caporaso, F.; Prakash, A. Radiat. Phys. Chem. 2002, *63*, 391.
84. Byun, M.-W.; Lee, J.W.; Yook, H.-S.; Lee, K.-H.; Kim, H.-Y. Radiat. Phys. Chem. 2002, *63*, 361.
85. Fu, A.H.; Sebranek, J.G.; Murano, E.A. J. Food Sci. 1995, *60*, 1001.
86. Alur, M.D.; Kamat, A.S.; Doke, S.N.; Nair, P.M. J. Food Sci. Technol. India 1998, *35*(1), 15.
87. Sommers, C.H.; Thayer, D.W. J. Food Saf. 2000, *20*, 127.
88. McAteer, N.J.; Grant, I.R.; Patterson, M.F.; Stevenson, M.H.; Weatherup, S.T.C. Int. J. Food Sci. Technol. 1995, *30*, 757.
89. Stevenson, M.H.; Stewart, E.M.; McAteer, N.J. Radiat. Phys. Chem. 1995, *46*, 785.
90. Patterson, M.F.; Stewart, E. Paper presented at the Second FAO/IAEA Research Coordination Meeting of the Coordinated Research Programme on Development of Shelf-Stable and Ready-to-Eat Food through High Dose Irradiation Processing, Beijing, May 3–8, 1998.
91. Foley, D.M.; Reher, E.; Caporaso, F.; Trimboli, S.; Musherraf, Z.; Prakash, A. Food Microbiol. 2001, *18*, 193.
92. Shamsuzzaman, K.N.; Chuaqui-Offermanns, N.; Lucht, L.; McDougall, T.; Borsa, J. J. Food Prot. 1992, *55*, 523.
93. Shamsuzzaman, K.N.; Lucht, L.; Chuaqui-Offermans, N. J. Food Prot. 1995, *58*, 497.
94. Farkas, J.; Polyák-Fehér, K.; Andrássy, É.; Mészáros, L. Radiat. Phys. Chem. 2002, *63*, 345.
95. Kiss, I.; Farkas, J. In *Combination Processes in Food Irradiation*; International Atomic Energy Agency: Vienna, 1981, 107 pp.
96. Farkas, J.; Andrássy, É. In *Fundamental and Applied Aspects of Bacterial Spores*. Dring, G.J., Ellar, D.J., Gould, G.W. Eds.; Academic Press: London, 1985; 397 pp.
97. Dickman, S. Nature 1991, *349*, 273.

98. Farkas, J. *Irradiation of Dry Food Ingredients*; CRC Press, Inc.: Boca Raton, FL, 1988.
99. Farkas, J. In *Food Irradiation: Principles and Applications*. Molins, R.A. Ed.; Wiley-Interscience: New York, 2001, 291 pp.
100. *Code of Good Irradiation Practice for the Control of Pathogens and Other Microflora in Spices, Herbs and Other Vegetable Seasonings*, International Consultative Group on Food Irradiation, ICGFI Document No. 5, International Atomic Energy Agency, Vienna, 1988.
101. *Facts About Food Irradiation*, A series of Fact Sheets from the International Consultative Group on Food Irradiation, International Atomic Energy Agency, Vienna, 1999.
102. *Food Irradiation. Available Research Indicates that Benefits Outweigh Risks*, Report to Congressional Requests, GAO/RCED-00-217. U.S. General Accounting Office, Washington, D.C., 2000.
103. Kim, M.-J.; Yook, H.-S.; Byun, M.-W. Radiat. Phys. Chem. 2000, *57*, 55.
104. Farkas, J. Food Control 1990, *1*, 223.
105. Thayer, D.W.; Songprasertchai, S.; Boyd, G. J. Food Prot. 1991, *54*, 718.
106. Grant, I.R.; Patterson, M.F. Int. J. Food Microbiol. 1995, *27*, 117.
107. Kempe, L.L. Appl. Microbiol. 1955, *3*, 346.
108. Stegeman, H.; Mossel, D.A.A.; Pilnik, W. In *Spore Research 1976*; Barker, A.N. Wolf, J., Ellar D.J., Dring, G.J., Gould, W., Eds.; Academic Press: New York, 1977; 565 pp.
109. Welt, B.A.; Teixeira, A.A.; Balaban, M.O.; Smerage, G.H.; Hintenlang, D.E.; Smittle, B.J. J. Food Sci. 2001, *66*,844.
110. Gagnon, M.; Lacroix, M.; Pringsulaka, V.; Latreille, B.; Jobin, M.; Nouchpramool, K.; Prachasitthisak, Y.; Charoen, S.; Abdulyatham, P.; Lettre, J.; Grad, B. Radiat. Phys. Chem. 1993, *42*, 283.
111. Lacroix, M.; Gagnon, M.; Pringsulaka, V.; Jobin, M.; Nouchpramool, K.; Prachsitthisak, V.; Charoen, S.; Adulyatham, P.; Lettre, J.; Grad, B. Radiat. Phys. Chem. 1993, *42*, 273.
112. Abdellaoui, S.; Lacroix, M.; Jobin, M.; Bonbekri, C.; Gagnon, M. Sci. Aliment. 1995, *15*, 217.
113. Mahrour, A.; Lacroix, M.; Nketsia-Tabiri, J.; Calderon, N.; Gagnon, M. Radiat. Phys. Chem. 1998, *52*, 77.
114. Thakur, B.R.; Singh, R.K. Trends Food Sci. Technol. 1995, *6*, 7.
115. Josephson, E.S.; Brynjolfsson, A.; Wierbicki, E.; Rowley, D.B.; Merritt, C. Jr.; Balleen, R.W.; Killoran, J.J.; Thomas, M.H. In *Radiation Preservation of Food*; International Atomic Energy Agency: Vienna, 1973; 471 pp.
116. Wierbicki, E. In *Proceedings of the 26th European Meeting of Meat Research Workers, Colorado Springs*, 1980; 194 pp.
117. deBruyn, I. In *Irradiation for Food Safety and Quality*. Loaharanu, P.; Thomas, P., Eds.; Technomic Publishing Co., Inc.: Lancaster-Basel, 2001; 206 pp.
118. Crawford, L.M.; Ruff, E.H. Food Control 1996, *7*, 87.
119. *Wholesomeness of Irradiated Food*, Report of a Joint FAO/IAEA/WHO Expert Committee, WHO Technical Report Series 659, World Health Organization, Geneva, 1981.
120. *Safety and Nutritional Adequacy of Irradiated Food*; World Health Organization: Geneva, 1994.
121. *High Dose Irradiation of Food*; World Health Organization: Geneva, 1999.
122. Fox, J.B.; Thayer, D.W.; Jenkins, R.U.; Phillips, J.G.; Aackerman, S.; Beecher, J.; Holden, M.; Morrow, F.D.; Quirback, D.M. Int. J. Radiat. Biol. 1989, *55*, 689.
123. Ingram, M.; Farkas, J. Acta Aliment. 1977, *6*, 123.
124. Farkas, J. Int. J. Food Microbiol. 1989, *9*, 1.
125. Loaharanu, P.; Thomas, P. In *Irradiation for Food Safety and Quality*; Loaharanu, P.; Thomas, P., Eds.; Technomic Publishing Co., Inc.: Lancaster-Basel, 2001; vii pp.
126. Molins, R.A. In *Food Irradiation: Principles and Applications*. Molins, R.A., Ed.; Wiley-Interscience: New York, 2001; 1 p.
127. *Directive 1999/2/EC of the European Parliament and of the Council of 22 February 1999*, Official Journal of the European Communities, 13/03/1999, L66/16 pp.
128. *Directive 1999/3/EC of the European Parliament and of the Council of 22 February 1999 on the Establishment of a Community List of Foods and Food Ingredients Treated with Ionising Radiation*, Official Journal of European Communities, 13.3.1999, L66/24 pp.

129. Vounakis, H. In *Irradiation for Food Safety and Quality*; Loaharanu, P., Thomas, P., Eds.; Technomic Publishing Co., Inc.: Lancaster-Basel, 2001; 26 pp.
130. *Comments on the DG SANCO Consultation of Consumer Organizations, Industry Concerned and Other Interested Parties on the Strategy for Completion of the Positive List of Food and Food Ingredients to be Authorised for Irradiation Treatment*, Annex of Directive 1999/3/EC, SANCO/4175/2000-rev.1, European Commission, Health and Consumer Protection Directorate-General, Directorate D - Food Safety: Production and Distribution Chain, Brussels, 2000.
131. Ehlermann, D.A.E. Radiat. Phys. Chem. 2002, *63*, 277.
132. Morehouse, K.M. Radiat. Phys. Chem. 2002, *63*, 281.
133. *Manual of Food Irradiation Dosimetry*, Technical report series No. 178, International Atomic Energy Agency, Vienna, 1977.
134. Borsa, J.; Chu, R.; Sun, J.; Linton, N.; Hunter, C. Radiat. Phys. Chem. 2002, *63*, 271.
135. Murray, C.H., Stewart, E.M., Gray, R., Pearce J. Eds.; *Detection Methods for Irradiated Foods—Current Status*; The Royal Society of Chemistry, Information Services: London, 1996.
136. Delincée, H. Radiat. Phys. Chem. 2002, *63*, 455.
137. Yang, J.-S.; Kim, C.-K.; Lee, H.J. Korean. J. Food Sci. Technol. 1999, *31*, 606.
138. Khan, H.M.; Bhatti, I.A.; Delincée, H. Radiat. Phys. Chem. 2002, *63*, 403.
139. Chabane, S.; Pouliquen-Sonaglia, I.; Raffi, J. Can. J. Physiol. Pharm. 2001, *79*, 103.
140. Horvatovich, P.; Miesch, M.; Hasselmann, C.; Marchioni, E. J. Chromatogr., A 2000, *897*, 259.
141. Delincée, H. Radiat. Phys. Chem. 2002, *63*, 443.
142. Heide, L.; Mohr, E.; Wichmann, G.; Bögl, K.W. In *Health Impact, Identification and Dosimetry of Irradiated Foods*, Institut für Strahlenhygiene des Bundesgesundheitsamtes: München, ISH-Heft, 1988, *125*, 176.
143. Farkas, J.; Sharif, M.M.; Koncz, Á. Radiat. Phys. Chem. 1990, *26*, 621.
144. Esteves, M.P.; Raymundo, A.; deSousa, I.; Andrade, M.E.; Empis, J. Radiat. Phys. Chem. 2002, *64*, 323.
145. Kunstadt, P. In *Irradiation for Food Safety and Quality*. Loaharanu, P.; Thomas, P., Eds.; Technomic Publishing Co., Inc.: Lancaster-Basel, 2001; 129 pp.
146. Kunstadt, P. In *Food Irradiation: Principles and Applications*; Molins, R.A., Ed.; Wiley-Interscience: New York, 2001; 415 pp.
147. Cleland, M.R.; Herer, A.S.; Cokragan, A. In *Irradiation for Food Safety and Quality*; Loaharanu, P., Thomas, P., Eds.; Technomic Publishing Co., Inc.: Lancaster-Basel, 2001; 158 pp.
148. Morrison, R.M. In *Food Irradiation Processing*; International Atomic Energy Agency: Vienna, 1985; 407 pp.
149. *Facts About Food Irradiation*; p. 33 International Consultative Group on Food Irradiation, International Atomic Energy Agency: Vienna, 1991; 33 pp.
150. Traegardh, C.; Hallström, B. *Energy Analysis of Selected Food Post-Harvest and Preservation Systems, Irradiators Compared to Conventional Food Preservation Methods*. Res. Contract No. 2850/TC, report to the International Atomic Energy Agency, Vienna, 1981.
151. Diehl, J.F. Food irradiation—past, present and future. IMRP-102, In *Conference Programme and Abstracts of the 12th International Meeting on Radiation Processing, 25–30 March 2001, Avignon, France*.
152. Bruhn, C. J. Food Prot. 1995, *58*, 175.
153. Resurreccion, A.V.A.; Galver, F.C.F. Food Technol. 1999, *59*(3), 52.
154. Frenzen, P.D.; DeBess, E.E.; Hechemy, K.E.; Kassenborg, H.; Kennedy, M.; McCombs, K.; McNees, A.the Foodnet Working Group. J. Food Prot. 2001, *64*, 2020.
155. Vickers, Z.M.; Wang, J. J. Food Sci. 2002, *67*, 380.
156. Nunes, K. Meat Poult. 2002, *34* (March Issue).
157. Anon. Food Environ. Prot. Newsl. 2001, *3*(1), 13.
158. Mermelstein, D.H. Food Technol. 2000, *54*(7), 88.
159. Anon. Dairy Food Environ. Sanitat. 2001, *49* (January Issue).
160. Moy, J.H.; Wong, L. Radiat. Phys. Chem. 2002, *63*, 397.

27

New Applications of Ion Beams to Material, Space, and Biological Science and Engineering

Mitsuhiro Fukuda, Hisayoshi Itoh, Takeshi Ohshima, Masahiro Saidoh, and Atsushi Tanaka
Japan Atomic Energy Research Institute, Takasaki, Gunma, Japan

1. INTRODUCTION

Ion accelerators first appeared around 1930 and began growing rapidly in the 1930s. At the beginning of the growth period, the ion accelerators were originally designed as experimental equipment required for nuclear physics and for radioisotope production. A variety of ion accelerators has so far been developed in succession to meet the demands from nuclear research. The beginning of the ion beam applications, except for nuclear research, was the medical application to the treatment of tumors with charged particles started in the 1950s. For industrial applications of ion accelerators, ion implantation began in the middle of the 1960s. The use of ion beams for material analysis became widespread in the 1970s. The full-scale applications of MeV ion beams to materials, bio-, medical, and life sciences have rapidly broadened in the 1990s.

So far, there is no ion accelerator facility adapted for biological or botanical research, ion beam processing research, etc., because of the lack of the strong incentive to the research using accelerator ion beams. The ion beam irradiation research facility, Takasaki Ion Accelerators for Advanced Radiation Applications (TIARA), may be the first facility to promote the research on a large scale. The facility was completed in 1993 at the Takasaki establishment of Japan Atomic Energy Research Institute (JAERI), and the extensive applications of ion beams have been conducted mainly for research and development in materials science and biotechnology under full-scale operation of the azimuthally varying field (AVF) cyclotron and three electrostatic accelerators [1]. In this chapter new applications of accelerated ion beams to materials, space, and biological science and engineering, carried out mainly at TIARA, have been reviewed. Various ion beam technologies pioneering ion beam applications have also been described.

2. CHARACTERISTICS OF ION BEAMS INTERACTING WITH MATTER

When energetic ions penetrate into matter, energy deposition from incident particles to the irradiated medium occurs through a variety of interactions between the incident particle and the constituent atoms or molecules of the irradiated medium. In general, energetic, heavy charged particles transfer their energy to the irradiated materials with high-density ionization and excitation along their nearly straight particle tracks (even for uniform irradiation) showing a maximum linear energy transfer (LET) in the curve of dose distribution against penetration depth until the particles are brought to rest. This characteristic results in microscopically nonuniform dose delivery having a specific dose distribution against depth, which are in contrast to relatively uniform dose delivery and monotonous dose distribution against depth in low-LET radiation irradiation such as gamma-ray and electron beam irradiation.

Another important characteristic is that ion beams can produce a variety of the secondary particles/photons such as secondary ions/atoms, electrons, positrons, X-rays, gamma rays, and so on, which enable us to use ion beams as analytical probes. Ion beam analyses are characterized by the respectively detected secondary species, such as secondary ion mass spectrometry (SIMS), sputtered neutral mass spectrometry (SNMS), electron spectroscopy, particle-induced X-ray emission (PIXE), nuclear reaction analyses (NRA), positron emission tomography (PET), and so on.

In addition to the above-mentioned features, the ion beam irradiation has functions of implantation of different atoms in the irradiated medium and of the nuclear transmutation of the irradiated medium atoms. So far, the underlying physics and the subsequent relaxation processes in the interaction between ion beams and matter have been extensively studied not only for purely scientific interests but also for practical purposes, such that a series of international conferences on these topics have been held on a worldwide scale [2].

The relationship between the properties of the ion beams interacting with matter and the parameters of ion beam applications is shown in Fig. 1. Recent progress of ion beam handling technologies such as microbeam and single-ion hit techniques enable us to apply ion beams for advanced research and development in materials science and in biotechnology, such as the analysis of the single-event effect in the semiconductor device for space, the radiation processing of microscopic region, cell surgery technique, simulation of radiation risk for human space flight, and so on.

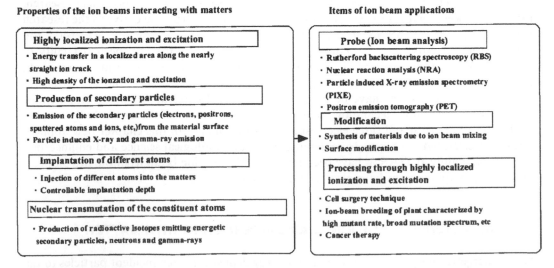

Figure 1   The relationship between properties of the ion beams interacting with matter and items of ion beam applications.

3. ION BEAM TECHNOLOGIES PIONEERING THE ION BEAM APPLICATIONS

3.1. Diversity of Ion Beam Irradiation Techniques

The most interesting parameters of the ion beams used for the nuclear research are mass number, atomic number, energy, and intensity of the accelerated particles. Special irradiation techniques are not always required for the nuclear research. In contrast, irradiation properties for the ion beam applications to materials science and biotechnology have been widely diversified to fulfill the requirements of the research. The optimum irradiation condition depends on how we want to bring about physical, chemical, or biological effects in a target substance, and how we investigate characteristics of the substance skillfully. The extent to which ion-induced effects are controllable and the analyzing techniques can be highly sophisticated by optimizing physical parameters of the ion beam, such as ion species, energy, beam intensity, irradiation area, particle density fluence, incident angle, and so on. The fundamental parameters of the ion beam irradiation are shown in Fig. 2. The response to the ion beam irradiation depends on the phase of the substance (gas, liquid, solid) and on temperature, atmospheric pressure, and so on.

Controllability and flexibility of the irradiation parameters are indispensable for the ion beam applications. The widely diverse requirements for the irradiation conditions have promoted the development of various ion beam technologies. Conversely, there are some cases that new ion beam technologies trigger a breakthrough in the ion beam applications. Nowadays, it is no exaggeration to say that the ion beam technologies are pioneering new research fields. The ion beam technologies classified according to the irradiation conditions are listed in Table 1. The present status of the representative ion beam technologies is described in the following sections.

3.2. Production of the Ions

Ion Source

A variety of ion species, from light to heavy ions, namely, from protons to uranium ions, are utilized for the ion beam applications. Wide diversity of ion species, charge state, and beam currents is required to meet the requirements for the research of the ion beam applications. Various types of ion sources [3], using different ionization methods, such as field ionization and surface ionization, have been developed to provide optimum ion beams. An ion source

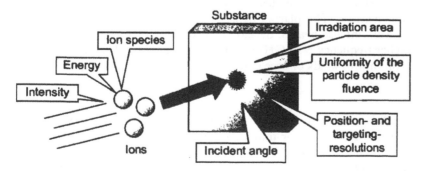

Figure 2 Typical parameters of the ion beam irradiation.

Table 1 Ion Beam Irradiation Parameters and Their Related Ion Beam Technologies

Parameter	Requirement	Ion beam technology
Ion species	Variety	Ion source
		Cluster ion source
	Multiplex ions	Dual/triple beam
Energy	Wide range from keV to GeV	Accelerator
	Quick change	Cocktail beam acceleration
Intensity	Wide range from a single	Ion source
	ion to mA	Accelerator
		Single-ion hit technique
Irradiation area and	Uniformity	Beam defocusing and a scatterer
particle density		Beam scanning: raster scan,
fluence		beam wobbling, spiral scan
Beam spot and	Minimization	Microbeam and nanobeam
targeting resolution	Precision	Micro-PIXE analysis
Time structure	Flexibility	Pulsated beam

of radio frequency (rf) plasma type produces positively or negatively charged particles by ionizing gaseous atoms and/or molecules in the plasma. The constituents of the plasma are the ions, electrons, and neutral atoms and/or molecules. The plasma remains stable by confining the ions and electrons with a magnetic field. The maximum ionization efficiency is obtained at electron energy about as high as 3 to 4 times the ionization potential of the constituent atoms and/or molecules. A gas or a vaporized metal is fed into the plasma generated by a discharge using microwaves or energetic electrons. The ions inside the plasma are extracted by an electric field formed between an ion reservoir and a perforated electrode on which a high voltage is applied. The beam intensity is fully dependent on the ion and electron densities in the plasma, the plasma electron temperature, the extraction voltage, and the geometry of the extraction system.

Higher charge states have a great advantage of increasing the particle energy, since the energy gain of the ions in the accelerator is proportional to the charge state. An electron cyclotron resonance (ECR) ion source [4], using the field ionization method, is substantially superior in the performance of heavy-ion production with higher charge state. A schematic drawing of a typical ECR ion source is shown in Fig. 3. The electrons in the plasma are heated by rf power. Energetic electrons and high-density plasmas are required to produce an intense beam of highly charged ions. A higher magnetic field enables one to achieve an efficient electron confinement [5]. The plasma density is related to the square of the rf frequency. To produce very intense beams, development of a superconducting ECR ion source working at a higher frequency ~28 GHz is in progress [6]. Nowadays, the ECR ion source is widely used for various kinds of accelerators.

Cluster Ion Beam

A cluster ion beam, which is composed of atomic particles such as C_n^+ and Au_n^+, where n is the number of atoms in the cluster [7], or molecular ions such as OH^- and AlO^+, has attracted a great deal of interest in atomic physics [8,9] and in materials science [10]. The cluster ion beam has distinctive properties of vicinity effects caused in substances, which is

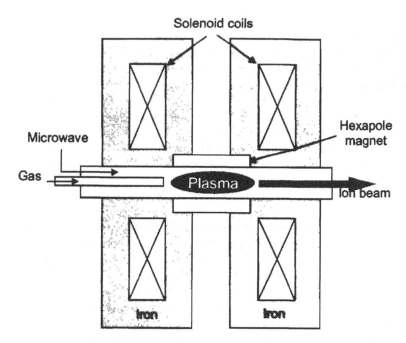

Figure 3 A schematic drawing of a typical ECR ion source. The ions and electrons are confined in the plasma by the magnetic field generated by the solenoid coils and the hexapole magnet.

rather different from those of a single ion. By irradiation of the high-energy cluster ion beam, a number of atoms are implanted simultaneously into a very small area, bringing high implantation density and large energy transfer. It is anticipated that the cluster ion applications will result in a breakthrough in the invention of new functional materials.

Cluster ion sources with different production techniques have been devised [10]. The cluster ions are produced, for instance, by vaporizing the materials using a heated oven, a laser ablation apparatus, or sputtering. Essential features of the cluster ion source are increased intensity and stability of the beam, and control of the mass distribution of the particles. Further development of cluster ion sources is in progress at several facilities [7,11].

3.3. Acceleration of Ions

Ion Accelerator

Ion beams are powerful tools for modification of specific material characteristics, investigation of material properties, and processing and creation of materials, bringing great benefit to industrial applications. Ion beams in the energy region of ~keV to several MeV have the advantage of enhancing the effectiveness of ion interaction with matter without significant damage. On the other hand, the ion beams of energy greater than several tens of MeV cause great damage to the substances along their trajectory. With larger range and higher LET in substances, the high-energy ion beams are typically utilized for investigating the radiation tolerance of materials, for creating new materials, and for generating new plant resources.

Ion accelerators can cover a wide range of energy from ~keV to ~GeV, depending entirely on the requirement of the application. A single accelerator can hardly provide various ion beams in such a broad dynamic range of energy due to limitations of the acceleration voltage, the magnetic field, beam quality, and other reasons. Practically, the required energy range may be covered by a combination of different accelerators like the accelerator complex at the TIARA facility of JAERI, Takasaki, as shown in Fig. 4 [12].

In principle, a charged particle is accelerated by traversing an electric field formed between electrodes. Acceleration methods are classified into two types: potential acceleration using static fields and sequential acceleration using alternating fields [13]. Electrostatic accelerators such as the Van de Graaff accelerator and a tandem accelerator, belonging to the former type, provide ion beams with energy of ~keV to ~MeV. The latter type of accelerators, such as the cyclotron, the linac, and the synchrotron, cover higher energy region from ~MeV to ~GeV.

Flexibility in ionic species, energy, intensity, time structure, beam spot size, etc., are the most essential characteristics common to all the accelerators used for ion beam applications. The TIARA facility is unique in its design specialized for the ion beam applications, especially to materials science and to biotechnology. A great variety of ionic species from proton through bismuth is available through four types of accelerators: a K110 AVF cyclotron, a 3-MV tandem accelerator, a 3-MV single-ended accelerator, and a 400-kV ion implanter [14,15]. The energy range covered by the four accelerators is shown in Fig. 5. A cyclotron [16] has the advantages of the diversity of accelerated particles from protons to heavy ions with mass number of greater than 200, a broad energy range from ~MeV to ~GeV, the ability to increase beam intensity up to mA, and compactness for saving space. An AVF cyclotron [17,18] has a radially increasing magnetic field to compensate for the

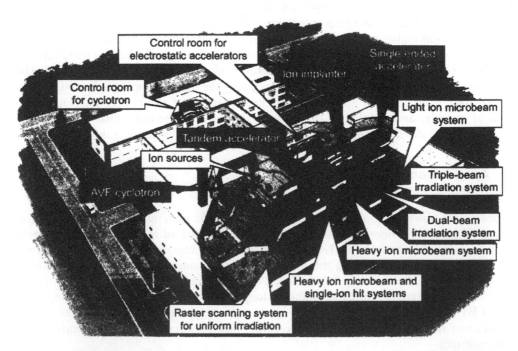

Figure 4   A bird's-eye view of the accelerator complex at the TIARA facility of JAERI, Takasaki.

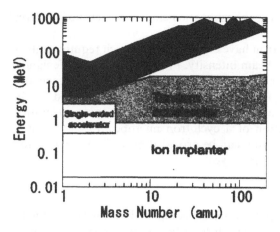

Figure 5 Energy range of the ion beams available at the TIARA facility of JAERI, Takasaki.

relativistic mass increase for isochronism, and an azimuthally varying field to provide the additional vertical focusing. A schematic drawing of the AVF cyclotron is shown in Fig. 6. Ions generated by the external ion source are axially injected at the center of the AVF cyclotron. The ions are led to the medium plane of the cyclotron magnet by the electric field generated between inflector electrodes in the shape of a spiral. The ions make revolutions periodically in the magnetic field generated by the cyclotron magnet. The revolution period of the particles is kept constant to accelerate charged particles using a fixed-rf voltage supplied to the electrodes in the frequency range of MHz. After the ions reach an extraction radius, the beam trajectory is shifted outward by using the electric field of a deflector,

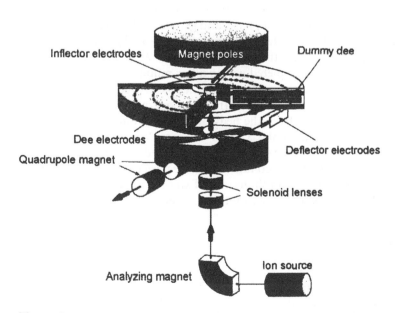

Figure 6 Schematic drawing of the ion acceleration in the AVF cyclotron.

followed by a magnetic channel for weakening the magnetic field, to extract the ions from the cyclotron.

Accelerator technologies for the cyclotron have evolved to meet such requirements of the ion beam as the increases in energy and beam intensity. Precise experiments demand a cyclotron beam of high quality, namely, high energy and time resolutions. Development of a flat-top acceleration system [19,20] and a beam cooling system [21] resulted in the minimization of the energy spread in the cyclotron beam. The flat-top acceleration technique is indispensable for the production of a cyclotron microbeam [22]. A highly stable beam required for the flat-top acceleration can be provided by a precise and flexible control of the temperature of the cyclotron magnet yoke.

Quick Change of Ion Species and Energy

Most of beam times allocated for ion beam applications are completed within 24 hr, which is comparatively shorter than that for nuclear physics. Ordinarily, ion species and energy have to be changed to collect the necessary data for investigation of dependence on LET in substances, for example. Such researches require several beam times for data acquisition, taking users almost a year or more. In case of a cyclotron, more than 2 hr of overall time is needed for reexcitation of the cyclotron magnet and optimization of the parameters of the whole systems. The change of the ion beam condition during the individual beam time is inefficient for increasing the rate of cyclotron operation. Quick change of the beam without loss of time is essential for efficient use of the beam time.

A "cocktail" beam acceleration method [23] is a smart technique, extremely useful for quickly changing ion species and energy in the cyclotron. Cocktail ions, consisting of different ion species, having almost identical mass-to-charge ratio (M/Q) produced by the ECR ion source, are simultaneously injected into the cyclotron. The cyclotron parameters are optimized for one of the cocktail ions. The selected ion is fully accelerated under the isochronous condition given by $2\pi f_{rf}/h = (Q/M)(B_0/m_0c^2)$, where f_{rf} is the rf frequency, h is an acceleration harmonic number, B_0 is the magnetic field for isochronism, m_0 is the unified atomic mass unit, and c is the speed of light. Other ions with slightly different M/Q values are gradually shifted an additional amount in rf phase, and get out of the accelerating phase of the rf cycle. Other ion species can be extracted from the cyclotron by slightly changing the frequency or the magnetic field to match the difference of the M/Q. Changing of the frequency or the magnetic field can be completed within a few minutes.

Ion beams are useful to simulate the environment in space, where semiconductor devices are exposed to high-energy heavy-ion impact. Incorrect operation of semiconductor devices such as single-event upset results from the heavy-ion irradiation. The cocktail ion families of $M/Q = 4$ and 5, available at the JAERI AVF cyclotron facility [24], are frequently utilized to investigate the tolerance of the semiconductor devices to the radiation, and to survey highly radiation-tolerant semiconductor devices appearing in the market. Efficiency of the radiation-tolerance testing for thousands of kinds of semiconductor devices has been totally improved by the cocktail acceleration technique.

3.4. Irradiation of Ions

Simultaneous Irradiation of Different Ion Beams

Structural materials for fission reactors (as well as for future fusion reactors) are being exposed to intense neutron flux for many years. In the case of fusion reactor, 14-MeV neutrons, produced by the fusion reaction of $d + t \rightarrow {}^4He + n$, induce nuclear reactions of (n, α) and (n, p), while helium and hydrogen gases are accumulated in the reactor materials. Constituent atoms of the reactor materials are recoiled by the neutrons. The neutron irradiation causes a degradation of the material properties through displacement of the constituent atoms, accumulation of the hydrogen and helium gases, and other nuclear transmutation products.

Development of advanced structural materials for blankets of the fusion reactor is particularly significant, because the efficiency and life expectancy of the blankets are limited, respectively, by the operating temperature and the radiation damage of the materials. Properties of candidate materials such as low-activation ferritic steel and silicon carbide fiber-reinforced silicon carbide (SiC/SiC) composite are being investigated under dual or triple ion beam irradiation at TIARA facility. A variety of the neutron irradiation conditions in the reactors can be simulated in a relatively short period by simultaneous irradiation of multiple ion beams. Especially, triple ion beam irradiation is extremely effective in simulating the synergistic effect of displacement damage and the helium and hydrogen accumulation in the reactor materials. It helps to develop new reactor materials.

Simulation of the neutron-induced damages using triple ion beams is schematically shown in Fig. 7. A proton and a helium ion are provided by the ion implanter and the single-ended accelerator, respectively. Heavy ions, such as iron or silicon, accelerated by the tandem accelerator, are injected into the target simultaneously. For example, the SiC/SiC composite was tested under triple ion beam irradiation consisting of a 380-keV proton, a 1.2-MeV helium ion, and a 7.8-MeV Si^{3+} ion. The triple irradiation system is equipped with an energy degrader and a beam scanner for uniform three-dimensional (3-D) irradiation.

Uniform Irradiation by Beam Scanning

A uniform-irradiation technique is commonly used for ion beam applications in materials science and biotechnology, and for biomedical application such as cancer therapy. Uniformity of the irradiated-particle density distribution is essential to bring about the same

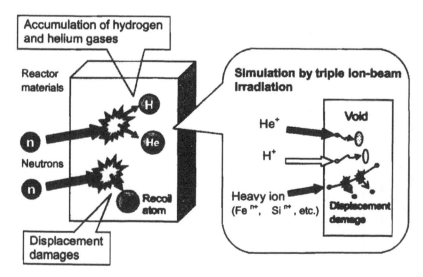

Figure 7 Schematic drawing of the damage in reactor materials induced by neutrons, and its simulation by the triple-ion-beam irradiation.

irradiation effects in a substance with a large area or with a number of target samples aligned in several lines. A relative dose variation of at most $\pm 10\%$ is maximally tolerable for the uniform irradiation. Several uniform-irradiation systems have been developed so far to meet individual demands of ion beam applications. The most economical method for uniform irradiation is a double-scattering beam delivery using two sets of scatterers [25]. In this system, we should pay attention to the energy loss in the scatterers, a change of a charge state of the ions, activation of the scatterer material, and production of nuclear fragments.

Formation of a beam spot with a homogeneous profile by using multipole magnetic fields is a good way to flatten the particle distribution [26]. Multipole fields, such as octupole, dodecapole, and 16-pole, generate a stronger focusing force for the particles traveling away from the axis of the multipole, while the focusing force acting on the particles close to the multipole axis is comparatively weak. The final particle distribution is highly dependent on initial beam size and angular spread. A very careful designing of the multipole magnet and the beam transport system is required to obtain a uniform particle distribution with a hard edge at an irradiation point.

A 2-D beam scanning system equipped with a set of magnets or electrostatic plates to deflect the beam horizontally and vertically is commonly used for uniform irradiation over a large area, e.g., $>10 \times 10$ cm^2. A schematic drawing of the 2-D beam scanning method is shown in Fig. 8. The scanning area may be varied by changing the excitation currents of the

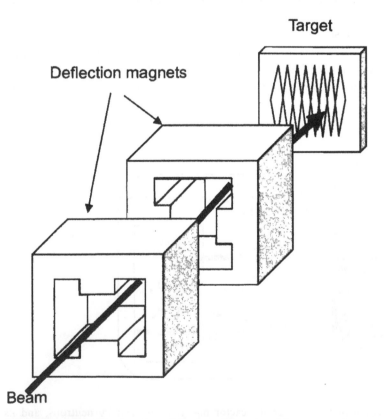

Figure 8 Schematic drawing of the two-dimensional beam scanning system using a set of magnets to deflect the beam horizontally and vertically.

deflection magnet or the voltages supplied to the electrostatic plates. By using a raster beam scanning system [27], a large rectangular area can be irradiated with continuously delivered particles. The beam-sweeping speed and the spacing between neighboring trajectories are kept constant. Uniformity of the particle distribution is determined by beam spot size and trajectory spacing. The raster beam scanning system is utilized for estimating damage in solar cells from exposure to high-energy protons and for inducing mutation in plants by heavy ions.

A wobbler beam delivery system [28], another 2-D beam scanning system, is used typically for biomedical applications. A beam is circularly wobbled at several different radii to form a uniform dose distribution in an area up to 30 cm in diameter. The beam-sweeping speed at the outer wobble radii increases, since the horizontal and vertical deflection frequencies are constant. When beam intensity remains constant, longer irradiation time makes up for a deficiency of particle density in outer wobbles to achieve a uniform radiation field. A continuous uniform irradiation over a round area can be realized by a spiral beam scanning method [29]. The beam-sweeping speed is kept constant even for larger radii, while the trajectory spacing in a radial direction is invariable. Precise control of the irradiation time is unnecessary for the spiral beam scanning system.

3.5. Microbeam

Superiority of the Microbeam

Spatial and targeting resolutions of the ion irradiation in a finite area are greatly enhanced by using a microbeam with a beam spot size of 1 µm in diameter. More precise microanalysis of elements included in substances is feasible by combining the nuclear microprobe with analytical techniques like PIXE, NRA, and ion channeling [30]. In recent years, the micro-analysis technique has become widespread in the field of life sciences [31]. The microbeam has also been applied to material processing and to elucidation of radiation effects caused by high-LET irradiation such as single-event upset of semiconductor devices used in space.

A high-energy ion has higher LET in substances than do electrons or gamma rays with the same energy. A heavy-ion microbeam with energy of hundreds of MeV is extremely useful for research in biology and biotechnology as a fine probe for investigation of cell response and as a specific tool for cell surgery. A single ion can cause great damage locally to a part of the DNA without destroying the whole cell. The microbeam technology is indispensable for research in the state-of-the-art bioscience, for example, in the investigation of cell-to-cell communications like bystander effects, the analysis of cellular spatial sensitivity, the interaction of damages caused by individual irradiation, the cellular repair dynamics, and in such an intracellular process as apoptosis.

Production of the Microbeam

The first microbeam was generated by using a Russian quadruplet lens configuration in the early 1970s at Harwell in Britain [32]. In recent years, microbeam technology has made remarkable progress to fulfill the requirements for several kinds of microprobe analyses. Nowadays, more than 200 facilities, at which microbeams with energy greater than several hundreds of keV/nucleon can be provided, are available worldwide for a variety of research fields.

Beam collimation using a microaperture foil with a hole of 5 to 10 µm in diameter or a precisely manufactured slit is an easy method for the microbeam production [33,34]. The minimum beam spot size was limited to around 5 µm due to the size of the microaperture and contamination of particles scattered at the edge of the microaperture.

Beam focusing using a multiplet of quadrupole lenses [30,31] is a sophisticated technique to reduce the beam size to less than 5 μm. A schematic diagram of the focusing system for microbeam production is shown in Fig. 9. The object of the primary beam is limited by the first slit, followed by a series of second slits used for divergence defining and for beam collimation. An image of these slits can be projected at a focal plane by means of the focusing lenses. The microbeam size is determined by the demagnification factor of the lens system. Spherical and chromatic aberrations in the lens system are also taken into account to minimize the beam size on the basis of the classical optical theory for charged particles [35].

Microbeam Systems at the TIARA Facility

Three microbeam systems were developed at the TIARA facility for application to materials science and biotechnology. A heavy-ion microbeam system installed on a beam line of the 3-MV tandem accelerator is the first one developed to study single-event upset (SEU) of semiconductor devices used for space [36]. The microbeam system can focus heavy-ion beams such as a 15-MeV nickel ion with a spot size of less than 1 μm. In order to observe the SEU phenomena at a specific position of the microdevice, the microbeam system is equipped with a single-ion hit system, consisting of single-ion detectors and a fast beam switcher.

A light-ion microbeam system connected with the 3-MV single-ended accelerator was developed for high-resolution ion beam microanalysis [37]. The highest spatial resolution of 0.25 μm was achieved for 2-MeV proton and helium ions. The beam spot size was estimated from the intensity distribution of the secondary electrons emitted from a silicon relief pattern irradiated with the 2-MeV helium ion microbeam as shown in Fig. 10.

PIXE analysis using the microbeam has an overwhelming advantage in visualizing very small quantities of elements with a very high sensitivity. For example, 2-D distribution of small amounts of elements included in cells can be obtained by the micro-PIXE technique. An in-air micro-PIXE system using a light-ion microbeam was developed at the 3-MV single-ended accelerator facility [38]. A schematic diagram of the in-air micro-PIXE system is shown in Fig. 11. A proton microbeam with spatial and targeting resolutions of 1 μm penetrates a 4-μm-thick Mylar film, used as a sample backing and a vacuum partition,

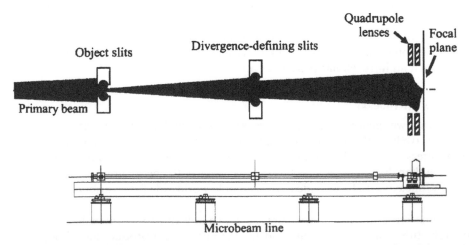

Figure 9   Schematic diagram of the focusing system for the microbeam formation.

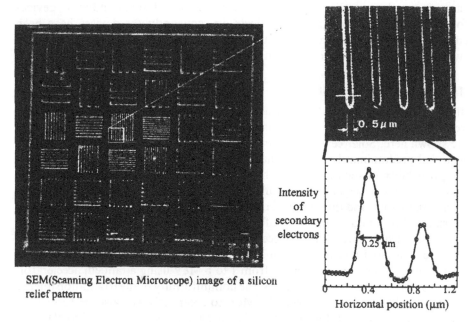

SEM(Scanning Electron Microscope) image of a silicon
relief pattern

Intensity
of
secondary
electrons

Horizontal position (μm)

Figure 10 Estimation of the spot size of the 2-MeV helium ion microbeam from the image of a
silicon relief pattern obtained by detecting secondary electrons.

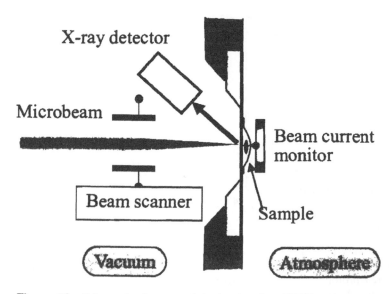

Figure 11 Schematic diagram of the in-air micro-PIXE system.

been examining and characterizing the radiation resistance of semiconductor devices intended for space applications. Radiation effects on semiconductor devices have been investigated to clarify the mechanisms of radiation-induced degradation, and as a result, to propose a way for developing better radiation-hard devices.

Radiation Degradation of Solar Cells

For an example of evaluating radiation degradation of semiconductor devices, the irradiation test results of silicon (Si) and gallium arsenide (GaAs) solar cells using protons and electrons are shown here. Solar cells, which are the main power sources for artificial satellites, are degraded predominantly by high-energy protons and electrons in space because the existence probability of such particles is high in their mission trajectories such as the geostationary orbit. Their degradation results from the introduction of displacement damage in semiconductors, and thus, control of irradiation parameters of charged particles such as uniformity of beam flux is important for detailed analysis of radiation tolerance of solar cells. In the tests using a proton beam accelerated at 10 MeV with AVF cyclotron, solar cell samples were irradiated uniformly in an area of 10×10 cm^2 with a magnetic beam scanner. In irradiation of protons at energies from 1 to 6 MeV using a tandem accelerator, the beam scan area was 4×4 cm^2. In both cases, the nonuniformity of the irradiation fluence was kept within 3%. Electron irradiation of 1 MeV to solar cells was also performed by using Cockcraft-Walton-type or Dynamitron Cascade-type accelerator at JAERI. The performance of solar cells was examined under illumination of pseudosunlight at zero air mass (AM0) condition before and after irradiation, or during irradiation by using simultaneous irradiation technique of charged particles and pseudosunlight [39].

Figure 13 shows typical results of the degradation of crystalline Si solar cells having a back surface field and reflector structure (Si-BSFR), which were qualified by National Space Development Agency of Japan (NASDA) for space usage, when irradiated by 10-MeV protons and 1-MeV electrons. The pn junction of the cell samples, with a size of 2 cm $\times$ 2 cm $\times$ 50 μm, was fabricated by phosphorus (P) doping to a depth of 0.15 μm into boron

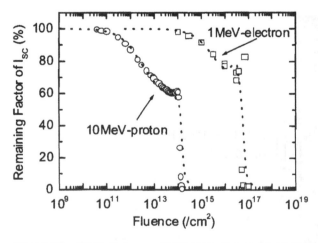

Figure 13 Dependence on 10-MeV-proton and 1-MeV-electron fluence of the remaining factor of short circuit current (I_{sc}) for Si-BSFR solar cells. Circles and squares represent the results for samples irradiated with 10-MeV protons and 1-MeV electrons, respectively. Broken lines represent the fitting results based on a model mentioned in the text.

(B)-doped, p-type substrate with a resistivity of 10 Ω cm. As shown in Fig. 13, which is a figure of the electrical performance of solar cells, normalized short-circuit current I_{sc} decreased with increasing fluence up to 1×10^{13}/cm^2 for proton irradiation and 1×10^{16}/cm^2 for electron irradiation. The logarithmic degradation observed in such fluence ranges is explained by the conventional model, i.e., a decrease in the minority carrier lifetime [40]. Over these fluences, we found an anomalous change in I_{sc}, i.e., I_{sc} recovered about 5% to 10%, and abruptly dropped to nearly zero at proton fluences around 1×10^{14}/cm^2 and electron fluences around 1×10^{17}/cm^2. The conventional model cannot account for this anomalous phenomenon. Based on an analysis of all the results obtained, the following model was proposed: The hole concentration in the p-type substrate decreases owing to the introduction of lattice defects, which act as hole traps. It raises the Fermi level of the substrate corresponding to a change in the electrical properties of the substrate from p-type to intrinsic. Thus, the depletion layer extends toward the backside of the solar cell, which enhances the efficiency of carrier collection and consequently the I_{sc} recovers. When the cells are irradiated further, minority-carrier mobility in the depletion region decreases and thus a rapid decrease in drift length takes place. As the drift length becomes shorter than the width of the depletion layer, electrons generated by light illumination do not reach the pn junction, leading to $I_{sc} = 0$. Simulation results using the model are also shown by broken lines in Fig. 13, indicating that the experimental data are well described by the proposed model [39].

Based on the model described above, the degradation of Si solar cells equipped in a Japanese artificial satellite ETS-6 has been predicted. The result is illustrated in Fig. 14. Actual flight data are also plotted in the figure. It should be noted that the obtained data are in good agreement with prediction, indicating high reliability of the method. The results obtained in this research have been summarized in *Handbook of Si Solar Cells for Space Application* [41], which is used as an important database on space Si solar cells developed in Japan. In addition, both junction-type Si solar cells with high radiation resistance have been developed from these results [42].

GaAs solar cells have also been evaluated at TIARA because in 1998 the artificial satellite COMETS, which was equipped with similar cells, orbited across a strong ionization

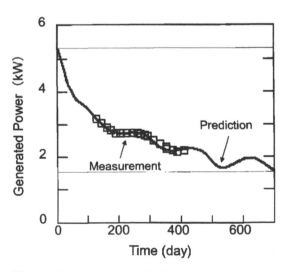

Figure 14 Prediction (solid line) and actual flight data (squares) of the degradation of Si solar cells equipped in ETS-6.

radiation environment called the van Allen belt due to its incorrect positioning. The characterization of proton and electron irradiated cells has been used to predict the total lifetime of the COMETS satellite. This has contributed to a revision of the COMETS initial mission program. On the basis of those experiences, at present we have been studying radiation effects on new types of solar cells, such as InGaP/GaAs/Ge triple-junction solar cells. It was found from those investigations that the improvement of radiation resistance of GaAs middle cells is a key issue for developing higher radiation resistant InGaP/GaAs/Ge solar cells [43].

Radiation-Induced Malfunction in Electronic Devices

In contrast to the radiation degradation of solar cells, malfunction called SEU is caused by irradiation of energetic particles into large-scale integrated circuits (LSIs) such as highly integrated memory devices. Because a malfunction of electronic devices is regarded as one of the most serious problems in space, the evaluation of single-event tolerance of LSIs is indispensable for their space applications. It is well known that SEUs are triggered by the generation of a high concentration of charge (highly dense electron–hole plasma) in semiconductors due to impingement of a heavy ion. The probability of SEU depends on the LET of an ion in a device substrate, typically Si, because the value of charge generated in semiconductors is proportional to LET. For accurate examination of SEUs taking place in an LSI by a single ion, a special apparatus has been designed and installed in TIARA. By using the apparatus, high-energy heavy ions accelerated by AVF cyclotron irradiate sample devices at low fluence rates, e.g., several hundreds of ions per second. We also used cocktail beams developed in TIARA to make SEU tests efficiently. Details of cocktail ion beams have been described in the previous section.

Typical results obtained for the 256-kbit static random access memories (SRAMs), which were developed for space application by NASDA, are shown in Fig. 15. The SEU cross section increased with LET in a range above a threshold value and reached a constant

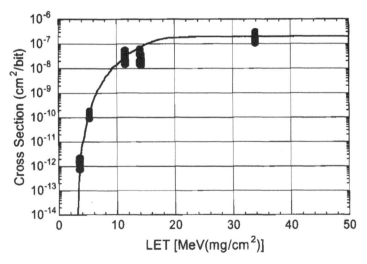

Figure 15 Single-event upset cross section for 256-kbit SRAM as a function of LET. Closed circles and solid line represent experimental and fitting results, respectively.

value called the saturated cross section. The threshold LET and the saturated cross section can be related to the critical charge necessary to cause SEU and to the size of memory cells, respectively. It was determined from those tests that the threshold LET and the saturated cross section were 6.4 MeV/mg/cm² and 3.2×10^{-8} cm², respectively, for the 256-kbit SRAM. Here the threshold LET is defined as the LET value corresponding to one hundredth of the saturated cross section. Using these values, the SEU tolerance for the devices in space have been estimated. As a result, it was deduced that SEU occurs approximately once a week in the geostationary orbit, showing its strong radiation resistance sufficient for space application. Based on results obtained at TIARA, radiation-hard devices like 1M gate arrays have been newly developed for components used in artificial satellites [44].

Because SEU occurs when the amount of electric charge collected in a memory cell exceeds a critical value, an understanding of the transportation of charges induced in memory devices by ion irradiation is indispensable to clarify the mechanism of SEU. For this purpose, a microbeam irradiation system was designed and installed in a beam line connected to the tandem accelerator. Details of the microbeam system have been described in the previous section. To measure transient currents in several nanoseconds, a wide-bandwidth measurement system with a charge sensitivity of 20 fC and a temporal resolution of 10 ps was developed. The developed system is schematically illustrated in Fig. 16. Using the system, charge collection behaviors for several kinds of devices such as Si pn junction diodes, GaAs Schottky diodes, etc., have been examined. Typical results obtained for silicon-on-insulator (SOI) devices are described below [45].

For the development of SEU-hardened memory devices, it is expedient to reduce charge collected in a memory cell. For this purpose, the formation of buried oxide in device structures, i.e., the fabrication of SOI structure, is considered a useful method because such a buried oxide layer can be expected to suppress the charge collection due to the drift and funneling processes. However, no experimental approach had been made for the charge collection in SOI devices. To investigate the charge collection in the SOI structure, transient currents induced in SOI pn junctions by heavy ions such as 15-MeV carbon (C) or oxygen (O) ions have been measured.

The samples were pn junction diodes formed on an SOI wafer fabricated by wafer bonding technique. The thicknesses of the n-type top Si layer with resistivity of 2–4 Ω cm, the oxide layer, and the n-type Si substrate with resistivity of 1–50 Ω cm are 5.7, 0.48, and 630 μm, respectively. The p region, which is 50 μm in diameter and 0.5 μm in depth, was

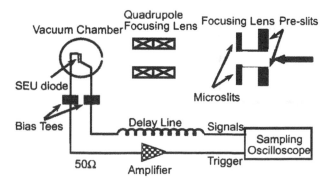

Figure 16 Schematic drawing of the system developed for transient current measurement using heavy-ion microbeam.

formed by diffusion of B. The sample diodes were mounted on a chip carrier with 50-Ω double-ended microstriplines. They were irradiated with 15-MeV C or 15-MeV O ions using a heavy-ion microbeam, and transient currents induced in the sample diodes by ion irradiation were monitored using a 50-GHz wideband sampling oscilloscope and a superconducting delay line, which were connected with the p contacts.

Figure 17 shows the transient currents induced in the SOI pn junction diode irradiated with 15-MeV C ions when the reverse bias applied to the junction was 10 V. The result obtained for the bulk Si pn junction diode when the reverse bias was 10 V is also displayed in Fig. 17 for comparison. The values of currents obtained for the SOI sample were shown to be lower than those for the bulk Si diode. The amount of charge collected in the p region can be derived from time integration of the obtained transient currents. For the SOI diode, the total amount of collected charges was estimated to be 210 fC, which was approximately one half of that obtained for the bulk Si diode (470 fC). The low values of collected charges in the SOI pn junction can be interpreted as the only charges generated in the top Si layer reach to the p region, i.e., the buried oxide suspends the charge conduction arising from the tunneling effect.

The amount of positive charge created in a depletion region can be estimated theoretically by using the length of a depletion layer, the value of LET for incident ions, and the ion energy necessary to produce an electron–hole pair in Si (3.6 eV). Assuming that all positive charges generated in the top Si layer of the SOI device by ion irradiation were transferred to the p contact, the amount of collected charge was estimated to be 260 fC. The total amount of collected charge obtained from transient currents in the SOI device was comparable with the estimated value. It indicates that the charge collection in the SOI pn junction is limited in the top Si layer, whereas the penetration depth of 15-MeV C ions (about 14 μm) is much deeper than the top Si layer. Similar results were also obtained for 15-MeV O ion irradiation. All the above findings demonstrate that the SOI structure has a strong SEU resistance, and therefore they are suitable for the application to electronic components used in space.

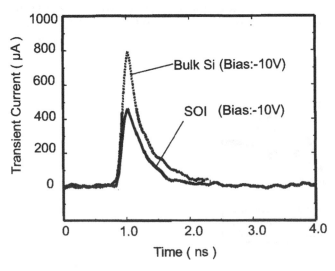

Figure 17   Transient current induced in SOI pn junction diode by 15-MeV C-ion irradiation (solid line). The result obtained for bulk Si pn diode is also shown as a dotted line in the figure for comparison. The reverse bias of 10 V was applied to pn diodes during measurements.

Development of Radiation-Resistant SiC Devices

Silicon carbide is regarded as a promising material for electronic devices used in space due to its outstanding electrical properties [46] and excellent radiation resistance [47]. For its application to highly integrated devices, it is indispensable to establish a selective doping technique of electrically active (donor and acceptor) impurities. Because diffusion technique cannot be employed due to the extremely low diffusion coefficients of impurities in SiC, ion implantation is the most promising way to dope impurities in selective regions of SiC.

For the electrical activation of implanted impurities, removal of defects by post-implantation annealing is required. Whereas doping of high concentration of donors or acceptors is necessary to produce low resistive regions, amorphous SiC layers subjected to high-dose implantation at room temperature (RT) can hardly be recovered by postimplantation annealing. In this case, hot implantation is expected to be effective for avoiding amorphization of ion-implanted SiC, because induced defects are partially removed during implantation. The influence of hot implantation on the electrical activation of P donors in SiC have been investigated [48]. Fourfold implantation of P ions at 80, 100, 150, and 200 keV was performed for p-type SiC with a net acceptor concentration of $4 \times 10^{15}/cm^3$ to form a box profile of P atoms with a mean concentration of 1×10^{18} to $5 \times 10^{19}/cm^3$ and a depth 0.25 μm. Implantation temperature was altered in the range from RT to 1200°C. After implantation, the samples were annealed in an Ar ambient up to 1500°C for 20 min. Ohmic contacts were formed by Al deposition on the P-implanted layers. The electrical properties of the samples were examined by Hall measurements using van der Pauw method.

Figure 18 shows the carrier (electron) concentration in SiC samples implanted with P at $5 \times 10^{19}/cm^3$ at RT or 1200°C as a function of annealing temperature. The carrier concentrations measured at RT are plotted in this figure. The carrier concentration for both samples increased with annealing temperature. This indicates an increase of electrical activation of P and a decrease of residual defects due to annealing. As for the comparison between RT and 1200°C implantations, samples implanted at RT showed lower carrier concentration as compared to samples implanted at 1200°C at every annealing temperature. Since, in this sample, an amorphous layer was produced by implantation, such a highly defective layer is thought to remain even after 1500°C annealing, which causes poor elec-

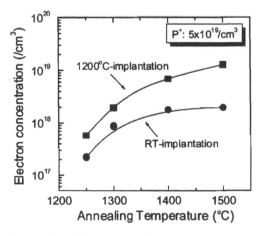

Figure 18 Free carrier (electron) concentration in SiC samples implanted with P ions at RT (circles) and 1200°C (squares) as a function of annealing temperature. Annealing was performed for 20 min in Ar atmosphere. The electron concentration was obtained from Hall effect measurement at RT.

trical activation of implanted P atoms. The relationship between implanted P concentration and implantation temperature has been investigated to clarify optimum condition of P doping. It was found that the effects on hot implantation enhanced with increasing implanted P concentration, and the optimum implantation temperature depends on implanted P concentration [48]. For example, for high-dose P implantation, the highest carrier concentration was obtained in samples implanted at 800°C. In addition, from positron annihilation spectroscopy (PAS) study, this result can be explained in terms of the migration and formation of vacancy defects due to annealing [49].

In fabrication of p-type SiC, doping of acceptors such as B and aluminum (Al) is one of the key issues because their electrical activation ratio (atoms activated electrically/ implanted atoms) was not high enough even when hot-implantation technique was adopted. It was reported that B and Al atoms located at Si sublattice sites act as shallow acceptors in SiC [50]. In addition, the incorporation of B and Al in SiC could be controlled by the change in Si/C ratio during epitaxial growth of SiC [51]. From these findings, the introduction of additional C atoms by implantation, called C coimplantation, is expected to enhance the electrical activation of acceptors. The effects of C coimplantation on the electrical activation of Al and B acceptors in SiC have been investigated. It was found that the electrical properties of B- and A-implanted SiC are improved by C co-implantation [52,53]. In this study, n-type 4H-SiC samples with a net donor concentration of $5.3 \times 10^{15}/\text{cm}^3$ were used. The samples were implanted with B ions at RT. A box profile of B with a mean concentration of $5 \times 10^{18}/\text{cm}^3$ and a depth of 0.5 μm was achieved by conducting a fivefold implantation. Prior to the implantation of B, a fivefold implantation of C or Si ions was performed at RT or 800°C to form a similar box profile to that of B. After implantation, samples were annealed at 1630°C for 30 min in a SiC crucible in pure Ar ambient. Ohmic contacts were formed using the deposition of Al and subsequent sintering at 950°C for 3 min in vacuum. Hall measurements were carried out in a temperature range from 180 to 700 K using a van der Pauw arrangement.

Figure 19 shows the free hole concentration as a function of reciprocal temperature for the samples implanted with B, co-implanted with C and B (C/B) and co-implanted with Si

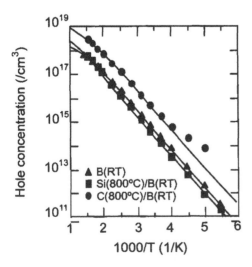

Figure 19   Free hole concentration as a function of reciprocal temperature for B- (triangles), C/B- (circles), and C/Si-implanted (squares) SiC. Solid lines represent the fitting results.

and B (Si/B). The ionization energy was determined from the slope of these curves (solid lines in the figure) to be equal to 285 meV, which agrees with the B ionization energy in 4H-SiC [54]. This indicates that implanted B species are activated electrically in all these samples. The free hole concentration was found to increase by the C coimplantation and to decrease by the Si coimplantaion. The hole concentration at RT in samples coimplanted with C at 800°C and B was approximately 7.1 times as high as that in the sample implanted with only B. Considering that hole concentration at RT in samples coimplanted with C at RT and B was approximately 4.1 times as high as that in the sample implanted with only B [52]; the results obtained indicate that hot coimplantation of C is quite effective for improving the electrical properties of B acceptors. A similar increase in the free hole concentration due to C coimplantation was also obtained for Al acceptors in SiC [53,55]. It was also observed that the intensity of the PL peak at 383.9 nm, which arises from shallow B acceptors, increased by C coimplantation. In addition, the value of the D_{II} center, which is related to C interstitials, increased by coimplantation of C. These results are interpreted in terms of a site competition effect and/or an alternation of the degree of compensation. In conclusion, the C coimplantation technique is quite effective to form p-type layers in SiC.

Oxidation for forming insulator on SiC is also one of the key issues for the fabrication process of SiC devices. The relationship between the oxidation process and the electrical characteristics of SiC metal–oxide–semiconductor field effect transistors (MOSFETs) have been studied [56,57]. A photograph of the enhancement-type SiC MOSFETs fabricated in this study is shown in Fig. 20. The source and drain regions of the MOSFETs were formed using P hot implantation mentioned above. In this study, oxide on SiC was formed by pyrogenic oxidation (a wet oxidation process using direct reaction between hydrogen and oxygen gases) and subsequent thermal treatment in steam or H_2 atmosphere. Details of the fabrication process of the MOSFET are described in Ref. [57]. From this study, it was found that the postoxidation annealing methods can improve the channel mobility of SiC MOSFETs, which is one of the most important characteristics, although the obtained value was only 30% of the ideal value. Furthermore, the radiation response of the SiC MOSFETs has been examined. It was found that SiC MOSFETs showed approximately 100 times higher radiation resistance than Si MOSFETs. [58,59].

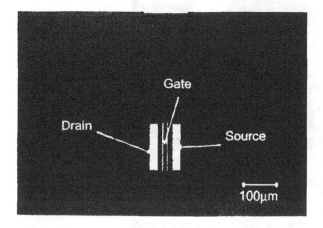

Figure 20 Photograph of the enhancement-type SiC MOSFET.

4.2. Characterization of Radiation Degradation of Nuclear Materials

The investigation of radiation degradation of materials used in nuclear fusion reactors as well as fission reactors is indispensable for their safety and long-term operation with a high reliability. In fusion reactors, structural materials, solid tritium breeder materials, and insulating ceramic materials will be exposed to fast (14 MeV) neutrons generated via DT reactions in high-density plasma. When reactor materials are irradiated with such fast neutrons, heavy displacement damage as well as He and H atoms, which are produced by transmutation reactions like (n, p) and (n) are introduced into the materials.

The Development of Triple-Beam Irradiation Apparatus

In order to develop materials for fusion reactors, it is important to investigate synergistic effects of displacement damage and transmutation-induced gas products on their physical and mechanical properties. For this purpose, a triple-beam irradiation apparatus has been designed and installed at TIARA. A photograph of the triple-beam irradiation apparatus is shown in Fig. 21. Using the apparatus connected with a 3-MV tandem accelerator, a 3-MV single-end accelerator, and a 400-kV ion implanter, three species of ions, e.g., heavy ions, He, and H can be simultaneously irradiated into samples for simulating the radiation environment of a fusion reactor. Each beam line has an angle of 15° from the other beam line, and three kinds of beams are focused on the target sample in a vacuum chamber. Besides, for wide-area irradiation with high uniformity, a beam scanning system is also installed in each line. Since Al foils are installed for energy degradation in the irradiation chamber, the penetration depth of irradiated ions can be adjusted. The temperature of samples is also controllable from 80 to 1300 K. Details of the apparatus were described elsewhere [60].

Austenitic Steel, Ferritic/Martensitic Steel, and Vanadium Alloy

Using the triple-ion beam irradiation apparatus, the microstructural evolution of austenitic stainless steel, which is considered as a structural material for water-cooled fusion reactors

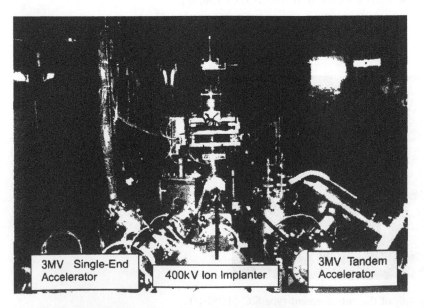

Figure 21 Photograph of the triple-beam irradiation apparatus.

such as the international thermonuclear experimental reactor (ITER) has been investigated [61,62]. The austenitic stainless steel samples with a composition of Fe–16Ni–14Cr–0.25Ti–2.5Mo in wt.% were irradiated simultaneously with 12-MeV Ni, 1-MeV He, and 350-keV H ions in a temperature range between 300° and 400°C. The irradiated samples were observed by transmission electron microscopy (TEM), and the number density of dislocation loops in the irradiated region was studied.

Figure 22 shows the depth distribution of the number density of dislocation loops for the sample irradiated with 12-MeV Ni, 1-MeV He, and 350-keV H triple-ion beams. The number density of dislocation loops decreases in a region between 1.0 and 1.5 μm. It was reported that hydrogen interacts with dislocation in steels at a relatively low temperature [63]. Considering that the estimated projection range of He and H using TRIM is around 1.5 μm, the results obtained in this study suggest that the generation of dislocation loops is suppressed by the interaction of defects and He and/or H. In addition, very small cavities whose average diameter and density were 1.4 nm and $1.5 \times 10^{24}/cm^3$ respectively were found in the triple-beam-irradiated region. Since such high-density small cavities were not observed in a region irradiated with only 12-MeV Ni ions, the creation of high-density small cavities is thought to be an effect of simultaneous irradiation of Ni, He, and H. Sekimura et al. [64] reported that high-density small cavities (diameter: 10 nm, density: $10^{21}/cm^3$) were created in austenitic stainless steel samples implanted with dual beams of Al and He ions. Therefore, the creation of small cavities is one of the synergistic effects of displacement damage and transmutation-induced gas products. The growth of the cavities in F82 was also obtained in the region irradiated with triple beams. The growth strongly depends on heat and mechanical treatment process [65]. Since such a growth relates to microhardness and leads to swelling of the steel, this finding is considered very important for the predicted lifetime of steels. Besides, the remarkable swelling for pure vanadium (V) and V alloys, which are also candidates for fusion materials due to triple-beam irradiation, was found [66]. Because this swelling could not be attributed to single-or and dual-beam irradiation, this also can be concluded to be "synergistic effects."

SiC/SiC Composites and Other Materials

Ceramic matrix composites are expected to be applied to structural materials for fusion reactors because of their excellent mechanical properties at high temperature and their

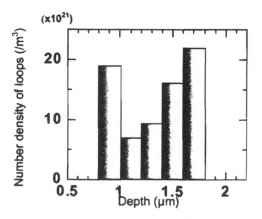

Figure 22 Depth distribution of the number density of dislocation loops for austenitic stainless steels irradiated with 12-MeV Ni, 1-MeV He, and 350-keV H triple beams.

noncatastrophic failure behavior. SiC/SiC composite materials are regarded as one of the attractive candidates for fusion materials. In addition to the excellent mechanical properties as ceramics, SiC/SiC materials have an advantage of low residual radioactivity. The effects of triple-beam irradiation on SiC/SiC materials are described below. SiC-fiber-reinforced SiC-matrix (SiCf/SiC) composites and a monolithic SiC (β-SiC) were used in this study [67]. The samples were irradiated with triple beams (6-MeV Si^{2+}, 1.2-MeV He^+ and 250-keV H^+ ions) at 1000°C. As a result of TEM observation, He bubbles and cracks were found in SiC matrix, and the number of He bubbles and cracks strongly depended on the fabrication process of SiC/SiC composites. The size of the bubbles also depended on the fabrication process of SiC/SiC composites. The turbostratic carbon structure was formed in the carbon interphase layer due to irradiation, although the interphase layer was the nongraphitic carbon structure before irradiation. Furthermore, a new crystalline SiC phase was observed in the carbon interphase irradiated with Si ions. Details of irradiation effects on SiC/SiC composites are described in Refs. [67–69].

In addition to SiC/SiC, irradiation effects on insulators such as Al_2O_3 were also examined [70]. As a result, synergistic effects such as the diffusion of H and the growth of cavities were observed in samples irradiated with triple beams. As for carbon fiber composites, the microhardness increased by triple-beam irradiation [71], which was a different result from that obtained from 14-MeV neutron irradiation (8×10^{19} n/cm^2) experiment.

4.3. Beam Analysis and Modification of Inorganic Materials

Applying ion beams, surface-sensitive analysis and modification in atomic and electronic structures of inorganic materials have been developed. Ion beam modification of titanium dioxide (TiO_2), carbon-based materials, and the analysis of Nb/Cu multilayers and VO_2 using ion beam are described as follows.

Modification of the Characteristics of TiO_2

Titanium dioxide has many attractive characteristics for photoactive applications, such as photocatalytic coatings and dye-sensitized solar cells. However, its large optical bandgap (3.0–3.2 eV) is a disadvantage taking into account the utilization of sunlight. Doping TiO_2 with different impurities using ion implantation is known to provide an effective modification of its electronic structures, thereby improving the separation of the photogenerated electron–hole pairs and/or extending the wavelength range of the TiO_2 photoresponse into the visible region [72]. Therefore, the effects of fluorine (F) and chromium (Cr) implantations on the characteristics of TiO_2 have been taken up [73–75].

First, the results of F implantation into TiO_2 will be described. Single-crystalline TiO_2 rutile with a $\langle 001 \rangle$ crystallographic axis was implanted with 200-keV F ions at a fluence range from 1×10^{16} to $1 \times 10^{17}/cm^2$ and subsequently annealed in air. The radiation damage and its annealing behavior were investigated by Rutherford backscattering spectrometry in channeling geometry (RBS/C) and PAS. X-ray photoelectron spectroscopy measurements confirmed the occupation of F atoms on O-lattice sites in the outermost region of the implanted surface layer. Secondary ion mass spectrometry (SIMS) was employed for probing F distribution in the as-implanted and post-annealed samples. The electronic structures of the F-doped TiO_2 were also evaluated by theoretical calculations to predict the doping effect on the spectral response of TiO_2 [74].

The disorder peaks in RBS/C spectra lessened during isochronal thermal treatment. This can be attributed to a columnar regrowth within the total thickness of the damaged layer [76]. The pronounced recovery in the Ti and O sublattices was achieved by such a recrystallization process due to annealing up to 1200°C. PAS study revealed that the crystal structure was recovered by the migration of vacancy-type defects to the surface. According to the SIMS results shown in Fig. 23, the F depth profile was shifted to a shallower region along with the damage recovery and this resulted in the formation of the F-doped layer where the impurity concentration steadily increased toward the surface. Importantly, this method enabled F atoms to be introduced in a controlled manner at specific locations to realize impurity concentration gradients in TiO_2. From the ab initio band calculations, it was found that the F doping caused a modification of the electronic structure around the lower edge of the conduction band. This probably leads to a reduction in the effective bandgap energy, in other words, to an improved optical response in the visible-light region. The fluorination of TiO_2 surfaces would enhance their chemical and optical stabilities and open avenues toward photoelectronic materials with various applications. In this sense, F ions have the possibility of acting as promising dopants in a TiO_2 photocatalyst.

The surface charge separation efficiency was characterized using photo-induced transient charge separation (PITCS) measurement proposed by JAERI [77]. By this technique, the total number of photocarriers transported toward the surface can be estimated as a function of illuminated photon energy without disturbing the spontaneous surface band bending, because of contactless observation under no-applied bias. The PITCS measurements were carried out for studying hole generation in TiO_2. The samples used in this study were TiO_2 rutile films. The samples were implanted with Cr ions at 150 keV at a dose of 5.0×10^{15} ions/cm^2 and subsequently annealed at 600°C for 5 hr in air. For comparison, uniformly Cr-doped TiO_2 rutile films, highly oriented undoped TiO_2 anatase, and TiO_2 rutile films were also examined. Figure 24 shows the incident-photon-energy dependence of the surface charge separation efficiency of the samples [75]. The vertical axis represents the ratios of the number of surface transported holes (h$^+$) to the number of induced photons. An increase in charge separation efficiency for samples implanted with Cr ions, and subsequently annealed, was observed, especially around 350 nm, as compared to other samples. The total number of h$^+$ for Cr implanted film was estimated to be approximately 15 times higher than that for TiO_2 anatase. The obtained result suggests that the charge separation properties are improved by making a gradient of Cr distribution.

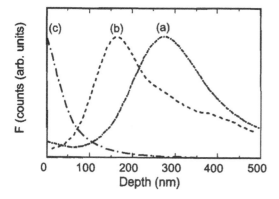

Figure 23 Depth profile of F in TiO_2 rutile single crystals implanted with 200-keV F ions. (a) As-implanted, (b) 600°C-annealed, and (c) 1200°C-annealed samples.

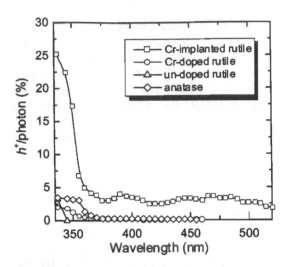

Figure 24 Incident photon energy dependence of the surface charge separation efficiency ($h^+/$ photon) measured by PITCS. Squares, circles, triangles, and diamonds represent the results obtained for Cr-implanted rutile, Cr-doped rutile, undoped rutile, and anatase TiO_2 films, respectively.

Carbon-Based Materials

In the research and development of inorganic materials, the formation mechanism and thermal stability of carbon-based materials have been intensively investigated. The formation processes of carbon onions (graphitic spheres) and nanocapsules in copper and gold have been elucidated through simultaneous TEM observation under C-ion irradiation [78]. This apparatus consists of two sets of accelerators and a TEM. One of the accelerators is a 400-kV ion implanter at TIARA, which can accelerate more than 30 kinds of elements including high-melting-point metals. The other one is a duoplasmatron ion source with accelerating voltage of 40 kV. In this apparatus, the following techniques are available as well as conventional transmission electron microscopy: high-resolution electron microscopy, scanning electron microscopy, scanning transmission electron microscopy, CCD camera and video-recording system and electron energy loss spectroscopy. For the observation of C onions, copper (Cu) plates were implanted with C ions at temperatures from 300° to 700°C. The high-magnification micrograph of C onions is shown in Fig. 25. As a result of TEM observation during C ion irradiation, it was found that the accumulation of C atoms above 10–20 at.% (i.e., $>1 \times 10^{17}$ C/cm^2) and high enough temperature for the migration of C atoms are needed for creating a nucleus of C onions in Cu plates. The production rate of C onions using ion implantation was 6 to 8 times higher than that using electron irradiation. Furthermore, the creation of C onions and nanocapsules was found to depend on the quality of substrate materials. Details of this study are described in Ref. [79].

In addition to C onions, C atoms condense into various kinds of chemically bonded forms, and they are known to have excellent physical properties depending on the bonding nature. This means that research and applications not only in the materials science but also in other scientific fields are expected. At JAERI, the optimum growth conditions have been successfully obtained for the preparation of high-quality C_{60}, diamondlike carbon, and nanocrystalline diamond by means of ion-beam-assisted deposition [80–82]. The susceptibility of Ni/C_{60} thin films to thermal treatment, the formation of nanocrystalline diamond and nanotubes due to codeposition of Co and C_{60}, and the surface modification of glassy

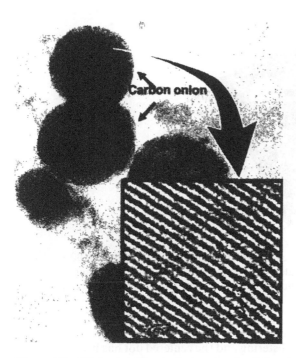

Figure 25 High-magnification micrograph of C onions.

carbon by hydrogen doping have also been characterized [83–86]. In the following, materials synthesized from C and C_{60} are described as examples.

Ion beam deposition (IBD) was carried out by employing an accelerator–decelerator system under ultrahigh vacuum. In this study, $^{12}C^+$ ions were used as source, and Si(111) and Ir(100)/MgO(100) single crystalline were used as substrates [87]. The typical ion beam conditions were in the energy range of 50–200 eV and in the beam current around 1 mA. On the other hand, as for the ion-beam-assisted deposition (IBAD) experiments, a special chamber was designed. In the chamber, three different sources (C_{60}, Ne ions, and transition metals) can be installed. In the case of evaporating C_{60} molecules, energetic Ne^+ ions were also introduced in an energy range up to 5 keV. Ne^+ ion beam was obliquely incident at 60° relative to the surface normal. In the case of using ions with high Z element, the sputtering effect should be considered during irradiation. The chamber can be evacuated up to 1.2×10^{-6} Pa, and the vacuum was kept at less than 5×10^{-5} Pa during IBAD experiments.

As a result of IBD and IBAD experiments [80], nanodiamonds immersed in the dominant sp^3 amorphous carbon films were found. Patterns were formed via the dynamical process between the sputtering and the deposition effects [82]. Hexagonal nanosized diamonds were also prepared from the C_{60} vapor with the simultaneous irradiation of 1.5-keV Ne^+ ions at a temperature of 700°C. Furthermore, although C and iridium (Ir) are immiscible, defects introduced by C implantation could favor the supersaturated C atoms in the subsurface region. Fig. 26 shows an atomic force microscope (AFM) micrograph of diamondlike carbon (DLC) film on Ir(100) prepared by IBD method (100-eV $^{12}C^+$) followed by thermal annealing at 600°C under He gas environment. After annealing, DLC film suffered from blistering because of immiscibility between C and Ir. One can recognize a symmetric, flowerlike pattern of blistering with a needlelike dot at the center.

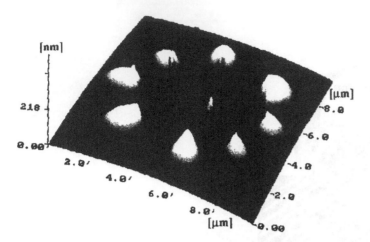

Figure 26 AFM micrograph of DLC film on Ir(100) prepared by IBD method (100 eV $^{12}C^+$), followed by thermal annealing at 600°C under He gas environment. After annealing, DLC film suffered from blistering because of immiscibility between C and Ir. Size of the micrograph is 9.85 × 9.85 μm^2.

The "petals" correspond to the bumps in blistering and the centered dot has been confirmed to be nanodiamond by micro-Raman spectroscopic analysis. The growth from the nucleated embryos was enhanced by a heat treatment. These results suggest that the nucleation sites can be controlled by using ion beam technique. Therefore, it is concluded that nanopattern as well as materials with interesting features may be synthesized by using ion beam irradiation technique with changing ion energies and substrate temperature.

Study of Materials by Ion Beam Analysis

Ion beam analysis performed on materials synthesized at TIARA is described in this section. Multilayers in which different thin metal films are stacked alternatively are considered as a mirror material reflecting short wavelength lights such as soft X-rays. To form high-quality mirror materials, it is indispensable to measure the thickness of films and the flatness of the film interfaces. Niobium (Nb) films deposited on single-crystal α-Al_2O^3, MgO substrates by electron beam evaporation technique has been investigated by RBS/C and X-ray diffraction analyses. Figure 27 shows 2.7-MeV $^4He^+$ RBS spectra from Nb on α-Al_2O_3 substrate measured under random conditions and the $\langle 110 \rangle$ aligned conditions. As a result, it was found that some amount of disorder exists around the interface region between Nb layer and α-Al_2O_3 substrate [88]. From the fact that the ratio of the $\langle 110 \rangle$ to random spectrum yield from Nb film is 0.044 at just behind surface peak, the excellent crystallinity of the film was confirmed. Multilayers consisting of Nd and Cu films have also been examined to understand the relationship between crystallinity and the fabrication process [89]. ^{15}N nuclear reaction analysis (^{15}N NRA) was conducted using ^{15}N ions accelerated by 3-MeV tandem accelerator in this study. Since ^{15}N NRA can detect small amounts of residual hydrogen existing at the interface between Nb and Cu layers, the thickness of each layer can be estimated. The results obtained from ^{15}N NRA are shown in Fig. 28. The calculated result is also shown as a solid line in the figure. From the comparison between experimental and calculated results, the optimum condition of the fabrication of the multilayers was determined. Details are described in Refs. [88,89].

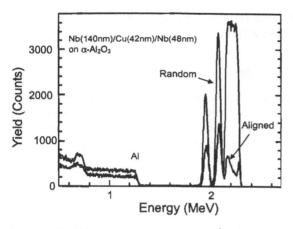

Figure 27 RBS/C spectra of 2.7-MeV-^{4}He$^+$ from Nb(110) epitaxial film on α-Al$_2$O$_3$ substrate [Nb (140 nm)/Cu (42 nm)/Nb (48 nm)/α-Al$_2$O$_3$]. The 100-nm-thick Nb films were fabricated at 750°C by electron beam evaporation.

The fabrication process of vanadium oxide (VO$_2$) has also been studied using RBS/C. Since optical and electrical properties of VO$_2$ are dramatically changed at 68°C due to phase transition, VO$_2$ is regarded as one of the candidates for thermally activated electronic or optical switching devices for optical fibers or sensors. To obtain the desired properties, the development of the fabrication process for very thin films, without crystalline defects on various substrates, is required. Single-crystalline VO$_2$ thin films on (0001) plane of a sapphire substrate have been synthesized by a laser ablation method. The quality of VO$_2$ was examined by X-ray diffraction and RBS/C method. The electrical resistance and the optical transmittance of the VO$_2$ film were measured under increasing and decreasing temperatures. At a temperature of 68°C, an abrupt transition of resistance from metal to

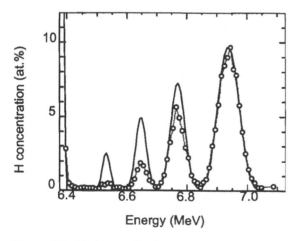

Figure 28 Hydrogen distribution observed in Nb/Cu multilayers by ^{15}N NRA (circles). The fitting result (solid line) is also shown in the figure. The structure of Nb/Cu multilayers is Nb (2 nm)/Cu/Nb (5 nm)/Cu/Nb (10 nm)/Cu/Nb (20 nm)/α-Al$_2$O$_3$, where the thickness of Cu layer is 20 nm.

insulator was observed [90]. In addition, as a result of the study of Mo doping effects, transition temperature was found to decrease with increasing Mo concentration [91].

4.4. Beam Processing of Organic Materials

Irradiation of heavy ions like Au and Kr deposits a heavy damage along the ion track in polymer films. When some polymer films are irradiated with heavy ions and dipped in an alkaline aqueous solution, micropores are produced along the ion track. Since the shape and size of pores of polymers can be controlled by etching conditions and by the species of heavy ions, these polymer films are expected to be used for high-performance filters and ion detectors [92]. The development of the polymer filters and ion detectors are described as follows.

Thermoresponsive Polymer Films

Polymer films that have useful functions such as selective transportation and absorption of specific molecules are expected to be applied for high-performance filters used in medicine and biology. The fabrication process of polymer films with pores using heavy-ion irradiation consists of the following four steps: (1) the preparation of a thin core film, (2) irradiation of heavy ions into the core film, (3) the treatment of the core film in an alkaline solution, and (4) grafting of a polymer, which changes the shape responsive to temperature. In this study [92], a copolymer consisting of diethylene glycol diallyl carbonate (DEBA) and acryloyl-L-proline methyl ester (A-ProOMe) was used as the core film. DEBA acts as a comonomer with high sensitivity to ion beam irradiation, and A-ProOMe enhances the affinity to the polymers that are subsequently grafted. The core films were irradiated with ^{84}Kr ions at 6.19 MeV/n at a fluence of $10^7/cm^2$. After irradiation, the films were treated in an aqueous 6 M NaOH solution at 60°C for 1 hr to make pores 1.3 μm in diameter. Then, the films were immersed in an aqueous A-ProOMe solution, and irradiated with gamma rays at a dose of 30 kGy for grafting. A-ProOMe is a cross-linkable polymer to give a temperature-responsive hydrogel. Thus, its volume increases by absorption of water below a critical temperature (14°C). The volume of swelled A-ProOMe is about 400 times larger than nonswelled one. The diameter of cylindrical pores was 0.7 μm at 30°C. Figure 29 shows the temperature dependence of the permeability of p-nitrophenol for A-ProOMe grafted films. No significant change of the permeability was observed in temperature ranges below 12°C. However, a remarkable increase in permeability was found above 14°C, which corresponds to the lower critical solution temperature. It is noteworthy that molecules can be selectively separated by changing the volume of the grafted polymers in the cylindrical pore, which are regulated by the change in temperature. Furthermore, when N-isopropylacrylamide (NIPAAm) was used as a responsive material grafted on films, polyethylene glycol molecules with the same mass could be separated from polyethylene glycol with a different mass by controlling temperatures, as shown in Fig. 30.

Ion Detectors Based on CR39

Polyallyl diglycol carbonate (CR39) is well known as a plastic track detector [93]. However, the sensitivity of a plastic detector to ion irradiation should be greater compared to that of CR39 for detection of lighter ion particles with higher energy, when the ion detectors are applied to the field of cosmic ray physics, nuclear physics, space radiation, and neutron dosimetries. Therefore, The development of a new plastic track detector based on CR39 that shows high sensitivity to proton beams are described next.

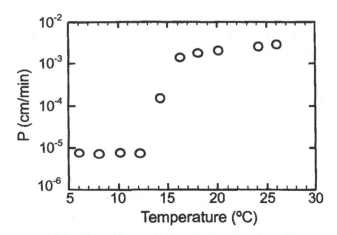

Figure 29 Temperature dependence of the permeability of p-nitrophenol for A-ProOMe grafted films. The permeability (P) was calculated from the equation, $P = \kappa V / AC$, where κ, V, A, and C are the slope of the permeation–time curve, the volume of the receiver chamber, the surface area of the membrane, and the concentration of p-nitropenol in the donor chamber, respectively.

Several types of CR39 copolymers containing 1–3 wt.% of N-isopropylacrylamide (NIPAAm) and 0.01 wt.% of Naugard 445 were prepared by cast polymerization. The copolymerizations were performed with 3% diisopropyl peroxydicarbonate (IPP) as an initiator at 70°C for 24 hr to give copolymer films (CR39/NIPAAm). Details of the fabrication process of CR39/NIPAAm are described in Ref. [94]. Protons at an energy of 10 MeV were incident onto CR39/NIPAAm. The effects of NaOH concentration and temperature of an etchant to ion beam sensitivity of CR39/NIPAAm were investigated. Results for CR39/NIPAAm show a broad maximum of sensitivity at 5 N NaOH at 70°C.

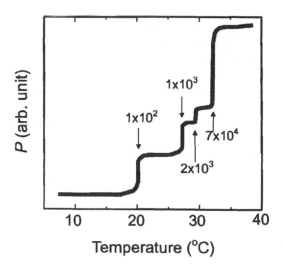

Figure 30 Temperature dependence of the permeability (P) of polyethylene glycol molecules for N-isopropylacrylamide (NIPAAm) grafted films. Polyethylene glycol with four kinds of masses (1×10^2, 1×10^3, 2×10^3, 7×10^4) are contained in the solution.

For the comparison between CR39 and CR39/NIPAAm, the sensitivity of CR/NIPAm to ion particles was much higher than that for CR39 in LET ranges below 10 keV/μm [94]. This indicates that CR/NIPAAm is very useful for an ion particle detector.

5. APPLICATIONS TO BIOLOGICAL SCIENCE

5.1. Mutagenesis in Plants

Introduction and Irradiation Methods

Since the discovery of mutation induction by X-rays in maize [95], numerous studies on plant mutagenesis by ionizing radiation have been carried out. The biological effect of ion beams was also investigated and it was found that ion beams show high RBE of lethality, mutation, and so on compared with low-LET radiation such as gamma rays, X-rays, and electrons [96,97]. Because ion beams deposit high energy on a target densely and locally, as opposed to low-LET radiation, it is suggested that ion beams predominantly induce single- or double-strand DNA breaks with damaged end groups whose reparability would be low [98]. Therefore, it seems plausible that ion beams can frequently produce large DNA alteration such as inversion, translocation, and large deletion rather than point mutation, resulting in producing characteristic mutants induced by ion beams. However, the characteristics of ion beams on the mutation induction have not yet been clearly elucidated.

As the first facility of ion beams for biological and material use, TIARA was established in 1991. The characteristics of ion beams for the biological effects and the induction of mutation have been investigated using model plants, *Arabidopsis*, chrysanthemum, tobacco, wheat, and so on. Several kinds of ion beams such as C, He, Ne, and Ar were experimented on in this study and mostly 220-MeV C ions were used. All ions were generated by the AVF cyclotron in TIARA. The physical properties of the 220-MeV carbon ions are as follows: Incident energy at the target surface was 17.4 MeV/u, mean LET in a target (0.25-mm thickness) was estimated to be 110 keV/μm as water equivalent, and the range of ions was ca. 1.0 mm. These physical properties were calculated by ELOSS code program, a modified OSCAR code program [99]. Particle fluences of the ions were determined using a diethyleneglycol-bis-allylcarbonate (CR39) film track detector. Electron beams (2 MeV; JAERI, Takasaki, Japan) were also used as a low-LET ionizing radiation control.

In general, ion beams were scanned at more than 50 × 50 mm, and exited the vacuum chamber through the beam window made of a 30-μm-thick titanium foil. The irradiation sample was placed in the air at a distance of 10 cm from the beam window. In the case of *Arabidopsis* or tobacco seeds, for example, 100–3000 seeds were sandwiched between kapton films (7.5-μm thickness) to make a monolayer of the seeds for homogeneous irradiation. In the case of rice or barley seeds, the embryo side was kept facing toward ion beams, whereas for calli or explants in a petri dish, the lid of the petri dish was replaced with a kapton-film cover to decrease the loss of the energy of ion beams. Samples were irradiated for less than 3 min for all doses.

Biological Effects of Ion Beams

Determination of biological effects of ion beams is an inevitable problem for the application of ion beams to mutagenesis in plants. In the previous investigations, it was shown that RBE for growth inhibition and so on reached a peak at around 72–174 [100] or 190 keV/μm [101], suggesting LET dependence of RBE of higher plants is similar to that of mammalian cells. However, in these experiments, the Bragg peak was adjusted on the target and/or only a few

LETs were investigated. To acquire more detailed data on biological effects by using penetrating ions with various kinds of LET, several investigations have been carried out in TIARA [102–106]. Most efforts were focused on LET dependence of survival and other endpoints of ion beams in the case of plant seeds. In general, all the experiments showed that the RBE peak of survival and other endpoints was around ca. 230 keV/μm and more (Fig. 31). By using carbon ions with different LET, Hase et al. [106] indicated that the RBE for survival and chromosome aberration in tobacco increased with increasing LET and showed the highest value at 230 keV/μm. Shikazono et al. [105] found that the highest LET of the peak RBE for lethality of *Arabidopsis* was around 221 keV/μm for carbon ions and around 350 keV/μm and more for neon and argon ions. Therefore, it is almost certain that the LET having a maximum of RBE for lethality is higher in plant seeds than in mammalian cells.

Mutation Rate

In order to compare mutation frequency induced by 220-MeV C ions with that by low-LET radiation (electrons), *Arabidopsis* visible phenotype loci were chosen as follows: transparent testa (*tt*) whose seed coat is transparent because of lack of pigment; glaborous (*gl*), which have no hair on their leaves and stems; and long hypocotyl (*hy*) whose hypocotyl is longer than that of the wild type in the light condition. Mutation frequencies of *tt*, *gl*, and *hy* induced by carbon ions were 8- to 34-fold higher than those by electrons (Table 2). In this study, irradiation doses for the induction of mutation were determined from the RBE of carbon ions compared with that of electrons on the survival of plants, which was approximately 5. Both doses are at three-quarters of the shoulder dose of each survival curve [104].

Mutation Spectrum

It is necessary to compare the mutation spectrum induced by ion beams with that by the low-LET radiation to figure out a special feature of ion beams as an effective mutagen. Mutation

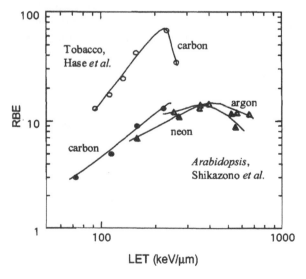

Figure 31 The relative biological effectiveness (RBE) of lethality as a function of linear energy transfer (LET). (From Refs. 105 and 106.)

Table 2 Mutation Frequency Induced by Carbon Ions and Electrons

Mutagen (dose)	Locus	Mutation frequency/locus/diploid cell/dose (Gy) ($\times$ 10^{-6})
Carbon ions (150 Gy)	*tt*	2.3 (34-fold)[a]
	gl	1.7 (8)
	Hy	2.0 (16)
Electrons (750 Gy)	*Tt*	0.068
	gl	0.21
	hy	0.12

[a] Compared to the frequency of the electrons.

spectrum was first investigated on the flower color of chrysanthemum cv. Taihei (pink color) by Nagatomi et al. [107]. The explants of leaf and floral petals incubated in agar medium were irradiated with carbon ions of 220 MeV. After irradiation, the culture was transferred to a new medium to induce callus. The mutation induction of the regenerated plants from the callus was investigated, comparing the results with those with gamma-ray irradiation. The mutation rates of flower color induced by ion beams were approximately half of those induced by gamma rays in both floral petal and leaf. Most flower color mutants induced by gamma rays were light pink, and a few were dark pink (Table 3). On the other hand, flower mutants induced by ion beams showed complex and stripe types other than single color (Fig. 32). The complex-type flower color mutants increased as the dose of ion beams rose. The color spectrum of the ion-beam-induced mutants shifted from pink to yellow. Those specific mutants such as complex and striped color types have never been obtained by gamma-ray irradiation in the cultivar. Thus, mutation spectrum of flower color induced by ion beams is wide and novel mutation phenotypes can be induced.

Recently, mutation spectrum of flower color and flower shape of carnation was also investigated by Okamura et al. [108]. In the carnation variety Vital (spray-type, cherry pink flowers with frilly petals) tested, flower color mutants such as pink, white, and red were obtained by X-ray irradiation, whereas the color spectrum of the mutants obtained by carbon ion irradiation was wide such as pink, light pink, salmon, red, yellow, and complex and striped types. Furthermore, many kinds of round shape of petals were induced in addition to flower colors. These indicate that ion beams can induce novel flower color and shape with high frequency.

Novel Mutants

Until now, several new mutants were induced by ion beams. Representative mutants will be described below.

Table 3 Mutation Spectrum of Flower Color in Chrysanthemum

Mutagen	Mutation frequency (%)					
	White	Light pink	Dark pink	Orange	Yellow	Complex/Stripe
Unirradiated	0	0.3	0	0	0	0
Gamma rays	0	27.7	2.1	0	0	0
Carbon ions	0.3	4.6	0.3	0.3	0.2	10.2

Original variety "Taihei" with pink flowers was used.

Figure 32 Novel mutants induced by ion beams. Top left: 1-month-old plants of *Arabidopsis* wild type (top) and UV-B-resistant mutants (bottom) under high UV-B (11–13 kJ/m²/day) condition. Top center and right: complex (center) and striped (right) flower-color varieties of chrysanthemum. Bottom left: flower of *Arabidopsis frl1* mutant. Bottom center and right: new carnation varieties with round petals (center) and Dianthus type petals (right).

Arabidopsis as a Model Plant. In an attempt to isolate novel mutants by means of ion beams, *Arabidopsis* [*Arabidopsis thaliana* (L.) Heynh.] was used as a model plant. This is because thousands of mutants were already induced in *Arabidopsis* by using chemical mutagens, X-rays, T-DNA mutagemesis, etc. Therefore, if a novel mutant phenotype is discovered by using ion beams, it would be well grounded that ion beams are a powerful tool as a new mutagen. As one of such a mutant phenotype, UV-B-resistant plants were chosen, because many UV-B-sensitive mutants have been isolated in *Arabidopsis* but there has been no report of mutants in any higher plants that are resistant to UV-B light. Four UV-B-resistant mutants [*ultraviolet-light insensitive (uvi) 1–4*] have been isolated in 5100 M_2 families derived from carbon-ion-irradiated 1280 M_1 seeds. The fresh weight of *uvi1*, *uvi2* and *uvi4*, after 1 month under UV-B radiation, was 2-fold higher than that of wild type, and 1.5-fold higher in the *uvi3* (Fig. 32). Especially, *uvi1* showed very great ability of both photoreactivation and dark repair. The reduction of CPD and (6-4) photoproducts that are two major DNA damages caused by UV-B are both faster in *uvi1* than in wild type, indicating that *uvi1* mutation would gain the DNA reparability [109].

New anthocyanin-accumulated or -defective mutants were also isolated. A new mutant, *anthocyanin spotted testa (ast)*, which has the spotted pigmentation on the seed coat, has been isolated from 11,960 M_2 plants derived from 1488 self-pollinated M_1. Anthocyanin content was about 6 times higher at 6 days after flowering than that of the wild-type seeds [110]. Two new loci of anthocyanin-defective *tt* phenotype (tentatively *ttA*, *ttB*) have been found along with the research of mutation frequency experiment (unpublished data). Both genes seem to encode enzymes of anthocyanin biosynthetic pathway. On the other hand, a novel flower mutant, *frl1*, has serrated petals and sepals, but the other floral and vegetative organs appear to be normal (Fig. 32). *FRL1* gene should act in petal and sepal development in an organ-identity-specific manner [111].

Other Plants. As already described in "Mutation Spectrum," complex and striped types of flower color have been obtained in chrysanthemum. A higher mutation frequency of complex flower color mutants were derived from floral petal irradiation than that from leaf irradiation [107]. Also, many kinds of flower color such as pink, light pink, salmon, red, yellow, complex and striped types, and various kinds of round petals were induced from the carnation variety Vital [108] (Fig. 32).

In order to develop an efficient procedure for obtaining a desired mutation, ion beams were applied to tobacco anthers, and potato virus Y (PVY)-resistant mutants have been selected. A high frequency (2.9–3.9%) of resistant mutants was obtained by the irradiation of C and He ions with a dose of 5–10 Gy [112,113].

Two mutant lines of yellow mosaic virus resistant barley were found in ca. 50,000 M_2 families. The character of the resistance in a field was not changed over three generations [114].

High efficiency of getting blast-resistant mutants of rice was also obtained using ion beams, although resistant mutants were already obtained by gamma rays and by thermal neutrons [115].

Molecular Mechanisms of Mutation

In order to investigate the DNA alteration of mutations induced by ion beams in plants, polymerase chain reaction (PCR) and sequencing analysis was performed to compare DNA fragments amplified from carbon-ion- and electron-induced *Arabidopsis* mutants [116,117]. Fourteen out of 30 loci possessed intragenic mutation ("small mutation") such as point mutation or deletion of several to hundreds of bases. For comparison, 16 out of 30 loci possessed intergenic mutation ("large mutation"), such as chromosomal inversion, translocation, and total deletion covering their own loci. These results imply that in the case of mutation induced by ion beams, half of the mutants have intragenic small mutation and the other half have large DNA alteration such as inversion, translocation, and large deletion (Table 4). In such an alteration, a common feature was that all the DNA strand breaks induced by carbon ions were found to be rejoined using short homologies [117]. These results suggest that the nonhomologous end joining pathway operates after plant cells are exposed to ion beams.

Conclusion

The *uvi1–uvi4* mutants were induced from 1280 M_1, and the *ast* and *frl1* mutants were obtained from the offspring of 1488 M_1 seeds. As ion beams showed high induction of mutation of the known loci such as *tt, gl, hy*, it would be true that ion beams can highly induce not only known mutants but also novel mutants. In chrysanthemum and/or carnation, complex and striped flower color and new flower-shape mutants that have never been induced by low-LET radiation have been produced, indicating that ion beams could induce various kinds of mutants on the similar phenotypes. In conclusion, the character-

Table 4 Classification of Mutation Induced by Carbon Ions

	Large DNA structural alteration	
Intragenic mutation (point mutation, small deletion, etc.)	Inversion translocation	Total deletion
47%	40%	13%

istics of ion beams for the mutation induction are (1) to induce mutants with high frequency, (2) to show broad mutation spectrum, and (3) to produce novel mutants. For these reasons, chemical mutagens such as EMS and low-LET ionizing radiation such as gamma rays and electrons will predominantly induce many but small modifications or DNA damages on the DNA strands, producing several pointlike mutations on the genome. On the contrary, ion beams as a high-LET ionizing radiation will cause few but large and irreparable DNA damages locally, producing a limited number of null mutations (Fig. 33).

5.2. Cross Incompatibility

In plant breeding, obtaining interspecific hybrid is important for introducing desirable genes from wild species to cultivated species. However, it is very difficult to obtain a viable hybrid plant between widely related species. These problems would predominantly come from sexual reproduction such as cross incompatibility, hybrid inviability, etc [118,119].

The wild species of tobacco, *Nicotiana gossei*, has been reported to be resistant to more diseases and insects than other *Nicotiana* species, several attempts to hybridize it with *N. tabacum* (cultivar) have been made. In *N. gossei* crossed with *N. tabacum*, it was easy to get hybrid seeds with conventional cross, but the hybrid plants cannot survive. To overcome the cross incompatibility, mature pollen from *N. tabacum* cv. Bright Yellow 4 were irradiated with helium ions from a tandem accelerator (6 MeV). Two viable hybrid plants were obtained at the rate of 1.1×10^{-3}. Both the hybrids were resistant to tobacco mosaic virus and aphid although the degree of resistance varied. With varying energy and kinds of ions, hybrid production rate ranged from 10^{-3} to 10^{-2} [120–122]. The highest rate, 3.6×10^{-2}, was obtained with 10 Gy of 100-MeV He ion beams. In the case of gamma irradiation, the rate of producing hybrids was 3.7×10^{-5}. Thus, ion beams are powerful tools to overcome the hybrid incompatibility.

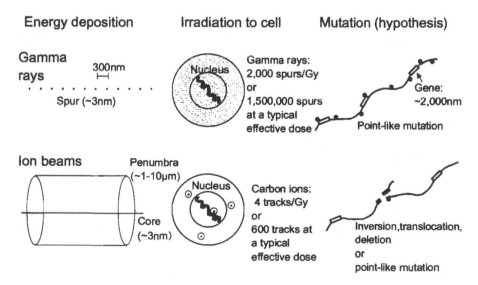

Figure 33 Characteristics of energy transfer of ion beams for mutation induction.

5.3. Gene Transfer

Agrobacterium-mediated DNA transfer technique and direct gene transfer into protoplasts on the basis of the electroporation technique have frequently been used for the gene transfer methods in plants. However, these methods were hardly adapted to monocotyledonous plants or to varieties in which regeneration system was not established. These deficiencies found in gene transfer techniques will be largely overcome if pollen could be used as a DNA vector.

Penetration-controlled irradiation with ion beams for biological study has been established [123]. When dry pollens were exposed to He or C ions with shallow penetration depth (e.g., 4-μm depth in the case of He ions), leaky pollen, which was thought to be a physical lesion in the pollen envelop induced by ion beams, was efficiently produced. Using this penetration-controlled irradiation system, the transient exogenous GUS expression was observed at a high frequency (55.1%) when tobacco pollen was incubated in a DNA solution, following ion beam irradiation of pollen at the penetration depth of 4 μm [124]. The GUS expression frequency increased with the irradiation dose, indicating that the foreign DNA can be efficiently introduced into pollen [125]. However, any transformant cannot be found in more than 10,000 seeds obtained from the crosses using the DNA introduced pollen. It is thought that the pollen transformation is difficult only by raising the transient GUS expression frequency and that the nature of the pollen, such as bimembrane structure and DNA incorporation capability, may be a fundamental problem in a pollen transformation system [126].

5.4. Application of Microbeams

A heavy-ion microbeam provides a unique way to control the number of particles traversing individual cells and localization of dose within the cell. A collimated heavy-ion microbeam apparatus has been installed in a beam line from the AVF cyclotron to develop a novel cell surgery technique [127] (Fig. 34) (Table 5).

Functional Analysis of Animal Development

As a functional analysis of animal development, an accurate fate map was carried out using the silkworm, *Bombyx mori* [128]. Among the beam sizes tested (60- to 250-μm φ), 250-μm φ was adopted for the fate mapping of egg at the cellular blastderm stage. When eggs were irradiated near the anterior, posterior, or dorsal periphery, no defective larvae were observed. A close correlation was observed between the site of irradiation and the site of the defect induced. Head defects were induced only by irradiation at the anterior half of the egg, whereas defects in thoracic and abdominal segments were induced by irradiation at the middle and the posterior part of the egg, respectively. Based on the correlations, a fate map of the *B. mori* egg was established [128].

Functional Analysis of Plant Physiology

Microbeam system is also very useful for analyzing plant physiology. Root gravitropism is a simple but wondrously complex phenomenon. Perception of gravity in plant roots has been well studied, but its signaling from gravity sensing to physical change of root bending is still unclear. When primary root apical tissues of *Arabidopsis* were irradiated by microbeams with 120-μm diameter, strong inhibition of root elongation and curvature were observed at root tip. Irradiation of cells that become the lower part of the root cap after gravistimulation

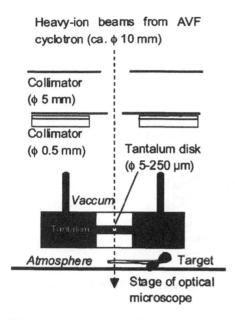

Heavy-ion beams from AVF
cyclotron (ca. φ 10 mm)

Collimator
(φ 5 mm)

Collimator
(φ 0.5 mm)

Tantalum disk
(φ 5-250 μm)

Vaccum

Tantalum

Atmosphere

Target

Stage of optical
microscope

Figure 34 Schematic diagram of microbeam system for biological study. An ion beam from an AVF cyclotron is collimated by collimators and a tantalum disk with microaperture with a diameter of 5–250 μm. The collimated ion beam passes through the room air to a target that is set on the micropositioning stage of an optical microscopic system.

showed strong inhibition of root curvature, whereas irradiation of cells that become the upper part of the root cap after gravistimulation did not show severe damage in either root curvature or root growth [129]. These results indicate not only that the root tip cells are the most sensitive sites for root gravity, but also that the signaling of root gravity would go through the lower part of the root tip cells after the perception.

5.5. Application of Positron-Emitting Radioisotopes

In plant researches, a real-time imaging system to visualize the movement of various molecules and ions has been desired for understanding the regulation of plant functions. The positron-emitting tracer imaging system (PETIS) makes use of positron-emitting nuclides and enables the investigator to trace the radioactivity in a plant in real time without touching the sample plant [130] (Fig. 35). This has been first realized at TIARA by use of

Table 5 Physical Parameters of Heavy-Ion Microbeams

Ions	Specific energy (MeV/u)	Beam size (φ, μm)	Projectile range (in water, mm)	LET[a] (keV/μm)
$^4He^{2+}$	12.5	40~250	1.8	14~50
$^{12}C^{5+}$	18.3	10~250	1.2	100~300
$^{20}Ne^{7+}$	13.0	5~250	0.7	400~800
$^{40}Ar^{13+}$	11.5	5~250	0.24	1200~2500

[a] LET value on the surface of the target.

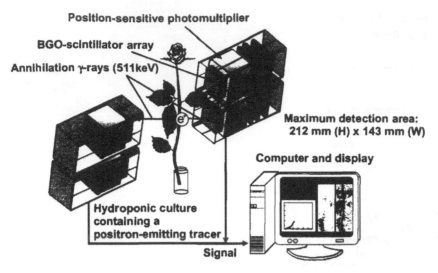

Figure 35 Schematic of positron-emitting tracer imaging system (PETIS) and an example of experimental setup.

cyclotron-produced positron-emitting radioisotopes from which various labeled compounds needed for plant research can be produced. As most positron-emitting radioisotopes are very short-lived, except for ^{22}Na, they are produced at the site for experiments using the PETIS and labeled compounds of nutrients and amino acids (Table 6).

Using the PETIS, real-time $[^{11}C]$methionine translocation was studied for barley. For the mechanism of Fe uptake in an Fe-deficient barley, it was found that leaf methionine does not participate in the reaction of mugineic acid synthesis, but the methionine produced in barley roots is used in the biosynthesis of mugineic acid phytosiderophores [131,132]. $^{13}NH_4^+$ and H_2 ^{15}O translocation in rice plants were also investigated [133,134]. On the study of root-nodule functions in common bean, the root site of nitrate uptake and assimilation was clarified by using ^{13}N-nitrate or ammonium [135]. Furthermore, use of PETIS and positron-emitting nuclides has given us deep insights into water movement in living plants [136] and nitrogen (N) distribution in soybean using N-deficient and/or N-sufficient plants [137].

Table 6 Production and Application of Positron Emitters in TIARA

Nuclide	Half-life	Reaction	Tracer for experiments on plants
^{11}C	20 min	$^{14}N(p, \alpha)^{11}C$	$^{11}CO_2$, $[^{11}C]$methionine
^{13}N	10 min	$^{16}O(p, \alpha)^{13}N$	$^{13}NO_2^-/^{13}NO_3^-$, $^{13}NH_4^+$, $[^{13}N]N_2$
^{18}F	110 min	$^{16}O(\alpha, pn)^{18}F$ $^{18}O(p, n)^{18}F$	Aqueous ^{18}F-, $[^{18}F]$-2-deoxy-2-fluoro-D-glucose, ^{18}F-proline
^{48}V	16 days	$^{45}Sc(\alpha, p)^{48}V$	$[^{48}V]H_2VO_4^-$ solution
^{52}Mn	5.6 days	$^{nat}Cr(p, xn)^{52}Mn$	$^{52}Mn^{2+}$ solution
^{52}Fe	8.3 hr	$^{nat}Cr(\alpha, xn)^{52}Fe$	$^{52}Fe^{3+}$ solution
^{62}Zn	9.2 hr	$^{63}Cr(p, 2n)^{62}Zn$	$^{62}Zn^{2+}$ solution

5.6. Future Prospects

Ion beams as new mutagen showed breakthrough for the next step of biotechnology by making new mutants that cannot be obtained by other means. One of the most important objectives of the postgenome age is to elucidate the function of genes. Unfortunately, it is very difficult in the plant kingdom to apply the gene-knockout system that was used in mammals or bacteria. Therefore, ion beams will give us thousands of novel mutants that inform the function of the mutated gene. On the other hand, microbeam irradiation system is very useful for research on a cell or cell-to-cell function. If the size of the microbeams are on the order of several microns or submicrons, accurate and effective irradiation for cell or chromosome inactivation will be carried out. In researches using a whole plant, behavior of molecules and ions in a living plant can be analyzed with PETIS. As the problems of elevated CO_2 concentrations and environmental pollution increase, real-time imaging for physiological understanding of plants is necessary for the security of food resources and conservation of the environment.

REFERENCES

1a. Tanaka, R. *Present Status and Future Development of Research for Advanced Radiation Application Using Fast Ions*, Oyo Buturi 2000, 69, 369 (in Japanese) (a monthly publication of The Japan Society of Applied Physics).

1b. Tanaka, R. *Application of Ion Beams*, Radioisotopes 2001, 50, 27 (in Japanese) (A monthly publication of Japan Radioisotope Association).

2. See the proceedings of the following International Conferences appeared in special issues of Nuclear Instruments and Methods in Physics Research Section B: Nucl. Instrum. Methods Phys. Res. B 2002, 193, Proc. of the 19th Intern. Conf. on Atomic Collisions in Solids; Nucl. Instrum. Methods Phys. Res. B 2002, 190, Proc. of the 15th Intern. Conf. on Ion Beam Analysis; Nucl. Instrum. Methods Phys. Res. B 2001, 175 177, Proc. of the 12th Intern. Conf. on Ion Beam Modification of Materials; Nucl. Instrum. Methods Phys. Res. B 2001, 181, Proc. of the 7th Intern. Conf. on Nuclear Microprobe Technology and Applications.

3. Brown, I.G. *The Physics and Technology of Ion Sources*; Wiley-Interscience: New York, 1989.

4. Geller, R. Electron Cyclotron Resonance Ion Sources and ECR Plasmas; Institute of Physics Publishing: Bristol, 1996.

5. Melin, G.; Bourg, F.; Briand, P.; Delaunay, M.; Gaudart, G.; Girard, A.; Hitz, D.; Klein, J.P.; Ludwig, P.; Nguyen, T.K.; Pontonnier, M.; Su, Y. Rev. Sci. Instrum. 1994, 65, 1051.

6. Gammino, S.; Ciavola, G.; Celona, L.; Romano, P.; Torrisi, L.; Hitz, D.; Girard, A.; Melin, G. In *Proceedings of the 16th International Conference on Cyclotrons and Their Applications 2001*; Marti, F., Ed.; American Institute of Physics: New York, 2001; 223 pp.

7. Della-Negra, S.; Brunelle, A.; Le Beyec, Y.; Curaudeau, J.M.; Mouffron, J.P.; Waast, B.; Hakansson, P.; Sundqvist, B.U.R.; Parilis, E. Nucl. Instrum. Methods 1993, *B74*, 453.

8. Tombrello, T.A. Nucl. Instrum. Methods 1995, *B99*, 225.

9. In *Proceedings of an International Workshop on the Structure of Small Molecules and Ions*; Naaman, R., Vager, Z., Eds.; Plenum Press: New York, 1988.

10. Milani, P.; Iannotta, S. *Cluster Beam Synthesis of Nanostructured Materials*; Springer: Berlin, 1999.

11. Saitoh, Y.; Mizuhashi, K.; Tajima, S. Nucl. Instrum. Methods 2000, *A452*, 61.

12. Tanaka, R.; Arakawa, K.; Yokota, W.; Nakamura, Y.; Kamiya, T.; Fukuda, M.; Agematsu, T.; Watanabe, H.; Akiyama, N.; Tanaka, S.; Nara, T.; Hagiwara, M.; Okada, S.; Maruyama, M. In *Proceedings of the 12th International Conference on Cyclotrons and Their Applications*; Martin, B., Ziegler, K., Eds.; World Scientific: Singapore, 1989; 566 pp.

13. In *Advances of Accelerator Physics and Technologies, Advanced Series on Directions in High Energy Physics*; Schopper, H., Ed. World Scientific: Singapore, 1993; Vol. 12.

14. Arakawa, K.; Nakamura, Y.; Yokota, W.; Fukuda, M.; Nara, T.; Agematsu, T.; Okumura, S.; Ishibori, I.; Karasawa, T.; Tanaka, R.; Shimizu, A.; Tachikawa, T.; Hayashi, Y.; Ishii, K.; Satoh, T. In *Proceedings of the 13th International Conference on Cyclotrons and Their Applications*; Dutto, G., Craddock, M.K., Eds.; World Scientific: Singapore, 1992; 119 pp.

15. Saitoh, Y.; Tajima, S.; Takada, I.; Mizuhashi, K.; Uno, S.; Ohkoshi, K.; Ishii, Y.; Kamiya, T.; Yotumoto, K.; Tanaka, R.; Iwamoto, E. Nucl. Instrum. Methods 1994, *B89*, 23.

16. Livingood, J.J. *Principles of Cyclic Particle Accelerators*; D. Van Nostrand: Princeton, 1961.

17. Thomas, L.H. Phys. Rev. 1938, *54*, 580.

18. Richardson, J.R. *Sector Focusing Cyclotrons, Progress in Nuclear Techniques and Instrumentation*; North-Holland Publishing Co.: Amsterdam, 1965.

19. Gordon, M.M. Part. Accel. 1971, *2*, 203.

20. Joho, W. In *Proceedings of the 11th International Conference on Cyclotrons and Their Applications*; Sekiguchi, M., Yano, Y., Hatanaka, K., Eds.; IONICS: Tokyo, 1986; 31 pp.

21. *Handbook of Accelerator Physics and Engineering*; Chao, A.W., Tigner, M., Eds.; World Scientific: Singapore, 1999.

22. Kurashima, S.; Fukuda, M.; Nakamura, Y.; Nara, T.; Agematsu, T.; Ishibori, I.; Tamura, H.; Yokota, W.; Okumura, S.; Arakawa, K.; Kumata, Y.; Fukumoto, Y. In *Proceedings of the 16th International Conference on Cyclotrons and Their Applications 2001*; Marti, F., Ed.; American Institute of Physics: New York, 2001; 303 pp.

23. McMahan, M.A.; Wozniak, G.J.; Lyneis, C.M.; Bowman, D.R.; Liu, Charity.Z.H.; Moretto, L.G.; Kehoe, W.L.; Mignerey, A.C.; Namboodiri, M.N. Nucl. Instrum. Methods 1986, *A253*, 1.

24. Fukuda, M.; Arakawa, K.; Okumura, S.; Nara, T.; Ishibori, I.; Nakamura, Y.; Yokota, W.; Agematsu, T.; Tamura, H. In *Proceedings of the 1999 Particle Accelerator Conference*; Luccio, A., MacKay, W., Eds. http://accelconf.web.cern.ch/AccelConf/, 1999; 2259 pp.

25. Koehler, A.M.; Schneider, R.J.; Sisterson, J.M. Med. Phys. 1977, *4*, 297.

26. Sherrill, B.; Bailey, J.; Kashy, E.; Leakeas, C. Nucl. Instrum. Methods 1989, *B40/41*, 1004.

27. Rogers, E.J. Nucl. Instrum. Methods 1981, *189*, 305.

28. Renner, T.R.; Chu, W.T. Med. Phys. 1987, *14*, 825.

29. Fukuda, M.; Okumura, S.; Arakawa, K. Nucl. Instrum. Methods 1997, *A396*, 45.

30. Breese, M.B.H.; Jamieson, D.N.; King, P.J.C. *Materials Analysis Using A Nuclear Microprobe*; Wiley-Interscience: New York, 1996.

31. Llabador, Y.; Moretto, P. *Applications of Nuclear Microprobes in the Life Sciences—An Efficient Analytical Technique for Research in Biology and Medicine*; World Scientific: Singapore, 1998.

32. Cookson, J.A.; Ferguson, A.T.G.; Pilling, F. J. Radioanal. Chem. 1972, *12*, 39.

33. Kobayashi, Y.; Taguchi, M.; Shimizu, T.; Okumura, S.; Watanabe, H. J. Radiat. Res. 1996, *36*(4), 290.

34. Kamiya, T.; Yokota, W.; Kobayashi, Y.; Cholewa, M.; Krochmal, M.S.; Laken, G.; Larsen, I.D.; Fiddes, L.; Parkhill, G.; Dowsey, K. Nucl. Instrum. Methods 2001, *B181*, 27.

35. Grime, G.W.; Watt, F. *Beam Optics of Quadrupole Probe-Forming Systems*; Adam Hilger: Bristol, 1984.

36. Kamiya, T.; Utsunomiya, N.; Minehara, E.; Tanaka, R. Nucl. Instrum. Methods 1992, *B64*, 362.

37. Kamiya, T.; Suda, T.; Tanaka, R. Nucl. Instrum. Methods 1995, *B104*, 43.

38. Sakai, T.; Kamiya, T.; Oikawa, M.; Sato, T.; Tanaka, A.; Ishii, K. Nucl. Instrum. Methods 2002, *B190*, 271.

39. Ohshima, T.; Morita, Y.; Nashiyama, I.; Kawasaki, O.; Hisamatsu, T.; Nakao, T.; Wakow, Y.; Matsuda, S. IEEE Trans. Nucl. Sci 1996, *43*, 2990.

40. Tada, H.Y.; Carter, J.R., Jr.; Anspaugh, B.E.; Downing, R.G. *Solar Cell Radiation Handbook*; 3rd Ed. JPL Publication, 1982; 82–86.

41. *Handbook of Si Solar Cells for Space Application*. Private communication by JAERI and NASDA, 1999.

42. Anzawa, O.; Aoyama, K.; Imaizumi, M.; Matsuda, S.; Ohshima, T.; Hirao, T.; Itoh, H.; Saito M.; Matsumoto, Y. JAERI-Review 2000, *2000-024*, 8.

43. Ohshima, T.; Imaizumi, M.; Takamoto, T.; Sumita, T.; Ohi, A.; Kawakita, S.; Itoh, H.; Mat-

suda, S. Proceedings of the 5th International Workshop on Radiation Effects on Semiconductor Devices for Space Application, Takasaki; Japan Atomic Energy Research Institute (Japan), Oct 9–11, 2002; 113 pp.

44. Nemoto, N.; Shindou, H.; Makihara, A.; Kuboyama, S.; Matsuda, S.; Sugimoto, K.; Oikawa, K.; Itoh, H. Proceedings of European Space Components Conference (ESCCON2000); European Space Agency (ESA): Noordwijk, 2000; 325 pp.

45. Hirao, T.; Hamano, T.; Sakai, T.; Nashiyama, I. Nucl. Instrum. Methods, 1999, *B158*, 260.

46. For example, *Properties of Silicon Carbide*; Harris, G.L., Ed.; EMIS Dataviews Series No. 13; INSPEC Publication, IEE: London, 1995.

47. Itoh, H.; Yoshikawa, M.; Nashiyama, I.; Misawa, S.; Okumura, H.; Yoshida, S. IEEE Trans. Nucl. Sci. 1990, *NS-37*, 1732.

48. Ohshima, T.; Abe, K.; Itoh, H.; Yoshikawa, M.; Kojima, K.; Nashiyama, I.; Okada, S. Appl. Phys. 2000, *A71*, 141.

49. Ohshima, T.; Uedono, A.; Itoh, H.; Yoshikawa, M.; Kojima, K.; Okada, S.; Nashiyama, I.; Abe, K.; Tanigawa, S.; Frank, T.; Pensl, G. Mater. Sci. Forum 2000, *338–342*, 857.

50. Fukumoto, A. Phys. Rev. B 1996, *53*, 4458.

51. Larkin, D.J. Inst. Phys. Conf. Ser. 1996, *142*, 23.

52. Itoh, H.; Troffer, T.; Peppermuller, C.; Pensl, G. Appl. Phys. Lett. 1998, *73*, 1427.

53. Itoh, H.; Troffer, T.; Pensl, G. Mater. Sci. Forum 1998, *264–268*, 685.

54. Troffer, T.; Schadt, M.; Frank, T.; Itoh, H.; Pensl, G.; Heindl, J.; Strunk, H.P.; Maier, M. Phys. Status Solidi A 1997, *162*, 277.

55. Ohshima, T.; Itoh, H.; Yoshikawa, M. Mater. Sci. Forum 2001, *353–356*, 575.

56. Ohshima, T.; Yoshikawa, M.; Itoh, H.; Aoki, Y.; Nashiyama, I. Mater. Sci. Eng. B 1999, *61–62*, 480.

57. Ohshima, T.; Yoshikawa, M.; Itoh, H.; Kojima, K.; Okada, S.; Nashiyama, I. Mater. Sci. Forum 2000, *338–342*, 1299.

58. Ohshima, T.; Itoh, H.; Yoshikawa, M. J. Appl. Phys. 2001, *90*, 3038.

59. Lee, K.; Ohshima, T.; Itoh, H. Mater. Sci. Forum 2002, *389–393*, 1097.

60. Hamada, S.; Miwa, Y.; Yamaki, D.; Katano, Y.; Nakazawa, T.; Noda, K. J. Nucl. Mater. 1998, *258–263*, 383.

61. Hamada, S.; Zhang, Y.C.; Miwa, Y.; Yamaki, D. Radiat. Phys. Chem. 1997, *50*, 555.

62. Ioka, I.; Naito, A.; Shiba, K.; Robertson, J.P.; Jitsukawa, S.; Hishimura, A. J. Nucl. Mater. 1998, *258–263*, 1664.

63. Hirsh, J.P. Metall. Trans. A 1980, *11A*, 861.

64. Sekimura, N.; Kawanishi, H.; Nodaka, M.; Ishino, S. J. Nucl. Mater. 1984, *122*, 322.

65. Sawai, T.; Wakai, E.; Tomia, T.; Naito, A.; Jitsukawa, S. J. Nucl. Mater. 2002, *307–311*, 312.

66. Sekimura, N.; Iwai, T.; Arai, Y.; Yonamine, S.; Naito, A.; Miwa, Y.; Hamada, S. J. Nucl. Mater. 2000, *283–287*, 224.

67. Taguchi, T.; Wakai, E.; Igawa, N.; Nogami, S.; Snead, L.L.; Hasegawa, A.; Jitsukawa, S. J. Nucl. Mater. 2002, 307–311, 1135.

68. Nogami, S.; Hasegawa, A.; Abe, K.; Taguchi, T.; Yamada, R. J. Nucl. Mater. 2000, *283–287*, 268.

69. Nogami, S.; Hasegawa, A.; Taguchi, T.; Abe, K.; Yamada, R. Mater. Trans. 2001, *42*, 171.

70. Katano, Y.; Aruga, T.; Yamamoto, S.; Narumi, K.; Nakazawa, T.; Yamaki, D.; Noda, K. J. Nucl. Mater. 2000, *283–287*, 942.

71. Eto, M.; Baba, S.; Ishihara, M.; Ugachi, H. J. Nucl. Mater. 1998, *258–263*, 843.

72. Anpo, M.; Ichihashi, Y.; Takeuchi, M.; Yamashita, H. Res. Chem. Intermed. 1998, *24*, 143.

73. Yamaki, T.; Sumita, T.; Yamamoto, S. J. Mater. Sci. Lett. 2002, *21*, 33.

74. Umebayashi, T.; Yamaki, T.; Itoh, H.; Asai, K. Appl. Phys. Lett. 2002, *81*, 454.

75. Sumita, T.; Yamaki, T.; Yamamoto, S.; Miyashita, A. Thin Solid Films. 2002, *416*, 80.

76. Fromknecht, R.; Khubeis, I.; Massing, S.; Meyer, O. Nucl. Instrum. Methods 1999, *B147*, 191.

77. Sumita, T.; Yamaki, T.; Yamamoto, T.; Miyashita, A. Jpn. J. Appl. Phys. 2001, *40*, 4007.

78. Abe, H. Diam. Relat. Mater. 2001, *10*, 1201.

79. Abe, H.; Naramoto, H.; Iwase, A.; Kinoshita, C. Nucl. Instrum. Methods 1997, *B127–128*, 681.

80. Naramoto, H.; Zhu, X.; Xu, Y.; Narumi, K.; Vacik, J.; Yamamoto, S.; Miyashita, K. Phys. Solid State 2002, *44*, 668.
81. Zhu, X.D.; Naramoto, H.; Xu, Y.; Narumi, K. J. Chem. Phys. 2002, *116*, 10458.
82. Zhu, X.D.; Naramoto, H.; Xu, Y.; Narumi, K.; Miyashita, K. Phys Rev. B 2002, *60*, 165426.
83. Vacik, J.; Naramoto, H.; Narumi, K.; Yamamoto, S.; Miyashita, K. Proc. Mater. Res. Soc. Symp. 2001, *648*, P3.50.1.
84. Zhang, Z.J.; Narumi, K.; Naramoto, H. Appl. Phys. Lett. 2001, *79*, 2934.
85. Vacik, J.; Naramoto, H.; Narumi, K.; Yamamoto, S.; Miyashita, K. J. Chem. Phys. 2001, *114*, 9115.
86. Zhu, X.D.; Xu, Y.H.; Naramoto, H.; Narumi, K.; Miyashita, K. J. Phys.: Condens. Matter 2002, *14*, 5083.
87. Naramoto, H.; Xu, Y.H.; Zhu, X.D.; Narumi, K.; Vacik, J.; Yamamoto, S.; Miyashita, K. Proc. Mater. Res. Soc. Symp. 2001, *647*, O5.18.1.
88. Yamamoto, S.; Naramoto, H.; Narumi, K.; Tsuchiya, B.; Aoki, Y.; Kudo, H. Nucl. Instrum. Methods, B 1998, *134*, 400.
89. Yamamoto, S.; Naramoto, H. Nucl. Instrum. Methods 2000, *B161–163*, 605.
90. Zhu, P.R.; Yamamoto, S.; Miyashita, A.; Wu, Z.P.; Narumi, K.; Naramoto, H. Philos. Mag. Lett. 1999, *79*, 603.
91. Wu, Z.P.; Miyashita, A.; Yamamoto, S.; Abe, H.; Nashiyama, I.; Narumi, K.; Naramoto, H. J. Appl. Phys. 1999, *86*, 5311.
92. Yoshida, M.; Asano, M.; Safranj, A.; Omichi, H.; Spohr, R.; Vetter, J.; Katakai, R. Macromolecules 1996, *29* 8987.
93. Cartwright, B.G.; Shirk, E.K.; Price, P.B. Nucl. Instrum. Methods 1978, *153*, 457.
94. Ogura, K.; Hattori, T.; Asano, M.; Yoshida, M.; Omichi, H.; Nagaoka, N.; Kubota, H.; Katakai, R.; Hasegawa, H. Radiat. Meas. 1997, *28*, 197.
95. Stadler, L.J. Science 1928, *68*, 186.
96. Blakely, E.A. Radiat. Environ. Biophys. 1992, *31*, 181.
97. Lett, J.T. Radiat. Environ. Biophys. 1992, *31*, 257.
98. Goodhead, D.T. Radiat. Environ. Biophys. 1995, *34*, 67.
99. Hata, K.; Baba, H.H. JAERI-M 1988, 88–184.
100. Hirono, Y.; Smith, H.H.; Lyman, J.T.; Thompson, K.H.; Baum, J.W. Radiat. Res. 1970, *44*, 204.
101. Mei, M.; Deng, H.; Lu, Y.; Zhuang, C.; Liu, C.; Qiu, Q.; Qiu, Y.; Yang, T.C. Adv. Space Res. 1994, *10*, 363.
102. Hase, Y.; Shimono, K.; Inoue, M.; Tanaka, A.; Watanabe, H. Radiat. Environ. Biophys. 1999, *38*, 111.
103. Shimono, K.; Shikazono, N.; Inoue, M.; Tanaka, A.; Watanabe, H. Radiat. Environ. Biophys. 2001, *40*, 221.
104. Tanaka, A.; Shikazono, N.; Yokota, Y.; Watanabe, H.; Tano, S. Int. J. Radiat. Biol. 1997, *72*, 121.
105. Shikazono, N.; Tanaka, A.; Kitayama, S.; Watanabe, H.; Tano, S. Radiat. Environ. Biophys. 2002, *41*, 159.
106. Hase, Y.; Yamaguchi, M.; Inoue, M.; Tanaka, A. Int. J. Radiat. Biol. 2002, *78*, 799.
107. Nagatomi, S.; Tanaka, A.; Kato, A.; Watanabe, H.; Tano, S. JAERI-Review 1997, *96-017*, 50.
108. Okamura, M.; Ohtsuka, M.; Yasuno, N.; Hirosawa, T.; Tanaka, A.; Shikazono, N.; Hase, Y.; Tanase, M. JAERI-Review 2001, *2001-039*, 52.
109. Tanaka, A.; Sakamoto, A.; Ishigaki, Y.; Nikaido, O.; Sun, G.; Hase, Y.; Shikazono, N.; Tano, S.; Watanabe, H. Plant Physiol. 2002, *129*, 64.
110. Tanaka, A.; Tano, S.; Chantes, T.; Yokota, Y.; Shikazono, N.; Watanabe, H. Genes Genet. Syst. 1997, *72*, 141.
111. Hase, Y.; Tanaka, A.; Baba, T.; Watanabe, H. Plant J. 2000, *24*, 21.
112. Hamada, K.; Inoue, M.; Tanaka, A.; Watanabe, H. Plant Biotechnol. 1999, *16*, 285.
113. Hamada, K.; Inoue, M.; Tanaka, A.; Watanabe, H. Plant Biotechnol. 2001, *18*, 251.

114. Kishinami, I.; Tanaka, A.; Watanabe, H. Abstracts of the Fifth TIARA Research Review Meeting; Japan Atomic Energy Research Institute, Tokai-mura, Japan, 1996; 165–166 (in Japanese).

115. Nakai, H.; Watanabe, H.; Kitayama, S.; Tanaka, A.; Kobayashi, Y.; Takahashi, T.; Asai, T.; Imada, T. JAERI-Review 1995, 95-019, 34.

116. Shikazono, N.; Yokota, Y.; Tanaka, A.; Watanabe, H.; Tano, S. Genes Genet. Syst. 1998, 73, 173.

117. Shikazono, N.; Tanaka, A.; Watanabe, H.; Tano, S. Genetics 2001, 157, 379.

118. Goodspeed, T.H. The Genus Nicotiana; Chronica Botanica Company: Waltham, MA, 1954.

119. Takenaka, Y. Bot. Mag. Tokyo 1962, 75, 237.

120. Inoue, M.; Watanabe, H.; Tanaka, A.; Nakamura, A. JAERI TIARA Annual Report 1993; Japan Atomic Energy Research Institute, Tokai-mura, Japan, 1994; 44 pp.

121. Yamashita, T.; Inoue, M.; Watanabe, H.; Tanaka, A.; Tano, S. JAERI-Review 1995, 95-019, 37.

122. Yamashita, T.; Inoue, M.; Watanabe, H.; Tanaka, A.; Tano, S. JAERI-Review 1997, 96-017, 44.

123. Tanaka, A.; Watanabe, H.; Shimizu, T.; Inoue, M.; Kikuchi, M.; Kobayashi, Y.; Tano, S. Nucl. Instrum. Methods Phys. Res. B. 1997, 129, 42.

124. Tanaka, A.; Watanabe, H.; Hase, Y.; Inoue, M.; Kikuchi, M.; Kobayashi, Y.; Tano, S. JAERI-Review 1997, 97-015, 21.

125. Inoue, M.; Kitamura, S.; Toda, Y.; Watanabe, H.; Tanaka, A.; Hase, Y. JAERI-Review 1999, 99-025, 80.

126. Hase, Y.; Sakamoto, A.; Wada, S.; Kitamura, S.; Tanaka, A. JAERI-Review 2001, 2001-039, 87.

127. Kobayashi, Y.; Taguchi, M.; Okumura, S.; Watanabe, H. JAERI-Review 1997, 96-017, 38.

128. Kiguchi, K.; Kinjoh, Y.; Masahashi, K.; Tu, Z.T.; Tamura, H.; Shirai, K.; Kanetatsu, K.; Kobayashi, Y.; Taguchi, M.; Watanabe, H. JAERI-Review 2000, 2000-024, 51.

129. Tanaka, A.; Kobayashi, Y.; Hase, Y.; Watanabe, H. J. Exp. Bot. 2002, 53, 683.

130. Keutgen, N.; Matsuhashi, S.; Mizuniwa, C.; Ito, T.; Fujimura, T.; Ishioka, N.S.; Watanabe, S.; Sekine, T.; Uchida, H.; Hashimoto, S. Appl. Radiat. Isotopes 2002, 57, 225.

131. Nakanishi, H.; Bughio, N.; Matsuhashi, S.; Ishioka, N.S.; Uchida, H.; Tsuji, A.; Osa, A.; Sekine, T.; Kume, T.; Mori, S. J. Exp. Bot. 1999, 50, 637.

132. Bughio, N.; Nakanishi, H.; Kiyomia, S.; Marsuhashi, S.; Ishioka, N.S.; Watanabe, S.; Uchida, H.; Tsuji, A.; Osa, A.; Kume, T.; Hashimoto, S.; Sekine, T.; Mori, S. Planta 2001, 213, 708.

133. Kiyomiya, S.; Nakanishi, H.; Uchida, H.; Tsuji, A.; Nishiyama, S.; Futatsubashi, M.; Tsukada, H.; Ishioka, N.S.; Watanabe, S.; Ito, T.; Mizuniwa, C.; Osa, A.; Matsuhashi, S.; Hashimoto, S.; Sekine, T.; Mori, S. Plant Physiol. 2001, 125, 1743.

134. Mori, S.; Kiyomiya, S.; Nakanishi, H.; Ishioka, N.S.; Watanabe, S.; Osa, A.; Matsuhashi, S.; Hashimoto, S.; Sekine, T.; Uchida, H.; Nishiyama, S.; Tsukada, H.; Tsuji, A. Soil Sci. Plant Nutr. 2000, 46, 975.

135. Matsunami, H.; Arima, Y.; Watanabe, K.; Ishioka, N.S.; Watanabe, S.; Osa, A.; Sekine, T.; Uchida, H.; Tsuji, A.; Matsuhashi, S.; Ito, T.; Kume, T. Soil Sci. Plant Nutr. 1999, 45, 955.

136. Nakanishi, T.M.; Furukawa, J.; Tanoi, K.; Yokota, H.; Ueoka, S.; Ishioka, N.S.; Watanabe, S.; Osa, A.; Sekine, T.; Itoh, T.; Mizuniwa, T.; Matsuhashi, S.; Hashimoto, S.; Uchida, H.; Tsuji, A. J. Radioanal. Nucl. Chem. 2001, 249, 503.

137. Ohtake, N.; Sato, T.; Fujikake, H.; Sueyoshi, K.; Ohyama, T.; Ishioka, N.S.; Watanabe, S.; Osa, A.; Sekine, T.; Matsuhashi, S.; Ito, T.; Mizuniwa, C.; Kume, T.; Hashimoto, S.; Uchida, H.; Tsuji, A. J. Exp. Bot. 2001, 52, 277.

Index

.

Printed and bound by CPI Group (UK) Ltd, Croydon, CR0 4YY

23/10/2024

01778254-0010